List of Tables

Basic Engineering Circuit Analysis

2ND EDITION

Basic Engineering Circuit Analysis

J. David Irwin

Professor and Head
Department of Electrical Engineering
Auburn University, Alabama

Macmillan Publishing Company
New York

Collier Macmillan Publishers
London

Printed in the United States of America

Macmillan Publishing Company
866 Third Avenue, New York, New York 10022

Collier Macmillan Canada, Inc.

Library of Congress Cataloging-in-Publication Data

Irwin, J. David
Basic engineering circuit analysis.

Bibliography: p.
Includes index.
1. Electric circuit analysis. I. Title.
TK454.I78 1987 621.319'2 86-9747
ISBN 0-02-359860-3

Printing: 4 5 6 7 8 Year: 8 9 0 1 2 3 4 5

ISBN 0-02-359860-3

To Edie

Geri Marie
John David, Jr.
Laura Lynne

Preface

Perspective

This book is designed specifically for introductory courses in linear circuit analysis, which appear early in an electrical engineering program. The material can easily be split into a two-semester or three-quarter course sequence. The recommended semester sequence is Chapters 1–10 and 11–18; and the recommended quarter course sequence is 1–7, 8–13, and 14–18. The book may also serve as a survey textbook, through the selective use of various chapters and sections throughout the book.

A real attempt has been made to present the material in an easy-to-read and understandable fashion. The book is designed to aid (1) the teacher in presenting the material in a clear and lucid manner, and (2) the student in grasping the concepts and techniques in order to quickly become proficient in their use. In fact, I have endeavored to present the material in such a way that sufficiently mature students, with the help of the supplements, can learn the material on their own.

Features

Prerequisite Material. In general, the reader should have a background in differential and integral calculus and an understanding of the techniques involved in the solution of differential equations with constant coefficients; some basic knowledge of matrices and computer programming would also be helpful.

Pedagogy. The basic philosophy employed in the development of this book can be stated simply as follows: "To *learn* circuit analysis, one must *do* circuit analysis";

therefore, the text is replete with both examples and problems. I have made every effort to make the discussion as clear as possible, and in so doing, I have employed the old Roman concept of ''Exemplum Docet—The Example Teaches.'' I have tried to keep the mathematics as simple as possible so that the concepts and principles are not obscured in a maze of calculations and numbers. In addition, where appropriate I have worked the same problems in a number of ways to facilitate an understanding and comparison of various techniques, and their relationship to one another. In general, I have tried to present the material in a way that helps both the teacher and the student.

Illustrations. Illustrations are used extensively in order to clarify all the details of a circuit analysis procedure. In addition, a second color is introduced to highlight variables under investigation in both circuits and graphs.

Computer Solutions. The computer-aided circuit analysis program SPICE is employed throughout the book. Wherever appropriate it is used to determine currents and voltages in a wide variety of networks. This is an extremely important feature of the book, as SPICE is extensively used in industry to analyze circuit designs before committing them to production. In addition, the criteria employed by the Accreditation Board for Engineering and Technology specifically state: ''Students must demonstrate knowledge of the application and use of digital computation techniques to specific engineering problems.'' The employment of SPICE in the context helps satisfy this standard and, in addition, provides the basis for its use in electronics.

Flexibility. The material is presented in a manner that permits considerable flexibility in its use. For example, one can completely omit all op-amp material, Chapter 4 on topology, and all SPICE material with no loss of generality. In addition, many other sections and chapters can be skipped while proceeding from beginning to end in the development of a coherent presentation of the material.

Summaries. At the end of every chapter the important concepts are summarized in a concise fashion. These summaries serve as a quick reminder for the reader of all the significant principles and techniques contained within the chapter.

Problem Sets. There are approximately 300 drill problems and 1000 problems at the end of the chapters. The problems have been carefully selected and strategically placed to test the reader's understanding of all the material.

Supplements. A student manual is available that provides not only solutions to the drill problems, but also hundreds of other problems that are similar to the ones that appear at the end of the chapters. In addition, a solution manual for all the problems at the end of the chapters is available for teachers.

Changes for This Edition

The second edition is a refinement of the first edition; throughout the book material has been rewritten to improve the presentation. A considerable number of additions has been made. Some of the specific ones that can be easily identified are as follows.

- Increased the number of examples.
- Added approximately 300 drill problems.
- Increased the number of problems at the end of the chapters so that the total is approximately 1000.
- Added dc circuit analysis using SPICE to Chapter 5.
- Added transient circuit analysis using SPICE to Chapters 7 and 8.
- Added ac circuit analysis using SPICE to Chapter 10.
- Added SPICE analysis of three phase circuits to Chapter 12.
- Added the delta-connected source to Chapter 12.
- Simplified and expanded the coverage of Bode plots in Chapter 13.
- Expanded the coverage of resonance in Chapter 13.
- Added frequency analysis using SPICE to Chapter 13.
- Added SPICE analysis of magnetically coupled circuits to Chapter 14.
- Expanded the Fourier series and Fourier transform coverage.
- Split Chapter 17 into two chapters and doubled the material on Laplace transform to include convolution and a significant amount of material on circuit analysis.

Acknowledgment

I am indebted to a number of colleagues and friends who have helped in the preparation of this book. A number of these people were mentioned in the first edition, and I want to reemphasize my grateful thanks for their contributions here also. They are Professors James L. Lowry, Charles L. Rogers, Charles A. Gross, Edward R. Graf, Richard C. Jaeger, Mr. George B. Lindsey, and Miss Betty Kelley. In addition, I want to express my appreciation to Kevin Driscoll, David Mack, and John Parr for all their help. I am also grateful to Keith A. Jones, who was a great help in checking the entire manuscript. I am especially indebted to Professor Emeritus M. A. Honnell for his very careful and meticulous review of the galley and page proofs. Finally, I wish to express my deepest appreciation to my wife, Edie, without whose help and support this book would not have been possible.

J.D.I.

Contents

Basic Engineering Circuit Analysis

CHAPTER 1 Basic Concepts

It is important to note that the title of this book does not imply that it will be useful only to electrical engineers. Rather, in our time we find that all types of engineers need to be aware of the implications of circuit analysis because they are encountering circuits in many ways: as an integral part of their systems, in their instrumentation, and so on. In addition, many of today's problems are so complicated that they are attacked by a team of specialists with backgrounds in engineering, mathematics, physics, and chemistry. In many instances the electrical components literally permeate an entire system through implications of power, control, instrumentation, monitoring, and the like. Therefore, it is important that as many technical personnel as possible be at least familiar with electric circuit analysis.

1.1 System of Units

The system of units we employ is the international system of units, the Système International des Unités, which is normally referred to as the SI standard system. This system, which is composed of the basic units meter (m), kilogram (kg), second (s), ampere (A), degree kelvin (K), and candela (cd), is defined in all modern physics texts and therefore will not be defined here. However, we will discuss the units in some detail as we encounter them in our subsequent analyses.

The standard prefixes that are employed in SI are shown in Fig. 1.1. Note the decimal relationship between these prefixes. These standard prefixes are employed throughout our study of electric circuits.

Only a few years ago a millisecond, 10^{-3} s, was considered to be a short time in

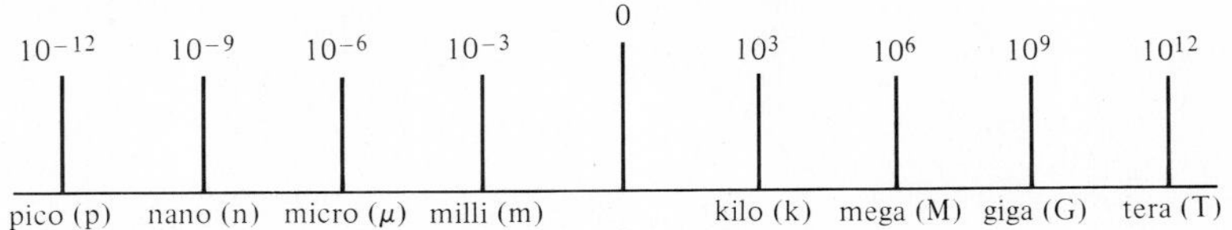

Figure 1.1 Standard SI prefixes.

the analysis of electric circuits. Advances in technology, however, have led to a state in which we now think of doing such things as performing calculations in nanoseconds or even picoseconds. At the same time, circuits have experienced a phenomenal decrease in size. For example, consider the integrated-circuit chip shown in Fig. 1.2. Such miniaturized circuits are commonplace in calculators, computers, and other electronic equipment.

1.2 Basic Quantities

Before we begin our analysis of electric circuits, we must define the basic terms that we will employ. However, in this chapter and throughout the text our definitions and explanations will be as simple as possible in order to foster an understanding of the use of the material. No attempt will be made to give complete definitions of many of the quantities because such definitions are not only unnecessary at this level but are often confusing.

Although most of us have an intuitive concept of what is meant by a circuit, we will simply refer to an *electric circuit* as an interconnection of electrical components.

It is very important at the outset that the reader understand the basic strategy that we will employ in our analysis of electric circuits. This strategy is outlined in Fig. 1.3 and we will be concerned only with the portion of the diagram to the right of the dashed line. In subsequent courses or further study the reader will learn how to model physical devices such as electronic components. Our procedure here will be to employ linear models for our circuit components, then define the variables that are used in the appropriate circuit equations to yield solution values, and finally, interpret the solution values for the variables in order to determine what is actually happening in the physical circuit. The variables may be time varying or constant depending on the nature of the physical parameters they represent.

The most elementary quantity in an analysis of electric circuits is the electric *charge*. We know from basic physics that the nature of charge is based on concepts of atomic theory. We view the atom as a fundamental building block of matter which is composed of a positively charged nucleus surrounded by negatively charged electrons. In the metric system, charge is measured in coulombs (C). The charge on an electron is negative and equal in magnitude to 1.602×10^{-19} C. However, our interest in electric charge is centered around its motion, since charge in motion results in an energy transfer. Of particular interest to us are those situations in which the motion is confined to a definite closed path.

An electric circuit is essentially a pipeline that facilitates the transfer of charge from one point to another. The time rate of change of charge constitutes an electric *current*.

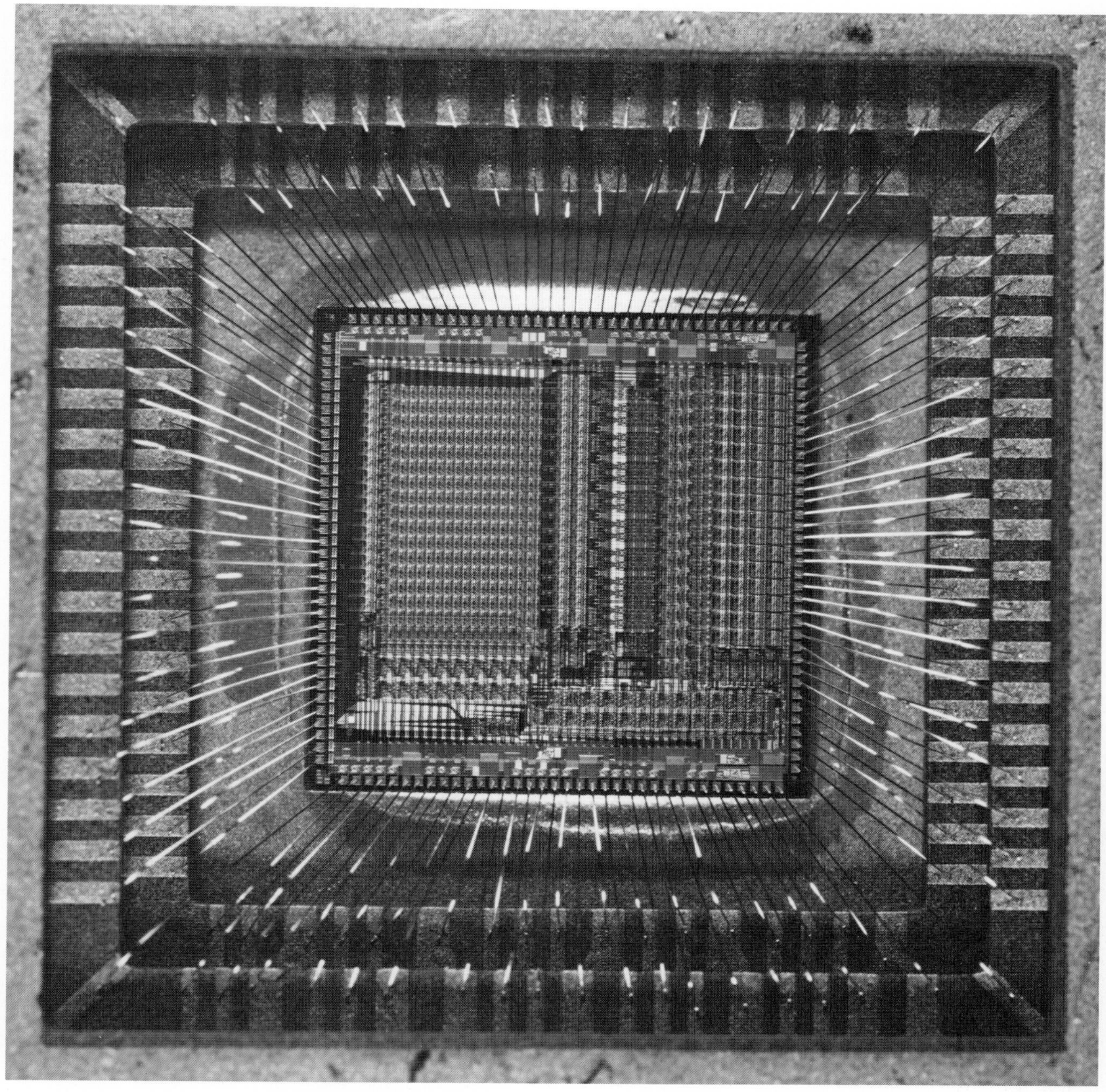

Figure 1.2 An example of a VLSI advanced bipolar chip with 30,000 transistors. This high-speed chip is approximately 0.25 inches on a side and uses a 144 pin-grid-array package (Courtesy of Honeywell Inc.).

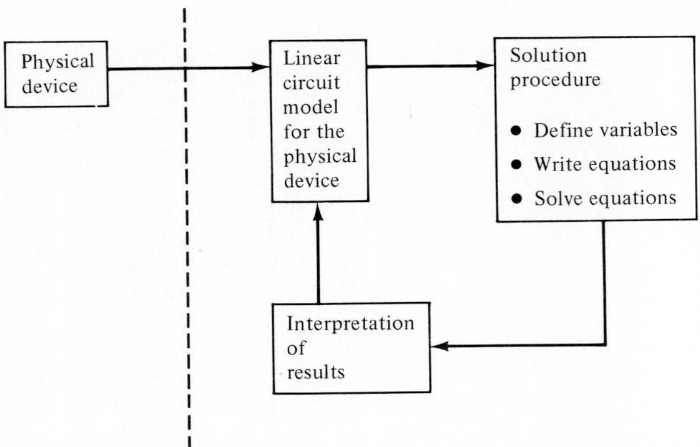

Figure 1.3 Basic strategy employed in circuit analysis.

Mathematically, the relationship is expressed as

$$i(t) = \frac{dq(t)}{dt} \tag{1.1}$$

where i and q represent current and charge, respectively (lowercase letters represent time dependency and capital letters are reserved for constant quantities). The basic unit of current is the ampere (A) and 1 ampere is 1 coulomb per second.

Although we know that current flow in metallic conductors results from electron motion, the conventional current flow, which is universally adopted, represents the movement of positive charges. It is important that the reader think of current flow as the movement of positive charge regardless of the physical phenomena that take place. The symbolism that will be used to represent current flow is shown in Fig. 1.4. The variable representing the current in the wire in Fig. 1.4a was defined as I_1 flowing in the wire from left to right in the figure. A set of equations was written for the circuit and a solution value of 2 A was obtained; that is, $I_1 = 2$ A. This means that the physical current in the wire is flowing from left to right, in the direction of our variable, and is 2 A. $I_1 = 2$ A in Fig. 1.4a indicates that at any point in the wire shown, 2 C of charge passes from left to right each second. The same procedure was followed for the wire shown in Fig. 1.4b, and the variable I_1 was defined in the same manner. A set of equations was written and a solution value of -2 A was obtained. An interpretation of this result is that the physical current in the wire is from right to left, opposite to our reference direction for I_1, and its value is 2 A. $I_1 = -2$ A in Fig. 1.4b indicates that at any point in the wire shown, 2 C of charge passes from right to left each second. Therefore, it is important

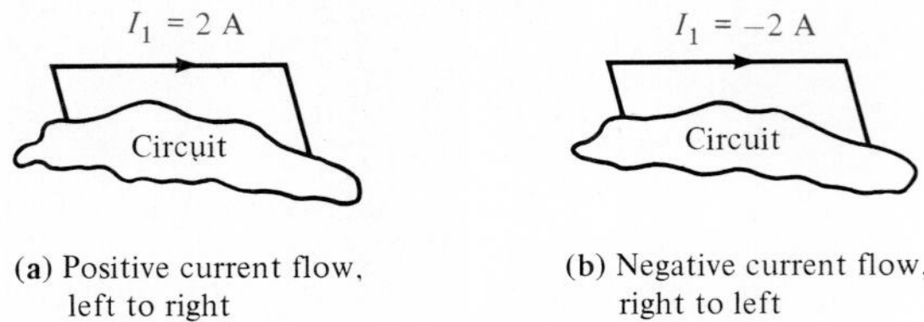

(a) Positive current flow, left to right

(b) Negative current flow, right to left

Figure 1.4 Conventional current flow.

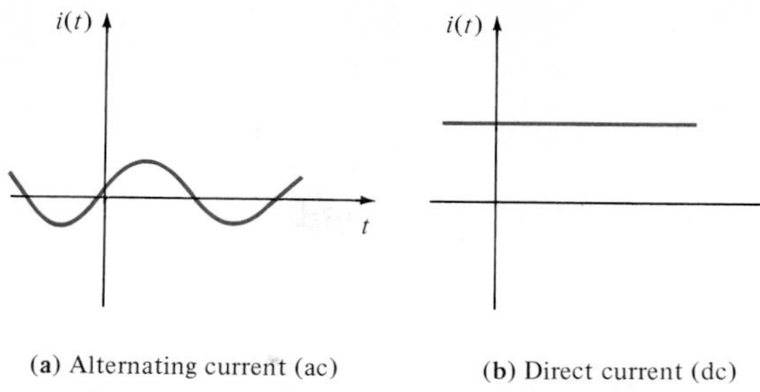

Figure 1.5 Two common types of current.

to specify not only the magnitude of the variable representing the current, but also its direction. After a solution is obtained for the current variable, the actual physical current is known.

There are two types of current that we encounter often in our daily lives, alternating current (ac) and direct current (dc), which are shown as a function of time in Fig. 1.5. *Alternating current* is the common current found in every household, used to run the refrigerator, stove, washing machine, and so on. Batteries, which are used in automobiles or flashlights, are one source of *direct current*. In addition to these two types of currents, which have a wide variety of uses, we can generate many other types of currents. We will examine some of these other types later in the book.

We have indicated that charges in motion yield an energy transfer. Now we define the *voltage* (also called the *electromotive force* or *potential*) between two points in a circuit as the difference in energy level of a unit positive charge located at each of the two points. Work or energy $w(t)$ or W is measured in joules (J) and 1 joule is 1 newton meter (N · m). Hence voltage [$v(t)$ or V] is measured in volts (V) and 1 volt is 1 joule per coulomb; that is, 1 volt = 1 joule per coulomb = 1 newton meter per coulomb.

If a unit positive charge is moved between two points, the energy required to move it is the difference in energy level between the two points and is the defined voltage. It is extremely important that the variables that are used to represent voltage between two points be defined in such a way that the solution will let us interpret which point is at the higher potential with respect to the other.

In Fig. 1.6a the variable that represents the voltage between points A and B has been defined as V_1, and it is assumed that point A is at a higher potential than point B, as indicated by the + and − signs associated with the variable and defined in the figure. Note that V_1 also has a reference direction associated with it. The equations for the circuit were written and a solution value of 2 V was obtained: that is, $V_1 = 2$ V, as shown on

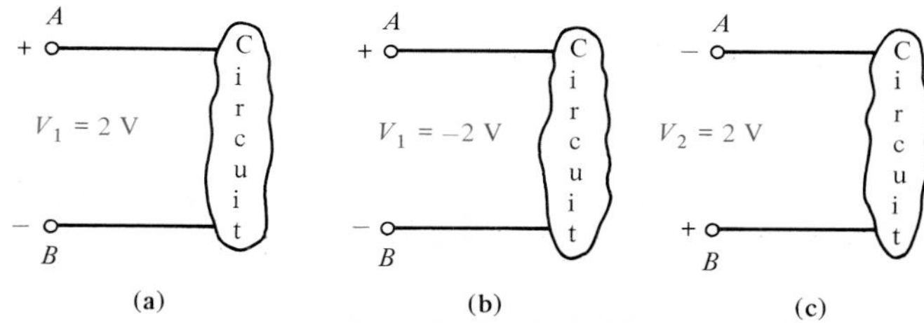

Figure 1.6 Voltage representations.

the figure. The physical interpretation of this is that the difference in potential of points A and B is 2 V and point A is at the higher potential. If a unit positive charge is moved from point A through the circuit to point B, it will give up energy to the circuit and have 2 J less energy when it reaches point B. If a positive charge is moved from point B to point A, extra energy must be added to the charge by the circuit, and hence the charge will end up with 2 J more energy at point A than it started with at point B.

The same procedure was followed for the circuit in Fig. 1.6b, and the variable V_1 was defined in the same manner. A set of equations was written and a solution value of -2 V was obtained: that is, $V_1 = -2\text{V}$. The physical interpretation of $V_1 = -2\text{V}$ is that the potential between points A and B is 2 V and point B is at the higher potential. If the circuit in Fig. 1.6b was reworked with the variable for the voltage defined as shown in Fig. 1.6c, the solution would have been $V_2 = 2$ V. The physical interpretation of this is that the difference in potential of points A and B is 2 V, with point B at the higher potential.

Note that it is important to define a variable with a reference direction so that the answer can be interpreted to give the physical condition in the circuit. We will find that it is not possible in many cases to define the variable so that the answer is positive, and we will also find that it is not necessary to do so. A negative number for a given variable gives exactly the same information as a positive number for a new variable that is the same as the old variable, except that it has an opposite reference direction. Hence, when we define either current or voltage, it is absolutely necessary that we specify both magnitude and direction. Therefore, it is incorrect to say that the voltage between two points is 10 V or the current in a line is 2 A, since only the magnitude and not the direction for the variables has been defined.

At this point we have presented the conventions that we employ in our discussions of current and voltage. Energy is yet another important term of basic significance. Figure 1.7 illustrates the voltage-current relationships for energy transfer. In Fig. 1.7a energy is being supplied *to* the element by whatever is attached to the terminals. Note that 2 A, that is, 2 C/s of charge, is moving from point A to point B through the element each second. Each coulomb loses 3 J of energy as it passes through the element from point A to point B. Therefore, the element is absorbing 6 J/s of energy. Note that when the element is *absorbing* energy, a positive current enters the positive terminal. In Fig. 1.7b energy is being supplied *by* the element to whatever is connected to terminals A-B. In this case, note that when the element is *supplying* energy, a positive current enters the

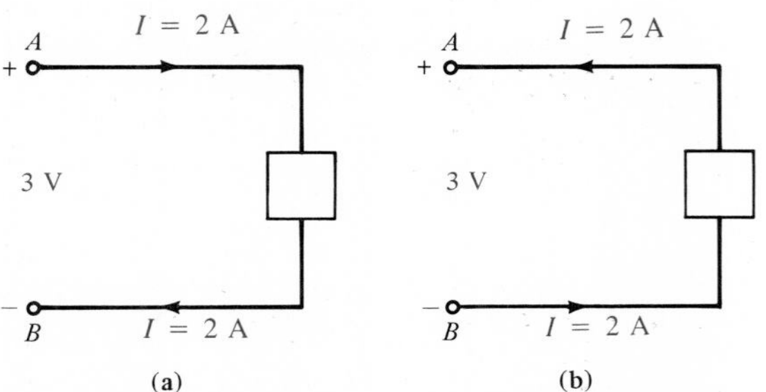

Figure 1.7 Voltage-current relationships for energy absorbed (a) and energy supplied (b).

negative terminal and leaves via the positive terminal. In this convention a negative current in one direction is equivalent to a positive current in the opposite direction, and vice versa.

EXAMPLE 1.1

Suppose that your car will not start. To determine if the battery is dead, you turn on the light switch and find that the lights are very dim, indicating a faulty battery. You borrow a friend's car and a set of jumper cables. However, how do you connect his car's battery to yours? What do you want his battery to do?

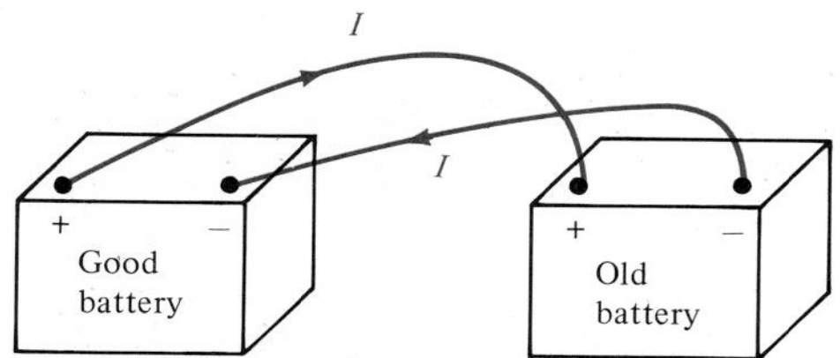

Figure 1.8 Diagram for Example 1.1.

Essentially, his car's battery must supply power to yours, and therefore it should be connected in the manner shown in Fig. 1.8. Note that the positive current leaves the positive terminal of the good battery (supplying energy) and enters the positive terminal of the old battery (absorbing energy). Note that the same connections are used when charging a battery. ■

We have defined voltage in joules per coulomb as the energy required to move a positive charge of 1 C through an element. If we assume that we are dealing with a differential amount of charge and energy, then

$$v = \frac{dw}{dq} \tag{1.2}$$

Multiplying this quantity by the current in the element yields

$$vi = \frac{dw}{dq} \times \frac{dq}{dt} = \frac{dw}{dt} = p \tag{1.3}$$

which is the time rate of change of energy or power measured in joules per second, or watts (W). Since, in general, both v and i are functions of time, p is also a time-varying quantity. Therefore, the change in energy from time t_1 to time t_2 can be found by integrating Eq. (1.3): that is,

$$w = \int_{t_1}^{t_2} p\,dt = \int_{t_1}^{t_2} vi\,dt \tag{1.4}$$

At this point, let us summarize our sign convention for power. To determine the sign of any of the quantities involved, the variables for the current and voltage should be arranged as shown in Fig. 1.9. The variable for the voltage $v(t)$ is defined as the voltage across the element with the positive reference at the same terminal that the current variable $i(t)$ is entering. This convention is called the *passive sign convention* and will be so noted in the remainder of this book. The product of v and i, with their attendant signs,

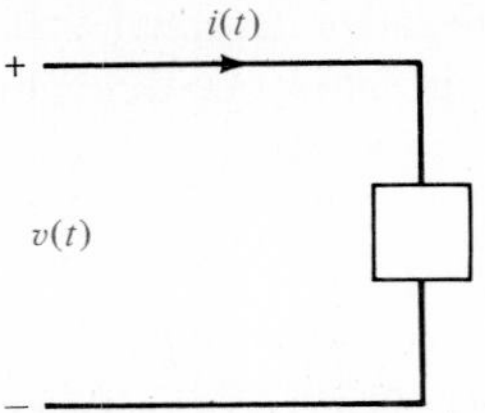

Figure 1.9 Sign convention for power.

will determine the magnitude and sign of the power. If the sign of the power is positive, power is being absorbed by the element; if the sign is negative, power is being supplied by the element.

EXAMPLE 1.2

We wish to determine the power absorbed by, or supplied by, the elements in Fig. 1.7.

In Fig. 1.7a, $P = VI = (3\text{ V})(2\text{ A}) = 6\text{ W}$ is absorbed by the element. In Fig. 1.7b, $P = VI = (3\text{ V})(-2\text{ A}) = -6\text{ W}$ is absorbed by the element, or $+6$ W is supplied by the element. ■

EXAMPLE 1.3

Given the two diagrams shown in Fig. 1.10, determine whether the element is absorbing or supplying power and by how much.

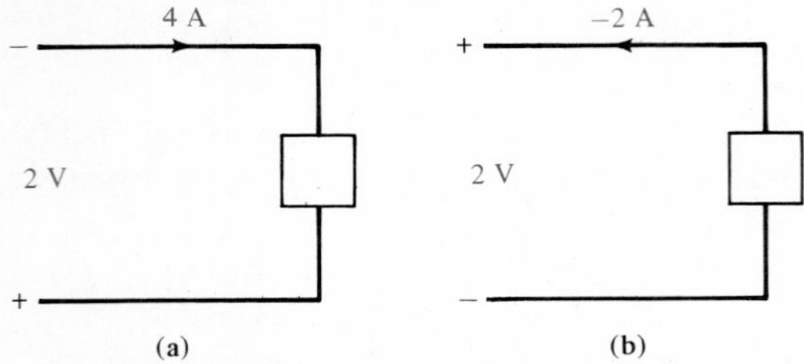

Figure 1.10 Elements for Example 1.3.

In Fig. 1.10a, the power is $P = (2\text{ V})(-4\text{ A}) = -8\text{ W}$. Therefore, the element is supplying power. In Fig. 1.10b, the power is $P = (2\text{ V})(2\text{ A}) = 4\text{ W}$. Therefore, the element is absorbing power. ■

DRILL EXERCISE

D1.1. Determine the amount of power absorbed or supplied by the elements in Fig. D1.1.

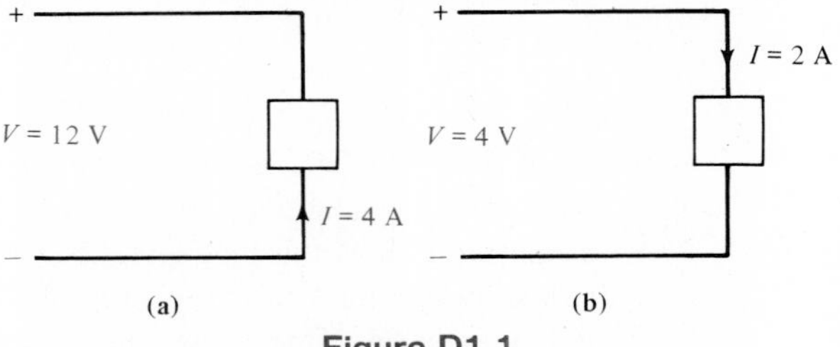

Figure D1.1

EXAMPLE 1.4

We wish to determine the unknown voltage or current in Fig. 1.11.

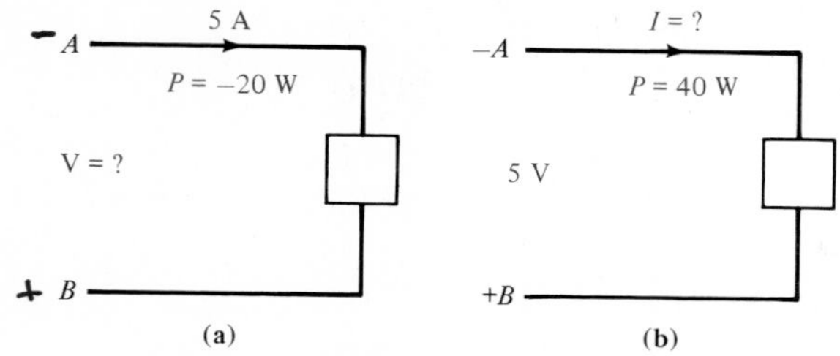

Figure 1.11 Elements for Example 1.4.

In Figure 1.11a, a power of −20 W indicates that the element is delivering power. Therefore, the current enters the negative terminal (terminal *A*), and from Eq. (1.3) the voltage is 4 V.

In Fig. 1.11b, a power of +40 W indicates that the element is absorbing power and therefore the current should enter the positive terminal *B*. The current thus has a value of −8 A, as shown in the figure. ■

DRILL EXERCISE

D1.2. Determine the unknown variables in Fig. D1.2.

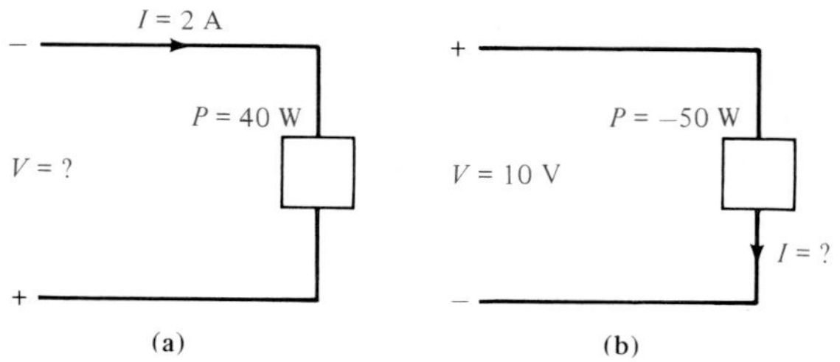

Figure D1.2

Finally, it is important to note that these electrical networks satisfy the principle of conservation of energy. For our present purposes this means that the power that is supplied in a network is exactly equal to the power absorbed.

1.3 Circuit Elements

It is important to realize at the outset that in our discussions we will be concerned with the behavior of a circuit element, and we will describe that behavior by a mathematical model. Thus, when we refer to a particular circuit element, we mean the *mathematical model* that describes its behavior.

Thus far we have defined voltage, current, and power. In the remainder of this chapter we will define both independent and dependent current and voltage sources. Although we will assume ideal elements, we will try to indicate the shortcomings of these assumptions as we proceed with the discussion.

In general, the elements we will define are in terminal devices that are completely characterized by the current through the element and/or the voltage across it. These elements, which we will employ in constructing electric circuits, will be broadly classified as being either active or passive. The distinction between these two classifications depends essentially upon one thing—whether they supply or absorb energy. As the words themselves imply, an *active* element is capable of generating energy and a *passive* element cannot generate energy.

However, we will show later that some passive elements are capable of storing energy. Typical active elements are batteries, generators, and transistor models. The three common passive elements are resistors, capacitors, and inductors.

In Chapter 2 we will launch an examination of passive elements by discussing the resistor in detail. However, before proceeding with that element, we first present some very important active elements.

1. Independent voltage source.
2. Independent current source.
3. Two dependent voltage sources.
4. Two dependent current sources.

Independent Sources

An *independent voltage source* is a two-terminal element that maintains a specified voltage between its terminals regardless of the current through it. The general symbol for an independent source, a circle, is shown in Fig. 1.12a. As the figure indicates, terminal A is $v(t)$ volts positive with respect to terminal B. The word "positive" may

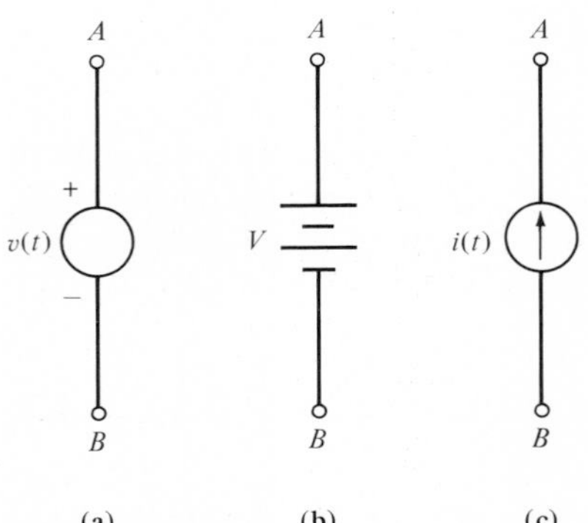

Figure 1.12 Symbols for (a) independent voltage source; (b) constant voltage source; (c) independent current source.

be somewhat misleading. What is meant in this case is that $v(t)$ is referenced positive at terminal A and that the physical voltage across the device must be interpreted from the numerical value of $v(t)$. That is, if $v(t)$ at $t = 2$ s is -10 V, point B is at a higher potential than point A at $t = 2$ s. The symbol $v(t)$ is normally employed for time-varying voltages. However, if the voltage is time invariant (i.e., constant), the symbol shown in Fig. 1.12b is sometimes used. This symbol, which is used to represent a battery, illustrates that terminal A is V volts positive with respect to terminal B, where the long line on the top and the short line on the bottom indicate the positive and negative terminals, respectively, and thus the polarity of the element.

In contrast to the independent voltage source, the *independent current source* is a two-terminal element that generates a specified current regardless of the voltage across its terminals. The general symbol for an independent current source is shown in Fig. 1.12c, where i is the specified current and the arrow indicates the direction of current flow.

It is important that we pause here to inject a comment concerning a shortcoming of the models. In general, mathematical models approximate actual physical systems only under a certain range of conditions. Rarely does a model accurately represent a physical system under every set of conditions. To illustrate this point, consider the model for the voltage source in Fig. 1.12a. We assume that the voltage source delivers v volts regardless of what is connected to its terminals. Theoretically, we could adjust the external circuit so that an infinite amount of current would flow, and therefore the voltage source would deliver an infinite amount of power. This is, of course, physically impossible. A similar argument could be made for the independent current source. Hence the reader is cautioned to keep in mind that models have limitations and are thus valid representations of physical systems only under certain conditions.

EXAMPLE 1.5

Determine the power absorbed or supplied by the elements in the network in Fig. 1.13.

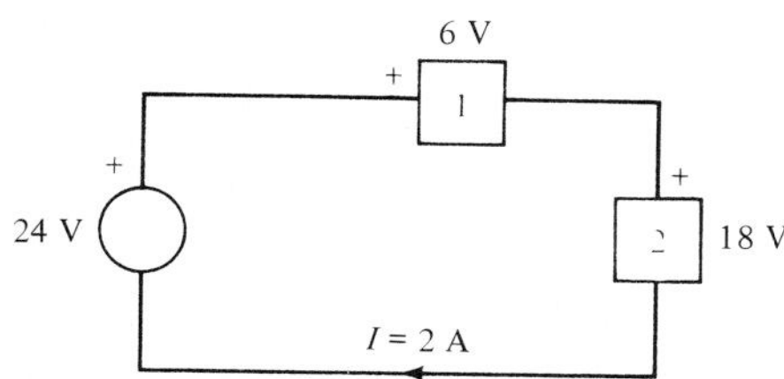

Figure 1.13 Network for Example 1.5.

The current flow is out of the positive terminal of the 24-V source, and therefore this element is supplying (2)(24) = 48 W of power. The current is into the positive terminals of elements 1 and 2, and therefore elements 1 and 2 are absorbing (2)(6) = 12 W and (2)(18) = 36 W, respectively. Note that the power supplied is equal to the power absorbed. ■

DRILL EXERCISE

D1.3. Find the power that is absorbed or supplied by the three elements in Fig. D1.3.

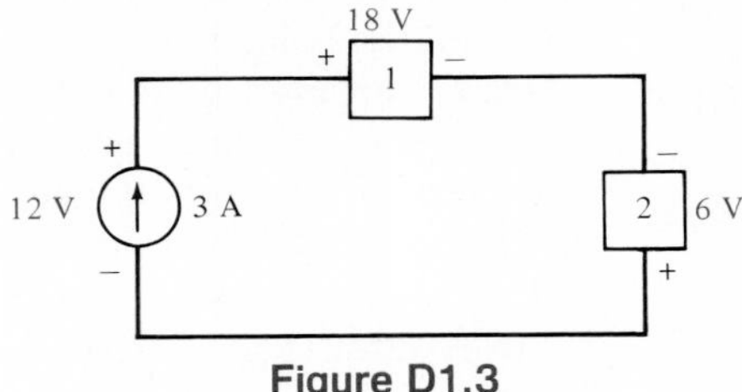

Figure D1.3

Dependent Sources

In contrast to the independent sources, which produce a particular voltage or current completely unaffected by what is happening in the remainder of the circuit, dependent sources generate a voltage or current that is determined by a voltage or current at a specified location in the circuit. These sources are very important because they are an integral part of the mathematical models used to describe the behavior of many electronic circuit elements (e.g., transistors).

In contrast to the circle used to represent independent sources, a diamond is used to represent a dependent or controlled source. Figure 1.14 illustrates the four types of dependent sources. The input terminals on the left represent the voltage or current that controls the dependent source, and the output terminals on the right represent the output current or voltage of the controlled source. Note that in Fig. 1.14a and d the quantities μ and β are dimensionless constants because we are transforming voltage to voltage and current to current. This is not the case in Fig. 1.14b and c; hence when we employ these elements a short time later, we must describe the units of the factors r and g.

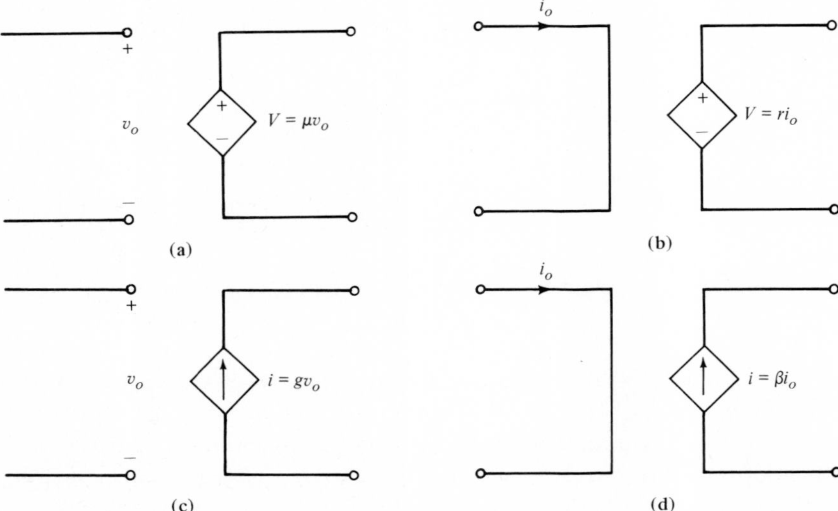

Figure 1.14 Four different types of dependent sources.

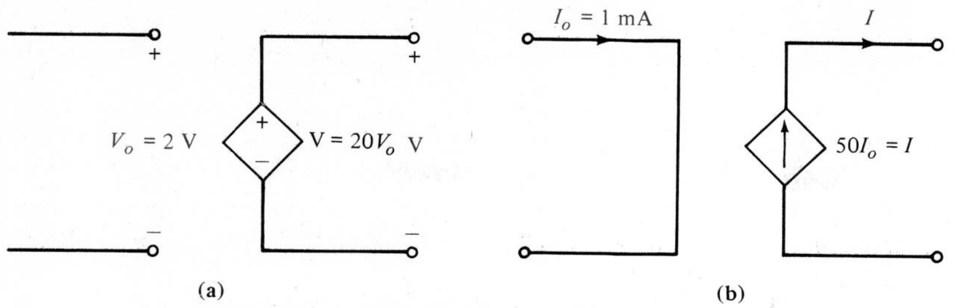

Figure 1.15 Circuits for Example 1.6.

EXAMPLE 1.6

Given the two-port networks shown in Fig. 1.15, we wish to determine the outputs.

In Fig. 1.15a the output voltage is $V = \mu V_o$ or $V = 20V_o = (20)(2\text{ V}) = 40\text{ V}$. Note that the output voltage has been amplified from 2 V at the input port to 40 V at the output port; that is, the circuit is an amplifier with an amplification factor of 20.

In Fig. 1.15b, the output current is $I = \beta I_o = (50)(1\text{ mA}) = 50\text{ mA}$; that is, the circuit has a current gain of 50, meaning that the output current is 50 times greater than the input current. ■

DRILL EXERCISE

D1.4. Determine the power supplied by the dependent sources in Fig. D1.4.

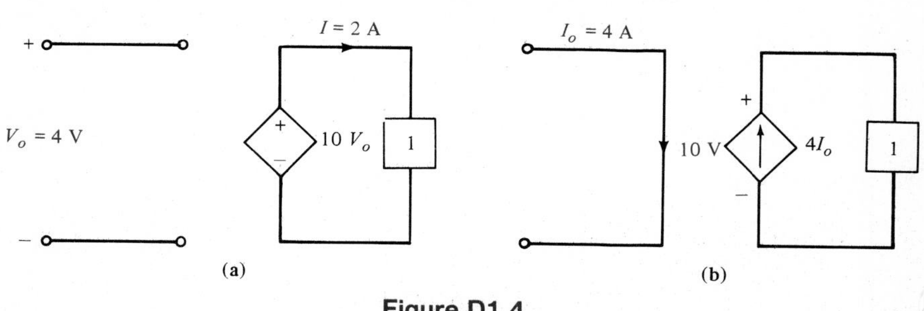

Figure D1.4

As we conclude this chapter, a number of important comments are in order. The reader must thoroughly understand our approach as outlined in Fig. 1.3 because in the following chapters we will simply define and refer to voltages and currents at specific locations within a circuit. Before any equations are written, readers should think carefully about what kinds of answers they expect. For example, if all the sources are constant, all the voltages and currents in the circuit will be constant, and therefore capital letters should be used for all the variables representing the voltages and currents. It is also useful for readers to think ahead about how variables are related to one another and define the variables so that a minimum number of minus signs are used. Although this is not important in the final solution, as indicated earlier, it does force readers to think about the problem from beginning to end before attacking it so that the solution will not come as a surprise.

1.4 Summary

We have introduced the basic strategy for an analysis of electric circuits. This strategy involves the use of linear models to represent the various circuit elements, the definition of variables used in the circuit equations, and the interpretation of the solution values of the variables to determine the actual values of the quantities which they represent in the physical network.

The system of units that has been adopted is the SI standard. Charge, current, voltage, power, and energy have been defined and the interrelationships among these quantities have been examined. Current and voltage sources, both dependent and independent, have been discussed, and the passive sign convention that will be used throughout the text has been presented.

KEY POINTS

- p = 10^{-12} k = 10^{3}
  n = 10^{-9} M = 10^{6}
  μ = 10^{-6} G = 10^{9}
  m = 10^{-3} T = 10^{12}
- The passive sign convention is defined in Fig. 1.9.
- Power is being absorbed by an element when the sign of the power is positive using the passive sign convention.
- Power is being supplied by an element when the sign of the power is negative using the passive sign convention.
- The electrical networks considered here satisfy the principle of conservation of energy.
- An ideal independent voltage (current) source is a two-terminal element that maintains a specified voltage (current) between its terminals regardless of the current (voltage) through (across) the element.
- Dependent or controlled sources generate a voltage or current that is determined by a voltage or current at a specified location in the circuit.

PROBLEMS

1.1. What charge is represented by 5,000,000 electrons?

1.2. The current entering the circuit in Fig. P1.2a is the periodic waveform shown in Fig. P1.2b. Determine the amount of charge that enters the circuit between $t = 0$ and $t = 4.5$ s.

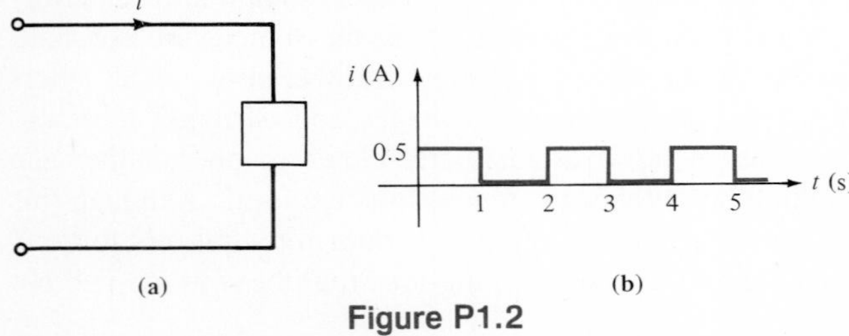

Figure P1.2

1.3. The current entering a circuit is the sawtooth pulse shown in Fig. P1.3. Determine the amount of charge that enters the circuit as a result of the current pulse.

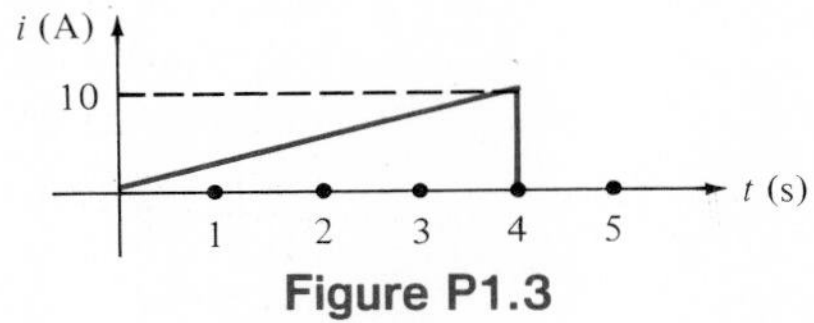

Figure P1.3

1.4. Determine how much power is absorbed by the circuit in Fig. P1.4 if
(a) $V = 10$ V and $I = 3$ A
(b) $V = 4$ V and $I = -4$ A

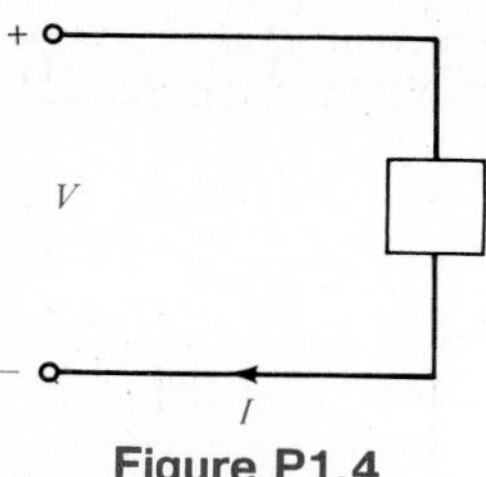

Figure P1.4

1.5. Determine how much power is supplied by the circuit in Fig. P1.4 if
(a) $V = 4$ V and $I = -2$ A
(b) $V = -6$ V and $I = -4$ A

1.6. Find the magnitude and direction of the current in the elements in Fig. P1.6.

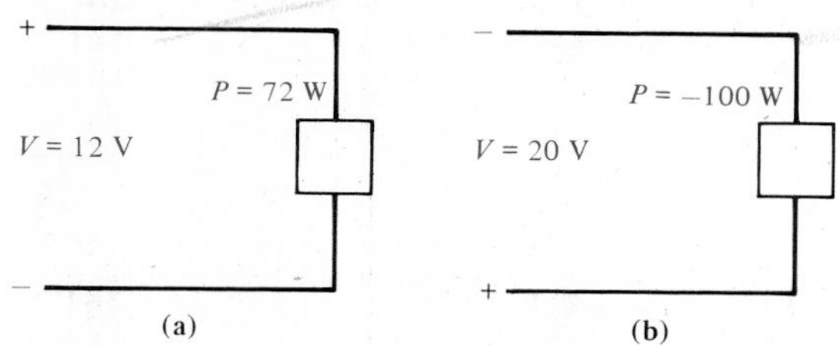

Figure P1.6

1.7. Determine the magnitude and direction of the voltage across the elements in Fig. P1.7.

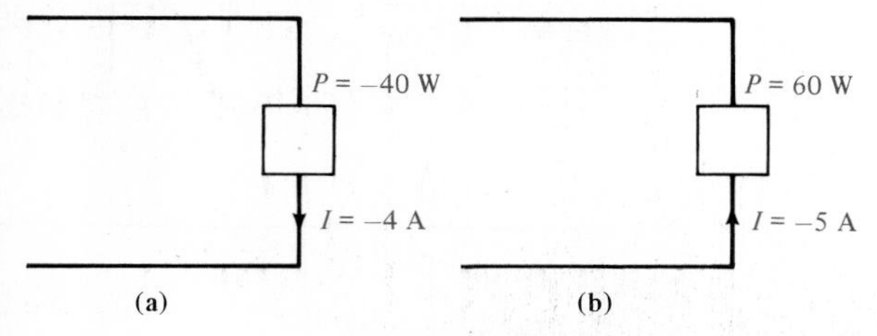

Figure P1.7

1.8. Determine the magnitude and direction of the current in the elements in Fig. P1.8.

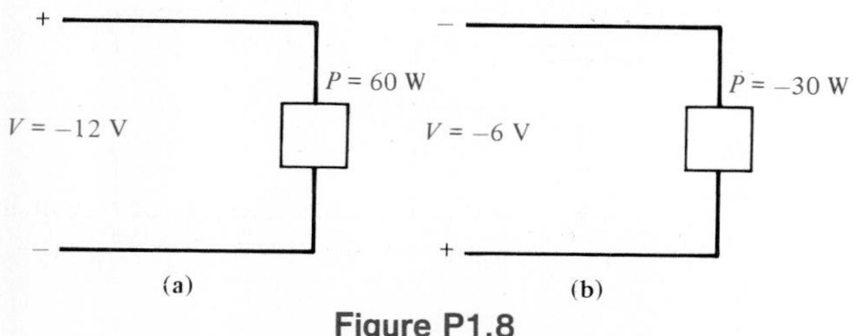

Figure P1.8

1.9. Suppose that a 12-V automobile battery that stores 6×10^6 J of energy is connected to a 5-A headlight. Determine the power delivered to the light and the total charge that has passed through the light during a 30-s interval.

1.10. The power absorbed by element 1 in the circuit in Fig. P1.10 is 20 W. Is power being absorbed or supplied by element 2, and how much?

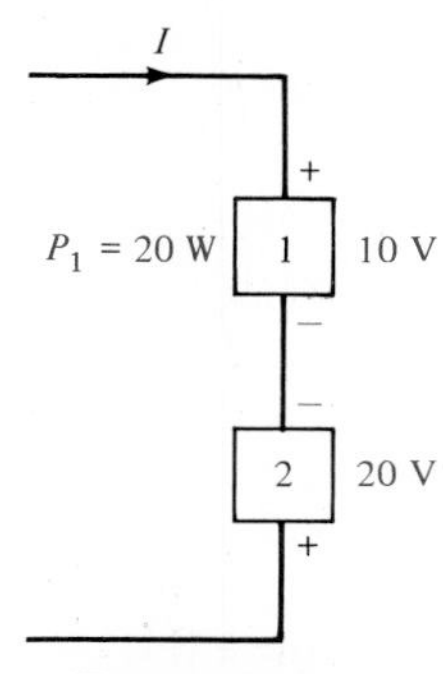

Figure P1.10

1.11. Find the power that is absorbed or supplied by the elements in the circuit in Fig. P1.11.

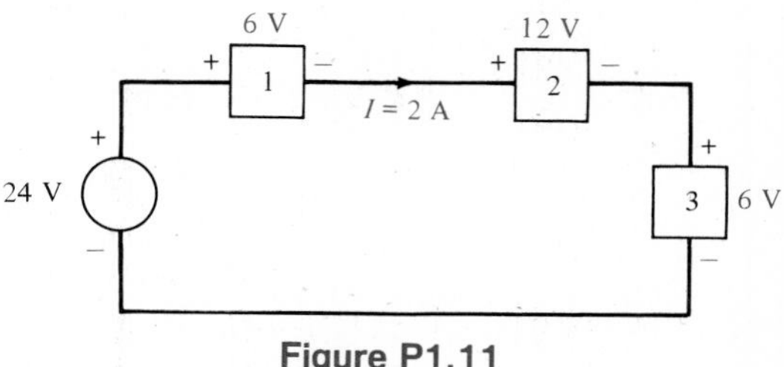

Figure P1.11

1.12. Determine the power absorbed or supplied by the elements in the circuit in Fig. P1.12.

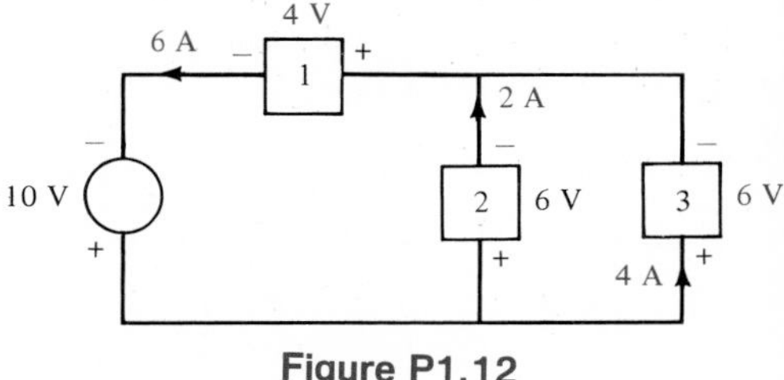

Figure P1.12

1.13. Find the power that is absorbed or supplied by the elements in the network in Fig. P1.13.

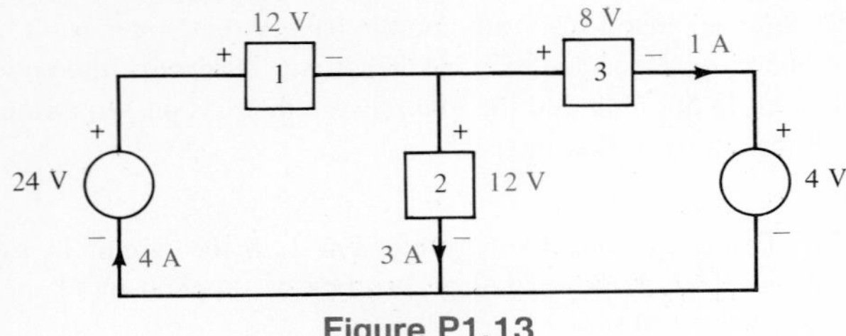

Figure P1.13

1.14. Find the power absorbed or supplied by the circuit elements in the network in Fig. P1.14.

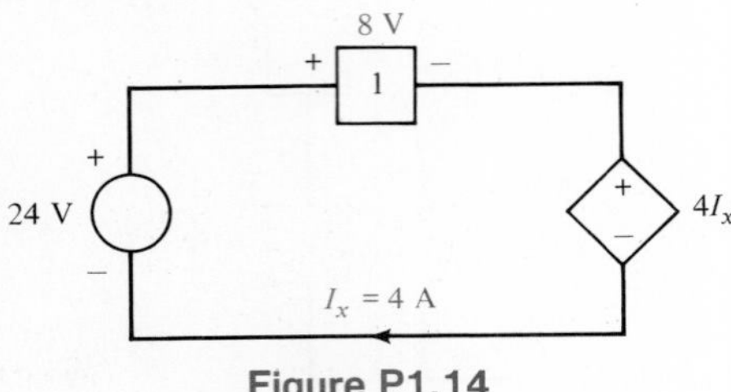

Figure P1.14

1.15. Find the power that is absorbed or supplied by the circuit elements in the network in Fig. P1.15.

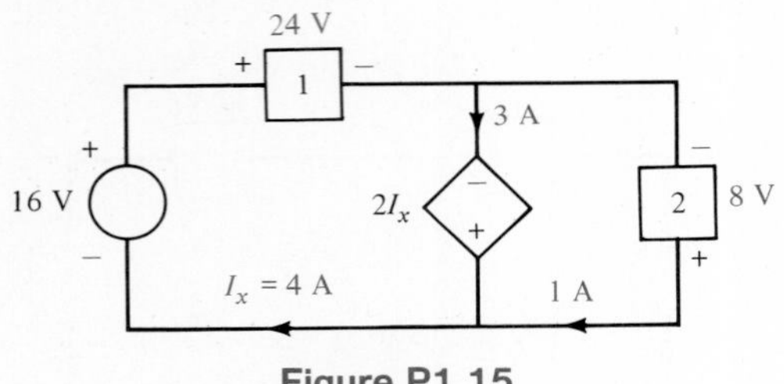

Figure P1.15

1.16. Determine the power supplied by each element of the circuit shown in Fig. P1.16.

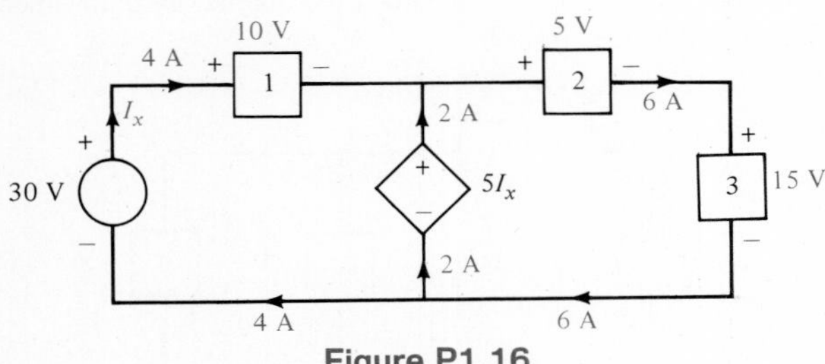

Figure P1.16

1.17. The output voltage of the dependent source shown in Fig. P1.17 is 24 V. Determine the control current, I_o.

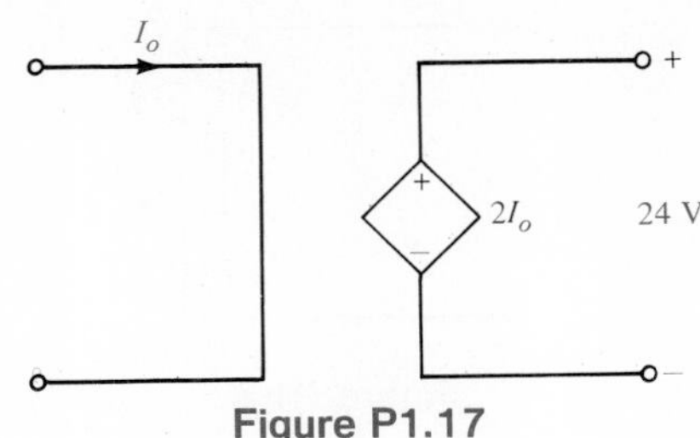

Figure P1.17

1.18. Compute the power that is absorbed or supplied by the elements in the network in Fig. P1.18.

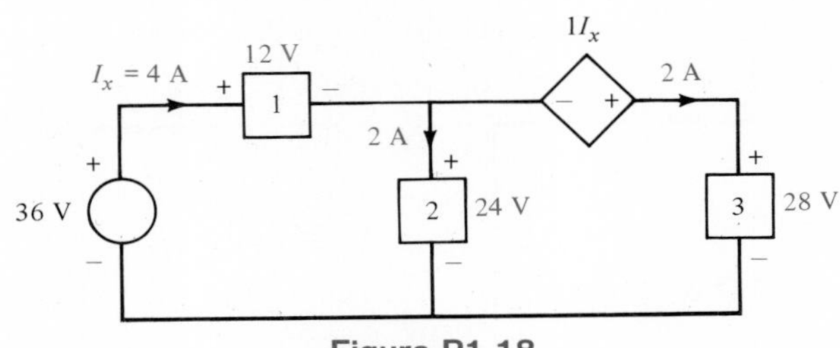

Figure P1.18

1.19. Find V_o in the network in Fig. P1.19.

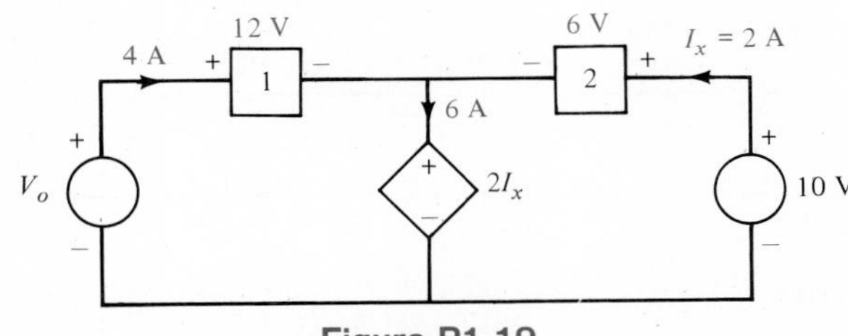

Figure P1.19

1.20. Compute I_o in the network in Fig. P1.20.

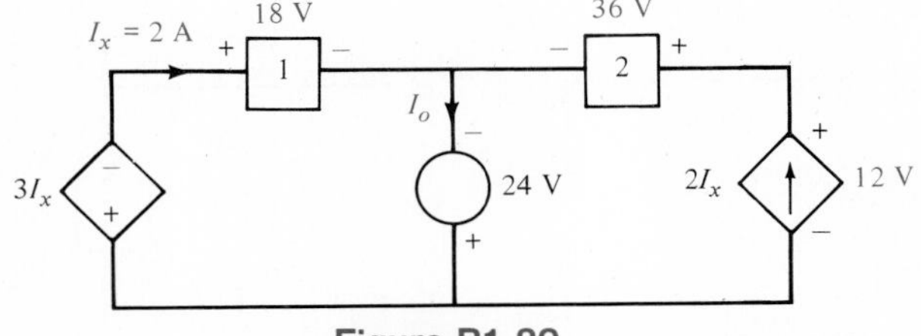

Figure P1.20

CHAPTER

2

Resistive Circuits

In this chapter we introduce some of the basic concepts and laws that are fundamental to circuit analysis. In general, we will restrict our activities to *analysis,* that is, to the determination of a specific voltage, current, or power somewhere in a network. The techniques we introduce have wide application in circuit analysis, even though we discuss them within the framework of simple networks.

2.1 Ohm's Law

Ohm's law is named for the German physicist Georg Simon Ohm, who is credited with establishing the voltage-current relationship for a resistor. As a result of his pioneering work, the unit of resistance bears his name.

Ohm's law states that *the voltage across a resistor is directly proportional to the current flowing through it*. The resistance, measured in ohms, is the constant of proportionality between the voltage and current. The symbol used to represent the resistive element is shown in Fig. 2.1, and the mathematical relationship of Ohm's law is illustrated by the equation

$$v(t) = Ri(t) \qquad \text{where } R \geq 0 \tag{2.1}$$

or equivalently, by the voltage-current characteristic shown in Fig. 2.2a. Note that we tacitly assume that the resistor has a constant value and therefore that the voltage–current characteristic is linear.

The symbol Ω is used to represent ohms, and therefore

$$1\ \Omega = 1\ \text{V/A}$$

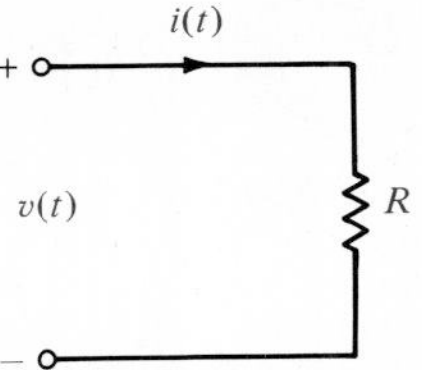

Figure 2.1 Symbol for the resistor.

Although in our analysis we will always assume that the resistors are *linear* and are thus described by a straight-line characteristic that passes through the origin, it is important that readers realize that some very useful and practical elements do exist that exhibit a *nonlinear* resistance characteristic; that is, the voltage-current relationship is not a straight line. Diodes, which are used extensively in electric circuits, are examples of nonlinear resistors. A typical diode characteristic is shown in Fig. 2.2b.

Since a resistor is a passive element, the proper current–voltage relationship is illustrated in Fig. 2.1. The power supplied to the terminals is absorbed by the resistor. Note that the charge moves from the higher to the lower potential as it passes through the resistor and the energy absorbed is dissipated by the resistor in the form of heat. As indicated in Chapter 1, the rate of energy dissipation is the instantaneous power, and therefore

$$p(t) = v(t)i(t) \tag{2.2}$$

which, using Eq. (2.1), can be written as

$$p(t) = Ri^2(t) = \frac{v^2(t)}{R} \tag{2.3}$$

This equation illustrates that the power is a nonlinear function of either current or voltage and that it is always a positive quantity.

Conductance, represented by the symbol G, is another quantity with wide application in circuit analysis. By definition, conductance is the inverse of resistance: that is,

$$G = \frac{1}{R} \tag{2.4}$$

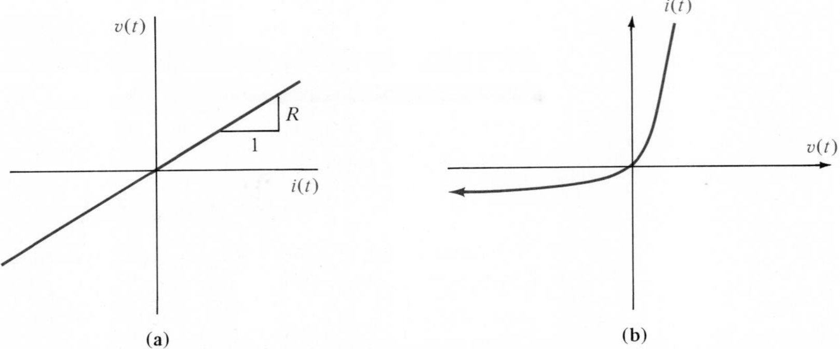

Figure 2.2 Graphical representation of the voltage-current relationship for (a) a linear resistor and (b) a diode.

The unit of conductance is the siemens, and the relationship between units

$$1 \text{ S} = 1 \text{ A/V}$$

Using Eq. (2.4), we can write two additional expressions,

$$i(t) = Gv(t) \tag{2.5}$$

and

$$p(t) = \frac{i^2(t)}{G} = Gv^2(t) \tag{2.6}$$

Equation (2.5) is another expression of Ohm's law.

Two specific values of resistance, and therefore conductance, are very important: $R = 0$ and $R = \infty$. If the resistance $R = 0$, we have what is called a *short circuit*. From Ohm's law,

$$\begin{aligned} v(t) &= Ri(t) \\ &= 0 \end{aligned}$$

Therefore, $v(t) = 0$, although the current could theoretically be any value. If the resistance $R = \infty$, we have what is called an *open circuit*, and from Ohm's law,

$$\begin{aligned} i(t) &= \frac{v(t)}{R} \\ &= 0 \end{aligned}$$

Therefore, the current is zero regardless of the value of the voltage across the open terminals.

EXAMPLE 2.1

In the circuit shown in Fig. 2.3a, determine the current and the power absorbed by the resistor.

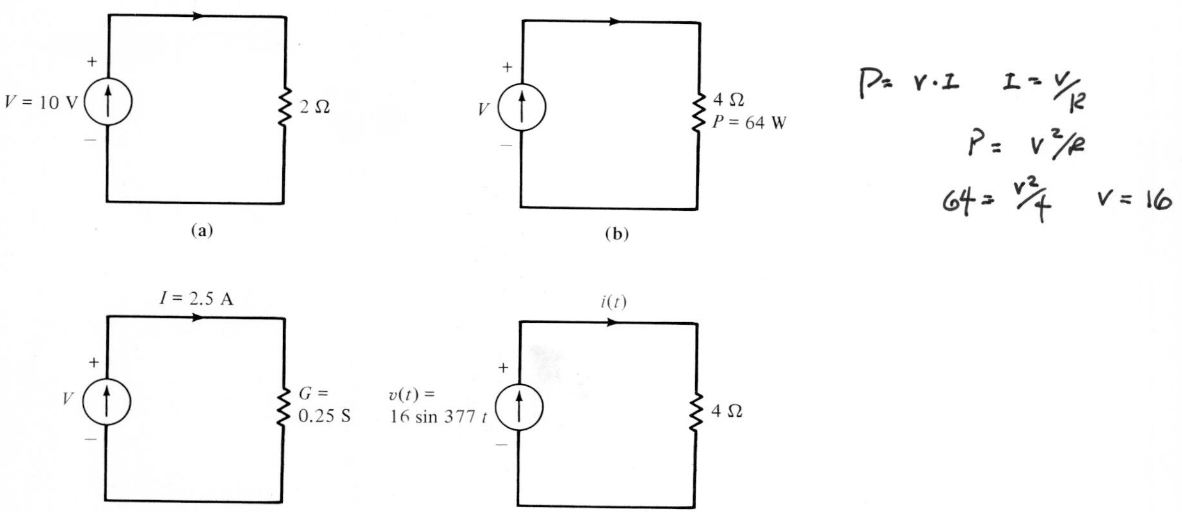

Figure 2.3 Circuits for Examples 2.1 to 2.4.

Using Eq. (2.1), we find the current to be

$$I = \frac{V}{R} = \frac{10}{2} = 5 \text{ A}$$

and from Eq. (2.2) or (2.3), the power absorbed by the resistor is

$$\begin{aligned} P &= VI = (10 \text{ V})(5 \text{ A}) = 50 \text{ W} \\ &= RI^2 = (2\ \Omega)(5 \text{ A})^2 = 50 \text{ W} \\ &= \frac{V^2}{R} = \frac{(10 \text{ V})^2}{2\ \Omega} = 50 \text{ W} \end{aligned}$$

■

EXAMPLE 2.2

The power absorbed by the 4-Ω resistor in Fig. 2.3b is 64 W. Determine the voltage and the current.

Using Eq. (2.3), we can immediately determine either of the unknowns.

$$64 \text{ W} = (4\ \Omega)I^2$$

$$4 \text{ A} = I$$

and

$$64 \text{ W} = \frac{V^2}{4\ \Omega}$$

$$16 \text{ V} = V$$

Note, however, that once I is determined, V could be derived from Ohm's law. Note carefully that $I = -4\text{A}$ and $V = -16$ V also satisfy the mathematical equations above.

■

EXAMPLE 2.3

Given the circuit in Fig. 2.3c, we wish to determine the voltage across the terminals and the power absorbed by the resistance.

From Eq. (2.5) the voltage is

$$\begin{aligned} V &= \frac{I}{G} \\ &= \frac{2.5}{0.25} = 10 \text{ V} \end{aligned}$$

The power is determined from Eq. (2.6) as

$$P = \frac{I^2}{G} = \frac{(2.5)^2}{0.25} = 25 \text{ W}$$

■

EXAMPLE 2.4

Given the circuit in Fig. 2.3d with a sinusoidal input, determine the resultant current and the power absorbed by the resistor.

From Ohm's law,

$$i(t) = \frac{v(t)}{R} = \frac{16 \sin 377t}{4} = 4 \sin 377t$$

Therefore, both the voltage and current are sinusoidal. The power is

$$p(t) = v(t)i(t) = (16 \sin 377t)(4 \sin 377t) \text{ A} = 64 \sin^2 377t \text{ W}$$

Note that although the voltage and current are negative during the intervals when the sine function is negative, the power is always a positive quantity. ■

DRILL EXERCISE

D2.1. Given the network in Fig. D2.1, determine the voltage V across the resistor, the power absorbed by the resistor, and the power supplied by the current source.

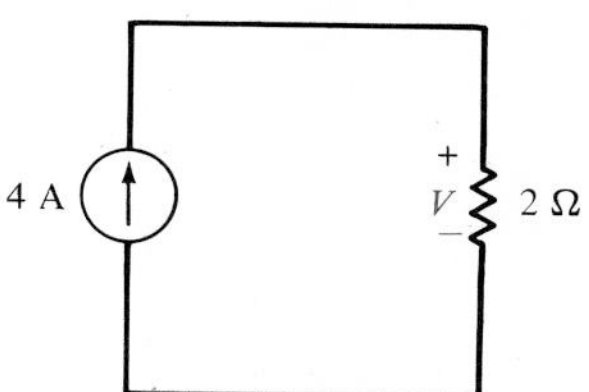

Figure D2.1

D2.2. Given the network in Fig. D2.2, determine the resistance R, the voltage V, and the power supplied by the current source.

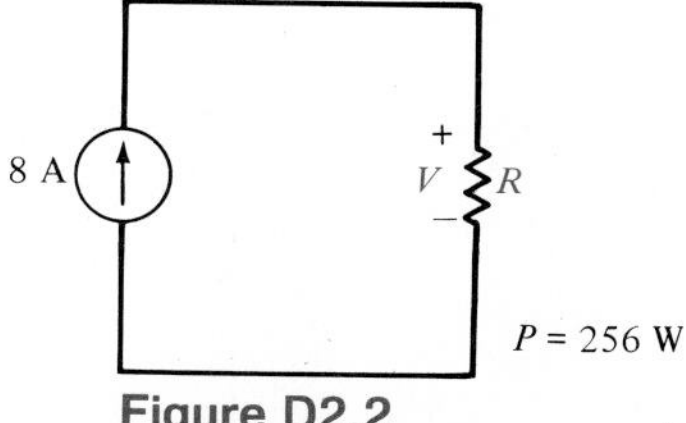

Figure D2.2

2.2 Kirchhoff's Laws

The previous circuits that we have considered have all contained a single resistor and were analyzed using Ohm's law. At this point we begin to expand our capabilities to handle more complicated networks which result from an interconnection of two or more of these simple elements. We will assume that the interconnection is performed by electrical conductors (wire) that have zero resistance: that is, perfect conductors. Because the wires have zero resistance, the energy in the circuit is in essence lumped in each element and we employ the term *lumped-parameter circuit* to describe the network.

To aid us in our discussion, we will define a number of terms that will be employed throughout our analysis. As will be our approach throughout this text, we will use examples to illustrate the concepts and define the appropriate terms. For example, the circuit shown in Fig. 2.4a will be used to describe the terms *node, loop,* and *branch.* A *node* is simply a point of connection of two or more circuit elements. The nodes in the circuit in Fig. 2.4a are exaggerated in Fig. 2.4b for clarity. The reader is cautioned to compare the two figures carefully and note that although one node can be spread out with perfect conductors, it is still only one node. For example, node 5 consists of the entire bottom connector of the circuit. In other words, if we start at some point in the circuit and move along perfect conductors in any direction until we encounter a circuit element, the total path we cover represents a single node. Therefore, we can assume that a node is one end of a circuit element together with all the perfect conductors that are attached to it. Examining the circuit, we note that there are numerous paths through it.

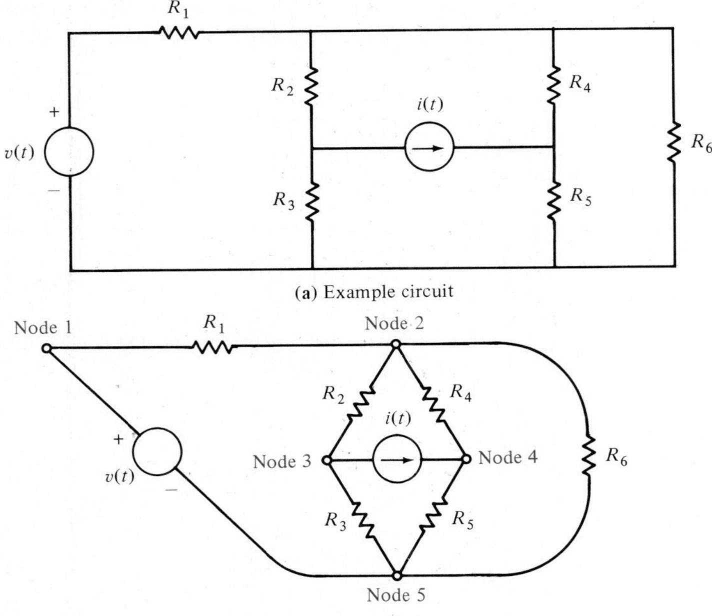

(b) Circuit in (a) with nodes illustrated

Figure 2.4 Circuits used to illustrate terms.

A *loop* is simply any *closed path* through the circuit in which no node is encountered more than once. For example, starting from node 1, one loop would contain the elements R_1, R_2, R_3, and $v(t)$; another loop would contain R_2, $i(t)$, R_5, and R_6, and so on. Finally, a *branch* is a portion of a circuit containing only a single element and the nodes at each end of the element. The circuit in Fig. 2.4 contains eight branches.

Given the previous definitions, we are now in a position to consider Kirchhoff's laws, named after the German scientist Gustav Robert Kirchhoff. These two laws are quite simple but extremely important. We will not attempt to prove them because the proofs are beyond our current level of understanding. However, we will demonstrate their usefulness and attempt to make the reader proficient in their use. The first law is *Kirchhoff's current law,* which states that *the algebraic sum of the currents entering any node is zero*. In mathematical form the law appears as

$$\sum_{j=1}^{N} i_j(t) = 0 \tag{2.7}$$

where $i_j(t)$ is the jth current entering the node through branch j and N is the number of branches connected to the node. To understand the use of this law, consider the node shown in Fig. 2.5. Applying Kirchhoff's current law to this node yields

$$i_1(t) + [-i_2(t)] + i_3(t) + i_4(t) + [-i_5(t)] = 0$$

We have assumed that the algebraic signs of the currents entering the node are positive and therefore that the signs of the currents leaving the node are negative.

If we multiply the foregoing equation by -1, we obtain the expression

$$-i_1(t) + i_2(t) - i_3(t) - i_4(t) + i_5(t) = 0$$

which simply states that *the algebraic sum of the currents leaving a node is zero*. Alternatively, we can write the equation as

$$i_1(t) + i_3(t) + i_4(t) = i_2(t) + i_5(t)$$

which states that *the sum of the currents entering a node is equal to the sum of the currents leaving the node*. Both of these italicized expressions are alternative forms of Kirchhoff's current law.

Note carefully that Kirchhoff's current law states that the *algebraic* sum of the currents either entering or leaving a node must be zero. We now begin to see why we stated

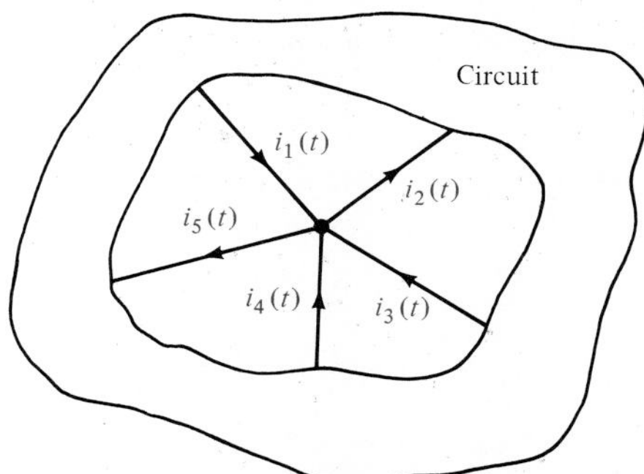

Figure 2.5 Currents at a node.

in Chapter 1 that it is critically important to specify both the magnitude and the direction of a current.

EXAMPLE 2.5

Using Kirchhoff's current law, we want to determine the value of I_1 in Fig. 2.6a.

The algebraic sum of the currents entering the node is

$$I_1 + 5 - 2 - 1 = 0$$

or

$$I_1 = -2 \text{ A}$$

This equation indicates that the magnitude of I_1 is 2 A, but the direction is opposite to that which we have assumed, and therefore I_1 is actually 2 A leaving the node. ■

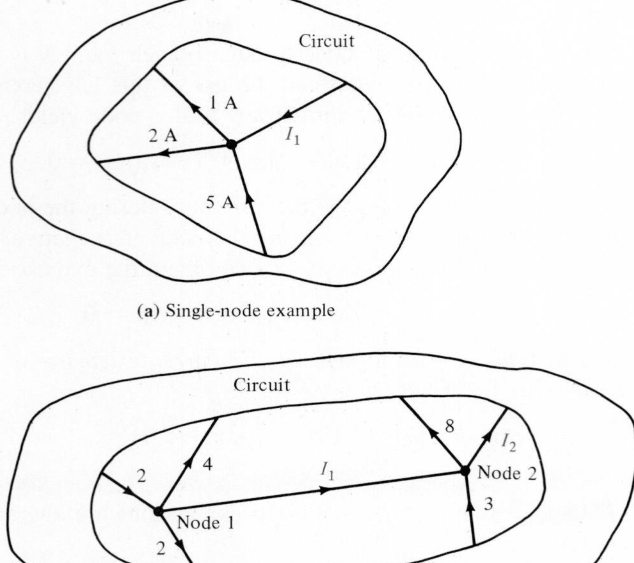

(a) Single-node example

(b) Multiple-node example

Figure 2.6 Illustration nodes for Kirchhoff's current law.

EXAMPLE 2.6

In Fig. 2.6b, we wish to determine the currents I_1 and I_2.

The equations for Kirchhoff's current law at nodes 1 and 2, respectively, are

$$2 - 4 - 2 - I_1 = 0$$

$$I_1 - 8 + 3 - I_2 = 0$$

From the first equation, $I_1 = -4$ A. Since it was assumed that I_1 was leaving node 1, the negative sign illustrates that the current is actually entering node 1. Using this value of I_1 in the second equation, we find that $I_2 = -9$ A. ■

Finally, it is possible to generalize Kirchhoff's current law to include a closed surface. By a closed surface we mean some set of elements completely contained within the surface that are interconnected. Since the current entering each element within the surface is equal to that leaving the element (i.e., the element stores no net charge), it follows that the current entering an interconnection of elements is equal to that leaving the surface. Therefore, Kirchhoff's current law can also be stated: *The algebraic sum of the currents entering any closed surface is zero.*

EXAMPLE 2.7

To illustrate this generalization of Kirchhoff's current law stated above, we need only consider the multiple-node arrangement discussed in Example 2.6 and illustrated in Fig. 2.6b.

Now we apply Kirchhoff's current law to the surface. Assuming that the currents entering the surface are positive and those leaving the surface are negative, we can write

$$2 - 4 - 8 - I_2 + 3 - 2 = 0$$

$$I_2 = -9 \text{ A}$$

which is, of course, what we obtained for I_2 in Example 2.6. Note however, that we did not need to solve for I_1 to determine I_2. ■

DRILL EXERCISE

D2.3. Find I_1 in the network in Fig. D2.3.

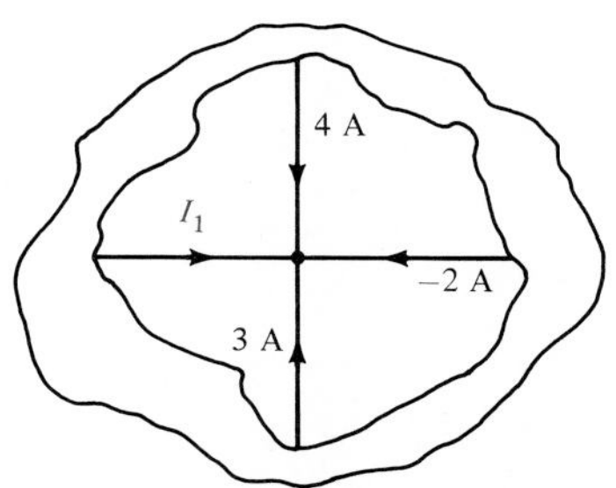

Figure D2.3

D2.4. Find the currents I_1 and I_2 in the network in Fig. D2.4.

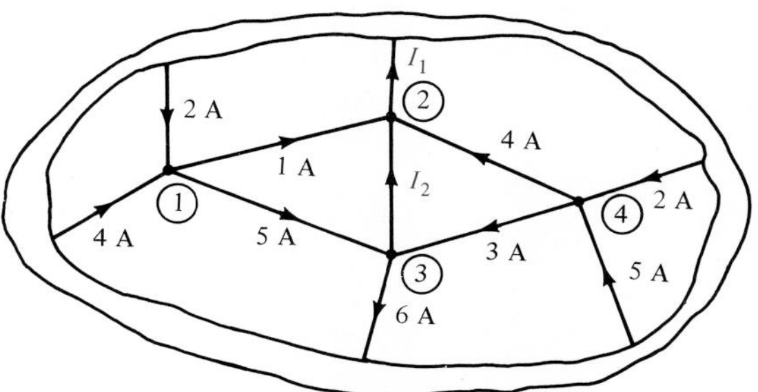

Figure D2.4

Kirchhoff's second law, called *Kirchhoff's voltage law,* states that *the algebraic sum of the voltages around any closed path is zero*. As was the case with Kirchhoff's current law, we will defer the proof of this law and concentrate on understanding how to apply it. Once again the reader is cautioned to remember that we are dealing only with lumped-parameter circuits. These circuits are conservative, meaning that the work required to move a unit charge around any closed path is zero.

Recall that in Kirchhoff's current law, the algebraic sign was required to keep track of whether the currents were entering or leaving a node. In Kirchhoff's voltage law the algebraic sign is used to keep track of the voltage polarity. In other words, as we traverse the circuit, it is necessary to sum to zero the increases and decreases in energy level. Therefore, it is important that we keep track of whether the energy level is increasing or decreasing as we go through each element. To begin, as we traverse the circuit we will consider an increase in potential (energy level) as positive and a decrease in potential (energy level) as negative and set the sum of them equal to zero. Later, for convenience in the solutions, we will put some of the terms on one side of the equal sign and some terms on the other side.

We will now employ several examples to illustrate the use of this law.

EXAMPLE 2.8

Consider the circuit shown in Fig. 2.7. If we start at point a and traverse the clockwise path $abcda$ summing the voltages, we obtain

$$-V_{R_1} - V_2 - V_{R_2} + V_1 = 0$$

or

$$-V_{R_1} - 6 - V_{R_2} + 12 = 0$$

which is Kirchhoff's voltage law for this circuit. Note that if we had taken the counter-clockwise direction, we would obtain

$$-V_1 + V_{R_2} + V_2 + V_{R_1} = 0$$

Multiplying this equation by -1 yields the same equation as that obtained in traversing the circuit in a clockwise direction. This illustrates that the application of Kirchhoff's voltage law is independent of the direction in which we traverse the circuit.

Now suppose that V_{R_1} is known to be 2 V; then the equations can be used to solve

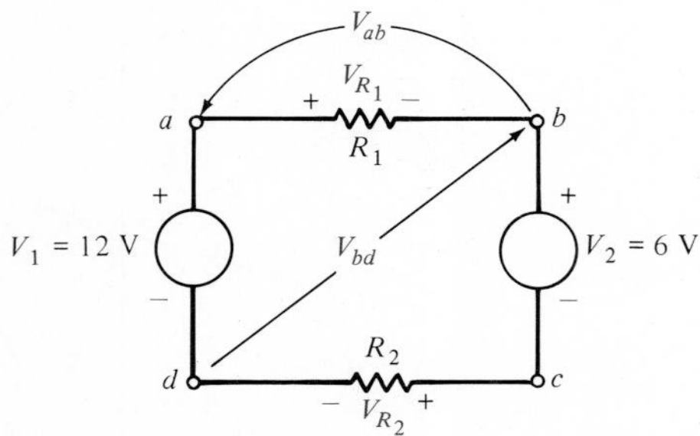

Figure 2.7 Circuit to illustrate Kirchhoff's voltage law.

for V_{R_2}. In the first equation above,

$$-2 - 6 - V_{R_2} + 12 = 0$$

$$V_{R_2} = 4 \text{ V}$$

Finally, we employ the convention V_{ab} to indicate the voltage of point a with respect to point b: that is, the variable for the voltage between point a and point b with point a as the positive reference. Since the potential is measured between two points, it is convenient to use an arrow between the two points with the head of the arrow located at the positive reference. Note that the double subscript notation, the + and − notation, and the single-headed arrow notation are all the same if the head of the arrow is pointing toward the positive terminal and the first subscript in the double subscript notation. Note that V_{bd} is shown as a single-headed arrow in Fig. 2.7, and in the same figure $V_{ab} = V_{R_1}$ is a voltage of 2 V, $V_{bc} = V_2$ is a voltage of 6 V, $V_{cd} = V_{R_2}$ is 4 V, and $V_{da} = -V_1$ is -12 V. Using this convention, we can apply Kirchhoff's voltage law to the circuit to determine the voltage between any two points. For example, suppose that we want to determine the voltage V_{bd} using the path $dabd$. We can write the equation

$$V_{da} + V_{ab} + V_{bd} = 0$$

$$-12 + 2 + V_{bd} = 0$$

$$V_{bd} = 10 \text{ V}$$

We could also use the path $dbcd$ and obtain

$$V_{db} + V_{bc} + V_{cd} = 0$$

$$-V_{bd} + 6 + 4 = 0$$

$$V_{bd} = 10 \text{ V}$$

■

EXAMPLE 2.9

Consider now the circuit shown in Fig. 2.8. In addition to the voltage and sources, we assume that we know the voltage across the resistors R_2 and R_3. Note that the voltage V_{R_1} is the only unknown voltage in the first loop and, therefore, applying Kirchhoff's voltage law to this loop will yield this unknown.

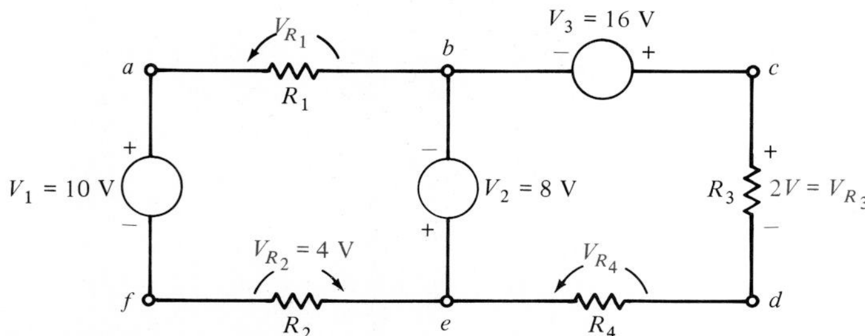

Figure 2.8 Another circuit to illustrate Kirchhoff's voltage law.

Traversing the path *abefa*, we obtain

$$-V_{R_1} + V_2 - V_{R_2} + V_1 = 0$$

$$-V_{R_1} + 8 - 4 + 10 = 0$$

$$V_{R_1} = 14 \text{ V}$$

Since V_{R_4} is the only unknown voltage in the second loop, we can use this loop to determine this unknown. Traversing the path *ebcde*, we obtain

$$-V_2 + V_3 - V_{R_3} + V_{R_4} = 0$$

$$-8 + 16 - 2 + V_{R_4} = 0$$

$$V_{R_4} = -6 \text{ V}$$

Note that although V_{ed} was assumed to be positive, it is actually negative, since using our definition, V_{R_4} was found to be negative. Since Kirchhoff's voltage law can be applied to any closed path, we can also use the path *abcdefa* to determine V_{R_4}.

Traversing this path, we obtain

$$-V_{R_1} + V_3 - V_{R_3} + V_{R_4} - V_{R_2} + V_1 = 0$$

$$-14 + 16 - 2 + V_{R_4} - 4 + 10 = 0$$

$$V_{R_4} = -6 \text{ V}$$ ■

In general, the mathematical representation of Kirchhoff's voltage law is

$$\sum_{j=1}^{N} v_j(t) = 0 \tag{2.8}$$

where $v_j(t)$ is the voltage across the jth branch (with the proper reference direction) in a loop containing N voltages. This expression is analogous to Eq. (2.7) for Kirchhoff's current law.

DRILL EXERCISE

D2.5. In the network in Fig. D2.5, V_{R_1} is known to be 4 V. Find V_{R_2} and V_{bd}.

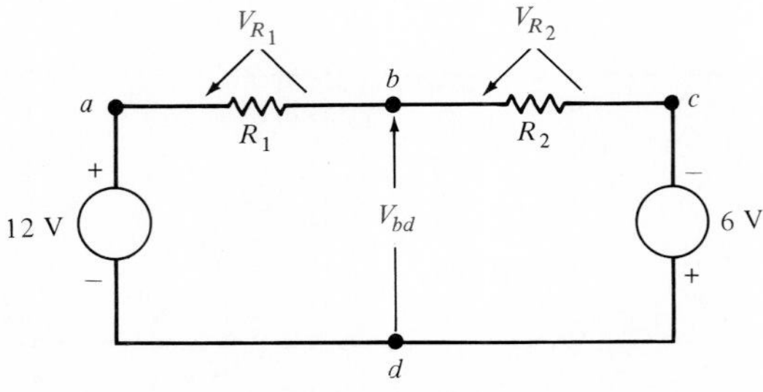

Figure D2.5

D2.6. In the network in Fig. D2.6, $V_{de} = 4$ V and $V_{bd} = 6$ V. Find V_{ab} and V_{cd}.

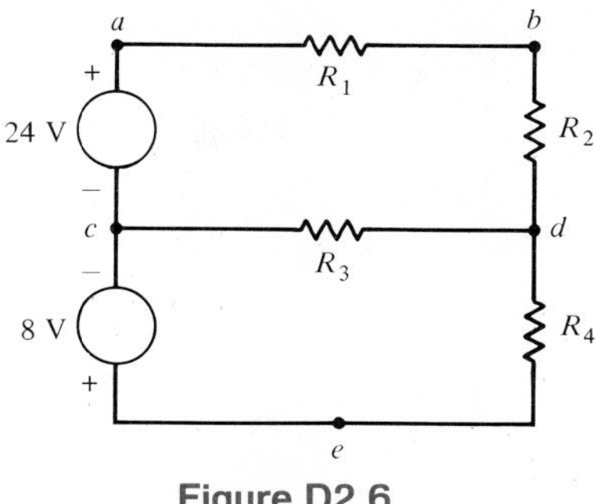

Figure D2.6

2.3 Single-Loop Circuits

At this point we can begin to apply the laws we have presented earlier to the analysis of simple circuits. To begin, we examine what is perhaps the simplest circuit—a simple closed path, or loop, of elements. The elements of a single loop carry the same current and therefore are said to be in *series*. However, we will apply Kirchhoff's voltage law and Ohm's law to the circuit to determine various quantities in the circuit.

Our approach will be to begin with a simple circuit and then generalize the analysis to more complicated ones. The circuit shown in Fig. 2.9 will serve as a basis for discussion. This circuit consists of an independent voltage source that is in series with two resistors. We have assumed that the current flows in a clockwise direction. If this assumption is correct, the solution of the equations that yields the current will produce a positive value. If the current is actually flowing in the opposite direction, the value of the current variable will simply be negative, indicating that the current is flowing in a direction opposite to that assumed. We have also made voltage polarity assignments for v_{R_1} and v_{R_2}. These assignments have been made using the convention employed in our discussion of Ohm's law and our choice for the direction of $i(t)$, that is, the convention shown in Fig. 2.1.

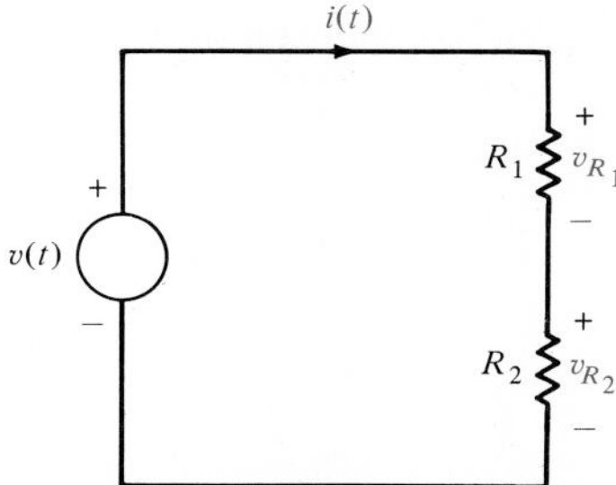

Figure 2.9 Single-loop circuit.

Applying Kirchhoff's voltage law to this circuit yields

$$+v(t) - v_{R_1} - v_{R_2} = 0$$

or

$$v(t) = v_{R_1} + v_{R_2}$$

However, from Ohm's law we know that

$$v_{R_1} = R_1 i(t)$$

$$v_{R_2} = R_2\, i(t)$$

Therefore,

$$v(t) = R_1 i(t) + R_2 i(t)$$

Solving the equation for $i(t)$ yields

$$i(t) = \frac{v(t)}{R_1 + R_2} \tag{2.9}$$

Knowing the current, we can now apply Ohm's law to determine the voltage across each resistor:

$$\begin{aligned} v_{R_1} &= R_1 i(t) \\ &= \frac{R_1}{R_1 + R_2}\, v(t) \end{aligned} \tag{2.10}$$

Similarly,

$$v_{R_2} = \frac{R_2}{R_1 + R_2}\, v(t) \tag{2.11}$$

Note that the equations satisfy Kirchhoff's voltage law, since

$$+v(t) - \frac{R_1}{R_1 + R_2}\, v(t) - \frac{R_2}{R_1 + R_2}\, v(t) = 0$$

Although simple, Eqs. (2.10) and (2.11) are very important because they describe the operation of what is called a *voltage divider*. In other words, *the source $v(t)$ is divided between the resistors R_1 and R_2 in direct proportion to their resistances.*

EXAMPLE 2.10

Consider the circuit shown in Fig. 2.10. The circuit is identical to Fig. 2.9 except that R_1 is a variable resistor such as the volume control for a radio or television set. Suppose that $V = 24$ V, $R_1 = 10\ \Omega$, and $R_2 = 2\ \Omega$.
Using Eq. (2.11), we find that the voltage V_1 is

$$\begin{aligned} V_1 &= \frac{R_2}{R_1 + R_2}\, V \\ &= \frac{2}{10 + 2}\, 24 \\ &= 4\ \text{V} \end{aligned}$$

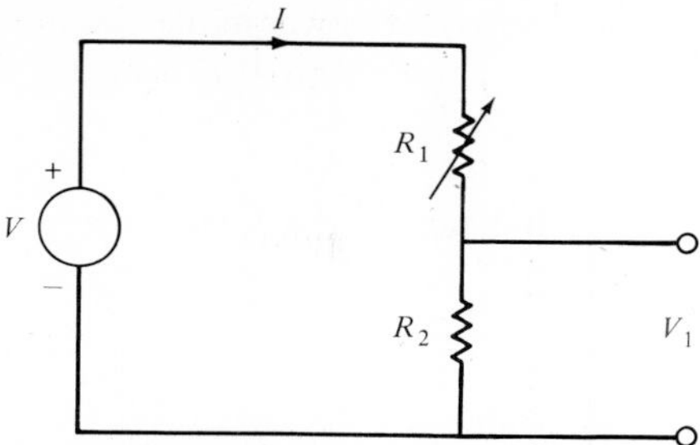

Figure 2.10 Voltage-divider circuit.

Now suppose that the variable resistor R_1 is changed from 10 to 0.4 Ω. Then

$$V_1 = \frac{2}{0.4 + 2} 24$$

$$= 20 \text{ V}$$

Note that the use of Eq. (2.11) is equivalent to determining the current I and then using Ohm's law to find V_1. Note that the largest voltage is across the largest resistor. This voltage-divider concept and the simple circuit we have employed to describe it are very useful because, as will be shown later, more complicated circuits can be reduced to this form. ■

Let us now consider the power relationship that exists in the circuit of Fig. 2.9. The instantaneous power delivered by the voltage source is

$$p(t) = v(t)i(t)$$

and the instantaneous power absorbed by resistors R_1 and R_2 is

$$p_1(t) = \frac{v_{R_1}^2(t)}{R_1} = \frac{R_1}{(R_1 + R_2)^2} v^2(t)$$

and

$$p_2(t) = \frac{v_{R_2}^2(t)}{R_2} = \frac{R_2}{(R_1 + R_2)^2} v^2(t)$$

respectively. Now note that

$$\begin{aligned} p_1(t) + p_2(t) &= \frac{R_1}{(R_1 + R_2)^2} v^2(t) + \frac{R_2}{(R_1 + R_3)^2} v^2(t) \\ &= \frac{v^2(t)}{R_1 + R_2} \\ &= \frac{v(t)}{R_1 + R_2} v(t) \\ &= v(t)i(t) \\ &= p(t) \end{aligned}$$

This analysis illustrates the conservation of power in the circuit, since the power supplied by the voltage source is completely absorbed by two resistors.

EXAMPLE 2.11

Determine the instantaneous power absorbed in the resistor R_2 of Example 2.10 when $R_1 = 10\ \Omega$ and when $R_1 = 0.4\ \Omega$.

$$P_2 = \frac{R_2}{(R_1 + R_2)^2} V^2$$

For $R_1 = 10\ \Omega$,

$$P_2 = \frac{2}{(10 + 2)^2} (24)^2$$
$$= 8 \text{ W}$$

and for $R_1 = 0.4\ \Omega$,

$$P_2 = \frac{2}{(0.4 + 2)^2} (24)^2$$
$$= 200 \text{ W}$$

■

At this point we wish to extend our analysis to include a multiplicity of voltage sources and resistors. For example, consider the circuit shown in Fig. 2.11a. Here we have assumed that the current flows in a clockwise direction and we have defined the variable $i(t)$ accordingly. This may or may not be the case, depending on the value of the various voltage sources. Kirchhoff's voltage law for this circuit is

$$-v_{R_1} - v_2(t) + v_3(t) - v_{R_2} - v_4(t) - v_5(t) + v_1(t) = 0$$

or using Ohm's law,

$$(R_1 + R_2)i(t) = v_1(t) - v_2(t) + v_3(t) - v_4(t) - v_5(t)$$

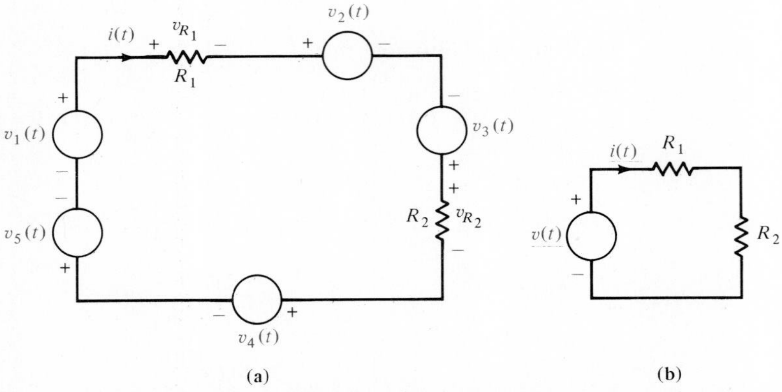

Figure 2.11 Equivalent circuits with multiple sources.

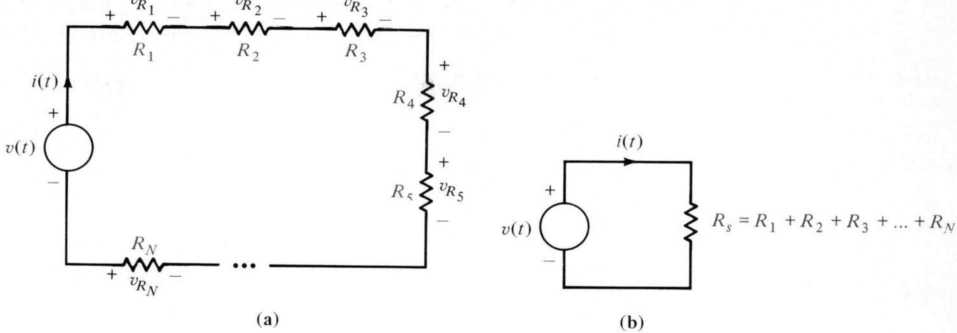

Figure 2.12 Equivalent circuits.

which can be written as

$$(R_1 + R_2)i(t) = v(t)$$

where

$$v(t) = v_1(t) + v_3(t) - [v_2(t) + v_4(t) + v_5(t)]$$

so that under the definitions above, Fig. 2.11a is equivalent to Fig. 2.11b. In other words, the sum of several voltage sources in series can be replaced by one source whose value is the algebraic sum of the individual sources. This analysis can, of course, be generalized to a circuit with N series sources.

Now consider the circuit with n resistors in series as shown in Fig. 2.12a. Applying Kirchhoff's voltage law to this circuit yields

$$\begin{aligned} v(t) &= v_{R_1} + v_{R_2} + \cdots + v_{R_N} \\ &= R_1 i(t) + R_2 i(t) + \cdots + R_N i(t) \end{aligned}$$

and therefore,

$$v(t) = R_s i(t) \tag{2.12}$$

where

$$R_s + R_1 + R_2 + \cdots + R_N \tag{2.13}$$

and hence

$$i(t) = \frac{v(t)}{R_s} \tag{2.14}$$

Note also that for any resistor R_i in the circuit, the voltage across R_i is given by the expression

$$v_{R_i} = \frac{R_i}{R_s} v(t) \tag{2.15}$$

which is the voltage-division property for multiple resistors in series.

Equation (2.13) illustrates that *the equivalent resistance of N resistors in series is simply the sum of the individual resistances*. Thus, using Eqs. (2.13) and (2.14), we can draw the circuit in Fig. 2.12b as an equivalent circuit for the one in Fig. 2.12a.

The instantaneous power absorbed by the series combination of N resistors is

$$\begin{aligned} p(t) &= \left[\frac{v(t)}{R_s}\right]^2 R_1 + \left[\frac{v(t)}{R_s}\right]^2 R_2 + \cdots + \left[\frac{v(t)}{R_s}\right]^2 R_N \\ &= \frac{R_1}{R_s^2} v^2(t) + \frac{R_2}{R_s^2} v^2(t) + \cdots + \frac{R_N}{R_s^2} v^2(t) \\ &= \frac{v^2(t)}{R_s} \\ &= v(t)i(t) \end{aligned}$$

which is the power supplied by the voltage source.

EXAMPLE 2.12

Given the circuit in Fig. 2.13, we wish to determine the current I, the power absorbed by the resistor R_2, and an equivalent circuit.

Assuming that the current flows in the counterclockwise direction and defining the current variable accordingly, Kirchhoff's voltage law yields the equation

$$-36 - 7I - 3I + 12 - 2I = 0$$

or

$$\begin{aligned} (7 + 3 + 2)I &= 12 - 36 \\ I &= -2 \text{ A} \end{aligned}$$

Therefore, the current is 2 A, but it flows in a clockwise direction. The power absorbed by the resistor R_2 is

$$\begin{aligned} P &= I^2R_2 \\ &= (2)^2(3) \\ &= 12 \text{ W} \end{aligned}$$

The equivalent circuit is as shown in Fig. 2.12b, where

$$V = 24 \text{ V}, \quad R_s = 12\ \Omega, \quad \text{and} \quad I = 2 \text{ A}.$$

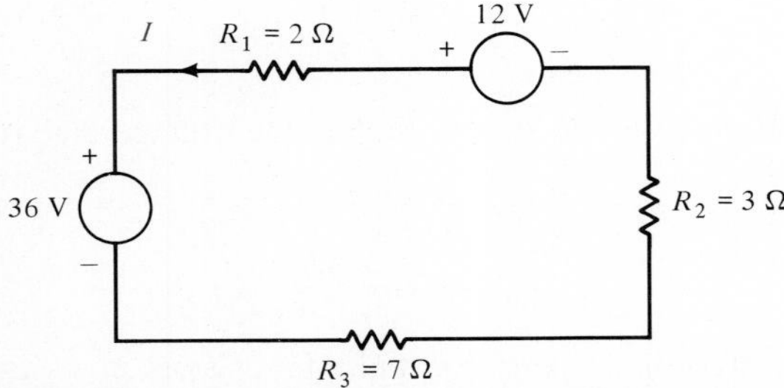

Figure 2.13 Example circuit with multiple sources and resistors. ■

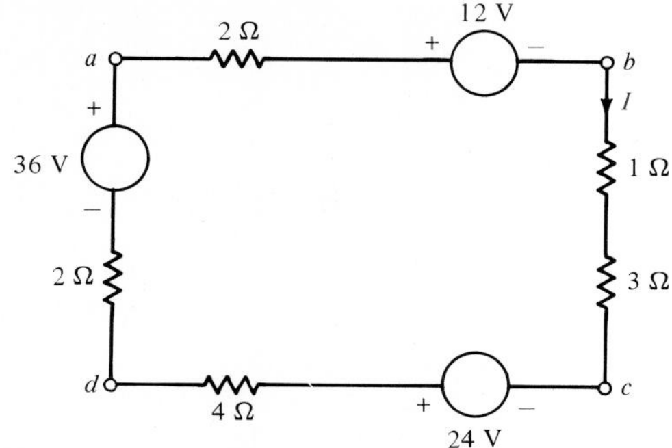

Figure 2.14 Another example circuit with multiple sources and resistors.

EXAMPLE 2.13

Given the circuit in Fig. 2.14, determine the current I and find the voltage V_{ac}.

We can write Kirchhoff's voltage law by inspection. In this case we put the voltage sources on the right-hand side of the equation and assign a positive sign to them if they aid the assumed direction of current flow and a negative sign if they oppose the assumed direction of the current. The left side of the equation is simply the voltage across the resistors. Therefore,

$$I(2 + 2 + 1 + 3 + 4) = 36 - 12 + 24$$

or

$$I = \frac{48}{12}$$

$$= 4 \text{ A}$$

The voltage V_{ac} can be computed using Kirchhoff's voltage law either via the route through b or d. Since the current is in the clockwise direction, the voltage V_{ac} via terminal b is

$$V_{ac} = +(2)(4) + 12 + (1)(4) + 3(4)$$

$$= 36 \text{ V}$$

V_{ac} can also be computed via the path through d. Paying close attention to the direction of the voltage across the resistors caused by the clockwise current flow, we obtain

$$V_{ac} = 36 - (2)(4) - (4)(4) + 24$$

$$= 36 \text{ V}$$

■

EXAMPLE 2.14

The voltage V_A across the 2-Ω resistor in Fig. 2.15 is known to be 8 V. Let us determine the voltages V_1 and V_o.

Upon using Ohm's law, the current in the 2-Ω resistor is found to be

$$V_A = I(2)$$

$$8 = I(2)$$

$$I = 4 \text{ A}$$

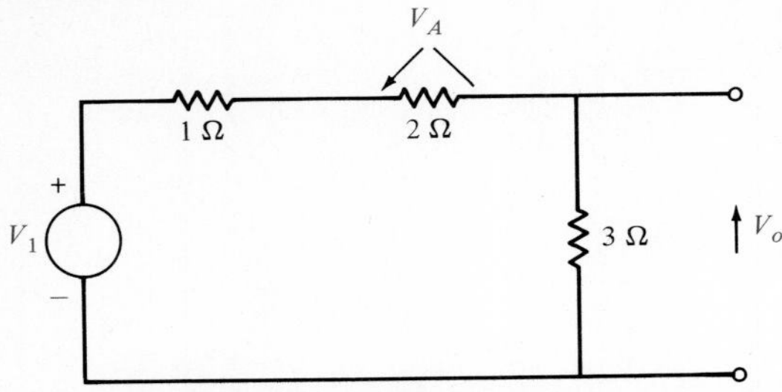

Figure 2.15 Simple series circuit.

Since V_A is positive on the left side of the 2-Ω resistor, the current flows from left to right in this resistor. The current I flows down through the 2-Ω resistor and hence

$$V_o = I(3)$$
$$= 12 \text{ V}$$

Applying Kirchhoff's voltage law around the entire loop yields

$$+V_1 - I(1) + I(2) - I(3) = 0$$
$$V_1 = (6)I$$
$$= 24 \text{ V}$$

or equivalently, $V_1 = I(1) + V_A + V_o$. ■

DRILL EXERCISE

D2.7. Find V_o in the network in Fig. D2.7.

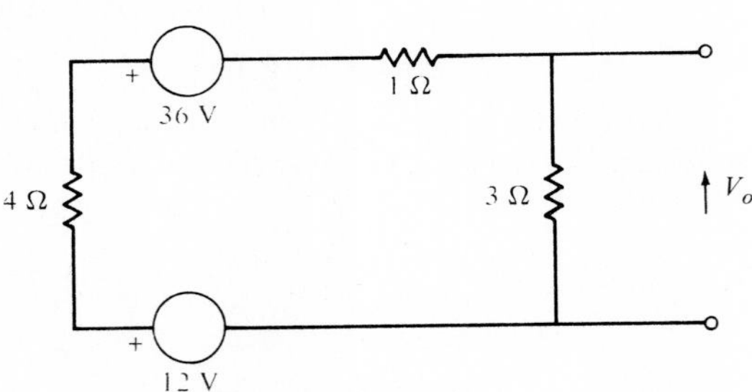

Figure D2.7

D2.8. Given the network in Fig. D2.8, find the current I, the power absorbed by the 5-Ω resistor, V_{bd}, and V_{be}.

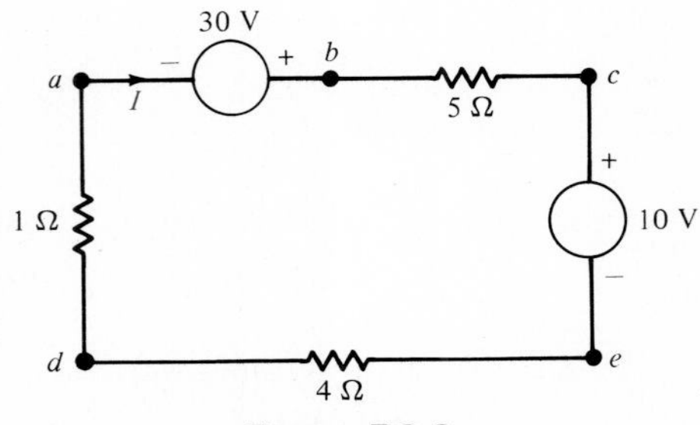

Figure D2.8

D2.9. Find V_o in the network in Fig. D2.9.

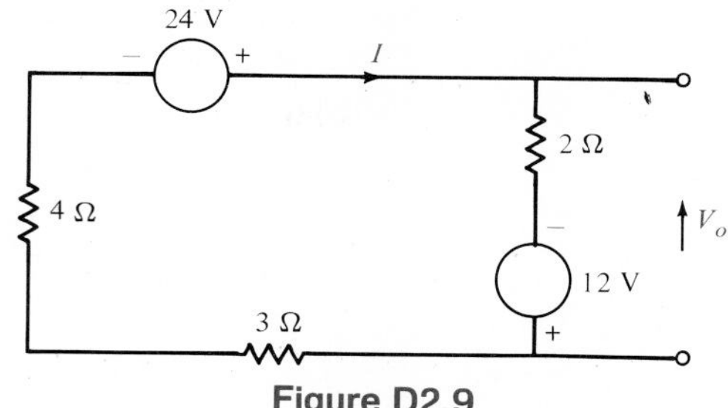

Figure D2.9

D2.10. V_o is known to be 4 V in the network in Fig. D2.10. Find V_2.

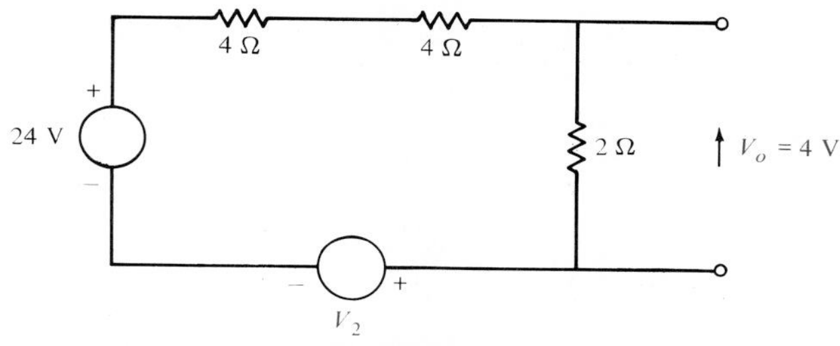

Figure D2.10

2.4 Single-Node-Pair Circuits

An important circuit is the single-node-pair circuit. In this case the elements have the same voltage across them, and therefore are in *parallel*. We will, however, apply Kirchhoff's current law and Ohm's law to determine various unknown quantities in the circuit.

Following our approach with the single-loop circuit, we will begin with the simplest case and then generalize our analysis. Consider the circuit shown in Fig. 2.16. Here we have an independent current source in parallel with two resistors. Although this circuit plainly has two nodes, one at the top and one at the bottom, the bottom terminal is simply a reference node—sometimes considered as ground. Therefore, we often refer to this circuit, in which all the elements are in parallel, as a single-node circuit.

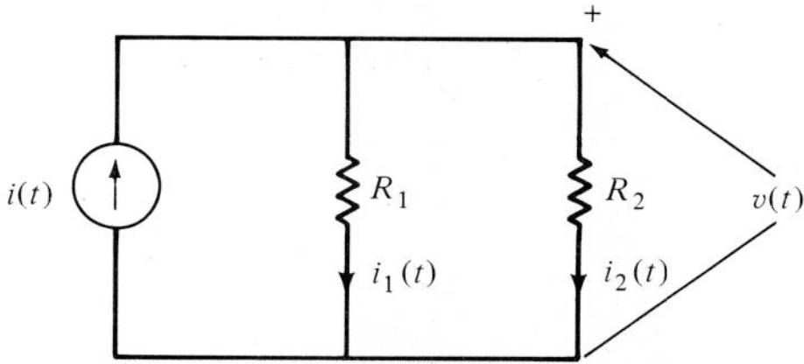

Figure 2.16 Simple parallel circuit.

Since all of the circuit elements are in parallel, the voltage $v(t)$ appears across all of them. The currents $i_1(t)$ and $i_2(t)$ are arbitrarily assigned as shown in the figure. Now applying Kirchhoff's current law to the upper node, we obtain

$$i(t) = i_1(t) + i_2(t)$$

Now, employing Ohm's law, we have

$$\begin{aligned} i(t) &= \frac{v(t)}{R_1} + \frac{v(t)}{R_2} \\ &= \left(\frac{1}{R_1} + \frac{1}{R_2}\right)v(t) \\ &= \frac{v(t)}{R_p} \end{aligned}$$

where

$$\frac{1}{R_p} = \frac{1}{R_1} + \frac{1}{R_2} \tag{2.16}$$

$$R_p = \frac{R_1 R_2}{R_1 + R_2} \tag{2.17}$$

Therefore, the equivalent resistance of two resistors connected in parallel is equal to the product of their resistances divided by their sum. Note also that this equivalent resistance R_p is always less than either R_1 or R_2. Hence by connecting resistors in parallel we reduce the overall resistance. In the special case when $R_1 = R_2$, the equivalent resistance is equal to half of the value of the individual resistors.

The manner in which the current $i(t)$ from the source divides between the two branches is called *current division* and can be found from the expressions above. For example,

$$\begin{aligned} v(t) &= R_p i(t) \\ &= \frac{R_1 R_2}{R_1 + R_2} i(t) \end{aligned} \tag{2.18}$$

and

$$\begin{aligned} i_1(t) &= \frac{v(t)}{R_1} \\ &= \frac{R_2}{R_1 + R_2} i(t) \end{aligned} \tag{2.19}$$

and

$$\begin{aligned} i_2(t) &= \frac{v(t)}{R_2} \\ &= \frac{R_1}{R_1 + R_2} i(t) \end{aligned} \tag{2.20}$$

Equations (2.19) and (2.20) are mathematical statements of the current-division rule. *Therefore, the current divides in inverse proportion to the resistances.* In other words, to determine the current in the branch containing R_1 we multiply the incoming current $i(t)$ by the opposite resistance R_2 and divide that product by the sum of the two resistors.

If we employ conductance $G_i = 1/R_i$ instead of resistance in the analysis, we can show that

$$i(t) = (G_1 + G_2)v(t)$$

$$G_p = G_1 + G_2 \tag{2.21}$$

and hence that

$$i_1(t) = \frac{G_1}{G_1 + G_2} i(t) \tag{2.22}$$

$$i_2(t) = \frac{G_2}{G_1 + G_2} i(t) \tag{2.23}$$

Note that in this case the current divides as the ratio of the path conductance to the total conductance.

EXAMPLE 2.15

Consider the circuit shown in Fig. 2.17a. The equivalent circuit is shown in Fig. 2.17b. Given the information on the circuit, we wish to determine the currents and equivalent resistance.

The equivalent resistance for the circuit is

$$\begin{aligned} R_p &= \frac{R_1R_2}{R_1 + R_2} \\ &= \frac{(3)(6)}{3 + 6} \\ &= 2\ \Omega \end{aligned}$$

Now V can be calculated as

$$\begin{aligned} V &= R_pI \\ &= (2)(12) \\ &= 24\ \text{V} \end{aligned}$$

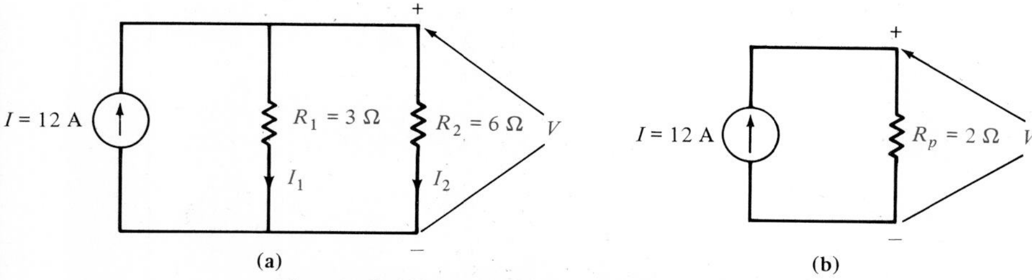

Figure 2.17 Example of a parallel circuit.

Once the voltage V is known, Ohm's law can be used to calculate the currents I_1 and I_2.

$$I_1 = \frac{V}{R_1} = \frac{24}{3} = 8 \text{ A}$$

and

$$I_2 = \frac{V}{R_2} = \frac{24}{6} = 4 \text{ A}$$

Note that these currents satisfy Kirchhoff's current law at both the upper and lower (reference) nodes.

$$I = I_1 + I_2$$

$$12 \text{ A} = 8 \text{ A} + 4 \text{ A}$$

We can also apply current division to determine I_1 and I_2. For example,

$$I_1 = \frac{R_2}{R_1 + R_2} I = \frac{6}{3 + 6}(12) = 8 \text{ A}$$

and

$$I_2 = \frac{3}{3 + 6}(12) = 4 \text{ A}$$

Note that the largest current flows through the smallest resistor, and vice versa. In addition, one should note that if R_1 and R_2 are equal, the current will divide equally between them. ■

The power delivered by the current source in Fig. 2.16 is

$$p(t) = v(t)i(t)$$

The power absorbed by the two resistors is

$$
\begin{aligned}
p_1(t) + p_2(t) &= i_1^2(t)R_1 + i_2^2(t)R_2 \\
&= \left[\frac{R_2 i(t)}{R_1 + R_2}\right]^2 R_1 + \left[\frac{R_1 i(t)}{R_1 + R_2}\right]^2 R_2 \\
&= \frac{R_2^2 R_1 + R_1^2 R_2}{(R_1 + R_2)^2}\, i^2(t) \\
&= \left[\frac{R_1 R_2}{R_1 + R_2}\, i(t)\right] i(t) \\
&= v(t)i(t)
\end{aligned}
$$

which is, of course, the power delivered by the current source.

EXAMPLE 2.16

Determine the instantaneous power supplied by the source and absorbed by each resistor in the circuit analyzed in Example 2.15.

The power supplied is

$$
\begin{aligned}
P &= VI \\
&= (24)(12) \\
&= 288 \text{ W}
\end{aligned}
$$

The power absorbed in R_1 and R_2 is, respectively,

$$
\begin{aligned}
P_1 &= I_1^2 R_1 \\
&= (8)^2(3) \\
&= 192 \text{ W}
\end{aligned}
$$

and

$$
\begin{aligned}
P_2 &= I_2^2 R_2 \\
&= (4)^2(6) \\
&= 96 \text{ W}
\end{aligned}
$$

and

$$
\begin{aligned}
P &= P_1 + P_2 \\
&= 192 + 96 \\
&= 288 \text{ W}
\end{aligned}
$$

■

Let us now extend our analysis to include a multiplicity of current sources and resistors. For example, consider the circuit shown in Fig. 2.18a. We have assumed that

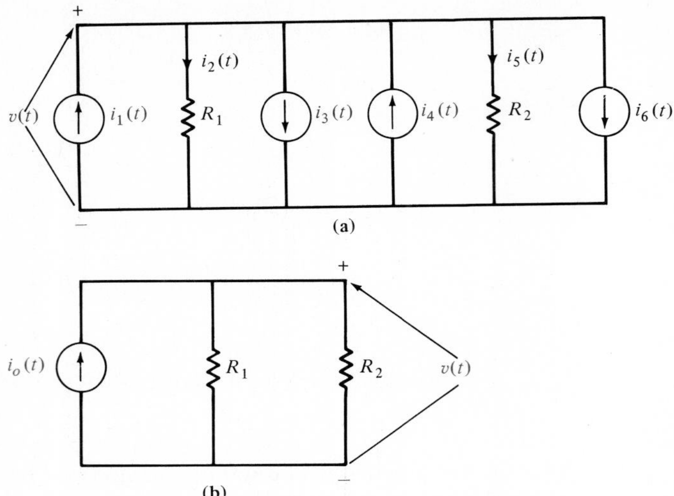

Figure 2.18 Equivalent circuits.

the upper node is $v(t)$ volts positive with respect to the lower node. Applying Kirchhoff's current law to the upper node yields

$$i_1(t) - i_2(t) - i_3(t) + i_4(t) - i_5(t) - i_6(t) = 0$$

or

$$i_1(t) - i_3(t) + i_4(t) - i_6(t) = i_2(t) + i_5(t)$$

which is equivalent to

$$\begin{aligned} i_o(t) &= \left(\frac{1}{R_1} + \frac{1}{R_2}\right)v(t) \\ &= \frac{R_1 R_2}{R_1 + R_2}\,v(t) \end{aligned}$$

where

$$i_o(t) = i_1(t) - i_3(t) + i_4(t) - i_6(t)$$

so that under the definitions above, the circuit in Fig. 2.18b is equivalent to that in Fig. 2.18a. We could, of course, generalize this analysis to a circuit with N current sources.

Now consider the circuit with N resistors in parallel as shown in Fig. 2.19a. Applying Kirchhoff's current law to the upper node yields

$$\begin{aligned} i_o(t) &= i_1(t) + i_2(t) + \cdots i_N(t) \\ &= \left(\frac{1}{R_1} + \frac{1}{R_2} + \cdots + \frac{1}{R_N}\right)v(t) \end{aligned} \tag{2.24}$$

or

$$= (G_1 + G_2 + \cdots + G_N)v(t) \tag{2.25}$$

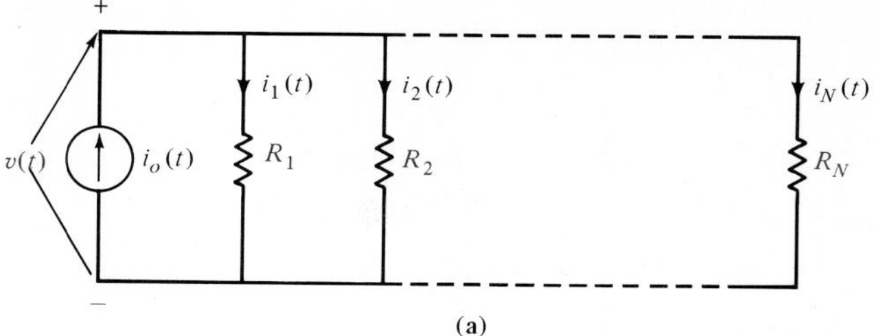

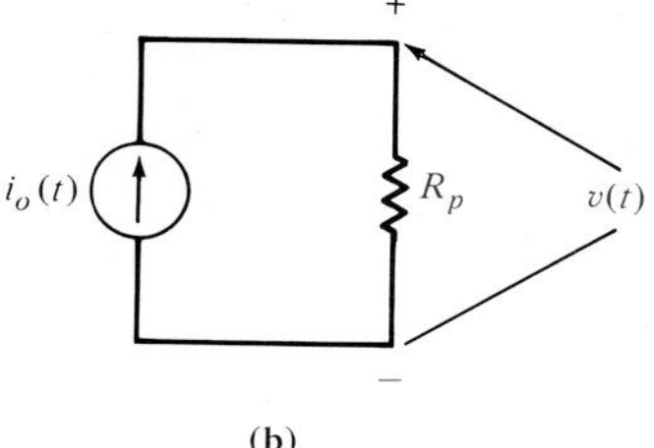

Figure 2.19 Equivalent circuits.

These two equations can be rewritten as

$$i_o(t) = \frac{v(t)}{R_p} \tag{2.26}$$

$$= G_p v(t) \tag{2.27}$$

where

$$\frac{1}{R_p} = \sum_{i=1}^{N} \frac{1}{R_i} \tag{2.28}$$

and

$$G_p = \sum_{i=1}^{N} G_i \tag{2.29}$$

so that Fig. 2.19a can be reduced to an equivalent circuit as shown in Fig. 2.19b.

The current division for any branch can be calculated using Ohm's law and the equations above. For example, for the jth branch in the network of Fig. 2.19a,

$$i_j(t) = \frac{v(t)}{R_j}$$

using Eq. (2.26), we obtain

$$i_j(t) = \frac{R_p}{R_j} i_o(t) \tag{2.30}$$

In a similar manner we find that

$$i_j(t) = \frac{G_j}{G_p} i_o(t) \tag{2.31}$$

Equations (2.30) and (2.31) define the current-division rule for the general case.

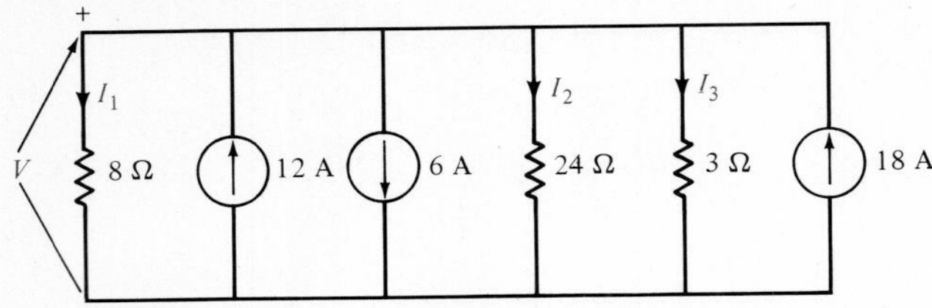

Figure 2.20 Single-node example circuit with multiple sources and resistors.

EXAMPLE 2.17

Given the circuit in Fig. 2.20, we wish to determine the voltage V, the currents in each resistor, and an equivalent circuit.

Employing Kirchhoff's current law, we obtain

$$\left(\frac{1}{8} + \frac{1}{24} + \frac{1}{3}\right)V = 12 - 6 + 18$$

$$\frac{1}{2}V = 24$$

$$V = 48 \text{ V}$$

Then

$$I_1 = \frac{48}{8}$$

$$= 6 \text{ A}$$

$$I_2 = \frac{48}{24}$$

$$= 2 \text{ A}$$

and

$$I_3 = \frac{48}{3}$$

$$= 16 \text{ A}$$

Now applying Kirchhoff's current law at the upper node yields

$$-6 + 12 - 6 - 2 - 16 + 18 = 0$$

Since the equivalent source is a current of 24 A entering the upper node and the equivalent resistance is $R_p = 2\ \Omega$, the equivalent circuit consists of a 24-A current source parallel with a 2-Ω resistor. ■

EXAMPLE 2.18

In the circuit in Fig. 2.21, the power absorbed by the 6-Ω resistor is 24 W. We wish to determine the value of the current source I_o.

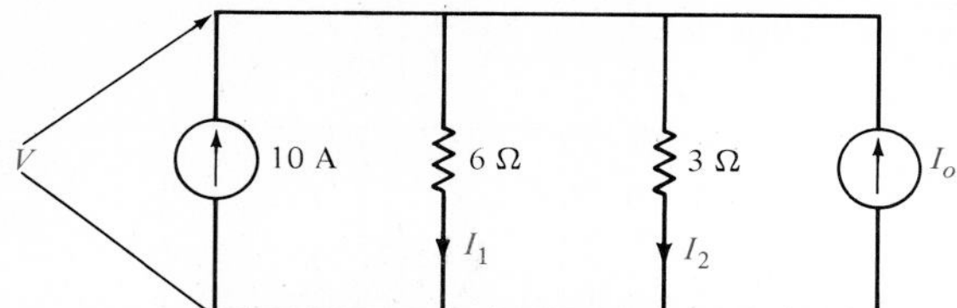

Figure 2.21 Another single-node circuit with multiple sources and resistors.

Since

$$P = I^2R$$

then

$$24 = I_1^2(6)$$

and hence

$$I_1 = \pm 2 \text{ A}$$

Therefore, the voltage V is

$$V = I_1(6) = \pm 12 \text{ V}$$

The current I_2 can now be found using Ohm's law as

$$I_2 = \frac{V}{3}$$

$$= \pm\frac{12}{3}$$

$$= \pm 4 \text{ A}$$

Now applying Kirchhoff's current law at the upper node yields

$$10 - 2 - 4 + I_o = 0 \qquad \text{or} \qquad 10 + 2 + 4 + I_o = 0$$

$$I_o = -4 \text{ A} \qquad\qquad I_o = -16 \text{ A}$$

■

DRILL EXERCISE

D2.11. Given the network in Fig. D2.11, determine the power absorbed by the 6-Ω resistor.

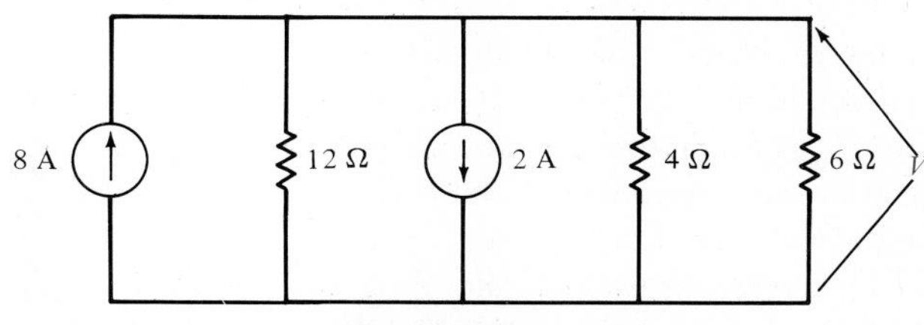

Figure D2.11

D2.12. In the network in Fig. D2.12, $V = 12$ V. Find I_o.

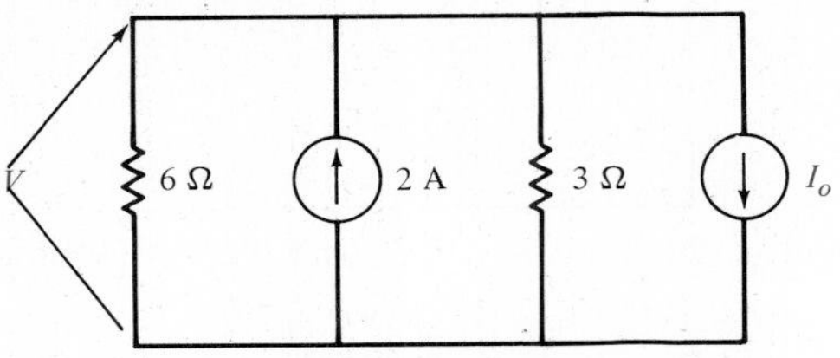

Figure D2.12

D2.13. In the network in Fig. D2.13, $I_2 = 4$ A. Find I_o.

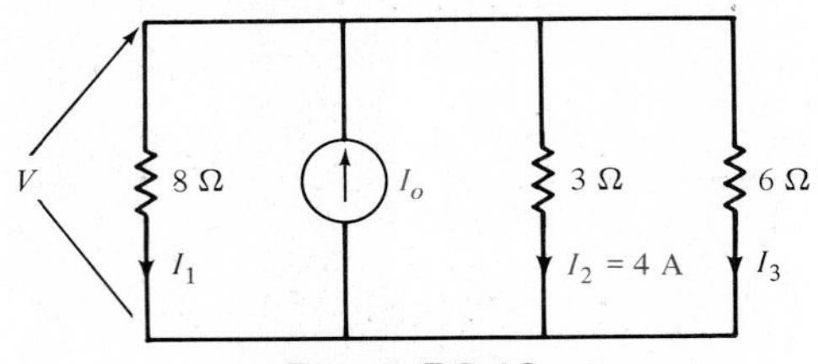

Figure D2.13

2.5 Series and Parallel Resistor Combinations

We have shown in our earlier developments that the equivalent resistance of N resistors in series is

$$R_s = R_1 + R_2 + \cdots + R_N$$

and the equivalent resistance of N resistors in parallel is

$$\frac{1}{R_p} = \frac{1}{R_1} + \frac{1}{R_2} + \cdots + \frac{1}{R_N}$$

Let us now examine some combinations of these two cases.

EXAMPLE 2.19

Let us determine the resistance at terminals A-B of the network shown in Fig. 2.22a. To determine the equivalent resistance at A-B, we begin at the opposite end of the network and combine resistors as we progress toward terminals A-B. The 1-, 2-, and 3-Ω resistors connected in series between terminals E and F are equivalent to one 6-Ω resistor, which in turn is in parallel with the 12-Ω resistor. This parallel combination is equivalent to one 4-Ω resistor connected between E and F as shown in Fig. 2.22b. The two 1-Ω resistors and the 4-Ω resistor are in series, and this combination is in parallel with the 3-Ω resistor. Combining these resistors reduces the network to that shown in Fig. 2.22c. Therefore, the resistance at terminals A-B is 14 Ω. ■

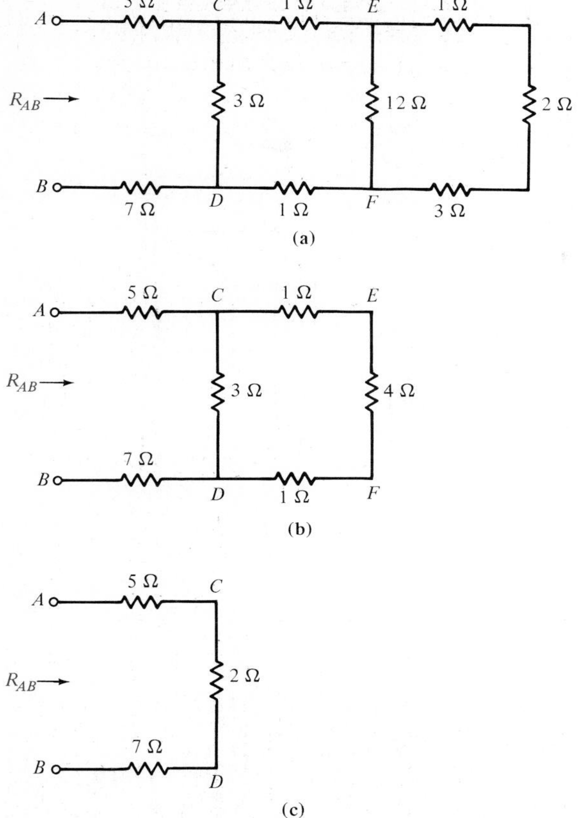

Figure 2.22 Simplification of a resistance network.

EXAMPLE 2.20

We wish to determine the resistance at terminals *A-B* in the network in Fig. 2.23a. Once again starting at the opposite end of the network from the terminals and combining resistors as shown in the sequence of circuits in Fig. 2.23, we find that the equivalent resistance at the terminals is 5 Ω. ■

DRILL EXERCISE

D2.14. Find the equivalent resistance at the terminals of the network in Fig. D2.14.

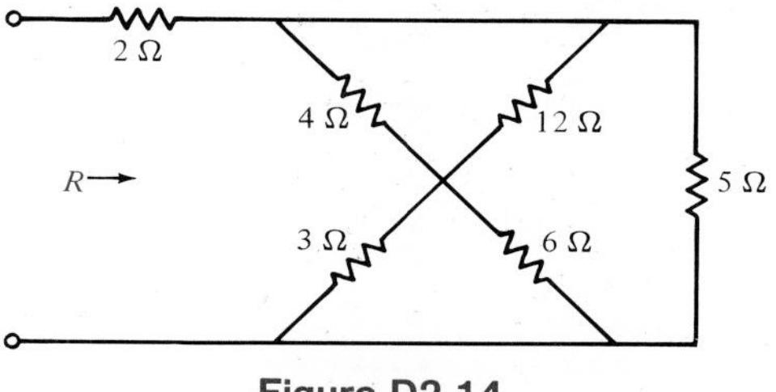

Figure D2.14

D2.15. Compute the resistance at the terminals of the network in Fig. D2.15.

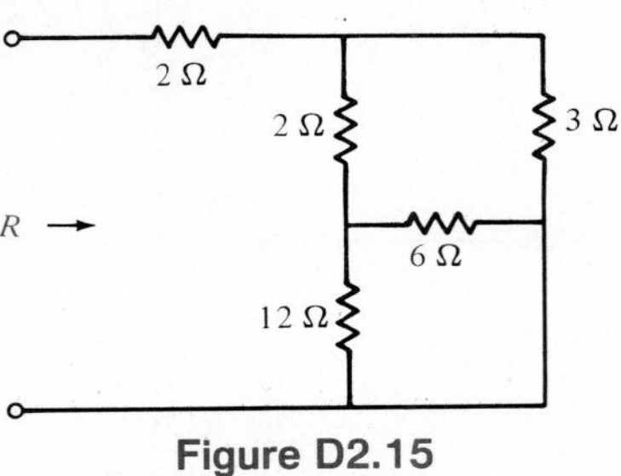

Figure D2.15

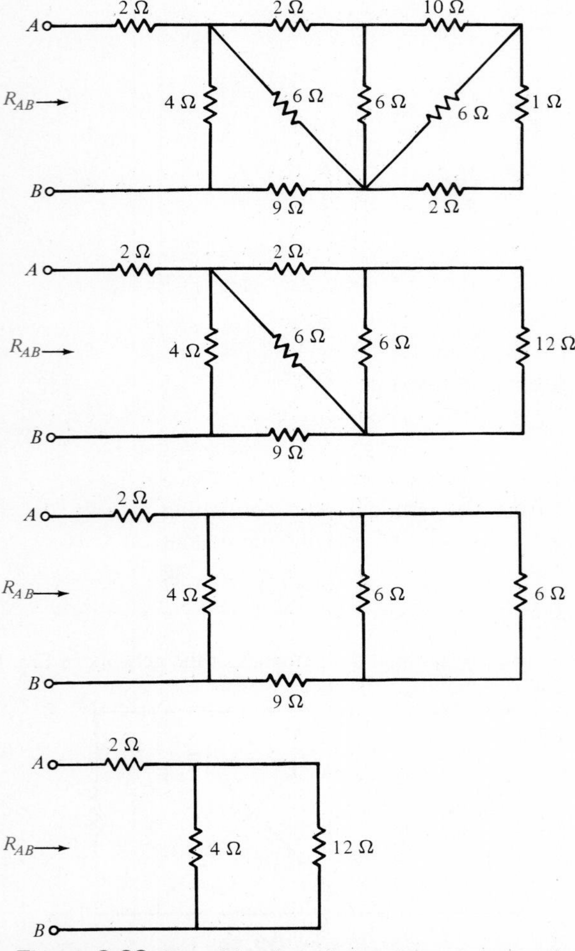

Figure 2.23 Simplification of a resistance network.

2.6

Circuits with Series-Parallel Combination of Resistors

At this point we have learned many techniques that are fundamental to circuit analysis. Now we wish to apply them and show how they can be used in concert to analyze circuits. We will illustrate their application through a number of examples that will be treated in some detail.

EXAMPLE 2.21

We wish to determine all the currents and voltages in the ladder network shown in Fig. 2.24a. To begin our analysis of the network, we start at the right end of the circuit and combine the resistors to determine the total resistance seen by the 64-V source. This will allow us to calculate the current I_1. Then employing Kirchhoff's current and voltage laws, Ohm's law, and/or current division, we will be able to calculate all currents and voltages in the circuit.

At the right end of the circuit the 9- and 3-Ω resistors are in series and thus can be combined into one equivalent 12-Ω resistor. This resistor is in parallel with the 4-Ω

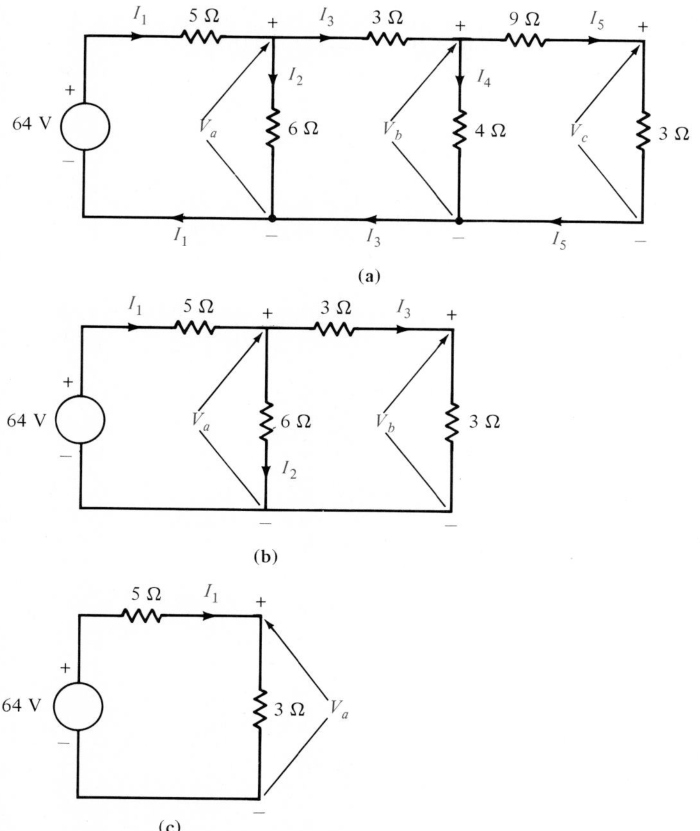

Figure 2.24 Analysis of a ladder network.

resistor, and their combination yields an equivalent 3-Ω resistor, shown at the right edge of the circuit in Fig. 2.24b. In Fig. 2.24b the two 3-Ω resistors are in series and their combination is in parallel with the 6-Ω resistor. Combining all three resistances yields the circuit shown in Fig. 2.24c.

Applying Kirchhoff's voltage law to the circuit in Fig. 2.24c yields

$$I_1(5 + 3) = 64$$

$$I_1 = 8 \text{ A}$$

V_a can be calculated from Ohm's law as

$$V_a = I_1(3)$$

$$= 24 \text{ V}$$

or using Kirchhoff's voltage law,

$$V_a = 64 - 5I_1$$

$$= 64 - 40$$

$$= 24 \text{ V}$$

Knowing I_1 and V_a, we can now determine all currents and voltages in Fig. 2.24b. For example, since $V_a = 24$ V, the current I_2 can be found using Ohm's law as

$$I_2 = \frac{24}{6}$$

$$= 4 \text{ A}$$

Then using Kirchhoff's law, we have

$$I_1 = I_2 + I_3$$

$$8 = 4 + I_3$$

$$I_3 = 4 \text{ A}$$

Note that I_3 could also be calculated using Ohm's law:

$$V_a = (3 + 3)I_3$$

$$I_3 = \frac{24}{6}$$

$$= 4 \text{ A}$$

Applying Kirchhoff's voltage law to the right-hand loop in Fig. 2.24b yields

$$V_a - V_b = 3I_3$$

$$24 - V_b = 12$$

$$V_b = 12 \text{ V}$$

or since V_b is equal to the voltage drop across the 3-Ω resistor, we could use Ohm's law as

$$\begin{aligned} V_b &= 3I_3 \\ &= 12 \text{ V} \end{aligned}$$

We are now in a position to calculate the final unknown currents and voltages in Fig. 2.24a. Knowing V_b, we can calculate I_4 using Ohm's law as

$$\begin{aligned} V_b &= 4I_4 \\ I_4 &= \frac{12}{4} \\ &= 3 \text{ A} \end{aligned}$$

Then from Kirchhoff's current law, we have

$$\begin{aligned} I_3 &= I_4 + I_5 \\ 4 \text{ A} &= 3 \text{ A} + I_5 \\ I_5 &= 1 \text{ A} \end{aligned}$$

We could also have calculated I_5 using the current-division rule. For example,

$$\begin{aligned} I_5 &= \frac{4}{4 + (9 + 3)} I_3 \\ &= 1 \text{ A} \end{aligned}$$

Finally, V_c can be computed as

$$\begin{aligned} V_c &= I_5(3) \\ &= 3 \text{ V} \end{aligned}$$

Note that Kirchhoff's current law is satisfied at every junction and Kirchhoff's voltage law is satisfied around every loop. ■

EXAMPLE 2.22

Given the circuit in Fig. 2.25 with $V_o = 72$ V, determine all the currents and voltages.

The circuit can be simplified as shown in the progression from Fig. 2.25a to Fig. 2.25b to Fig. 2.25c. Then I_1 can be computed from Ohm's law as

$$\begin{aligned} I_1 &= \frac{V_o}{6 + 2 + 4} \\ &= 6 \text{ A} \end{aligned}$$

Now using Kirchhoff's voltage law in Fig. 2.25c yields

$$72 = 6I_1 + V_a + V_b + 4I_1$$

$$72 - 6I_1 - 4I_1 = V_a + V_b$$

$$12\ V = V_a + V_b$$

This value could also be calculated from Ohm's law by multiplying the current of 6 A by the 2-Ω resistance. From Fig. 2.25b we can obtain I_2 as

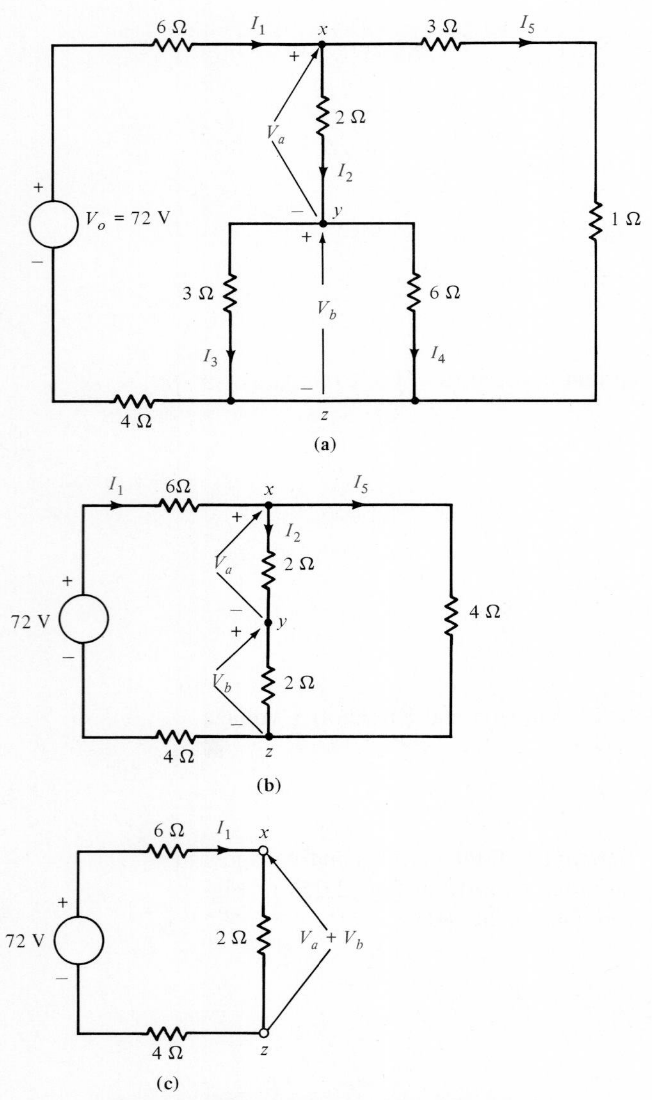

Figure 2.25 Example circuit for analysis.

$$I_2 = \frac{V_a + V_b}{2 + 2}$$

$$= 3 \text{ A}$$

Then using Ohm's law, we obtain

$$V_a = I_2(2)$$

$$= 6 \text{ V}$$

and

$$V_b = I_2(2)$$

$$= 6 \text{ V}$$

Therefore, point y is 6 V positive with respect to point z, and point x is 6 V positive with respect to point y, or 12 V positive with respect to point z. From Kirchhoff's current law,

$$I_1 = I_2 + I_5$$

$$6 = 3 + I_5$$

$$I_5 = 3 \text{ A}$$

Since V_b is known, currents I_3 and I_4 can be obtained from Ohm's law as

$$I_3 = \frac{V_b}{3}$$

$$= 2 \text{ A}$$

and

$$I_4 = \frac{V_b}{6}$$

$$= 1 \text{ A}$$

Either current could also be calculated using current division on the current I_2. For example, I_3 can be obtained as

$$I_3 = \frac{6}{3 + 6} I_2$$

$$= 2 \text{ A}$$

■

EXAMPLE 2.23

Suppose that we are given the circuit in Fig. 2.25a and $I_4 = \frac{1}{2}$ A, and we want to find the source voltage V_o.

If $I_4 = \frac{1}{2}$ A, then from Ohm's law, $V_b = 3$ V. V_b can now be used to calculate $I_3 = 1$ A. Kirchhoff's current law applied at node y yields

$$I_2 = I_3 + I_4$$
$$= 1.5 \text{ A}$$

Then from Ohm's law, we have

$$V_a = (1.5 \text{ A})(2\ \Omega)$$
$$= 3 \text{ V}$$

Since $V_a + V_b$ is now known, I_5 can be obtained:

$$I_5 = \frac{V_a + V_b}{3 + 1}$$
$$= 1.5 \text{ A}$$

Applying Kirchhoff's current law at node x yields

$$I_1 = I_2 + I_5$$
$$= 3 \text{ A}$$

Now using Fig. 2.25c, we can employ Kirchhoff's voltage law since we now know I_1.

$$V_o = 6I_1 + 2I_1 + 4I_1$$
$$= 12I_1$$
$$= 36 \text{ V}$$

■

EXAMPLE 2.24

Given the circuit in Fig. 2.26a, we wish to determine the current I_3 and the voltage V_o across the current source.

Current division may be applied directly to obtain I_3 as

$$I_3 = \frac{4}{4 + (4 + 4 + 4)} I_1$$
$$= \frac{4}{16} \times 12$$
$$= 3 \text{ A}$$

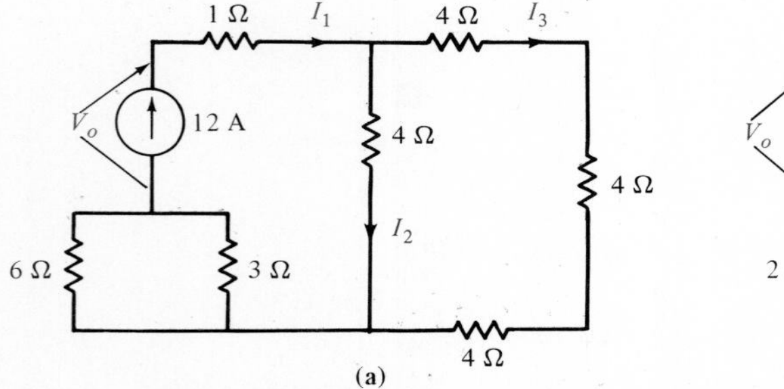

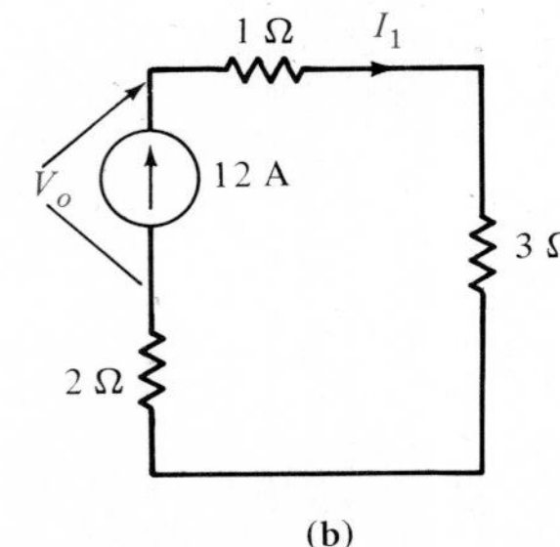

Figure 2.26 Example circuit containing a current source.

Combining resistances, we can reduce the circuit in Fig. 2.26a to that in Fig. 2.26b. Then using Kirchhoff's voltage law, we can compute V_o, since I_1 is by definition 12 A.

$$V_o = 1I_1 + 3I_1 + 2I_1$$
$$= 72 \text{ V}$$

■

DRILL EXERCISE

D2.16. Find all the currents in the network in Fig. D2.16.

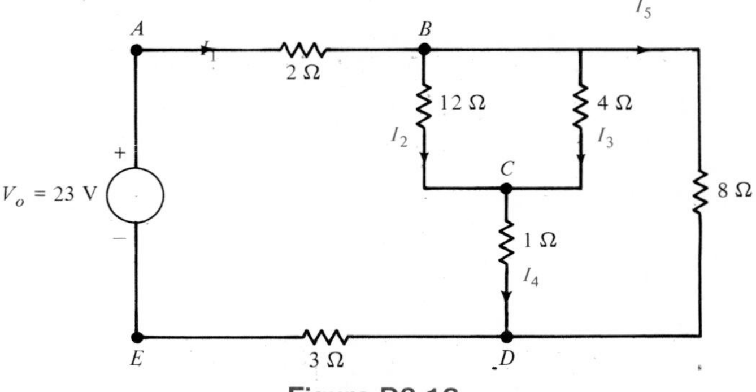

Figure D2.16

D2.17. Find V_o in the network in Fig. D2.16 if it is known that $I_3 = 6$ A.

2.7 Circuits with Dependent Sources

In Chapter 1 we outlined the different kinds of dependent sources. We will now show how to solve simple one-loop and one-node circuits that contain these dependent sources. Although the following examples are fairly simple, they will serve to illustrate the basic concepts.

The following example employs a current-controlled voltage source.

EXAMPLE 2.25

Consider the circuit shown in Fig. 2.27. To determine the voltage V_o across the 5-Ω resistor, we employ Kirchhoff's voltage law:

$$V - 5I_1 + 2I_1 - 3I_1 = 0$$
$$24 = I_1(5 + 3 - 2)$$
$$I_1 = 4 \text{ A}$$

Therefore,

$$V_o = (5)(4)$$
$$= 20 \text{ V}$$

■

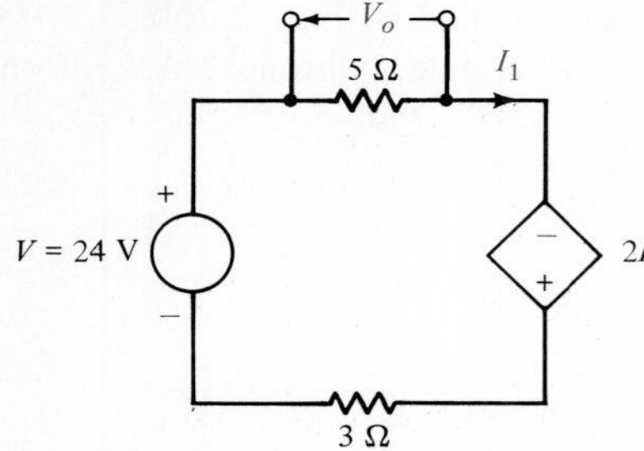

Figure 2.27 Example circuit containing a current-controlled voltage source.

EXAMPLE 2.26

Given the circuit in Fig. 2.28 containing a current-controlled current source, we wish to determine the voltage V_o.

Employing Kirchhoff's current law yields

$$-10 + 4I_o = \frac{V}{6} + \frac{V}{3}$$

but

$$I_o = \frac{V}{3}$$

Therefore, the equation above can be rewritten as

$$-10 + \frac{4V}{3} = \frac{V}{6} + \frac{V}{3}$$

Solving this equation, we obtain

$$V = 12 \text{ V}$$

Then

$$I_o = \frac{12}{3} = 4 \text{ A}$$

and hence

$$V_o = 2I_o$$
$$= 8 \text{ V}$$

■

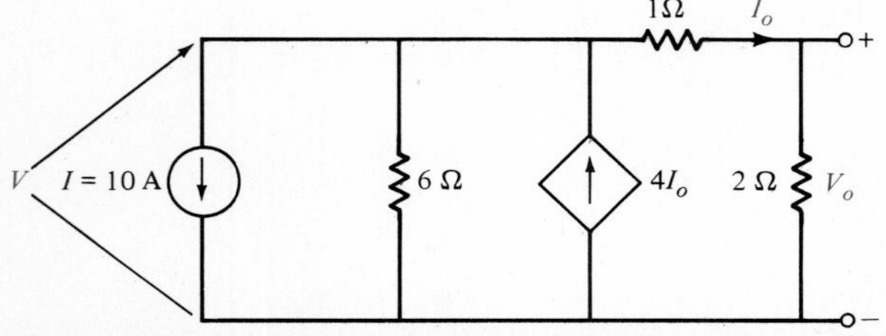

Figure 2.28 Example circuit containing a current-controlled current source.

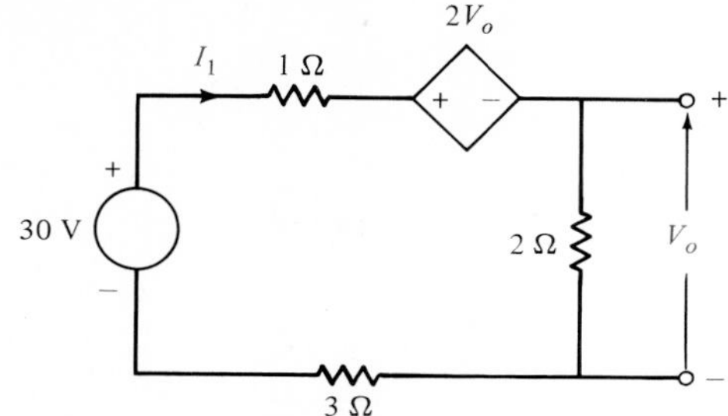

Figure 2.29 Example circuit containing a voltage-controlled voltage source.

EXAMPLE 2.27

We wish to solve the circuit in Fig. 2.29 containing a voltage-controlled voltage source to determine the voltage V_o.

Applying Kirchhoff's voltage law to this single-loop circuit yields

$$30 - 2V_o = (1 + 2 + 3)I_1$$

where

$$V_o = 2I_1$$

Therefore, the equation above becomes

$$30 - 4I_1 = 6I_1$$

or

$$I_1 = 3 \text{ A}$$

Then

$$V_o = 2I_1 = 6 \text{ V}$$

■

At this point it is perhaps helpful to point out that when analyzing circuits with dependent sources, we first treat the dependent source as though it were an independent source when we write a Kirchhoff's current or voltage law equation. Once the equation is written, we then write the controlling equation that specifies the relationship of the dependent source to the unknown parameter. For instance, the first equation in Example 2.27 treats the dependent source like an independent source. The second equation in the example specifies the relationship of the dependent source to the current, which is the unknown in the first equation.

DRILL EXERCISE

D2.18. Find V_o in the network in Fig. D2.18.

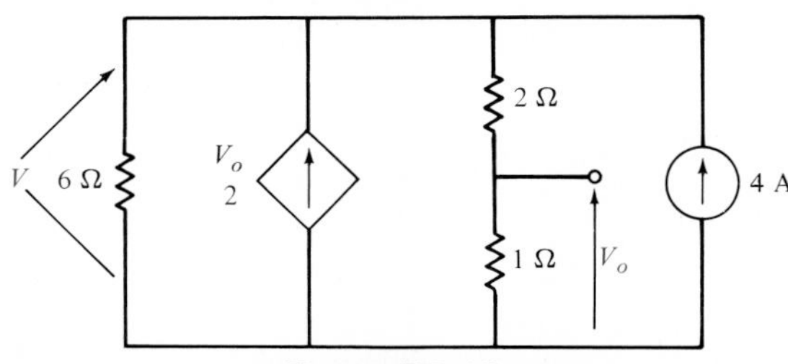

Figure D2.18

D2.19. Find V_o in the network in Fig. D2.19.

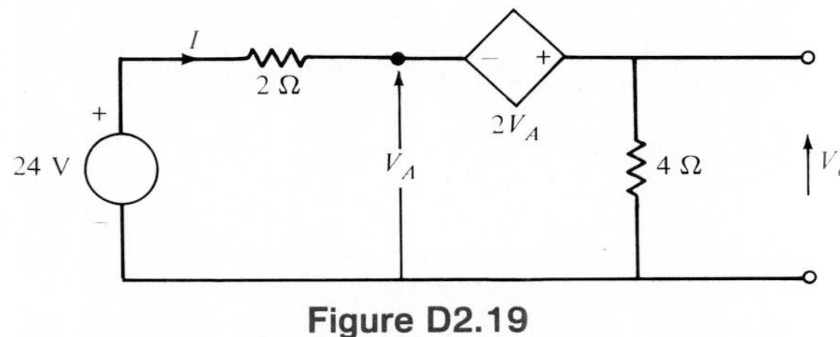

Figure D2.19

EXAMPLE 2.28

The midband small-signal equivalent circuit for a field-effect transistor (FET) common-source amplifier or bipolar junction transistor (BJT) common-emitter amplifier can be given by the circuit shown in Fig. 2.30a. We wish to determine an expression for the gain of the amplifier, which is the ratio of the output voltage to the input voltage. Note that although this circuit, which contains a voltage-controlled current source, appears to be somewhat complicated, we are actually in a position now to solve it with techniques we have studied up to this point. The loop on the left, or input to the amplifier, is essentially detached from the output portion of the amplifier on the right. The voltage across R_2 is $v_g(t)$, which controls the dependent current source.

To simplify the analysis, let us replace the resistors R_3, R_4, and R_5 with R_L such that

$$\frac{1}{R_L} = \frac{1}{R_3} + \frac{1}{R_4} + \frac{1}{R_5}$$

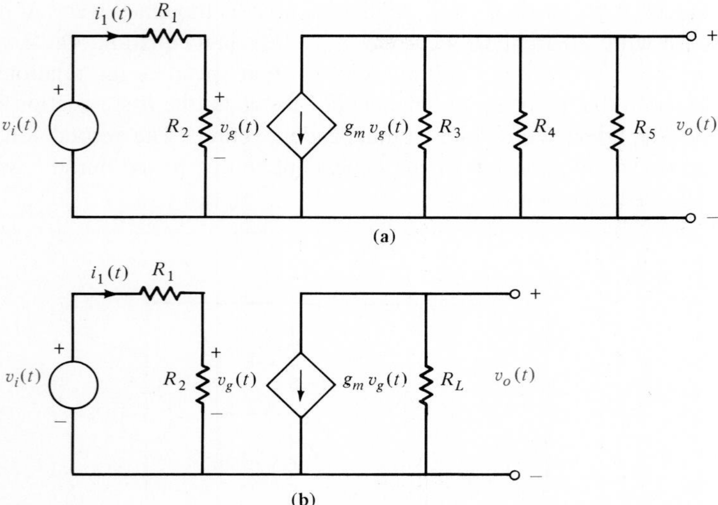

Figure 2.30 Example circuit containing a voltage-controlled current source.

Then the circuit reduces to that shown in Fig. 2.30b. Applying Kirchhoff's voltage law to the input portion of the amplifier yields

$$v_i(t) = i_1(t)(R_1 + R_2)$$

and

$$v_g(t) = i_1(t)R_2$$

Solving these equations for $v_g(t)$ yields

$$v_g(t) = \frac{R_2}{R_1 + R_2} v_i(t)$$

From the output circuit, note that the voltage $v_o(t)$ is given by the expression

$$v_o(t) = -g_m v_g(t) R_L$$

Combining this equation with the one above yields

$$v_o(t) = \frac{-g_m R_L R_2}{R_1 + R_2} v_i(t)$$

Therefore, the amplifier gain, which is the ratio of the output voltage to the input voltage, is given by

$$\frac{v_o(t)}{v_i(t)} = -\frac{g_m R_L R_2}{R_1 + R_2}$$

Reasonable values for the circuit parameters in Fig. 2.30a are $R_1 = 100\ \Omega$, $R_2 = 1\ \text{k}\Omega$, $g_m = 0.04$ S, $R_3 = 50\ \text{k}\Omega$, and $R_4 = R_5 = 10\ \text{k}\Omega$. Hence the gain of the amplifier under these conditions is

$$\frac{v_o(t)}{v_i(t)} = \frac{-(0.04)(4.545 \times 10^3)(1 \times 10^3)}{1.1 \times 10^3}$$

$$= -165.29$$

Thus the magnitude of the gain is 165.29. ■

2.8

Circuits with Operational Amplifiers

The operational amplifier, or op-amp as it is commonly known, is an extremely important circuit. It is a versatile interconnection of devices that vastly expands our capabilities in circuit design.

In sharp contrast to the circuit elements we have introduced thus far, the op-amp is modeled as a multiterminal device. It was first introduced in the 1940s for use in analog computers, and now finds wide application in circuit design as a result of the advances made in integrated-circuit technology. Although we will model the op-amp as a fairly simple device, its actual construction involves the use of numerous components, including resistors, capacitors, and transistors interconnected in a rather complicated manner.

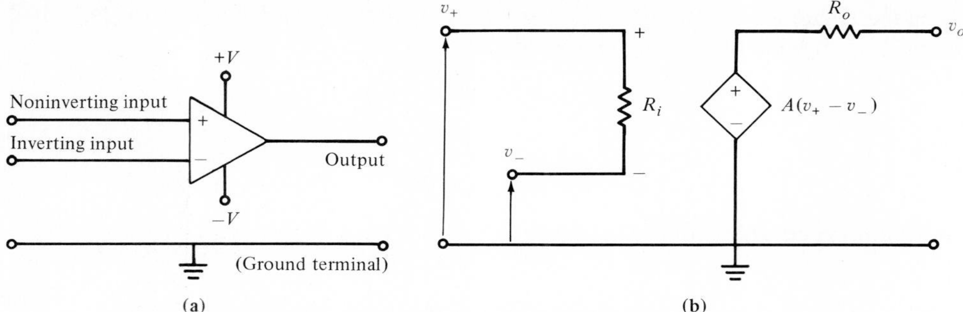

Figure 2.31 Operational amplifier representations.

The circuit symbol for the op-amp is shown in Fig. 2.31a. Only the five principal terminals of the device are shown. The device has other terminals; however, a discussion of them would only serve to complicate the presentation unnecessarily. The two voltage terminals are for dc power supply (typically 10 to 15 V); and, in addition, they provide a return path to ground for ac current. The + and − signs for the two input terminals indicate the polarity of the output as compared with the input; that is, a positive input on the noninverting input terminal produces a positive output, whereas a positive input on the inverting terminal produces a negative output.

Our interest in the op-amp is confined to its input/output characteristics, and therefore our model ignores the supply voltages. Functionally, the op-amp operates like the equivalent circuit shown in Fig. 2.31b. This equivalent circuit implies a unilateral device: that is, the output is determined by the difference in input voltages, but the input is unaffected by voltages applied at the output. The resistance R_i is very large, the gain A is very high, and the output resistance R_o is very low; typical values for these parameters are $10^5\ \Omega$, 10^5, and $10\ \Omega$, respectively. The values of these parameters in this model suggest an even simpler model. R_i is very large and therefore can be assumed to be an open circuit. R_o is very low with respect to the external recommended load connection, and therefore can be assumed to be a short circuit. Finally, the gain A is an extremely large amplification factor and would appear to represent essentially an *infinite* (∞) *gain*. At this point in our education, however, the term ''infinite gain'' causes some concern. Does this mean that if the input is 1 μV, the output is unlimited? Our common sense leads us to question such a condition. Actually, the high gain is used to guarantee accuracy in the device, and we will show that when the op-amp is connected into a *circuit,* the external resistors will actually control the operation of the device, and the voltage gain of the *circuit* will not be dependent on the amplifier gain A.

Under the assumptions stated above, the ideal model for the op-amp is reduced to that shown in Fig. 2.32. The important characteristics of this model are: (1) the input currents to the op-amp are approximately zero, and therefore $i_+ \simeq i_- \simeq 0$; and (2) the voltage between the input terminals is approximately zero (i.e., $v_+ \simeq v_-$). Hence we will assume that $i_+ = i_- = 0$ and $v_+ - v_- = 0$. The ground terminal ($\doteqdot$) shown on the op-amp is necessary for signal current return, and it guarantees that Kirchhoff's current law will be satisfied at both the op-amp and the ground node ($\doteqdot$) in the circuit.

Having described some of the fundamental properties of op-amps, let us now turn our attention to their application.

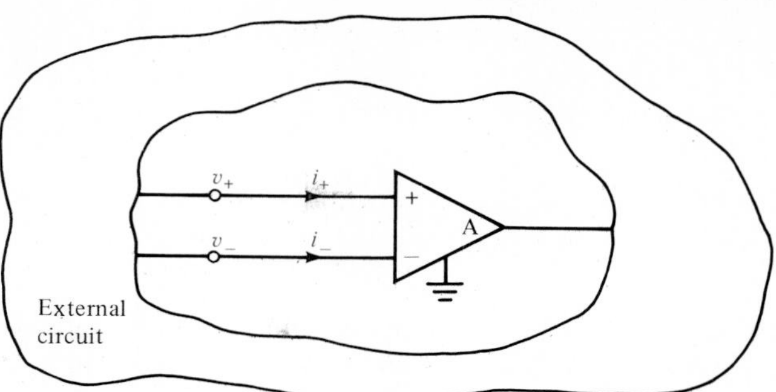

Figure 2.32 Ideal model for an operational amplifier. Model parameters: $i_+ = i_- = 0$, $v_+ = v_-$.

EXAMPLE 2.29

Consider the circuit shown in Fig. 2.33a. From the previous discussion we know that $i_- = i_+ = 0$ and that $v_- - v_+ = 0$. Applying Kirchhoff's current law at the inverting terminal produces the equation

$$i_1 + i_2 = i_-$$

or

$$\frac{v_s - v_-}{R_1} + \frac{v_o - v_-}{R_2} = i_-$$

However, $i_- = 0$ and $v_- = v_+$ and $v_+ = 0$ because the terminal is tied to ground. Hence

$$\frac{v_s}{R_1} + \frac{v_o}{R_2} = 0$$

or

$$v_o = \frac{-R_2}{R_1} v_s$$

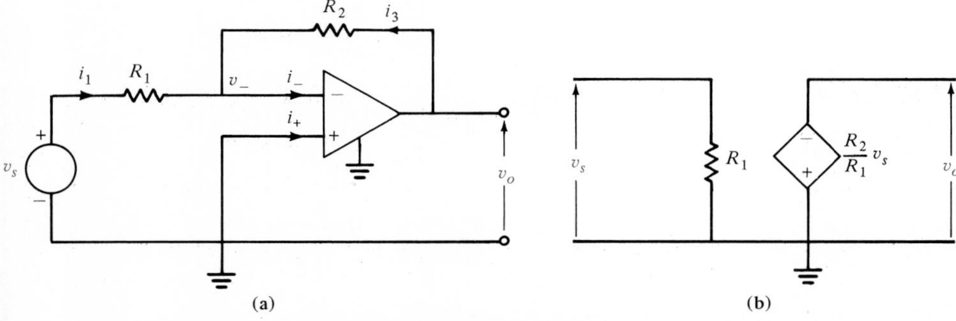

Figure 2.33 Inverting operational amplifier equivalent circuit.

Note that the equation indicates that this circuit is an inverting amplifier; that is, a positive input at v_s yields a negative output at v_o, and vice versa. The gain of this amplifier circuit is R_2/R_1, which is independent of the basic op-amp's gain. Therefore, the output can be written as

$$v_o = -\frac{R_2}{R_1} v_s$$

and hence an equivalent circuit for this amplifier can be drawn as shown in Fig. 2.33b. Note that this equivalent circuit is a voltage-controlled voltage source. ■

EXAMPLE 2.30

Let us examine the amplifier configuration shown in Fig. 2.34a. Once again we recall the basic properties of the op-amp: that is $i_+ = i_- = 0$, and $v_+ - v_- = 0$. Since the voltage across the op-amp input terminals is zero, then from Kirchhoff's voltage law $v_s = v_-$, which is the voltage across R_1. Applying Kirchhoff's current law to the node between R_1 and R_2 yields

$$i_- + \frac{v_s}{R_1} + \frac{v_s - v_o}{R_2} = 0$$

But since $i_- = 0$,

$$v_o = \left(1 + \frac{R_2}{R_1}\right) v_s$$

Hence the circuit configuration in Fig. 2.34a represents a noninverting amplifier with a gain of $1 + R_2/R_1$. Once again we note that this gain is independent of the op-amp's gain. The amplifier is noninverting since the sign of the output voltage is the same as that of the input. The equivalent circuit for this amplifier is shown in Fig. 2.34b. Note that this equivalent circuit is a voltage-controlled voltage source. ■

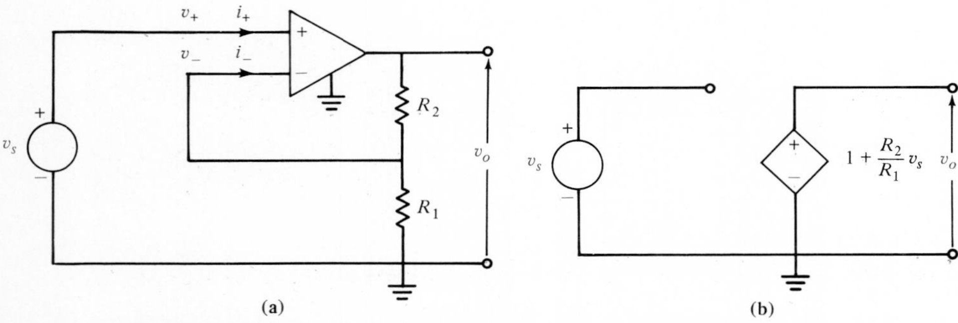

Figure 2.34 Noninverting operational amplifier and its equivalent circuit.

DRILL EXERCISE

D2.20. Determine both the gain and the output voltage of the op-amp configuration shown in Fig. D2.20.

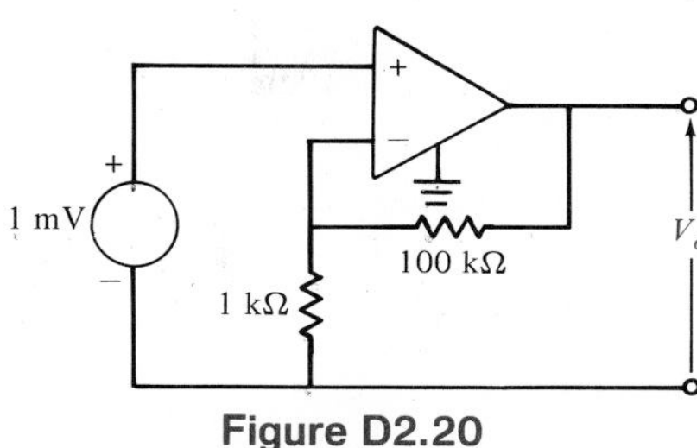

Figure D2.20

An important special case of the noninverting amplifier is the configuration shown in Fig. 2.35. Note carefully that this circuit is equivalent to that shown in Fig. 2.34a with $R_1 = \infty$ (i.e., open circuited) and $R_2 = 0$ (i.e., short circuited). Under these conditions we note that the gain of the circuit is $1 + R_2/R_1 = 1$ (i.e., $v_o = v_s$). Since v_o follows v_s, the circuit is called a *voltage follower*.

An obvious question at this point is: If $v_o = v_s$, why not just connect v_s to v_o via two parallel connecting wires; why do we need to place an op-amp between them? The answer to this question is fundamental and provides us with some insight that will aid us in circuit analysis and design.

Consider the circuit shown in Fig. 2.36a. In this case v_o is not equal to v_s because of the voltage drop across R_s:

$$v_o = v_s - iR_s$$

However, in Fig. 2.36b, the input current to the op-amp is zero and therefore v_s appears at the op-amp input. Since the gain of the op-amp configuration is 1, $v_o = v_s$. In Fig. 2.36a, the resistive network's interaction with the source caused the voltage v_o to be less than v_s. In other words, the resistive network loads the source voltage. However, in Fig. 2.36b the op-amp isolates the source from the resistive network, and therefore the voltage follower is referred to as a *buffer amplifier* because it can be used to isolate one circuit from another. The energy supplied to the resistive network in the first case must come from the source v_s, whereas in the second case it comes from the power supplies that supply the amplifier, and little or no energy is drawn from v_s.

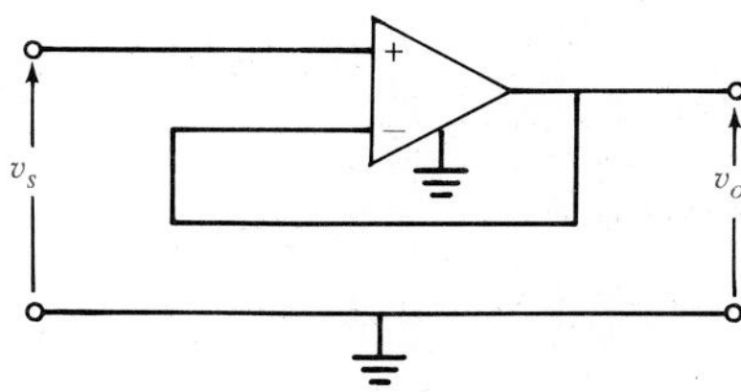

Figure 2.35 Voltage-follower operational amplifier configuration.

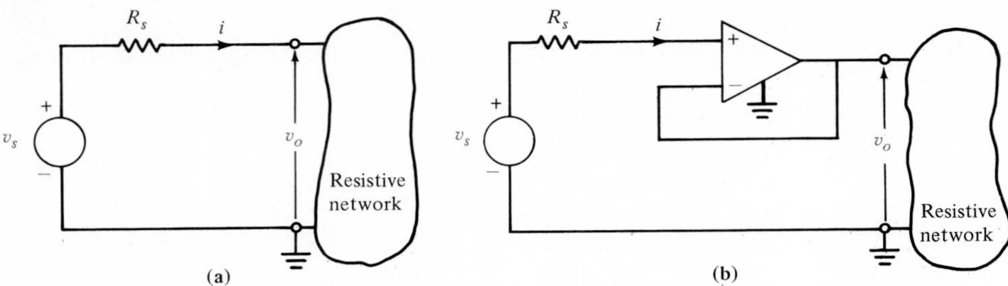

Figure 2.36 Illustration of the isolation capability of a voltage follower.

2.9 Summary

This chapter has dealt primarily with the fundamental laws that we employ in our analysis of circuits: Ohm's law and Kirchhoff's laws. The resistive circuit element has been introduced and we have shown how to compute an equivalent resistance when these elements are placed in series or parallel. Both voltage division and current division have been presented. We have shown that by employing the fundamental laws, we can analyze some fairly complicated circuits that contain only a single source. Finally, we have introduced the operational amplifier and its equivalent circuit, which we will use in a variety of circuits throughout this book.

KEY POINTS

- For a short circuit, the resistance is zero, the voltage across the short is zero, and the current in the short is determined by the rest of the circuit.
- For an open circuit, the resistance is infinite, the current is zero, and the voltage across the open terminals is determined by the rest of the circuit.
- A node is a point of interconnection of two or more circuit elements.
- A loop is any closed path through the circuit in which no node is encountered more than once.
- A branch is a portion of a circuit containing only a single element and the nodes at each end of the element.
- Kirchhoff's current law states that the algebraic sum of the currents entering a node is zero.
- Kirchhoff's voltage law states that the algebraic sum of the voltages around any closed path is zero.
- Current division shows that the current divides in inverse proportion to the resistance paths.
- Voltage division shows that the voltage is divided among resistors in direct proportion to their resistances.
- For an ideal op-amp $i_+ = i_- = 0$ and $v_+ = v_-$.

PROBLEMS

2.1. Determine the number of nodes and branches in the network in Fig. P2.1.

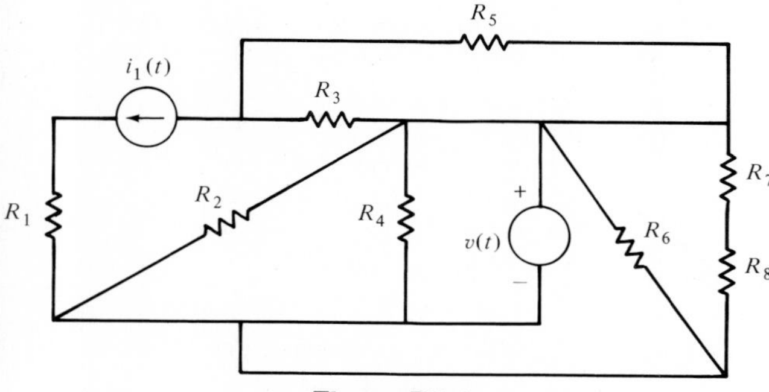

Figure P2.1

2.2. Find the power absorbed in the 1-Ω resistor in the circuit shown in Fig. P2.2.

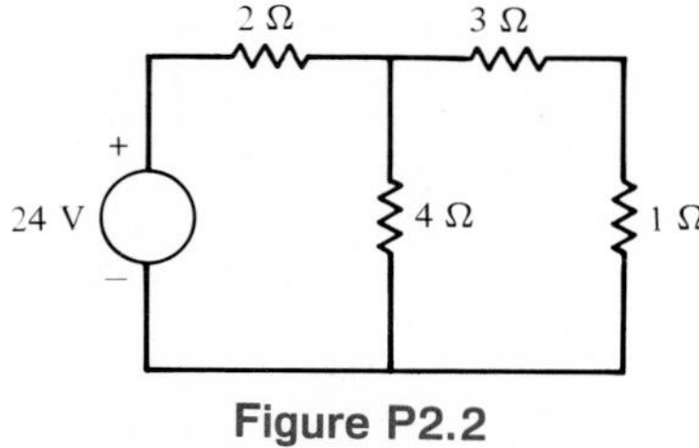

Figure P2.2

2.3. Find I and the power dissipated in the 4-Ω resistor in the network shown in Fig. P2.3.

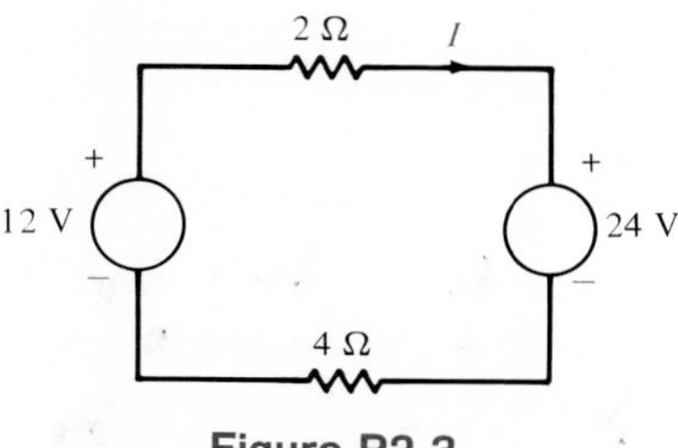

Figure P2.3

2.4. Find the power dissipated in the 2-Ω resistor in the network shown in Fig. P2.4.

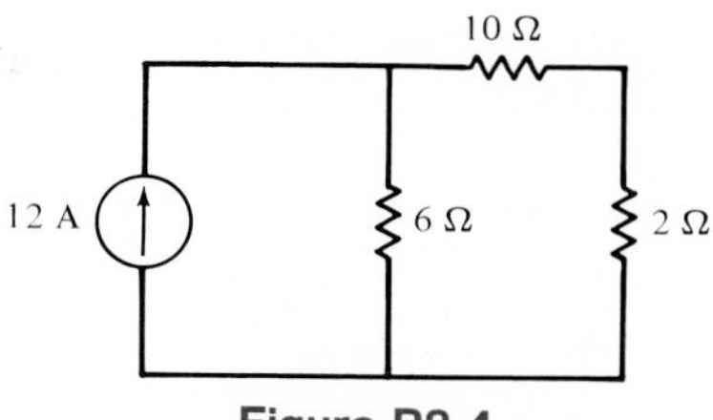

Figure P2.4

2.5. The power dissipated by the resistor in the circuit in Fig. P2.5 is 72 W. Determine the value of the voltage source V and the power generated by the source.

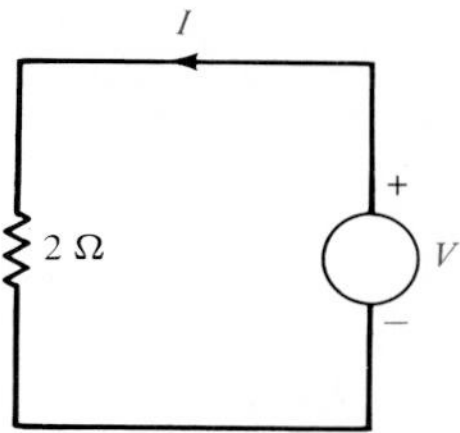

Figure P2.5

2.6. Given the circuit in Fig. P2.6, find I and the power dissipated in the 4-Ω resistor.

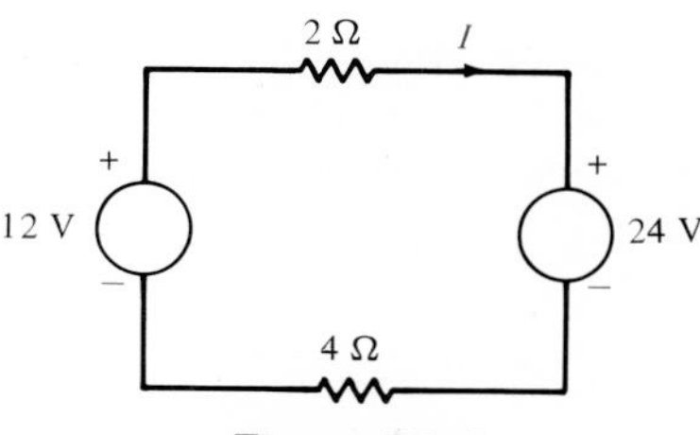

Figure P2.6

2.7. Find I and V_o in the network in Fig. P2.7.

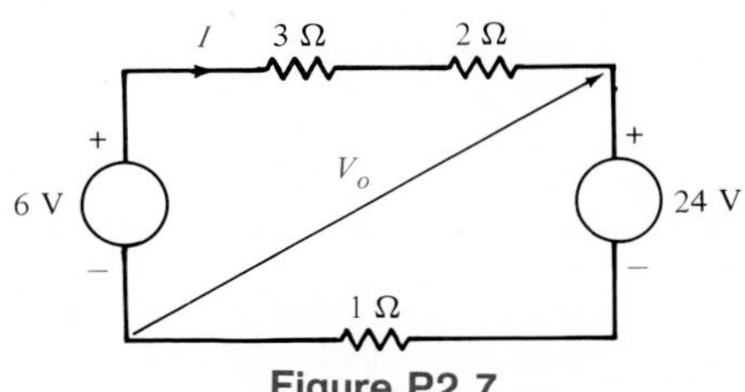

Figure P2.7

2.8. Use Kirchhoff's current law to determine the currents I_1 and I_2 in the circuit in Fig. P2.8.

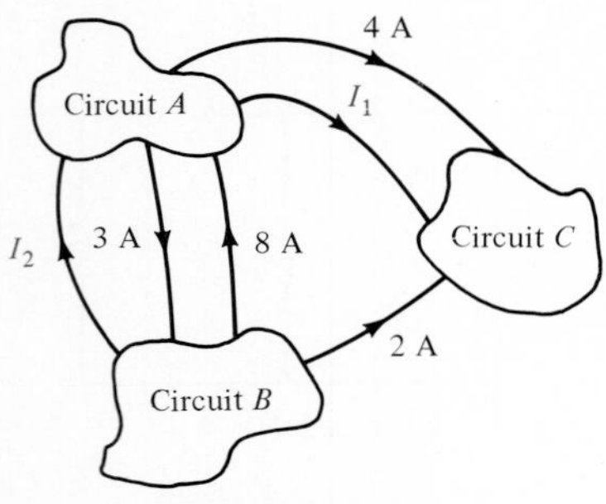

Figure P2.8

2.9. Use Kirchhoff's current law to determine the unknown currents in the circuit in Fig. P2.9.

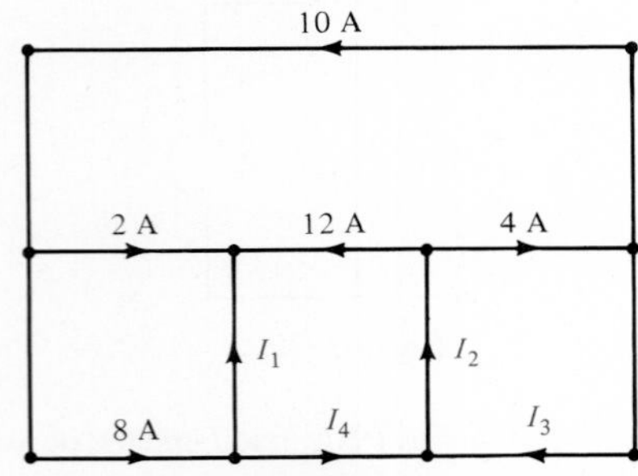

Figure P2.9

2.10. Since Kirchhoff's current law must be satisfied at every partition that cuts the circuit in two pieces, use the partitions to find the unknown currents in the circuit in Fig. P2.10.

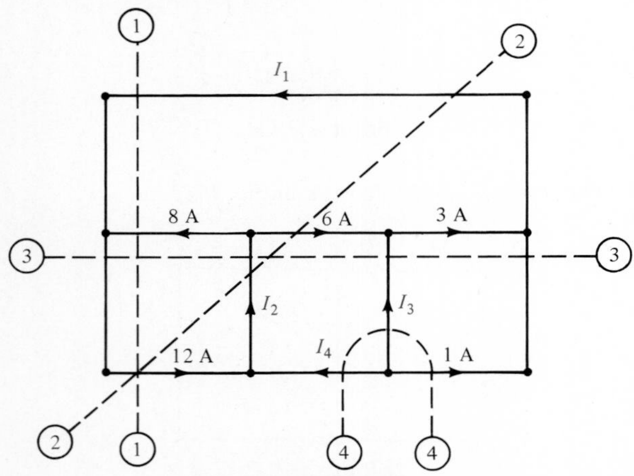

Figure P2.10

2.11. In the circuit in Fig. P2.11 $V_A = 21$ V and $V_{R_2} = 6$ V. Find V_{R_1}, V_{R_3}, and V_B.

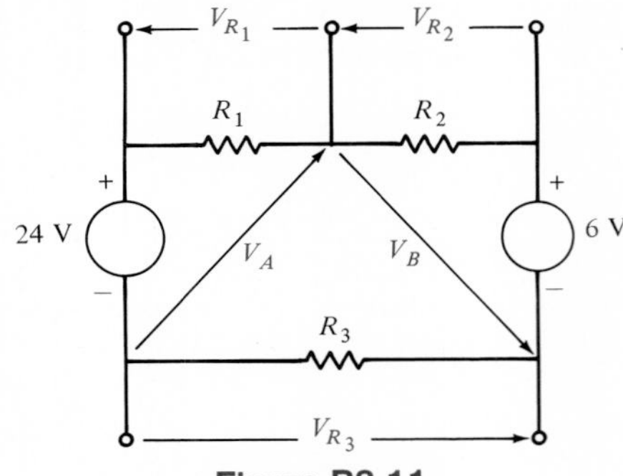

Figure P2.11

2.12. Determine V_o in the circuit shown in Fig. P2.12.

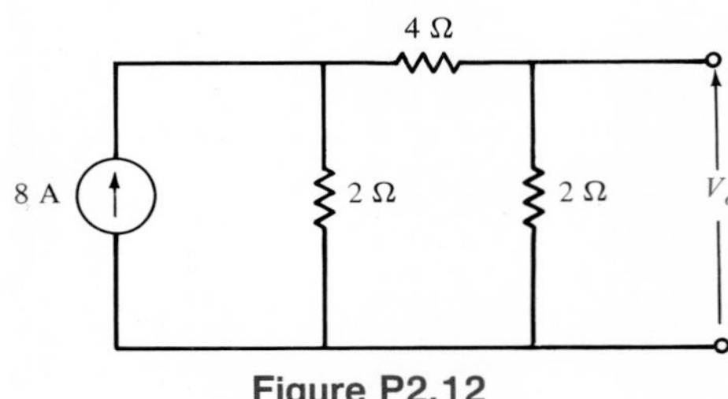

Figure P2.12

2.13. Find the current in each resistor shown in Fig. P2.13.

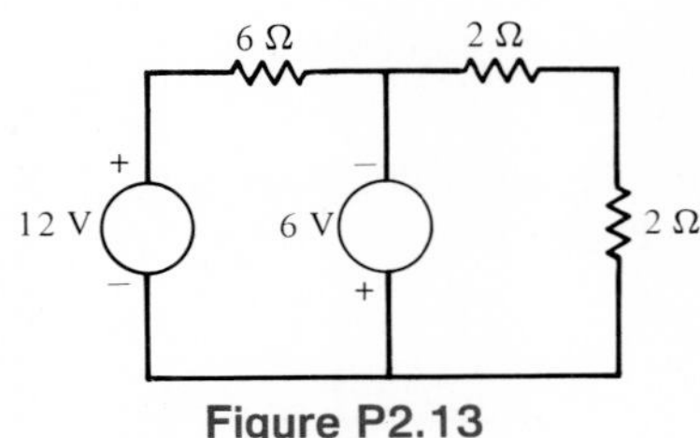

Figure P2.13

2.14. Find I_1 in the circuit shown in Fig. P2.14.

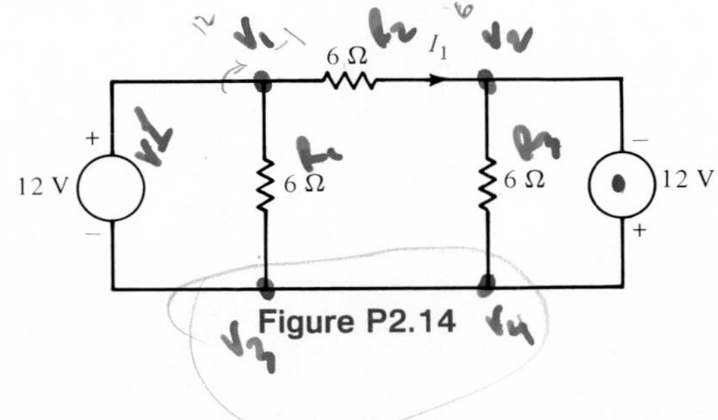

Figure P2.14

2.15. Given $I_o = 4$ A, find V in Fig. P2.15.

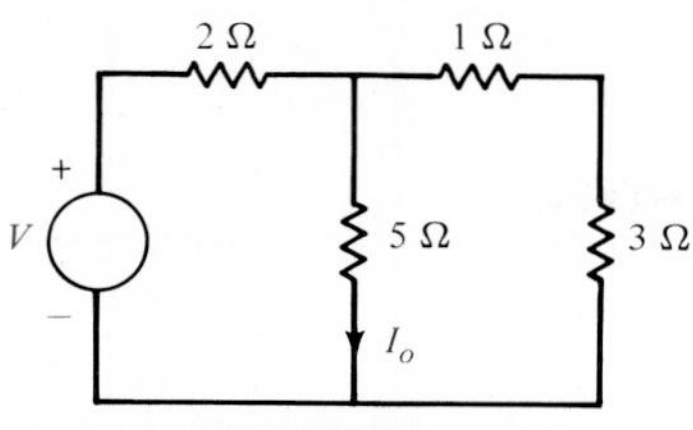

Figure P2.15

2.16. Find v_o in the network shown in Fig. P2.16.

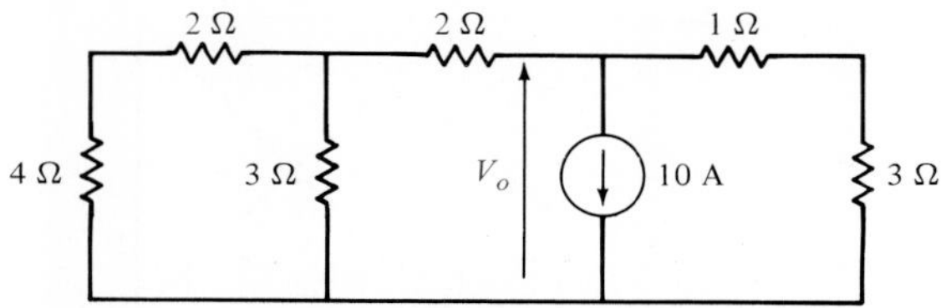

Figure P2.16

2.17. Given the network in Fig. P2.17, use voltage division to determine the voltages V_1, V_2, and V_3.

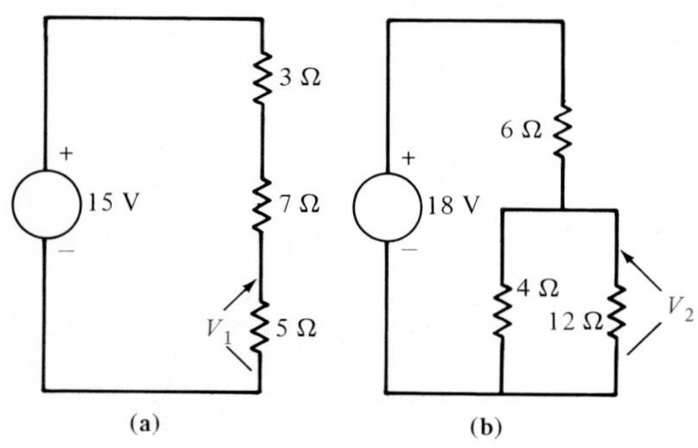

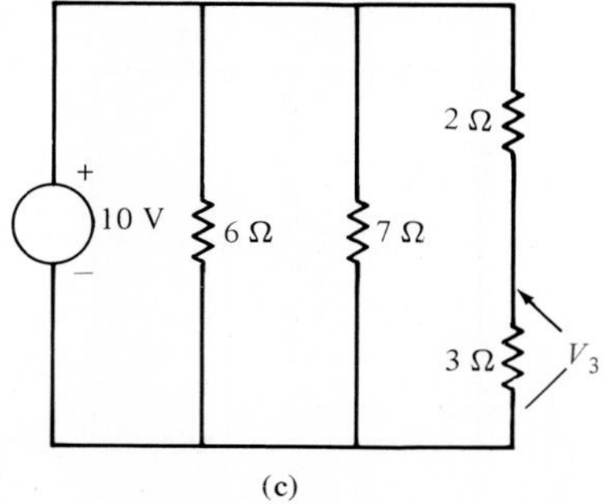

Figure P2.17

2.18. Determine the voltages V_1, V_2, and V_3 in the networks shown in Fig. P2.18 using voltage division.

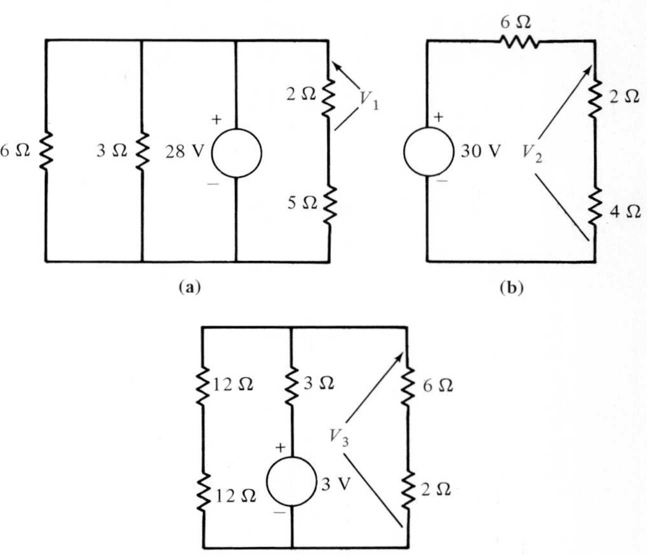

Figure P2.18

2.19. Determine the value of the current source in Fig. P2.19.

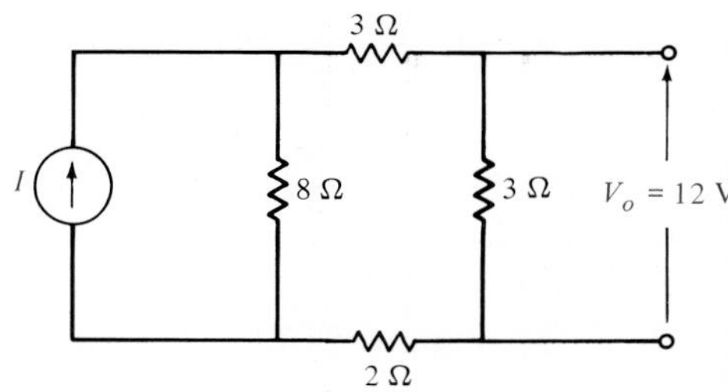

Figure P2.19

2.20. In the network in Fig. P2.20, $I_o = 2$ A, find V.

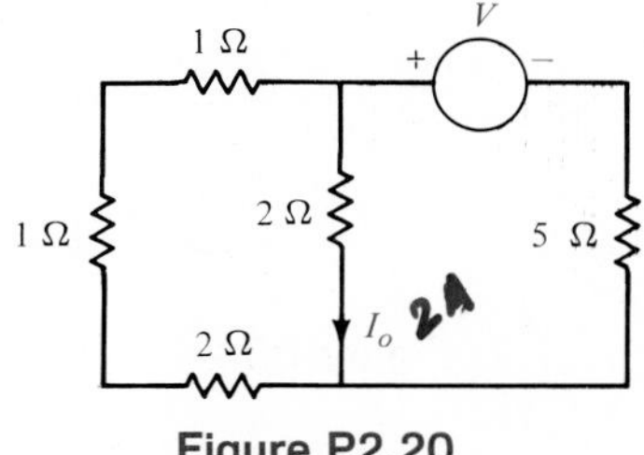

Figure P2.20

2.21. Find I_o and V in the network in Fig. P2.21.

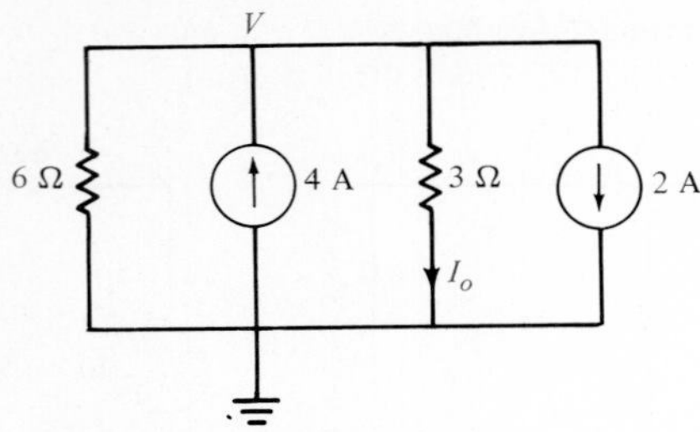

Figure P2.21

2.22. Given the network in Fig. P2.22, find V and I_o.

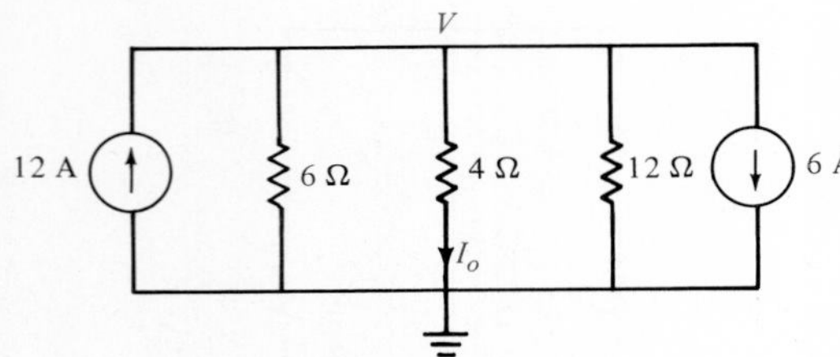

Figure P2.22

2.23. Find R_{eq} in the circuits in Fig. P2.23.

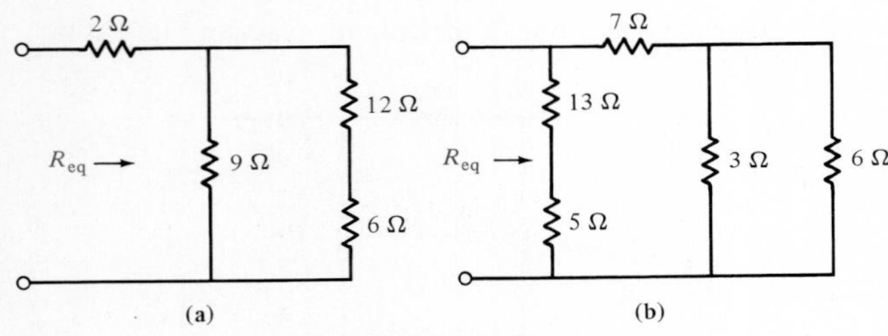

Figure P2.23

2.24. Find G_{eq} in the circuits in Fig. P2.24.

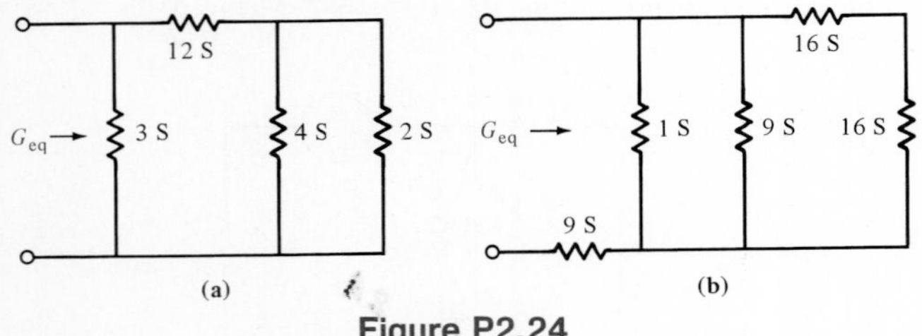

Figure P2.24

2.25. Determine the equivalent conductance of the circuit in Fig. P2.25.

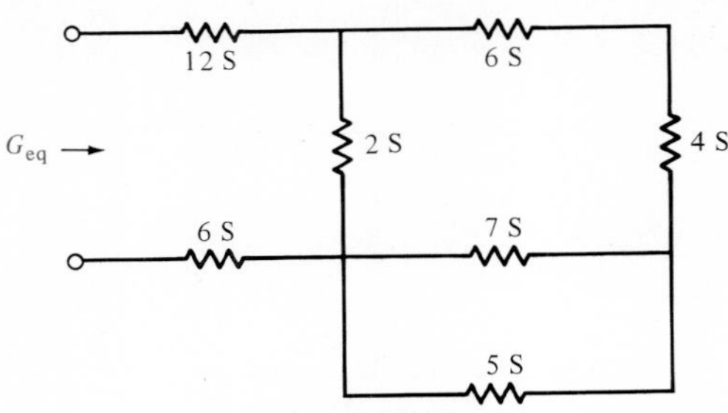

Figure P2.25

2.26. Find the equivalent resistance in Fig. P2.26.

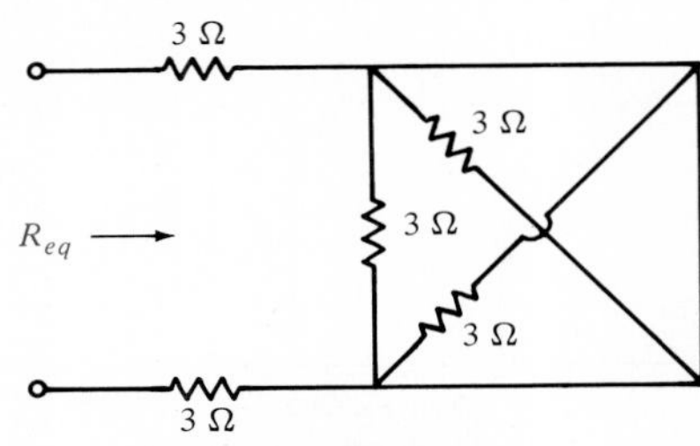

Figure P2.26

2.27. Find the equivalent resistance R_{eq} in Fig. P2.27.

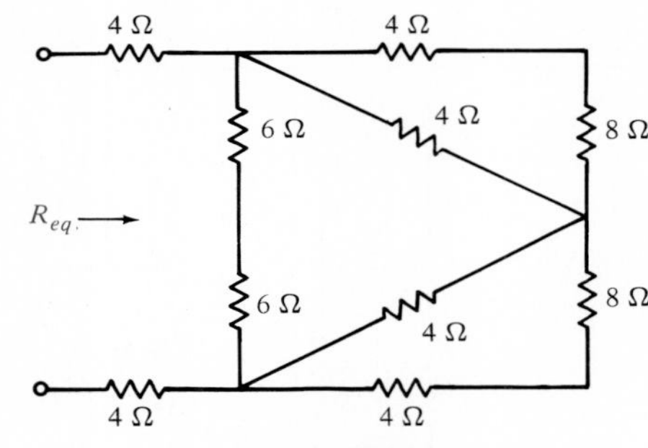

Figure P2.27

2.28. Find R_{eq} in the network shown in Fig. P2.28.

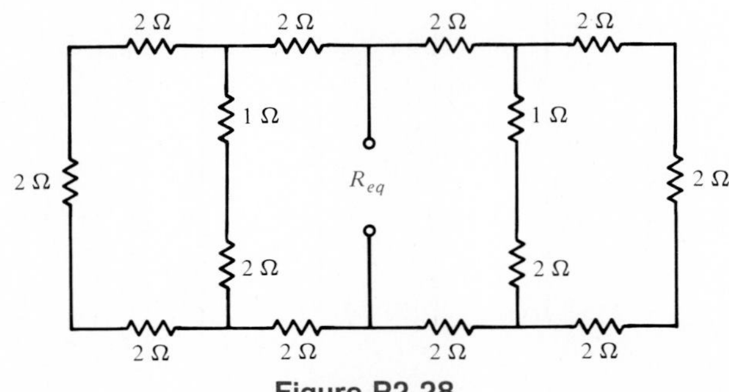

Figure P2.28

2.29. Find R_{eq} in the circuit shown in Fig. P2.29.

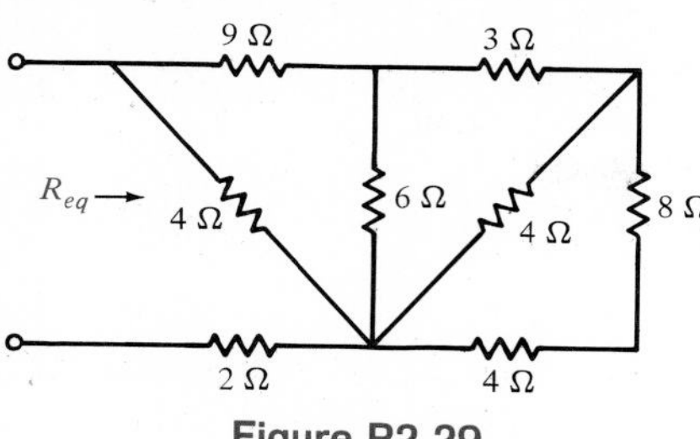

Figure P2.29

2.30. Find the equivalent resistance R_{eq} of the circuit in Fig. P2.30.

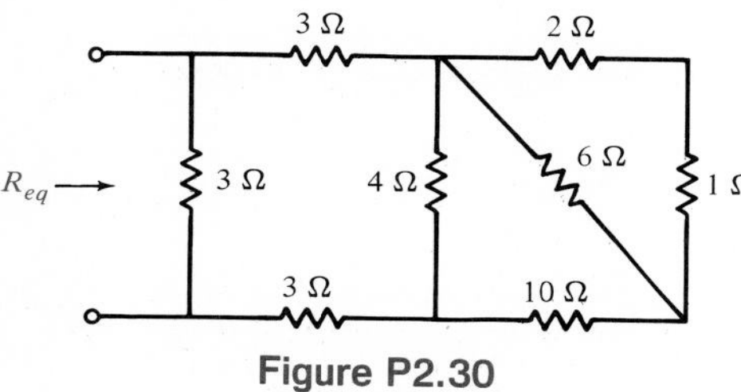

Figure P2.30

2.31. Find the equivalent resistance R_{eq} in the circuit shown in Fig. P2.31.

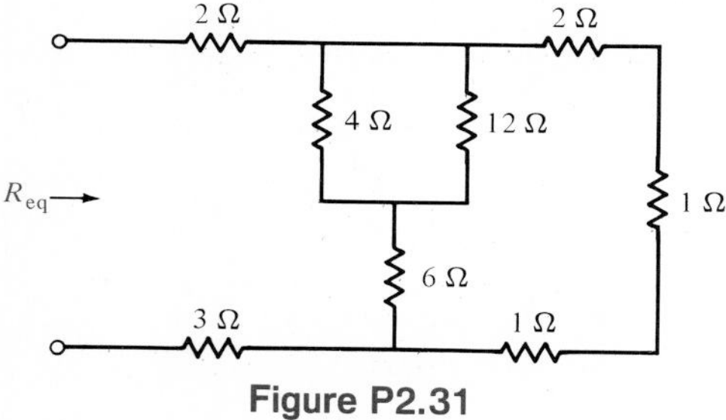

Figure P2.31

2.32. Find the equivalent resistance of the network in Fig. P2.32.

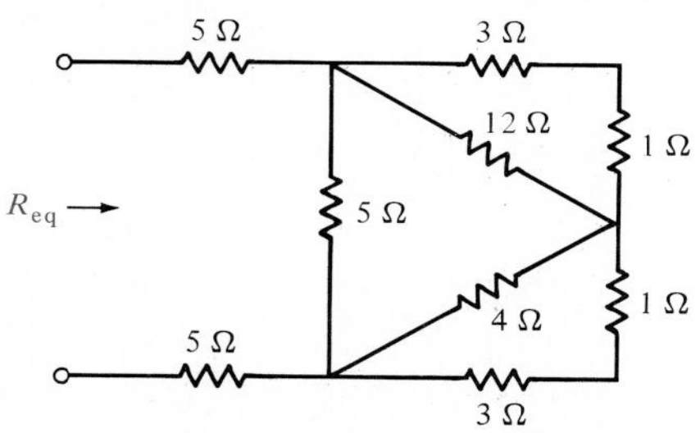

Figure P2.32

2.33. Given the resistor network in Fig. P2.33,

(a) Find R_{eq}.

(b) Find R_{eq} if the 2-Ω resistor is open-circuited.

(c) Find R_{eq} if the 12-Ω resistor is short-circuited.

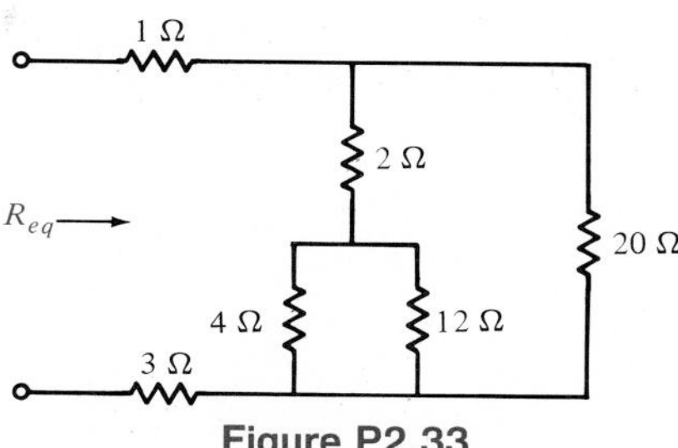

Figure P2.33

2.34. Find the current I in the network in Fig. P2.34.

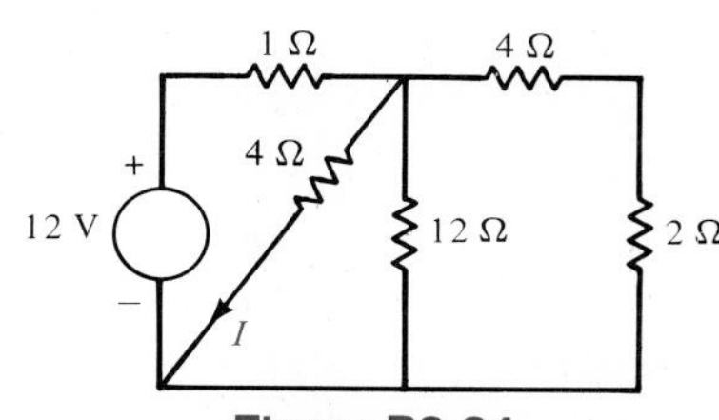

Figure P2.34

2.35. Find the voltage across the 1-Ω resistor in the circuit in Fig. P2.35.

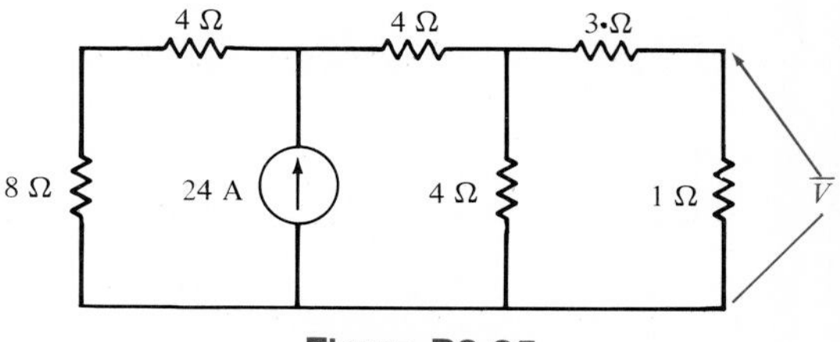

Figure P2.35

2.36. Find I in the network shown in Fig. P2.36.

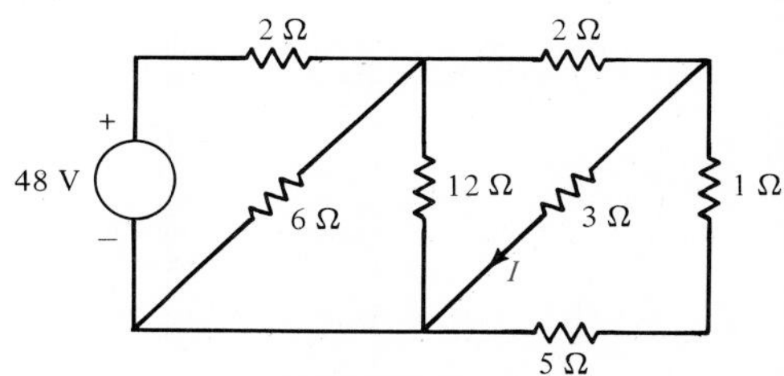

Figure P2.36

2.37. Given $I_o = 4$ A, find V in the circuit shown in Fig. P2.37.

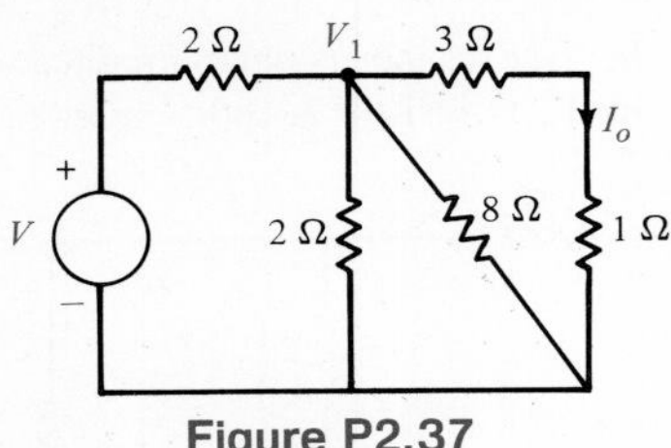

Figure P2.37

2.38. Find V in the circuit in Fig. P2.38.

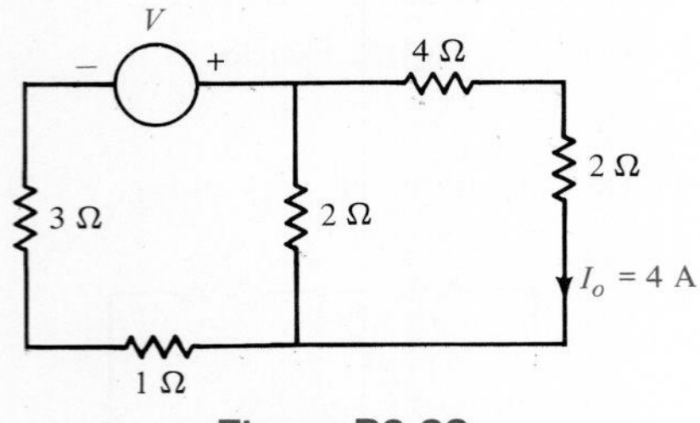

Figure P2.38

2.39. The power absorbed in the 2-Ω resistor is 32 W. Determine the value of the current source in Fig. P2.39.

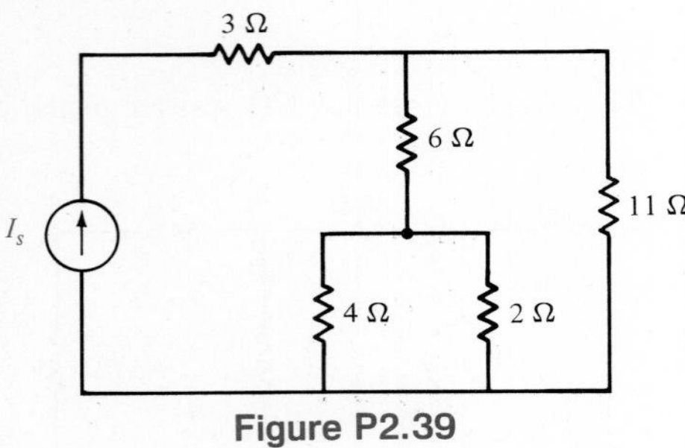

Figure P2.39

2.40. Find the voltage across the current source in Problem 2.39.

2.41. The power produced by the 48-V source in the circuit in Fig. P2.41 is 288 W. Find the value of R_o.

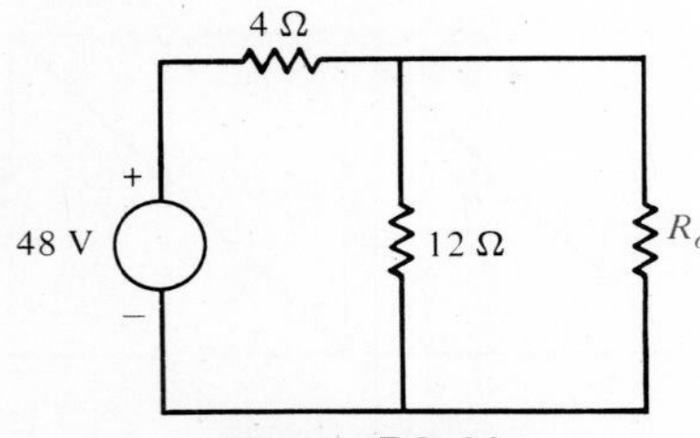

Figure P2.41

2.42. Determine the value of I_o in the network in Fig. P2.42 given that the power absorbed by the 6-Ω resistor is 24 W.

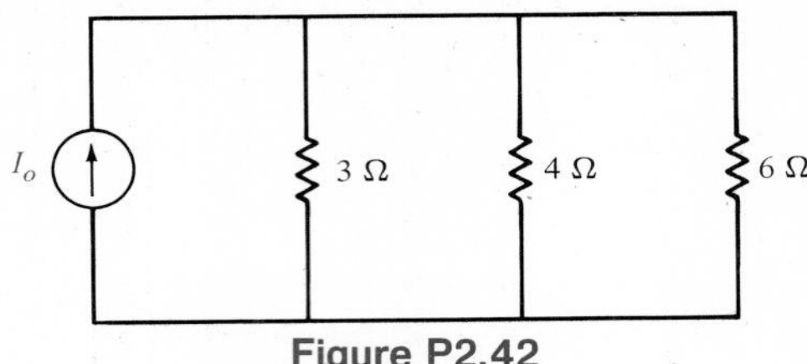

Figure P2.42

2.43. Find the power absorbed by the 8-Ω resistor in the network in Fig. P2.43.

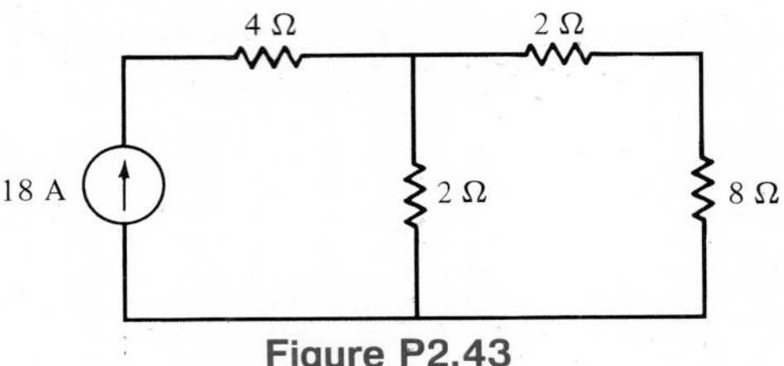

Figure P2.43

2.44. Given that the 3-Ω resistor in the network in Fig. P2.44 absorbs 27 W of power, determine the value of the current source I_1.

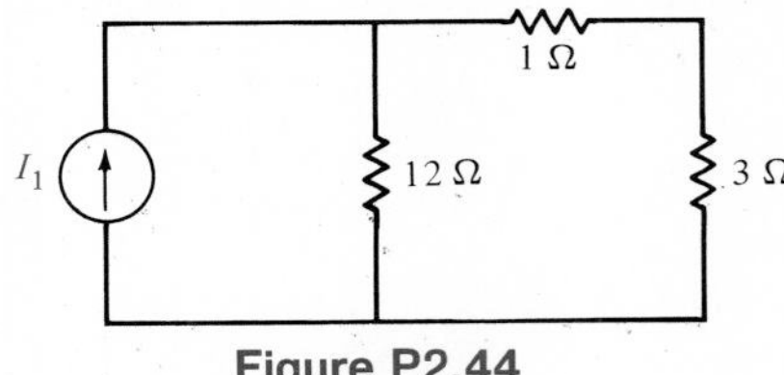

Figure P2.44

2.45. If $V_o = 4$ V in the circuit in Fig. P2.45, find I.

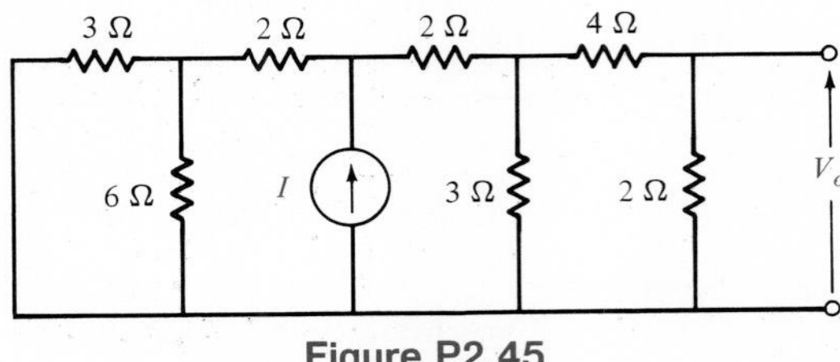

Figure P2.45

2.46. Find I_o in the network in Fig. P2.46.

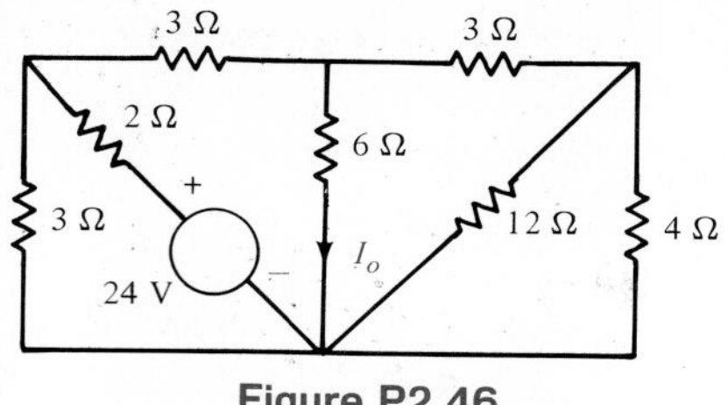

Figure P2.46

2.47. Find the current I_o in the circuit shown in Fig. P2.47.

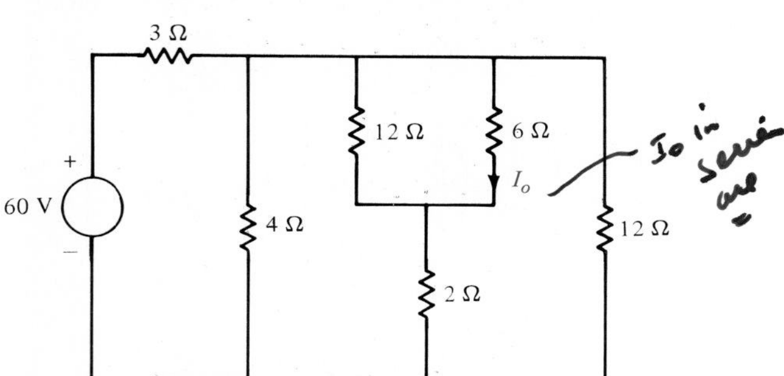

Figure P2.47

2.48. Find the voltage across the current source in Fig. P2.48.

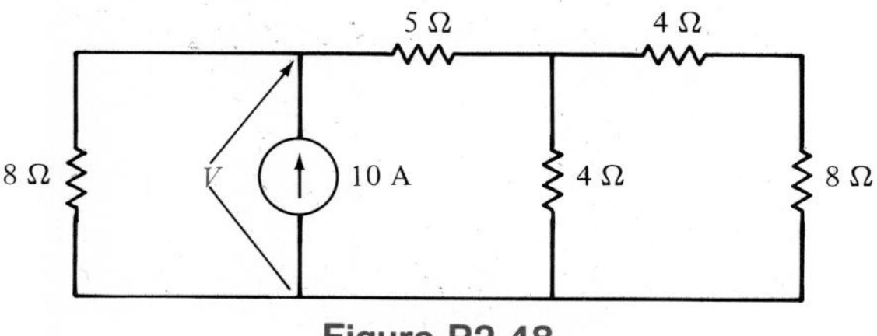

Figure P2.48

2.49. Determine the current I_o in the circuit shown in Fig. P2.49.

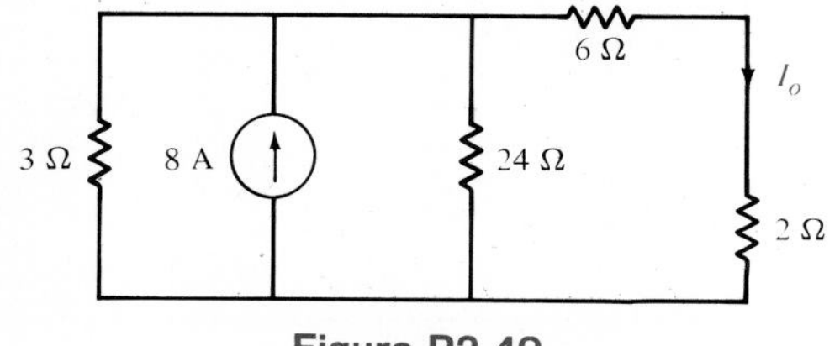

Figure P2.49

2.50. If $I_o = 4$ A, find V in the network in Fig. P2.50.

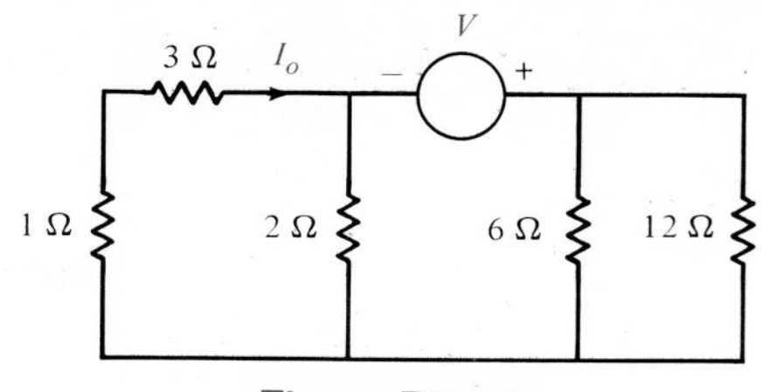

Figure P2.50

2.51. Determine I_o in the circuit shown in Fig. P2.51.

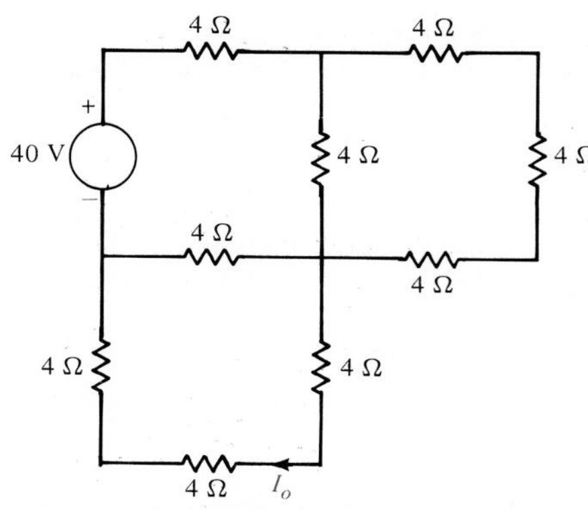

Figure P2.51

2.52. In Fig. P2.52, the current I is known to be 4 A. Determine the voltage source V.

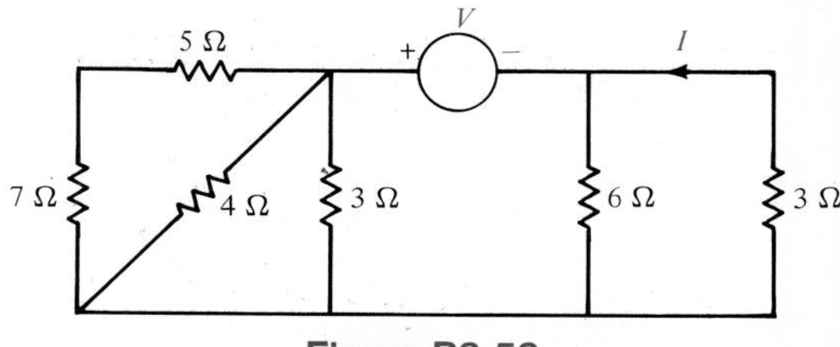

Figure P2.52

2.53. Find I_o in the network in Fig. P2.53 given that the voltage across the 3-Ω resistor is 12 V.

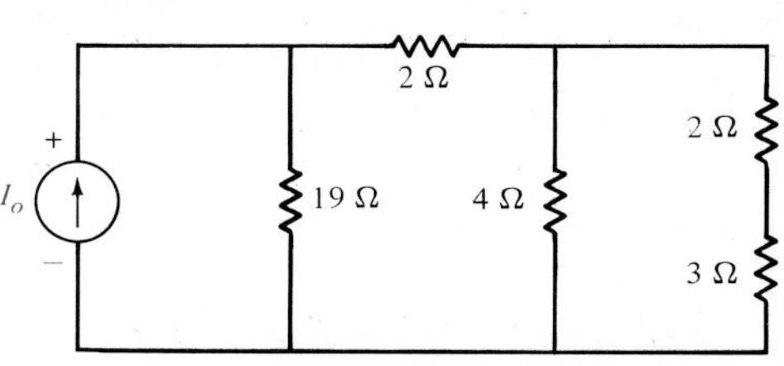

Figure P2.53

2.54. If $I = 1$ A, find V_o in the circuit in Fig. P2.54.

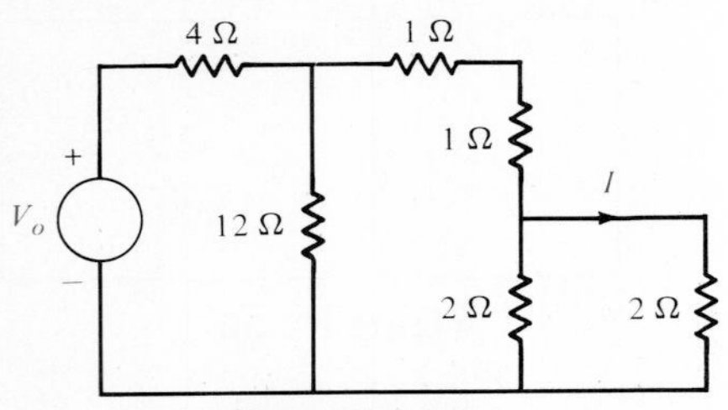

Figure P2.54

2.55. Given the network in Fig. P2.55,
(a) Find the current I_o if the 5-Ω resistor is short-circuited.
(b) Find the voltage V_o if the 3-Ω resistor is open-circuited.

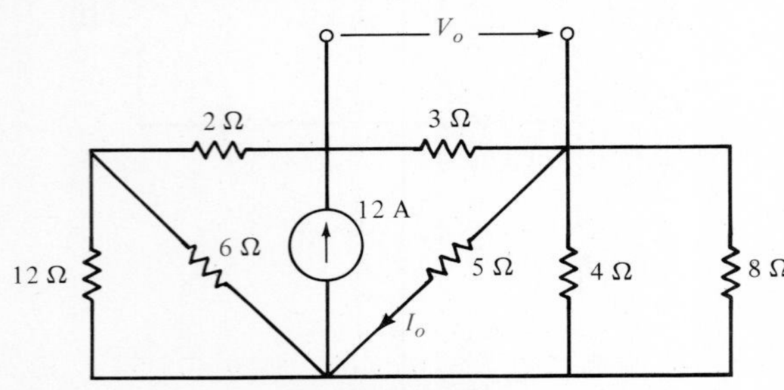

Figure P2.55

2.56. Given the network in Fig. P2.56,
(a) Find the current I_o if the 12-Ω resistor is short-circuited.
(b) Find the voltage V_o if the 2-Ω resistor is open-circuited.

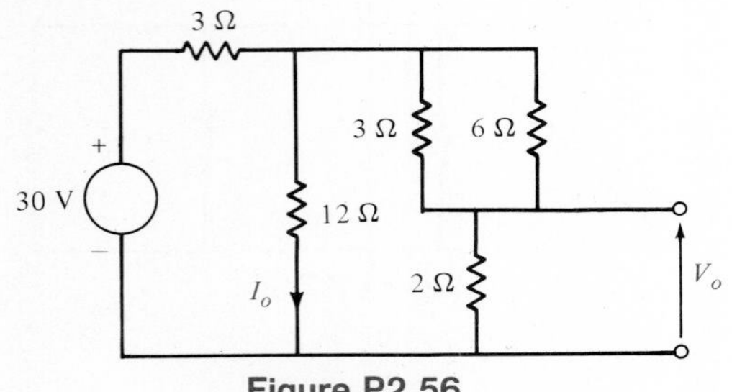

Figure P2.56

2.57. If V_1 is 20 V in the network in Fig. P2.57, compute V_o.

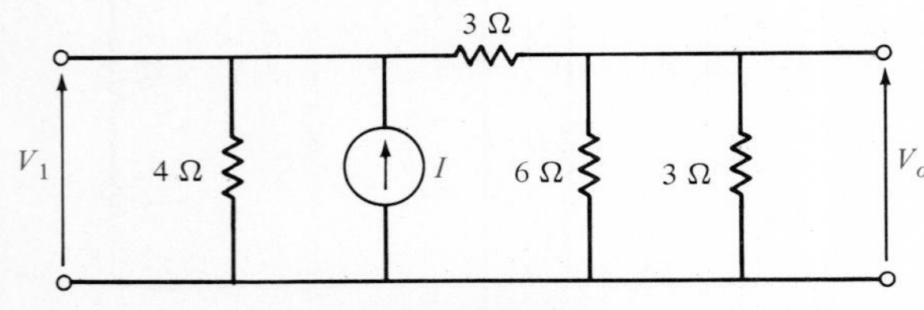

Figure P2.57

2.58. Find V in the network in Fig. P2.58 if $I_o = 12$ A.

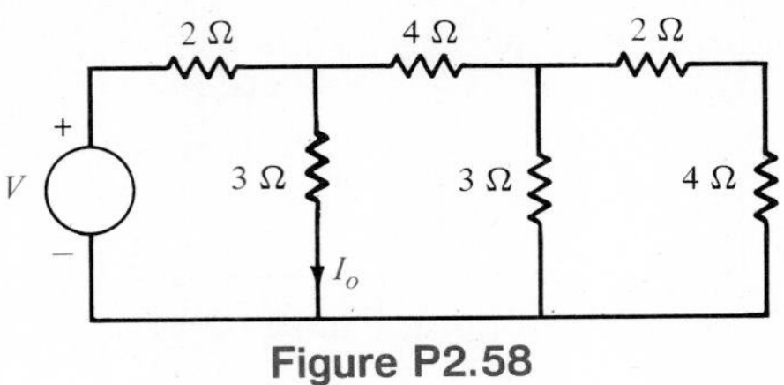

Figure P2.58

2.59. If V_o in the network in Fig. P2.59 is 12 V, find I.

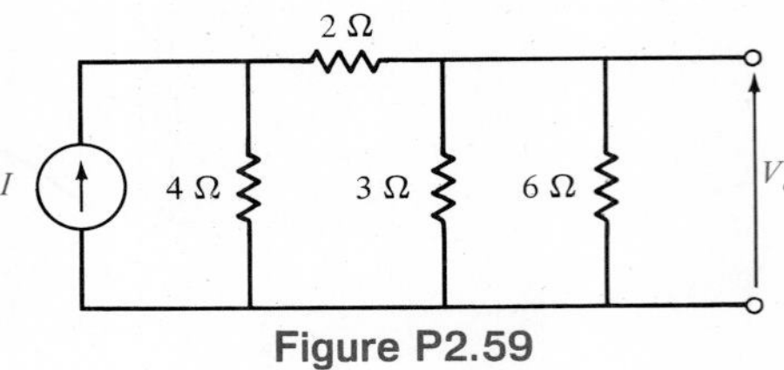

Figure P2.59

2.60. If the power dissipated in the 4-Ω resistor in the network in Fig. P2.60 is $P_{4\Omega} = 144$ W, find the power supplied by the voltage source.

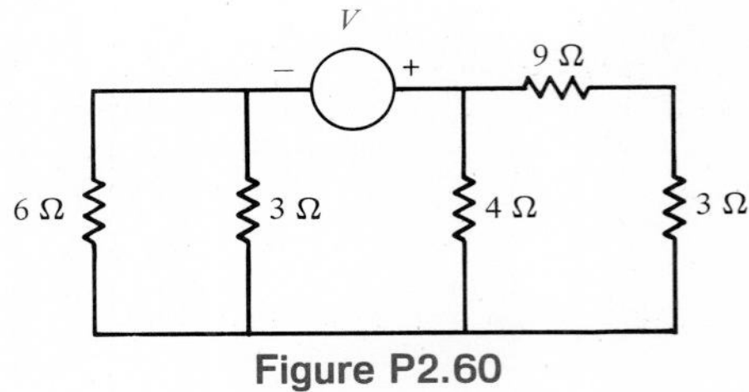

Figure P2.60

2.61. Find V_A in the network in Fig. P2.61.

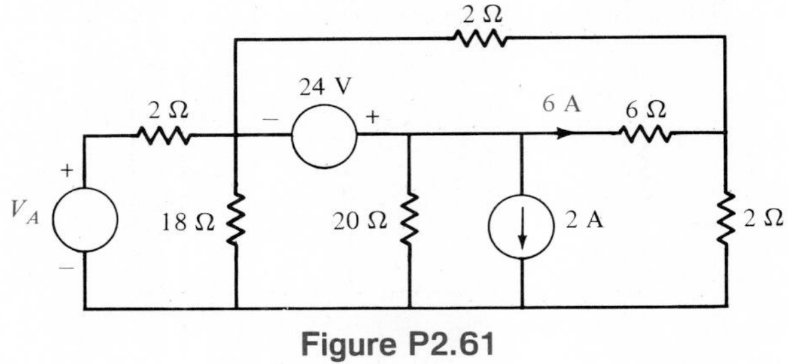

Figure P2.61

2.62. If V_o in the network in Fig. P2.62 is 12 V, find V_A.

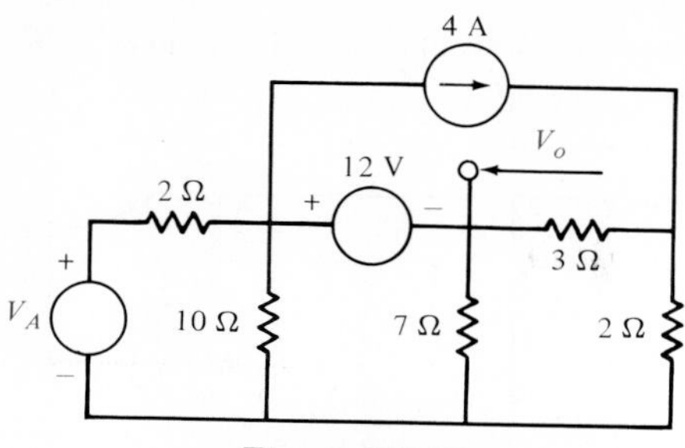

Figure P2.62

2.63. In the network in Fig. P2.63, find V_A.

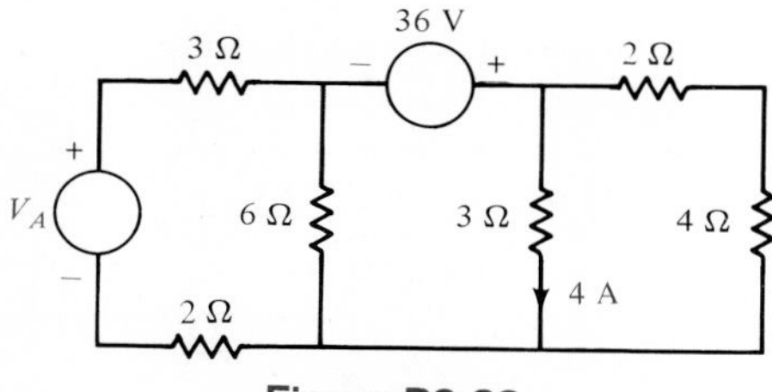

Figure P2.63

2.64. In the network in Fig. P2.64, find V_A.

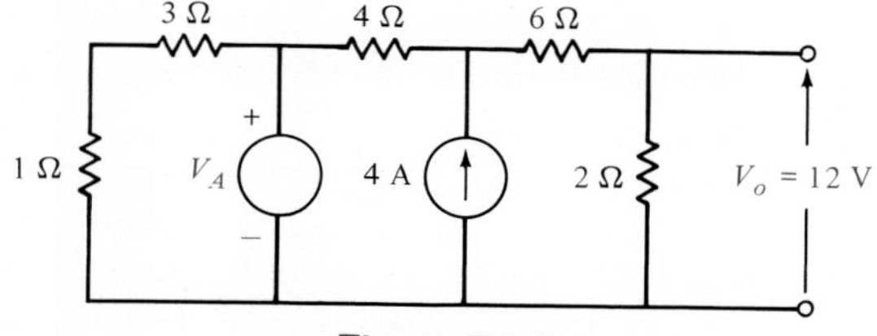

Figure P2.64

2.65. In the circuit shown in Fig. P2.65, V_o is known to be 12 V. Find the value of R.

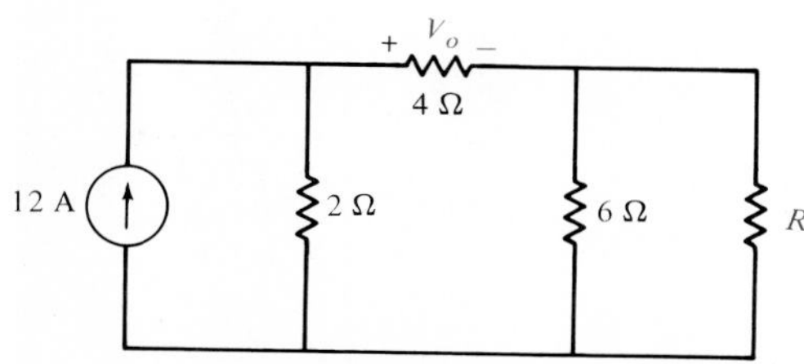

Figure P2.65

2.66. Determine the voltage V_o in the network in Fig. P2.66.

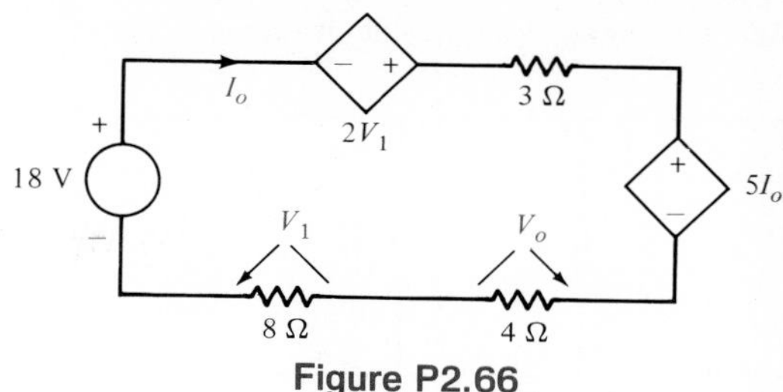

Figure P2.66

2.67. What must be the value of R_o in the network in Fig. P2.67 if the current I_o is 4 A?

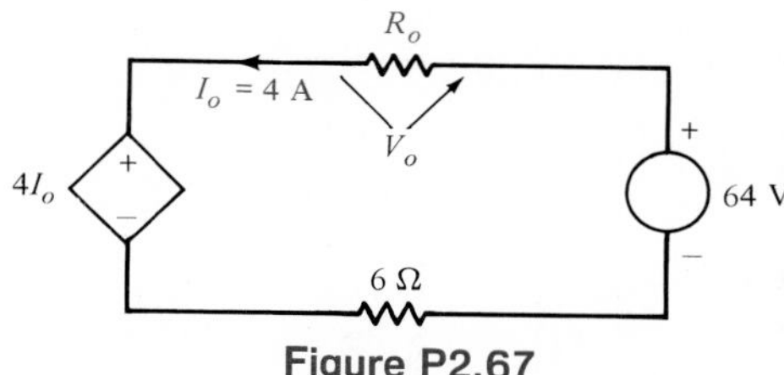

Figure P2.67

2.68. In the circuit shown in Fig. P2.68, find the current I_o.

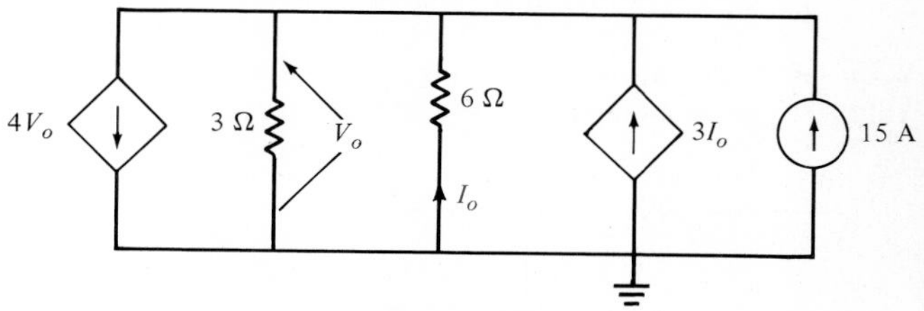

Figure P2.68

2.69. Find the voltage V in the network in Fig. P2.69.

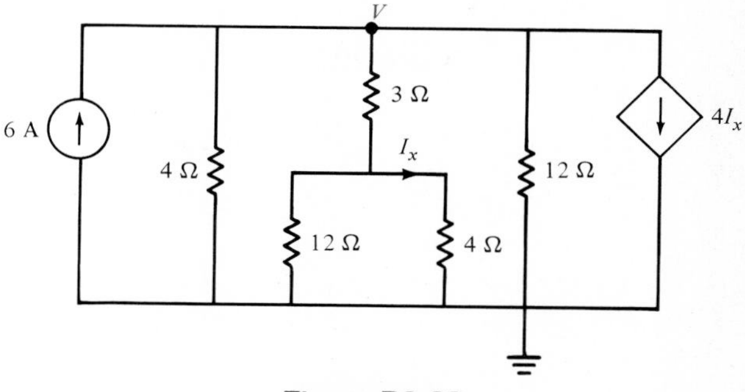

Figure P2.69

2.70. Find the voltage V in the circuit shown in Fig. P2.70.

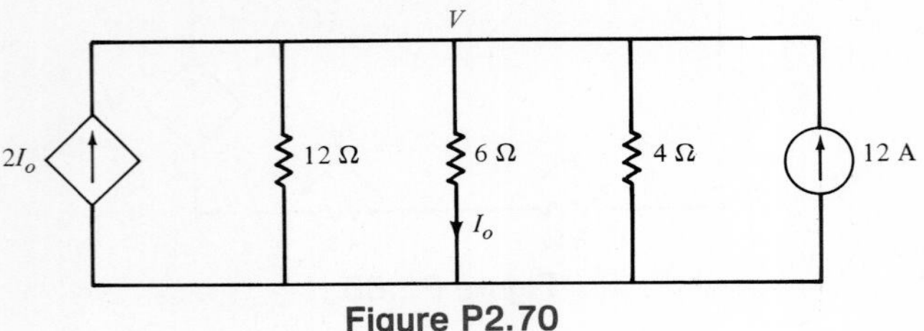

Figure P2.70

2.71. Determine the voltage V in Fig. P2.71.

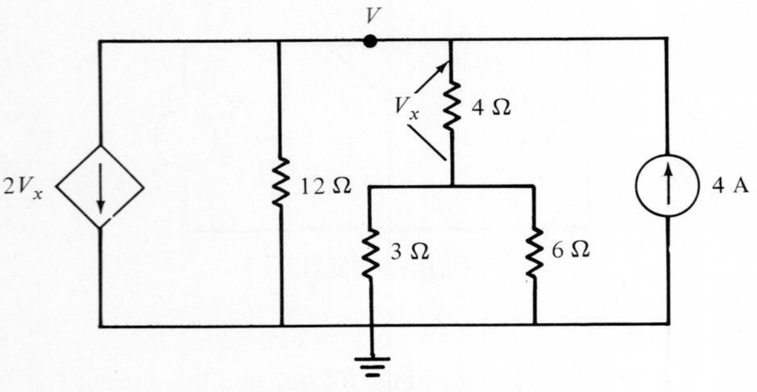

Figure P2.71

2.72. Given the circuit shown in Fig. P2.72, determine the output voltage V_o given an input voltage of $V_1 = 10$ mV.

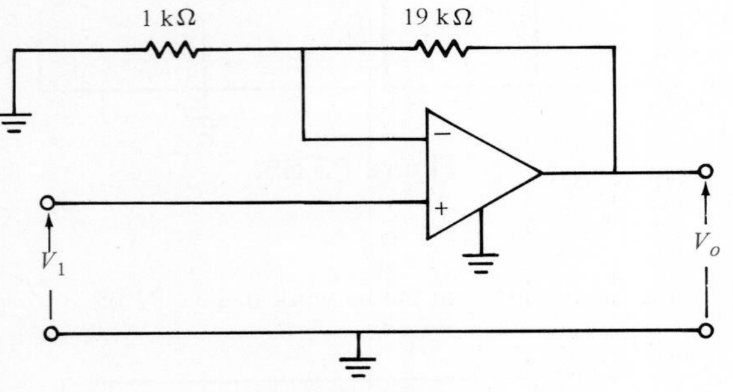

Figure P2.72

2.73. Determine the voltage gain of the op-amp circuit shown in Fig. P2.73.

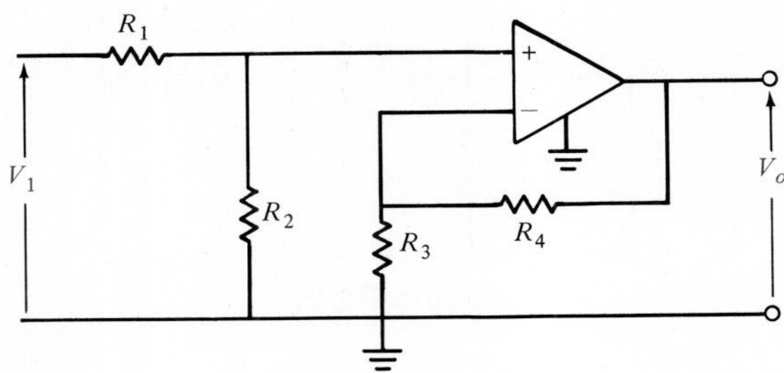

Figure P2.73

2.74. Find V_o in the circuit shown in Fig. P2.74.

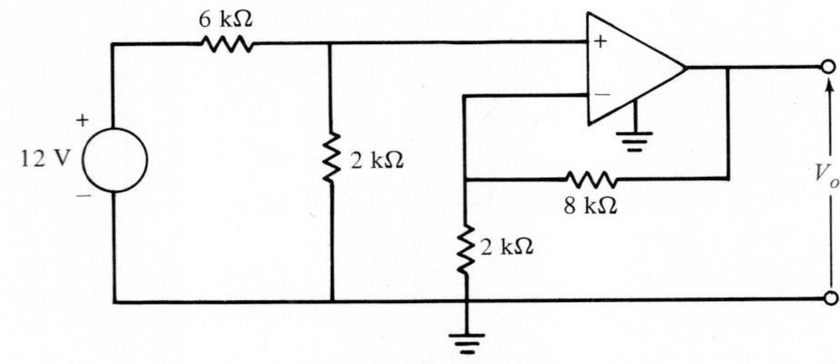

Figure P2.74

2.75. In the circuit shown in Fig. P2.75, select R_2 so that the output voltage $V_o = -24$ V.

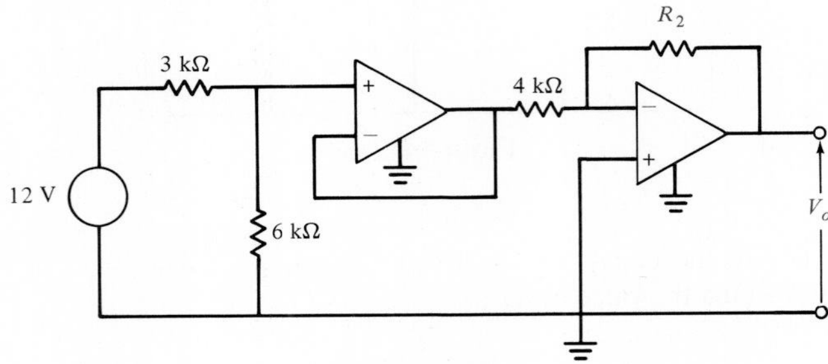

Figure P2.75

CHAPTER

Nodal and Loop Analysis Techniques

In Chapter 2 we analyzed the simplest possible circuits, those containing only a single node pair or a single loop. We found that these circuits can be completely analyzed via a single algebraic equation. In the case of the single-node pair circuit (i.e., one containing two nodes, one of which is a reference node), once the node voltage is known, we can calculate all the currents. In a single-loop circuit, once the loop current is known, we can calculate all the voltages.

In this chapter we extend our capabilities in a systematic manner so that we can calculate all currents and voltages in circuits that contain multiple nodes and loops. Our analyses are based primarily on two laws with which we are already familiar: Kirchhoff's current law (KCL) and Kirchhoff's voltage law (KVL). In a nodal analysis we employ KCL to determine the node voltages, and in a loop analysis we use KVL to determine the loop currents.

3.1 Nodal Analysis

In a nodal analysis the variables in the circuit are selected to be the node voltages. The node voltages are defined with respect to a common point in the circuit. One node is selected as the reference node and all other node voltages are defined with respect to that node. Quite often this node is the one to which the largest number of branches are connected. It is commonly called *ground* because it is said to be at ground-zero potential and it sometimes represents the chassis or ground line in a practical circuit.

We will select our variables as being positive with respect to the reference node. If

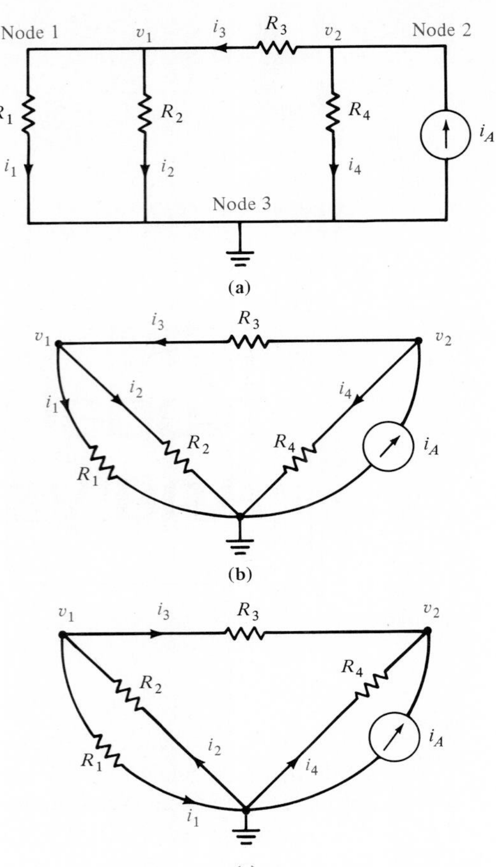

Figure 3.1 Three-node circuit.

one or more of the node voltages is actually negative with respect to the reference node, the analysis will indicate it.

Suppose for a moment that we know all the node voltages in a given network. Each resistive element in the circuit is connected between two nodes, one of which may be the reference node. Nevertheless, the voltage across each resistive element is known. Hence, using Ohm's law, we can calculate the current through the element. In this manner we can determine every voltage and every current in the circuit.

In a double-node circuit (i.e., one containing two nodes, one of which is the reference node), a single equation is required to solve for the unknown node voltage. In the case of an N-node circuit, $N - 1$ linearly independent simultaneous equations are required to determine the $N - 1$ unknown node voltages. These equations are written by employing KCL at $N - 1$ of the N nodes.

Consider, for example, the circuit shown in Fig. 3.1a. This circuit has three nodes. The node at the bottom is selected as the reference node and is so labeled using the ground symbol, $\overline{\underline{\underline{+}}}$. This reference node is assumed to be at zero potential and the node voltages v_1 and v_2 are defined with respect to this node; that is, if $v_1 = 4$ V and $v_2 =$

16 V, the voltage across R_1 and R_2 is 4 V, the voltage across R_4 and the current source is 16 V, and the voltage across R_3 is $16 - 4 = 12$ V. The right terminal of R_3 is positive with respect to the left terminal, and therefore the current will flow from right to left through R_3 as illustrated in Fig. 3.1. Thus, in applying Ohm's law to find the current in a resistor, we take the difference in potential between the two terminals of the resistor and divide it by the value of the resistor. As shown in Fig. 3.1, the current will flow from the terminal of higher potential to the terminal of lower potential.

The circuit is redrawn in Fig. 3.1b to indicate the nodes clearly. The branch currents are assumed to flow in the directions indicated in the figures. If one or more of the branch currents are actually flowing in a direction opposite to that assumed, the analysis will simply produce a branch current that is negative.

Applying KCL at node 1 yields

$$i_1 + i_2 - i_3 = 0$$

Using Ohm's law and noting that the reference node is at zero potential, we obtain

$$\frac{v_1 - 0}{R_1} + \frac{v_1 - 0}{R_2} - \frac{v_2 - v_1}{R_3} = 0$$

or

$$v_1\left(\frac{1}{R_1} + \frac{1}{R_2} + \frac{1}{R_3}\right) - v_2\frac{1}{R_3} = 0$$

which can, of course, be written

$$v_1(G_1 + G_2 + G_3) - v_2(G_3) = 0$$

At node 2, KCL for the currents leaving the node is

$$+i_3 + i_4 - i_A = 0$$

Note that i_A is negative because this current is entering the node. This equation can be written

$$+\frac{v_2 - v_1}{R_3} + \frac{v_2 - 0}{R_4} - i_A = 0$$

or

$$-v_1\frac{1}{R_3} + v_2\left(\frac{1}{R_3} + \frac{1}{R_4}\right) = i_A$$

which can also be written

$$-v_1(G_3) + v_2(G_3 + G_4) = i_A$$

Therefore, the two equations, which when solved yield the node voltages, are

$$\begin{aligned} v_1\left(\frac{1}{R_1} + \frac{1}{R_2} + \frac{1}{R_3}\right) - v_2\frac{1}{R_3} &= 0 \\ -v_1\frac{1}{R_3} + v_2\left(\frac{1}{R_3} + \frac{1}{R_4}\right) &= i_A \end{aligned} \tag{3.1}$$

Note that the analysis has produced two simultaneous equations in the unknowns v_1 and v_2. They can be solved using any convenient technique. In Appendix A two methods for solving simultaneous equations are presented. The first involves the use of determinants with Cramer's rule, and the second involves the use of matrices. As shown in the appendix, Eq. (3.1) may be rewritten in matrix form as

$$\mathbf{A}\mathbf{v} = \mathbf{i} \tag{3.2}$$

where

$$\mathbf{A} = \begin{bmatrix} \frac{1}{R_1} + \frac{1}{R_2} + \frac{1}{R_3} & -\frac{1}{R_3} \\ -\frac{1}{R_3} & \frac{1}{R_3} + \frac{1}{R_4} \end{bmatrix}, \quad \mathbf{v} = \begin{bmatrix} v_1 \\ v_2 \end{bmatrix}, \quad \text{and} \quad \mathbf{i} = \begin{bmatrix} 0 \\ i_A \end{bmatrix}$$

Thus

$$\begin{bmatrix} \frac{1}{R_1} + \frac{1}{R_2} + \frac{1}{R_3} & -\frac{1}{R_3} \\ -\frac{1}{R_3} & \frac{1}{R_3} + \frac{1}{R_4} \end{bmatrix} \begin{bmatrix} v_1 \\ v_2 \end{bmatrix} = \begin{bmatrix} 0 \\ i_A \end{bmatrix} \tag{3.3}$$

In general, the solution of Eq. (3.2) is

$$\mathbf{v} = \mathbf{A}^{-1}\mathbf{i} \tag{3.4}$$

where $\mathbf{A}^{-1}$ is the inverse of matrix $\mathbf{A}$. As a rule, matrix methods used in conjunction with a digital computer are extremely useful in circuit analysis.

Note that a nodal analysis employs KCL in conjunction with Ohm's law. Once the direction of the branch currents has been *assumed,* then Ohm's law, as illustrated by Fig. 2.1 and expressed by Eq. (2.1), is used to express the branch currents in terms of the unknown node voltages. We can assume the currents to be in any direction. However, once we assume a particular direction, we must be very careful to write the currents correctly in terms of the node voltages using Ohm's law. For example, suppose we had assumed the direction of the branch currents as shown in Fig. 3.1(c). Then noting that when we employ Ohm's law, as illustrated in Fig. 2.1 and expressed in Eq. (2.1), the current in a resistor is equal to the difference in potential between the two terminals of the resistor divided by the value of the resistor, and that the current flows from the terminal of higher potential toward the terminal of lower potential, the branch currents for the circuit in Fig. 3.1c are

$$i_1 = \frac{v_1 - 0}{R_1} \qquad i_3 = \frac{v_1 - v_2}{R_3}$$

$$i_2 = \frac{0 - v_1}{R_2} \qquad i_4 = \frac{0 - v_2}{R_4}$$

KCL applied at the two nodes of the circuit in Fig. 3.1c yields

$$i_1 - i_2 + i_3 = 0$$

and

$$i_3 + i_4 + i_A = 0$$

When we combine these equations for the circuit in Fig. 3.1c, we obtain the same equations for the node voltage that we derived for the network in Fig. 3.1a and b.

EXAMPLE 3.1

Suppose that the network shown in Fig. 3.1 has the following parameters: $R_1 = 0.5\ \Omega$, $R_1 = 1\ \Omega$, $R_3 = 1\ \Omega$, $R_4 = 0.5\ \Omega$, and $i_A = I_A = 11$ A. Then Eq. (3.1) becomes

$$4V_1 - V_2 = 0$$

$$-V_1 + 3V_2 = 11$$

where as stated in Chapter 1, we are employing capital letters because the voltages are constant. Multiplying the first equation by 3 and adding it to the second equation yields

$$11V_1 = 11$$

or

$$V_1 = 1\text{ V}$$

Substituting this value into either equation above yields

$$V_2 = 4\text{ V}$$

The two circuit equations can also be solved using Cramer's rule. The determinant for the two equations is

$$\Delta = \begin{vmatrix} 4 & -1 \\ -1 & 3 \end{vmatrix}$$

Using Cramer's rule gives us

$$V_1 = \frac{\begin{vmatrix} 0 & -1 \\ 11 & 3 \end{vmatrix}}{\begin{vmatrix} 4 & -1 \\ -1 & 3 \end{vmatrix}} = \frac{11}{12 - 1} = 1\text{ V}$$

and

$$V_2 = \frac{\begin{vmatrix} 4 & 0 \\ -1 & 11 \end{vmatrix}}{\begin{vmatrix} 4 & -1 \\ -1 & 3 \end{vmatrix}} = \frac{44}{11} = 4\text{ V}$$

The circuit equations can also be solved using matrix analysis. In matrix form the equations are

$$\begin{bmatrix} 4 & -1 \\ -1 & 3 \end{bmatrix} \begin{bmatrix} V_1 \\ V_2 \end{bmatrix} = \begin{bmatrix} 0 \\ 11 \end{bmatrix}$$

and therefore

$$\begin{bmatrix} V_1 \\ V_2 \end{bmatrix} = \begin{bmatrix} 4 & -1 \\ -1 & 3 \end{bmatrix}^{-1} \begin{bmatrix} 0 \\ 11 \end{bmatrix}$$

To calculate the inverse of $\mathbf{A}$, we need the adjoint and the determinant. The adjoint is

$$\text{adj } \mathbf{A} = \begin{bmatrix} 3 & 1 \\ 1 & 4 \end{bmatrix}$$

and the determinant is

$$|\mathbf{A}| = (3)(4) - (-1)(-1) = 11$$

Therefore,

$$\begin{bmatrix} V_1 \\ V_2 \end{bmatrix} = \frac{1}{11}\begin{bmatrix} 3 & 1 \\ 1 & 4 \end{bmatrix}\begin{bmatrix} 0 \\ 11 \end{bmatrix} = \begin{bmatrix} 1 \\ 4 \end{bmatrix}$$

Thus $V_1 = 1$ V and $V_2 = 4$ V.

Knowing the node voltages, we can determine all the currents from Ohm's law:

$$I_1 = \frac{V_1}{R_1} = \frac{1 \text{ V}}{0.5} = 2 \text{ A}$$

$$I_2 = \frac{V_1}{R_2} = \frac{1 \text{ V}}{1} = 1 \text{ A}$$

and

$$I_3 = \frac{V_2 - V_1}{R_3}$$

$$= \frac{4 - 1}{1}$$

$$= 3 \text{ A}$$

Note that the positive value of I_3 simply means that the current is actually flowing in the circuit from right to left as indicated by the values of I. The sum of the currents leaving node 1,

$$I_1 + I_2 - I_3 = 2 + 1 - 3 = 0$$

The current

$$I_4 = \frac{V_2}{R_4} = \frac{4 \text{ V}}{0.5} = 8 \text{ A}$$

and the sum of the currents leaving node 2 is

$$+I_3 + I_4 - I_A = 3 + 8 - 11 = 0$$

Since KCL is satisfied at each node, the answers check. ■

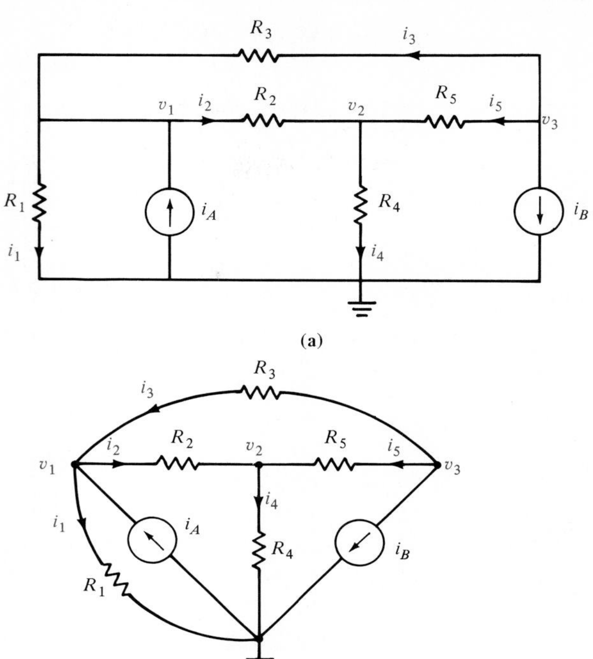

Figure 3.2 Four-node circuit.

Let us now examine the circuit in Fig. 3.2a and redrawn in Fig. 3.2b. The current directions are assumed as shown in the figures.

At node 1 KCL yields

$$i_1 - i_A + i_2 - i_3 = 0$$

or

$$\frac{v_1}{R_1} - i_A + \frac{v_1 - v_2}{R_2} - \frac{v_3 - v_1}{R_3} = 0$$

$$v_1\left(\frac{1}{R_1} + \frac{1}{R_2} + \frac{1}{R_3}\right) - v_2\frac{1}{R_2} - v_3\frac{1}{R_3} = i_A$$

At node 2 KCL yields

$$-i_2 + i_4 - i_5 = 0$$

or

$$-\frac{v_1 - v_2}{R_2} + \frac{v_2}{R_4} - \frac{v_3 - v_2}{R_5} = 0$$

$$-v_1\frac{1}{R_2} + v_2\left(\frac{1}{R_2} + \frac{1}{R_4} + \frac{1}{R_5}\right) - v_3\frac{1}{R_5} = 0$$

At node 3 the equation is

$$i_3 + i_5 + i_B = 0$$

or

$$\frac{v_3 - v_1}{R_3} + \frac{v_3 - v_2}{R_5} + i_B = 0$$

$$-v_1 \frac{1}{R_3} - v_2 \frac{1}{R_5} + v_3\left(\frac{1}{R_3} + \frac{1}{R_5}\right) = -i_B$$

Grouping the node equations together, we obtain

$$\begin{aligned} v_1\left(\frac{1}{R_1} + \frac{1}{R_2} + \frac{1}{R_3}\right) - v_2 \frac{1}{R_2} - v_3 \frac{1}{R_3} &= i_A \\ -v_1 \frac{1}{R_2} + v_2\left(\frac{1}{R_2} + \frac{1}{R_4} + \frac{1}{R_5}\right) - v_3 \frac{1}{R_5} &= 0 \\ -v_1 \frac{1}{R_3} - v_2 \frac{1}{R_5} + v_3\left(\frac{1}{R_3} + \frac{1}{R_5}\right) &= -i_B \end{aligned} \tag{3.5}$$

Note that our analysis has produced three simultaneous equations in the three unknown node voltages v_1, v_2, and v_3. The equations can also be written in matrix form as

$$\begin{bmatrix} \frac{1}{R_1} + \frac{1}{R_2} + \frac{1}{R_3} & -\frac{1}{R_2} & -\frac{1}{R_3} \\ -\frac{1}{R_2} & \frac{1}{R_2} + \frac{1}{R_4} + \frac{1}{R_5} & -\frac{1}{R_5} \\ -\frac{1}{R_3} & -\frac{1}{R_5} & \frac{1}{R_3} + \frac{1}{R_5} \end{bmatrix} \begin{bmatrix} v_1 \\ v_2 \\ v_3 \end{bmatrix} = \begin{bmatrix} i_A \\ 0 \\ -i_B \end{bmatrix} \tag{3.6}$$

EXAMPLE 3.2

Given that the network shown in Fig. 3.2 has the following parameters, determine the node voltages.

$$R_1 = 0.5\ \Omega \qquad R_3 = 1\ \Omega \qquad R_5 = 0.5\ \Omega \qquad i_B = I_B = 1\ \text{A}$$

$$R_2 = 0.25\ \Omega \qquad R_4 = 1\ \Omega \qquad i_A = I_A = 4\ \text{A}$$

The data indicate that

$$G_1 = 2 \qquad G_3 = 1 \qquad G_5 = 2 \qquad G_2 = 4 \qquad G_4 = 1$$

The node equations are

$$\begin{aligned} 7V_1 - 4V_2 - V_3 &= 4 \\ -4V_1 + 7V_2 - 2V_3 &= 0 \\ -V_1 - 2V_2 + 3V_3 &= -1 \end{aligned}$$

The determinant for this set of equations is

$$\Delta = \begin{vmatrix} 7 & -4 & -1 \\ -4 & 7 & -2 \\ -1 & -2 & 3 \end{vmatrix}$$

Expanding this determinant using the first row yields

$$\Delta = 7\begin{vmatrix} 7 & -2 \\ -2 & 3 \end{vmatrix} - (-4)\begin{vmatrix} -4 & -2 \\ -1 & 3 \end{vmatrix} - 1\begin{vmatrix} -4 & 7 \\ -1 & -2 \end{vmatrix}$$

$$= 7(17) + 4(-14) - 1(15)$$

$$= 48$$

Employing Cramer's rule gives us

$$V_1 = \frac{\Delta_1}{\Delta} = \frac{\begin{vmatrix} 4 & -4 & -1 \\ 0 & 7 & -2 \\ -1 & -2 & 3 \end{vmatrix}}{48}$$

Expanding this determinant using the first column, we obtain

$$\Delta_1 = 4\begin{vmatrix} 7 & -2 \\ -2 & 3 \end{vmatrix} - 1\begin{vmatrix} -4 & -1 \\ 7 & -2 \end{vmatrix}$$

$$= 53$$

Therefore,

$$V_1 = \frac{\Delta_1}{\Delta} = \frac{53}{48} = 1.104 \text{ V}$$

Similarly,

$$V_2 = \frac{\Delta_2}{\Delta} = \frac{\begin{vmatrix} 7 & 4 & -1 \\ -4 & 0 & -2 \\ -1 & -1 & 3 \end{vmatrix}}{48}$$

Expansion using the second column yields

$$V_2 = \frac{1}{48}[4(14) - 1(18)]$$

$$= \frac{38}{48} = 0.792 \text{ V}$$

Finally,

$$V_3 = \frac{\Delta_3}{\Delta} = \frac{\begin{vmatrix} 7 & -4 & 4 \\ -4 & 7 & 0 \\ -1 & -2 & -1 \end{vmatrix}}{48}$$

Expanding Δ_3 using the third column produces

$$V_3 = \frac{4(15) - 1(33)}{48}$$

$$= \frac{27}{48} = 0.563 \text{ V}$$

The matrix equation for the circuit is

$$\begin{bmatrix} 7 & -4 & -1 \\ -4 & 7 & -2 \\ -1 & -2 & 3 \end{bmatrix} \begin{bmatrix} V_1 \\ V_2 \\ V_3 \end{bmatrix} = \begin{bmatrix} 4 \\ 0 \\ -1 \end{bmatrix}$$

and therefore

$$\begin{bmatrix} V_1 \\ V_2 \\ V_3 \end{bmatrix} = \begin{bmatrix} 7 & -4 & -1 \\ -4 & 7 & -2 \\ -1 & -2 & 3 \end{bmatrix}^{-1} \begin{bmatrix} 4 \\ 0 \\ -1 \end{bmatrix}$$

The cofactors that produce the adjoint of the matrix are

$$A_{11} = \begin{vmatrix} 7 & -2 \\ -2 & 3 \end{vmatrix} = 17, \qquad A_{12} = (-1)\begin{vmatrix} -4 & -2 \\ -1 & 3 \end{vmatrix} = 14,$$

$$A_{13} = \begin{vmatrix} -4 & 7 \\ -1 & -2 \end{vmatrix} = 15$$

The remaining cofactors are calculated as

$$A_{21} = 14 \qquad A_{22} = 20 \qquad A_{23} = 18$$

$$A_{31} = 15 \qquad A_{32} = 18 \qquad A_{33} = 33$$

The determinant of the matrix was calculated earlier to be

$$\Delta = |A| = 48$$

Therefore,

$$\begin{bmatrix} V_1 \\ V_2 \\ V_3 \end{bmatrix} = \frac{1}{48}\begin{bmatrix} 17 & 14 & 15 \\ 14 & 20 & 18 \\ 15 & 18 & 33 \end{bmatrix} \begin{bmatrix} 4 \\ 0 \\ -1 \end{bmatrix}$$

Hence

$$V_1 = \frac{1}{48}[(4)(17) - (1)(15)] = +1.104 \text{ V}$$

$$V_2 = \frac{1}{48}[(4)(14) - (1)(18)] = +0.792 \text{ V}$$

$$V_3 = \frac{1}{48}[(4)(15) - (1)(33)] = +0.563 \text{ V}$$

We can now calculate all the currents.

$$I_1 = \frac{V_1}{R_1} = \frac{1.10}{0.5} = 2.21 \text{ A}$$

$$I_2 = \frac{V_1 - V_2}{R_2} = \frac{1.10 - 0.79}{0.25} = 1.25 \text{ A}$$

$$I_3 = \frac{V_3 - V_1}{R_3} = \frac{0.56 - 1.10}{1} = -0.54 \text{ A}$$

$$I_4 = \frac{V_2}{R_4} = \frac{0.79}{1} = 0.79 \text{ A}$$

$$I_5 = \frac{V_3 - V_2}{R_2} = \frac{0.56 - 0.79}{0.5} = -0.46 \text{ A}$$

We can now check our results using KCL and KVL. At node 1 the currents entering the node are

$$-I_1 + I_A - I_2 + I_3 = -2.21 + 4 - 1.25 - 0.54 = 0$$

At node 2 the currents entering the node are

$$I_2 - I_4 + I_5 = 1.25 - 0.79 - 0.46 = 0$$

At node 3 the currents entering the node are

$$-I_3 - I_5 - I_B = 0.54 + 0.46 - 1 = 0$$

■

At this point it is important that we note the symmetrical form of the equations that describe the two previous networks. Equations (3.1) and (3.3) and (3.5) and (3.6) exhibit the same type of symmetrical form. The A matrix for each network (3.3) and (3.6) is a symmetrical matrix. This symmetry is not accidental. Networks containing only resistors and independent current sources will always exhibit this symmetrical form. We can take advantage of this fact and learn to write the equations by inspection. Note in the first equation of (3.1) that the coefficient of v_1 is the sum of all the conductances connected to node 1 and the coefficient of v_2 is the negative of the conductances connected between node 1 and node 2. The right-hand side of the equation is the sum of the currents entering node 1 through current sources. This equation is KCL at node 1. In the second equation in (3.1), the coefficient of v_2 is the sum of all the conductances connected to node 2, the coefficient of v_1 is the negative of the conductance connected between node 2 and node 1, and the right-hand side of the equation is the sum of the currents entering node 2 through current sources. This equation is KCL at node 2. Similarly, in the first equation in (3.5) the coefficient of v_1 is the sum of the conductances connected to node 1, the coefficient of v_2 is the negative of the conductance connected between node 1 and node 2, the coefficient of v_3 is the negative of the conductance connected between node 1 and node 3, and the right-hand side of the equation is the sum of the currents entering node 1 through current sources. The other two equations in (3.5) are obtained in a similar manner. In general, if KCL is applied to node j with node voltage v_j, the coefficient of v_j is the sum of all the conductances connected to node j and the coefficients of the other node voltages (e.g., v_{j-1}, v_{j+1}) are the negative of the sum of the conductances connected directly between these nodes and node j. The right-hand side of the equation is equal to the sum of the currents entering the node via current sources. Therefore, the left-hand side of the equation represents the sum of the currents leaving node j and the right-hand side of the equation represents the currents entering node j.

EXAMPLE 3.3

Let us apply what we have just learned and write down the equations for the network in Fig. 3.3 by inspection.

The equations are

$$v_1\left(\frac{1}{R_1 + R_2} + \frac{1}{R_3}\right) - v_2\frac{1}{R_1 + R_2} = i_A$$

$$-v_1\frac{1}{R_1 + R_2} + v_2\left(\frac{1}{R_1 + R_2} + \frac{1}{R_5}\right) - v_4\frac{1}{R_5} = -i_B$$

$$v_3\left(\frac{1}{R_4} + \frac{1}{R_7}\right) - v_5\frac{1}{R_7} = -i_A$$

$$-v_2\frac{1}{R_5} + v_4\left(\frac{1}{R_5} + \frac{1}{R_6}\right) = -i_c$$

$$-v_3\frac{1}{R_7} + v_5\frac{1}{R_7} = i_c$$

or in matrix form the equations become

$$\begin{bmatrix} \frac{1}{R_1 + R_2} + \frac{1}{R_3} & -\frac{1}{R_1 + R_2} & 0 & 0 & 0 \\ \frac{-1}{R_1 + R_2} & \frac{1}{R_1 + R_2} + \frac{1}{R_5} & 0 & -\frac{1}{R_5} & 0 \\ 0 & 0 & \frac{1}{R_4} + \frac{1}{R_7} & 0 & -\frac{1}{R_7} \\ 0 & -\frac{1}{R_5} & 0 & \frac{1}{R_5} + \frac{1}{R_6} & 0 \\ 0 & 0 & -\frac{1}{R_7} & 0 & \frac{1}{R_7} \end{bmatrix} \begin{bmatrix} v_1 \\ v_2 \\ v_3 \\ v_4 \\ v_5 \end{bmatrix} = \begin{bmatrix} i_A \\ -i_B \\ -i_A \\ -i_C \\ i_C \end{bmatrix}$$

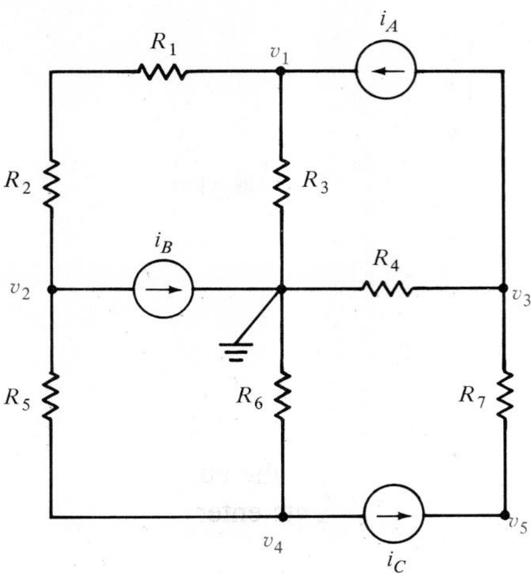

Figure 3.3 Example six-node circuit.

In this example note that no node was chosen between R_1 and R_2, since they can simply be grouped together to form another resister equal to $R_1 + R_2$. Once again, note the symmetry of the $\boldsymbol{A}$ matrix. ■

DRILL EXERCISE

D3.1. Use nodal analysis to find all the branch currents in the circuit in Fig. D3.1.

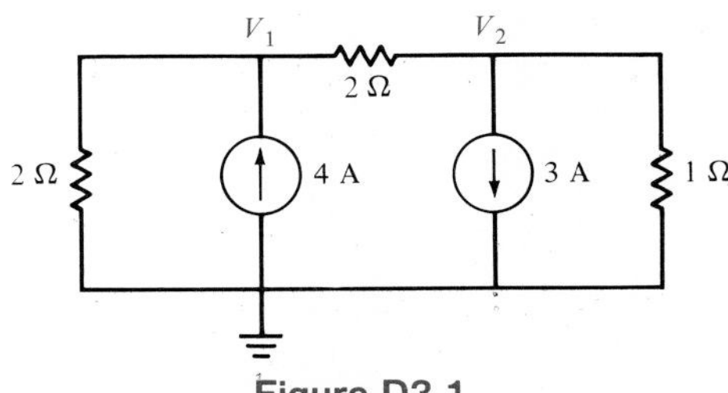

Figure D3.1

D3.2. Find all the branch currents in the network in Fig. D3.2 using nodal analysis.

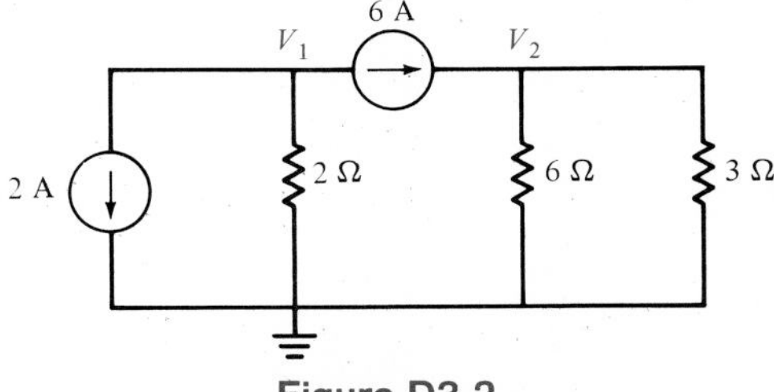

Figure D3.2

D3.3 Write the nodal equations for the circuit in Fig. D3.3.

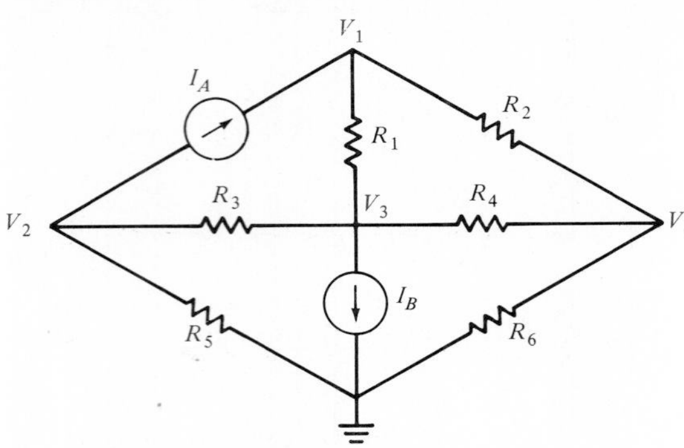

Figure D3.3

Node Equations for Circuits Containing Independent Voltage Sources

To introduce this topic, let us consider the circuit shown in Fig. 3.4a. It might appear at first glance that there are two unknown node voltages, v_A and v_B. However, let us redraw the circuit as shown in Fig. 3.4b. A close examination of this network shows that the independent voltage sources exist between the nodes, and thus the node voltages are completely defined. For example, $v_A = v_1$ because the source v_1 is tied directly across the terminals A to ground. In addition, note that $v_{R_5} = v_B$. KVL around the loop containing v_1, v_2, and v_{R_5} yields

$$+v_1 - v_2 - v_B = 0$$

Therefore,

$$v_B = v_1 - v_2$$

and since v_1 and v_2 are known quantities, all node voltages are known. The currents can be immediately calculated as

$$i_{12} = \frac{v_2}{R_1 + R_2}$$

$$i_3 = \frac{v_1}{R_3}$$

$$i_4 = \frac{v_1}{R_4}$$

$$i_5 = \frac{v_B}{R_5} = \frac{v_1 - v_2}{R_5}$$

Although this is a very simple example, it is clear that the presence of the voltage

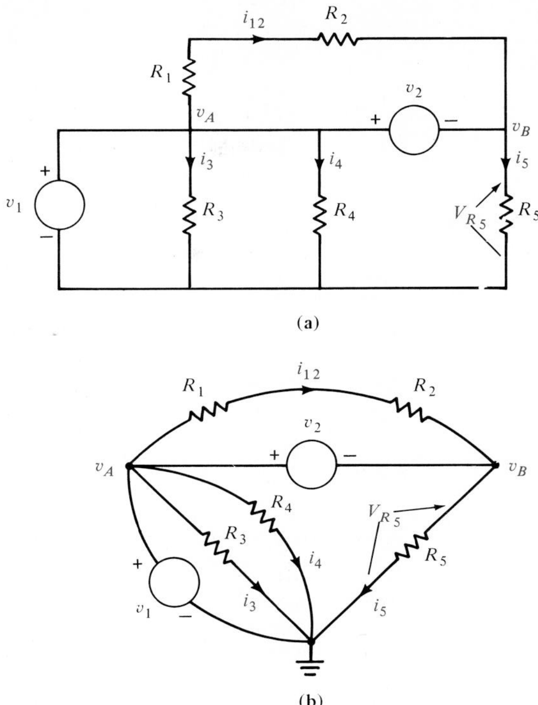

Figure 3.4 Network containing voltage sources between nodes.

sources has simplified the analysis. As a general rule, any time that voltage sources exist between nodes, the node voltage equations that describe the network will be simpler.

The network in Fig. 3.5 has three nodes, and therefore two linearly independent equations are required to solve for the two unknown node voltages, V_A and V_B. Note, however, that if V_A or V_B is known, then the other node voltage V_B or V_A is also known, since the difference in potential between the two nodes is *constrained* by the voltage source V_1. Hence one of the two linearly independent equations required to solve for the node voltages is

$$V_A - V_B = V_1$$

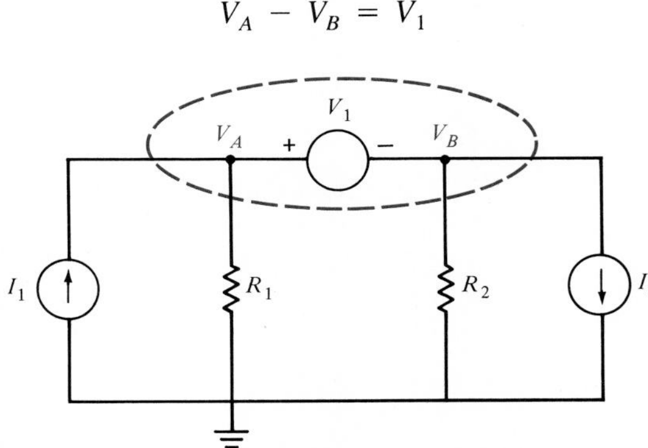

Figure 3.5 Simple circuit containing a supernode.

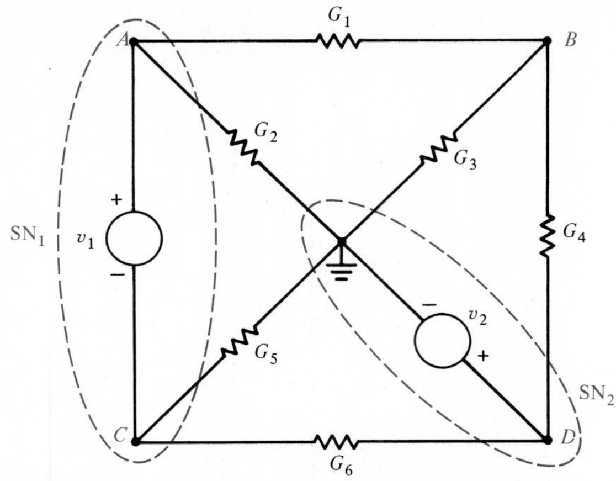

Figure 3.6 Circuit containing voltage sources, illustrating supernodes.

The other linearly independent equation can be obtained simply by enclosing the voltage source in a dashed surface, which we refer to as a *supernode,* and writing KCL for this supernode. We demonstrated in Chapter 2 that KCL must hold for such a surface, and this technique also eliminates the problem of dealing with a current through a voltage source. KCL for the supernode is

$$-I_1 + \frac{V_A}{R_1} + \frac{V_B}{R_2} + I_2 = 0$$

This equation, together with the constraint equation above, will yield the two node voltages, V_A and V_B.

Let us now consider a more complicated circuit such as the one shown in Fig. 3.6. From our previous discussion we recognize immediately that the voltage v_D is known, and that the voltage between nodes A and C is known. These two conditions are thus *constraints* on the circuit and represent two of the basic equations necessary to solve for all the node voltages. Therefore, two of the independent equations required to solve for the four node voltages are

$$\begin{aligned} v_D &= v_2 \\ v_A - v_C &= v_1 \end{aligned} \tag{3.7}$$

The supernodes are labeled SN_1 and SN_2.

If we now examine the circuit in Fig. 3.6 it appears that we have three nodes, two supernodes and node B. Since the node voltage v_D is known, KCL applied at nodes B and SN_1 yields the two remaining required equations. For these nodes the equations are

$$(V_B - V_A)G_1 + V_BG_3 + (V_B - V_D)G_4 = 0$$

$$(V_A - V_B)G_1 + V_AG_2 + V_CG_5 + (V_C - V_D)G_6 = 0$$

which when simplified become

$$
\begin{aligned}
-V_AG_1 + V_B(G_1 + G_3 + G_4) - V_DG_4 &= 0 \\
V_A(G_1 + G_2) - V_B(G_1) + V_C(G_5 + G_6) - V_DG_6 &= 0
\end{aligned}
\tag{3.8}
$$

The four equations in (3.7) and (3.8) can be used to determine all the unknown voltages.

EXAMPLE 3.4

The circuit shown in Fig. 3.6 has the following parameters:

$$
\begin{array}{llll}
G_1 = 0.5\ \text{S} & G_3 = 0.25\ \text{S} & G_5 = 0.25\ \text{S} & v_1 = V_1 = 12\ \text{V} \\
G_2 = 0.5\ \text{S} & G_4 = 1\ \text{S} & G_6 = 1\ \text{S} & v_2 = V_2 = 24\ \text{V}
\end{array}
$$

Equations (3.7) and (3.8) become

$$
\begin{aligned}
V_D &= 24 \\
V_A - V_C &= 12 \\
-V_A(0.5) + V_B(1.75) &= V_D \\
V_A(1) - V_B(0.5) + V_C(1.25) &= V_D
\end{aligned}
$$

Eliminating V_C and V_D yields

$$
\begin{aligned}
-0.5V_A + 1.75V_B &= 24 \\
2.25V_A - 0.5V_B &= 39
\end{aligned}
$$

Using Cramer's rule, V_A and V_B can be determined as

$$
V_A = \frac{\begin{vmatrix} 24 & 1.75 \\ 39 & -0.5 \end{vmatrix}}{\begin{vmatrix} -0.5 & 1.75 \\ 2.25 & -0.5 \end{vmatrix}} = \frac{-80.25}{-3.6875} = 21.7627\ \text{V}
$$

$$
V_B = \frac{\begin{vmatrix} -0.5 & 24 \\ 2.25 & 39 \end{vmatrix}}{\begin{vmatrix} -0.5 & 1.75 \\ 2.25 & -0.5 \end{vmatrix}} = \frac{-73.5}{-3.6875} = 19.9322\ \text{V}
$$

In matrix form,

$$
\begin{bmatrix} -0.5 & 1.75 \\ 2.25 & -0.5 \end{bmatrix} \begin{bmatrix} V_A \\ V_B \end{bmatrix} = \begin{bmatrix} 24 \\ 39 \end{bmatrix}
$$

Inverting the coefficient matrix yields

$$
\begin{bmatrix} V_A \\ V_B \end{bmatrix} = \frac{-1}{3.6875} \begin{bmatrix} -0.5 & -1.75 \\ -2.25 & -0.5 \end{bmatrix} \begin{bmatrix} 24 \\ 39 \end{bmatrix} = \begin{bmatrix} 21.7627 \\ 19.9322 \end{bmatrix}
$$

As a quick check, we sum the currents leaving node B:

$$\frac{V_B - V_A}{R_1} + \frac{V_B}{R_3} + \frac{V_B - V_D}{R_4} = \frac{19.9322 - 21.7627}{2} + \frac{19.9322}{4} + \frac{19.9322 - 24}{1} = 0$$

Note that in the solution of this circuit we had to solve only two equations after several simple substitutions. ■

DRILL EXERCISE

D3.4. Use nodal analysis to find the currents in the two resistors in the network in Fig. D3.4.

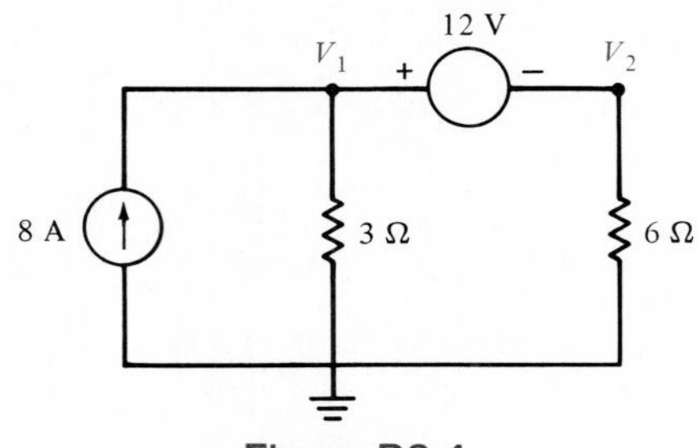

Figure D3.4

D3.5. Use nodal analysis to find the current I in the 24-V source for the network shown in Fig. D3.5.

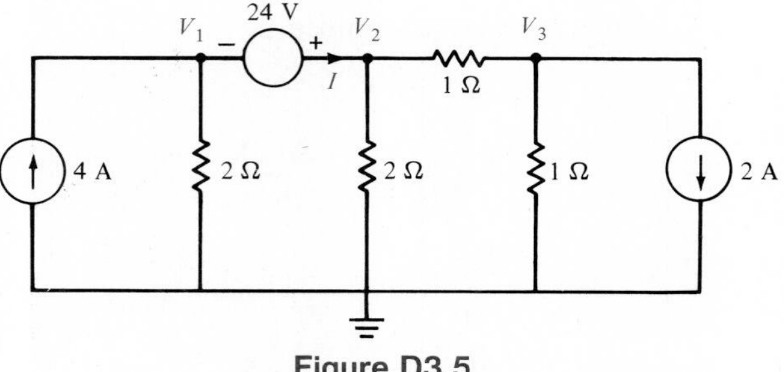

Figure D3.5

At this point it seems appropriate to give some guidelines for solving circuits of this type. When using node equations to solve circuits containing voltage sources, the following rules govern the formulation of the equations required.

(1) number of equations = number of topological nodes − 1 (i.e., nodes remaining with current sources open and voltage sources shorted − 1)

(2) number of constraints = number of voltage sources

(3) number of unknowns = number of physical nodes − 1

As a final point in this discussion, consider the dashed portion of the network shown in Fig. 3.7. Note that R_5 and node A form a supernode. This supernode exists because the current in R_5 is known and thus the voltage across R_5 is also known.

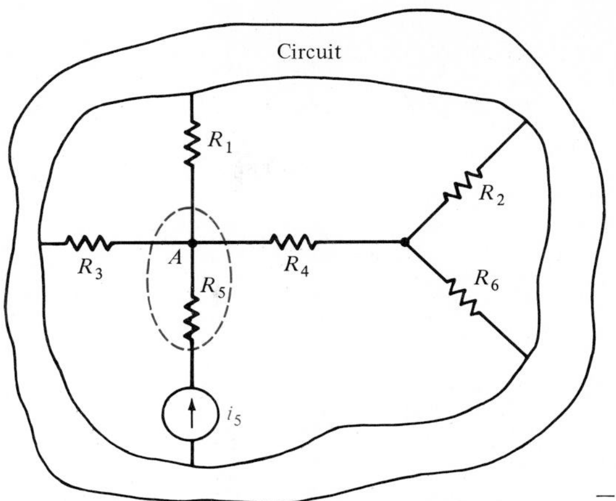

Figure 3.7 Network containing a supernode.

Node Equations for Circuits Containing Dependent Voltage Sources

Networks containing dependent or controlled sources are treated in the same manner as described above. There is, however, one important difference in the form of the resulting equations: The presence of a dependent source may destroy the symmetrical form of the nodal equations that define the circuit.

Consider the network shown in Fig. 3.8. Note that $v_4 = v_B$ and $v_A = v_2 - v_3$. Therefore, the equations that describe the network are

$$v_4 = v_B$$

$$v_1(G_1 + G_3) - v_2(G_1) - v_4(G_3) = i_A$$

$$-v_1(G_1) + v_2(G_1 + G_2) - v_3(G_2) = -i_A - 3(v_2 - v_3)$$

$$-v_2(G_2) + v_3(G_2 + G_4) = i_B$$

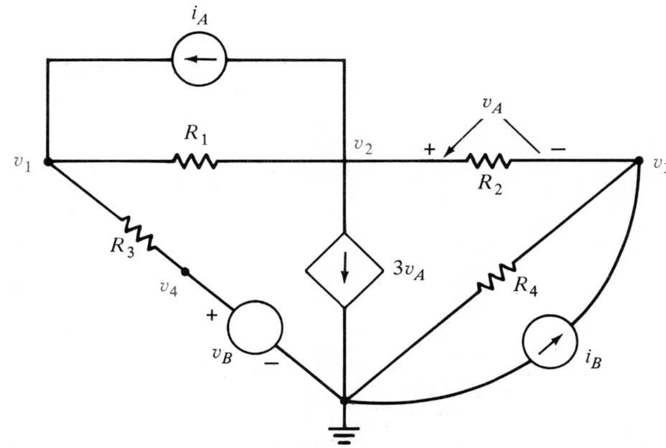

Figure 3.8 Network containing a controlled current source.

EXAMPLE 3.5

Given the following parameters for the network in Fig. 3.8, determine the node voltages.

$$R_1 = 1\ \Omega \qquad R_3 = 2\ \Omega \qquad i_A = I_A = 4\text{ A} \qquad v_B = V_B = 10\text{ V}$$

$$R_2 = 0.5\ \Omega \qquad R_4 = 1\ \Omega \qquad i_B = I_B = 5\text{ A}$$

Substituting these values into the equations above yields

$$\begin{aligned} 1.5V_1 - V_2 \qquad &= 9 \\ -V_1 + 6V_2 - 5V_3 &= -4 \\ -2V_2 + 3V_3 &= 5 \end{aligned}$$

Using Cramer's rule, we find that

$$\Delta = |A| = \begin{vmatrix} 1.5 & -1 & 0 \\ -1 & 6 & -5 \\ 0 & -2 & 3 \end{vmatrix} = 9$$

and therefore

$$V_1 = \frac{\Delta_1}{\Delta} = \frac{\begin{vmatrix} 9 & -1 & 0 \\ -4 & 6 & -5 \\ 5 & -2 & 3 \end{vmatrix}}{9} = 9.4444\text{ V}$$

$$V_2 = \frac{\Delta_2}{\Delta} = \frac{\begin{vmatrix} 1.5 & 9 & 0 \\ -1 & -4 & -5 \\ 0 & 5 & 3 \end{vmatrix}}{9} = 5.1666\text{ V}$$

and

$$V_3 = \frac{\Delta_3}{\Delta} = \frac{\begin{vmatrix} 1.5 & -1 & 9 \\ -1 & 6 & -4 \\ 0 & -2 & 5 \end{vmatrix}}{9} = 5.1111\text{ V}$$

or in matrix form,

$$\begin{bmatrix} 1.5 & -1 & 0 \\ -1 & 6 & -5 \\ 0 & -2 & 3 \end{bmatrix} \begin{bmatrix} V_1 \\ V_2 \\ V_3 \end{bmatrix} = \begin{bmatrix} 9 \\ -4 \\ 5 \end{bmatrix}$$

The cofactors of the coefficient matrix (i.e., the $\boldsymbol{A}$ matrix) are

$$A_{11} = 8 \qquad A_{12} = 3 \qquad A_{13} = 2$$

$$A_{21} = 3 \qquad A_{22} = 4.5 \qquad A_{23} = 3$$

$$A_{31} = 5 \qquad A_{32} = 7.5 \qquad A_{33} = 8$$

The determinant is $|\boldsymbol{A}| = 9$, and therefore the solution of the equation is

$$\begin{bmatrix} V_1 \\ V_2 \\ V_3 \end{bmatrix} = \frac{1}{9}\begin{bmatrix} 8 & 3 & 5 \\ 3 & 4.5 & 7.5 \\ 2 & 3 & 8 \end{bmatrix}\begin{bmatrix} 9 \\ -4 \\ 5 \end{bmatrix}$$

Therefore,

$$\begin{bmatrix} V_1 \\ V_2 \\ V_3 \end{bmatrix} = \begin{bmatrix} 9.4444 \\ 5.1666 \\ 5.1111 \end{bmatrix}$$

As a quick check, KCL at node 3 yields

$$\frac{V_3 - V_2}{0.5} + \frac{V_3}{1} - 5 = \frac{5.1111 - 5.1666}{0.5} + \frac{5.1111}{1} - 5 = 0$$

■

Next, let us consider the circuit shown in Fig. 3.9, which contains a dependent voltage source. The nodal equations for this circuit are

$$\frac{v_1 - v_2}{R_2} + \frac{v_1 - v_3}{R_1} = i_A$$

$$v_2 = 10i_4 = \frac{10v_3}{R_4}$$

$$\frac{v_3 - v_1}{R_1} + \frac{v_3 - v_2}{R_3} + \frac{v_3}{R_4} = -i_B$$

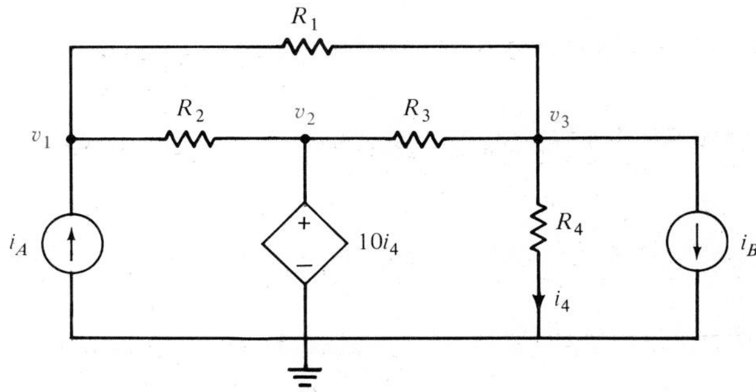

Figure 3.9 Circuit containing a controlled voltage source.

Grouping the various terms in the equations and substituting for v_2 yields

$$v_1\left(\frac{1}{R_1}+\frac{1}{R_2}\right) - v_3\left(\frac{1}{R_1}+\frac{10}{R_2R_4}\right) = i_A$$

$$-v_1\frac{1}{R_1} + v_3\left(\frac{1}{R_1}+\frac{1}{R_3}+\frac{1}{R_4}-\frac{10}{R_3R_4}\right) = -i_B$$

Once again a simple substitution changes the problem from a three-variable problem to a two-variable problem, since once v_3 is known, v_2 can be calculated directly. A dependent voltage source will reduce the order of the equations in the same manner as an independent voltage source.

EXAMPLE 3.6

Suppose that the parameters in the circuit shown in Fig. 3.9 are

$$R_1 = 0.5\ \Omega \qquad R_3 = 2\ \Omega \qquad i_A = I_A = 10\text{ A}$$

$$R_2 = 1\ \Omega \qquad R_4 = 10\ \Omega \qquad i_B = I_B = 2\text{ A}$$

Substituting these values into the equations developed above yields

$$3V_1 - 3V_3 = 10$$

$$-2V_1 + 2.1\,V_3 = -2$$

or

$$\begin{bmatrix} 3 & -3 \\ -2 & 2.1 \end{bmatrix}\begin{bmatrix} V_1 \\ V_3 \end{bmatrix} = \begin{bmatrix} 10 \\ -2 \end{bmatrix}$$

and therefore,

$$\begin{bmatrix} V_1 \\ V_3 \end{bmatrix} = \begin{bmatrix} 3 & -3 \\ -2 & 2.1 \end{bmatrix}^{-1}\begin{bmatrix} 10 \\ -2 \end{bmatrix} = \frac{1}{0.3}\begin{bmatrix} 2.1 & 3 \\ 2 & 3 \end{bmatrix}\begin{bmatrix} 10 \\ -2 \end{bmatrix}$$

$$= \begin{bmatrix} 50.0000 \\ 46.6666 \end{bmatrix}$$

Since

$$V_2 = \frac{10}{R_4}V_3$$

then

$$V_2 = V_3$$

This means, of course, that no current flows in R_3. Let us check this answer by applying KCL at node 3.

$$\frac{V_3 - V_2}{R_3} + \frac{V_3 - V_1}{R_1} + \frac{V_3}{R_4} + I_B = 0 + \frac{46.6666 - 50.0000}{0.5} + \frac{46.6666}{10} + 2$$

$$= 0$$

■

DRILL EXERCISE

D3.6. Find the node voltages for the network in Fig. D3.6.

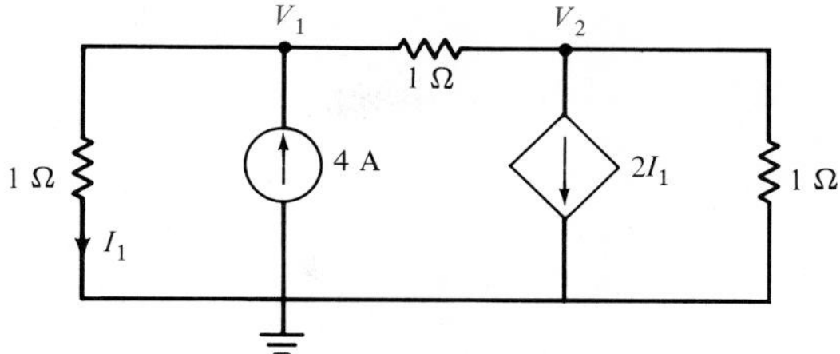

Figure D3.6

D3.7. Compute the node voltages for the network in Fig. D3.7.

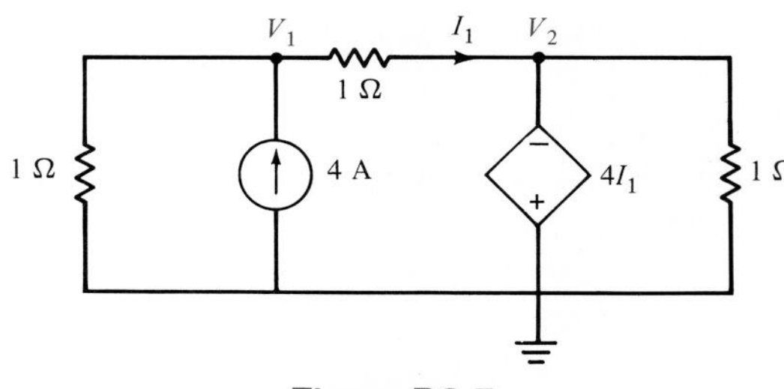

Figure D3.7

Node Equations for Operational Amplifier Circuits

The following examples illustrate the use of nodal equations in the solution of circuits containing operational amplifiers.

EXAMPLE 3.7

As another example of the use of node equations, consider the op-amp circuit shown in Fig. 3.10. Recall from our earlier analysis that $i_- = 0$, and therefore the node equation

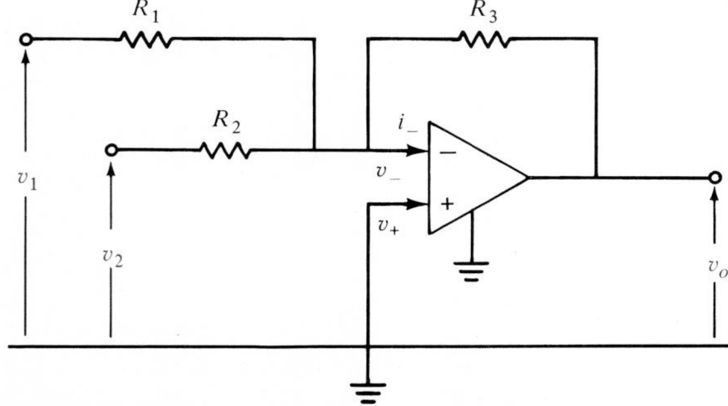

Figure 3.10 Voltage-adder operational amplifier circuit.

at the inverting terminal of the op-amp is

$$\frac{v_1 - v_-}{R_1} + \frac{v_2 - v_-}{R_2} + \frac{v_o - v_-}{R_3} = 0$$

However, $v_- - v_+ = 0$; v_+ is grounded, and therefore zero. Hence v_- is zero, so the equation above reduces to

$$\frac{v_1}{R_1} + \frac{v_2}{R_2} + \frac{v_o}{R_3} = 0$$

Solving for v_o yields

$$v_o = -\left(\frac{R_3}{R_1} v_1 + \frac{R_3}{R_2} v_2\right)$$

Note that if R_1 is chosen equal to R_2, the equation above becomes

$$v_o = -K_1(v_1 + v_2)$$

where $K_1 = R_3/R_1$. Note that this is an adder circuit in that the output is proportional to the addition of input voltages v_1 and v_2. ■

EXAMPLE 3.8

Consider the op-amp circuit shown in Fig. 3.11. The node equation at the inverting terminal is

$$\frac{v_1 - v_-}{R_1} + \frac{v_o - v_-}{R_2} = i_-$$

At the noninverting terminal KCL yields

$$\frac{v_2 - v_+}{R_3} = \frac{v_+}{R_4} + i_+$$

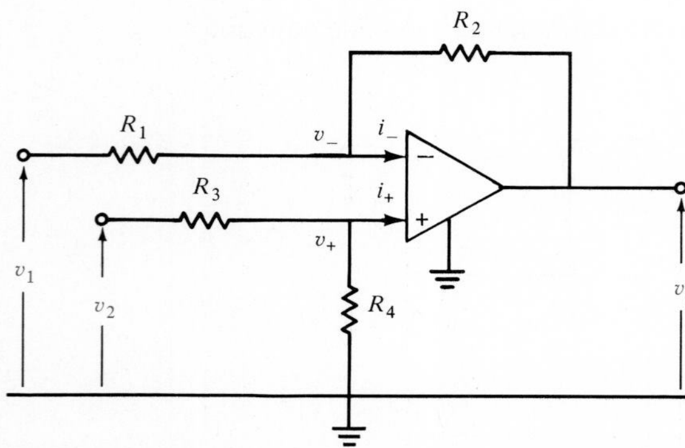

Figure 3.11 Differential amplifier operational amplifier circuit.

However, $i_+ = i_- = 0$ and $v_+ = v_-$. Substituting these values into the two equations above yields

$$\frac{v_1 - v_-}{R_1} + \frac{v_o - v_-}{R_2} = 0$$

and

$$\frac{v_2 - v_-}{R_3} = \frac{v_-}{R_4}$$

Solving these two equations for v_o results in the expression

$$v_o = \frac{R_2}{R_1}\left(1 + \frac{R_1}{R_2}\right)\frac{R_4}{R_3 + R_4} v_2 - \frac{R_2}{R_1} v_1$$

Note that if $R_4 = R_2$ and $R_3 = R_1$, the expression reduces to

$$v_o = \frac{R_2}{R_1}(v_2 - v_1)$$

Therefore, this op-amp circuit can be employed to subtract two input voltages. ■

EXAMPLE 3.9

The circuit shown in Fig. 3.12a is a precision differential voltage-gain device. It is used to provide a single-ended input for an analog-to-digital converter. We wish to derive an expression for the output of the circuit in terms of the two inputs. To accomplish this we draw the equivalent circuit shown in Fig. 3.12b. Recall that the voltage across the input terminals of the op-amp is approximately zero and the currents into the op-amp input terminals are approximately zero. Note that we can write node equations for node voltages v_1 and v_2 in terms of v_o and v_a. Since we are interested in an expression for v_o

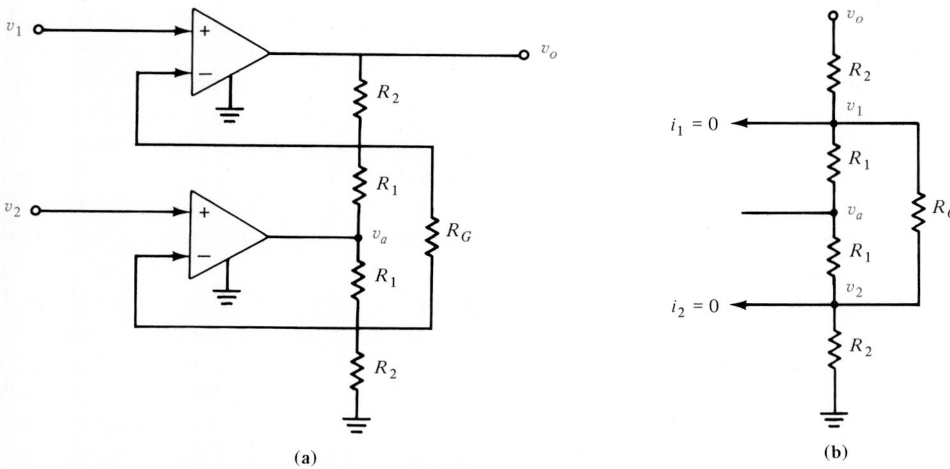

Figure 3.12 Instrumentation amplifier circuit.

in terms of the voltages v_1 and v_2, we simply eliminate the v_a terms from the two node equations. The node equations are

$$\frac{v_1 - v_o}{R_2} + \frac{v_1 - v_a}{R_1} + \frac{v_1 - v_2}{R_G} = 0$$

$$\frac{v_2 - v_a}{R_1} + \frac{v_2 - v_1}{R_G} + \frac{v_2}{R_2} = 0$$

Adding the two equations will eliminate v_a, and hence if v_o is written in terms of v_1 and v_2, we obtain

$$v_o = (v_1 - v_2)\left(1 + \frac{R_2}{R_1} + \frac{2R_2}{R_G}\right)$$

■

DRILL EXERCISE

D3.8. Use nodal analysis to find I_o in the network in Fig. D3.8.

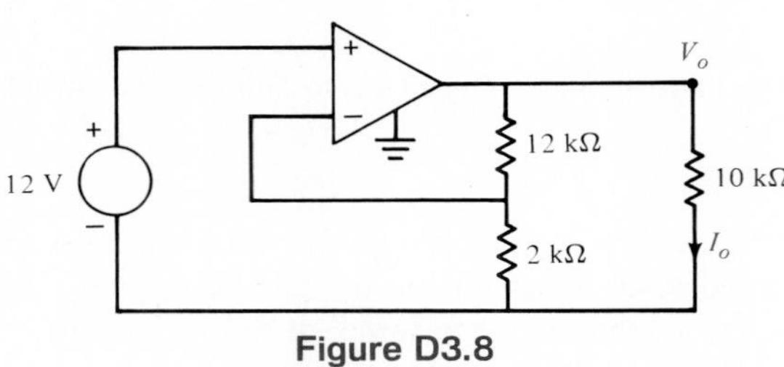

Figure D3.8

3.2 Loop Analysis

In a nodal analysis the unknown parameters are the node voltages, and KCL is employed to determine them. In contrast to this approach, a loop analysis uses KVL to determine currents in the circuit. Once the currents are known, Ohm's law can be used to calculate the voltages. Recall that, in Chapter 2, we found that a single equation was sufficient to determine the current in a circuit containing a single loop. If the circuit contains N independent loops, we will show that N independent simultaneous equations will be required to describe the network. For now we will assume that the circuits are *planar,* which simply means that we can draw the circuit on a sheet of paper in a way such that no conductor crosses another conductor.

To begin our analysis, consider the circuit shown in Fig. 3.13. Let us also identify two loops, *A-B-E-F-A* and *B-C-D-E-B*. We now define a new set of current variables, called *loop currents,* which can be used to find the physical currents in the circuit. Let us assume that current i_1 flows in the first loop and that current i_2 flows in the second loop. Then the current flowing from B to E through R_3 is $i_1 - i_2$. The directions of the currents have been assumed. As was the case in the nodal analysis, if the actual currents are not in the direction indicated, the values calculated will be negative.

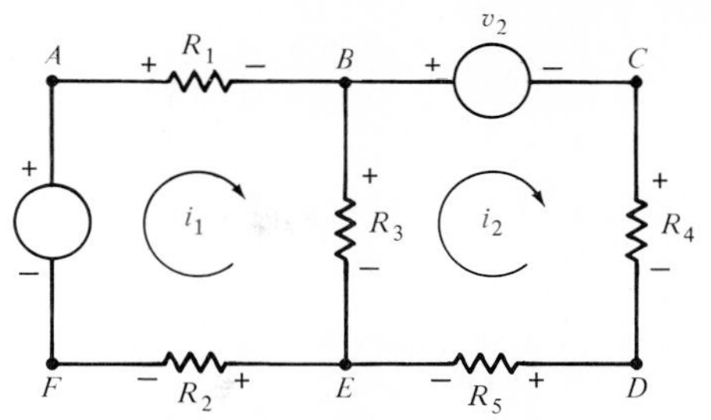

Figure 3.13 Two-loop circuit.

Applying KVL to the first loop yields

$$-i_1R_1 - (i_1 - i_2)R_3 - i_1R_2 + v_1 = 0$$

or

$$i_1(R_1 + R_2 + R_3) - i_2R_3 = v_1$$

KVL applied to loop 2 yields

$$-v_2 - i_2R_4 - i_2R_5 + (i_1 - i_2)R_3 = 0$$

or

$$-i_1R_3 + i_2(R_3 + R_4 + R_5) = -v_2$$

Therefore, the two simultaneous equations required to solve this two-loop circuit are

$$\begin{aligned} i_1(R_1 + R_2 + R_3) - i_2(R_3) &= v_1 \\ -i_1(R_3) + i_2(R_3 + R_4 + R_5) &= -v_2 \end{aligned} \tag{3.9}$$

or in matrix form,

$$\begin{bmatrix} R_1 + R_2 + R_3 & -R_3 \\ -R_3 & R_3 + R_4 + R_5 \end{bmatrix} \begin{bmatrix} i_1 \\ i_2 \end{bmatrix} = \begin{bmatrix} v_1 \\ -v_2 \end{bmatrix}$$

EXAMPLE 3.10

If the circuit in Fig. 3.13 has the following parameters, determine the loop currents.

$$R_1 = 2\ \Omega \quad R_3 = 2\ \Omega \quad R_5 = 2\ \Omega \quad v_2 = V_2 = 36\text{ V}$$

$$R_2 = 4\ \Omega \quad R_4 = 5\ \Omega \quad v_1 = V_1 = 24\text{ V}$$

The equations for the circuit are

$$\begin{aligned} 8I_1 - 2I_2 &= 24 \\ -2I_1 + 9I_2 &= -36 \end{aligned}$$

or

$$\begin{bmatrix} 8 & -2 \\ -2 & 9 \end{bmatrix} \begin{bmatrix} I_1 \\ I_2 \end{bmatrix} = \begin{bmatrix} 24 \\ -36 \end{bmatrix}$$

Solving for the currents yields

$$\begin{bmatrix} I_1 \\ I_2 \end{bmatrix} = \frac{1}{68}\begin{bmatrix} 9 & 2 \\ 2 & 8 \end{bmatrix}\begin{bmatrix} 24 \\ -36 \end{bmatrix} = \begin{bmatrix} \dfrac{144}{68} \\ -\dfrac{240}{68} \end{bmatrix} = \begin{bmatrix} 2.1176 \\ -3.5294 \end{bmatrix}$$

Note that since I_2 is negative, it is actually flowing in the direction opposite to that which was assumed.

We can check the validity of our answers via KVL. In loop 1 KVL requires that

$$\begin{aligned} -I_1R_1 - (I_1 - I_2)R_3 - I_1R_2 + V_1 \\ = (2.1176)(2) - (2.1176 + 3.5294)(2) - (2.1176)(4) + 24 = 0 \end{aligned}$$

In loop 2, KVL requires that

$$\begin{aligned} -V_2 - I_2R_4 - I_2R_5 + (I_1 - I_2)R_3 \\ = -36 + (3.5294)(5) + (3.5294)(2) + (2.1176 + 3.5294)(2) = 0 \end{aligned}$$

The reader is encouraged to check KVL around the entire outside loop *A-B-C-D-E-F-A*.

Once the loop currents are known, the voltages at any point in the circuit can easily be determined. For example, the voltage at node *A* with respect to node $E = V_1 - I_1R_2 = I_1R_1 + (I_1 - I_2)R_3$. The voltage at node *C* with respect to node $E = -V_2 + (I_1 - I_2)R_3 = I_2R_4 + I_2R_5$. The voltage at any point in the circuit with respect to any other point can be obtained in a similar manner. ■

Let us now consider the circuit shown in Fig. 3.14. KVL applied to the closed path *A-B-E-D-C-A* yields

$$+v_1 - i_1R_2 + v_4 - R_4(i_1 - i_3) - R_3(i_1 - i_2) - i_1R_1 = 0$$

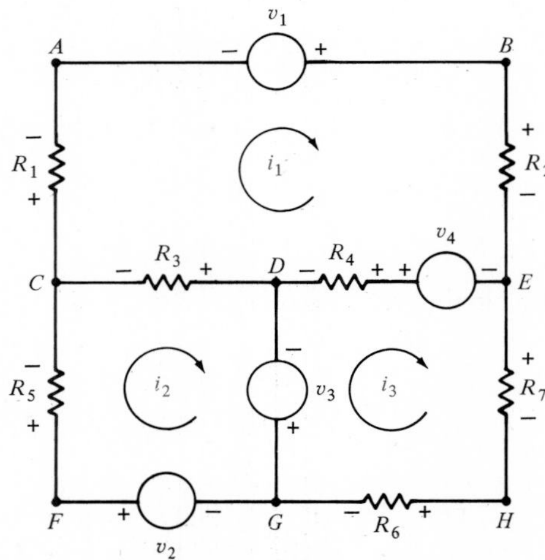

Figure 3.14 Three-loop circuit.

or

$$i_1(R_1 + R_2 + R_3 + R_4) - i_2(R_3) - i_3(R_4) = v_1 + v_4$$

KVL for loop 2 around the path *C-D-G-F-C* is

$$+R_3(i_1 - i_2) + v_3 + v_2 - i_2R_5 = 0$$

or

$$-i_1(R_3) + i_2(R_3 + R_5) = v_2 + v_3$$

KVL for loop 3 is

$$+R_4(i_1 - i_3) - v_4 - R_7i_3 - R_6i_3 - v_3 = 0$$

or

$$-i_1(R_4) + i_3(R_4 + R_6 + R_7) = -v_3 - v_4$$

Therefore, the three equations necessary to solve for the three unknown loop currents are

$$\begin{aligned} i_1(R_1 + R_2 + R_3 + R_4) \quad -i_2(R_3) \quad -i_3(R_4) &= v_1 + v_4 \\ -i_1(R_3) \quad +i_2(R_3 + R_5) &= v_2 + v_3 \\ -i_1(R_4) \quad +i_3(R_4 + R_6 + R_7) &= -v_3 - v_4 \end{aligned} \qquad (3.10)$$

or in matrix form,

$$\begin{bmatrix} R_1 + R_2 + R_3 + R_4 & -R_3 & -R_4 \\ -R_3 & R_3 + R_5 & 0 \\ -R_4 & 0 & R_4 + R_6 + R_7 \end{bmatrix} \begin{bmatrix} i_1 \\ i_2 \\ i_3 \end{bmatrix} = \begin{bmatrix} v_1 + v_4 \\ v_2 + v_3 \\ -v_3 - v_4 \end{bmatrix}$$

EXAMPLE 3.11

Given the following parameters for the network shown in Fig. 3.14, determine the loop currents.

$R_1 = 2\ \Omega \quad R_3 = 3\ \Omega \quad R_5 = 4\ \Omega \quad R_7 = 1\ \Omega$

$R_2 = 1\ \Omega \quad R_4 = 1\ \Omega \quad R_6 = 2\ \Omega \quad v_1 = V_1 = 12\text{ V}$

$v_2 = V_2 = 6\text{ V} \quad v_4 = V_4 = 24\text{ V}$

$v_3 = V_3 = 18\text{ V}$

The matrix equation above becomes

$$\begin{bmatrix} 7 & -3 & -1 \\ -3 & 7 & 0 \\ -1 & 0 & 4 \end{bmatrix} \begin{bmatrix} I_1 \\ I_2 \\ I_3 \end{bmatrix} = \begin{bmatrix} 36 \\ 24 \\ -42 \end{bmatrix}$$

The cofactors of the matrix are $A_{11} = 28$, $A_{12} = 12$, $A_{13} = 7$, $A_{21} = 12$, $A_{22} = 27$, $A_{23} = 3$, $A_{31} = 7$, $A_{32} = 3$, and $A_{33} = 40$. The determinant of $\mathbf{A}$, $|\mathbf{A}| = 153$. Therefore,

$$\begin{bmatrix} I_1 \\ I_2 \\ I_3 \end{bmatrix} = \frac{1}{153}\begin{bmatrix} 28 & 12 & 7 \\ 12 & 27 & 3 \\ 7 & 3 & 40 \end{bmatrix}\begin{bmatrix} 36 \\ 24 \\ -42 \end{bmatrix} = \begin{bmatrix} 6.5490 \\ 6.2353 \\ -8.8627 \end{bmatrix}$$

Using these data and summing the voltages around loop 1 yields

$$+V_1 - I_1R_2 + V_4 - R_4(I_1 - I_3) - R_3(I_1 - I_2) - I_1R_1$$
$$= 12 - (6.5490)(1) + 24 - 3(6.5490 - 6.2353) - 1(6.5490 + 8.8627) - 2(6.5490) = 0$$

As we saw in Example 3.10, once the loop currents are known, we can determine all the branch currents and the voltage at any point with respect to any other point. ■

Once again we are compelled to note the symmetrical form of the loop equations that describe the previous networks. Note also that the **A** matrix for each circuit is symmetrical. Since this symmetry is generally exhibited by networks containing resistors and independent voltage sources, we can learn to write the loop equations by inspection. In the first equation of (3.9) the coefficient of i_1 is the sum of the resistances through which loop current 1 flows, and the coefficient of i_2 is the negative of the sum of the resistances common to loop current 1 and loop current 2. The right-hand side of the equation is the sum of the voltage sources in loop 1. The sign of the voltage source is positive if it aids the assumed direction of the current flow and negative if it opposes the assumed flow. The first equation is KVL for loop 1. In the second equation in (3.9), the coefficient of i_2 is the sum of all the resistances in loop 2, the coefficient of i_1 is the negative of the sum of the resistances common to loop 1 and loop 2, and the right-hand side of the equation is the sum of the voltage sources in loop 2. The equations in (3.10) are obtained in a similar manner. In general, if we assume all of the loop currents to be in the same direction, then if KVL is applied to loop j with loop current i_j, the coefficient of i_j is the sum of the resistances in loop j and the coefficients of the other loop currents (e.g., i_{j-1}, i_{j+1}) are the negatives of the resistances common to these loops and loop j. The right-hand side of the equation is equal to the sum of the voltage sources in loop j. These voltage sources have a positive sign if they aid the current flow i_j and a negative sign if they oppose it.

EXAMPLE 3.12

Let us apply this inspection technique which we have just presented to the circuit shown in Fig. 3.15.

By inspection, KVL yields

$$(R_1 + R_2)i_1 - R_2i_2 = v_1 + v_3$$
$$-R_2i_1 + (R_2 + R_3 + R_4 + R_5)i_2 - R_5i_4 = -v_2$$
$$(R_6 + R_7 + R_8)i_3 - R_8i_4 = -v_3 - v_4$$
$$-R_5i_2 - R_8i_3 + (R_5 + R_8 + R_9)i_4 = v_4 - v_5$$

or in matrix form,

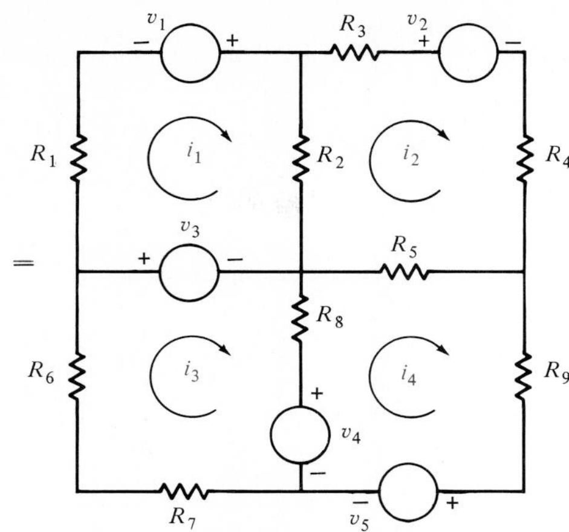

Figure 3.15 Example four-loop circuit.

$$\begin{bmatrix} R_1 + R_2 & -R_2 & 0 & 0 \\ -R_2 & R_2 + R_3 + R_4 + R_5 & 0 & -R_5 \\ 0 & 0 & R_6 + R_7 + R_8 & -R_8 \\ 0 & -R_5 & -R_8 & R_5 + R_8 + R_9 \end{bmatrix} \begin{bmatrix} i_1 \\ i_2 \\ i_3 \\ i_4 \end{bmatrix} = \begin{bmatrix} v_1 + v_3 \\ -v_2 \\ -v_3 - v_4 \\ v_4 - v_5 \end{bmatrix}$$

■

For planar networks we can define what is commonly called a *mesh*. Basically, a mesh is a special loop that does not contain any other loops. Therefore, as we traverse the path of a mesh, we do not encircle any circuit elements. For example, the circuit of Fig. 3.15 has four meshes. Mesh 1 contains the elements v_1, R_2, v_3, and R_1; mesh 2 contains elements R_3, v_2, R_4, R_5, and R_2; and the other two meshes are defined in a similar manner. Therefore, the loop currents defined for this circuit are also *mesh currents*. However, the path defined by the circuit elements v_1, R_3, v_2, R_4, R_5, v_3, and R_1 is a loop, but not a mesh. Note that all the loop currents we have defined thus far are also mesh currents, and therefore our analysis could be called a *mesh analysis*.

DRILL EXERCISE

D3.9. Use loop analysis to compute v_o in the network in Fig. D3.9.

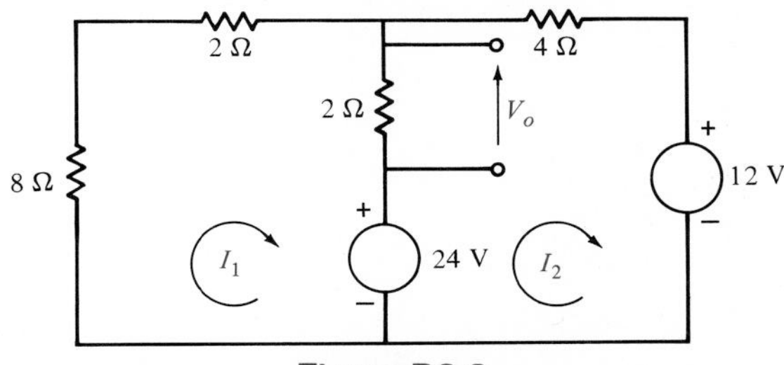

Figure D3.9

D3.10. Write the loop equations in matrix form for the network shown in Fig. D3.10.

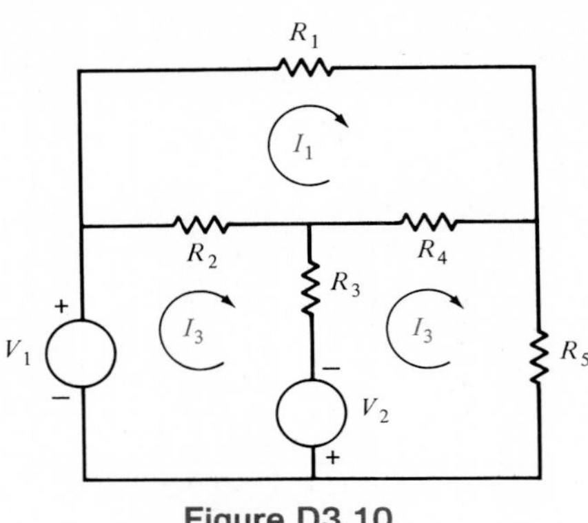

Figure D3.10

Loop Equations for Circuits Containing Independent Current Sources

Consider the circuit shown in Fig. 3.16. It appears that there are two unknown currents i_1 and i_2. However, since i_1 is assumed to go through the branch containing i_A, i_1 must be equal to i_A. Therefore, since $i_1 = i_A$, the only unknown current is i_2. Hence KVL for the second loop is

$$-i_1R_2 + i_2(R_2 + R_3 + R_4) = v_1 + v_2$$

or since $i_1 = i_A$,

$$i_2(R_2 + R_3 + R_4) = v_1 + v_2 + i_AR_2$$

Thus i_2 can be computed immediately. Once the loop currents are known, all voltages in the network can be calculated. For example,

$$v_C = -v_2 + i_2R_4$$

$$v_B = v_C + i_2R_3 = -v_2 + i_2R_4 + i_2R_3$$

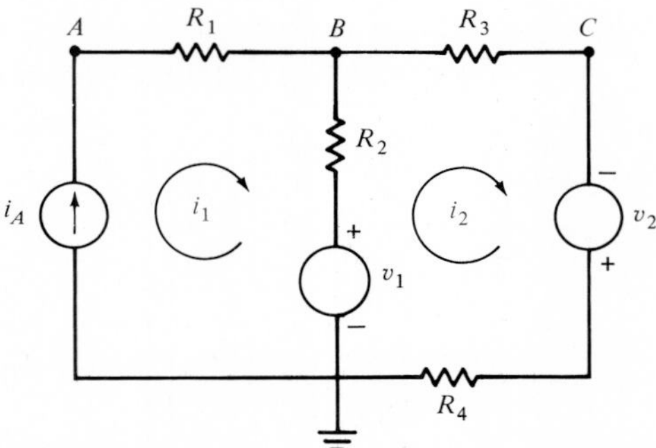

Figure 3.16 Circuit containing an independent current source.

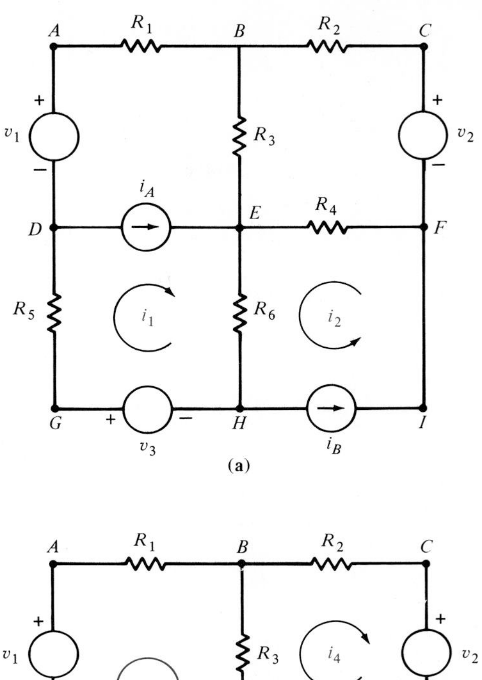

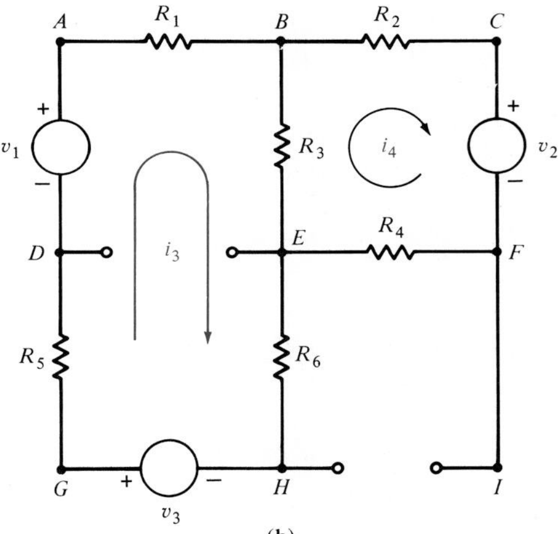

Figure 3.17 Circuit containing multiple independent current sources.

or

$$v_B = v_1 + (i_1 - i_2)R_2$$

and

$$v_A = v_B + i_A R_1$$

Clearly, the presence of a current source has simplified the loop equations. As a general rule, KVL equations are simpler whenever current sources are present.

Consider now the more complicated circuit shown in Fig. 3.17. There are obviously four loops in this circuit, and therefore four independent equations are needed to determine the currents. Note that i_1 and i_2 are chosen so that they coincide with i_A and i_B, respectively. Therefore, this selection places two *constraints* on the circuit equations: that is,

$$\begin{aligned} i_1 &= i_A \\ i_2 &= i_B \end{aligned} \tag{3.11}$$

We need two additional equations. We can determine a proper choice by open-circuiting the current sources as shown in Fig. 3.17b. The additional equations should come from this modified circuit. For example, we could use the loops *A-B-E-H-G-D-A* and *B-C-F-E-B*. Assuming the currents i_3 and i_4 as shown in Fig. 3.17b, together with the currents i_1 and i_2 shown in Fig. 3.17a, we derive the following equations:

$$\begin{aligned} R_1 i_3 + R_3(i_3 - i_4) + R_6(i_1 + i_2 + i_3) + R_5(i_1 + i_3) &= v_1 + v_3 \\ R_2 i_4 + R_4(i_2 + i_4) + R_3(i_4 - i_3) &= -v_2 \end{aligned} \tag{3.12}$$

These two equations, together with the constraint equations in (3.11), will yield the currents and voltages in the circuit.

EXAMPLE 3.13

Using the approach outlined above, determine the currents in the circuit of Fig. 3.17 if

$$\begin{array}{lllll} R_1 = 2\ \Omega & R_3 = 2\ \Omega & R_5 = 1\ \Omega & i_A = I_A = 2\text{ A} & v_1 = V_1 = 12\text{ V} \\ R_2 = 2\ \Omega & R_4 = 2\ \Omega & R_6 = 1\ \Omega & i_B = I_B = 4\text{ A} & v_2 = V_2 = 6\text{ V} \\ & & & & v_3 = V_3 = 24\text{ V} \end{array}$$

Rearranging the equations in (3.12) gives us

$$I_3(R_1 + R_3 + R_6 + R_5) - R_3 I_4 = V_1 + V_3 - R_6(I_1 + I_2) - R_5 I_1$$

$$-R_3 I_3 + I_4(R_2 + R_4 + R_3) = -V_2 - R_4 I_2$$

where $I_1 = I_A$ and $I_2 = I_B$. Using the constants above, we find that the equations become

$$\begin{aligned} 6I_3 - 2I_4 &= 28 \\ -2I_3 + 6I_4 &= -14 \end{aligned}$$

These equations yield

$$\begin{aligned} I_3 &= 4.375\text{ A} \\ I_4 &= -0.875\text{ A} \end{aligned}$$

A quick check of the answers can be obtained by substituting them into the first equation in (3.12). ■

When using loop equations to solve circuits containing current sources, the following rules govern the formulation of the equations required:

(1) number of equations = number of topological loops (i.e., loops remaining with current sources open and voltage sources shorted)

(2) number of constraints = number of current sources

(3) number of unknowns = number of physical loops

DRILL EXERCISE

D3.11. Use loop analysis to find V_o in the network in Fig. D3.11.

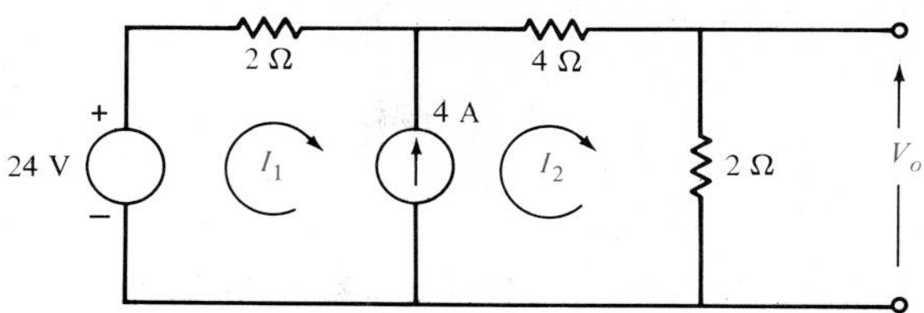

Figure D3.11

D3.12. Find V_o in the network in Fig. D3.12 using loop analysis.

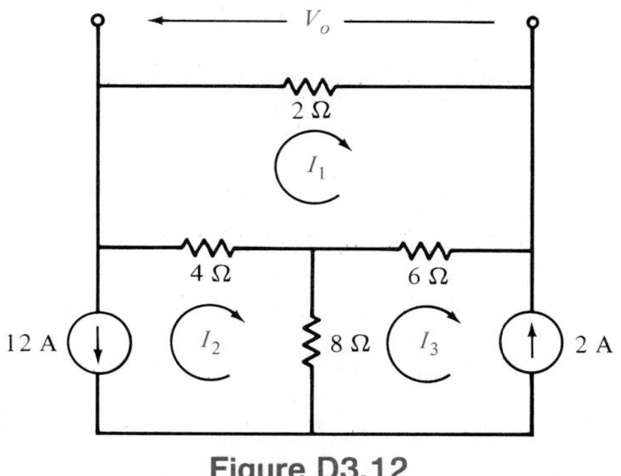

Figure D3.12

Loop Equations for Circuits Containing Dependent Sources

Although we deal with circuits containing dependent sources in the same manner as the circuits described above, we must recall that the dependent source may destroy the symmetry of the resulting equations. The following simple example will illustrate the application of loop equations in circuits with dependent sources.

EXAMPLE 3.14

We want to find the output voltage in the circuit shown in Fig. 3.18a.

First the circuit is redrawn as shown in Fig. 3.18b. Now loop current I_1 goes through the 12-A current source. The circuit equations are then

$$6(I_2 - I_1) + 4I_2 = 24 + 2V_a$$

$$V_a = 4(I_2 - I_1)$$

$$I_1 = 12$$

Solving these equations, we find that $I_2 = 0$ and therefore $V_o = 0$. ■

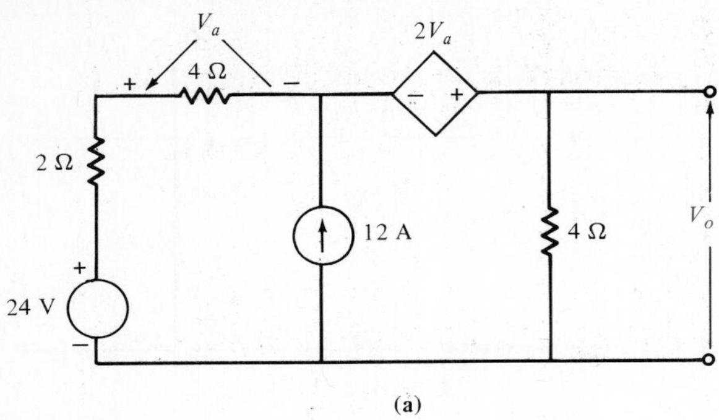

(a)

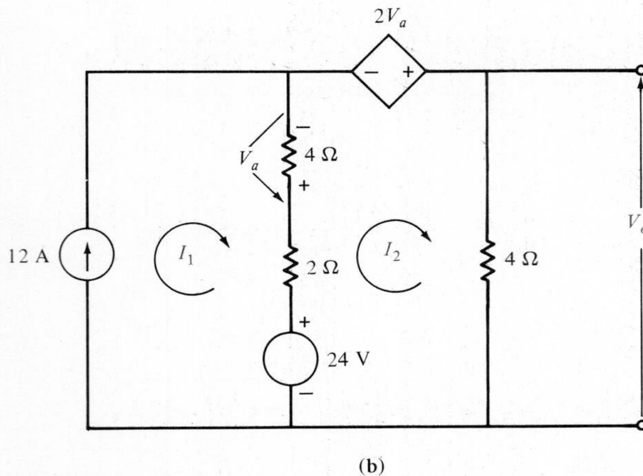

(b)

Figure 3.18 Circuit containing a dependent source.

EXAMPLE 3.15

We wish to calculate the input resistance of the op-amp circuit shown in Fig. 3.19a.

The input resistance is the resistance seen by the input voltage source v_1. Using the op-amp model shown in Fig. 2.31b, we obtain the circuit shown in Fig. 3.19b. The loop equations for this circuit are

$$v_1 = i_1(R_i + R_1) - i_2 R_1$$

$$-Av_i = -i_1 R_1 + i_2(R_1 + R_2 + R_o)$$

$$v_i = R_i i_1$$

Solving these equations for i_1 yields

$$i_1 = \frac{v_1(R_1 + R_2 + R_o)}{(R_i + R_1)(R_1 + R_2 + R_o) + R_1(AR_i - R_1)}$$

The input resistance is

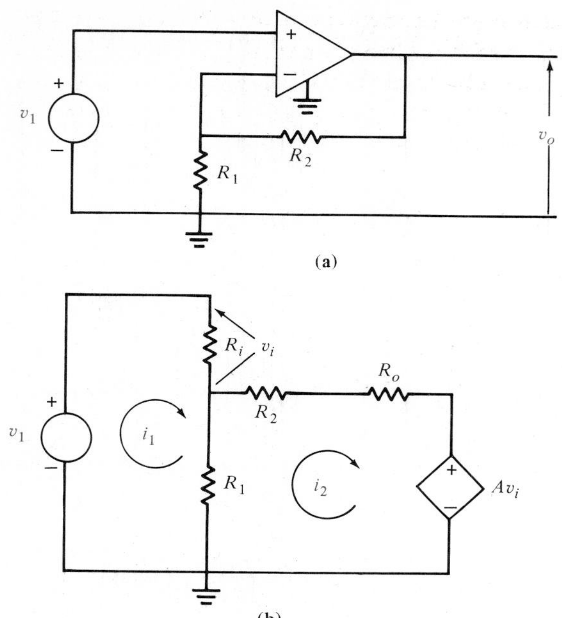

Figure 3.19 Circuit and model for a noninverting operational amplifier.

$$R_{\text{in}} = \frac{v_1}{i_1} = \frac{(R_i + R_1)(R_1 + R_2 + R_o) + R_1(AR_i - R_1)}{R_1 + R_2 + R_o}$$

$$= R_i + \frac{R_1(R_2 + R_o + AR_i)}{R_1 + R_2 + R_o}$$

Since A and R_i are normally very large and R_o is very small, the expression for R_{in} can be approximated by the equation

$$R_{\text{in}} \simeq \frac{AR_i}{1 + R_2/R_1}$$

■

DRILL EXERCISE

D3.13. Use loop analysis to find V_o in the network in Fig. D3.13.

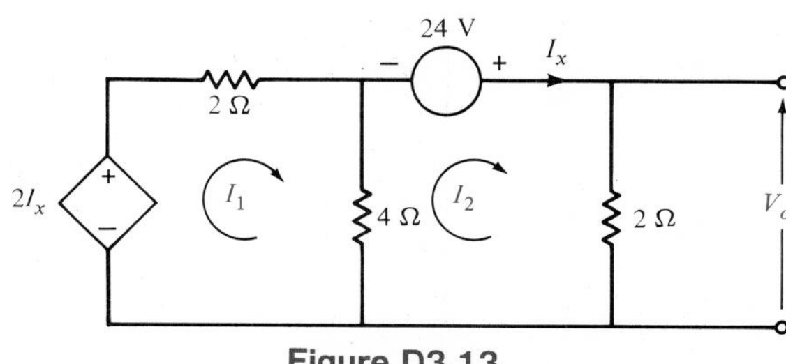

Figure D3.13

D3.14. Find V_o in the network in Fig. D3.14 using loop analysis.

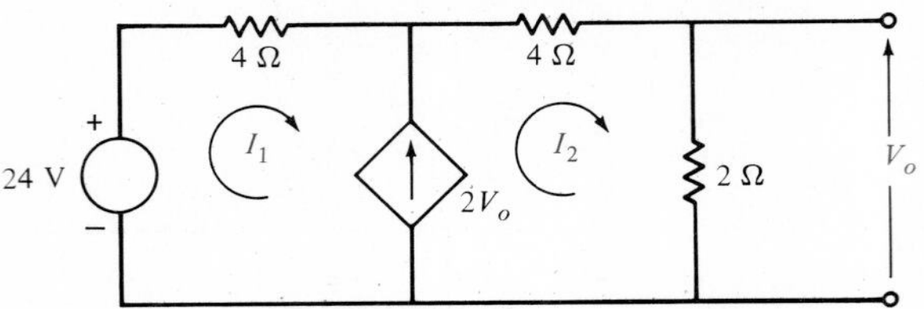

Figure D3.14

3.3

Review of Solution Techniques

As a quick review of some of the techniques we have presented, consider this very simple example.

EXAMPLE 3.16

Consider the circuit shown in Fig. 3.20a. Let us determine the output voltage V_o using several of the methods we presented earlier.

Note that the circuit in Fig. 3.20b is equivalent to that in Fig. 3.20a from the standpoint of the current I_1. From this single loop we find that

$$I_1 = \frac{60}{4 + 8 + 3} = 4 \text{ A}$$

Using KVL, we have

$$V_1 = 60 - 4(4 + 8)$$
$$= 12 \text{ V}$$

Therefore, the current I_2 is

$$I_2 = \frac{12 \text{ V}}{5 + 7} = 1 \text{ A}$$

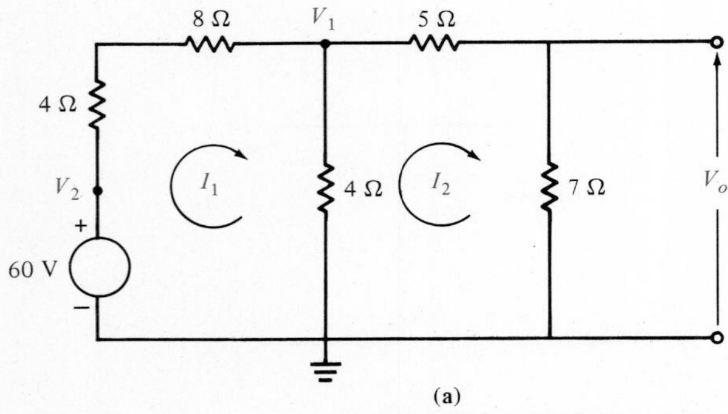

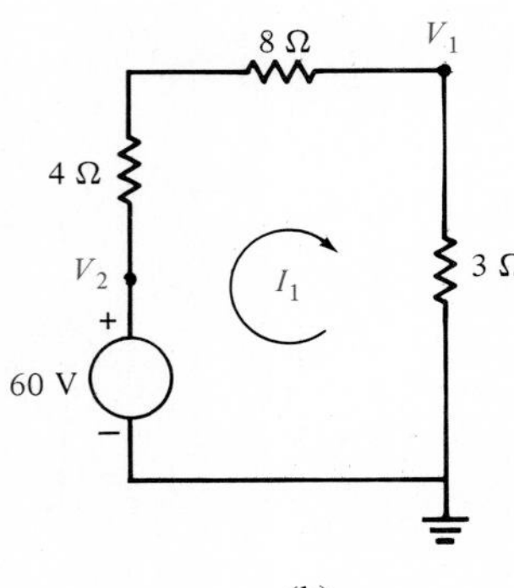

Figure 3.20 Example circuit.

and hence

$$V_o = (1)(7)$$
$$= 7 \text{ V}$$

Once V_1 is known, we could also use a voltage divider to determine V_o:

$$V_o = V_1\left(\frac{7}{7 + 5}\right)$$
$$= 12\left(\frac{7}{12}\right)$$
$$= 7 \text{ V}$$

Current division can also be employed:

$$I_2 = I_1\left(\frac{4}{4 + 5 + 7}\right)$$
$$= 4\left(\frac{4}{16}\right)$$
$$= 1 \text{ A}$$

and then

$$V_o = (I_2)(7)$$
$$= 7 \text{ V}$$

Let us now use node equations. Note that we have three nodes, V_1, V_2, and V_o, in Fig. 3.20a. V_2 is known, however. Note that once we determine V_1, we can calculate V_o immediately using a voltage divider or simply calculate I_2 and then employ Ohm's law. The node equations are

$$V_2 = 60 \text{ V}$$
$$V_1\left(\frac{1}{12} + \frac{1}{4} + \frac{1}{12}\right) - \frac{V_2}{12} = 0$$

Solving these equations yields

$$V_1 = 12 \text{ V}$$

V_o can now be immediately calculated as shown above.

The loop equations for the circuit in Fig. 3.20a are

$$16I_1 - 4I_2 = 60$$
$$-4I_1 + 16I_2 = 0$$

Using Cramer's rule, matrix analysis, or elimination of variables in this simple case yields

$$I_1 = 4 \text{ A}$$

$$I_2 = 1 \text{ A}$$

Therefore, $V_o = (I_2)(7) = 7$ V. ■

Hopefully, this type of example will help to tie things together.

3.4 Summary

Two very important and extremely useful techniques for circuit analysis have been introduced. Both the node voltage and loop current methods produce a set of simultaneous equations which when solved enable us to determine any voltage or current in the network. Procedures for simplifying the analysis were presented and techniques for solving the equations using both Cramer's rule and matrices were examined. In addition, the node voltage and loop current methods were applied to circuits containing op-amps.

KEY POINTS

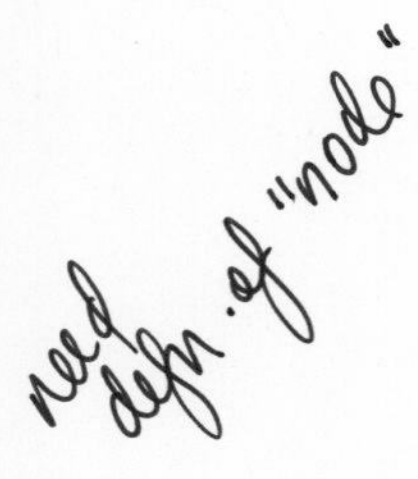

- When solving an N-node circuit using nodal analysis:
  1. Label all nodes, selecting one as the reference node.
  2. If the circuit contains only independent current sources, write $N - 1$ linearly independent simultaneous equations using KCL to determine the $N - 1$ unknown node voltages.
  3. If the circuit contains an independent voltage source, a supernode enclosing this source is created and a constraint equation for this source is written in addition to the KCL equations for the supernode and the remaining nodes.
  4. If the circuit contains a dependent source, treat the dependent source as though it were an independent source when writing the $N - 1$ independent node equations, and then write one additional equation for the controlling parameter of the dependent source.
  5. Solve the equations using any convenient method, such as Cramer's rule, matrices, or substitution.
- When solving an N-loop circuit using loop analysis:
  1. Label N loop currents for N distinct closed paths through the network.
  2. If the circuit contains only independent voltage sources, write N linearly independent simultaneous equations using KVL to determine the N unknown loop currents.
  3. If the circuit contains an independent current source, a constraint equation is written for the current that flows through it; the current source is then open-circuited and another equation is written for the newly created loop.
  4. If the circuit contains a dependent source, treat the dependent source as though it were an independent source when writing the N-loop equations, and then write one additional equation for the controlling parameter of the dependent source.
  5. Solve the equations using any convenient method, such as Cramer's rule, matrices, or substitution.

PROBLEMS

3.1. Using nodal equations, determine the voltage V in Fig. P3.1.

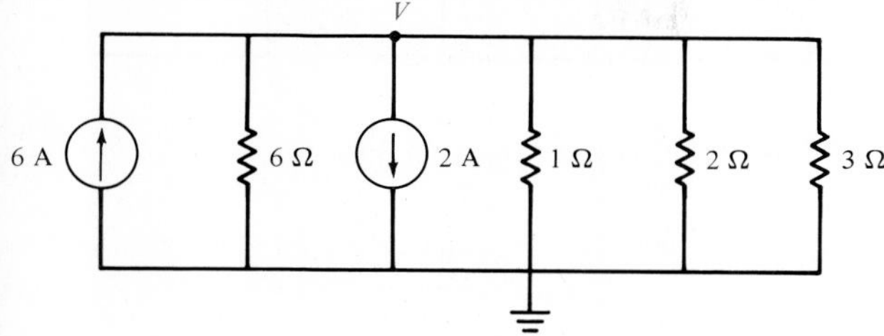

Figure P3.1

3.2. Using nodal equations, find I_o in Fig. P3.2.

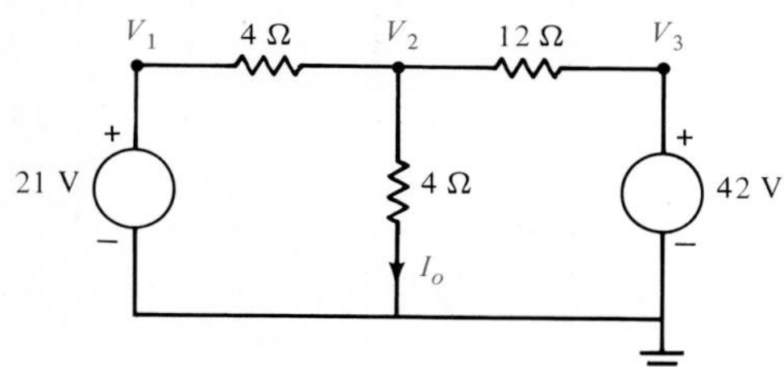

Figure P3.2

3.3. Use nodal equations to determine V_o in the circuit shown in Fig. P3.3.

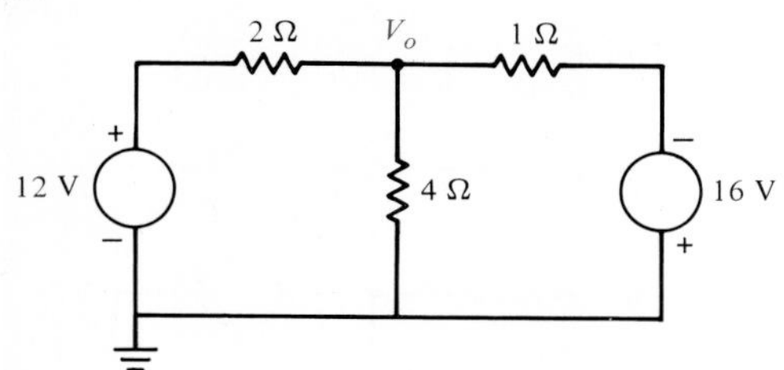

Figure P3.3

3.4. Use nodal equations to find V_2 in the circuit shown in Fig. P3.4.

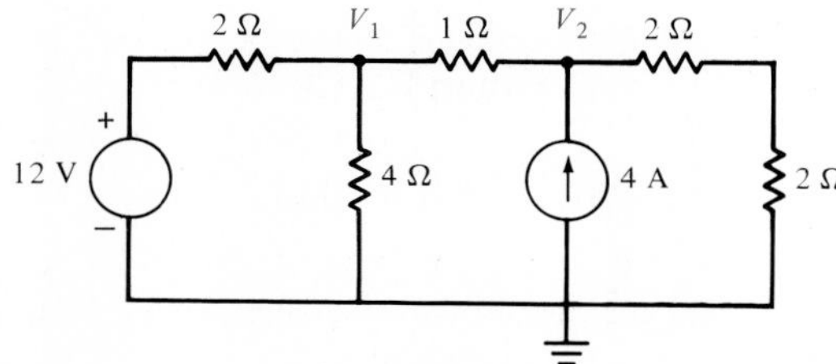

Figure P3.4

3.5. Using nodal equations, find V_o in the circuit shown in Fig. P3.5.

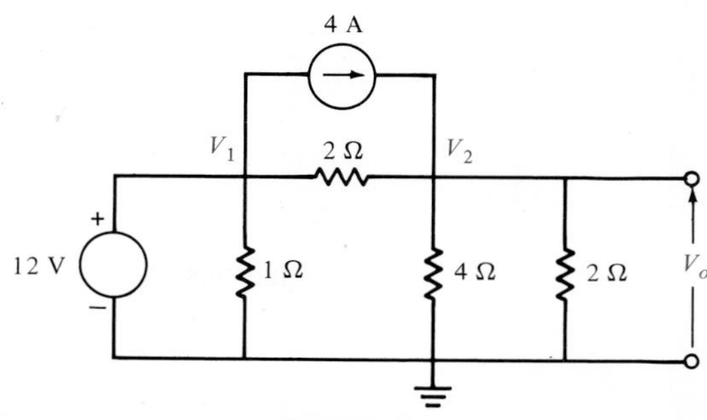

Figure P3.5

3.6. Use nodal analysis to find I_o in the network in Fig. P3.6.

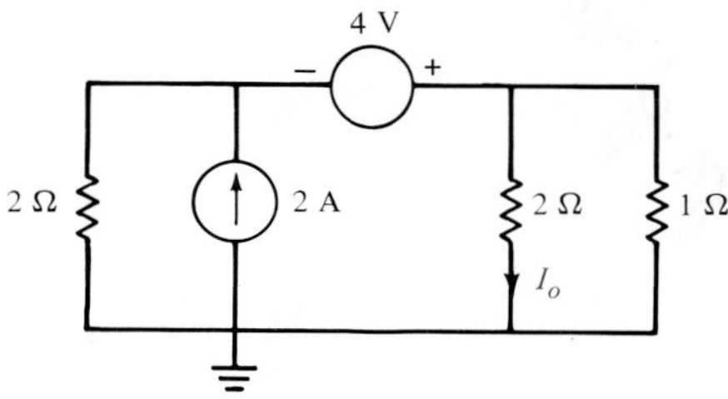

Figure P3.6

3.7. Use node equations to find V_2 in the circuit shown in Fig. P3.7.

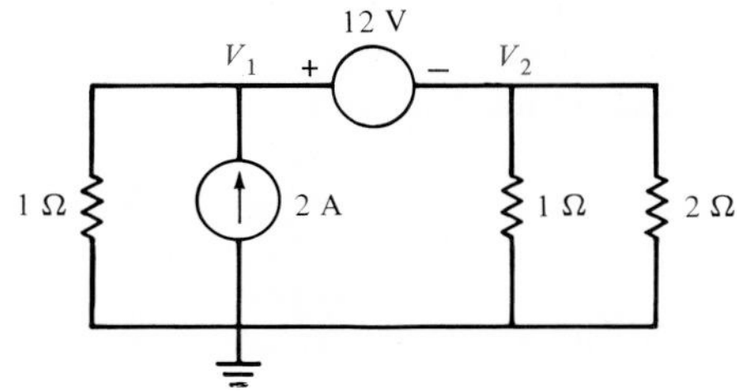

Figure P3.7

3.8. Find I_o in the network in Fig. P3.8 using node equations.

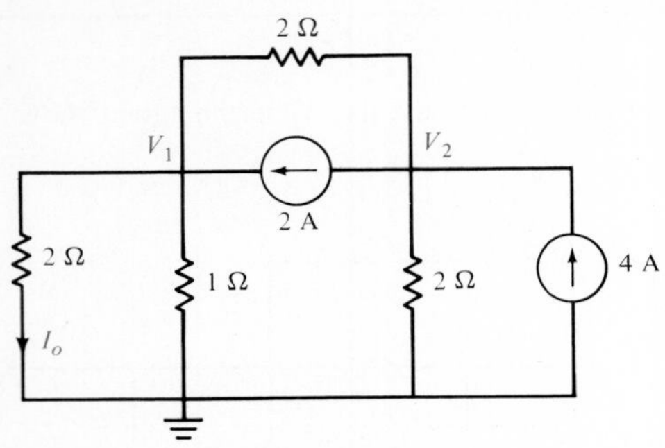

Figure P3.8

3.9. Use node equations to find I_o in the circuit in Fig. P3.9.

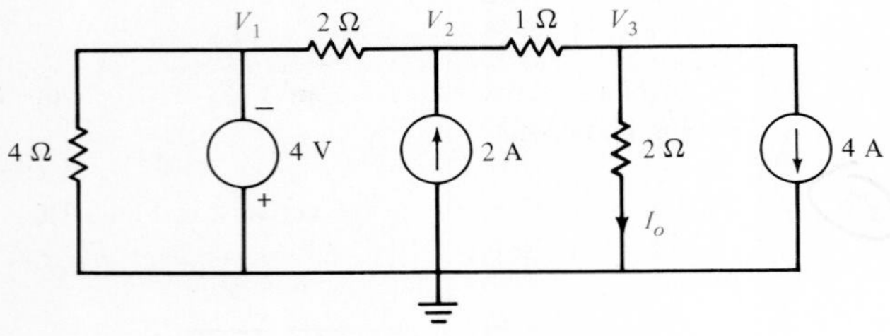

Figure P3.9

3.10. Use node equations to find V_3 in the network in Fig. P3.10.

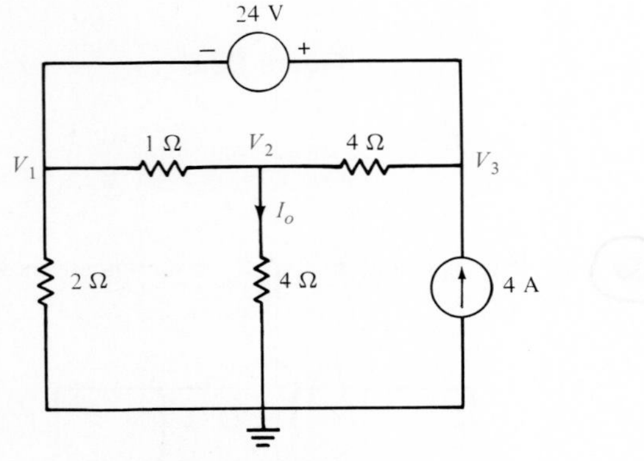

Figure P3.10

3.11. Use node equations to find I_o in the network shown in Fig. P3.11.

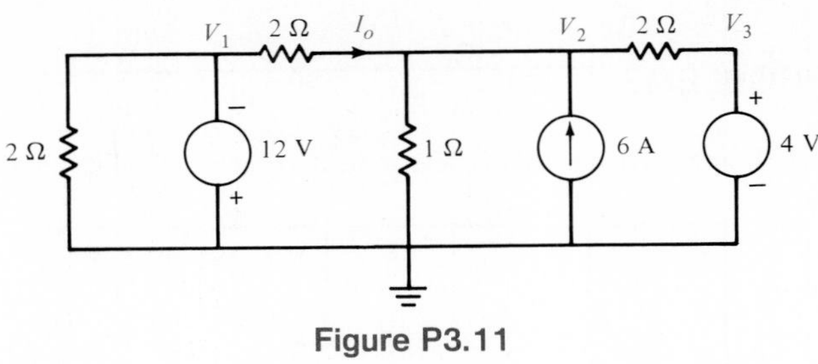

Figure P3.11

3.12. Find I_o in Fig. P3.12 using nodal equations.

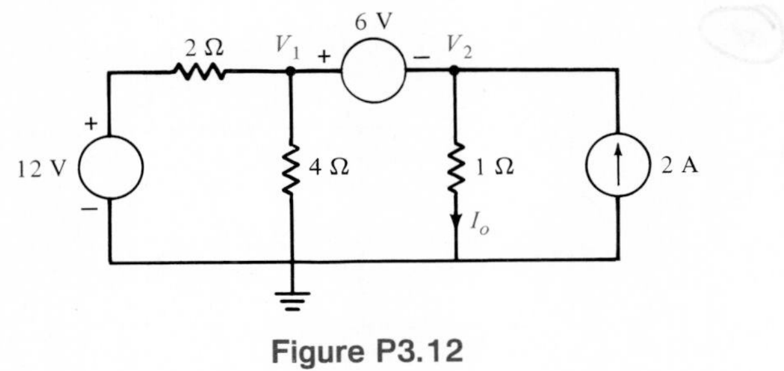

Figure P3.12

3.13. Find all node voltages in the circuit shown in Fig. P3.13 using nodal equations.

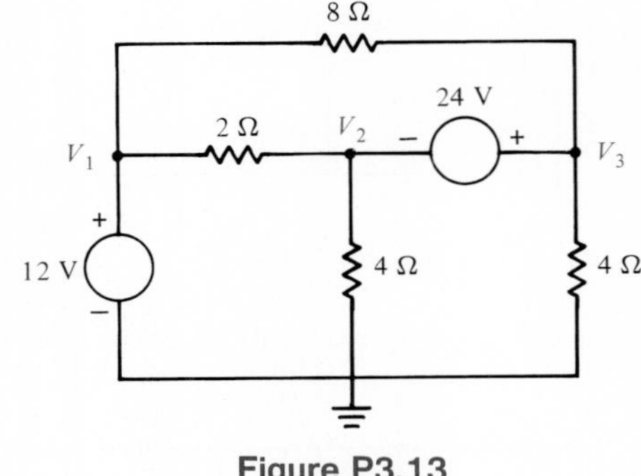

Figure P3.13

3.14. Find I_o in the circuit shown in Fig. P3.14 using nodal equations.

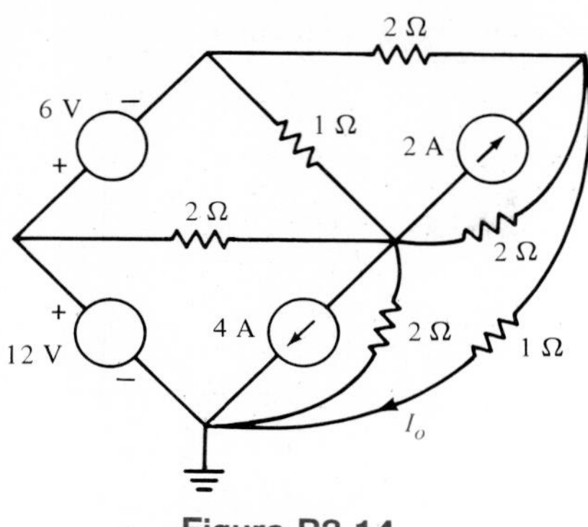

Figure P3.14

3.15. Use nodal equations to find V_o in Fig. P3.15.

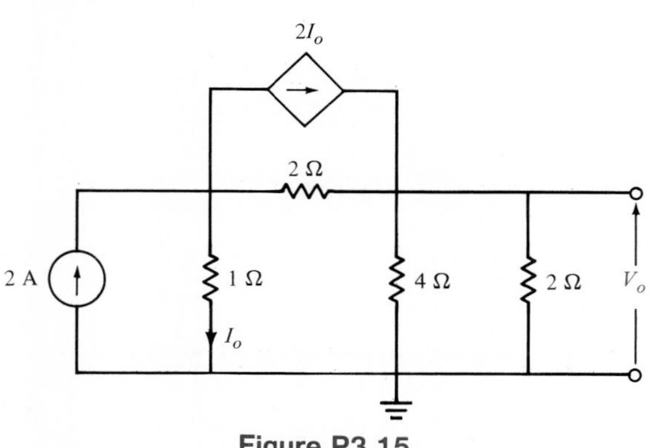

Figure P3.15

3.16. Find the node voltages in the circuit shown in Fig. P3.16 using nodal equations.

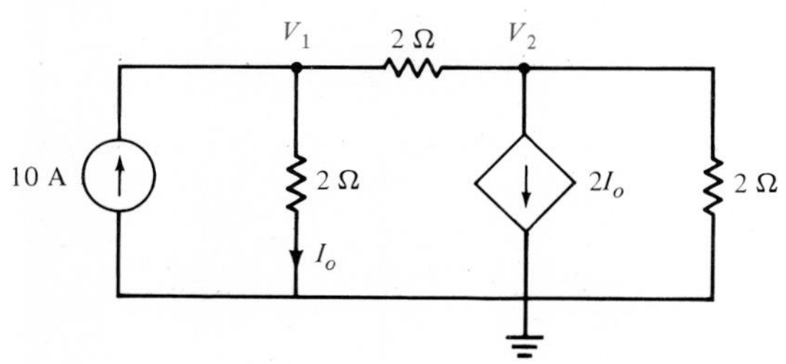

Figure P3.16

3.17. Use node equations to find I_o in the circuit in Fig. P3.17.

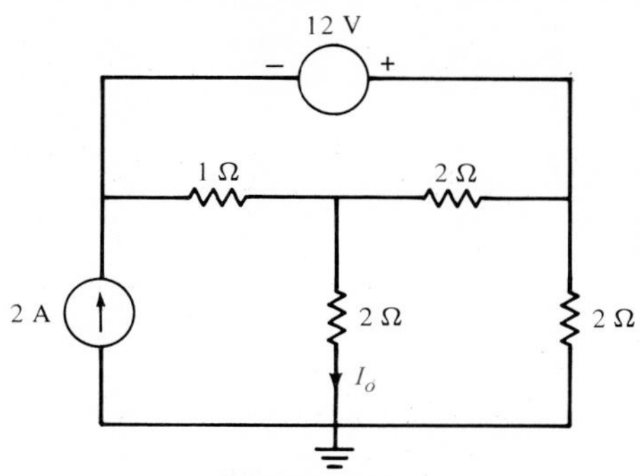

Figure P3.17

3.18. Find the voltage across the 24-Ω resistor in the network in Fig. P3.18 using nodal analysis.

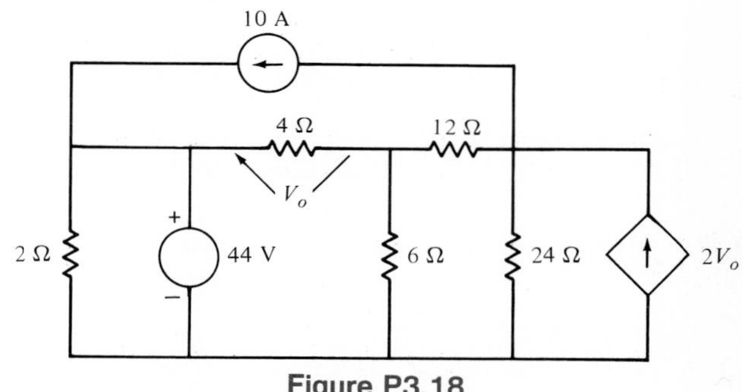

Figure P3.18

3.19. Find the node voltages in the circuit in Fig. P3.19.

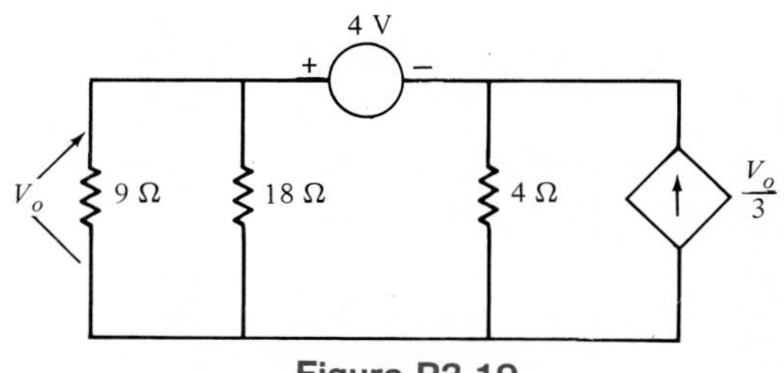

Figure P3.19

3.20. Determine the node voltages in the network in Fig. P3.20.

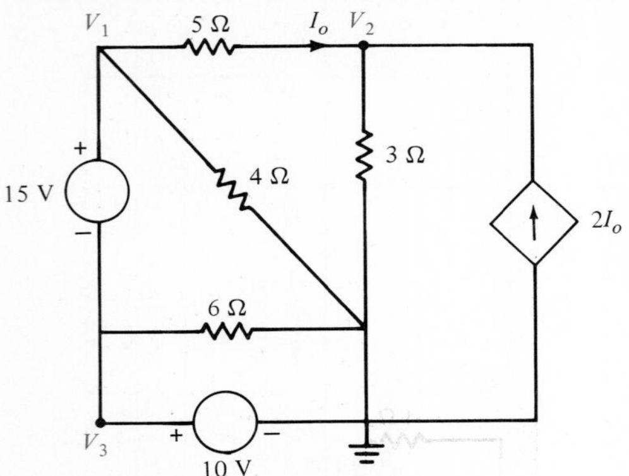

Figure P3.20

3.21. Find the voltage V_o in the network shown in Fig. P3.21 using nodal analysis.

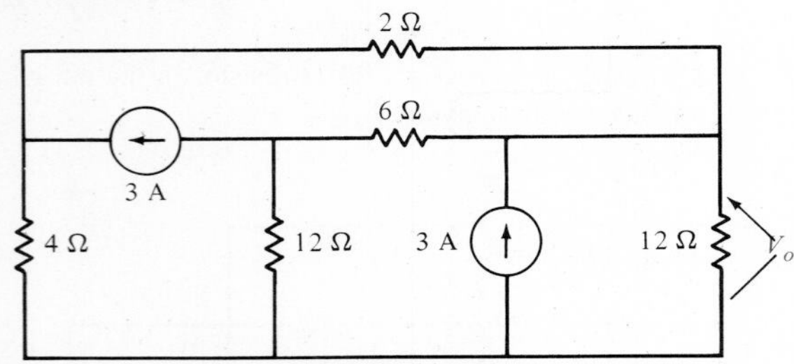

Figure P3.21

3.22. Determine the node voltages in the network in Fig. P3.22.

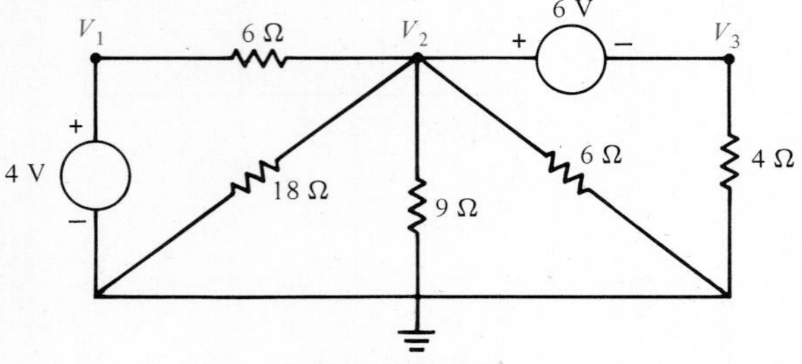

Figure P3.22

3.23. Find the current I_o in the network in Fig. P3.23 using nodal analysis.

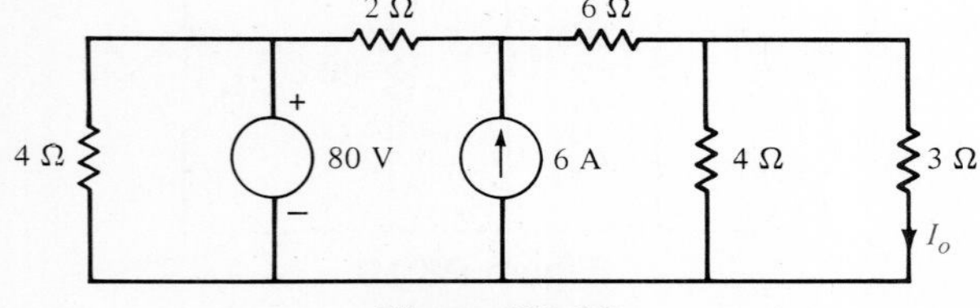

Figure P3.23

3.24. Calculate the voltage V_2 in Fig. P3.24, using nodal equations.

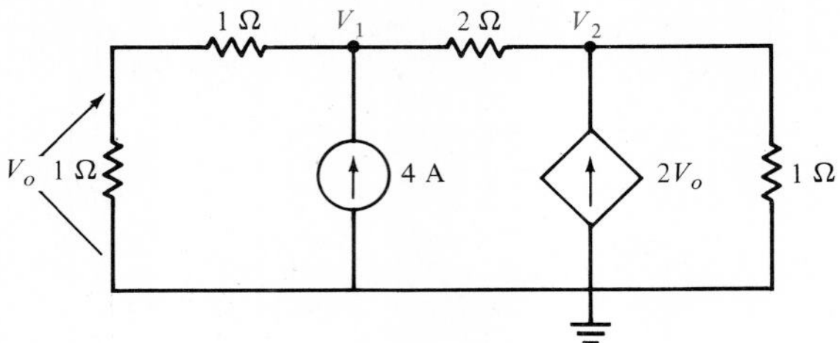

Figure P3.24

3.25. Find V_o in Fig. P3.25, using nodal equations.

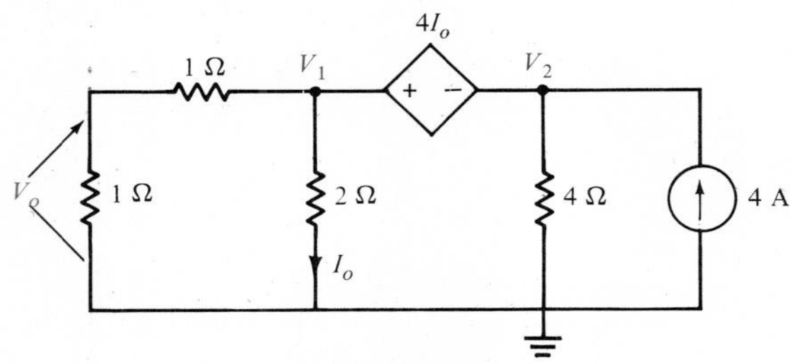

Figure P3.25

3.26. Find V_o in the circuit shown in Fig. P3.26, using nodal equations.

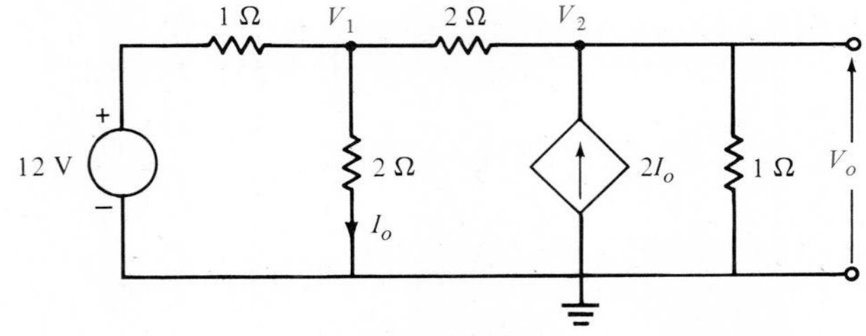

Figure P3.26

3.27. Write the node equations for the network in Fig. P3.27 in standard form.

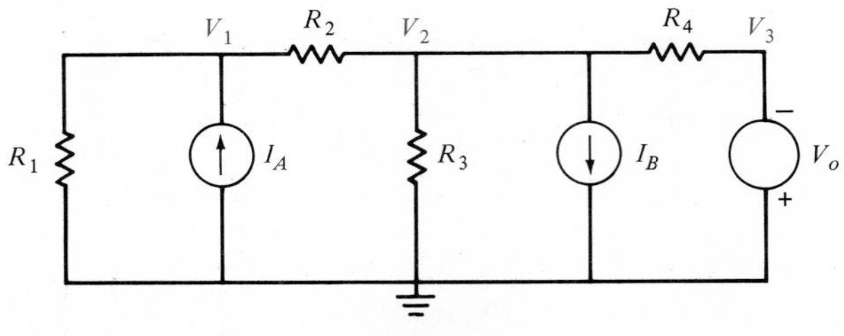

Figure P3.27

3.28. Write the nodal equations for the network in Fig. P3.28 in standard form.

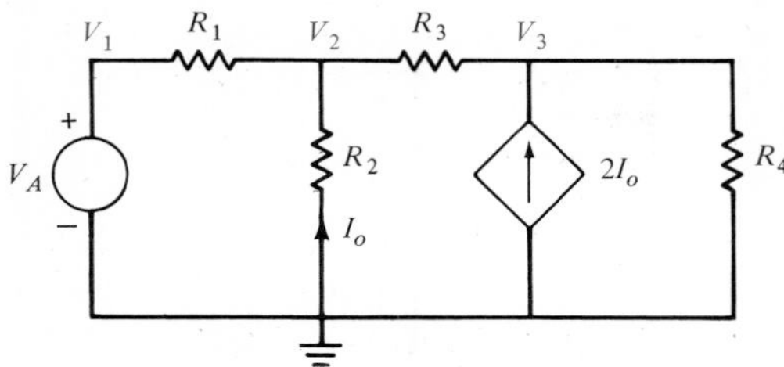

Figure P3.28

3.29. Write the node equations for the network shown in Fig. P3.29 in standard form.

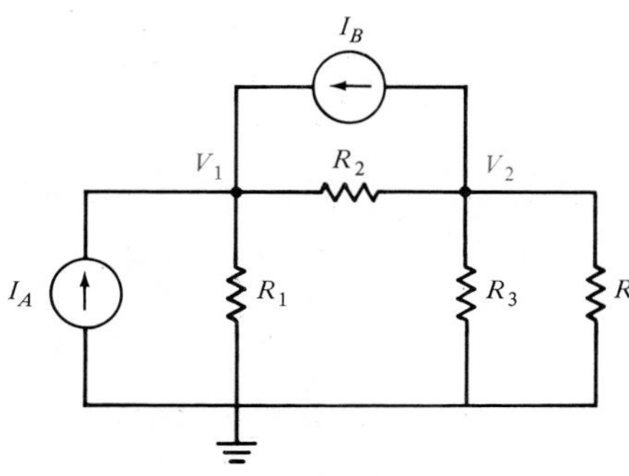

Figure P3.29

3.30. Write the node equations for the network in Fig. P3.30.

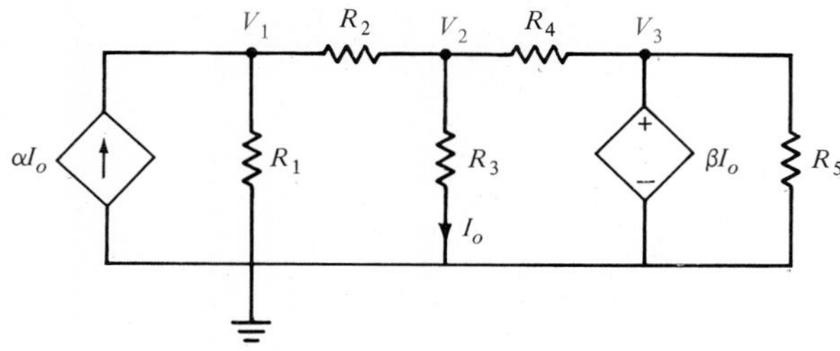

Figure P3.30

3.31. Use loop equations to find I_o in the circuit shown in Fig. P3.31.

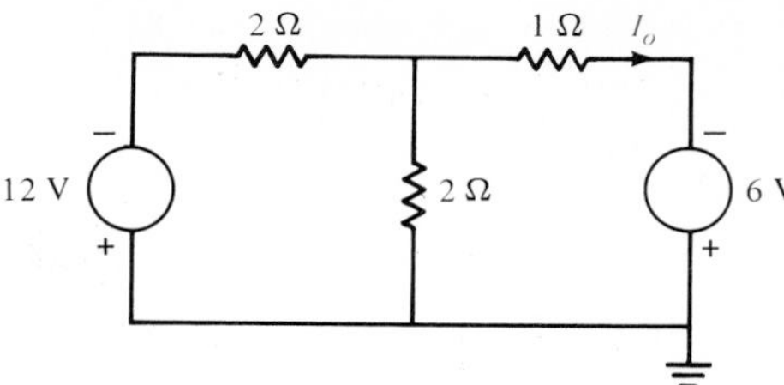

Figure P3.31

3.32. Use loop equations to calculate the power absorbed by the 5-Ω resistor in Fig. P3.32.

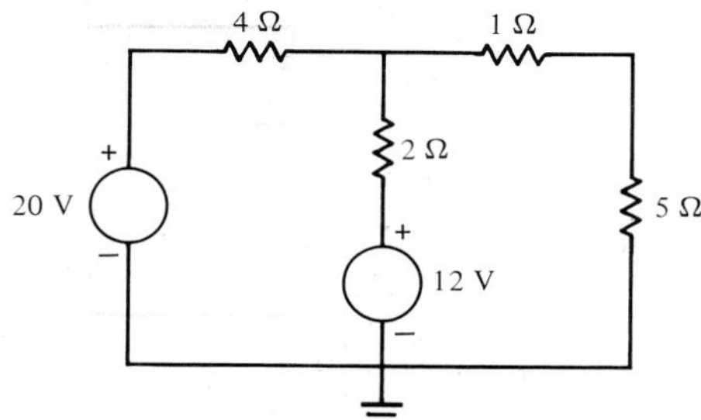

Figure P3.32

3.33. Use loop equations to find V_o in the circuit shown in Fig. P3.33.

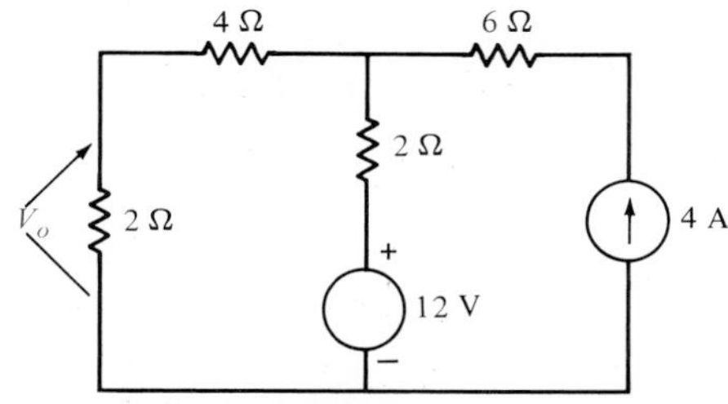

Figure P3.33

3.34. Use loop equations to find V_o in the circuit shown in Fig. P3.34.

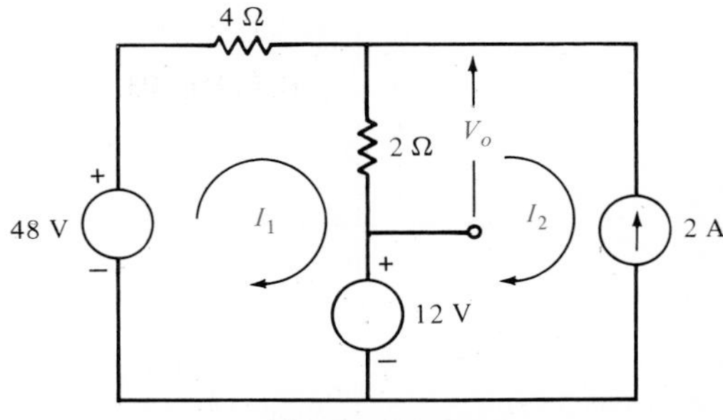

Figure P3.34

3.35. Use loop analysis to determine V_o in the network shown in Fig. P3.35.

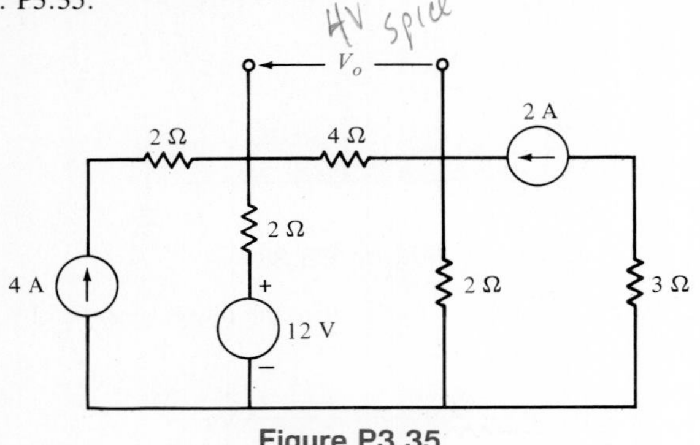

Figure P3.35

3.36. Use loop equations to find V_4 in the network in Fig. P3.36.

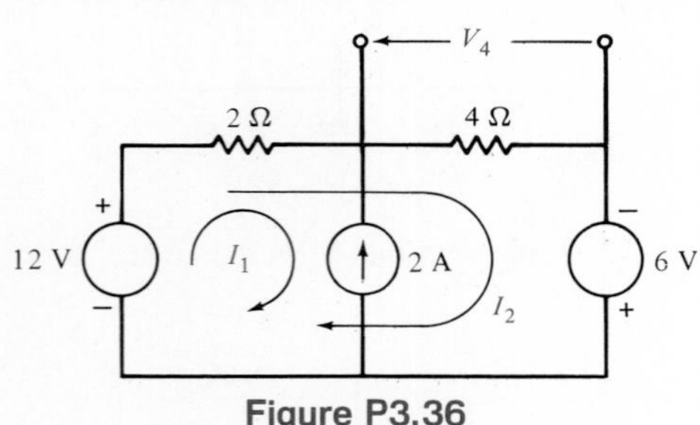

Figure P3.36

3.37. Use loop equations to find V_o in the network shown in Fig. P3.37.

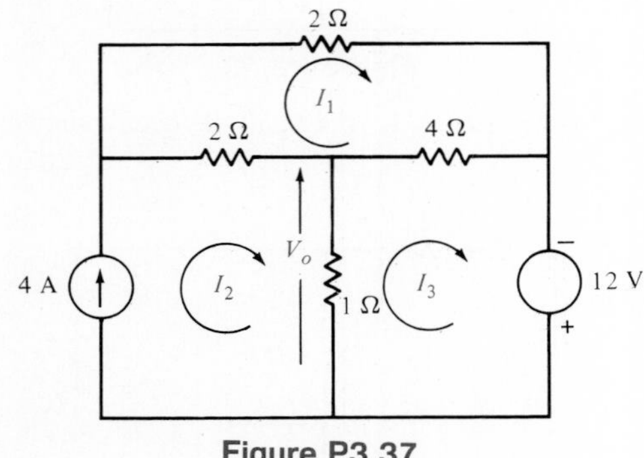

Figure P3.37

3.38. Use loop equations to find V_o in the network shown in Fig. P3.38.

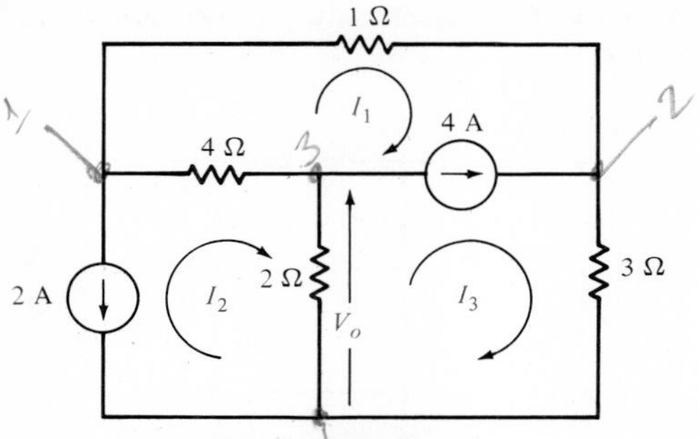

1 Ω I_1 4 Ω 4 A 2 A I_2 2 Ω V_o I_3 3 Ω

Figure P3.38

3.39. Use loop equations to find the power absorbed in the 1-Ω resistor in the network in Fig. P3.39.

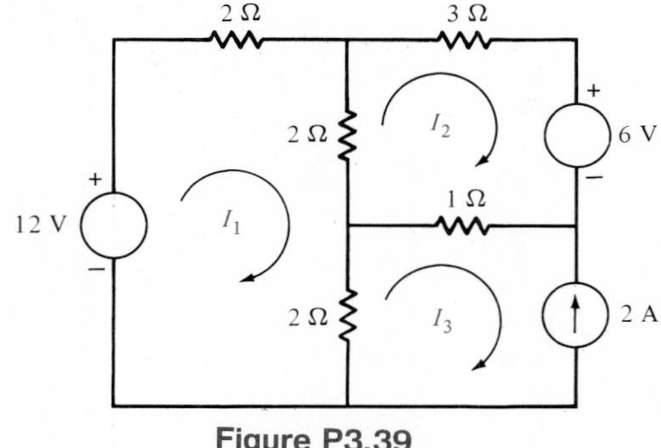

Figure P3.39

3.40. Determine V_o in the network shown in Fig. P3.40.

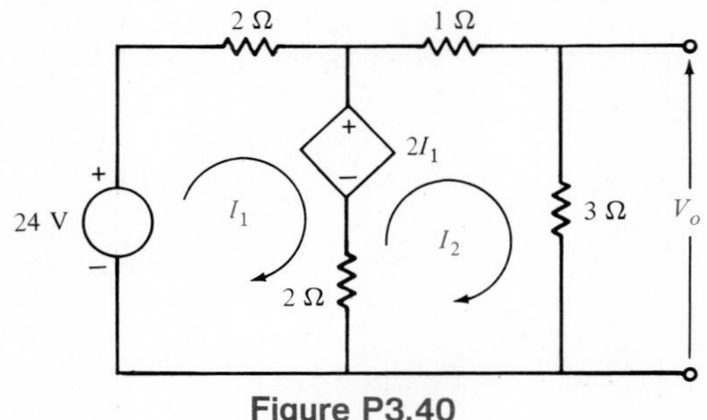

Figure P3.40

3.41. Write the loop equations for the circuit shown in Fig. P3.41.

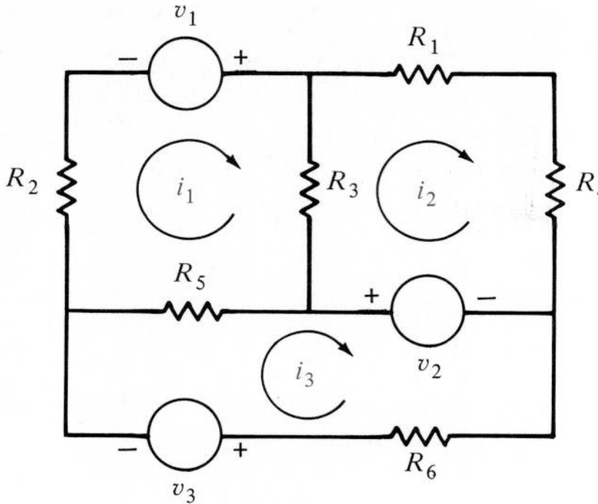

Figure P3.41

3.42. Write the loop equations for the circuit in Fig. P3.42 in standard form.

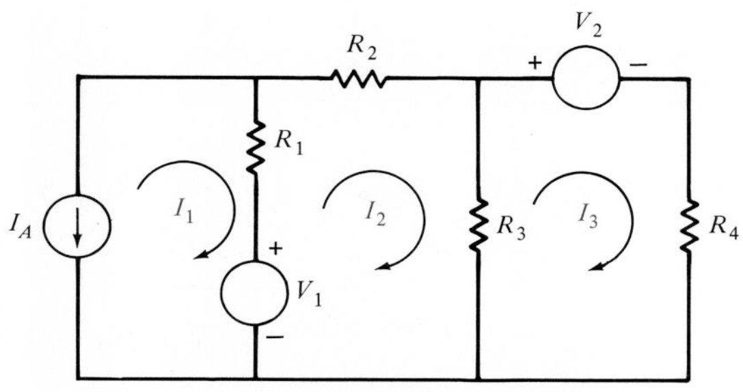

Figure P3.42

3.43. Write the mesh equations for the network in Fig. P3.43.

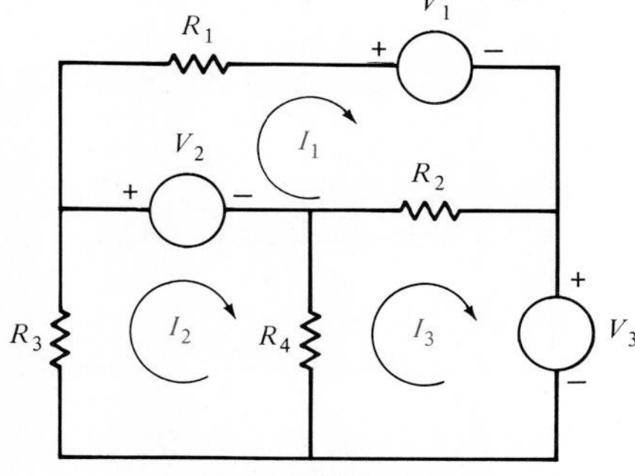

Figure P3.43

3.44. Write the loop equations for the network shown in Fig. P3.44.

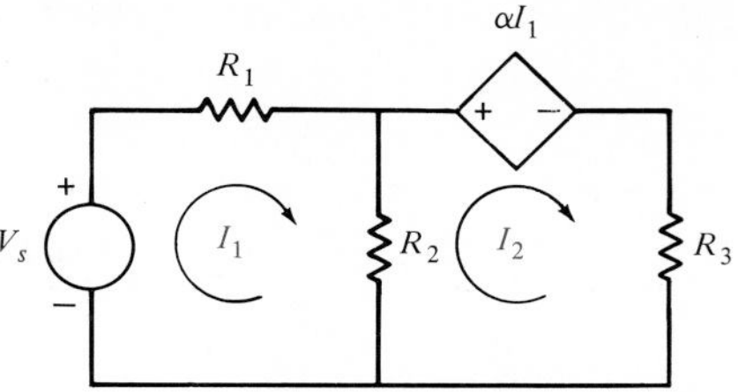

Figure P3.44

3.45. Write the mesh equations for the network shown in Fig. P3.45 in standard form.

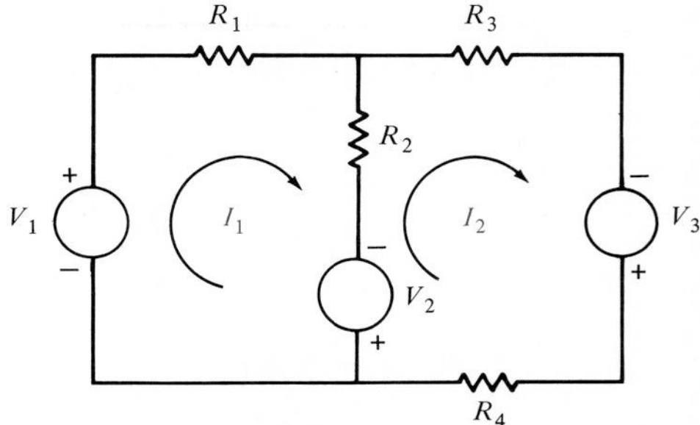

Figure D3.45

3.46. Determine I_o in the circuit shown in Fig. P3.46.

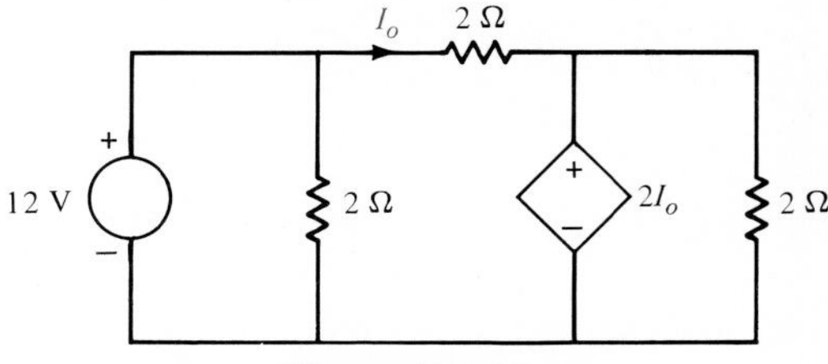

Figure P3.46

3.47. Given the network in Fig. P3.47 write the loop equations in standard form which, when solved, will yield the loop currents.

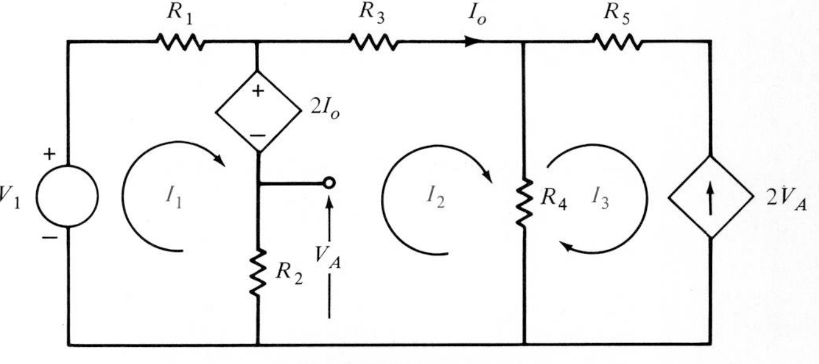

Figure P3.47

3.48. Determine the current flowing through the 4-Ω resistor in the network in Fig. P3.48.

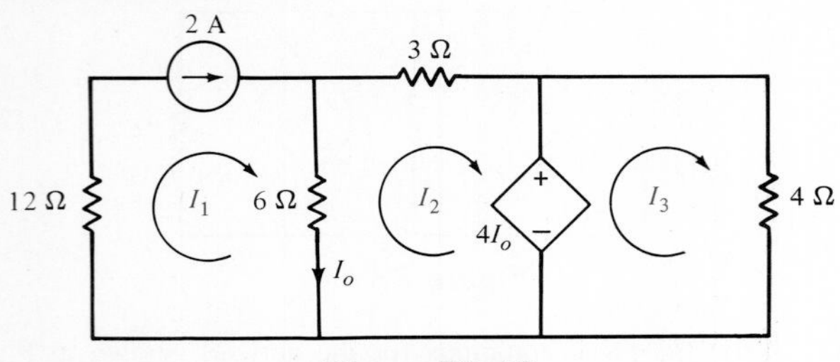

Figure P3.48

3.49. For the circuit in Fig. P3.49, determine the loop currents.

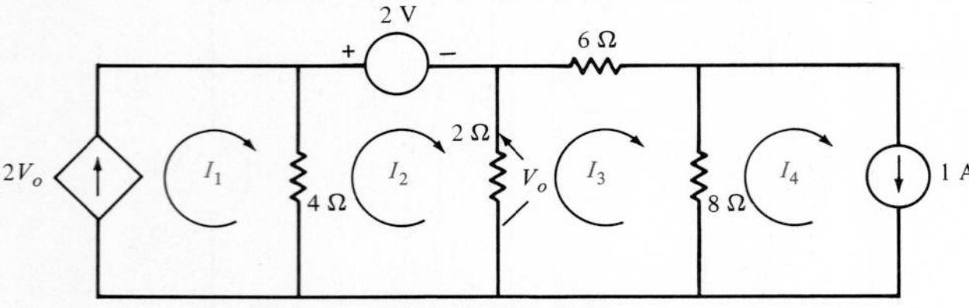

Figure P3.49

3.50. Determine the voltage across the 3-Ω resistor in the network in Fig. P3.50 using loop analysis.

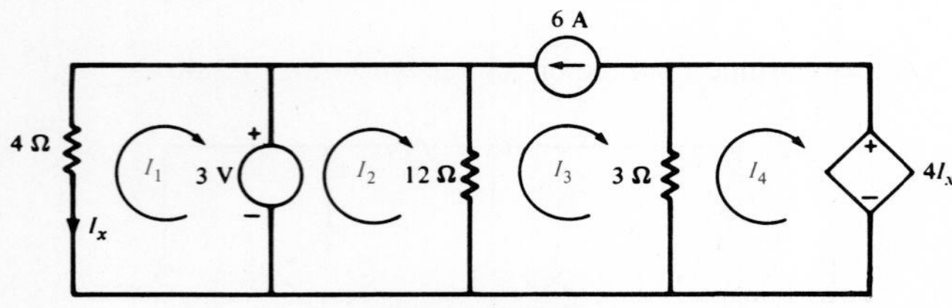

Figure P3.50

3.51. Find the loop currents I_1, I_2, and I_3 in the circuit in Fig. P3.51.

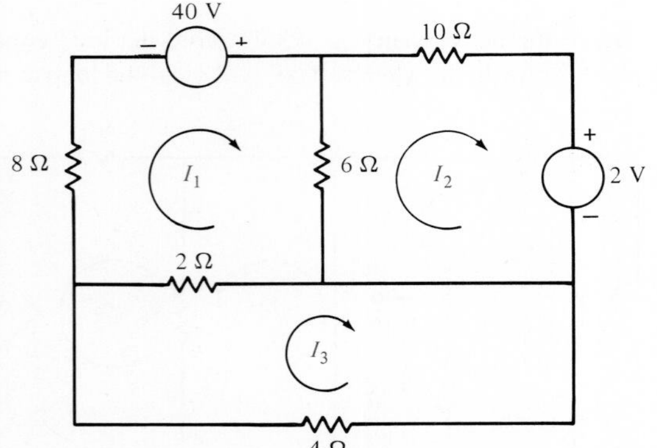

Figure P3.51

3.52. Find the currents I_1, I_2, and I_3 in the circuit in Fig. P3.52.

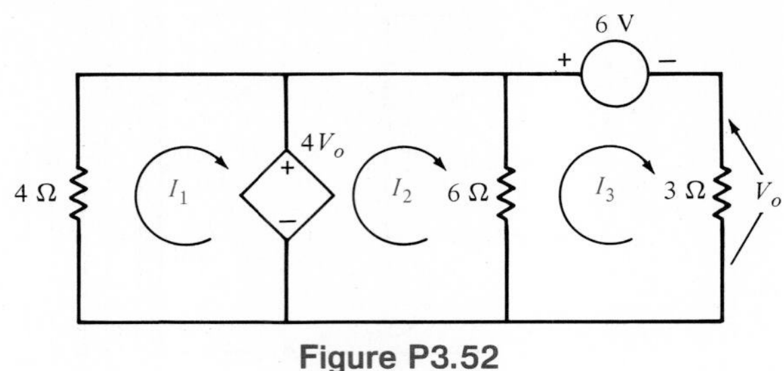

Figure P3.52

3.53. Find V_o in the network in Fig. P3.53.

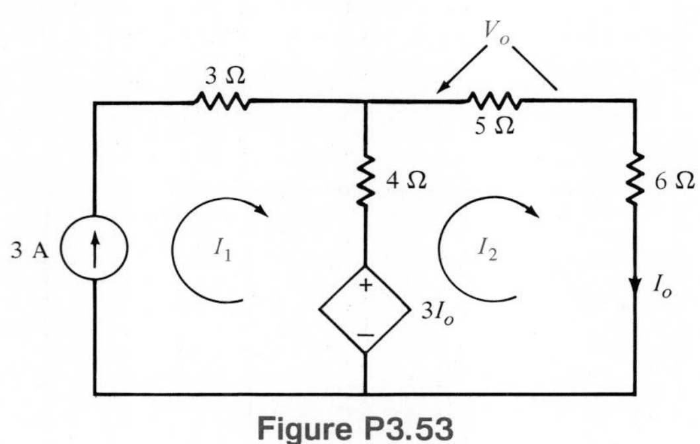

Figure P3.53

3.54. Write the loop equations for the circuit shown in Fig. P3.54.

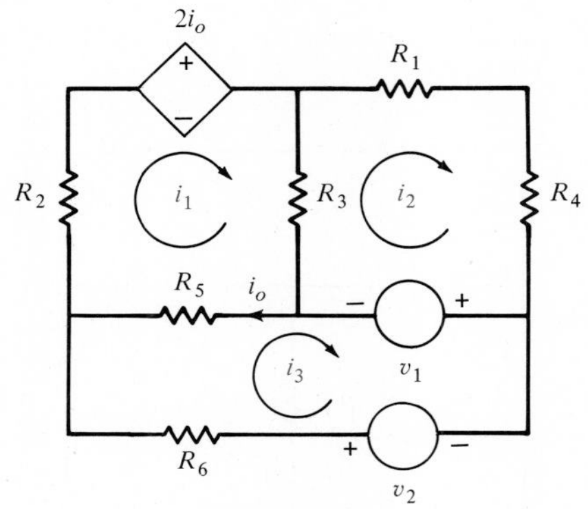

Figure P3.54

3.55. Find I_o in the circuit shown in Fig. P3.55.

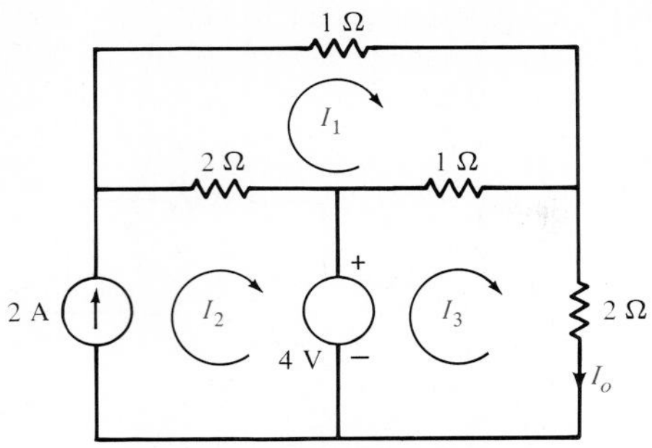

Figure P3.55

3.56. Write the loop equations necessary to solve for the currents in the circuit shown in Fig. P3.56.

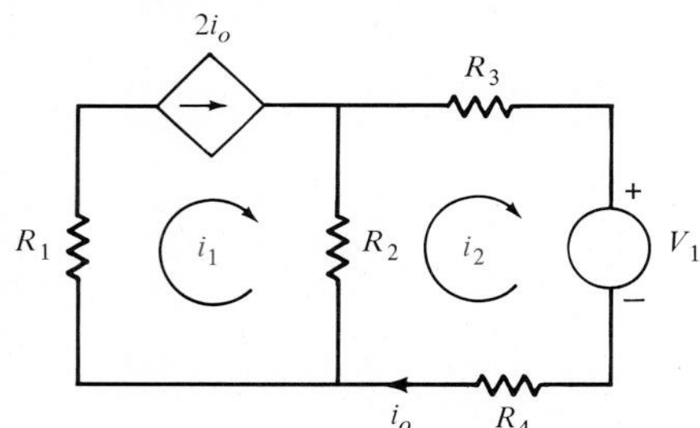

Figure P3.56

3.57. Find V_o in Fig. P3.57.

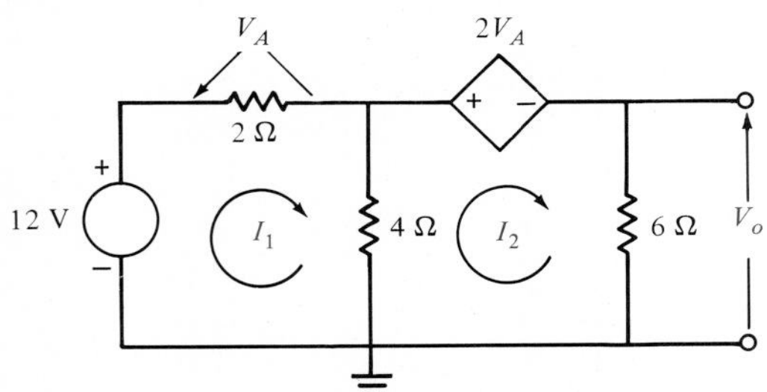

Figure P3.57

3.58. Use loop equations to find V_{out} in the circuit shown in Fig. P3.58.

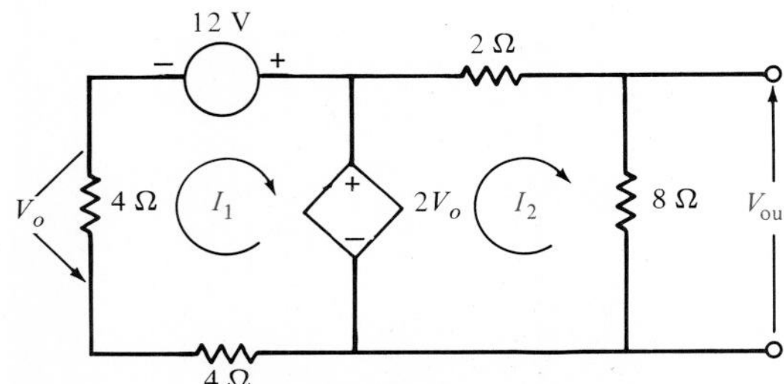

Figure P3.58

3.59. Use loop equations to find I_o in Fig. P3.59.

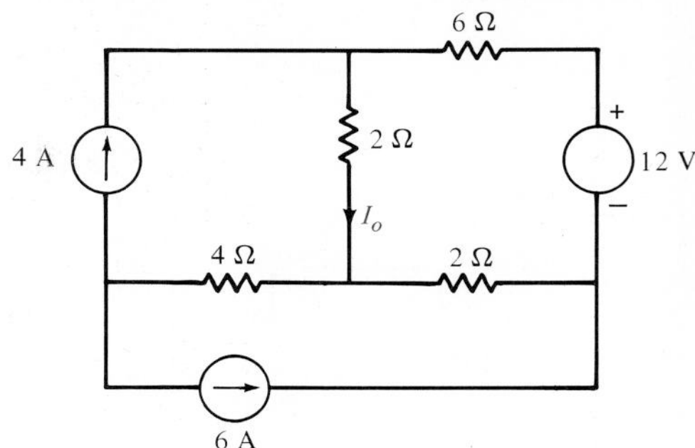

Figure P3.59

3.60. Determine I_o in Fig. P3.60 using loop equations.

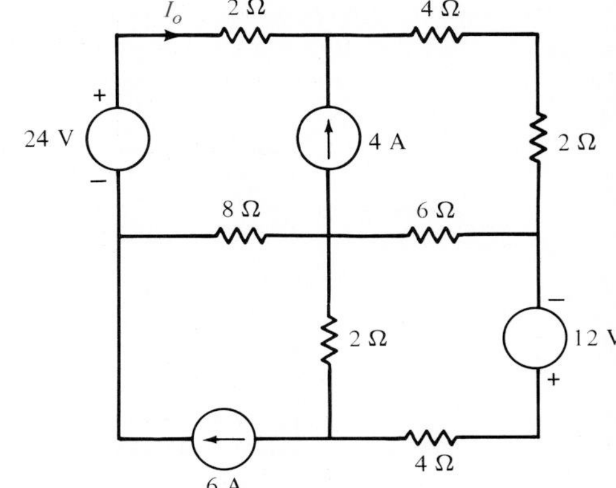

Figure P3.60

3.61. Given a box of 10-kΩ resistors and an op-amp, design a circuit that will have an output voltage of

$$v_o = -2v_1 - 4v_2$$

3.62. Find the expression for the output voltage in the op-amp circuit shown in Fig. P3.62.

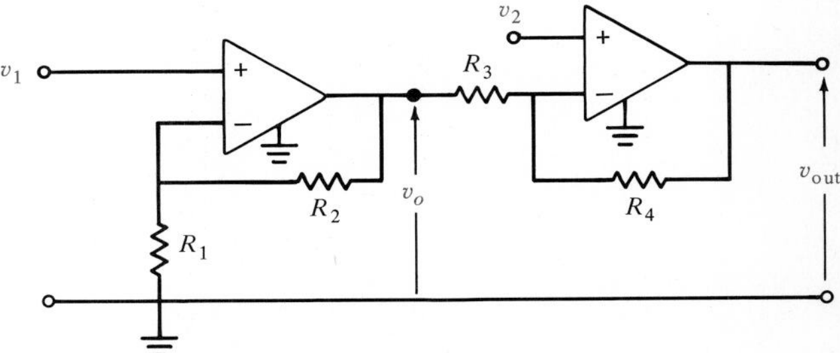

Figure P3.62

3.63. Using nodal equations, determine the gain v_o/v_1 of the circuit shown in Fig. P3.63.

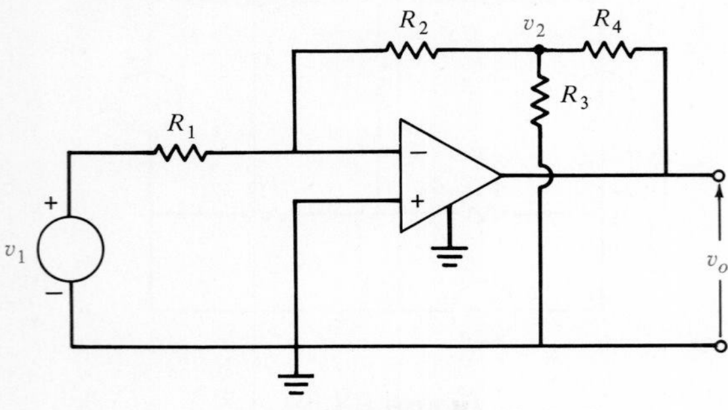

Figure P3.63

3.64. Given the amplifier circuit in Fig. P3.64, show that the output voltage is related directly to any small change ΔR in the resistor which connects the output to the inverting terminal.

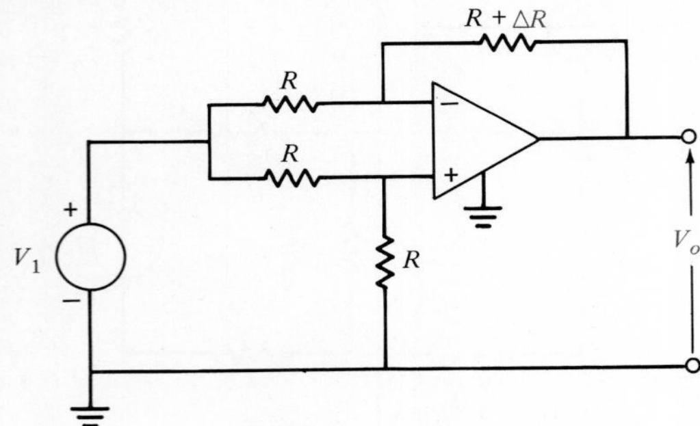

Figure P3.64

3.65. Using the equivalent circuit in Fig. 2.31, find the input resistance $R_i = v_1/i_1$ for the op-amp circuit shown in Fig. P3.65. Show that for R_o small and A large R_i reduces to R_1.

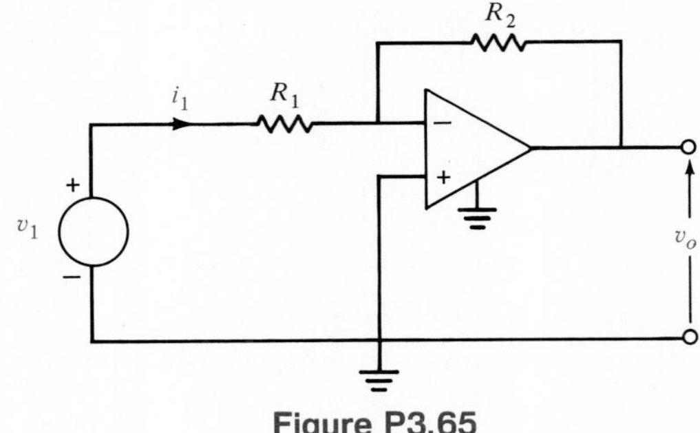

Figure P3.65

CHAPTER

4

Circuit Equations via Network Topology

In this chapter we generalize the methods of nodal and loop analyses that we have presented in Chapter 3. The vehicle that we will employ to accomplish this is *topology*. For our purposes topology refers to the properties that relate to the geometry of the circuit. These properties remain unchanged even if the circuit is bent into any other shape provided that no parts are cut and no new connections are made.

4.1 Basic Definitions

At this point readers should feel fairly comfortable about their ability to solve for the currents and voltages in a circuit via node and loop equations. And indeed, if they understand the material in Chapter 3, this feeling is well justified. However, to provide motivation for the concepts to be discussed next, consider the network shown in Fig. 4.1. Suppose that we are required to find all the currents and voltages in this circuit. This circuit seems to be much more complicated than those we considered in Chapter 3. It is more complicated because it is what is called a *nonplanar* circuit. A *planar* circuit is one that can be drawn on a plane surface with no crossovers; that is, no branch passes over any other branch. And, of course, a nonplanar circuit is one that is not planar. All the circuits we have considered thus far have been planar.

Let us now present some topological concepts to aid us in our discussion. Since the geometrical properties of a circuit are independent of the circuit elements that are contained in each branch, we will simply represent each network branch by a line segment. Such a drawing of the circuit is called a *graph*.

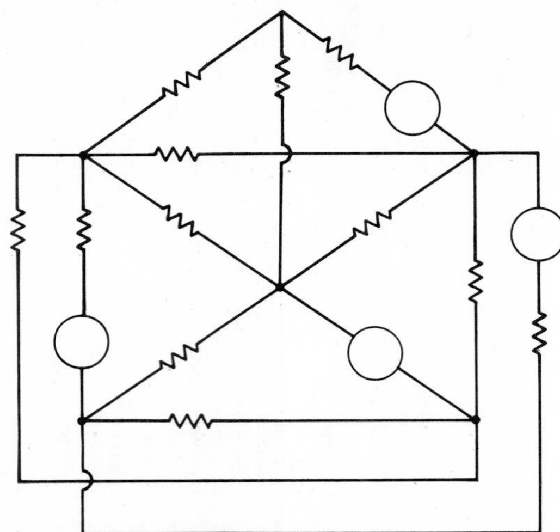

Figure 4.1 Examples of a nonplanar circuit.

EXAMPLE 4.1

The graph associated with the circuit shown in Fig. 4.2a is given in Fig. 4.2b.

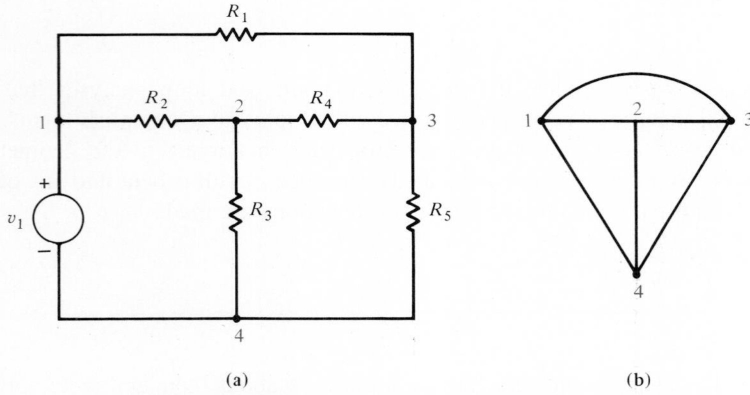

Figure 4.2 Circuit and its associated graph. ■

As can be seen from the example, the graph consists of a number of interconnected nodes. If a path exists from any node to every other node, the graph is said to be *connected*.

EXAMPLE 4.2

The graph in Fig. 4.3a is connected; however, the graph in Fig. 4.3b is not, because no path exists from node 4 to node 5. ■

A *tree* of the graph is defined as a set of branches that connects every node to every other node via some path without forming a closed path. In general, there exist a number of different trees for any given circuit.

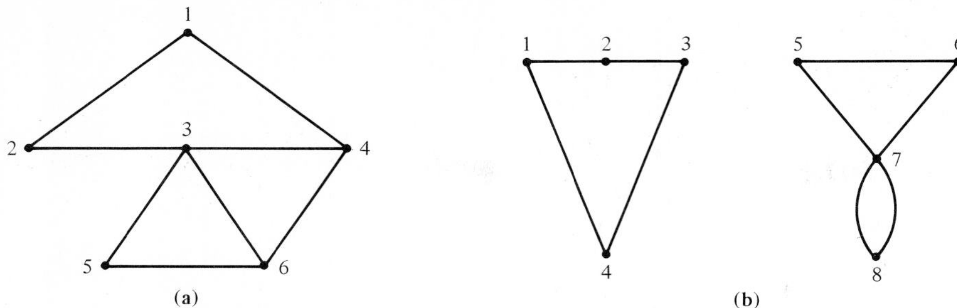

Figure 4.3 Examples of (a) connected and (b) unconnected graphs.

EXAMPLE 4.3

Three possible trees for the graph shown in Fig. 4.2b are given in Fig. 4.4a. The set of branches shown in Fig. 4.4b are not trees for the graph in Fig. 4.2b. ■

If a graph for a network is known and a particular tree is specified, those branches of the graph that are not part of the tree form what is called the *cotree*. The cotree consists of what we call *links*.

EXAMPLE 4.4

The links that belong to the cotrees that correspond to the trees shown in Fig. 4.4a are shown dashed in Fig. 4.5. ■

Note that it is impossible to specify categorically a particular branch of a graph as a link, since it may be a link for one choice of tree and not for another.

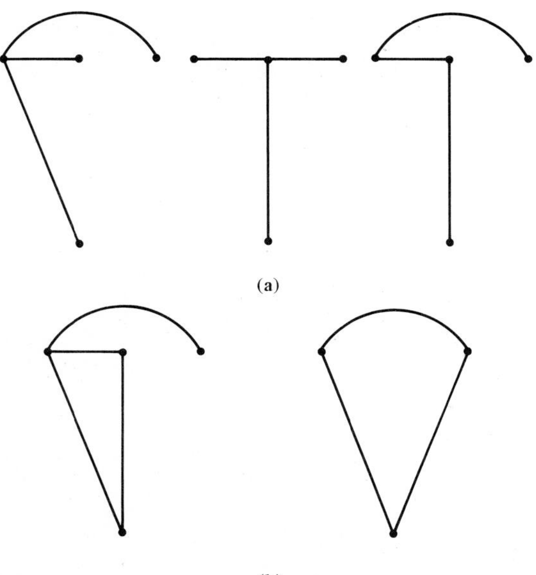

Figure 4.4 Trees and sets of branches that are not trees for the graph in Fig. 4.2b.

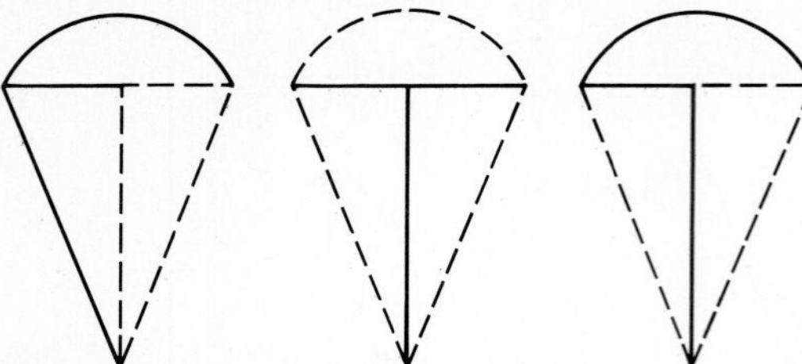

Figure 4.5 Cotrees consisting of links for the trees shown in Fig. 4.4a.

Finally, we wish to define what is called a *cut set*. A cut set is simply a minimum set of branches which, when cut, will divide the graph into two separate parts. Hence it is impossible to go from a node in one part of the graph to a node in another part without passing through a branch of the cut set.

EXAMPLE 4.5

Two cut sets for the graph shown in Fig. 4.2b are shown in Fig. 4.6. Note that in one case node 4 is separated from nodes 1, 2, and 3; in the other cases nodes 1 and 2 are separated from nodes 3 and 4. ■

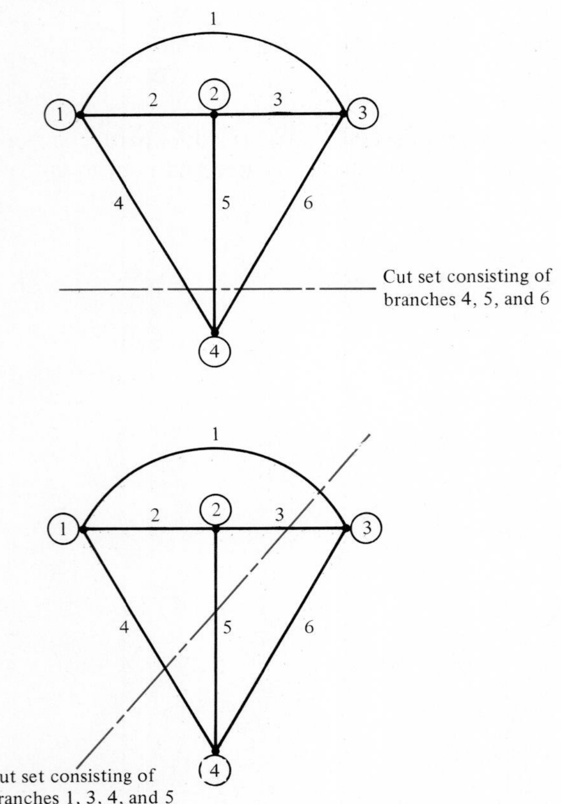

Figure 4.6 Two cut sets for the graph in Fig. 4.2b.

DRILL EXERCISE

D4.1. Given the graph for a network in Fig. D4.1a, state whether the tree–cotree combinations in Fig. D4.1b to f form a valid set of tree branches and links.

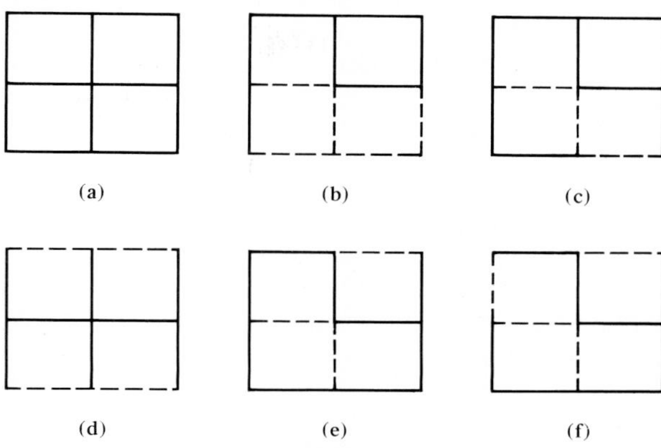

Figure D4.1

D4.2. Given the graph for the network in Fig. D4.2a, state whether the tree–cotree combinations in Fig. D4.2b to f form a valid set of tree branches and links.

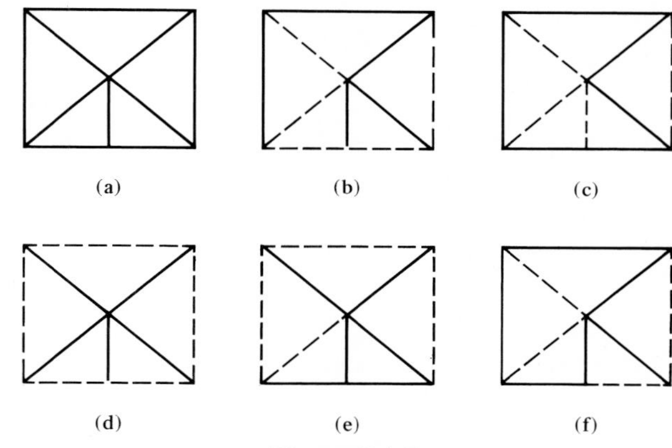

Figure D4.2

4.2 Relationships Among Links, Nodes, and Tree Branches

Suppose that we have a graph that represents a circuit, and let us define the following parameters:

$$L = \text{number of links}$$

$$B = \text{number of branches in the graph}$$

$$N = \text{number of nodes in the graph}$$

Now suppose that we remove all the branches from the graph so that only nodes remain. In order to construct a tree, we begin by placing one branch between two nodes. We continue adding branches to the tree without forming any loops. Each time we add an additional branch we add one more node. Note that the addition of every successive branch connects one node, except for the first branch, which connected two nodes. Hence there is one more node than there are branches on the tree. Therefore, in general, a tree consists of $N - 1$ branches. Since the entire graph contains B branches, the number of links is given by the expression

$$L = B - (N - 1) = B - N + 1 \tag{4.1}$$

EXAMPLE 4.6

A specific tree for the network shown in Fig. 4.1 is illustrated in Fig. 4.7. Note that the graph of the network satisfies the relationship in Eq. (4.1).

$$L = B - N + 1$$

$$8 = 13 - 6 + 1$$

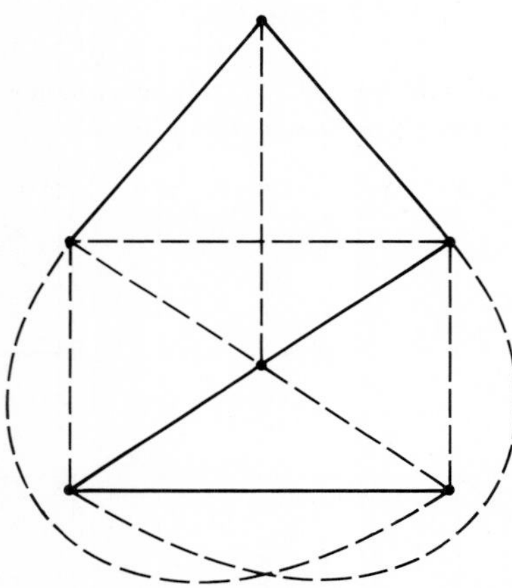

Figure 4.7 Tree and its corresponding links for the network in Fig. 4.1. ■

DRILL EXERCISE

D4.3. Compute the proper number of links for the graphs in Figs. D4.1a and D4.2a.

4.3 General Nodal Analysis Equations

Our objective here will be to determine in general the proper number of nodal equations necessary to solve for all the voltages in a network. In Chapter 3 we showed that if we had a *linearly independent* set of $N - 1$ nodal equations for an N-node network, we could determine the necessary node voltages. We will now present an argument, based on network topology, which illustrates that for an N-node circuit, the number of independent KCL equations is $N - 1$.

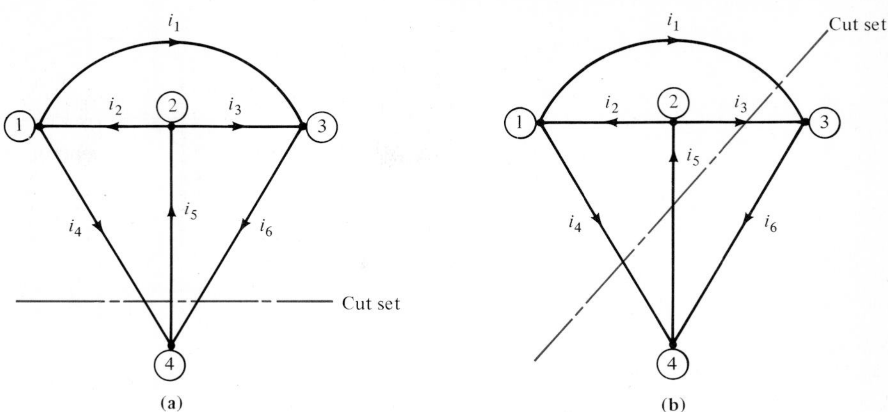

Figure 4.8 Example graph for illustrating cut-set equations.

To begin, let us redraw the graphs in Fig. 4.6 with assumed directions of currents as shown in Fig. 4.8. KCL for either part of the circuit divided by a cut set is called a *cut-set equation*. Note that each cut-set equation is nothing more than a sum of KCL equations written for all nodes in either part of the circuit. For example, the cut-set equation for Fig. 4.8a is

$$i_4 - i_5 + i_6 = 0$$

which is KCL at node 4, where $i_k(t)$ represents the kth branch as defined in Fig. 4.8a. Similarly, the cut-set equation for Fig. 4.8b is

$$i_4 - i_5 + i_3 + i_1 = 0$$

which is the sum of the KCL equations at nodes 3 and 4, or equivalently at nodes 1 and 2:

$$\text{For node 3:} \qquad i_1 + i_3 - i_6 = 0$$

$$\text{For node 4:} \qquad i_4 - i_5 + i_6 = 0$$

Adding these two equations yields

$$i_4 - i_5 + i_3 + i_1 = 0$$

Because of this relationship between KCL equations and the cut-set equations, if we can determine the number of independent cut-set equations, we will also have determined the number of independent KCL equations.

The cut-set equations are derived from a tree which, as we have shown, contains $N - 1$ branches. If any branch of the tree is cut, the tree will be divided into two parts. Each tree branch that is cut, together with the links that connect the two separate parts of the graph, form a cut set. For example, the cut sets for a specific tree for the graph shown in Fig. 4.2 are shown in Fig. 4.9. Note that cut set 1 includes tree branch 5 and links 4 and 6. Cut set 2 includes tree branch 2 and links 4, 3, and 6. Cut set 3 includes tree branch 1 and links 3 and 6. We call these cut sets the *fundamental* cut sets. There are $N - 1$ fundamental cut sets because there are $N - 1$ tree branches. As shown above, each fundamental cut set produces a cut-set equation. Since each cut-set equation contains

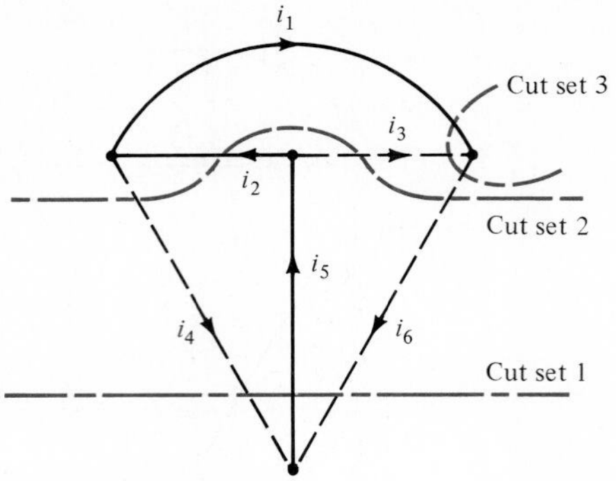

Figure 4.9 Illustration of fundamental cut sets.

a tree branch current that does not appear in any other cut-set equation, each equation is independent; that is, it cannot be derived as a linear combination of the other equations. Therefore, at least $N - 1$ cut-set equations are independent.

Let us now see if other cut-set equations can be added to the fundamental cut-set equations and remain independent. If they cannot, no other independent cut-set equations exist, and the number of independent cut-set equations, and therefore KCL equations, is $N - 1$.

Since by definition a cut set must contain at least one tree branch, a cut set that is not a fundamental cut set must contain at least two tree branches. Let us suppose, then, that we have a cut set with two or more tree branches. The cut-set equation resulting from this cut set can be expressed completely in terms of link currents by expressing each tree branch current in terms of link currents using the fundamental cut-set equations. If this is done, the coefficient of each link current in the resulting equation must be zero, and hence the equation reduces to $0 = 0$. To see this, consider the fundamental cut sets in Fig. 4.9 and a new cut set that cuts tree branches 1 and 2 and link 4. For the assumed directions of current the fundamental cut-set equations are

$$\text{Cut set 1:} \quad i_5 = i_4 + i_6$$

$$\text{Cut set 2:} \quad i_2 = i_4 - i_3 + i_6$$

$$\text{Cut set 3:} \quad i_1 = -i_3 + i_6$$

The equation for the new cut set is

$$i_4 - i_2 + i_1 = 0$$

Using the fundamental cut-set equations for i_1 and i_2 yields

$$i_4 - (i_4 - i_3 + i_6) + (-i_3 + i_6) = 0$$

$$0 = 0$$

Note what this condition means. If the foregoing condition were not true, we could express one link current as a linear combination of other link currents: for example, i_6 in terms of i_4 and i_3 in Fig. 4.9. However, in this case, if i_4 and i_3 are cut, thus forcing them to zero, i_6 is also forced to zero. But this is impossible since i_6 together with the tree branches form a closed path that contains a current which could exist even if $i_3 = i_4 = 0$. Therefore, cut-set equations that are not a linear combination of the fundamental cut-set equations are not independent, and hence the number of independent cut-set equations, and thus KCL equations, is $N - 1$. If there are voltage sources present in the network, then, in general, the number of independent KCL equations is $N - 1 -$ the number of voltage sources.

In our general nodal analysis technique we begin with an electrical network and for this circuit we draw a graph. The variables for the graph are the tree branch voltages and hence the number of variables is $N - 1 -$ the number of voltage sources. The voltage across each link is then a sum of the tree branch voltages. Therefore, all the branch voltages in the network can be expressed in terms of the tree branch voltages. Substituting the branch voltages into the cut-set equations yields a set of independent equations that number $N - 1 -$ number of voltage sources.

It is important to note that in a circuit which contains no voltage sources, not only do the $N - 1$ tree branch voltages form an independent set of equations, but as shown in Chapter 3, the $N - 1$ nonreference node voltages also constitute an independent set, since we can determine the node voltages from the tree branch voltages, and vice versa. In fact, any independent set of $N - 1$ voltages can be used to solve the network.

EXAMPLE 4.7

Given the graph of a network in Fig. 4.10a with the assumed direction of currents and the specified tree, find an independent set of KCL equations.

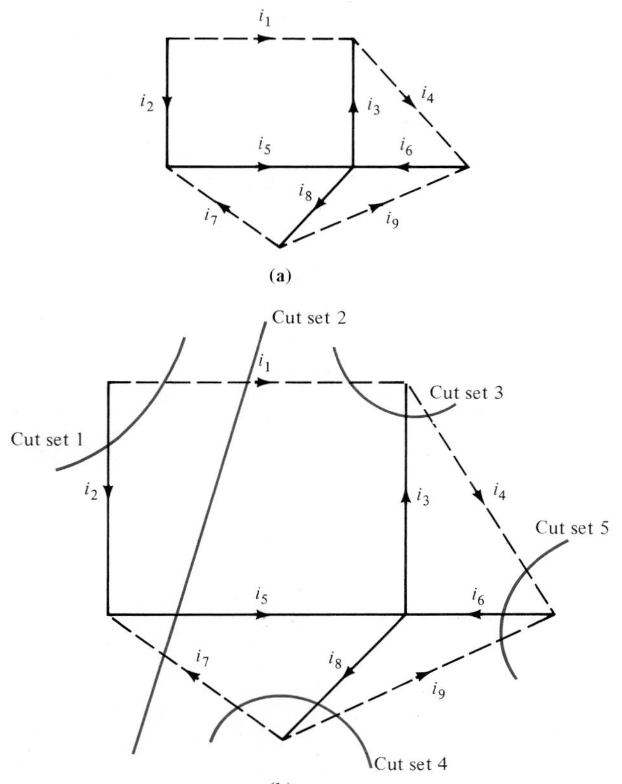

Figure 4.10 Example illustrating fundamental cut-set equations.

Since the tree is defined, the independent set of KCL equations are equivalent to the fundamental cut-set equations. The fundamental cut sets are shown in Fig. 4.10b. Therefore, the independent set of KCL equations are

$$i_1 + i_2 = 0$$

$$i_1 + i_5 - i_7 = 0$$

$$i_1 + i_3 - i_4 = 0$$

$$i_7 - i_8 + i_9 = 0$$

$$i_4 - i_6 + i_9 = 0$$

■

EXAMPLE 4.8

Using the topology methods we have just described, we wish to derive the nodal equations for the network in Fig. 3.1. A graph for the network shown in Fig. 3.1 is given in Fig. 4.11.

The cut-set equations for the graph are

$$i_1 + i_2 + i_3 = 0$$

and

$$-i_3 + i_4 - i_A = 0$$

Expressing the currents in terms of the tree branch voltages, we obtain

$$\frac{v_1}{R_1} + \frac{v_1}{R_2} + \frac{v_1 - v_2}{R_3} = 0$$

$$\frac{-(v_1 - v_2)}{R_3} + \frac{v_2}{R_4} - i_A = 0$$

or

$$v_1\left(\frac{1}{R_1} + \frac{1}{R_2} + \frac{1}{R_3}\right) - v_2\frac{1}{R_3} = 0$$

$$-v_1\frac{1}{R_3} + v_2\left(\frac{1}{R_3} + \frac{1}{R_4}\right) = i_A$$

■

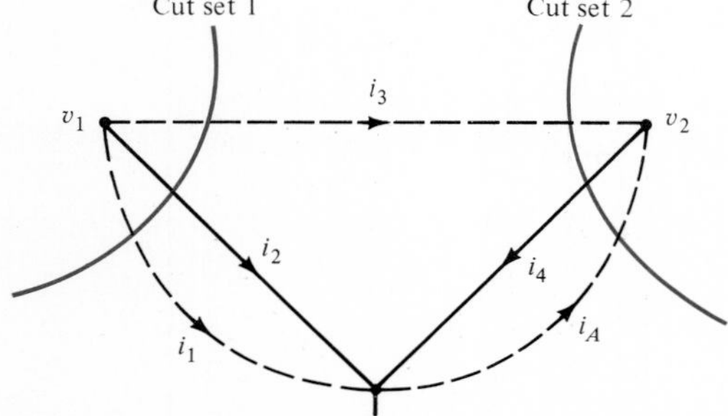

Figure 4.11 Graph and cut sets for the network in Fig. 3.1.

EXAMPLE 4.9

Repeat Example 4.8 for the network in Fig. 3.2. A graph for this network is shown in Fig. 4.12.

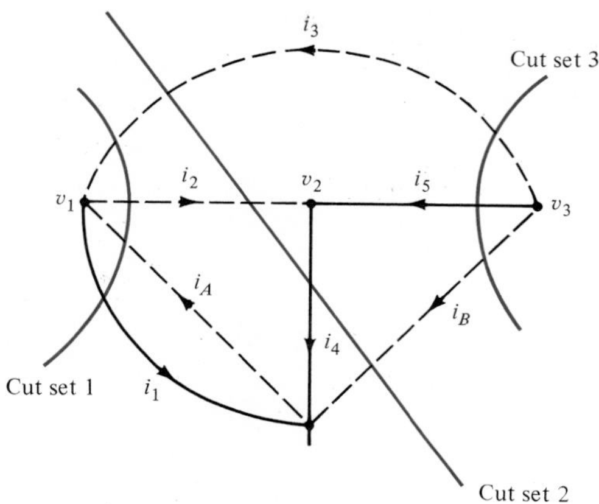

Figure 4.12 Graph and cut sets for the network shown in Fig. 3.2.

The cut-set equations for this graph are

$$i_1 - i_A + i_2 - i_3 = 0$$

$$i_3 - i_2 + i_4 + i_B = 0$$

$$i_3 + i_5 + i_B = 0$$

Subtracting the third equation from the second equation and using the resultant equation with the first and third equations yields

$$i_1 + i_2 - i_3 = i_A$$

$$-i_2 + i_4 - i_5 = 0$$

$$i_3 + i_5 = -i_B$$

Expressing these currents in terms of the node voltages, we obtain

$$\frac{v_1}{R_1} + \frac{v_1 - v_2}{R_2} - \frac{v_3 - v_1}{R_3} = i_A$$

$$\frac{-(v_1 - v_2)}{R_2} + \frac{v_2}{R_4} - \frac{v_3 - v_2}{R_5} = 0$$

$$\frac{v_3 - v_1}{R_3} + \frac{v_3 - v_2}{R_5} = -i_B$$

which are the same as the equations in (3.5). ■

DRILL EXERCISE

D4.4. Given the network in Fig. D4.4a, use the specified graph for the network in Fig. D4.4b to write a proper set of nodal equations for the network.

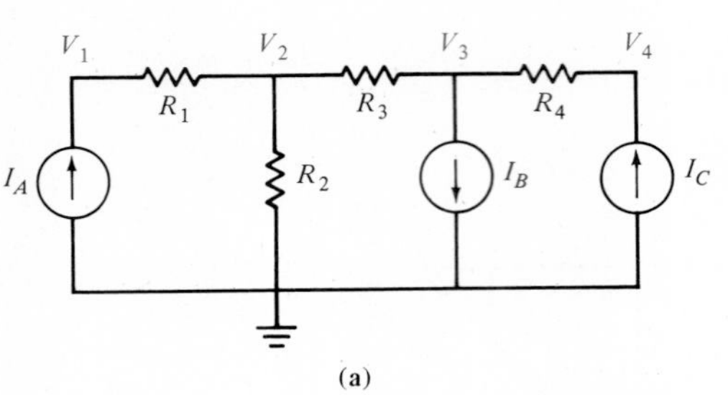

(a)

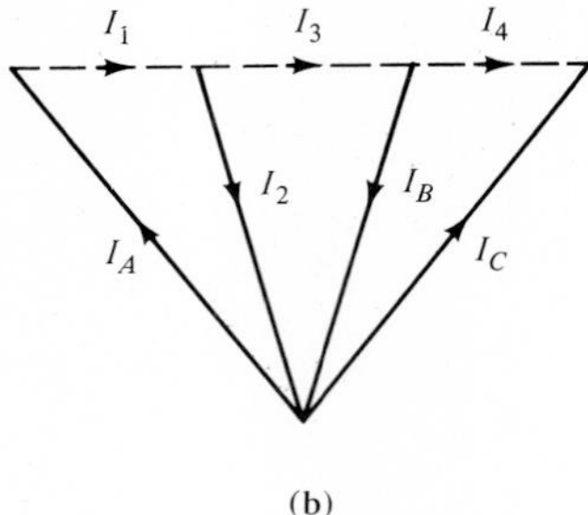

(b)

Figure D4.4

D4.5. Given the network in Fig. D4.5a, use the specified graph for the network in Fig. D4.5b to write a proper set of nodal equations for the network.

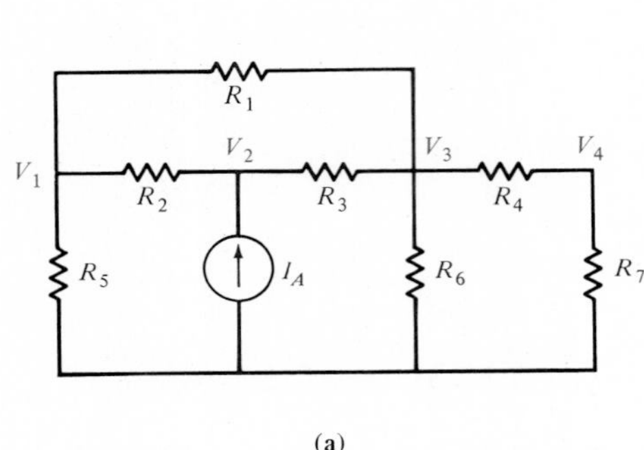

(a)

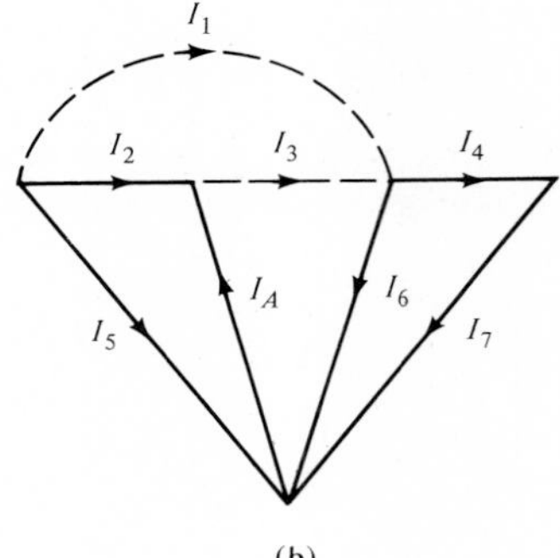

(b)

Figure D4.5

4.4 General Loop Analysis Equations

Our objective in this section will be to determine, in general, the proper number of KVL equations necessary to solve a circuit. The approach will be analogous and essentially parallel to our previous discussion concerning the number of independent KCL equations.

Consider once again the graph shown in Fig. 4.9. Imagine that we begin only with the tree branches 1, 2, and 5. If we now add one link at a time to the tree, we create a new loop with each link. For example, adding link 3 creates the loop consisting of that link and tree branches 1 and 2. The other links added one at a time create similar loops. Since we have shown that the number of links is equal to $B - N + 1$, we will construct this same number of loops, each one of which contains only one link. The loops constructed in this manner are called the *fundamental loops*. If KVL equations are written for these fundamental loops, these equations will be independent since each one contains

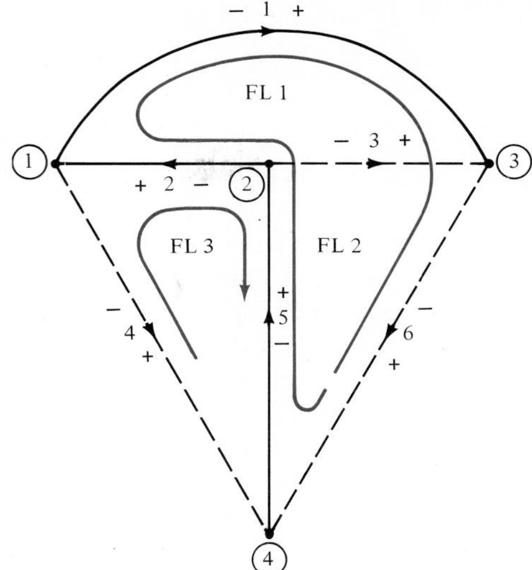

Figure 4.13 Illustration of fundamental loops.

a link voltage that is not present in any other equation. Hence there are at least $B - N + 1$ independent KVL equations.

Let us now see if there are other KVL equations that are independent of the fundamental loop equations. If there are none, the number of independent KVL equations is $B - N + 1$.

As has just been shown, every closed path, or loop, must contain at least one link since by definition a tree alone contains no closed paths. Therefore, any loop other than a fundamental loop contains at least two or more links. Consider the fundamental loops (FLs) shown in Fig. 4.13, and let us select the closed path containing tree branches 1 and 5 and links 3 and 4. Using the fundamental loops, the link voltages can be written in terms of the tree branch voltages. If the link voltages (3 and 4) are eliminated from the equation for the closed path, the coefficient for each tree branch voltage in the resulting equation must be zero. Hence, the equation becomes $0 = 0$. This fact can easily be seen using Fig. 4.13. Assume a generator sign convention for each branch so that the arrowheads in the graph point to the positive voltage terminal for each tree branch or link. The fundamental loop equations are therefore

$$v_3 = v_1 + v_2$$

$$v_4 = -v_2 - v_5$$

$$v_6 = -v_5 - v_2 - v_1$$

The equation for the closed path ①-③-②-④-① is

$$v_1 - v_3 - v_5 - v_4 = 0$$

Substituting for the link voltages v_3 and v_4 yields

$$v_1 - (v_1 + v_2) - v_5 - (v_2 - v_5) = 0$$
$$0 = 0$$

If this condition were not true, it would be possible to express one tree branch voltage in terms of only other tree branch voltages. For example, if one tree branch voltage, say v_1, could be expressed only in terms of others, say v_2 and v_5, and if v_2 and v_5 are short-circuited so that they are reduced to zero, v_1 would also become zero. However, as can be seen in Fig. 4.13, this condition is impossible, since v_1 could have a value other than zero even if $v_2 = v_5 = 0$. Therefore, KVL equations that are not a linear combination of the fundamental loop equations are not independent and hence the number of independent KVL equations is $B - N + 1$.

If there are current sources present in the network, then in general, the number of independent KVL equations is $B - N + 1 -$ the number of current sources.

In our general loop analysis technique we begin with an electrical network and for this network we draw a graph. Branch voltages are selected and the fundamental loops identified. The fundamental loop current variables together with the equation $v = iR$ yield a set of independent loop equations that number $N - B + 1 -$ the number of current sources.

Thus far our analysis has been very general. In the special case of planar networks, ''windowpanes'' can be employed to write the necessary KVL equations. Figure 4.14, which is a redrawn version of Fig. 4.13 with a different tree, illustrates this approach. If all the mesh currents are assumed to be in the same direction and no controlled sources are present, the KVL equations for the windowpanes can be written by inspection, as was shown in Chapter 3.

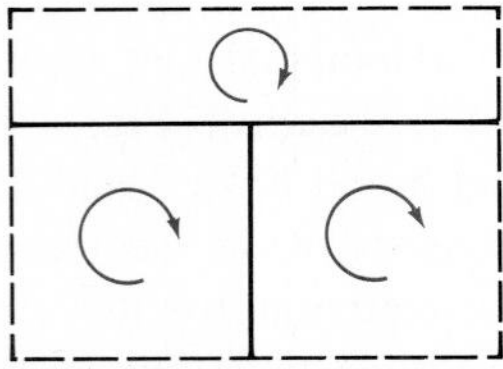

Figure 4.14 Graph illustrating windowpanes for circuit in Fig. 4.2.

EXAMPLE 4.10

Using the topology methods we have just described, we wish to derive the loop equations for the network in Fig. 3.13. A graph of this network is shown in Fig. 4.15a. Branch voltages have been selected and the links together with the appropriate tree branches form the fundamental loops.

The fundamental loop current variables are i_1 and i_2 and the fundamental loop equations are

$$v_1 - v_{R_1} - v_{R_3} - v_{R_2} = 0$$
$$v_{R_3} - v_2 - v_{R_4} - v_{R_5} = 0$$

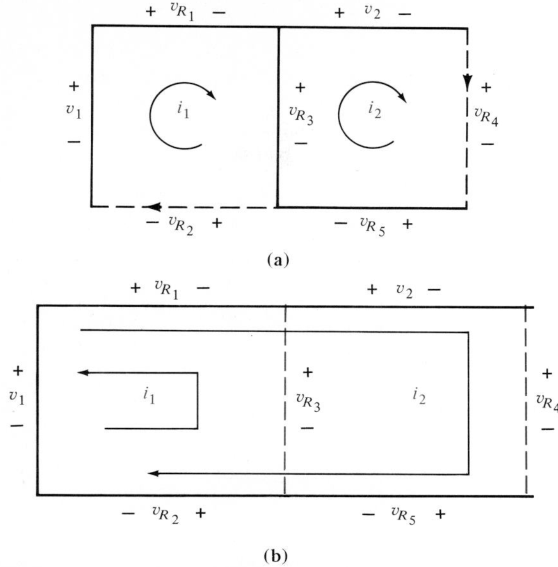

Figure 4.15 Graph for the network in Fig. 3.12 illustrating fundamental loops.

Using the relationship $v = iR$, we obtain

$$i_1R_1 + (i_1 - i_2)R_3 + i_1R_2 = v_1$$

$$-(i_1 - i_2)R_3 + i_2R_4 + i_2R_5 = -v_2$$

or

$$i_1(R_1 + R_2 + R_3) - i_2R_3 = v_1$$

$$-i_1R_3 + i_2(R_3 + R_4 + R_5) = -v_2$$

which are the same as the equations in (3.9).

Suppose, however, that we had selected tree branches and corresponding links for the network as shown in Fig. 4.15b. The fundamental loop currents are i_1 and i_2, and the fundamental loop equations in this case are

$$v_1 - v_{R_1} - v_{R_3} - v_{R_2} = 0$$

$$v_1 - v_{R_1} - v_2 - v_{R_4} - v_{R_5} - v_{R_2} = 0$$

Since $v_{R_1} = (i_1 + i_2)R_1$, and so on, the equations above can be written in terms of the loop currents as

$$(R_1 + R_2 + R_3)i_1 + (R_1 + R_2)i_2 = v_1$$

$$(R_1 + R_2)i_1 + (R_1 + R_2 + R_4 + R_5)i_2 = v_1 - v_2$$

These two linearly independent equations will also yield all the currents and voltages in the network. ■

DRILL EXERCISE

D4.6. Given the network in Fig. D4.6a and the graph for the network in Fig. D4.6b, write the proper set of loop equations that will yield all the currents in the network.

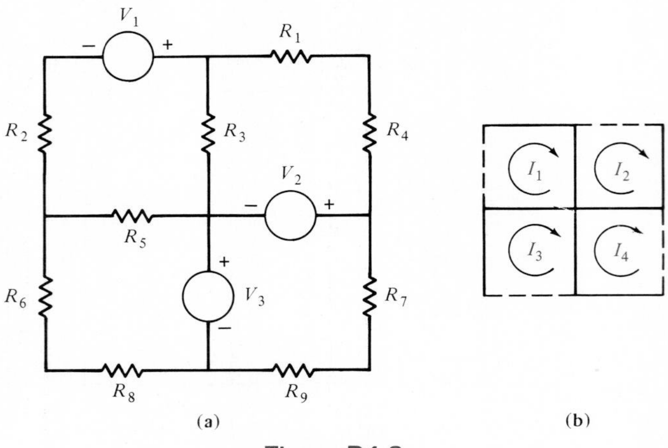

Figure D4.6

D4.7. Given the network in Fig. D4.7a and the graph for the network in Fig. D4.7b, write the proper set of loop equations that will yield all the currents in the network.

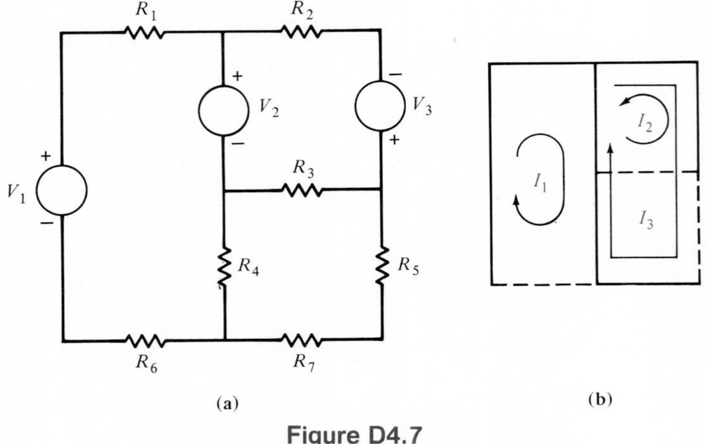

Figure D4.7

4.5 General Circuit Analysis Techniques

We have shown that ($N - 1 -$ the number of voltage sources) independent KCL equations and ($B - N + 1 -$ the number of current sources) independent KVL equations are required to solve a circuit, that is, determine all currents and voltages. We have also illustrated the manner in which to choose a tree and cotree. However, since there is more than one tree for a given network, perhaps one choice may be better than another. The following guidelines for selecting a tree and a cotree are presented to facilitate a solution for the network.

1. When constructing a tree and cotree for a network, select the tree so that it includes all voltage sources and select the cotree so that it includes all current sources.
2. If possible, place all voltage control branches for voltage-controlled dependent sources in the tree and all current control branches for current-controlled dependent sources in the cotree.

The remainder of this section contains a number of examples that have been chosen to illustrate the manner in which to apply the techniques we have discussed. The reader should note how simply some problems can be solved by faithfully using the guidelines outlined above.

EXAMPLE 4.11

We wish to solve the circuit shown in Fig. 4.16a. We note that $B = 6$, $N = 4$, the number of voltage sources is 2, and the number of current sources is 1. Therefore, $B - N + 1 -$ the number of current sources $= 2$ and $N - 1 -$ the number of voltage sources $= 1$.

If we begin constructing a tree and cotree using the guidelines above, we develop the graph shown in Fig. 4.16b. One way to complete the tree and cotree is shown in Fig. 4.16c. At this point we find that we can use one nodal equation and two constraints

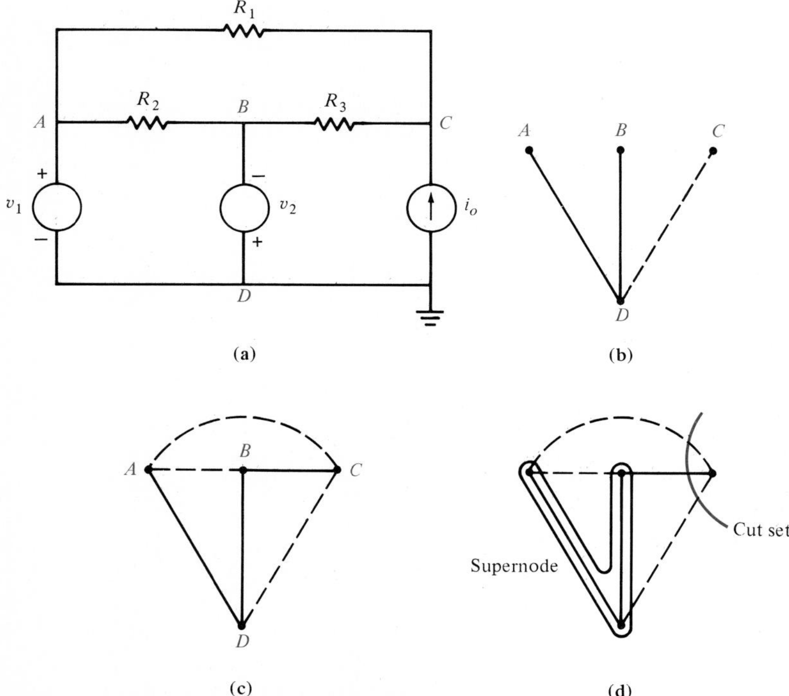

Figure 4.16 Formation of a tree and a cotree for a network.

or two loop equations with one constraint. The loop equations are simplified by the presence of the current source (i.e., one loop current is known). However, the voltage sources define two node voltages (i.e., two node voltages are known). Because of v_1 and v_2, *A-D-B* forms a supernode, and therefore we need only write one node equation. The supernode is shown in Fig. 4.16d together with the only cut set. If we write this equation at node *C*, we obtain

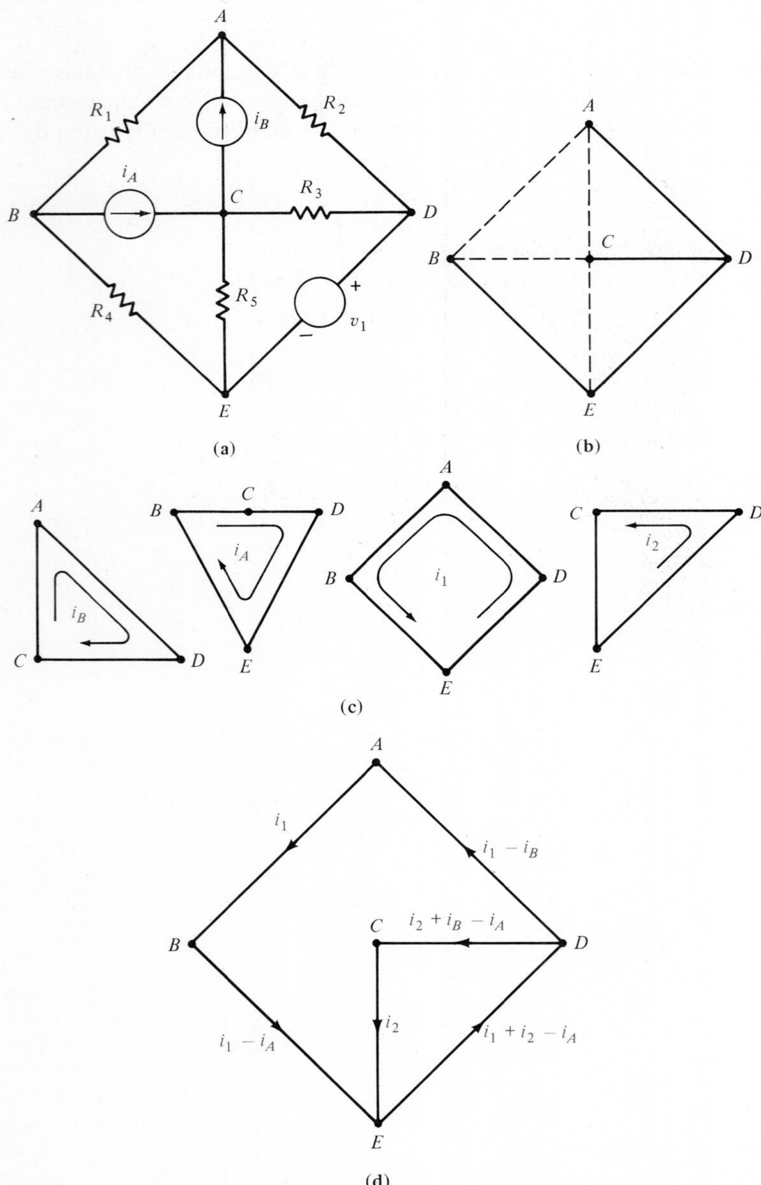

Figure 4.17 Development of the proper loop equations to solve the network in (a).

$$\frac{v_C - v_1}{R_1} + \frac{v_C - (-v_2)}{R_3} = i_o$$

v_C is the only unknown in the equation. If $V_1 = 10$ V, $V_2 = 12$ V, $I_o = 2$ A, $R_1 = 1\ \Omega$, $R_2 = 2\ \Omega$, and $R_3 = 2\ \Omega$, then $V_C = 4$ V. Now all currents can be immediately calculated. ■

EXAMPLE 4.12

Consider the circuit in Fig. 4.17a. Following the guidelines, we select the tree and cotree shown in Fig. 4.17b. We note that $B = 8$, $N = 5$, the number of voltage sources is 1, and the number of current sources is 2. Therefore, the number of independent node equations is 3 and the number of independent loop equations is 2. Each link, together with a portion of the tree, forms a loop. These loops are shown in Fig. 4.17c for the two current sources and the two unknown current variables i_1 and i_2. The current information in Fig. 4.17c is transferred to the two loops in Fig. 4.17d, which are the loops for the current variables i_1 and i_2.

Note that we do not know the voltage drop across the current source and this procedure does not require us to write a KVL equation around a loop containing a current source. Using Fig. 4.17d and a, we can write the two loop equations as

$$(i_1 - i_B)R_2 + i_1R_1 + (i_1 - i_A)R_4 = v_1$$

$$(i_2 + i_B - i_A)R_3 + i_2R_5 = v_1$$

Note that in this case the two equations are not coupled and therefore can be solved individually for the variables i_1 and i_2. In general, however, this will not be the case.

If $R_1 = 1\ \Omega$, $R_2 = 2\ \Omega$, $R_3 = 2\ \Omega$, $R_4 = 1\ \Omega$, $R_5 = 2\ \Omega$, $I_A = 2$ A, $I_B = 4$ A, and $V_1 = 12$ V, the equations above can be solved to yield $I_1 = 5.5$ A and $I_2 = 2$ A. Using these values and the branch currents shown in Fig. 4.17d, one can easily show that KVL is satisfied around the closed path *A-D-C-E-B-A*. Similar checks can also be made. ■

EXAMPLE 4.13

Consider the network shown in Fig. 4.18a, which contains a voltage-controlled voltage source. The tree for this circuit is shown in Fig. 4.18b. A careful analysis of the circuit indicates that all node voltages would be known if v_2 were known. Note also that v_1 and the controlled source $4v_2$ form a supernode. Therefore, if we select node D as the reference node and write a single-node equation at node B, we obtain

$$\frac{v_2 - v_1}{R_1} + \frac{v_2}{R_3} + \frac{v_2 - (v_1 - 4v_2)}{R_2} = 0$$

If $R_1 = 1\ \Omega$, $R_2 = 2\ \Omega$, $R_3 = 2\ \Omega$, $V_1 = 8$ V, and $I_o = 4$ A, the equation above can be used to compute $V_2 = 3$ V. Hence $V_A = 8$ V, $V_B = 3$ V, and $V_C = -4$ V, and therefore all currents can be immediately calculated. ■

EXAMPLE 4.14

A circuit containing a current-controlled current source is shown in Fig. 4.19a. Using the guidelines for constructing a tree and cotree, we obtain the graph shown in Fig.

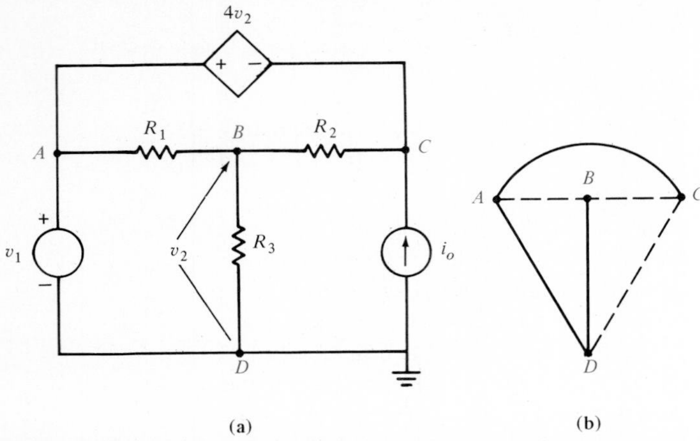

Figure 4.18 Circuit containing a voltage-controlled voltage source.

4.19b. The tree is completed as shown in Fig. 4.19c. Note that $N - 1 -$ the number of voltage sources $= 3$ and $B - N + 1 -$ the number of current sources $= 1$. Since there are three current sources in the circuit, only one KVL equation around the loop that does not contain a current source is required to solve this circuit. The four loops defined by the links and their currents are shown in Fig. 4.19d. The current information in Fig. 4.19d is placed on the loop in Fig. 4.19e, which contains no current sources. A single KVL equation for this loop is

$$v_1 - R_3(i_1 - i_B) - i_1 R_4 - (i_1 + i_A)R_2 - (i_1 + i_A - 2i_1)R_1 = 0$$

The only unknown in this equation is the current i_1. ■

EXAMPLE 4.15

The circuit in Fig. 4.20a is a nonplanar network that contains a current-controlled voltage source. The graph in Fig. 4.20b is obtained following the guidelines for constructing a tree and cotree. One particular tree and cotree are shown in Fig. 4.20c. Note that $N = 5$, $B = 10$, the number of current sources $= 1$, and the number of voltage sources $= 3$. Therefore, $N - 1 -$ the number of voltage sources is 1, and $B - N + 1 -$ the number of current sources is 5. In addition, note that nodes A-B-C-E form a supernode. Therefore, the only two unknowns in the network are v_D and i_1. If we write a KCL equation at node D, we obtain

$$\frac{v_D - (v_2 + 4i_1)}{R_3} + \frac{v_D - (v_1 + v_2)}{R_4} + \frac{v_D - v_2}{R_5} + \frac{v_D}{R_6} = 0$$

where

$$\frac{v_D}{R_6} = -i_1$$

■

EXAMPLE 4.16

A nonplanar network containing a voltage-controlled current source is shown in Fig. 4.21a. For this network, $B = 9$, $N = 6$, the number of current sources $= 3$, and the number of voltage sources $= 1$. Therefore, $N - 1 -$ the number of voltage sources is

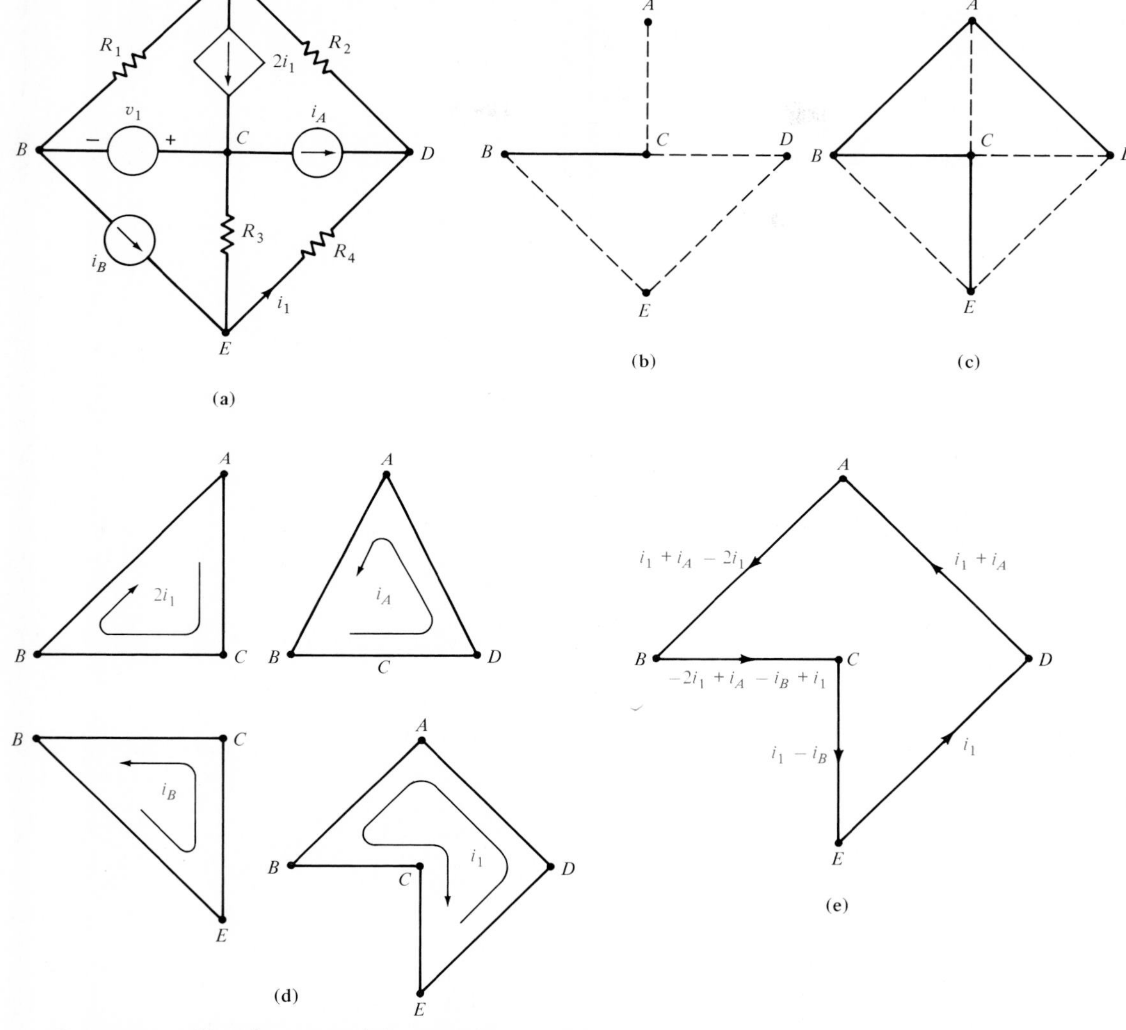

Figure 4.19 Development of the equation necessary to solve the network in (a).

4, and $B - N + 1 -$ the number of current sources is 1. A tree and cotree for the network are shown in Fig. 4.21b. The various currents are shown in Fig. 4.21c and all this current information is placed on the loop in Fig. 4.21d, which contains no current sources. KVL for the loop in Fig. 4.21d is

$$R_1(i+4v_o) + R_2(i+4v_o+i_A) + R_4(i+4v_o+i_A+i_B) + v_1 + R_5(i+i_B) + R_3 i = 0$$

where

$$v_o = R_2(i + 4v_o + i_A)$$

This single-loop equation, together with its constraint equation, which results from the dependent source, is all that is needed to solve this fairly complicated circuit. ■

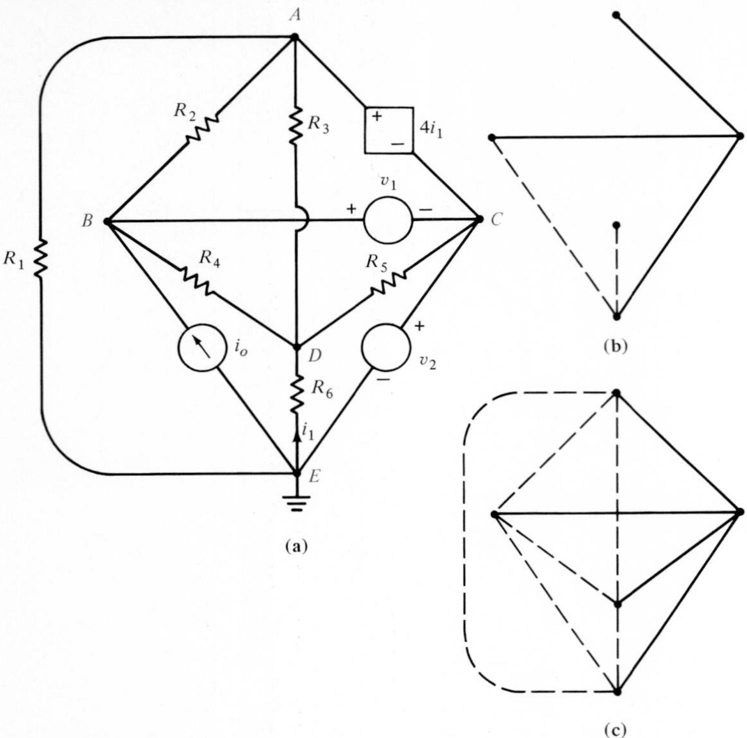

Figure 4.20 Nonplanar network containing a current-controlled voltage source.

DRILL EXERCISE

D4.8. Given the network in Fig. D4.8a and the specified tree in Fig. D4.8b, find the current I_1.

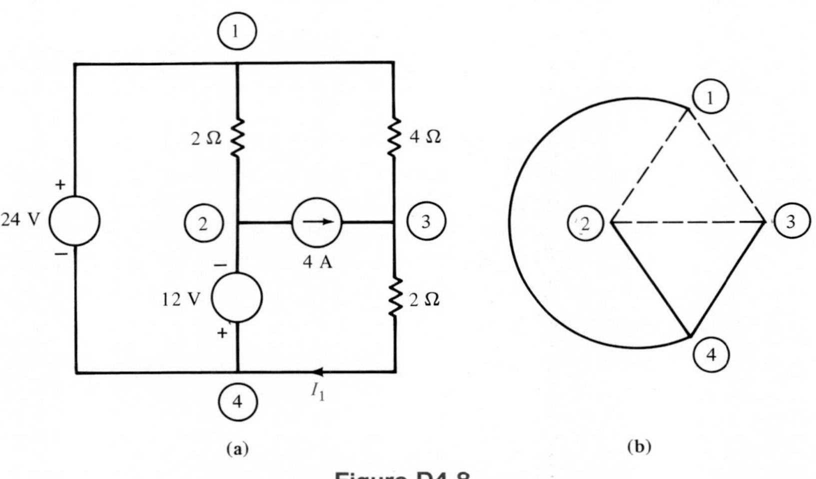

Figure D4.8

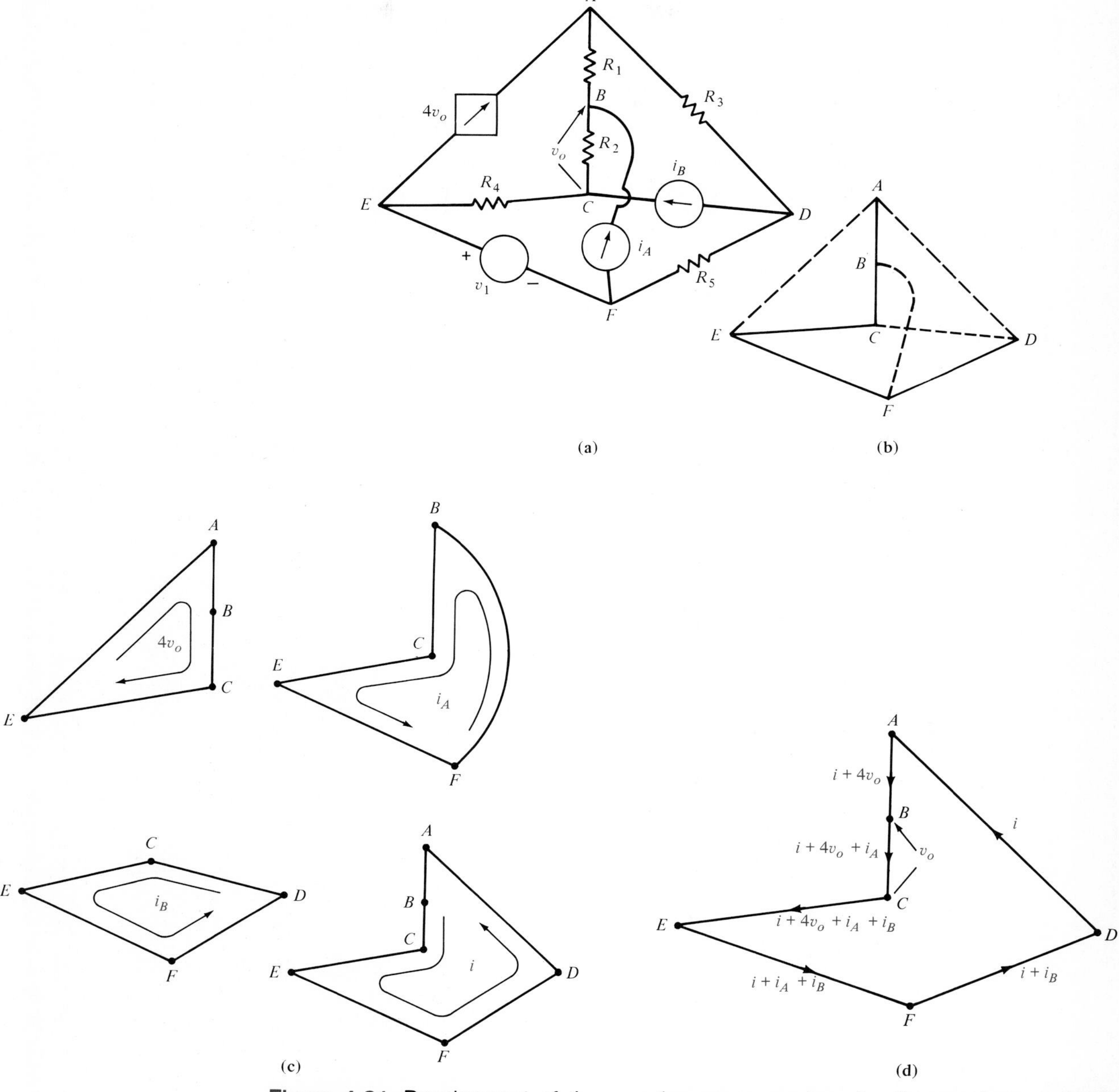

Figure 4.21 Development of the equation necessary to solve the nonplanar network in (a).

D4.9. Given the network in Fig. D4.9a and the specified tree in Fig. D4.9b, find the current I_o.

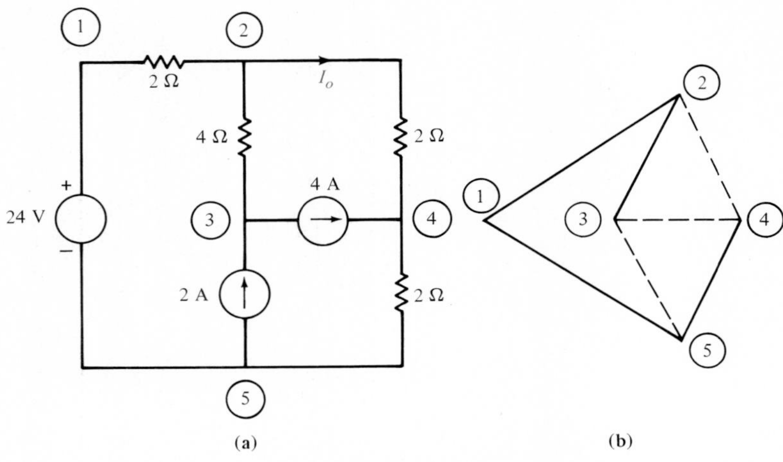

Figure D4.9

D4.10. Given the network in Fig. D4.10a and the specified tree in Fig. D4.10b, find the current I_o.

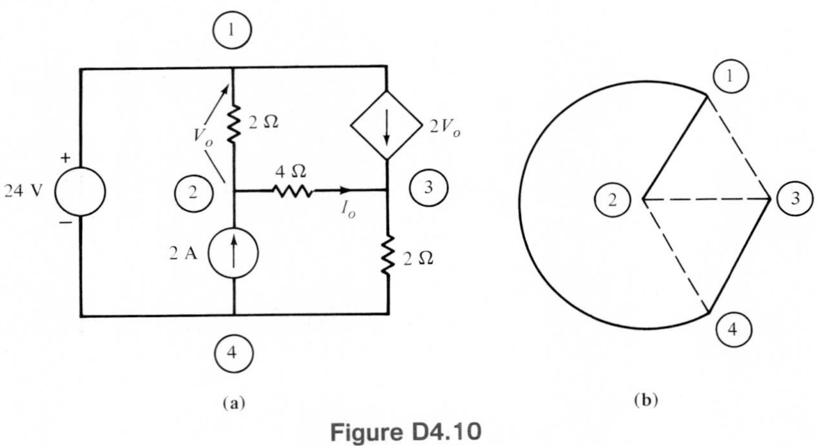

Figure D4.10

We have tried to demonstrate via the examples the topological techniques that can generally be applied to solve circuits. The reader must keep in mind, however, that these techniques are essentially the same as those proposed in Chapter 3. To illustrate the similarities, let us consider again the circuits analyzed in Examples 3.11 and 4.12.

EXAMPLE 4.17

The circuit shown in Fig. 4.16a is redrawn in Fig. 4.22 with all voltage sources shorted and current sources opened. The number of topological nodes is 2: the reference node and node C. The equation for node C is

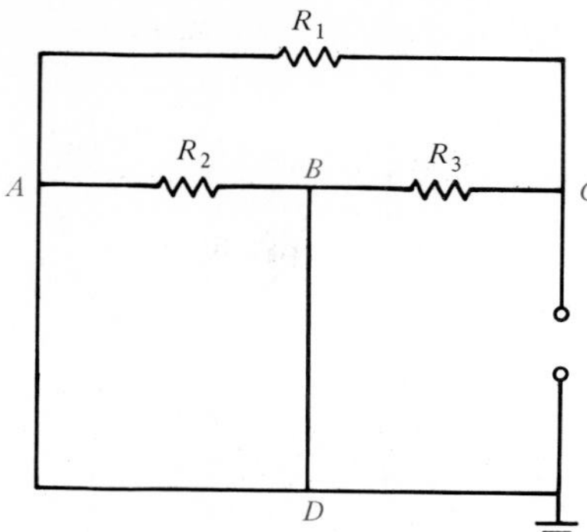

Figure 4.22 Circuit in Fig. 4.16a with all voltage sources shorted and all current sources opened.

$$\frac{v_C - v_B}{R_3} + \frac{v_C - v_A}{R_1} = i_o$$

The two constraints are

$$v_B = -v_2$$

$$v_A = v_1$$

Therefore, the node equation becomes

$$\frac{v_C + v_2}{R_3} + \frac{v_C - v_1}{R_1} = i_o$$

which is, of course, the equation obtained in Example 4.11. ■

EXAMPLE 4.18

The circuit in Fig. 4.17a is redrawn in Fig. 4.23a with all voltage sources shorted and all current sources opened. Note that the number of topological loops is 2. Writing the loop equations for these two loops using the current assignments shown in Fig. 4.23b, we obtain

$$R_2(i_1 - i_3) + R_1 i_1 + R_4(i_1 - i_4) = v_1$$

$$R_3(i_2 + i_3 - i_4) + R_5(i_2) = v_1$$

The constraints are

$$i_3 = i_B$$

$$i_4 = i_A$$

Therefore, the two loop equations become

$$R_2(i_1 - i_B) + R_1 i_1 + R_4(i_1 - i_A) = v_1$$

$$R_3(i_2 + i_B - i_A) + R_5 i_2 = v_1$$

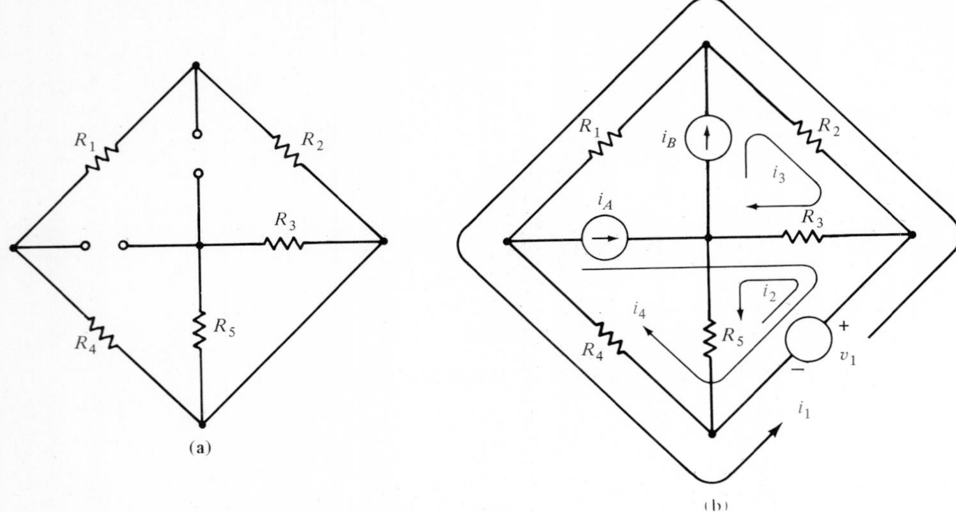

Figure 4.23 Reexamination of the circuit in Fig. 4.17a.

These equations are exactly the same as those obtained in Example 4.12. Other current assignments could have been chosen, but the ones selected above were employed so that the reader could compare the results directly with those of Example 4.12. ■

4.6 Summary

This chapter has dealt with the geometrical properties of networks. The relationship among branches, links, and nodes has been employed in order to determine the proper number of equations required to yield a solution. We have shown that we can use topological concepts involving graph theory to provide a systematic approach for solving circuit problems.

KEY POINTS

- A planar circuit is one that can be drawn on a plane surface with no crossovers.
- A tree of a graph is a set of branches that connects every node to every other node via some path without forming a closed loop. In general, there are numerous trees for a given graph.
- The branches of a graph that are not part of the tree form a cotree. The branches of the cotree are called links.
- There are $N - 1$ branches in a tree and $L = B - N + 1$ links in a cotree, where N is the number of nodes, L the number of links, and B the number of branches.
- For a network, the number of independent cut-set equations (KCL equations) is $N - 1 -$ the number of voltage sources.
- For a network, the number of independent fundamental loop equations (KVL equations) is $B - N + 1 -$ the number of current sources.

PROBLEMS

4.1. Are the graphs shown in Fig. P4.1 planar or nonplanar?

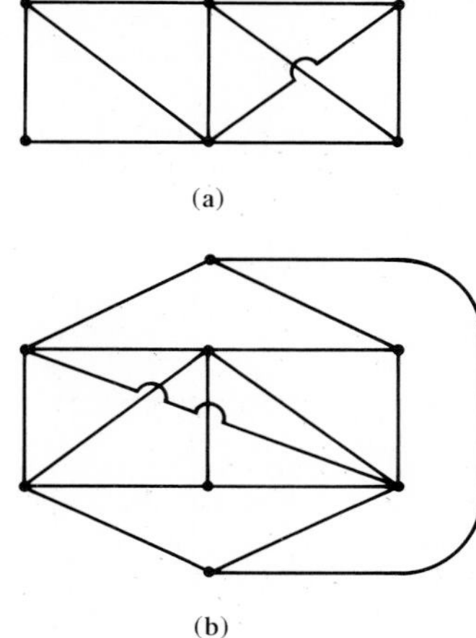
(a)

(b)

Figure P4.1

4.2. For the graphs in Fig. P4.2, determine the number of tree branches and the number of links.

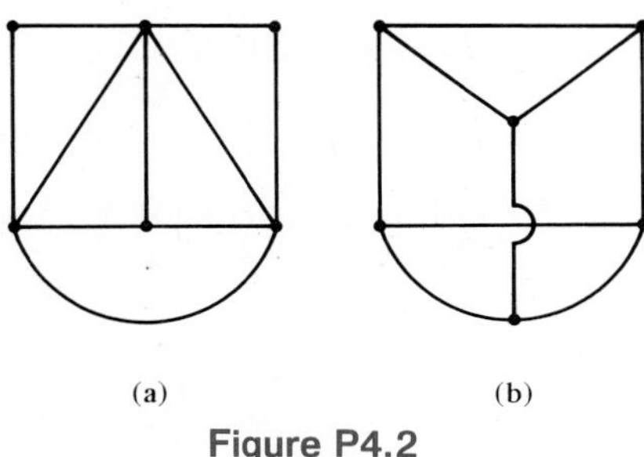
(a) (b)

Figure P4.2

4.3. For the graph in Fig. P4.3, can a tree be selected so that each loop corresponds to a mesh?

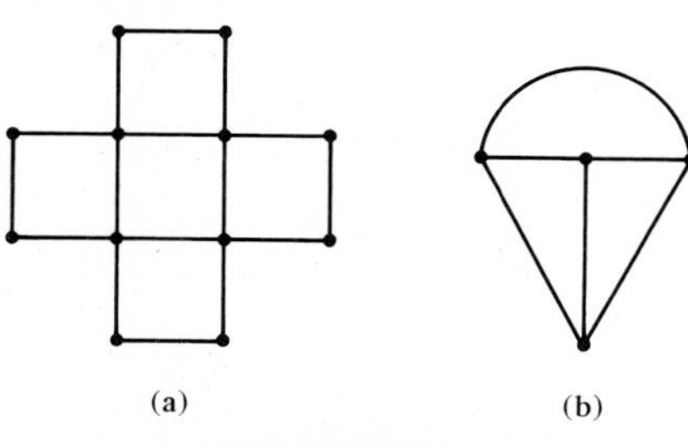
(a) (b)

Figure P4.3

4.4. For the graphs shown in Fig. P4.4, determine the number of tree branches and the number of links.

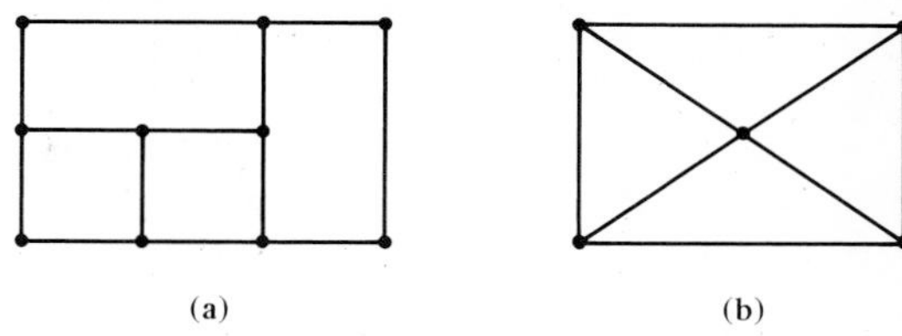
(a) (b)

Figure P4.4

4.5. For the graphs shown in Fig. P4.5, determine the number of tree branches and the number of links.

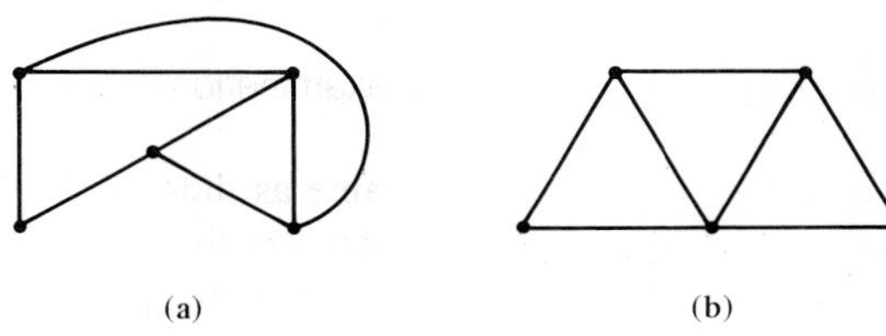
(a) (b)

Figure P4.5

4.6. For the graphs shown in Fig. P4.6, determine the number of tree branches and the number of links.

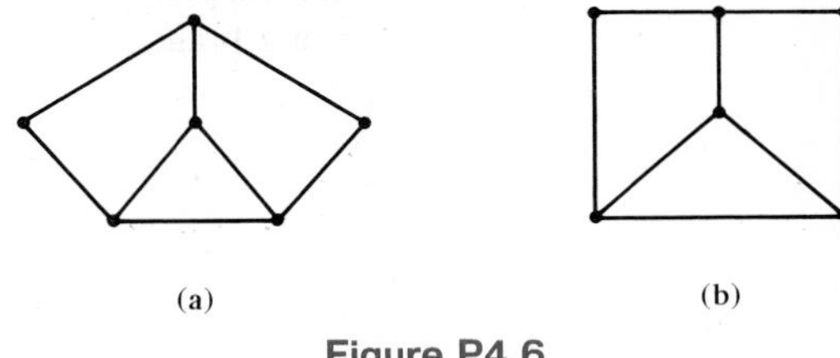
(a) (b)

Figure P4.6

4.7. For the graphs shown in Fig. P4.7, with a specified tree, find an independent set of KCL equations using cut sets.

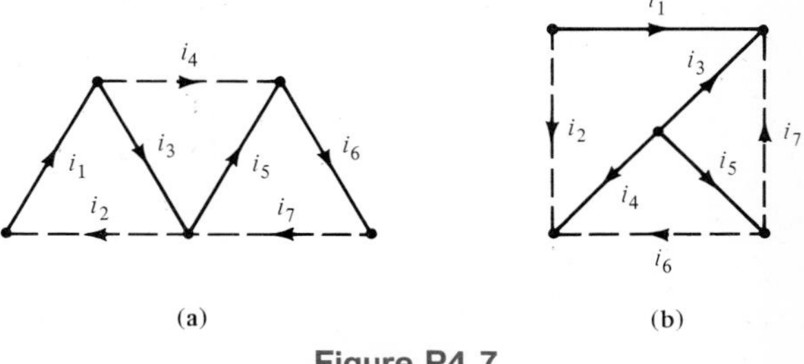

(a) (b)

Figure P4.7

4.8. For the graphs shown in Fig. P4.8 with a specified tree, find an independent set of KCL equations using cut sets.

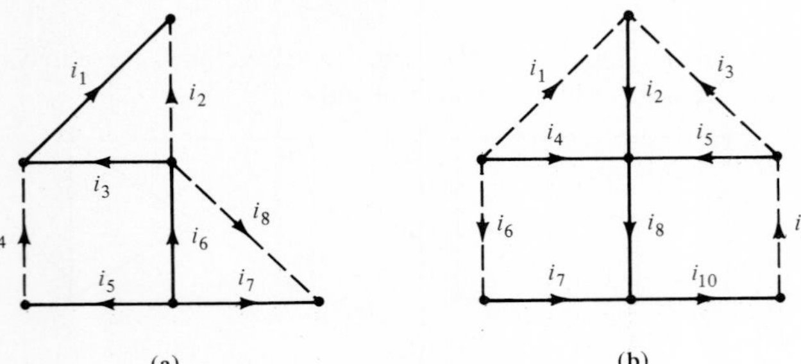

Figure P4.8

4.9. For the graphs shown in Fig. P4.9 with a specified tree, find an independent set of KCL equations using using cut sets.

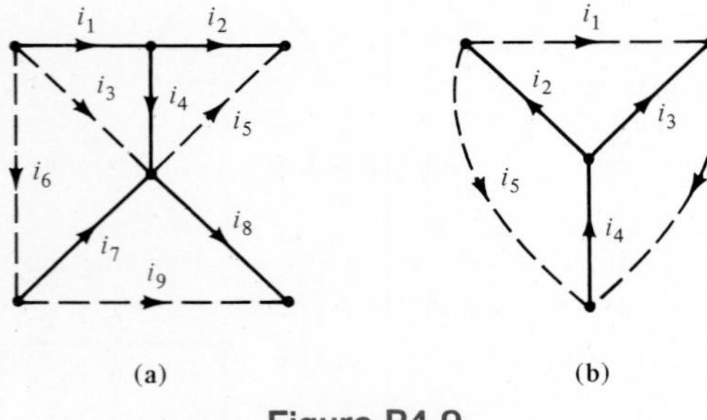

Figure P4.9

4.10. For the graph shown in Fig. P4.10 with a specified tree, find an independent set of KCL equations using cut sets.

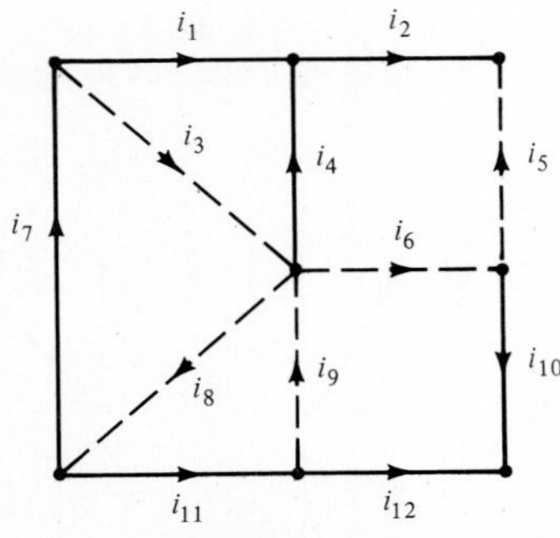

Figure P4.10

4.11. For the graphs shown in Fig. P4.11 with a specified tree, find an independent set of KCL equations using cut sets.

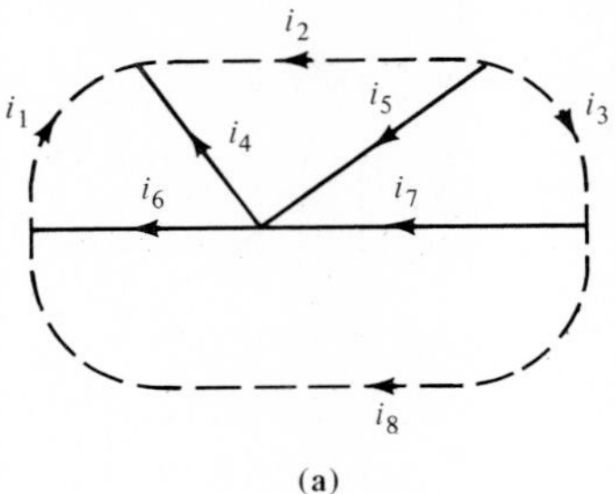

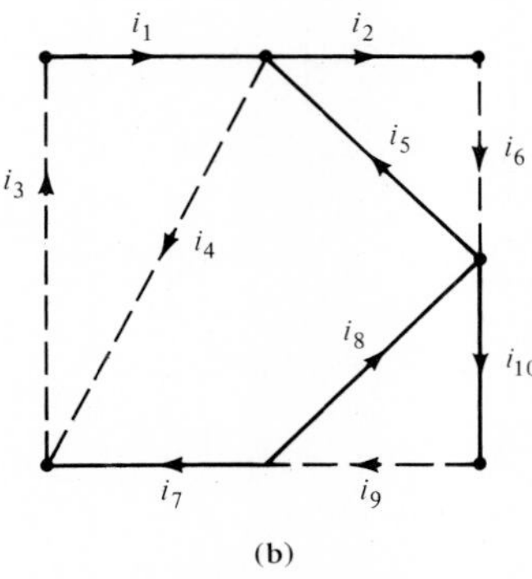

Figure P4.11

4.12. Determine the independent set of KCL equations for the circuit shown in Fig. P4.12, using cut sets.

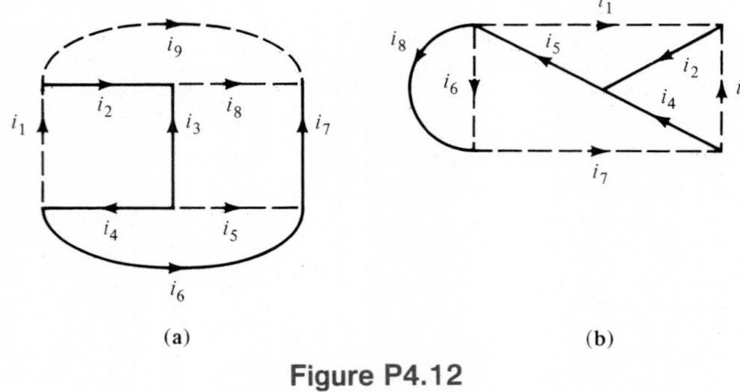

Figure P4.12

4.13. For the graphs shown in Fig. P4.13 with a specified tree, find an independent set of KVL equations using fundamental loops.

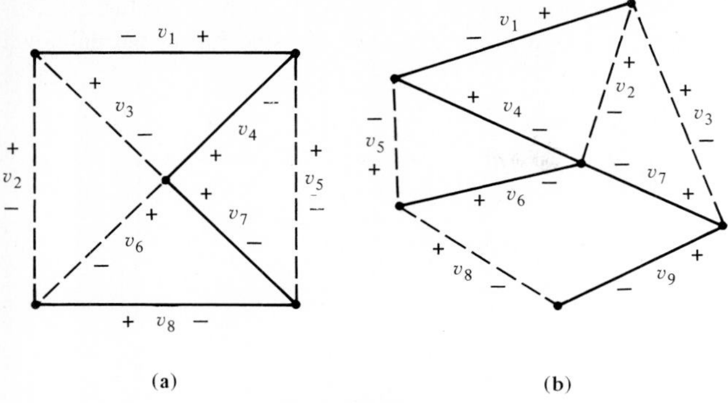

Figure P4.13

4.14. For the graphs shown in Fig. P4.14 with a specified tree, find an independent set of KVL equations using fundamental loops.

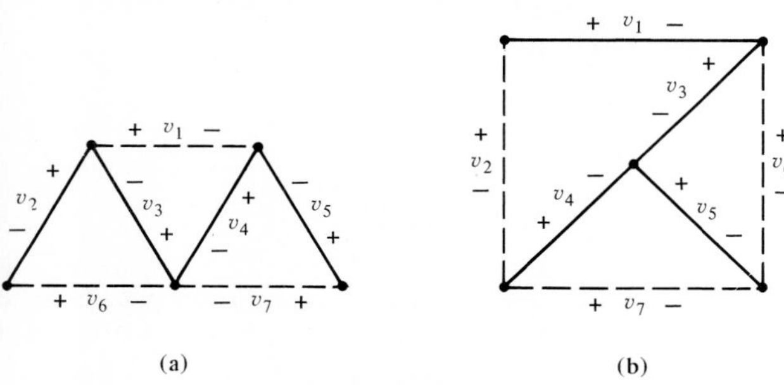

Figure P4.14

4.15. For the graphs shown in Fig. P4.15 with a specified tree, find an independent set of KVL equations using fundamental loops.

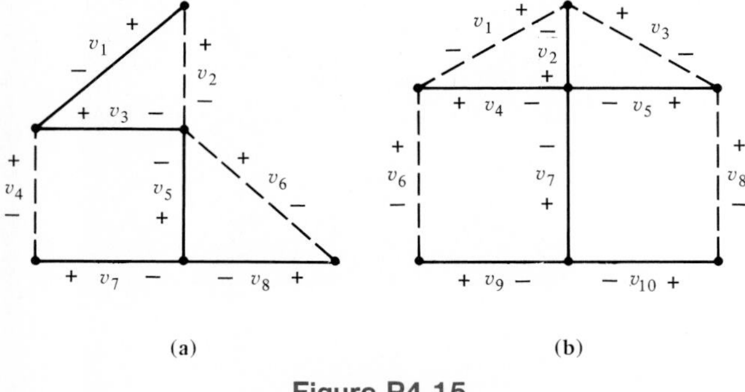

Figure P4.15

4.16. For the graphs shown in Fig. P4.16 with a specified tree, find an independent set of KVL equations using fundamental loops.

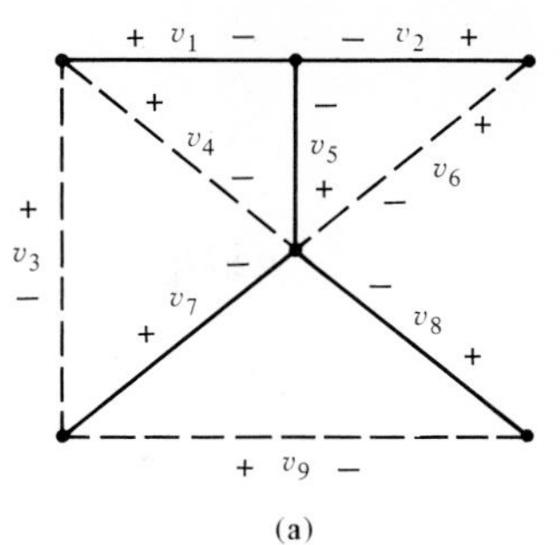

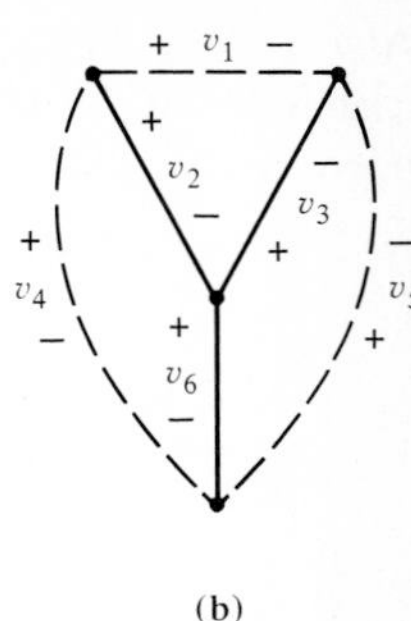

Figure P4.16

4.17. In Fig. P4.17 the graph is given for a network and the known currents are specified. Find the unknown currents I_1, I_2, I_3, I_4, and I_5.

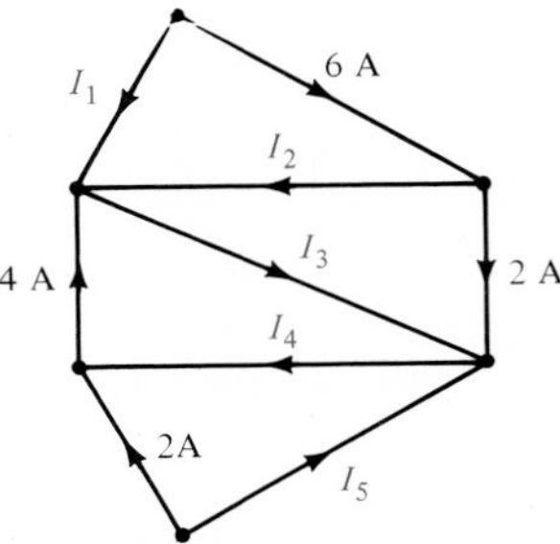

Figure P4.17

4.18. The tree for the network in Fig. P4.18 is given in Fig. 4.18b. Write the KCL equations in terms of the unknown node voltages.

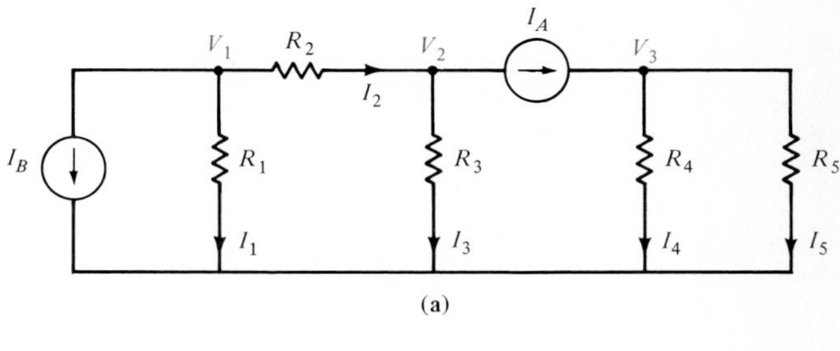

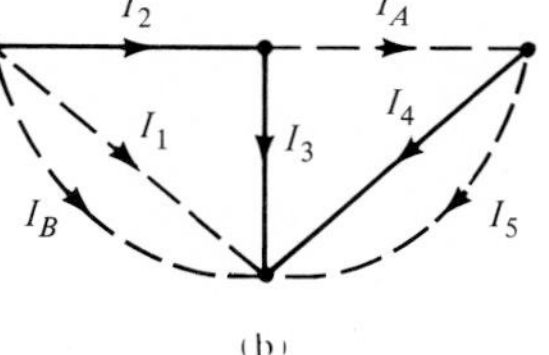

Figure P4.18

4.19 Use the specified tree in Fig. P4.19b to write the KCL equations for the network in Fig. 4.19a in terms of the unknown node voltages.

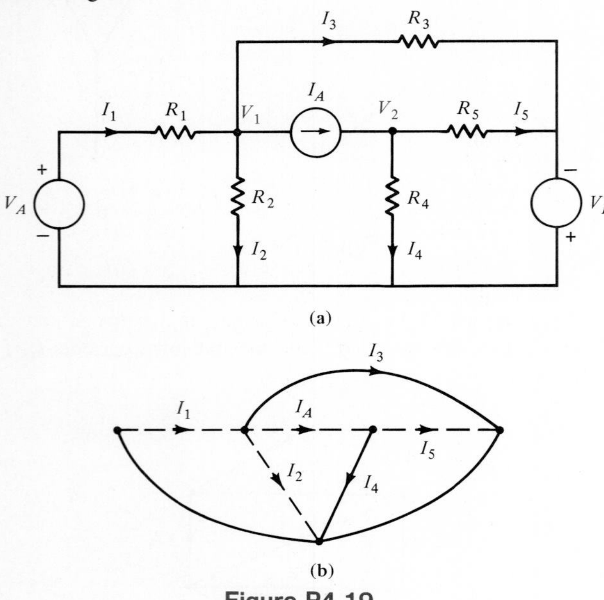

Figure P4.19

4.20. Given the tree in Fig. P4.20b for the network in Fig. P4.20a, write the loop equations necessary to solve for all node voltages.

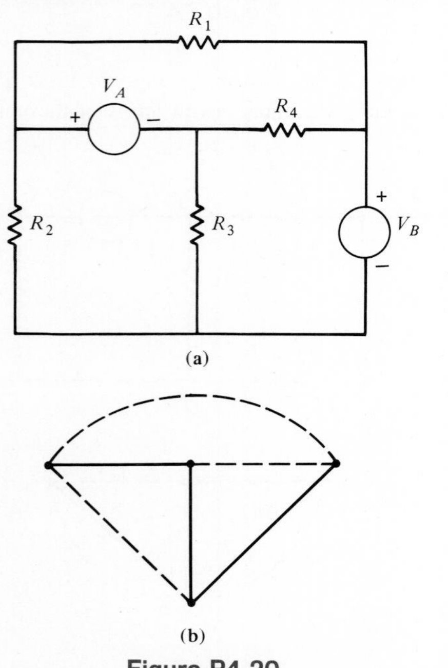

Figure P4.20

4.21. Given the tree in Fig. P4.21b for the network in Fig. P4.21a, write the loop equations necessary to solve for all the unknown voltages.

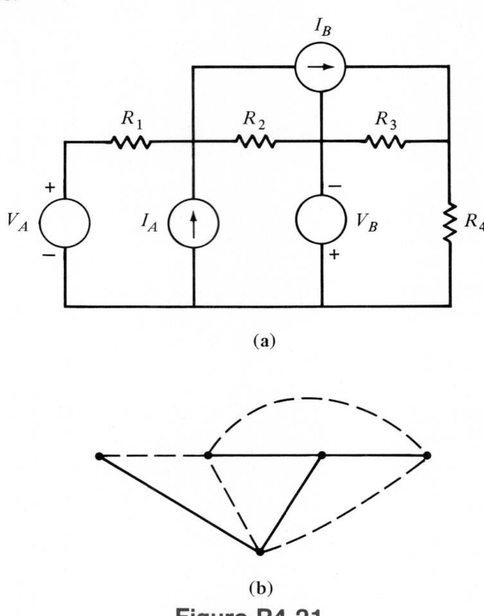

Figure P4.21

4.22. Given the network in Fig. P4.22a and the specified tree and links shown in Fig. P4.22b, write the cut-set equations in standard form using the specified tree that will yield the node voltages.

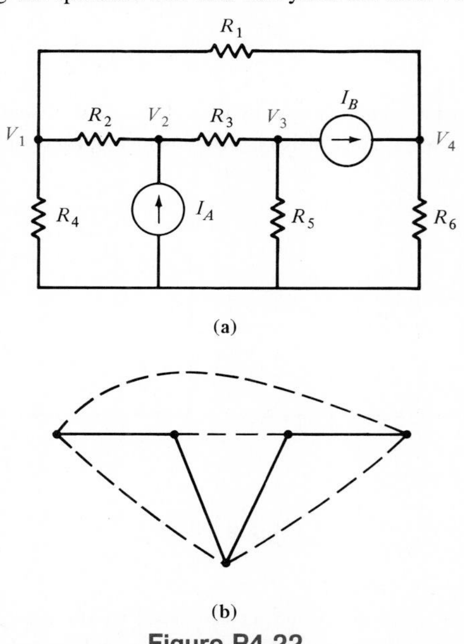

Figure P4.22

4.23. Show that by using two supernodes, one KCL equation is all that is required to determine all node voltages in the network shown in Fig. P4.23.

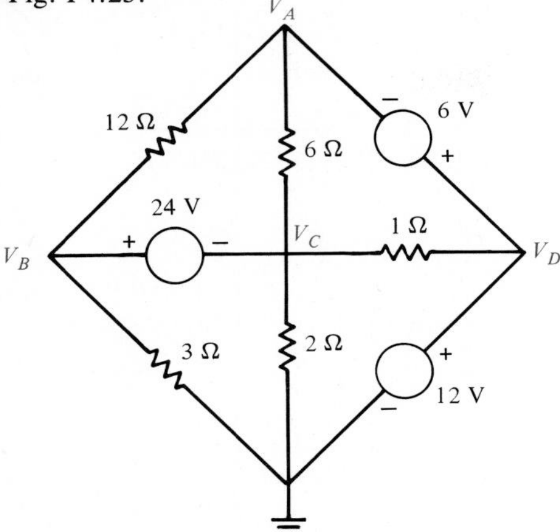

Figure P4.23

4.24. The network in Fig. P4.24 has four loops and four nodes. However, if the tree is chosen as shown and the supernode identified, show that the node voltages can be obtained by means of the cut sets with only two equations.

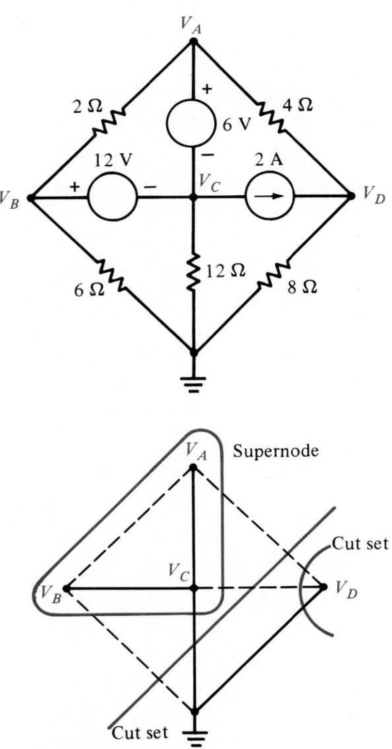

Figure P4.24

4.25. Using the guidelines stated in Section 4.5, develop a tree for the circuit shown in Fig. P4.25. Write a single equation to determine V_C and use this to find I_o.

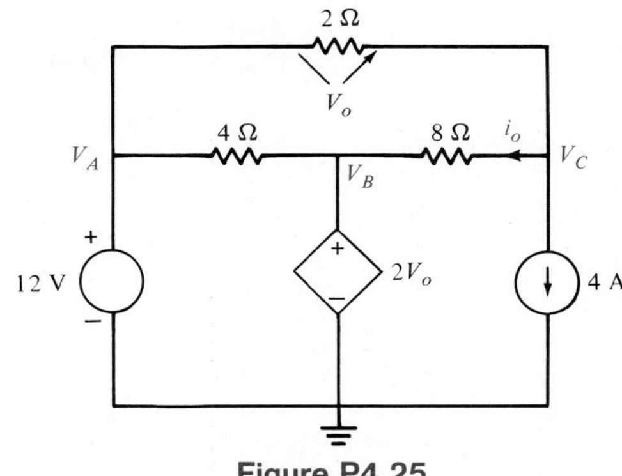

Figure P4.25

4.26. Find the current I in Fig. P4.26.

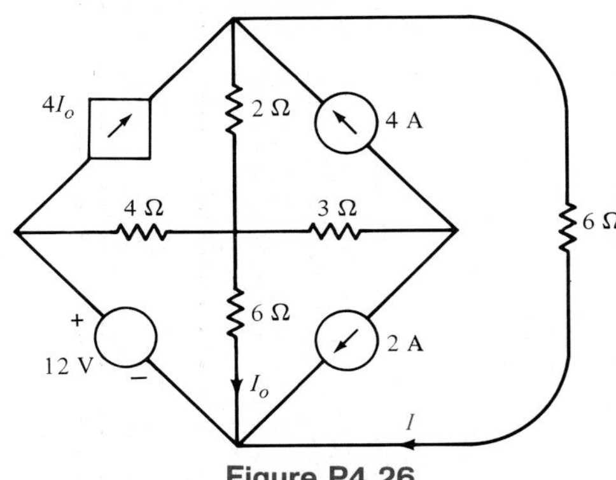

Figure P4.26

4.27. Find the voltage V_A in Fig. P4.27.

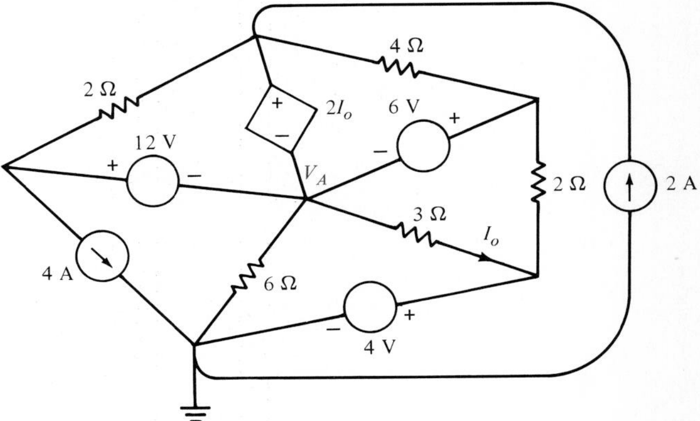

Figure P4.27

4.28. Find the voltage across the 8-Ω resistor in the circuit in Fig. P4.28.

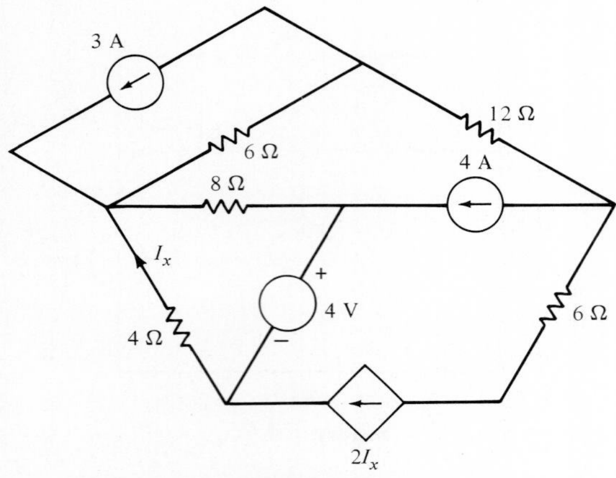

Figure P4.28

4.29. Determine the current I_o in the network in Fig. P4.29.

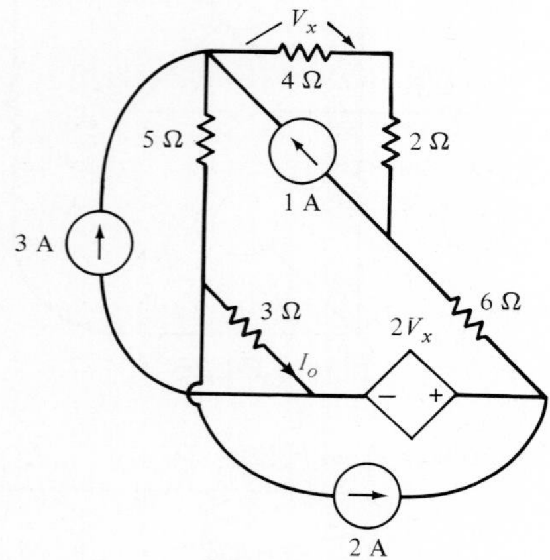

Figure P4.29

4.30. Find the value of the current I_o in the network in Fig. P4.30 by using only one KVL equation.

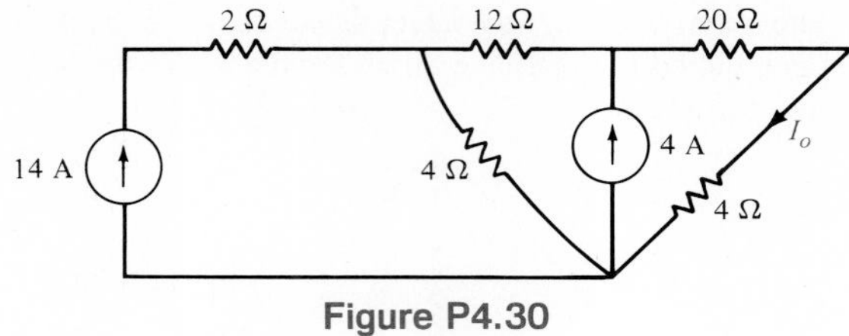

Figure P4.30

4.31. Determine the current I_x in the network in Fig. P4.31 by following the guidelines in Section 4.5 and writing a single equation.

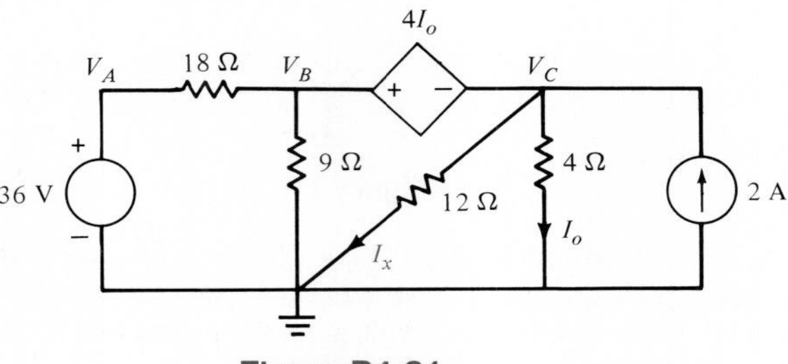

Figure P4.31

4.32. Show that the node voltages in the network in Fig. P4.32 can be determined using only one KCL equation.

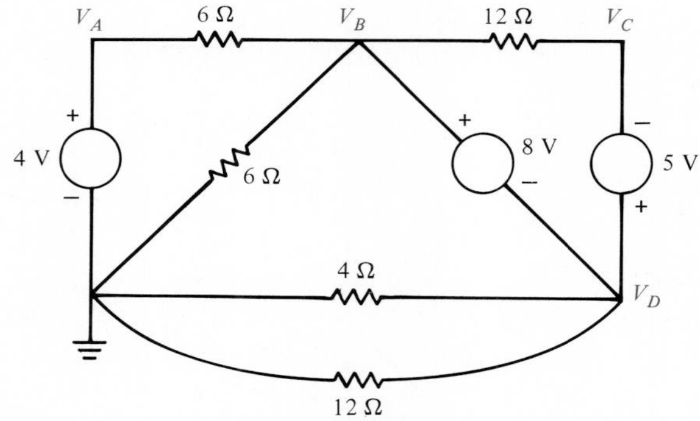

Figure P4.32

CHAPTER

5

Useful Theorems for Circuit Analysis

In this chapter we present a number of theorems that will be very useful to us in our analysis of circuits. The reader is encouraged to compare their use with some of the more brute-force techniques we have presented thus far. In addition, it is in this chapter that we introduce SPICE, a computer-aided circuit analysis program. Although SPICE is an extremely powerful technique, we will confine our discussion to the solution of some relatively simple dc circuits. SPICE will, however, be revisited numerous times throughout this book as we apply it to a wide variety of circuit analysis problems.

5.1 Linearity

All the circuits we have analyzed thus far and all the circuits we will treat in this book are linear circuits. We have shown earlier that the resistor is a linear element because its current–voltage relationship is a linear characteristic; that is,

$$v(t) = Ri(t) \tag{5.1}$$

Linearity requires both additivity and homogeneity (scaling). In the case of a resistive element, if $i_1(t)$ is applied, the voltage across the resistor is

$$v_1(t) = Ri_1(t) \tag{5.2}$$

Similarly, if $i_2(t)$ is applied, then

$$v_2(t) = Ri_2(t) \tag{5.3}$$

However, if $i_1(t) + i_2(t)$ is applied, the voltage across the resistor is

$$v(t) = R[i_1(t) + i_2(t)] = Ri_1(t) + Ri_2(t) = v_1(t) + v_2(t) \tag{5.4}$$

This demonstrates the additive property. In addition, if the current is scaled by a constant K_1, the voltage is also scaled by the constant K_1, since

$$RK_1 i(t) = K_1 Ri(t) = K_1 v(t) \tag{5.5}$$

This demonstrates homogeneity.

Dependent sources are linear if their output current or voltage is proportional to the sum of the first power of current or voltage variables present in the circuit. In other words, $v(t)$ or $i(t)$ may be of the form $a_1 i_1(t)$ or $a_2 v_2(t)$ or $a_1 i_1(t) + a_2 v_2(t)$. However, if either $v(t)$ or $i(t)$ is given by an expression of the form $a_1 i_1^2(t)$ or $a_2 i_1(t) v_2(t)$, the dependent source is nonlinear.

We have shown in the preceding chapters that a circuit that contains only independent sources, linear dependent sources, and resistors is described by equations of the form

$$a_1 v_1(t) + a_2 v_2(t) + \cdots + a_n v_n(t) = i(t) \tag{5.6}$$

or

$$b_1 i_1(t) + b_2 i_2(t) + \cdots + b_n i_n(t) = v(t)$$

Note that if the independent sources are multiplied by a constant, the node voltages or loop currents are also multiplied by the same constant. Thus we define a linear circuit as one composed of only independent sources, linear dependent sources, and linear elements. Capacitors and inductors, which we will examine in Chapter 6, are also circuit elements that have a linear input/output relationship provided that their initial stored energy is zero.

EXAMPLE 5.1

Given the circuit in Fig. 5.1a, we wish to determine the currents if the voltage source $v_1(t) = V_1 = 12$ V, and knowing these values, calculate the currents for any other value of V_1.

The 2- and 4-Ω resistors are in series and their combination is in parallel with the 3-Ω resistor, resulting in the circuit shown in Fig. 5.1b. The current I_1 can now easily be calculated as 3 A. The currents I_2 and I_3 can now be obtained in a variety of ways. Using current division, we can immediately determine that I_2 is equal to 2 A and I_3 is equal to 1 A. Once these currents are known, we can calculate their values for any $v_1(t)$

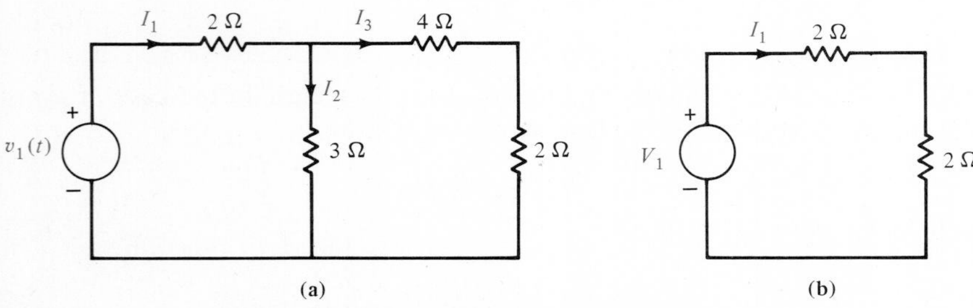

Figure 5.1 Circuits for Example 5.1.

using linearity. For example, if $v_1(t)$ is changed to $V_1 = 48$ V, then because this value is four times the 12 V we assumed above, all the currents are four times the value we have previously calculated. Therefore, $I_1 = 12$ A, $I_2 = 8$ A, and $I_3 = 4$ A. ■

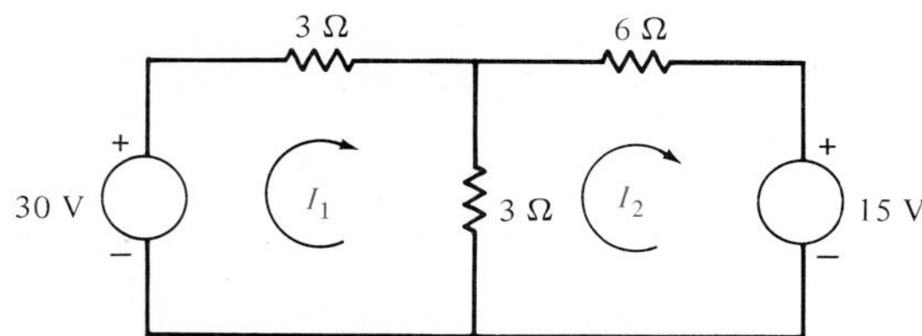

Figure 5.2 Circuit for Example 5.2.

EXAMPLE 5.2

We wish to illustrate the property of linearity for the circuit given in Fig. 5.2.

Suppose that we wish to calculate I_1 for the voltage sources given and also for the case in which they are doubled in value. The two loop equations are

$$6I_1 - 3I_2 = 30$$

$$-3I_1 + 9I_2 = -15$$

Using determinants, we find that

$$I_1 = \frac{\begin{vmatrix} 30 & -3 \\ -15 & 9 \end{vmatrix}}{\begin{vmatrix} 6 & -3 \\ -3 & 9 \end{vmatrix}}$$

and using a matrix approach, we have

$$\begin{bmatrix} I_1 \\ I_2 \end{bmatrix} = \begin{bmatrix} 6 & -3 \\ -3 & 9 \end{bmatrix}^{-1} \begin{bmatrix} 30 \\ -15 \end{bmatrix}$$

Using either approach to solve for I_1 yields $I_1 = 5$ A. If the sources are doubled, then

$$I_1 = \frac{\begin{vmatrix} 60 & -3 \\ -30 & 9 \end{vmatrix}}{\begin{vmatrix} 6 & -3 \\ -3 & 9 \end{vmatrix}}$$

Using the fact that if any row or column of a determinant is multiplied by a constant, the entire determinant is multiplied by the same constant, we obtain

$$I_1 = 2\frac{\begin{vmatrix} 30 & -3 \\ -15 & 9 \end{vmatrix}}{\begin{vmatrix} 6 & -3 \\ -3 & 9 \end{vmatrix}}$$

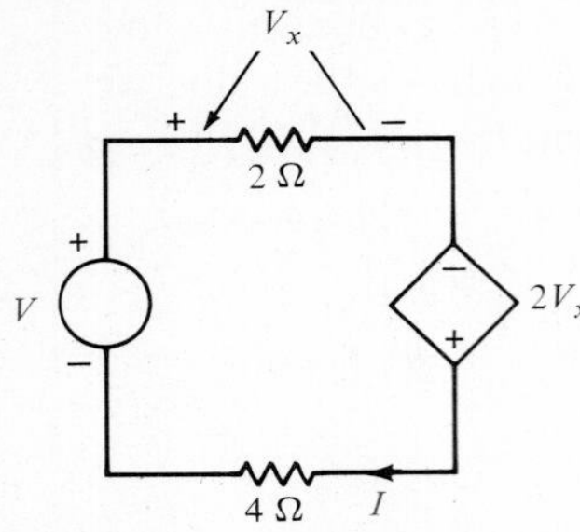

Figure 5.3 Circuit for Example 5.3.

which, of course, is double the value calculated above. As shown in Appendix A.

$$\begin{bmatrix} I_1 \\ I_2 \end{bmatrix} = \begin{bmatrix} 6 & -3 \\ -3 & 9 \end{bmatrix}^{-1} \begin{bmatrix} 60 \\ -30 \end{bmatrix} = 2 \begin{bmatrix} 6 & -3 \\ -3 & 9 \end{bmatrix}^{-1} \begin{bmatrix} 30 \\ -15 \end{bmatrix}$$

and therefore, again, I_1 is twice the value calculated earlier. ■

EXAMPLE 5.3

Calculate the value of the current I in Fig. 5.3 if $V = 6$ V and $V = 30$ V. Applying KVL yields

$$V + 2V_x = 6I$$

where

$$V_x = 2I$$

Therefore,

$$\frac{V}{2} = I$$

If $V = 6$ V, then $I = 3$ A; and if $V = 30 \text{ V} = 5 \times 6$ V, then $I = 5 \times 3 \text{ A} = 15$ A. ■

EXAMPLE 5.4

For the circuit shown in Fig. 5.4, we wish to determine the output voltage V_{out}. However, rather than approach the problem in a straightforward manner and calculate I_o, then I_1,

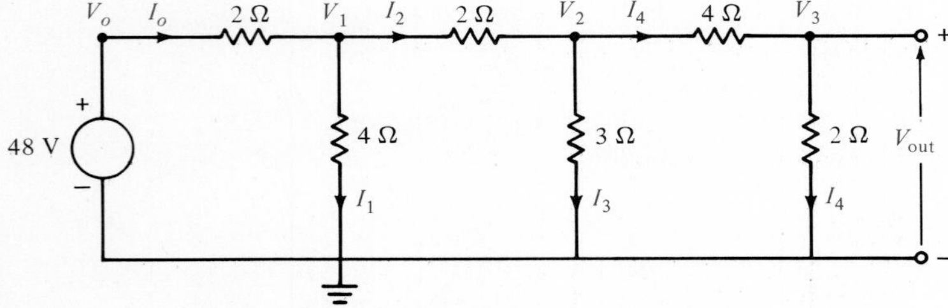

Figure 5.4 Circuit for Example 5.4.

then I_2, and so on, we will use linearity and simply assume that the output voltage is $V_{out} = 1$ V. This assumption will yield a value for the source voltage. We will then use the actual value of the source voltage and linearity to compute the actual value of V_{out}.

If we assume that $V_{out} = V_3 = 1$ V, then

$$I_4 = \frac{V_3}{2} = 0.5 \text{ A}$$

V_2 can then be calculated as

$$\begin{aligned} V_2 &= 4I_4 + V_3 \\ &= 3 \text{ V} \end{aligned}$$

Hence

$$I = \frac{V_2}{3} = 1 \text{ A}$$

Using KCL gives us

$$\begin{aligned} I_2 &= I_3 + I_4 \\ &= 1.5 \text{ A} \end{aligned}$$

Then

$$\begin{aligned} V_1 &= 2I_2 + V_2 \\ &= 6 \text{ V} \end{aligned}$$

I_1 is then computed as

$$\begin{aligned} I_1 &= \frac{V_1}{4} \\ &= 1.5 \text{ A} \end{aligned}$$

Applying KCL again, we have

$$\begin{aligned} I_o &= I_1 + I_2 \\ &= 3 \text{ A} \end{aligned}$$

and finally,

$$\begin{aligned} V_o &= 2I_o + V_1 \\ &= 12 \text{ V} \end{aligned}$$

Therefore, assuming that $V_{out} = 1$ V yields a source voltage of 12 V. However, the actual source voltage is 48 V and hence the actual output voltage is $1 \text{ V } (\frac{48}{12}) = 4$ V.■

DRILL EXERCISE

D5.1. Use linearity and the assumption that $V_o = 1$ V to compute the correct voltage V_o in the network in Fig. D5.1 if $I_o = 4$ A.

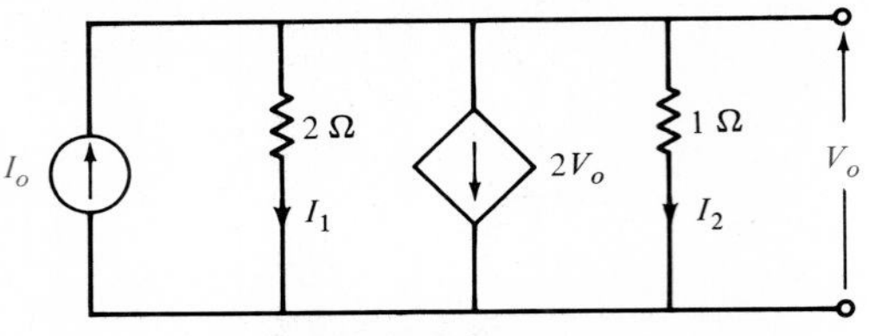

Figure D5.1

D5.2. Use linearity and the assumption that $I_o = 1$ A to compute the correct current I_o in the network in Fig. D5.2 if $I = 12$ A.

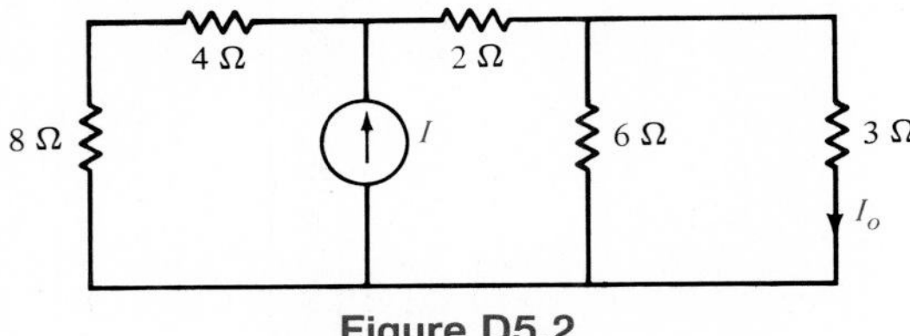

Figure D5.2

D5.3. Use linearity and the assumption that $I_o = 1$ A to determine the actual value of I_o in the network in Fig. D5.3 if $V = 24$ V.

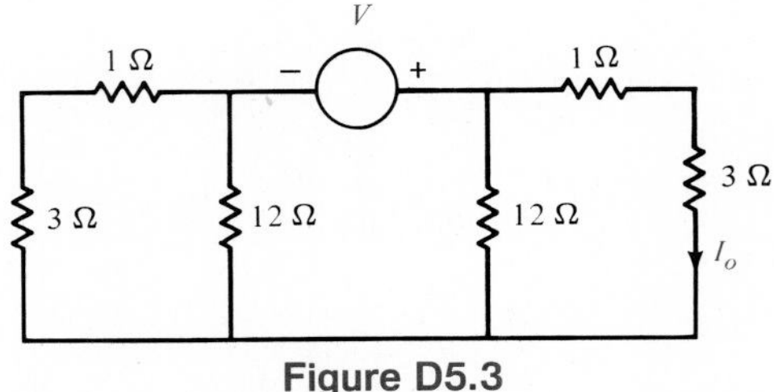

Figure D5.3

5.2 Superposition

In order to provide motivation for this subject, it is instructive to reexamine a problem that we have just considered in Section 5.1.

EXAMPLE 5.5

The circuit given in Fig. 5.2 is redrawn in Fig. 5.5, where the values of the voltage sources are unspecified. The value of the current $i_1(t)$ can be computed using KVL as

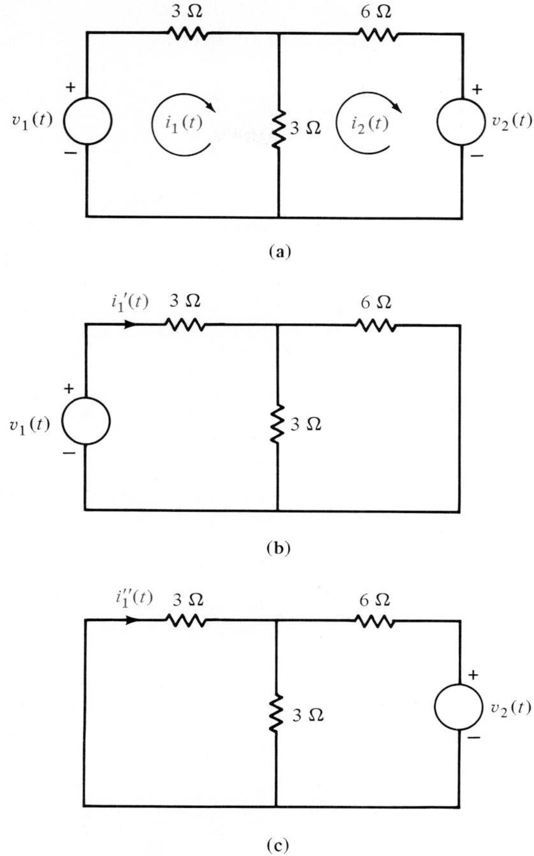

Figure 5.5 Illustration of superposition via an example.

shown in Example 5.2 as

$$\begin{bmatrix} i_1(t) \\ i_2(t) \end{bmatrix} = \frac{1}{45} \begin{bmatrix} 9 & 3 \\ 3 & 6 \end{bmatrix} \begin{bmatrix} v_1(t) \\ -v_2(t) \end{bmatrix}$$

$$i_1(t) = \frac{1}{45} [9v_1(t) - 3v_2(t)]$$

$$= \frac{v_1(t)}{5} - \frac{v_2(t)}{15}$$

In other words, the current $i_1(t)$ has a component due to $v_1(t)$ and a component due to $v_2(t)$. In view of the fact that $i_1(t)$ has two components, one due to each independent source, it would be interesting to examine what each source acting alone would contribute to $i_1(t)$. For $v_1(t)$ to act alone, $v_2(t)$ must be zero. As we pointed out in Chapter 1,

$v_2(t) = 0$ means that $v_2(t)$ is short-circuited. Therefore, to determine the value of $i_1(t)$ due to $v_1(t)$ only, we employ the circuit in Fig. 5.5b and refer to this value of $i_1(t)$ as $i_1'(t)$.

$$i_1'(t) = \frac{v_1(t)}{3 + \frac{(3)(6)}{3 + 6}}$$

$$= \frac{v_1(t)}{5}$$

Let us now determine the value of $i_1(t)$ due to $v_2(t)$ acting alone and refer to this value as $i_1''(t)$. We employ the circuit shown in Fig. 5.5c and compute this value as

$$i_1''(t) = -\frac{v_2(t)}{6 + \frac{(3)(3)}{3 + 3}}\left(\frac{3}{3 + 3}\right)$$

$$= \frac{-v_2(t)}{15}$$

Now if we add the values $i_1'(t)$ and $i_1''(t)$, we obtain the value computed directly:

$$i_1(t) = i_1'(t) + i_1''(t)$$

$$= \frac{v_1(t)}{5} - \frac{v_2(t)}{15}$$

Note that we have *superposed* the value of $i_1'(t)$ on $i_1''(t)$, or vice versa, to determine the total value of the unknown current. ■

What we have demonstrated in Example 5.5 is true in general for linear circuits and is a direct result of the property of linearity. The *principle of superposition,* which provides us with this ability to reduce a complicated problem to several easier problems—each containing only a single independent source—states that

> In any linear circuit containing multiple independent sources, the current or voltage at any point in the network may be calculated as the algebraic sum of the individual contributions of each source acting alone.

When determining the contribution due to an independent source, any remaining voltage sources are made zero by short-circuiting them, and any remaining current sources are made zero by open-circuiting them; however, dependent sources are not made zero and remain in the circuit. The following examples will illustrate the technique.

EXAMPLE 5.6

We wish to determine the current I_o shown in Fig. 5.6a. For comparison the problem will be solved using nodal equations, loop equations, and superposition.

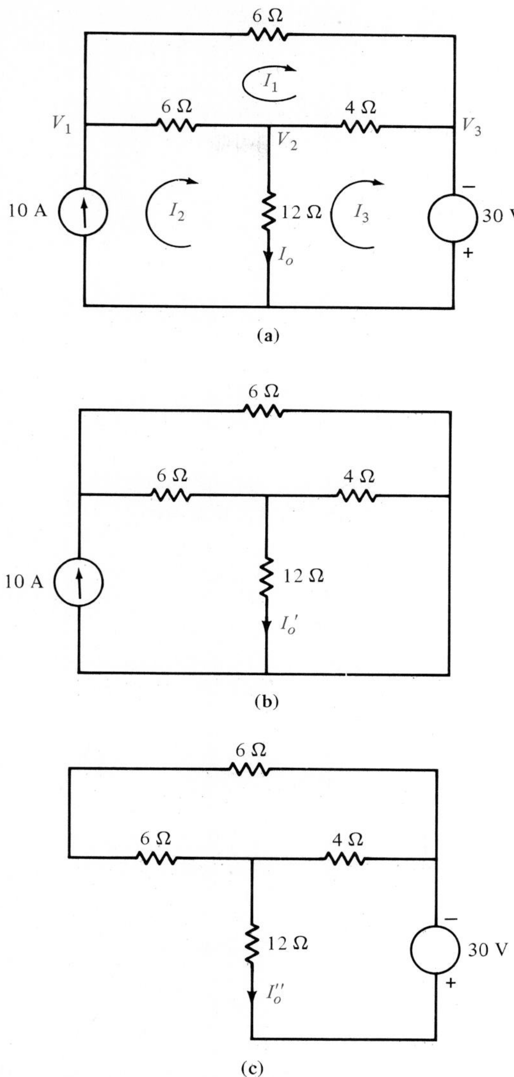

Figure 5.6 Example circuit, illustrating superposition.

The nodal equations are

$$
\begin{aligned}
V_1\left(\frac{1}{6} + \frac{1}{6}\right) - V_2\left(\frac{1}{6}\right) \qquad\qquad - V_3\left(\frac{1}{6}\right) &= 10 \\
-V_1\left(\frac{1}{6}\right) \qquad + V_2\left(\frac{1}{6} + \frac{1}{12} + \frac{1}{4}\right) - V_3\left(\frac{1}{4}\right) &= 0 \\
V_3 &= -30 \text{ V}
\end{aligned}
$$

If these equations are solved for V_2, we obtain $V_2 = -12$ V, and therefore $I_o = -1$ A.

The loop equations for this circuit are

$$16I_1 - 6I_2 - 4I_3 = 0$$

$$-4I_1 - 12I_2 + 16I_3 = 30$$

$$I_2 = 10 \text{ A}$$

Solving these equations for I_3 yields $I_3 = 11$ A, and hence

$$\begin{aligned} I_o &= I_2 - I_3 \\ &= 10 - 11 \\ &= -1 \text{ A} \end{aligned}$$

I_o can be determined via superposition by determining I_o' in Fig. 5.6b and I_o'' in Fig. 5.6c and algebraically adding the results. Note that these two circuits are very simple because they contain only one source. Therefore, I_o' and I_o'' can be immediately calculated as shown below. Using current division, we have

$$\begin{aligned} I_o' &= 10\left(\frac{6}{6 + 9}\right)\frac{4}{4 + 12} \\ &= 1 \text{ A} \end{aligned}$$

Similarly,

$$\begin{aligned} I_0'' &= -\frac{30}{12 + \dfrac{(4)(12)}{4 + 12}} \\ &= -2 \text{ A} \end{aligned}$$

Hence

$$\begin{aligned} I_o &= I_o' + I_o'' \\ &= 1 - 2 \\ &= -1 \text{ A} \end{aligned}$$

■

EXAMPLE 5.7

Consider the circuit shown in Fig. 5.7a. We want to determine the voltage V_o.

The loop equations that we will use for this circuit are

$$24 + 2V_x = 4(I_1 - I_2) + 2I_1$$

$$V_x = -4(I_1 - I_2)$$

$$I_2 = -3 \text{ A}$$

Solving these equations yields $I_1 = -12/14$ A, and therefore

$$V_o = \frac{24}{14} \text{ V}$$

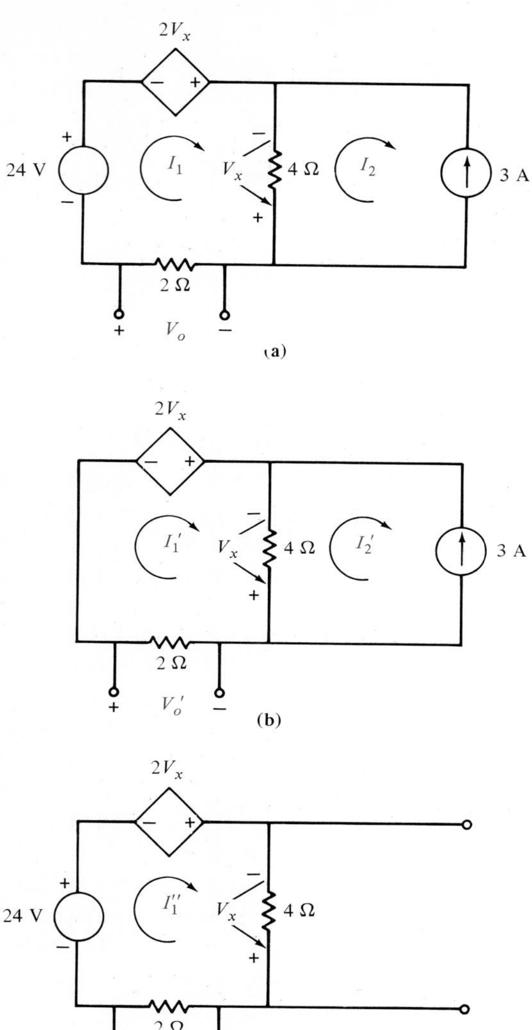

Figure 5.7 Illustration of superposition for a circuit containing a dependent source.

Employing the principle of superposition, we first solve the circuit in Fig. 5.7b. The equations for this circuit are

$$2V_x = 4(I_1' - I_2') + 2I_1'$$

$$V_x = -4(I_1' - I_2')$$

$$I_2' = -3 \text{ A}$$

These equations yield $I_1' = -36/14$ A, and therefore

$$V_o' = \frac{72}{14} \text{ V}$$

For the circuit in Fig. 5.7c the equations are

$$24 + 2V_x = 6I_1''$$

$$V_x = -4I_1''$$

From these equations we obtain $I_1'' = 24/14$ A, and hence

$$V_o'' = -\frac{48}{14} \text{ V}$$

Therefore,

$$V_o = V_o' + V_o''$$

$$= \frac{24}{14} \text{ V}$$

■

Superposition can be applied to a circuit with any number of dependent and independent sources. In fact, superposition can be applied to such a network in a variety of ways. For example, a circuit with three independent sources can be solved using each source acting alone, as we have demonstrated above, or we could use two at a time and sum the result with that obtained from the third acting alone. In addition, the independent sources do not have to assume their actual value or zero. However, it is mandatory that the sum of the different values chosen add to the total value of the source.

Example 5.7 illustrates that applying superposition to circuits with dependent sources is not always a viable scheme. Unfortunately, this seems to be true in general, since the application of superposition does not generally reduce the number of equations that must be solved when dependent sources are present.

Superposition is a fundamental property of linear equations, and therefore can be applied to any effect that is linearly related to its cause. In this regard it is important to point out that although superposition applies to the current and voltage in a linear circuit, it cannot be used to determine power, because power is a nonlinear function. As a quick illustration, note that the power in the 2-Ω resistor in Example 5.7 is $P = I^2R = (-6/7)^2(2) = 1.47$ W. However, superposition would have yielded $P = (-18/7)^2(2) + (12/7)^2(2) = 19.1$ W, which is incorrect.

DRILL EXERCISE

D5.4. Compute V_o in the network in Fig. D5.4 using superposition.

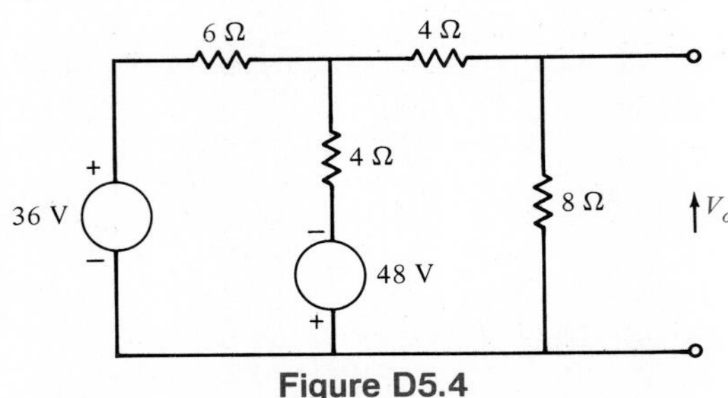

Figure D5.4

D5.5. Compute V_o in the network in Fig. D5.5 using superposition.

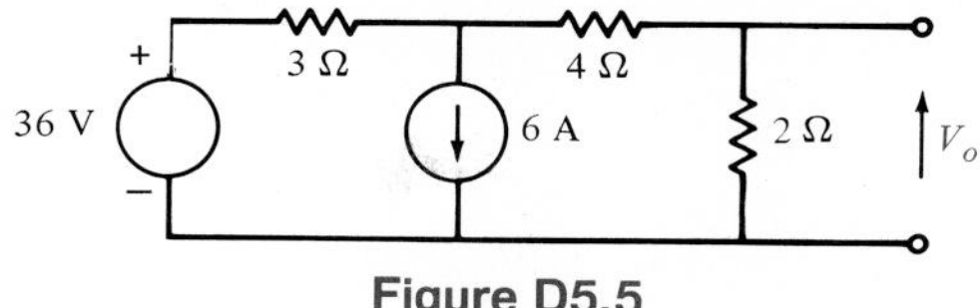

Figure D5.5

D5.6. Find V_o in the network in Fig. D5.6 using superposition.

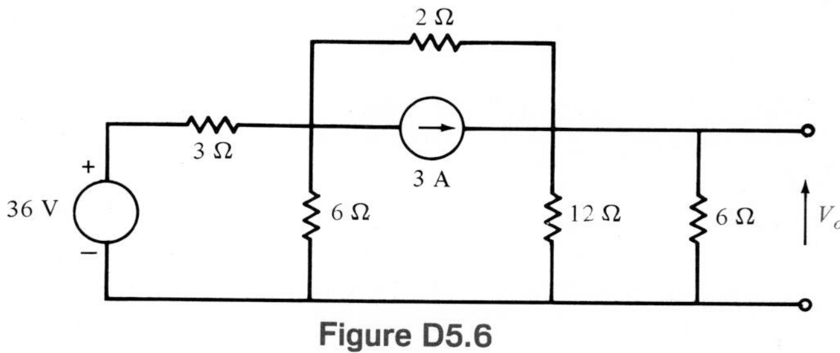

Figure D5.6

EXAMPLE 5.8

As another example of the use of superposition, consider the operational amplifier circuit shown in Fig. 5.8. The configuration has three input sources and one output.

Employing superposition, we short out v_2 and v_3 and determine the output as a function of v_1. Note that one end of R_2 and one end of R_3 are tied to ground as a result of shorting the sources, and the other ends are shorted to ground because $v_+ = 0$ and $v_- = v_+$. Therefore, R_2 and R_3 are out of the circuit and v_0 due to the source v_1 is

$$v_o' = \frac{-R_4}{R_1} v_1$$

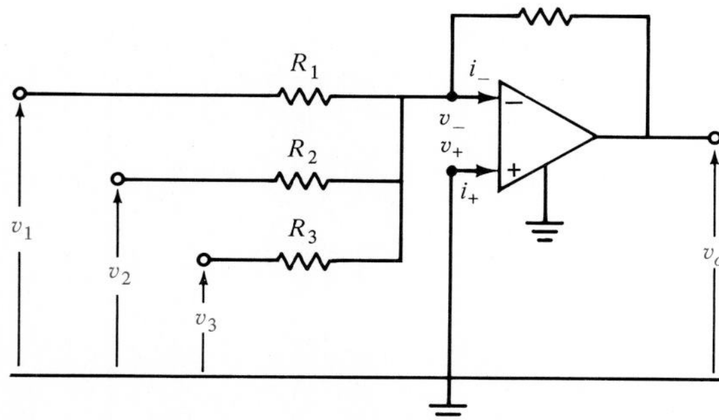

Figure 5.8 Operational amplifier adder circuit.

Similarly,

$$v_o'' = \frac{-R_4}{R_2} v_2$$

and

$$v_o''' = \frac{-R_4}{R_3} v_3$$

Therefore, employing superposition, we obtain

$$v_o = -\left(\frac{R_4}{R_1} v_1 + \frac{R_4}{R_2} v_2 + \frac{R_4}{R_3} v_3\right)$$

Note that this is the same result that we obtained in Chapter 3. ■

EXAMPLE 5.9

Consider the differential amplifier op-amp circuit shown in Fig. 5.9. Employing superposition, we first set $v_2 = 0$. We note that since $i_+ = 0$, there can be no current in R_1 or R_2 and therefore $v_+ = 0$. The circuit is therefore an inverting amplifier and hence

$$v_o' = -\frac{R_2}{R_1} v_1$$

Then setting, $v_1 = 0$, the circuit equations are

$$\frac{v_2 - v_+}{R_1} = \frac{v_+}{R_2} + i_+$$

$$\frac{v_o'' - v_-}{R_2} = \frac{v_-}{R_1} + i_-$$

However, $i_- = i_+ = 0$ and $v_+ = v_-$. Substituting these values into the equations yields

$$v_o'' = \frac{R_2}{R_1} v_2$$

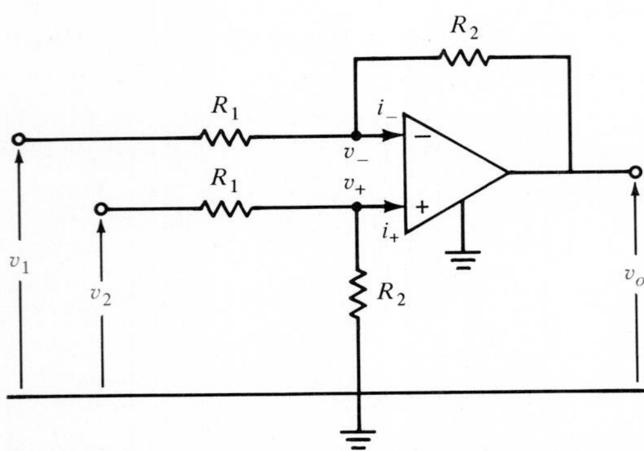

Figure 5.9 Operational amplifier circuit for a differential amplifier.

Therefore,

$$v_o = v_o' + v_o''$$
$$= \frac{R_2}{R_1}(v_2 - v_1)$$

which is the same value as that obtained in Chapter 3. ■

5.3 Source Transformation

Before we begin discussing source transformations, it is necessary that we point out that real sources differ from the ideal models we have presented thus far. In general, a practical voltage source does not produce a constant voltage regardless of the load resistance or the current it delivers, and a practical current source does not deliver a constant current regardless of the load resistance or the voltage across its terminals. Practical sources contain internal resistance, and therefore the models shown in Fig. 5.10a and b more closely represent actual sources. Note that the power delivered by the practical voltage source is given by the expression

$$P_L = \frac{v^2}{R_L}\left[\frac{1}{1 + (R_v/R_L)}\right]^2$$

and therefore if $R_L >> R_v$, then

$$P_L = \frac{v^2}{R_L}$$

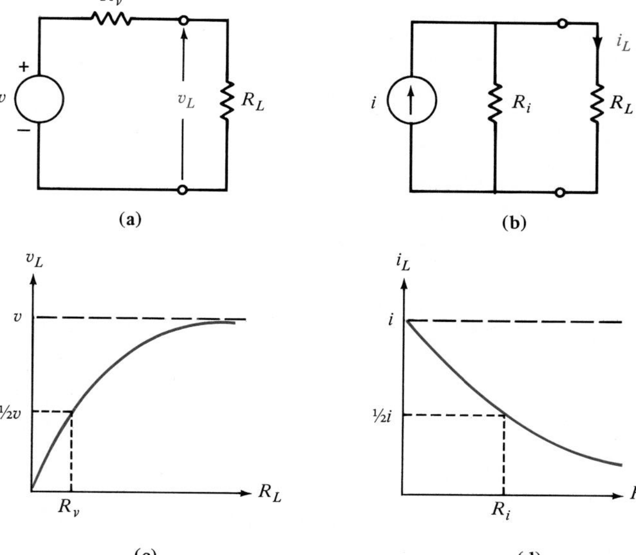

Figure 5.10 Effect of internal resistance on actual voltage and current sources.

which is the power delivered by the ideal voltage source. Similarly, the power delivered by the practical current source is

$$P_L = i^2 R_L \left[\frac{1}{1 + (R_L/R_i)} \right]^2$$

and hence if $R_i >> R_L$, then

$$P_L = i^2 R_L$$

which is the power delivered by the ideal current source.

The graphs shown in Fig. 5.10c and d illustrate the effect of the internal resistance on the voltage source and current source, respectively. The graphs indicate that the output voltage approaches the ideal source voltage only for large values of load resistance and, therefore, small values of current. Similarly, the load current is approximately equal to the ideal source current only for values of the load resistance R_L that are small in comparison to the internal resistance R_i.

With this material on practical sources as background information, we now ask if it is possible to exchange one source model for another: that is, exchange a voltage source model for a current source model, or vice versa. We could exchange one source for another provided that they are equivalent: that is, each source produces exactly the same voltage and current for any load that is connected across its terminals.

Let us examine the two circuits shown in Fig. 5.11. In order to determine the conditions required for the two sources to be equivalent, the load currents are set equal:

$$\frac{R_i i}{R_i + R_L} = \frac{v}{R_v + R_L}$$

These two expressions will be equal for any load R_L if

$$R_i = R_v = R \qquad v = Ri \tag{5.7}$$

It is interesting to note that we would arrive at exactly the same conditions if we required the load voltages to be equal for any value of R_L.

The relationships specified in Eq. (5.7) and Fig. 5.11 are extremely important and the reader should not fail to grasp their significance. What these relationships tell us is that if we have embedded within a network a current source i in parallel with a resistor R, we can replace this combination with a voltage source of value $v = iR$ in series with the resistor R. The reverse is also true; that is, a voltage source v in series with a resistor

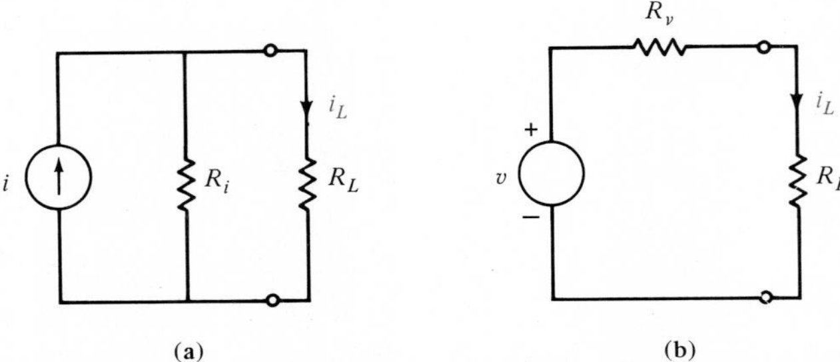

Figure 5.11 Circuits used to determine conditions for a source exchange.

R can be replaced with a current source of value $i = v/R$ in parallel with the resistor R. Parameters within the circuit (e.g., an output voltage) are unchanged under these transformations.

The following examples will demonstrate the utility of a source exchange.

EXAMPLE 5.10

Given the voltage source in Fig. 5.12a, determine the equivalent current source.

The current source can be obtained using the expressions in Eq. (5.7) and is shown in Fig. 5.12b.

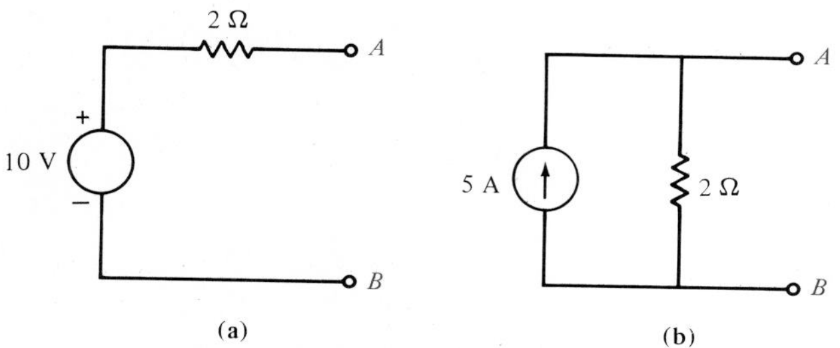

Figure 5.12 Two equivalent sources. ■

The reader is cautioned to keep the polarity of the voltage source and the direction of the current source in agreement, as shown in Fig. 5.12.

EXAMPLE 5.11

Consider the circuit shown in Fig. 5.13a. For comparison we will solve this circuit using a variety of techniques, including source transformation.

The loop equations for the circuit are

$$9I_1 - 3I_2 = 60$$

$$-3I_1 + 6I_2 = 15$$

These equations yield $I_1 = 9$ A and $I_2 = 7$ A, so that $I_o = I_1 - I_2 = 2$ A. Note that we could have used different loops so that only one current would have to be found.

The nodal equations for the circuit are

$$-V_1\left(\frac{1}{6}\right) + V_2\left(\frac{1}{6} + \frac{1}{3} + \frac{1}{3}\right) - V_3\left(\frac{1}{3}\right) = 0$$

$$V_1 = 60 \text{ V}$$

$$V_3 = -15 \text{ V}$$

Solving for V_2 yields $V_2 = 6$ V, and therefore $I_o = 2$ A. Employing superposition and using Fig. 5.13b and c yields

$$I_o' = \frac{60}{6 + \dfrac{(3)(3)}{3 + 3}}\left(\frac{3}{3 + 3}\right)$$

$$= 4 \text{ A}$$

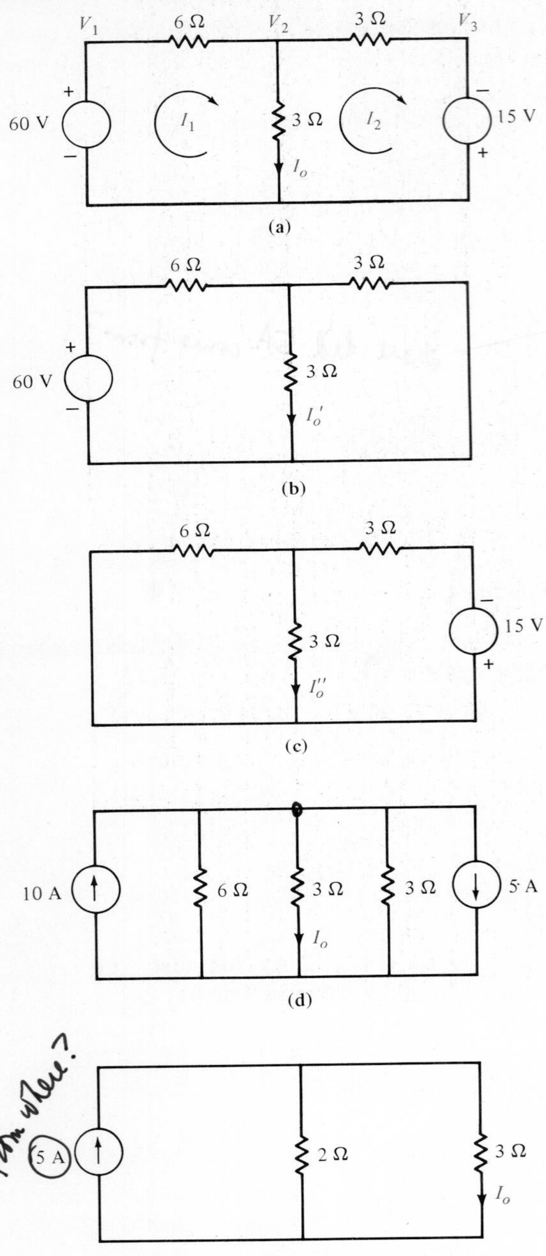

Figure 5.13 Variety of circuits employed in Example 5.11.

and

$$I_o'' = -\frac{15}{3 + \dfrac{(6)(3)}{6 + 3}}\left(\frac{6}{6 + 3}\right)$$

$$= -2 \text{ A}$$

Hence $I_o = I_o' + I_o'' = 2$ A.

Using source transformations, we can immediately convert the circuit in Fig. 5.13a to the circuit in Fig. 5.13d. Combining resistors and sources produces the circuit in Fig. 5.13e. Now employing current division, we obtain

$$I_o = 5\left(\frac{2}{2 + 3}\right) = 2 \text{ A}$$

Note the simplicity with which the problem was solved using source exchange. ■

EXAMPLE 5.12

Suppose that we are given the circuit in Fig. 5.14a and asked to determine the current I_o. We could, of course, solve for I_o using loop equations, nodal equations, or superposition. However, we will not do that. Instead, we will solve for I_o rather quickly through repeated application of the source transformation.

If we begin at the left end of the network, the 60-V source is converted to a current source and the resultant 6-Ω and 3-Ω resistors in parallel are combined to produce the circuit in Fig. 5.14b. The parallel combination of the 10-A current source and 2-Ω resistor are converted to a 20-V source in series with a 2-Ω resistor, as shown in Fig. 5.14c. The 20-V source in series with the two 2-Ω resistors is now converted into a current source and parallel resistor, as shown in Fig. 5.14d. The 5-A and 4-A current sources are combined and converted to the voltage source shown in Fig. 5.14e. Using the source transformation again, we convert the circuit in Fig. 5.14e to that shown in Fig. 5.14f. Combining resistors in this figure produces the circuit shown in Fig. 5.14g. When we employ current division, we find that $I_o = 1$ A.

DRILL EXERCISE

D5.7. Find V_o in the circuit in Fig. D5.4 using source transformation.

D5.8. Find V_o in the network in Fig. D5.5 using source transformation.

D5.9. Find V_o in the circuit in Fig. D5.6 using source transformation.

5.4 Thévenin's and Norton's Theorems

Thus far we have presented a number of techniques for circuit analysis. At this point we will add two theorems to our collection of tools that will prove to be extremely useful. The theorems are named after their authors, M. L. Thévenin, a French engineer, and E. L. Norton, a scientist formerly with Bell Telephone Laboratories.

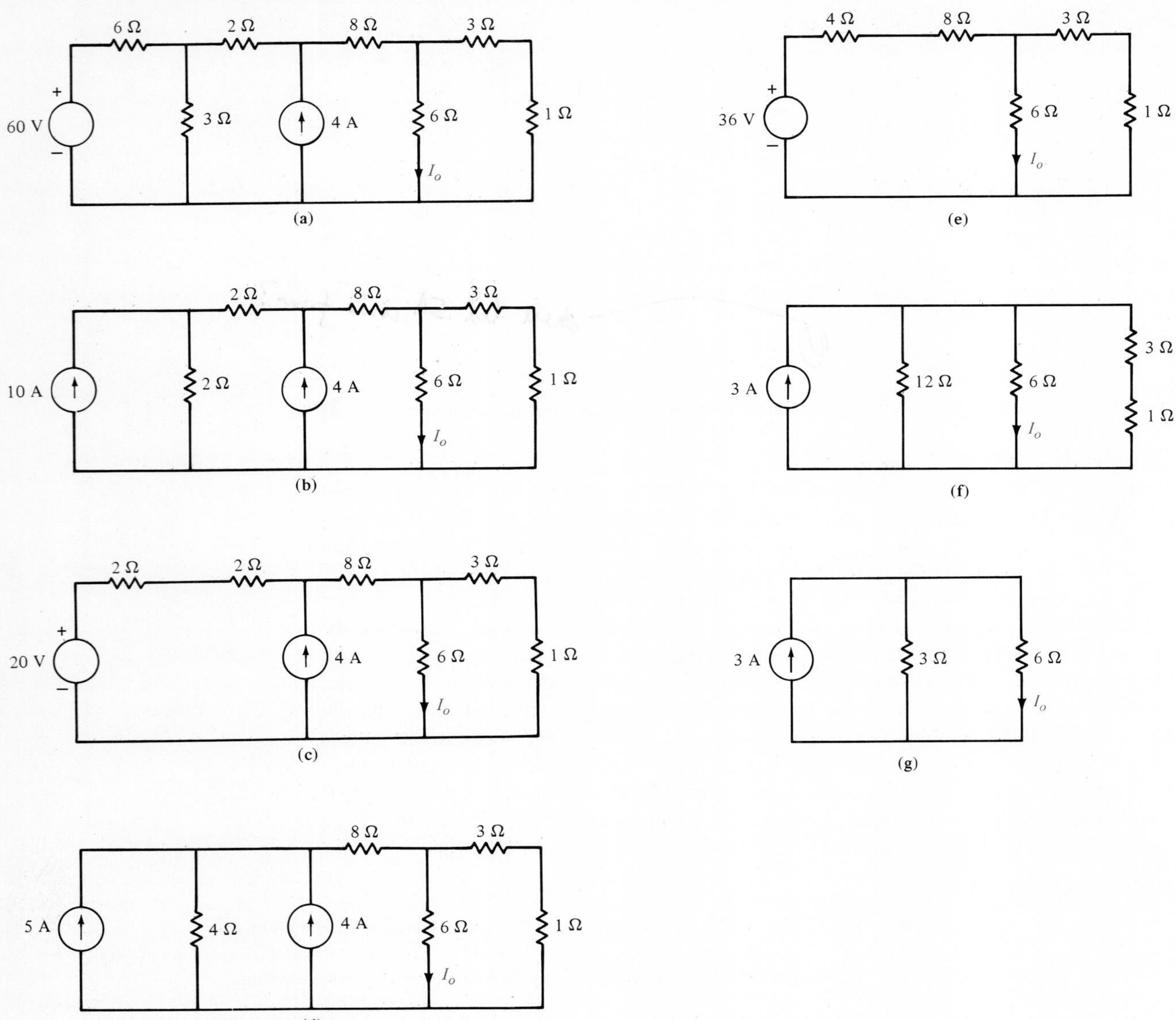

Figure 5.14 Application of the source transformation technique to an example problem.

Suppose that we are given a circuit and that we wish to find the current, voltage, or power that is delivered to some resistor of the network which we will call the load. *Thévenin's theorem* tells us that we can replace the entire network, exclusive of the load, by an equivalent circuit that contains only an independent voltage source in series with a resistor in such a way that the current–voltage relationship at the load is unchanged. *Norton's theorem* is identical to the statement above except that the equivalent circuit is an independent current source in parallel with a resistor.

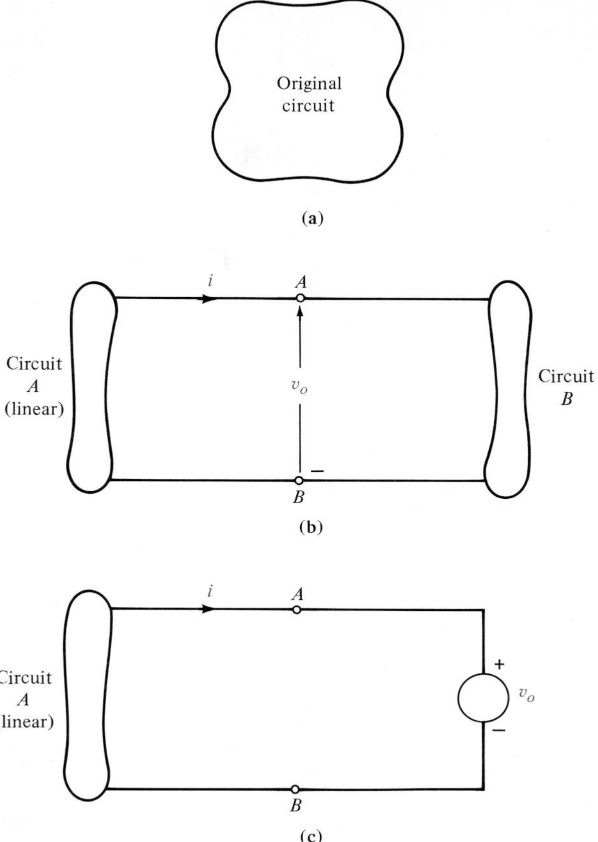

Figure 5.15 Concepts used to develop Thévenin's theorem.

In developing the theorems, we will assume that the circuit shown in Fig. 5.15a can be split into two parts, as shown in Fig. 5.15b. In general, circuit B is the load and may be linear or nonlinear. Circuit A is the balance of the original network exclusive of the load and must be linear. As such, circuit A may contain independent sources, dependent sources and resistors, or any other linear element. We require, however, that a dependent source and its control variable appear in the same circuit.

Circuit A delivers a current i to circuit B and produces a voltage v_o across the input terminals of circuit B. From the standpoint of the terminal relations of circuit A, we can replace circuit B by a voltage source of v_o volts (with the proper polarity), as shown in Fig. 5.15c. Since the terminal voltage is unchanged and circuit A is unchanged, the terminal current i is unchanged.

Now applying the principle of superposition to the network shown in Fig. 5.15c, the total current i shown in the figure is the sum of the currents caused by all the sources in circuit A and the source v_o which we have just added. Therefore, via superposition the current i can be written

$$i = i_o + i_{sc} \tag{5.8}$$

where i_o is the current due to v_o with all independent sources in circuit A made zero (i.e., voltage sources short-circuited and current sources open-circuited), and i_{sc} is the short-circuit current due to all sources in circuit A with v_o short-circuited.

The terms i_o and v_o are related by the equation

$$i_o = \frac{v_o}{R_{Th}} \tag{5.9}$$

where R_{Th} is the equivalent resistance looking back into circuit A from terminals A-B with all independent sources in circuit A made zero.

Substituting Eq. (5.9) into Eq. (5.8) yields

$$i = -\frac{v_o}{R_{Th}} + i_{sc} \tag{5.10}$$

This is a general relationship and, therefore, must hold for any specific condition at terminals A-B. As a specific case, suppose that the terminals are open-circuited. For this condition $i = 0$ and v_o is equal to the open-circuit voltage v_{oc}. Thus Eq. (5.10) becomes

$$i = 0 = \frac{-v_{oc}}{R_{Th}} + i_{sc} \tag{5.11}$$

or

$$v_{oc} = R_{Th} i_{sc} \tag{5.12}$$

This equation states that the open-circuit voltage is equal to the short-circuit current times the equivalent resistance looking back into circuit A with all independent sources made zero. We refer to R_{Th} as the Thévenin equivalent resistance.

Substituting Eq. (5.12) into Eq. (5.10) yields

$$i = \frac{-v_o}{R_{Th}} + \frac{v_{oc}}{R_{Th}}$$

or

$$v_o = v_{oc} - R_{Th} i \tag{5.13}$$

Let us now examine the circuits that are described by these equations. The circuit represented by Eq. (5.13) is shown in Fig. 5.16a. The fact that this circuit is equivalent at the terminals to circuit A in Fig. 5.15 is a statement of *Thévenin's theorem*. The circuit represented by Eq. (5.10) is shown in Fig. 5.16b. The fact that this circuit is equivalent at the terminals to circuit A in Fig. 5.15 is a statement of *Norton's theorem*.

Note carefully that the circuits in Fig. 5.16 together with the relationship in Eq. (5.12) represent a source transformation.

The manner in which these theorems are applied depends on the structure of the original network under investigation. For example, if only independent sources are present, we can calculate the open-circuit voltage or short-circuit current and the Thévenin equivalent resistance. However, if dependent sources are also present, the Thévenin equivalent will be determined by calculating v_{oc} and i_{sc}, since this is normally the best approach for determining R_{Th} in a network containing dependent sources. Finally, if only dependent

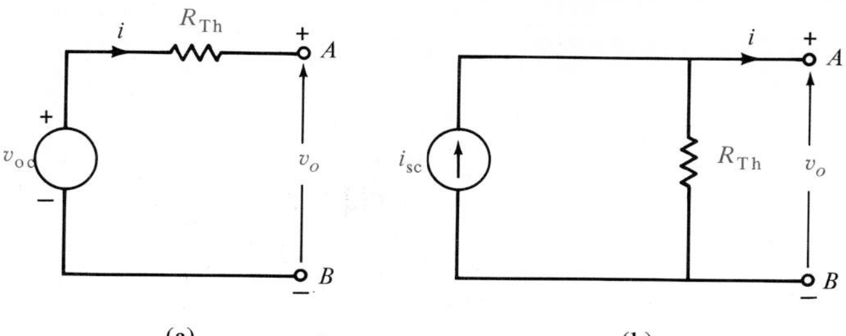

Figure 5.16 Thévenin and Norton equivalent circuits.

sources are present in the network, we can either apply a 1-A source at the terminals, calculate the terminal voltage, and use $v_o = 1 \times R_{Th}$, or we can apply a 1-V source at the terminals, calculate the terminal current, and use $i_o = 1/R_{Th}$.

A variety of examples are now presented to demonstrate the utility of these theorems.

EXAMPLE 5.13

Suppose that we wish to find the output voltage V_o, as shown in Fig. 5.17a. Using Thévenin's theorem, we break the circuit at points *A*-*B* and determine the open-circuit voltage as shown in Fig. 5.17b. Note that no current flows in the 8-Ω resistor, and therefore, using voltage division, $V_{oc} = 24$ V. The Thévenin equivalent resistance determined from Fig. 5.17c is 10 Ω. Now attaching the Thévenin equivalent circuit to the load, we can determine the output voltage as shown in Fig. 5.17d as $V_o = 4$ V.

The circuit could also be solved using Norton's theorem. In this case we will break the circuit just to the right of the 6-Ω resistor and determine the short-circuit current and the Thévenin equivalent resistance as shown in Fig. 5.18a and b. $I_{sc} = 12$ A and $R_{Th} = 2$ Ω. Using this equivalent circuit in the original network yields the circuit shown in Fig. 5.18c. Now we apply Norton's theorem again and break the circuit to the right of the 8-Ω resistor and once again determine I_{sc} and R_{Th} using Fig. 5.18d and e. Using current division, I_{sc} in Fig. 5.18d is 24/10 A and R_{Th} is 10 Ω. Now attaching the load to this equivalent circuit yields the circuit in Fig. 5.18f, from which we can calculate $V_o = 4$ V. ■

Let us examine for a moment some of the salient features of Example 5.13. Note that in applying Thévenin's theorem there is no point in breaking the network to the left of the 6-Ω resistor, since the 36-V source and 3-Ω resistor is a Thévenin equivalent as is. In addition we could have broken the network to the left of the 8-Ω resistor. Breaking the network at this point would not change the open circuit voltage, but in this case $R_{Th} = 2$-Ω. However, when the Thévenin equivalent is connected to the 8- and 2-Ω resistors, the final circuit would again be that shown in Fig. 5.17d. Similar arguments could be made for the analysis using Norton's theorem. Finally, let us take note of the repeated application of Norton's theorem in the solution of this problem. This is an extremely important point and very useful in the application of both theorems. For example, in analyzing a ladder-type network such as that shown in Fig. 2.24a, we may simply work our way down the network from left to right by forming a sequence of either Thévenin or Norton equivalents.

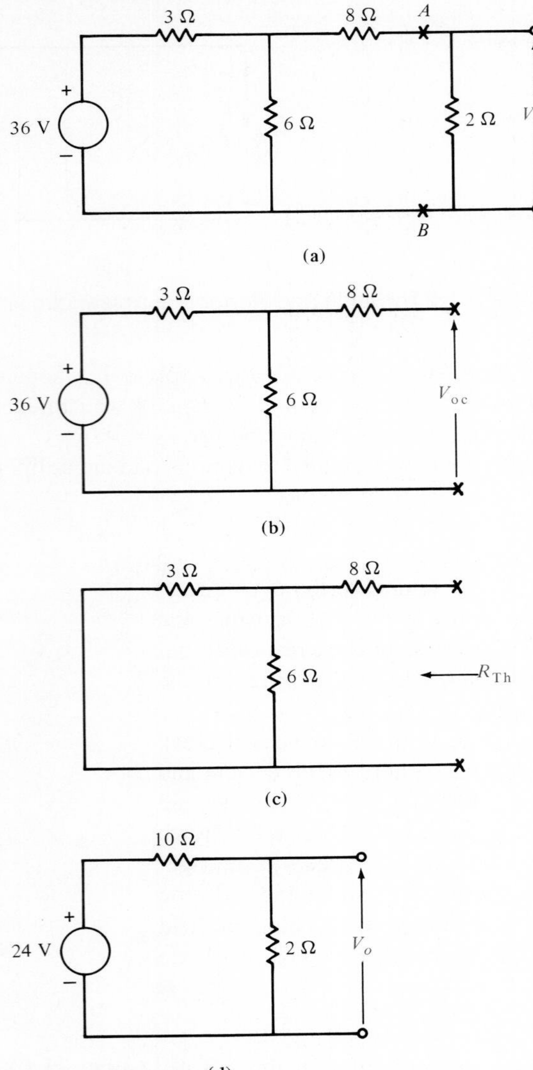

Figure 5.17 Use of Thévenin's theorem.

EXAMPLE 5.14

Let us determine the output voltage V_o as shown in Fig. 5.19a. The loop equations for this circuit are

$$12 = 4(I_2 - I_1) + 4I_2$$

$$I_1 = 3\ A$$

Solving these equations yields $I_2 = 3$ A and therefore $V_o = 6$ V.

Upon using Thévenin's theorem, the open-circuit voltage is determined from Fig. 5.19b. The 3-A source produces a 12-V drop across the 4-Ω resistor, and therefore

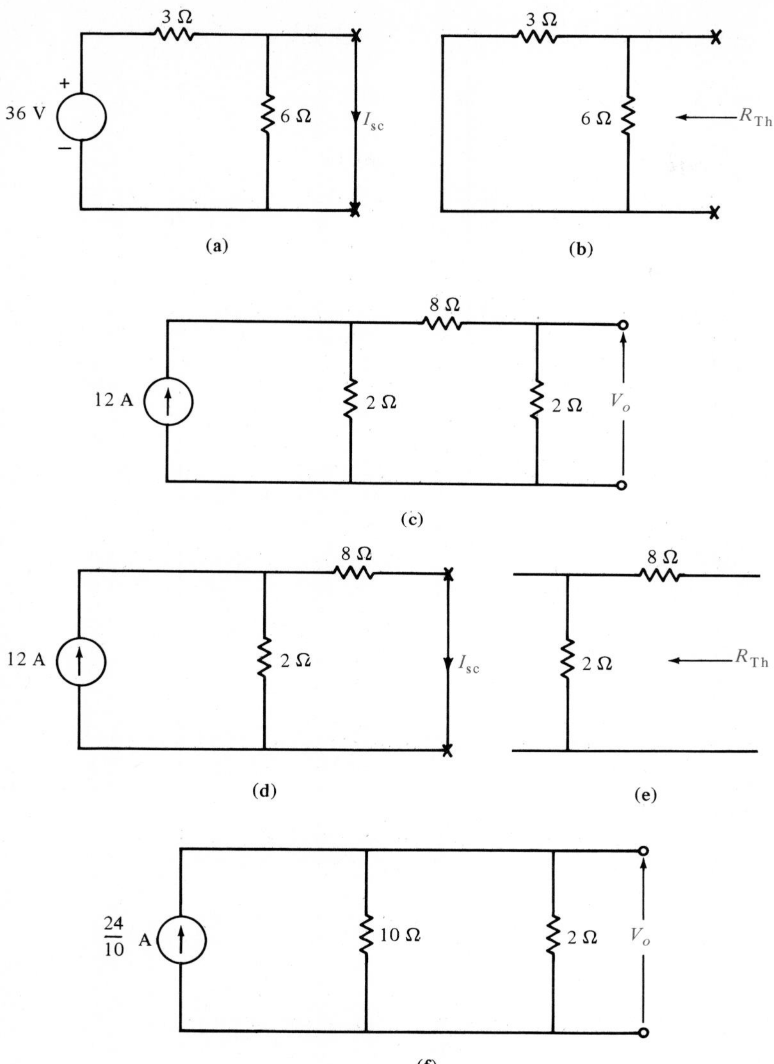

Figure 5.18 Use of Norton's theorem.

$V_{oc} = 12\text{ V} + 12\text{ V} = 24\text{ V}$. R_{Th} obtained from Fig. 5.19c is $R_{Th} = 6\ \Omega$. If the equivalent circuit is now attached to the 2-Ω load as shown in Fig. 5.19d, the output voltage is found to be $V_o = 6\text{ V}$.

DRILL EXERCISE

D5.10. Use Thévenin's theorem to find V_o in the circuit in Fig. D5.4.

D5.11. Use Thévenin's theorem to find V_o in the circuit in Fig. D5.5.

D5.12. Use Thévenin's theorem to find V_o in the network in Fig. D5.6.

D5.13. Use Norton's theorem to find V_o in the circuit in Fig. D5.4.

D5.14. Use Norton's theorem to find V_o in the network in Fig. D5.5.

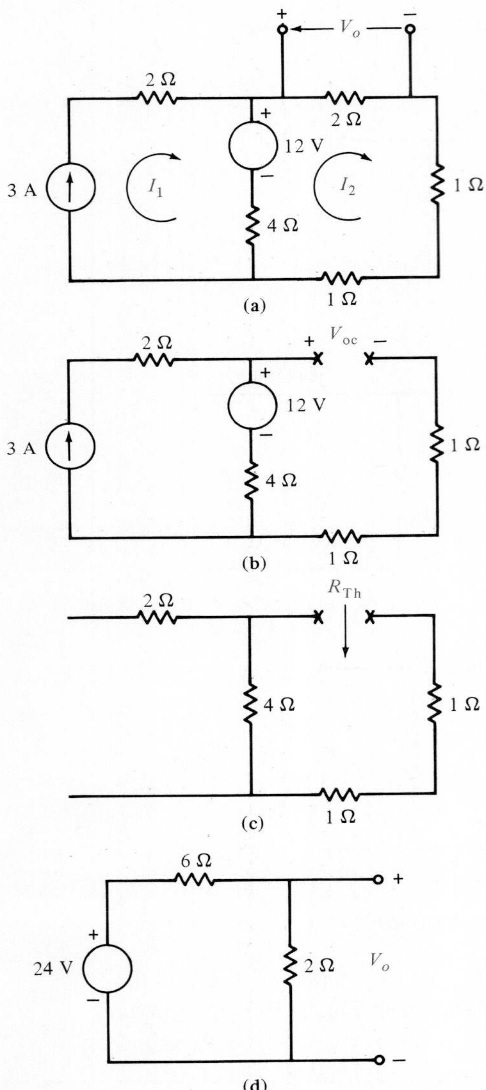

Figure 5.19 Application of Thévenin's theorem.

EXAMPLE 5.15

Let us solve the circuit in Example 5.7 using Thévenin's theorem.

The open-circuit voltage is determined using the circuit shown in Fig. 5.20a. Note

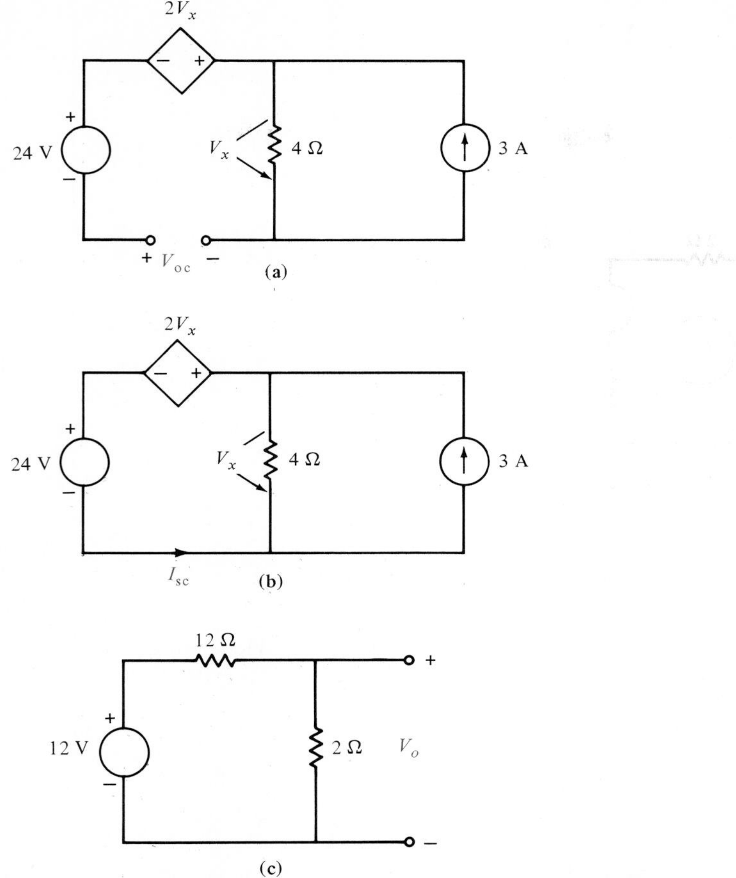

Figure 5.20 Application of Thévenin's theorem to a circuit containing a dependent source.

that the 3-A source produces $V_x = -12$ V, and therefore applying KVL to the open-circuit loop yields

$$V_{oc} + 24 + 2V_x + V_x = 0$$

But $V_x = -12$, and therefore

$$V_{oc} = 12 \text{ V}$$

Since a dependent source is present, we must use I_{sc} to determine R_{Th}. The short-circuit current is shown in Fig. 5.20b. To determine the short-circuit current, we note that KVL must hold around the left-hand loop. For this loop KVL is

$$24 + 2V_x + V_x = 0$$

and therefore $V_x = -8$ V. By employing Ohm's law, we find that 2 A must flow down through the 4-Ω resistor. Hence applying KCL at the node at the top of the 4-Ω resistor yields

$$I_{sc} + 2 - 3 = 0$$

or

$$I_{sc} = 1 \text{ A}$$

R_{Th} can now be computed as

$$R_{Th} = \frac{V_{oc}}{I_{sc}} = \frac{12}{1} = 12 \ \Omega$$

Therefore, attaching the equivalent circuit to the load as shown in Fig. 5.20c yields

$$V_o = \frac{24}{14} \text{ V}$$

■

EXAMPLE 5.16

Consider the circuit shown in Fig. 5.21a. The loop equations for this network are

$$36 + 2V_a = 2(I_1 + I_2) + 2I_2$$

$$V_a = -2(I_1 + I_2)$$

$$I_1 = -6 \text{ A}$$

Solving these equations yields $I_2 = 9$ A and hence $V_o = 18$ V. If we use Thévenin's theorem, the open-circuit voltage can be calculated using Fig. 5.21b. Applying KVL to the outer loop yields

$$36 + V_a + 2V_a - V_{oc} = 0$$

However, the 6-A source produces a V_a across the 2-Ω resistor of 12 V. Therefore, $V_{oc} = 72$ V.

I_{sc} is computed from Fig. 5.21c. Using KCL, the current in the 36-V source and 2-Ω resistor are found to be $(I_{sc} - 6)$ A. Applying KVL to the outer loop yields

$$36 - 2(I_{sc} - 6) + 2V_a = 0$$

$$V_a = -2(I_{sc} - 6)$$

and hence

$$I_{sc} = 12 \text{ A}$$

Therefore,

$$R_{Th} = \frac{V_{oc}}{I_{sc}} = \frac{72}{12} = 6 \ \Omega$$

The equivalent circuit attached to the load is shown in Fig. 5.21d. From this circuit we can calculate

$$V_o = 18 \text{ V}$$

■

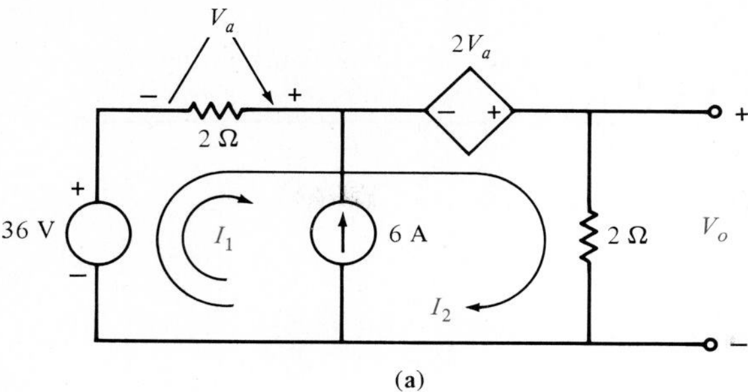

(a)

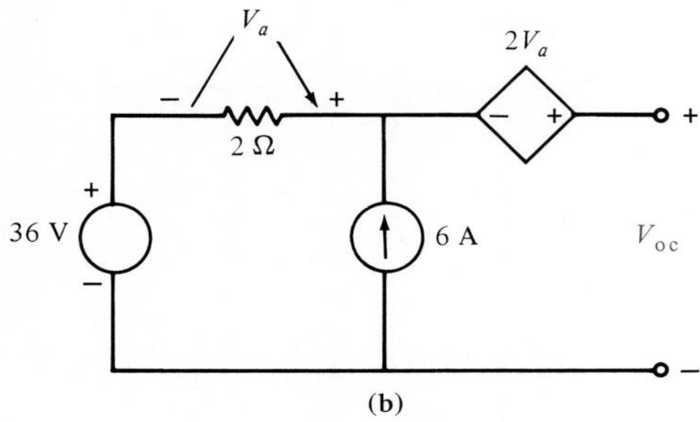

(b)

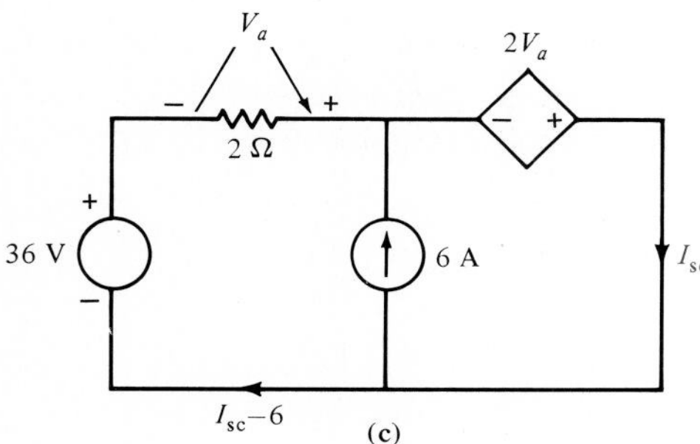

(c)

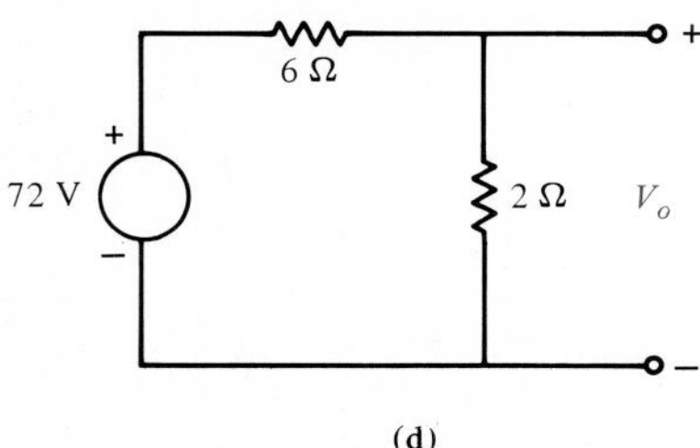

(d)

Figure 5.21 Circuit solved using Thévenin's theorem.

DRILL EXERCISE

D5.15. Use Thévenin's theorem to find V_o in the network in Fig. D5.15.

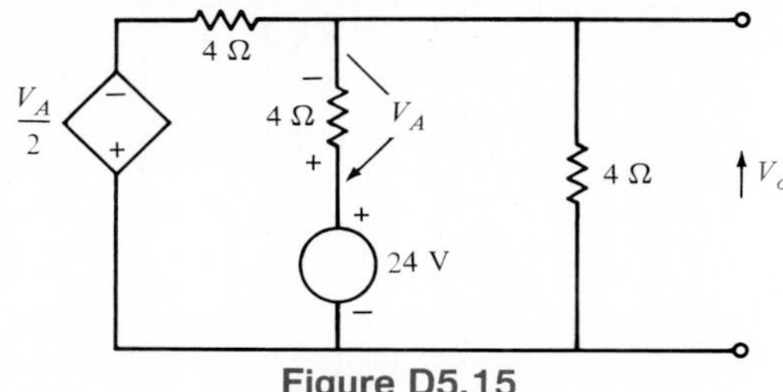

Figure D5.15

D5.16. Use Thévenin's theorem to find V_o in the circuit in Fig. D5.16.

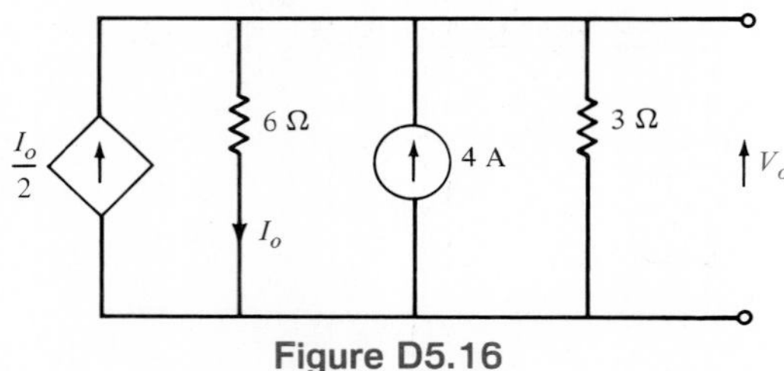

Figure D5.16

EXAMPLE 5.17

As another example of the application of Thévenin's theorem, consider the circuit shown in Fig. 5.22a. Note that this circuit contains no independent sources. Hence we will assume that a voltage source of 1 V is applied at the terminals and calculate the current I as shown in Fig. 5.22b. From these we can determine R_{Th}.

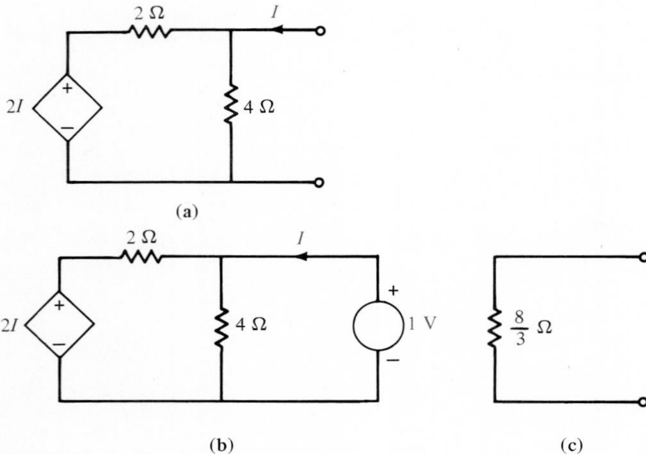

Figure 5.22 Application of Thévenin's theorem to a circuit containing a dependent source but no independent sources.

Applying KCL at the node at the top of the 4-Ω resistor yields

$$\frac{1 - 2I}{2} + \frac{1}{4} - I = 0$$

Solving this equation yields

$$I = \frac{3}{8}\text{ A}$$

and therefore

$$R_{\text{Th}} = \frac{1}{3/8} = \frac{8}{3}\ \Omega$$

The Thévenin equivalent circuit is then as shown in Fig. 5.22c. ■

EXAMPLE 5.18

Using the operational amplifier model shown in Fig. 2.31b, find the output resistance of the circuit shown in Fig. 5.23a. The output resistance is the Thévenin equivalent re-

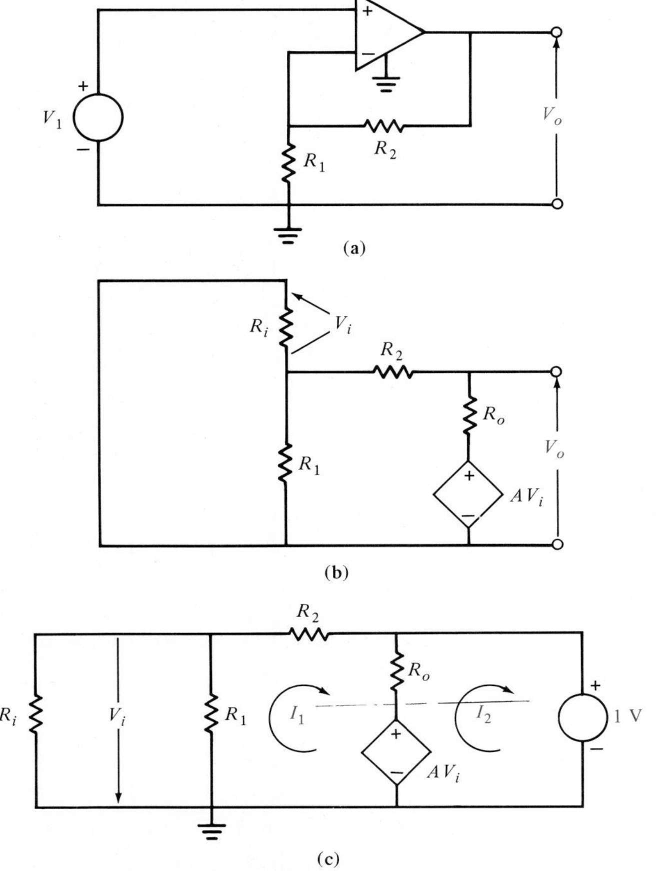

Figure 5.23 Circuits for determining the output resistance of an operational amplifier circuit.

sistance at the output terminals. Using the model, the equivalent circuit is shown in Fig. 5.23b.

The Thévenin equivalent resistance at the output terminals can be obtained by assuming a voltage source of 1 V connected to the terminals and determining the resultant current it produces. The circuit in Fig. 5.23b has been redrawn in Fig. 5.23c with this connection.

The circuit equations for this configuration are

$$-AV_i = I_1(R_{i1} + R_2 + R_o) - I_2R_o$$

$$AV_i - 1 = -I_1R_o + I_2R_o$$

$$V_i = R_{i1}I_1$$

where R_{i1} is the parallel combination of R_i and R_1. Solving these equations for I_2 yields

$$I_2 = \frac{-(R_{i1} + AR_{i1} + R_2 + R_o)}{R_o(R_{i1} + R_2)}$$

Then

$$R_{out} = \frac{1}{-I_2} = \frac{R_o(R_{i1} + R_2)}{R_{i1} + AR_{i1} + R_2 + R_o}$$

Since R_i is normally much greater than R_1, $R_{i1} \approx R_1$ and thus

$$R_{out} = \frac{R_o(R_1 + R_2)}{R_1 + AR_1 + R_2 + R_o}$$

Since A is very large, AR_1 is normally much greater than $R_1 + R_2 + R_o$. Therefore, R_{out} can be approximated by the expression

$$R_{out} = \frac{R_o(R_1 + R_2)}{AR_1}$$

$$= \frac{R_o(1 + R_2/R_1)}{A}$$

Note that this equation indicates that the output resistance of this configuration is much lower than the basic op-amp whose output resistance is R_o. ■

As a final point, the reader should note that we can apply Thévenin's or Norton's theorem over and over again if necessary to a large network to calculate some current, voltage, or power within the network.

5.5 Maximum Power Transfer

In circuit analysis we are sometimes interested in determining the maximum power that can be delivered to a load. By employing Thévenin's theorem, we can determine the maximum power that a source can supply and the manner in which to adjust the load to effect maximum power transfer.

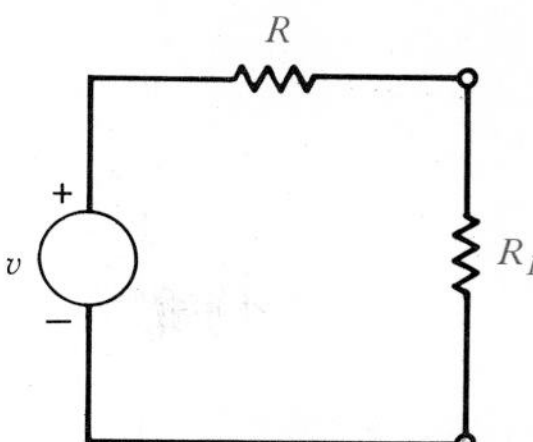

Figure 5.24 Equivalent circuit for examining maximum power transfer.

Suppose that we are given the circuit shown in Fig. 5.24. The power that is delivered to the load is given by the expression

$$\begin{aligned} P_{\text{load}} &= i^2 R_L \\ &= \left(\frac{v}{R + R_L}\right)^2 R_L \end{aligned}$$

We want to determine the value of R_L that maximizes this quantity. Hence we differentiate this expression with respect to R_L and equate the derivative to zero.

$$\frac{dP_{\text{load}}}{dR_L} = \frac{(R + R_L)^2 v^2 - 2v^2 R_L (R + R_L)}{(R + R_L)^4}$$

which yields

$$R_L = R$$

In other words, maximum power transfer takes place when the load resistance $R_L = R$. Although this is a very important result, we have derived it using the simple network in Fig. 5.24. However, we should recall that v and R in Fig. 5.24 could represent anything from the Thévenin equivalent circuit for a very complicated network to a single source v and its resistance R.

EXAMPLE 5.19

Let us determine the value of R_L for maximum power transfer in the network in Fig. 5.25a. After finding this value of R_L, we will find the maximum power that can be transferred to this load.

If we break the network at the load, we obtain the circuit shown in Fig. 5.25b. At this point it is important to note that we simply have another network, and therefore all the techniques that we have learned can be applied to solve it. Note that we can form a supernode around the dependent source. The voltage with respect to the reference node at the positive terminal of the dependent source is V_{oc}, and therefore the voltage with respect to the reference node at the negative terminal of the dependent source is $V_{oc} - 2I$. KCL applied to the supernode is

$$\frac{V_{oc} - 2I}{4} - 12 + \frac{V_{oc}}{2} = 0$$

where

$$I = \frac{V_{oc}}{2}$$

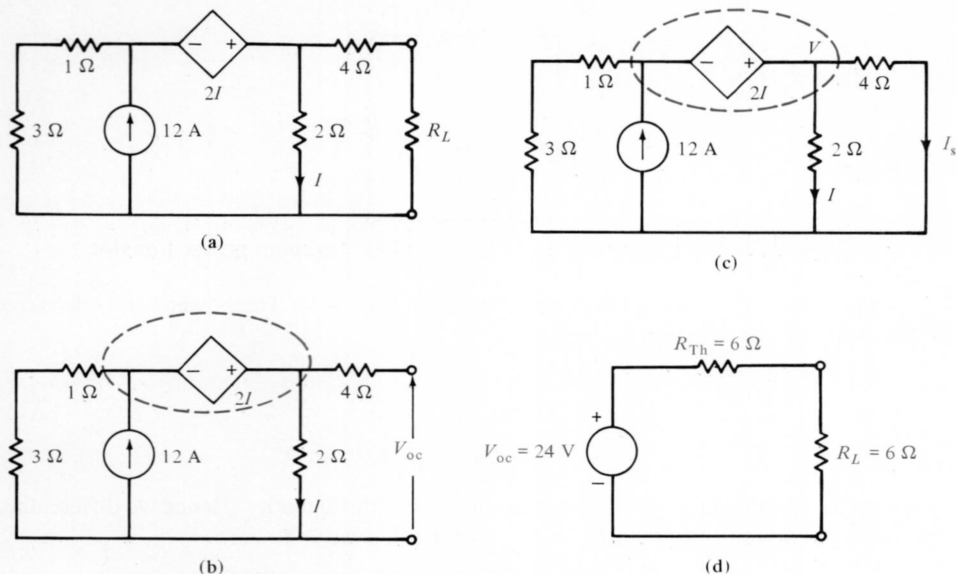

Figure 5.25 Example circuit for determining maximum power transfer.

Solving these equations yields

$$V_{oc} = 24 \text{ V}$$

I_{sc} can be derived in a similar manner using the network in Fig. 5.25c. KCL for the supernode is

$$\frac{V - 2I}{4} - 12 + \frac{V}{2} + \frac{V}{4} = 0$$

where

$$I = \frac{V}{2}$$

Therefore,

$$V = 16 \text{ V}$$

and hence

$$I_{sc} = \frac{V}{4} = 4 \text{ A}$$

The Thévenin equivalent resistance is then

$$R_{Th} = \frac{V_{oc}}{I_{sc}} = \frac{24}{4} = 6\ \Omega$$

The Thévenin equivalent network is shown in Fig. 5.25d, and hence the maximum power transferred to the load is

$$P_L = \left[\frac{24}{6 + 6}\right]^2 (6) = 24 \text{ W}$$

■

DRILL EXERCISE

D5.17. In the network in Fig. D5.17, find R_L for maximum power transfer and the maximum power absorbed by the load.

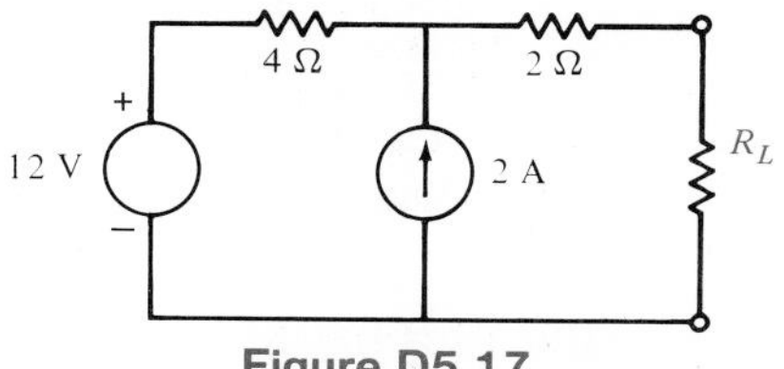

Figure D5.17

D5.18. In the network in Fig. D5.18, find R_L for maximum power transfer and the maximum power absorbed by the load.

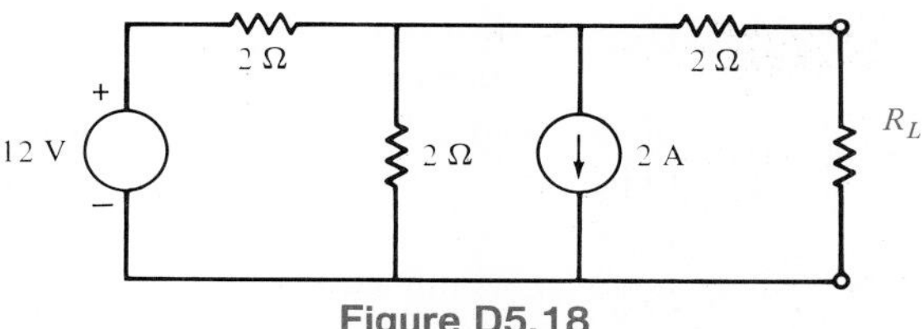

Figure D5.18

5.6 DC Circuit Analysis Using SPICE

SPICE was developed in the Department of Electrical Engineering and Computer Science at the University of California at Berkeley. This program applies the computational power of the digital computer to the analysis of complicated circuits.

Although we demonstrate its use here with simple examples that the reader can easily check using techniques presented in this book, the real power of the program lies in the ease with which it can be applied to very complicated circuits that would appear intractable if they had to be solved by hand calculations.

Unfortunately, SPICE is no panacea or substitute for a thorough understanding of circuit analysis, and therefore we cannot just turn the problem over to the computer. It is perhaps this "let the computer figure it out" philosophy that resulted in the acronym GIGO, which stands for "garbage in, garbage out."

SPICE is capable of performing dc, ac, or transient analysis. In the dc analysis case, a set of simultaneous linear or nonlinear equations with real coefficients is solved; in the transient analysis case, the solution is obtained by solving a set of simultaneous nonlinear integrodifferential equations.

In general, the use of SPICE in the analysis of an electric circuit requires the following steps.

1. Select the type of analysis to be performed: dc, ac, or transient analysis.
2. Draw the circuit in SPICE format using only the building blocks that are allowed in the type of analysis desired.
3. Write the SPICE statements describing the circuit.
4. Write the SPICE statements that tell the program how to analyze the circuit and describe the type of solution desired.

The SPICE statements may be subdivided into the following five categories:

1. Title and comment statements.
2. Data statements.
3. Solution control statements.
4. Output specification statements.
5. End statement.

Title and Comment Statements

The *title statement* is the first statement in any SPICE program and typically contains a title descriptive enough to identify the particular problem under investigation. The computer will print this description as a heading for the output. The title statement *must* be present, since SPICE always uses the first line as a title.

The *comment statement* is basically an aid to the programmer and appears as a part of the output listing. It is characterized by an asterisk in the first column, with the remaining columns containing information describing what the program is doing at that point. These statements are extremely useful when referring back to a problem that has not been examined in some time.

Data Statements

Describing a circuit to someone verbally, without the use of a circuit diagram, would appear to be a difficult task at best and an insurmountable one if the circuit is very complicated. Hence, in describing the circuit to the computer, a systematic approach must be employed that clearly itemizes the exact manner in which the circuit elements are interconnected.

The fundamental building block in a SPICE analysis is a circuit branch, and the standard branch in SPICE is shown in Fig. 5.26. Note that a double subscript is employed

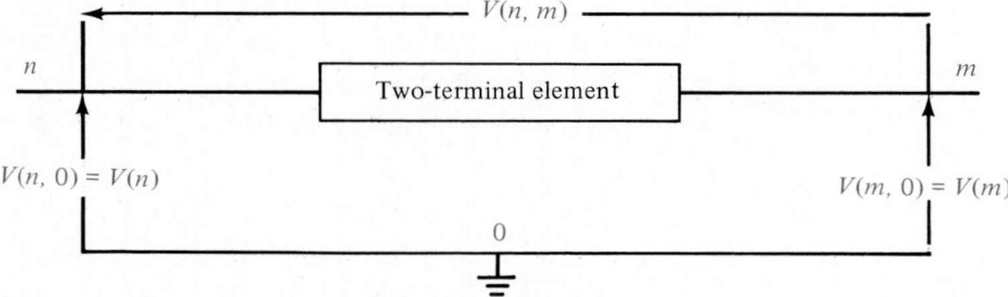

Figure 5.26 Standard SPICE circuit branch.

to describe the branch voltages. The first subscript is assumed to be the positive node for the branch voltage. The node voltages are denoted by $V(n, 0)$ or, for simplicity, $V(n)$. The branch current is referenced so that a positive current in a resistor produces a positive branch voltage, and therefore the reference direction for the current is from the first listed subscript on the branch voltage to the second.

Each branch may contain any two-terminal element, such as a resistor, current source, or voltage source. The values of the current or voltage sources may be zero, but the resistor must be nonzero.

The branches are connected together at nodes which are identified to SPICE with numbers. The numbers of the nodes must be nonnegative but need not be sequential. The reference ground node must be numbered zero. The circuit cannot contain a loop of voltage sources or a cut set of current sources. Each node in the circuit must have a dc path to ground, and every node must have at least two connections.

One of the main problems encountered in describing a circuit and the desired solutions is keeping track of the reference directions of all voltages and currents. All reference directions in SPICE are fixed by the programmer and specified by the order in which the node connections are listed in the data statements.

A free format is used on all data statements. The fields must be separated by one or more blanks, an equal sign, or an opening or closing parenthesis. A statement may be continued by entering a plus sign in column 1 of the continuation statement, and SPICE will continue reading beginning with column 2.

Each data segment is composed of three fields:

1. The element name.
2. The circuit nodes to which the element is connected.
3. The value of the parameters that determine the electrical characteristic of the element.

The element name field must begin with a letter of the alphabet and cannot contain any delimiters. The first letter of the name specifies the element type:

R	resistor
V	independent voltage source
I	independent current source
G	voltage-controlled current source
E	voltage-controlled voltage source
F	current-controlled current source
H	current-controlled voltage source

A name can contain from one to eight characters. For example, a resistor name must begin with the letter R and can be followed by one to seven additional characters. Hence R, R1348, RINPUT, ROUT3, and RAIB66DG are all valid resistor names. In the discussion that follows, XXXXXXX, YYYYYYY, and ZZZZZZZ will denote arbitrary alphanumeric strings.

The number field may be an integer field such as 12 or -44, a floating-point field such as 3.142 or 1.4146, an integer or floating-point followed by an exponent such as 1E-14 or 2.65E3, or an integer or floating-point followed by one of the following scale factors:

T = 1E12	G = 1E9	MEG = 1E6
K = 1E3	M = 1E-3	U = 1E-6
N = 1E-9	P = 1E-12	F = 1E-15

Letters immediately following a number that are not scale factors are ignored, and letters immediately following a scale factor are ignored. Hence 10, 10V, 10VOLTS, and 10HZ all represent the same scale factor. Note also that 1000, 1000.0, 1000HZ, 1E3, 1.OE3, 1KHZ, and 1K all represent the same number. Note, however, that:

1. Commas are not allowed.
2. Missing signs are assumed + by the computer.
3. A missing decimal point is placed at the end of the series of numbers.

The following is a detailed description of the format of the data statements that are used to describe each type of branch in a network.

Branch Statements for Resistive Elements. The general form for resistive elements is

```
RXXXXXXX N1 N2 VALUE
```

Here XXXXXXX denotes an arbitrary alphanumeric string that uniquely identifies the particular element. N1 and N2 are the circuit nodes to which the element is connected. VALUE is the value of the resistance in ohms. The value of a resistor can be positive or negative but cannot be zero.

Branch Statements for Independent Sources. Branch statements for independent sources have the form

```
BXXXXXXX N+ N- DC (DC VALUE)
```

B is the letter V for voltage sources or I for current sources. N+ and N− are the positive and negative nodes, respectively. Positive current, for either voltage or current sources, is assumed to flow from the positive node, through the source, and into the negative node. DC VALUE is the dc value of the source. If the dc value is zero, this value may be omitted.

Branch Statements for Linear Dependent Sources. A linear dependent source in SPICE is defined as a voltage source or current source whose voltage or current is a constant times either a specified branch voltage or the current flowing through a voltage source. The latter point introduces an interesting idiosyncracy of SPICE. In order to employ the current in a particular branch, we must introduce a dummy voltage source in series with this particular branch so that the program can calculate the current in this voltage source.

Branch statements for linear voltage-controlled sources have the general form

```
BXXXXXXX N+ N- NC+ NC- VALUE
```

B is either the letter G or E, where I = GV for voltage-controlled current sources and

V = EV for voltage-controlled voltage sources. N+ and N− are the positive and negative nodes, respectively. Current flow is from the positive node, through the source, and into the negative node. NC+ and NC− are the positive and negative controlling nodes, respectively. VALUE is either the value of the voltage gain E or the transconductance G.

Branch statements for linear current-controlled sources have the form

```
BXXXXXXX N+ N- VNAME VALUE
```

B is either the letter F or H, where I = FI for current-controlled current sources and V = HI for current-controlled voltage sources. N+ and N− are the positive and negative nodes, respectively. Current flow is from the positive node, through the source, to the negative node. VNAME is the name of the voltage source through which the controlling current flows. Positive controlling current flows into the positive node, through the source VNAME, and out of the negative node. VALUE is either the value of current gain F or the transresistance H.

Solution Control Statements

Three of the dc analysis options are specified by the statements .OP, .DC, and .TF. The .OP statement will cause all the dc node voltages and currents in voltage sources to be listed as part of the output.

The dc analysis may be performed for a range of source voltages or currents using the .DC statement. The general form of this statement is

```
.DC SCRNAM VSTART VSTOP VINCR
```

SCRNAM is the name of an independent voltage or current source to be varied. VSTART, VSTOP, and VINCR are the starting, final, and incremental values, respectively. A second source may optionally be specified with associated sweep parameters. The form for this statement is

```
.DC SCRNAM VSTART VSTOP VINCR SCR2 START2 STOP2
INCR2
```

In this case, the first source will be swept over its range for each value of the second source. Some typical examples of the .DC statement are

```
.DC V1 1.25 10.50 0.25
.DC V0 0 50 1 I1 1M 10M 1M
```

The dc analysis option may also be used to evaluate dc small-signal transfer functions, that is, the ratio of an output variable to an input variable. Input resistance and output resistance will also be computed automatically as part of the solution:

```
.TF OUTVAR INSRC
```

OUTVAR is the small-signal output variable. INSRC is the small-signal input source. Two typical example statements are

```
.TF V(4, 1) VIN
.TF I(VOUT) VIN
```

Recall that the voltage between any two nodes n and m is referred to as $V(n, m)$. If the second node is zero, it may be omitted. For the equations above, SPICE would compute the dc small-signal value of the ratio of $V(4, 1)$ to VIN, the input resistance at VIN, and the output resistance at VOUT. If no analysis is specified, SPICE will perform an .OP analysis by default.

Output Specification Statements

Printed output is requested using the .PRINT statement with the format

```
.PRINT DC OV1 OV2 . . . OV8
```

OV1, OV2, . . . , OV8 are the current or voltage output variables desired with a maximum of eight output variables per statement.

The branch voltage between nodes N1 and N2 is specified by V(N1, N2). If N2 and the comma are omitted, ground (0) is assumed. So the voltage at node N1 may be requested simply as V(N1).

The current flowing in the independent voltage source VXXXXXXX is specified by I(VXXXXXXX). Positive current flows from the positive node, through the source, to the negative node, as shown in Fig. 5.27. A typical PRINT statement is

```
.PRINT DC V(5) V(24) I(VCC) V(22, 17)
```

Note that when the .OP option is employed, no .PRINT statement is required, since .OP causes all node voltages and currents through voltage sources to be printed automatically.

Plotted output is obtained using the .PLOT statement. One PLOT statement is required for each graph to be plotted with a maximum of eight output variables per statement.

```
.PLOT DC OV1(PLO1, PHI1) OV2(PLO2, PHI2) . . .
OV8(PLO8, PHI8)
```

The plot limits (PLO, PHI) are optional and may be specified after any of the output variables. All output variables to the left of a pair of plot limits will be plotted using the same upper and lower bounds, PHI and PLO, respectively. If plot limits are not specified, SPICE will automatically determine the plot limits so that all variables may be plotted on the same graph. When more than one output variable appears on the same plot, the first

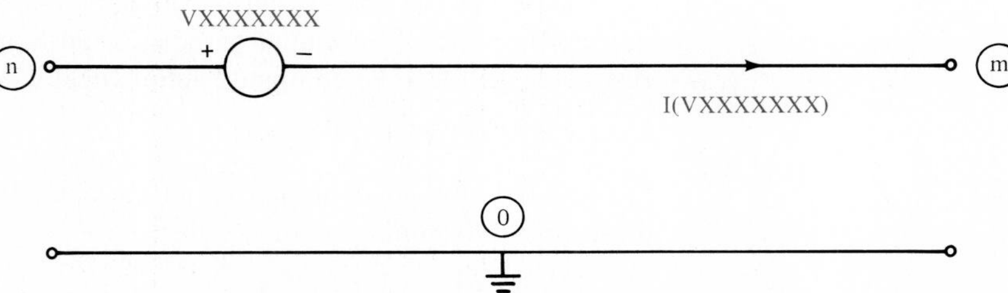

Figure 5.27 Current referenced to an independent voltage source.

variable specified will be printed as well as plotted. If printed values of the other variables are desired, PRINT statements should be included in the SPICE program. There is no limit on the number of plot statements that may be used. An example .PLOT statement is

```
.PLOT DC V(4) V(23, 2) I(VCC) (0, 20)
```

The width of the output can be controlled by using the .WIDTH OUT = XX statement, where XX is the number of columns desired. In many cases where the output is to a CRT and not a line printer, .WIDTH OUT = 80 is used so that the output will not "wrap around" on the CRT. If the .WIDTH statement is not used, SPICE defaults to a width of 132 columns.

End Statement

The programmer must tell SPICE when the end of the input data has been reached. This is done by placing a statement at the end of the program with the format .END.

The following examples will serve to illustrate the techniques we have described.

EXAMPLE 5.20

Let us determine the node voltages in the circuit shown in Fig. 5.28. The SPICE program for solving this circuit follows.

```
DC ANALYSIS OF A SIMPLE CIRCUIT

V1  1  0  DC  36
R1  1  2  3
R2  2  0  6
R3  2  3  8
R4  3  0  2

*PRINT AS OUTPUT NODE VOLTAGES V1, V2 AND V3 =
*VOUT
.PRINT DC V(2) V(3)
.END
```

The output listing from the computer includes the following:

```
NODE  VOLTAGE   NODE  VOLTAGE   NODE  VOLTAGE
(1)   36.0000   (2)   20.0000   (3)   4.0000
```

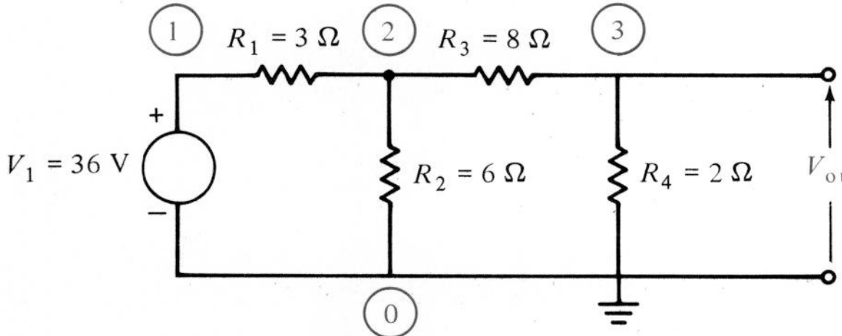

Figure 5.28 Example 5.20 dc circuit.

Output voltage is always listed in volts. Note that the .OP analysis is performed by default, since no analysis was specified in the program, and therefore all node voltages are printed automatically. ■

EXAMPLE 5.21

Let us determine the node voltages and the current in the 1-Ω resistor in the circuit shown in Fig. 2.25a. The circuit is redrawn in Fig. 5.29 and labeled to facilitate the SPICE analysis.

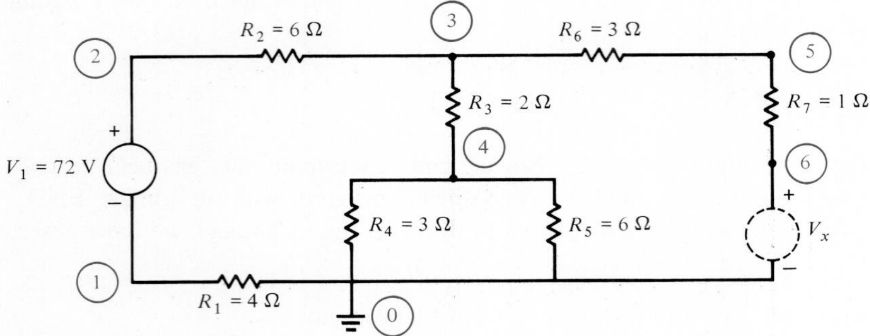

Figure 5.29 Circuit in Fig. 2.25a redrawn for SPICE analysis.

Note that since the current in R7 is to be determined directly, the dummy voltage source VX has been introduced. Therefore, the current in this source can be used in the analysis and, in this case, printed as part of the output.

The SPICE program for this circuit follows.

```
CALCULATE THE NODE VOLTAGES AND THE CURRENT IN
*R7.
R1   1   0   4
V1   2   1   DC   72
R2   2   3   6
R3   3   4   2
R4   4   0   3
R5   4   0   6
R6   3   5   3
R7   5   6   1
VX   6   0   DC
.OP
.END
```

The output listing from the computer includes the following:

```
NODE   VOLTAGE    NODE   VOLTAGE    NODE   VOLTAGE
(1)    -24.0000   (2)    48.0000    (3)    12.0000
(4)    6.0000     (5)    3.0000     (6)    0.0000
```

```
VOLTAGE SOURCE CURRENTS
NAME    CURRENT
V1      -6.000D+00
VX       3.000D+00
```

■

DRILL EXERCISE

D5.19. Using SPICE, determine all node voltages and the currents I_x and I_o in the network in Fig. D5.19.

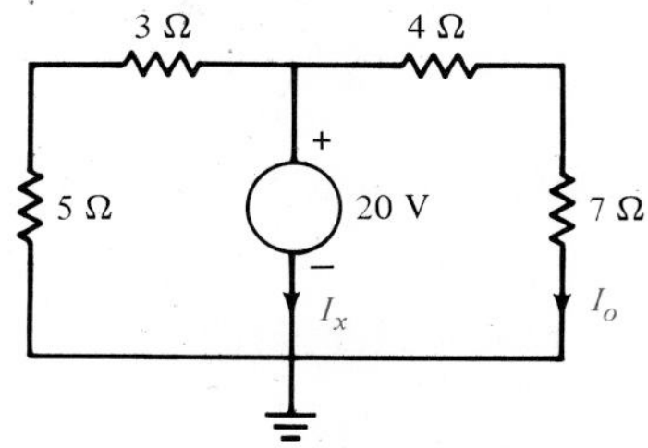

Figure D5.19

D5.20. Using SPICE, determine all node voltages and the currents I_x, I_y, and I_o in the network in Fig. D5.20.

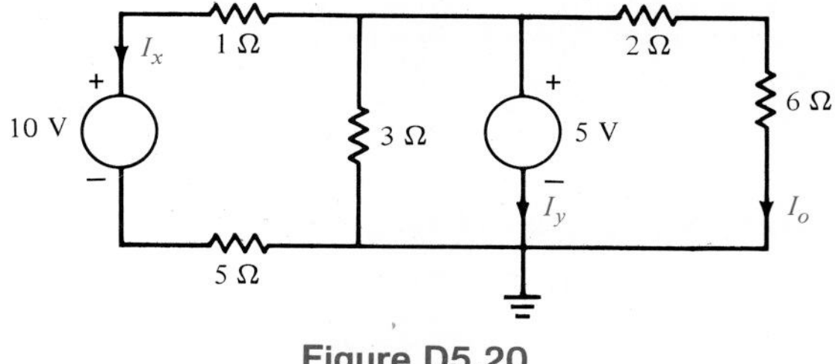

Figure D5.20

EXAMPLE 5.22

We wish to determine the node voltages for the circuit in Fig. 3.8 using the values in Example 3.5. The SPICE program for the circuit follows.

```
CALCULATE THE NODE VOLTAGES

VB    4    0    DC    10
R3    1    4    2
R1    1    2    1
IA    2    1    DC    4
G1    2    0    2     3    3
R2    2    3    0.5
R4    3    0    1
IB    0    3    DC    5
.DC   VB   10   10    1
.PRINT  DC  V(1)  V(2)  V(3)  V(4)  I(VB)
.END
```

The output listing from the computer includes the following:

```
NODE VOLTAGE NODE VOLTAGE NODE VOLTAGE NODE VOLTAGE
(1)  9.4444  (2)  5.1667  (3)  5.1111  (4)  10.0000

   VOLTAGE SOURCE CURRENTS
   NAME    CURRENT
   VB      -2.778D-01
```

■

EXAMPLE 5.23

The circuit in Fig. 5.30 was solved using other techniques in Example 5.16. Let us now write a SPICE program to determine the node voltages in this network. The program is as follows:

```
CALCULATE THE NODE VOLTAGES
V1    1    0    DC    36
E1    3    2    2     1     2
R1    1    2    2
I1    0    2    DC    6
R2    3    0    2
.DC   V1   36   36    1
*PRINT AS OUTPUT NODE VOLTAGES WHERE V3 = VOUT
.PRINT DC V(1)  V(2)  V(3)
.END
```

The node voltages listed by the computer are

```
NODE   VOLTAGE    NODE   VOLTAGE    NODE   VOLTAGE
(1)    36.0000    (2)    30.0000    (3)    18.0000
```

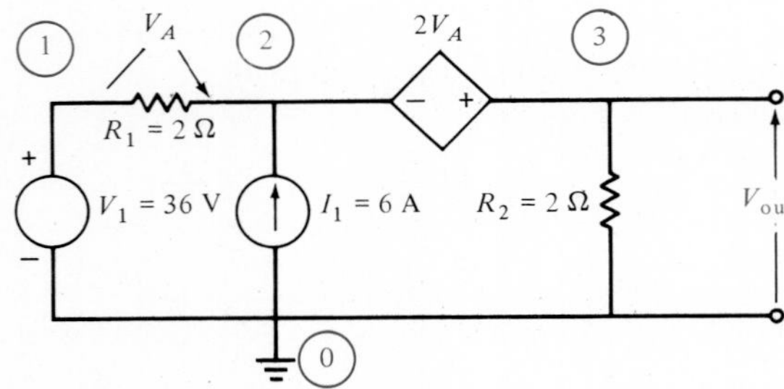

Figure 5.30 Circuit in Fig. 5.21a redrawn for SPICE analysis. ■

EXAMPLE 5.24

The network in Fig. P3.15 will be solved using SPICE. The circuit is redrawn in Fig. 5.31. Once again, note that a dummy voltage source is introduced because the current in R2 controls the dependent source. The SPICE program for the network follows.

```
CALCULATE THE NODE VOLTAGES
I1     0    1    DC     2
R2     1    3    1
VX     3    0    DC
F1     1    2    VX     2
R1     1    2    2
R3     2    0    4
R4     2    0    2
.DC    I1   2    2    1
.PRINT  DC  V(1)   V(2)   V(3)
.END
```

The node voltages computed by the machine are

```
NODE  VOLTAGE    NODE  VOLTAGE    NODE  VOLTAGE
(1)   0.8000     (2)   1.6000     (3)   0.0000
```

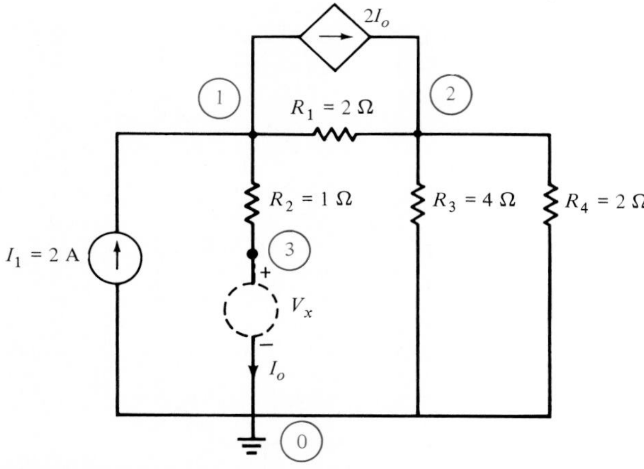

Figure 5.31 Circuit in Fig. P3.15 redrawn for SPICE analysis. ■

EXAMPLE 5.25

Let us determine the node voltages and the current in the 10-Ω resistor in the circuit in Fig. 3.9 using the values in Example 3.6. To aid a SPICE analysis the circuit is redrawn as shown in Fig. 5.32. The SPICE program for the network follows.

```
CALCULATE THE NODE VOLTAGES

IA     0    1    DC     10
R1     1    3    0.5
R2     1    2    1
H1     2    0    VX     10
VX     4    0    DC
R3     2    3    2
R4     3    4    10
IB     3    0    DC     2
.END
```

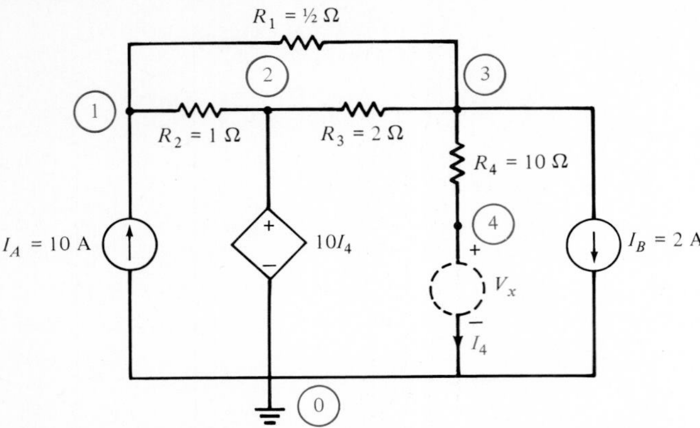

Figure 5.32 Circuit in Fig. 3.9 redrawn for SPICE analysis.

The computer output lists the following:

```
NODE VOLTAGE NODE VOLTAGE NODE VOLTAGE NODE VOLTAGE
(1)  50.0000 (2)   46.6667 (3)   46.6667 (4)    0.0000
   VOLTAGE SOURCE CURRENTS
   NAME      CURRENT
   VX        4.667D+00
```

Once again no analysis was specified, and therefore .OP is run by default. ■

DRILL EXERCISE

D5.21. Using SPICE, compute all the node voltages, the current I_x, and the current in the 10-V source in the network in Fig. D5.21.

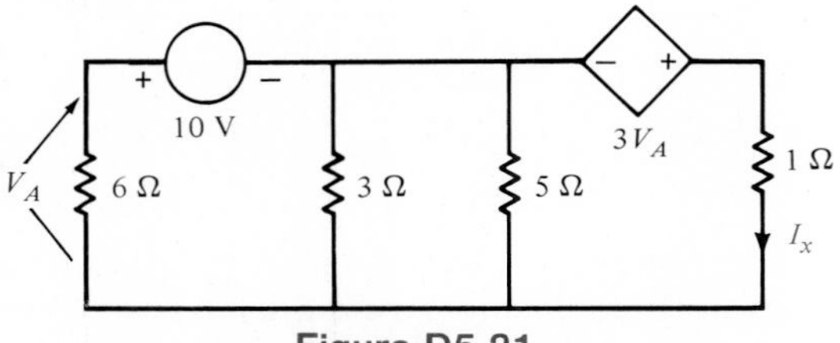

Figure D5.21

D5.22. Using SPICE, compute all the node voltages, the currents I_x and I_a, and the current in the 10-V source in the network in Fig. D5.22.

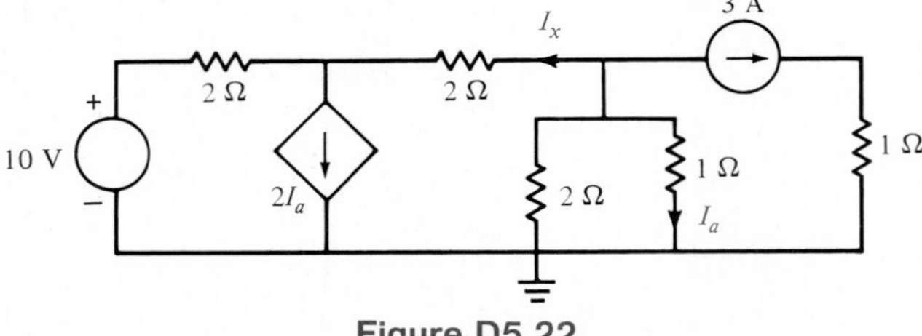

Figure D5.22

5.7 Summary

In this chapter we have presented a number of very powerful tools that have wide applications in circuit analysis. Specifically, we have introduced the principle of superposition, which allows us to treat each source independently and then algebraically add the response due to each to determine the total response. We have also shown that source transformation and Thévenin's and Norton's theorems quite often allow us to simplify circuit analysis by solving a series of simple problems rather than a more complicated one. The concept of maximum power transfer from a network to a resistive load was presented. Finally, we have demonstrated how to apply SPICE to dc circuit problems.

KEY POINTS

- In a linear network containing multiple independent sources, the principle of superposition allows us to compute any current or voltage in the network as the algebraic sum of the individual contributions of each source acting alone.
- Superposition is a linear property and does not apply to nonlinear functions such as power.
- Source transformation permits us to replace a voltage source V in series with a resistance R by a current source $I = V/R$ in parallel with the resistance R. The reverse is also true.
- Using Thévenin's theorem, we can replace some portion of a network at a pair of terminals with a voltage source V_{oc} in series with a resistor R_{Th}. V_{oc} is the open-circuit voltage at the terminals, and R_{Th} is the Thévenin equivalent resistance obtained by looking into the terminals with all independent sources made zero.
- Using Norton's theorem, we can replace some portion of a network at a pair of terminals with a current source I_{sc} in parallel with a resistor R_{Th}. I_{sc} is the short-circuit current at the terminals and R_{Th} is the Thévenin equivalent resistance.
- In a dc SPICE analysis a set of simultaneous linear equations with real coefficients is solved.
- The SPICE program consists of the following five categories of statements:
  1. Title and comment statements.
  2. Data statements.
  3. Solution control statements.
  4. Output specification statements.
  5. End statement.

PROBLEMS

5.1. Given the circuit in Fig. P5.1, use linearity and assume that $I_o = 1$ A to determine the actual value of I_o.

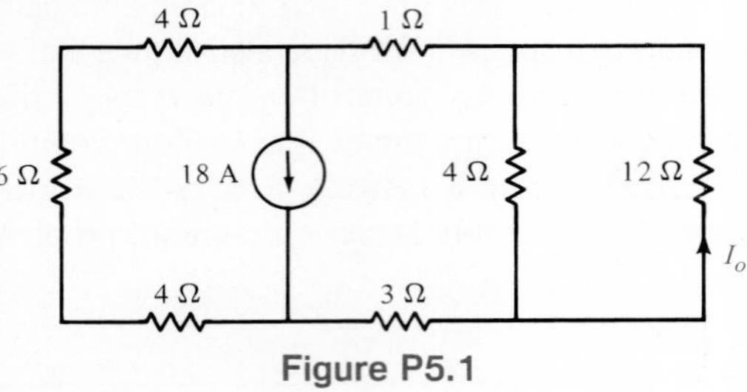

Figure P5.1

5.2. Determine I_o in the network in Fig. P5.2. Use linearity and assume that $I_o = 1$ A.

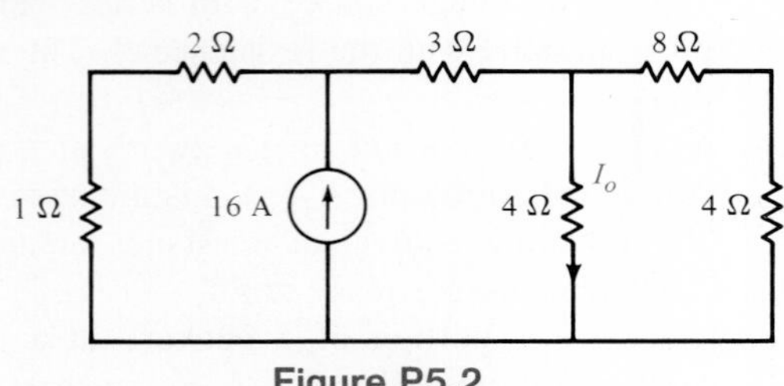

Figure P5.2

5.3. Use superposition to find the output voltage in Fig. P5.3.

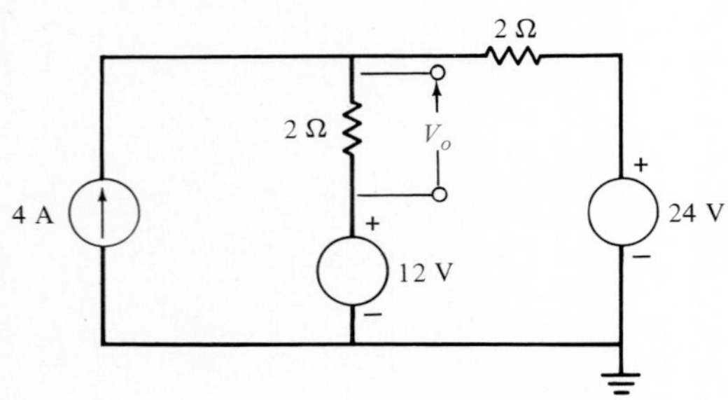

Figure P5.3

5.4. Determine V_o in the network shown in Fig. P5.4 using superposition.

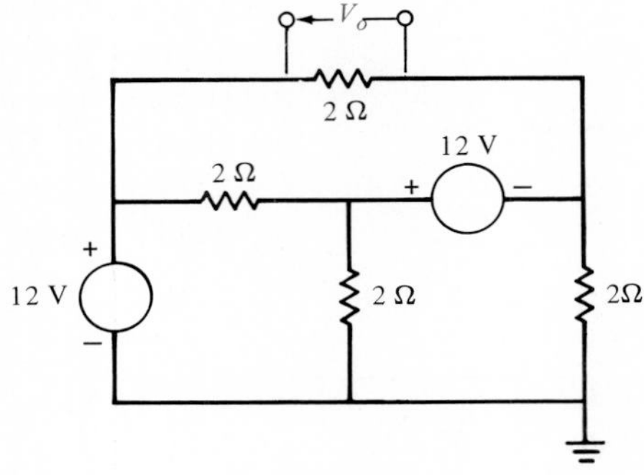

Figure P5.4

5.5. Use superposition to find I_o in Fig. P5.5.

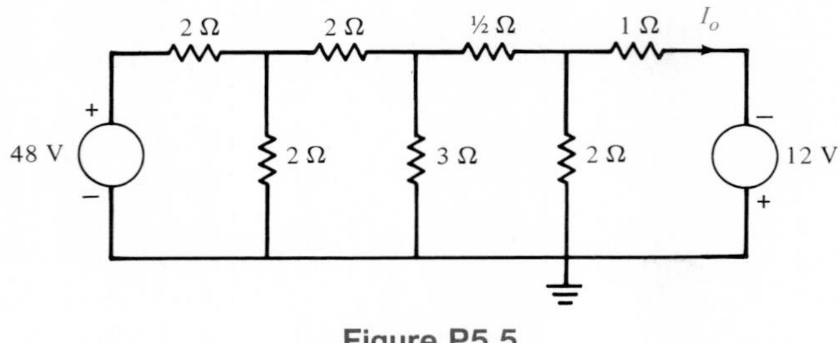

Figure P5.5

5.6. Determine I_o in the circuit shown in Fig. P5.6 using superposition.

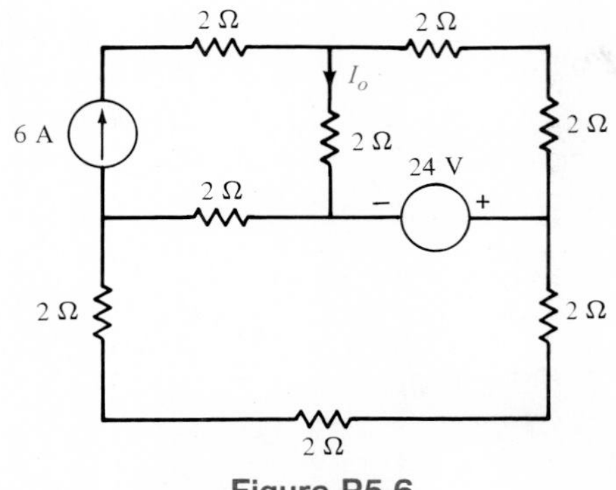

Figure P5.6

5.7. Use superposition to find V_o in Fig. P5.7.

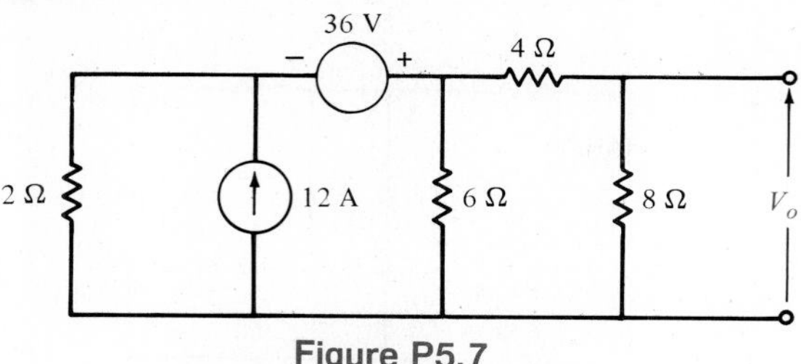

Figure P5.7

5.8. Find the voltage across the 4-Ω resistor by using superposition in the circuit in Fig. P5.8.

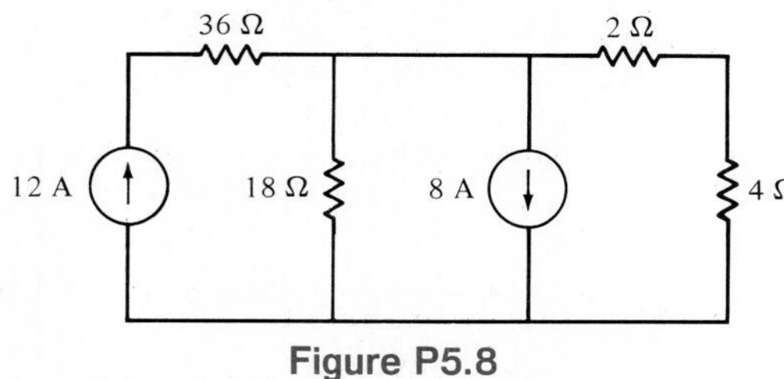

Figure P5.8

5.9. Use superposition to find I_o in the network in Fig. P5.9.

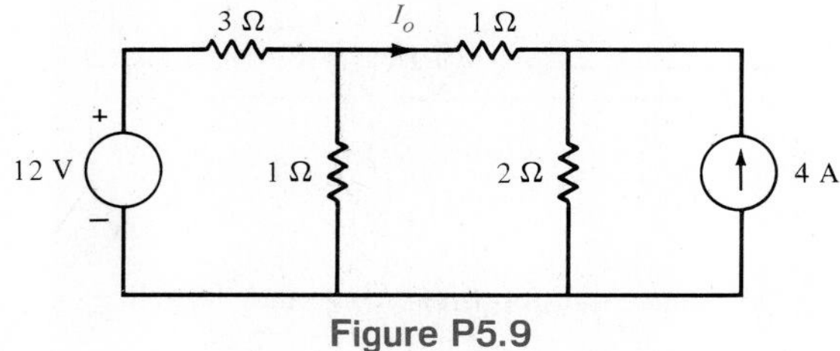

Figure P5.9

5.10. Use superposition to find I_o in the circuit in Fig. P5.10.

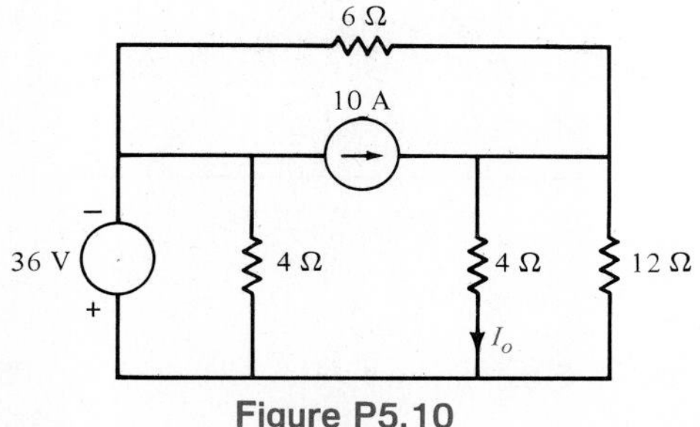

Figure P5.10

5.11. Use superposition to find I_o in the circuit in Fig. P5.11.

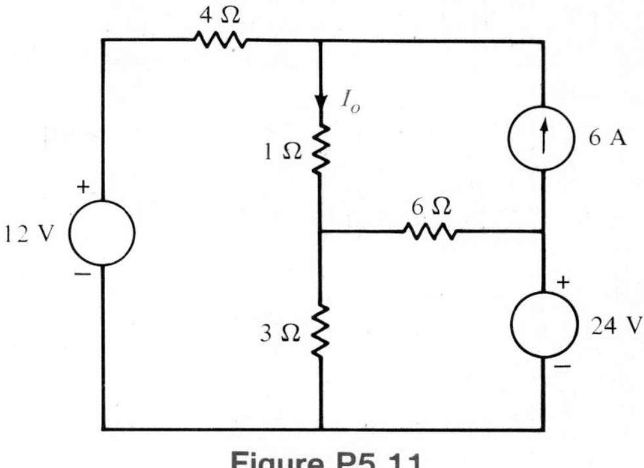

Figure P5.11

5.12. Use superposition to find I_o in the network in Fig. P5.12.

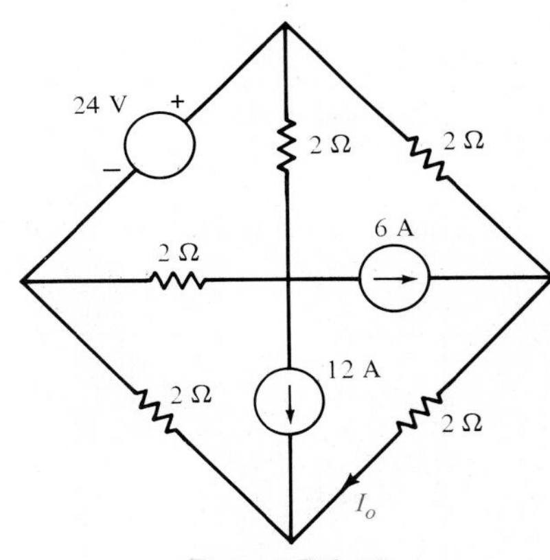

Figure P5.12

5.13. Use superposition to find I_o in the network in Fig. P5.13.

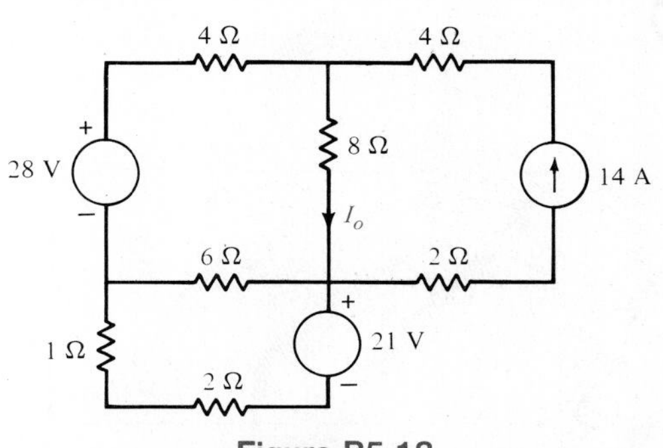

Figure P5.13

5.14. Use superposition to find I_o in the circuit in Fig. P5.14.

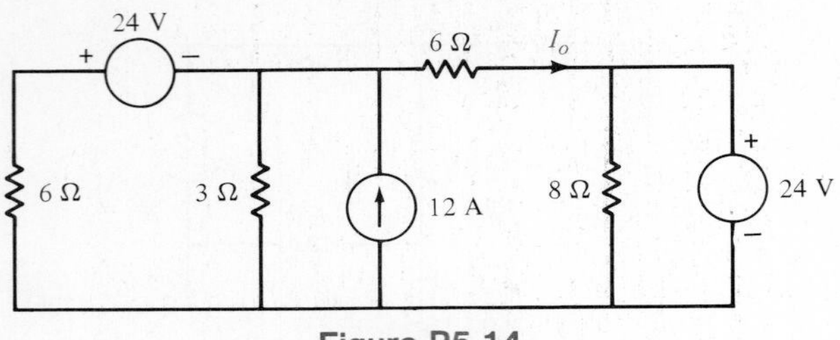

Figure P5.14

5.15. Use source exchange to find I_o in the circuit in Fig. P5.15.

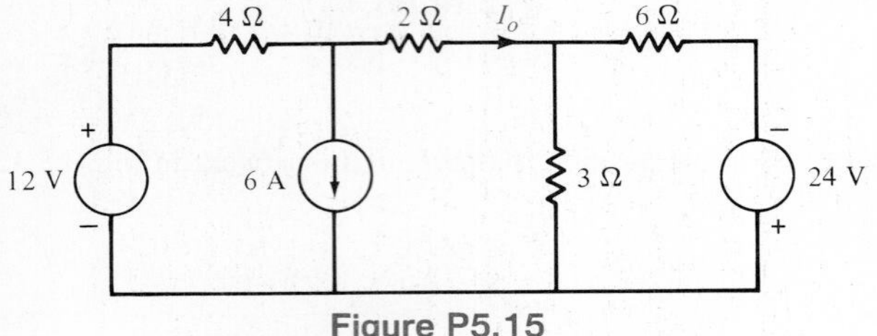

Figure P5.15

5.16. Use source transformation to find I_o in the circuit shown in Fig. P5.16.

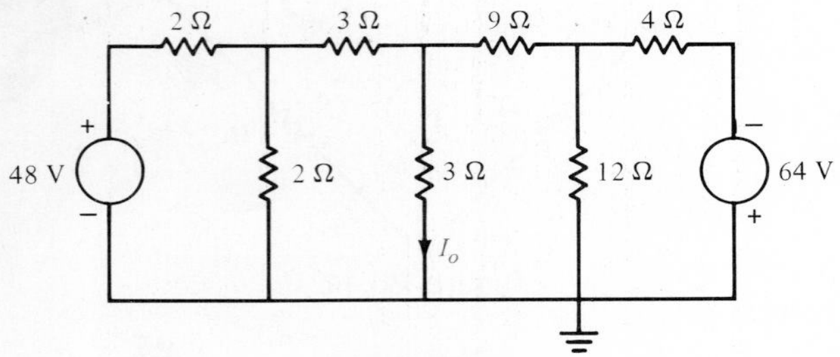

Figure P5.16

5.17. Use source transformation to find I_o in the network in Fig. P5.17.

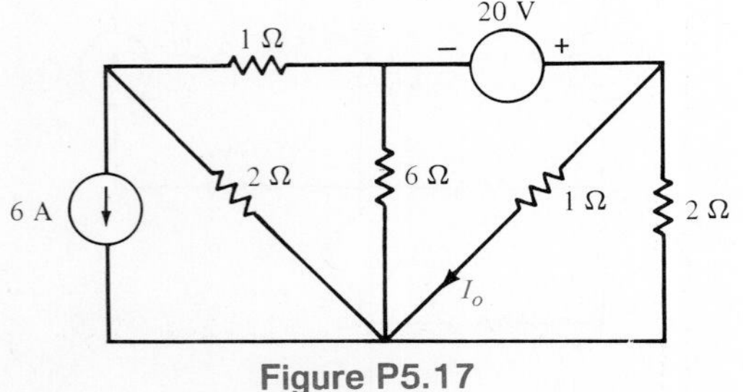

Figure P5.17

5.18. Use source transformation to find I_o in the network in Fig. P5.18.

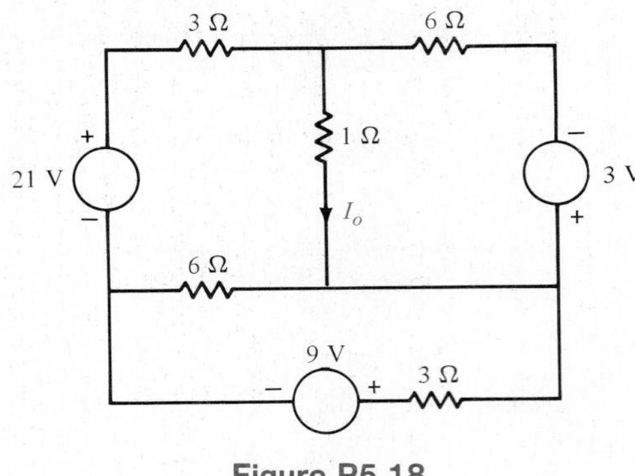

Figure P5.18

5.19. Use source transformation to find I_o in the circuit in Fig. P5.19.

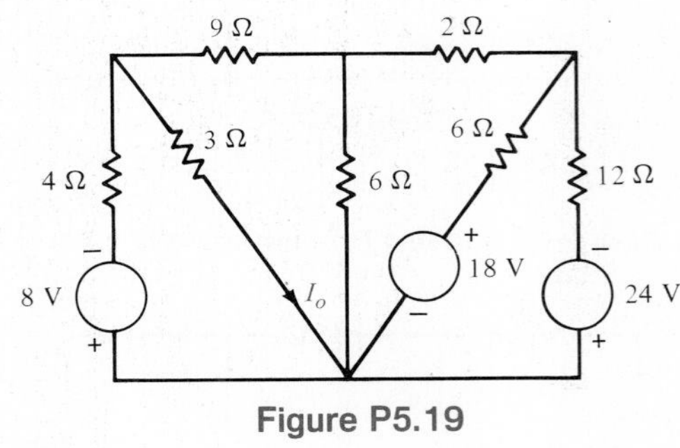

Figure P5.19

5.20. Using source transformation, determine the voltage V_o across the 4-Ω resistor in the circuit in Fig. P5.20.

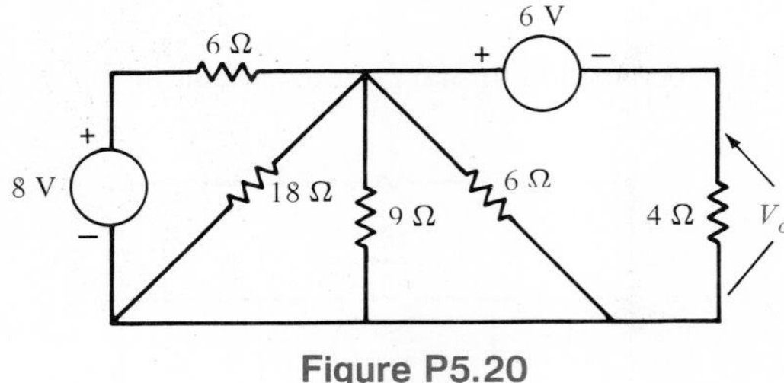

Figure P5.20

5.21. Determine the current I_o in the circuit in Fig. P5.21 using superposition.

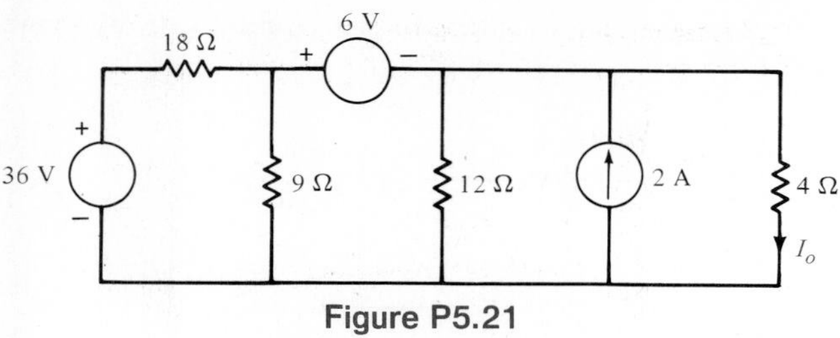

Figure P5.21

5.22. Use source transformation to determine I_o in the network in Fig. P5.21.

5.23. Find I_o in Fig. P5.23 using source transformation.

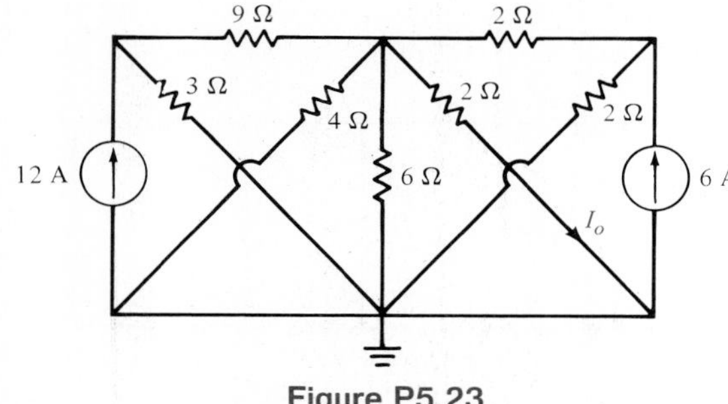

Figure P5.23

5.24. Use source transformation to find V_o in Problem 5.3.

5.25. Use Thévenin's theorem to solve Problem 5.3.

5.26. Solve Problem 5.4 using Thévenin's theorem.

5.27. Use repeated application of Thévenin's theorem to find I_o in Problem 5.5.

5.28. Use Thévenin's theorem to find V_o in the circuit shown in Fig. P5.28.

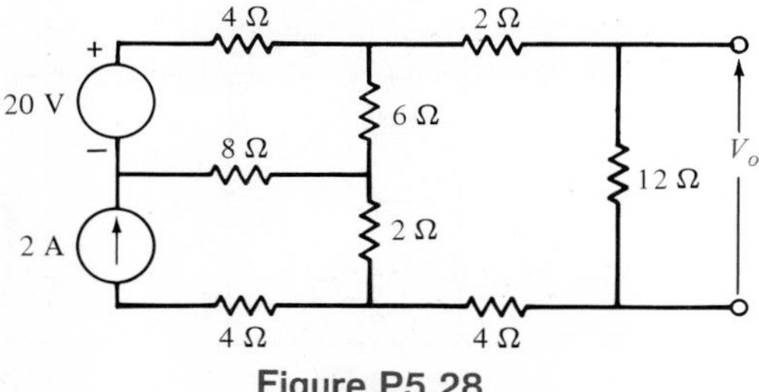

Figure P5.28

5.29. Find I_o in the circuit shown in Fig. P5.29 using Norton's theorem.

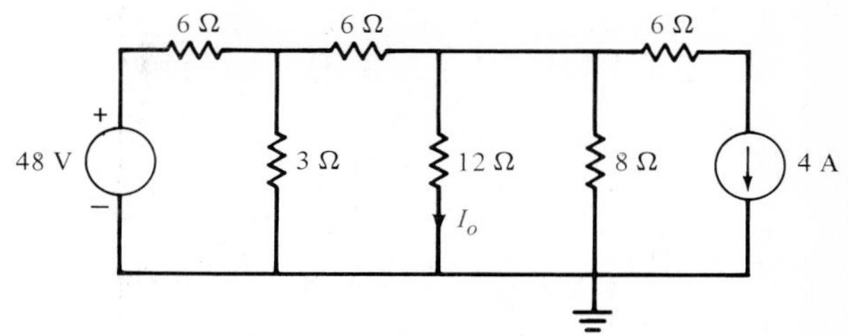

Figure P5.29

5.30. Use Thévenin's theorem to find the voltage across the 8-Ω load in Fig. P5.30.

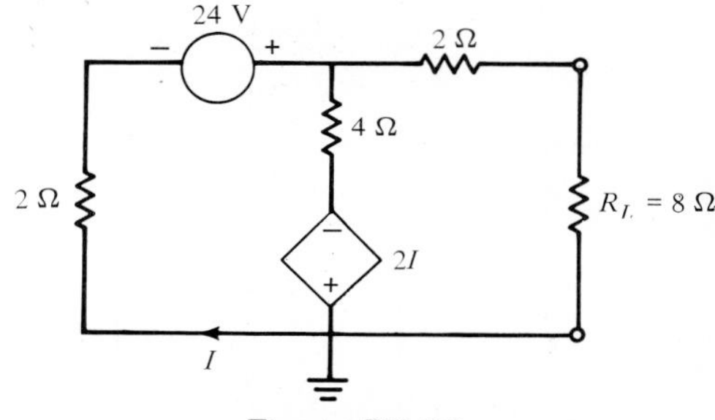

Figure P5.30

5.31. Use Norton's theorem to find I_o in the network in Fig. P5.31.

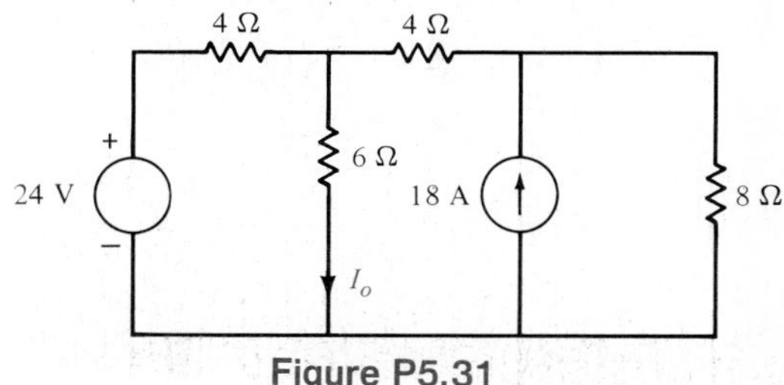

Figure P5.31

5.32. Use Norton's theorem to find I_o in the network in Fig. P5.32.

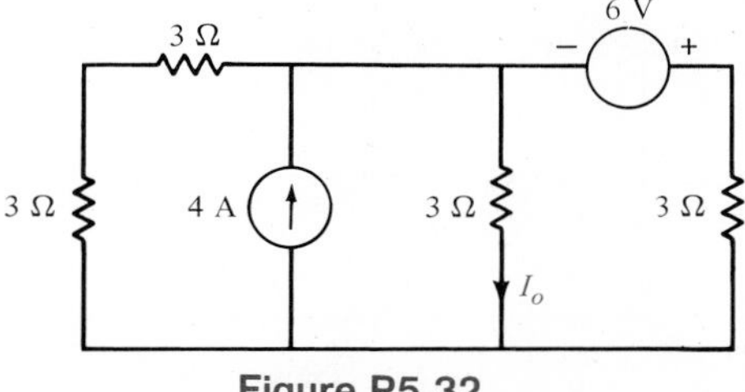

Figure P5.32

5.33. Use Thévenin's theorem to find V_o in the network shown in Fig. P5.33.

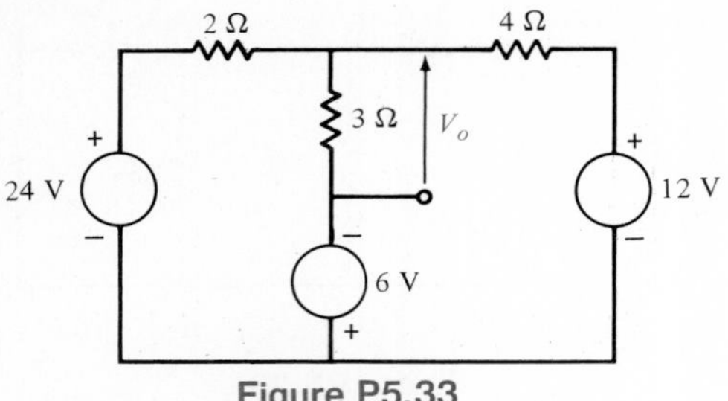

Figure P5.33

5.34. Find V_o in the network in Fig. P5.34 using Thévenin's theorem.

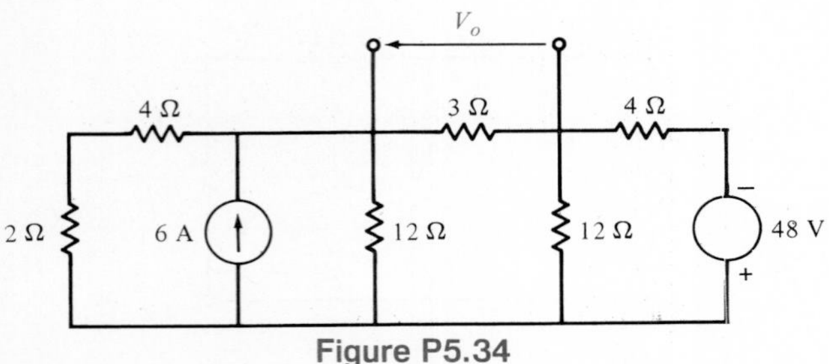

Figure P5.34

5.35. Apply Thévenin's theorem twice to determine V_o in the circuit in Fig. P5.35.

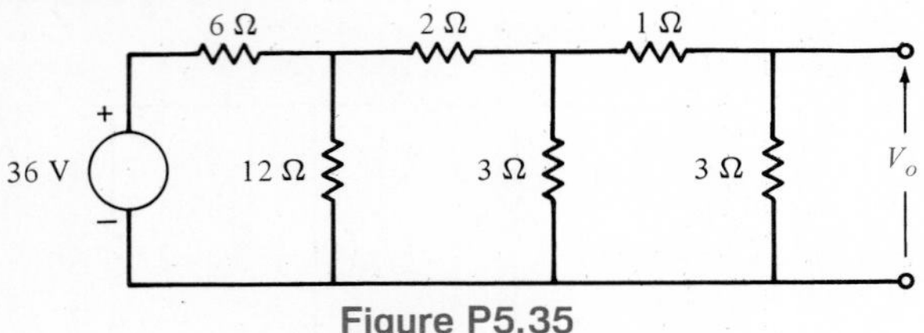

Figure P5.35

5.36. Use Thévenin's theorem to find V_o in the network in Fig. P5.36.

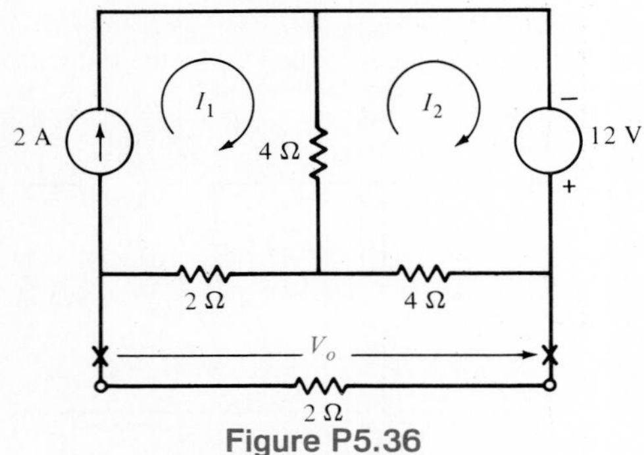

Figure P5.36

5.37. Use Thévenin's theorem to find V_o in the network shown in Fig. P5.37.

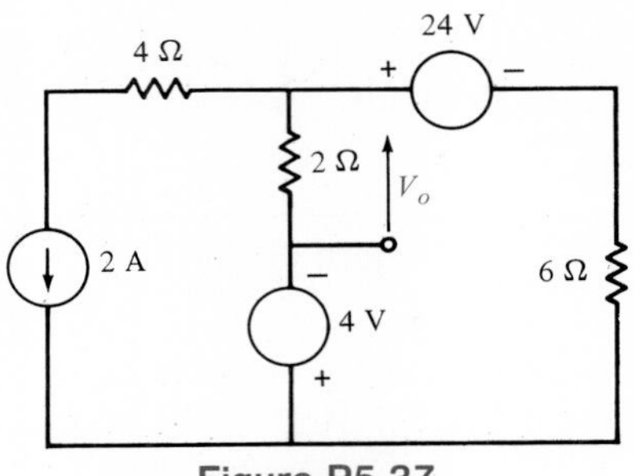

Figure P5.37

5.38. Apply Norton's theorem to find I_o in the network in Fig. P5.38.

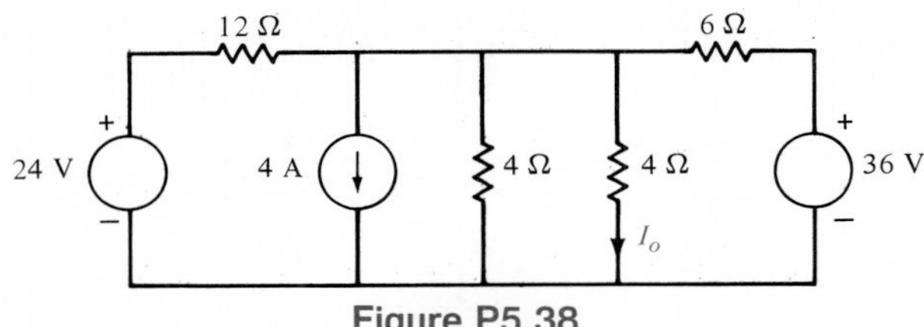

Figure P5.38

5.39. Apply Thévenin's theorem to find V_o in the circuit in Fig. P5.39.

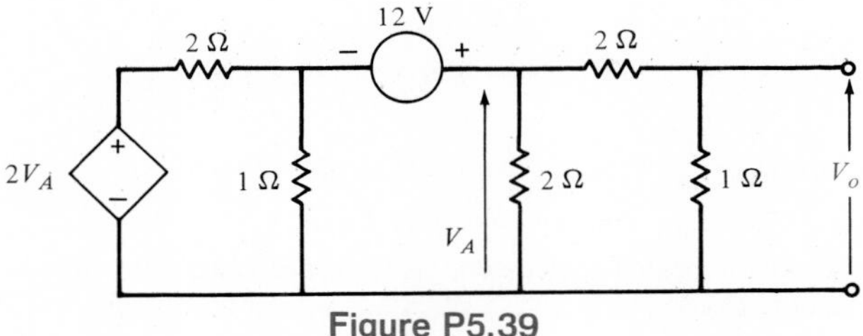

Figure P5.39

5.40. Use Thévenin's theorem to calculate V_o in the network in Fig. P5.40.

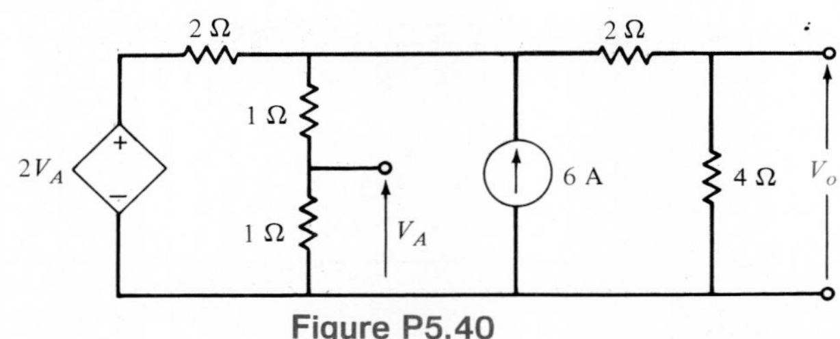

Figure P5.40

5.41. Find the current flowing through the 18-Ω resistor using Thévenin's theorem in the network shown in Fig. P5.41.

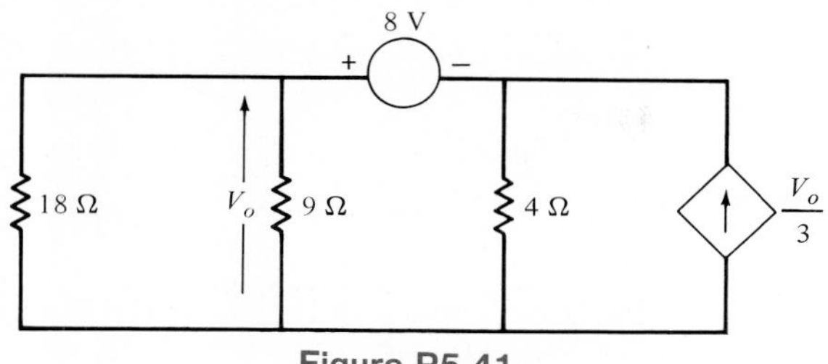

Figure P5.41

5.42. Using Thévenin's theorem, (a) determine the current flowing through the 6-Ω resistor in the network in Fig. P5.42. (b) Using the results obtained above (V_{oc}, I_{sc}, R_{Th}), replace the 6-Ω resistor with a 12-Ω resistor and determine the current flowing through it.

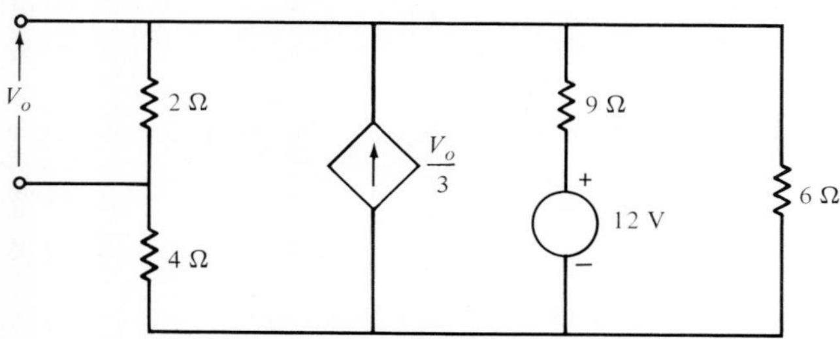

Figure P5.42

5.43. Determine the voltage developed across the 10-Ω resistor in the network in Fig. P5.43 using Norton's theorem.

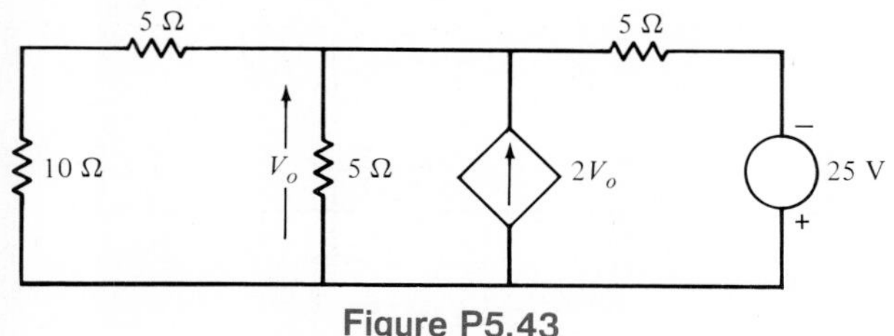

Figure P5.43

5.44. **(a)** Find the current I_x in the network in Fig. P5.44, using Norton's theorem. **(b)** Using the results obtained from part (a)—(V_{oc}, I_{sc}, R_{Th}), replace the 9-Ω resistor with a 3-Ω resistor and compute the current I_x under this condition.

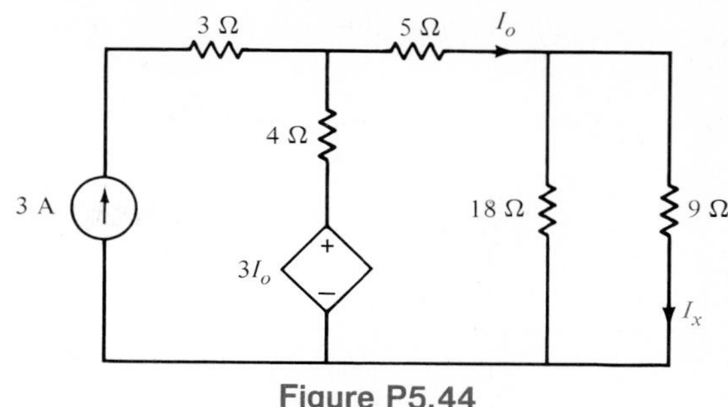

Figure P5.44

5.45. Use Thévenin's theorem to compute the current in R_L in Fig. P5.45.

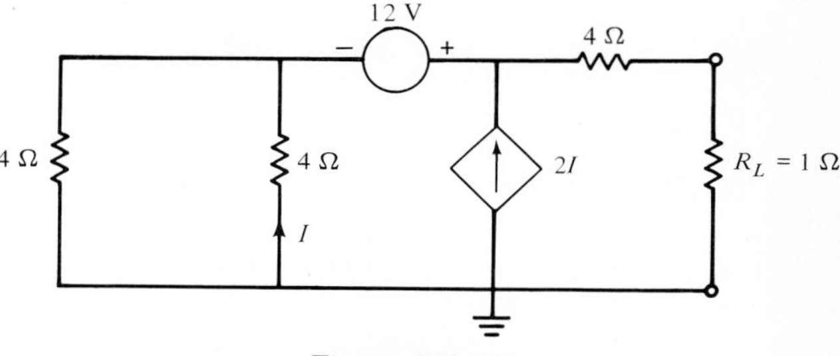

Figure P5.45

5.46. Find V_o in Fig. P5.46 using Thévenin's theorem.

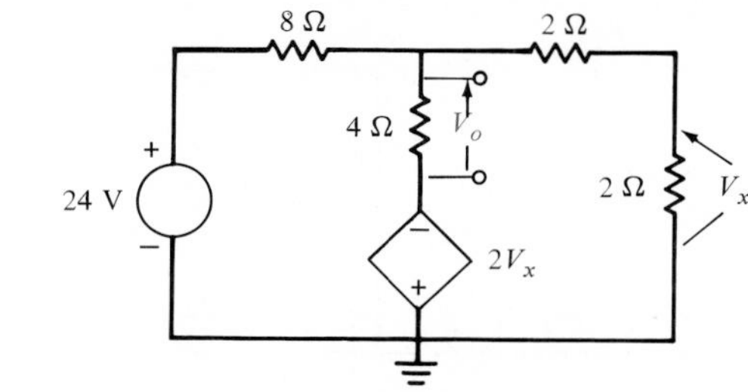

Figure P5.46

5.47. Use Thévenin's theorem to find V_o in the circuit shown in Fig. P5.47.

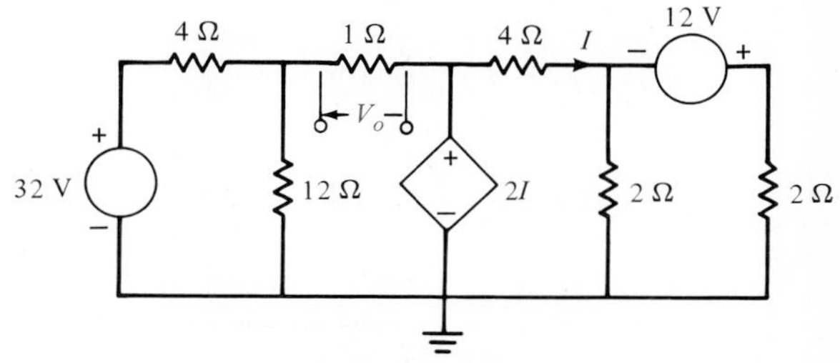

Figure P5.47

5.48. Use Thévenin's theorem to find V_o in Fig. P5.48.

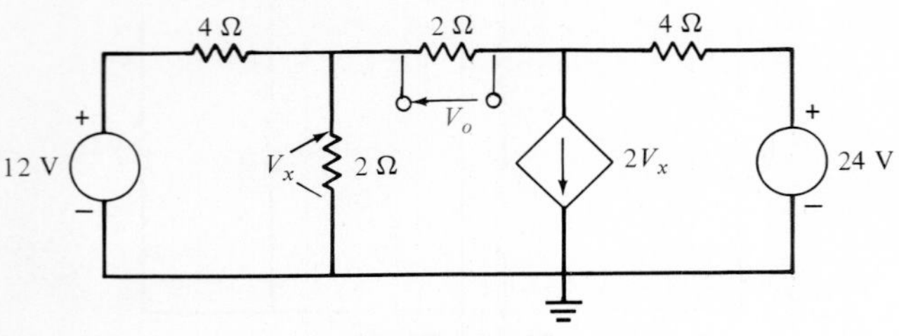

Figure P5.48

5.49. Determine V_o in Fig. P5.49 using Thévenin's theorem.

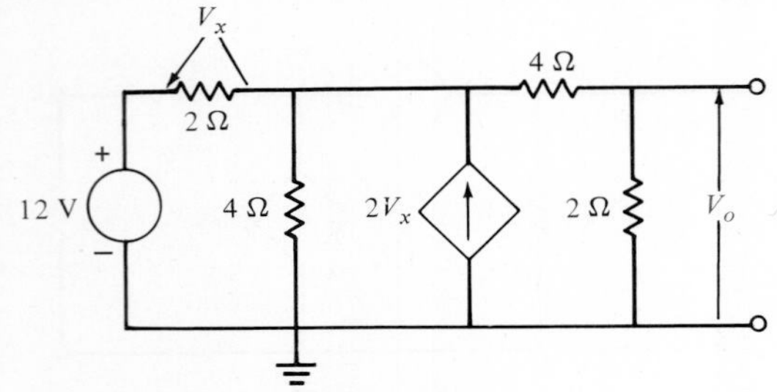

Figure P5.49

5.50. Determine the value of R_L in Fig. P5.50 for maximum power transfer.

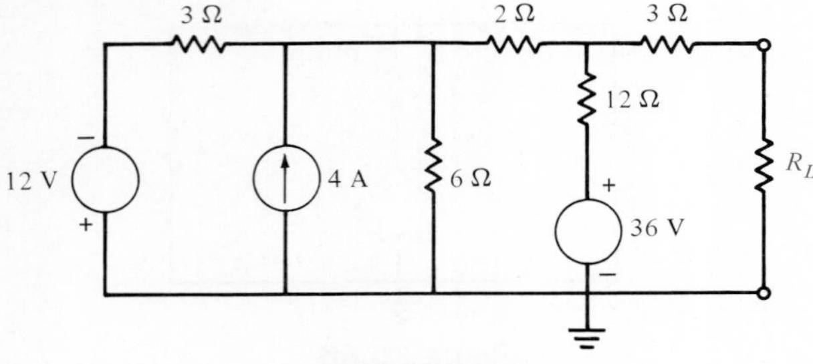

Figure P5.50

5.51. Determine the value of R_i in Fig. P5.51 so that maximum power is transferred to the 6-Ω load.

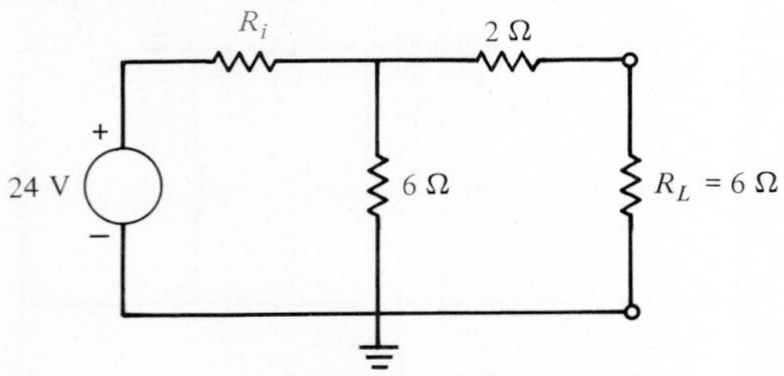

Figure P5.51

5.52. Determine the value of R_L in Problem 5.45 for maximum power transfer.

5.53. In Fig. P5.53, find the value of R_L for maximum power transfer to this load.

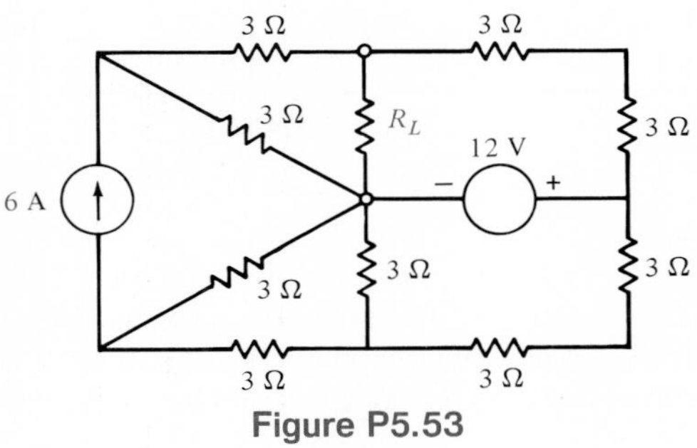

Figure P5.53

5.54. Find the value of R_L in Problem 5.30 for maximum power transfer.

5.55. Using superposition, find an expression for the output voltage v_o in the circuit shown in Fig. P5.55.

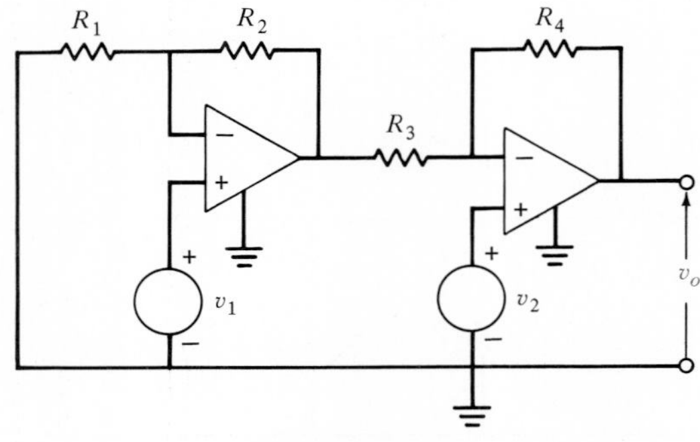

Figure P5.55

5.56. Use SPICE to determine all node voltages and currents I_x, I_y, and I_z in the network in Fig. P5.56.

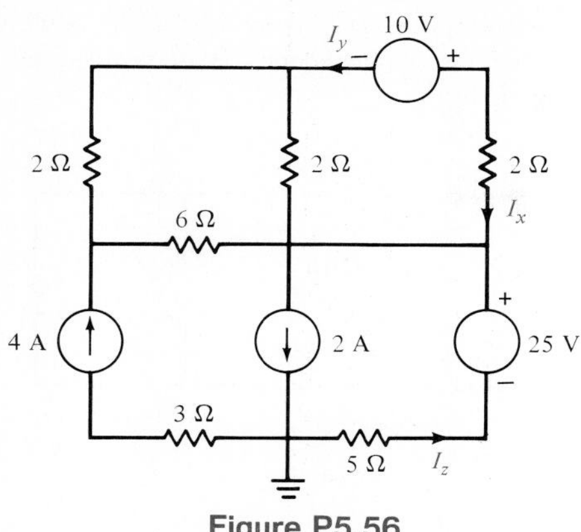

Figure P5.56

5.57. Use SPICE to determine I_x and all node voltages in the network in Fig. P5.57.

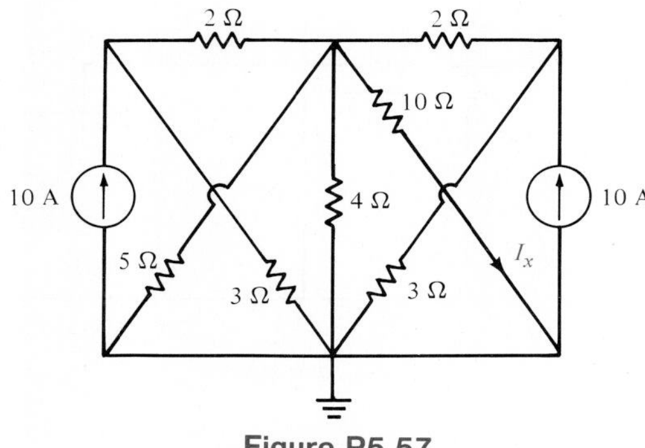

Figure P5.57

5.58. Write a SPICE program to determine all node voltages, I_x, V_x, and the current through both voltage sources in the network in Fig. P5.58.

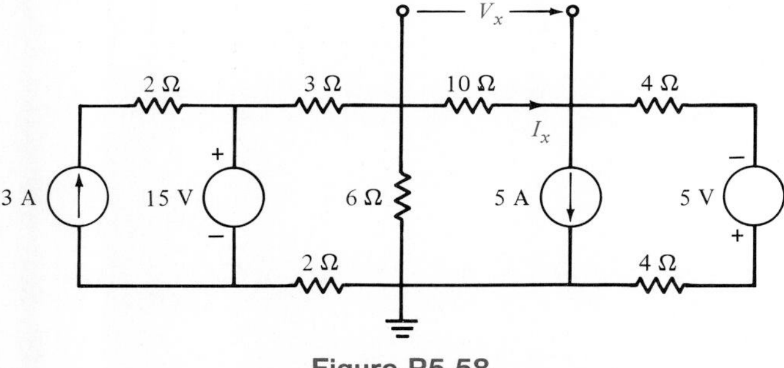

Figure P5.58

5.59. Given the circuit in Fig. P5.59, determine I_a, I_x, all node voltages, and the current through the 50-V source using a SPICE analysis.

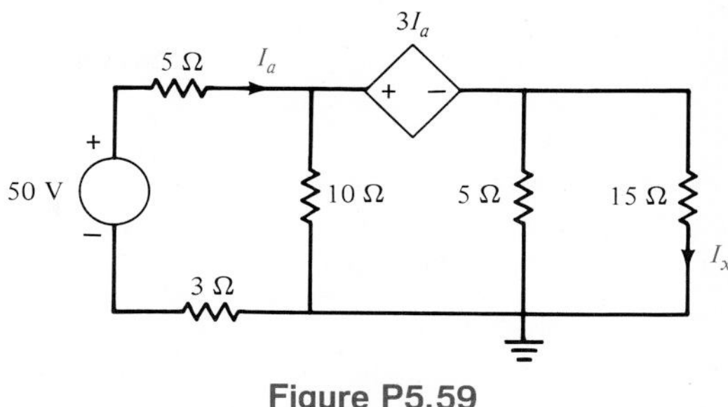

Figure P5.59

5.60. Write a SPICE program to calculate I_x and the node voltages in the network in Fig. P5.60.

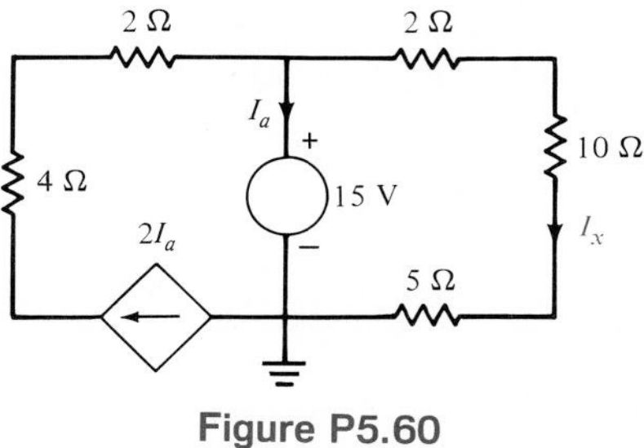

Figure P5.60

5.61. Given the network in Fig. P5.61, determine the node voltages, I_x, and the current in all the sources using SPICE.

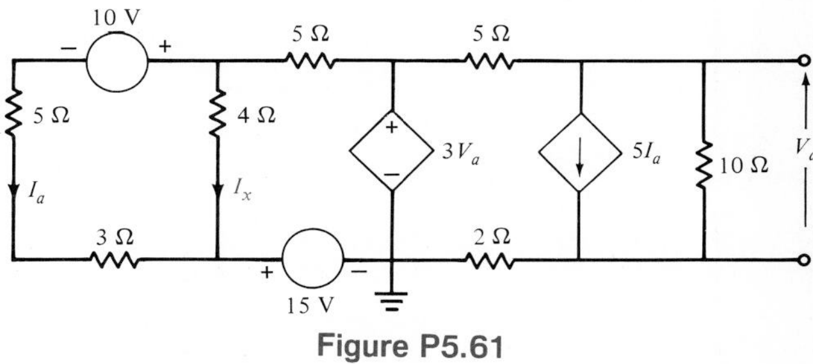

Figure P5.61

5.62. Write a SPICE program to determine all the node voltages and I_a in the network in Fig. P5.62.

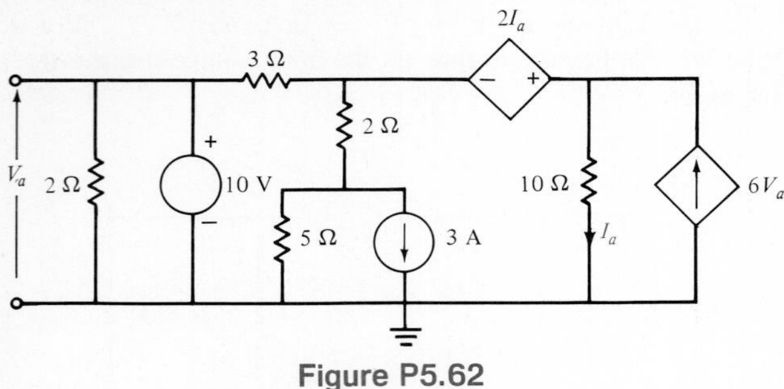

Figure P5.62

5.63. Given the network in Fig. P5.63, determine I_x and the voltage V_x using SPICE.

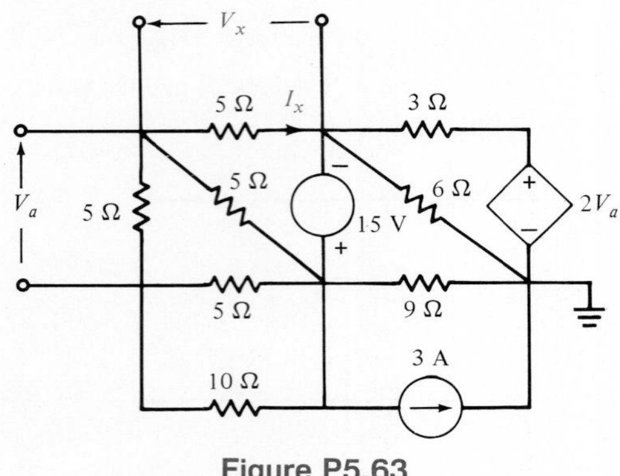

Figure P5.63

5.64. Given the circuit in Fig. P5.64, write a SPICE program to calculate V_o and I_o.

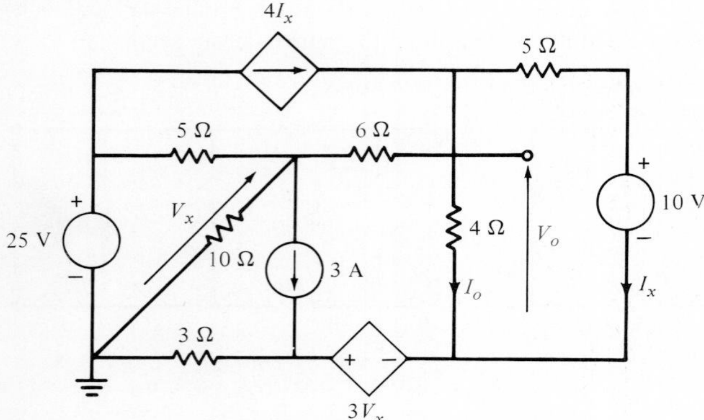

Figure P5.64

5.65. Write a SPICE program to determine all the node voltages, the current through the independent voltage source, and I_a in the network in Fig. P5.65.

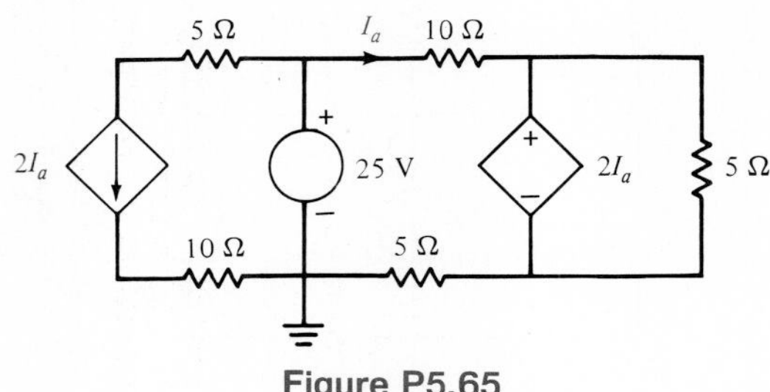

Figure P5.65

5.66. Write a SPICE program to determine all node voltages and currents through all the sources in the network in Fig. P5.66.

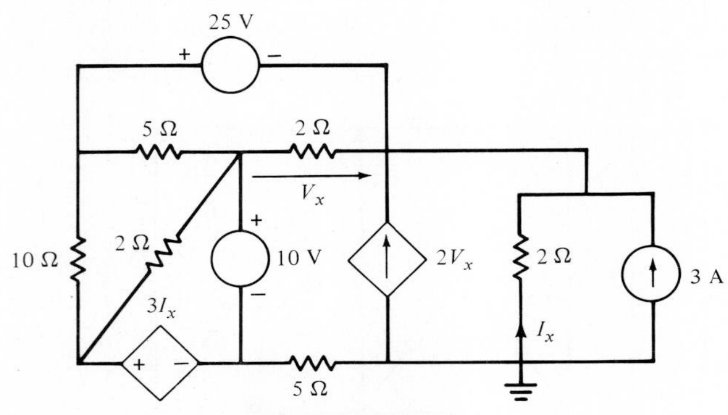

Figure P5.66

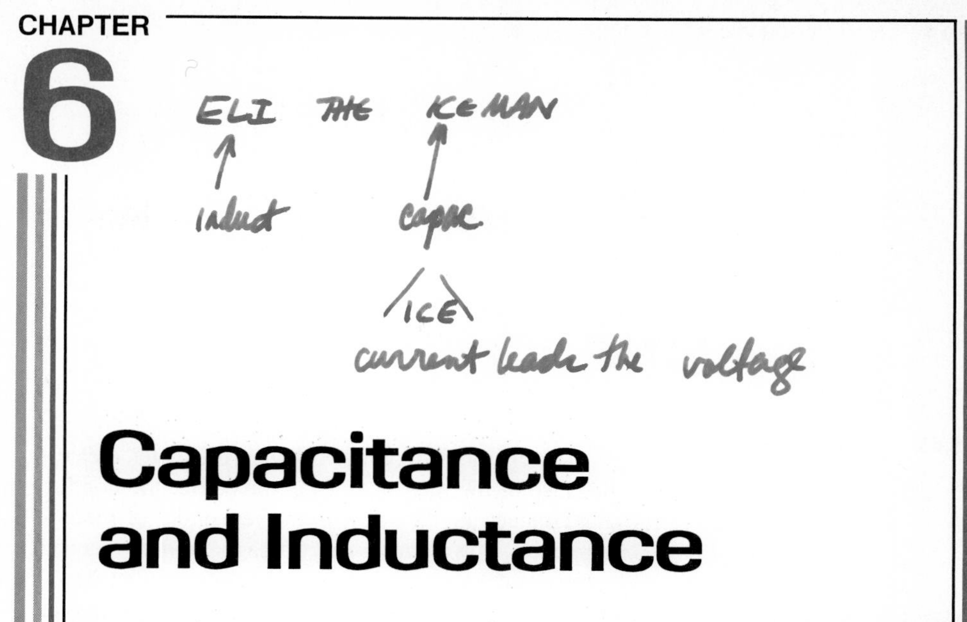

CHAPTER 6 Capacitance and Inductance

Our analysis of electric circuits thus far has been confined to circuits that contain only independent sources, dependent sources, and resistors. At this point we introduce two additional circuit elements: the capacitor and the inductor. They are linear elements; however, in contrast to the resistor, their terminal characteristics are described by linear differential equations. Another distinctive feature of these new elements is their ability to absorb energy from the circuit, store it temporarily, and later, return it. Elements that possess this energy storage capability are referred to simply as *storage elements*.

Finally, the principle of duality is introduced, which illustrates the parallelism that exists between loop and nodal analysis.

6.1 Capacitors

A *capacitor* is a circuit element that consists of two conducting surfaces separated by a nonconducting, or *dielectric,* material. A simplified capacitor and its electrical symbol are shown in Fig. 6.1. If a source is connected to the capacitor, positive charges will be transferred to one plate and negative charges to the other. The charge on the capacitor is proportional to the voltage across it such that

$$q = Cv \tag{6.1}$$

where C is the proportionality factor known as the capacitance of the element in coulombs per volt or farads (F). The unit ''farad'' is named after Michael Faraday, a famous English physicist.

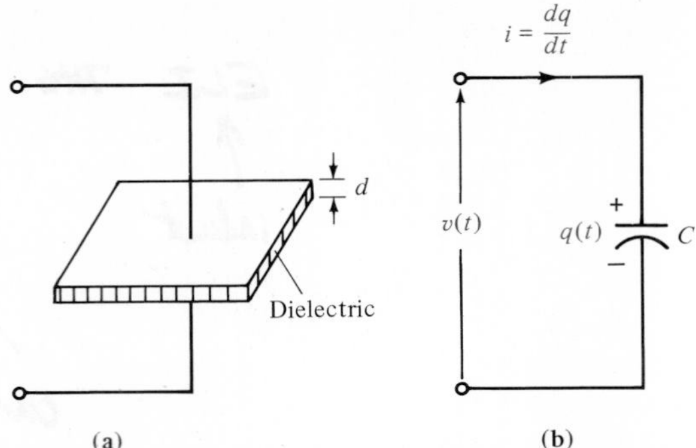

Figure 6.1 Capacitor and its electrical symbol.

The charge differential between the plates creates an electric field that stores energy. Because of the presence of the dielectric, the conduction current that flows in the wires that connect the capacitor to the remainder of the circuit cannot flow internally between the plates. However, via electromagnetic field theory it can be shown that this conduction current is equal to the displacement current that flows between the plates of the capacitor and is present any time that an electric field or voltage varies with time.

Our primary interest is in the current–voltage terminal characteristics of the capacitor. Since the current is

$$i = \frac{dq}{dt}$$

then for a capacitor

$$i = \frac{d}{dt} Cv$$

which for constant capacitance is

$$i = C\frac{dv}{dt} \tag{6.2}$$

Equation (6.2) can be rewritten as

$$dv = \frac{1}{C} i \, dt$$

Now integrating this expression from $t = -\infty$ to some time t and assuming $v(-\infty) = 0$ yields

$$v(t) = \frac{1}{C}\int_{-\infty}^{t} i(x)\, dx \tag{6.3}$$

where $v(t)$ indicates the time dependence of the voltage. Equation (6.3) can be expressed as two integrals, so that

$$\begin{aligned} v(t) &= \frac{1}{C}\int_{-\infty}^{t_0} i(x)\, dx + \frac{1}{C}\int_{t_0}^{t} i(x)\, dx \\ &= v(t_0) + \frac{1}{C}\int_{t_0}^{t} i(x)\, dx \end{aligned} \tag{6.4}$$

where $v(t_0)$ is the voltage due to the charge that accumulates on the capacitor from time $t = -\infty$ to time $t = t_0$.

The energy stored in the capacitor can be derived from the power that is delivered to the element. This power is given by the expression

$$p(t) = v(t)i(t) = Cv(t)\frac{dv(t)}{dt} \tag{6.5}$$

and hence the energy stored in the electric field is

$$\begin{aligned} w_c(t) &= \int_{-\infty}^{t} Cv(x)\frac{dv(x)}{dx}\,dx = C\int_{-\infty}^{t} v(x)\frac{dv(x)}{dx}\,dx \\ &= C\int_{v(-\infty)}^{v(t)} v(x)\,dv(x) = \tfrac{1}{2}Cv^2(x)\Big|_{v(-\infty)}^{v(t)} \\ &= \tfrac{1}{2}Cv^2(t) \qquad \text{joules} \end{aligned} \tag{6.6}$$

since $v(t = -\infty) = 0$. The expression for the energy can also be written using Eq. (6.1) as

$$w_c(t) = \frac{1}{2}\frac{q^2(t)}{C} \tag{6.7}$$

Equations (6.6) and (6.7) represent the energy stored by the capacitor, which, in turn, is equal to the work done by the source to charge the capacitor.

The polarity of the voltage across a capacitor being charged is shown in Fig. 6.1b. In the ideal case the capacitor will hold the charge for an indefinite period of time if the source is removed. If at some later time an energy-absorbing device (e.g., a flash bulb) is connected across the capacitor, a discharge current will flow from the capacitor, and therefore the capacitor will supply the energy stored to the device.

Although we will model the capacitor as an ideal device, in practice there is normally a very large leakage resistance in parallel which provides a conduction path between the plates. It is through this parallel resistance that the realistic capacitor slowly discharges itself.

Capacitors may be fixed or variable and typically range from thousands of microfarads (μF) to a few picofarads (pF). We employ them in numerous situations to obtain specific circuit performance characteristics. Unfortunately, they can also have a detrimental effect on circuit behavior when they are inherently present in the form of stray or parasitic capacitance, which can occur between any two conducting surfaces in a circuit.

EXAMPLE 6.1

If the charge accumulated on two parallel conductors charged to 12 V is 600 pC, what is the capacitance of the parallel conductors?

Using Eq. (6.1), we find that

$$C = \frac{Q}{V} = \frac{600 \times 10^{-12}}{12} = 50 \text{ pF}$$

■

EXAMPLE 6.2

Given a capacitor that is initially uncharged, determine the voltage across the capacitor as a function of time if it is subjected to the current pulse shown in Fig. 6.2a.

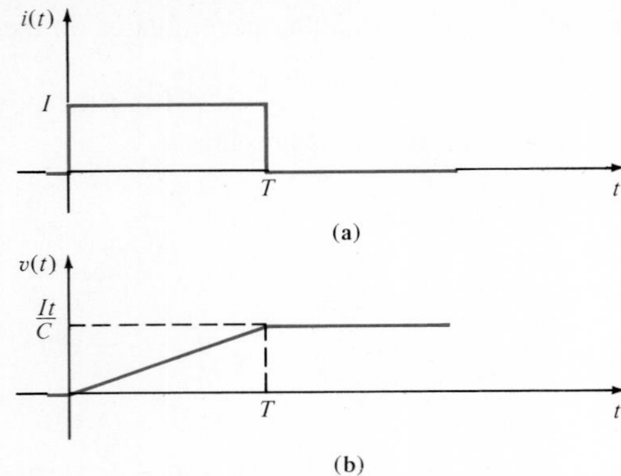

Figure 6.2 Current and voltage waveforms for an initially uncharged capacitor.

Using Eq. (6.4) and the fact that $v(t_0) = V(0) = 0$, we have

$$v(t) = 0 + \frac{1}{C}\int_0^t I\,dx \qquad 0 \le t \le T$$

$$= \frac{It}{C} \qquad 0 \le t \le T$$

Therefore, the voltage waveform is shown in Fig. 6.2b. ■

EXAMPLE 6.3

The voltage across a 5-μF capactior has the waveform shown in Fig. 6.3a. Determine the current waveform.

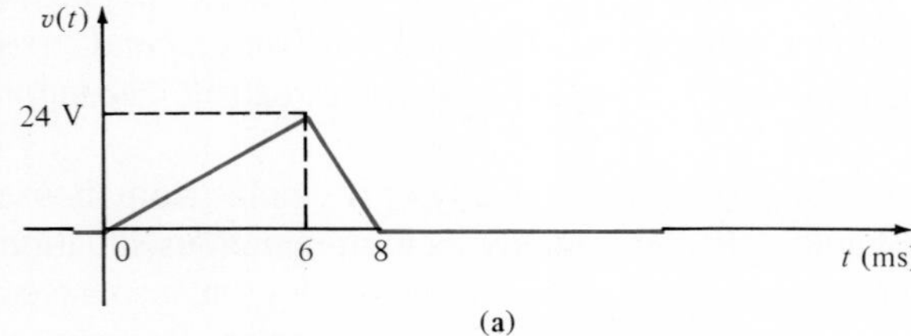

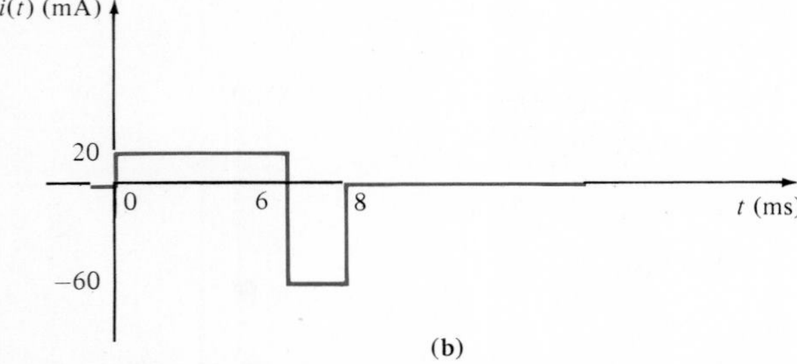

Figure 6.3 Voltage and current waveforms for a 5-μF capacitor.

Using Eq. (6.2) and noting that

$$v(t) = \frac{24}{6 \times 10^{-3}} t \qquad 0 \le t \le 6 \text{ ms}$$

$$= \frac{-24}{2 \times 10^{-3}} t + 96 \qquad 6 \le t \le 8 \text{ ms}$$

we find that

$$i(t) = C \frac{dv(t)}{dt}$$

$$= 5 \times 10^{-6} \times 4 \times 10^{3} \qquad 0 \le t \le 6 \text{ ms}$$

$$= 20 \text{ mA} \qquad 0 \le t \le 6 \text{ ms}$$

and

$$i(t) = 5 \times 10^{-6}(-12 \times 10^{3}) \qquad 6 \le t \le 8 \text{ ms}$$

$$= -60 \text{ mA} \qquad 6 \le t \le 8 \text{ ms}$$

Therefore, the current waveform is as shown in Fig. 6.3b and $i(t) = 0$ for $t > 8$ ms. ■

EXAMPLE 6.4

Determine the energy stored in the electric field of the capacitor in Example 6.3 at $t = 6$ ms.

Using Eq. (6.6), we have

$$w(t) = \tfrac{1}{2} C v^2(t)$$

At $t = 6$ ms,

$$w(6 \text{ ms}) = \tfrac{1}{2}(5 \times 10^{-6})(24)^2$$

$$= 1440 \ \mu\text{J}$$

■

The equations and examples above illustrate a number of salient features of the capacitor. An ideal capacitor only stores energy; it does not dissipate energy as a resistor does. If the voltage across a capacitor is constant (i.e., non-time varying), the current through it is zero, and therefore to direct current, a capacitor looks like an open circuit. Although the capacitor current is zero, the capacitor can still store a finite amount of energy. An instantaneous jump in the voltage across a capacitor is not physically realizable, because it requires the movement of a finite amount of charge in zero time, which is an infinite current. Although we can fabricate situations where the voltage across a capacitor is discontinuous, these do not occur in practical situations; however, such an approximation is normally a good one.

DRILL EXERCISE

D6.1. A 10-μF capacitor has an accumulated charge of 500 nC. Determine the voltage across the capacitor.

D6.2. The voltage across a 4-μF capacitor is shown in Fig. D6.2. Determine the waveform for the capacitor current.

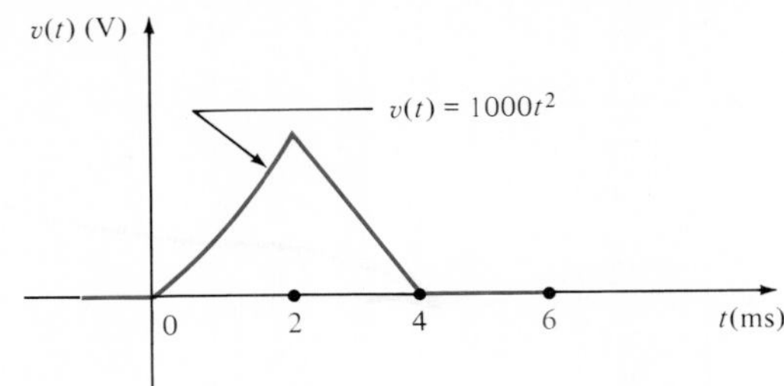

Figure D6.2

D6.3. Compute the energy stored in the electric field of the capacitor in Drill Exercise D6.2 at $t = 2$ ms.

6.2 Inductors

An *inductor* is a circuit element that consists of a conducting wire usually in the form of a coil. A typical inductor and its electrical symbol are shown in Fig. 6.4. From a historical standpoint, developments that lead to the mathematical model we employ to represent the inductor are as follows. It was first shown that a current-carrying conductor would produce a magnetic field. It was later found that the magnetic field and the current that produced it were linearly related. Finally, it was shown that a changing magnetic field produced a voltage that was proportional to the time rate of change of the current that produced the magnetic field; that is,

$$v(t) = L\frac{di(t)}{dt} \tag{6.8}$$

The constant of proportionality L is called the inductance, and is measured in the unit ''henry,'' named after the American inventor Joseph Henry, who discovered the relationship. As seen in Eq. (6.8), 1 henry (H) is dimensionally equal to 1 volt-second per ampere.

Following the development of the mathematical equations for the capacitor, we find that the expression for the current in an inductor is

$$i(t) = \frac{1}{L}\int_{-\infty}^{t} v(x)\,dx \tag{6.9}$$

which can also be written as

$$i(t) = i(t_0) + \frac{1}{L}\int_{t_0}^{t} v(x)\,dx \tag{6.10}$$

The power delivered to the inductor can be used to derive the energy stored in the element.

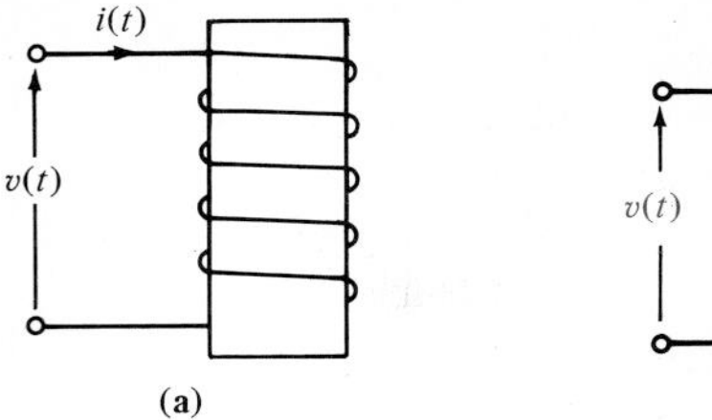

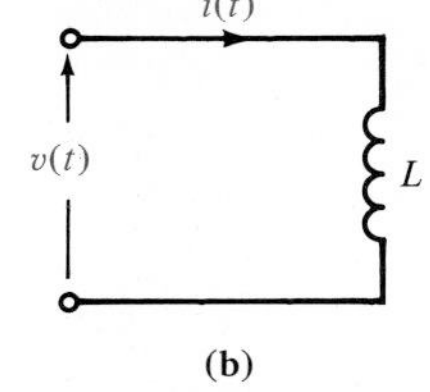

Figure 6.4 Inductor model and its electrical symbol.

This power is equal to

$$p(t) = v(t)i(t) = \left[L\frac{di(t)}{dt}\right]i(t) \tag{6.11}$$

Therefore, the energy stored in the magnetic field is

$$w_L(t) = \int_{-\infty}^{t}\left[L\frac{di(x)}{dx}\right]i(x)\,dx$$

Following the development of Eq. (6.6), we obtain

$$w_L(t) = \tfrac{1}{2}Li^2(t) \qquad \text{joules} \tag{6.12}$$

The inductor, like the resistor and capacitor, is a passive element. The polarity of the voltage across the inductor is shown in Fig. 6.4.

Like the capacitor, the inductor is an energy storage device. However, a practical inductor cannot store energy as well as can a practical capacitor. The inductor always has some winding resistance in the coil, which quickly dissipates energy.

Practical inductors typically range from a few microhenrys to tens of henrys. From a circuit design standpoint it is important to note that inductors cannot be easily fabricated on an integrated-circuit chip, and therefore chip designs typically employ only active electronic devices, resistors and capacitors, which can be easily fabricated in microcircuit form.

EXAMPLE 6.5

The current in a 10-mH inductor has the waveform shown in Fig. 6.5a. Determine the voltage waveform.

Using Eq. (6.8) and noting that

$$i(t) = \frac{20 \times 10^{-3}t}{2 \times 10^{-3}} \qquad 0 \le t \le 2 \text{ ms}$$

and

$$i(t) = \frac{-20 \times 10^{-3}t}{2 \times 10^{-3}} + 40 \times 10^{-3} \qquad 2 \le t \le 4 \text{ ms}$$

we find that

$$v(t) = (10 \times 10^{-3})\frac{20 \times 10^{-3}}{2 \times 10^{-3}} \qquad 0 \le t \le 2 \text{ ms}$$

$$= 100 \text{ mV}$$

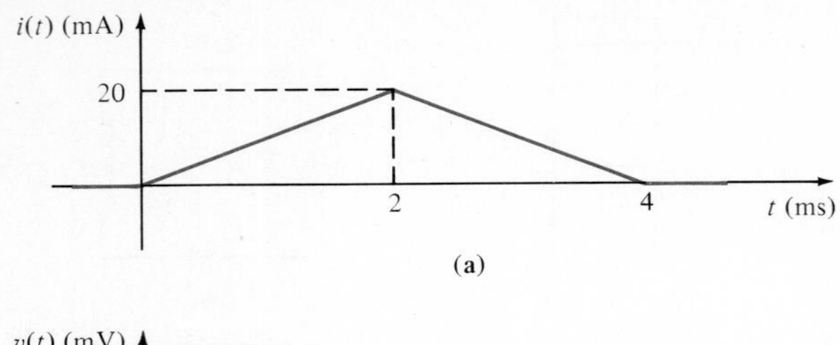

(a)

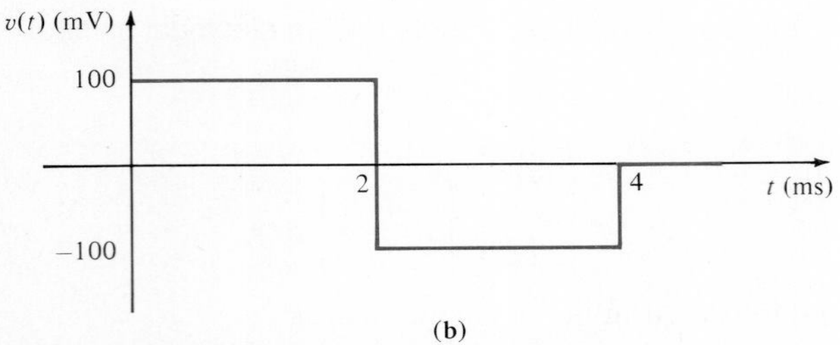

(b)

Figure 6.5 Current and voltage waveforms for a 10-mH inductor.

and

$$v(t) = (10 \times 10^{-3}) \frac{-20 \times 10^{-3}}{2 \times 10^{-3}} \qquad 2 \le t \le 4 \text{ ms}$$
$$= -100 \text{ mV}$$

and $v(t) = 0$ for $t > 4$ ms. Therefore, the voltage waveform is shown in Fig. 6.5b. ■

EXAMPLE 6.6

The current in a 2-mH inductor is

$$i(t) = 2 \sin 377t$$

Determine the voltage across the inductor and the energy stored in the inductor.

From Eq. (6.8), we have

$$v(t) = L \frac{di(t)}{dt}$$
$$= (2 \times 10^{-3}) \frac{d}{dt} (2 \sin 377t)$$
$$= 1.508 \cos 377t$$

and from Eq. (6.12),

$$w_L(t) = \tfrac{1}{2} L i^2(t)$$
$$= \tfrac{1}{2}(2 \times 10^{-3})(2 \sin 377t)^2$$
$$= 0.004 \sin^2 377t$$

■

The previous material illustrates a number of important features of the inductor. An ideal inductor only stores energy; it does not dissipate any energy. As has been indicated, this is not the case for physically realizable inductors. From Eq. (6.8) we note that if the current is constant, the voltage across an inductor is zero. Hence to direct current, the inductor looks like a short circuit. The equation also indicates that an instantaneous change in current would require an infinite voltage. Therefore, it is not possible to change instantaneously the current in an inductor.

DRILL EXERCISE

D6.4. The current in a 5-mH inductor has the waveform shown in Fig. D6.4. Compute the waveform for the inductor voltage.

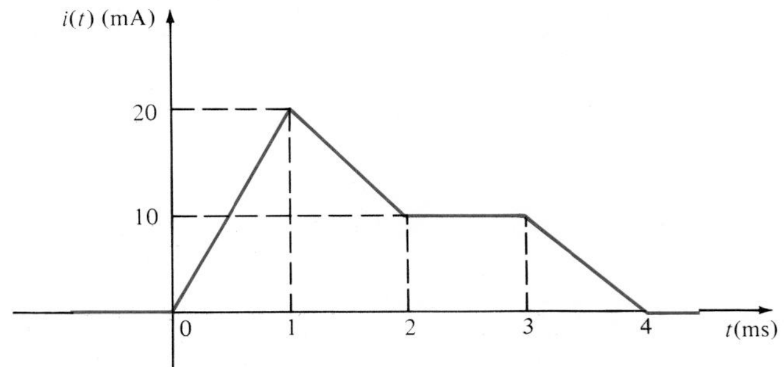

Figure D6.4

D6.5. Compute the energy stored in the magnetic field of the inductor in Drill Exercise D6.4 at $t = 1.5$ ms.

6.3 Capacitor and Inductor Combinations

Series Capacitors

If a number of capacitors are connected in series, their equivalent capacitance can be calculated using KVL. Consider the circuit shown in Fig. 6.6a. For this circuit

$$v(t) = v_1(t) + v_2(t) + v_3(t) + \cdots + v_N(t) \tag{6.13}$$

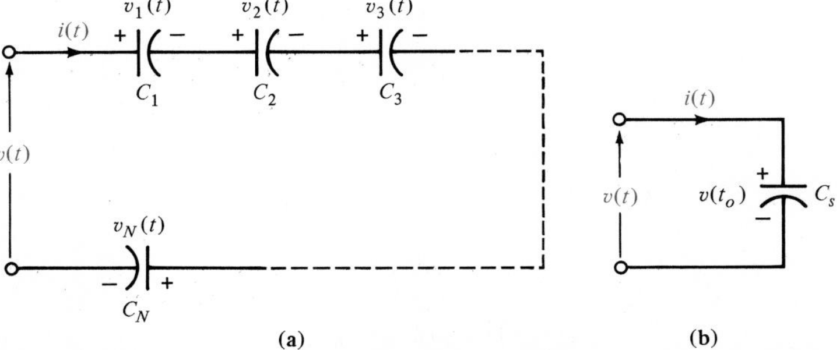

Figure 6.6 Equivalent circuit for N series-connected capacitors.

but

$$v_i(t) = \frac{1}{C_i} \int_{t_0}^{t} i(t)\, dt + v_i(t_0) \quad (6.14)$$

Therefore, Eq. (6.13) can be written as follows using Eq. (6.14):

$$v(t) = \left(\sum_{i=1}^{N} \frac{1}{C_i} \right) \int_{t_0}^{t} i(t) dt + \sum_{i=1}^{N} v_i(t_0) \quad (6.15)$$

$$= \frac{1}{C_s} \int_{t_0}^{t} i(t)\, dt + v(t_0) \quad (6.16)$$

where

$$\frac{1}{C_s} = \sum_{i=1}^{N} \frac{1}{C_i} = \frac{1}{C_1} + \frac{1}{C_2} + \cdots + \frac{1}{C_N} \quad (6.17)$$

and $v(t_0)$ is equal to the voltage on C_s at $t = t_0$. Thus the circuit in Fig. 6.6b is equivalent to that in Fig. 6.6a under the conditions stated above.

It is also important to note that since the same current flows in each of the series capacitors, each capacitor gains the same charge in the same time period. The voltage drop across each capacitor will depend on this charge and the capacitance of the element.

EXAMPLE 6.7

Determine the equivalent capacitance and the initial voltage for the circuit shown in Fig. 6.7.

The equivalent capacitance is

$$\frac{1}{C_s} = \frac{1}{2\ \mu\text{F}} + \frac{1}{3\ \mu\text{F}} + \frac{1}{6\ \mu\text{F}} = \frac{1}{1\ \mu\text{F}}$$

Therefore, $C_s = 1\ \mu\text{F}$ and, as seen from the figure, $v(t_0) = -3$ V. Note that the energy stored is $w(t_0) = \frac{1}{2}(8 + 48 + 6)10^{-6}$ J, not $\frac{1}{2}(3)^2 10^{-6}$ J.

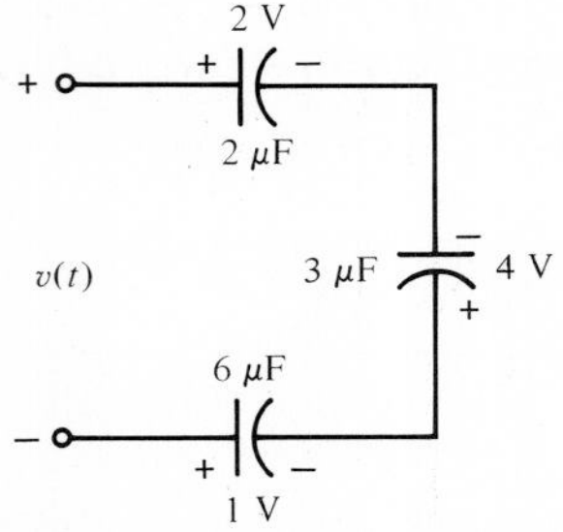

Figure 6.7 Circuit containing multiple capacitors with initial voltages. ■

EXAMPLE 6.8

Two capacitors in series are charged so that the voltage across the combination is 12 V. One capacitor is 30 μF and the other is unknown. If the voltage across the 30-μF capacitor is 8 V, find the capacitance of the unknown capacitor.

The charge on the 30-μF capacitor is

$$Q = CV = (30\ \mu\text{F})(8\ \text{V}) = 240\ \mu C$$

The unknown capacitor has the same charge, and therefore

$$C = \frac{Q}{V} = \frac{240\ \mu C}{4\ \text{V}} = 60\ \mu\text{F}$$

■

Parallel Capacitors

In order to determine the equivalent capacitance of N capacitors connected in parallel, we employ KCL. As can be seen from Fig. 6.8a,

$$\begin{aligned} i(t) &= i_1(t) + i_2(t) + i_3(t) + \cdots + i_N(t) \qquad (6.18) \\ &= C_1 \frac{dv(t)}{dt} + C_2 \frac{dv(t)}{dt} + C_3 \frac{dv(t)}{dt} + \cdots + C_N \frac{dv(t)}{dt} \\ &= \left(\sum_{i=1}^{N} C_i\right) \frac{dv}{dt} \\ &= C_p \frac{dv(t)}{dt} \qquad (6.19) \end{aligned}$$

where

$$C_p = C_1 + C_2 + C_3 + \cdots + C_N \qquad (6.20)$$

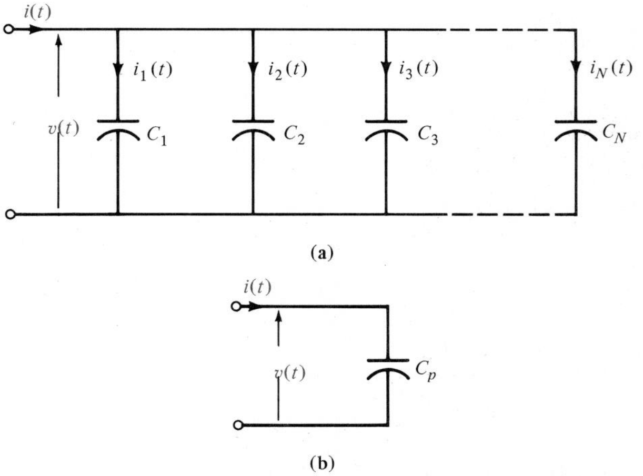

(a)

(b)

Figure 6.8 Equivalent circuits for N capacitors connected in parallel.

EXAMPLE 6.9

Determine the equivalent capacitance at terminals A-B of the circuit shown in Fig. 6.9.

$$\begin{aligned} C_p &= 4\ \mu\text{F} + 6\ \mu\text{F} + 2\ \mu\text{F} + 3\ \mu\text{F} \\ &= 15\ \mu\text{F} \end{aligned}$$

■

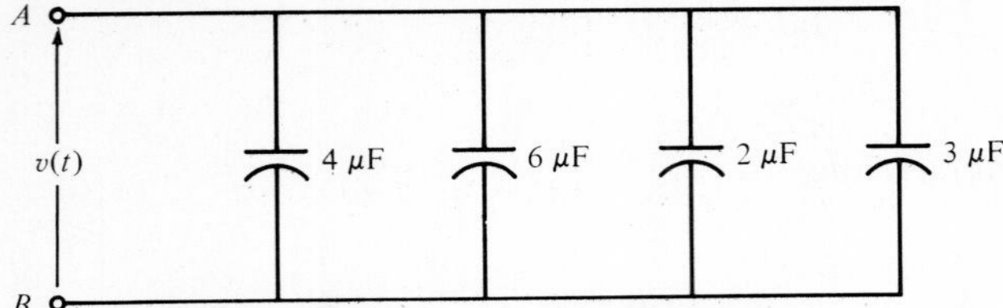

Figure 6.9 Circuit containing multiple capacitors in parallel.

DRILL EXERCISE

D6.6. Two capacitors are connected as shown in Fig. D6.6. Determine the value of C_1.

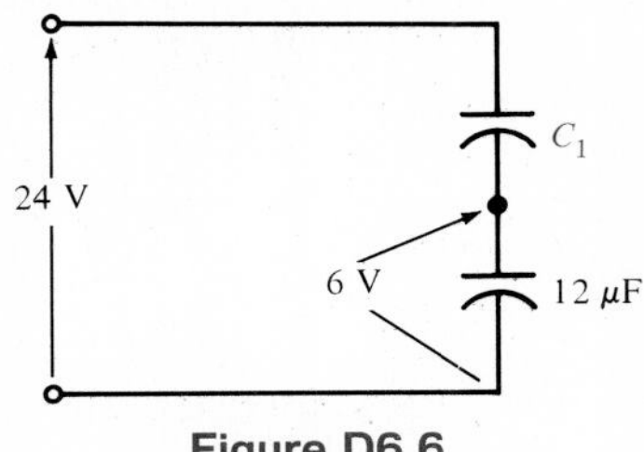

Figure D6.6

D6.7. Compute the equivalent capacitance of the network in Fig. D6.7.

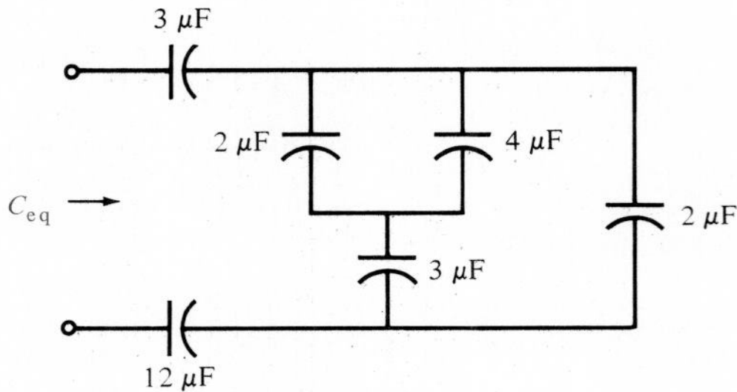

Figure D6.7

D6.8. Compute the equivalent capacitance of the network in Fig. D6.8 if all the capacitors are 4 μF.

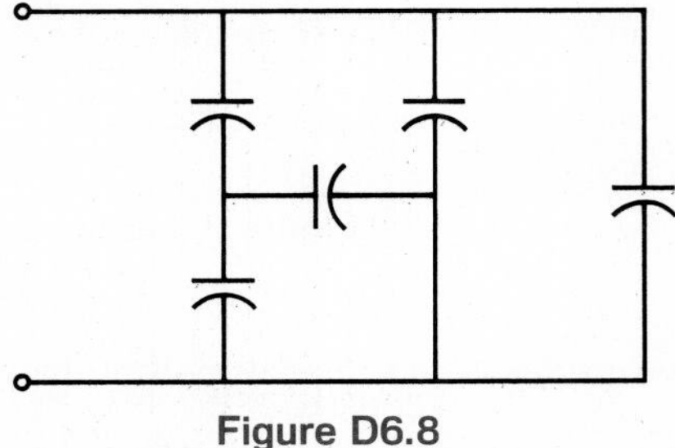
Figure D6.8

(a)

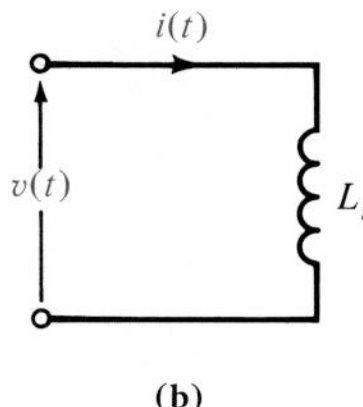

(b)

Figure 6.10 Equivalent circuit for N series-connected inductors.

Series Inductors

If N inductors are connected in series, the equivalent inductance of the combination can be determined as follows. Referring to Fig. 6.10a and using KVL, we see that

$$v(t) = v_1(t) + v_2(t) + v_3(t) + \cdots + v_N(t) \tag{6.21}$$

and therefore,

$$v(t) = L_1 \frac{di(t)}{dt} + L_2 \frac{di(t)}{dt} + L_3 \frac{di(t)}{dt} + \cdots + L_N \frac{di(t)}{dt} \tag{6.22}$$

$$= \left(\sum_{i=1}^{N} L_i \right) \frac{di(t)}{dt}$$

$$= L_s \frac{di(t)}{dt} \tag{6.23}$$

where

$$L_s = \sum_{i=1}^{N} L_i \tag{6.24}$$

EXAMPLE 6.10

The equivalent inductance of the circuit shown in Fig. 6.11 is

$$L_s = 1 \text{ H} + 2 \text{ H} + 4 \text{ H}$$

$$= 7 \text{ H}$$

■

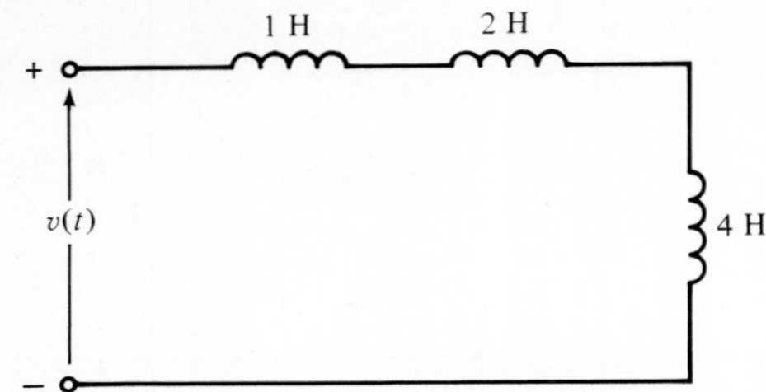

Figure 6.11 Circuit containing multiple inductors.

Parallel Inductors

Consider the circuit shown in Fig. 6.12a, which contains N parallel inductors. Using KCL, we can write

$$i(t) = i_1(t) + i_2(t) + i_3(t) + \cdots + i_N(t) \tag{6.25}$$

However,

$$i_j(t) = \frac{1}{L_j}\int_{t_0}^{t} v(x)\,dx + i_j(t_0) \tag{6.26}$$

Substituting this expression into Eq. (6.25) yields

$$i(t) = \left(\sum_{j=1}^{N}\frac{1}{L_j}\right)\int_{t_0}^{t} v(x)\,dx + \sum_{j=1}^{N} i_j(t_0) \tag{6.27}$$

$$= \frac{1}{L_p}\int_{t_0}^{t} v(x)\,dx + i(t_0) \tag{6.28}$$

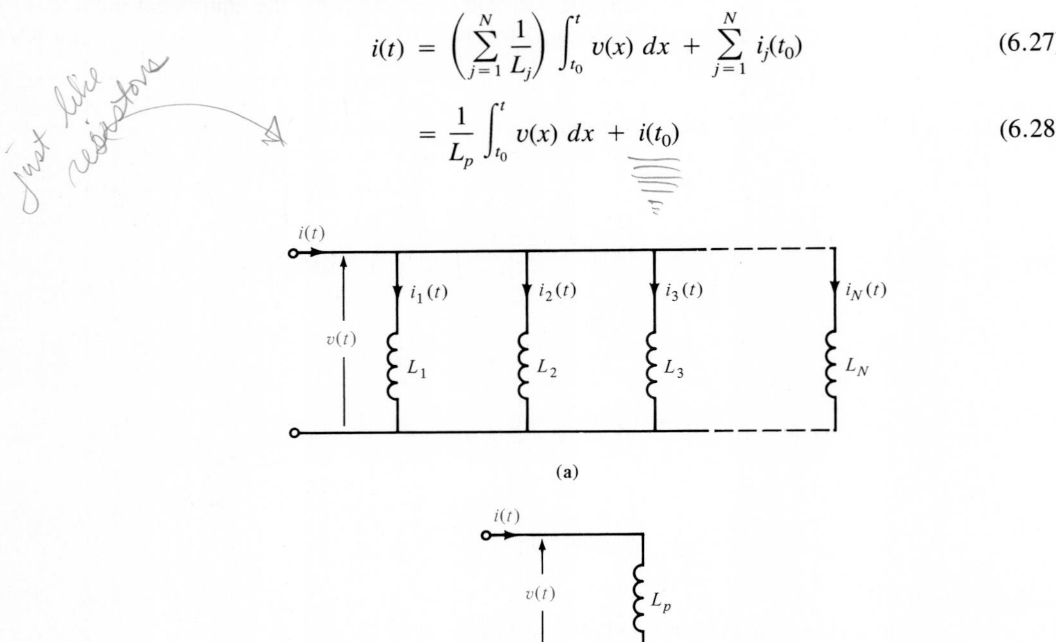

Figure 6.12 Equivalent circuits for N inductors connected in parallel.

where

$$\frac{1}{L_p} = \frac{1}{L_1} + \frac{1}{L_2} + \frac{1}{L_3} + \cdots + \frac{1}{L_N} \tag{6.29}$$

and $i(t_0)$ is equal to the current in L_p at $t = t_0$. Thus the circuit in Fig. 6.12b is equivalent to that in Fig. 6.12a under the conditions stated above.

EXAMPLE 6.11

Determine the equivalent inductance and the initial current for the circuit shown in Fig. 6.13.

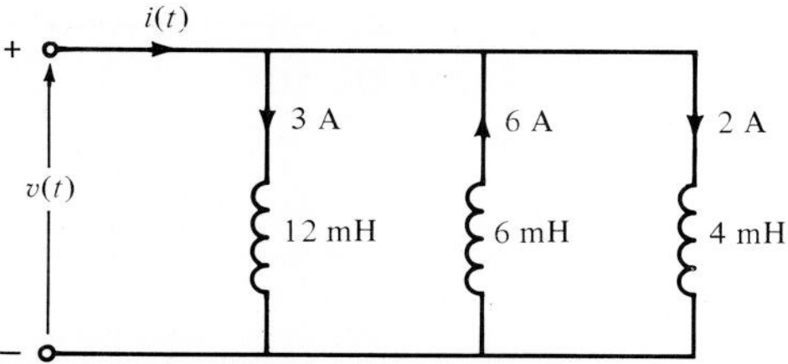

Figure 6.13 Circuit containing multiple inductors with initial currents.

The equivalent inductance is

$$\frac{1}{L_p} = \frac{1}{12\text{ mH}} + \frac{1}{6\text{ mH}} + \frac{1}{4\text{ mH}}$$

$$L_p = 2\text{ mH}$$

and the initial current is $i(t_0) = -1$ A. ■

The previous material indicates that capacitors combine like conductances, whereas inductances combine like resistances.

DRILL EXERCISE

D6.9. Compute the equivalent inductance of the network in Fig. D6.9 if all inductors are 4 mH.

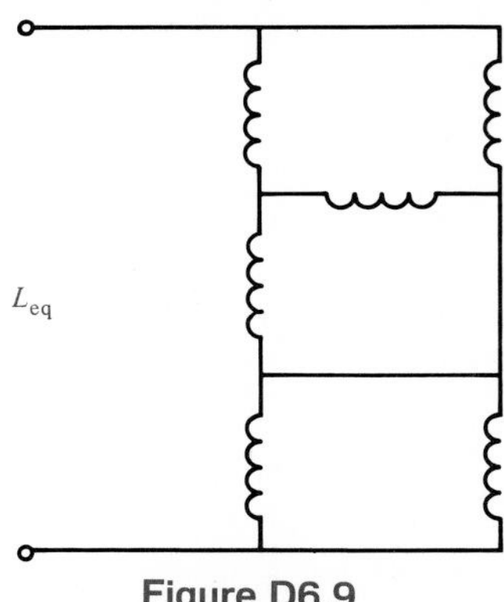

Figure D6.9

D6.10. Determine the equivalent inductance of the network in Fig. D6.10 if all inductors are 6 mH.

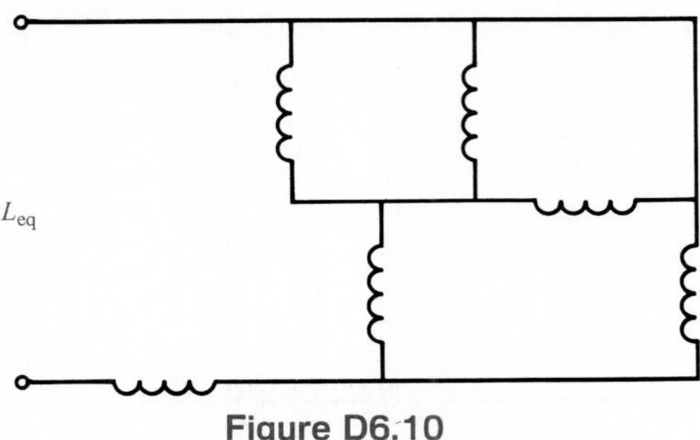

Figure D6.10

6.4 Duality

We have noted a striking similarity between the defining equations for a capacitor and an inductor. We recall that this similarity extends to the equations for resistance and conductance. In view of these relationships among the various components, we are naturally led to ask if there exists a more general analogy of which each one of these relations is simply a part. The answer to this question lies in what we call *duality*. From a circuit theory standpoint we say that one circuit is the *dual* of the other if the mesh equations that define one of them are of the same mathematical form as the nodal equations that define the other. To illustrate this concept, consider the two circuits shown in Fig. 6.14. The mesh equation for the circuit in Fig. 6.14a is

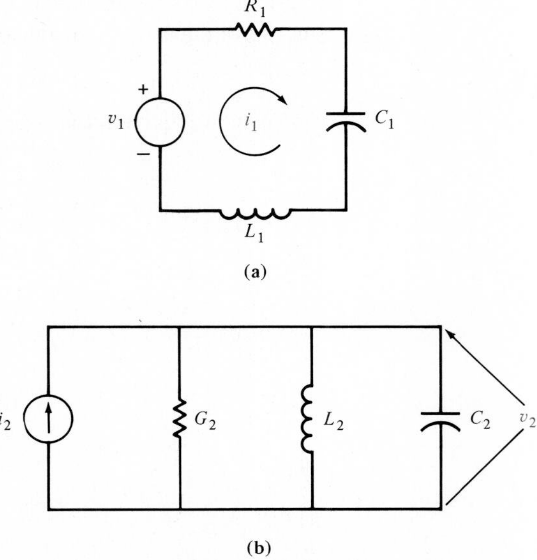

Figure 6.14 Pair of dual circuits.

$$v_1(t) = R_1 i_1(t) + \frac{1}{C_1}\int_0^t i_1(x)\,dx + v_{C_1}(0) + L_1 \frac{di_1(t)}{dt}$$

and the nodal equation for the circuit in Fig. 6.14b is

$$i_2(t) = G_2 v_2(t) + \frac{1}{L_2}\int_0^t v_2(x)\,dx + i_{L_2}(0) + C_2 \frac{dv_2(t)}{dt}$$

The similarity between the two equations is obvious. The series circuit in Fig. 6.14a is the dual of the parallel circuit in Fig. 6.14b, with the following symbol changes.

$$\begin{array}{ll} v_1 \rightarrow i_2 & C_1 \rightarrow L_2 \\ R_1 \rightarrow G_2 & v_{C_1} \rightarrow i_{L2} \\ i_1 \rightarrow v_2 & L_1 \rightarrow C_2 \end{array}$$

What is extremely important here is that the reader note that the solutions of the two circuits are identical under the symbol changes above, and therefore there is no need to analyze both circuits.

In general, it is not necessary that the circuit elements in two circuits have the same numerical values; to be duals, it is sufficient that their equations have the same mathematical form.

Our concept of duality extends well beyond the simple circuit shown in Fig. 6.14. Let us now consider the circuits shown in Fig. 6.15. The loop equations for the circuit in Fig. 6.15a are

$$v_o(t) = \frac{1}{C}\int_0^t i_1(x)\,dx + R_1 i_1(t) - R_1 i_2(t)$$

$$0 = -R_1 i_1(t) + R_1 i_2(t) + L\frac{di_2(t)}{dt} + R_2 i_2(t)$$

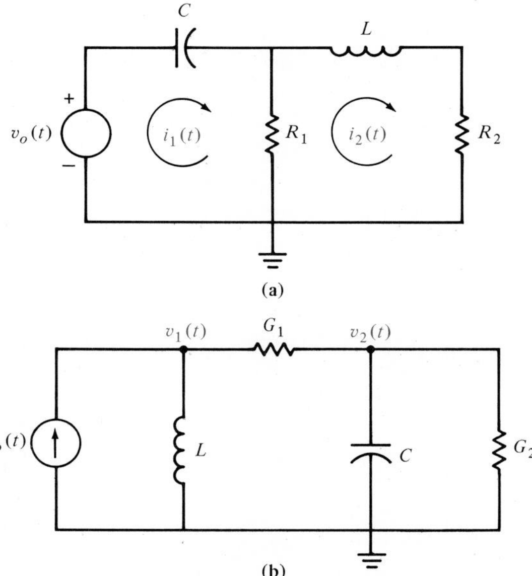

Figure 6.15 Dual circuits.

Table 6.1 Dual Conversions for Electrical Circuits

Loop Analysis	Node Analysis
Independent or dependent voltage sources	Independent or dependent current sources
Resistance	Conductance
Inductance	Capacitance
Mesh currents	Nonreference node voltages
A mesh containing B branches	A node joining B branches
Open or short circuit	Short or open circuit

and the node equations for the circuit in Fig. 6.15b are

$$i_o(t) = \frac{1}{L}\int_0^t v_1(x)\,dx + G_1 v_1(t) - G_2 v_2(t)$$

$$0 = -G_1 v_1(t) + G_1 v_2(t) + C\frac{dv_2(t)}{dt} + G_2 v_2(t)$$

Once again, the equations are identical under the following symbol changes: $v_o \to i_o$, $i_1 \to v_1$, $i_2 \to v_2$, $C \to L$, $R_1 \to G_1$, $L \to C$, and $R_2 \to G_2$. Some of the more important dual relationships for electrical circuits are shown in Table 6.1.

The dual of a given circuit can be found by the following procedure. First, the circuit equations for the original network are formulated on either a loop or a node basis. Second, the equations are rewritten by inserting the dual quantities shown in Table 6.1. Finally, the dual equations are interpreted and the network they describe is drawn. Although this technique will always produce a dual circuit, a more convenient method of deriving a dual circuit is the following graphical approach.

The graphical method for finding the dual of a planar network involves the following steps.

1. Place a dot inside each mesh of the original network. These dots correspond to nonreference nodes in the dual network.
2. Place a dot outside the entire original network. This dot corresponds to the reference node in the dual network.
3. Connect the inside dots between adjacent meshes by drawing a dashed line through the circuit elements that are common to both meshes. The circuit elements traversed are replaced by their duals and therefore form the branches that are common to the independent nodes in the dual network.
4. Connect the inside dots to the outside dot by drawing a dashed line through the circuit elements in all external branches. The elements crossed are replaced by their duals and form the branches that connect the independent nodes with the reference node.
5. The convention for voltage polarity and current direction is
   (a) A clockwise mesh current corresponds to a positive polarity for a nonreference node.
   (b) A voltage source that aids the clockwise mesh current corresponds to a current source flowing into the dual nonreference node.

The following examples will illustrate the graphical method.

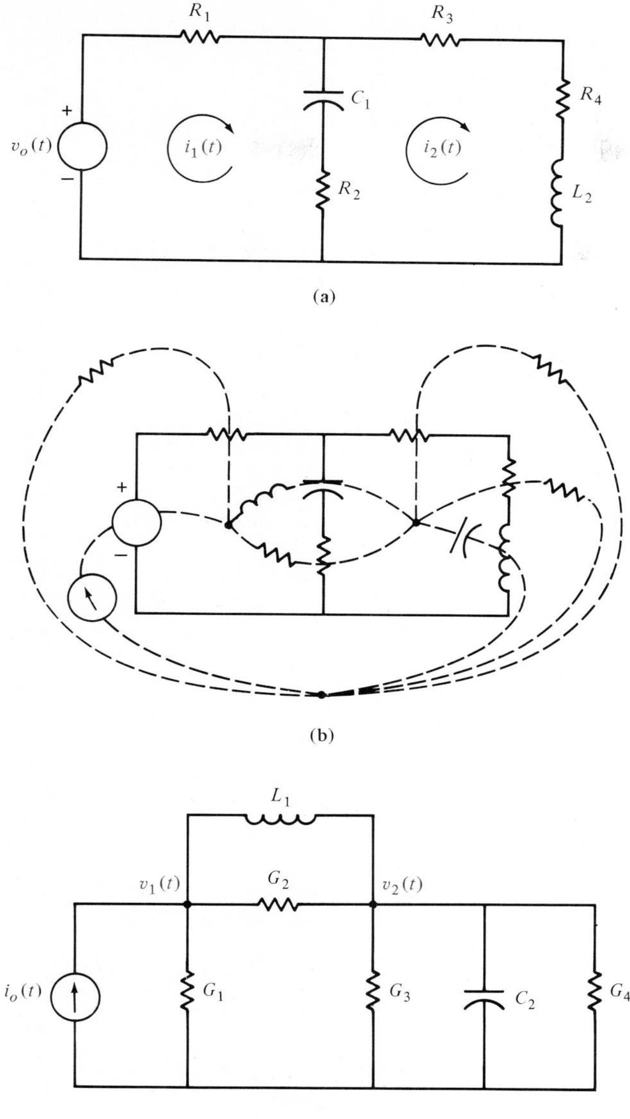

Figure 6.16 Development of a dual network.

EXAMPLE 6.12

We wish to draw the dual of the circuit shown in Fig. 6.16a using the graphical approach.

By following the procedure described above, the inside and outside dots are shown in Fig. 6.16b together with the dashed lines connecting the dots via the circuit elements. Since there are two independent loops in the original network, there will be two independent nodes in the dual network. To illustrate the dual replacement the dual elements are drawn into the dashed lines. The final network is shown in Fig. 6.16c. ■

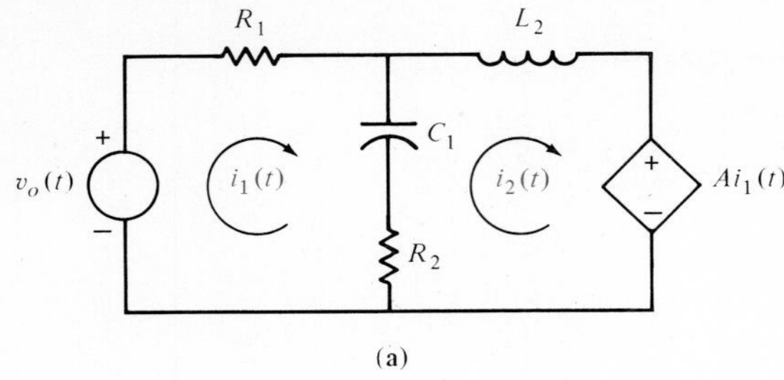

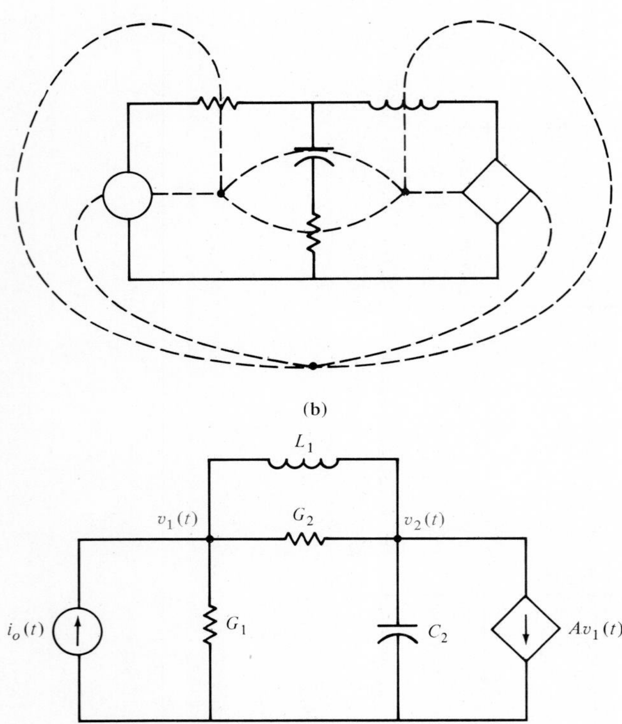

Figure 6.17 Development of a dual network with a dependent source.

EXAMPLE 6.13

The dual of the network shown in Fig. 6.17a is derived via Fig. 6.17b and shown in Fig. 6.17c. For comparison the loop equations for Fig. 6.17a are

$$v_o(t) = R_1 i_1(t) + \frac{1}{C_1}\int_0^t [i_1(x) - i_2(x)]\,dx + R_2[i_1(t) - i_2(t)]$$

$$-Ai_1(t) = R_2[i_2(t) - i_1(t)] + \frac{1}{C_1}\int_0^t [i_2(x) - i_1(x)]\,dx + L_2\frac{di_2(t)}{dt}$$

and the node equations for Fig. 6.17c are

$$i_o(t) = G_1 v_1(t) + \frac{1}{L_1}\int_0^t [v_1(x) - v_2(x)]\,dx + G_2[v_1(t) - v_2(t)]$$

$$-Av_1(t) = G_2[v_2(t) - v_1(t)] + \frac{1}{L_1}\int_0^t [v_2(x) - v_1(x)]\,dx + C_2\frac{dv_2(t)}{dt}$$

■

EXAMPLE 6.14

The dual of the circuit shown in Fig. 6.18a is given in Fig. 6.18b. Note the difference in the switch positions in the two networks.

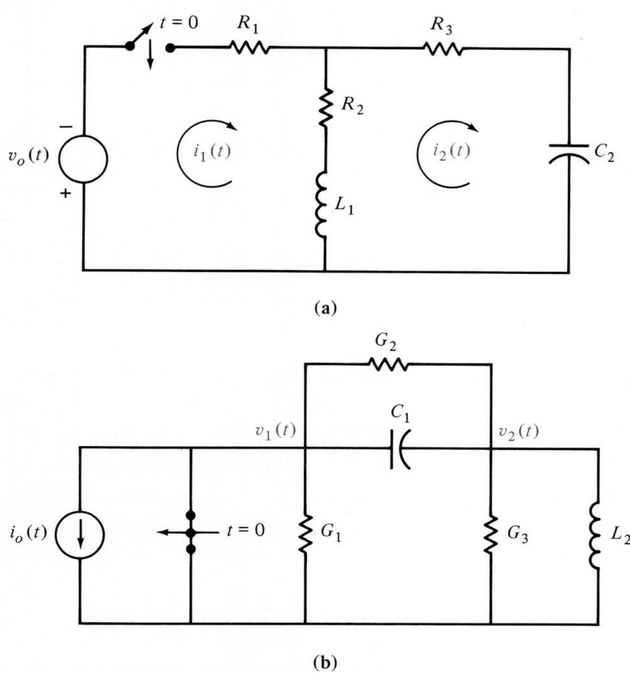

Figure 6.18 Example of dual networks. ■

DRILL EXERCISE

D6.11. Determine the dual of the network shown in Fig. D6.11.

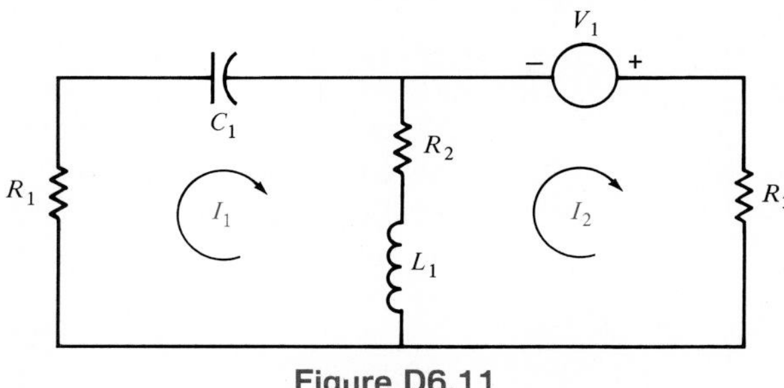

Figure D6.11

D6.12. Find the dual of the network shown in Fig. D6.12.

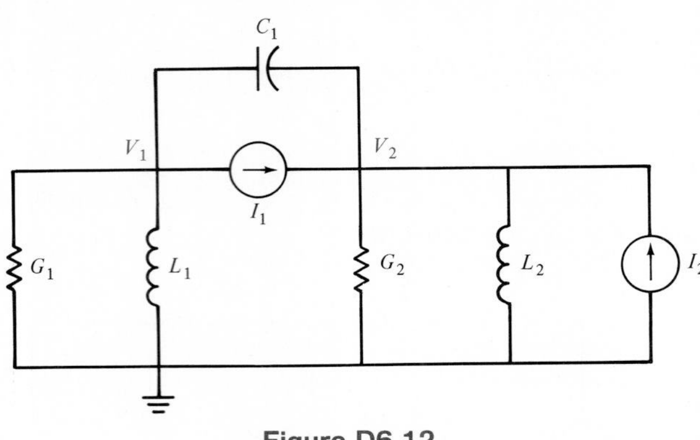

Figure D6.12

6.5 Summary

Two additional ideal circuit elements were introduced in this chapter: the capacitor and the inductor. We showed that in sharp contrast to the resistor, both of these new elements are capable of storing energy, and the energy relationships for these elements were determined. We illustrated how to compute equivalent capacitance when capacitors are interconnected in series or parallel, and how to determine equivalent inductance when these elements are interconnected. Finally, the concept of duality was presented and discussed.

KEY POINTS

- Capacitors and inductors are capable of storing energy.
- A capacitor looks like an open circuit to dc.
- An inductor looks like a short circuit to dc.
- When determining the equivalent capacitance of a number of interconnected capacitors, capacitors in series combine like resistors in parallel and capacitors in parallel combine like resistors in series.
- When determining the equivalent inductance of a number of interconnected inductors, inductors combine like resistors whether they are connected in series or parallel.
- Two circuits are said to be duals of one another if the mesh equations for one network have the same mathematical form as the nodal equations for the other.

PROBLEMS

6.1. Given a 3-μF capacitor and a 6-μF capacitor, what values of capacitance can be obtained by interconnecting them?

6.2. The two capacitors in Problem 6.1 are connected in parallel and charged to 12 V. Find

(a) The charge stored by each capacitor.

(b) The total energy stored.

6.3. Find the equivalent capacitance of the network shown in Fig. P6.3.

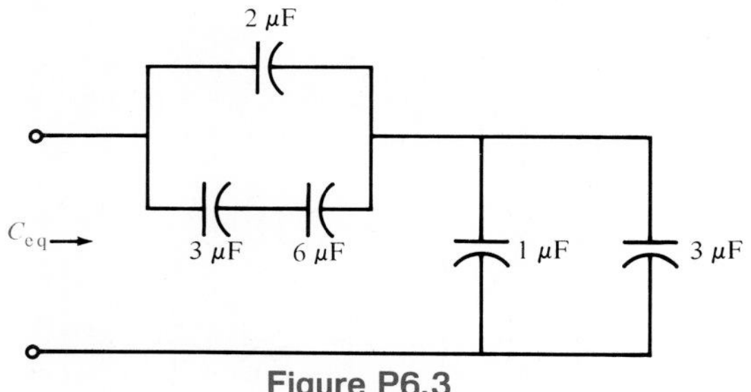

Figure P6.3

6.4. Find the equivalent capacitance of the network shown in Fig. P6.4.

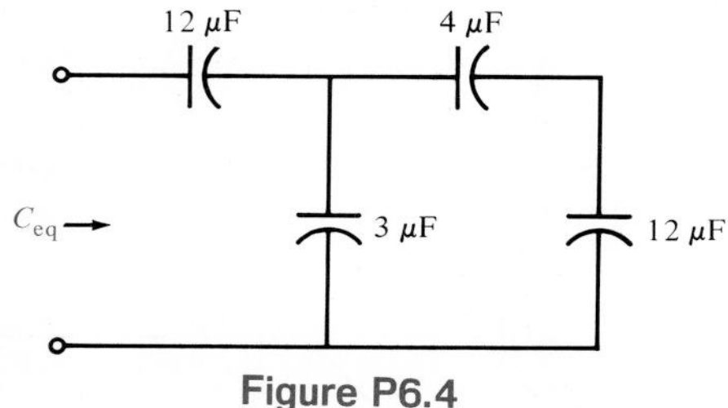

Figure P6.4

6.5. Find C_{eq} for the network in Fig. P6.5.

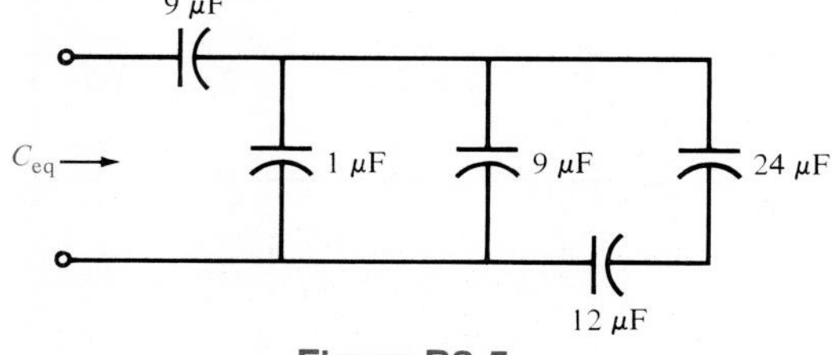

Figure P6.5

6.6. What values of capacitance can be obtained by interconnecting a 4-μF capacitor, a 6-μF capacitor, and a 12-μF capacitor?

6.7 Find the voltage across each of the capacitors if the open-circuit voltage in Fig. P6.7 is 30 V.

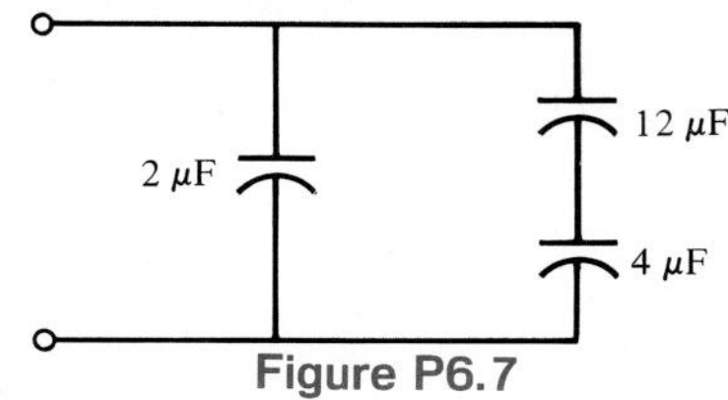

Figure P6.7

6.8. If the open-circuit voltage in the network in Fig. P6.8 is 28 V, find the voltage across each capacitor.

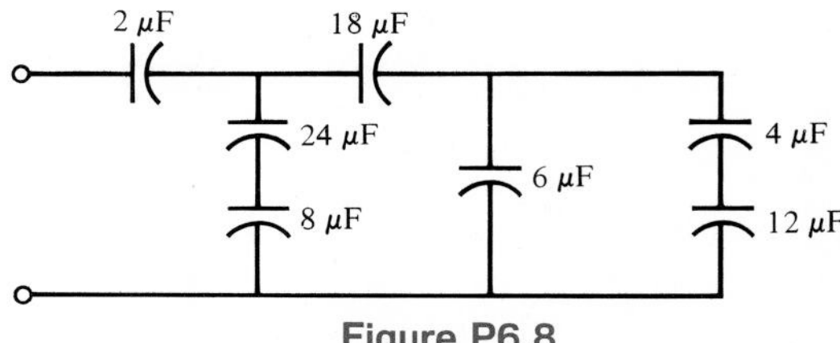

Figure P6.8

6.9. Find the equivalent capacitance of the network in Fig. P6.9 if

(a) *C-D* is short circuited—measured at *A-B*.

(b) *A-B* is open circuited—measured at *C-D*.

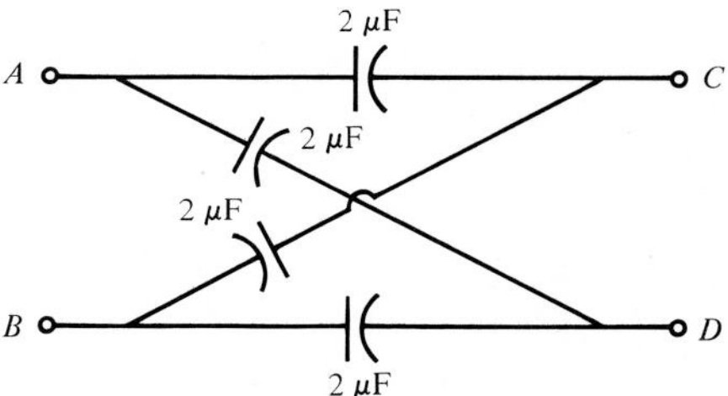

Figure P6.9

6.10. Determine the total capacitance of the network in Fig. P6.10.

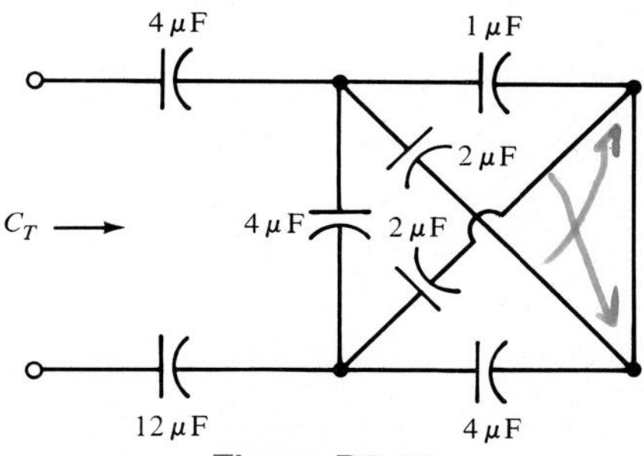

Figure P6.10

6.11. Find the total capacitance of the network in Fig. P6.11.

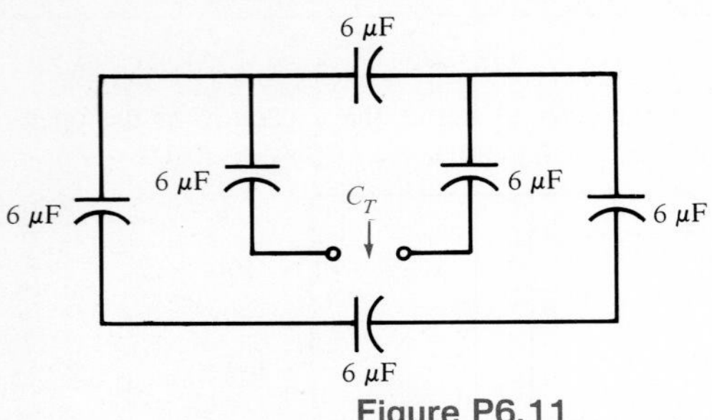

Figure P6.11

6.12. Find the total capacitance of the network in Fig. P6.12.

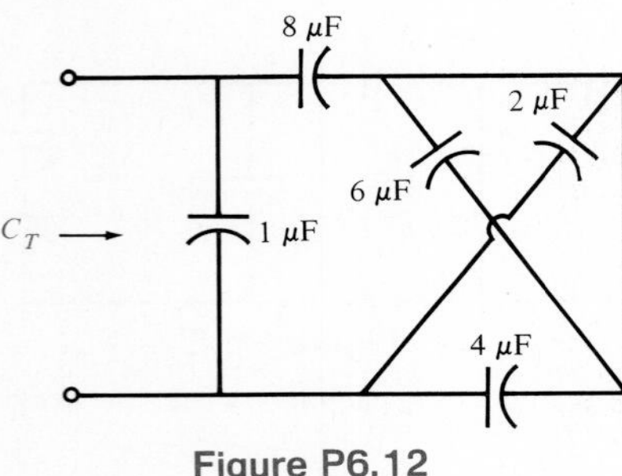

Figure P6.12

6.13. Find the total capacitance of the network in Fig. P6.13.

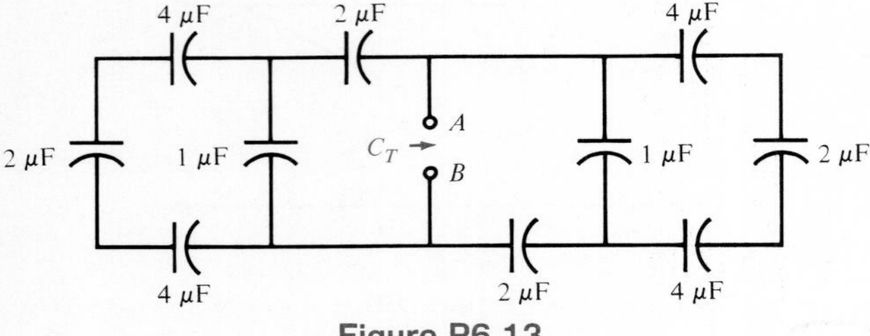

Figure P6.13

6.14. Determine the total capacitance of the network in Fig. P6.14.

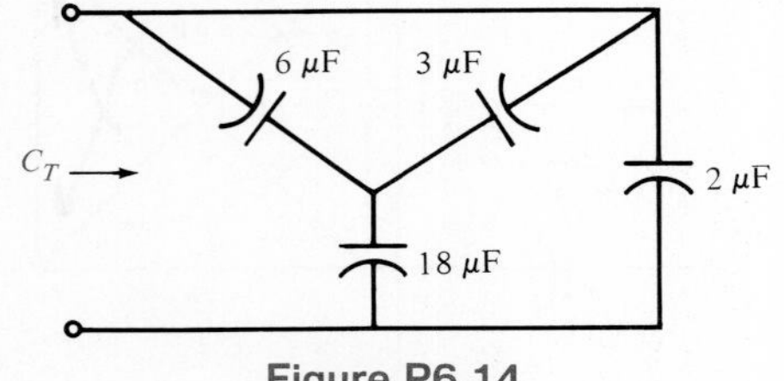

Figure P6.14

6.15. Find the total capacitance of the network in Fig. P6.15.

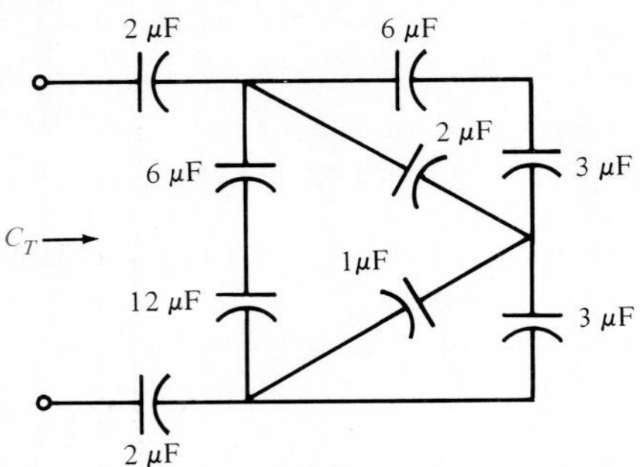

Figure P6.15

6.16. Determine the capacitance at terminals *A-B* in the network in Fig. P6.16.

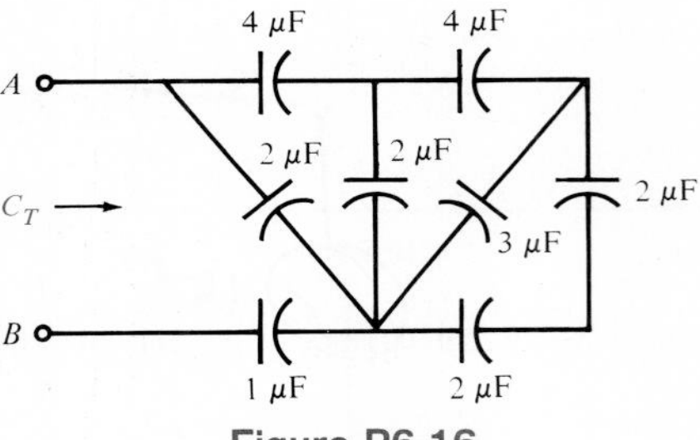

Figure P6.16

6.17. Determine the capacitance at terminals *A-B* in the network in Fig. P6.17.

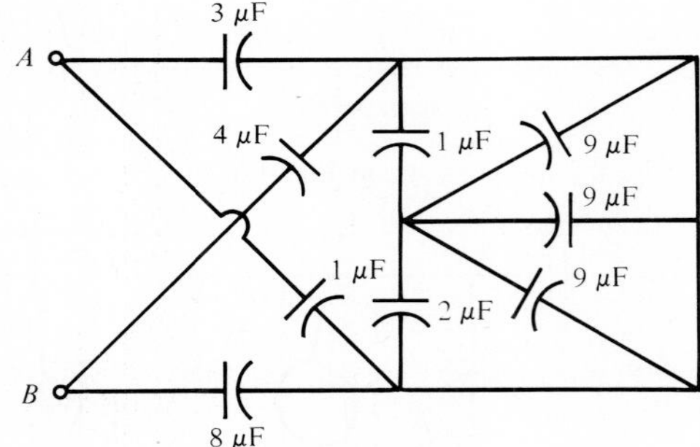

Figure P6.17

6.18. Two capacitors C_1 and C_2 are connected in series. If the voltage across the combination is $v_o(t)$, show that the voltage division between the capacitors is

$$v_1 = \frac{C_2}{C_1 + C_2} v_o \quad \text{and} \quad v_2 = \frac{C_1}{C_1 + C_2} v_o$$

where v_1 is the voltage across C_1 and v_2 is the voltage across C_2.

6.19. The voltage across the capacitance network in Fig. P6.4 is 24 V. Find the voltage across each capacitor.

6.20. Find the equivalent capacitance at terminals *A-B* in Fig. P6.20.

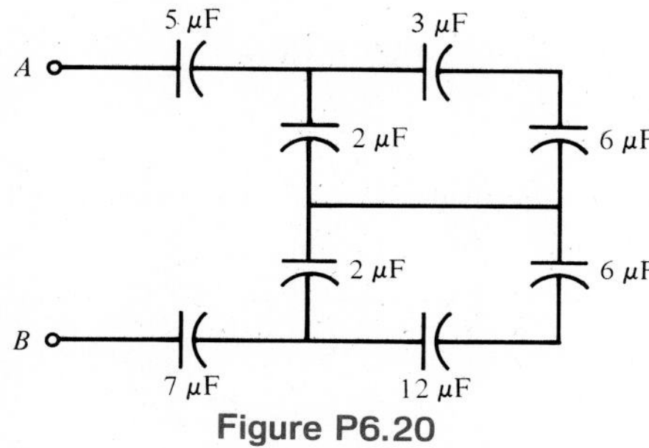

Figure P6.20

6.21. Draw the waveform for the current in a 5-μF capacitor when the voltage across the capacitor is as given in Fig. P6.21.

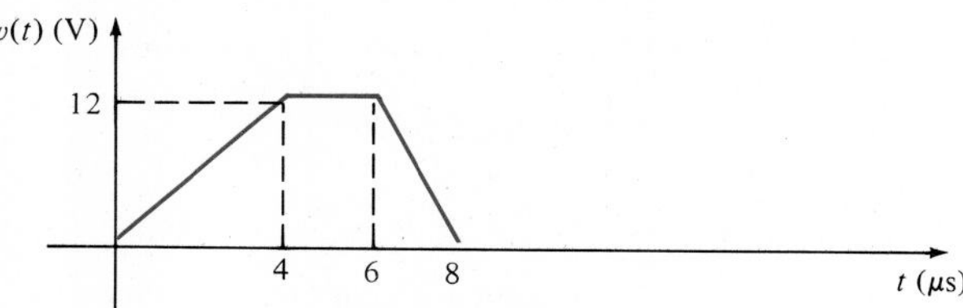

Figure P6.21

6.22. The voltage across a 10-μF capacitor is given in Fig. P6.22. Calculate the current waveform.

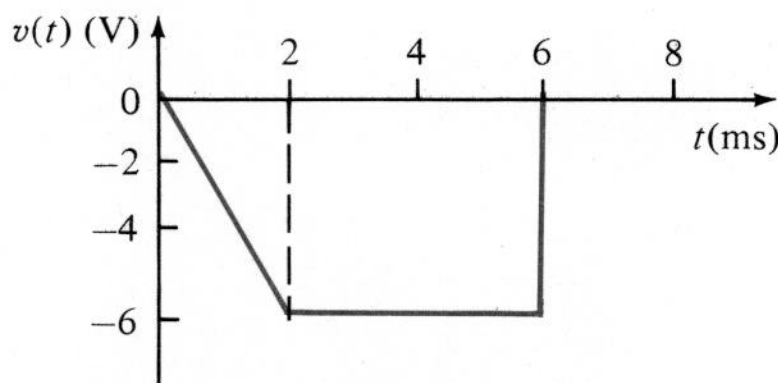

Figure P6.22

6.23. Draw the waveform for the current in a 3-μF capacitor when the voltage across the capacitor is given in Fig. P6.23.

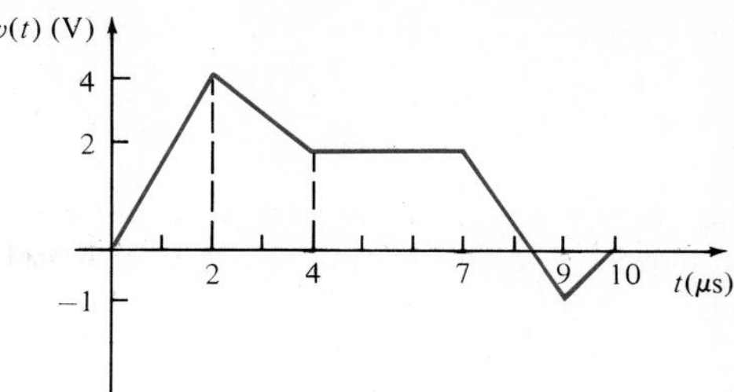

Figure P6.23

6.24. If the current in a 10-μF capacitor is given by the expression $i(t) = 30e^{-t}$, find the expression for the voltage and sketch the waveform. Assume that the initial voltage is 2 V.

6.25. The voltage across a 50-μF capacitor is shown in Fig. P6.25. Determine the current waveform.

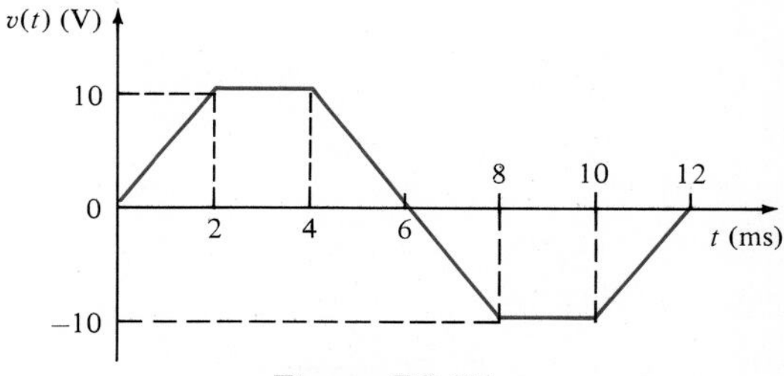

Figure P6.25

6.26. The voltage across a 2-μF capacitor is shown in Fig. P6.26. What is the current waveform?

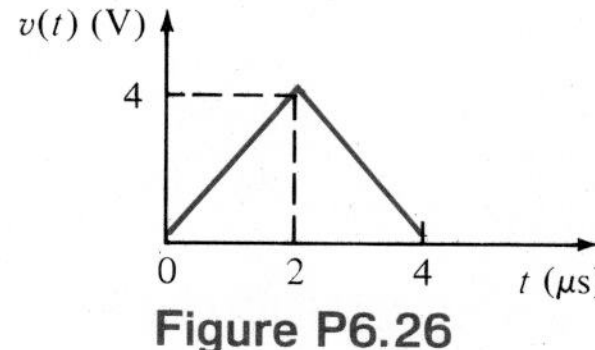

Figure P6.26

6.27. Draw the waveform for the current in a 12-μF capacitor when the capacitor voltage is as described in Fig. P6.27.

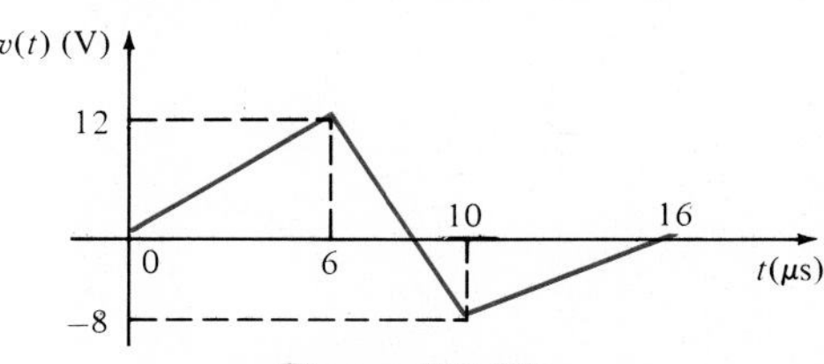

Figure P6.27

6.28. The voltage across a 20-μF capacitor is given by the expression $v(t) = 120 \sin 377t$. Find the expression for the current and sketch the waveform.

6.29. The current in an inductor changes from 0 to 200 mA in 4 ms and induces a voltage of 100 mV. What is the value of the inductor?

6.30. If the current $i(t) = 1.5t$ A flows through a 2-H inductor, find the energy stored at $t = 2$ s.

6.31. Find the equivalent inductance in Fig. P6.31.

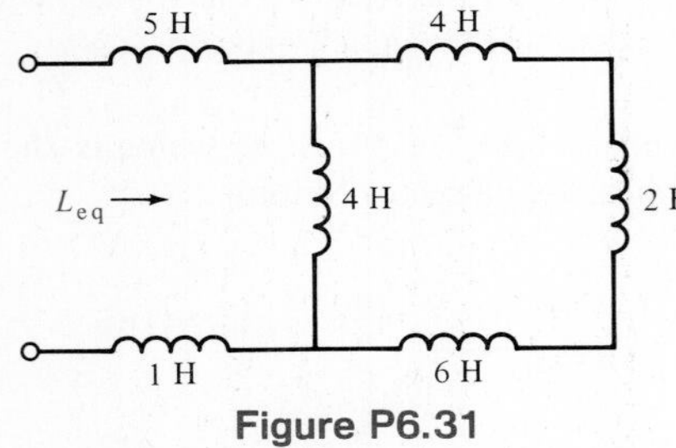

Figure P6.31

6.32. If the voltage across the network in Fig. P6.32 is 20 V, determine the voltage across each inductor.

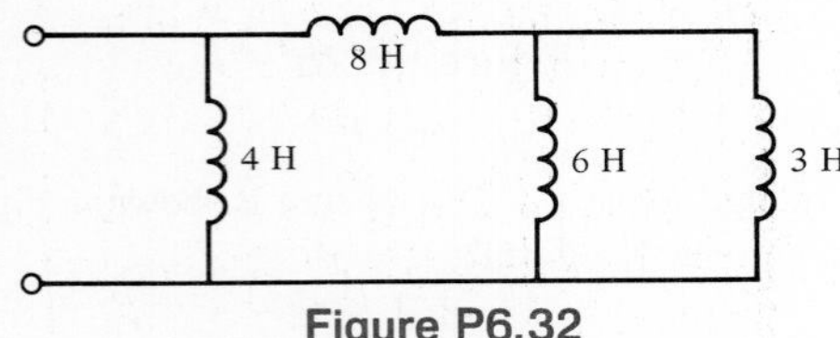

Figure P6.32

6.33. If the current flowing into the circuit in Fig. P6.32 is 7A, determine the current passing through each inductor.

6.34. Determine the inductance at terminals *A-B* in the network in Fig. P6.34.

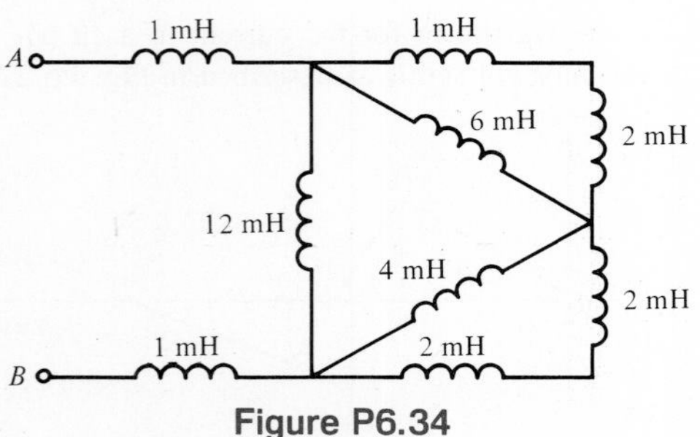

Figure P6.34

6.35. Determine the inductance at terminals *A-B* in the network in Fig. P6.35.

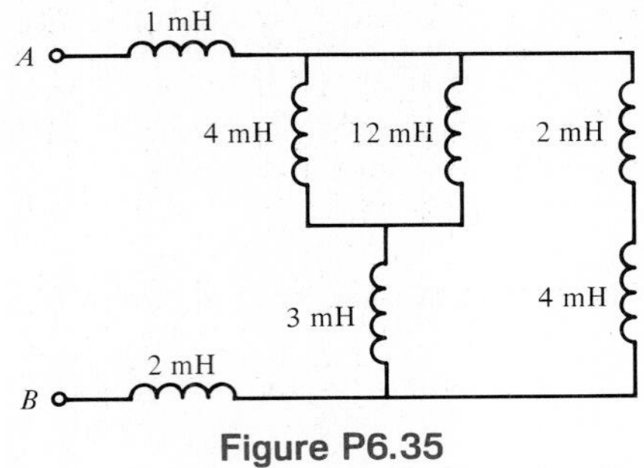

Figure P6.35

6.36. Find the equivalent inductance in the network in Fig. P6.36.

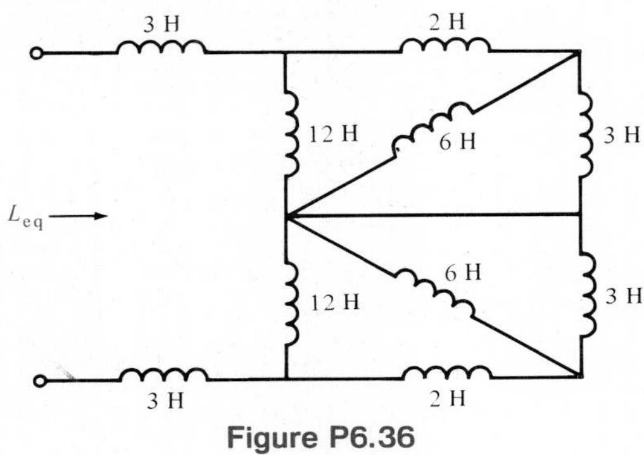

Figure P6.36

6.37. Determine the equivalent inductance at terminals *A-B* in Fig. P6.37.

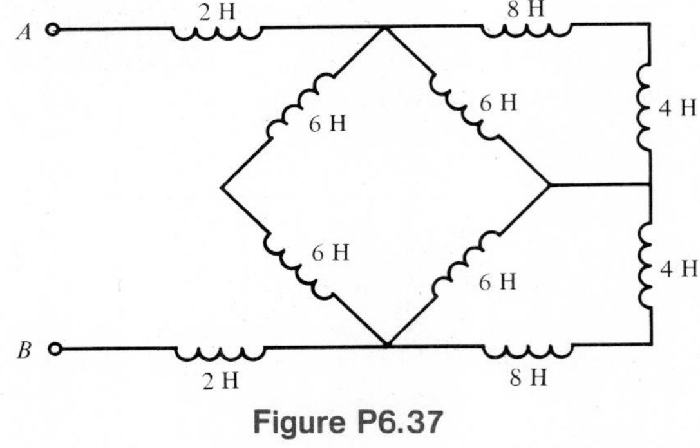

Figure P6.37

6.38. Given the two-port network shown in Fig. P6.38, find

(a) The equivalent inductance at terminals A-B with terminals C-D short-circuited.

(b) The equivalent inductance at terminals C-D with terminals A-B open-circuited.

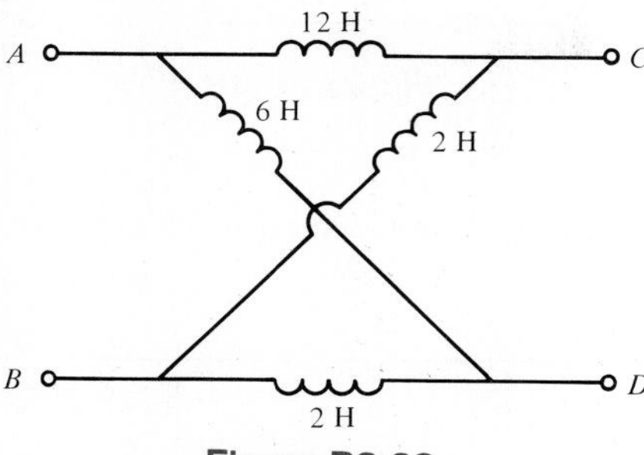

Figure P6.38

6.39. Find the total inductance for the network in Fig. P6.39.

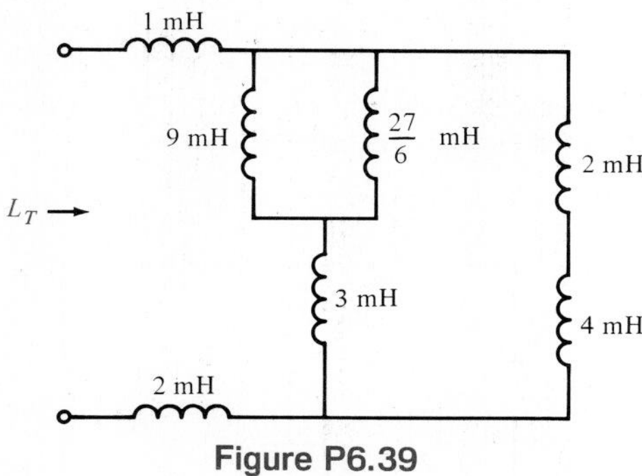

Figure P6.39

6.40. Find the total inductance for the network in Fig. P6.40.

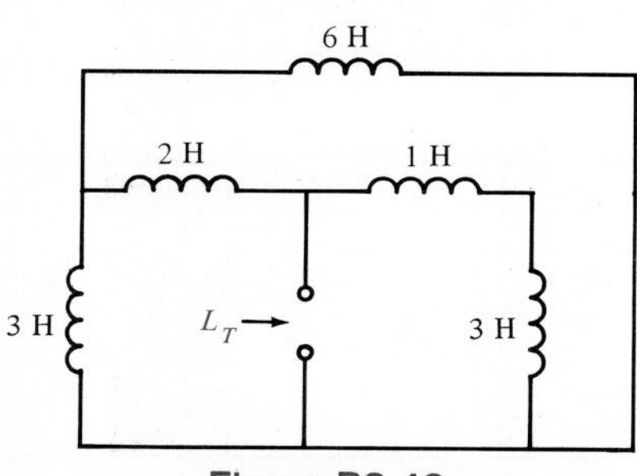

Figure P6.40

6.41. Find the equivalent inductance for the network in Fig. P6.41.

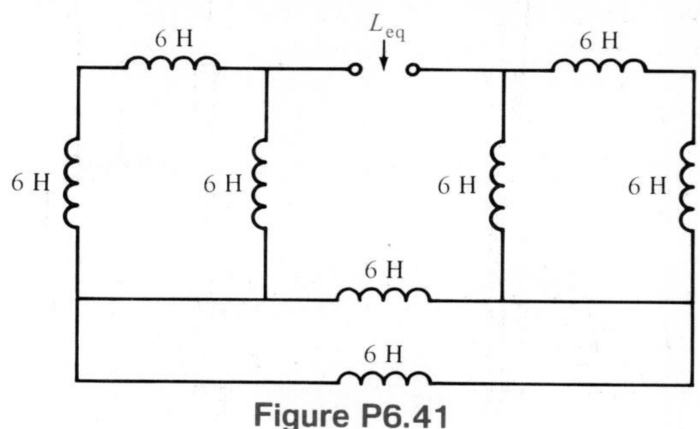

Figure P6.41

6.42. Two inductors L_1 and L_2 are connected in parallel. The current entering the combination is $i_o(t)$. Derive the expressions for the current division between the inductors.

6.43. The current in a 3-H inductor is shown in Fig. P6.43. Determine the waveform for the inductor voltage.

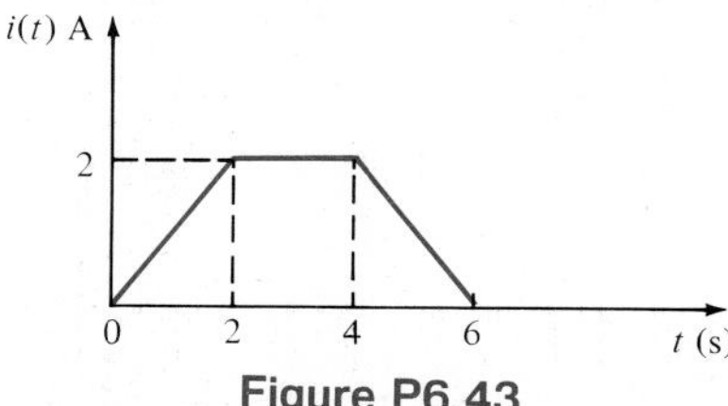

Figure P6.43

6.44. The current in a 50-mH inductor is given in Fig. P6.44. Sketch the inductor voltage.

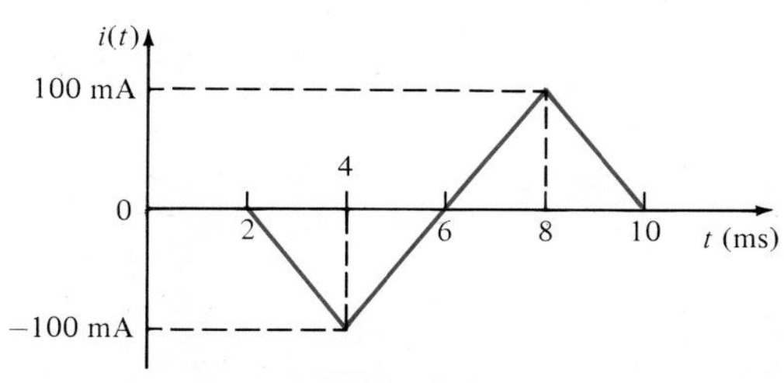

Figure P6.44

6.45. Draw the waveform for the voltage across a 10-mH inductor when the inductor current is given by the waveform shown in Fig. P6.45.

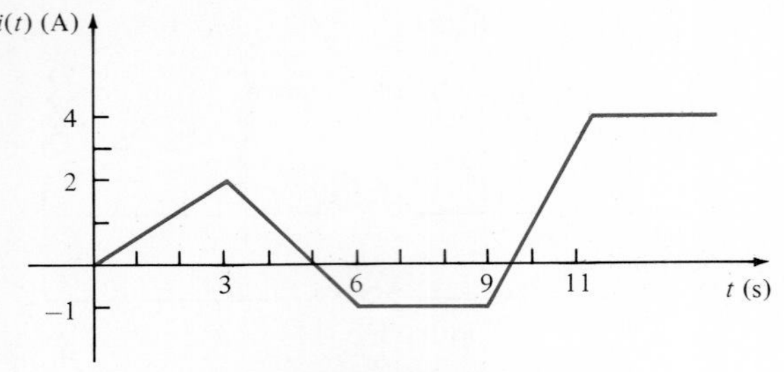

Figure P6.45

6.46. Determine the waveform for the voltage developed across a 30-mH inductor whose current waveform is shown in Fig. P6.46. The expression for the current in the interval $0 > t > 4$ s is $5e^{-t}$ A.

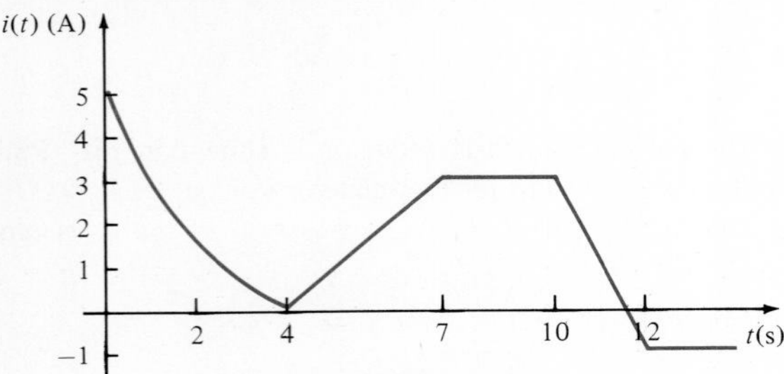

Figure P6.46

6.47. The current in a 2-H inductor is given by the waveform shown in Fig. P6.47. Determine the waveform for the inductor voltage.

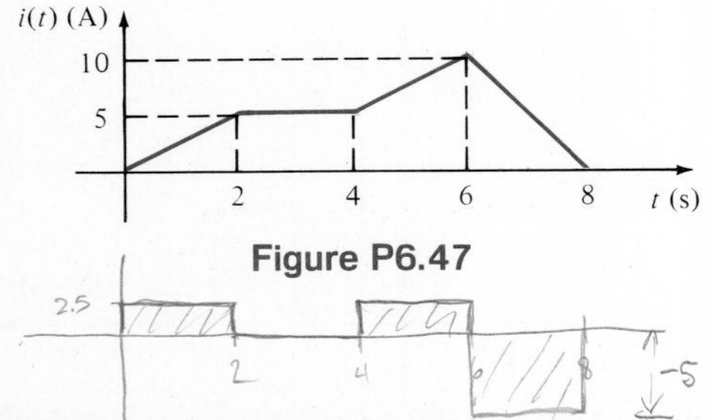

Figure P6.47

6.48. The voltage across a 2-H inductor is given by the waveform shown in Fig. P6.48. Find the waveform for the current in the inductor.

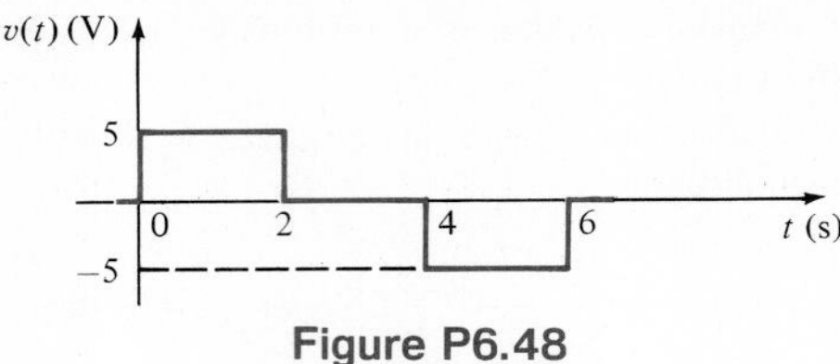

Figure P6.48

6.49. Draw the dual network for the network shown in Fig. P6.49.

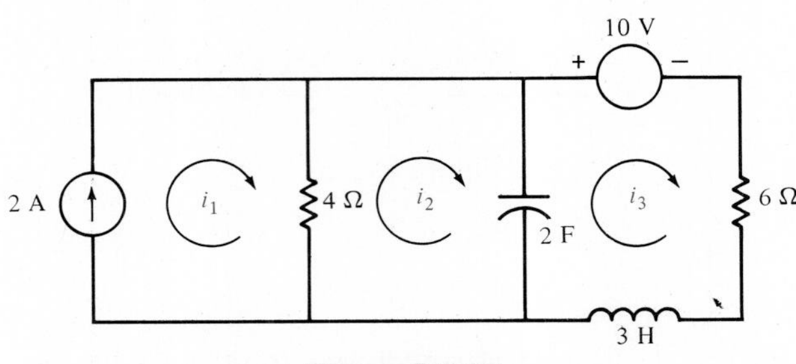

Figure P6.49

6.50. In the network shown in Fig. P6.50, the switch closes at $t = 0$. Draw the dual for this network.

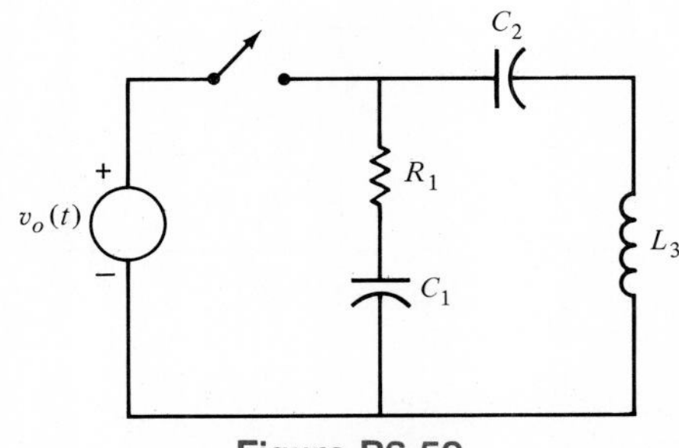

Figure P6.50

6.51. Determine the dual of the circuit shown in Fig. P6.51.

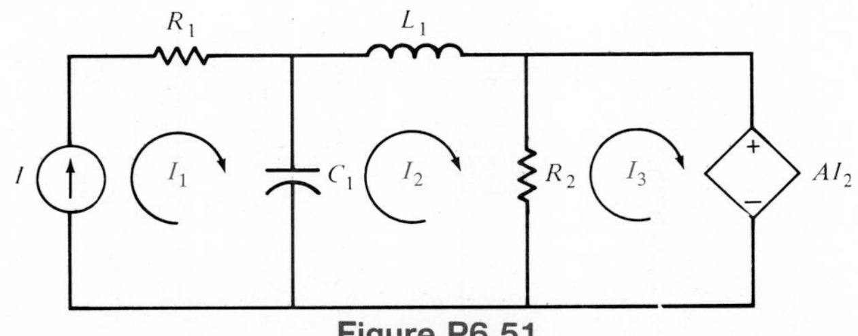

Figure P6.51

CHAPTER

RC and *RL* Circuits

In this chapter we perform what is normally referred to as a transient analysis of a network that contains a single storage element. We examine the behavior of a circuit as a function of time after a sudden change in the network occurs due to switches opening or closing. The time constant of a network is introduced, and a step procedure for obtaining a transient solution of a network is presented. Finally, we present the techniques required to perform a SPICE transient analysis.

7.1 Development of the Fundamental Procedures

In our study of *RC* and *RL* circuits we will show that the solution of these circuits, that is, finding a voltage or current, requires us to solve a first-order differential equation of the form

$$\frac{dx(t)}{dt} + ax(t) = f(t) \tag{7.1}$$

Although there are a number of techniques for solving an equation of this type—for example, separation of variables, integrating factor method, or simply guessing a solution—we will employ the procedure outlined below to obtain a general solution that is applicable to the *RC* and *RL* circuit problems we analyze in this chapter.

A fundamental theorem of differential equuations states that if $x(t) = x_p(t)$ is any solution to Eq. (7.1), and $x(t) = x_c(t)$ is any solution to the homogeneous equation

$$\frac{dx(t)}{dt} + ax(t) = 0 \tag{7.2}$$

then

$$x(t) = x_p(t) + x_c(t) \tag{7.3}$$

is a solution to the original Eq. (7.1). The term $x_p(t)$ is called the *particular integral solution* and $x_c(t)$ is called the *complementary solution*.

At the present time we confine ourselves to the situation in which $f(t) = A$ (i.e., some constant). The general solution of the differential equation then consists of two parts that are obtained by solving the two equations

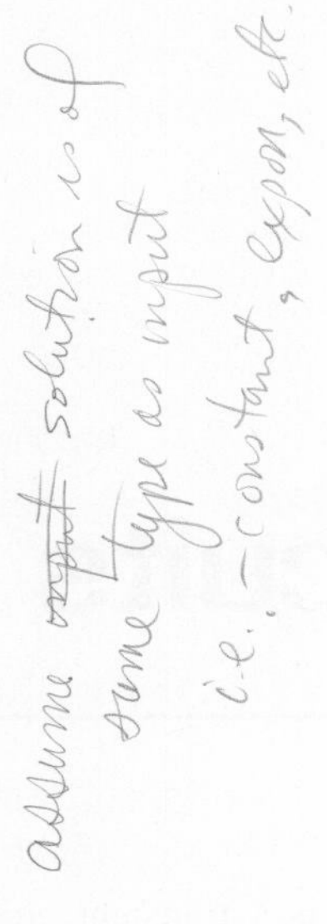

$$\frac{dx_p(t)}{dt} + ax_p(t) = A \tag{7.4}$$

$$\frac{dx_c(t)}{dt} + ax_c(t) = 0 \tag{7.5}$$

Since the right-hand side of Eq. (7.4) is a constant, it is reasonable to assume that the solution $x_p(t)$ must also be a constant. Therefore, we assume that

$$x_p(t) = K_1 \tag{7.6}$$

Substituting this constant into Eq. (7.4) yields

$$K_1 = \frac{A}{a} \tag{7.7}$$

Examining Eq. (7.5), we note that

$$\frac{dx_c(t)/dt}{x_c(t)} = -a \tag{7.8}$$

This equation is equivalent to

$$\frac{d}{dt}[\ln x_c(t)] = -a$$

Hence

$$\ln x_c(t) = -at + c$$

and therefore

$$x_c(t) = K_2 e^{-at} \tag{7.9}$$

Therefore, a solution of Eq. (7.1) is

$$x(t) = x_p(t) + x_c(t)$$

$$= \frac{A}{a} + K_2 e^{-at} \tag{7.10}$$

The constant K_2 can be found if the value of the independent variable $x(t)$ is known at one instant of time.

Once the solution in Eq. (7.10) is obtained, certain elements of the equation are given names that are commonly employed in electrical engineering. For example, the term A/a is referred to as the *steady-state solution*: the value of the variable $x(t)$ as

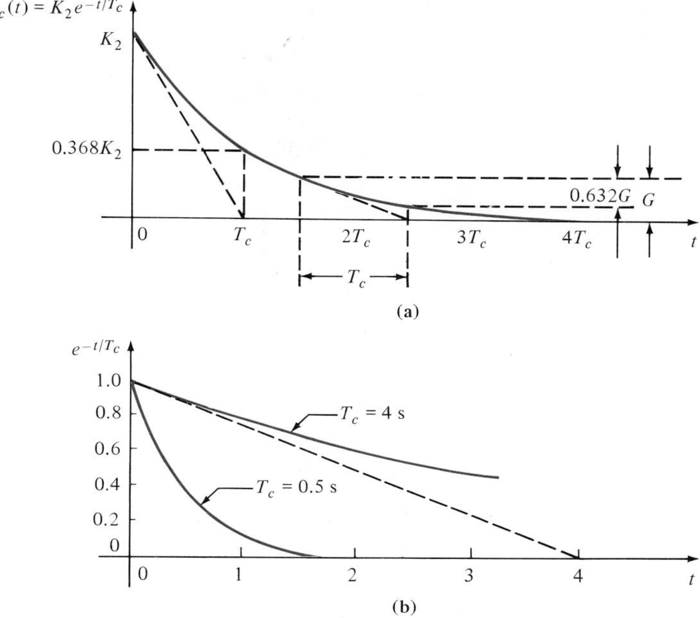

Figure 7.1 Time-constant illustrations.

$t \to \infty$ when the second term becomes negligible. The constant $1/a$ is called the *time constant* of the circuit. Note that the second term in Eq. (7.10) is a decaying exponential that has a value, if $a > 0$, of K_2 for $t = 0$ and a value of 0 for $t = \infty$. The rate at which this exponential decays is determined by the time constant $T_c = 1/a$. A graphical picture of this effect is shown in Fig. 7.1a. As can be seen from the figure, the value of $x_c(t)$ has fallen from K_2 to a value of $0.368K_2$ in one time constant, a drop of 63.2%. In two time constants the value of $x_c(t)$ has fallen to $0.135K_2$, a drop of 63.2% from the value at time $t = T_c$. This means that the gap between a point on the curve and the final value of the curve is closed by 63.2% each time constant. Finally, after five time constants, $x_c(t) = 0.0067K_2$, which is less than 1%.

An interesting property of the exponential function shown in Fig. 7.1a is that the initial slope of the curve intersects the time axis at a value of $t = T_c$. In fact, we can take any point on the curve, not just the initial value, and find the time constant by finding the time required to close the gap by 63.2%. Finally, the difference between a small time constant (i.e., fast response) and a large time constant (i.e., slow response) is shown in Fig. 7.1b. These curves indicate that if the circuit has a small time constant, it settles down quickly to a steady-state value. Conversely, if the time constant is large, more time is required for the circuit to settle down or reach steady state. In any case, note that the circuit response essentially reaches steady state within five time constants (i.e., $5T_c$).

Note that the previous discussion has been very general, in that no particular form of the circuit has been assumed—only that it results in a first-order differential equation.

Next, we will study two specific circuits and then outline a method for handling these circuits in general. We will then look at more specific cases using the general method.

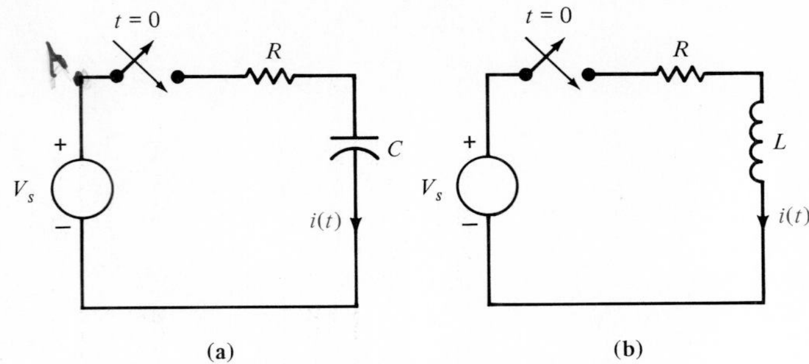

(a) (b)

Figure 7.2 *RC* and *RL* circuits.

Consider the circuit shown in Fig. 7.2a. At time $t = 0$ the switch closes. The equation that describes the circuit for time $t > 0$ is

$$\frac{1}{C}\int i(x)\,dx + Ri(t) = V_S$$

Taking the derivative of this equation with respect to t yields

$$\frac{i(t)}{C} + R\frac{di(t)}{dt} = 0$$

or

$$\frac{di(t)}{dt} + \frac{1}{RC}i(t) = 0$$

Following our previous development, we assume that the solution of this first-order differential equation is of the form

$$i(t) = K_2 e^{-t/T_c}$$

Substituting this solution into the differential equation yields

$$-\frac{K_2 e^{-t/T_c}}{T_c} + \frac{1}{RC}K_2 e^{-t/T_c} = 0$$

$$\left(-\frac{1}{T_c} + \frac{1}{RC}\right)K_2 e^{-t/T_c} = 0$$

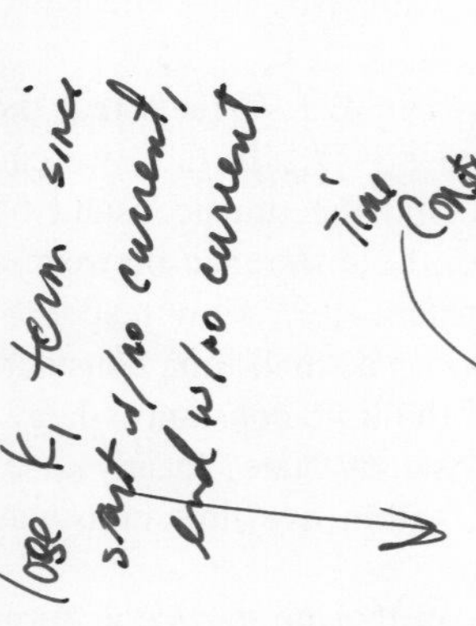

Our assumed solution is valid if $K_2 e^{-t/T_c} = 0$ or $T_c = RC$. The first case implies that $i(t) = 0$ for all t and therefore is disregarded. Therefore,

$$T_c = RC$$

and hence the solution is

$$i(t) = K_2 e^{-t/RC}$$

The constant K_2 is chosen so that the general solution satisfies the particular conditions of the circuit.

The circuit in Fig. 7.2b can be examined in a manner similar to that employed for the circuit in Fig. 7.2a. The equation that describes the circuit for $t > 0$ is

$$L\frac{di(t)}{dt} + Ri(t) = V_S$$

or

$$\frac{di(t)}{dt} + \frac{R}{L}i(t) = \frac{V_S}{L}$$

From our earlier discussions we assume a solution of the form

$$i(t) = K_1 + K_2 e^{-t/T_c}$$

Substituting this expression into the differential equation, we obtain

$$-\frac{1}{T_c}K_2 e^{-t/T_c} + \frac{R}{L}(K_1 + K_2 e^{-t/T_c}) = \frac{V_S}{L}$$

Equating the constant and exponential terms yields

$$\frac{R}{L}K_1 = \frac{V_S}{L}$$

$$\left(-\frac{1}{T_c} + \frac{R}{L}\right)K_2 e^{-t/T_c} = 0$$

Our previous analysis indicates that

$$K_1 = \frac{V_S}{R}$$

and

$$T_c = \frac{L}{R}$$

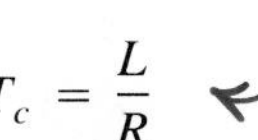

Therefore, the solution is

$$i(t) = \frac{V_S}{R} + K_2 e^{-(R/L)t}$$

where once again the constant K_2 is chosen so that the general solution will satisfy the particular conditions of the circuit.

The importance of the two circuits in Fig. 7.2 stems from the fact that we will find that by employing Thévenin's theorem, we can reduce complicated circuits to these forms, which yield the circuit time constant upon which the circuit response is based.

In general, when the input to an *RC* or *RL* circuit, which contains only a single storage element, is a dc voltage or current, the solution of the differential equation that describes an unknown current or voltage *anywhere in the network* can be written as

$$x(t) = K_1 + K_2 e^{-t/T_c} \tag{7.11}$$

Note that this was the case in the development above and the value of K_1, the steady-state solution, was obtained directly from the differential equation. In our analysis of electrical circuits it is more convenient to determine the constants from an analysis of a modified circuit, as will be shown later.

From Eq. (7.11) we note that as $t \rightarrow \infty$, $e^{-at} \rightarrow 0$ and $x(t) = K_1$. Therefore, if the circuit is solved for the variable $x(t)$ in steady state (i.e., $t \rightarrow \infty$) with the capacitor replaced by an open circuit [v is constant and therefore $i = C(dv/dt) = 0$] or the inductor replaced by a short circuit [i is constant and therefore $v = L(di/dt) = 0$], then the variable $x(t) = K_1$. Note that since the capacitor or inductor has been removed, the circuit is a dc circuit with constant sources and resistors and therefore only dc analysis is required in the steady-state solution.

The constant K_2 in Eq. (7.11) can also be obtained via the solution of a dc circuit in which a capacitor is replaced by a voltage source or an inductor is replaced by a current source. The value of the voltage source for the capacitor or the current source for the inductor is a known value at one instant of time. In general, we will use the initial condition value since it is generally the one known, but the value at any instant could be used. This value can be obtained in numerous ways and is often specified as input data in a statement of the problem. However, a more likely situation is one in which a switch is thrown in the circuit and the initial value of the capacitor voltage or inductor current is determined from the previous circuit (i.e., the circuit before the switch is thrown). It is normally assumed that the previous circuit has reached steady state, and therefore the voltage across the capacitor or the current through the inductor can be found in exactly the same manner as was used to find K_1. Therefore, K_2 is found by solving the circuit as it was connected before the switch was thrown with the capacitor replaced by an open circuit or the inductor replaced by a short circuit.

Finally, the value of the time constant can be found by determining the Thévenin equivalent resistance at the terminals of the storage element. Then $T_c = R_{\text{Th}} C$ for an RC circuit, and $T_c = L/R_{\text{Th}}$ for an RL circuit.

Let us now reiterate this procedure in a step-by-step fashion.

Step 1. We assume a solution for the variable $x(t)$ of the form $x(t) = K_1 + K_2 e^{-t/T_c}$.

Step 2. Assuming that the original circuit has reached steady state before a switch was thrown (thereby producing a new circuit), draw this previous circuit with the capacitor replaced by an open circuit or the inductor replaced by a short circuit. Solve for the voltage across the capacitor, $v_C(0-)$, or the current through the inductor, $i_L(0-)$, prior to switch action.

Step 3. Assuming that the energy in the storage element cannot change in zero time, draw the circuit, valid only at $t = 0+$. The switches are in their new positions and the capacitor is replaced by a voltage source with a value of $v_C(0+) = v_C(0-)$ or the inductor is replaced by a current source with value $i_L(0+) = i_L(0-)$. Solve for the initial value of the variable $x(0+)$.

Step 4. Assuming that steady state has been reached after the switches are thrown, draw the equivalent circuit, valid for $t > 5T_c$, by replacing the capacitor by an open circuit or the inductor by a short circuit. Solve for the steady-state value of the variable

$$x(t)\Big|_{t>5T_c} \doteq x(\infty)$$

Step 5. Since the time constant for all voltages and currents in the circuit will be the same, it can be obtained by reducing the entire circuit to a simple series circuit containing a voltage source, resistor, and a storage element (i.e., capacitor or inductor) by forming a simple Thévenin equivalent circuit at the terminals of the storage element. This Thévenin equivalent circuit is obtained by looking into the circuit from the terminals of the storage element. The time constant for a circuit containing a capacitor is $T_c = R_{Th} C$, and for a circuit containing an inductor it is $T_c = L/R_{Th}$.

Step 6. Using the results of steps 3, 4, and 5, we can evaluate the constants in step 1 as

$$x(0) = K_1 + K_2$$

$$x(\infty) = K_1$$

and therefore $K_1 = x(\infty)$, $K_2 = x(0) - x(\infty)$, and hence the solution is

$$x(t) = x(\infty) + [x(0) - x(\infty)]e^{-t/T_c}$$

7.2 Source-Free Circuits

We begin our analysis of source-free networks by examining the circuit, which consists of an initially charged capacitor in series with a resistor as shown in Fig. 7-3a. It is assumed that the capacitor is charged to a voltage $v(t) = V_o$ at $t = 0$. Since no independent sources are present in the network, the circuit response is dependent only on the passive circuit elements and the initial voltage. The currents and voltages in the circuit for $t > 0$ can be obtained via KCL. Employing KCL at the top node yields

$$C\frac{dv(t)}{dt} + \frac{v(t)}{R} = 0$$

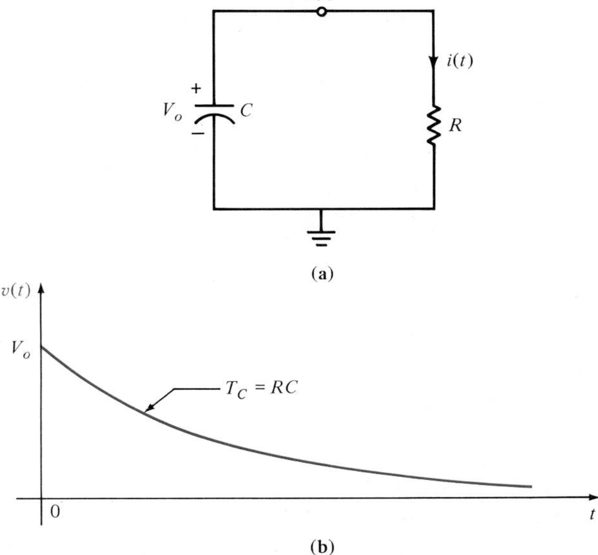

Figure 7.3 Source-free *RC* circuit and its response to an initial condition.

or

$$\frac{dv(t)}{dt} + \frac{1}{RC} v(t) = 0$$

Let us now apply the step-by-step procedure above to the analysis of this circuit.

Step 1. $v(t)$ is of the form

$$v(t) = K_1 + K_2 e^{-t/T_c}$$

Step 2. The initial voltage across the capacitor is given as $v_C(0-) = V_o$.

Step 3. $v_C(0-) = v_C(0+) = V_o$. Since our unknown is $v(t)$, which is equal to the voltage across the capacitor $v_C(t)$, $v(0+) = V_o$.

Step 4. If the capacitor is replaced by an open circuit, the voltage $v(t)$ across the resistor will be zero. Hence for $t > 5T_c$, $v(\infty) = 0$.

Step 5. The Thévenin equivalent resistance looking into the circuit from the capacitor terminals is obviously $R_{\text{Th}} = R$. Therefore, the circuit time constant is $T_c = RC$.

Step 6. Using the results above, we find that

$$v(\infty) = 0 = K_1$$

$$v(0) = V_o = K_1 + K_2$$

and therefore,

$$v(t) = V_o e^{-t/RC}$$

Note that the current in the resistor has the same form as the voltage and is given by the expression

$$i(t) = \frac{v(t)}{R} = \frac{V_o}{R} e^{-t/RC}$$

At time $t = 0$ the energy stored in the capacitor is

$$w_C(0) = \frac{1}{2} CV_o^2 \qquad \text{joules}$$

As time increases, the voltage decreases as shown in Fig. 7.3b, and hence the energy stored in the capacitor also decreases. As this energy decreases, it is absorbed by the resistor. The total energy absorbed by the resistor is

$$\begin{aligned} w_R(\infty) &= \int_0^\infty p_R(x)\, dx \\ &= \int_0^\infty \left(\frac{V_o}{R} e^{-x/RC}\right)^2 R\, dx \\ &= \frac{V_o^2}{R} \int_0^\infty e^{-2x/RC}\, dx \\ &= \frac{1}{2} CV_o^2 \qquad \text{joules} \end{aligned}$$

This is, of course, equal to the initial energy stored in the capacitor.

It is interesting to note that if KVL had been employed to determine the describing equation for the circuit, the result would have been

$$\frac{1}{C}\int_0^t i(x) - v_C(0) + i(t)R = 0$$

Differentiating the equation yields

$$\frac{i(t)}{C} + R\frac{di(t)}{dt} = 0$$

since

$$i(t) = \frac{v(t)}{R}$$

Then

$$\frac{v(t)}{C} + \frac{dv(t)}{dt} = 0$$

which is, of course, the original equation in our analysis.

EXAMPLE 7.1

Consider the circuit shown in Fig. 7.4a. Assuming that the switch has been in position 1 for a long time, at time $t = 0$ the switch is moved to position 2. We wish to calculate the current $i(t)$ for $t > 0$.

Step 1. We assume the current to be of the form

$$i(t) = K_1 + K_2e^{-t/T_c}$$

Step 2. In steady state prior to switch action, the initial capacitor voltage is found from Fig. 7.4b as

$$v_C(0-) = \frac{(12)(3\ k\Omega)}{6\ \text{k}\Omega + 3\ \text{k}\Omega} = 4\ \text{V}$$

Step 3. The new circuit, valid only for $t = 0+$, is shown in Fig. 7.4c. The value of the voltage source that replaces the capacitor is $v_C(0-) = v_C(0+) = 4$ V. Therefore,

$$i(0+) = \frac{4}{3} \times 10^{-3}\ \text{A}$$

Step 4. The equivalent circuit, valid for $t > 5T_c$, is shown in Fig. 7.4d. Since there are no sources present, the value of $i(\infty) = 0$.

Step 5. The Thévenin equivalent resistance, obtained by looking into the circuit from the open-circuit terminals of the capacitor in Fig. 7.4d, is

$$R_{\text{Th}} = \frac{(6\ \text{k}\Omega)(3\ \text{k}\Omega)}{(6\ \text{k}\Omega) + (3\ \text{k}\Omega)} = 2\ \text{k}\Omega$$

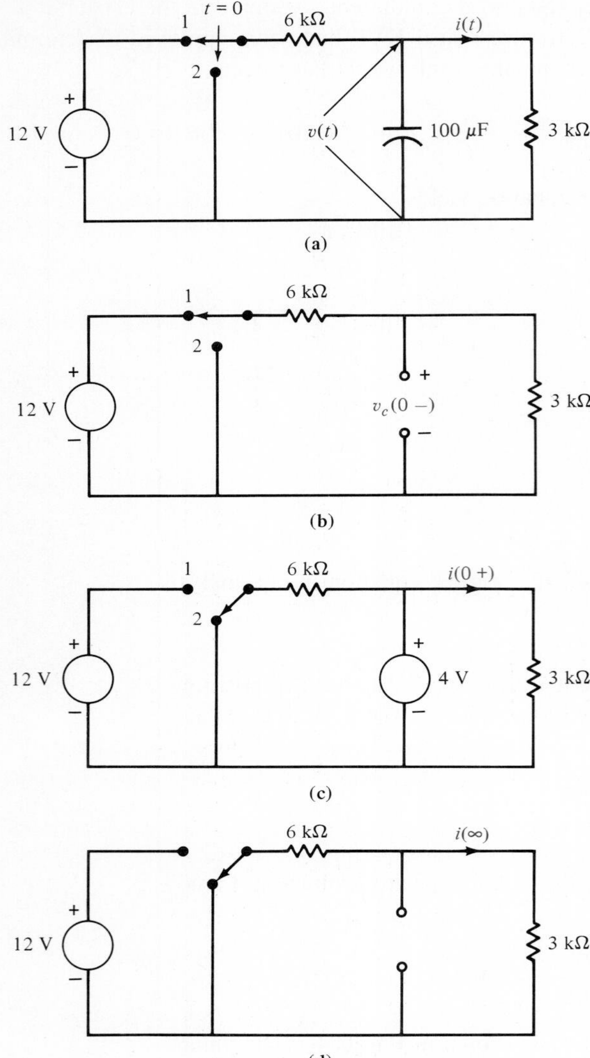

Figure 7.4 Analysis of an *RC* transient circuit.

Therefore, the circuit time constant is

$$\begin{aligned} T_c &= R_{\text{Th}}\, C \\ &= (2 \times 10^3)(100 \times 10^{-6}) \\ &= 0.2 \text{ s} \end{aligned}$$

Step 6. From the previous analysis

$$K_1 = i(\infty) = 0$$

$$K_2 = i(0) - i(\infty) = \tfrac{4}{3} \times 10^{-3}$$

and hence

$$i(t) = \tfrac{4}{3} e^{-t/0.2}$$
$$= 1.33\, e^{-5t} \text{ mA} \qquad \blacksquare$$

DRILL EXERCISE

D7.1. Consider the network in Fig. D7.1. At $t = 0$ the switch opens. Find $v_o(t)$ for $t > 0$.

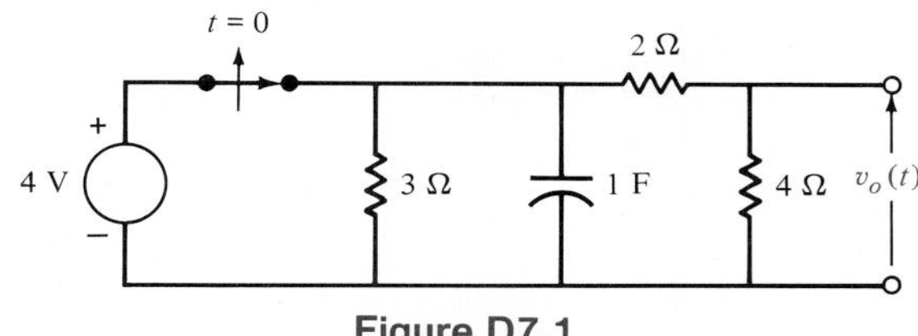

Figure D7.1

D7.2. Consider the network in Fig. D7.2. The switch closes at $t = 0$. Find $v_o(t)$ for $t > 0$.

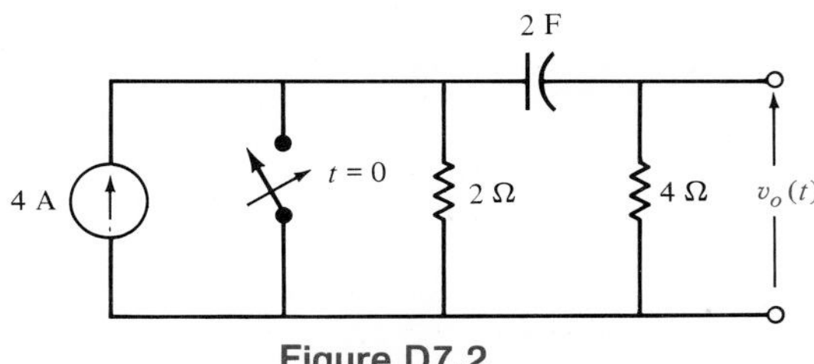

Figure D7.2

Let us next consider a source-free *RL* circuit as shown in Fig. 7.5a. It is assumed that the inductor has an initial current of $i(t) = I_o$ at $t = 0$. Applying KVL around the loop yields

$$L\frac{di(t)}{dt} + Ri(t) = 0$$

or

$$\frac{di(t)}{dt} + \frac{R}{L}i(t) = 0$$

Once again we employ the step-by-step procedure to analyze this circuit:

Step 1. $i(t)$ is of the form $K_1 + K_2 e^{-t/T_c}$.

Step 2. The initial current through the inductor is $i_L(0-) = I_o$.

Step 3. $i(0+) = i_L(0+) = i_L(0-) = I_o$.

Step 4. If the inductor is replaced by a short circuit, $v(t)$ for $t > 5T_c$ will be zero and hence $i(\infty) = 0$.

Step 5. The Thévenin equivalent resistance looking into the circuit from the inductor terminals is $R_{\text{Th}} = R$; therefore, the circuit time constant is $T_c = L/R$.

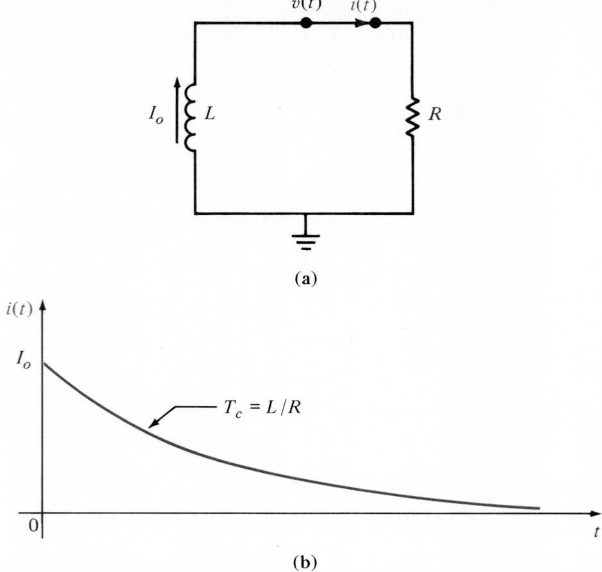

Figure 7.5 Source-free *RL* circuit and its response to an initial condition.

Step 6. The results above show that

$$i(\infty) = K_1 = 0$$

Thus

$$i(0) = K_1 + K_2 = I_o$$

and hence

$$i(t) = I_o e^{-Rt/L}$$

At time $t = 0$ the energy stored in the inductor is

$$w_L(0) = \tfrac{1}{2} L I_o^2 \qquad \text{joules}$$

As time increases the current decreases, as shown in Fig. 7.5b, and hence the energy stored in the inductor will also decrease. This energy is absorbed by the resistor. The energy that the resistor absorbs is given by the expression

$$\begin{aligned} w_R(\infty) &= \int_0^\infty p_R(t)\, dt \\ &= R I_o^2 \int_0^\infty e^{-2Rt/L}\, dt \\ &= \frac{1}{2} L I_o^2 \qquad \text{joules} \end{aligned}$$

This value coincides with that for the energy initially stored in the inductor.

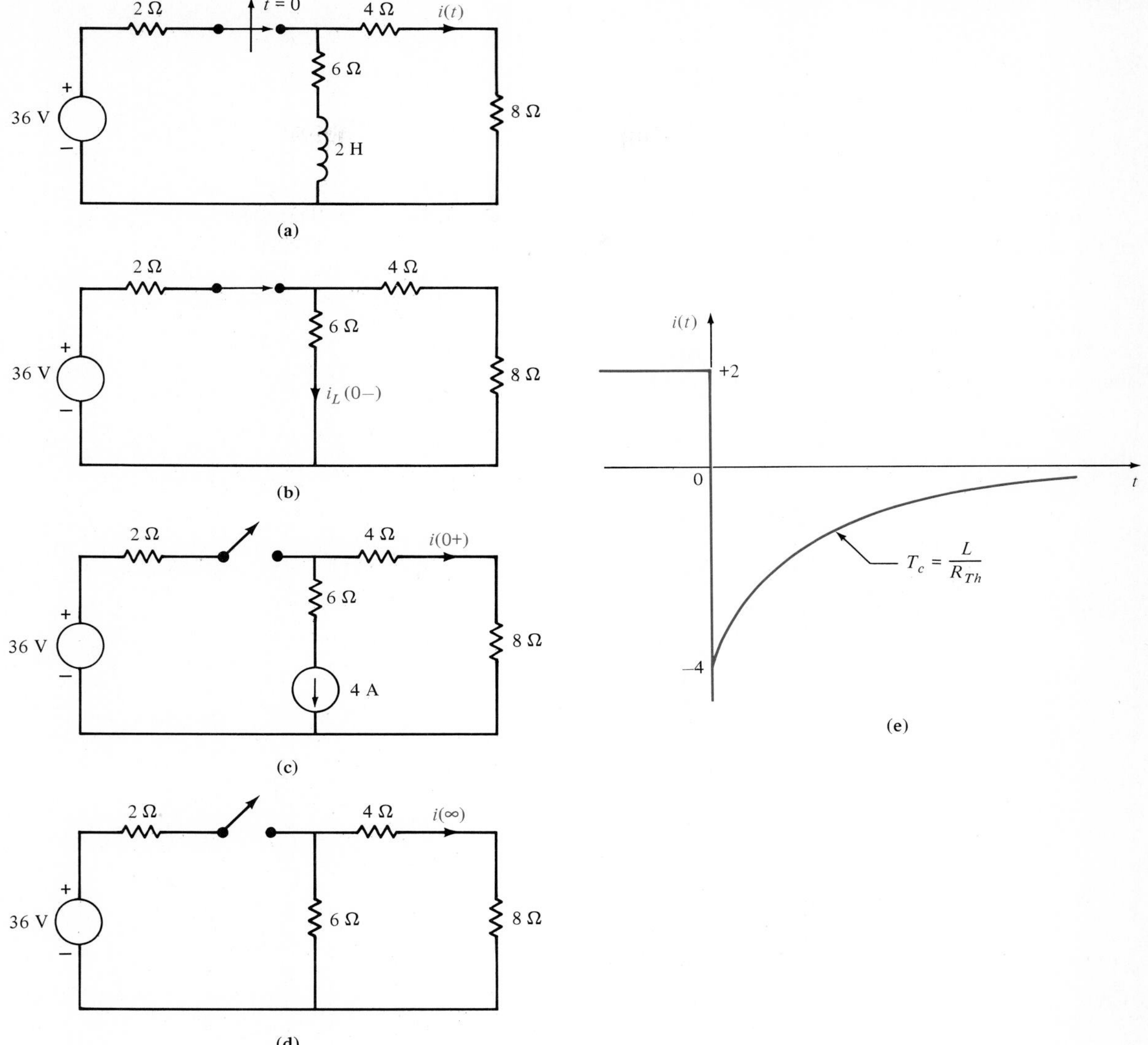

Figure 7.6 Analysis of an *RL* circuit.

EXAMPLE 7.2

Let us examine the circuit shown in Fig. 7.6a. At $t = 0$ the switch is opened and we wish to calculate the current $i(t)$ for $t > 0$.

Step 1. The current is assumed to be of the form

$$i(t) = K_1 + K_2 e^{-t/T_c}$$

Step 2. In the steady-state mode prior to the time the switch opens, the initial inductor current can be found from Fig. 7.6b as

$$i_L(0-) = \frac{36}{2 + \dfrac{(6)(12)}{6 + 12}} \left(\frac{12}{18}\right)$$

$$= 4 \text{ A}$$

Step 3. The new circuit, valid only for $t = 0+$, is shown in Fig. 7.6c. The value of the current source that replaces the inductor is $i_L(0-) = i_L(0+) = 4$ A. Therefore,

$$i(0+) = -4 \text{ A}$$

Step 4. The equivalent circuit, valid only for $t > 5T_c$, is given in Fig. 7.6d. Since there are no sources present, the value of $i(\infty) = 0$.

Step 5. The Thévenin equivalent resistance found by looking into the circuit from the inductor terminals is

$$R_{\text{Th}} = 6 + 4 + 8$$

$$= 18\ \Omega$$

Therefore, the circuit time constant is

$$T_c = \frac{L}{R_{\text{Th}}}$$

$$= \frac{2}{18}$$

$$= \frac{1}{9} \text{ s}$$

Step 6. From the foregoing analysis,

$$K_1 = i(\infty) = 0$$

Then

$$K_2 = i(0) - i(\infty) = -4$$

and hence

$$i(t) = -4e^{-9t} \text{ A}$$

Note that although the current in the inductor is continuous at $t = 0$, the current $i(t)$ in the 4- and 8-Ω resistors has jumped abruptly at $t = 0$, as shown in Fig. 7.6e. ■

DRILL EXERCISE

D7.3. The switch in the network in Fig. D7.3 opens at $t = 0$. Find $v_o(t)$ for $t > 0$.

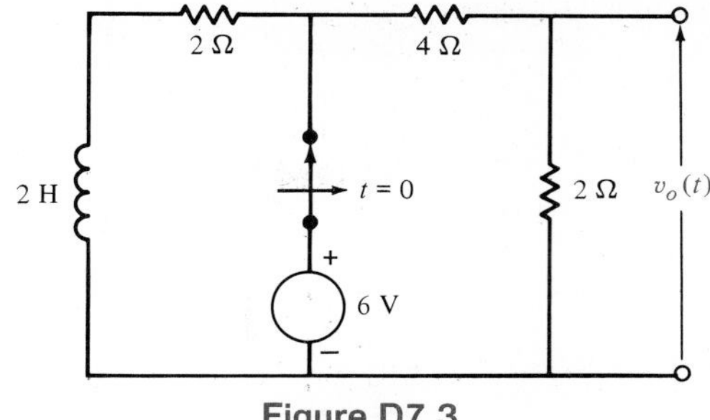

Figure D7.3

D7.4. The switch in the network in Fig. D7.4 closes at $t = 0$. Find $i_o(t)$ for $t > 0$.

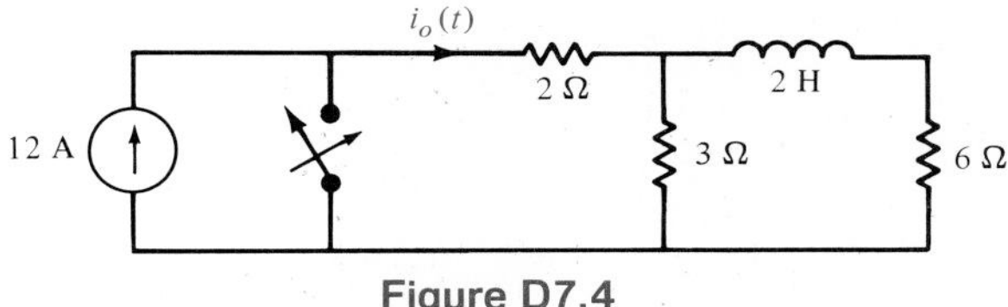

Figure D7.4

7.3 Circuits with Constant Forcing Functions

In Section 7.2 we were concerned with the natural response of circuits to certain initial conditions. In every case the independent variable $x(t) \to 0$ as $t \to \infty$. In this section we examine circuits that contain constant forcing functions in addition to stored initial conditions. To solve these circuits, we will simply employ the same general procedure outlined above. In this case, however, we find that $K_1 = x(\infty) \neq 0$; that is, the steady-state value is some constant other than zero.

EXAMPLE 7.3

Consider the circuit shown in Fig. 7.7a. The circuit is in steady state prior to time $t = 0$, when the switch is closed. Let us calculate the current $i(t)$ for $t > 0$.

Step 1. $i(t)$ is of the form $K_1 + K_2 e^{-t/T_c}$.

Step 2. The initial voltage across the capacitor is calculated from Fig. 7.7b as

$$\begin{aligned} v_C(0-) &= 36 \text{ V} - i(0-)2 \text{ k}\Omega \\ &= 32 \text{ V} \end{aligned}$$

Step 3. The new circuit, valid only for $t = 0+$, is shown in Fig. 7.7c.

The value of the voltage source that replaces the capacitor is $v_C(0-) = v_C(0+) = 32$.

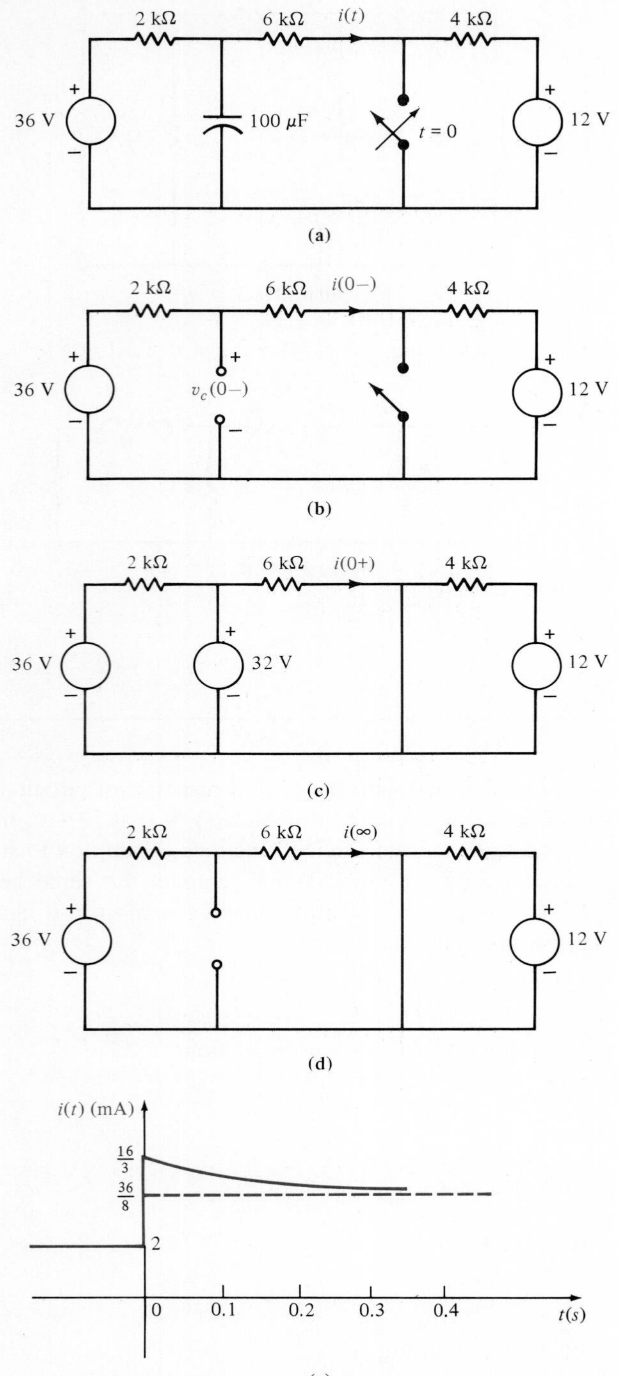

Figure 7.7 Analysis of an *RC* transient circuit with a constant forcing function.

Hence

$$i(0+) = \frac{32 \text{ V}}{6 \text{ k}\Omega} = \frac{16}{3} \times 10^{-3} \text{ A}$$

Step 4. The equivalent circuit, valid for $t > 5T_c$, is shown in Fig. 7.7d. The current $i(\infty)$ caused by the 36-V source is

$$i(\infty) = \frac{36 \text{ V}}{2 \text{ k}\Omega + 6 \text{ k}\Omega} = \frac{36}{8} \times 10^{-3} \text{ A}$$

Step 5. The Thévenin equivalent resistance, obtained by looking into the open-circuit terminals of the capacitor in Fig. 7.7d, is

$$R_{\text{Th}} = \frac{(2 \text{ k}\Omega)(6 \text{ k}\Omega)}{2 \text{ k}\Omega + 6 \text{ k}\Omega} = \frac{3}{2} \text{ k}\Omega$$

Therefore, the circuit time constant is

$$\begin{aligned} T_c &= R_{\text{Th}}C \\ &= \left(\frac{3}{2} \times 10^3\right)\left(100 \times 10^{-6}\right) \\ &= 0.15 \text{ s} \end{aligned}$$

Step 6

$$K_1 = i(\infty) = \frac{36}{8} \text{ mA}$$

$$\begin{aligned} K_2 &= i(0) - i(\infty) = i(0) - K_1 \\ &= \frac{16}{3} \text{ mA} - \frac{36}{8} \text{ mA} \\ &= \frac{5}{6} \text{ mA} \end{aligned}$$

Therefore,

$$i(t) = \frac{36}{8} + \frac{5}{6} e^{-t/0.15} \text{ mA}$$

Once again we see that although the voltage across the capacitor is continuous at $t = 0$, the current $i(t)$ in the 6-kΩ resistor jumps at $t = 0$ from 2 mA to $5\frac{1}{3}$ mA, and finally decays to $\frac{36}{8}$ mA. ■

EXAMPLE 7.4

The circuit shown in Fig. 7.8a is assumed to have been in a steady-state condition prior to switch closure at $t = 0$. We wish to calculate the voltage $v(t)$ for $t > 0$.

Step 1. $v(t)$ is of the form $K_1 + K_2 e^{-t/T_c}$.

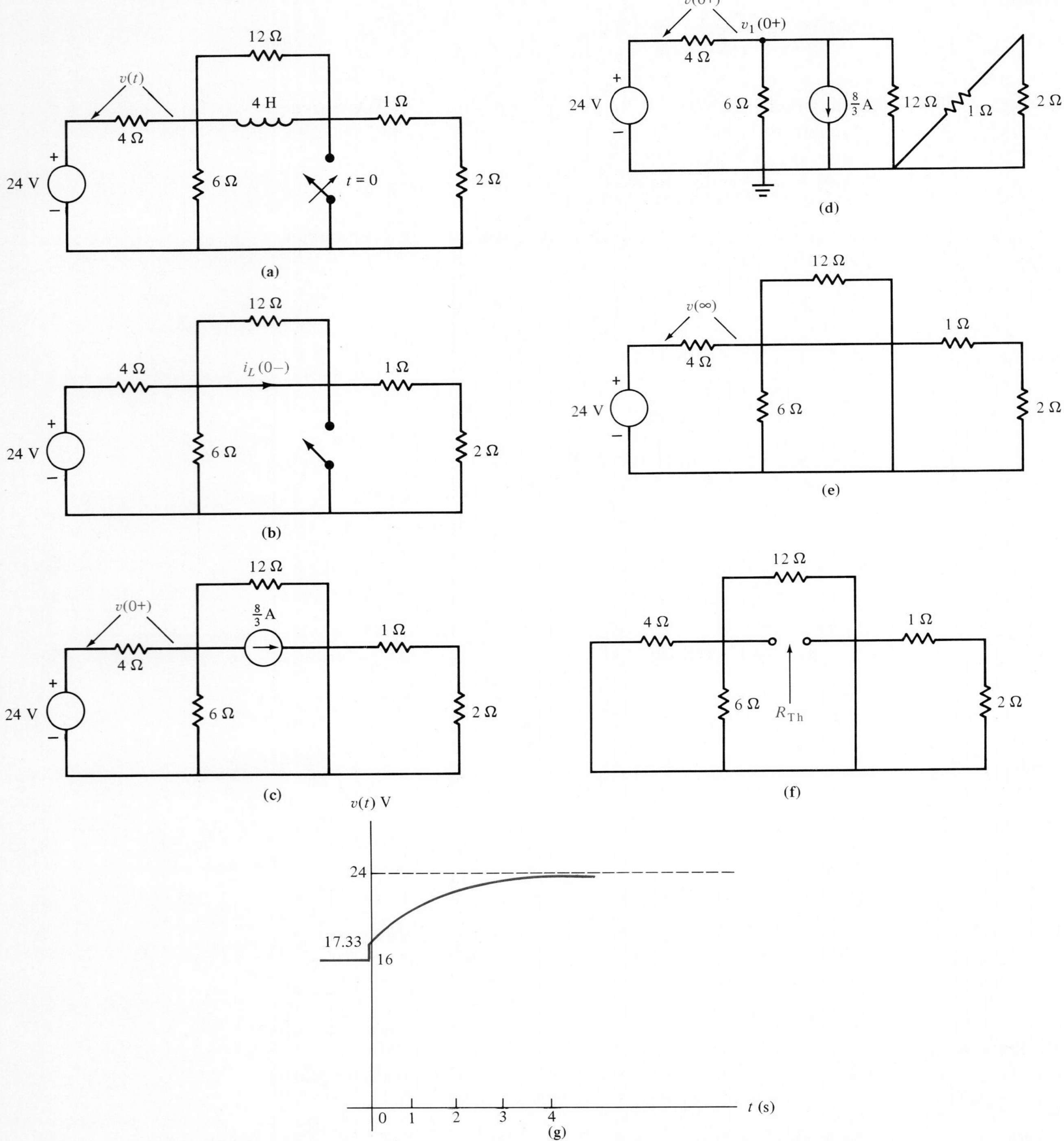

Figure 7.8 Analysis of an *RL* transient circuit with a constant forcing function.

Step 2. In Fig. 7-8b we see that

$$i_L(0-) = \frac{24}{4 + \frac{(6)(3)}{6+3}}\left(\frac{6}{6+3}\right)$$

$$= \frac{8}{3} \text{ A}$$

Step 3. The new circuit, valid only for $t = 0+$, is shown in Fig. 7.8c, which is equivalent to the circuit shown in Fig. 7.8d. The value of the current source that replaces the inductor is $i_L(0-) = i_L(0+) = \frac{8}{3}$ A. The node voltage $v_1(0+)$ can be determined from the circuit in Fig. 7.8d using a single node equation, and $v(0+)$ is equal to the difference between the source voltage and $v_1(0+)$. The equation for $v_1(0+)$ is

$$\frac{v_1(0+) - 24}{4} + \frac{v_1(0+)}{6} + \frac{8}{3} + \frac{v_1(0+)}{12} = 0$$

or

$$v_1(0+) = \frac{20}{3} \text{ V}$$

Then

$$v(0+) = 24 - v_1(0+)$$

$$= \frac{52}{3} \text{ V}$$

Step 4. The equivalent circuit for the steady-state condition after switch closure is given in Fig. 7.8e. Note that the 6-, 12-, 1-, and 2-Ω resistors are shorted, and therefore $v(\infty) = 24$ V.

Step 5. The Thévenin equivalent resistance is found by looking into the circuit from the inductor terminals. This circuit is shown in Fig. 7.8f. Note carefully that R_{Th} is equal to the 4-, 6-, and 12-Ω resistors in parallel. Therefore, $R_{Th} = 2\ \Omega$, and the circuit time constant is

$$T_c = \frac{L}{R_{Th}} = \frac{4 \text{ H}}{2\ \Omega} = 2 \text{ s}$$

Step 6. From the previous analysis we find that

$$K_1 = v(\infty) = 24$$

$$K_2 = v(0) - v(\infty) = -\frac{20}{3}$$

and hence that

$$v(t) = 24 - \frac{20}{3} e^{-t/2} \text{ V}$$

From Fig. 7.8b we see that the value of $v(t)$ before switch closure is 16 V. Therefore, the circuit response, $v(t)$ as a function of time is shown in Fig. 7.8g. ■

DRILL EXERCISE

D7.5. Consider the network in Fig. D7.5. The switch opens at $t = 0$. Find $v_o(t)$ for $t > 0$.

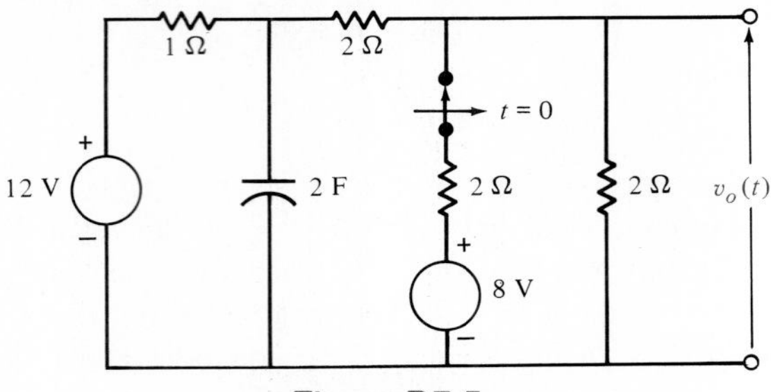

Figure D7.5

D7.6. Consider the network shown in Fig. D7.6. If the switch opens at $t = 0$, find the output voltage $v_o(t)$ for $t > 0$.

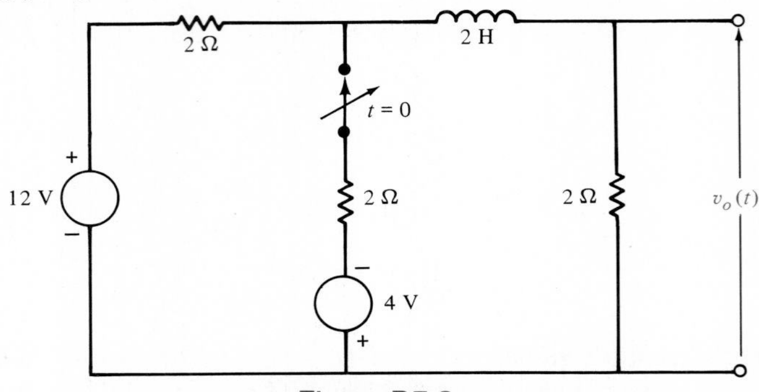

Figure D7.6

EXAMPLE 7.5

The circuit shown in Fig. 7.9a has reached steady state with the switch in position 1. At time $t = 0$ the switch moves from position 1 to position 2. We want to calculate $v_o(t)$ for $t > 0$.

Step 1. $v_o(t)$ is of the form $K_1 + K_2 e^{-t/T_c}$.

Step 2. Using the circuit in Fig. 7.9b, we can calculate $i_L(0-)$.

$$i_A = \frac{12}{4} = 3 \text{ A}$$

Then

$$i_L(0-) = \frac{12 + 2i_A}{6} = \frac{18}{6} = 3 \text{ A}$$

Step 3. The new circuit, valid only for $t = 0+$, is shown in Fig. 7.9c. The value of the current source that replaces the inductor is $i_L(0-) = i_L(0+) = 3$ A. Because of the current source

$$v_o(0+) = (3)(6) = 18 \text{ V}$$

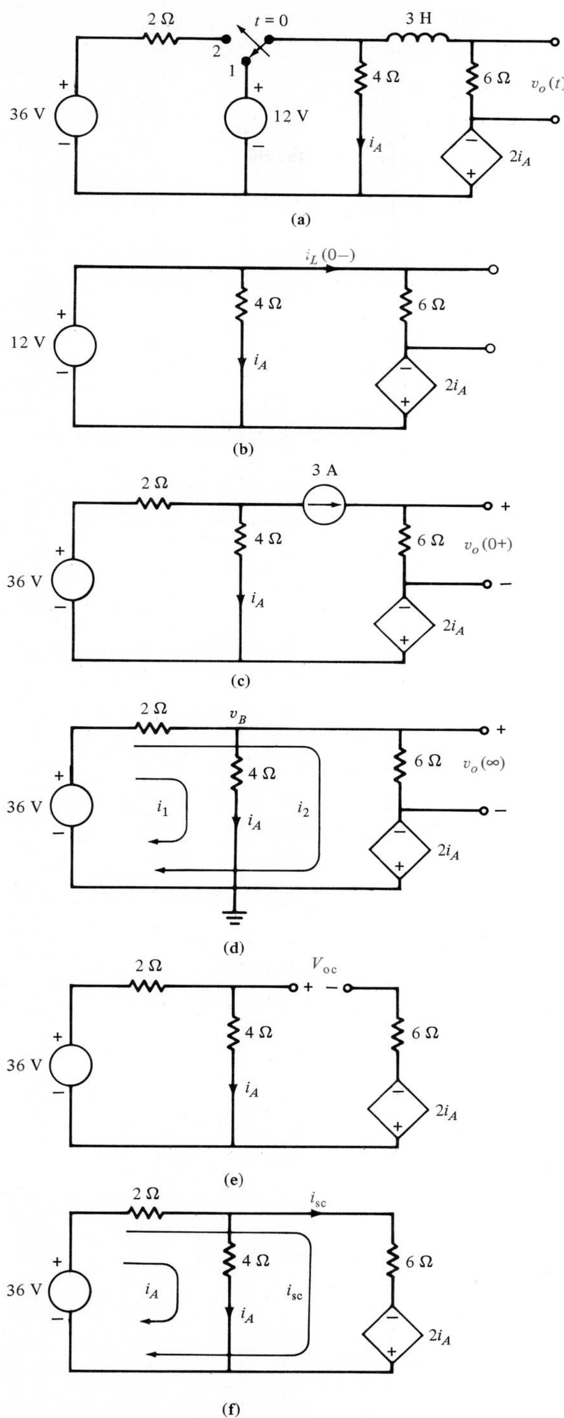

Figure 7.9 Analysis of an *RL* transient circuit containing a dependent source.

Step 4. The equivalent circuit for the steady-state condition after switch closure is given in Fig. 7.9d. Using the voltages and currents defined in the figure, we can compute $v_o(\infty)$ in a variety of ways. For example, using node equations we can find $v_o(\infty)$ from

$$\frac{v_B - 36}{2} + \frac{v_B}{4} + \frac{v_B + \frac{2v_B}{4}}{6} = 0$$

$$i_A = \frac{v_B}{4}$$

$$V_o(\infty) = v_B + 2i_A$$

or using loop equations,

$$36 = 2(i_1 + i_2) + 4i_1$$

$$36 = 2(i_1 + i_2) + 6i_2 - 2i_1$$

$$v_o(\infty) = 6i_2$$

Using either approach, we find that $v_o(\infty) = 27$ V.

Step 5. The Thévenin equivalent resistance can be obtained via V_{oc} and i_{sc}, because of the presence of the dependent source. From Fig. 7.9e we note that

$$i_A = \frac{36}{2 + 4} = 6 \text{ A}$$

Therefore,

$$V_{oc} = (4)(6) + 2(6)$$

$$= 36 \text{ V}$$

From Fig. 7.9f we can write the following loop equations:

$$36 = 2(i_A + i_{sc}) + 4i_A$$

$$36 = 2(i_A + i_{sc}) + 6i_{sc} - 2i_A$$

Solving these equations for i_{sc} yields

$$i_{sc} = \frac{36}{8} \text{ A}$$

Therefore,

$$R_{Th} = \frac{V_{oc}}{i_{sc}} = \frac{36}{36/8} = 8\ \Omega$$

Hence the circuit time constant is

$$T_c = \frac{L}{R_{Th}} = \frac{3}{8} \text{ s}$$

Step 6. Using the information computed above, we can derive the final equation for $v_o(t)$:

$$K_1 = v_o(\infty) = 27$$

$$K_2 = v_o(0) - v_o(\infty) = 18 - 27 = -9$$

Therefore,

$$v_o(t) = 27 - 9e^{-t/(3/8)} \text{ V}$$

■

DRILL EXERCISE

D7.7. If the switch in the network in Fig. D7.7 closes at $t = 0$, find $v_o(t)$ for $t > 0$.

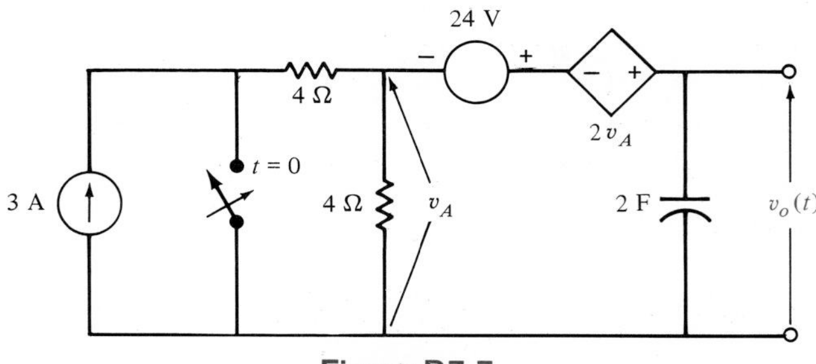

Figure D7.7

At this point it is appropriate to state that not all switch action will always occur at time $t = 0$. If may occur at any time t_0. In this case the results of the step-by-step analysis yield the following equations:

$$x(t_0) = K_1 + K_2$$

$$x(\infty) = K_1$$

and

$$x(t) = x(\infty) + [x(t_0) - x(\infty)]e^{-(t-t_0)/T_c} \qquad t > t_0$$

The function is essentially time shifted by t_0 seconds.

Finally, it should be noted that if more than one independent source is present in the network, we can simply employ superposition to obtain the total response.

7.4 Pulse Response

Thus far we have examined networks in which a voltage or current source is suddenly applied. As a result of this sudden application of a source, voltages or currents in the circuit are forced to change abruptly. A forcing function whose value changes in a discontinuous manner or has a discontinuous derivative is called a *singular function*.

Two such singular functions that are very important in circuit analysis are the unit impulse function and the unit step function. We will defer a discussion of the former until a later chapter and concentrate on the latter.

The *unit step function* is defined by the following mathematical relationship:

$$u(t) = \begin{cases} 0 & t < 0 \\ 1 & t > 0 \end{cases}$$

In other words, this function, which is dimensionless, is equal to zero for negative values of the argument and equal to 1 for positive values of the argument. It is undefined for a zero argument where the function is discontinuous. A graph of the unit step is shown in Fig. 7.10a. The unit step is dimensionless, and therefore a voltage step of V_o volts or a current step of I_o amperes is written as $V_o u(t)$ and $I_o u(t)$, respectively. Equivalent circuits for a voltage step are shown in Fig. 7.10b and c. Equivalent circuits for a current step are shown in Fig. 7.10d and e. If we use the definition of the unit step, it is easy to

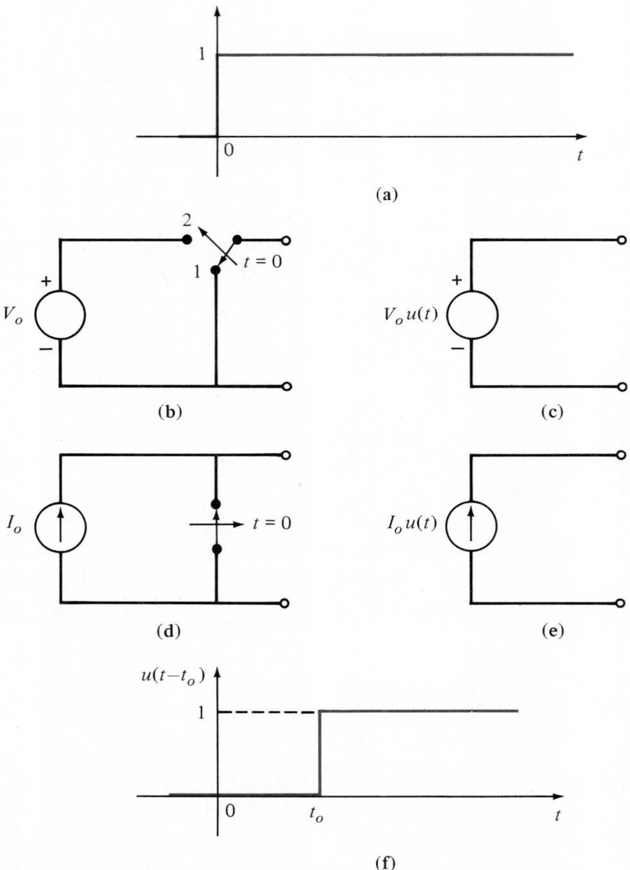

Figure 7.10 Graphs and models of the unit step function.

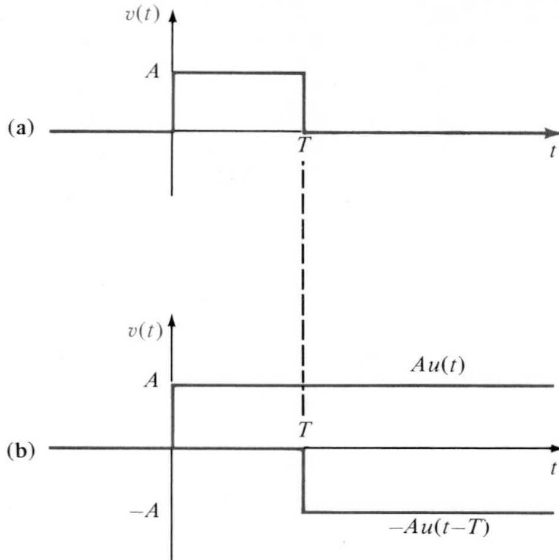

Figure 7.11 Construction of a pulse via two step functions.

generalize this function by replacing the argument t by $t - t_0$. In this case

$$u(t - t_0) = \begin{cases} 0 & t < t_0 \\ 1 & t > t_0 \end{cases}$$

A graph of this function is shown in Fig. 7.10f. Note that $u(t - t_0)$ is equivalent to delaying $u(t)$ by t_0 seconds, so that the abrupt change occurs at time $t = t_0$.

Step functions can be used to construct one or more pulses. For example, the voltage pulse shown in Fig. 7.11a can be formulated by initiating a unit step at $t = 0$ and subtracting one that starts at $t = T$ as shown in Fig. 7.11b. The equation for the pulse is

$$v(t) = A[u(t) - u(t - T)]$$

If the pulse is to start at $t = t_0$ and have width T, the equation would be

$$v(t) = A[u(t - t_0) - u(t - (t_0 + T))]$$

Using this approach, we can write the equation for a pulse starting at any time and ending at any time. Similarly, using this approach, we could write the equation for a series of pulses, called a *pulse train*, by simply forming a summation of pulses constructed in the manner illustrated above.

The following example will serve to illustrate many of the concepts we have just presented.

EXAMPLE 7.6

Consider the circuit shown in Fig. 7.12a. The input function is the voltage pulse shown in Fig. 7.12b. The initial conditions for the network are zero [i.e., $v_C(0-) = 0$] and therefore the response $v_o(t)$ for $0 < t < 0.3$ s is simply the step response. At $t = 0.3$ s the forcing function becomes zero and therefore $v_o(t)$ for $t > 0.3$ s is the source-free or natural response of the network.

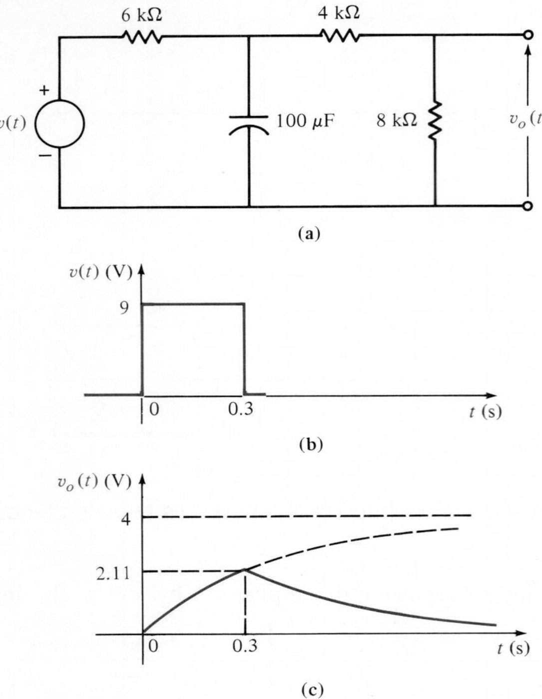

Figure 7.12 Pulse response of a network.

Since the output voltage $v_o(t)$ is a voltage division of the capacitor voltage, and the initial voltage across the capacitor is zero, we know that $v_o(0+) = 0$.

The steady-state value of $v_o(t)$ [i.e., $v_o(\infty)$] due to the application of the unit step is

$$v_o(\infty) = \frac{9 \text{ V}}{6 \text{ k}\Omega + 4 \text{ k}\Omega + 8 \text{ k}\Omega}(8 \text{ k}\Omega)$$
$$= 4 \text{ V}$$

The Thévenin equivalent resistance is

$$R_{\text{Th}} = \frac{(6 \text{ k}\Omega)(12 \text{ k}\Omega)}{6 \text{ k}\Omega + 12 \text{ k}\Omega}$$
$$= 4 \text{ k}\Omega$$

Therefore, the circuit time constant T_c is

$$T_c = R_{\text{Th}} C$$
$$= (4 \times 10^3)(100 \times 10^{-6})$$
$$= 0.4 \text{ s}$$

Therefore, the response $v_o(t)$ for the period $0 < t < 0.3$ s is

$$v_o(t) = 4 - 4e^{-t/0.4} \qquad 0 < t < 0.3 \text{ s}$$

The capacitor voltage can be calculated by realizing that $v_o(t) = \frac{2}{3}\, v_C(t)$. Therefore,

$$v_C(t) = \tfrac{3}{2}(4 - 4e^{-t/0.4})$$

Since the capacitor voltage is continuous,

$$v_C(0.3-) = v_C(0.3+)$$

and therefore

$$\begin{aligned} v_o(0.3+) &= \tfrac{2}{3} v_C(0.3+) \\ &= 4(1 - e^{-0.3/0.4}) \\ &= 2.11 \text{ V} \end{aligned}$$

The final value for $v_o(t)$ as $t \to \infty$ is zero, since there are no sources present. Therefore, the expression for $v_o(t)$ for $t > 0.3$ s is

$$v_o(t) = 2.11e^{-(t-0.3)/0.4} \qquad t > 0.3 \text{ s}$$

The term $e^{-(t-0.3)/0.4}$ indicates that the exponential decay starts at $t = 0.3$ s. The complete solution can be written by means of superposition as

$$v_o(t) = 4(1 - e^{-t/0.4})u(t) - 4(1 - e^{-(t-0.3)/0.4})u(t - 0.3)$$

or equivalently, the complete solution is

$$v_o(t) = \begin{cases} 0 & t < 0 \\ 4(1 - e^{-t/0.4}) & 0 < t < 0.3 \text{ s} \\ 2.11e^{-(t-0.3)/0.4} & 0.3 < t \end{cases}$$

which in mathematical form is

$$v_o(t) = 4(1 - e^{-t/0.4})[u(t) - u(t - 0.3)] + 2.11e^{-(t-0.3)/0.4}u(t - 0.3) \text{ V}$$

Note that the term $[u(t) - u(t - 0.3)]$ acts like a gating function that captures only the part of the step response that exists in the time interval $0 < t < 0.3$ *s*. The output as a function of time is shown in Fig. 7.12c. ■

DRILL EXERCISE

D7.8. The voltage source in the network in Fig. D7.8a is shown in Fig. D7.8b. The initial current in the inductor is zero. Determine the output voltage $v_o(t)$ for $t > 0$.

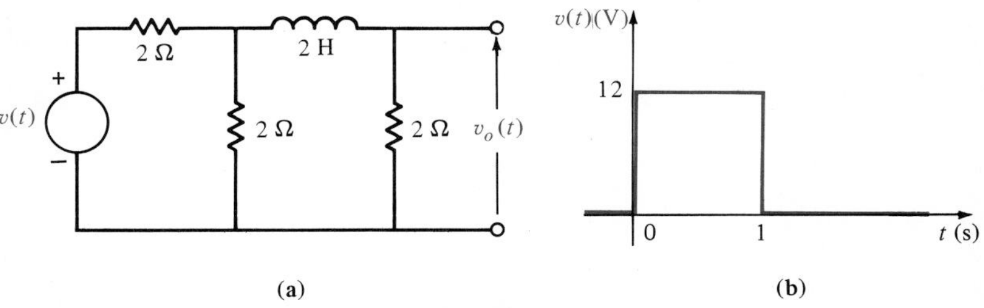

Figure D7.8

D7.9. The current source in the network in Fig. D7.9a is defined in Fig. D7.9b. The initial voltage across the capacitor is zero. Determine the current $i_o(t)$ for $t > 0$.

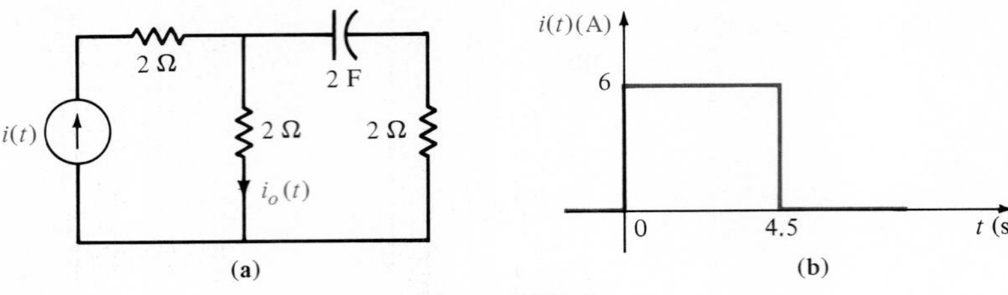

Figure D7.9

7.5 *RC* Operational Amplifier Circuits

Two very important *RC* op-amp circuits are the differentiator and the integrator. These circuits are derived from the circuit for an inverting op-amp by replacing the resistors R_1 and R_2, respectively, by a capacitor. Consider, for example, the circuit shown in Fig. 7.13a. The circuit equations are

$$C_1 \frac{d}{dt}(v_1 - v_-) + \frac{v_o - v_-}{R_2} = i_-$$

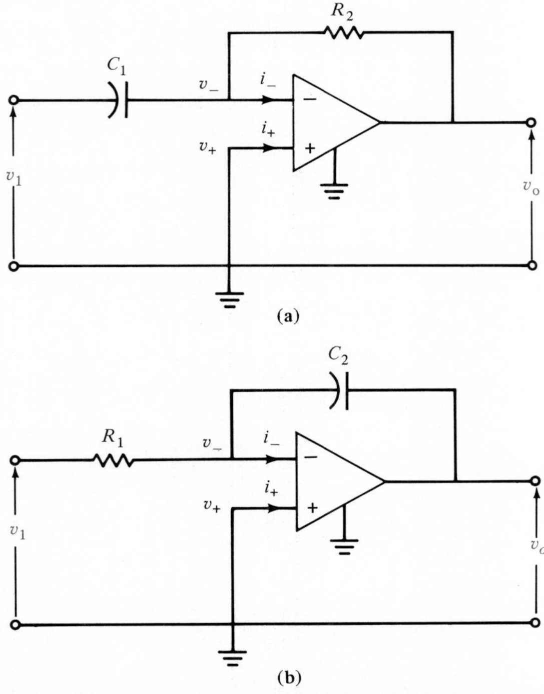

Figure 7.13 Differentiator and integrator operational amplifier circuits.

However, $v_- = 0$ and $i_- = 0$. Therefore,

$$v_o(t) = -R_2C_1 \frac{dv_1(t)}{dt} \qquad (7.12)$$

and thus the output of the op-amp circuit is proportional to the derivative of the input.

The circuit equations for the op-amp configuration in Fig. 7.13b are

$$\frac{v_1 - v_-}{R_1} + C_2 \frac{d}{dt}(v_o - v_-) = i_-$$

but since $v_- = 0$ and $i_- = 0$, the equation reduces to

$$\frac{v_1}{R_1} = -C_2 \frac{dv_o}{dt}$$

or

$$\begin{aligned} v_o(t) &= \frac{-1}{R_1C_2} \int_{-\infty}^{t} v_1(x)\,dx \\ &= \frac{-1}{R_1C_2} \int_{0}^{t} v_1(x)\,dx + v_o(0) \end{aligned} \qquad (7.13)$$

If the capacitor is initially discharged, then $v_o(0) = 0$ and hence

$$v_o(t) = \frac{-1}{R_1C_2} \int_{0}^{t} v_1(x)\,dx \qquad (7.14)$$

and thus the output voltage of the op-amp circuit is proportional to the integral of the input voltage.

EXAMPLE 7.7

The waveform in Fig. 7.14a is applied at the input of the differentiator circuit shown in Fig. 7.13a. If $R_2 = 1\ \text{k}\Omega$ and $C_1 = 2\ \mu\text{F}$, determine the waveform at the output of the op-amp.

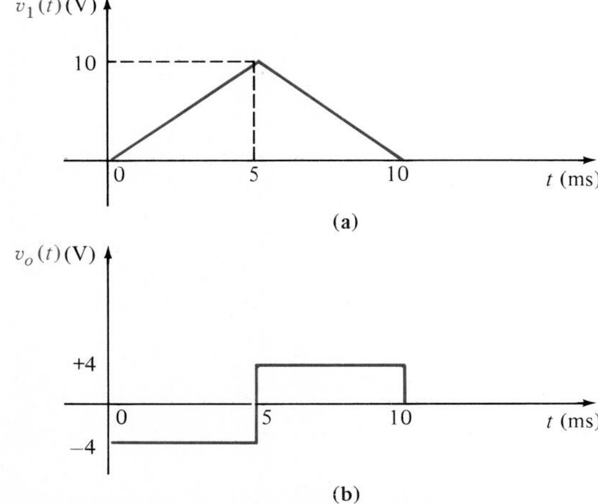

Figure 7.14 Input and output waveforms for a differentiator circuit.

Using Eq. (7.12), we find that the op-amp output is

$$v_o(t) = -R_2C_1 \frac{dv_1(t)}{dt}$$

$$= -2 \times 10^{-3} \frac{dv_1(t)}{dt}$$

$dv_1(t)/dt = 2 \times 10^3$ for $0 \leq t < 5$ ms and therefore

$$v_o(t) = -4 \text{ V} \qquad 0 \leq t < 5 \text{ ms}$$

$dv_1(t)/dt = -2 \times 10^3$ for $5 \leq t < 10$ ms and therefore

$$v_o(t) = 4 \text{ V} \qquad 5 \leq t < 10 \text{ ms}$$

Hence the output waveform of the differentiator is shown in Fig. 7.14b. ■

EXAMPLE 7.8

If the integrator shown in Fig. 7.13b has the parameters $R_1 = 5$ kΩ and $C_2 = 0.2$ μF, determine the waveform at the op-amp output if the input waveform is given in Fig. 7.15a and the capacitor is initially discharged.

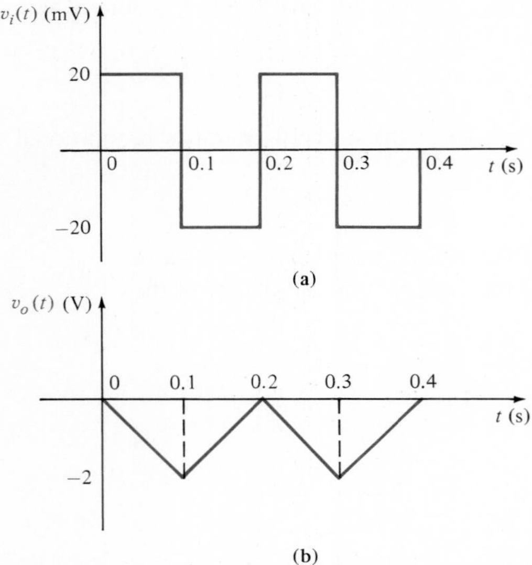

Figure 7.15 Input and output waveforms for an integrator circuit.

The integrator output is given by the expression

$$v_o(t) = \frac{-1}{R_1C_2} \int_0^t v_1(x)\, dx$$

which with the given circuit parameters is

$$v_o(t) = -10^3 \int_0^t v_1(x)\, dx$$

In the interval $0 \leq t < 0.1$ s, $v_1(t) = 20$ mV. Hence

$$v_o(t) = -10^3(20 \times 10^{-3}t) \qquad 0 \leq t < 0.1 \text{ s}$$
$$= -20t$$

At $t = 0.1$ s, $v_o(t) = -2$ V. In the interval from 0.1 to 0.2 s, the integrator produces a positive slope output of $20t$ from $v_o(0.1) = -2$ V to $v_o(0.2) = 0$ V. This waveform from $t = 0$ to $t = 0.2$ s is repeated in the interval $t = 0.2$ to $t = 0.4$ s and therefore the output waveform is shown in Fig. 7.15b. ■

DRILL EXERCISE

D7.10. The waveform in Fig. D7.10 is applied to the input terminals of the op-amp differentiator circuit. Determine the differentiator output waveform if the op-amp circuit parameters are $C_1 = 2$ F and $R_2 = 2\ \Omega$.

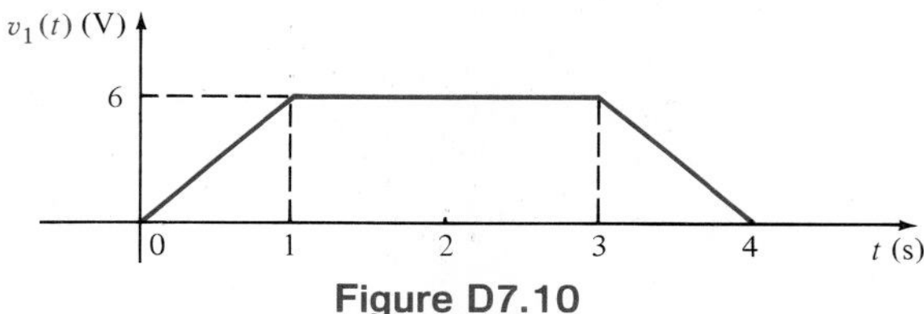

Figure D7.10

7.6 Transient Circuit Analysis Using SPICE

As we begin our discussion of this topic, it is important to point out that all the material in Section 5.6 applies here also. The circuit is described in exactly the same manner and, in fact, a dc analysis is performed automatically prior to transient analysis to determine the circuit's initial conditions. The independent variable in transient analysis is always time, and the circuit variables are tabulated and/or plotted as functions of time. It is also important to note that the circuit under analysis cannot contain a loop of voltage sources and/or inductors and cannot contain a cut set of current sources and/or capacitors.

We will now introduce some new data, solution control, and output specification statements which will be useful in transient analysis. The examples will indicate the ease with which we can perform analyses that would be difficult to treat mathematically.

Branch Statements for Inductors and Capacitors

The general statement form for these elements is

```
BXXXXXXX N1 N2 VALUE
```

B is the letter L or C, depending on the type of element. XXXXXXX denotes an arbitrary alphanumeric string that uniquely identifies the particular element. N1 and N2 are the circuit nodes to which the element is connected. VALUE is the value of the element. Capacitance is in farads, and inductance is in henrys.

Branch Statements for Time-Varying Sources

Any independent source can be assigned a time-dependent value for transient analysis. If a source is assigned a time-dependent value, the time-zero value will be used for any dc analysis that is requested. The time-varying sources may be of four different types: pulse, exponential, sinuoidal, or piecewise linear. The type of source is specified on the data statement that defines the source.

The data statement for pulse sources is of the form

```
VXXXXXXX N+ N- PULSE (V1 V2 TD TR TF PW PER)
```

V1 is the initial value of the source in volts or amperes, and V2 is the value of the source during the pulse. TD is the delay time between time zero and the start of the pulse in seconds. TR, TF, and PW are the rise time, fall time, and width of the pulse in seconds. PER is the period in seconds for a periodic pulse train. Intermediate points on the pulse are determined by linear interpolation. Therefore, the following table describes the single pulse shown in Fig. 7.16.

Time	Value
0	V1
TD	V1
TD + TR	V2
TD + TR + PW	V2
TD + TR + PW + TF	V1

Exponential sources are described using the format

```
VXXXXXXX N+ N- EXP(V1 V2 TD1 TAU1 TD2 TAU2)
```

V1 is the initial value in volts or amperes. V2 is the peak value in volts or amperes. TD1 and TAU1 are the rise-delay-time and rise-time constants, respectively. TD2 and TAU2 are the fall-delay-time and fall-time constants, respectively. All times are ex-

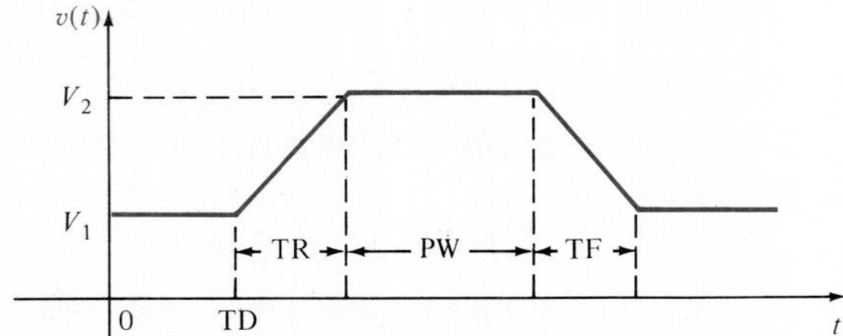

Figure 7.16 General form of a pulse.

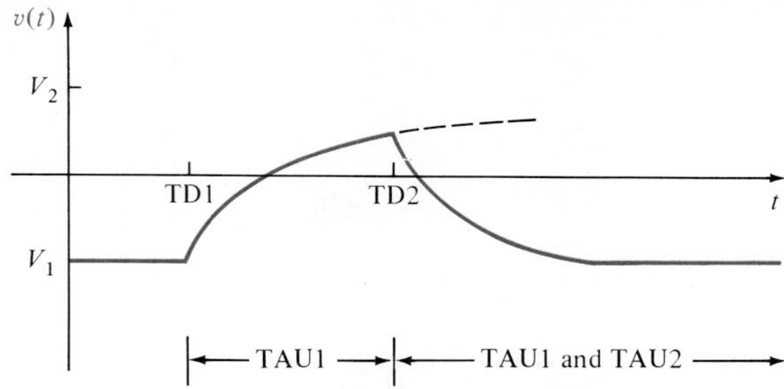

Figure 7.17 General form of an exponential source.

pressed in seconds. The waveform shown in Fig. 7.17 is described by the expressions

```
Time 0 to TD1:       V1
Time TD1 to TD2:     V1 +(V2 - V1)*
                        (1 - EXP(-(TIME - TD1)/TAU1))
Time TD2 to TSTOP:   V1 +(V2 - V1)*
                        (1 - EXP(-(TIME - TD1)/TAU1))
                       +(V1 - V2)*
                        (1 - EXP(-(TIME - TD2)/TAU2))
```

The data statement for sinusoidal sources is of the form

```
VXXXXXXX N + N - SIN(VO VA FREQ TD THETA)
```

The shape of this waveform is described by the equation

```
Time 0 to TD:        VO
Time TD to TSTOP:    VO + VA*EXP(-(TIME-TD)*THETA)*
                     SINE(2PI*FREQ*(TIME + TD))
```

VO is the offset in volts or amperes. VA is the amplitude in volts or amperes. FREQ is the frequency in hertz. TD is the delay in seconds, and THETA is the damping factor in $(\text{seconds})^{-1}$.

Piecewise linear sources are described using the format

```
VXXXXXXX N+ N- PWL (T1 V1 T2 V2 T3 V3 T4
V4 . . .)
```

Each pair of values (TI, VI) specifies that the value of the source is VI volts or amperes at time TI. The value of the source at intermediate values of time is determined by linear interpolation. Values of TI must always be increasing in time.

Solution Control Statements

Transient analysis in SPICE is invoked using a .TRAN statement of the form

```
.TRAN TSTEP TSTOP TSTART TMAX
```

TSTEP is the printing or plotting increment for line printer output. TSTART and TSTOP are the start and stop times for the analysis. If TSTART is omitted, it is assumed to be zero. The transient analysis always begins at time zero. In the interval [0, TSTART], the circuit is analyzed, but no outputs are stored. In the interval [TSTART, TSTOP], analysis continues and the outputs are stored. TMAX is the maximum time step that SPICE will use and defaults to a value of (TSTOP-TSTART)/50 or TSTEP, whichever is smaller.

If initial conditions for the capacitive and inductive elements are known and are to be employed in the SPICE analysis, the .TRAN statement includes the keyword UIC (use initial conditions) as shown below.

```
.TRAN TSTEP TSTOP TSTART TMAX UIC
```

If this keyword is specified, SPICE will use the initial values that can be added to the data statements in the form

```
CXXXXXXX N+ N- VALUE IC= INCOND
LYYYYYYY N+ N- VALUE IC= INCOND
```

For the capacitor, the initial condition is the initial (time-zero) value of the capacitor voltage in volts. For the inductor, the initial condition is the initial (time-zero) value of inductor current in amperes that flows from N+, through the inductor, to N−. When employing the UIC feature, SPICE uses the initial values on the data statements as the initial transient condition and proceeds with the analysis.

Output Specification Statements

The .PRINT and .PLOT statements employed in the dc analysis are useful here also. In a transient analysis they take the form

```
.PRINT TRAN OV1 . . .OV8
.PLOT TRAN OV1(PLO1, PHI1) . . . . .OV8(PLO8,
*PHI8)
```

EXAMPLE 7.9

Consider the network shown in Fig. 7.18a. At $t = 0$ the switch is opened. We wish to determine the response $v_o(t)$ for $t > 0$ using SPICE.

In the steady-state mode prior to the time the switch opens, the initial inductor current can be found using Thévenin's theorem. Breaking the circuit at the two x marks in Fig. 7.18a yields the network in Fig. 7.18b. The open-circuit voltage is

$$V_{oc} = \frac{24 - 12}{16} \times 4 + 12 = 15 \text{ V}$$

The Thévenin equivalent resistance is

$$R_{Th} = \frac{4 \times 12}{4 + 12} = 3\ \Omega$$

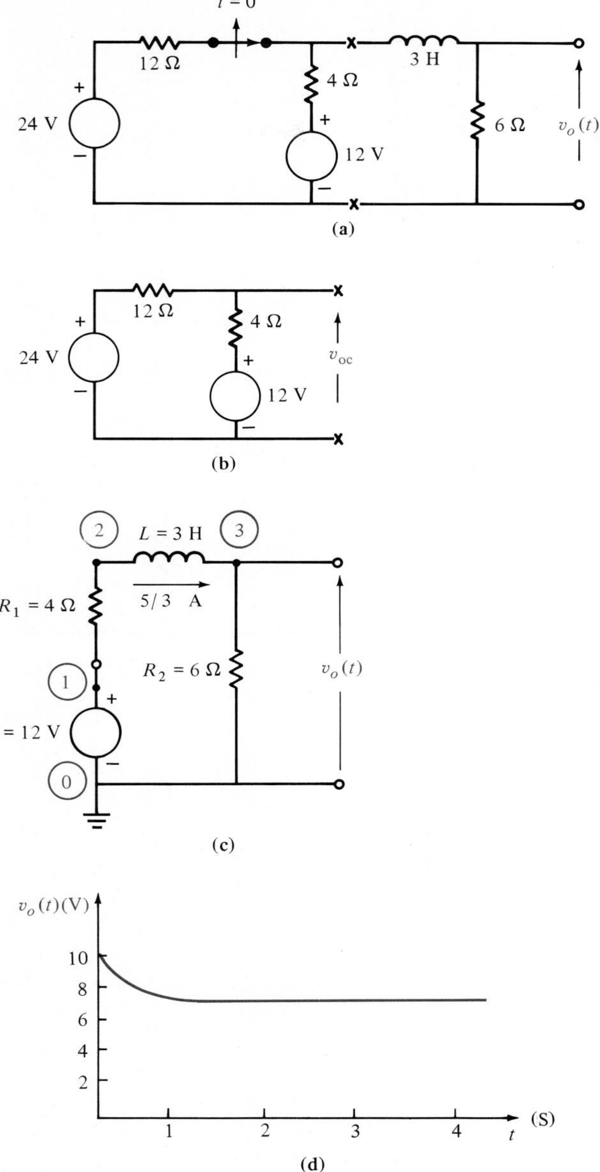

Figure 7.18 Analysis of an *RL* circuit.

In the steady state the inductor is a short circuit to dc and therefore the current in the 6-Ω resistor from top to bottom, which is the initial current, is $\frac{15}{9}$ or $\frac{5}{3}$ A.

At $t = 0+$ the network, labeled for a SPICE analysis, is shown in Fig. 7.18c. The SPICE program for plotting $v_o(t)$ in 0.05-s intervals over a 4-s time frame follows.

```
EXAMPLE 7.9   TRANSIENT ANALYSIS
V1   1     0     DC     12
R1   2     1     4
L1   2     3     3      IC = 1.66667
R2   3     0     6
.TRAN      0.05  4      UIC
.PLOT      TRAN  V(3)
.END
```

The waveform generated by the program is shown in Fig. 7.18d. ■

DRILL EXERCISE

D7.11. In the network in Fig. D7.11 the switch opens at $t = 0$. Using SPICE, plot the output voltage $v_o(t)$ from 0 to 2 s in steps of 50 ms.

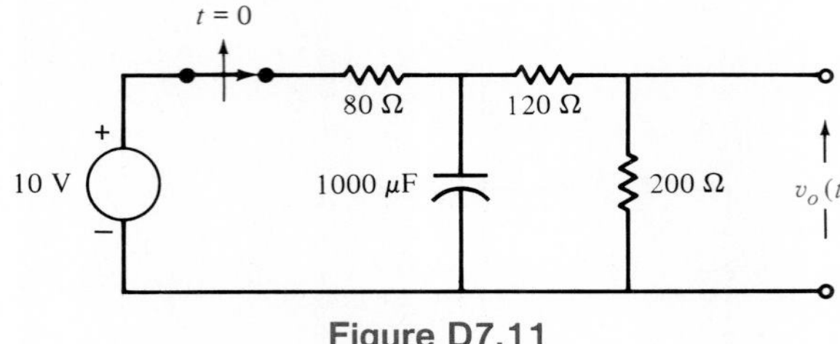

Figure D7.11

D7.12. In the network in Fig. D7.12 the switch opens at $t = 0$. Using SPICE, plot the output voltage from 0 to 3 s in steps of 0.05 s.

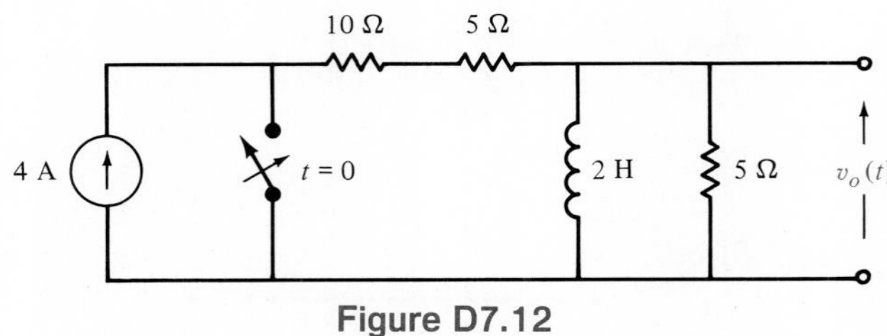

Figure D7.12

EXAMPLE 7.10

Let us consider again the problem outlined in Example 7.6. The circuit is shown in Fig. 7.12a and the input is given in Fig. 7.12b. A SPICE program for plotting the output voltage follows.

```
EXAMPLE 7.10  SINGLE PULSE RESPONSE
VIN   1   0   PULSE  ( 0 9 0 0 0 0.3 10)
R1   1     2     6K
C    2     0     100UF
R2   2     3     4K
R3   3     0     8K
```

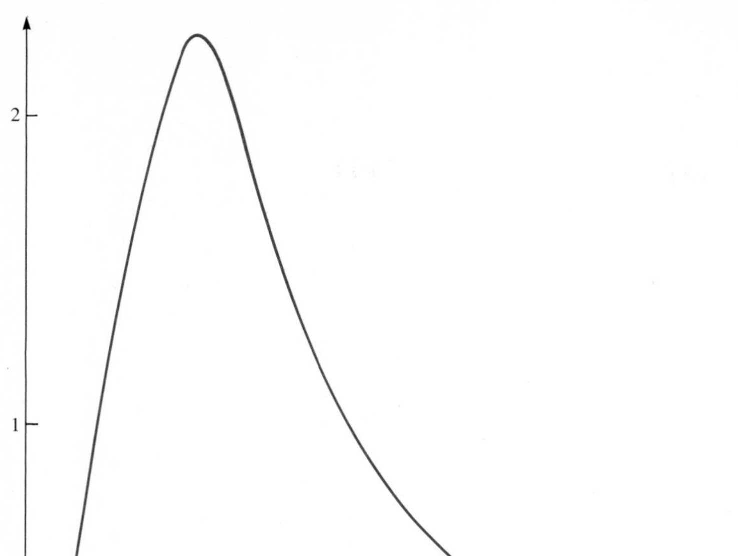

Figure 7.19 Pulse response of the network in Fig. 7.12a.

```
.TRAN     .05  1.5
.PLOT     TRAN  V(3)
.END
```

Note that the interval over which the response is examined is 1.5 s, as specified on the .TRAN statement. Since the period of the pulse train is 10 s, the program provides a single pulse response.

The computer printout of the pulse response is shown in Fig. 7.19. ■

EXAMPLE 7.11

Consider the network shown in Fig. 7.20a. Let us write a SPICE program for plotting the output voltage together with the input voltage if the input is an exponential source described by the following equations.

Time 0 to time 1 s: $v_1(t) = 0$

Time 1 to time 2 s: $v_1(t) = 2[1 - e^{-(t-1)/2}]$ V

Time 2 to time 8 s: $v_1(t) = 2[1 - e^{-(t-1)/2}] - 2[1 - e^{-(t-2)/2}]$ V

The SPICE program that will plot the input and output voltages in 0.1-s intervals over an 8-s range is shown on p. 278.

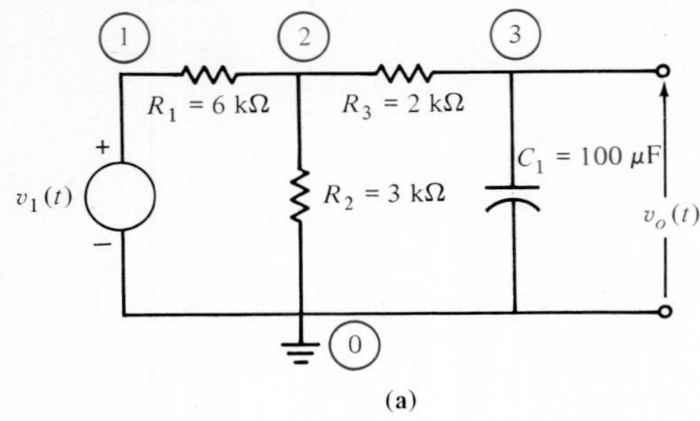

(a)

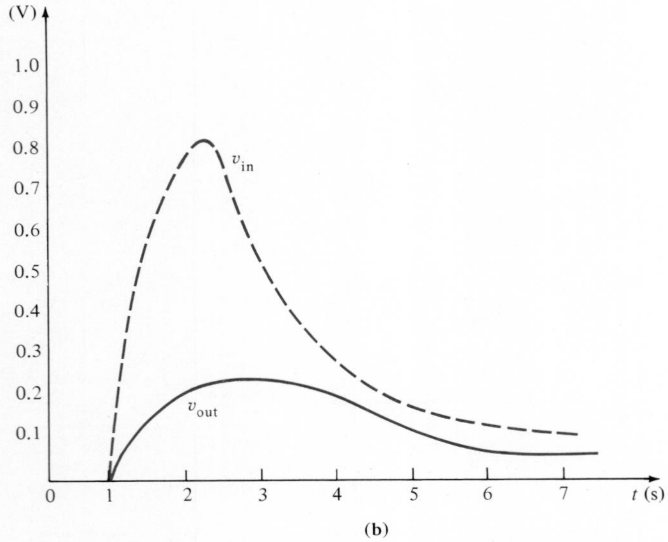

(b)

Figure 7.20 Analysis of an *RC* circuit with an exponential input.

```
EXAMPLE 7.11   EXPONENTIAL SOURCE RESPONSE
V1   1     0   EXP  ( 0 2 1 2 2 2 )
R1   1     2   6K
R2   2     0   3K
R3   2     3   2K
C1   3     0   100UF
.TRAN      0.1  8
.PLOT      TRAN   V(3)   V(1)
.END
```

The input and output voltages are shown in Fig. 7.20b. ■

EXAMPLE 7.12

Suppose that the input voltage to the network in Fig. 7.20a is a damped sinusoid of the form

$$v_1(t) = 10e^{-0.8t} \sin [2\pi(2)t]$$

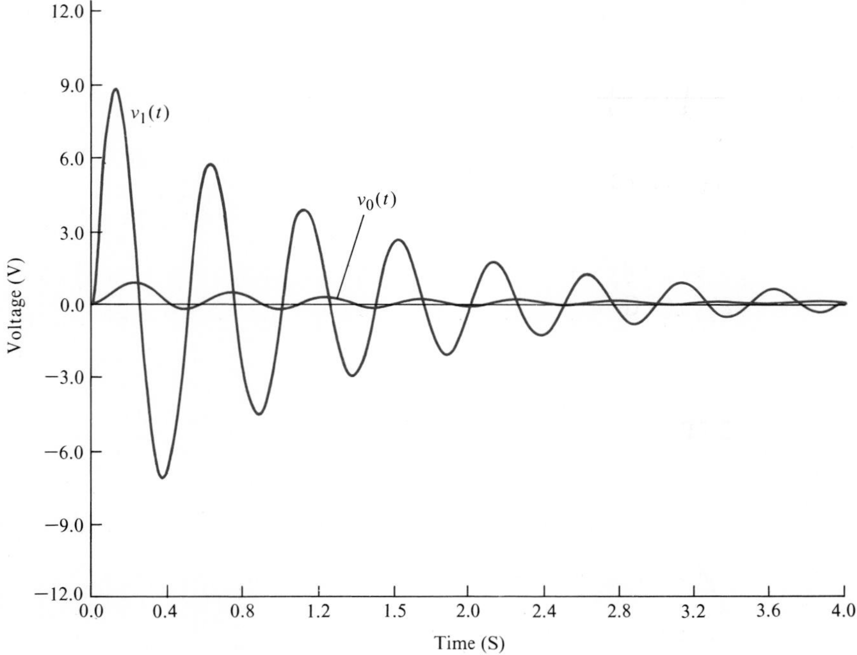

Figure 7.21 Input/output of an *RC* circuit with damped sinusoidal forcing function.

The SPICE program that will plot the input and output voltages in 0.25-s intervals over a 4-s range is shown below.

```
EXAMPLE 7.12  DAMPED SINUSOIDAL SOURCE RESPONSE
V1  1  0  sin  (  0  10  2  0  0.8)
R1  1  2  6K
R2  2  0  3K
R3  2  3  2K
C1  3  0  100UF

.TRAN    0.025 4
.PLOT    TRAN V(1) V(3)
.END
```

The input and output voltages are shown in Fig. 7.21. ■

EXAMPLE 7.13

Consider the network in Fig. 7.22a. The input voltage is shown in Fig. 7.22b. The SPICE program that will plot the input and the output response of the network in 0.05-s intervals over a 4-s range is listed below.

```
EXAMPLE 7.13  PIECEWISE LINEAR SOURCE RESPONSE
V1  1  0  PWL (0 0 0.2 0 0.4 4 0.8 4 1 0 4 0)
R1  1  2  6K
R2  2  3  4K
C1  2  0  100UF
R3  3  0  8K
.TRAN   0.05 4
.PLOT   TRAN V(1) V(3)
.END
```

The input and output voltages are shown in Fig. 7.23. ■

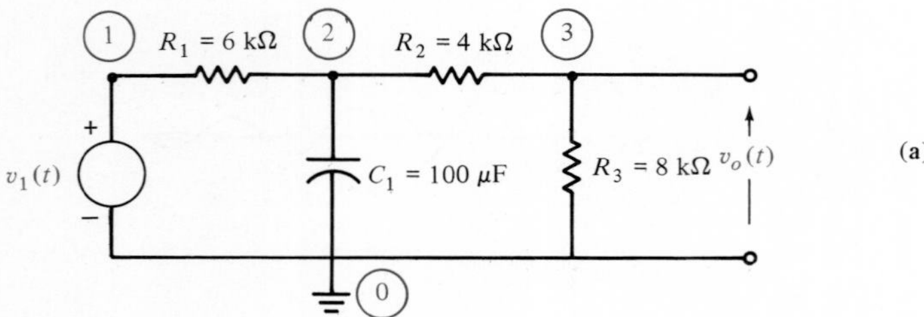

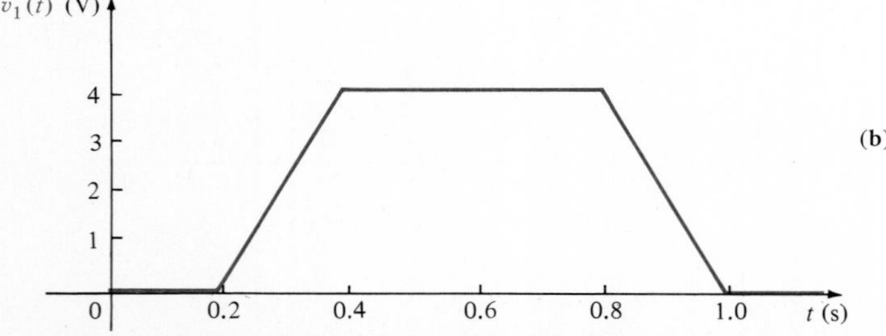

Figure 7.22 *RC* network with a piecewise linear input.

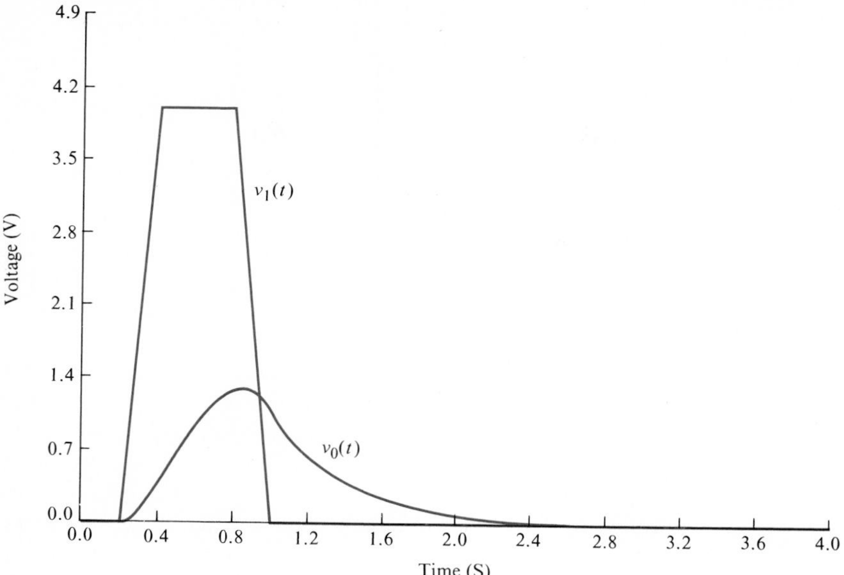

Figure 7.23 Input and output voltages for the network in Fig. 7.22a.

DRILL EXERCISE

D7.13. In the network in Fig. D7.13a the voltage source is specified by the waveform shown in Fig. D7.13b. Use SPICE to plot the output voltage $v_o(t)$ over the interval 0 to 1 s in increments of 10 ms.

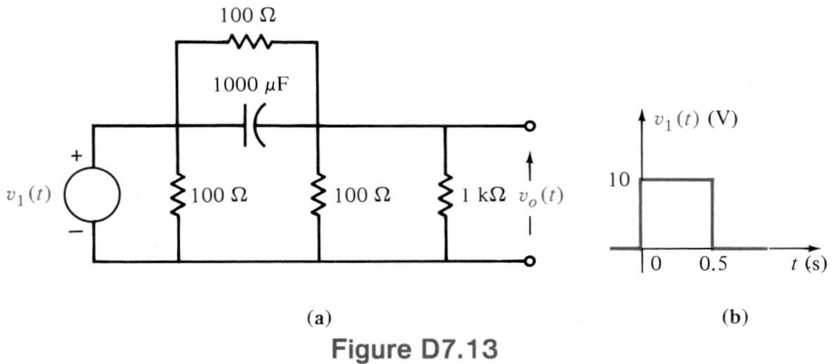

Figure D7.13

D7.14. Consider the network in Fig. D7.14. Use SPICE to plot the output voltage $v_o(t)$ over the interval 0 to 7 s in increments of 200 ms if the voltage source $v_s(t)$ is given by the expression

$$v_s(t) = 10(1 - e^{-t/1}) - 10(1 - e^{-(t-2)/2}) \text{ V}$$

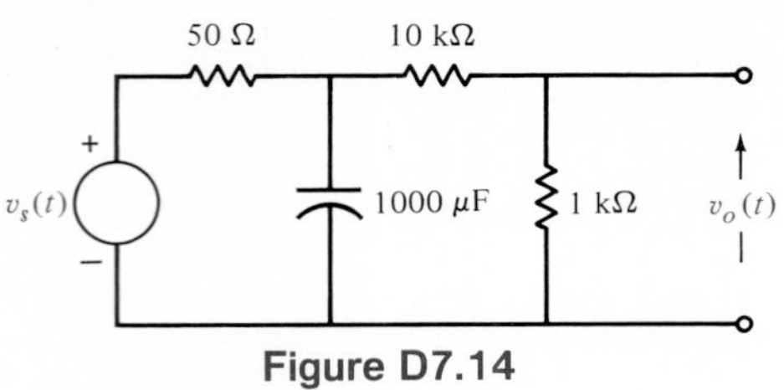

Figure D7.14

7.7 Summary

We have shown that the solution of *RC* and *RL* circuits which contain only a single energy storage element (i.e., *C* or *L*) involve the solution of a first-order differential equation. A general solution method was developed and a step-by-step procedure outlined for determining a solution. Both the natural and forced response of these networks were examined. It was found that the circuit response in the form of a decaying exponential was controlled by the circuit time constant, and that the response essentially decays to its final value in five time constants. The time constants for *RC* and *RL* circuits are $R_{Th}C$ and L/R_{Th}, respectively, where R_{Th} is the Thévenin equivalent resistance seen from the terminals of the storage element.

Finally, the application of SPICE to *RC* and *RL* problems was presented.

KEY POINTS

- The voltage or current anywhere in an *RC* or *RL* circuit is obtained by solving a first-order differential equation.
- A solution to the equation

$$\frac{dx(t)}{dt} + ax(t) = A \text{ is } x(t) = \frac{A}{a} + k_2e^{-at}$$

  where A/a is referred to as the steady-state solution and $1/a$ is called the time constant.
- The function e^{-at} decays to within 1% if its final value (which is zero) after a period of $5/a$ seconds.
- If a network has a small time constant, its response to some input will quickly settle to its steady-state value; however, if the network time constant is large, a long time is required for the network to reach steady state.
- An *RC* circuit has a time constant of $R_{Th}C$ seconds and an *RL* circuit has a time constant of L/R_{Th} seconds.
- The unit step function $u(t - t_o)$ has a value of zero for $t < t_o$ and a value of 1 for $t > t_o$.
- In a transient SPICE analysis a set of simultaneous integrodifferential equations is solved.
- The dc SPICE analysis specifications apply in a transient analysis.
- The independent variable in a transient analysis is always time.
- A circuit under a SPICE analysis cannot contain either a loop of voltage sources and/or inductors or a cut set of current sources and/or capacitors.

PROBLEMS

In the following problems it is assumed that the circuit is in steady state prior to any switch action.

7.1. Find $v_C(t)$ for $t > 0$ in the circuit shown in Fig. P7.1.

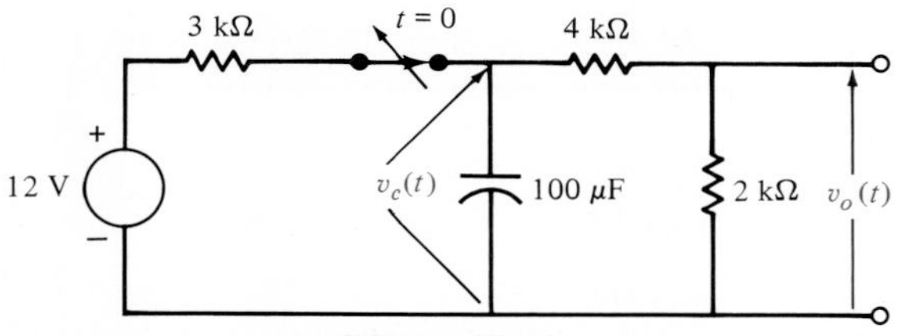

Figure P7.1

7.2. Find $v_o(t)$ for $t > 0$ in Fig. P7.1.

7.3. Find $i_o(t)$ for $t > 0$ in Fig. P7.3.

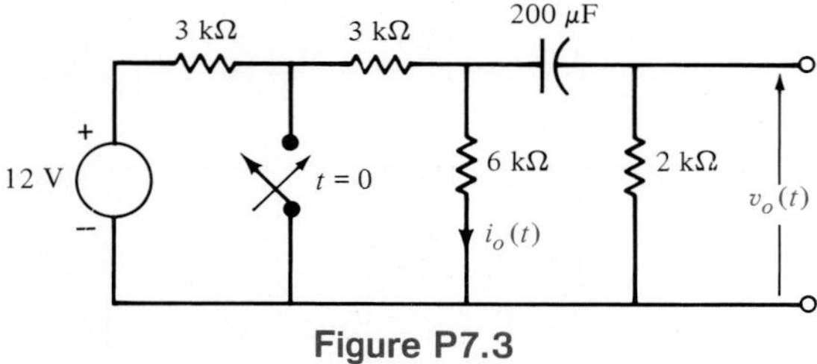

Figure P7.3

7.4. Find $v_o(t)$ for $t > 0$ in Fig. P7.3.

7.5. Find $i_o(t)$ for $t > 0$ in Fig. P7.5.

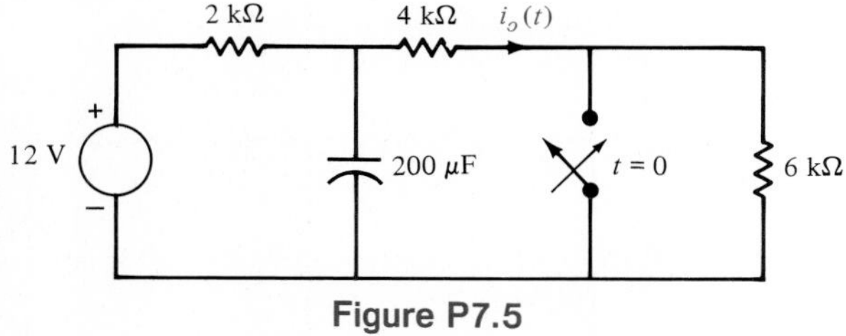

Figure P7.5

7.6. Given the circuit shown in Fig. P7.6, find $i_o(t)$ for $t > 0$.

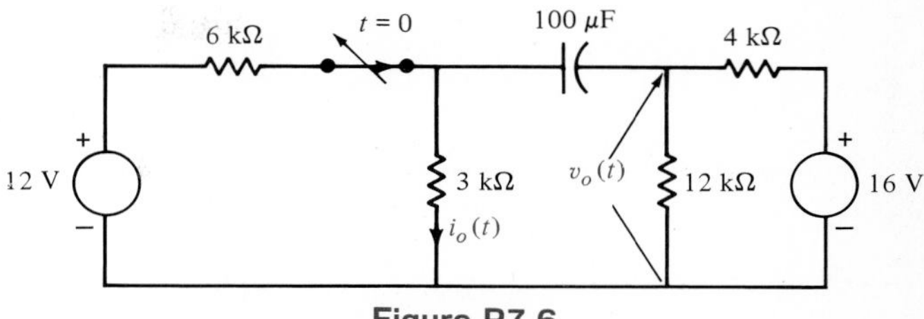

Figure P7.6

7.7. Find $v_o(t)$ for $t > 0$ in Fig. P7.6.

7.8. In Fig. P7.8, find $i_o(t)$ for $t > 0$.

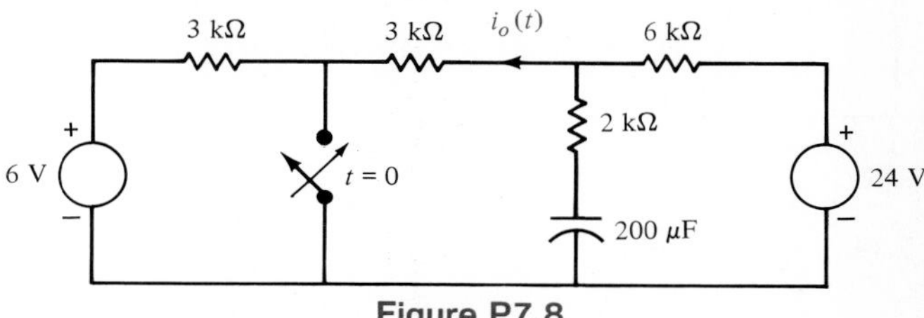

Figure P7.8

7.9. Find $i_o(t)$ for $t > 0$ in the circuit shown in Fig. P7.9.

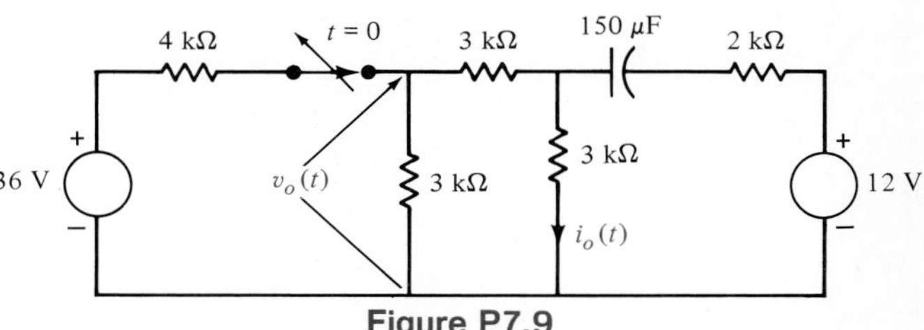

Figure P7.9

7.10. Find $v_o(t)$ for $t > 0$ for the circuit in Fig. P7.9.

7.11. Find $v_o(t)$ for $t > 0$ in the network in Fig. P7.11.

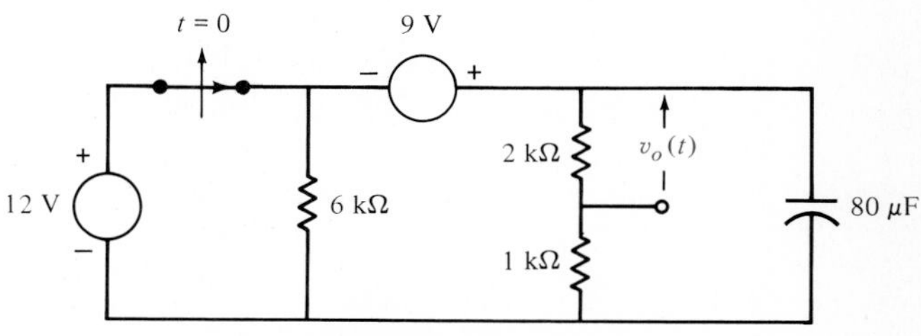

Figure P7.11

7.12. Find $v_o(t)$ and $i_o(t)$ for $t > 0$ in the network in Fig. P7.12.

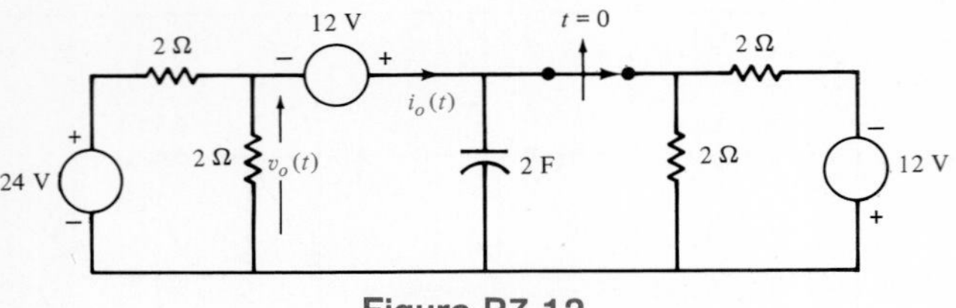

Figure P7.12

7.13. Find $v_o(t)$ for $t > 0$ in the circuit in Fig. P7.13.

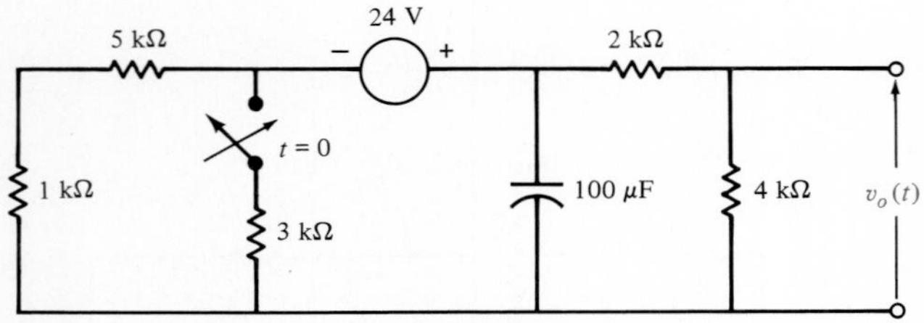

Figure P7.13

7.14. Find $v_o(t)$ for $t > 0$ in the circuit in Fig. P7.14.

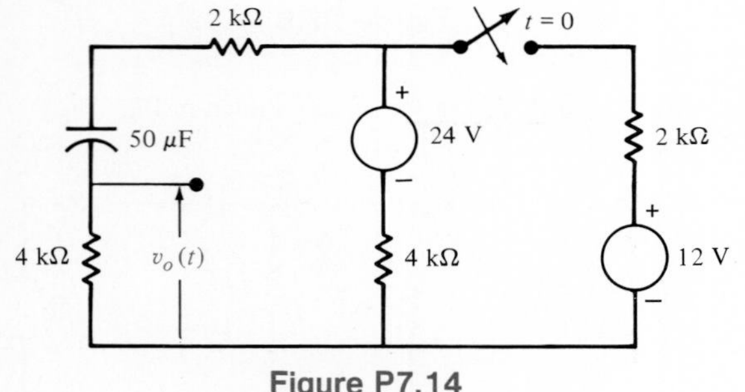

Figure P7.14

7.15. Find $v_o(t)$ and $i_o(t)$ for $t > 0$ in the network in Fig. P7.15.

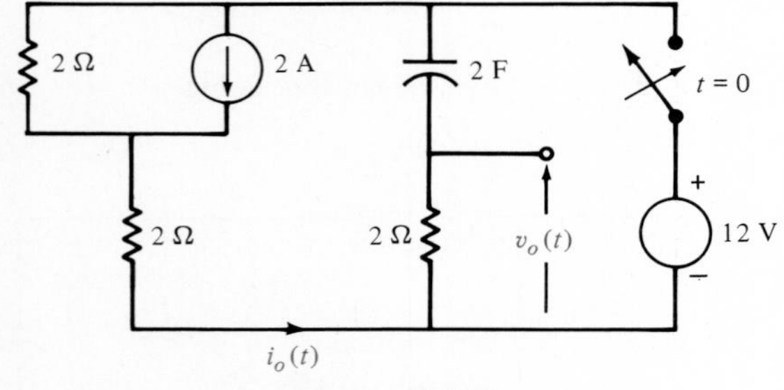

Figure P7.15

7.16. Find $v_o(t)$ and $i_o(t)$ for $t > 0$ in the network in Fig. P7.16.

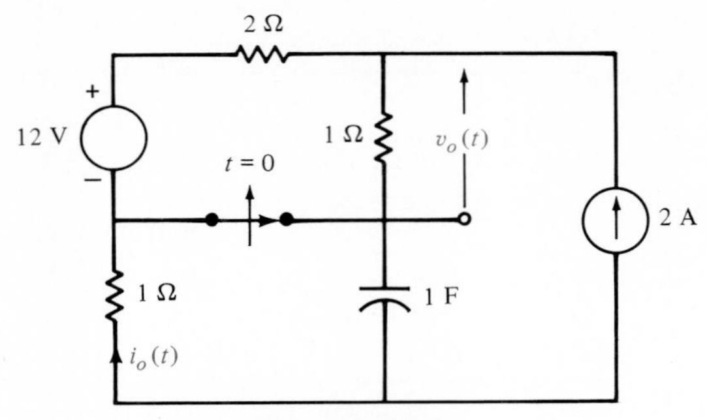

Figure P7.16

7.17. Find $v_o(t)$ and $i_o(t)$ for $t > 0$ in the network in Fig. P7.17.

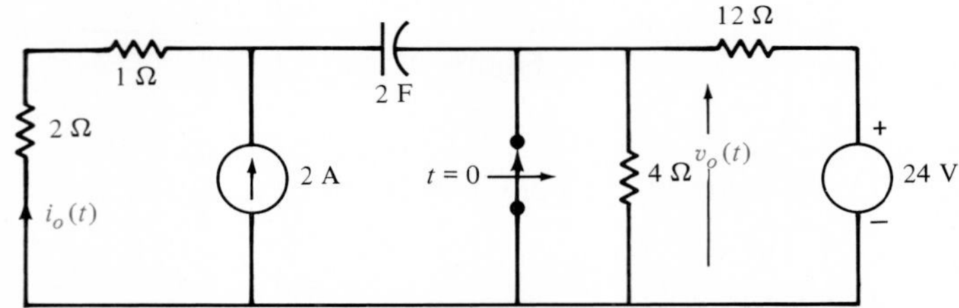

Figure P7.17

7.18. Find $v_o(t)$ for $t > 0$ in the circuit in Fig. P7.18.

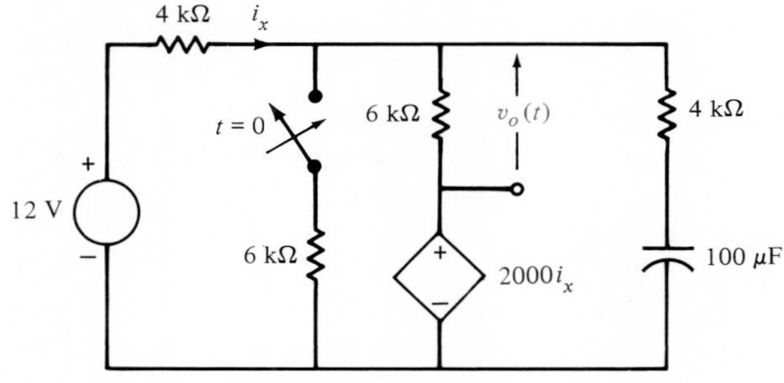

Figure P7.18

7.19. In Fig. P7.19, find $v_o(t)$ for $t > 0$.

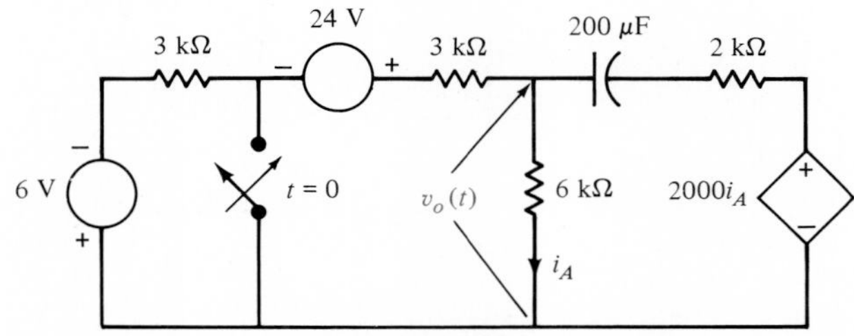

Figure P7.19

7.20. In the circuit shown in Fig. P7.20, the switch opens at $t = 0$. Find $i_1(t)$ for $t > 0$.

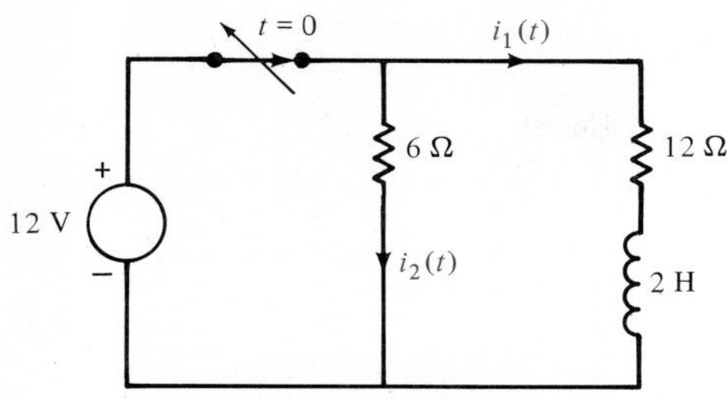

Figure P7.20

7.21. Determine the expression for $i_2(t)$, $t > 0$, in Fig. P7.20. Sketch the response.

7.22. In Fig. P7.20, if the switch was open for $t < 0$ and then closes at $t = 0$, find $i_1(t)$ for $t > 0$.

7.23. In Figure P7.23, determine $i_o(0-)$ before the switch is open. Find $i_o(t)$ for $t > 0$ and plot it.

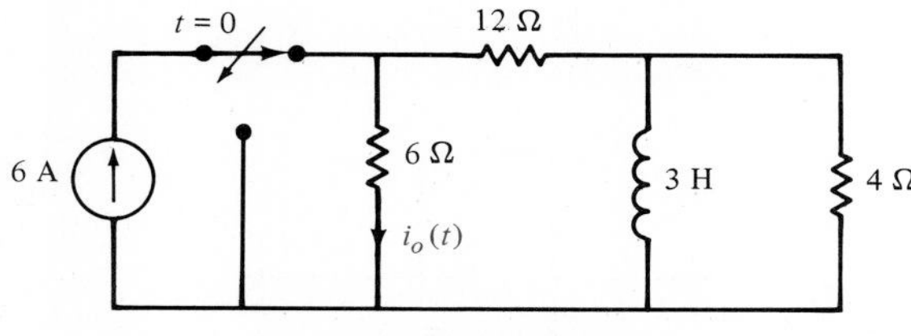

Figure P7.23

7.24. Find $v_o(t)$ for $t > 0$ in Fig. P7.24.

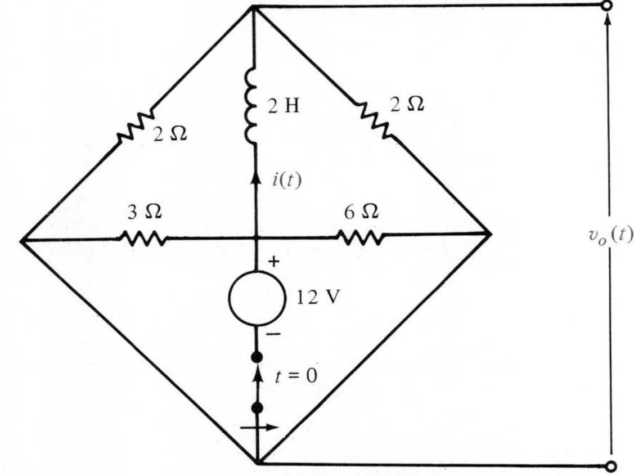

Figure P7.24

7.25. Find $i(t)$ for $t > 0$ in the network in Problem 7.24.

7.26. In the circuit shown in Fig. P7.26, both switches close at $t = 0$. Find $i_o(t)$ for $t > 0$.

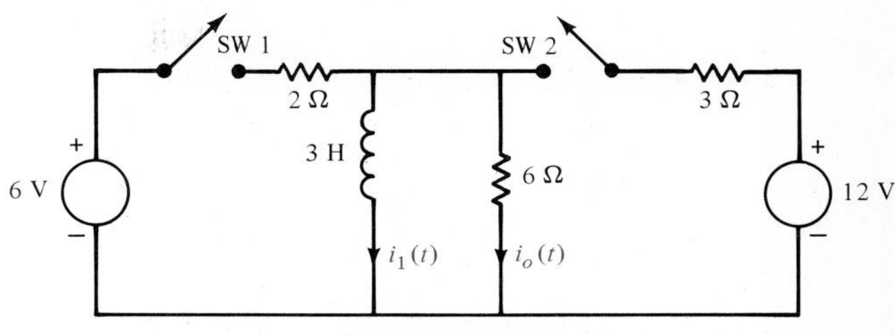

Figure P7.26

7.27. Under the conditions stated in Problem 7.26, find $i_1(t)$, $t > 0$.

7.28. The circuit in Fig. P7.26 has the following initial conditions: SW 1 is open and SW 2 is closed. At $t = 0$, SW 1 closes and SW 2 opens. Find $i_1(t)$ for $t > 0$.

7.29. Find $i_o(t)$ for $t > 0$ in Problem 7.28.

7.30. Determine $i_o(t)$ for $t > 0$ in Fig. P7.30.

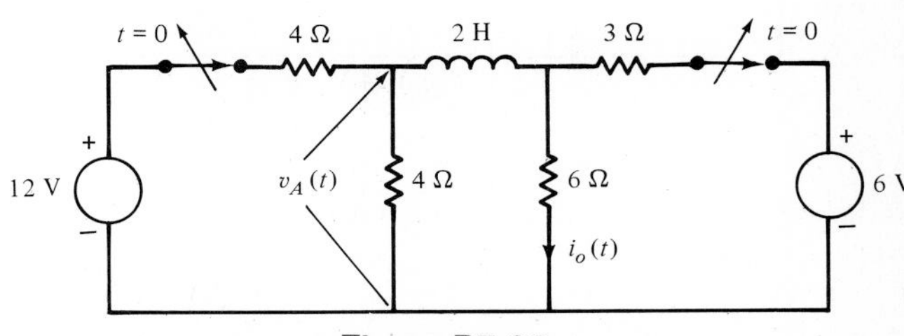

Figure P7.30

7.31. Find $v_A(t)$ for $t > 0$ in Fig. P7.30.

7.32. Find $i_o(t)$ for $t > 0$ in Fig. P7.32.

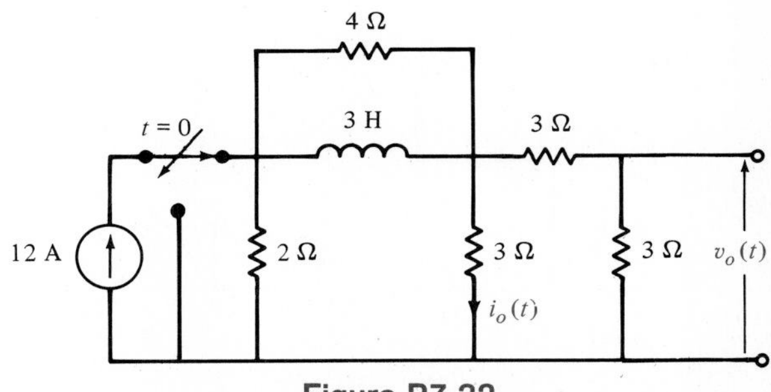

Figure P7.32

7.33. Determine the expression for $v_o(t)$, $t > 0$, in Fig. P7.32.

7.34. Determine the expression for $v_o(t)$, $t > 0$ in the network in Fig. P7.34.

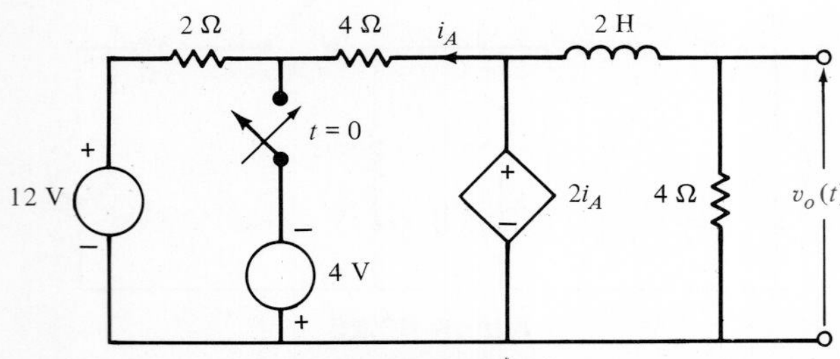

Figure P7.34

7.35. In Fig. P7.35, determine
(a) $i_1(t)$, $t > 0$
(b) $i_2(t)$, $t > 0$

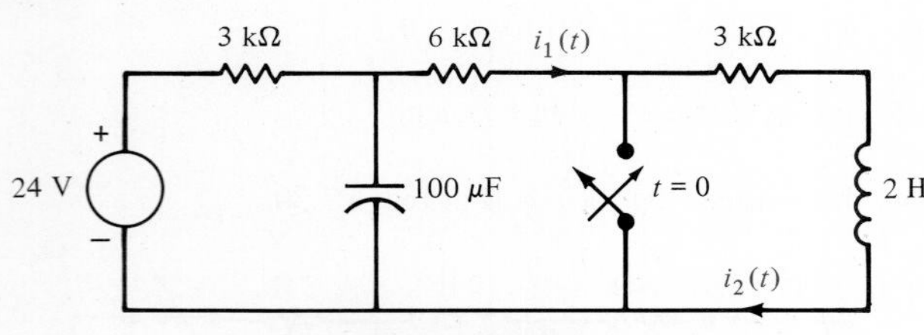

Figure P7.35

7.36. In Problem 7.2, if the switch opens at $t = 1$ s, find the voltage $v_o(t)$, $t > 1$, and plot the response.

7.37. The switch in Problem 7.23 opens at $t = 0.5$ s. Find $i_o(t)$ for $t > 0.5$ s and plot the response.

7.38. Determine the equations for the combination of pulses using the unit step function for the waveforms shown in Fig. P7.38.

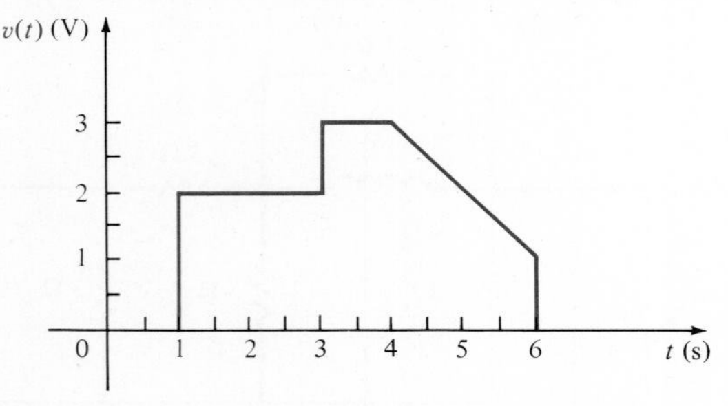

(a)

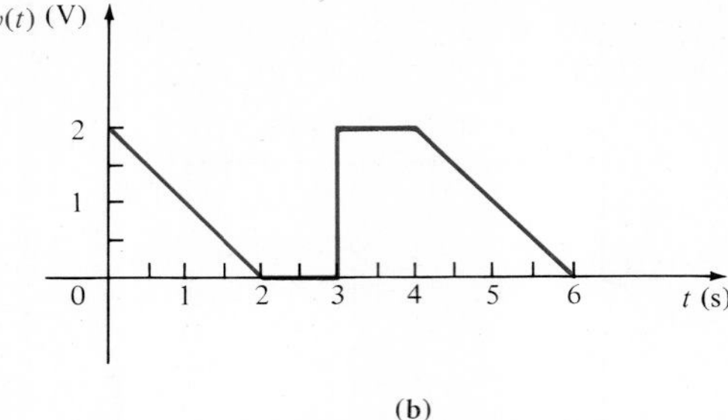

(b)

Figure P7.38

7.39. Determine the equation for the voltage $v_o(t)$, $t > 0$, in Fig. P7.39a when subjected to the input pulse shown in Fig. P7.39b.

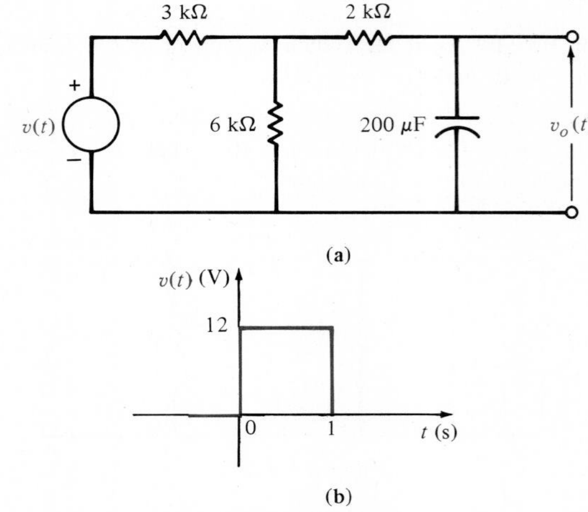

Figure P7.39

7.40. Find $v_o(t)$, $t > 0$, in the network in Fig. P7.40.

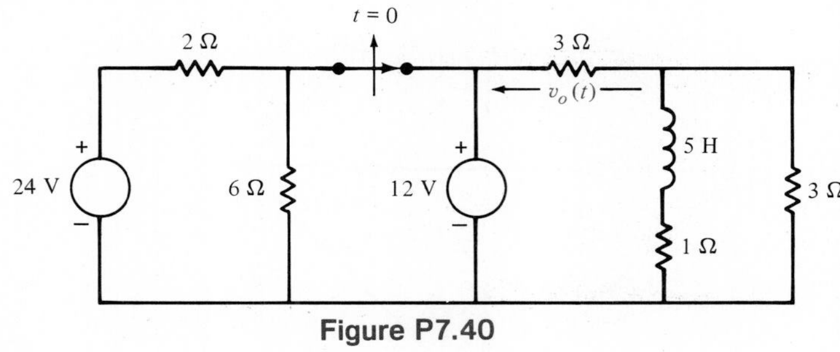

Figure P7.40

7.41. Find $v_o(t)$ for $t > 0$ in the network in Fig. P7.41.

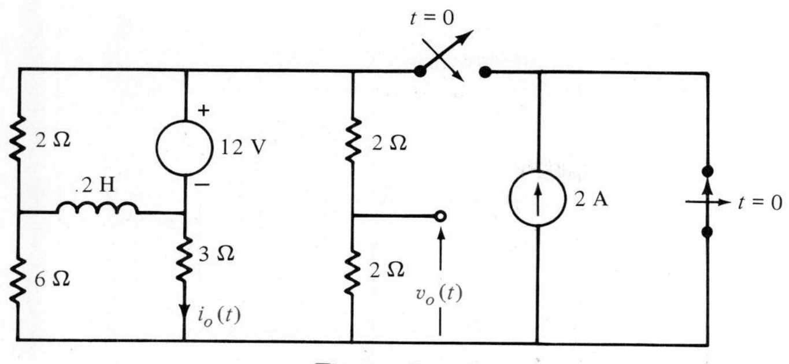

Figure P7.41

7.42. Find $i_o(t)$ for $t > 0$ in the network in Problem P7.41.

7.43. Find $v_o(t)$ for $t > 0$ in the network in Fig. P7.43.

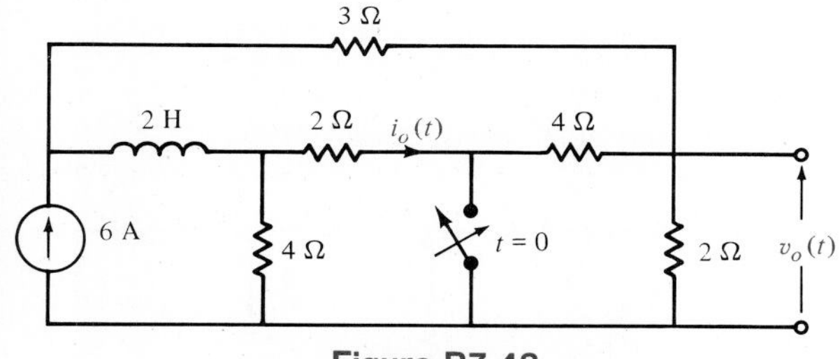

Figure P7.43

7.44. Find $i_o(t)$ for $t > 0$ in the network in Fig. P7.43.

7.45. Find $v_o(t)$ for $t > 0$ in the network in Fig. P7.45.

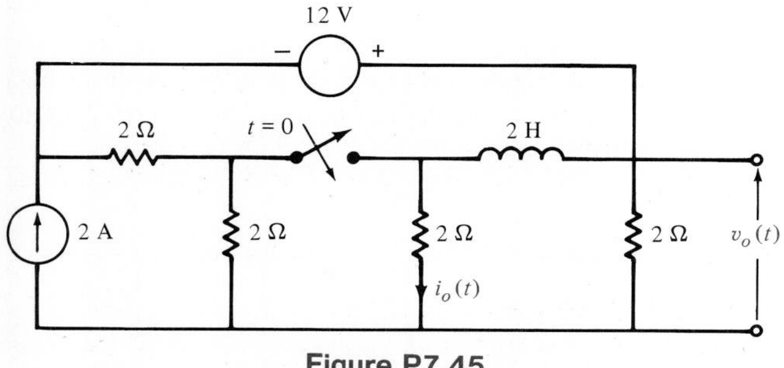

Figure P7.45

7.46. Find $i_o(t)$ for $t > 0$ in the network in Fig. P7.45.

7.47. Find $v_o(t)$ for $t > 0$ in the network in Fig. P7.47.

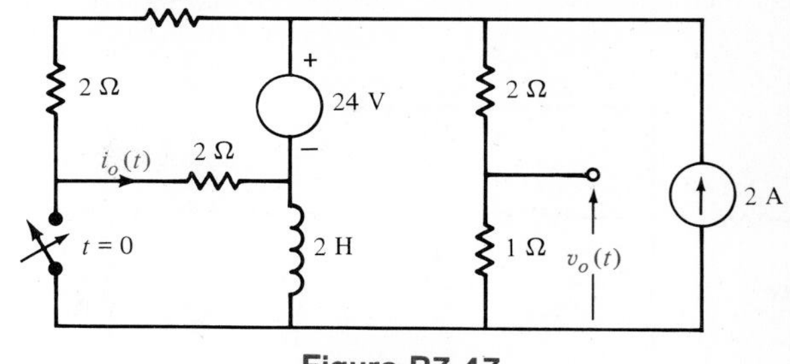

Figure P7.47

7.48. Find $i_o(t)$ for $t > 0$ in the network in Fig. P7.47.

7.49. Find $v_o(t)$ for $t > 0$ in the network in Fig. P7.49.

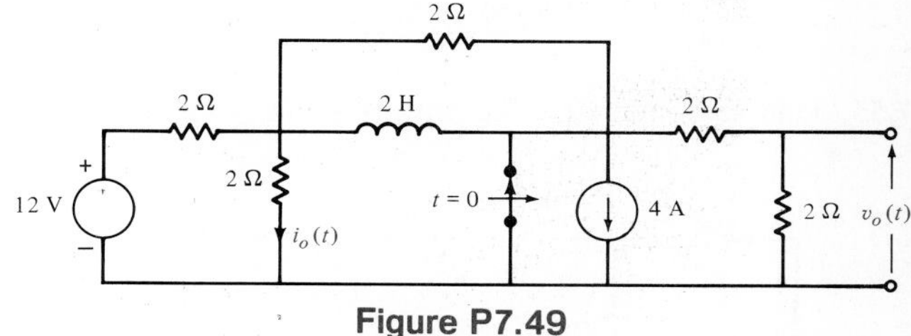

Figure P7.49

7.50. Find $i_o(t)$ for $t > 0$ in the network in Fig. P7.49.

7.51. Find $v_o(t)$ for $t > 0$ in the network in Fig. P7.51.

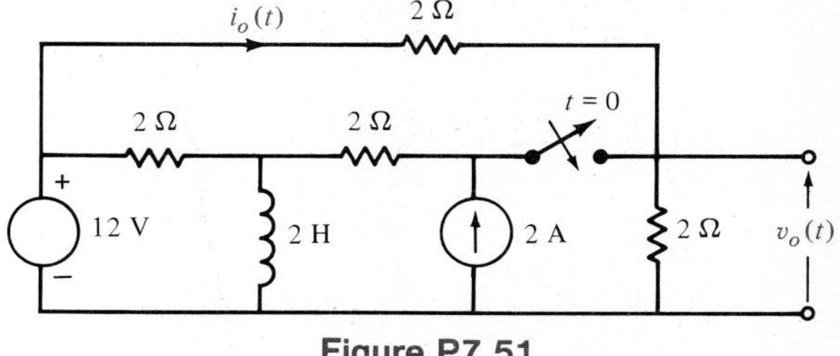

Figure P7.51

7.52. Find $i_o(t)$ for $t > 0$ in the network in Problem 7.51.

7.53. Find $v_o(t)$ for $t > 0$ in the network in Fig. P7.53.

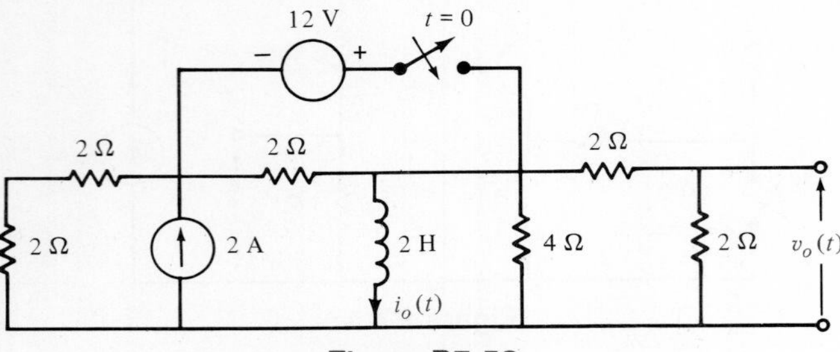

Figure P7.53

7.54. Find $i_o(t)$ for $t > 0$ in the network in Fig. P7.53.

7.55. Find $i(t)$ for $t > 0$ in the network in Fig. P7.55.

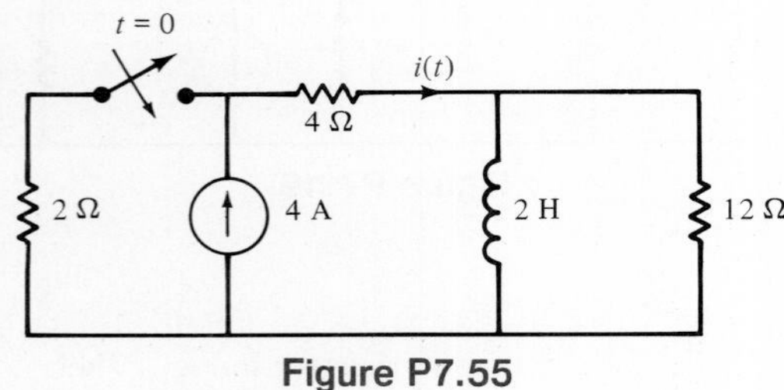

Figure P7.55

7.56. Find $i(t)$ for $t > 0$ for the network in Fig. P7.56.

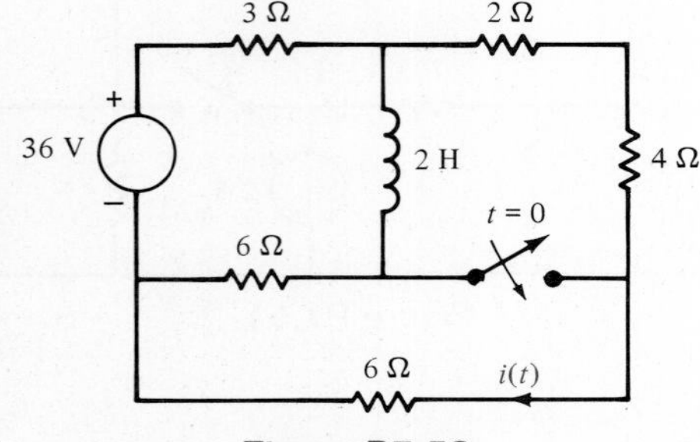

Figure P7.56

7.57. Determine the relationship between the input and output voltages for the circuit shown in Fig. P7.57.

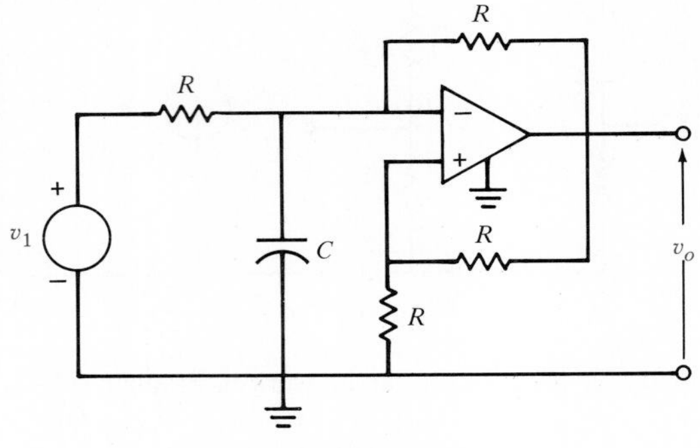

Figure P7.57

7.58. Given the network in Fig. P7.58, plot $v_R(t)$ from 0 to 0.6 s in increments of 50 ms using SPICE.

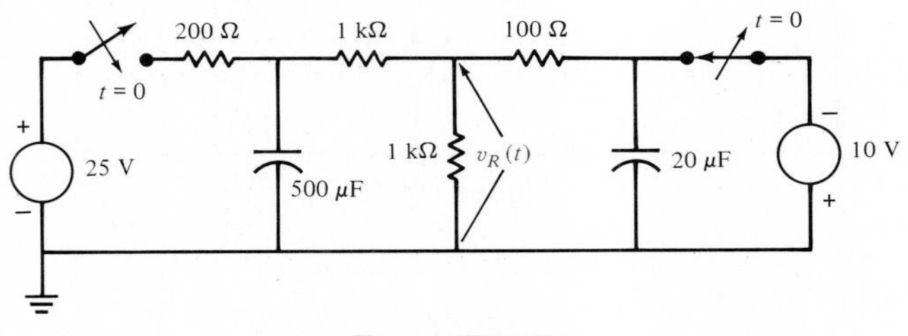

Figure P7.58

7.59. Write a SPICE program to plot $v_o(t)$ in the circuit in Fig. P7.59 from 0 to 0.02 s in steps of 0.5 ms. (Hint: Use SPICE to determine initial conditions.)

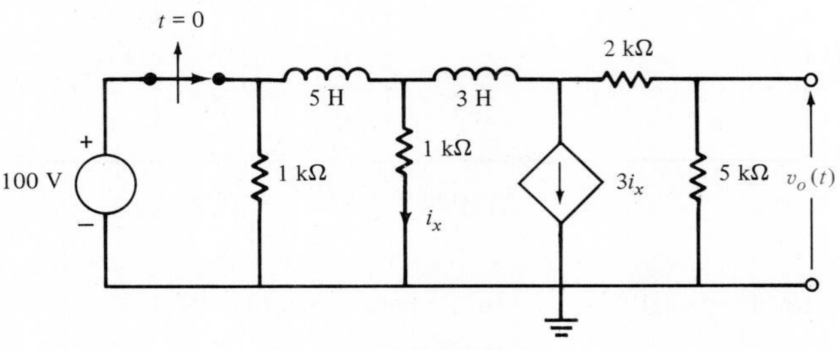

Figure P7.59

7.60. Plot $v_o(t)$ from 0 to 0.5 s with a step size of 10 ms using SPICE for the network in Fig. P7.60.

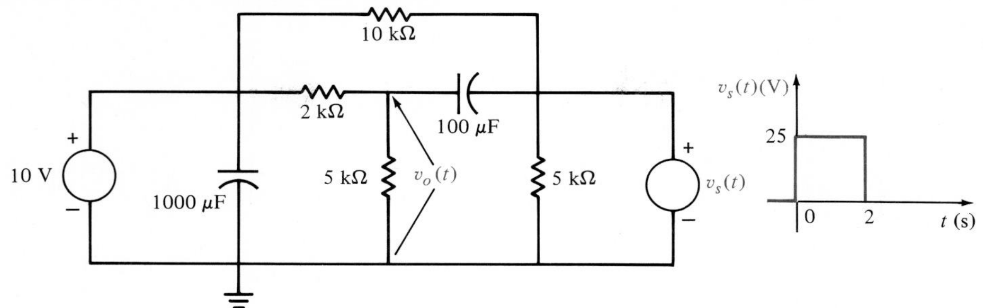

Figure P7.60

7.61. In Fig. P7.61, use SPICE to plot $v_o(t)$ over the interval 0 to 7 s using a step size of 0.2 s, where $v_s(t) = 10(1 - e^{-t/0.5}) - 10[1 - e^{-(t-2)/2}$ V].

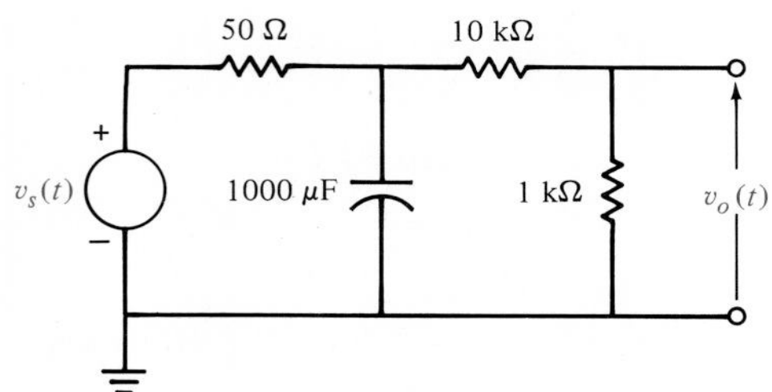

Figure P7.61

7.62. Given the circuit in Fig. P7.62, plot $v_o(t)$ from 0 to 8 s in steps of 100 ms, using SPICE analysis.

$v_{s1}(t)$: $0 \rightarrow 2$ s $\quad v_{s1} = 25(1 - e^{-t/3})$ V

$2 \rightarrow 8$ s $\quad v_{s1} = 25(1 - e^{-t/3}) - 25\,(1 - e^{-t/2})$ V

$v_{s2}(t)$: $0 \rightarrow 1$ s $\quad v_{s2} = 0$ V

$1 \rightarrow 8$s $\quad v_{s2} = 15(1 - e^{-t/4}) - 15[1 - e^{-(t-3)/4}]$ V

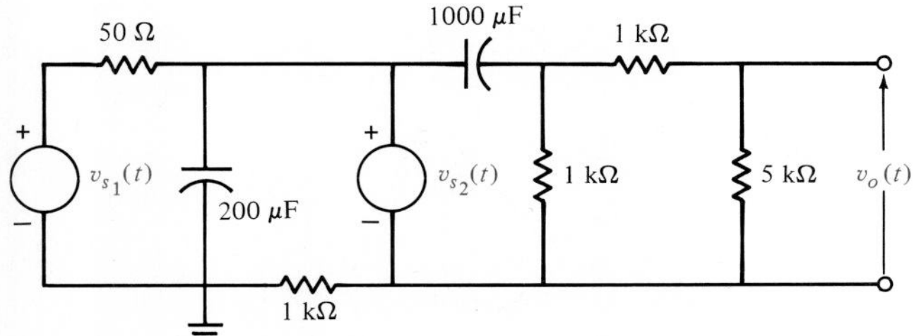

Figure P7.62

7.63. Using SPICE plot the voltage across the 1-Ω resistor in the network in Fig. P7.63. Use a step size of 100 ms from 0 to 5 s.

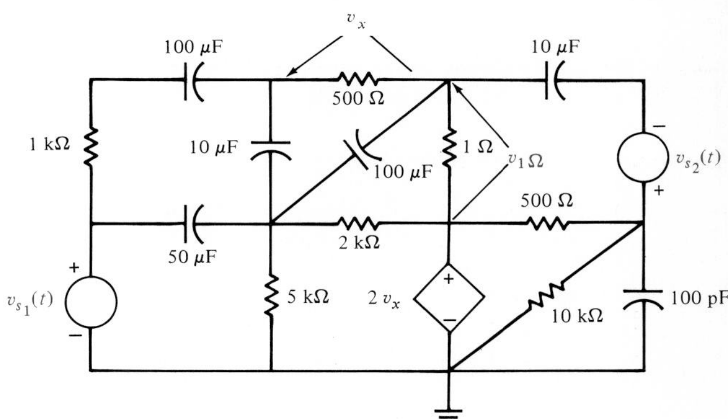

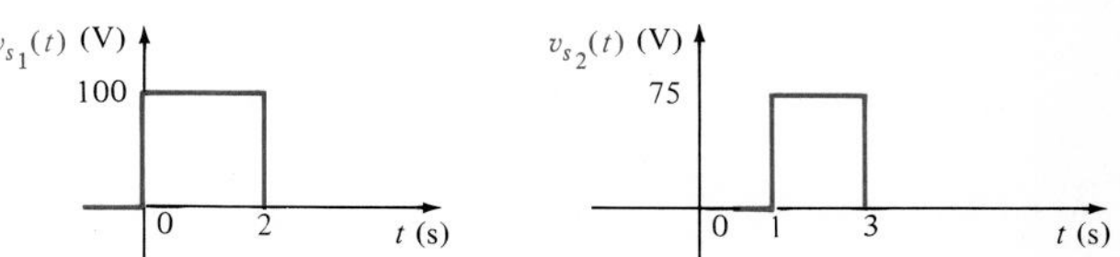

Figure P7.63

7.64. Given the network in Fig. P7.64, plot $v_L(t)$ using an increment of 20 ms over the range 0 to 0.6 s.

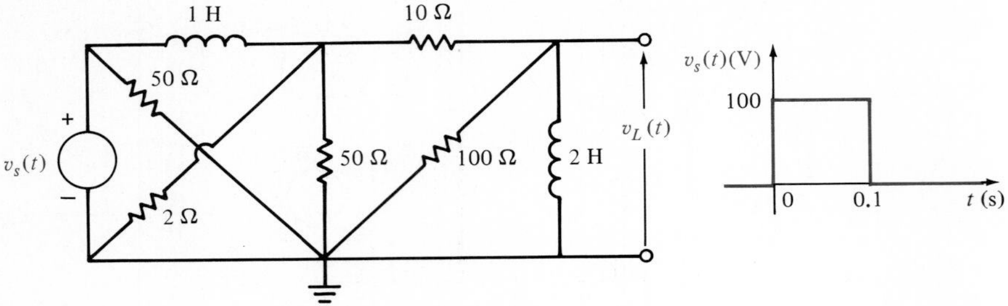

Figure P7.64

7.65. Plot the voltage across the 4-Ω resistor in the circuit in Fig. P7.65. Use a step size of $\frac{1}{8}$ s over the interval 0 to 5 s.

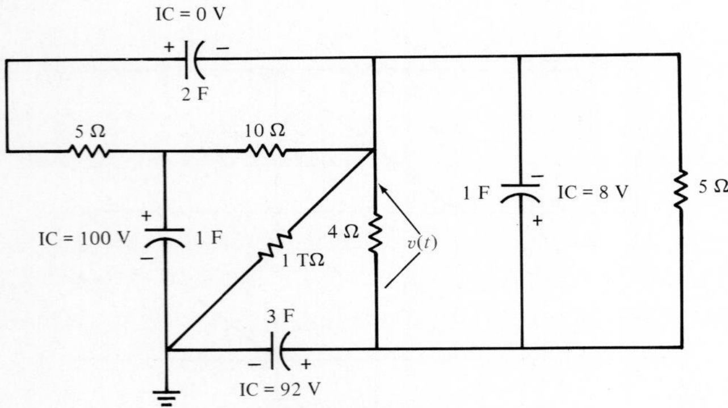

Figure P7.65

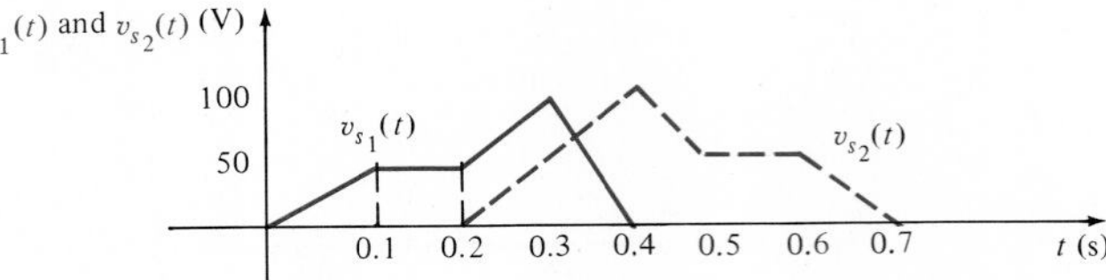

Figure P7.66

7.66. Given the network in Fig. P7.66, plot $v_R(t)$ from 0 to 5 s using an increment of 50 ms.

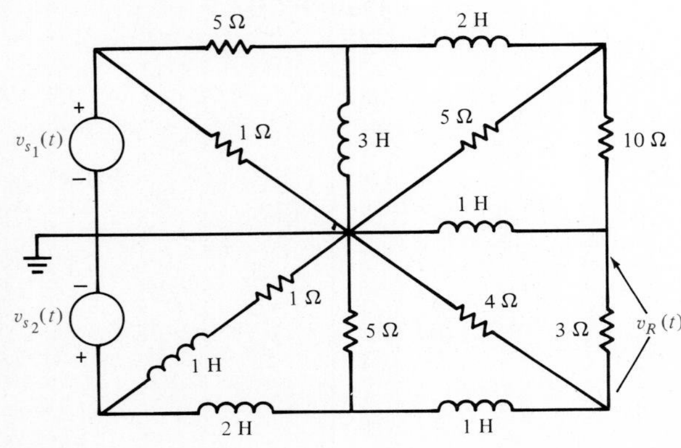

7.67. Plot $v_x(t)$ in the network in Fig. P7.67 using SPICE analysis for 20 ms in steps of 1 ms.

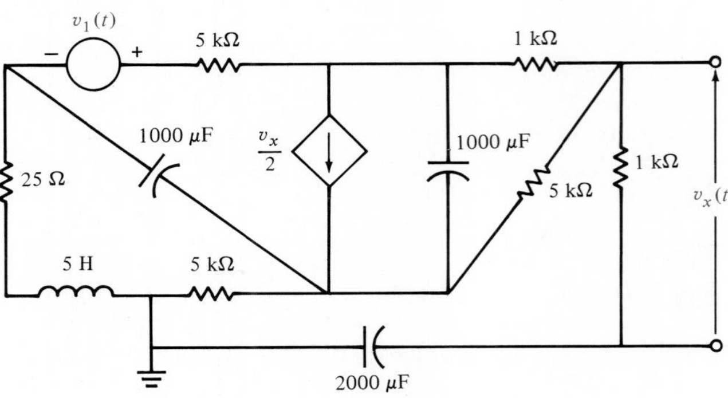

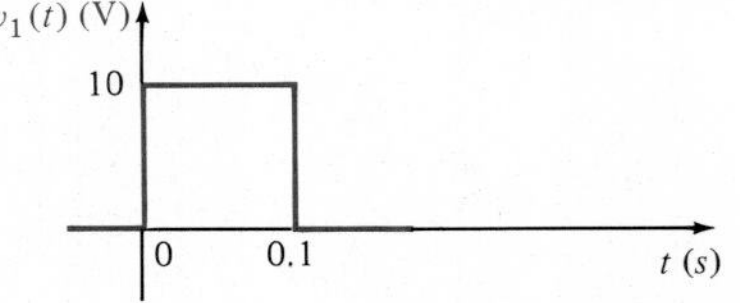

Figure P7.67

CHAPTER

RLC Circuits

In this chapter we extend our analysis of circuits that contain storage elements to the case where an inductor and a capacitor are present simultaneously. Although these *RLC* circuits are more complicated than those we have analyzed, we will follow a development similar to that used earlier in obtaining a solution. In addition, we extend our SPICE analysis to networks that contain both an inductor and a capacitor.

8.1 The Basic Circuit Equation

To begin our development, let us consider the two basic *RLC* circuits shown in Fig. 8.1. We assume that energy may be initially stored in both the inductor and capacitor. The node equation for the parallel *RLC* circuit is

$$\frac{v}{R} + \frac{1}{L}\int_{t_0}^{t} v(x)\,dx - i_L(t_0) + C\frac{dv}{dt} = i_S(t)$$

Similarly, the loop equation for the series *RLC* circuit is

$$Ri + \frac{1}{C}\int_{t_0}^{t} i(x)\,dx - v_C(t_0) + L\frac{di}{dt} = v_S(t)$$

The similarity between the two equations is obvious. We expect this close relationship to exist since, as shown in Chapter 6, the two circuits in Fig. 8.1 are *duals* of one another. This means that the equation for the node voltage in the parallel circuit is of the same form as that for the loop current in the series circuit. Therefore, the solution

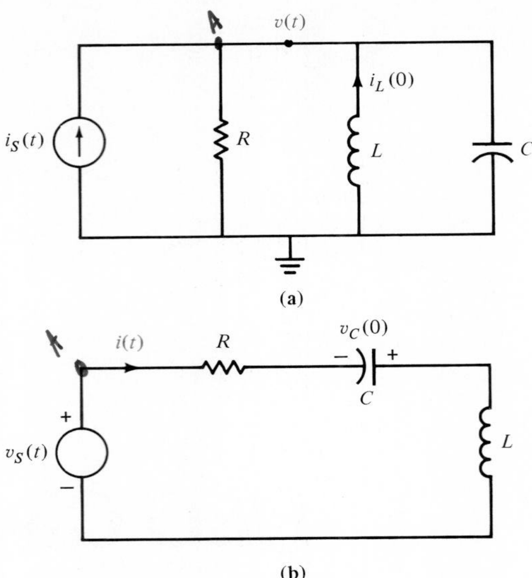

Figure 8.1 Parallel and series *RLC* circuits.

of these two circuits is dependent on solving one equation. If the two equations above are differentiated with respect to time, we obtain

$$C\frac{d^2v}{dt^2} + \frac{1}{R}\frac{dv}{dt} + \frac{v}{L} = \frac{di_s}{dt}$$

and

$$L\frac{d^2i}{dt^2} + R\frac{di}{dt} + \frac{i}{C} = \frac{dv_s}{dt}$$

Since both circuits lead to a second-order differential equation with constant coefficients, we will concentrate our analysis on this type of equation.

8.2 Mathematical Development of the Response Equations

In concert with our development of the solution of a first-order differential equation that results from the analysis of either an *RL* or an *RC* circuit as outlined in Chapter 7, we will now employ the same approach here to obtain the solution of a second-order differential equation that results from the analysis of *RLC* circuits. As a general rule, for this case we are confronted with an equation of the form

$$\frac{d^2x(t)}{dt^2} + a_1\frac{dx(t)}{dt} + a_2x(t) = \mathrm{f}(t) \tag{8.1}$$

Once again we use the fact that if $x(t) = x_p(t)$ is a solution to Eq. (8.1), and if $x(t) = x_c(t)$ is a solution to the homogeneous equation

$$\frac{d^2x(t)}{dt^2} + a_1\frac{dx(t)}{dt} + a_2x(t) = 0$$

then

$$x(t) = x_p(t) + x_c(t)$$

is a solution to the original Eq. (8.1). If we again confine ourselves to a constant forcing function [i.e., $f(t) = A$], the development at the beginning of Chapter 7 shows that the solution of Eq. (8.1) will be of the form

$$x(t) = \frac{A}{a_2} + x_c(t) \tag{8.2}$$

Let us now turn our attention to the solution of the homogeneous equation

$$\frac{d^2x(t)}{dt^2} + a_1 \frac{dx(t)}{dt} + a_2 x(t) = 0$$

where a_1 and a_2 are constants. For simplicity we will rewrite the equation in the form

$$\frac{d^2x(t)}{dt^2} + 2\alpha \frac{dx(t)}{dt} + \omega_0^2 x(t) = 0 \tag{8.3}$$

where we have made the following simple substitutions for the constants $a_1 = 2\alpha$ and $a_2 = \omega_0^2$. Following the development of a solution for the first-order homogeneous differential equation in Chapter 7, the solution of Eq. (8.3) must be a function whose first- and second-order derivatives have the same form, so that the left-hand side of Eq. (8.3) will become identically zero for all t. Again we assume that

$$x(t) = Ke^{st}$$

Substituting this expression into Eq. (8.3) yields

$$s^2Ke^{st} + 2\alpha sKe^{st} + \omega_0^2 Ke^{st} = 0$$

Dividing both sides of the equation by Ke^{st} yields

$$s^2 + 2\alpha s + \omega_0^2 = 0 \tag{8.4}$$

This equation is commonly called the *characteristic equation*. α is called the exponential *damping coefficient* and ω_0 is referred to as the *undamped resonant frequency*. The importance of this terminology will become clear as we proceed with the development. If this equation is satisfied, our assumed solution $x(t) = Ke^{st}$ is correct. Employing the quadratic formula, we find that Eq. (8.4) is satisfied if

$$\begin{aligned} s &= \frac{-2\alpha \pm \sqrt{4\alpha^2 - 4\omega_0^2}}{2} \\ &= -\alpha \pm \sqrt{\alpha^2 - \omega_0^2} \end{aligned} \tag{8.5}$$

Therefore, there are two values of s, s_1 and s_2, that satisfy Eq. (8.4).

$$\begin{aligned} s_1 &= -\alpha + \sqrt{\alpha^2 - \omega_0^2} \\ s_2 &= -\alpha - \sqrt{\alpha^2 - \omega_0^2} \end{aligned} \tag{8.6}$$

This means that $x_1(t) = K_1e^{s_1t}$ is a solution of Eq. (8.3) and that $x_2(t) = K_2e^{s_2t}$ is also a solution of Eq. (8.3); that is,

$$\frac{d^2}{dt^2}(K_1e^{s_1t}) + 2\alpha\frac{d}{dt}(K_1e^{s_1t}) + \omega_0^2K_1e^{s_1t} = 0$$

and

$$\frac{d^2}{dt^2}(K_2e^{s_2t}) + 2\alpha\frac{d}{dt}(K_2e^{s_2t}) + \omega_0^2K_2e^{s_2t} = 0$$

The addition of these two equations produces the equality

$$\frac{d^2}{dt^2}(K_1e^{s_1t} + K_2e^{s_2t}) + 2\alpha\frac{d}{dt}(K_1e^{s_1t} + K_2e^{s_2t}) + \omega_0^2(K_1e^{s_1t} + K_2e^{s_2t}) = 0$$

Note that the sum of the two solutions is also a solution. Therefore, in general, the complementary solution of Eq. (8.1) is of the form

$$x_c(t) = K_1e^{s_1t} + K_2e^{s_2t} \tag{8.7}$$

K_1 and K_2 are arbitrary constants that can be evaluated via the initial conditions $x(0)$ and $dx(0)/dt$. For example, since

$$x(t) = K_1e^{s_1t} + K_2e^{s_2t}$$

then

$$x(0) = K_1 + K_2$$

and

$$\frac{dx(0)}{dt} = s_1K_1 + s_2K_2$$

Hence $x(0)$ and $dx(0)/dt$ produce two simultaneous equations, which when solved yield the constants K_1 and K_2.

Close examination of Eqs. (8.6) and (8.7) indicates that the form of the solution of the homogeneous equation is dependent on the relative magnitude of the values of α and ω_0. For example, if $\alpha > \omega_0$, the roots of the characteristic equation, s_1 and s_2, also called the *natural frequencies* because they determine the natural (unforced) response of the network, are real and unequal; if $\alpha < \omega_0$, the roots are complex numbers; and finally if $\alpha = \omega_0$, the roots are real and equal. Each of these cases is very important; hence we will now examine each one in some detail.

Case 1, $\alpha > \omega_0$. This case is commonly called *overdamped*. The natural frequencies s_1 and s_2 are real and unequal, and therefore the natural response of the network described by the second-order differential equation is of the form

$$x_c(t) = K_1e^{-(\alpha-\sqrt{\alpha^2-\omega_0^2})t} + K_2e^{-(\alpha+\sqrt{\alpha^2-\omega_0^2})t} \tag{8.8}$$

where K_1 and K_2 are found from the initial conditions. This indicates that the natural response is the sum of two decaying exponentials.

Case 2, $\alpha < \omega_0$. This case is called *underdamped*. Since $\omega_0 > \alpha$, the roots of the characteristic equation given in Eq. (8.6) can be written as

$$s_1 = -\alpha + \sqrt{-(\omega_0^2 - \alpha^2)} = -\alpha + j\omega_n$$

$$s_2 = -\alpha - \sqrt{-(\omega_0^2 - \alpha^2)} = -\alpha - j\omega_n$$

where $j = \sqrt{-1}$ and $\omega_n = \sqrt{\omega_0^2 - \alpha^2}$. Thus the natural frequencies are complex numbers. The natural response is then

$$x_c(t) = K_1 e^{-(\alpha - j\omega_n)t} + K_2 e^{-(\alpha + j\omega_n)t}$$

This equation can be simplified in the following manner. First the equation is rewritten as

$$x_c(t) = e^{-\alpha t}(K_1 e^{j\omega_n t} + K_2 e^{-j\omega_n t})$$

Then by using Euler's identities,

$$e^{j\theta} = \cos\theta + j\sin\theta$$

and

$$e^{-j\theta} = \cos\theta - j\sin\theta$$

we obtain

$$\begin{aligned} x_c(t) &= e^{-\alpha t}[K_1(\cos\omega_n t + j\sin\omega_n t) + K_2(\cos\omega_n t - j\sin\omega_n t)] \\ &= e^{-\alpha t}[(K_1 + K_2)\cos\omega_n t + (jK_1 - jK_2)\sin\omega_n t] \\ &= e^{-\alpha t}(A_1\cos\omega_n t + A_2\sin\omega_n t) \end{aligned} \tag{8.9}$$

where A_1 and A_2, like K_1 and K_2, are constants, which are evaluated using the initial conditions $x(0)$ and $dx(0)/dt$. If $x_c(t)$ is real, K_1 and K_2 will be complex and $K_2 = K_1^*$. $A_1 = K_1 + K_2$ is therefore two times the real part of K_1, and $A_2 = jK_1 - jK_2$ is -2 times the imaginary part of K_1. A_1 and A_2 are real numbers. This illustrates that the natural response is an exponentially damped oscillatory response.

Case 3, $\alpha = \omega_0$. This case, called *critically damped*, results in

$$s_1 = s_2 = -\alpha$$

as shown in Eq. (8.6). Therefore, Eq. (8.7) reduces to

$$x_c(t) = K_3 e^{-\alpha t}$$

where $K_3 = K_1 + K_2$. However, this cannot be a solution to the second-order differential Eq. (8.3) because in general it is not possible to satisfy the two initial conditions $x(0)$ and $dx(0)/dt$ with the single constant K_3.

In the case where the characteristic equation has repeated roots, a solution can be obtained in the following manner. If $x_1(t)$ is known to be a solution of the second-order homogeneous equation, then via the substitution $x(t) = x_1(t)y(t)$ we can transform the given differential equation into a first-order equation in $dy(t)/dt$. Since this resulting equation is only a function of $y(t)$, it can be solved to find the general solution $x(t) = x_1(t)y(t)$.

For the present case, $s_1 = s_2 = -\alpha$, and hence the basic equation is

$$\frac{d^2x(t)}{dt^2} + 2\alpha\frac{dx(t)}{dt} + \alpha^2 x(t) = 0 \tag{8.10}$$

and one known solution is

$$x_1(t) = K_3 e^{-\alpha t}$$

By employing the substitution

$$x_2(t) = x_1(t)y(t) = K_3 e^{-\alpha t} y(t)$$

Eq. (8.10) becomes

$$\frac{d^2}{dt^2}[K_3 e^{-\alpha t} y(t)] + 2\alpha \frac{d}{dt}[K_3 e^{-\alpha t} y(t)] + \alpha^2 K_3 e^{-\alpha t} y(t) = 0$$

Evaluating the derivatives, we obtain

$$\frac{d}{dt}[K_3 e^{-\alpha t} y(t)] = -K_3 \alpha e^{-\alpha t} y(t) + K_3 e^{-\alpha t} \frac{dy(t)}{dt}$$

$$\frac{d^2}{dt^2}[K_3 e^{-\alpha t} y(t)] = K_3 \alpha^2 e^{-\alpha t} y(t) - 2K_3 \alpha e^{-\alpha t} \frac{dy(t)}{dt} + K_3 e^{-\alpha t} \frac{d^2 y(t)}{dt^2}$$

Substituting these expressions into the equation above yields

$$K_3 e^{-\alpha t} \frac{d^2 y(t)}{dt^2} = 0$$

Therefore,

$$\frac{d^2 y(t)}{dt^2} = 0$$

and hence

$$y(t) = A_1 + A_2 t$$

Therefore, the general solution is

$$\begin{aligned} x_2(t) &= x_1(t)y(t) \\ &= K_3 e^{-\alpha t}(A_1 + A_2 t) \end{aligned}$$

which can be written as

$$x_2(t) = B_1 e^{-\alpha t} + B_2 t e^{-\alpha t} \tag{8.11}$$

where B_1 and B_2 are constants derived from the initial conditions.

It is informative to sketch the natural response for the three cases we have discussed: overdamped, Eq. (8.8); underdamped, Eq. (8.9); and critically damped, Eq. (8.11). Figure 8.2 graphically illustrates the three cases for the situations in which $x_C(0) = 0$. Note that the critically damped response peaks and decays faster than the overdamped response. The underdamped response is an exponentially damped sinusoid whose rate of decay is dependent on the factor α. Actually, the terms $\pm e^{-\alpha t}$ define what is called the *envelope* of the response, and the damped oscillations (i.e, the oscillations of decreasing amplitude) exhibited by the waveform in Fig. 8.2b are called *ringing*.

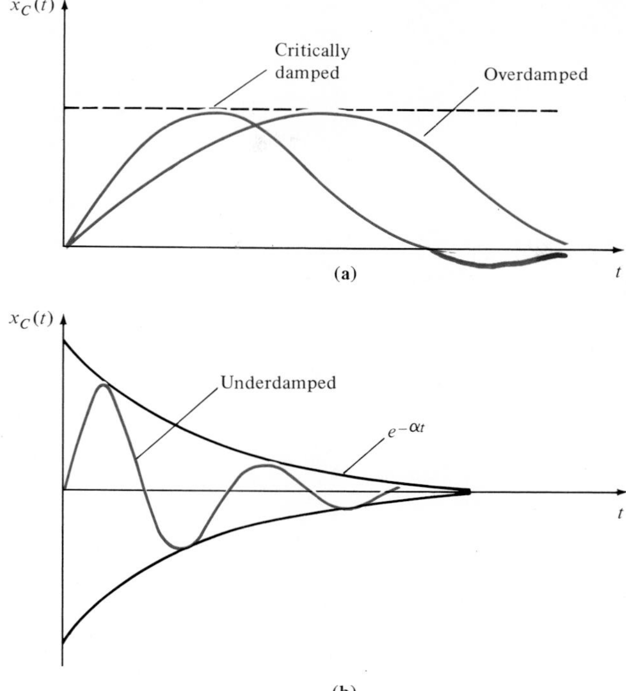

Figure 8.2 Comparison of overdamped, critically damped, and underdamped responses.

DRILL EXERCISE

D8.1. A parallel *RLC* circuit has the following circuit parameters: $R = 1\ \Omega$, $L = 2$ H, and $C = 2$ F. Compute the damping ratio and the undamped natural frequency of this network.

D8.2. A series *RLC* circuit consists of $R = 2\ \Omega$, $L = 1$ H, and a capacitor. Determine the type of response exhibited by the network if (a) $C = \frac{1}{2}$ F, (b) $C = 1$ F, and (c) $C = 2$ F.

8.3 Source-Free Response

Let us now examine the source-free response of a number of simple *RLC* networks. We will analyze circuits that exhibit overdamped, underdamped, and critically damped responses. The following examples will serve as a vehicle for this analysis.

EXAMPLE 8.1

Consider the parallel *RLC* circuit shown in Fig. 8.3. The second-order differential equation that describes the voltage $v(t)$ is

$$\frac{d^2v}{dt^2} + \frac{1}{RC}\frac{dv}{dt} + \frac{v}{LC} = 0$$

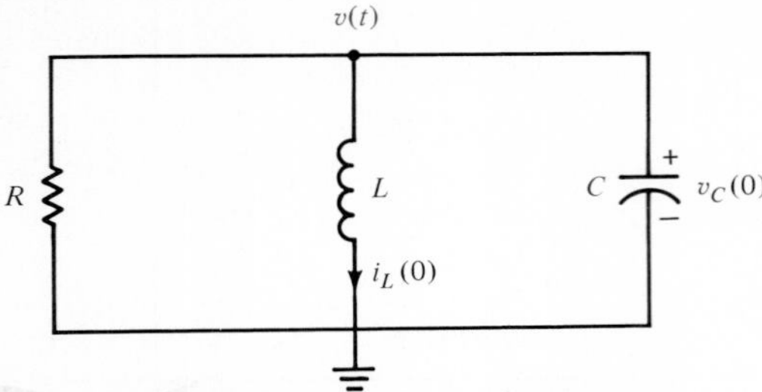

Figure 8.3 Parallel *RLC* circuit.

A comparison of this equation with Eq. (8.3) and (8.4) indicates that for the parallel *RLC* circuit the damping coefficient is $1/2RC$ and the resonant frequency is $1/\sqrt{LC}$. If the circuit parameters are $R = 2\ \Omega$, $C = \frac{1}{5}$ F, and $L = 5$ H, the equation becomes

$$\frac{d^2v}{dt^2} + 2.5\frac{dv}{dt} + v = 0$$

Let us assume that the initial conditions on the storage elements are $i_L(0) = -1\ A$, and $v_C(0) = 4$ V.

The characteristic equation for the network is

$$s^2 + 2.5s + 1 = 0$$

and the roots are

$$s_1 = -2$$

$$s_2 = -0.5$$

Since the roots are real and unequal, the circuit is overdamped, and $v(t)$ is of the form

$$v(t) = K_1e^{-2t} + K_2e^{-0.5t}$$

The initial conditions are now employed to determine the constants K_1 and K_2. Since $v(t) = v_c(t)$,

$$v_C(0) = v(0) = 4 = K_1 + K_2$$

The second equation needed to determine K_1 and K_2 is normally obtained from the expression

$$\frac{dv(t)}{dt} = -2K_1e^{-2t} - 0.5K_2e^{-0.5t}$$

However, the second initial condition is not $dv(0)/dt$. If this were the case, we would simply evaluate the equation above at $t = 0$. This would produce a second equation in the unknowns K_1 and K_2. We can, however, circumvent this problem by noting that the node equation for the circuit can be written as

$$C\frac{dv(t)}{dt} + \frac{v(t)}{R} + i_L(t) = 0$$

or

$$\frac{dv(t)}{dt} = \frac{-1}{RC} v(t) - \frac{i_L(t)}{C}$$

At $t = 0$,

$$\frac{dv(0)}{dt} = \frac{-1}{RC} v(0) - \frac{1}{C} i_L(0)$$

$$= -2.5(4) - 5(-1)$$

$$= -5$$

But since

$$\frac{dv(t)}{dt} = -2K_1 e^{-2t} - 0.5K_2 e^{-0.5t}$$

then when $t = 0$

$$-5 = -2K_1 - 0.5K_2$$

This equation, together with the equation

$$4 = K_1 + K_2$$

produces the constants $K_1 = 2$ and $K_2 = 2$. Therefore, the final equation for the voltage is

$$v(t) = 2e^{-2t} + 2e^{-0.5t} \text{ V}$$

Note that the voltage equation satisfies the initial condition $v(0) = 4$ V.

Before leaving this example, let us examine this problem from another viewpoint. Since the circuit is linear, the characteristic equation for any current or voltage in the network will be the same. Therefore, the current in the inductor must be of the form

$$i_L(t) = K_3 e^{-2t} + K_4 e^{-0.5t}$$

Using the initial conditions, we can evaluate the constants K_3 and K_4, and thus obtain an exact expression for $i_L(t)$.

$$i_L(0) = -1 = K_3 + K_4$$

and

$$\frac{di_L(0)}{dt} = -2K_3 - 0.5K_4$$

However, once again we do not know $di_L(0)/dt$. We do know that

$$v(0) = L \frac{di_L(0)}{dt}$$

and therefore

$$\frac{di_L(0)}{dt} = \frac{v(0)}{L} = \frac{4}{5}$$

Hence

$$\tfrac{4}{5} = -2K_3 - 0.5K_4$$

This equation, together with the equation

$$-1 = K_3 + K_4$$

comprise the two simultaneous linearly independent equations in K_3 and K_4. Solving these two equations yields $K_3 = -\frac{1}{5}$ and $K_4 = -\frac{4}{5}$. Therefore, $i_L(t)$ is

$$i_L(t) = -\tfrac{1}{5}e^{-2t} - \tfrac{4}{5}e^{-0.5t} \text{ A}$$

The voltage across the inductor is related to the inductor current by the expression

$$\begin{aligned} v(t) &= L\frac{di_L(t)}{dt} \\ &= 5\frac{d}{dt}\left(-\frac{1}{5}e^{-2t} - \frac{4}{5}e^{-0.5t}\right) \\ &= 2e^{-2t} + 2e^{-0.5t} \text{ V} \end{aligned}$$

which is, of course, the expression we derived earlier. The response curve for this voltage $v(t)$ is shown in Fig. 8.4.

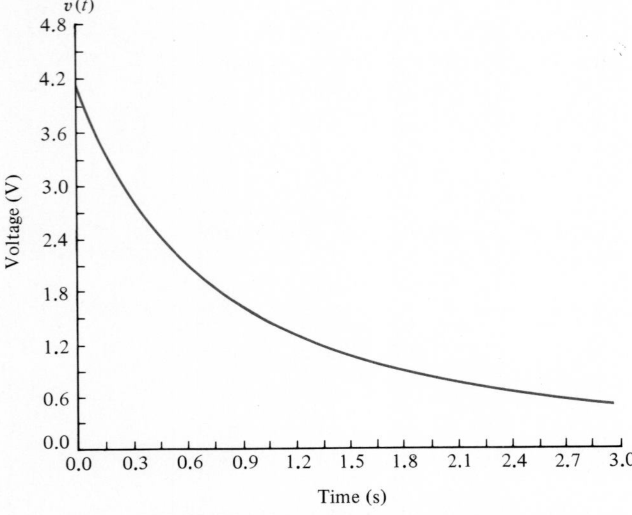

Figure 8.4 Overdamped response.

Note that in comparison with the *RL* and *RC* circuits we analyzed in Chapter 7, the response of this *RLC* circuit is controlled by two time constants. The first term has a time constant of $\frac{1}{2}$, and the second term has a time constant of 2. ■

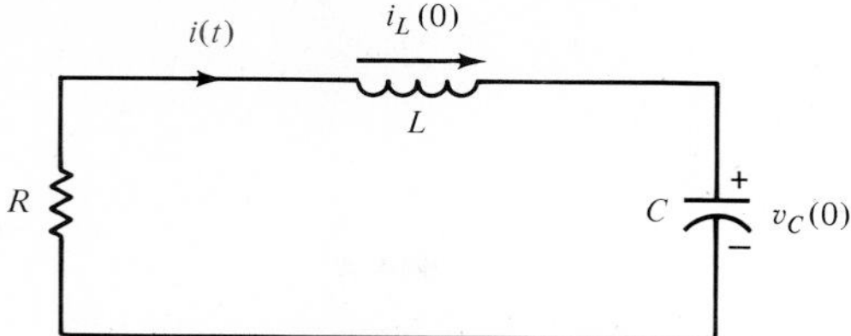

Figure 8.5 Series *RLC* circuit.

EXAMPLE 8.2

The series *RLC* circuit shown in Fig. 8.5 has the following parameters: $C = 0.04$ F, $L = 1$ H, $R = 6\ \Omega$, $i(0) = 4$ A, and $v_C(0) = -4$ V. The equation for the current in the circuit is given by the expression

$$\frac{d^2i}{dt^2} + \frac{R}{L}\frac{di}{dt} + \frac{i}{LC} = 0$$

A comparison of this equation with Eq. (8.3) and (8.4) illustrates that for a series *RLC* circuit the damping coefficient is $R/2L$ and the resonant frequency is $1/\sqrt{LC}$. Substituting the circuit element values into the equation above yields

$$\frac{d^2i}{dt^2} + 6\frac{di}{dt} + 25i = 0$$

The characteristic equation is then

$$s^2 + 6s + 25 = 0$$

and the roots are

$$s_1 = -3 + j4$$

$$s_2 = -3 - j4$$

Since the roots are complex, the circuit is underdamped, and the expression for $i(t)$ is

$$i(t) = K_1e^{-3t}\cos 4t + K_2e^{-3t}\sin 4t$$

Using the initial conditions, we find that

$$i(0) = 4 = K_1$$

and

$$\frac{di}{dt} = -4K_1e^{-3t}\sin 4t - 3K_1e^{-3t}\cos 4t + 4K_2e^{-3t}\cos 4t - 3K_2e^{-3t}\sin 4t$$

and thus

$$\frac{di(0)}{dt} = -3K_1 + 4K_2$$

Although we do not know $di(0)/dt$, we can find it via KVL. From the circuit we note that

$$Ri(0) + L\frac{di(0)}{dt} + v_C(0) = 0$$

or

$$\frac{di(0)}{dt} = -\frac{R}{L}i(0) - \frac{v_C(0)}{L}$$

$$= -\frac{6}{1}(4) + \frac{4}{1}$$

$$= -20$$

Therefore,

$$-3K_1 + 4K_2 = -20$$

and since $K_1 = 4$, $K_2 = -2$. The expression then for $i(t)$ is

$$i(t) = 4e^{-3t}\cos 4t - 2e^{-3t}\sin 4t \text{ A}$$

Note that this expression satisfies the initial condition $i(0) = 4$.

The voltage across the capacitor could be determined via KVL using this current:

$$Ri(t) + L\frac{di(t)}{dt} + v_C(t) = 0$$

or

$$v_C(t) = -Ri(t) - L\frac{di(t)}{dt}$$

Substituting the expression above for $i(t)$ into this equation yields

$$v_C(t) = -4e^{-3t}\cos 4t + 22e^{-3t}\sin 4t \text{ V}$$

Note that this expression satisfies the initial condition $v_C(0) = -4$ V.

Once the characteristic equation was known, we could have simply assumed that $v_C(t)$ was of the form

$$v_C(t) = K_3e^{-3t}\cos 4t + K_4e^{-3t}\sin 4t$$

The constants K_3 and K_4 can be obtained via the initial conditions

$$v_C(0) = -4 = K_3$$

and

$$\frac{dv_C(t)}{dt} = -3K_3e^{-3t}\cos 4t - 4K_3e^{-3t}\sin 4t - 3K_4e^{-3t}\sin 4t + 4K_4e^{-3t}\cos 4t$$

This expression can be related to known initial conditions using the expression

$$i(t) = C\frac{dv_C(t)}{dt}$$

or

$$\frac{dv_C(t)}{dt} = \frac{i(t)}{C}$$

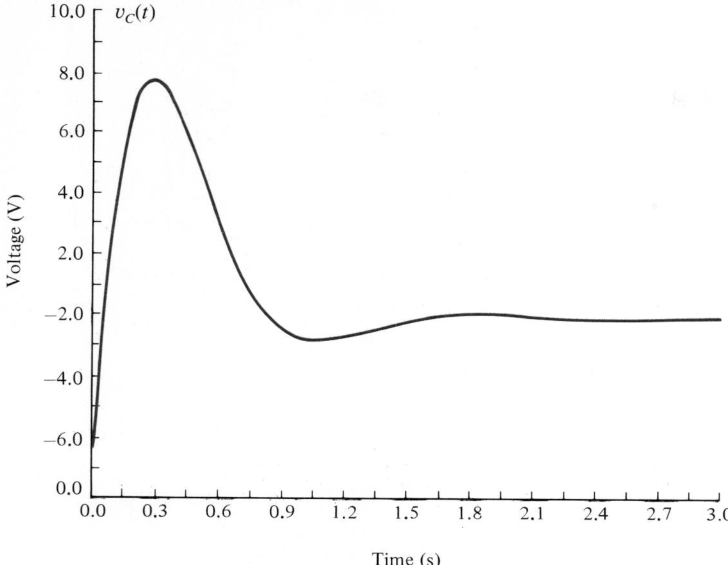

Figure 8.6 Underdamped response.

Setting the two expressions above equal to one another and evaluating the resultant equation at $t = 0$ yields

$$\frac{i(0)}{C} = -3K_3 + 4K_4$$

$$100 = -3K_3 + 4K_4$$

Since $K_3 = -4$, $K_4 = 22$, and therefore

$$v_C(t) = -4e^{-3t} \cos 4t + 22e^{-3t} \sin 4t \text{ V}$$

which is equivalent to the expression derived earlier. A plot of the function $v_C(t)$ is shown in Fig. 8.6. ■

EXAMPLE 8.3

Let us examine the circuit in Fig. 8.7, which is slightly more complicated than the two we have considered earlier. The two equations that describe the network are

$$L\frac{di(t)}{dt} + R_1 i(t) + v(t) = 0$$

$$i(t) = C\frac{dv(t)}{dt} + \frac{v(t)}{R_2}$$

Substituting the second equation into the first yields

$$\frac{d^2v}{dt^2} + \left(\frac{1}{R_2C} + \frac{R_1}{L}\right)\frac{dv}{dt} + \frac{R_1 + R_2}{R_2LC}v = 0$$

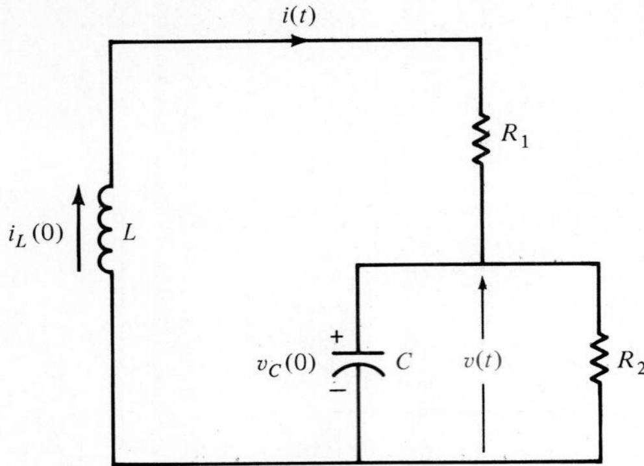

Figure 8.7 Series/parallel *RLC* circuit.

If the circuit parameters and initial conditions are

$$R_1 = 10\ \Omega \qquad C = \tfrac{1}{8}\text{ F} \qquad v_C(0) = 1\text{ V}$$

$$R_2 = 8\ \Omega \qquad L = 2\text{ H} \qquad i_L(0) = \tfrac{1}{2}\text{ A}$$

the differential equation becomes

$$\frac{d^2v}{dt^2} + 6\frac{dv}{dt} + 9v = 0$$

The characteristic equation is then

$$s^2 + 6s + 9 = 0$$

and hence the roots are

$$s_1 = -3$$

$$s_2 = -3$$

Since the roots are real and equal, the circuit is critically damped. The term $v(t)$ is then given by the expression

$$v(t) = K_1e^{-3t} + K_2te^{-3t}$$

Since $v(t) = v_C(t)$,

$$v(0) = v_C(0) = 1 = K_1$$

In addition,

$$\frac{dv(t)}{dt} = -3K_1e^{-3t} + K_2e^{-3t} - 3K_2te^{-3t}$$

However,

$$\frac{dv(t)}{dt} = \frac{i(t)}{C} - \frac{v(t)}{R_2C}$$

Setting these two expressions equal to one another and evaluating the resultant equation at $t = 0$ yields

$$\frac{1/2}{1/8} - \frac{1}{1} = -3K_1 + K_2$$

$$3 = -3K_1 + K_2$$

Since $K_1 = 1$, $K_2 = 6$ and the expression for $v(t)$ is

$$v(t) = e^{-3t} + 6te^{-3t} \text{ V}$$

Note that the expression satisfies the initial condition $v(0) = 1$.

As demonstrated in the preceding examples, once the roots of the characteristic equation are known, the current $i(t)$ could have been expressed as

$$i(t) = K_3e^{-3t} + K_4te^{-3t}$$

The constants can be evaluated as follows:

$$i(0) = \tfrac{1}{2} = K_3$$

and

$$\frac{di(t)}{dt} = -3K_3e^{-3t} + K_4e^{-3t} - 3K_4te^{-3t}$$

but

$$\frac{di(t)}{dt} = -\frac{R_1}{L}i(t) - \frac{v(t)}{L}$$

Equating the two expressions and evaluating the resultant equation at $t = 0$ yields

$$-3 = -3K_3 + K_4$$

Hence

$$K_4 = -\tfrac{3}{2}$$

Therefore,

$$i(t) = \tfrac{1}{2}e^{-3t} - \tfrac{3}{2}te^{-3t} \text{ A}$$

If this expression for the current is employed in the circuit equation,

$$v(t) = -L\frac{di(t)}{dt} - R_1i(t)$$

we obtain

$$v(t) = e^{-3t} + 6te^{-3t} \text{ V}$$

which is identical to the expression derived earlier. This function is shown in Fig. 8.8. ■

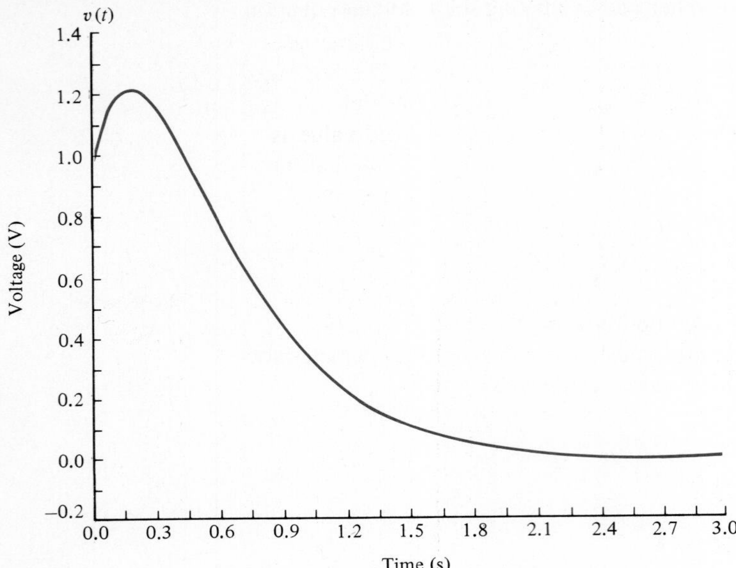

Figure 8.8 Critically damped response.

DRILL EXERCISE

D8.3. The switch in the network in Fig. D8.3 opens at $t = 0$. Find $i(t)$ for $t > 0$.

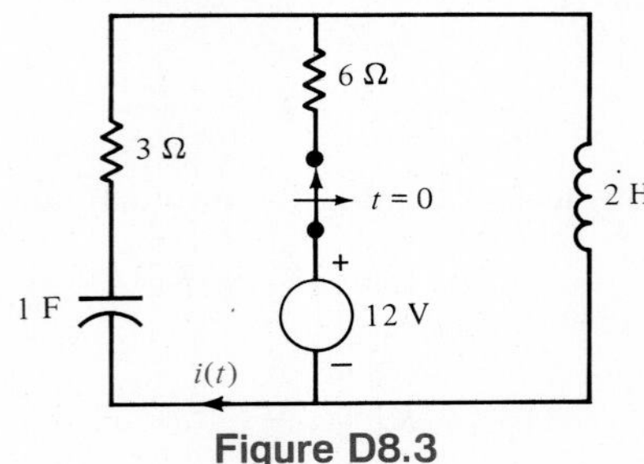

Figure D8.3

D8.4. The switch in the network in Fig. D8.4 moves from position 1 to position 2 at $t = 0$. Find $v_o(t)$ for $t > 0$.

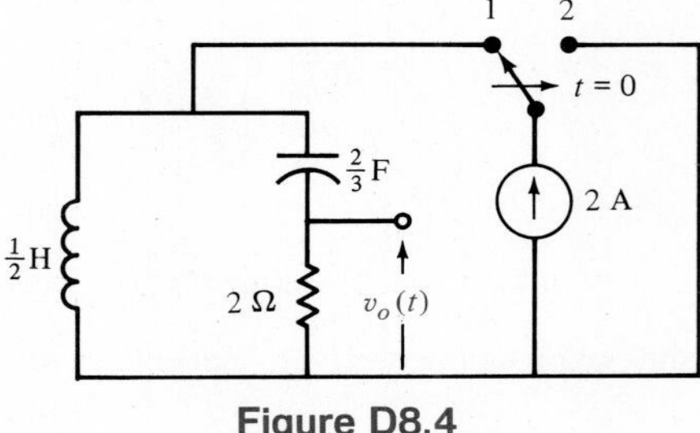

Figure D8.4

As we have shown in Fig. 8.2, the response function for a source-free *RLC* circuit with zero initial value for the response variable rises to some maximum and then exponentially decays to zero. Quite often it is advantageous for this transient response to decay, or be damped out, as rapidly as possible. A measure of the time required for the transient response to decay to some selected value is what we term the *settling time* t_s. Since the decay rate is exponential, we know that the response will be zero only at $t = \infty$. However, the response will reach a very small value in much less time.

One definition of settling time for a source-free circuit is the time required for the response to decay to 1% of its maximum value. Therefore, the settling time for a circuit can be calculated by determining the maximum value of the response, equating 1% of this value to the response function, and finally solving the resultant equation to find the time t_s.

EXAMPLE 8.4

Let us determine the settling time, t_s, for the parallel *RLC* circuit whose voltage equation is given by the expression

$$v(t) = 24(e^{-t} - e^{-4t}) \text{ V}$$

The time at which the response reaches a maximum can be determined by setting the derivative equal to zero, that is,

$$\frac{dv}{dt} = 0 = 24(-e^{-t} + 4e^{-4t})$$

Hence

$$e^{-t_{\max}} = 4e^{-4t_{\max}}$$

or

$$e^{3t_{\max}} = 4$$

and hence

$$t_{\max} = 0.462 \text{ s}$$

The maximum voltage is then

$$v_{\max} = 24(e^{-0.462} - e^{-4(0.462)})$$

$$= 11.339 \text{ V}$$

The settling time is then obtained by solving the equation

$$0.11339 = 24(e^{-t_s} - e^{-4t_s})$$

Since e^{-4t_s} is very small, this term can essentially be ignored and the settling time calculated to be $t_s = 5.355$ s. ■

DRILL EXERCISE

D8.5. The output voltage for a network is given by the expression

$$v_o(t) = 2e^{-t} - 6e^{-3t} \text{ V}$$

Determine the settling time of the output voltage.

8.4 Response of Circuits with Constant Forcing Functions

In an attempt to make our analysis more general, we will now analyze circuits that contain, in addition to nonzero initial conditions, constant forcing functions. We will illustrate not only the differential equation techniques, but also the manner in which equivalent circuits, valid at $t = 0+$ and $t = \infty$, can be employed in the analysis. The two examples that follow will illustrate the basic procedure, and then we will present a general method that can always be used when the circuits are more complex.

EXAMPLE 8.5

Consider the circuit shown in Fig. 8.9. This circuit is the same as that analyzed in Example 8.2, except that a constant forcing function is present. The circuit parameters are the same as those used in Example 8.2:

$$C = 0.04\ \text{F} \qquad i_L(0) = 4\ \text{A}$$

$$L = 1\ \text{H} \qquad v_C(0) = -4\ \text{V}$$

$$R = 6\ \Omega$$

We want to find an expression for $v_C(t)$ for $t > 0$.

From our earlier mathematical development we know that the general solution of this problem will consist of a particular solution plus a complementary solution. From Example 8.2 we know that the complementary solution is of the form $K_3e^{-3t} \cos 4t + K_4e^{-3t} \sin 4t$. The particular solution is a constant, since the input is a constant and therefore the general solution is

$$v_C(t) = K_3e^{-3t} \cos 4t + K_4e^{-3t} \sin 4t + K_5$$

An examination of the circuit shows that in the steady state the final value of $v_C(t)$ is 12 V, since in the steady-state condition, the inductor is a short circuit and the capacitor is an open circuit. Thus $K_5 = 12$. The steady-state value could also be immediately calculated from the differential equation. The form of the general solution is then

$$v_C(t) = K_3e^{-3t} \cos 4t + K_4e^{-3t} \sin 4t + 12$$

The initial conditions can now be used to evaluate the constants K_3 and K_4.

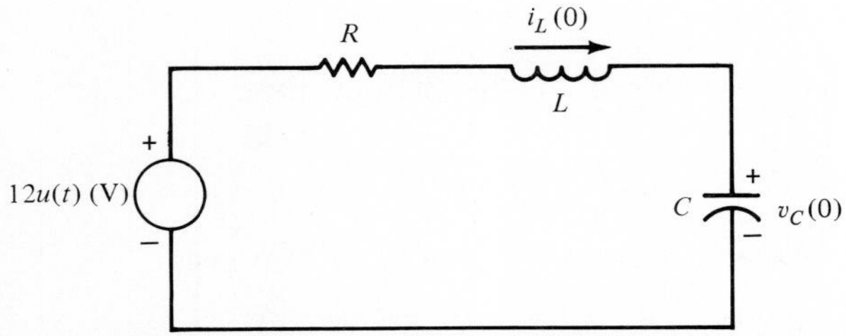

Figure 8.9 Series *RLC* circuit with a step function input.

$$v_C(0) = -4 = K_3 + 12$$

$$-16 = K_3$$

Since the derivative of a constant is zero, the results of Example 8.2 show that

$$\frac{dv_C(0)}{dt} = \frac{i(0)}{C} = 100 = -3K_3 + 4K_4$$

and since $K_3 = -16$, $K_4 = 13$. Therefore, the general solution for $v_C(t)$ is

$$v_C(t) = 12 - 16e^{-3t} \cos 4t + 13e^{-3t} \sin 4t \text{ V}$$

Note that this equation satisfies the initial condition $v_C(0) = -4$, and the final condition $v_c(\infty) = 12$ V. ■

EXAMPLE 8.6

Let us examine the circuit shown in Fig. 8.10. A close examination of this circuit will indicate that it is identical to that shown in Example 8.3 except that a constant forcing function is present. We assume the circuit is in steady state at $t = 0-$. The equations that describe the circuit for $t > 0$ are

$$L\frac{di(t)}{dt} + R_1 i(t) + v(t) = 24$$

$$i(t) = C\frac{dv(t)}{dt} + \frac{v(t)}{R_2}$$

Combining these equations, we obtain

$$\frac{d^2v(t)}{dt^2} + \left(\frac{1}{R_2C} + \frac{R_1}{L}\right)\frac{dv(t)}{dt} + \frac{R_1 + R_2}{R_2LC}v(t) = \frac{24}{LC}$$

If the circuit parameters are $R_1 = 10\ \Omega$, $R_2 = 2\ \Omega$, $L = 2$ H, and $C = \frac{1}{4}$ F, the differential equation for the output voltage reduces to

$$\frac{d^2v(t)}{dt^2} + 7\frac{dv(t)}{dt} + 12v(t) = 48$$

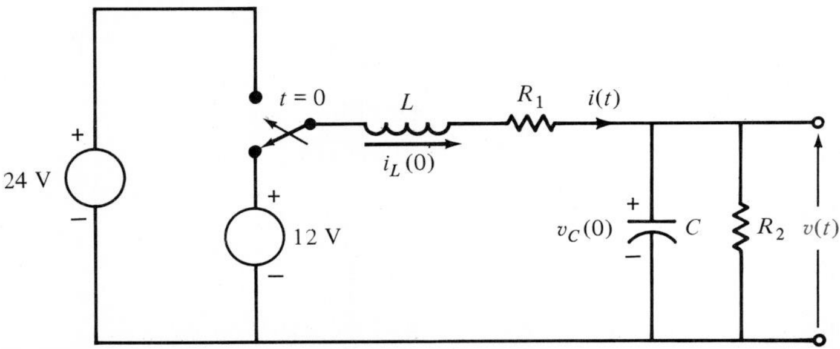

Figure 8.10 Series/parallel *RLC* circuit with a constant forcing function.

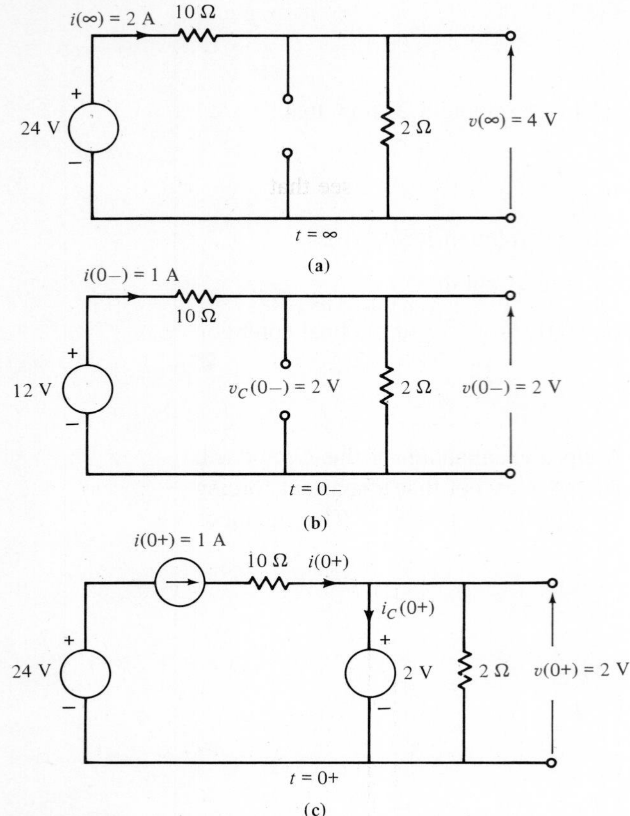

Figure 8.11 Equivalent circuits at $t = \infty$, $t = 0-$, and $t = 0+$ for the circuit in Fig. 8.10.

The characteristic equation is

$$s^2 + 7s + 12 = 0$$

and hence the roots are

$$s_1 = -3$$

$$s_2 = -4$$

The circuit response is overdamped, and therefore the general solution is of the form

$$v(t) = K_1 e^{-3t} + K_2 e^{-4t} + K_3$$

The steady-state value of the voltage, K_3, can be computed from Fig. 8.11a. Note that

$$v(\infty) = 4 \text{ V} = K_3$$

The initial conditions can be calculated from Fig. 8.11b and c, which are valid at $t =$

0− and $t = 0+$, respectively. Note that $v(0+) = 2$ V and hence from the response equation

$$v(0+) = 2 \text{ V} = K_1 + K_2 + 4$$

$$-2 = K_1 + K_2$$

Figure 8.11c illustrates that $i(0+) = 1$. From the response equation we see that

$$\frac{dv(0)}{dt} = -3K_1 - 4K_2$$

and since

$$\begin{aligned}\frac{dv(0)}{dt} &= \frac{i(0)}{C} - \frac{v(0)}{R_2C}\\ &= 4 - 4\\ &= 0\end{aligned}$$

therefore

$$0 = -3K_1 - 4K_2$$

Solving the two equations for K_1 and K_2 yields $K_1 = -8$ and $K_2 = 6$. Therefore, the general solution for the voltage response is

$$v(t) = 4 - 8e^{-3t} + 6e^{-4t} \text{ V}$$

Note that this equation satisfies both the initial and final values of $v(t)$. ■

DRILL EXERCISE

D8.6. The switch in the network in Fig. D8.6 moves from position 1 to position 2 at $t = 0$. Compute $i_o(t)$ for $t > 0$ and use this current to determine $v_o(t)$ for $t > 0$.

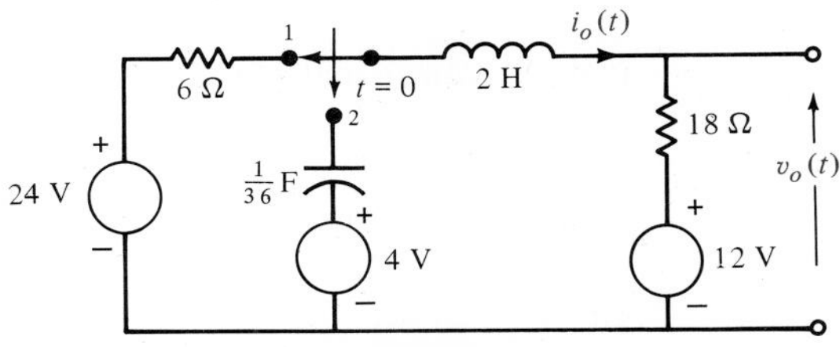

Figure D8.6

The circuits we have analyzed thus far have been very simple. We have encountered little or no difficulty in obtaining a solution. We would indeed be most fortunate if this were true in general; however, it is not. As we will illustrate in our subsequent development, a slight rearrangement of the circuit elements can cause us considerable difficulty. Therefore, we will present a technique that can be applied in general to obtain a solution.

8.5 General Analysis Technique for Circuit Response

This technique is based on the graph theory we employed in Chapter 4. Our unknowns in this case are not node voltages or loop currents, but rather *capacitor voltages* and *inductor currents*, which we will refer to as *state variables*.

The procedure we will employ in this case is based on the following guidelines. The suggestions given in Chapter 4 for selecting a tree and cotree also apply here, with the added stipulation that the capacitors are placed in the tree and inductors are placed in the links. If the capacitors are assigned a tree branch voltage and the inductors are assigned link currents, then KVL around a loop containing an inductor and KCL at nodes or supernodes to which a capacitor is connected form a partial or complete set of equations that will yield a solution to the network. If the KVL equations involving the inductor currents and the KCL equations involving the capacitor voltages are not sufficient to solve the circuit, additional equations can be employed. These additional equations are normally derived by assigning a voltage variable to the resistors that appear in a tree branch or a current variable to a resistor that appears in a link. KVL for a fundamental loop containing the resistor with the assigned current variable or KCL for the fuundamental cut set containing the resistor with the assigned voltage variable provides the necessary additional equations.

These equations can be used to derive the differential equation for the unknown voltage or current. The solution of this equation produces the system reponse for the variable in question.

This technique could also be used simply to obtain the characteristic equation for the network. Once the roots of this equation are known, the form of the solution can be written as

$$x(t) = K_1e^{-\alpha_1 t} + K_2e^{-\alpha_2 t} + K_3$$

if the circuit is overdamped,

$$x(t) = e^{-\alpha t}(K_1 \cos \beta t + K_2 \sin \beta t) + K_3$$

if the circuit is underdamped, and

$$x(t) = K_1e^{-\alpha t} + K_2te^{-\alpha t} + K_3$$

if the circuit is critically damped, where $x(t)$ may be any $v(t)$ or $i(t)$ in the circuit. The unknown constants in the equations above can be evaluated by examining the equivalent circuits that are valid at $t = 0+$ and $t = \infty$ and using this information in conjunction with the basic circuit equations, as illustrated in Example 8.6. The examples that follow will illustrate these concepts.

EXAMPLE 8.7

Let us analyze the circuit in Fig. 8.12a and determine the expression for the voltage $v_C(t)$ for $t > 0$. Our approach will be first to derive the differential equation for $v_C(t)$ and then to assign circuit parameters and solve for the voltage expression.

In the time frame $t > 0$ the circuit shown in Fig. 8.12a reduces to that shown in Fig. 8.12b. The tree and corresponding links selected for this circuit, according to the guide-

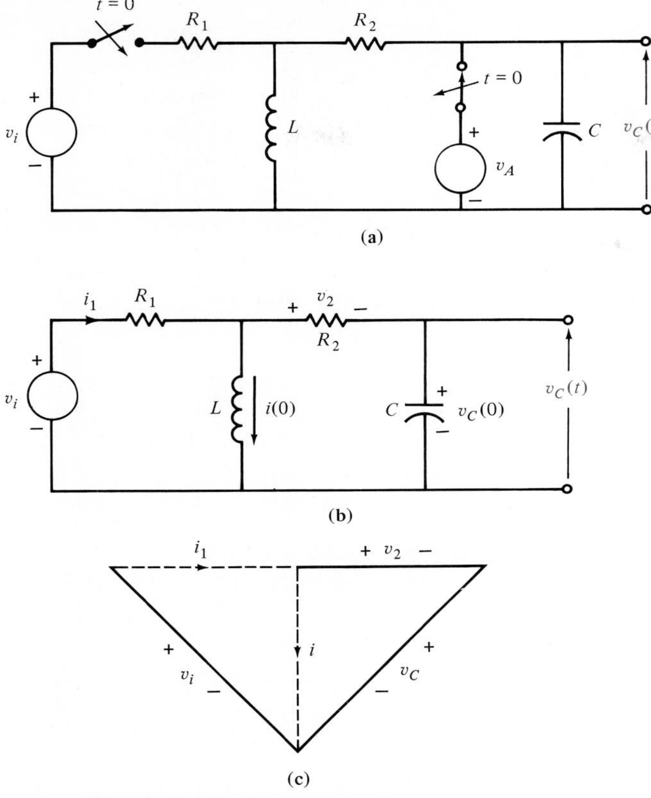

Figure 8.12 Circuit diagrams for Example 8.7.

lines stated earlier, are shown in Fig. 8.12c. The selection of R_1 as a link and R_2 as a tree branch was arbitrary (i.e., they could easily be reversed).

The equations involving the state variables are obtained from the figures as follows. KVL around the loop containing the inductor is

$$L\frac{di}{dt} - v_C - v_2 = 0 \tag{8.12}$$

KCL at the node to which the capacitor is connected is

$$i_1 - i - C\frac{dv_C}{dt} = 0 \tag{8.13}$$

Note that we have introduced two new variables v_2 and i_1. KVL for the fundamental loop containing R_1 is

$$v_i = i_1R_1 + v_2 + v_C \tag{8.14}$$

KCL for the fundamental cut set containing R_2 is

$$i_1 = i + \frac{v_2}{R_2} \tag{8.15}$$

Equations (8.14) and (8.15) are linearly independent simultaneous equations in the variables v_2 and i_1. Solving these equations yields

$$v_2 = \frac{(v_i - v_C)R_2 - R_1R_2i}{R_1 + R_2} \tag{8.16}$$

and

$$i_1 = \frac{v_i - v_C + iR_2}{R_1 + R_2} \tag{8.17}$$

Substituting these expressions for v_2 and i_1 into the two equations for the state variables yields

$$L\frac{di}{dt} = \frac{R_1}{R_1 + R_2}v_C + \frac{R_2}{R_1 + R_2}v_i - \frac{R_1R_2}{R_1 + R_2}i \tag{8.18}$$

$$C\frac{dv_C}{dt} = \frac{-1}{R_1 + R_2}v_C + \frac{1}{R_1 + R_2}v_i - \frac{R_1}{R_1 + R_2}i \tag{8.19}$$

The differential equation for the voltage $v_C(t)$ can now be easily obtained by solving Eq. (8.19) for the current i and substituting this expression into Eq. (8.18). The expression for i is

$$i = \frac{-C(R_1 + R_2)}{R_1}\frac{dv_C}{dt} - \frac{v_C}{R_1} + \frac{v_i}{R_1} \tag{8.20}$$

Substituting this into Eq. (8.18) yields

$$\frac{d^2v_C}{dt^2} + \frac{R_1R_2C + L}{(R_1 + R_2)LC}\frac{dv_C}{dt} + \frac{R_1}{(R_1 + R_2)LC}v_C = 0 \tag{8.21}$$

Let us assume now that the circuit parameters are

$$R_1 = 1\ \Omega \qquad C = \tfrac{1}{2}\ \text{F} \qquad v_i = V_i = 4\ \text{V}$$

$$R_2 = 1\ \Omega \qquad L = 1\ \text{H} \qquad v_A = V_A = 1\ \text{V}$$

The differential equation for the output voltage is then

$$\frac{d^2v_C}{dt^2} + 1.5\frac{dv_C}{dt} + v_C = 0 \tag{8.22}$$

The characteristic equation is

$$s^2 + 1.5s + 1 = 0$$

and

$$s_1 = -\frac{3}{4} + j\frac{\sqrt{7}}{4}$$

$$s_2 = -\frac{3}{4} - j\frac{\sqrt{7}}{4}$$

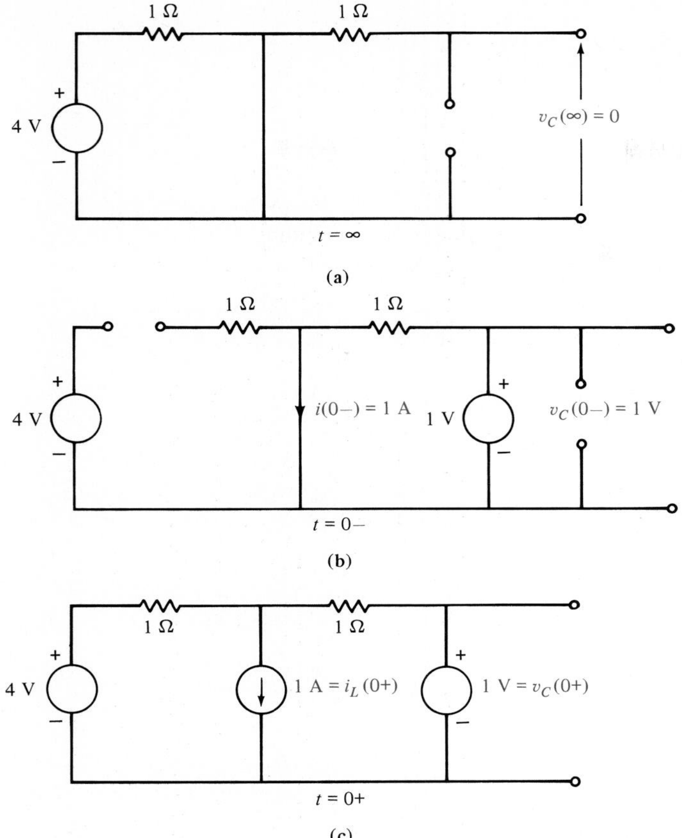

Figure 8.13 Equivalent circuits at $t = \infty$, $t = 0-$, and $t = 0+$ for the circuit in Fig. 8.12a.

The circuit is therefore underdamped and the equation for $v_C(t)$ is of the form

$$v_C(t) = K_1 e^{-3/4t} \cos \frac{\sqrt{7}}{4} t + K_2 e^{-3/4t} \sin \frac{\sqrt{7}}{4} t + K_3 \tag{8.23}$$

Since there is no constant forcing function in the differential Eq. (8.22), the constant $K_3 = 0$. K_3 is the steady-state response $v_C(\infty)$, and as can be seen in Fig. 8.13a, this value is zero. The initial conditions are derived from the equivalent circuits in Fig. 8.13b and c. The values for the inductor current and capacitor voltage are

$$i(0) = 1 \text{ A}$$

$$v_C(0) = 1 \text{ V}$$

Since $K_3 = 0$, $v_C(t)$ reduces to

$$v_C(t) = K_1 e^{-3/4t} \cos \frac{\sqrt{7}}{4} t + K_2 e^{-3/4t} \sin \frac{\sqrt{7}}{4} t \tag{8.24}$$

Using the initial conditions, we have

$$v_C(0) = 1 \text{ V} = K_1 \tag{8.25}$$

In addition,

$$\frac{dv_C(0)}{dt} = -\frac{3}{4}K_1 + \frac{\sqrt{7}}{4}K_2 \tag{8.26}$$

Although we do not have a value for $dv_C(0)/dt$, we can compute one by combining Eq. (8.13) and (8.17). Substituting Eq. (8.17) into Eq. (8.13) and evaluating the resultant expression at $t = 0$ yields

$$\begin{aligned} \frac{dv_C(0)}{dt} &= \frac{1}{C}\left(\frac{v_i - v_C(0) + i(0)R_2}{R_1 + R_2}\right) - \frac{i(0)}{C} \\ &= 2\left(\frac{4 - 1 + 1}{2}\right) - 2 \\ &= 2 \end{aligned} \tag{8.27}$$

Substituting this value into Eq. (8.26) and using Eq. (8.25) yields

$$K_2 = \frac{11}{\sqrt{7}} \tag{8.28}$$

If the values for K_1 and K_2 are now used in Eq. (8.24), we obtain the final expression for $v_C(t)$:

$$v_C(t) = e^{-3/4t}\cos\frac{\sqrt{7}}{4}t + \frac{11}{\sqrt{7}}e^{-3/4t}\sin\frac{\sqrt{7}}{4}t \text{ V} \tag{8.29}$$

Note that this expression satisfies both the initial and final values for $v_C(t)$. ■

EXAMPLE 8.8

Consider the circuit shown in Fig. 8.14a. This circuit is more complicated than the circuit in Example 8.7. In addition, we wish to find the voltage $v_o(t)$ for $t > 0$, which is not a capacitor voltage. We will follow the development outlined in Example 8.7.

For $t > 0$, the circuit shown in Fig. 8.14a reduces to that of Fig. 8.14b. An appropriate tree with the corresponding links is shown in Fig. 8.14c. The equations involving the state variables are

$$v_{i2} = L\frac{di}{dt} + v_o \tag{8.30}$$

$$i_A = C\frac{dv_C}{dt} + i_B \tag{8.31}$$

In addition, KVL for the fundamental loops containing i_A and i_B are

$$v_{i2} = i_A R_1 + v_C \tag{8.32}$$

$$v_C = i_B R_2 + v_o \tag{8.33}$$

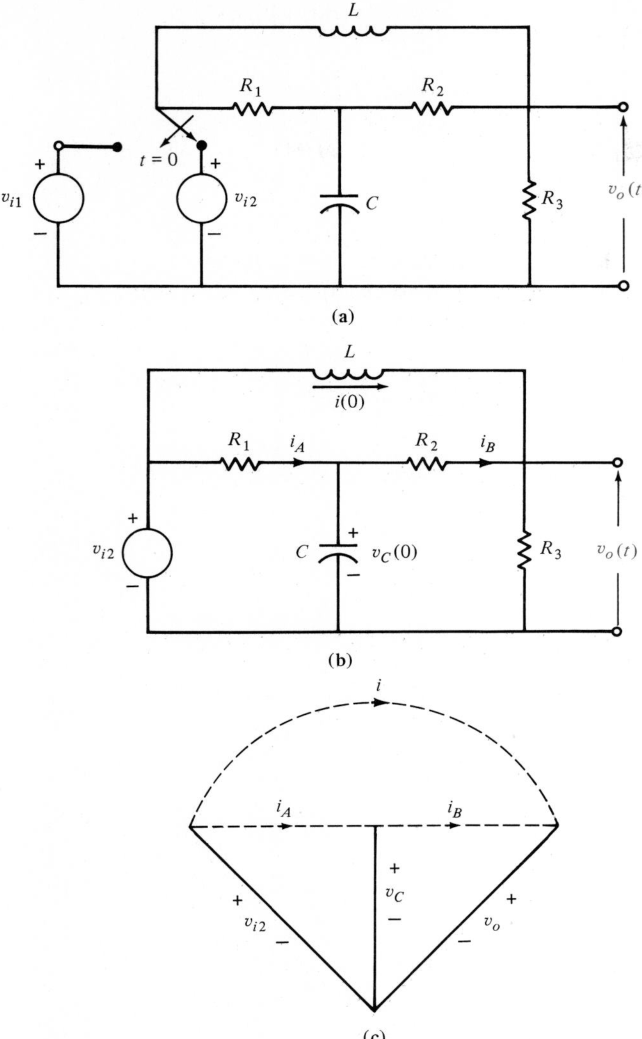

Figure 8.14 Circuit diagrams for Example 8.8.

and KCL for the fundamental cut set containing R_3 is

$$i + i_B = \frac{v_o}{R_3} \tag{8.34}$$

Equations (8.33) and (8.34) are linearly independent simultaneous equations in the variables i_B and v_o. Solving these equations yields

$$i_B = \frac{v_C - iR_3}{R_2 + R_3} \tag{8.35}$$

$$v_o = \frac{iR_2R_3 + v_CR_3}{R_2 + R_3} \tag{8.36}$$

Substituting Eqs. (8.32), (8.35), and (8.36) into Eqs. (8.30) and (8.31) yields

$$L\frac{di}{dt} + i\frac{R_2R_3}{R_2 + R_3} + v_C\frac{R_3}{R_2 + R_3} = v_{i2} \tag{8.37}$$

$$C\frac{dv_C}{dt} + v_C\left(\frac{1}{R_2 + R_3} + \frac{1}{R_1}\right) - \frac{iR_3}{R_2 + R_3} = \frac{v_{i2}}{R_1} \tag{8.38}$$

We could, of course, obtain the characteristic equation for the circuit by deriving either the differential equation for the inductor current or the capacitor voltage. If we choose to find the differential equation for the inductor current, we would solve Eq. (8.37) for v_C and substitute this expression into Eq. (8.38). Performing this operation yields the following equation.

$$\frac{d^2i}{dt^2} + \frac{R_1R_2R_3C + (R_1 + R_2 + R_3)L}{R_1LC(R_2 + R_3)}\frac{di}{dt} + \frac{R_3^2R_1 + R_2R_3(R_1 + R_2 + R_3)}{R_1LC(R_2 + R_3)^2}i = \frac{(R_1 + R_2)}{R_1LC(R_2 + R_3)}v_{i2} \tag{8.39}$$

Let us assume now that the circuit parameters are

$$R_1 = 1\ \Omega \qquad R_3 = 2\ \Omega \qquad C = 1\ \text{F} \qquad v_{i1} = V_{i1} = 1\ \text{V}$$

$$R_2 = 1\ \Omega \qquad L = \tfrac{1}{2}\ \text{H} \qquad v_{i2} = V_{i2} = 4\ \text{V}$$

Equation (8.39) now becomes

$$\frac{d^2i}{dt^2} + \frac{8}{3}\frac{di}{dt} + \frac{24}{9}i = \frac{4}{3}v_{i2}$$

The characteristic equation for the circuit is then

$$s^2 + \frac{8}{3}s + \frac{24}{9} = 0$$

and the roots are

$$s_1 = -\frac{4}{3} + j\frac{2\sqrt{2}}{3}$$

$$s_2 = -\frac{4}{3} - j\frac{2\sqrt{2}}{3}$$

Hence the circuit is underdamped and the expression for $v_o(t)$ must be of the form

$$v_o(t) = K_1e^{-4t/3}\cos\frac{2\sqrt{2}}{3}t + K_2e^{-4t/3}\sin\frac{2\sqrt{2}}{3}t + K_3 \tag{8.40}$$

As shown in Fig. 8.15a, the final value of $v_o(t)$ [i.e., $v_o(\infty)$] is 4 V and hence $K_3 = 4$. Therefore,

$$v_o(t) = 4 + K_1e^{-4t/3}\cos\frac{2\sqrt{2}}{3}t + K_2e^{-4t/3}\sin\frac{2\sqrt{2}}{3}t \tag{8.41}$$

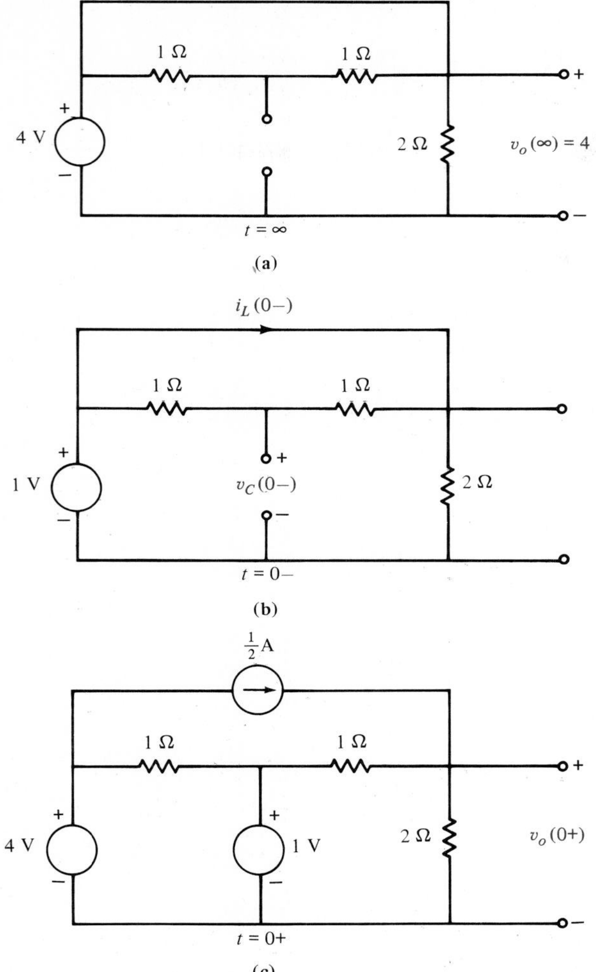

Figure 8.15 Equivalent circuits at $t=\infty$, $t=0-$, and $t=0+$ for the circuit in Fig. 8.14a.

Figure 8.15b and c are used to calculate the initial conditions. At $t = 0-$ there is no current in the 1-Ω resistors, and therefore $i(0-) = \frac{1}{2}$ A and $v_o(0-) = 1$ V. At $t = 0+$ the node voltage $v_o(0+)$ can be computed by applying KCL at that node:

$$\frac{v_o(0+) - 1}{1} + \frac{v_o(0+)}{2} = \frac{1}{2}$$

or

$$v_o(0+) = 1 \text{ V}$$

Using this condition in conjunction with Eq. (8.41) yields

$$v_o(0) = 1 \text{ V} = 4 + K_1 \qquad (8.42)$$
$$-3 = K_1$$

From Eq. (8.41) we can show that

$$\frac{dv_o(0)}{dt} = -\frac{4}{3}K_1 + \frac{2\sqrt{2}}{3}K_2 \tag{8.43}$$

Now we need an expression that relates $dv_o(0)/dt$ to $i(0)$, $v_C(0)$, and v_{i2} so that the constant K_2 can be evaluated. From Eq. (8.36),

$$v_o = \frac{R_2R_3}{R_2 + R_3}i + \frac{R_3}{R_2 + R_3}v_C$$

and then

$$\frac{dv_o}{dt} = \frac{R_2R_3}{R_2 + R_3}\frac{di}{dt} + \frac{R_3}{R_2 + R_3}\frac{dv_C}{dt}$$

Substituting Eqs. (8.30) and (8.31) into the equation above yields

$$\frac{dv_o}{dt} = \frac{R_2R_3}{R_2 + R_3}\frac{v_{i2} - v_o}{L} + \frac{R_3}{R_2 + R_3}\frac{i_A - i_B}{C}$$

Using the expressions for i_A and i_B in Eqs. (8.32) and (8.33), we obtain

$$\frac{dv_o}{dt} = \frac{R_2R_3}{L(R_2 + R_3)}(v_{i2} - v_o) + \frac{R_3}{R_2 + R_3}\left(\frac{v_{i2} - v_c}{CR_1} - \frac{v_c - v_o}{CR_2}\right)$$

Evaluating this expression at $t = 0$ yields

$$\frac{dv_o(0)}{dt} = \frac{4}{3}(4 - 1) + \frac{2}{3}\left(\frac{4 - 1}{1} - \frac{1 - 1}{1}\right) = 6$$

Using this value in Eq. (8.43), we obtain

$$6 = \frac{-4}{3}K_1 + \frac{2\sqrt{2}}{3}K_2$$

But $K_1 = -3$, and therefore $K_2 = 3/\sqrt{2}$. Hence the expression for $v_o(t)$ is

$$v_o(t) = 4 - 3e^{-4/3t}\cos\frac{2\sqrt{2}}{3}t + \frac{3}{\sqrt{2}}e^{-4/3t}\sin\frac{2\sqrt{2}}{3}t \text{ V}$$

A quick check indicates that this expression satisfies the initial and final values for $v_o(t)$. ■

DRILL EXERCISE

D8.7. Given the network in Fig. D8.7,

(a) Derive the differential equation for $v_c(t)$ for $t > 0$.

(b) Determine the type of damping the network exhibits if $L = 1\text{H}$, $C = 1$ F, and $R_1 = R_2 = R_3 = 1\ \Omega$.

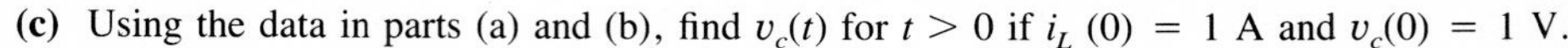

(c) Using the data in parts (a) and (b), find $v_c(t)$ for $t > 0$ if $i_L(0) = 1$ A and $v_c(0) = 1$ V.

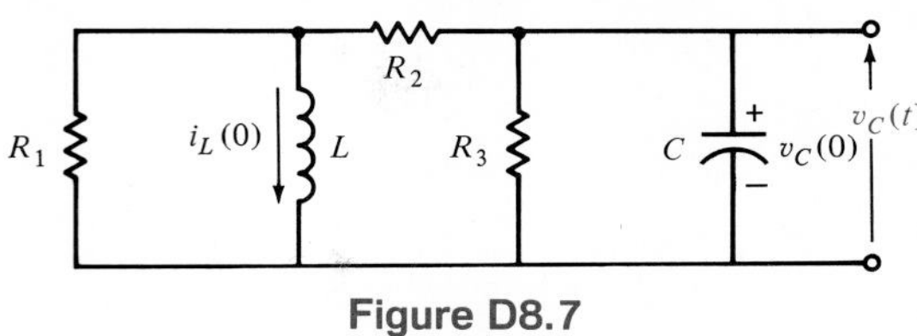

Figure D8.7

8.6 SPICE Analysis of *RLC* Circuits

In the previous sections of this chapter, our analysis techniques have been mathematical in nature. Although the mathematical approach does provide certain insight into the salient features of a circuit, it does not lead quickly to a solution. In contrast, a SPICE program consisting of only a few statements is sufficient to yield results that easily display the circuit behavior. Now, using SPICE, we reexamine some of the circuits that we treated in earlier examples.

EXAMPLE 8.9

The *RLC* circuit in Example 8.2 is redrawn for a SPICE analysis in Fig. 8.16. Let us write a SPICE program that will plot the capacitor voltage in intervals of 0.1 s over a 3-s range. The following program plots the capacitor voltage.

```
EXAMPLE 8.9  PLOT OF CAPACITOR VOLTAGE IN
*EXAMPLE 8.2
R     1     0     6
L     1     2     1     IC=4
C     2     0     0.04  IC=-4
.TRAN       0.1   3     UIC
.PLOT       TRAN  V(2)
.END
```

A plot of the capacitor voltage is shown in Fig. 8.6. ■

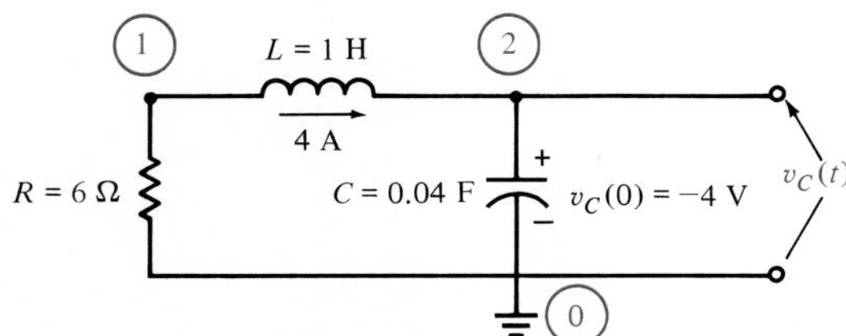

Figure 8.16 Circuit in Fig. 8.5 redrawn for SPICE analysis.

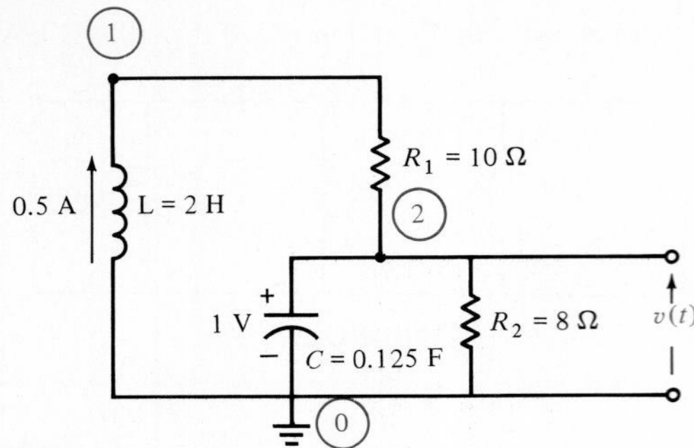

Figure 8.17 Circuit in Example 8.3 redrawn for SPICE analysis.

EXAMPLE 8.10

The *RLC* circuit in Example 8.3 is redrawn in Fig. 8.17 for a SPICE analysis. Let us write a SPICE program that will plot the voltage $v(t)$ in intervals of 0.1 s over a 3-s range. With reference to Fig. 8.17, the SPICE program is

```
EXAMPLE 8.10    PLOT OF V(T) IN EXAMPLE 8.3
L    0    1    2    IC=0.5
R1   1    2    10
C    2    0    0.125      IC=1
R2   2    0    8
.TRAN     0.1  3    UIC
.PLOT     TRAN V(2)
.END
```

A plot of the voltage $v(t)$ is shown in Fig. 8.8. ■

EXAMPLE 8.11

Consider the circuit in Example 8.7. This circuit, for $t > 0$, is redrawn in Fig. 8.18a with the appropriate initial conditions. A SPICE program that will plot the output voltage in 0.025-s intervals over a 4-s range is listed below.

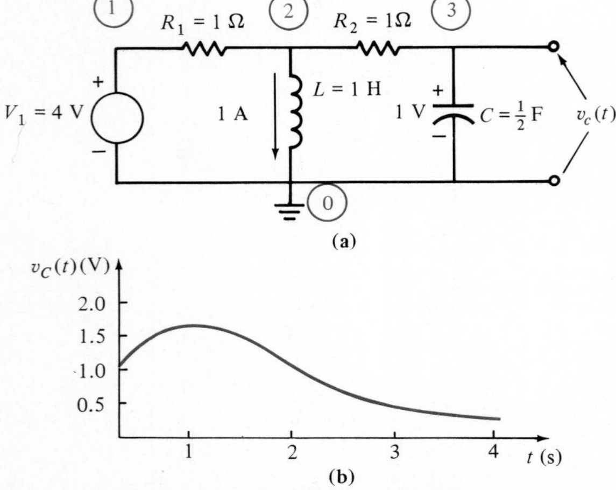

Figure 8.18 SPICE analysis of Example 8.7.

```
EXAMPLE 8.11   PLOT OF THE OUTPUT VOLTAGE IN
*EXAMPLE 8.7
V1   1     0     DC     4
R1   1     2     1
L    2     0     1      IC=1
R2   2     3     1
C    3     0     0.5    IC=1
.TRAN      0.025        4      UIC
.PLOT      TRAN V(3)
.END
```

A plot of the output voltage is shown in Fig. 8.18b. ■

EXAMPLE 8.12

Let us reexamine the network in Example 8.8. This circuit, for $t > 0$, is shown in Fig. 8.19a with the appropriate initial conditions. A SPICE program that will plot the output voltage in 0.05-s intervals over a 4-s range is listed below.

```
EXAMPLE 8.12   PLOT OF THE OUTPUT VOLTAGE IN
*EXAMPLE 8.8
VI2  1     0     DC     4
R1   1     2     1
L    1     3     0.5    IC=0.5
C    2     0     1      IC=1
R2   2     3     1
R3   3     0     2
.TRAN      0.05  4      UIC
.PLOT      TRAN V(3)
.END
```

A plot of the output voltage is shown in Fig. 8.19b. ■

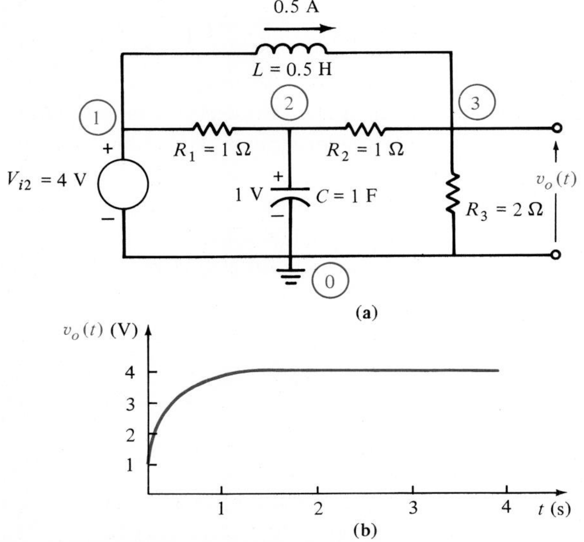

Figure 8.19 SPICE analysis of Example 8.8.

DRILL EXERCISE

D8.8. For the network in Fig. D8.8a and the source in Fig. D8.8b, plot the voltage $v(t)$ from 0 to 2.5 ms in increments of 50 μs using SPICE.

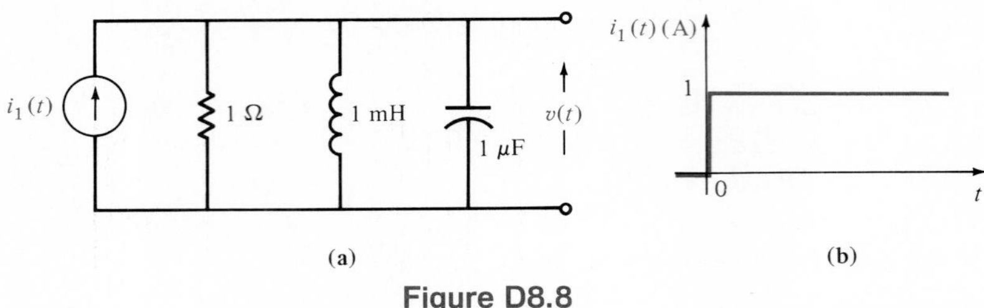

Figure D8.8

D8.9. For the network in Fig. D8.9, use SPICE to plot the transient response of $i(t)$ from 0 to 0.5 s in increments of 10 ms.

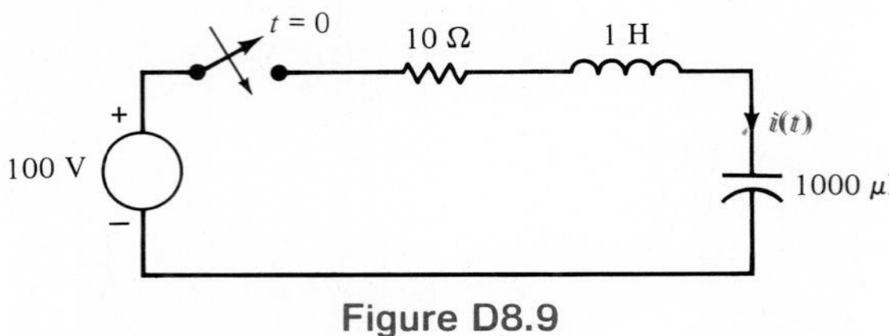

Figure D8.9

8.7 Summary

We have shown that the analysis of *RLC* circuits leads to a second-order differential equation. The roots of the circuit's characteristic equation control the type of response. These roots are a function of the circuit's damping ratio and undamped resonant frequency. If the roots are real and unequal, the response is overdamped; if the roots are complex conjugates, the response is underdamped; and if the roots are real and equal the response is critically damped.

A general analysis technique based on the graph theory introduced in Chapter 4 is presented. In this technique it is the capacitor voltages and inductor currents, which we call state variables, that are employed to obtain a solution. Finally, the SPICE analysis of *RLC* circuits is presented.

KEY POINTS

- The characteristic equation for an *RLC* circuit is $s^2 + 2\alpha s + \omega_0^2 = 0$, where α is the damping coefficient and ω_0 is the undamped resonant frequency.
- The two roots of the characteristic equation for an *RLC* circuit are real and unequal if $\alpha > \omega_0$; they are complex numbers if $\alpha < \omega_0$, and they are real and equal if $\alpha = \omega_0$.

- The response of an *RLC* circuit is said to be overdamped if the roots of the network's characteristic equation are real and unequal, underdamped if they are complex numbers, and critically damped if they are real and equal.
- The settling time for a source-free network is the time required for the response to decay to 1% of its maximum value.
- The SPICE analysis of networks containing both an inductor and a capacitor is performed in exactly the same manner as outlined in Chapter 7.

PROBLEMS

8.1. The parameters for a parallel *RLC* circuit are $R = 1\ \Omega$, $L = \frac{1}{5}$ H, and $C = \frac{1}{4}$ F. Determine the type of damping exhibited by the circuit.

8.2. If the inductor in Problem 8.1 is changed from $L = \frac{1}{5}$ H to $L = \frac{4}{3}$ H, what effect does this have on the circuit response?

8.3. The parallel *RLC* circuit shown in Fig. P8.3 has the following parameters: $R = 1\ \Omega$, $L = \frac{1}{8}$ H, and $C = \frac{1}{12}$ F. Is it possible to select a positive value of R_1 that will produce a voltage response that is critically damped?

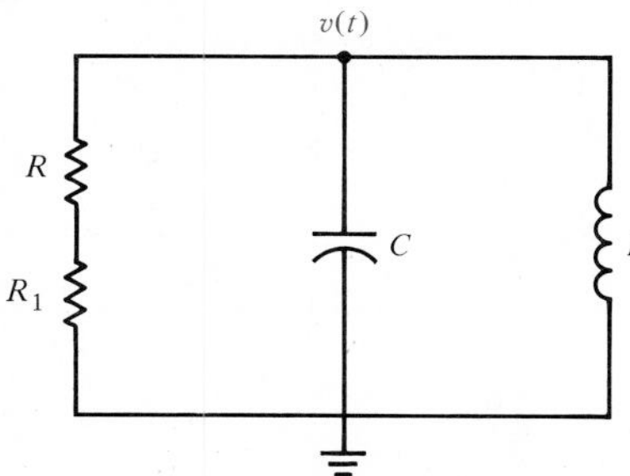

Figure P8.3

8.4. A series *RLC* circuit contains a resistor $R = 2\ \Omega$ and a capacitor $C = \frac{1}{8}$ F. Select the value of the inductor so that the circuit is critically damped.

8.5. Find a value in the network of R_x shown in Fig. P8.5 that will produce a voltage response that is critically damped.

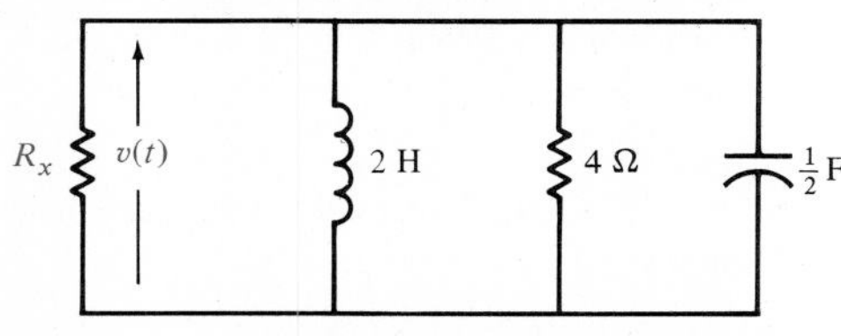

Figure P8.5

8.6. The response of the circuit given in Example 8.3 is

$$v(t) = e^{-3t} + 6te^{-3t}$$

Determine the settling time for the response.

8.7. In the circuit shown in Fig. P8.7, the switch opens at $t = 0$. Find the equation for the voltage $v_o(t)$ for $t > 0$.

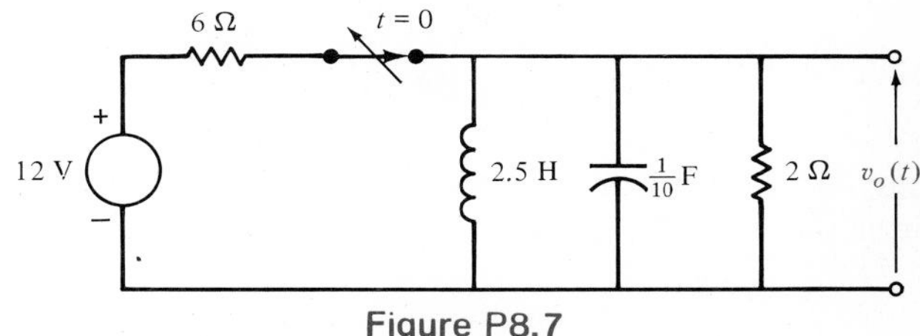

Figure P8.7

8.8. For the underdamped circuit shown in Fig. P8.8, determine the voltage $v(t)$ if the initial conditions on the storage elements are $i_L(0) = 1$ A and $v_c(0) = 10$ V.

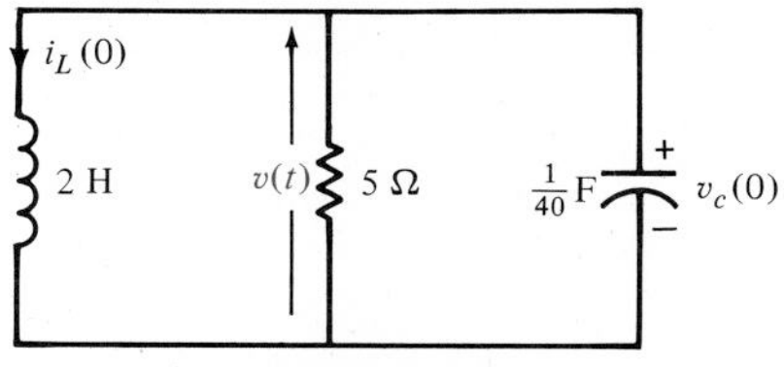

Figure P8.8

8.9. Given the circuit and the initial conditions of Problem 8.8, determine the current through the inductor.

8.10. In the critically damped circuit shown in Fig. P8.10, the initial conditions on the storage elements are $i_L(0) = 2$ A and $v_c(0) = 5$ V. Determine the voltage $v(t)$.

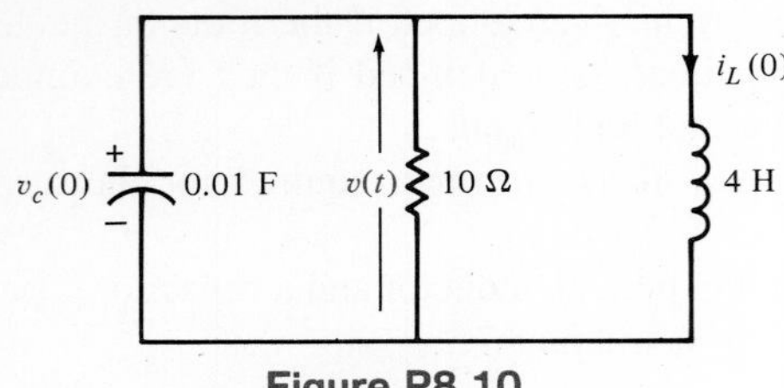

Figure P8.10

8.11. Given the circuit and the initial conditions from Problem 8.10, determine the current $i_L(t)$ that is flowing through the inductor.

8.12. Find the equation for $v(t)$, $t > 0$, in Fig. P8.12.

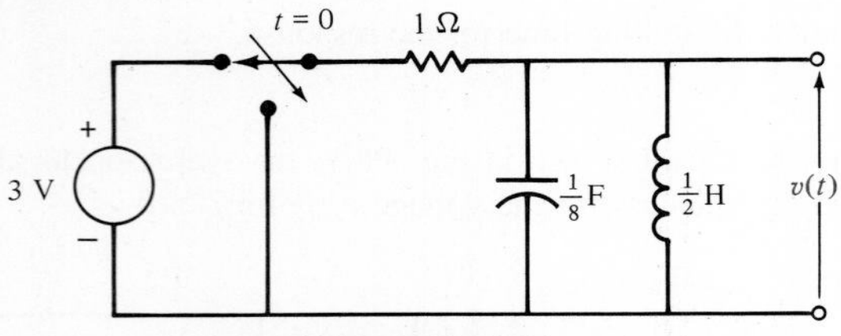

Figure P8.12

8.13. Determine the equation for the current $i(t)$, $t > 0$, in the circuit shown in Fig. P8.13.

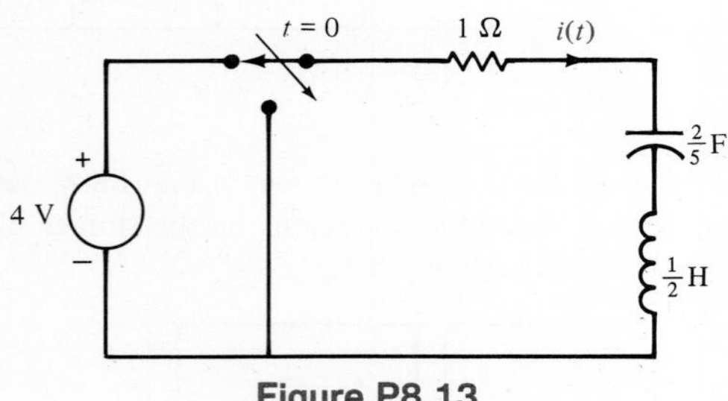

Figure P8.13

8.14. Given the circuit in Fig. P8.14, find the equation for $i(t)$, $t > 0$.

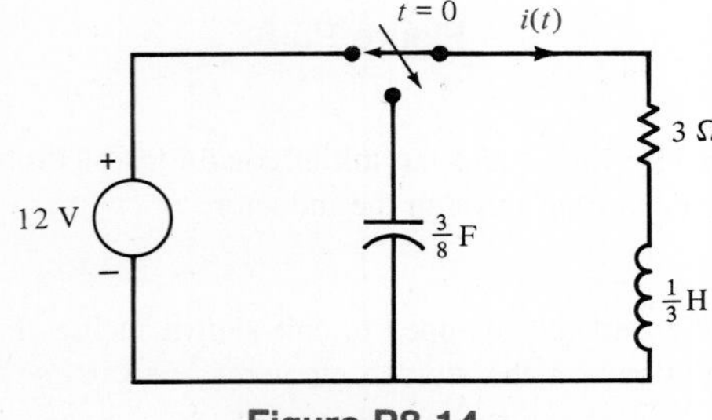

Figure P8.14

8.15. In the circuit shown in Fig. P8.15, switch action occurs at $t = 0$. Determine the voltage $v_o(t)$, $t > 0$.

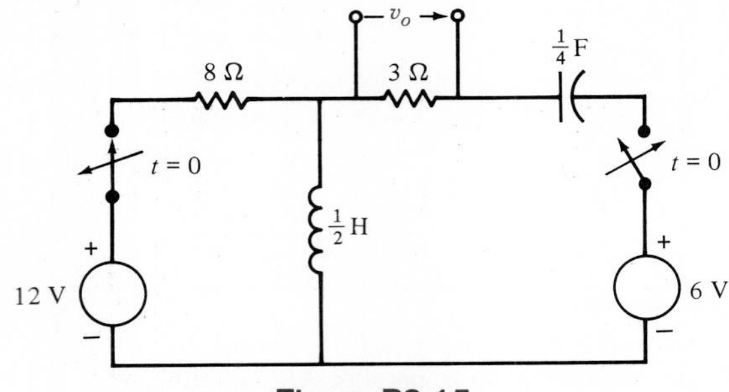

Figure P8.15

8.16. In the circuit shown in Fig. P8.16, find $v(t)$, $t > 0$.

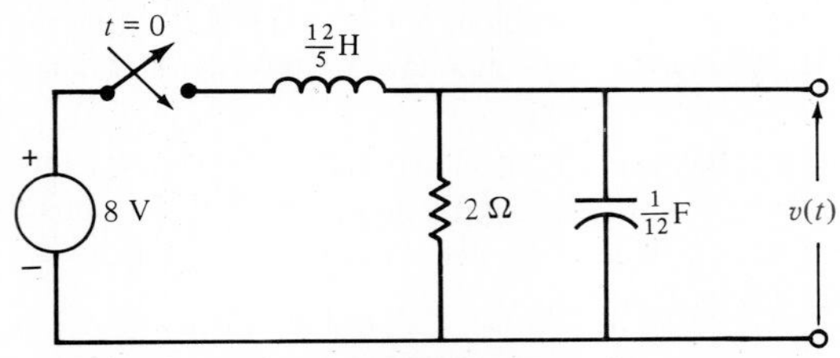

Figure P8.16

8.17. For the circuit in Fig. P8.17, find the current $i_L(t)$ for $t > 0$ if $R_x = 15\ \Omega$.

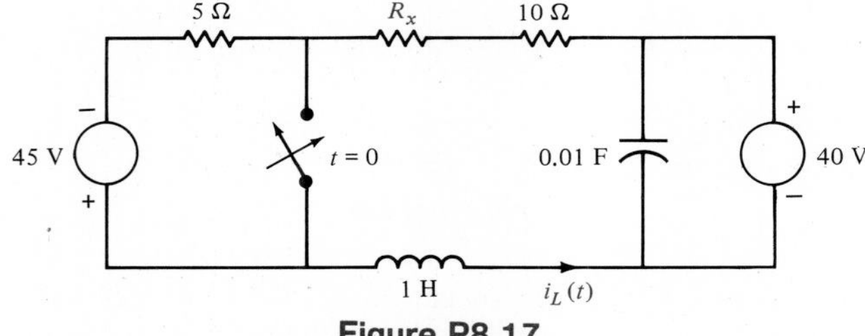

Figure P8.17

8.18. If the resistor R_x in Problem 8.17 is changed to 2 Ω, find $i_L(t)$ for $t > 0$.

8.19. If the resistor R_x in Problem 8.17 is changed to 10 Ω, determine $i_L(t)$ for $t > 0$.

8.20. Derive the differential equation for the current $i(t)$ in the inductor in Example 8.7.

8.21. Find $v_o(t)$ for $t > 0$ in Fig. P8.21.

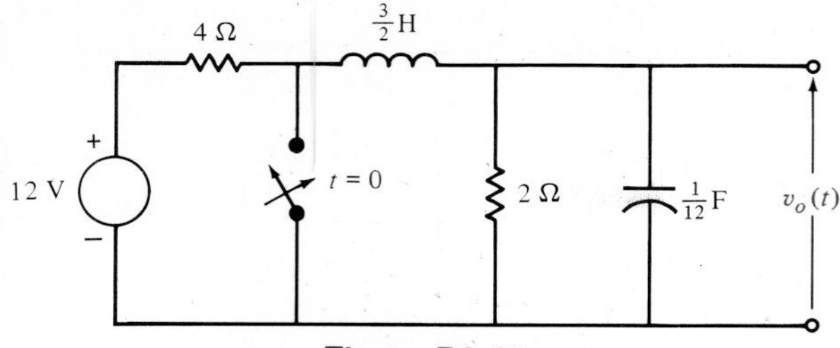

Figure P8.21

8.22. For the circuit shown in Fig. P8.22,

(a) Derive the differential equation for $v_c(t)$ for $t > 0$.

(b) Determine the type of damping the network exhibits if $L = 1$ H, $R_1 = 2\ \Omega$, $R_2 = \frac{1}{2}\ \Omega$, $C = 1$ F.

(c) if $V_s = 2$ V and $I_s = 2$ A, find $v_c(t)$ for $t > 0$.

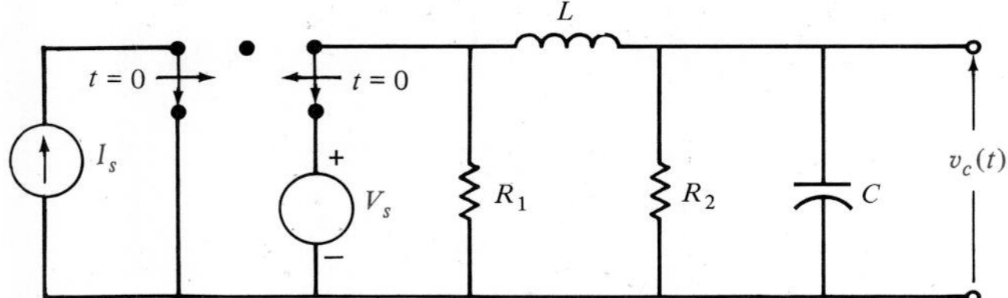

Figure P8.22

8.23. For the network shown in Fig. P8.23,

(a) Derive the general form of the characteristic equation for the network.

(b) Determine the type of damping the network exhibits if $L = 1$ H, $C = \frac{1}{2}$ F, $R_1 = R_3 = 1\ \Omega$ and $R_2 = \frac{1}{2}\ \Omega$.

(c) If $V_1 = 4$V and $V_2 = 2$V, find $v_o(t)$ for $t > 0$.

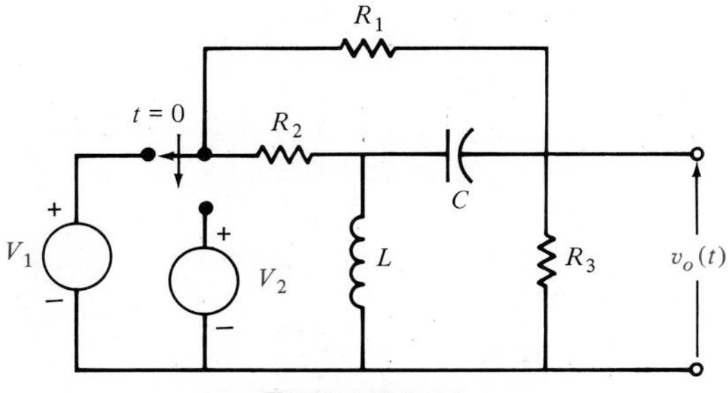

Figure P8.23

8.24. For the network shown in Fig. P8.24,

(a) Determine the general expression for the characteristic equation.

(b) Find the equation for the voltage $v_o(t)$, $t > 0$, if $R_1 = 6\ \Omega$, $R_2 = 12\ \Omega$, $L = 1$ H, $C = \frac{1}{3}$ F, and $V_i = 12$ V.

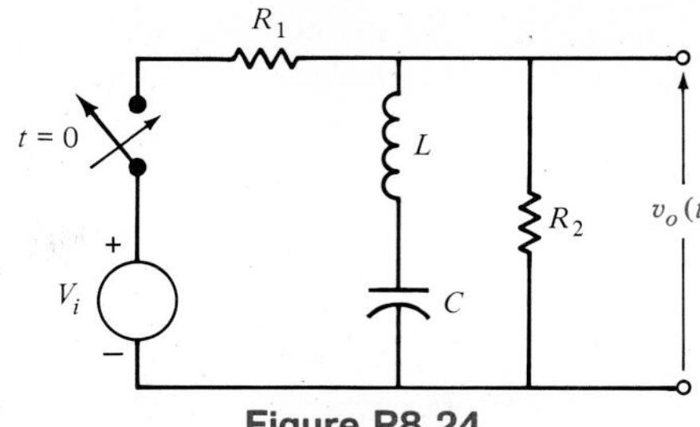

Figure P8.24

8.25. Given the circuit shown in Fig. P8.25, find an expression for

(a) The initial conditions at $t = 0$: $v_C(0)$, $i_L(0)$, and $v_o(0)$.

(b) The steady-state response.

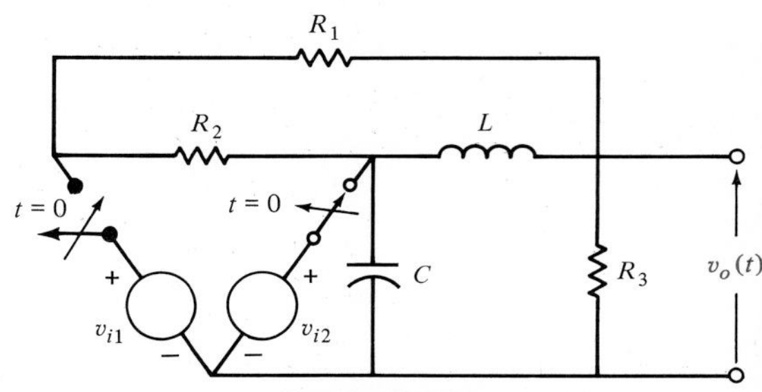

Figure P8.25

8.26. Derive a general expression for the characteristic equation of the circuit in Problem 8.25.

8.27. Following the development of Problems 8.25 and 8.26, find the equation for $v_o(t)$, $t > 0$, if the circuit parameters are R_1 and $R_2 = R_3 = 1\ \Omega$, $L = 1$ H, $C = 1$ F, $v_{i2} = 2$ V, and $v_{i1} = V_{i1} = 4$ V.

8.28. Given the network in Fig. P8.28, use SPICE to plot $v_L(t)$. Plot over the interval 0 to 50 ms with a step size of 0.5 ms.

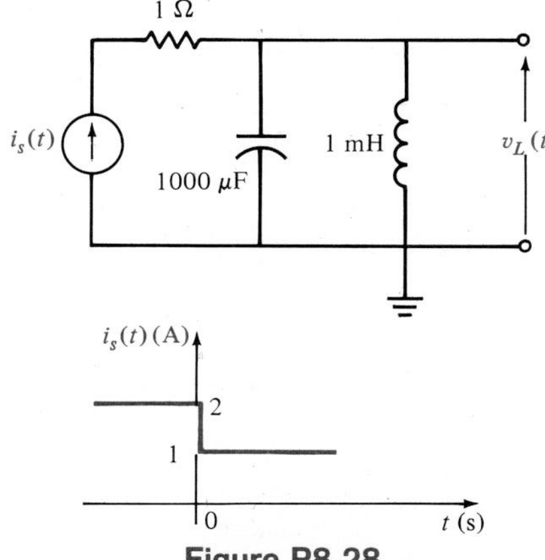

Figure P8.28

8.29. Using SPICE, plot the transient response of $v_c(t)$ in the network in Fig. P8.29. Use a step size of 0.02 s and plot over the interval 0-1 s.

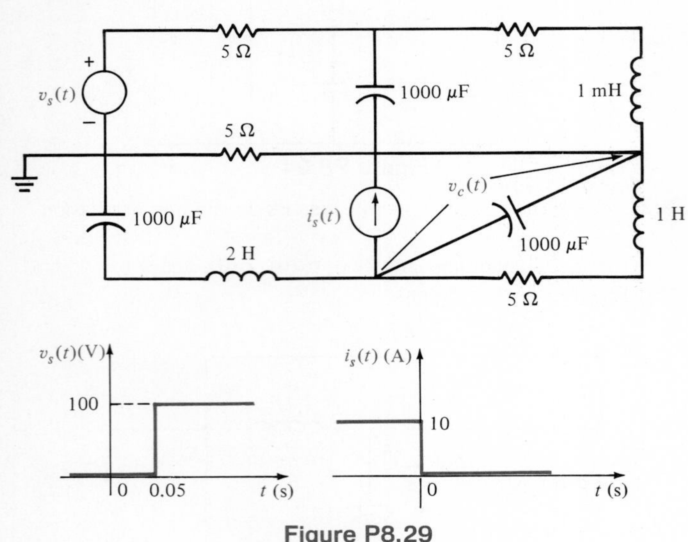

Figure P8.29

8.30. Write a SPICE program to plot $v_R(t)$ and $v_c(t)$ in the network in Fig. P8.30 from 0 to 1 s.

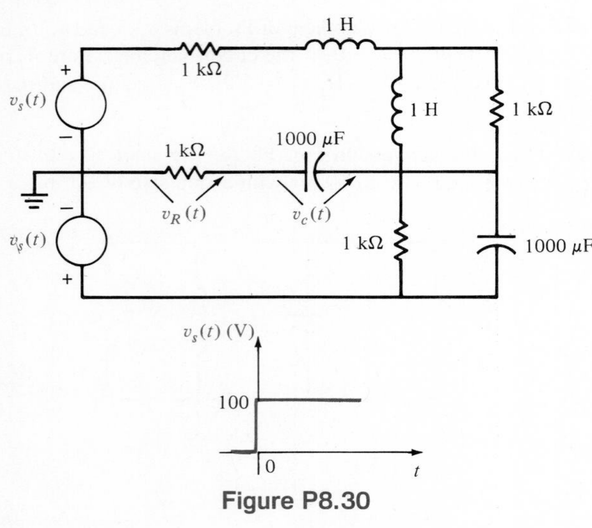

Figure P8.30

8.31. Use SPICE to plot the transient analysis of $v_o(t)$ in the network shown in Fig. P8.31. Plot over the interval from 0 to 5 ms.

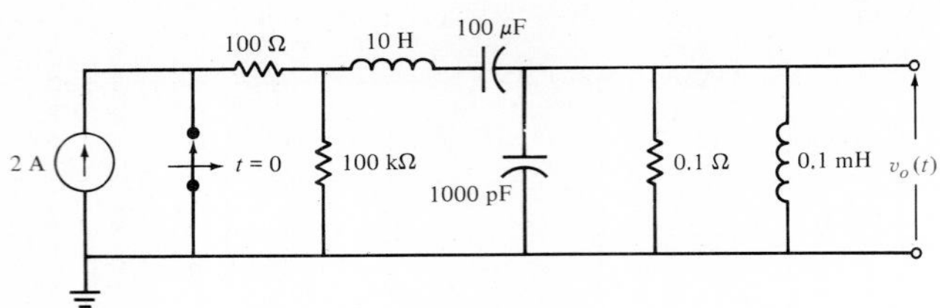

Figure P8.31

8.32. Write a SPICE program to plot $i_R(t)$ and $v_C(t)$ in the network shown in Fig. P8.32 over the range 0 to 0.3 s.

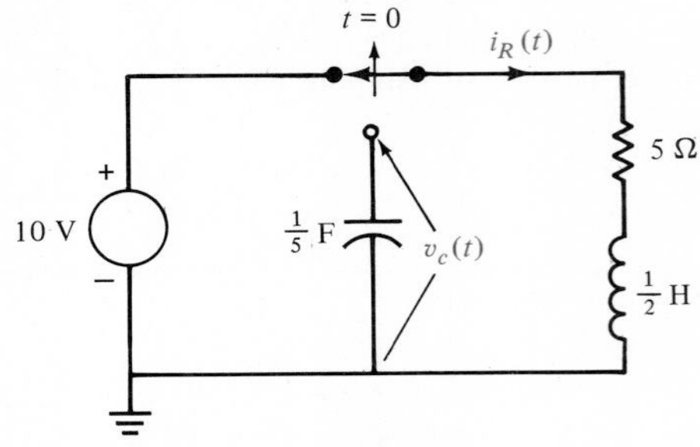

Figure P8.32

8.33. Given the network in Fig. P8.33, plot $i(t)$ and $v_c(t)$ from 0 to 5 s.

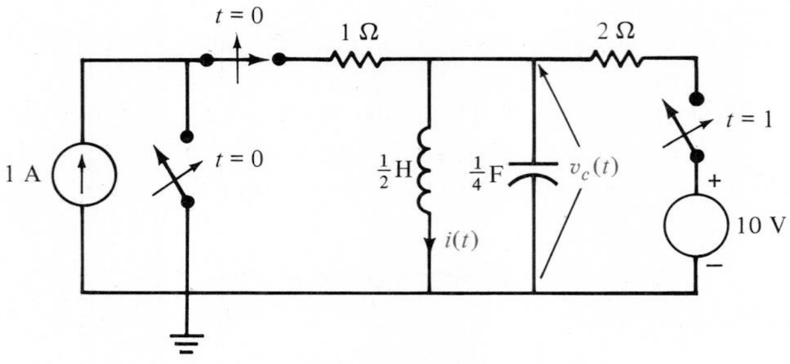

Figure P8.33

8.34. Write a SPICE program to plot $v_L(t)$ in the network in Fig. P8.34 over the interval 0 to 4 s.

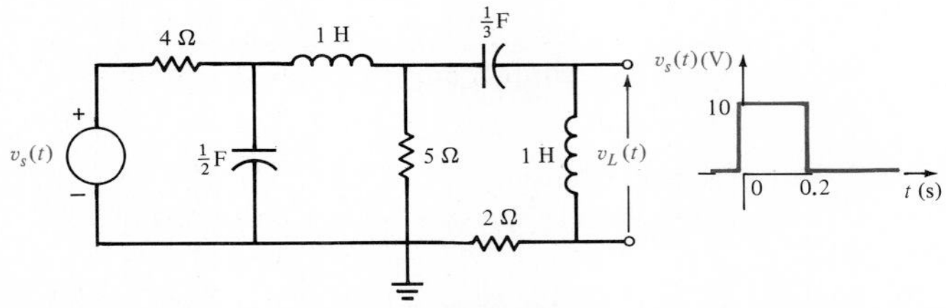

Figure P8.34

8.35. Use SPICE to plot the transient response of $V_R(t)$ in the network in Fig. P8.35. Use a step size of 0.1 s over the interval 0 to 3 s.

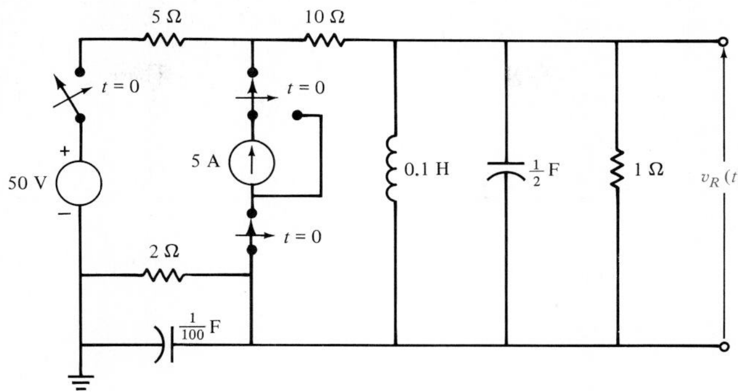

Figure P8.35

8.36. Given the network in Fig. P8.36, plot the transient response of $v(t)$ from 0 to 1.5 s.

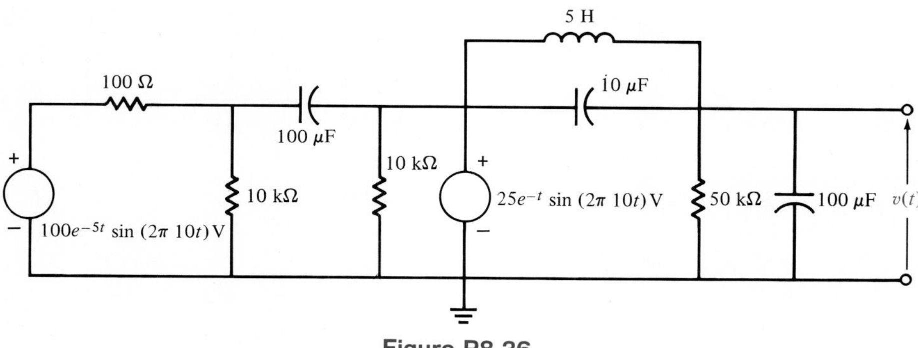

Figure P8.36

8.37. Write a SPICE program to plot the response of $v(t)$ from 0 to 200 ms for the network in Fig. P8.37.

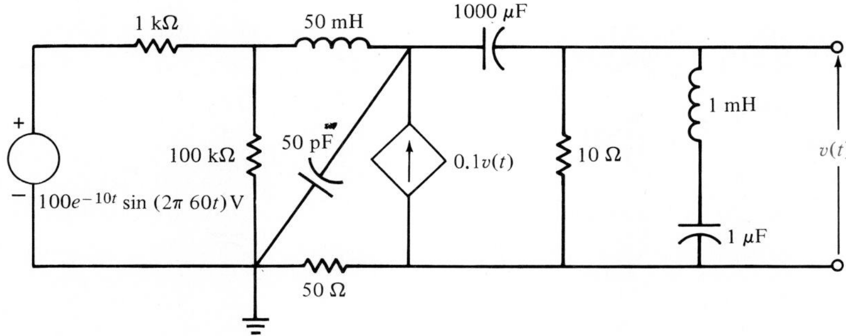

Figure P8.37

8.38. Plot the transient response of $v_o(t)$ in the network in Fig. P8.38 from 0.5 to 1.1 s.

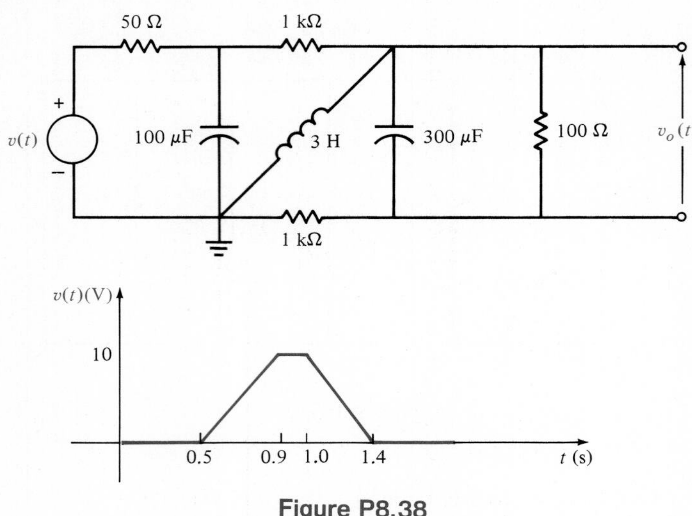

Figure P8.38

8.39. Plot $v_o(t)$ in the network in Fig. P8.39 over the time interval $0 \leq t \leq 0.1$ s.

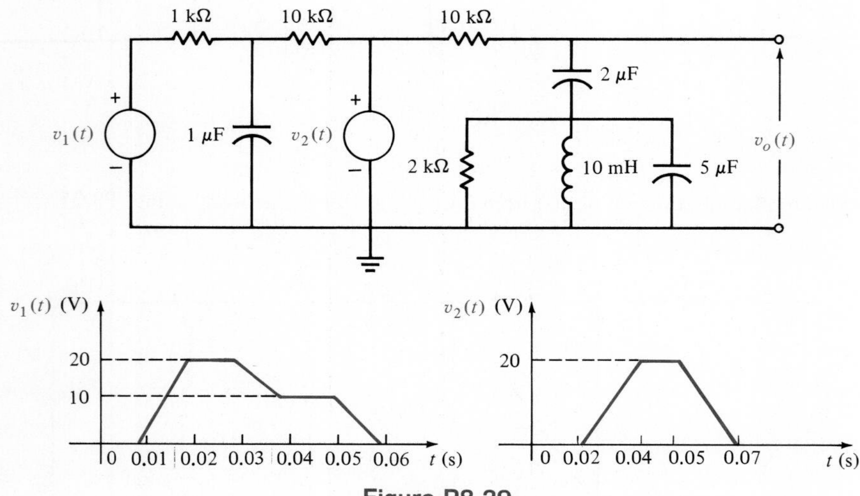

Figure P8.39

8.40. Plot the transient response of $v_o(t)$ in the network in Fig. P8.40 over the interval from 0 to 5 s.

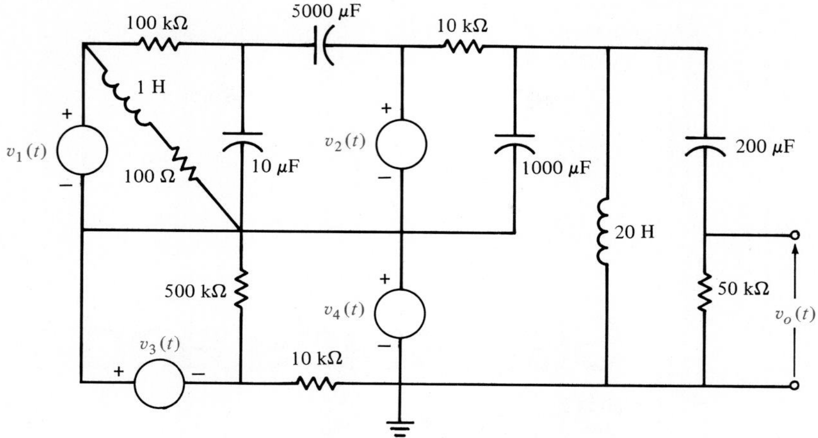

$v_2(t) = 10e^{-2t} \sin(2\pi\, 60t)$

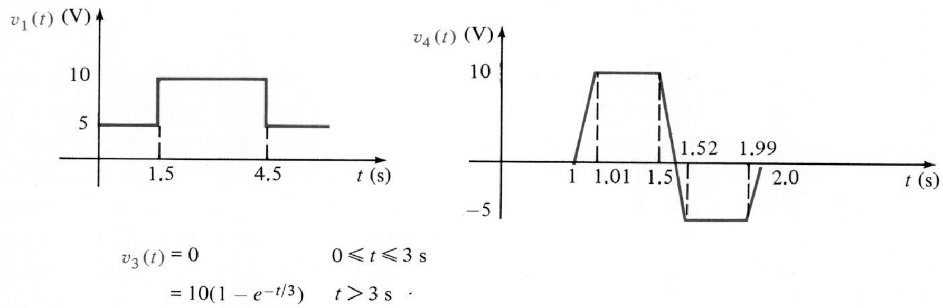

$v_3(t) = 0 \qquad 0 \leq t \leq 3$ s

$= 10(1 - e^{-t/3}) \qquad t > 3$ s

Figure P8.40

CHAPTER

Sinusoids and Phasors

In the preceding chapters we have considered in some detail both the natural and forced response of a network. We found that the natural response was a characteristic of the network and was independent of the forcing function. The forced response, however, depends directly on the type of forcing function, which up until now has always been a constant. At this point we will diverge from this tack to consider an extremely important excitation: the *sinusoidal forcing function*. Nature is replete with examples of sinusoidal phenomena, and although this is important to us as we examine many different types of physical systems, one reason that we can appreciate at this point for studying this forcing function is that it is the dominant waveform in the electric power industry. The signal present at the ac outlets in our home, office, laboratory, and so on, is sinusoidal. In addition, it is interesting to note that via Fourier analysis we can represent any periodic electrical signal by a sum of sinusoids.

In this chapter we concentrate on the steady-state forced response of networks with sinusoidal driving functions. We will ignore the initial conditions and the transient or natural response, which will eventually vanish for the type of circuits with which we will be dealing. We refer to this as an *ac steady-state analysis*. Before proceeding with this analysis, however, we must thoroughly understand the nature of the sinusoidal function.

9.1 Sinusoids

Let us begin our discussion of sinusoidal functions by considering the sine wave

$$x(\omega t) = X_M \sin \omega t \tag{9.1}$$

where $x(t)$ could represent either $v(t)$ or $i(t)$. X_M is the *amplitude* or *maximum value*, ω

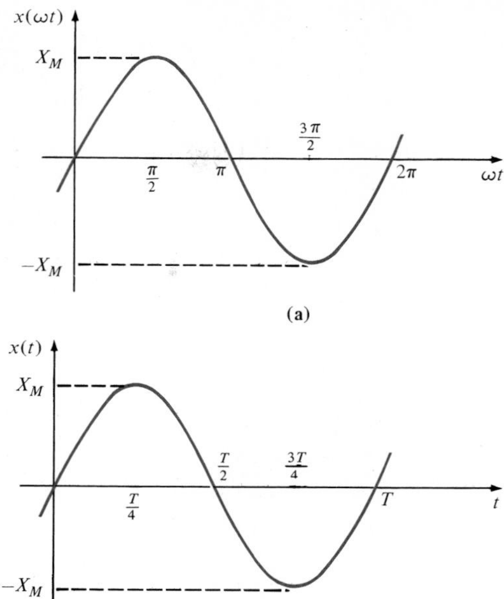

Figure 9.1 Plots of a sine wave as a function of both ωt and t.

is the *radian* or *angular frequency*, and ωt is the *argument* of the sine function. A plot of the function in Eq. (9.1) as a function of its argument is shown in Fig. 9.1a. Obviously, the function repeats itself every 2π radians and therefore the *period* of the function is 2π radians. This condition is described mathematically as $x(\omega t + 2\pi) = x(\omega t)$ or in general for period T as

$$x[\omega(t + T)] = x(\omega t) \tag{9.2}$$

meaning that the function has the same value at time $t + T$ as it does at time t.

The waveform can also be plotted as a function of time as shown in Fig. 9.1b. Note that this function goes through one period every T seconds, or in other words, in 1 second it goes through $1/T$ periods or cycles. The number of cycles per second, called hertz, is the frequency f, where

$$f = \frac{1}{T} \tag{9.3}$$

Now since $\omega T = 2\pi$ as shown in Fig. 9.1a, we find that

$$\omega = \frac{2\pi}{T} = 2\pi f \tag{9.4}$$

which is of course the general relationship among period in seconds, frequency in hertz, and radian frequency.

Now that we have discussed some of the basic properties of a sine wave, let us consider the following general expression for a sinusoidal function:

$$x(t) = X_M \sin(\omega t + \theta) \tag{9.5}$$

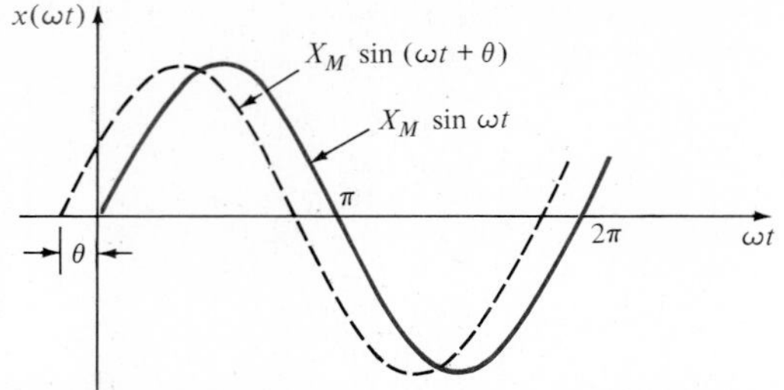

Figure 9.2 Graphical illustration of X_M sin (ωt + θ) leading X_M sin ωt by θ radians.

In this case $(\omega t + \theta)$ is the argument of the sine function, and θ is called the *phase angle*. A plot of this function is shown in Fig. 9.2, together with the original function in Eq. (9.1) for comparison. Because of the presence of the phase angle, any point on the waveform $X_M \sin(\omega t + \theta)$ occurs θ radians earlier in time than the corresponding point on the waveform $X_M \sin \omega t$. Therefore, we say that $X_M \sin \omega t$ *lags* $X_M \sin(\omega t + \theta)$ by θ radians. In the more general situation, if

$$x_1(t) = X_{M1} \sin(\omega t + \theta)$$

and

$$x_2(t) = X_{M2} \sin(\omega t + \phi)$$

then $x_1(t)$ leads $x_2(t)$ by $\theta - \phi$ radians and $x_2(t)$ lags $x_1(t)$ by $\theta - \phi$ radians. If $\theta = \phi$, the waveforms are identical and the functions are said to be *in phase*. If $\theta \neq \phi$, the functions are *out of phase*.

The phase angle is normally expressed in degrees rather than radians, and therefore we will simply state at this point that we will use the two forms interchangeably, that is,

$$x(t) = X_M \sin\left(\omega t + \frac{\pi}{2}\right) = X_M \sin(\omega t + 90°) \tag{9.6}$$

In addition, it should be noted that adding to the argument integer multiples of either 2π radians or 360° does not change the original function. This can easily be shown mathematically, but is visibly evident when examining the waveform, as shown in Fig. 9.2.

Although our discussion has centered about the sine function, we could just as easily have used the cosine function, since the two waveforms differ only by a phase angle; that is,

$$\cos \omega t = \sin\left(\omega t + \frac{\pi}{2}\right) \tag{9.7}$$

$$\sin \omega t = \cos\left(\omega t - \frac{\pi}{2}\right) \tag{9.8}$$

It should be noted that when comparing one sinusoidal function with another *of the same frequency* to determine the phase difference, it is necessary to express both functions as either sines or cosines with positive amplitudes. Once in this format, the phase angle between the functions can be computed as outlined above. Two other trigonometric identities that normally prove useful in phase angle determination are

$$-\cos(\omega t) = \cos(\omega t \pm 180^\circ) \tag{9.9}$$

$$-\sin(\omega t) = \sin(\omega t \pm 180^\circ) \tag{9.10}$$

Finally, the angle-sum and angle-difference relationships for sines and cosines may be useful in the manipulation of sinusoidal functions.

These relations are

$$\begin{aligned} \sin(\alpha + \beta) &= \sin\alpha\cos\beta + \cos\alpha\sin\beta \\ \cos(\alpha + \beta) &= \cos\alpha\cos\beta - \sin\alpha\sin\beta \\ \sin(\alpha - \beta) &= \sin\alpha\cos\beta - \cos\alpha\sin\beta \\ \cos(\alpha - \beta) &= \cos\alpha\cos\beta + \sin\alpha\sin\beta \end{aligned} \tag{9.11}$$

Our interest in these relations will revolve about the cases where $\alpha = \omega t$ and $\beta = \theta$. Since θ is a constant, the equations above indicate that either a sinusoidal or cosinusoidal function with a phase angle can be written as the sum of a sine function and a cosine function with the same arguments. For example, using Eq. (9.11), Eq. (9.5) can be written as

$$\begin{aligned} x(t) &= X_M \sin(\omega t + \theta) \\ &= X_M(\sin\omega t \cos\theta + \cos\omega t \sin\theta) \\ &= A \sin\omega t + B\cos\omega t \end{aligned}$$

where

$$A = X_M \cos\theta$$

$$B = X_M \sin\theta$$

EXAMPLE 9.1

Two voltages $v_1(t)$ and $v_2(t)$ are given by the equations

$$v_1(t) = 12\sin(377t + 45^\circ)\text{ V}$$

$$v_2(t) = 6\sin(377t - 15^\circ)\text{ V}$$

We wish to determine the frequency of the voltages and the phase angle between $v_1(t)$ and $v_2(t)$.

The frequency f in hertz (Hz) is given by the expression

$$f = \frac{\omega}{2\pi} = \frac{377}{2\pi} = 60\text{ Hz}$$

The phase angle between the voltages is $45^\circ - (-15^\circ) = 60^\circ$; that is, $v_1(t)$ leads $v_2(t)$ by 60° or $v_2(t)$ lags $v_1(t)$ by 60°. ■

EXAMPLE 9.2

We wish to plot the waveforms for the following functions: (a) $v(t) = 1\cos(\omega t + 45^\circ)$, (b) $v(t) = 1\cos(\omega t + 225^\circ)$, and (c) $v(t) = 1\cos(\omega t - 315^\circ)$.

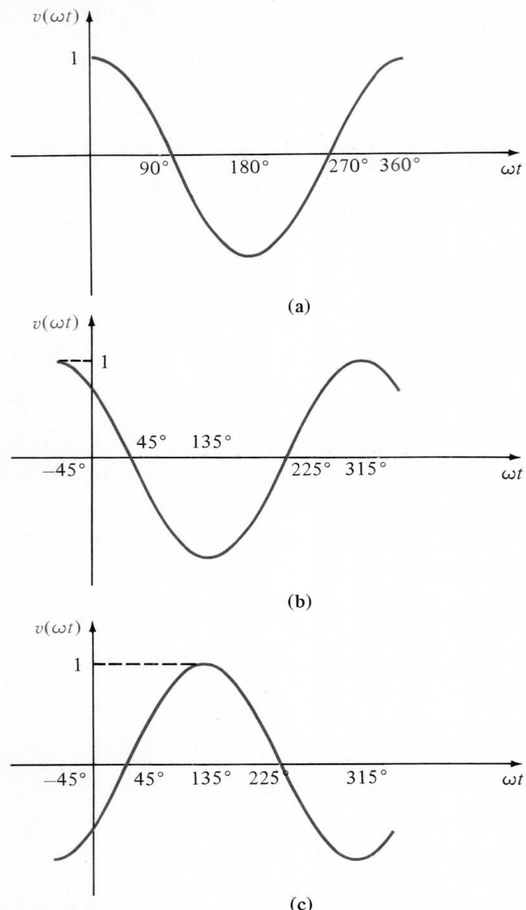

Figure 9.3 Cosine waveforms with various phase angles.

In Fig. 9.3a is shown a plot of the function $v(t) = 1 \cos \omega t$. Figure 9.3b is a plot of the function $v(t) = 1 \cos (\omega t + 45°)$. Figure 9.3c is a plot of the function $v(t) = 1 \cos (\omega t + 225°)$. Note that since

$$v(t) = 1 \cos (\omega t + 225°) = 1 \cos (\omega t + 45° + 180°)$$

this waveform is 180° out of phase with the waveform in Fig. 9.3b; that is, $\cos (\omega t + 225°) = -\cos (\omega t + 45°)$ and Fig. 9.3c is the negative of Fig. 9.3b. Finally, since the function

$$v(t) = 1 \cos (\omega t - 315°) = 1 \cos (\omega t - 315° + 360°) = 1 \cos (\omega t + 45°)$$

this function is identical to that shown in Fig. 9.3b. ■

EXAMPLE 9.3

Determine the phase angle between the two voltages $v_1(t) = 12 \sin (\omega t + 60°)$ V and $v_2(t) = -6 \cos (\omega t + 30°)$ V.

Using Eq. (9.9), $v_2(t)$ can be written as

$$v_2(t) = -6 \cos(\omega t + 30°) = 6 \cos(\omega t + 210°) \text{ V}$$

Then employing Eq. (9.7), we obtain

$$v_2(t) = 6 \cos(\omega t + 210°)$$

$$= 6 \sin(\omega t + 300°) \text{ V}$$

Now that both voltages of the same frequency are expressed as sine waves with positive amplitudes, the phase angle between $v_1(t)$ and $v_2(t)$ is $60° - (300°) = -240°$ [i.e., $v_1(t)$ leads $v_2(t)$ by $+120°$]. ■

DRILL EXERCISE

D9.1. Given the voltages $v_1(t)$ and $v_2(t)$, determine their frequency and the phase angle between them.

$$v_1(t) = 12 \cos(2513t - 70°) \text{ V}$$

$$v_2(t) = 4 \cos(2513t - 30°) \text{ V}$$

D9.2. Determine the relative position of the two sine waves.

$$v_1(t) = 12 \sin(377t - 45°)$$

$$v_2(t) = 6 \sin(377t + 675°)$$

D9.3. Determine the phase angle between the two currents.

$$i_1(t) = -4 \sin(377t - 60°) \text{ A}$$

$$i_2(t) = 10 \cos(377t + 45°) \text{ A}$$

9.2 Sinusoidal and Complex Forcing Functions

In the preceding chapters we applied a constant forcing function to a network and found that the steady-state response was also constant.

In a similar manner, if we apply a sinusoidal forcing function to a linear network, the steady-state voltages and currents in the network will also be sinusoidal. This should also be clear from the KVL and KCL equations. For example, if one branch voltage is a sinusoid of some frequency, the other branch voltages must be sinusoids of the same frequency if KVL is to apply around any closed path. This means, of course, that the forced solutions of the differential equations that describe a network with a sinusoidal forcing function are sinusoidal functions of time. For example, if we assume that our input function is a voltage $v(t)$ and our output response is a current $i(t)$ as shown in Fig. 9.4, then if $v(t) = A \sin(\omega t + \theta)$, $i(t)$ will be of the form $i(t) = B \sin(\omega t + \phi)$. The critical point here is that we know the form of the output response, and therefore the solution involves simply determining the values of the two parameters B and ϕ.

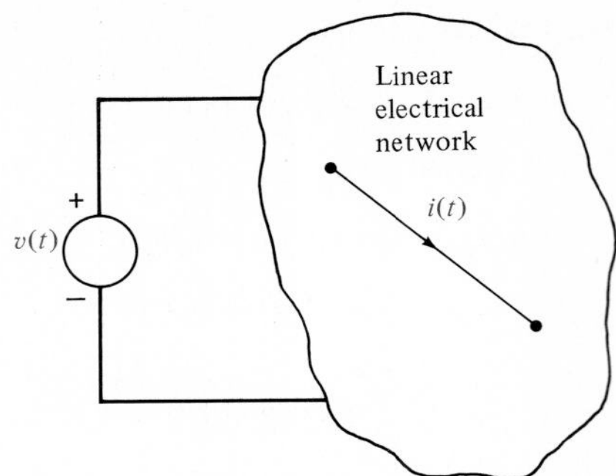

Figure 9.4 Current response to an applied voltage in an electrical network.

EXAMPLE 9.4

Consider the circuit in Fig. 9.5. The KVL equation for this circuit is

$$L\frac{di(t)}{dt} + Ri(t) = V_M \cos \omega t$$

Since the input forcing function is $V_M \cos \omega t$, we assume that the current $i(t)$ is of the form

$$i(t) = A \cos (\omega t + \phi)$$

which can be written using Eq. (9.11) as

$$\begin{aligned} i(t) &= A \cos \phi \cos \omega t - A \sin \phi \sin \omega t \\ &= A_1 \cos \omega t + A_2 \sin \omega t \end{aligned}$$

Substituting this form for $i(t)$ into the differential equation above yields

$$L\frac{d}{dt}(A_1 \cos \omega t + A_2 \sin \omega t) + R(A_1 \cos \omega t + A_2 \sin \omega t) = V_M \cos \omega t$$

Evaluating the indicated derivative produces

$$-A_1\omega L \sin \omega t + A_2\omega L \cos \omega t + RA_1 \cos \omega t + RA_2 \sin \omega t = V_M \cos \omega t$$

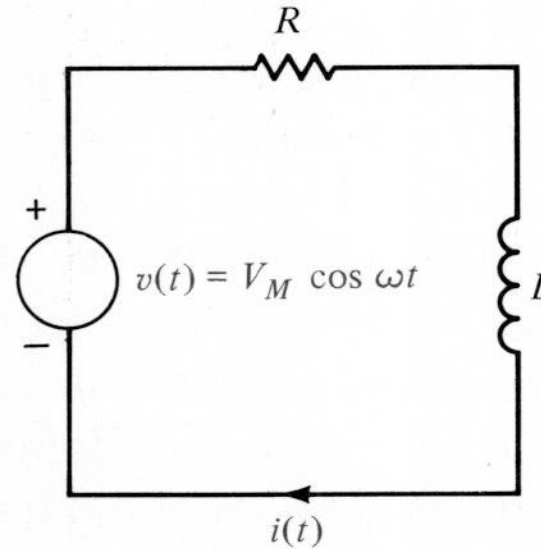

Figure 9.5 *RL* circuit.

By equating coefficients of the sine and cosine functions, we obtain

$$-A_1\omega L + RA_2 = 0$$

$$A_1R + A_2\omega L = V_M$$

Our two simultaneous equations in the unknowns A_1 and A_2 are

$$-\omega LA_1 + RA_2 = 0$$

$$RA_1 + \omega LA_2 = V_M$$

Solving these two equations for A_1 and A_2 yields

$$A_1 = \frac{RV_M}{R^2 + \omega^2L^2}$$

$$A_2 = \frac{\omega LV_M}{R^2 + \omega^2L^2}$$

Therefore,

$$i(t) = \frac{RV_M}{R^2 + \omega^2L^2}\cos\omega t + \frac{\omega LV_M}{R^2 + \omega^2L^2}\sin\omega t$$

which using the last identity in Eq. (9.11) can be written as

$$i(t) = A\cos(\omega t + \phi)$$

where A and ϕ are determined as follows:

$$A\cos\phi = \frac{RV_M}{R^2 + \omega^2L^2}$$

$$A\sin\phi = \frac{-\omega LV_M}{R^2 + \omega^2L^2}$$

Hence

$$\tan\phi = \frac{A\sin\phi}{A\cos\phi} = -\frac{\omega L}{R}$$

and therefore

$$\phi = -\tan^{-1}\frac{\omega L}{R}$$

since

$$(A\cos\phi)^2 + (A\sin\phi)^2 = A^2(\cos^2\phi + \sin^2\phi) = A^2$$

$$A^2 = \frac{R^2V_M^2}{(R^2 + \omega^2L^2)^2} + \frac{(\omega L)^2V_M^2}{(R^2 + \omega^2L^2)^2}$$

$$A = \frac{V_M}{\sqrt{R^2 + \omega^2L^2}}$$ = amplitude

Hence the final expression for $i(t)$ is

$$i(t) = \frac{V_M}{\sqrt{R^2 + \omega^2 L^2}} \cos\left(\omega t - \tan^{-1}\frac{\omega L}{R}\right)$$

The analysis above indicates that ϕ is zero if either $\omega = 0$ or $L = 0$ and hence $i(t)$ is in phase with $v(t)$. If $R = 0$, $\phi = 90°$ and the current lags the voltage by 90°. If R and L are both present, the current lags the voltage by some angle between 0° and 90°. ■

This example illustrates an important point—solving even a simple one-loop circuit containing one resistor and one inductor is very complicated when compared to the solution of a single-loop circuit containing only two resistors. Imagine for a moment how laborious it would be to solve a more complicated circuit using the procedure we have employed in Example 9.4. To circumvent this approach we will establish a correspondence between sinusoidal time functions and complex numbers. We will then show that this relationship leads to a set of algebraic equations for currents and voltages in a network (e.g., loop currents or node voltages) in which the coefficients of the variables are complex numbers. Hence, once again we will find that determining the currents or voltages in a circuit can be accomplished by solving a set of algebraic equations; however, in the sinusoidal steady-state case, their solution is complicated by the fact that variables in the equations have complex, rather than real, coefficients.

The vehicle we will employ to establish a relationship between time-varying sinusoidal functions and complex numbers is Euler's equation, which for our purposes is written as

$$e^{j\omega t} = \cos \omega t + j \sin \omega t \tag{9.12}$$

This complex function has a real part and an imaginary part:

$$\begin{aligned} \text{Re}\,(e^{j\omega t}) &= \cos \omega t \\ \text{Im}\,(e^{j\omega t}) &= \sin \omega t \end{aligned} \tag{9.13}$$

where Re $(\cdot)$ and Im $(\cdot)$ represent the real part and the imaginary part, respectively, of the function in the parenthesis.

Now suppose that we select as our forcing function in Fig. 9.4 the nonrealizable voltage

$$v(t) = V_M e^{j\omega t} \tag{9.14}$$

which because of Euler's identity can be written as

$$v(t) = V_M \cos \omega t + jV_M \sin \omega t \tag{9.15}$$

The real and imaginary parts of this function are each realizable. We think of this complex forcing function as two forcing functions, a real one and an imaginary one, and as a consequence of linearity, the principle of superposition applies and thus the current response can be written as

$$i(t) = I_M \cos(\omega t + \phi) + jI_M \sin(\omega t + \phi) \tag{9.16}$$

Where $I_M \cos(\omega t + \phi)$ is the response due to $V_M \cos \omega t$ and $jI_M \sin(\omega t + \phi)$ is the response due to $jV_M \sin \omega t$. This expression for the current containing both a real and an imaginary term can be written via Euler's equation as

$$i(t) = I_M e^{j(\omega t + \phi)} \tag{9.17}$$

Because of the relationships above we find that rather than applying the forcing function $V_M \cos \omega t$ and calculating the response $I_M \cos(\omega t + \phi)$, we can apply the complex forcing function $V_M e^{j\omega t}$ and calculate the response $I_M e^{j(\omega t + \phi)}$, the real part of which is the desired response $I_M \cos(\omega t + \phi)$. Although this procedure may initially appear to be more complicated, it is not. It is via this technique that we will convert the differential equation to an algebraic equation which is much easier to solve.

EXAMPLE 9.5

Once again, let us determine the current in the RL circuit examined in Example 9.4. However, rather than applying $V_M \cos \omega t$ we will apply $V_M e^{j\omega t}$. The response will be of the form

$$i(t) = I_M e^{j(\omega t + \phi)}$$

where only I_M and ϕ are unknown. Substituting $v(t)$ and $i(t)$ into the differential equation for the circuit, we obtain

$$RI_M e^{j(\omega t + \phi)} + L \frac{d}{dt}(I_M e^{j(\omega t + \phi)}) = V_M e^{j\omega t}$$

Taking the indicated derivative, we obtain

$$RI_M e^{j(\omega t + \phi)} + j\omega L I_M e^{j(\omega t + \phi)} = V_M e^{j\omega t}$$

Dividing each term of the equation by the common factor $e^{j\omega t}$ yields

$$RI_M e^{j\phi} + j\omega L I_M e^{j\phi} = V_M$$

which is an algebraic equation with complex coefficients. This equation can be written as

$$I_M e^{j\phi} = \frac{V_M}{R + j\omega L}$$

Converting the right-hand side of the equation to exponential or polar form produces the equation

$$I_M e^{j\phi} = \frac{V_M}{\sqrt{R^2 + \omega^2 L^2}} e^{j[-\tan^{-1}(\omega L/R)]}$$

(A quick refresher on complex numbers is given in Appendix B for readers who need to sharpen their skills in this area.) The form above clearly indicates that the magnitude and phase of the resulting current are

$$I_M = \frac{V_M}{\sqrt{R^2 + \omega^2 L^2}}$$

RL-circuit

and

$$\phi = -\tan^{-1}\frac{\omega L}{R}$$

However, since our actual forcing function was $V_M \cos \omega t$ rather than $V_M e^{j\omega t}$, our actual response is the real part of the complex response:

$$\begin{aligned} i(t) &= I_M \cos(\omega t + \phi) \\ &= \frac{V_M}{\sqrt{R^2 + \omega^2 L^2}} \cos\left(\omega t - \tan^{-1}\frac{\omega L}{R}\right) \end{aligned}$$

Note that this is identical to the response obtained in the previous example by solving the differential equation for the current $i(t)$. ■

9.3 Phasors

Once again let us assume that the forcing function for a linear network is of the form

$$v(t) = V_M e^{j\omega t}$$

Then every steady-state voltage or current in the network will have the same form and the same frequency ω; for example, a current $i(t)$ will be of the form $i(t) = I_M e^{j(\omega t + \phi)}$.

As we proceed in our subsequent circuit analyses, we will simply note the frequency and then drop the factor $e^{j\omega t}$ since it is common to every term in the describing equations. Dropping the term $e^{j\omega t}$ indicates that every voltage or current can be fully described by a magnitude and phase. For example, a voltage $v(t)$ can be written in exponential form as

$$v(t) = V_m \cos(\omega t + \theta) = \text{Re}\,[V_m e^{j(\omega t + \theta)}] \tag{9.18}$$

or as a complex number

$$v(t) = \text{Re}\,(V_M \angle \theta\; e^{j\omega t}) \tag{9.19}$$

Since we are working with a complex forcing function, the real part of which is the desired answer, and each term in the equation will contain $e^{j\omega t}$, we can drop Re $(\cdot)$ and $e^{j\omega t}$ and work only with the complex number $V_M \angle \theta$. This complex representation is commonly called a *phasor*. As a distinguishing feature, phasors will be written in boldface type. In a completely identical manner a voltage $v(t) = V_M \cos(\omega t + \theta) = \text{Re}\,[V_M e^{j(\omega t + \theta)}]$ and a current $i(t) = I_M \cos(\omega t + \phi) = \text{Re}\,[I_M e^{j(\omega t + \phi)}]$ are written in phasor notation as $\mathbf{V} = V_M \angle \theta$ and $\mathbf{I} = I_M \angle \phi$, respectively.

EXAMPLE 9.6

Again, we consider the *RL* circuit in Example 9.4. The differential equation is

$$L\frac{di(t)}{dt} + Ri(t) = V_M \cos \omega t$$

The forcing function can be replaced by a complex forcing function that is written as

$\mathbf{V}e^{j\omega t}$ with phasor $\mathbf{V} = V_M \angle 0°$. Similarly, the current $i(t)$ can be replaced by a complex function that is written as $\mathbf{I}e^{j\omega t}$ with phasor $\mathbf{I} = I_M \angle \phi$. From our previous discussions we recall that the solution of the differential equation is the real part of this current.

Using the complex forcing function, the differential equation becomes

$$L\frac{d}{dt}(\mathbf{I}e^{j\omega t}) + R\mathbf{I}e^{j\omega t} = \mathbf{V}e^{j\omega t}$$

$$j\omega L\mathbf{I}e^{j\omega t} + R\mathbf{I}e^{j\omega t} = \mathbf{V}e^{j\omega t}$$

Note that $e^{j\omega t}$ is a common factor and, as we have already indicated, can be eliminated leaving the phasors, that is,

$$j\omega L\mathbf{I} + R\mathbf{I} = \mathbf{V}$$

Therefore,

$$\mathbf{I} = \frac{\mathbf{V}}{R + j\omega L} = I_M \angle \phi = \frac{V_M}{\sqrt{R^2 + \omega^2 L^2}} \angle -\tan^{-1}\frac{\omega L}{R}$$

The real part of this function is the solution

$$i(t) = \frac{V_M}{\sqrt{R^2 + \omega^2 L^2}} \cos\left(\omega t - \tan^{-1}\frac{\omega L}{R}\right)$$

which once again is the function we obtained earlier. ■

For convenience we define the relation between the phasors after the $e^{j\omega t}$ has been eliminated as the frequency domain. Thus we have transformed a set of differential equations with sinusoidal forcing functions in the time domain to a set of algebraic equations containing complex numbers in the frequency domain. In effect, we are now faced with solving a set of algebraic equations for the unknown phasors. The phasors are then simply transformed back to the time domain to yield the solution of the original set of differential equations. In addition, we note that the solution of sinusoidal steady-state circuits would be relatively simple if we could write the phasor equation directly from the circuit discription. In Section 9.4 we will lay the groundwork for doing just that.

Note that in our discussions we have tacitly assumed that sinusoidal functions would be represented as phasors with a phase angle based on a cosine function. Therefore, if sine functions are used, we will simply employ the relationship in Eq. (9.7) to obtain the proper phase angle.

Table 9.1 Phasor Representation

Time Domain	Frequency Domain
$A\cos(\omega t \pm \theta)$	$A \angle \pm\theta$
$A\sin(\omega t \pm \theta)$	$A \angle \pm\theta - 90°$

In summary, while $v(t)$ represents a voltage in the time domain, the phasor **V** represents the voltage in the frequency domain. The phasor contains only magnitude and phase information, and the frequency is implicit in this representation. The transformation from the time domain to the frequency domain, as well as the reverse transformation, is shown in Table 9.1. Recall that the phase angle is based on a cosine function, and therefore if a sine function is involved, a 90° shift factor must be employed as shown in the table.

The following examples illustrate the use of the phasor transformation.

EXAMPLE 9.7

Convert the time functions $v(t) = 24 \cos(377t - 45°)$ and $i(t) = 12 \sin(377t + 120°)$ to phasors.

Using the phasor transformation shown above, we have

$$\mathbf{V} = 24 \angle -45°$$

$$\mathbf{I} = 12 \angle 120° - 90° = 12 \angle 30°$$

■

EXAMPLE 9.8

Convert the following phasors $\mathbf{V} = 16 \angle 20°$ and $\mathbf{I} = 10 \angle -75°$ from the frequency domain to the time domain if the frequency is 60 Hz. Write **V** as a cosine function and **I** as a sine function.

Employing the reverse transformation for phasors, we find that

$$v(t) = 16 \cos(377t + 20°)$$

$$i(t) = 10 \sin(377t + 15°)$$

■

DRILL EXERCISE

D9.4. Convert the following voltage functions to phasors.

$$v_1(t) = 12 \cos(377t - 425°) \text{ V}$$

$$v_2(t) = 18 \sin(2513t + 4.2°) \text{ V}$$

D9.5. Convert the following phasors to the time domain if the frequency is 400 Hz.

$$\mathbf{V}_1 = 10 \angle 20°$$

$$\mathbf{V}_2 = 12 \angle -60°$$

9.4 Phasor Relationships for Circuit Elements

As we proceed in our development of the techniques required to analyze circuits in the sinusoidal steady state, we are now in a position to establish the phasor relationships between voltage and current for the three passive elements R, L, and C.

In the case of a resistor as shown in Fig. 9.6a, the voltage-current relationship is known to be

$$v(t) = Ri(t) \tag{9.20}$$

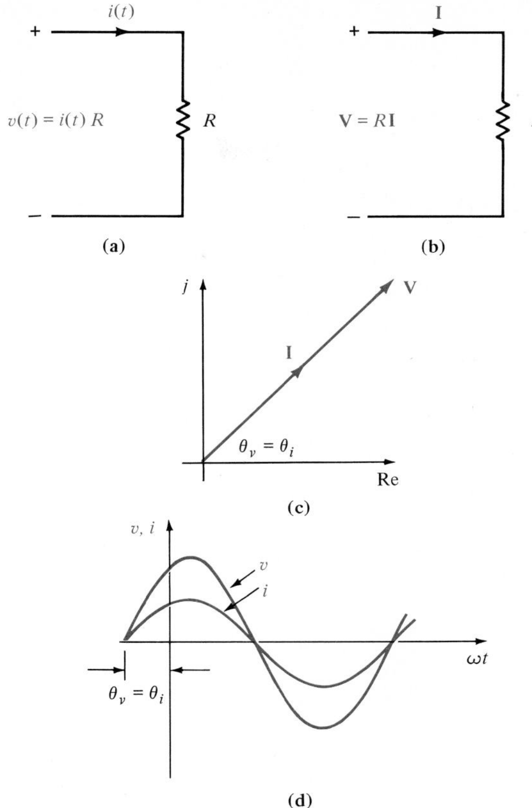

Figure 9.6 Voltage-current relationships for a resistor.

Applying the complex voltage $V_M e^{j(\omega t + \theta_v)}$ results in the complex current $I_M e^{j(\omega t + \theta_i)}$, and therefore Eq. (9.20) becomes

$$V_M e^{j(\omega t + \theta_v)} = RI_M e^{j(\omega t + \theta_i)}$$

which reduces to

$$V_M e^{j\theta_v} = RI_M e^{j\theta_i} \tag{9.21}$$

Equation (9.21) can be written in phasor form as

$$\mathbf{V} = R\mathbf{I} \tag{9.22}$$

where

$$\mathbf{V} = V_M e^{j\theta_v} = V_M \angle \theta_v \quad \text{and} \quad \mathbf{I} = I_M e^{j\theta_i} = I_M \angle \theta_i$$

From Eq. (9.21) we see that $\theta_v = \theta_i$ and thus the current and voltage for this circuit are *in phase*.

Historically, complex numbers have been represented as points on a graph in which the x-axis represents the real axis and the y-axis the imaginary axis. The line segment connecting the origin with the point provides a convenient representation of the magnitude and angle when the complex number is written in polar form. A review of Appendix B will indicate how these complex numbers or line segments can be added, subtracted, and so on. Since phasors are complex numbers it is convenient to represent the phasor voltage and current graphically as line segments. A plot of the line segments representing the phasors is called a *phasor diagram*. This pictorial representation of phasors provides immediate information on the relative magnitude of one phasor with another, the angle between two phasors, and the relative position of one phasor with respect to another (i.e., leading or lagging). A phasor diagram and the sinusoidal waveforms for the resistor are shown in Fig. 9.6c and d, respectively. A phasor diagram will be drawn for each of the other circuit elements in the remainder of this section.

EXAMPLE 9.9

If the voltage $v(t) = 24 \cos (377t + 75°)$ V is applied to a 6-Ω resistor as shown in Fig. 9.6a, we wish to determine the resultant current.

Since the phasor voltage is

$$\mathbf{V} = 24 \underline{/75°} \text{ V}$$

the phasor current from equation (9.22) is

$$\mathbf{I} = \frac{24 \underline{/75°}}{6} = 4 \underline{/75°} \text{ A}$$

which in the time domain is

$$i(t) = 4 \cos (377t + 75°) \text{ A}$$ ■

DRILL EXERCISE

D9.6. The current in a 4-Ω resistor is known to be $\mathbf{I} = 12 \underline{/60°}$ A. Express the voltage across the resistor as a time function if the frequency of the current is 60 Hz.

The voltage-current relationship for an inductor, as shown in Fig. 9.7a, is

$$v(t) = L \frac{di(t)}{dt} \tag{9.23}$$

Substituting the complex voltage and current into this equation yields

$$V_M e^{j(\omega t + \theta_v)} = L \frac{d}{dt} I_M e^{j(\omega t + \theta_i)}$$

which reduces to

$$V_M e^{j\theta_v} = j\omega L I_M e^{j\theta_i} \tag{9.24}$$

Equation (9.24) in phasor notation is

$$\mathbf{V} = j\omega L \mathbf{I} \tag{9.25}$$

Note that the differential equation in the time domain (9.23) has been converted to an algebraic equation with complex coefficients in the frequency domain. This relationship

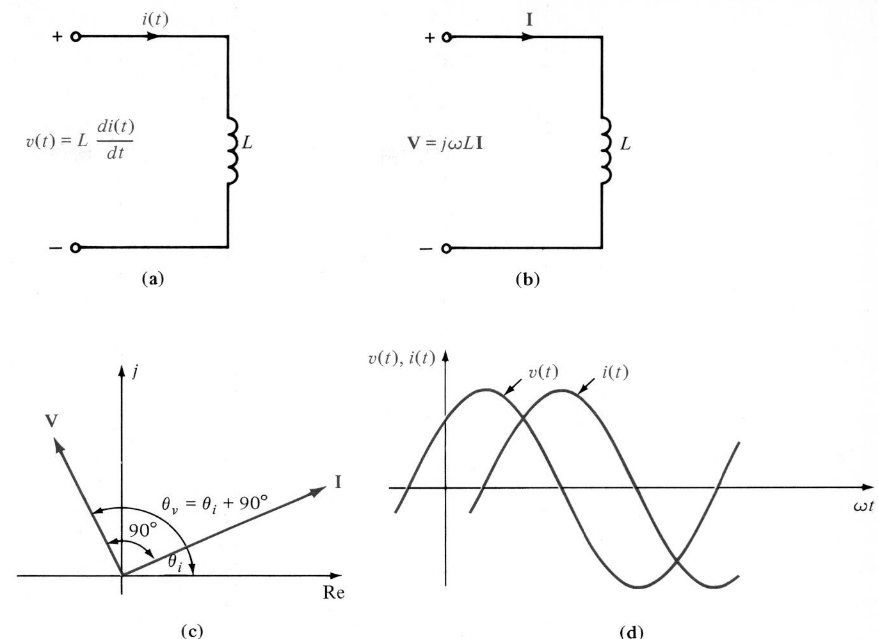

Figure 9.7 Voltage-current relationships for an inductor.

is shown in Fig. 9.7b. Since the imaginary operator $j = 1e^{j90°} = 1\underline{/90°} = \sqrt{-1}$, Eq. (9.24) can be written as

$$V_M e^{j\theta_v} = \omega L I_M e^{j(\theta_i + 90°)} \tag{9.26}$$

Therefore, the voltage and current are *90° out of phase* and in particular the voltage leads the current by 90° or the current lags the voltage by 90°. The phasor diagram and the sinusoidal waveforms for the inductor circuit are shown in Fig. 9.7c and d, respectively.

EXAMPLE 9.10

The voltage $v(t) = 12 \cos(377t + 20°)$ V is applied to a 20-mH inductor as shown in Fig. 9.7a. Find the resultant current.

The phasor current is

$$\mathbf{I} = \frac{\mathbf{V}}{j\omega L} = \frac{12\underline{/20°}}{\omega L\underline{/90°}}$$

$$= \frac{12\underline{/20°}}{(377)(20 \times 10^{-3})\underline{/90°}}$$

$$= 1.59\underline{/-70°} \text{ A}$$

or

$$i(t) = 1.59 \cos(377t - 70°) \text{ A}$$

■

DRILL EXERCISE

D9.7. The current in a 0.05-H inductor is $\mathbf{I} = 4 \angle -30°$ A. If the frequency of the current is 60 Hz, determine the voltage across the inductor.

The voltage-current relationship for our last passive element, the capacitor, as shown in Fig. 9.8a, is

$$i(t) = C\frac{dv(t)}{dt} \tag{9.27}$$

Once again employing the complex voltage and current, we obtain

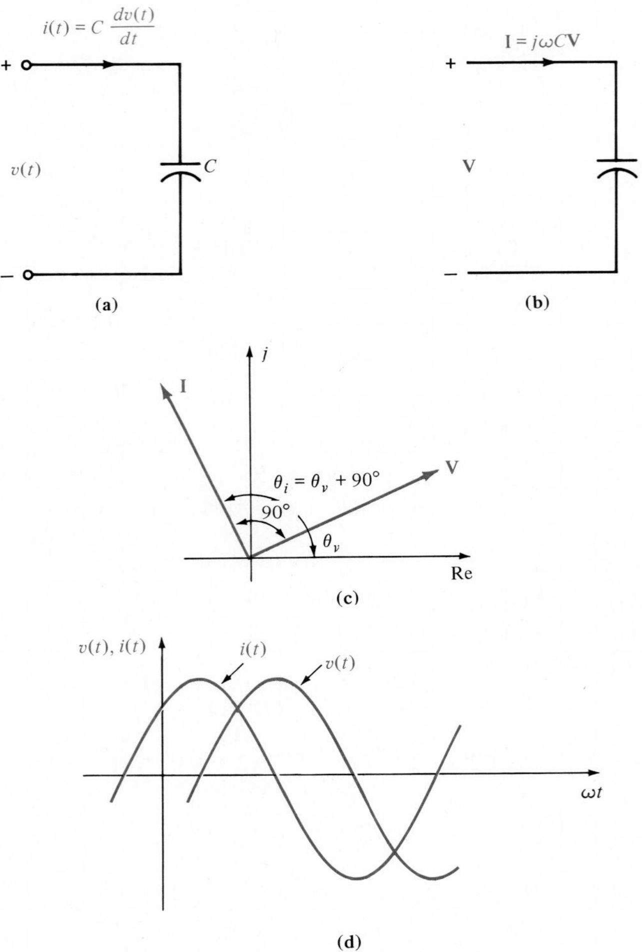

Figure 9.8 Voltage-current relationships for a capacitor.

$$I_M e^{j(\omega t + \theta_i)} = C \frac{d}{dt} V_M e^{j(\omega t + \theta_v)}$$

which reduces to

$$I_M e^{j\theta_i} = j\omega C V_M e^{j\theta_v} \tag{9.28}$$

In phasor notation this equation becomes

$$\mathbf{I} = j\omega C \mathbf{V} \tag{9.29}$$

Equation (9.27), a differential equation in the time domain, has been transformed into Eq. (9.29), an algebraic equation with complex coefficients in the frequency domain. The phasor relationship is shown in Fig. 9.8b. Substituting $j = 1e^{j90°}$ into Eq. (9.28) yields

$$I_M e^{j\theta_i} = \omega C V_M e^{j(\theta_v + 90°)} \tag{9.30}$$

Note that the voltage and current are *90° out of phase*. Equation (9.30) states that the current leads the voltage by 90° or the voltage lags the current by 90°. The phasor diagram and the sinusoidal waveforms for the capacitor circuit are shown in Fig. 9.8c and d, respectively.

EXAMPLE 9.11

The voltage $v(t) = 100 \cos (377t + 15°)$ V is applied to a 100-μF capacitor as shown in Fig. 9.8a. The resultant phasor current is

$$\begin{aligned}\mathbf{I} &= j\omega C(100 \angle 15°) \\ &= (377)(100 \times 10^{-6} \angle 90°)(100 \angle 15°) \\ &= 3.77 \angle 105° \text{ A}\end{aligned}$$

Therefore, the current written as a time function is

$$i(t) = 3.77 \cos (377t + 105°) \text{ A}$$ ■

DRILL EXERCISE

D9.8. The current in a 150-μF capacitor is $\mathbf{I} = 3.6 \angle -145°$ A. If the frequency of the current is 60 Hz, determine the voltage across the capacitor.

It is interesting to note that the phasor concept can be used to derive trigonometric identities. For example, consider the function $A \cos \omega t + B \sin \omega t$. Since

$$e^{j\omega t} = \cos \omega t + j \sin \omega t$$

and

$$-je^{j\omega t} = -j \cos \omega t + \sin \omega t$$

Therefore,

$$\text{Re}\,(-je^{j\omega t}) = \sin \omega t$$

and hence

$$\begin{aligned} A\cos\omega t + B\sin\omega t &= \text{Re}\,(Ae^{j\omega t} - jBe^{j\omega t}) \\ &= \text{Re}\,[(A - jB)e^{j\omega t}] \\ &= \text{Re}\,(D\,\underline{/\theta}\; e^{j\omega t}) \\ &= D\cos(\omega t + \theta) \end{aligned}$$

where the phasor $A - jB$ is written in polar form as

$$A - jB = D\,\underline{/\theta}$$

EXAMPLE 9.12

We wish to convert the function

$$v(t) = 10\cos(377t + 30°) + 5\sin(377t - 20°)$$

to a single sinusoidal function using the phasor concept.

From the discussion above we note that

$$\begin{aligned} D\,\underline{/\theta} &= A - jB \\ &= 10\,\underline{/30°} - j(5\,\underline{/-20°}) \\ &= 10\,\underline{/30°} - 5\,\underline{/70°} \\ &= 6.96\,\underline{/2.47°} \end{aligned}$$

and therefore

$$v(t) = 6.96\cos(377t + 2.47°)$$ ■

EXAMPLE 9.13

We wish to prove the trigonometric identity

$$\cos(\alpha + \beta) = \cos\alpha\cos\beta - \sin\alpha\sin\beta$$

using the phasor concept.

We can write the expression $\cos(\alpha + \beta)$ as

$$\begin{aligned} \cos(\alpha + \beta) &= \text{Re}\,[e^{j(\alpha+\beta)}] \\ &= \text{Re}\,(e^{j\alpha}e^{j\beta}) \\ &= \text{Re}\,[(\cos\alpha + j\sin\alpha)(\cos\beta + j\sin\beta)] \\ &= \cos\alpha\cos\beta - \sin\alpha\sin\beta \end{aligned}$$ ■

This discussion illustrates that our newly found technique of phasors has many uses beyond ac circuit analysis.

9.5 Impedance and Admittance

We have examined each of the circuit elements in the frequency domain on an individual basis. We now wish to treat these passive circuit elements in a more general fashion. We now define *impedance*, which we denote by **Z**, in exactly the same manner in which we defined resistance earlier. Impedance is defined as the ratio of the phasor voltage **V** to the phasor current **I**:

$$\mathbf{Z} = \frac{\mathbf{V}}{\mathbf{I}} \tag{9.31}$$

at the two terminals of the element related to one another by the passive sign convention, as illustrated in Fig. 9.9. Since **V** and **I** are complex, the impedance **Z** is complex and

$$\mathbf{Z} = \frac{V_M \angle \theta_v}{I_M \angle \theta_i} = \frac{V_M}{I_M} \angle \theta_v - \theta_i = Z_M \angle \theta_z \tag{9.32}$$

Since **Z** is the ratio of **V** to **I**, the units of **Z** are ohms. Thus impedance in an ac circuit is analogous to resistance in a dc circuit. In rectangular form, impedance is expressed as

$$\mathbf{Z}(j\omega) = R(\omega) + jX(\omega) \tag{9.33}$$

where $R(\omega)$ is the real, or resistive, component and $X(\omega)$ is the imaginary, or reactive, component. In general, we simply refer to R as the resistance and X as the reactance. It is important to note that R and X are real functions of ω and therefore $\mathbf{Z}(j\omega)$ is frequency dependent. Equation (9.33) clearly indicates that **Z** is a complex number; however, it is not a phasor.

Equations (9.32) and (9.33) indicate that

$$Z_M \angle \theta_z = R + jX \tag{9.34}$$

Therefore,

$$\begin{aligned} Z_M &= \sqrt{R^2 + X^2} \\ \theta_z &= \tan^{-1}\frac{X}{R} \end{aligned} \tag{9.35}$$

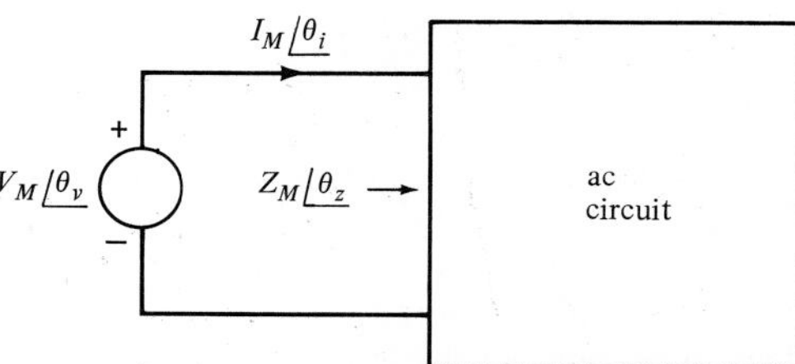

Figure 9.9 General impedance relationship.

Table 9.2 Passive Element Impedances

Passive Element	Impedance
R	$\mathbf{Z} = R$
L	$\mathbf{Z} = j\omega L = jX_L = \omega L \angle 90°,\ X_L = \omega L$
C	$\mathbf{Z} = \dfrac{1}{j\omega C} = jX_C = -\dfrac{1}{\omega C} \angle 90°,\ X_C = -\dfrac{1}{\omega C}$

where

$$R = Z_M \cos \theta_z$$

$$X = Z_M \sin \theta_z$$

For the individual passive elements the impedance is as shown in Table 9.2. However, just as it was advantageous to know how to determine the equivalent resistance of a series and/or parallel combination of resistors in a dc circuit, we now want to learn how to determine the equivalent impedance in an ac circuit when the passive elements are interconnected. The determination of equivalent impedance is based on KCL and KVL. Therefore, we must see if these laws are valid in the frequency domain. Suppose, for example, that a circuit is driven by voltage sources of the form $V_{M_m} \cos (\omega t + \theta_m)$. Then every steady-state current in the network will be of the form $I_{M_k} \cos (\omega t + \phi_k)$. At any node in the circuit, KCL, written in the time domain, is

$$i_1(t) + i_2(t) + \cdots + i_n(t) = 0$$

or

$$I_{M_1} \cos (\omega t + \phi_1) + I_{M_2} \cos (\omega t + \phi_2) + \cdots + I_{M_n} \cos (\omega t + \phi_n) = 0$$

From our previous work we can immediately apply the phasor transformation to the equation above to obtain

$$\mathbf{I}_1 + \mathbf{I}_2 + \cdots + \mathbf{I}_n = 0$$

However, this equation is simply KCL in the frequency domain. In a similar manner we can show that KVL applies in the frequency domain. Using the fact that KCL and KVL are valid in the frequency domain, we can show, as was done in Chapter 2 for resistors, that impedances can be combined using the same rules that we established for resistor combinations, that is, if $\mathbf{Z}_1, \mathbf{Z}_2, \mathbf{Z}_3, \ldots, \mathbf{Z}_n$ are connected in series, the equivalent impedance $\mathbf{Z}_s$ is

$$\mathbf{Z}_s = \mathbf{Z}_1 + \mathbf{Z}_2 + \mathbf{Z}_3 + \cdots + \mathbf{Z}_n \tag{9.36}$$

and if $\mathbf{Z}_1, \mathbf{Z}_2, \mathbf{Z}_3, \ldots, \mathbf{Z}_n$ are connected in parallel, the equivalent impedance is

$$\frac{1}{\mathbf{Z}_p} = \frac{1}{\mathbf{Z}_1} + \frac{1}{\mathbf{Z}_2} + \frac{1}{\mathbf{Z}_3} + \cdots + \frac{1}{\mathbf{Z}_n} \tag{9.37}$$

EXAMPLE 9.14

Determine the equivalent impedance of the network shown in Fig. 9.10 if the frequency is $f = 60$ Hz. Then compute the current $i(t)$ if the voltage source is $v(t) = 50 \cos (\omega t + 30°)$ V. Finally, calculate the equivalent impedance if the frequency is $f = 400$ Hz.

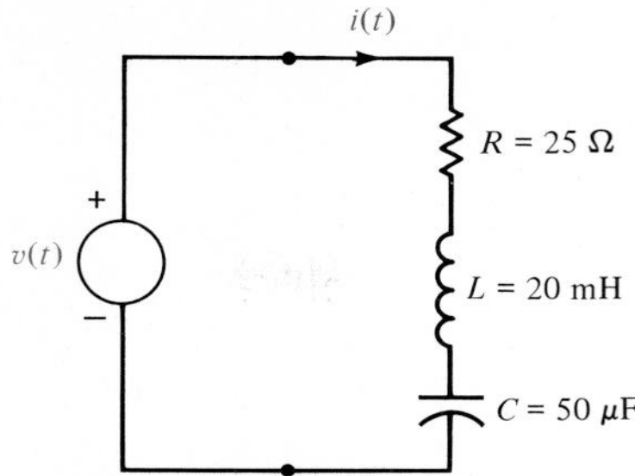

Figure 9.10 Series ac circuit.

The impedances of the individual elements at 60 Hz are

$$\mathbf{Z}_R = 25\ \Omega$$

$$\mathbf{Z}_L = j\omega L = j(2\pi \times 60)(20 \times 10^{-3}) = j7.54\ \Omega$$

$$\mathbf{Z}_C = \frac{-j}{\omega C} = \frac{-j}{(2\pi \times 60)(50 \times 10^{-6})} = -j53.05\ \Omega$$

Since the elements are in series,

$$\mathbf{Z} = \mathbf{Z}_R + \mathbf{Z}_L + \mathbf{Z}_C$$

$$= 25 - j45.51\ \Omega$$

the current in the circuit is given by

$$\mathbf{I} = \frac{\mathbf{V}}{\mathbf{Z}} = \frac{50\,\underline{/30^\circ}}{25 - j45.51} = \frac{50\,\underline{/30^\circ}}{51.92\,\underline{/-61.22^\circ}} = 0.96\,\underline{/91.22^\circ}\ \text{A}$$

or in the time domain, $i(t) = 0.96 \cos (377t + 91.22^\circ)$ A.

If the frequency is 400 Hz, the impedance of each element is

$$\mathbf{Z}_R = 25\ \Omega$$

$$\mathbf{Z}_L = j\omega L = j50.27\ \Omega$$

$$\mathbf{Z}_C = \frac{-j}{\omega C} = -j7.96\ \Omega$$

The total impedance is then

$$\mathbf{Z} = 25 + j42.31 = 49.14\,\underline{/59.42^\circ}\ \Omega$$

It is important to note that at the frequency $f = 60$ Hz, the reactance of the circuit is capacitive, that is, if the impedance is written as $R + jX$, $X < 0$; however, at $f = 400$ Hz the reactance is inductive since $X > 0$. ■

EXAMPLE 9.15

Determine the equivalent impedance of the circuit in Fig. 9.11 at 60 Hz. Compute the voltage $\mathbf{V}_s$ if the current $\mathbf{I}$ is known to be $\mathbf{I} = 0.5\,\underline{/-22.98^\circ}$ A.

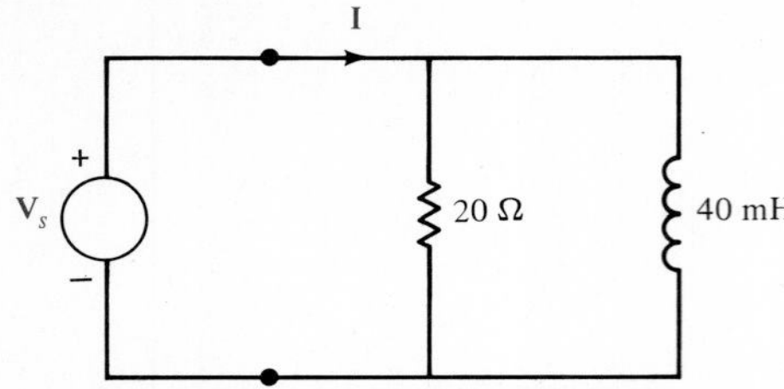

Figure 9.11 Parallel ac circuit.

The individual impedances are

$$\mathbf{Z}_R = 20\ \Omega$$

$$\mathbf{Z}_L = j\omega L = j(2\pi \times 60)(40 \times 10^{-3}) = j15.08\ \Omega$$

The equivalent impedance of the parallel elements is

$$\mathbf{Z} = \frac{\mathbf{Z}_R \mathbf{Z}_L}{\mathbf{Z}_R + \mathbf{Z}_L}$$

Using the values for $\mathbf{Z}_R$ and $\mathbf{Z}_L$, we obtain

$$\begin{aligned}\mathbf{Z} &= \frac{(20)(j15.08)}{20 + j15.08} \\ &= \frac{301.60\ \underline{/90^\circ}}{25.05\ \underline{/37.02^\circ}} = 7.25 + j9.61\ \Omega\end{aligned}$$

The reader should note carefully that the resistive component of the impedance is *not* the 20-Ω resistor. Recall that $\mathbf{Z}(j\omega) = R(\omega) + jx(\omega)$, and therefore in general the resistive component of an equivalent impedance is a function of ω.

If the current $\mathbf{I} = 0.5\ \underline{/-22.98^\circ}$ A, the voltage $\mathbf{V}_s$ is

$$\begin{aligned}\mathbf{V}_s &= \mathbf{IZ} \\ &= (0.5\ \underline{/-22.98^\circ})(12.04\ \underline{/52.98^\circ}) \\ &= 6.02\ \underline{/30^\circ}\ \text{V}\end{aligned}$$

■

DRILL EXERCISE

D9.9. Compute the voltage $v(t)$ in the network in Fig. D9.9.

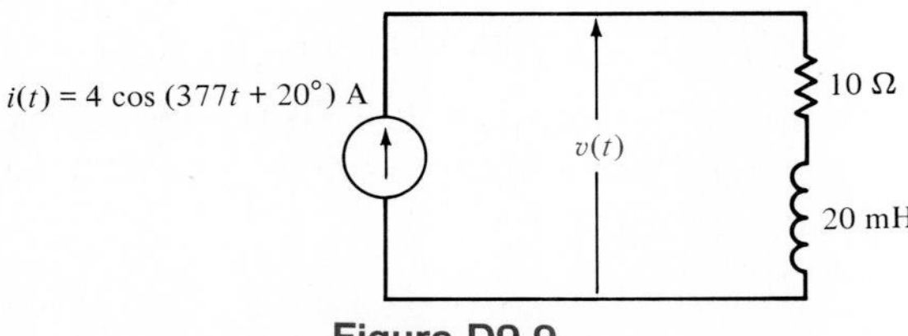

Figure D9.9

D9.10. Find the current $i(t)$ in the network in Fig. D9.10.

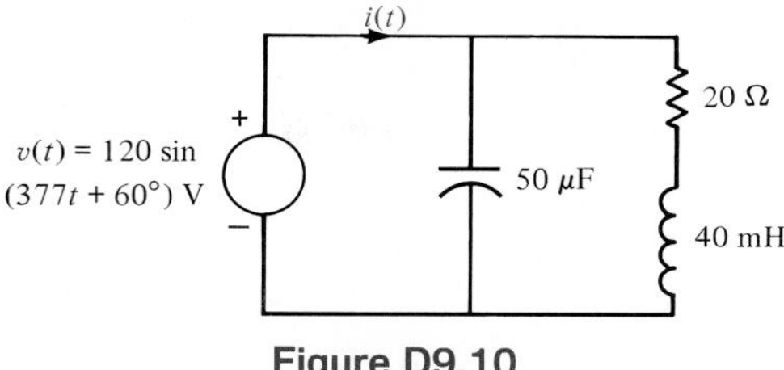

Figure D9.10

Another quantity that is very useful in the analysis of ac circuits is *admittance*, which is the reciprocal of impedance, that is,

$$\mathbf{Y} = \frac{1}{\mathbf{Z}} = \frac{\mathbf{I}}{\mathbf{V}} \tag{9.38}$$

The units of **Y** are siemens, and this quantity is analogous to conductance in resistive dc circuits. Since **Z** is a complex number, **Y** is also a complex number.

$$\mathbf{Y} = Y_M \angle \theta_y \tag{9.39}$$

which is written in rectangular form as

$$\mathbf{Y} = G + jB \tag{9.40}$$

where G and B are called *conductance* and *susceptance*, respectively. Because of the relationship between **Y** and **Z**, we can express the components of one quantity as a function of the components of the other.

$$G + jB = \frac{1}{R + jX} \tag{9.41}$$

Rationalizing the right-hand side of this equation yields

$$G + jB = \frac{R - jX}{R^2 + X^2}$$

and therefore

$$G = \frac{R}{R^2 + X^2}, \qquad B = \frac{-X}{R^2 + X^2} \tag{9.42}$$

In a similar manner we can show that

$$R = \frac{G}{G^2 + B^2}$$
$$X = \frac{-B}{G^2 + B^2} \tag{9.43}$$

It is very important to note that in general R and G are *not* reciprocals of one another.

The same is true for X and B. The purely resistive case is an exception. In the purely reactive case the quantities are negative reciprocals of one another.

The admittance of the individual passive elements is

$$\begin{aligned} \mathbf{Y}_R &= \frac{1}{R} = G \\ \mathbf{Y}_L &= \frac{1}{j\omega L} = -\frac{1}{\omega L}\angle 90^\circ \\ \mathbf{Y}_C &= j\omega C = \omega C \angle 90^\circ \end{aligned} \tag{9.44}$$

Once again, since KCL and KVL are valid in the frequency domain, we can show, using the same approach outlined in Chapter 2 for conductance in resistive circuits, that the rules for combining admittances are the same as those for combining conductances; that is, if $\mathbf{Y}_1, \mathbf{Y}_2, \mathbf{Y}_3, \ldots, \mathbf{Y}_n$ are connected in parallel, the equivalent admittance is

$$\mathbf{Y}_p = \mathbf{Y}_1 + \mathbf{Y}_2 + \cdots + \mathbf{Y}_n \tag{9.45}$$

and if $\mathbf{Y}_1, \mathbf{Y}_2, \ldots, \mathbf{Y}_n$ are connected in series, the equivalent admittance is

$$\frac{1}{\mathbf{Y}_s} = \frac{1}{\mathbf{Y}_1} + \frac{1}{\mathbf{Y}_2} + \cdots + \frac{1}{\mathbf{Y}_n} \tag{9.46}$$

EXAMPLE 9.16

If the equivalent impedance of a network is $\mathbf{Z} = 10\angle 30^\circ\ \Omega$, compute the equivalent admittance and draw the equivalent circuits.

Converting $\mathbf{Z}$ to rectangular form gives us

$$\begin{aligned} \mathbf{Z} &= 10\angle 30^\circ \\ &= 8.66 + j5.0\ \Omega \end{aligned}$$

The admittance is then

$$\begin{aligned} \mathbf{Y} &= \frac{1}{\mathbf{Z}} \\ &= 0.1\angle -30^\circ \\ &= 0.0866 - j0.05\ \text{S} \end{aligned}$$

The equivalent circuits are shown in Fig. 9.12. A quick check of the calculations can be made by converting $\mathbf{Y}_R$ to $\mathbf{Z}_R$ and $\mathbf{Y}_L$ to $\mathbf{Z}_L$, and then determining the impedance of the parallel combination; for example,

$$\mathbf{Z}_R = \frac{1}{\mathbf{Y}_R} = \frac{1}{0.0866} = 11.55\ \Omega$$

$$\mathbf{Z}_L = \frac{1}{\mathbf{Y}_L} = \frac{1}{-j0.05} = j20\ \Omega$$

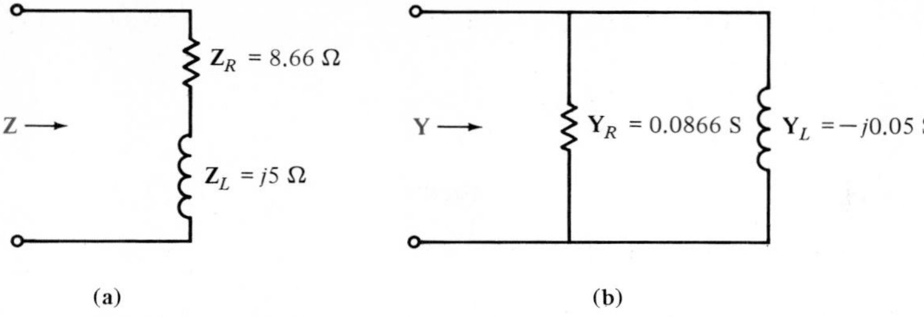

Figure 9.12 Equivalent circuits for impedance and admittance.

and then

$$\mathbf{Z} = \frac{\mathbf{Z}_R\mathbf{Z}_L}{\mathbf{Z}_R + \mathbf{Z}_L}$$

$$= \frac{(11.55)(j20)}{11.55 + j20}$$

$$= \frac{231 \angle 90^\circ}{23.1 \angle 60^\circ} = 10 \angle 30^\circ\ \Omega$$

which is our original impedance. ■

EXAMPLE 9.17

Calculate the equivalent admittance $\mathbf{Y}_p$ for the network in Fig. 9.13 and use it to determine the current $\mathbf{I}$ if $\mathbf{V}_s = 60 \angle 45^\circ$ V.

From Fig. 9.13 we note that

$$\mathbf{Y}_R = \frac{1}{\mathbf{Z}_R} = \frac{1}{2}\ \text{S}$$

$$\mathbf{Y}_L = \frac{1}{\mathbf{Z}_L} = \frac{-j}{4}\ \text{S}$$

Therefore,

$$\mathbf{Y}_p = \frac{1}{2} - j\frac{1}{4}\ \text{S}$$

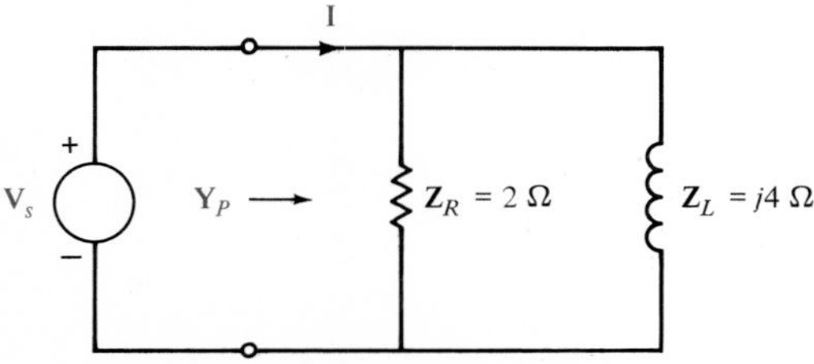

Figure 9.13 Example parallel circuit.

and hence

$$\begin{aligned}\mathbf{I} &= \mathbf{Y}_p\mathbf{V}_s \\ &= \left(\frac{1}{2} - j\frac{1}{4}\right)(60 \angle 45^\circ) \\ &= 33.6 \angle 18.43^\circ \text{ A}\end{aligned}$$

A quick check of this result can be made using the equivalent impedance

$$\begin{aligned}\mathbf{Z}_p &= \frac{(2)(j4)}{2 + j4} \\ &= \frac{8j}{2 + j4} \\ &= \frac{8 \angle 90^\circ}{4.47 \angle 63.43^\circ} \\ &= 1.79 \angle 26.57^\circ \ \Omega\end{aligned}$$

$$\begin{aligned}\mathbf{I} &= \frac{\mathbf{V}_s}{\mathbf{Z}_p} \\ &= \frac{60 \angle 45^\circ}{1.79 \angle 26.57^\circ} \\ &= 33.6 \angle 18.43^\circ \text{ A}\end{aligned}$$

which of course checks with our previous result. ■

EXAMPLE 9.18

Calculate the equivalent impedance for the circuit in Fig. 9.14 and use this value to determine the voltage $\mathbf{V}_s$ if the current is $\mathbf{I} = 10 \angle 30^\circ$ A.

As shown in Fig. 9.14,

$$\begin{aligned}\mathbf{Z}_s &= \mathbf{Z}_R + \mathbf{Z}_L + \mathbf{Z}_C \\ &= 4 - j2 \ \Omega\end{aligned}$$

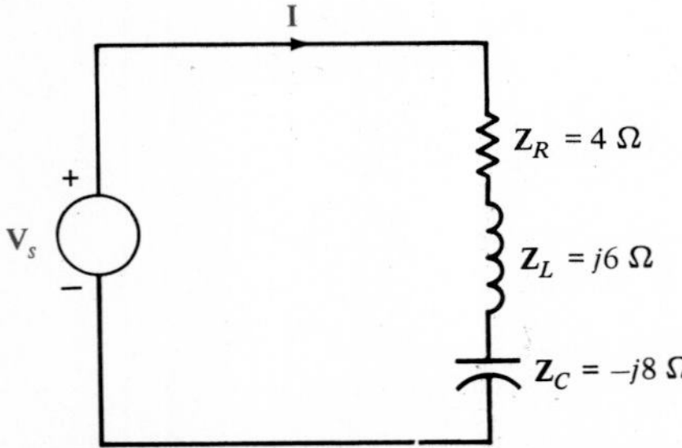

Figure 9.14 Example series circuit.

Then

$$\begin{aligned}\mathbf{V}_s &= \mathbf{I}\mathbf{Z}_s \\ &= (4 - j2)(10\,\underline{/30^\circ}) \\ &= (4.47\,\underline{/-26.57^\circ})(10\,\underline{/30^\circ}) \\ &= 44.7\,\underline{/3.43^\circ}\text{ V}\end{aligned}$$

■

DRILL EXERCISE

D9.11. Find the current **I** in the network in Fig. D9.11.

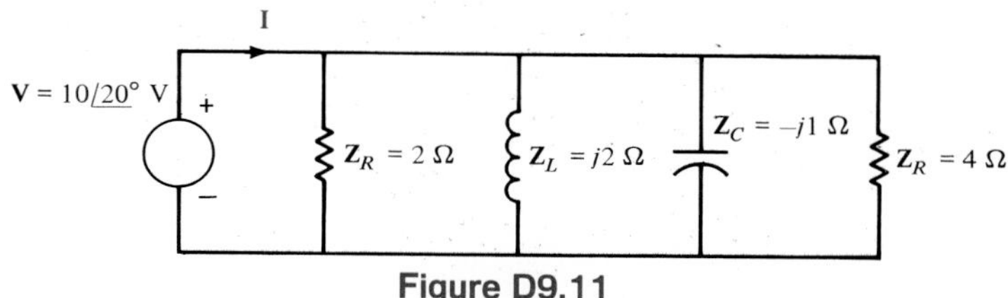

Figure D9.11

D9.12. Determine the voltage across the current source in the network in Fig. D9.12.

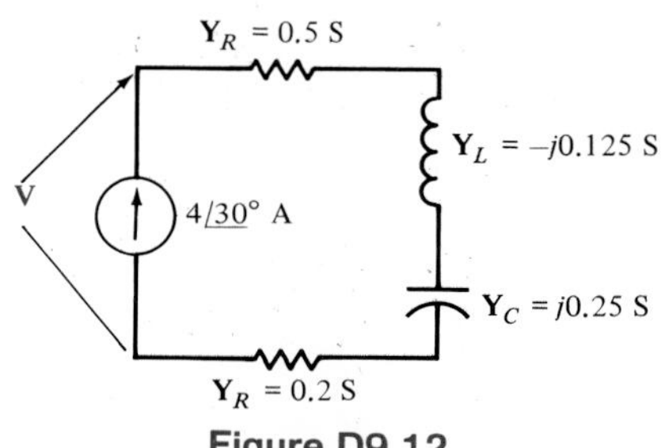

Figure D9.12

As a prelude to our analysis of more general ac circuits, let us examine the techniques for computing the impedance or admittance of circuits in which numerous passive elements are interconnected. The following example illustrates that our technique is based simply on the repeated application of Eq. (9.36), (9.37), (9.45), and (9.46), and is analogous to our earlier computations of equivalent resistance.

EXAMPLE 9.19

Consider the network shown in Fig. 9.15a. The impedance of each element is given in the figure. We wish to calculate the equivalent impedance of the network $\mathbf{Z}_{eq}$ at terminals *A-B*.

The equivalent impedance $\mathbf{Z}_{eq}$ could be calculated in a variety of ways; we could use only impedances, or only admittances, or a combination of the two. We will use two approaches to illustrate the various techniques involved. We begin by noting that the

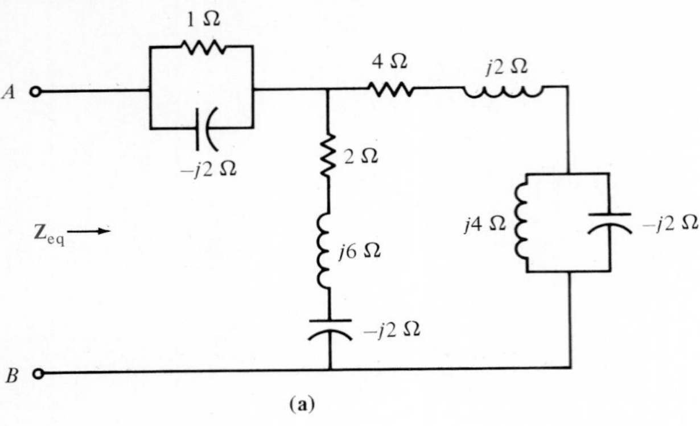

(a)

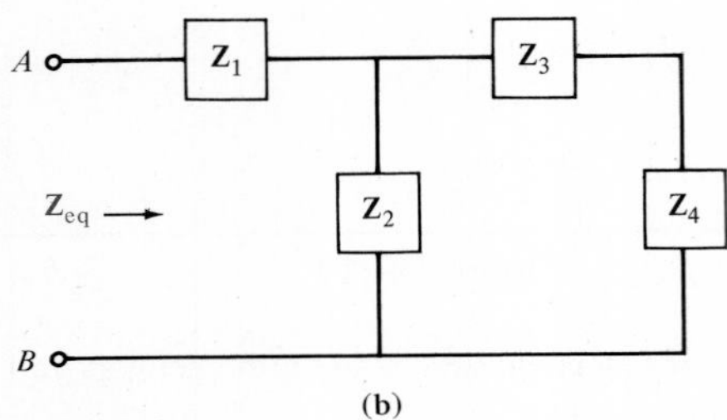

(b)

Figure 9.15 Example circuits for determining equivalent impedance.

circuit in Fig. 9.15a can be represented by the circuit in Fig. 9.15b. Using strictly an impedance approach, we note that

$$\begin{aligned}\mathbf{Z}_4 &= \frac{(j4)(-j2)}{j4 - j2}\\ &= -j4\ \Omega\end{aligned}$$

Since $\mathbf{Z}_3 = 4 + j2\ \Omega$, then $\mathbf{Z}_{34}$, which is the combined impedance of $\mathbf{Z}_3$ and $\mathbf{Z}_4$, is

$$\begin{aligned}\mathbf{Z}_{34} &= 4 + j2 - j4\\ &= 4 - j2\ \Omega\end{aligned}$$

The figure indicates that

$$\begin{aligned}\mathbf{Z}_2 &= 2 + j6 - j2\\ &= 2 + j4\ \Omega\end{aligned}$$

and therefore $\mathbf{Z}_{234}$, which is the combined impedance of $\mathbf{Z}_2$, $\mathbf{Z}_3$, and $\mathbf{Z}_4$, is

$$\begin{aligned}\mathbf{Z}_{234} &= \frac{(2 + j4)(4 - j2)}{(2 + j4) + (4 - j2)}\\ &= \frac{16 + j12}{6 + j2}\ \Omega\end{aligned}$$

If we multiply numerator and denominator by $6 - j2$ and perform the indicated algebra, we obtain

$$\mathbf{Z}_{234} = 3 + j1\ \Omega$$

From the figure

$$\mathbf{Z}_1 = \frac{(1)(-j2)}{1 - j2}$$

$$= \frac{4}{5} - j\frac{2}{5}\ \Omega$$

Therefore, if $\mathbf{Z}_{1234}$ is the combined impedance of $\mathbf{Z}_1$, $\mathbf{Z}_2$, $\mathbf{Z}_3$, and $\mathbf{Z}_4$, then

$$\mathbf{Z}_{\text{eq}} = \mathbf{Z}_{1234} = 3 + j1 + \frac{4}{5} - j\frac{2}{5}$$

$$= 3.8 + j0.6\ \Omega$$

We could obtain the same result by using both impedances and admittances, as illustrated below.

$$\mathbf{Y}_4 = \mathbf{Y}_L + \mathbf{Y}_C$$

$$= \frac{1}{j4} + \frac{1}{-j2}$$

$$= j\frac{1}{4}\ \text{S}$$

Therefore,

$$\mathbf{Z}_4 = -j4\ \Omega$$

Now

$$\mathbf{Z}_{34} = \mathbf{Z}_3 + \mathbf{Z}_4$$

$$= (4 + j2) + (-j4)$$

$$= 4 - j2\ \Omega$$

and hence

$$\mathbf{Y}_{34} = \frac{1}{\mathbf{Z}_{34}}$$

$$= \frac{1}{4 - j2}$$

$$= 0.20 + j0.10\ \text{S}$$

Since

$$\mathbf{Z}_2 = 2 + j6 - j2$$

$$= 2 + j4\ \Omega$$

then

$$\mathbf{Y}_2 = \frac{1}{2 + j4}$$
$$= 0.10 - j0.20 \text{ S}$$
$$\mathbf{Y}_{234} = \mathbf{Y}_2 + \mathbf{Y}_{34}$$
$$= 0.30 - j0.10 \text{ S}$$

The reader should note carefully our approach—we are adding impedances in series and adding admittances in parallel.

From $\mathbf{Y}_{234}$ we can compute $\mathbf{Z}_{234}$ as

$$\mathbf{Z}_{234} = \frac{1}{\mathbf{Y}_{234}}$$
$$= \frac{1}{0.30 - j0.10}$$
$$= 3 + j1 \ \Omega$$

Now

$$\mathbf{Y}_1 = \mathbf{Y}_R + \mathbf{Y}_C$$
$$= \frac{1}{1} + \frac{1}{-j2}$$
$$= 1 + j\frac{1}{2} \text{ S}$$

and then

$$\mathbf{Z}_1 = \frac{1}{1 + j\frac{1}{2}}$$
$$= 0.8 - j0.4 \ \Omega$$

Therefore,

$$\mathbf{Z}_{\text{eq}} = \mathbf{Z}_1 + \mathbf{Z}_{234}$$
$$= 0.8 - j0.4 + 3 + j1$$
$$= 3.8 + j0.6 \ \Omega$$

which is exactly what we obtained using an impedance approach. ■

DRILL EXERCISE

D9.13. Compute the impedance $\mathbf{Z}_T$ in the network in Fig. D9.13.

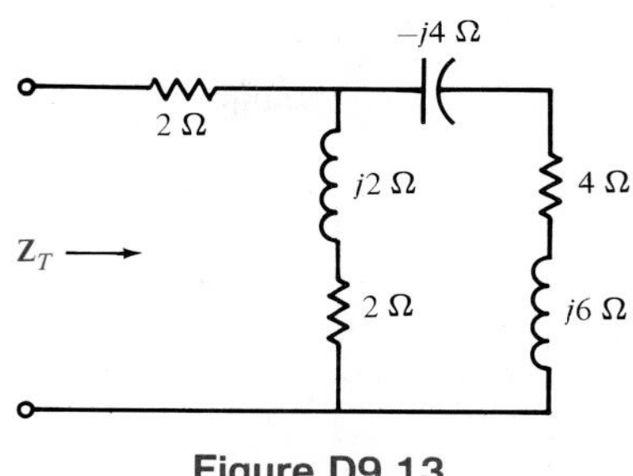

Figure D9.13

Impedance and admittance are functions of frequency, and therefore their values change as the frequency changes. These changes in $\mathbf{Z}$ and $\mathbf{Y}$ have a resultant effect on the current-voltage relationships in a network. This impact of changes in frequency on circuit parameters can be easily seen via a phasor diagram. The following examples will serve to illustrate these points.

EXAMPLE 9.20

Consider the circuit shown in Fig. 9.16. The pertinent variables are labeled on the figure. For convenience in forming a phasor diagram we select $\mathbf{V}$ as a reference phasor and arbitrarily assign it a 0° phase angle. We will, therefore, measure all currents with respect to this phasor. We suffer no loss of generality by assigning $\mathbf{V}$ a 0° phase angle, since if it is actually, for example, 30°, we will simply rotate the entire phasor diagram by 30° because all the currents are measured with respect to this phasor.

KCL at the upper node in the circuit is

$$\mathbf{I}_s = \mathbf{I}_R + \mathbf{I}_L + \mathbf{I}_C = \frac{\mathbf{V}}{R} + \frac{\mathbf{V}}{j\omega L} + \frac{\mathbf{V}}{1/j\omega C}$$

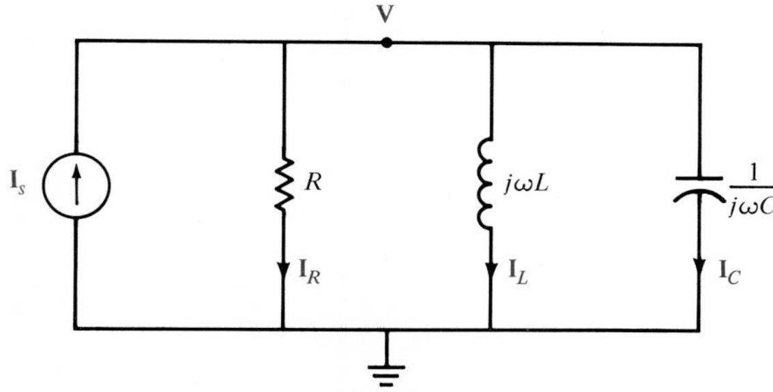

Figure 9.16 Example parallel circuit.

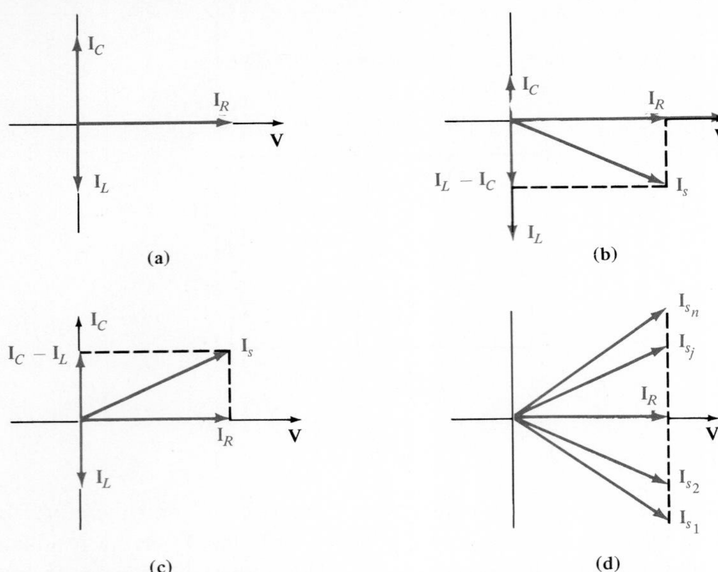

Figure 9.17 Phasor diagrams for the circuit in Fig. 9.16.

Since $\mathbf{V} = V_M \underline{/0°}$, then

$$\mathbf{I}_s = \frac{V_M \underline{/0°}}{R} + \frac{V_M \underline{/-90°}}{\omega L} + V_M \omega C \underline{/90°}$$

The phasor diagram that illustrates the phase relationship between $\mathbf{V}$, $\mathbf{I}_R$, $\mathbf{I}_L$, and $\mathbf{I}_C$ is shown in Fig. 9.17a. For small values of ω such that the magnitude of $\mathbf{I}_L$ is greater than that of $\mathbf{I}_C$, the phasor diagram for the currents is shown in Fig. 9.17b. In the case of large values of ω, that is, those for which $\mathbf{I}_C$ is greater than $\mathbf{I}_L$, the phasor diagram for the currents is shown in Fig. 9.17c. Note that as ω increases, the phasor $\mathbf{I}_s$ moves from $\mathbf{I}_{s_1}$ to $\mathbf{I}_{s_n}$ along a locus of points specified by the dashed line shown in Fig. 9.17d.

Note that $\mathbf{I}_s$ is in phase with $\mathbf{V}$ when $\mathbf{I}_C = \mathbf{I}_L$ or, in other words, when $\omega L = 1/\omega C$. Hence the node voltage $\mathbf{V}$ is in phase with the current source $\mathbf{I}_s$ when

$$\omega = \frac{1}{\sqrt{LC}}$$

This can also be seen from the KCL equation

$$\mathbf{I}_s = \left[\frac{1}{R} + j\left(\omega C - \frac{1}{\omega L}\right)\right]\mathbf{V}$$

■

EXAMPLE 9.21

Let us examine the series circuit shown in Fig. 9.18a. KVL for this circuit is of the form

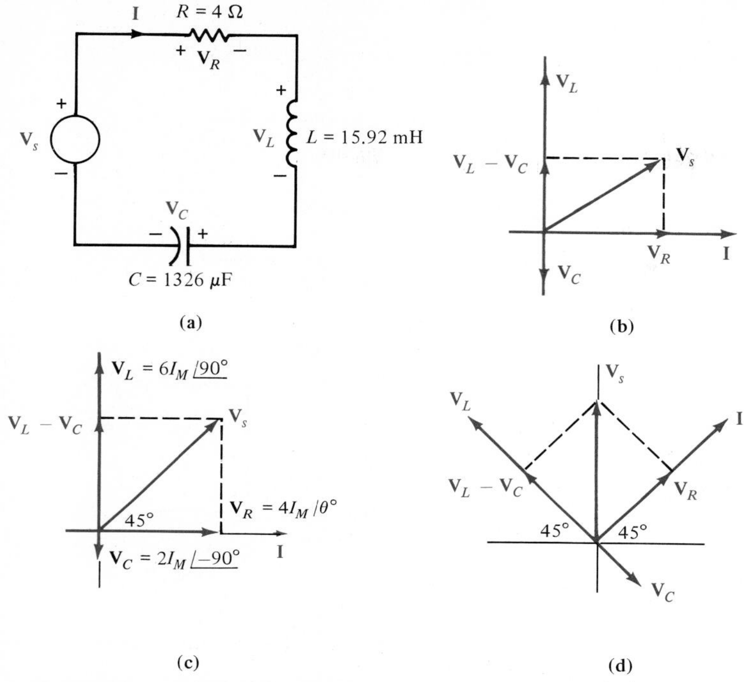

Figure 9.18 Series circuit and certain specific phasor diagrams.

$$\mathbf{V}_s = \mathbf{V}_R + \mathbf{V}_L + \mathbf{V}_C$$

$$= \mathbf{I}R + \omega L\mathbf{I} \angle 90^\circ + \frac{\mathbf{I}}{\omega C} \angle -90^\circ$$

If we select **I** as a reference phasor so that $\mathbf{I} = I_M \angle 0^\circ$, then if $\omega L I_M > I_M/\omega C$, the phasor diagram will be of the form shown in Fig. 9.18b. Specifically, if $\omega = 377$ rad/sec or r/s (i.e., $f = 60$ Hz), then $\omega L = 6$ and $1/\omega C = 2$. Under these conditions the phasor diagram is as shown in Fig. 9.18c. If, however,

$$\mathbf{V}_s(t) = 12\sqrt{2} \cos (377t + 90^\circ) \text{ V}$$

then

$$\mathbf{I} = \frac{\mathbf{V}}{\mathbf{Z}} = \frac{12\sqrt{2} \angle 90^\circ}{4 + j6 - j2}$$

$$= \frac{12\sqrt{2} \angle 90^\circ}{4\sqrt{2} \angle 45^\circ}$$

$$= 3 \angle 45^\circ \text{ A}$$

and the entire phasor diagram, as shown in Fig. 9.18b and c, is rotated 45° as shown in Fig. 9.18d. ■

DRILL EXERCISE

D9.14. Draw a phasor diagram illustrating all currents and voltages for the network in Fig. D9.14.

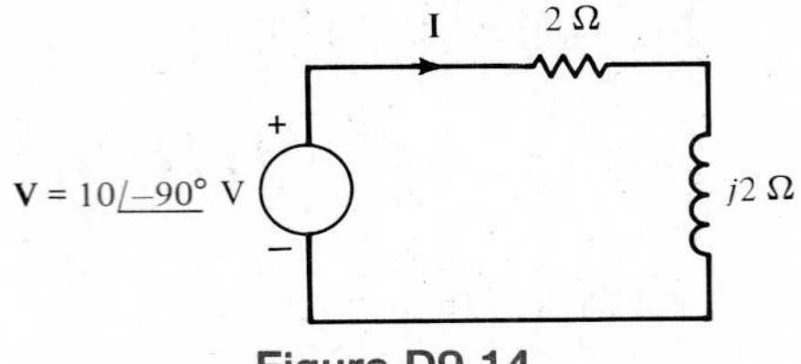

Figure D9.14

D9.15. Draw a phasor diagram illustrating all currents and voltages for the network in Fig. D9.15.

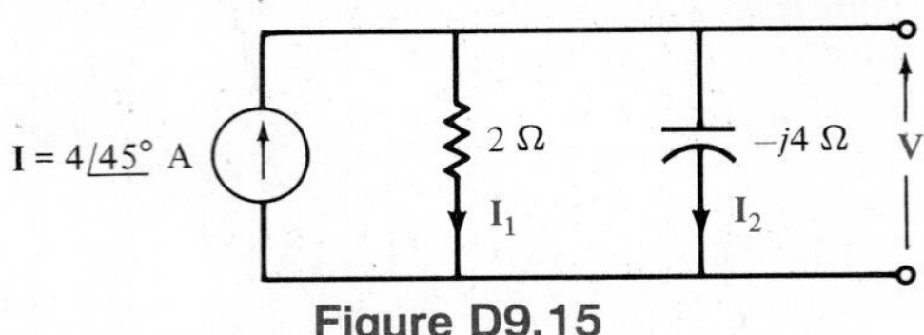

Figure D9.15

9.6

Basic Analysis Using Kirchhoff's Laws

We have shown that Kirchhoff's laws apply in the frequency domain, and therefore they can be used to compute steady-state voltages and currents in ac circuits. This approach involves expressing these voltages and currents as phasors, and once this is done, the ac steady-state analysis employing phasor equations is performed in an identical fashion to that used in the dc analysis of resistive circuits. Complex number algebra is the vehicle that is used for the mathematical manipulation of the phasor equations, which, of course, have complex coefficients. We will begin by illustrating that the techniques we have applied in the solution of dc resistive circuits are valid in ac circuit analysis also—the only difference being that in steady-state ac circuit analysis the algebraic phasor equations have complex coefficients. The SPICE circuit analysis program can also be applied to analyze the ac circuits which we will present here and in Chapter 10.

EXAMPLE 9.22

We wish to calculate all the voltages and currents in the circuit shown in Fig. 9.19. Our approach will be as follows. We will calculate the total impedance seen by the source

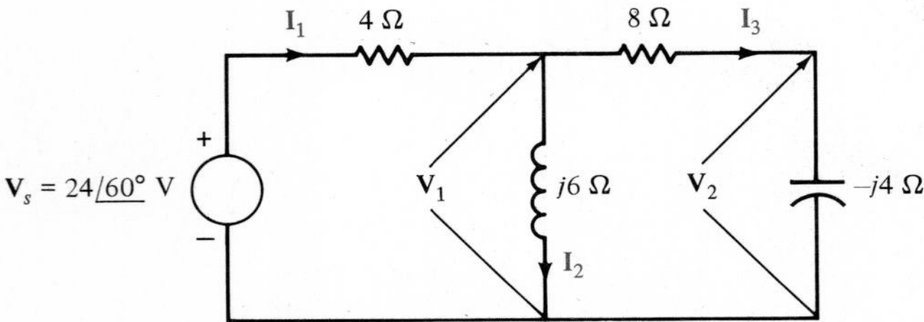

Figure 9.19 Example ac circuit.

$\mathbf{V}_s$. Then we will use this to determine $\mathbf{I}_1$. Knowing $\mathbf{I}_1$, we can compute $\mathbf{V}_1$ using KVL. Knowing $\mathbf{V}_1$, we can compute $\mathbf{I}_2$ and $\mathbf{I}_3$, and so on.

The total impedance seen by the source $\mathbf{V}_s$ is

$$\begin{aligned}\mathbf{Z}_{eq} &= 4 + \frac{(j6)(8 - j4)}{j6 + 8 - j4} \\ &= 4 + \frac{24 + j48}{8 + j2} \\ &= 4 + 4.24 + j4.94 \\ &= 9.61 \underline{/30.94^\circ}\ \Omega\end{aligned}$$

Then

$$\begin{aligned}\mathbf{I}_1 &= \frac{\mathbf{V}_s}{\mathbf{Z}_{eq}} = \frac{24 \underline{/60^\circ}}{9.61 \underline{/30.94^\circ}} \\ &= 2.5 \underline{/29.06^\circ}\ \text{A}\end{aligned}$$

$\mathbf{V}_1$ can be determined using KVL:

$$\begin{aligned}\mathbf{V}_1 &= \mathbf{V}_s - 4\mathbf{I}_1 \\ &= 24 \underline{/60^\circ} - 10 \underline{/29.06^\circ} \\ &= 3.26 + j15.92 = 16.25 \underline{/78.43^\circ}\ \text{V}\end{aligned}$$

Note that $\mathbf{V}_1$ could also be computed via voltage division:

$$\mathbf{V}_1 = \frac{\mathbf{V}_s \dfrac{(j6)(8 - j4)}{j6 + 8 - j4}}{4 + \dfrac{(j6)(8 - j4)}{j6 + 8 - j4}}\ \text{V}$$

which from our previous calculation is

$$\begin{aligned}\mathbf{V}_1 &= \frac{(24 \underline{/60^\circ})(6.51 \underline{/49.39^\circ})}{9.61 \underline{/30.49^\circ}} \\ &= 16.25 \underline{/78.43^\circ}\ \text{V}\end{aligned}$$

Knowing $\mathbf{V}_1$, we can calculate both $\mathbf{I}_2$ and $\mathbf{I}_3$:

$$\begin{aligned}\mathbf{I}_2 &= \frac{\mathbf{V}_1}{j6} = \frac{16.25 \underline{/78.43^\circ}}{6 \underline{/90^\circ}} \\ &= 2.71 \underline{/-11.56^\circ}\ \text{A}\end{aligned}$$

and

$$\begin{aligned}\mathbf{I}_3 &= \frac{\mathbf{V}_1}{8 - j4} \\ &= 1.82 \underline{/105^\circ}\ \text{A}\end{aligned}$$

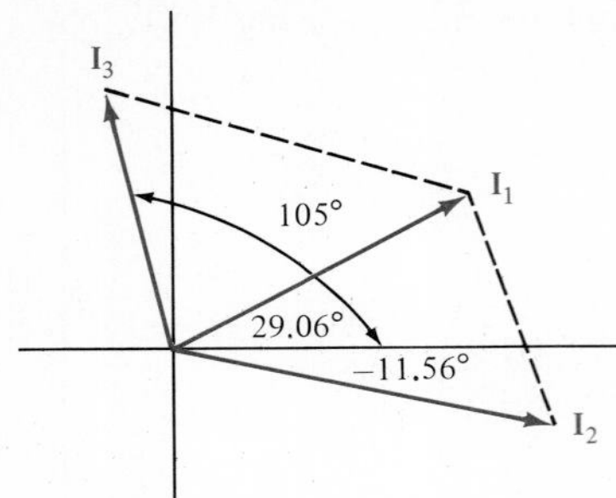

Figure 9.20 Phasor diagram for the currents in Example 9.22

$\mathbf{I}_2$ and $\mathbf{I}_3$ could have been calculated by current division. For example, $\mathbf{I}_2$ could be determined by

$$\mathbf{I}_2 = \frac{\mathbf{I}_1(8 - j4)}{8 - j4 + j6}$$

$$= \frac{(2.5 \angle 29.06°)(8.94 \angle -26.57°)}{8 + j2}$$

$$= 2.71 \angle -11.56° \text{ A}$$

Finally, $\mathbf{V}_2$ can be computed as

$$\mathbf{V}_2 = \mathbf{I}_3(-j4)$$

$$= 7.28 \angle 15° \text{ V}$$

This value could also have been computed by voltage division. The phasor diagram for the currents $\mathbf{I}_1$, $\mathbf{I}_2$, and $\mathbf{I}_3$ is shown in Fig. 9.20 and is an illustration of KCL. ■

EXAMPLE 9.23

If the current in the 4-Ω resistor in Fig. 9.21 is known to be $\mathbf{I}_4 = 3 \angle 45°$ A, calculate the value of the source voltage $\mathbf{V}_s$.

Since $\mathbf{I}_4 = 3 \angle 45°$ A, then

$$\mathbf{V}_2 = 4\mathbf{I}_4$$

$$= 12 \angle 45° \text{ V}$$

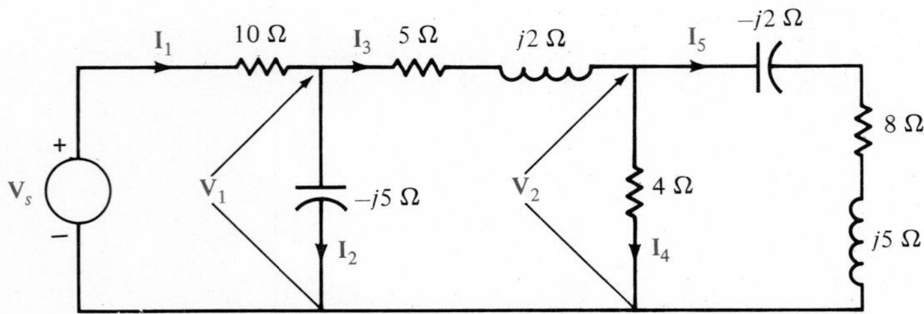

Figure 9.21 Example ac circuit.

Hence

$$\begin{aligned}\mathbf{I}_5 &= \frac{\mathbf{V}_2}{8 + j5 - j2}\\ &= 1.41 \angle 24.44^\circ = 1.28 + j0.58 \text{ A}\end{aligned}$$

Employing KCL, we get

$$\begin{aligned}\mathbf{I}_3 &= \mathbf{I}_4 + \mathbf{I}_5\\ &= 2.12 + j2.12 + 1.28 + j0.58\\ &= 3.40 + j2.70 = 4.34 \angle 38.45^\circ \text{ A}\end{aligned}$$

$\mathbf{V}_1$ can then be calculated using KVL:

$$\begin{aligned}\mathbf{V}_1 &= \mathbf{I}_3(5 + j2) + \mathbf{V}_2\\ &= (4.34) \angle 38.45^\circ)(5.39 \angle 21.8^\circ) + 12 \angle 45^\circ\\ &= 20.10 + j28.80 = 35.12 \angle 55.09^\circ \text{ V}\end{aligned}$$

$\mathbf{I}_2$ can now be calculated as

$$\begin{aligned}\mathbf{I}_2 &= \frac{\mathbf{V}_1}{-j5}\\ &= 7.02 \angle 145.09^\circ \text{ A}\end{aligned}$$

$\mathbf{I}_1$ can be computed via KCL:

$$\begin{aligned}\mathbf{I}_1 &= \mathbf{I}_2 + \mathbf{I}_3\\ &= 7.02 \angle 145.09^\circ + 4.34 \angle 38.45^\circ\\ &= -2.36 + j6.72 = 7.12 \angle 109.35^\circ \text{ A}\end{aligned}$$

$\mathbf{V}_s$ is then computed using KVL:

$$\begin{aligned}\mathbf{V}_s &= 10\mathbf{I}_1 + \mathbf{V}_1\\ &= 71.2 \angle 109.35^\circ + 35.12 \angle 55.09^\circ\\ &= -3.5 + j96 = 96.06 \angle 92.09^\circ \text{ V}\end{aligned}$$

Note that these are exactly the same kind of calculations made earlier in the analysis of resistive dc circuits without the use of complex numbers.

As a quick check on our calculations, let us examine KVL around the entire outer loop:

$$\mathbf{V}_s - 10\mathbf{I}_1 - (5 + j2)\mathbf{I}_3 - (8 + j3)\mathbf{I}_5 = 0$$

$$-3.5 + j96 + 23.60 - j67.20 - 11.61 - j20.31 - 8.49 - j8.49 = 0$$

$$0 = 0$$

which indicates that our calculations are correct. ■

DRILL EXERCISE

D9.16. Find the currents $\mathbf{I}_1$, $\mathbf{I}_2$, and $\mathbf{I}_3$ in the network in Fig. D9.16.

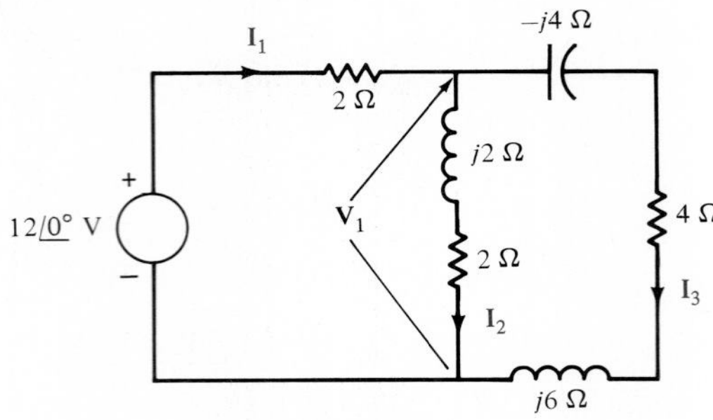

Figure D9.16

D9.17. In the network in Fig. D9.17, $\mathbf{V}_o$ is known to be $8\angle 45°$ V. Compute $\mathbf{V}_s$.

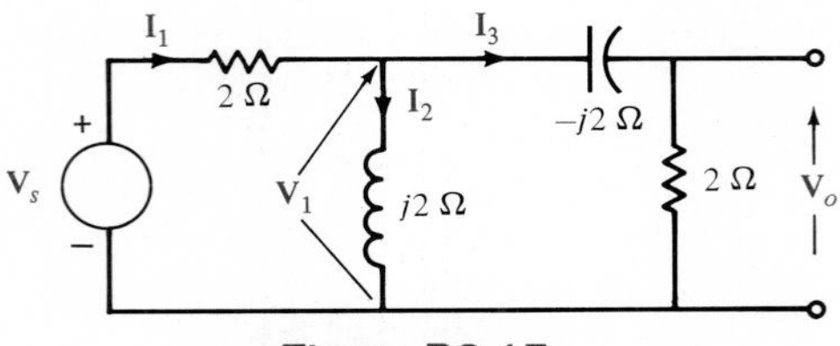

Figure D9.17

9.7

Use of Phasors in Operational Amplifier Circuits

The phasor equations that govern the operation of the op-amp are identical to those in the time domain. Therefore, the analysis techniques that we have presented in earlier chapters are also valid in ac op-amp circuits.

EXAMPLE 9.24

We want to derive the input/output relationship in the frequency domain for the op-amp shown in Fig. 9.22, and then determine the output voltage when the circuit is a differentiator (i.e., $\mathbf{Z}_i = 1/j\omega C_i$ and $\mathbf{Z}_F = R_F$) or an integrator (i.e., $Z_i = R_i$ and $\mathbf{Z}_F = 1/j\omega C_F$).

Using the op-amp characteristics described in Chapter 2, the KCL phasor equation at the op-amp input is

$$\frac{\mathbf{V}_i}{\mathbf{Z}_i} + \frac{\mathbf{V}_o}{\mathbf{Z}_F} = 0$$

Therefore,

$$\frac{\mathbf{V}_o}{\mathbf{V}_i} = -\frac{\mathbf{Z}_F}{\mathbf{Z}_i}$$

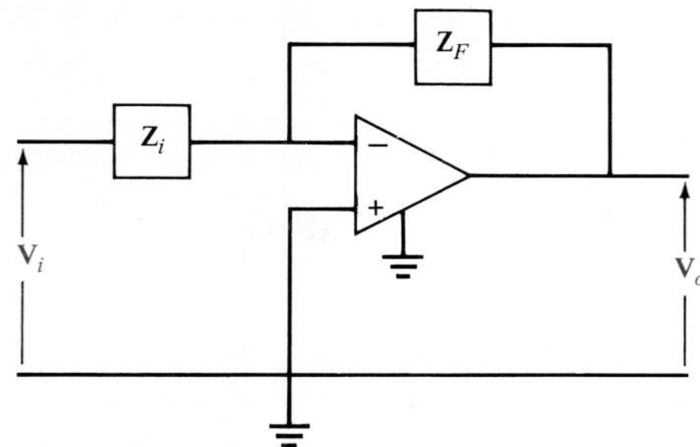

Figure 9.22 ac operational amplifier circuit.

For a differentiator configuration, $\mathbf{Z}_F = R_F$ and $\mathbf{Z}_i = 1/j\omega C_i$ and hence

$$\mathbf{V}_o = j\omega C_i R_F \mathbf{V}_i$$

In the integrator case $\mathbf{Z}_F = 1/j\omega C_F$ and $\mathbf{Z}_i = R_i$, and therefore

$$\mathbf{V}_o = \frac{-\mathbf{V}_i}{j\omega C_F R_i}$$

■

These relationships are very useful in phasor circuits containing differentiators and integrators. In addition, the reader will soon learn that differentiation in the time domain is equivalent to multiplying by $j\omega$ in the frequency domain, and integration in the time domain is equivalent to dividing by $j\omega$ in the frequency domain.

9.8 Summary

In this chapter it was shown that our solution approach for ac circuits involves an analysis in the frequency domain. A set of differential equations with sinusoidal forcing functions in the time domain was transformed into a set of algebraic equations with complex coefficients in the frequency domain. The phasor method was introduced and phasor relationships were established for the circuit elements. Algebraic equations in the frequency domain involving the unknown circuit quantities were solved using the phasor technique.

Impedance and admittance were introduced and used in conjunction with phasors to solve ac circuits containing a single source. The solution technique is based on the following facts: (1) Kirchhoff's laws hold for phasor currents and voltages; and (2) for an impedance $\mathbf{Z}$, its voltage and current are related by $\mathbf{V} = \mathbf{IZ}$.

KEY POINTS

- The sinusoidal function $x(\omega t) = X_M \sin(\omega t + \theta)$ has an amplitude of X_M, a radian frequency of ω, a period of $2\pi/\omega$, and a phase angle of θ.
- If $x_1(t) = X_{M_1} \sin(\omega t + \theta)$ and $x_2(t) = X_{M_2} \sin(\omega t + \phi)$, $x_1(t)$ leads $x_2(t)$ by $\theta - \phi$ radians and $x_2(t)$ lags $x_1(t)$ by $\theta - \phi$ radians.

- When comparing one sinusoidal function with another of the same frequency to determine the phase difference, it is necessary to express both functions as either sines or cosines with positive amplitudes.
- The sinusoidal voltage $v(t) = V_M \cos(\omega t + \theta)$ can be written in exponential form as $v(t) = R_e[V_M e^{j(\omega t + \theta)}]$ and in phasor form as $\mathbf{V} = V_M \angle \theta$.
- Since it is assumed that sinusoidal functions are represented as phasors with a phase angle based on a cosine function, sine functions must be converted to cosine functions using a 90° shift factor to obtain the correct phasor representation.
- If θ_V and θ_i represent the phase angles of the voltage across and the current through a circuit element, then $\theta_i = \theta_V$ if the element is a resistor, θ_i lags θ_V by 90° if the element is an inductor, θ_i leads θ_V by 90° if the element is a capacitor.
- Impedance, $\mathbf{Z}$, is defined as the ratio of the phasor voltage, $\mathbf{V}$, to the phasor current, $\mathbf{I}$, where $\mathbf{Z} = R$ for a resistor, $\mathbf{Z} = j\omega L$ for an inductor, and $\mathbf{Z} = 1/j\omega C$ for a capacitor.
- Admittance $\mathbf{Y}$ is the reciprocal of impedance (i.e., $\mathbf{Y} = 1/\mathbf{Z}$).
- When impedance and admittance are written in rectangular form, $\mathbf{Z} = R + jX$ and $\mathbf{Y} = G + jB$ where R is the resistance, X is the reactance, G is the conductance, and B is the susceptance.
- $\mathbf{Z}$ and $\mathbf{Y}$ are functions of frequency, and therefore their values change as frequency changes.
- KCL and KVL apply in the frequency domain.
- Impedance in the ac case combines like resistance in the dc case, and admittance in the ac case combines like conductance in the dc case.

PROBLEMS

9.1. Given the voltage $v(t) = 120 \cos(314t + \pi/4)$ V, determine the frequency of the voltage in hertz and the phase angle in degrees.

9.2. Express the following functions as sine functions with positive amplitudes.
(a) $A \cos(\omega t - 25°)$
(b) $-B \cos(\omega t + 60°)$
(c) $-C \sin(\omega t + 20°)$

9.3. Draw the waveforms for the following voltage functions.
(a) $v_1(t) = 12 \sin(\omega t + 135°)$ V
(b) $v_2(t) = 6 \sin(\omega t - 45°)$ V

9.4. Three branch currents in a network are known to be

$$i_1(t) = 2 \sin(377t + 45°) \text{ A}$$

$$i_2(t) = 0.5 \cos(377t + 10°) \text{ A}$$

$$i_3(t) = -0.25 \sin(377t + 60°) \text{ A}$$

Determine the phase angles by which $i_1(t)$ leads $i_2(t)$ and $i_1(t)$ leads $i_3(t)$.

9.5. Determine the phase angles by which $v_1(t)$ leads $i_1(t)$, and $v_1(t)$ leads $i_2(t)$, where

$$v_1(t) = 4 \sin(377t + 25°)$$

$$i_1(t) = 0.05 \cos(377t - 10°)$$

$$i_2(t) = -0.1 \sin(377t + 75°)$$

9.6. Express the function $12 \sin \omega t - 8 \cos \omega t$ as a sine function with an angle.

9.7. Convert the following voltages to phasors in the frequency domain.
(a) $v_1(t) = 12 \cos(2\pi 400t + 60°)$ V
(b) $v_2(t) = 6 \sin(2\pi 400t - 20°)$ V

9.8. Express the following phasors as cosine functions with a frequency of 60 Hz.
(a) $\mathbf{V}_1 = 24 \angle -45°$
(b) $\mathbf{V}_2 = 10 \angle 120°$

9.9. Express the phasors below as sine functions with a frequency of 400 Hz.

(a) $\mathbf{V}_1 = 8\angle -60°$

(b) $\mathbf{V}_2 = 2\angle 30°$

9.10. Express the function in Problem 9.6 in phasor notation.

9.11. Convert the following expressions in exponential form to phasor notation and sinusoidal form.

(a) $5e^{j(377t)}$

(b) $6e^{j(40t+20°)}$

(c) $2e^{j(2\pi 60t-180°)}$

(d) $-4e^{j(20t+30°)}$

9.12. Add the following phasors.

(a) $3\angle 30° + 4\angle 30°$

(b) $2\angle 45° - 4\angle -45°$

(c) $6\angle 20° + 5\angle 40°$

(d) $1\angle -20° + 1\angle 180°$

9.13. Find the magnitude and phase angle of the following impedances, which are written in rectangular form. Draw the result on a complex plane.

(a) $-1.5 + j2\ \Omega$

(b) $2.82 - j1.03\ \Omega$

(c) $1.32 + j1.5\ \Omega$

(d) $-1 - j1\ \Omega$

9.14. Calculate the current in the resistor in Fig. P9.14 if the voltage input is

(a) $v_1(t) = 10 \cos (377t + 180°)$ V

(b) $v_2(t) = 12 \sin (377t + 45°)$ V

Give the answers in both the time and frequency domains.

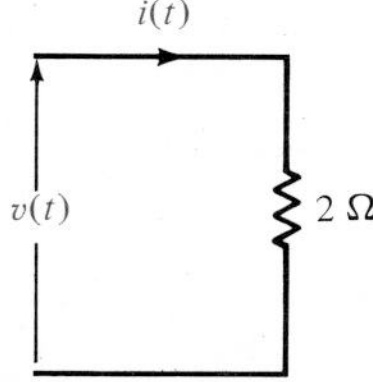

Figure P9.14

9.15. Calculate the current in Problem 9.14 in both the time and frequency domains if the voltage input is

$$v(t) = 6 \cos 377t + 2 \sin 377t \text{ V}$$

9.16. Calculate the current in the inductor shown in Fig. P9.16 if the voltage input is

(a) $v_1(t) = 24 \cos (377t + 12°)$ V

(b) $v_2(t) = 18 \sin (377t - 48°)$ V

Give the answers in both the time and frequency domains.

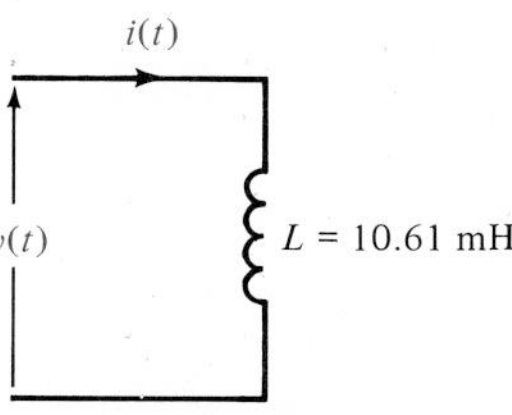

Figure P9.16

9.17. Calculate the current in Problem 9.16 in both the time and frequency domains if the voltage input is

$$v(t) = 2 \cos 377t + 4 \sin 377t \text{ V}$$

9.18. Calculate the current in the capacitor shown in Fig. P9.18 if the voltage input is

(a) $v_1(t) = 16 \cos (377t - 22°)$ V

(b) $v_2(t) = 8 \sin (377t + 64°)$ V

Give the answers in both the time and frequency domains.

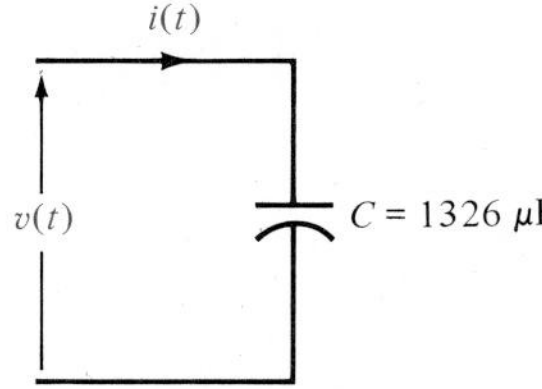

Figure P9.18

9.19. Calculate the current in Problem 9.18 in both the time and frequency domains if the input voltage is

$$v(t) = 6 \sin 377t - 4 \cos 377t \text{ V}$$

9.20. Find the following impedances and write the result in polar form.

(a) A 2-H inductor at 100 Hz.

(b) A 25-μF capacitor at 50 Hz.

(c) A 25-μF capacitor at 50 rad/s.

(d) A 75-Ω resistor at 377 rad/s.

9.21. Determine the impedance, in polar form, for the following elements.
(a) A 0.2-H inductor and a 300-Ω resistor in parallel at 377 rad/s.
(b) A 125-μF capacitor and a 40-Ω resistor in parallel at 60 Hz.
(c) A 1-H inductor and a 50-μF capacitor in parallel at 100 rad/s.

9.22. Calculate the equivalent impedance at terminals *A-B* in the circuit shown in Fig. P9.22.

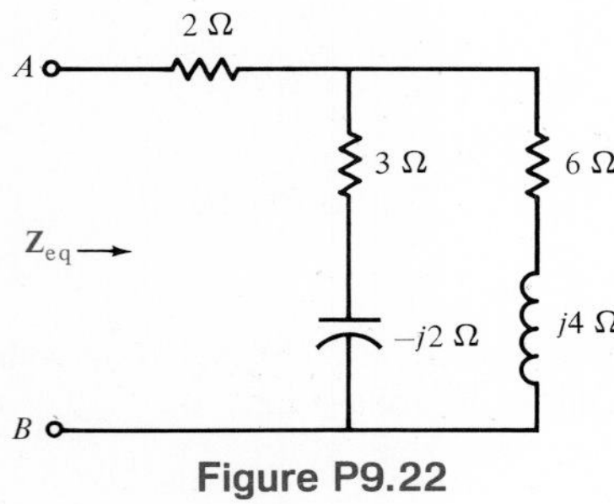

Figure P9.22

9.23. Calculate the equivalent impedance of the network shown in Fig. P9.23.

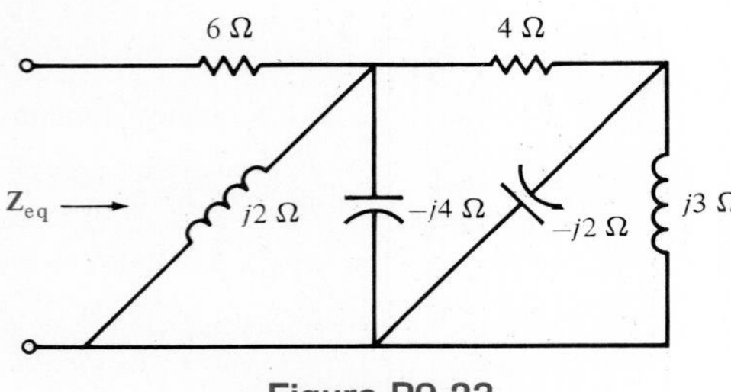

Figure P9.23

9.24. Given the following values of $\mathbf{Z}_1$, $\mathbf{Z}_2$, and $\mathbf{Z}_3$, determine the equivalent impedance of the circuit in Fig. P9.24.

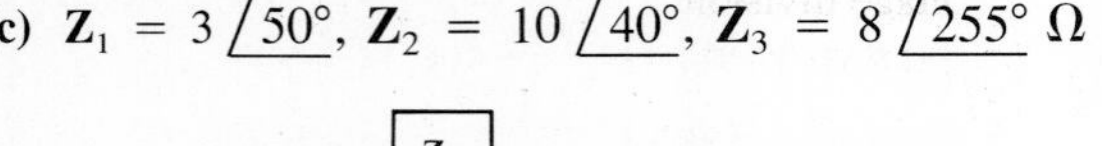
(a) $\mathbf{Z}_1 = 5\angle 40°$, $\mathbf{Z}_2 = 3\angle -35°$, $\mathbf{Z}_3 = 7\angle 135°\ \Omega$
(b) $\mathbf{Z}_1 = 2\angle -110°$, $\mathbf{Z}_2 = 4\angle 180°$, $\mathbf{Z}_3 = 5\angle 20°\ \Omega$
(c) $\mathbf{Z}_1 = 3\angle 50°$, $\mathbf{Z}_2 = 10\angle 40°$, $\mathbf{Z}_3 = 8\angle 255°\ \Omega$

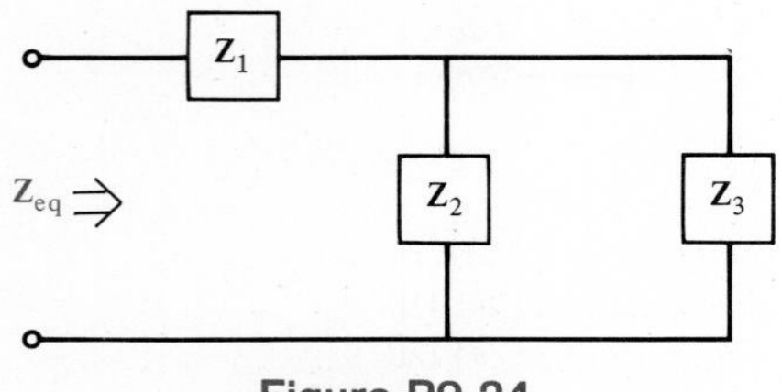

Figure P9.24

9.25. Find $\mathbf{Z}_T$ in Fig. P9.25.

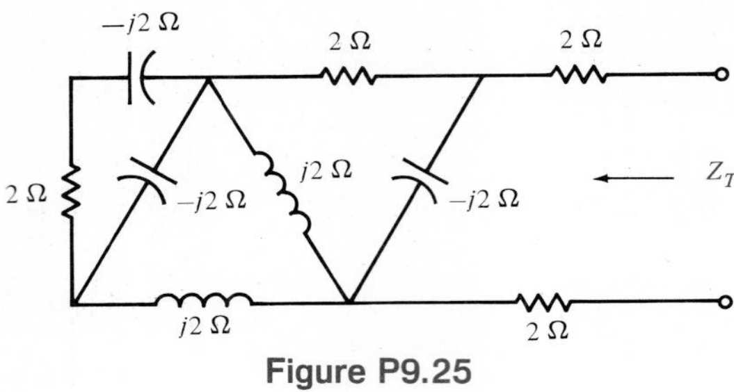

Figure P9.25

9.26. Calculate the equivalent admittance in the circuit shown in Fig. P9.26.

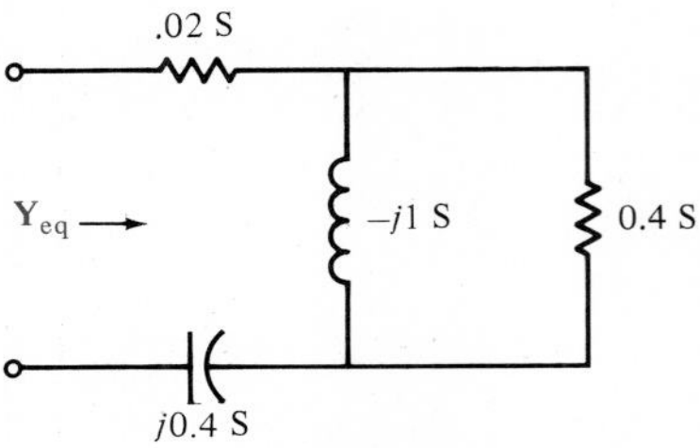

Figure P9.26

9.27. Determine the equivalent admittance for the circuit shown in Fig. P9.27.

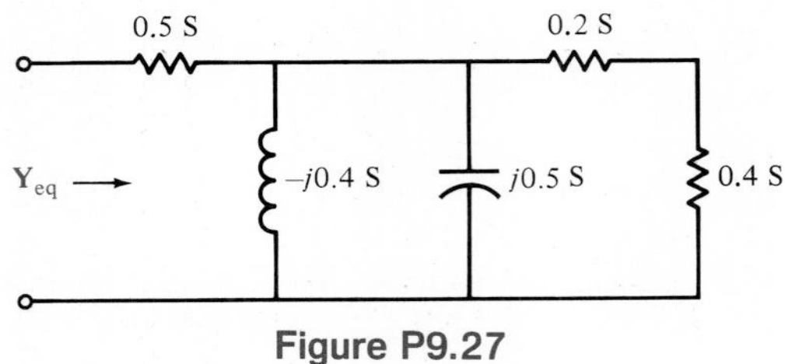

Figure P9.27

9.28. Calculate $\mathbf{Y}_{eq}$ as shown in Fig. P9.28.

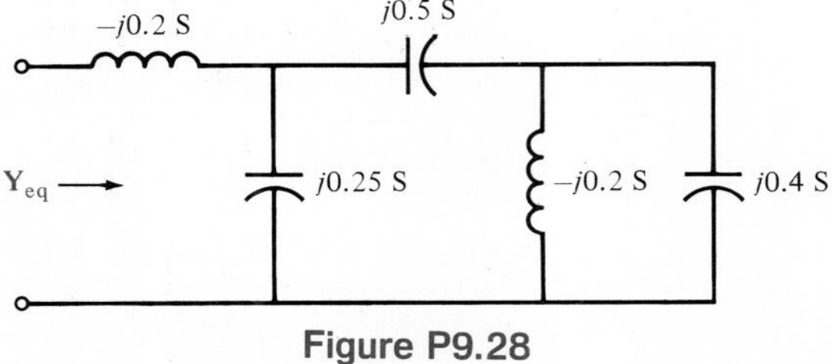

Figure P9.28

9.29. Find the magnitude and phase of the voltage across the inductor in the network in Fig. P9.29.

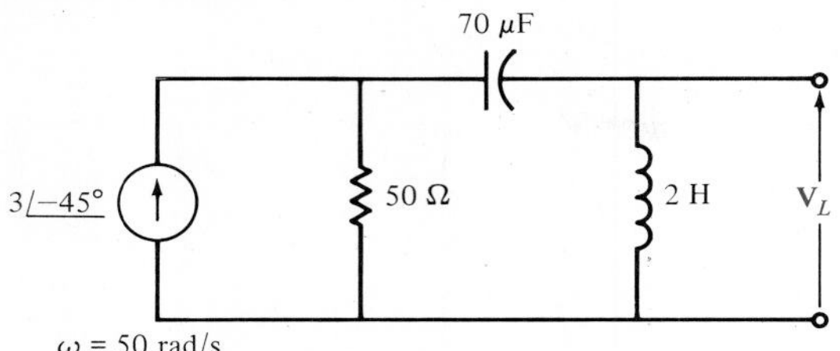

Figure P9.29

9.30. Using $\mathbf{I}_s$ as a reference, plot the phasor diagram for all voltages and currents shown in Fig. P9.30.

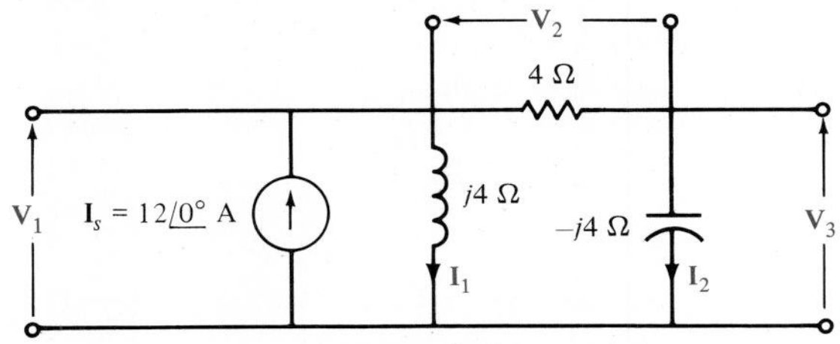

Figure P9.30

9.31. Using $\mathbf{V}_s$ as a reference, plot the phasor diagram for all voltages and currents in Fig. P9.31.

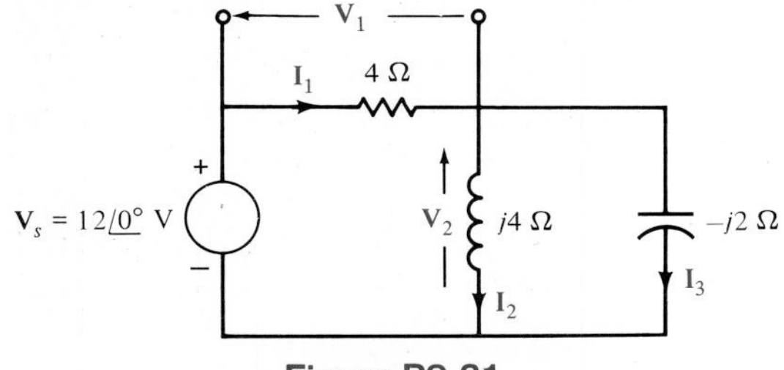

Figure P9.31

9.32. In the circuit shown in Fig. P9.32, determine the value of the inductance such that the current is in phase with the source voltage.

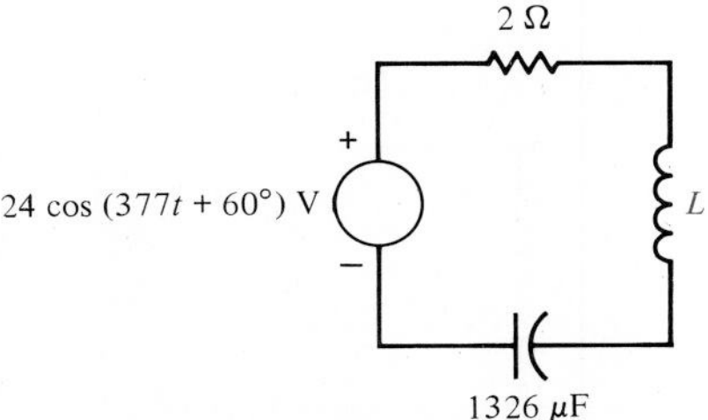

Figure P9.32

9.33. At 60 Hz the impedances of the elements in the circuit shown in Fig. P9.33 are $\mathbf{Z}_R = 4\ \Omega$, $\mathbf{Z}_L = j2\ \Omega$, and $\mathbf{Z}_C = -J8\ \Omega$. Determine the current in the circuit if the input voltage is $v(t) = 36\cos(2513.3t + 60°)$ V.

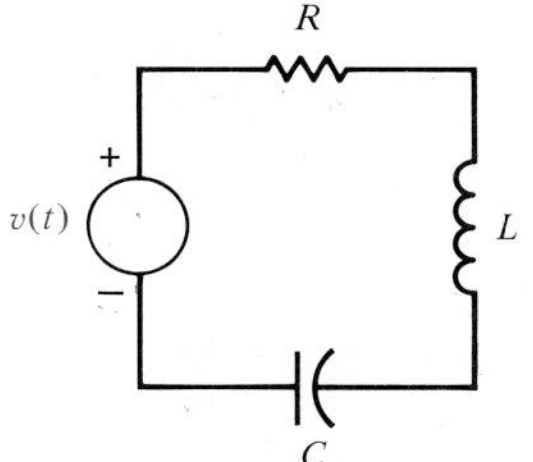

Figure P9.33

9.34. Find the value of the inductor in the network in Fig. P9.34 that will create a totally real equivalent impedance. The circuit operates at 50 Hz.

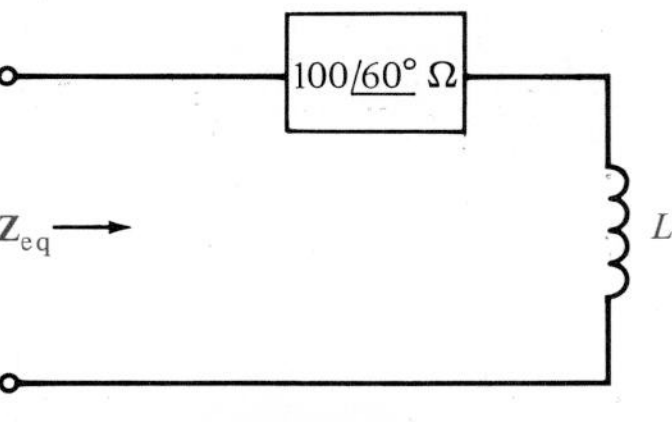

Figure P9.34

9.35. Determine the value of the inductor in the network in Fig. P9.35 that will create a totally real equivalent impedance if the circuit operates at 120 Hz. (*Hint*: Determine the admittance first.)

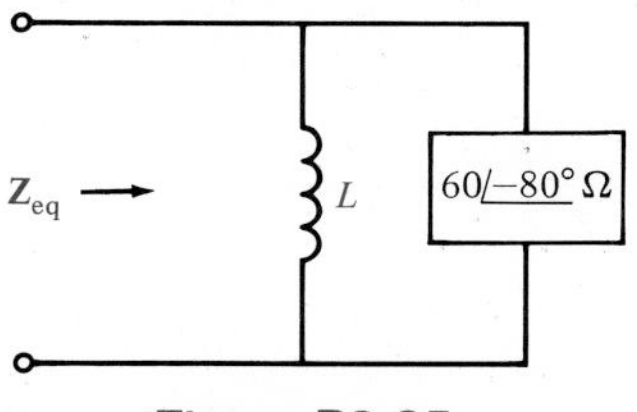

Figure P9.35

9.36. Using voltage division, compute the voltage across the parallel *LC* circuit shown in Fig. P9.36.

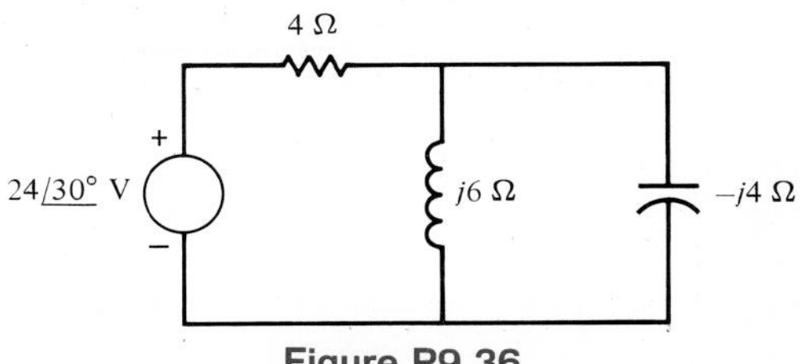

Figure P9.36

9.37. Compute the voltage $\mathbf{V}_o$ in the circuit shown in Fig. P9.37.

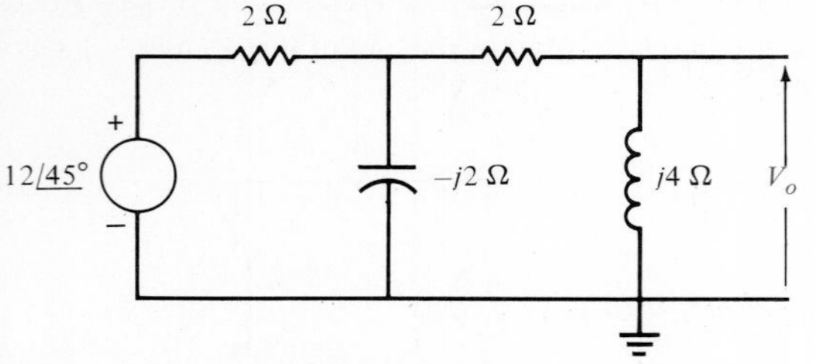

Figure P9.37

9.38 In the circuit shown in Fig. P9.38, determine the current in the 3-Ω resistor.

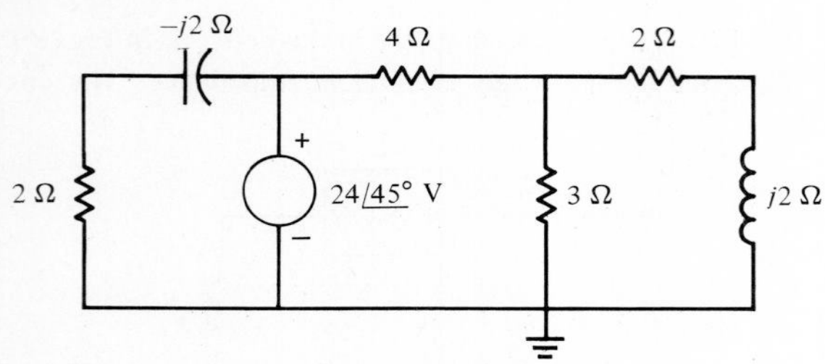

Figure P9.38

9.39. Calculate all the currents in the circuit shown in Fig. P9.39.

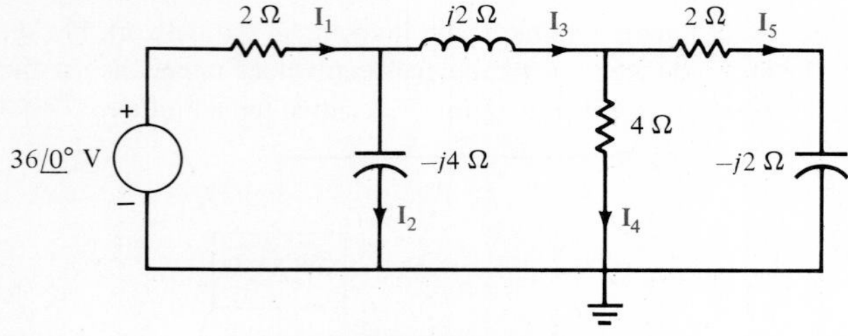

Figure P9.39

9.40. The voltage $\mathbf{V}_C$ in the circuit shown in Fig. P9.40 is known to be $\mathbf{V}_C = 8\,\underline{/-45°}$ V. Determine the value of $\mathbf{V}_s$.

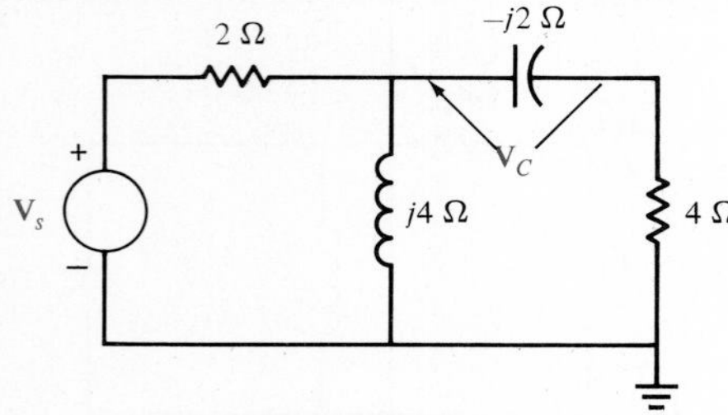

Figure P9.40

9.41. Compute the value of the source current in the circuit shown in Fig. P9.41 if the value of the output voltage is $\mathbf{V}_o = 10\,\underline{/0°}$ V.

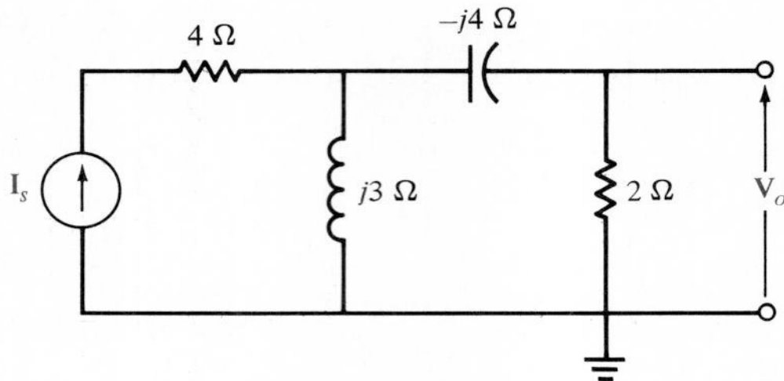

Figure P9.41

9.42. The current in the 2-Ω resistor in the circuit shown in Fig. P9.42 is $\mathbf{I}_o = 4\,\underline{/30°}$ A. Compute $\mathbf{V}_s$.

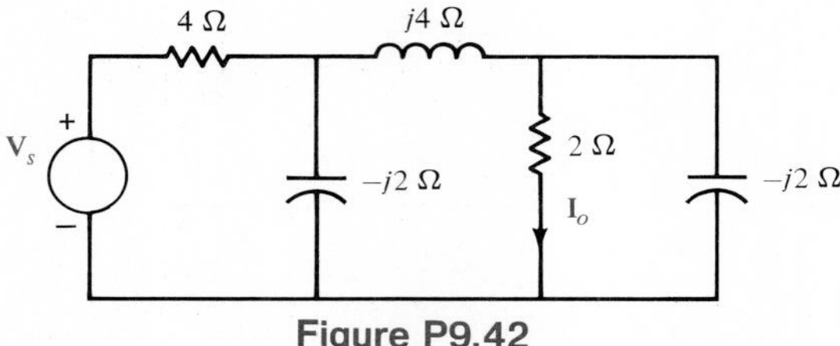

Figure P9.42

9.43. Find $\mathbf{I}_s$ in the network in Fig. P9.43 if $\mathbf{I}_o = 2\,\underline{/0°}$ A.

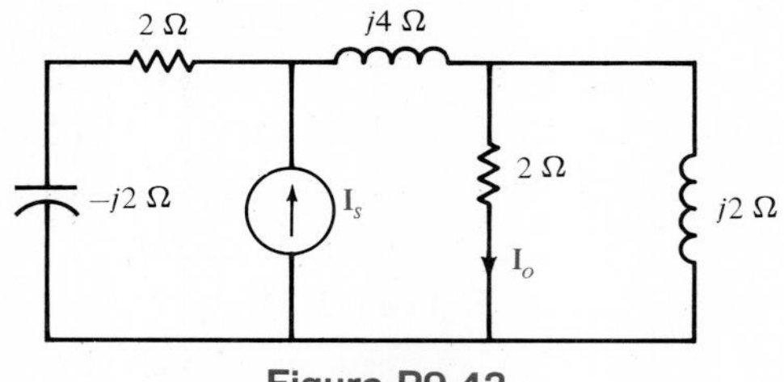

Figure P9.43

9.44. Find $\mathbf{I}_s$ in the network in Fig. P9.44 if $\mathbf{V}_o = 12\,\underline{/0°}$ V.

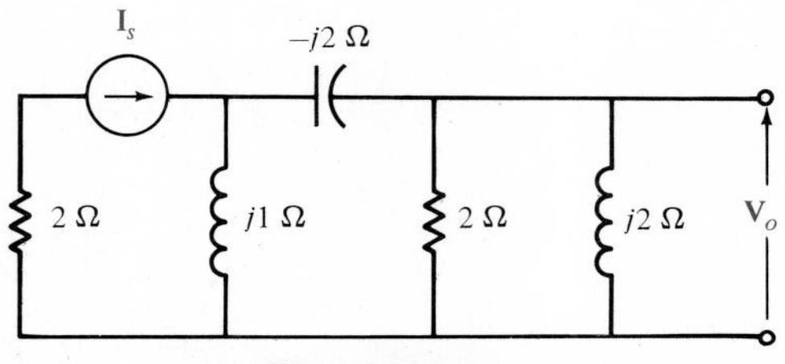

Figure P9.44

9.45. Find $\mathbf{V}_o$ in the network in Fig. P9.45.

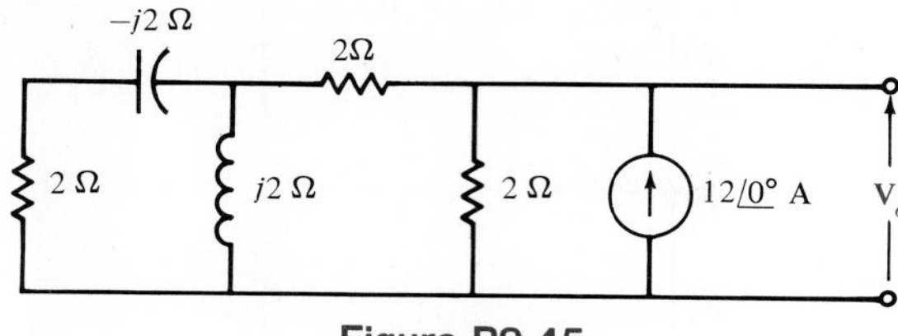

Figure P9.45

9.46. Find $\mathbf{I}_o$ in the network in Fig. P9.46.

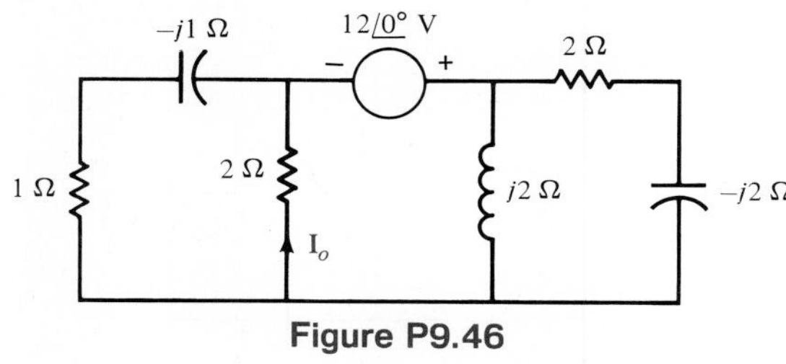

Figure P9.46

9.47. Find $\mathbf{I}_o$ in the network in Fig. P9.47.

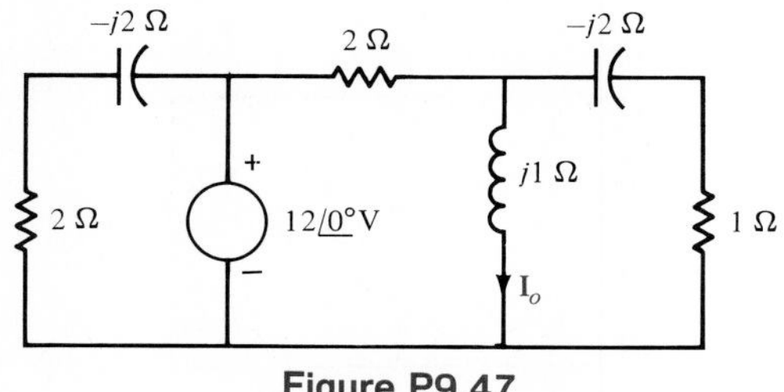

Figure P9.47

9.48. Find $\mathbf{V}_o$ in the network in Fig. P9.48.

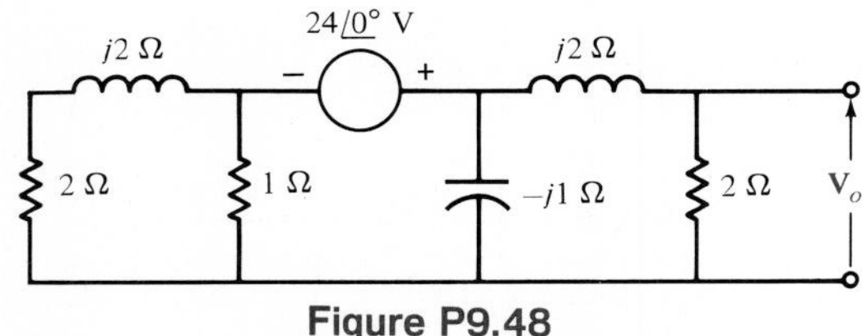

Figure P9.48

9.49. Find $\mathbf{V}_o$ in the network in Fig. P9.49.

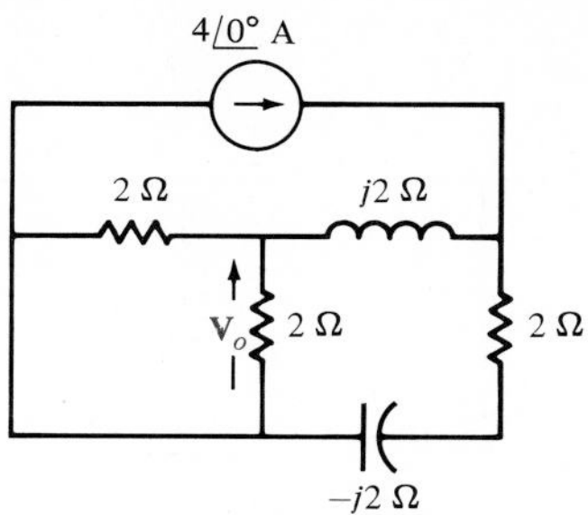

Figure P9.49

9.50. Find $\mathbf{V}_s$ in the network in Fig. P9.50 if $\mathbf{I}_o = 2\,\underline{/0°}$ A.

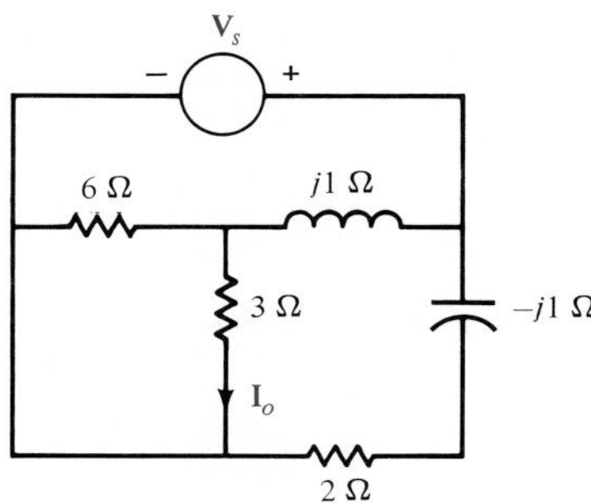

Figure P9.50

9.51. Find $\mathbf{V}_s$ in the network in Fig. P9.51 if $\mathbf{I}_1 = 2\,\underline{/0°}$ A.

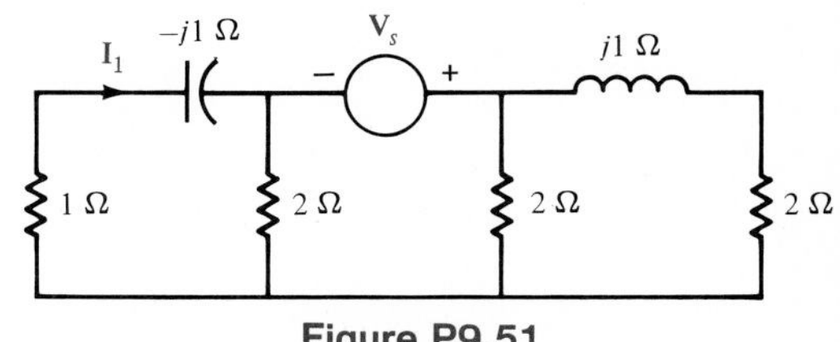

Figure P9.51

9.52. Find $\mathbf{V}_s$ in the network in Fig. P9.52 if $\mathbf{V}_1 = 4\,\underline{/0°}$ V.

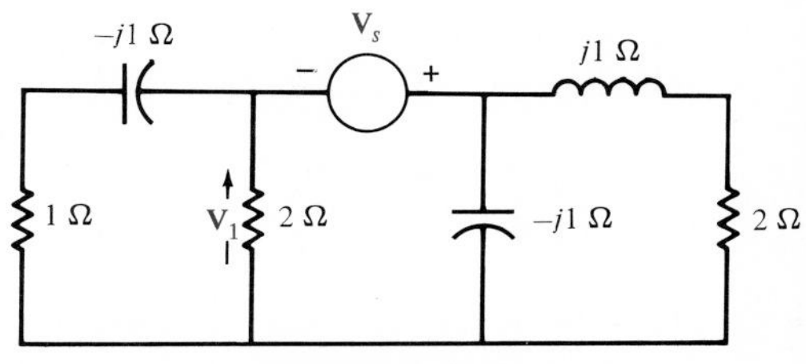

Figure P9.52

9.53. Find $\mathbf{V}_o$ in the network in Fig. P9.53.

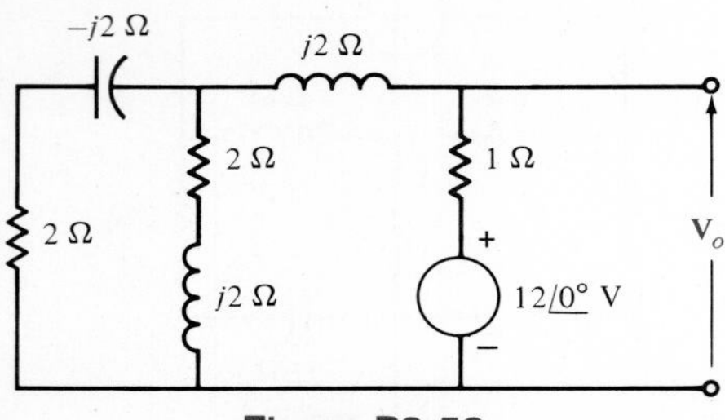

Figure P9.53

9.54. Find $\mathbf{I}_o$ in the network in Fig. P9.54.

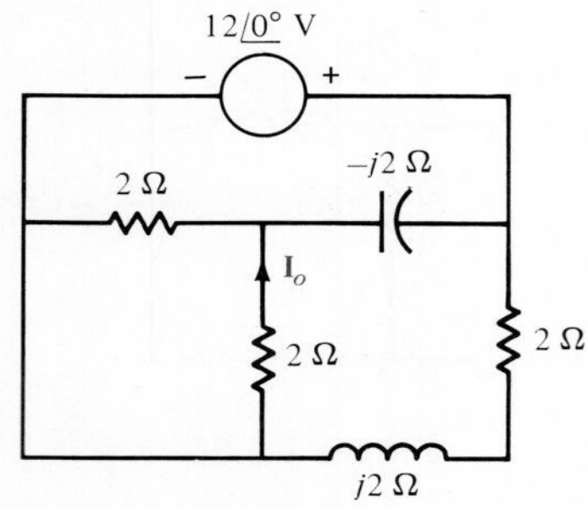

Figure P9.54

9.55. Find $\mathbf{I}_s$ in the network in Fig. P9.55 if $\mathbf{V}_1 = 8\,\underline{/0^\circ}$ V.

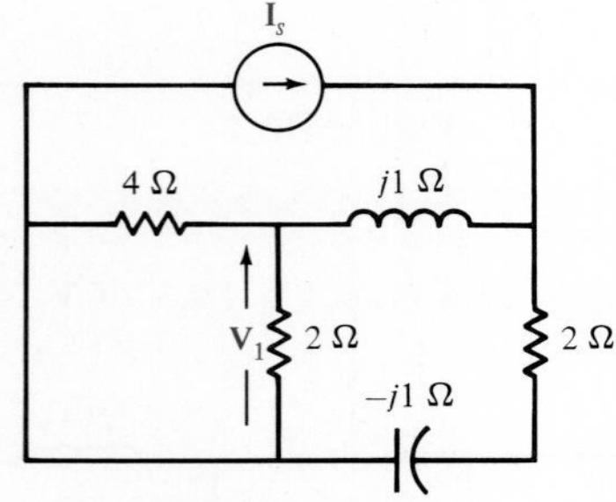

Figure P9.55

9.56. If $\mathbf{V}_1 = 4\,\underline{/0^\circ}$ V, find $\mathbf{I}_o$ in Fig. P9.56.

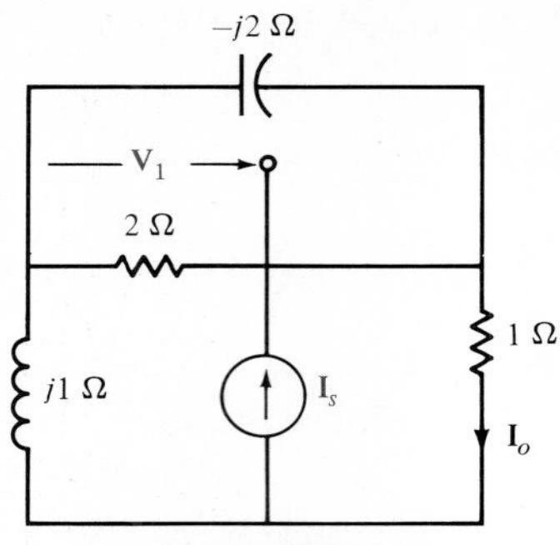

Figure P9.56

9.57. Given $\mathbf{I}_L = 4\,\underline{/0^\circ}$ A, find $\mathbf{V}_s$ in Fig. P9.57.

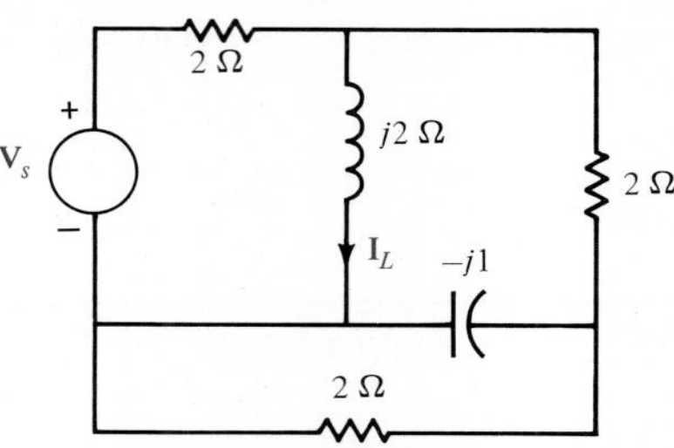

Figure P9.57

9.58. Given **I**, find $\mathbf{V}_s$ in Fig. P9.58.

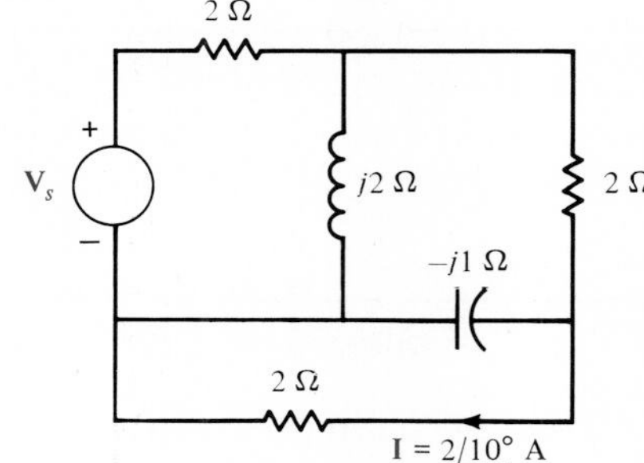

Figure P9.58

9.59. If $\mathbf{V}_1 = 10\,\underline{/0^\circ}$ V, find $\mathbf{V}_s$ in Fig. P9.59.

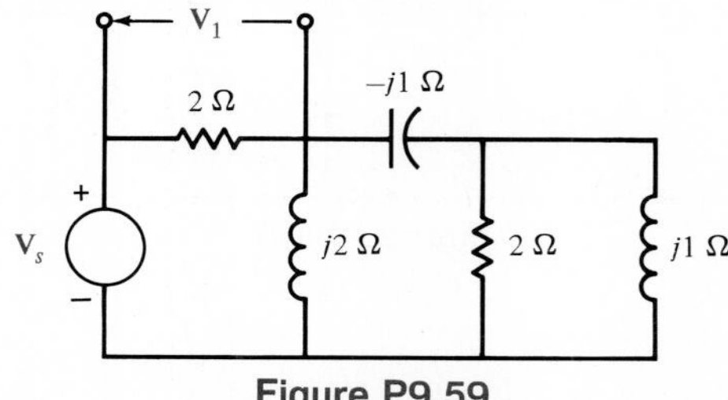

Figure P9.59

9.60. Given $\mathbf{V}_s = 12\angle 0°$ V, find $\mathbf{I}_s$ in Fig. P9.60.

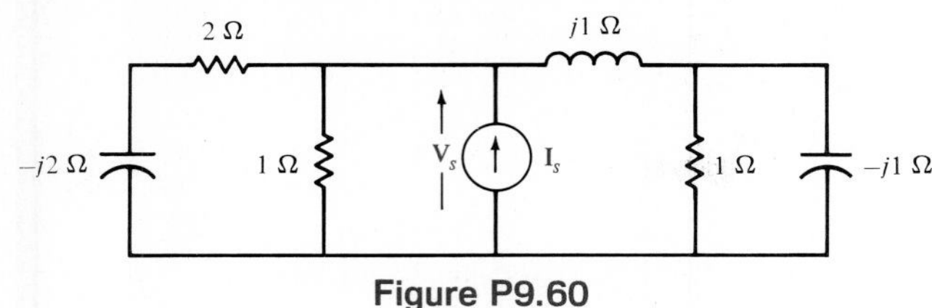

Figure P9.60

9.61. If $\mathbf{I}_o = 4\angle 0°$ A, find $\mathbf{V}_s$ in Fig. P9.61.

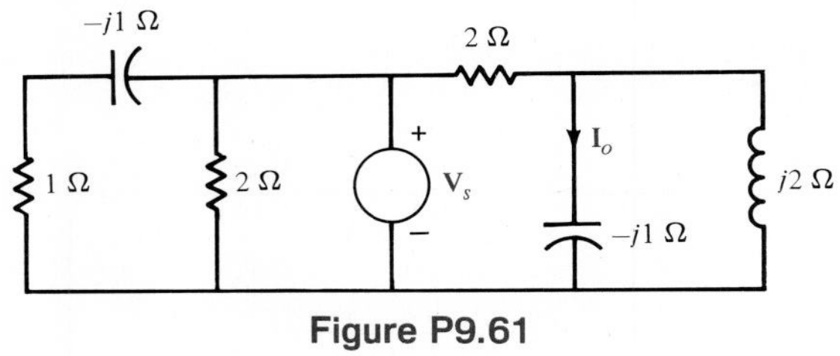

Figure P9.61

9.62. In the circuit shown in Fig. P9.62 $\mathbf{V}_4 = 12\angle 30°$ V. Calculate $\mathbf{V}_o$.

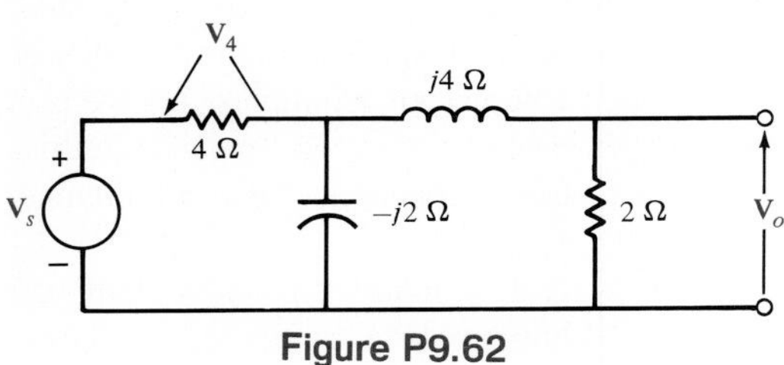

Figure P9.62

9.63. Derive the input/output relationship $\mathbf{V}_o/\mathbf{V}_i$ in the frequency domain for the op-amp circuit in Fig. P9.63.

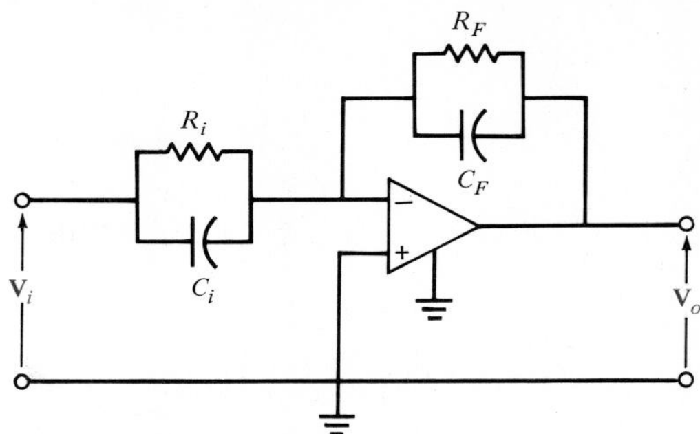

Figure P9.63

9.64. Determine the expression for the output voltage in the frequency domain for the circuit shown in Fig. P9.64.

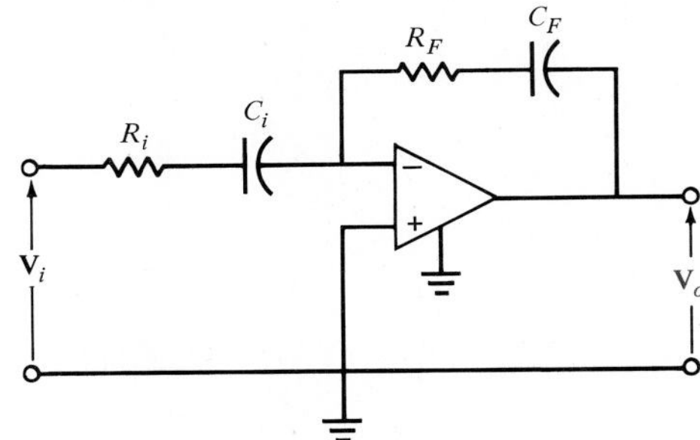

Figure P9.64

9.65

(a) Determine the input/output relationship in the frequency domain for the op-amp circuit shown in Fig. P9.65.

(b) Compute the transfer function $\mathbf{V}_o/\mathbf{V}_i$ if $\mathbf{Z}_i = R_i + 1/j\omega C_i$ and $\mathbf{Z}_F = 1/j\omega C_F$.

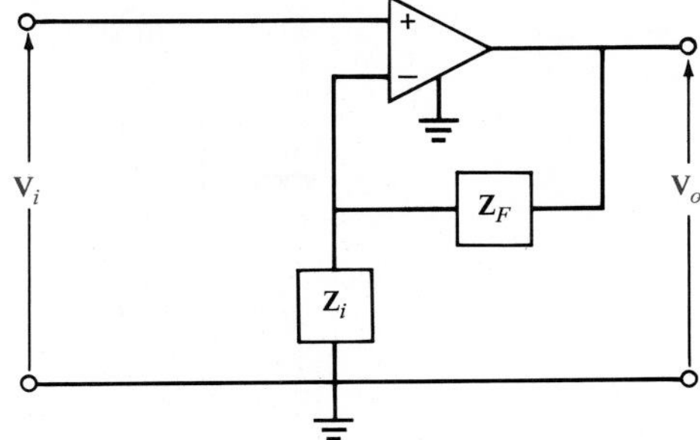

Figure P9.65

CHAPTER

10

Sinusoidal Steady-State Analysis

We have shown in Chapter 9 that the R, L, and C passive circuit elements are linear components. We have also demonstrated in Chapter 9 the use of KCL, KVL, and the relationship $\mathbf{V} = \mathbf{IZ}$ in the solution of ac steady-state circuit problems. Since the network theorems and other circuit analysis techniques, which we successfully employed to solve resistive dc circuits, are applicable to linear circuits, it would appear that these circuit analysis tools could be applied to networks containing R, L, and C elements. In this chapter we demonstrate the applicability of these various techniques to the steady-state analysis of ac circuits. Since we have discussed the analysis techniques in earlier chapters, we concentrate here on illustrating their use via numerous examples. As we proceed, it will become obvious to the reader that the difference between the material in this chapter and that in preceding chapters is our use of phasors, impedance, and admittance, which lead to equations involving complex numbers.

The examples employed in this chapter were specifically chosen to demonstrate the applicability of the various circuit analysis techniques to a variety of problems. In addition, an attempt has been made to help readers understand the various ways in which a circuit analysis problem can be approached by using more than one method to solve a particular problem.

In this chapter ac circuits will also be analyzed using SPICE.

10.1 Linearity

In order to illustrate the principle of linearity and its use in the solution of ac steady-state circuit problems, consider the following example.

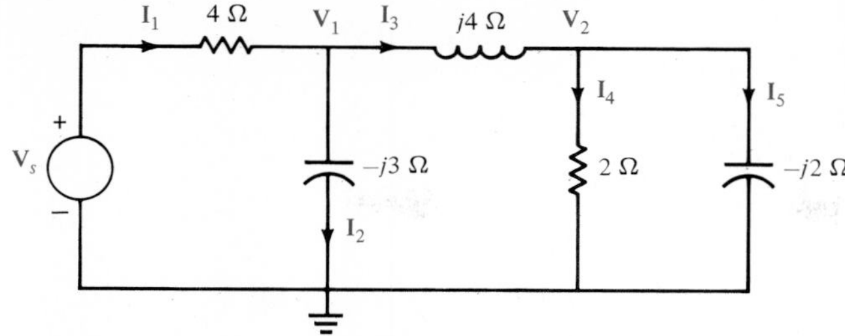

Figure 10.1 Example circuit used to illustrate linearity.

EXAMPLE 10.1

The input voltage to the circuit shown in Fig. 10.1 is $v_s(t) = 12 \cos(377t + 30°)$ V. We wish to determine the current $\mathbf{I}_4$ using linearity. Therefore, we will assume that $\mathbf{I}_4 = 1 \underline{/0°}$ A, compute the required value of $\mathbf{V}_s$, and then use the actual value of $\mathbf{V}_s$ to determine the actual value of $\mathbf{I}_4$.

If $\mathbf{I}_4 = 1 \underline{/0°}$ A, then $\mathbf{V}_2 = (2)(1 \underline{/0°}) = 2 \underline{/0°}$ V. Hence

$$\mathbf{I}_5 = \frac{2 \underline{/0°}}{2 \underline{/-90°}} = 1 \underline{/90°} \text{ A}$$

Applying KCL, we have

$$\begin{aligned}\mathbf{I}_3 &= \mathbf{I}_4 + \mathbf{I}_5 \\ &= 1 + j1 \text{ A}\end{aligned}$$

Now applying KVL gives us

$$\begin{aligned}\mathbf{V}_1 &= (1 + j1)(j4) + \mathbf{V}_2 \\ &= -2 + j4 \text{ V}\end{aligned}$$

$\mathbf{I}_2$ is then

$$\mathbf{I}_2 = \frac{-2 + j4}{-j3} = \frac{-4}{3} - j\frac{2}{3} \text{ A}$$

Again using KCL, we obtain

$$\begin{aligned}\mathbf{I}_1 &= \mathbf{I}_2 + \mathbf{I}_3 \\ &= \frac{-4}{3} - \frac{j2}{3} + 1 + j1 = \frac{-1}{3} + j\frac{1}{3} \text{ A}\end{aligned}$$

KVL around the loop at the far left of the circuit gives us

$$\begin{aligned}\mathbf{V}_s &= 4(\mathbf{I}_1) + \mathbf{V}_1 \\ &= -\frac{4}{3} + j\frac{4}{3} - 2 + j4 \\ &= -\frac{10}{3} + j\frac{16}{3} = 6.29 \underline{/122°} \text{ V}\end{aligned}$$

Now since $\mathbf{V}_s = 6.29 \angle 122°$ V produces an $I_4 = 1 \angle 0°$ A, then $\mathbf{V}_s = 12 \angle 30°$ V produces an $\mathbf{I}_4$ of

$$\mathbf{I}_4 = \frac{(12 \angle 30°)(1 \angle 0°)}{6.29 \angle 122°} = 1.91 \angle -92° \text{ A}$$

Therefore,

$$i_4(t) = 1.91 \cos (377t - 92°) \text{ A}$$ ■

DRILL EXERCISE

D10.1. Compute $\mathbf{V}_o$ in the network in Fig. D10.1 using linearity by first assuming that $\mathbf{V}_o = 1 \angle 0°$ V.

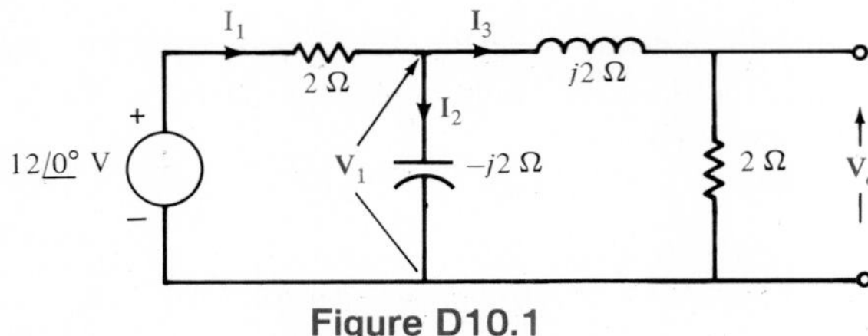

Figure D10.1

10.2

Nodal Analysis

The following examples illustrate that nodal analysis is performed in exactly the same manner as it was in dc resistive circuits. In this section nodal analysis will be employed in the frequency domain using the phasor technique.

EXAMPLE 10.2

We wish to calculate the node voltages in the circuit in Fig. 10.2. To begin, we note that $\mathbf{V}_1$ and $\mathbf{V}_2$ form a supernode (i.e., $\mathbf{V}_1 = \mathbf{V}_2 + 12 \angle 30°$). Therefore, only one equation, together with this constraint equation, is needed to determine the node voltages.

Applying KCL at the supernode yields the equation

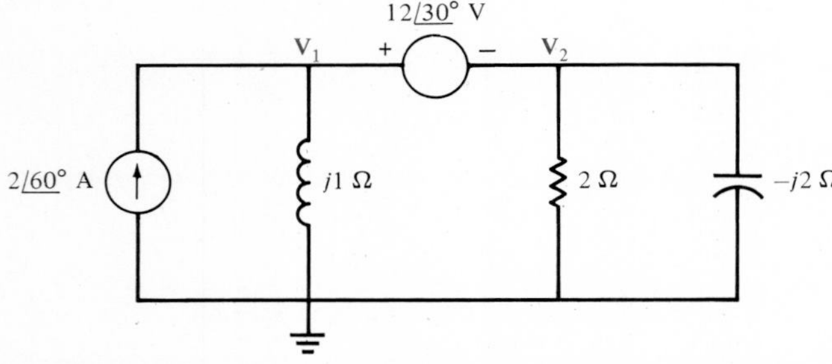

Figure 10.2 Example circuit used to illustrate nodal analysis.

$$\frac{\mathbf{V}_1}{j1} + \frac{\mathbf{V}_1 - 12 \underline{/30^\circ}}{2} + \frac{\mathbf{V}_1 - 12 \underline{/30^\circ}}{-j2} = 2 \underline{/60^\circ}$$

$$\mathbf{V}_1\left(\frac{1}{j1} + \frac{1}{2} + \frac{1}{-j2}\right) = 2 \underline{/60^\circ} + \frac{12 \underline{/30^\circ}}{2} - \frac{12 \underline{/30^\circ}}{j2}$$

Simplifying this equation, we obtain

$$(0.71 \underline{/-45^\circ})\, \mathbf{V}_1 = 3.2 + j9.93$$

or

$$\mathbf{V}_1 = 14.69 \underline{/117.14^\circ} \text{ V}$$

Then using the constraint equation gives us

$$\begin{aligned} \mathbf{V}_2 &= \mathbf{V}_1 - 12 \underline{/30^\circ} \\ &= -6.70 + j13.07 - (10.4 + j6) \\ &= -17.1 + j7.07 = 18.50 \underline{/157.54^\circ} \text{ V} \end{aligned}$$

■

DRILL EXERCISE

D10.2. Use nodal analysis to find $\mathbf{V}_o$ in the network in Fig. D10.2.

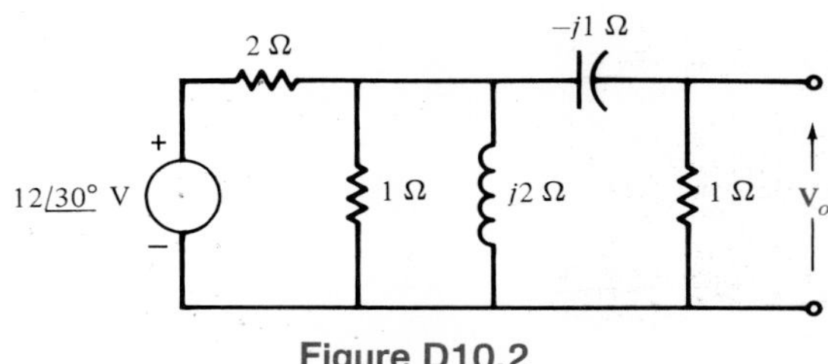

Figure D10.2

The following example illustrates the use of nodal analysis for a circuit containing a dependent source.

EXAMPLE 10.3

Let us compute the node voltages for the circuit shown in Fig. 10.3. There are three nodes, and therefore two equations are required to solve the circuit. The nodal equations are

$$\frac{\mathbf{V}_1}{2} + \frac{\mathbf{V}_1 - \mathbf{V}_2}{-j1} = 2\mathbf{V}_2$$

$$\frac{\mathbf{V}_2 - \mathbf{V}_1}{-j1} + \frac{\mathbf{V}_2}{j2} = 2 \underline{/30^\circ}$$

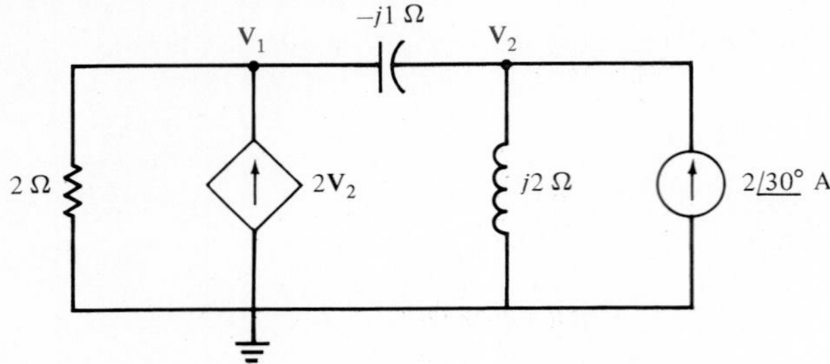

Figure 10.3 Example circuit used to illustrate nodal analysis.

These equations can be rewritten as

$$\left(\tfrac{1}{2} + j1\right)\mathbf{V}_1 - (2 + j1)\mathbf{V}_2 = 0$$

$$-(j1)\mathbf{V}_1 + \left(j\tfrac{1}{2}\right)\mathbf{V}_2 = 2\underline{/30^\circ}$$

The node voltages can be determined via matrix analysis. The two simultaneous equations for the node voltages in matrix form are

$$\boldsymbol{AV} = \boldsymbol{I}$$

$$\begin{bmatrix} \frac{1}{2} + j1 & -(2 + j1) \\ -j1 & j\frac{1}{2} \end{bmatrix} \begin{bmatrix} \mathbf{V}_1 \\ \mathbf{V}_2 \end{bmatrix} = \begin{bmatrix} 0 \\ 2\underline{/30^\circ} \end{bmatrix}$$

The determinant of $\boldsymbol{A}$ is

$$|\boldsymbol{A}| = \left(\tfrac{1}{2} + j1\right)\left(j\tfrac{1}{2}\right) - (-j1)(-2 - j1) = \tfrac{1}{2} - j1.75$$

The adjoint of $\boldsymbol{A}$ is

$$\text{adj}\ \boldsymbol{A} = \begin{bmatrix} j\frac{1}{2} & 2 + j1 \\ j1 & \frac{1}{2} + j1 \end{bmatrix}$$

Therefore, the inverse of $\boldsymbol{A}$ is

$$\boldsymbol{A}^{-1} = \frac{1}{\frac{1}{2} - j1.75} \begin{bmatrix} j\frac{1}{2} & 2 + j1 \\ j1 & \frac{1}{2} + j1 \end{bmatrix}$$

The solution of the matrix equation is

$$\boldsymbol{V} = \boldsymbol{A}^{-1}\boldsymbol{I}$$

or

$$\begin{bmatrix} \mathbf{V}_1 \\ \mathbf{V}_2 \end{bmatrix} = \frac{1}{1.82\underline{/-74.05^\circ}} \begin{bmatrix} j\frac{1}{2} & 2 + j1 \\ j1 & \frac{1}{2} + j1 \end{bmatrix} \begin{bmatrix} 0 \\ 2\underline{/30^\circ} \end{bmatrix}$$

Hence

$$\mathbf{V}_1 = \frac{(2\underline{/30^\circ})(2 + j1)}{1.82\underline{/-74.05^\circ}}\ \text{V} = 2.46\underline{/130.62^\circ}\ \text{V}$$

and

$$\mathbf{V}_2 = \frac{(2\,\underline{/30°})(\frac{1}{2} + j1)}{1.82\,\underline{/-74.05°}}\text{ V} = 1.23\,\underline{/167.48°}\text{ V}$$

DRILL EXERCISE

D10.3. Use nodal analysis to determine the node voltages in the network in Fig. D10.3.

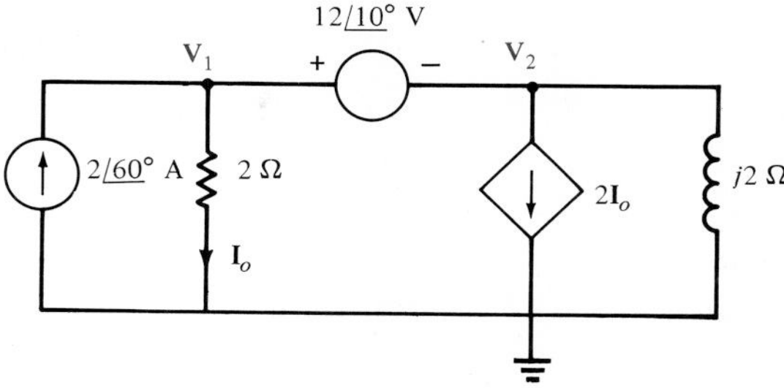

Figure D10.3

EXAMPLE 10.4

Calculate the output voltage for the op-amp circuit shown in Fig. 10.4a. Use the equivalent circuit for the op-amp in Fig. 2.31b and assume that $R_i = \infty$ and $R_o = 0$.

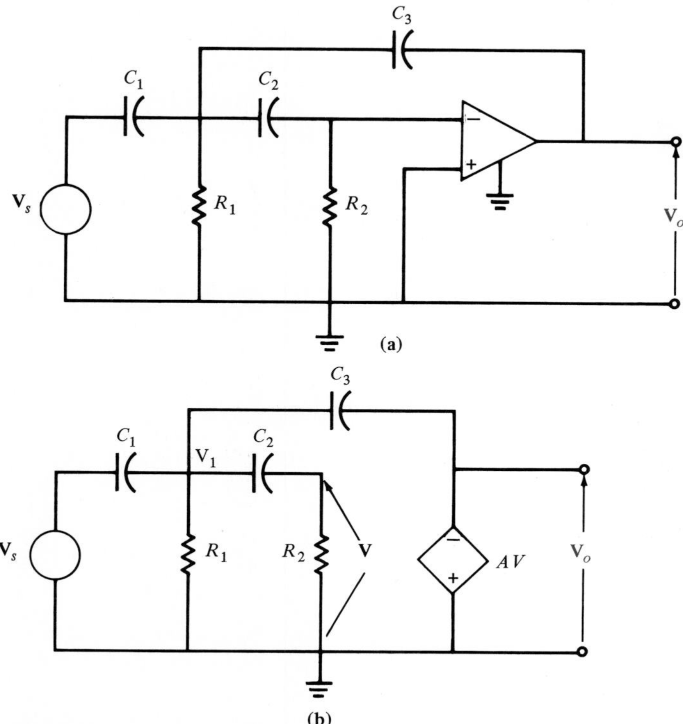

Figure 10.4 Operational amplifier circuit used to illustrate nodal analysis.

The original network containing the equivalent op-amp circuit is shown in Fig. 10.4b. The dependent source creates a supernode, and therefore only two equations are needed to compute the output voltage. These equations are

$$(\mathbf{V}_1 - \mathbf{V}_s)j\omega C_1 + \frac{\mathbf{V}_1}{R_1} + (\mathbf{V}_1 - \mathbf{V})j\omega C_2 + (\mathbf{V}_1 + A\mathbf{V})j\omega C_3 = 0$$

$$(\mathbf{V} - \mathbf{V}_1)j\omega C_2 + \frac{\mathbf{V}}{R_2} = 0$$

or

$$\mathbf{V}_1\left(j\omega C_1 + \frac{1}{R_1} + j\omega C_2 + j\omega C_3\right) - \mathbf{V}(j\omega C_2 - Aj\omega C_3) = j\omega C_1\mathbf{V}_s$$

$$-\mathbf{V}_1(j\omega C_2) + \mathbf{V}\left(j\omega C_2 + \frac{1}{R_2}\right) = 0$$

Solving for the voltage $\mathbf{V}$, we obtain

$$\mathbf{V} = \frac{-\omega^2 C_1 C_2 \mathbf{V}_s}{[1/R_1 + j\omega(C_1 + C_2 + C_3)](j\omega C_2 + 1/R_2) - j\omega C_2(j\omega C_2 - j\omega C_3 A)}$$

Simplifying this function yields

$$\mathbf{V} = \frac{-\omega^2 C_1 C_2 R_1 R_2 \mathbf{V}_s}{1 + j\omega[R_2C_2 + R_1(C_1 + C_2 + C_3)] - \omega^2 R_1 R_2 C_2(C_1 + C_3 + C_3 A)}$$

The output voltage $\mathbf{V}_0$ is then

$$\mathbf{V}_o = \frac{+A\omega^2 C_1 C_2 R_1 R_2 \mathbf{V}_s}{1 + j\omega[R_2C_2 + R_1(C_1 + C_2 + C_3)] - \omega^2 R_1 R_2 C_2(C_1 + C_3 + C_3 A)}$$

Note that as $A \rightarrow \infty$ the input/output relationship becomes

$$\mathbf{V}_o = -\frac{C_1}{C_3}\mathbf{V}_s$$

■

EXAMPLE 10.5

The circuit shown in Fig. 10.5 represents a portion of an equivalent circuit for a transistor amplifier. We wish to calculate the input admittance. The two node equations for the circuit are

$$-\mathbf{I}_1 + j\omega C_1\mathbf{V} + j\omega C_2(\mathbf{V} - \mathbf{V}_o) = 0$$

$$j\omega C_2(\mathbf{V}_o - \mathbf{V}) + g_m\mathbf{V} + \frac{\mathbf{V}_o}{R_L} = 0$$

or

$$(j\omega C_1 + j\omega C_2)\mathbf{V} - j\omega C_2\,\mathbf{V}_o = I_1$$

$$(g_m - j\omega C_2)\mathbf{V} + \left(j\omega C_2 + \frac{1}{R_L}\right)\mathbf{V}_o = 0$$

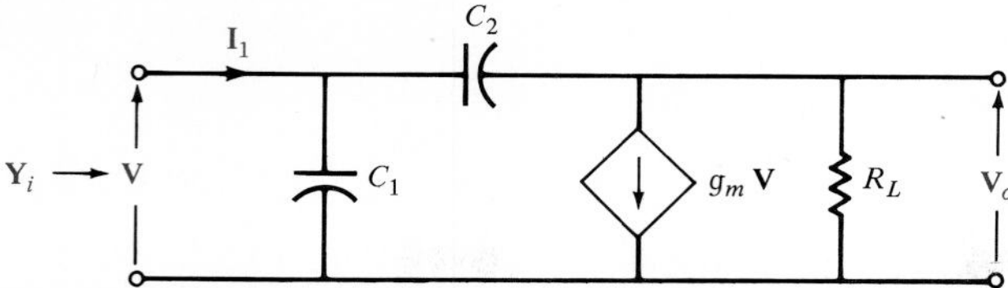

Figure 10.5 Portion of a transistor amplifier used to illustrate nodal analysis.

Solving these equations for **V** yields

$$\mathbf{V} = \frac{\mathbf{I}_1(j\omega C_2 + 1/R_L)}{(j\omega C_1 + j\omega C_2)(j\omega C_2 + 1/R_L) + j\omega C_2(g_m - j\omega C_2)}$$

The input admittance is

$$\mathbf{Y}_i = \frac{\mathbf{I}_1}{\mathbf{V}} = j\omega(C_1 + C_2) + \frac{j\omega C_2(g_m - j\omega C_2)}{j\omega C_2 + 1/R_L}$$

In electronics we find that for an important range of frequencies $1/R_L \gg \omega C_2$ and $g_m \gg \omega C_2$. Under these conditions the equation above reduces to

$$\begin{aligned}\mathbf{Y}_i &= j\omega C_1 + j\omega C_2 + j\omega C_2 g_m R_L \\ &= j\omega[C_1 + C_2(1 + g_m R_L)]\end{aligned}$$

The quantity $g_m R_L$ is generally large, and therefore the equation above indicates that the input capacitance of the amplifier is significantly increased due to the capacitor C_2, which links the input and the output of the amplifier. This increase in the input capacitance of an amplifier stage is known in electronics as the *Miller effect*. ■

10.3 Mesh and Loop Analysis

The following examples illustrate the use of mesh and loop analysis in ac steady-state circuits.

EXAMPLE 10.6

Let us determine the voltage $\mathbf{V}_o$ in the circuit in Fig. 10.6. Although there are two meshes, we need only one equation to determine all the mesh currents, because of the

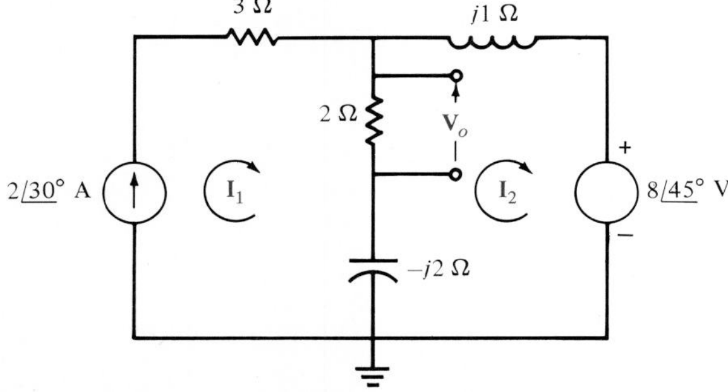

Figure 10.6 Circuit used to illustrate mesh equations.

presence of the current source. KVL for the right mesh is

$$(\mathbf{I}_2 - 2\,\underline{/30^\circ})(2 - j2) + \mathbf{I}_2(j1) = -8\,\underline{/45^\circ}$$

or

$$\mathbf{I}_2 = \frac{(2\,\underline{/-30^\circ})(2 - j2) - 8\,\underline{/45^\circ}}{2 - j1} = 3.18\,\underline{/-64.96^\circ}\text{ A}$$

To solve for V_o, we need the current through the 2-Ω resistor.

$$\mathbf{I}_1 - \mathbf{I}_2 = 2\,\underline{/30^\circ} - 3.18\,\underline{/-64.96^\circ}$$

$$= 0.38 + j3.88\text{ A}$$

and hence

$$\mathbf{V}_o = 2(\mathbf{I}_1 - \mathbf{I}_2)$$

$$= 0.76 + j7.76\text{ V}$$

■

EXAMPLE 10.7

Determine $\mathbf{V}_o$ for the circuit shown in Fig. 10.7 using mesh analysis. Note from the circuit that $\mathbf{I}_x = \mathbf{I}_1$. The equations are

$$2\mathbf{I}_1 + [\mathbf{I}_1 - \mathbf{I}_2](-j1) = 6\underline{/30^\circ} - 2\mathbf{I}_1$$

$$(\mathbf{I}_2 - \mathbf{I}_1)(-j1) + \mathbf{I}_2(j2) + \mathbf{I}_2(4) = 2\mathbf{I}_1$$

Simplifying these equations, we obtain

$$\mathbf{I}_1(4 - j1) + \mathbf{I}_2(j1) = 6\,\underline{/30^\circ}$$

$$\mathbf{I}_1(-2 + j1) + \mathbf{I}_2(4 + j1) = 0$$

The matrix equation is then

$$\begin{bmatrix} 4 - j1 & j1 \\ -2 + j1 & 4 + j1 \end{bmatrix}\begin{bmatrix} \mathbf{I}_1 \\ \mathbf{I}_2 \end{bmatrix} = \begin{bmatrix} 6\,\underline{/30^\circ} \\ 0 \end{bmatrix}$$

The determinant of the coefficient matrix is $18 + j2$. Hence

$$\begin{bmatrix} \mathbf{I}_1 \\ \mathbf{I}_2 \end{bmatrix} = \frac{1}{18 + j2}\begin{bmatrix} 4 + j1 & -j1 \\ 2 - j1 & 4 - j1 \end{bmatrix}\begin{bmatrix} 6\,\underline{/30^\circ} \\ 0 \end{bmatrix}$$

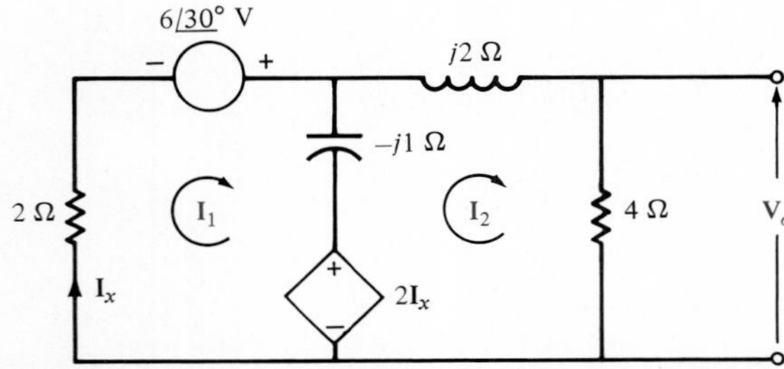

Figure 10.7 Example circuit used to illustrate mesh equations.

and hence

$$\mathbf{I}_1 = \frac{(4 + j1)(6 \underline{/30^\circ})}{18 + j2} = 1.36 \underline{/37.70^\circ} \text{ A}$$

and

$$\mathbf{I}_2 = \frac{(2 - j1)(6 \underline{/30^\circ})}{18 + j2} = 0.74 \underline{/-2.91^\circ} \text{ A}$$

Then $\mathbf{V}_o$ is

$$\mathbf{V}_o = 4\mathbf{I}_2 = 2.96 \underline{/-2.91^\circ} \text{ V}$$

■

DRILL EXERCISE

D10.4. Use mesh equations to find $\mathbf{V}_o$ in the network in Fig. D10.4.

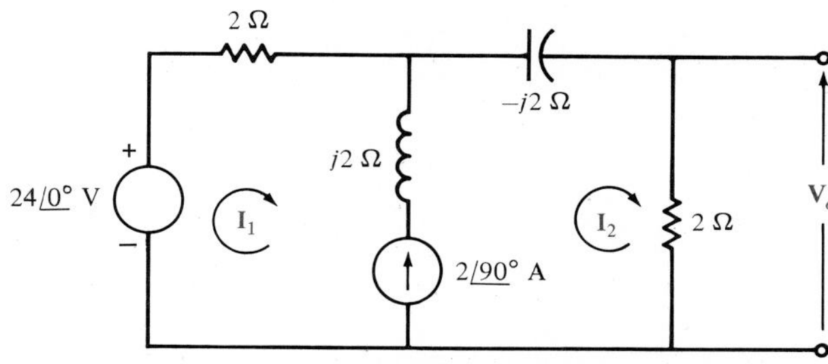

Figure D10.4

D10.5. Find $\mathbf{V}_o$ in the network in Fig. D10.5 using mesh equations.

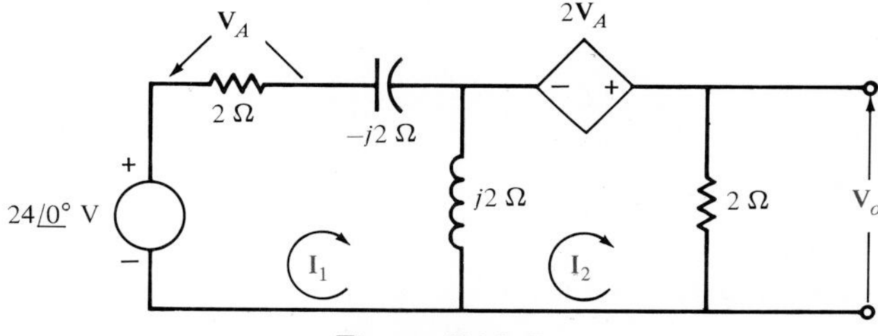

Figure D10.5

10.4 Superposition

We will now illustrate the principle of superposition in ac steady-state circuit analysis. Our vehicle for illustrating this principle will be the circuit analyzed in Example 10.6 and shown in Fig. 10.6.

EXAMPLE 10.8

Let us use superposition to determine $\mathbf{V}_o$ in the circuit in Fig. 10.6. We first replace the voltage source with a short circuit and compute $\mathbf{V}_o'$, which is the component of $\mathbf{V}_o$ due

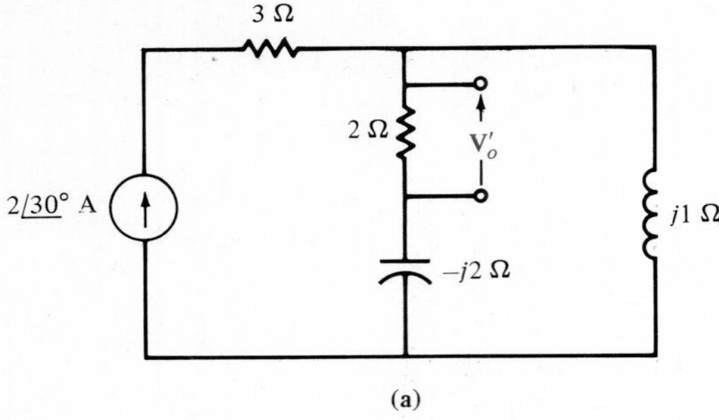

(a)

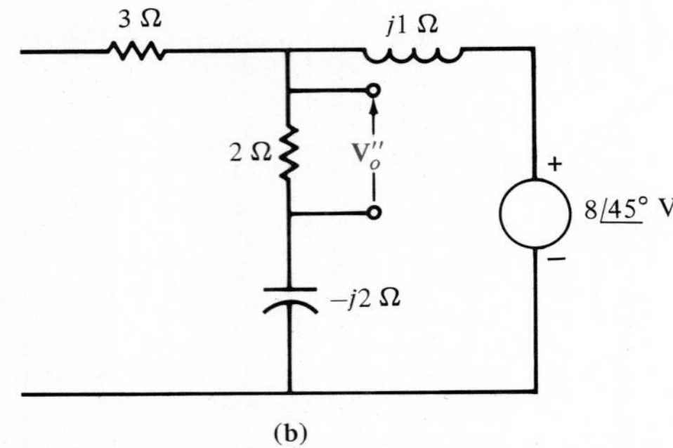

(b)

Figure 10.8 Circuits used to determine $\mathbf{V}_o$ in Fig. 10.6 via superposition.

to the current source, as shown in Fig. 10.8a. Using current division, we obtain

$$\mathbf{V}_o' = \frac{(2\,\underline{/30^\circ})(j1)}{2 - j2 + j1}(2)$$
$$= 1.79\,\underline{/146.57^\circ} \text{ V}$$

Now replacing the current source with an open circuit, we can determine $\mathbf{V}_o''$, which is the component of $\mathbf{V}_o$ due to the voltage source, as shown in Fig. 10.8b. Using voltage division, we have

$$\mathbf{V}_o'' = \frac{8\,\underline{/45^\circ}}{2 - j2 + j1}(2)$$
$$= 7.14\,\underline{/71.57^\circ} \text{ V}$$

Finally,

$$\mathbf{V}_o = \mathbf{V}_o' + \mathbf{V}_o''$$
$$= 1.79\,\underline{/146.57^\circ} + 7.14\,\underline{/71.57^\circ}$$
$$= 0.77 + j7.76 \text{ V}$$

■

DRILL EXERCISE

D10.6. Using superposition, find $\mathbf{V}_o$ in the network in Fig. D10.6.

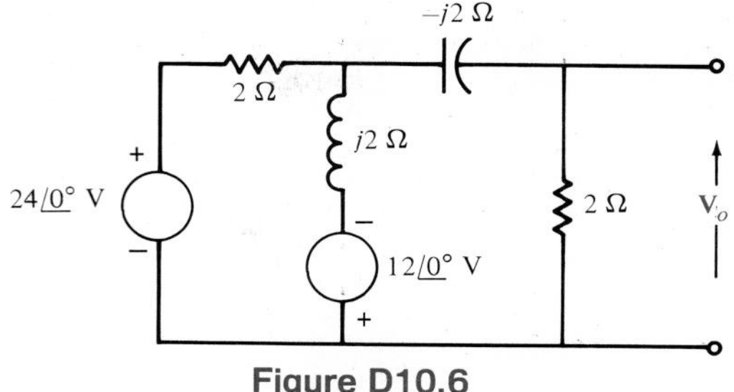

Figure D10.6

D10.7. Find $\mathbf{V}_o$ in the network in Fig. D10.7 using superposition.

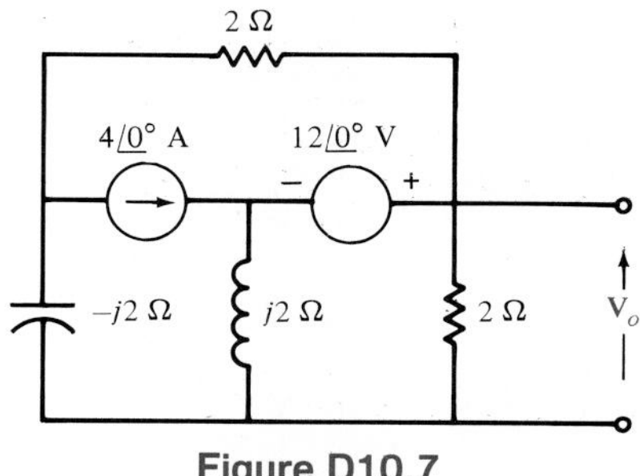

Figure D10.7

10.5 Source Transformation

The application of source transformation to ac steady-state circuit analysis will be illustrated by the following example.

EXAMPLE 10.9

We wish to determine $\mathbf{I}_4$ in the circuit in Fig. 10.1 using source transformation.

First we convert the voltage source and series 4-Ω resistor to a current source and parallel 4-Ω resistor as shown in Fig. 10.9a. The current source and parallel resistor and capacitor are now converted into an equivalent impedance in series with an equivalent voltage source as shown in Fig. 10.9b, where

$$\mathbf{Z}_1 = \frac{(4)(-j3)}{4 - j3}$$

$$= \frac{-j12}{4 - j3}\ \Omega$$

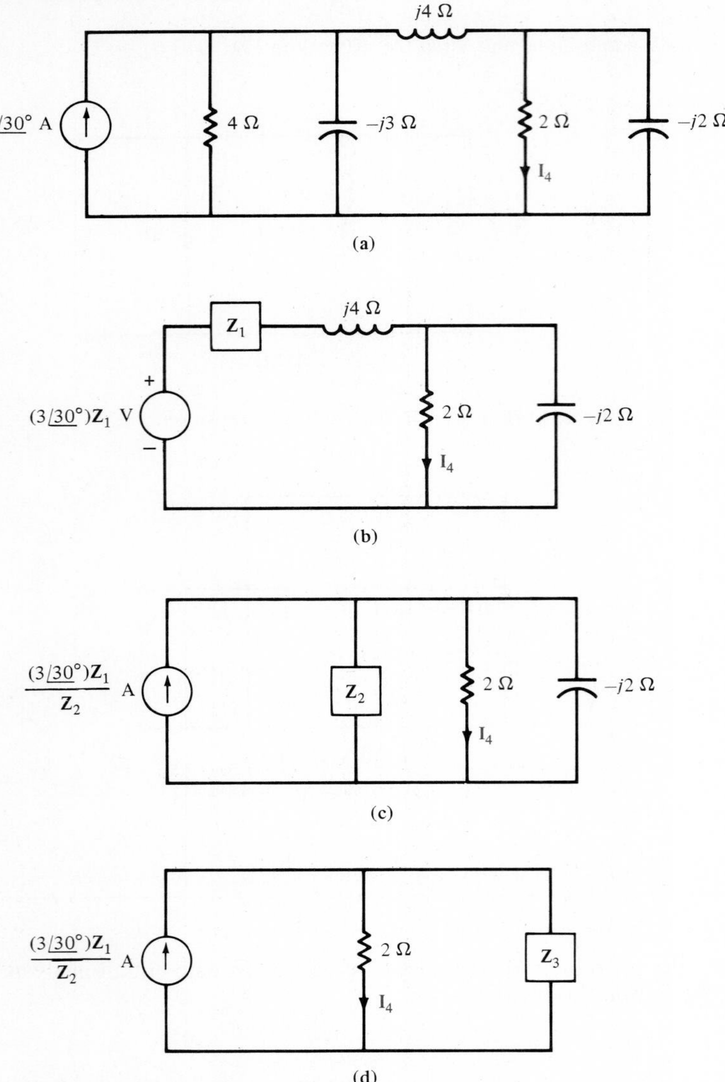

Figure 10.9 Circuits used in solving Example 10.1 via source transformation.

Combining $\mathbf{Z}_1$ with the inductor yields $\mathbf{Z}_2$, where

$$\mathbf{Z}_2 = \mathbf{Z}_1 + j4 = \frac{12 + j4}{4 - j3}\ \Omega$$

Converting the circuit in Fig. 10.9b back to a current source in parallel with $\mathbf{Z}_2$ produces the circuit in Fig. 10.9c. Combining $\mathbf{Z}_2$ with the parallel capacitor produces the circuit in Fig. 10.9d, where

$$\mathbf{Z}_3 = \frac{(-j2)(\mathbf{Z}_2)}{-j2 + \mathbf{Z}_2}$$

$$= \frac{4 - j12}{3 - j2}\ \Omega$$

The value of the current source is

$$\frac{(3\ \underline{/30^\circ})\mathbf{Z}_1}{\mathbf{Z}_2} = \frac{(3\underline{/30^\circ})[-j12/(4 - j3)]}{\dfrac{12 + j4}{4 - j3}} = 2.85\ \underline{/-78.43^\circ}\ \text{A}$$

Now employing current division

$$\mathbf{I}_4 = \frac{(2.85\ \underline{/-78.43^\circ})[(4 - j12)/(3 - j2)]}{2 + \dfrac{4 - j12}{3 - j2}}$$

$$= 1.91\ \underline{/-92^\circ}\ \text{A}$$

■

DRILL EXERCISE

D10.8. Find $\mathbf{V}_o$ in the network in Fig. D10.6 using source transformation.

10.6

Thévenin's and Norton's Theorems

In the following examples we will demonstrate the use of Thévenin's and Norton's theorems in the solution of a variety of problems including those in which these theorems are combined with other analysis techniques. We will illustrate the use of Thévenin's and Norton's theorems in the frequency domain and show that the open-circuit voltage and short-circuit current are phasors, and therefore their ratio is a complex quantity which we call the Thévenin equivalent impedance.

EXAMPLE 10.10

We will determine the voltage $\mathbf{V}_o$ in the circuit in Fig. 10.6 using Thévenin's theorem. The open-circuit voltage $\mathbf{V}_{oc}$ is obtained from the circuit in Fig. 10.10a. Since the current in the outer loop is $2\ \underline{/30^\circ}$ A, then

$$\mathbf{V}_{oc} = (2\ \underline{/30^\circ})(j1) + 8\ \underline{/45^\circ}$$

$$= 8.73\ \underline{/57.75^\circ}\ \text{V}$$

The Thévenin equivalent impedance is found by open-circuiting the current source, short-circuiting the voltage source, and computing the impedance at the terminals, where $\mathbf{V}_{oc}$ is determined as shown in Fig. 10.10b. Note that

$$\mathbf{Z}_{Th} = -j2 + j1 = -j1\ \Omega$$

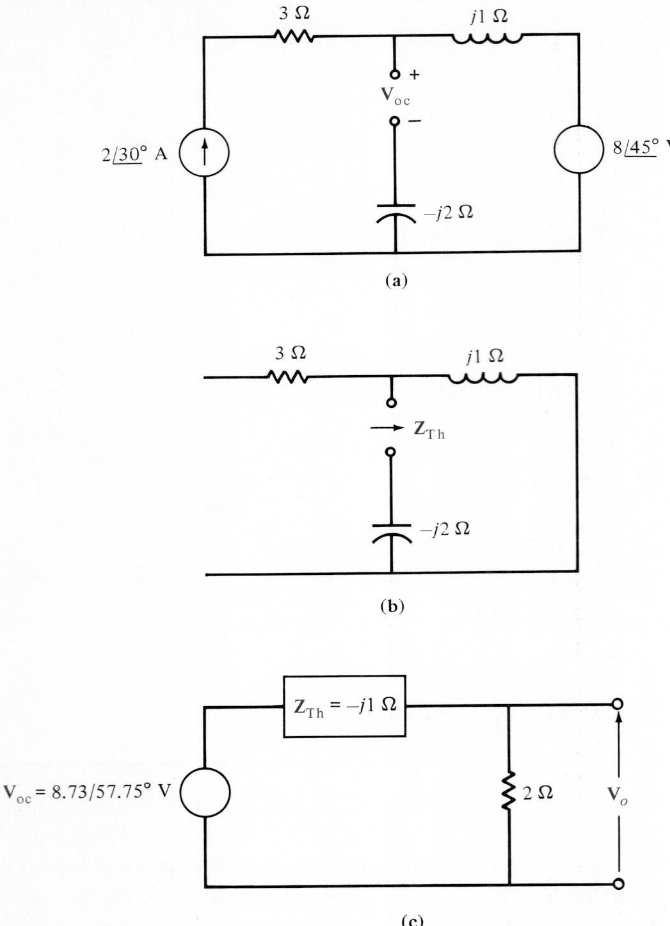

Figure 10.10 Circuits used to compute $\mathbf{V}_o$ in Fig. 10.6 using Thévenin's theorem.

The Thévenin equivalent circuit is then shown in Fig. 10.10c together with the 2-Ω load. The voltage $\mathbf{V}_o$ is then calculated as

$$\mathbf{V}_o = \frac{(8.73\,\underline{/57.75^\circ})(2)}{2 - j1}$$

$$= 0.76 + j7.76 \text{ V}$$

■

As another example of the use of Thévenin's theorem, let us determine the voltage $\mathbf{V}_o$ in the circuit in Example 10.7. This will indicate the use of Thévenin's theorem with a dependent source.

EXAMPLE 10.11

We wish to compute $\mathbf{V}_o$ in Fig. 10.7 using Thévenin's theorem. The open-circuit voltage $\mathbf{V}_{oc}$ is determined from the circuit in Fig. 10.11a. KVL for the loop is

$$2\mathbf{I}_x - 6\,\underline{/30^\circ} - (j1)\mathbf{I}_x + 2\mathbf{I}_x = 0$$

$$\mathbf{I}_x = \frac{6\,\underline{/30^\circ}}{4 - j1}\text{ A}$$

Therefore,

$$\mathbf{V}_{oc} = (2 - j1)\mathbf{I}_x$$

$$= \frac{(6\,\underline{/30^\circ})(2 - j1)}{4 - j1}\text{ V}$$

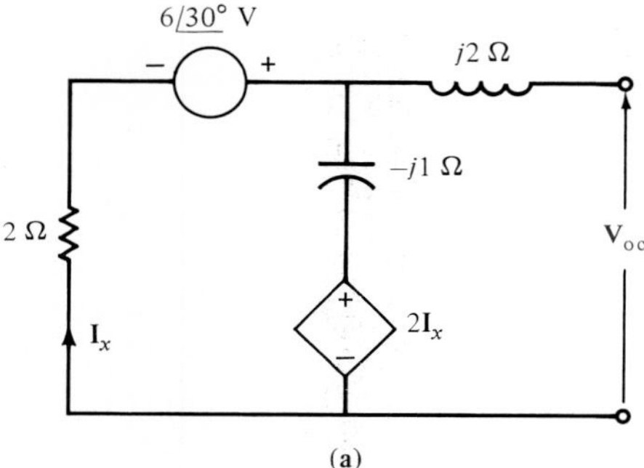

(a)

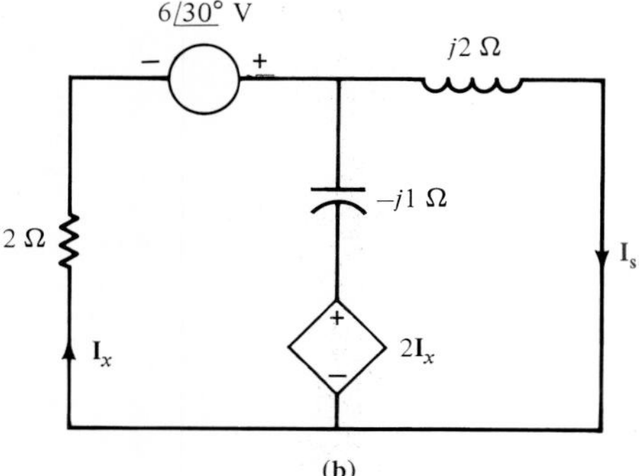

(b)

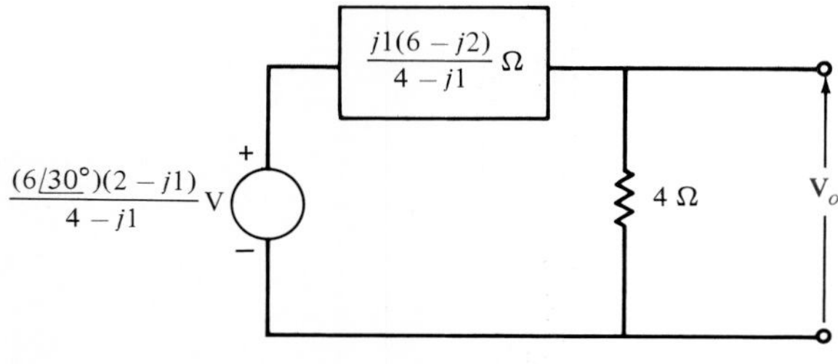

(c)

Figure 10.11 Circuits used to compute $\mathbf{V}_o$ in Fig. 10.7 using Thévenin's theorem.

The short-circuit current is derived from the circuit in Fig. 10.11b. The two mesh equations for the circuit are

$$-2\mathbf{I}_z + 6\,\underline{/30^\circ} = (2 - j1)\mathbf{I}_x - \mathbf{I}_{sc}(-j1)$$

$$2\mathbf{I}_x = -\mathbf{I}_x(-j1) + \mathbf{I}_{sc}(j2 - j1)$$

or

$$(4 - j1)\mathbf{I}_x + (j1)\mathbf{I}_{sc} = 6\,\underline{/30^\circ}$$

$$(-2 + j1)\mathbf{I}_x + (j1)\mathbf{I}_{sc} = 0$$

Solving these equations for $\mathbf{I}_{sc}$, we obtain

$$\mathbf{I}_{sc} = \frac{(2 - j1)(6\,\underline{/30^\circ})}{j1(6 - j2)}\ \text{A}$$

The Thévenin equivalent impedance is then

$$\mathbf{Z}_{\text{Th}} = \frac{\mathbf{V}_{oc}}{\mathbf{I}_{sc}} = \frac{(j1)(6 - j2)}{(4 - j1)}\ \Omega$$

The Thévenin equivalent circuit, together with the 4-Ω load, is shown in Fig. 10.11c. $\mathbf{V}_o$ is then computed as

$$\mathbf{V}_o = \frac{\dfrac{(6\,\underline{/30^\circ})(2 - j1)}{4 - j1}}{4 + \dfrac{(j1)(6 - j2)}{4 - j1}}(4)$$

$$= 2.96\,\underline{/-2.91^\circ}\ \text{V}$$

This is the same value obtained in Example 10.7. ■

The following example illustrates the use of Norton's theorem. Once again the problem has been solved in an earlier example, and therefore we can check our results and compare the methods.

EXAMPLE 10.12

Let us employ Norton's theorem to find the current in the 2-Ω resistor in Fig. 10.2, which flows from the node labeled $\mathbf{V}_2$ to ground.

The short-circuit current is computed from the circuit in Fig. 10.12a. Because the short circuit places the voltage source directly across the inductor, the inductor current is

$$\mathbf{I}_L = \frac{12\,\underline{/30^\circ}}{j1} = 12\,\underline{/-60^\circ}\ \text{A}$$

KCL applied at node 1 then indicates that

$$\mathbf{I}_{sc} = 2\,\underline{/60^\circ} - 12\,\underline{/-60^\circ}$$

$$= 13.11\,\underline{/112.42^\circ}\ \text{A}$$

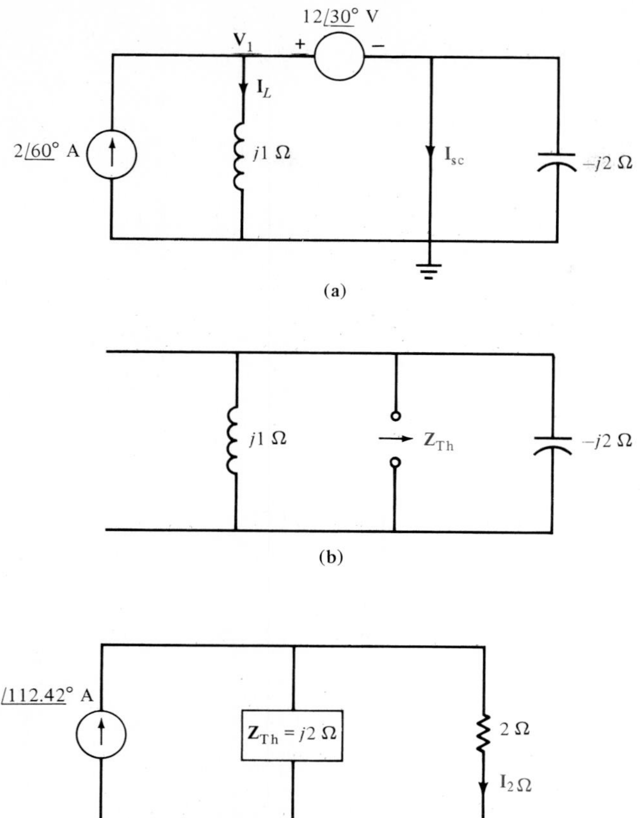

Figure 10.12 Circuits used to compute the current in the 2-Ω resistor in Fig. 10.2 using Norton's theorem.

The Thévenin equivalent impedance is calculated from Fig. 10.12b as

$$\mathbf{Z}_{\text{Th}} = \frac{(j1)(-j2)}{j1 - j2} = j2 \ \Omega$$

Therefore, the current in the 2-Ω resistor in Fig. 10.2 is computed by employing the Norton equivalent circuit as shown in Fig. 10.12c.

$$\mathbf{I}_{2\Omega} = \frac{(13.11 \angle 112.42^\circ)(j2)}{2 + j2}$$

$$= \frac{18.50 \angle 157.54^\circ}{2} \text{ A}$$

Since the voltage $\mathbf{V}_2$ in Fig. 10.2 was computed to be $\mathbf{V}_2 = 18.50 \angle 157.54^\circ$ V in Example 10.2, the value of $\mathbf{I}_{2\Omega}$ checks with previous results. ■

DRILL EXERCISE

D10.9. Use Thévenin's theorem to compute $\mathbf{V}_o$ in the network in Fig. D10.6.

D10.10. Use Norton's theorem to find $\mathbf{V}_o$ in the network in Fig. D10.6.

D10.11. Use Thévenin's theorem to find $\mathbf{V}_o$ in the network in Fig. D10.7.

D10.12. Find $\mathbf{V}_o$ in the network in Fig. D10.5 using Thévenin's theorem.

EXAMPLE 10.13

We wish to find the output voltage $\mathbf{V}_o$ in the network shown in Fig. 10.13a.

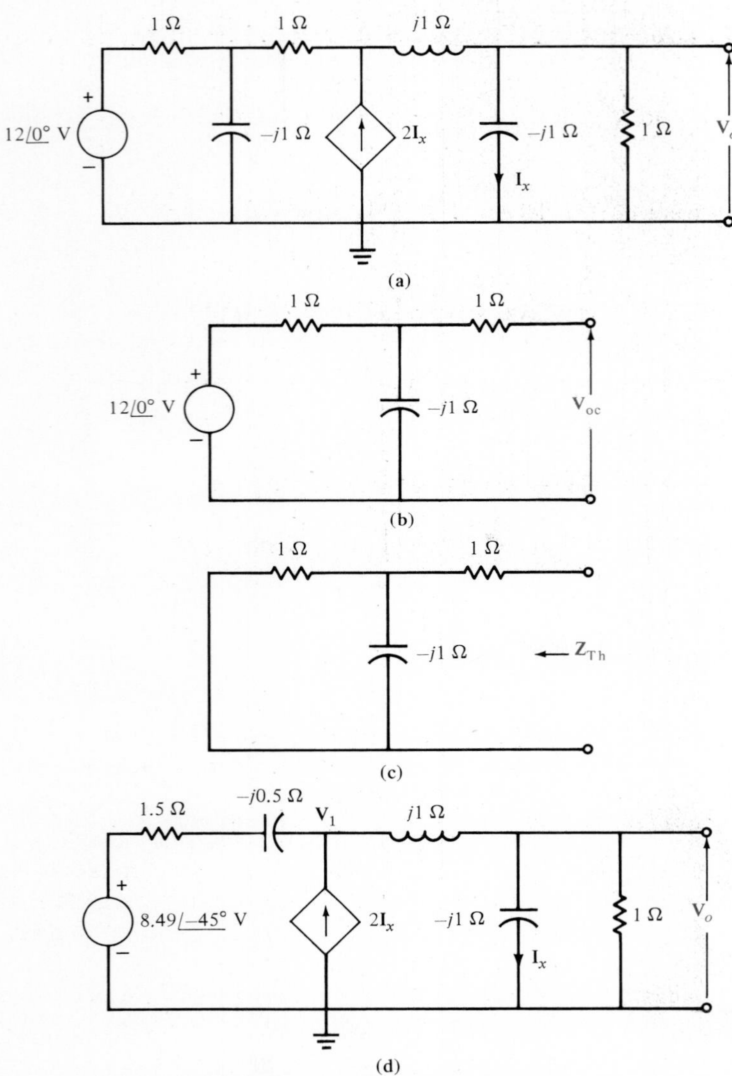

Figure 10.13 Example network.

The network is fairly complicated, and therefore it would appear that a direct attack might be time consuming. Since it is the output voltage at the right of the network that is required, we can begin by simplifying the left portion of the network. Note that if we form a simple Thévenin equivalent for the portion of the network to the left of the dependent source, we can reduce the circuit to one which has two unknown node voltages, one of which is the output voltage $\mathbf{V}_o$.

The portion of the network to the left of the current source is shown in Fig. 10.13b. The open-circuit voltage is

$$\begin{aligned}\mathbf{V}_{oc} &= \frac{(12 \angle 0^\circ)(-j1)}{1 - j1} \\ &= 8.49 \angle -45^\circ \text{ V}\end{aligned}$$

The Thévenin equivalent impedance derived from Fig. 10.13c is

$$\begin{aligned}\mathbf{Z}_{Th} &= 1 + \frac{-j1}{1 - j1} \\ &= 1.5 - j0.5\ \Omega\end{aligned}$$

The simplified network employing the Thévenin equivalent is shown in Fig. 10.13d. The node equations for this network are

$$\frac{\mathbf{V}_1 - 8.49 \angle -45^\circ}{1.5 - j0.5} + \frac{\mathbf{V}_1 - \mathbf{V}_o}{j1} = 2\mathbf{I}_x$$

$$\frac{\mathbf{V}_o - \mathbf{V}_1}{j1} + \frac{\mathbf{V}_o}{-j1} + \frac{\mathbf{V}_o}{1} = 0$$

$$\mathbf{I}_x = \frac{\mathbf{V}_o}{-j1}$$

Solving these equations yields

$$\mathbf{V}_o = 6.00 \angle 0^\circ \text{ V}$$

■

10.7 SPICE Analysis Techniques

The ac-analysis portion of SPICE is similar to the dc-analysis part except that a steady-state frequency-domain solution is obtained. The circuit solved by SPICE is a frequency-domain circuit. Since the analysis is in the frequency domain, all ac sources must be sinusoidal and have the same frequency. The computer calculates the impedances of all branches from the input values of the elements (and the frequency), and the programmer furnishes the magnitudes and phase angles of all sources. Dependent sources are also phasors and are always a constant times a branch voltage or current.

We have already presented, at the end of Chapters 5 and 7, all the necessary tools for a SPICE analysis in this case, with the exception of the proper solution control statement. For an ac-analysis the solution control statement is

```
.AC XXX NO FSTART FSTOP
```

XXX is DEC for decade variation, OCT for octave variation, or LIN for linear variation. NO is number of points per decade, octave, or just points, depending on whether DEC, OCT, or LIN was specified. FSTART is the starting frequency, and FSTOP is the final frequency. Note that the AC statement is not meaningful if no independent ac sources are specified.

For the ac analysis, five additional output variables can be used by replacing V by

VR	real part
VI	imaginary part
VM	magnitude
VP	phase
VDB	20 log10 (magnitude)

Note that in this chapter we will use only the linear variation. In fact, we will compute the unknown voltages and currents at only one frequency—the frequency of the sources. Decade and octave variations are explained and used in Chapter 13.

The following examples will serve to illustrate the use of SPICE in ac-circuit analysis.

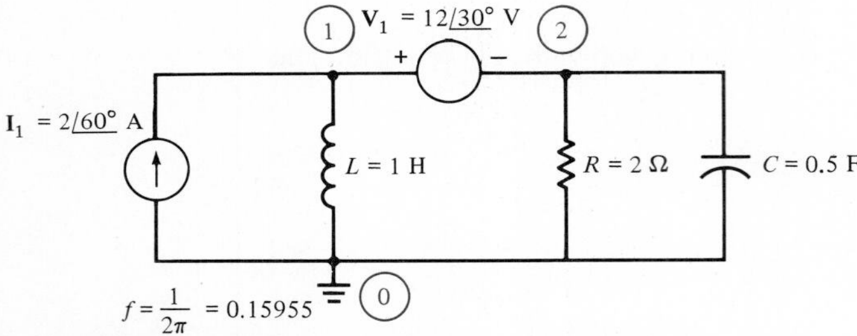

Figure 10.14 Circuit analyzed in Example 10.14

EXAMPLE 10.14

Let us write a program that will determine the node voltage of the network in Fig. 10.14. The program that will compute and print the node voltages is listed below.

```
EXAMPLE 10-14  AC ANALYSIS
I1   0   1   AC  2    60
L    1   0   1
VI   1   2   AC  12   30
R    2   0   2
C    2   0   .5
.AC  LIN 1   .159155   .159155
.PRINT AC VM(1) VP(1) VM(2) VP(2)
.END
```

Note that the analysis is requested at only one frequency, $f = 1/2\pi$ hertz. The output from the computer includes the following.

```
  FREQ        VM(1)        VP(1)        VM(2)        VP(2)
1.592D-01   1.475D+01    1.172D+02    1.855D+01    1.574D+02
```

which indicates that V1 = 14.74 $\angle 117.2°$ V and V2 = 18.55 $\angle 157.4°$ V. ■

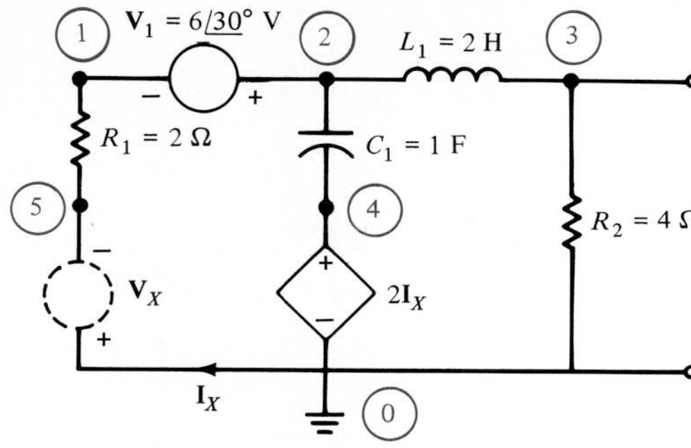

Figure 10.15 Figure 10.7 redrawn for SPICE analysis.

EXAMPLE 10.15

Consider once again the problem presented in Example 10.7. The circuit in Fig. 10.7 is redrawn for a SPICE analysis in Fig. 10.15. The program for computing the output voltage is

```
EXAMPLE 10.15   AC ANALYSIS
V1   2   1   AC   6   30
R1   1   5   2
VX   0   5   AC   0   0
C1   2   4   1
H1   4   0   VX   2
L1   2   3   2
R2   3   0   4
.AC LIN 1    0.159155   0.159155
.PRINT  AC   VM(3)      VP(3)
.END
```

The program yields the following computer output:

```
   FREQ             VM(3)              VP(3)
 1.592E-10        2.963E+00        -2.905E+00
```

■

EXAMPLE 10.16

Let us consider again the problem in Example 10.13. The network in Fig. 10.13a is redrawn for a SPICE analysis in Fig. 10.16. The program for calculating the output voltage is listed below.

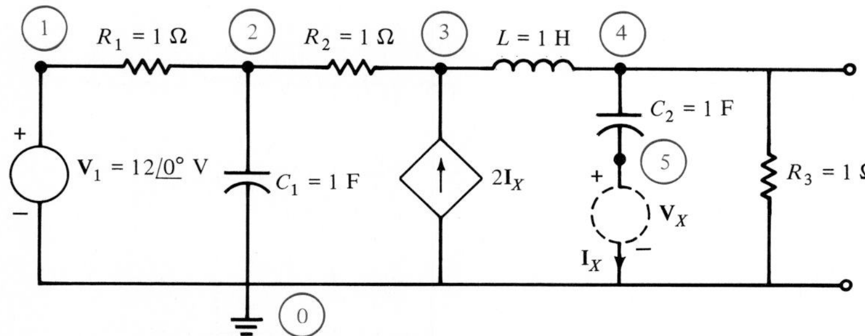

Figure 10.16 Network in Fig. 10.13a redrawn for SPICE analysis.

```
EXAMPLE 10.16   AC ANALYSIS
V1   1   0    AC   12   0
R1   1   2    1
C1   2   0    1
R2   2   3    1
F1   0   3    VX   2
L1   3   4    1
C2   4   5    1
VX   5   0    AC   0    0
R3   4   0    1
.AC LIN 1     0.159155     0.159155
.PRINT   AC   VM(4)        VP(4)
.END
```

The computer output for the program is

```
    FREQ              VM(4)            VP(4)
  1.592E-01        6.000E+00        2.049E-05
```

■

Since our analysis approach in this chapter has been confined to a single frequency, the SPICE examples have maintained that format. However, the use of the plot routines for analysis over a range of frequencies is presented in Chapter 13.

DRILL EXERCISE

D10.13. Given the network in Fig. D10.13, compute $\mathbf{V}_o$ and $\mathbf{I}_o$ using SPICE if the frequency is $f = 1/2\pi$ Hz.

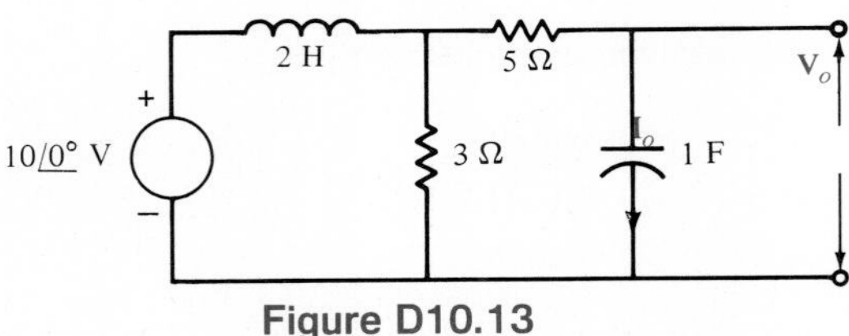

Figure D10.13

D10.14. Given the network in Fig. D10.14, determine $\mathbf{V}_o$ and $\mathbf{I}_o$ using SPICE if the frequency is $f = 1/2\pi$ Hz.

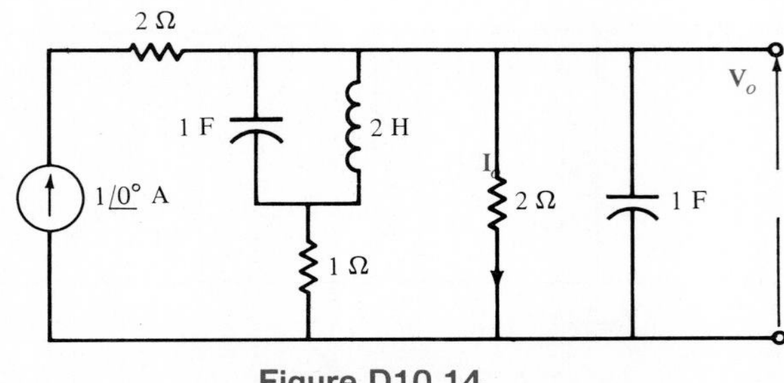

Figure D10.14

D10.15. Use SPICE to determine $\mathbf{V}_o$ and $\mathbf{I}_o$ in the network in Fig. D10.15 if the frequency is $f = 1/2\pi$ Hz.

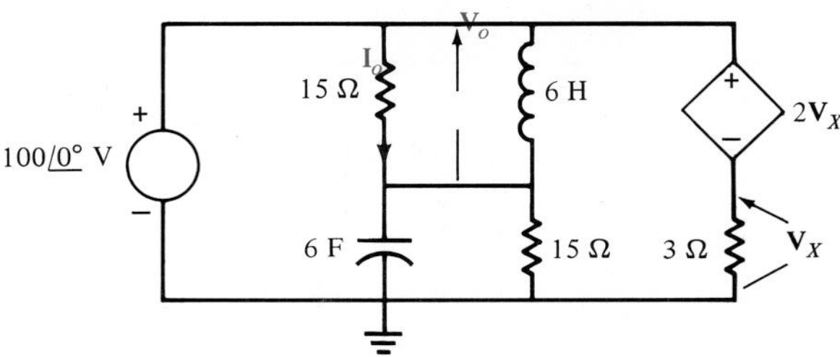

Figure D10.15

10.8 Summary

This chapter has illustrated that through the use of phasors, impedance, and admittance, we can determine the steady-state response of an ac circuit to a sinusoidal input using such techniques as nodal analysis, loop analysis, superposition, source transformation, and Thévenin's and Norton's theorems.

We have also demonstrated the use of SPICE in the solution of ac steady-state circuit problems. Finally, it has been shown that the ac-analysis portion of the SPICE circuit-analysis program is a simple and efficient technique for solving ac circuit analysis problems.

KEY POINTS

- The principle of linearity can be applied to ac steady-state circuit problems.
- All the solution techniques applied to dc circuits—nodal analysis, loop analysis, superposition, source transformation, Thévenin's theorem, and Norton's theorem—are all applicable in the solution of ac steady-state circuit problems.
- With the addition of a solution control statement and five output variables, the SPICE techniques presented in earlier chapters can be applied to the solution of ac steady-state circuit problems.

PROBLEMS

10.1. Find $\mathbf{I}_o$ in Fig. P10.1 by assuming that $\mathbf{I}_o = 1\,\underline{/0^\circ}$ A and using linearity.

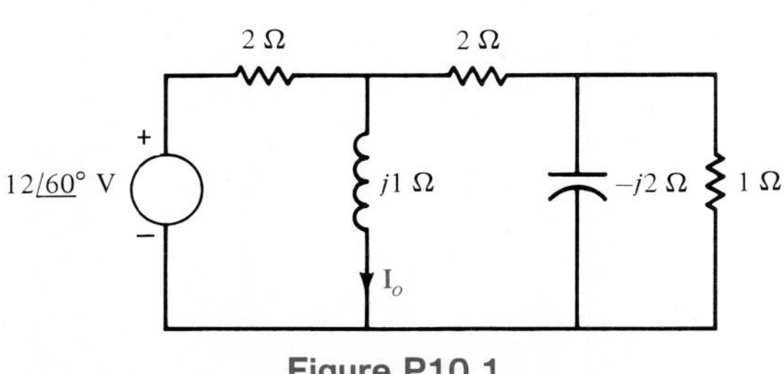

Figure P10.1

10.2. Find the current in the inductor in the circuit shown in Fig. P10.2, using nodal analysis.

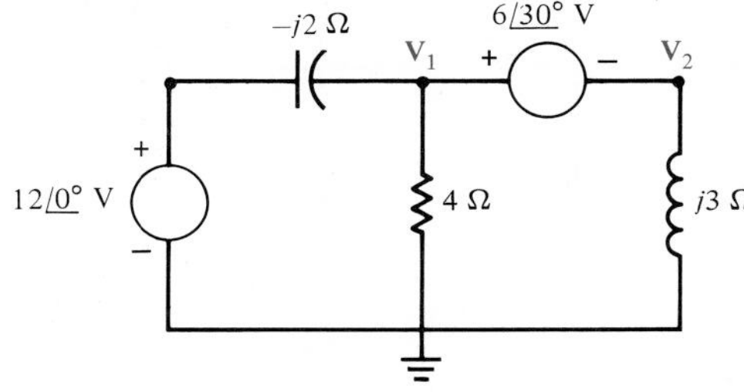

Figure P10.2

10.3. Use nodal analysis to find $\mathbf{I}_o$ in the network in Fig. P10.3.

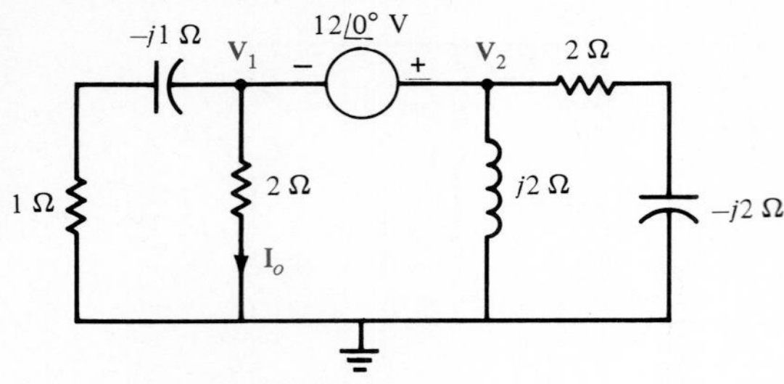

Figure P10.3

10.4. Use nodal analysis to find $\mathbf{V}_o$ in the network in Fig. P10.4.

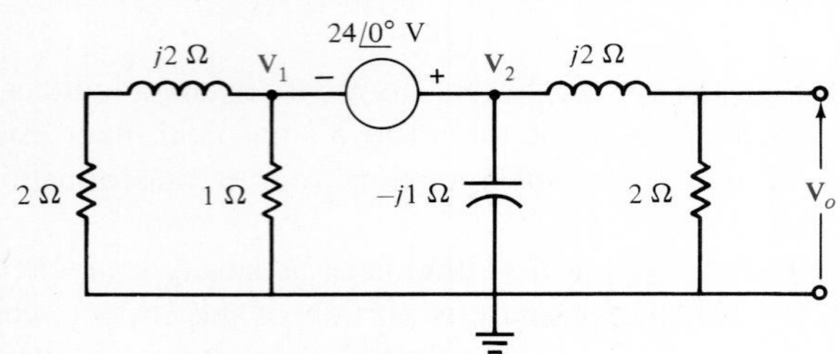

Figure P10.4

10.5. Find $\mathbf{I}_o$ in Fig. P10.5 using nodal analysis.

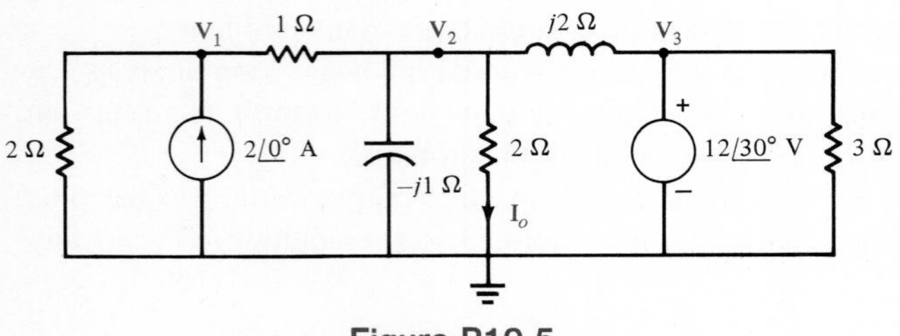

Figure P10.5

10.6. Find $\mathbf{I}_o$ in the circuit shown in Fig. P10.6 using nodal analysis.

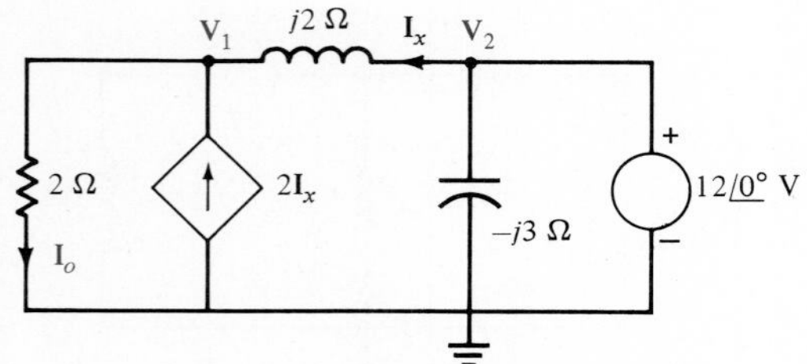

Figure P10.6

10.7. Determine $\mathbf{V}_x$ in Fig. P10.7 using nodal analysis.

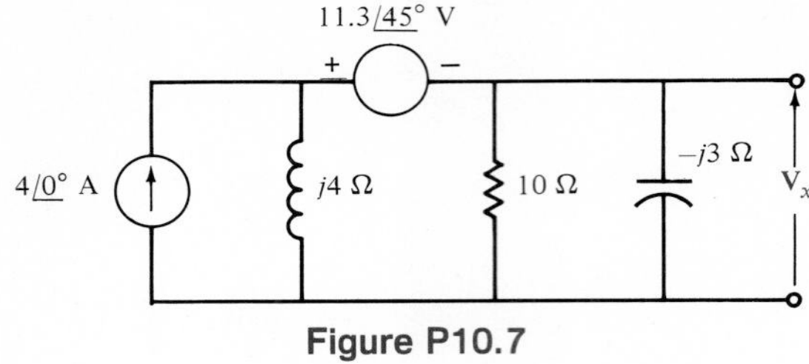

Figure P10.7

10.8. Calculate $\mathbf{I}_o$ in Fig. P10.8 using nodal analysis.

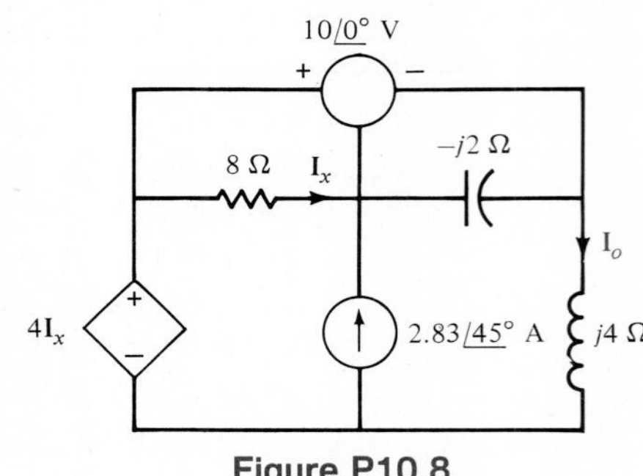

Figure P10.8

10.9. Use nodal analysis to determine $\mathbf{I}_o$ in Fig. P10.9.

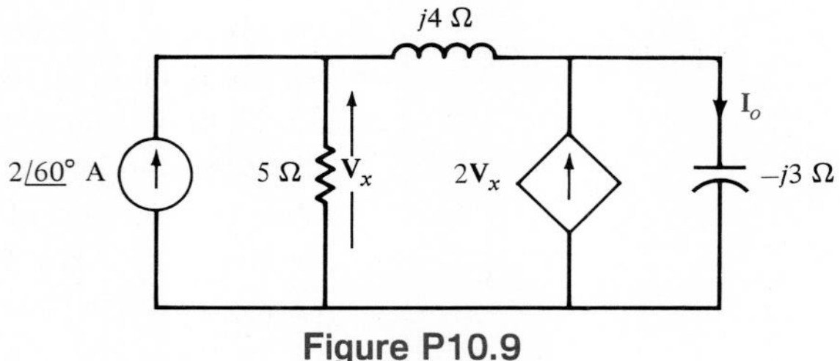

Figure P10.9

10.10. Find the voltage across the inductor in the circuit shown in Fig. P10.10 using nodal analysis.

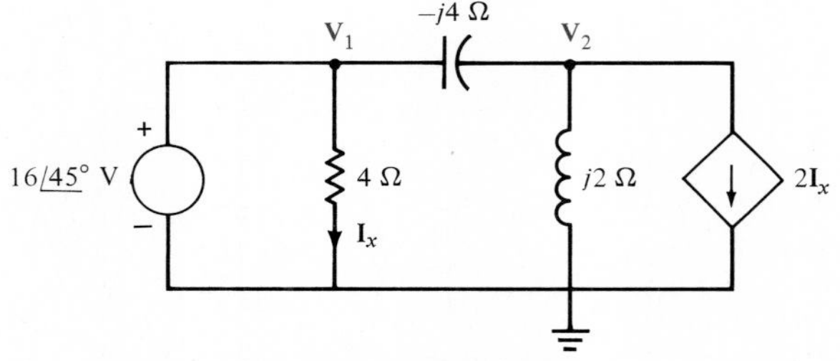

Figure P10.10

10.11. Determine $\mathbf{I}_o$ in the circuit shown in Fig. P10.11 using nodal analysis.

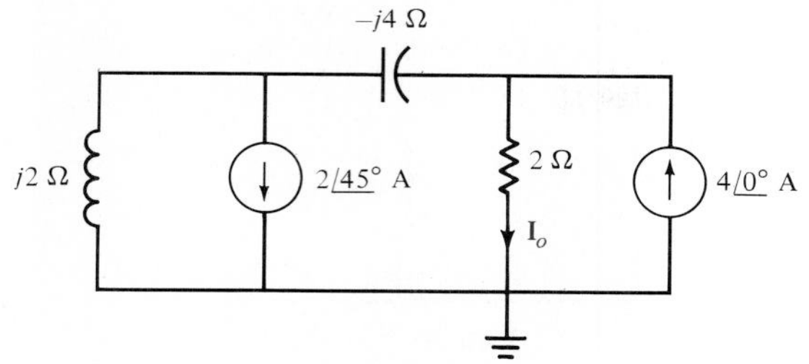

Figure P10.11

10.12. Find the voltage gain $\mathbf{V}_o/\mathbf{V}_s$ for the low-pass filter shown in Fig. P10.12 (and described in Chapter 13) using nodal analysis. Assume that the op amp is ideal so that the op amp together with resistors R_3 and R_4 can be represented by a gain of $A = (1 + R_4/R_3)$, using the representation of Fig. 2.31b with $R_i = \infty$ and $R_o = 0$.

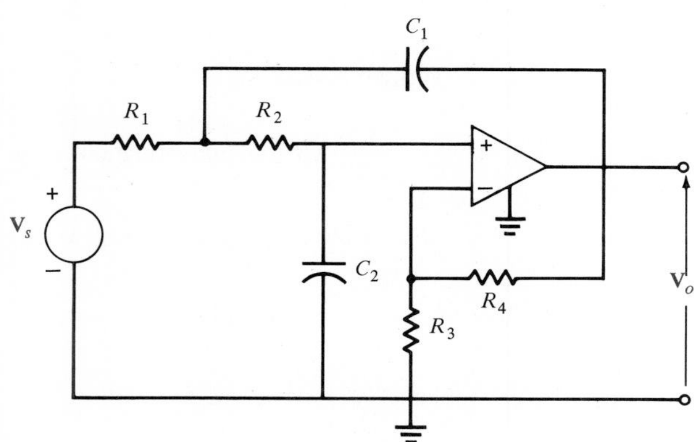

Figure P10.12

10.13. Find the voltage gain $\mathbf{V}_o/\mathbf{V}_s$ for the high-pass filter shown in Fig. P10.13 (and described in Chapter 13) using nodal analysis. Assume that the op amp is ideal so that the op amp together with resistors R_3 and R_4 can be represented by a gain of $A = (1 + R_4/R_3)$, using the representation in Fig. 2.31b with $R_i = \infty$ and $R_o = 0$.

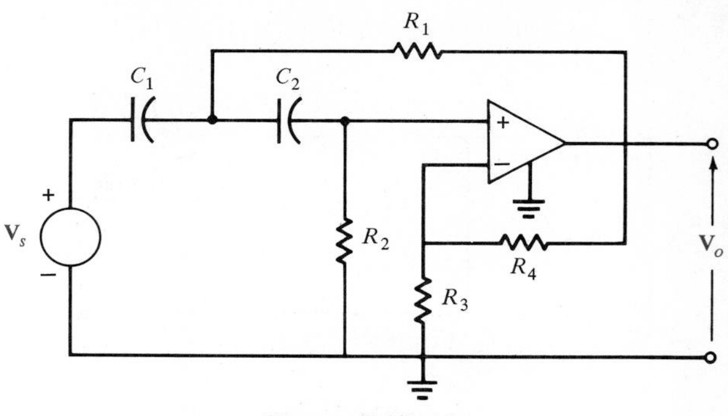

Figure P10.13

10.14. The low-frequency equivalent circuit for a common-emitter transistor amplifier is shown in Fig. P10.14. Compute the voltage gain $\mathbf{V}_o/\mathbf{V}_s$.

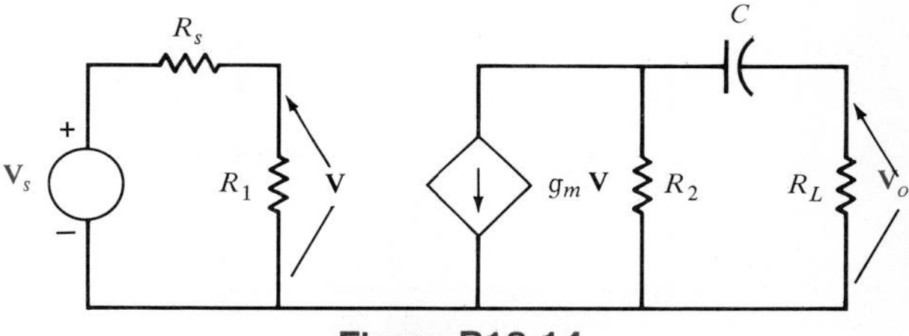

Figure P10.14

10.15. Given the small-signal high-frequency equivalent circuit for a shunt-peaked amplifier as shown in Fig. P10.15, calculate the current gain $\mathbf{I}_o/\mathbf{I}_s$.

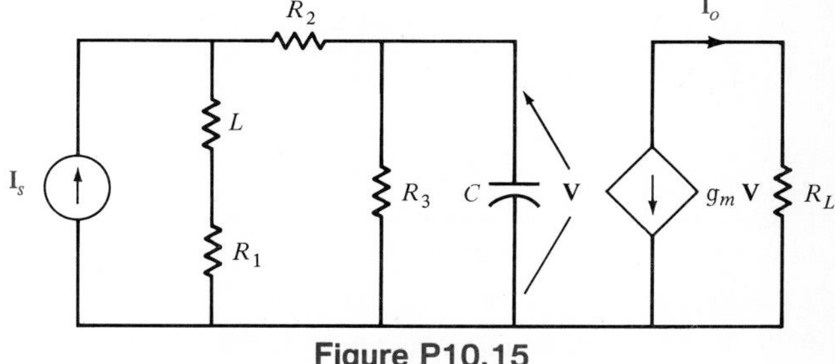

Figure P10.15

10.16. Solve Problem 10.2 using mesh analysis.

10.17. Solve for $\mathbf{V}_x$ in Problem 10.7 using loop analysis.

10.18. Find $\mathbf{I}_o$ in Problem 10.8 using loop analysis.

10.19. Determine $\mathbf{I}_o$ in Problem 10.9 using loop analysis.

10.20. Use loop analysis to find $\mathbf{I}_o$ in Fig. P10.20.

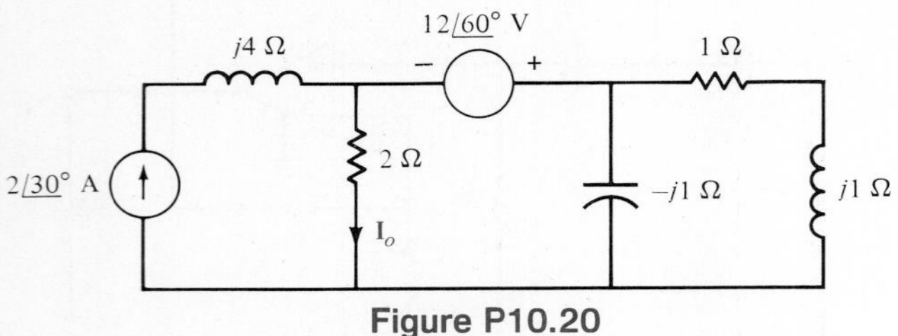

Figure P10.20

10.21. Use loop analysis to find $\mathbf{V}_o$ in Fig. P10.21.

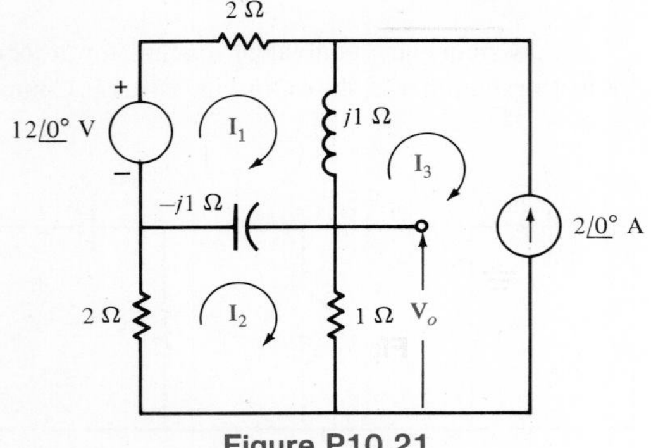

Figure P10.21

10.22. Find $\mathbf{V}_o$ in the circuit shown in Fig. P10.22 using mesh equations.

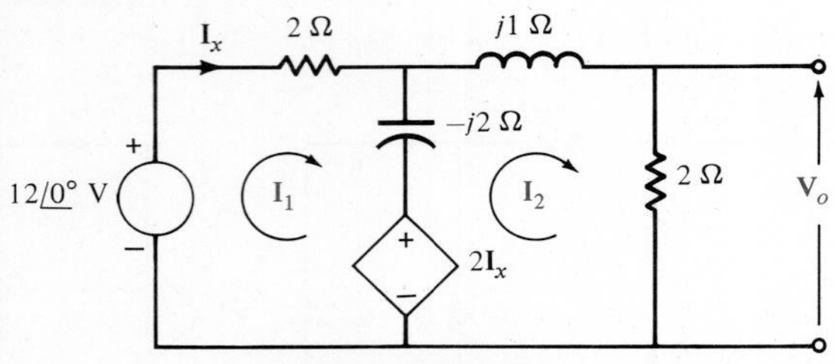

Figure P10.22

10.23. Use mesh analysis to find $\mathbf{V}_o$ in the circuit shown in Fig. P10.23.

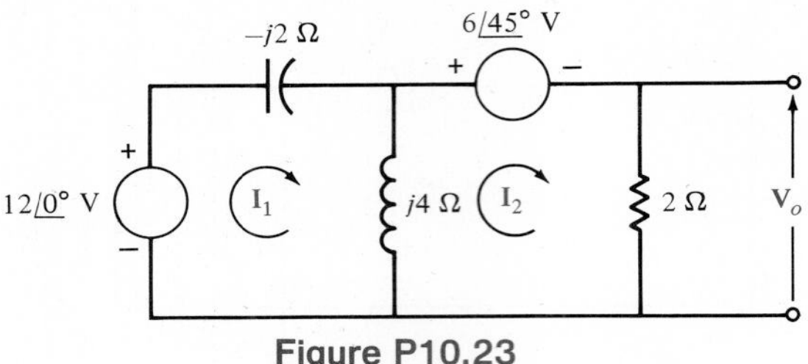

Figure P10.23

10.24. Use mesh analysis to find $\mathbf{V}_o$ in the circuit shown in Fig. P10.24.

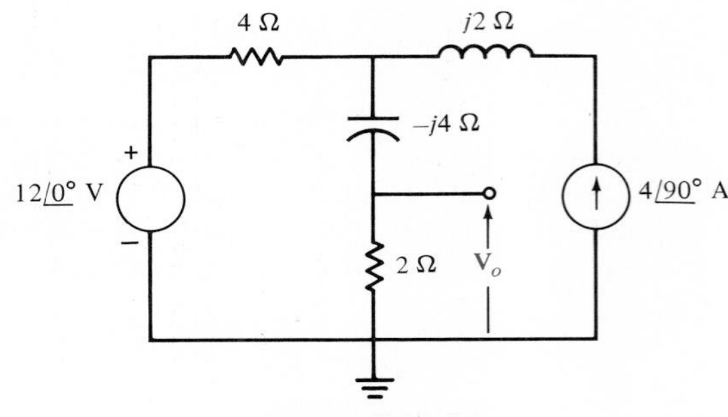

Figure P10.24

10.25. Solve Problem 10.2 using source transformation.

10.26. Solve Problem 10.11 using source transformation.

10.27. Solve Problem 10.23 using source transformation.

10.28. Find $\mathbf{V}_o$ in Problem 10.24 using source transformation.

10.29. Solve Problem 10.2 using superposition.

10.30. Use superposition to solve for $\mathbf{V}_o$ in Fig. P10.30.

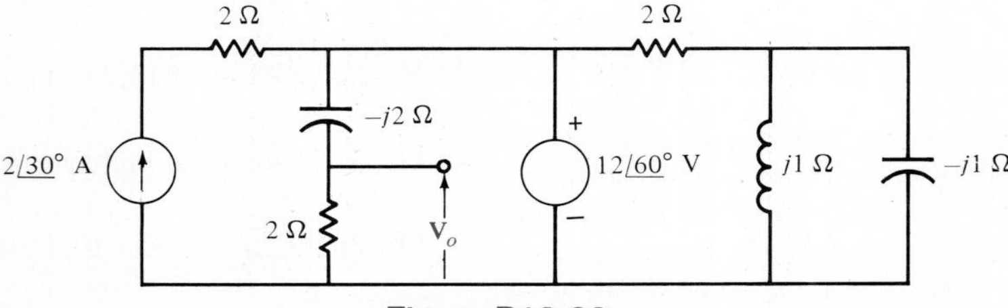

Figure P10.30

10.31. Solve for $\mathbf{V}_x$ in Problem 10.7 using superposition.

10.32. Solve for $\mathbf{I}_o$ in Problem 10.8 using superposition.

10.33. Use superposition to solve Problem 10.11.

10.34. Find $\mathbf{V}_o$ in Problem 10.23 using superposition.

10.35. Solve Problem 10.24 using superposition.

10.36. Solve Problem 10.2 using Thévenin's theorem.

10.37. Use Thévenin's theorem to find $\mathbf{V}_o$ in Fig. P10.37.

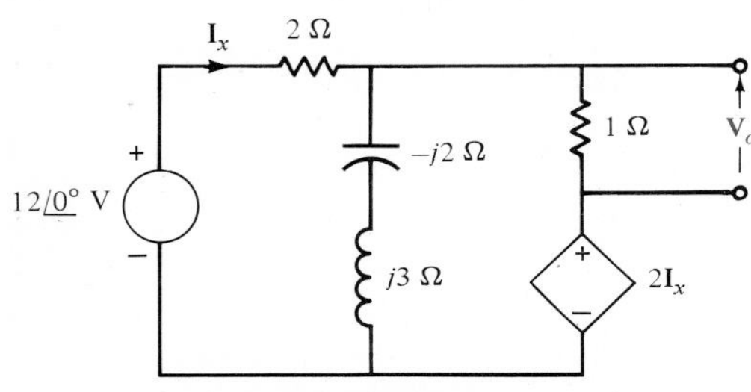

Figure P10.37

10.38. Solve for $\mathbf{V}_x$ in Problem 10.7 using Norton's theorem.

10.39. Solve for $\mathbf{V}_x$ in Problem 10.7 using Thévenin's theorem.

10.40. Solve Problem 10.10 using Thévenin's theorem.

10.41. Find $\mathbf{V}_o$ in Problem 10.22 using Thévenin's theorem.

10.42. Solve Problem 10.23 using Thévenin's theorem.

10.43. Use Thévenin's theorem to solve Problem 10.24.

10.44. Solve Problem 10.2 using Norton's theorem.

10.45. Use Norton's theorem to find $\mathbf{I}_o$ in Problem 10.11.

10.46. The high-frequency small-signal equivalent circuit for a single-stage common-emitter transistor amplifier is shown in Fig. P10.46. Determine the voltage gain $\mathbf{V}_o/\mathbf{V}_s$. (*Hint:* Apply Thévenin's theorem to the resistive network at the amplifier input to simplify the problem.)

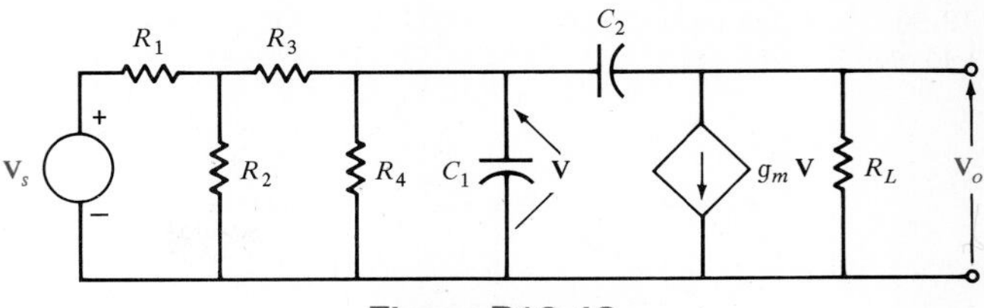

Figure P10.46

10.47. Use SPICE to calculate **I** and **V** in the network shown in Fig. P10.47. $f = \frac{1}{2}\pi$ Hz.

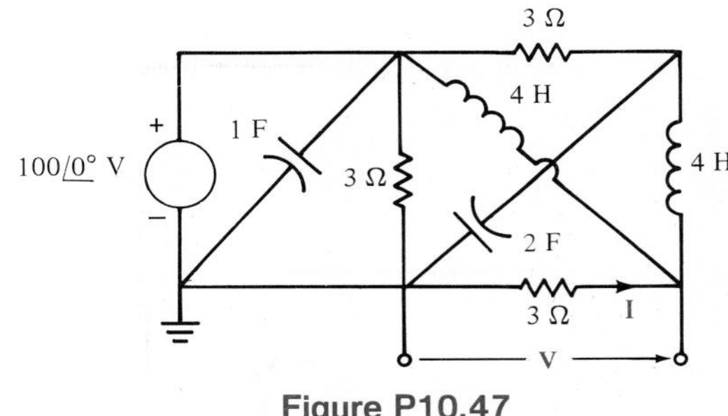

Figure P10.47

10.48. Use SPICE to determine $\mathbf{I}_o$ and $\mathbf{V}_o$ in the network in Fig. P10.48. $f = \frac{1}{2}\pi$ Hz.

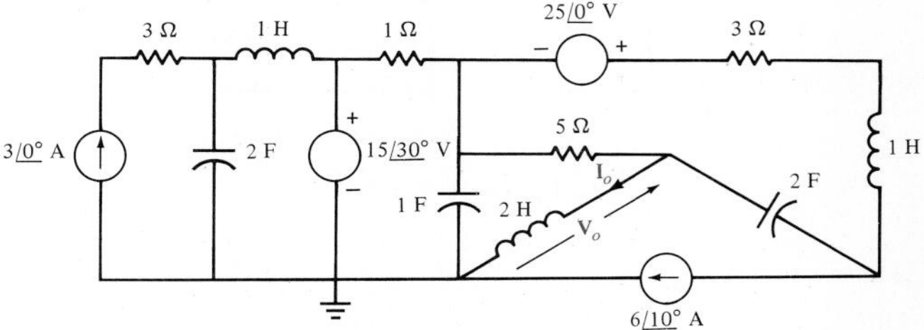

Figure P10.48

10.49. Write a SPICE program to calculate **V** and **I** in the network in Fig. P10.49. $f = \frac{1}{2}\pi$ Hz.

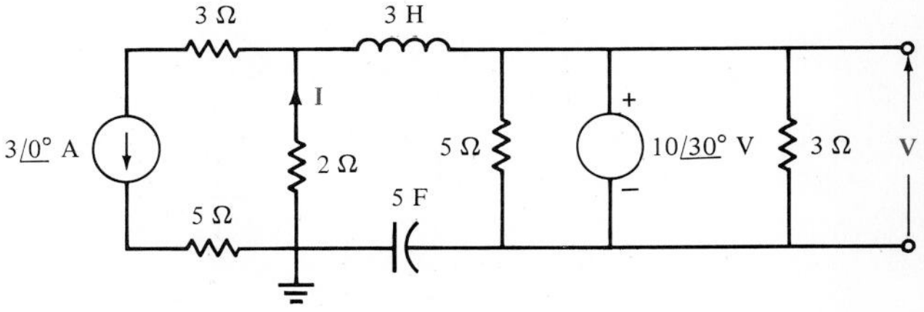

Figure P10.49

10.50. Use SPICE to calculate $\mathbf{V}$ and $\mathbf{I}_L$ in the circuit in Fig. P10.50. $f = \frac{1}{2}\pi$ Hz.

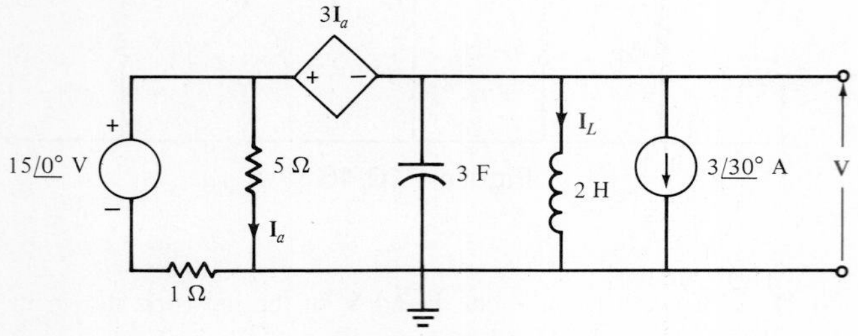

Figure P10.50

10.51. Given the circuit in Fig. P10.51, determine $\mathbf{V}_C$ using SPICE analysis. $f = \frac{1}{2}\pi$ Hz.

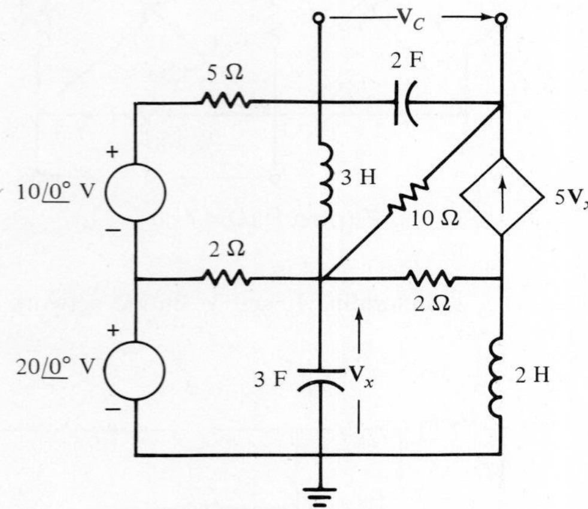

Figure P10.51

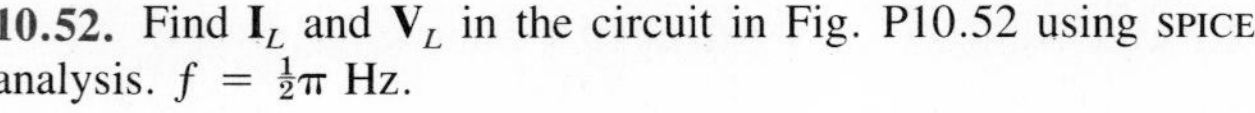

10.52. Find $\mathbf{I}_L$ and $\mathbf{V}_L$ in the circuit in Fig. P10.52 using SPICE analysis. $f = \frac{1}{2}\pi$ Hz.

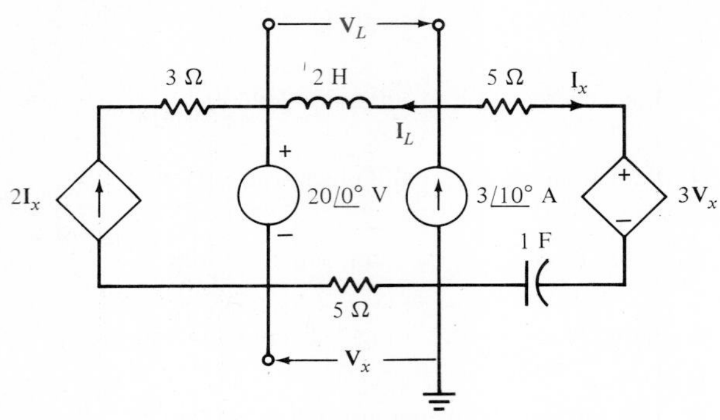

Figure P10.52

10.53. Determine $\mathbf{V}_R$ and $\mathbf{I}_R$ in the circuit in Fig. P10.53 using SPICE. $f = \frac{1}{2}\pi$ Hz.

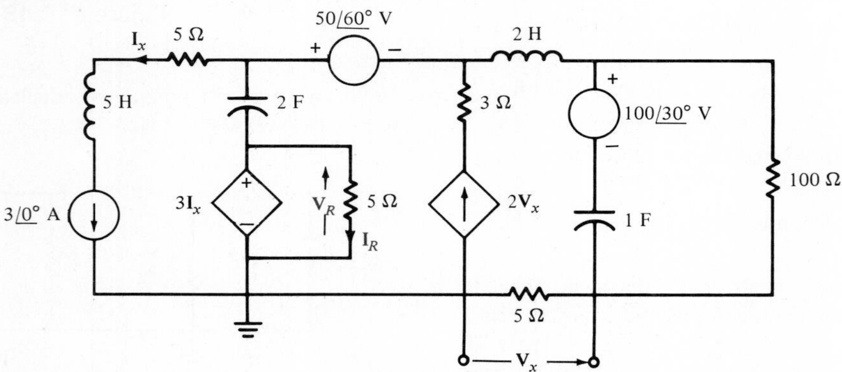

Figure P10.53

10.54. Write a SPICE program to calculate $\mathbf{V}_L$ and $\mathbf{I}_L$ in the network in Fig. P10.54. $f = \frac{1}{2}\pi$ Hz.

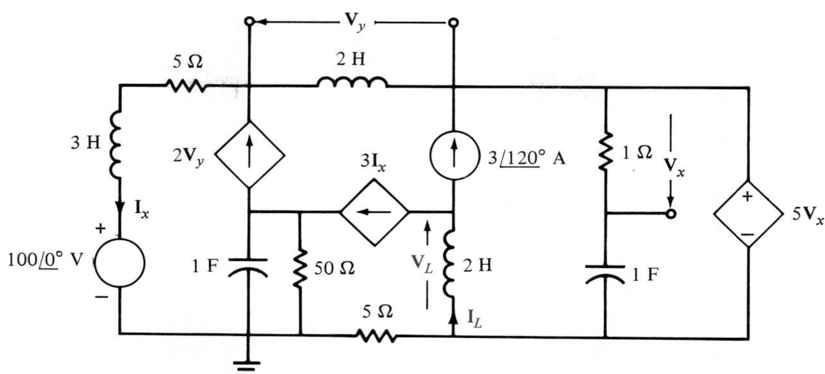

Figure P10.54

10.55. Calculate $\mathbf{V}_x$ and $\mathbf{I}_x$ in the circuit in Fig. P10.55 using SPICE techniques. $f = \frac{1}{2}\pi$ Hz.

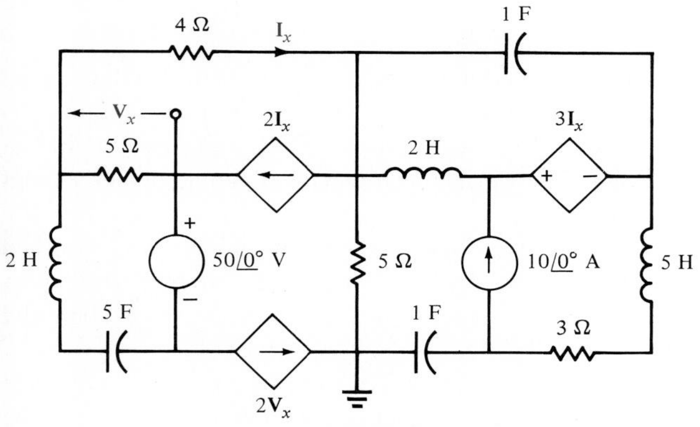

Figure P10.55

10.56. Calculate $\mathbf{V}_C$ and the current through the 40 $\underline{/0°}$-V source shown in the network in Fig. P10.56 using SPICE. $f = \frac{1}{2}\pi$ Hz.

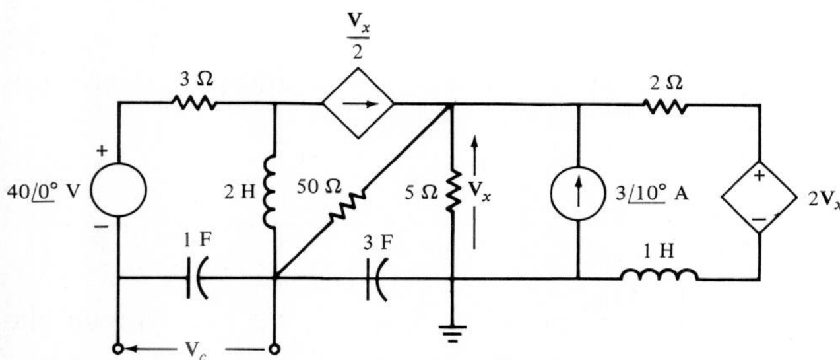

Figure P10.56

10.57. Given the network in Fig. P10.57, use SPICE to calculate $\mathbf{V}_o$. $f = \frac{1}{2}\pi$ Hz.

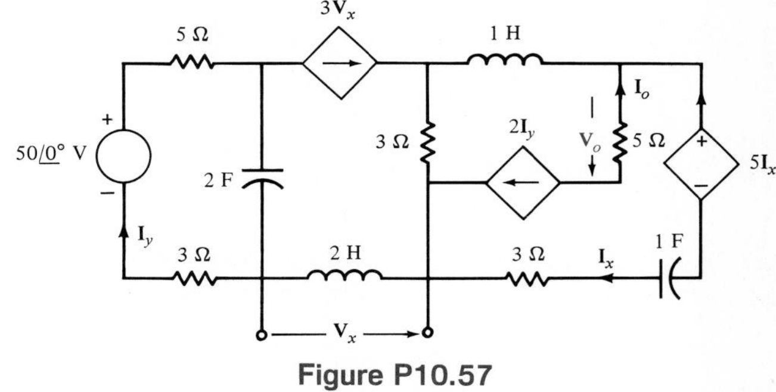

Figure P10.57

CHAPTER

Steady-State Power Analysis

We have been primarily concerned in the preceding chapters with determining the voltage or current at some point in a network. Of equal importance to us in many situations is the power that is supplied or absorbed by some element. Typically, electrical and electronic devices have peak power or maximum instantaneous power ratings which cannot be exceeded without damaging the devices.

In electrical and electronic systems, power comes in all sizes. The power absorbed by some device on an integrated-circuit chip may be in picowatts, whereas the power supplied by a large generating station may be in megawatts. Note that the range between these two examples is phenomenally large (10^{18}).

In our previous work we defined instantaneous power to be the product of voltage and current. Average power, obtained by averaging the instantaneous power, is the average rate at which energy is absorbed or supplied by a device. In the dc case, where both current and voltage are constant, the instantaneous power is equal to the average power. However, as we will demonstrate, this is not the case when the currents and voltages are sinusoidal functions of time.

In this chapter we explore the many ramifications of power in ac circuits. We examine instantaneous power, average power, maximum power transfer, average power for periodic nonsinuoidal waveforms, the power factor, complex power, and power measurement.

11.1 Instantaneous Power

By employing the sign convention adopted in the earlier chapters, we can compute the instantaneous power supplied or absorbed by any device as the product of the instantaneous voltage across the device and the instantaneous current through it.

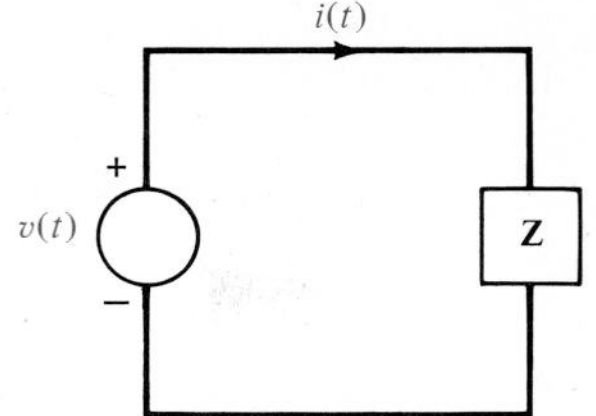

Figure 11.1 Simple ac network.

Consider the circuit shown in Fig. 11.1. In general, the steady-state voltage and current for the network can be written as

$$v(t) = V_M \cos(\omega t + \theta_v) \tag{11.1}$$

$$i(t) = I_M \cos(\omega t + \theta_i) \tag{11.2}$$

The instantaneous power is then

$$\begin{aligned} p(t) &= v(t)i(t) \\ &= V_M I_M \cos(\omega t + \theta_v) \cos(\omega t + \theta_i) \end{aligned} \tag{11.3}$$

Employing the following trigonometric identity

$$\cos\phi_1 \cos\phi_2 = \tfrac{1}{2}[\cos(\phi_1 - \phi_2) + \cos(\phi_1 + \phi_2)] \tag{11.4}$$

the instantaneous power can be written as

$$p(t) = \frac{V_M I_M}{2}[\cos(\theta_v - \theta_i) + \cos(2\omega t + \theta_v + \theta_i)] \tag{11.5}$$

Note that the instanteneous power consists of two terms. The first term is a constant (i.e., it is time independent); and the second term is a cosine wave of twice the excitation frequency. We will examine this equation in more detail in Section 11.2.

EXAMPLE 11.1

The circuit in Fig. 11.1 has the following parameters; $v(t) = 4\cos(\omega t + 60°)$ V and $\mathbf{Z} = 2\,\underline{/30°}\ \Omega$. We wish to determine equations for the current and the instantaneous power as a function of time, and plot these functions with the voltage on a single graph for comparison.

Since

$$\begin{aligned} \mathbf{I} &= \frac{4\,\underline{/60°}}{2\,\underline{/30°}} \\ &= 2\,\underline{/30°}\text{ A} \end{aligned}$$

then

$$i(t) = 2\cos(\omega t + 30°)\text{ A}$$

From equation (11.5),

$$\begin{aligned} p(t) &= 4[\cos(30°) + \cos(2\omega t + 90°)] \\ &= 3.46 + 4\cos(2\omega t + 90°)\text{ W} \end{aligned}$$

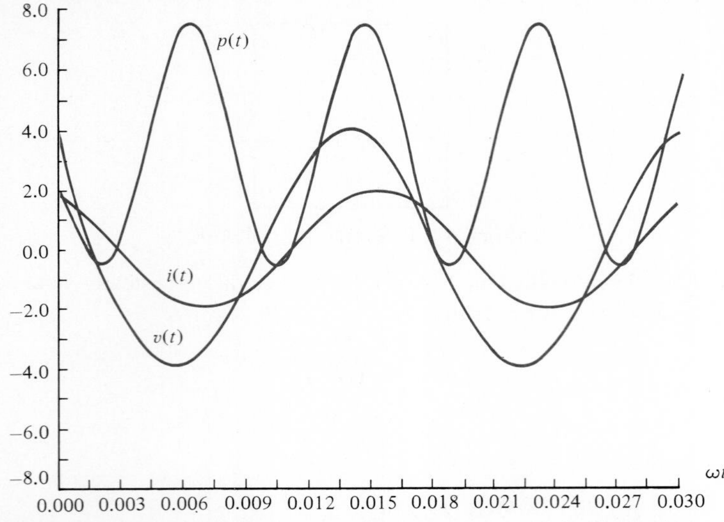

Figure 11.2 Plots of *v(t)*, *i(t)*, and *p(t)* for the circuit in Example 11.1

A plot of this function, together with plots of the voltage and current, are shown in Fig. 11.2. As can be seen in the figure, the instantaneous power has an average value, and the frequency is twice that of the voltage or current. ■

11.2 Average Power

From a basic mathematical perspective, the average value of a periodic waveform (e.g., a sinusoidal function) can be derived by integrating the function over a complete period and dividing this result by the period. Therefore, if the voltage and current are given by Eq. (11.1) and (11.2), respectively, the average power can be computed via the expression

$$\begin{aligned} P &= \frac{1}{T}\int_{t_0}^{t_0+T} p(t)\,dt \\ &= \frac{1}{T}\int_{t_0}^{t_0+T} V_m I_M \cos(\omega t + \theta_v)\cos(\omega t + \theta_i)\,dt \end{aligned} \tag{11.6}$$

where t_0 is arbitrary, $T = 2\pi/\omega$ is the period of the voltage or current, and P is measured in watts. Actually, we may average the waveform over any integral number of periods so that Eq. (11.6) can also be written as

$$P = \frac{1}{nT}\int_{t_0}^{t_0+nT} V_M I_M \cos(\omega t + \theta_v)\cos(\omega t + \theta_i)\,dt \tag{11.7}$$

where n is a positive integer.

Employing Eq. (11.5) in the expression (11.6), we obtain

$$P = \frac{1}{T}\int_{t_0}^{t_0+T} \frac{V_M I_M}{2}[\cos(\theta_v - \theta_i) + \cos(2\omega t + \theta_v + \theta_i)]\,dt \tag{11.8}$$

We could, of course, blindly perform the indicated integration; however, with a little forethought we can determine the result by inspection. The first term is independent of t, and therefore a constant in the integration. Integrating the constant over the period and dividing by the period simply results in the original constant. The second term is a cosine wave. It is well known that the average value of a cosine wave over one complete period or an integral number of periods is zero, and therefore the second term in Eq. (11.8) vanishes. In view of this discussion, Eq. (11.8) reduces to

$$P = \tfrac{1}{2} V_M I_M \cos(\theta_v - \theta_i) \tag{11.9}$$

Note that since $\cos(-\theta) = \cos(\theta)$, the argument for the cosine function can be either $\theta_v - \theta_i$ or $\theta_i - \theta_v$. In addition, note that $\theta_v - \theta_i$ is the angle of the circuit impedance as shown in Fig. 11.1. Therefore, for a purely resistive circuit,

$$P = \tfrac{1}{2} V_M I_M \tag{11.10}$$

and for a purely reactive circuit,

$$\begin{aligned} P &= \tfrac{1}{2} V_M I_M \cos(90°) \\ &= 0 \end{aligned}$$

Because purely reactive impedances absorb no average power, they are often called *lossless elements*. The purely reactive network operates in a mode in which it stores energy over one part of the period and releases it over another.

EXAMPLE 11.2

We wish to determine the average power absorbed by the impedance shown in Fig. 11.3.

From the figure we note that

$$\mathbf{I} = \frac{\mathbf{V}}{\mathbf{Z}} = \frac{V_M \underline{/\theta_v}}{2 + j2} = \frac{10 \underline{/60°}}{2.83 \underline{/45°}} = 3.53 \underline{/15°} \text{ A}$$

Therefore,

$$I_M = 3.53 \text{ A} \qquad \text{and} \qquad \theta_i = 15°$$

Hence

$$\begin{aligned} P &= \tfrac{1}{2} V_M I_M \cos(\theta_v - \theta_i) \\ &= \tfrac{1}{2}(10)(3.53)\cos(60° - 15°) \\ &= 12.5 \text{ W} \end{aligned}$$

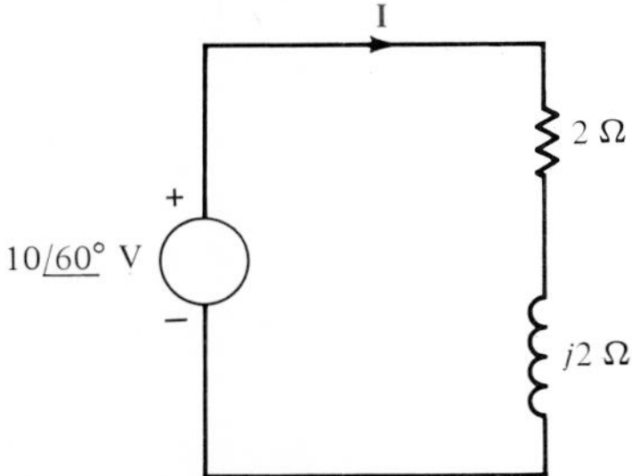

Figure 11.3 Example *RL* circuit.

Since the inductor absorbs no power, we can employ Eq. (11.10) provided that V_M in that equation is the voltage across the resistor. Using voltage division, we obtain

$$\mathbf{V}_R = \frac{(10 \underline{/60^\circ})(2)}{2 + 2j} = 7.07 \underline{/15^\circ} \text{ V}$$

and therefore

$$\begin{aligned} P &= \tfrac{1}{2}(7.07)(3.53) \\ &= 12.5 \text{ W} \end{aligned}$$

In addition, using Ohm's law, we could also employ the expressions

$$P = \frac{1}{2}\frac{V_M^2}{R}$$

or

$$P = \tfrac{1}{2} I_M^2 R$$

where once again we must be careful that the V_M and I_M in these equations refer to the voltage across the resistor and the current through it, respectively. ■

EXAMPLE 11.3

For the circuit shown in Fig. 11.4 we wish to determine both the total average power absorbed and the total average power supplied.

From the figure we note that

$$\mathbf{I}_1 = \frac{12 \underline{/45^\circ}}{4} = 3 \underline{/45^\circ} \text{ A}$$

$$\mathbf{I}_2 = \frac{12 \underline{/45^\circ}}{2 - j1} = \frac{12 \underline{/45^\circ}}{2.24 \underline{/-26.57^\circ}} = 5.37 \underline{/71.57^\circ} \text{ A}$$

and therefore

$$\begin{aligned} \mathbf{I} &= \mathbf{I}_1 + \mathbf{I}_2 \\ &= 3 \underline{/45^\circ} + 5.37 \underline{/71.57^\circ} \\ &= 8.16 \underline{/62.08^\circ} \text{ A} \end{aligned}$$

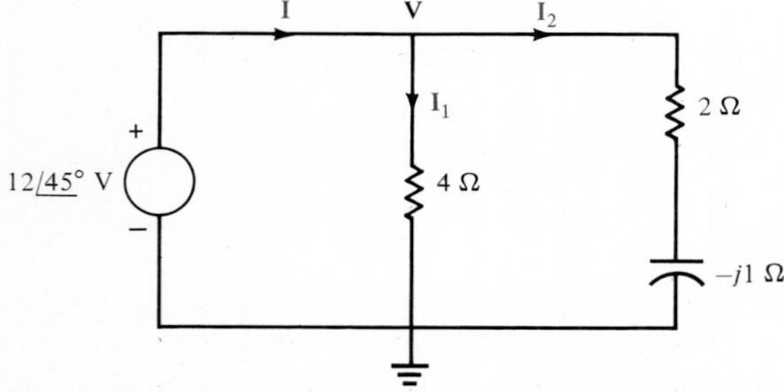

Figure 11.4 Example circuit for illustrating a power balance.

The power absorbed in the 4-Ω resistor is

$$P_4 = \tfrac{1}{2}V_M I_M = \tfrac{1}{2}(12)(3) = 18 \text{ W}$$

The power absorbed in the 2-Ω resistor is

$$P_2 = \tfrac{1}{2}I_M^2 R = \tfrac{1}{2}(5.37)^2(2) = 28.8 \text{ W}$$

Therefore, the total power absorbed is

$$P_A = 18 + 28.8 = 46.8 \text{ W}$$

Note that we could have calculated the power absorbed in the 2-Ω resistor using $\frac{1}{2}V_M^2/R$ if we had first calculated the voltage across the 2-Ω resistor.

The total power supplied by the source is

$$\begin{aligned} P_s &= +\tfrac{1}{2}V_M I_M \cos(\theta_v - \theta_i) \\ &= +\tfrac{1}{2}(12)(8.16)\cos(45° - 62.08°) \\ &= +46.8 \text{ W} \end{aligned}$$

Thus the total power supplied is, of course, equal to the total power absorbed. ■

DRILL EXERCISE

D11.1. Find the average power absorbed by each resistor in the network in Fig. D11.1.

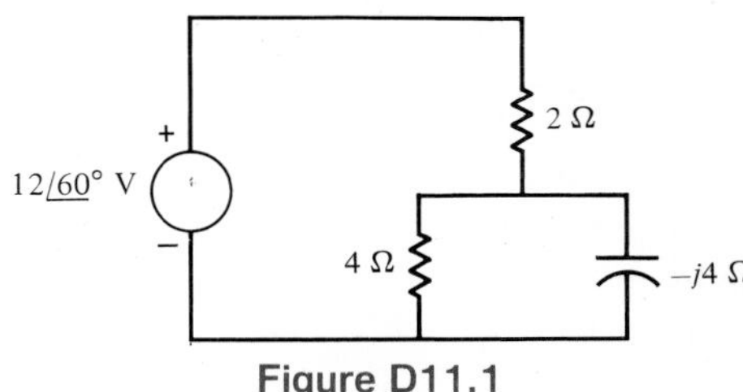

Figure D11.1

D11.2. Given the network in Fig. D11.2, find the average power absorbed by each passive circuit element and the total average power supplied by the current source.

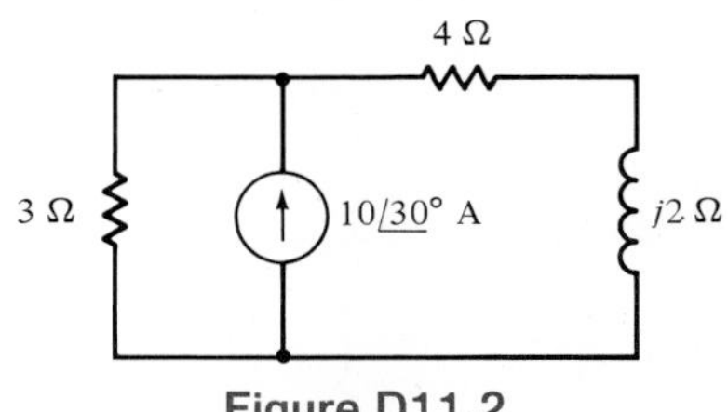

Figure D11.2

When determining average power, if more than one source is present in a network, we can use any of our network analysis techniques to find the necessary voltage and/or current to compute the power. However, we must remember that in general we cannot apply superposition to power. We should, however, mention that there is a special case in which superposition does apply to power. If the current is of the form

$$i(t) = I_1 \cos(\omega_1 t + \theta_1) + I_2 \cos(\omega_2 t + \theta_2)$$

where $\omega_1 \neq \omega_2$, the average power that is absorbed by a resistor R is given by the expression

$$P = \frac{1}{T}\int_0^T [I_1 \cos(\omega_1 t + \theta_1) + I_2 \cos(\omega_2 t + \theta_2)]^2 R\, dt$$

where both ω_1 and ω_2 are periodic in T. The equation above can be written as

$$\begin{aligned} P &= \frac{1}{T}\int_0^T [I_1^2 \cos^2(\omega_1 t + \theta_1) + I_2^2 \cos^2(\omega_2 t + \theta_2) \\ &\quad + 2I_1 I_2 \cos(\omega_1 t + \theta_1)\cos(\omega_2 t + \theta_2)]R\, dt \\ &= \frac{I_1^2}{2}R + \frac{I_2^2}{2}R + \frac{1}{T}\int_0^T \{I_1 I_2 \cos[(\omega_1 + \omega_2)t + \theta_1 + \theta_2] \\ &\quad + I_1 I_2 \cos[(\omega_1 - \omega_2)t + \theta_1 - \theta_2]\}R\, dt \end{aligned}$$

The last term will be zero if $\omega_1 - \omega_2 \neq 0$. Therefore, superposition will hold if the terms are of different frequencies and thus

$$P = \frac{I_1^2}{2}R + \frac{I_2^2}{2}R$$

Superposition also applies if one, and only one, of the sources is dc.

EXAMPLE 11.4

Consider the network shown in Fig. 11.5. We wish to determine the total average power absorbed and supplied by each element.

From the figure we note that

$$\mathbf{I}_2 = \frac{12\,\underline{/30^\circ}}{2} = 6\,\underline{/30^\circ}\text{ A}$$

and

$$\mathbf{I}_3 = \frac{12\,\underline{/30^\circ} - 6\,\underline{/0^\circ}}{j1} = \frac{4.39 + j6}{j1} = 7.43\,\underline{/-36.19^\circ}\text{ A}$$

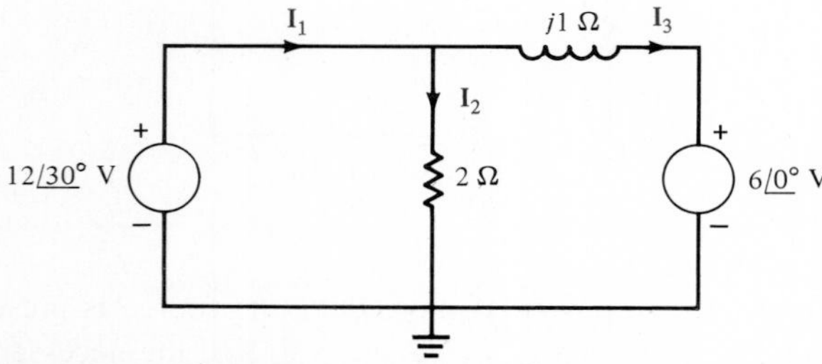

Figure 11.5 Example *RL* circuit with two sources.

The power absorbed by the 2-Ω resistor is

$$P_2 = \tfrac{1}{2}V_M I_M = \tfrac{1}{2}(12)(6) = 36 \text{ W}$$

According to the direction of $\mathbf{I}_3$ the $6 \underline{/0^\circ}$-V source is absorbing power. The power it absorbs is given by

$$P_{6\underline{/0^\circ}} = \tfrac{1}{2}V_M I_M \cos(\theta_v - \theta_i)$$

$$= \tfrac{1}{2}(6)(7.43)\cos[0^\circ - (-36.19^\circ)]$$

$$= 18 \text{ W}$$

At this point an obvious question arises: How do we know if the $6 \underline{/0^\circ}$-V source is supplying power to the remainder of the network or absorbing it? The answer to this question is actually straightforward. If we employ our passive sign convention that was adopted in the earlier chapters, that is, if the current reference direction enters the positive terminal of the source and the answer is positive, the source is absorbing power. If the answer is negative, the source is supplying power to the remainder of the circuit. A generator sign convention could have been used and under this condition the interpretation of the sign of the answer would be reversed. Note that once the sign convention is adopted and used, the sign for average power will be negative only if the angle difference is greater than 90° (i.e., $|\theta_v - \theta_i| > 90^\circ$).

To obtain the power supplied to the network, we compute $\mathbf{I}_1$ as

$$\mathbf{I}_1 = \mathbf{I}_2 + \mathbf{I}_3$$

$$= 6 \underline{/30^\circ} + 7.43 \underline{/-36.19^\circ}$$

$$= 11.29 \underline{/-7.07^\circ} \text{ A}$$

Therefore, the power supplied by the $12 \underline{/30^\circ}$-V source using the generator sign convention is

$$P_s = +\tfrac{1}{2}(12)(11.29)\cos(30^\circ + 7.07^\circ)$$

$$= +54 \text{ W}$$

and hence the power absorbed is equal to the power supplied. ■

EXAMPLE 11.5

The current in a 2-Ω resistor is of the form

$$i(t) = 4\cos(377t + 30^\circ) + 2\cos(754t + 60^\circ)$$

We wish to find the average power absorbed by the resistor.

Since $\omega_1 = 377$ and $\omega_2 = 754$ (i.e., $\omega_1 \neq \omega_2$), then

$$P = \tfrac{1}{2}(4^2 + 2^2)2 = 20 \text{ W}$$ ■

DRILL EXERCISE

D11.3. Determine the total average power absorbed and supplied by each element in the network in Fig. D11.3.

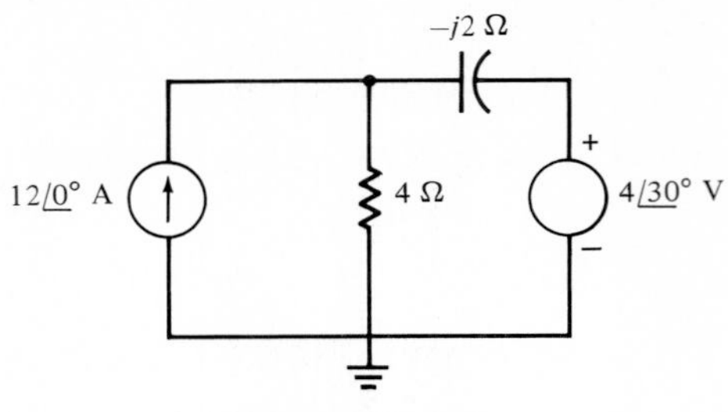

Figure D11.3

D11.4. Given the network in Fig. D11.4, determine the total average power absorbed or supplied by each element.

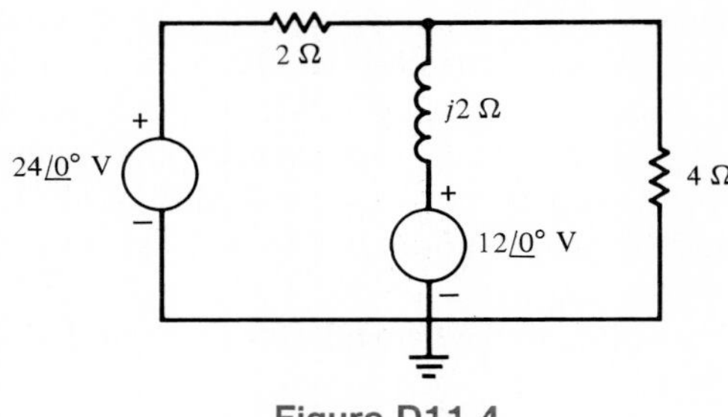

Figure D11.4

11.3 Maximum Average Power Transfer

In our study of resistive networks we addressed the problem of maximum power transfer to a resistive load. We showed that if the network excluding the load was represented by a Thévenin equivalent circuit, maximum power transfer would result if the value of the load resistor was equal to the Thévenin equivalent resistance (i.e., $R_L = R_{Th}$). We will now reexamine this issue within the present context to determine the load impedance for the network shown in Fig. 11.6 that will result in maximum average power being absorbed by the load impedance $\mathbf{Z}_L$.

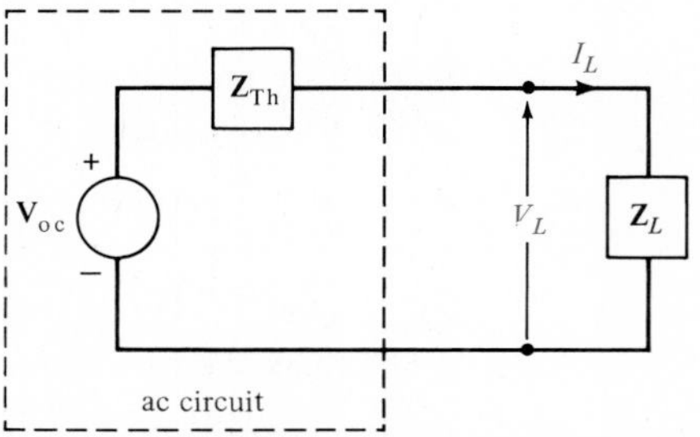

Figure 11.6 Circuit used to examine maximum power transfer.

The equation for average power at the load is

$$P_L = \tfrac{1}{2} V_L I_L \cos (\theta_{V_L} - \theta_{i_L}) \tag{11.11}$$

The phasor current and voltage at the load are given by the expressions

$$\mathbf{I}_L = \frac{\mathbf{V}_{\text{oc}}}{\mathbf{Z}_{\text{Th}} + \mathbf{Z}_L} \tag{11.12}$$

$$\mathbf{V}_L = \frac{\mathbf{V}_{\text{oc}}\mathbf{Z}_L}{\mathbf{Z}_{\text{Th}} + \mathbf{Z}_L} \tag{11.13}$$

where

$$\mathbf{Z}_{\text{Th}} = R_{\text{Th}} + jX_{\text{Th}} \tag{11.14}$$

and

$$\mathbf{Z}_L = R_L + jX_L \tag{11.15}$$

The magnitude of the phasor current and voltage are given by the expressions

$$I_L = \frac{V_{\text{oc}}}{[(R_{\text{Th}} + R_L)^2 + (X_{\text{Th}} + X_L)^2]^{1/2}} \tag{11.16}$$

$$V_L = \frac{V_{\text{oc}}(R_L^2 + X_L^2)^{1/2}}{[(R_{\text{Th}} + R_L)^2 + (X_{\text{Th}} + X_L)^2]^{1/2}} \tag{11.17}$$

The phase angles for the phasor current and voltage are contained in the quantity $(\theta_{V_L} - \theta_{i_L})$. Note also that $\theta_{V_L} - \theta_{i_L} = \theta_{\mathbf{Z}_L}$ and in addition

$$\cos \theta_{\mathbf{Z}_L} = \frac{R_L}{[R_L^2 + X_L^2]^{1/2}} \tag{11.18}$$

Substituting Eqs. (11.16) to (11.18) into Eq. (11.11) yields

$$P_L = \frac{1}{2} \frac{V_{\text{oc}}^2 R_L}{(R_{\text{Th}} + R_L)^2 + (X_{\text{Th}} + X_L)^2} \tag{11.19}$$

which could, of course, be obtained directly from Eq. (11.16) using $\frac{1}{2} I_L^2 R_L$. Once again, a little forethought will save us some calculus. From the standpoint of maximizing P_L, V_{oc} is a constant. In addition, the quantity $(X_{\text{Th}} + X_L)$ absorbs no power and therefore any value of this quantity only serves to reduce P_L. Hence we can eliminate this term by selecting $X_L = -X_{\text{Th}}$. Our problem then reduces to maximizing

$$P_L = \frac{1}{2} \frac{V_{\text{oc}}^2 R_L}{(R_L + R_{\text{Th}})^2} \tag{11.20}$$

However, this is the same quantity we maximized in the purely resistive case by selecting $R_L = R_{\text{Th}}$. Therefore, for maximum average power transfer to the load shown in Fig. 11.6, $\mathbf{Z}_L$ should be chosen so that

$$\mathbf{Z}_L = R_L + jX_L = R_{\text{Th}} - jX_{\text{Th}} = Z_{\text{Th}}^* \tag{11.21}$$

Finally, if the load impedance is purely resistive (i.e., $X_L = 0$), the condition for maximum average power transfer can be derived via the expression

$$\frac{dP_L}{dR_L} = 0$$

where P_L is the expression in (11.19) with $X_L = 0$. The value of R_L that maximizes P_L under the condition $X_L = 0$ is

$$R_L = \sqrt{R_{Th}^2 + X_{Th}^2} \tag{11.22}$$

EXAMPLE 11.6

Given the circuit in Fig. 11.7a we wish to find the value of $\mathbf{Z}_L$ for maximum average power transfer. In addition, we wish to find the value of the maximum average power delivered to the load.

In order to solve the problem, we form a Thévenin equivalent at the load. The circuit in Fig. 11.7b is used to compute the open-circuit voltage

$$\mathbf{V}_{oc} = \frac{4\,\underline{/0^\circ}\,(2)}{6 + j1}(4) = 5.28\,\underline{/-9.46^\circ}\text{ V}$$

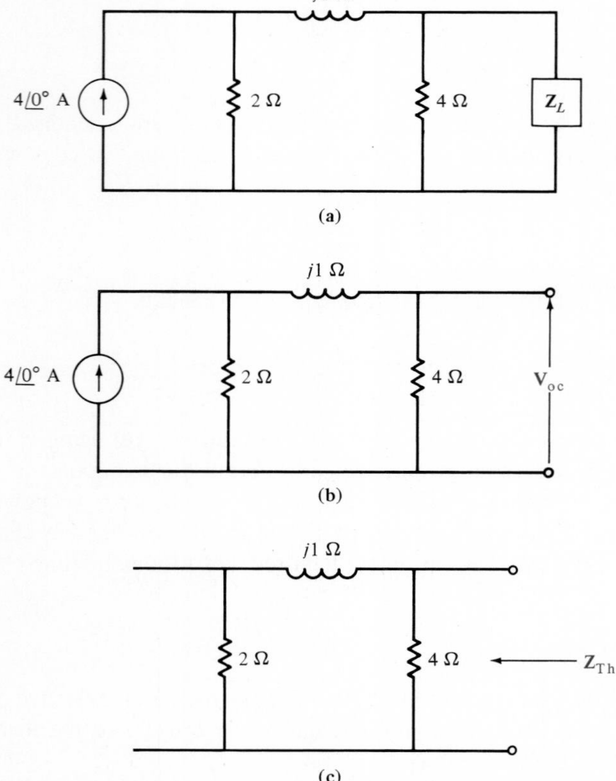

Figure 11.7 Circuits for illustrating maximum average power transfer.

The Thévenin equivalent impedance can be derived from the circuit in Fig. 11.7c. As shown in the figure,

$$\mathbf{Z}_{Th} = \frac{4(2 + j1)}{6 + j1} = 1.40 + j0.43\ \Omega$$

Therefore, $\mathbf{Z}_L$ for maximum average power transfer is

$$\mathbf{Z}_L = 1.40 - j0.43\ \Omega$$

With $\mathbf{Z}_L$ as given above, the current in the load is

$$\mathbf{I} = \frac{5.28\ \underline{/-9.46^\circ}}{2.8} = 1.89\ \underline{/-9.46^\circ}\ \text{A}$$

Therefore, the maximum average power transferred to the load is

$$P_L = \tfrac{1}{2}I_M^2R_L = \tfrac{1}{2}(1.89)^2(1.4) = 2.50\ \text{W}$$

■

EXAMPLE 11.7

For the circuit shown in Fig. 11.8a, we wish to find the value of $\mathbf{Z}_L$ for maximum average power transfer. In addition, let us determine the value of the maximum average power delivered to the load.

We will first reduce the circuit, with the exception of the load, to a Thévenin equivalent circuit. The open-circuit voltage can be computed from Fig. 11.8b. The equations for the circuit are

$$\mathbf{V}X + 4 = (2 + j4)\mathbf{I}_1$$
$$VX = -2\mathbf{I}_1$$

Solving for $\mathbf{I}_1$, we obtain

$$\mathbf{I}_1 = \frac{1\ \underline{/-45^\circ}}{\sqrt{2}}$$

The open-circuit voltage is then

$$\begin{aligned}\mathbf{V}_{oc} &= 2\mathbf{I}_1 - 4\\ &= \sqrt{2}\ \underline{/-45^\circ} - 4\\ &= -3 - j1\\ &= -3.16\ \underline{/18.43^\circ}\ \text{V}\end{aligned}$$

The short-circuit current can be derived from Fig. 11.8c. The equations for this circuit are

$$\mathbf{V}X + 4 = (2 + j4)\mathbf{I} - 2\mathbf{I}_{sc}$$
$$-4 = -2\mathbf{I} + (2 - j2)\mathbf{I}_{sc}$$
$$\mathbf{V}X = -2(\mathbf{I} - \mathbf{I}_{sc})$$

Solving these equations for $\mathbf{I}_{sc}$ yields

$$\mathbf{I}_{sc} = -(1 + j2)\ \text{A}$$

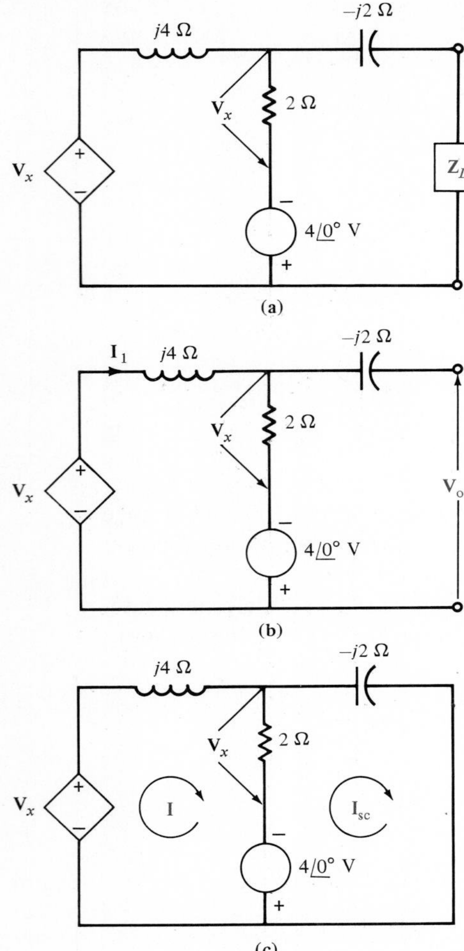

Figure 11.8 Circuits for illustrating maximum average power transfer.

The Thévenin equivalent impedance is then

$$\mathbf{Z}_{Th} = \frac{\mathbf{V}_{oc}}{\mathbf{I}_{sc}} = \frac{3 + j1}{1 + j2} = 1 - j1\ \Omega$$

Therefore, for maximum average power transfer the load impedance should be

$$\mathbf{Z}_L = 1 + j1\ \Omega$$

The current in this load $\mathbf{Z}_L$ is then

$$\mathbf{I}_L = \frac{\mathbf{V}_{oc}}{\mathbf{Z}_{Th} + \mathbf{Z}_L} = \frac{-3 - j1}{2} = -1.58\ \underline{/18.43^\circ}\ \text{A}$$

Hence the maximum average power transferred to the load is

$$P_L = \tfrac{1}{2}(1.58)^2(1)$$

$$= 1.25 \text{ W}$$

DRILL EXERCISE

D11.5. Given the network in Fig. D11.5, find $\mathbf{Z}_L$ for maximum average power transfer and the maximum average power transferred to the load.

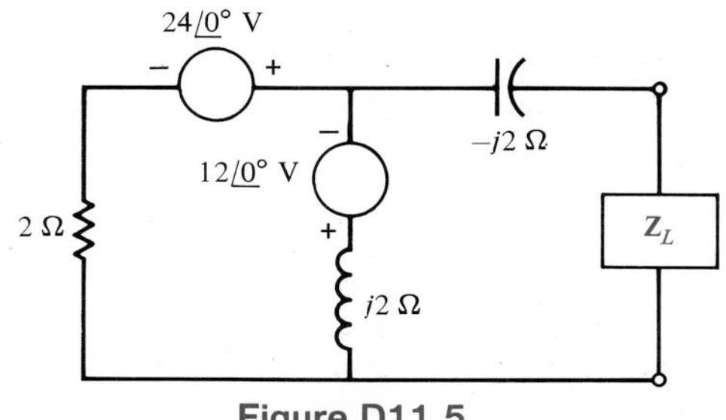

Figure D11.5

D11.6. Find $\mathbf{Z}_L$ for maximum average power transfer and the maximum average power transferred to the load in the network in Fig. D11.6.

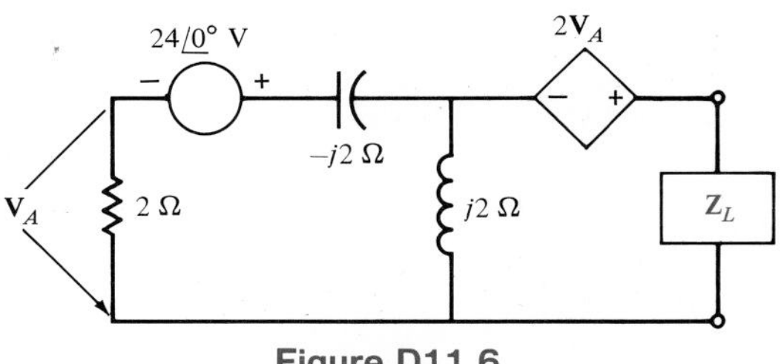

Figure D11.6

11.4 Effective or RMS Values

In the preceding sections of this chapter we have shown that the average power absorbed by a resistive load is directly dependent on the type, or types, of sources that are delivering power to the load. For example, if the source was dc, the average power absorbed was I^2R; and if the source was sinusoidal, the average power was $\frac{1}{2}I_M^2 R$. Although these two types of waveforms are extremely important, they are by no means the only waveforms we will encounter in circuit analysis. Therefore, a technique by which we can compare the *effectiveness* of different sources in delivering power to a resistive load would be quite useful.

In order to accomplish this comparison, we define what is called the *effective value of a periodic waveform*, representing either voltage or current. Although either quantity could be used, we will employ current in the definition. Hence we define the effective value of a periodic current as a constant value that is equal to the dc value which delivers the same average power to a resistor R. Let us call the constant value of the current that is equal to the dc value I_{eff}. Then the average power delivered to a resistor as a result of this current is

$$P = I_{\text{eff}}^2 R$$

Similarly, the average power delivered to a resistor by a periodic current $i(t)$ is

$$P = \frac{1}{T}\int_{t_0}^{t_0+T} i^2(t)R\, dt$$

Equating these two expressions, we find that

$$I_{\text{eff}} = \sqrt{\frac{1}{T}\int_{t_0}^{t_0+T} i^2(t)\, dt} \tag{11.23}$$

Note that this effective value is found by first determining the *square* of the current, then computing the average or *mean* value, and finally taking the square *root*. Thus in ''reading'' the mathematical Eq. (11.23), we are determining the root-mean-square which we abreviate as rms, and therefore I_{eff} is called I_{rms}.

Since dc is a constant, the rms value of dc is simply the constant value. Let us now determine the rms value of other waveforms. The most important waveform is the sinusoid, and therefore we address this particular one in the following example.

EXAMPLE 11.8

We wish to compute the rms value of the waveform $i(t) = I_M \cos(\omega t - \theta)$, which has a period of $T = 2\pi/\omega$.

Substituting these expressions into Eq. (11.23) yields

$$I_{\text{rms}} = \left[\frac{1}{T}\int_0^T I_M^2 \cos^2(\omega t - \theta)\, dt\right]^{1/2}$$

Using the trigonometric identity

$$\cos^2\phi = \tfrac{1}{2} + \tfrac{1}{2}\cos 2\phi$$

the equation above can be expressed as

$$I_{\text{rms}} = I_M\left[\frac{\omega}{2\pi}\int_0^{2\pi/\omega} [\tfrac{1}{2} + \tfrac{1}{2}\cos(2\omega t - 2\theta)]\, dt\right]^{1/2}$$

Since we know that the average or mean value of a cosine wave is zero,

$$\begin{aligned} I_{\text{rms}} &= I_M\left[\frac{\omega}{2\pi}\int_0^{2\pi/\omega} \frac{1}{2}\, dt\right]^{1/2} \\ &= I_M\left[\frac{\omega}{2\pi}\left(\frac{t}{2}\right)\Bigg|_0^{2\pi/\omega}\right]^{1/2} = \frac{I_M}{\sqrt{2}} \end{aligned} \tag{11.24}$$

Therefore, the rms value of a sinusoid is equal to the maximum value divided by the $\sqrt{2}$. Hence a sinusoidal current with a maximum value of I_M delivers the same average power to a resistor R as a dc current with a value of $I_M/\sqrt{2}$. ■

Using the rms values for voltage and current, the average power can be written in general as

$$P = V_{rms} I_{rms} \cos(\theta_v - \theta_i) \tag{11.25}$$

The power absorbed by a resistor R is

$$P = I_{rms}^2 R = \frac{V_{rms}^2}{R} \tag{11.26}$$

In dealing with voltages and currents in numerous electrical applications, it is important to know whether the values quoted are maximum, average, rms, or what. For example, the normal 120-V ac electrical outlets have an rms value of 120 V, an average value of 0 V, and a maximum value of $120\sqrt{2}$ V.

Finally, if the current in a resistor R is composed of a sum of sinusoids of different frequencies, the power absorbed by the resistor can be expressed as

$$P = (I_{1rms}^2 + I_{2rms}^2 + \cdots + I_{nrms}^2)R \tag{11.27}$$

where the rms value of the total current is

$$I_{rms} = \sqrt{I_{1rms}^2 + I_{2rms}^2 + \cdots + I_{nrms}^2} \tag{11.28}$$

and each component represents a current of different frequency.

EXAMPLE 11.9

We wish to compute the rms value of the voltage waveform shown in Fig. 11.9.

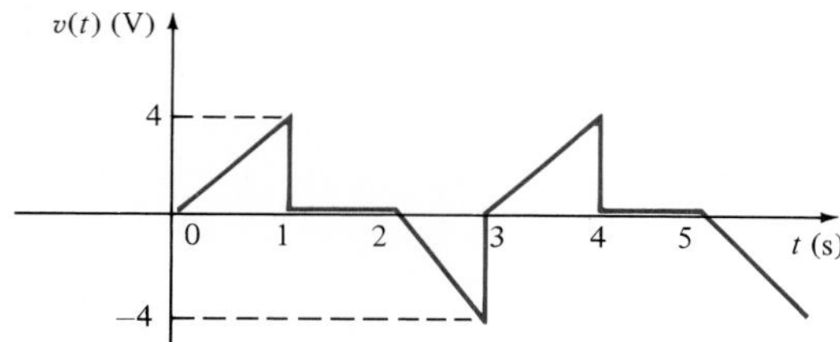

Figure 11.9 Waveform used to illustrate rms values.

The waveform is periodic with period $T = 3$ s. The equation for the voltage in the time frame $0 \leq t \leq 3$ s is

$$v(t) = \begin{cases} 4t \text{ V} & 0 < t \leq 1 \\ 0 \text{ V} & 1 < t \leq 2 \\ -4t + 8 \text{ V} & 2 < t \leq 3 \end{cases}$$

The rms value is

$$V_{rms} = \left\{ \frac{1}{3} \left[\int_0^1 (4t)^2 \, dt + \int_1^2 (0)^2 \, dt + \int_2^3 (8 - 4t)^2 \, dt \right] \right\}^{1/2}$$

$$= \left\{ \frac{1}{3} \left[\frac{16t^3}{3} \Big|_0^1 + \left[64t - \frac{64t^2}{2} + \frac{16t^3}{3} \right] \Big|_2^3 \right] \right\}^{1/2}$$

$$= 1.89 \text{ V}$$

■

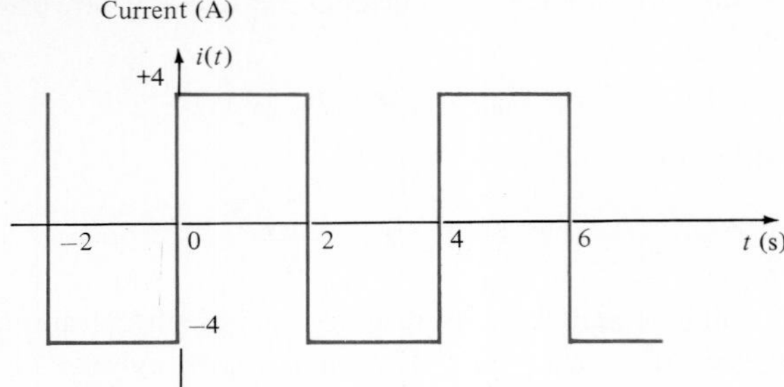

Figure 11.10 Waveform used to illustrate rms values.

EXAMPLE 11.10

Determine the rms value of the current waveform in Fig. 11.10 and use this value to compute the average power delivered to a 2-Ω resistor through which this current is flowing.

The current waveform is periodic with a period of $T = 4$ s. The rms value is

$$\begin{aligned} I_{\text{rms}} &= \left\{\frac{1}{4}\left[\int_0^2 (4)^2\,dt + \int_2^4 (-4)^2\,dt\right]\right\}^{1/2} \\ &= \left[\frac{1}{4}\left(16t\,\Big|_0^2 + 16t\,\Big|_2^4\right)\right]^{1/2} \\ &= 4\text{ A} \end{aligned}$$

The average power delivered to a 2-Ω resistor with this current is

$$P = I_{\text{rms}}^2 R = (4)^2(2) = 32\text{ W}$$

■

EXAMPLE 11.11

We wish to find the rms value of the current:

$$i(t) = 12 \sin 377t + 6 \sin (754t + 30°)\text{ A}$$

Since the frequencies of the two sinusoidal waves are different,

$$\begin{aligned} I_{\text{rms}}^2 &= I_{1\text{rms}}^2 + I_{2\text{rms}}^2 \\ &= \left(\frac{12}{\sqrt{2}}\right)^2 + \left(\frac{6}{\sqrt{2}}\right)^2 \end{aligned}$$

Therefore,

$$I_{\text{rms}} = 9.49\text{ A}$$

■

DRILL EXERCISE

D11.7. Compute the rms value of the voltage waveform shown in Fig. D11.7.

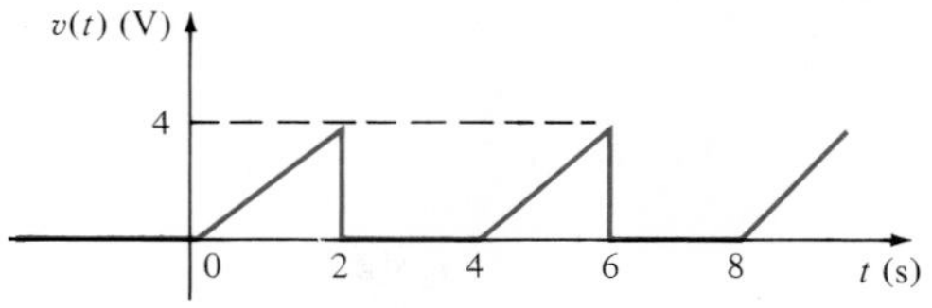

Figure D11.7

D11.8. The current waveform in Fig. D11.8 is flowing through a 4-Ω resistor. Compute the average power delivered to the resistor.

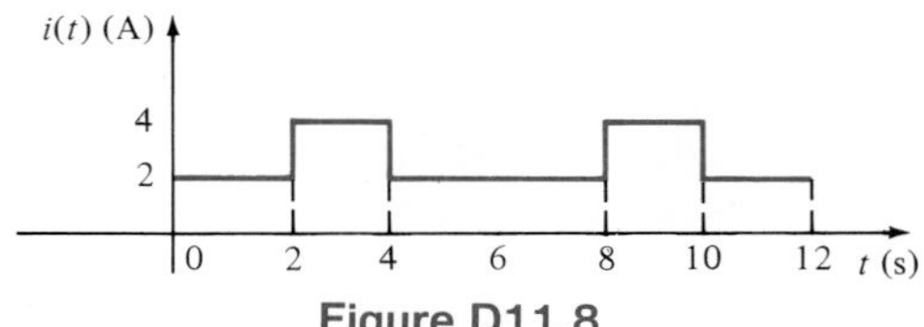

Figure D11.8

D11.9. The current waveform in Fig. D11.9 is flowing through a 10-Ω resistor. Determine the average power delivered to the resistor.

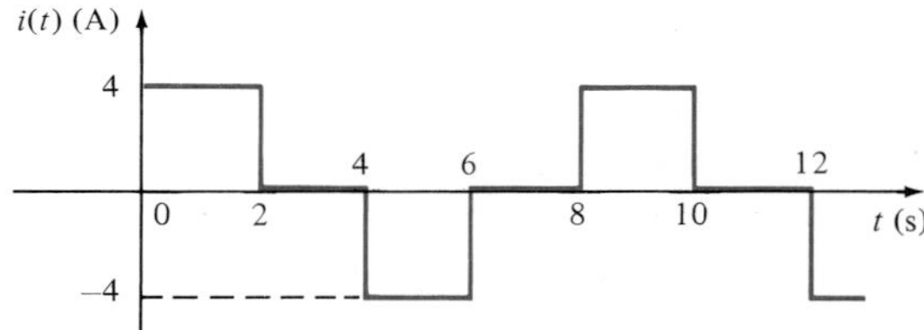

Figure D11.9

D11.10. The current $i(t)$ is flowing in a 4-Ω resistor. Determine the power absorbed by the resistor if

$$i(t) = 4 \sin (377t - 30°) + 6 \cos (754t - 45°) \text{ A}$$

11.5 The Power Factor

The power factor is a very important quantity. Its importance stems in part from the economic impact it has on industrial users of large amounts of power. In this section we carefully define this term and then illustrate its significance via some practical examples.

In Section 11.4 we showed that a load operating in the ac steady state is delivered an average power of

$$P = V_{rms} I_{rms} \cos (\theta_v - \theta_i)$$

We will now further define the terms in this important equation. The product $V_{rms}I_{rms}$ is referred to as the *apparent power*. Although the term $\cos (\theta_v - \theta_i)$ is a dimensionless quantity, and the units of P are watts, apparent power is normally stated in volt-amperes (VA) or kilovolt-amperes (kVA) in order to distinguish it from average power.

We now define the *power factor* (PF) as the ratio of the average power to the apparent power, that is,

$$\text{PF} = \frac{P}{V_{\text{rms}} I_{\text{rms}}} = \cos(\theta_v - \theta_i) \tag{11.29}$$

where

$$\cos(\theta_v - \theta_i) = \cos\theta_{\mathbf{Z}_L} \tag{11.30}$$

The angle $\theta_v - \theta_i = \theta_{\mathbf{Z}_L}$ is the phase angle of the load impedance and is often referred to as the *power factor angle*. The two extreme positions for this angle correspond to a purely resistive load where $\theta_{\mathbf{Z}_L} = 0$, the PF is 1, and the purely reactive load where $\theta_{\mathbf{Z}_L} = \pm 90°$ and the PF is 0. It is, of course, possible to have a unity PF for a load containing R, L, and C elements if the values of the circuit elements are such that a zero phase angle is obtained at the particular operating frequency.

There is, of course, a whole range of power factor angles between $\pm 90°$ and $0°$. If the load is an equivalent RC combination (i.e., $R - jX_C$), then the PF angle lies between the limits $-90° < \theta_{\mathbf{Z}_L} < 0°$. On the other hand, if the load is an equivalent RL combination (i.e., $R + jX_L$), then the PF angle lies between the limits $0 < \theta_{\mathbf{Z}_L} < 90°$. Obviously, confusion in identifying the type of load could result, due to the fact that $\cos\theta_{\mathbf{Z}_L} = \cos(-\theta_{\mathbf{Z}_L})$. To circumvent this problem, the PF is said to be either *leading* or *lagging*, where these two terms *refer to the phase of the current with respect to the voltage*. Since the current leads the voltage in an RC load, the load has a leading PF. In a similar manner, an RL load has a lagging PF; therefore, load impedances of $\mathbf{Z}_L = 1 - j1$ and $\mathbf{Z}_L = 2 + j1$ have power factors of $\cos(-45°) = 0.707$ leading and $\cos(26.59°) = 0.894$ lagging, respectively.

EXAMPLE 11.12

An industrial load consumes 88 kW at a PF of 0.707 lagging from a 480-V rms line. The transmission line resistance from the power company's transformer to the plant is 0.08 Ω. Let us determine the power that must be supplied by the power company (a) under present conditions, and (b) if the PF is somehow changed to 0.90 lagging.

(a) The equivalent circuit for these conditions is shown in Fig. 11.11. Using Eq.

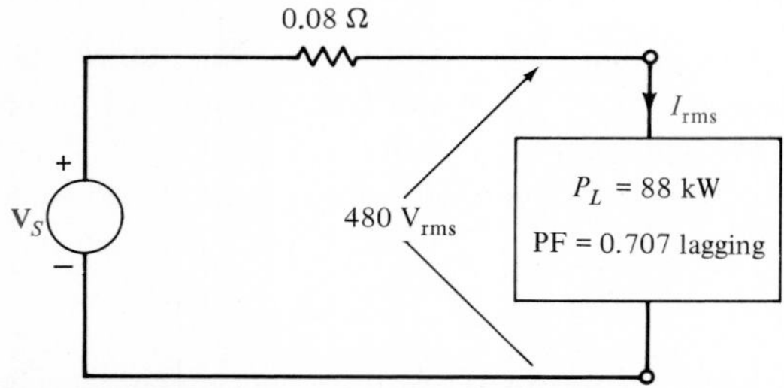

Figure 11.11 Example circuit for examining changes in power factor.

(11.29), the magnitude of the rms current into the plant is

$$I_{\text{rms}} = \frac{P_L}{(\text{PF})(V_{\text{rms}})}$$

$$= \frac{88 \times 10^3}{(0.707)(480)}$$

$$= 259.3 \text{ A}$$

The power that must be supplied by the power company is

$$P_S = P_L + (0.08)I_{\text{rms}}^2$$

$$= 88{,}000 + (0.08)(259.3)^2$$

$$= 93.38 \text{ kW}$$

(b) Suppose now that the PF is somehow changed to 0.90 lagging but the voltage remains constant at 480 V. The rms load current for this condition is

$$I_{\text{rms}} = \frac{P_L}{(\text{PF})(V_{\text{rms}})}$$

$$= \frac{88 \times 10^3}{(0.90)(480)}$$

$$= 203.7 \text{ A}$$

Under these conditions the power company must generate

$$P_S = P_L + (0.08)I_{\text{rms}}^2$$

$$= 88{,}000 + (0.08)(203.7)^2$$

$$= 91.32 \text{ kW}$$

Note carefully the difference between the two cases. A simple change in the PF of the load from 0.707 lagging to 0.90 lagging has had an interesting effect. Note that in the first case the power company must generate 93.38 kW in order to supply the plant with 88 kW of power because the low power factor means that the line losses will be high—5.38 kW. However, in the second case the power company need only generate 91.32 kW in order to supply the plant with its required power, and the corresponding line losses are only 3.32 kW. ■

The example clearly indicates the economic impact of the load's power factor. A low power factor at the load means that the power company generators must be capable of carrying more current at constant voltage, and they must also supply power for higher $I_{\text{rms}}^2 R$ line losses than would be required if the load's power factor were high. Since line losses represent energy expended in heat and benefit no one, the power company will insist that a plant maintain a high PF, typically 0.90 lagging, and adjust their rate schedule to penalize plants that do not conform to this requirement.

DRILL EXERCISE

D11.11. An industrial load consumes 100 kW at 0.707 PF lagging. The 60-Hz line voltage at the load is $480 \angle 0°$ V rms. The transmisison-line resistance between the power company's transformer and the load is 0.1 Ω. Determine the power savings that could be obtained if the PF is changed to 0.94 lagging.

Industrial plants that require large amounts of power have a wide variety of loads. However, by nature the loads normally have a lagging power factor. In view of our discussion above, we are naturally led to ask if there is any convenient technique for raising the power factor of a load. Since a typical load may be a bank of induction motors or other expensive machinery, the technique for raising the PF should be an economical one in order to be feasible.

To answer the question we pose, let us examine the circuit shown in Fig. 11.12. The circuit illustrates a typical industrial load. In parallel with this load we have placed a capacitor. Let us assume that the load voltage $\mathbf{V} = V \angle 0°$. The 0° phase angle causes no loss of generality but does simplify the following development.

The load current is

$$\mathbf{I}_L = \frac{V \angle 0°}{R_L + jX_L} = \frac{V}{Z_L \angle \theta_{\mathbf{Z}_L}} = I_L \angle -\theta_{\mathbf{Z}_L}$$

The capacitor current is

$$\mathbf{I}_C = \frac{V \angle 0°}{-jX_C} = I_C \angle 90° = \omega C V \angle 90°$$

Therefore, the total current is given by the expression

$$\begin{aligned}\mathbf{I}_T &= \mathbf{I}_L + \mathbf{I}_C \\ &= I_L \angle -\theta_{\mathbf{Z}_L} + I_C \angle 90°\end{aligned}$$

Without the capacitor present, the PF angle, the angle between the voltage V and the current $\mathbf{I}_T$, is $\theta_{\mathbf{Z}_L}$. Since $\theta_{\mathbf{Z}_L}$ is positive, the PF is lagging and the smaller the value of $\theta_{\mathbf{Z}_L}$, the higher the power factor cos $\theta_{\mathbf{Z}_L}$. If we now assume that the capacitor is present and express the total current in rectangluar form, we have

$$\mathbf{I}_T = I_{L\,\text{Re}} - jI_{L\,\text{Im}} + jI_C$$

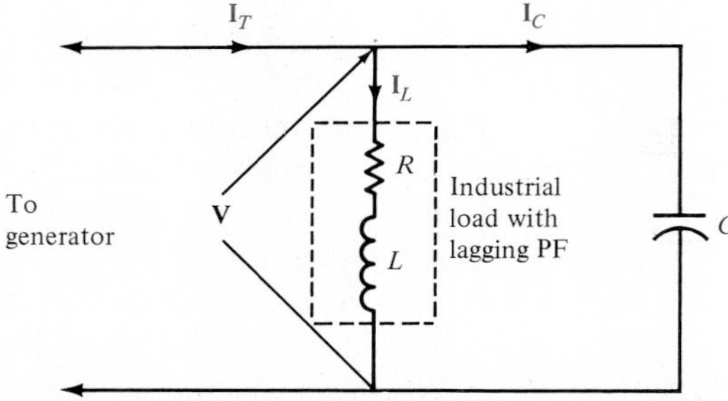

Figure 11.12 Circuit for power factor correction.

where the subscripts Re and Im refer to real and imaginary parts, and $I_C = \omega CV$. Grouping the imaginary terms in the equation above yields

$$\mathbf{I}_T = I_{L\,\mathrm{Re}} - j(I_{L\,\mathrm{Im}} - I_C)$$

Recall that I_C is dependent on the capacitor value; that is, by selecting the value of the capacitor, we can make this quantity any reasonable number. Note also that since the voltage angle is 0°, the lagging PF angle for the total load (i.e., the industrial load plus the capacitor) is

$$\theta = \tan^{-1} \frac{I_{L\,\mathrm{Im}} - I_c}{I_{L\,\mathrm{Re}}} \tag{11.31}$$

and therefore since $I_C = \omega CV$, Eq. (11.31) can be written as

$$\tan \theta = \frac{I_{L\,\mathrm{Im}} - \omega CV}{I_{L\,\mathrm{Re}}}$$

or rearranging the terms,

$$C = \frac{I_{L\,\mathrm{Im}} - I_{L\,\mathrm{Re}} \tan \theta}{\omega V} \tag{11.32}$$

Note carefully that in applying Eq. (11.32), the fact that $I_{L\,\mathrm{Im}}$ is negative has been accounted for in developing Eq. (11.31); therefore, if, for example, $I_L = 3.07 - j2.52$, then $I_{L\,\mathrm{Im}}$ in Eq. (11.32) is $+2.52$. For a given PF, C is the only unknown in this equation. Hence we can obtain a particular power factor for the total load by simply judiciously selecting a capacitor and placing it in parallel with the load. In general, we want the PF angle to be small and therefore we select the capacitor so that it reduces the quantity $I_{L\,\mathrm{Im}} - I_C$ by some prescribed amount. The following example will illustrate the approach.

EXAMPLE 11.13

An industrial load consisting of a bank of induction motors consumes 50 kW at a PF of 0.8 lagging from a 220 $\underline{/0°}$-V rms, 60-Hz line. We wish to raise the PF to 0.95 by placing a bank of capacitors in parallel with the load.

The circuit diagram for this problem is shown in Fig. 11.13. The magnitude of the rms load current is

$$I_L = \frac{50 \times 10^3}{(0.8)(220)}$$
$$= 284.09 \text{ A}$$

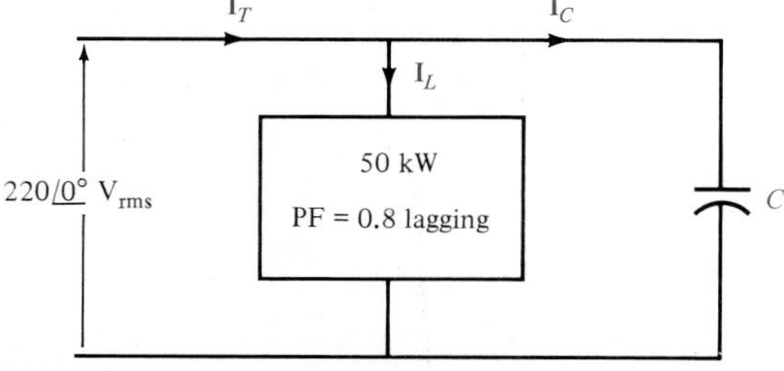

Figure 11.13 Example circuit for power factor correction.

Therefore, since $\cos^{-1} 0.8 = 36.87°$, the phasor load current is

$$\mathbf{I}_L = 284.09 \underline{/-36.87°}$$
$$= 227.27 - j170.45 \text{ A}$$

A PF = 0.95 corresponds to

$$\theta = \cos^{-1} 0.95$$
$$= 18.19°$$

Hence using Eq. (11.32), we have

$$C = \frac{170.45 - 227.27 \tan 18.19°}{(377)(220)}$$

Therefore,

$$C = 1155\ \mu\text{F}$$

By using a capacitor of this magnitude in parallel with the industrial load, we create, from the power company's perspective, a load PF of 0.95. However, since the capacitor is in parallel and absorbs no power, the parameters of the actual load remain unchanged. ■

DRILL EXERCISE

D11.12. Compute the value of the capacitor necessary to change the power factor in Drill Exercise D11.11 to 0.95 lagging.

11.6

Complex Power

In our study of ac steady-state power, it is convenient to introduce another quantity, which is commonly called *complex power*. To develop the relationship between this quantity and others we have presented in the preceding sections, consider the circuit shown in Fig. 11.14.

The complex power is defined to be

$$\mathbf{S} = \mathbf{V}_{\text{rms}}\mathbf{I}^*_{\text{rms}} \tag{11.33}$$

where $\mathbf{I}^*_{\text{rms}}$ refers to the complex conjugate of $\mathbf{I}_{\text{rms}}$; that is, if $\mathbf{I}_{\text{rms}} = I_{\text{rms}} \underline{/\theta_i} = I_R + jI_I$,

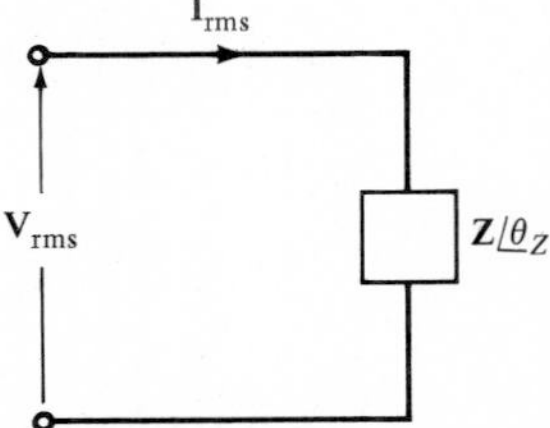

Figure 11.14 Circuit used to explain power relationships.

then $\mathbf{I}^*_{\text{rms}} \underline{/-\theta_i} = I_R - jI_I$. Since

$$\mathbf{V}_{\text{rms}} = V_{\text{rms}} \underline{/\theta_v}$$

and

$$\mathbf{I}_{\text{rms}} = I_{\text{rms}} \underline{/\theta_i}$$

Complex power can be expressed in the form

$$\mathbf{S} = V_{\text{rms}} I_{\text{rms}} \underline{/\theta_v - \theta_i} \tag{11.34}$$

In this form it is clearly a complex quantity with a real part

$$P = V_{\text{rms}} I_{\text{rms}} \cos (\theta_v - \theta_i) \tag{11.35}$$

and an imaginary part

$$Q = V_{\text{rms}} I_{\text{rms}} \sin (\theta_v - \theta_i) \tag{11.36}$$

where, of course, $\theta_v - \theta_i = \theta_{\mathbf{Z}}$. We note from Eq. (11.35) that the real part of the complex power is simply the *real* or *average power*. The imaginary part of **S**, as defined in Eq. (11.36) we call the *reactive* or *quadrature power*. Therefore, complex power can be expressed in the form

$$\mathbf{S} = P + jQ \tag{11.37}$$

where

$$P = \text{Re}\,(\mathbf{S}) = V_{\text{rms}} I_{\text{rms}} \cos (\theta_v - \theta_i) \tag{11.38}$$

$$Q = \text{Im}\,(\mathbf{S}) = V_{\text{rms}} I_{\text{rms}} \sin (\theta_v - \theta_i) \tag{11.39}$$

and as shown in Eq. (11.34), the magnitude of the complex power is what we have called the *apparent power*, and the phase angle for complex power is simply the power factor angle. Complex power like apparent power is measured in volt-amperes, real power is measured in watts, and in order to distinguish Q from the other quantities, which in fact have the same dimensions, it is measured in var, volt-amperes reactive.

In addition to the relationships expressed above, we note that since

$$\cos (\theta_v - \theta_i) = \frac{\text{Re}\,(\mathbf{Z})}{|\mathbf{Z}|}$$

$$\sin (\theta_v - \theta_i) = \frac{\text{Im}\,(\mathbf{Z})}{|\mathbf{Z}|}$$

and

$$I_{\text{rms}} = \frac{V_{\text{rms}}}{|\mathbf{Z}|}$$

Eqs. (11.38) and (11.39) can be written as

$$P = I^2_{\text{rms}} \text{Re}\,(\mathbf{Z}) \tag{11.40}$$

$$Q = I^2_{\text{rms}} \text{Im}\,(\mathbf{Z}) \tag{11.41}$$

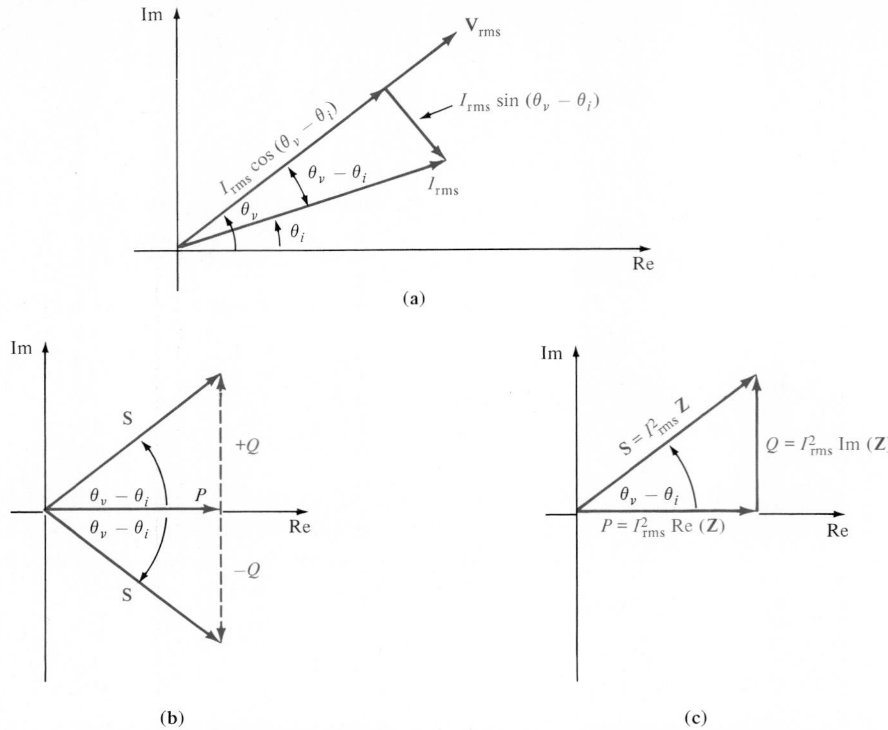

Figure 11.15 Diagram for illustrating power relationships.

and therefore Eq. (11.37) can be expressed as

$$\mathbf{S} = I^2_{rms}\,\mathbf{Z} \tag{11.42}$$

The diagrams in Fig. 11.15 serve to explain further the relationships among the various quantities of power. As shown in Fig. 11.15a, the phasor current can be split into two components: one that is in phase with $\mathbf{V}_{rms}$ and one that is 90° out of phase with $\mathbf{V}_{rms}$. Equations (11.38) and (11.39) illustrate that the in-phase component produces the real power, and the 90° component, called the *quadrature component*, produces the reactive or quadrature power. In addition, Eqs. (11.37) to (11.39) indicate that

$$\tan(\theta_v - \theta_i) = \frac{Q}{P} \tag{11.43}$$

which relates the PF angle to P and Q in what is called the *power triangle*.

The relationships among **S**, P, and Q can be expressed via the diagrams shown in Fig. 11.15b and c. In Fig. 11.15b we note the following conditions. If Q is positive, the load is inductive, the power factor is lagging, and the complex number **S** lies in the first quadrant. If Q is negative, the load is capacitive, the power factor is leading, and the complex number **S** lies in the fourth quadrant. If Q is zero, the load is resistive, the power factor is unity, and the complex number **S** lies along the positive real axis. Figure

11.15c illustrates the relationships expressed by Eq. (11.40) to (11.42) for an inductive load.

Finally, it is important to state that complex power, like energy, is conserved; that is, the total complex power delivered to any number of individual loads is equal to the sum of the complex powers delivered to each individual load, regardless of how the loads are interconnected.

EXAMPLE 11.14

A load operates at 20 kW, 0.8 PF lagging. The load voltage is $220 \underline{/0°}$ V rms at 60 Hz. The impedance of the line is $0.09 + j0.3\ \Omega$. We wish to determine the voltage and power factor at the input to the line.

The circuit diagram for this problem is shown in Fig. 11.16. The magnitude of the rms load current is

$$I_L = \frac{20 \times 10^3}{(220)(0.8)}$$
$$= 113.64 \text{ A rms}$$

This current lags the voltage by $\theta = \cos^{-1} 0.8 = 36.87°$. The reactive power at the load is

$$Q_L = P_L \tan \theta$$
$$= (20 \times 10^3)(0.75)$$
$$= 15{,}000 \text{ var}$$

The real power loss in the line is

$$P_{\text{line}} = (113.64)^2(0.09)$$
$$= 1162.26 \text{ W}$$

The reactive power loss in the line is

$$Q_{\text{line}} = (113.64)^2(0.3)$$
$$= 3874.21 \text{ var}$$

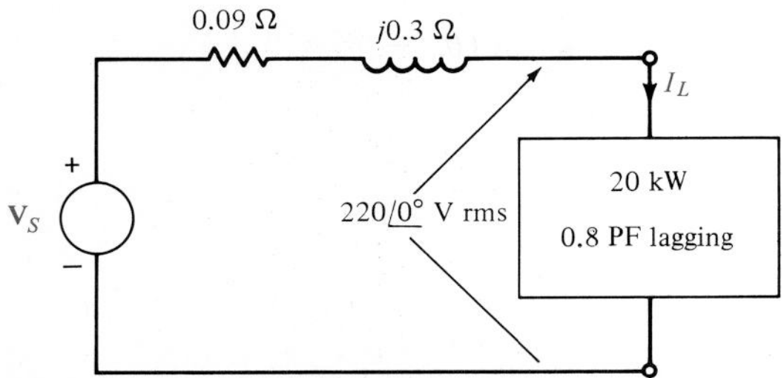

Figure 11.16 Example circuit for power analysis.

The power supplied at the generator must equal that consumed in the system and hence

$$\begin{aligned} P_S &= P_L + P_{\text{line}} \\ &= 20{,}000 + 1162.26 \\ &= 21{,}162.26 \text{ W} \end{aligned}$$

and

$$\begin{aligned} Q_S &= Q_L + Q_{\text{line}} \\ &= 15{,}000 + 3874.21 \\ &= 18{,}874.21 \text{ var} \end{aligned}$$

Therefore, the complex power at the generator is

$$\begin{aligned} \mathbf{S}_S &= P_S + jQ_S \\ &= 21{,}162.26 + j18{,}874.21 \\ &= 28{,}356.25 \,\underline{/41.73^\circ} \text{ VA} \end{aligned}$$

Hence the generator voltage is

$$\begin{aligned} V_S &= \frac{28{,}356.25}{113.64} \\ &= 249.53 \text{ V rms} \end{aligned}$$

and the generator power factor is

$$\cos(41.73^\circ) = 0.75 \text{ lagging}$$

We could have solved this problem using KVL. For example, we calculated the load current as

$$\mathbf{I}_L = 113.64 \,\underline{/-36.87^\circ} \text{ A rms}$$

Hence the voltage drop in the transmission line is

$$\begin{aligned} \mathbf{V}_{\text{line}} &= (113.64 \,\underline{/-36.87^\circ})(0.09 + j0.3) \\ &= 35.59 \,\underline{/36.43^\circ} \text{ V rms} \end{aligned}$$

Therefore, the generator voltage is

$$\begin{aligned} \mathbf{V}_S &= 220 \,\underline{/0^\circ} + 35.59 \,\underline{/36.43^\circ} \\ &= 249.53 \,\underline{/4.86^\circ} \text{ V rms} \end{aligned}$$

Hence the generator voltage is 249.54 V rms. In addition,

$$\theta_v - \theta_i = 4.86^\circ - (-36.87^\circ) = 41.73^\circ$$

and therefore,

$$\text{PF} = \cos(41.73^\circ) = 0.75 \text{ lagging}$$ ■

DRILL EXERCISE

D11.13. An industrial load requires 40 kW at 0.84 PF lagging. The load voltage is 220 $\underline{/0°}$ V rms at 60 Hz. The transmission-line impedance is $0.1 + j0.25\ \Omega$. Determine the real and reactive power losses in the line and the real and reactive power required at the input to the transmission line.

D11.14. A load requires 60 kW at 0.85 PF lagging. The 60-Hz line voltage at the load is 220 $\underline{/0°}$ V rms. If the transmission-line impedance is $0.12 + j0.18\ \Omega$, determine the line voltage and power factor at the input.

In Section 11.5 we illustrated a technique for load power factor correction by placing a capacitor in parallel with the load. We will briefly reexamine this technique using the quantities we have just defined.

Consider once again the circuit in Fig. 11.12. The original complex power for the load $\mathbf{Z}_L$, which we will denote as $\mathbf{S}_{\text{old}}$, is

$$\mathbf{S}_{\text{old}} = P_{\text{old}} + jQ_{\text{old}} = |\mathbf{S}_{\text{old}}| \underline{/\theta_{\text{old}}}$$

The new complex power that results from adding a capacitor is

$$\mathbf{S}_{\text{new}} = P_{\text{old}} + jQ_{\text{new}} = |\mathbf{S}_{\text{new}}| \underline{/\theta_{\text{new}}}$$

where θ_{new} is specified by the required power factor. The difference between the new and old complex powers is caused by the addition of the capacitor. Hence

$$\mathbf{S}_{\text{cap}} = \mathbf{S}_{\text{new}} - \mathbf{S}_{\text{old}} \tag{11.44}$$

and since the capacitor is purely reactive,

$$\mathbf{S}_{\text{cap}} = +jQ_c = -j\omega C V_{\text{rms}}^2 \tag{11.45}$$

Equation (11.45) can be used to find the required value of C. The procedure we have described is illustrated in Fig. 11.17. Remember that the smaller the angle $\theta_v - \theta_i$, the larger the power factor.

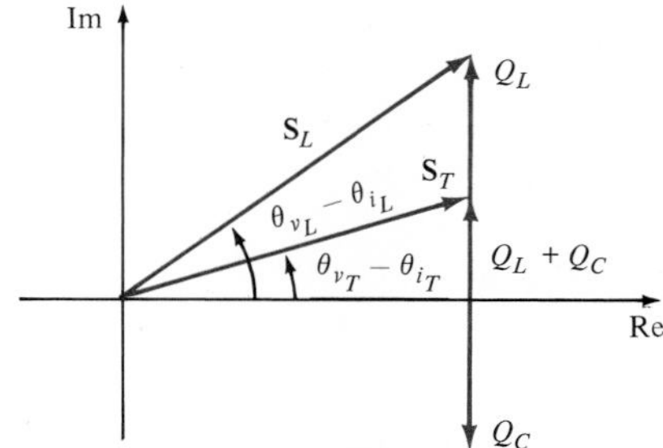

Figure 11.17 Illustration of the technique for power factor correction.

EXAMPLE 11.15

Let us repeat Example 11.13 using the complex number technique we have developed in this section. Recall that $P_L = 50$ kW and $\theta_{\text{old}} = 36.87°$. Therefore,

$$Q_{\text{old}} = P_{\text{old}} \tan \theta_{\text{old}} = (50)(10^3)(0.75) = 37.5 \text{ kvar}$$

Hence

$$\mathbf{S}_{\text{old}} = 50{,}000 + j37500$$

Since the required power factor is 0.95, $\theta_{\text{new}} = 18.19°$. Then

$$\begin{aligned}\mathbf{S}_{\text{new}} &= 50{,}000 + j50{,}000 \tan^{-1} 18.19° \\ &= 50{,}000 + j16{,}430\end{aligned}$$

and therefore

$$\begin{aligned}\mathbf{S}_{\text{cap}} &= \mathbf{S}_{\text{new}} - \mathbf{S}_{\text{old}} \\ &= -j21{,}070\end{aligned}$$

Hence

$$\begin{aligned}C &= \frac{21{,}070}{(377)(220)^2} \\ &= 1155\ \mu\text{F}\end{aligned}$$

■

DRILL EXERCISE

D11.15. Use equation (11.45) to solve Drill Exercise D11.12.

11.7 Summary

The basic power relationships which apply in ac steady-state circuits have been presented. Instantaneous power and average power were defined. Techniques for maximum average power transfer, which is analogous to maximum power transfer for dc circuits, was presented for various load conditions.

The effective, or rms, value of a periodic waveform was introduced as a means of measuring the effectiveness of a source in delivering power to a resistive load.

The power factor angle was introduced, together with a scheme for correcting it if necessary. Complex power and its relationship to real and reactive power was also presented.

KEY POINTS

- If the current and voltage are sinusoidal functions of time, the instantaneous power is equal to a time-independent average value plus a sinusoidal term which has a frequency twice that of the voltage or current.
- Capacitors and inductors are lossless elements and absorb no average power.
- When multiple sources are present in a network, superposition cannot be used in computing power unless each source is of a different frequency.
- To obtain the maximum average power transfer to a load, the load impedance should

be chosen equal to the conjugate of the Thévenin equivalent impedance representing the remainder of the network.

- The effective value of a periodic waveform is found by determining the root-mean-square (rms) value of the waveform.
- The rms value of a sinusoidal function is equal to the maximum value of the sinusoid divided by $\sqrt{2}$.
- Apparent power is defined as the product $V_{rms}I_{rms}$.
- The power factor is defined as the ratio of the average power to the apparent power and is said to be leading when the phase of the current leads the voltage, and lagging when the phase of the current lags the voltage.
- The power factor of a load with a lagging power factor can be corrected by placing a capacitor in parallel with the load.
- Complex power, **S**, is defined as the product $V_{rms}I^*_{rms}$.
- The complex power **S** can be written as $\mathbf{S} = P + jQ$, where P is the real or average power and Q is the imaginary or quadrature power.

PROBLEMS

11.1. The voltage and current at the input of a network are given by the expressions

$$v(t) = 6 \cos \omega t \text{ V}$$

$$i(t) = 4 \sin \omega t \text{ A}$$

Determine the average power absorbed by the network.

11.2. Find the average power absorbed by the resistor in the circuit shown in Fig. P11.2. Let $i_1(t) = 4 \cos (377t + 60°)$ A, $i_2(t) = 6 \cos (754t + 10°)$ A, and $i_3(t) = 4 \cos (377t - 30°)$ A.

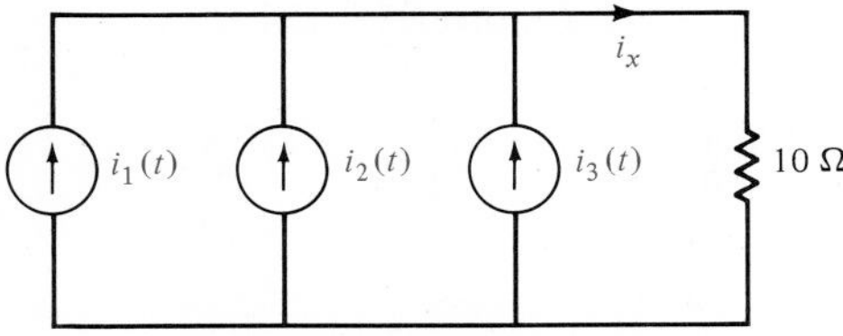

Figure P11.2

11.3. Find the average power absorbed by the network shown in Fig. P11.3.

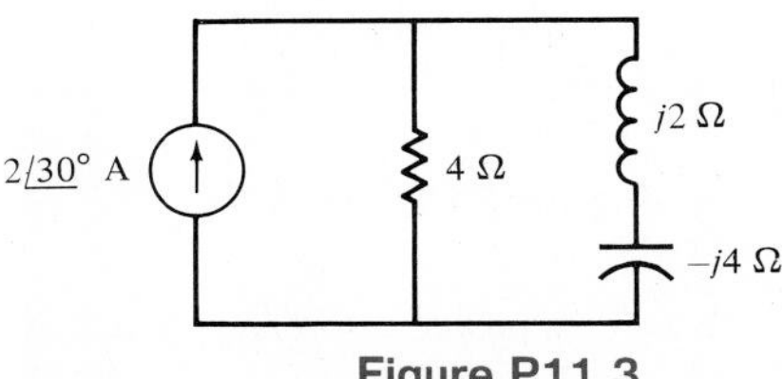

Figure P11.3

11.4. Compute the average power absorbed by the elements at the right of the dashed line in the network shown in Fig. P11.4.

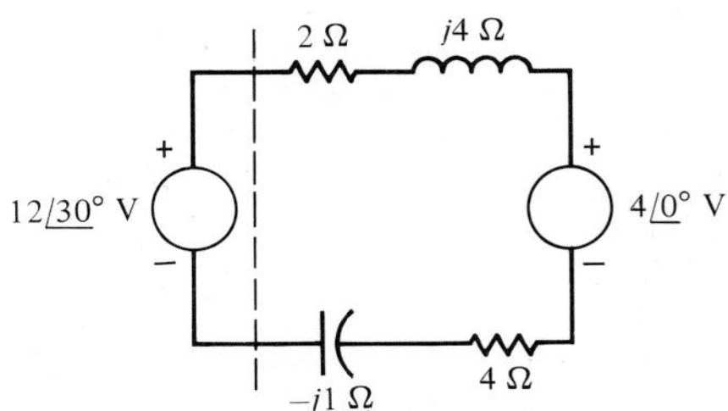

Figure P11.4

11.5. Determine the average power supplied by each source in Problem 11.4.

11.6. Given the network in Fig. P11.6, show that the power supplied by the sources is equal to the power absorbed by the passive elements.

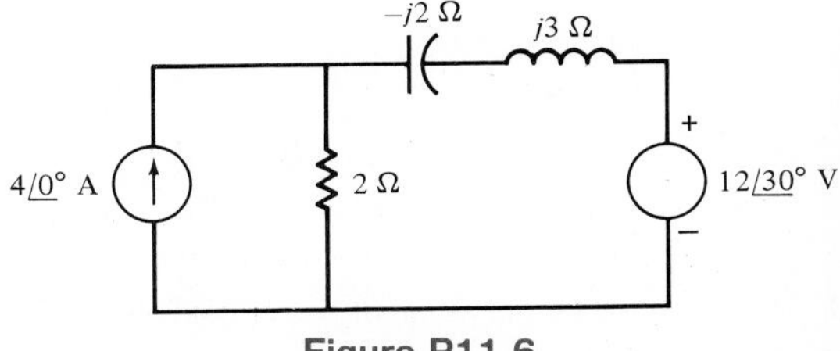

Figure P11.6

11.7. Show that an alternative expression for the average power absorbed by an impedance **Z** is

$$P = \tfrac{1}{2} I_m^2 \text{ Re } (\mathbf{Z})$$

where $\mathbf{Z} = \text{Re } (\mathbf{Z}) + j \text{ Im } (\mathbf{Z})$ and Re (**Z**) and Im (**Z**) represent the real and imaginary parts of the impedance, respectively.

11.8. Find the average power absorbed by the 2-Ω resistor in the circuit shown in Fig. P11.8.

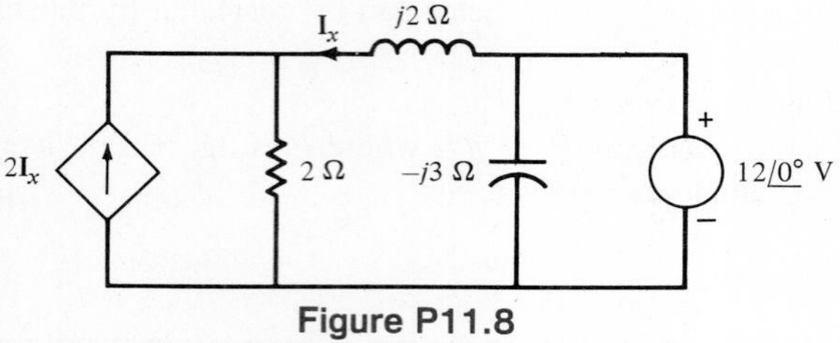

Figure P11.8

11.9. Determine the average power absorbed by the 4-Ω resistor in the network shown in Fig. P11.9.

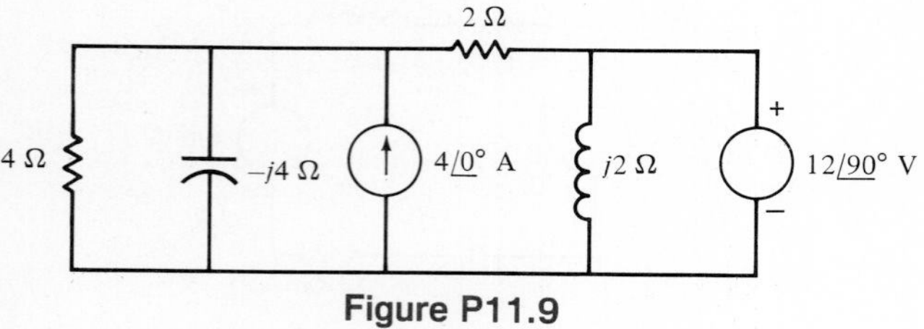

Figure P11.9

11.10. Calculate the average power absorbed by the 1-Ω resistor in the network shown in Fig. P11.10.

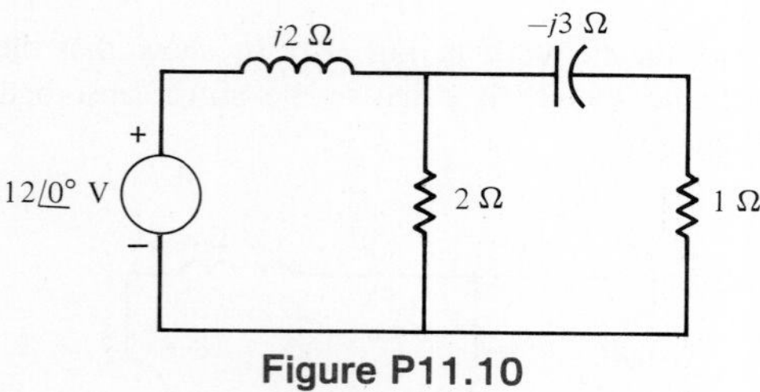

Figure P11.10

11.11. Show that the conservation of power holds for the network shown in Fig. P11.11.

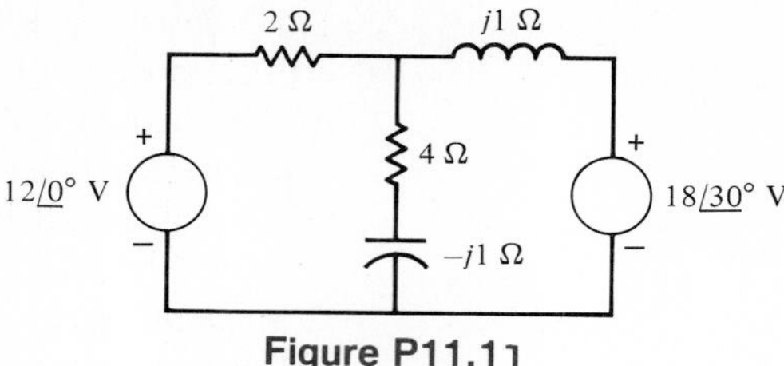

Figure P11.11

11.12. Show that the conservation of power holds for the network shown in Fig. P11.12.

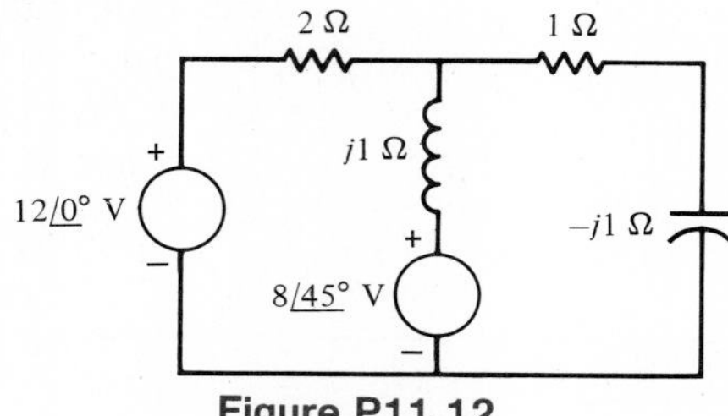

Figure P11.12

11.13. Show that the conservation of power holds for the network shown in Fig. P11.13.

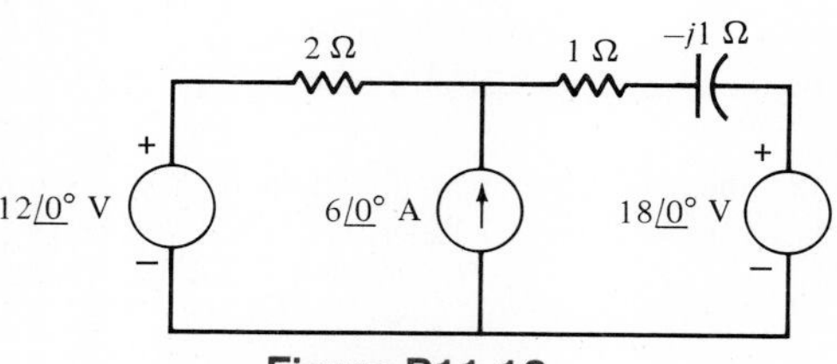

Figure P11.13

11.14. Determine the average power absorbed by the 2-kΩ output resistor in Fig. P11.14.

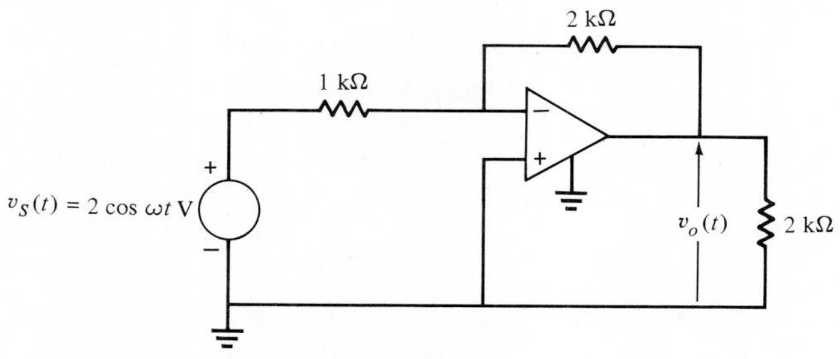

Figure P11.14

11.15. Determine the average power absorbed by the 4-kΩ resistor in Fig. P11.15.

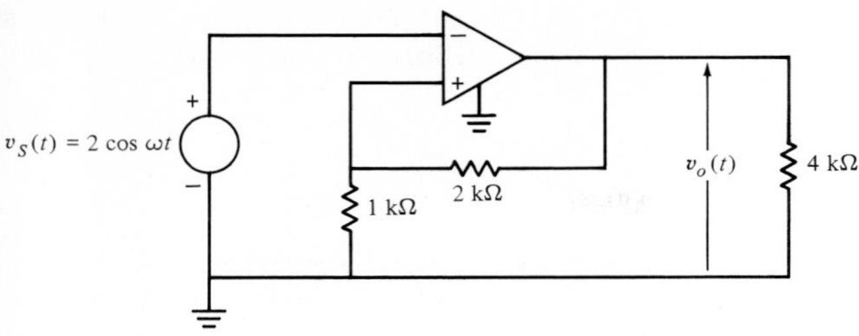

Figure P11.15

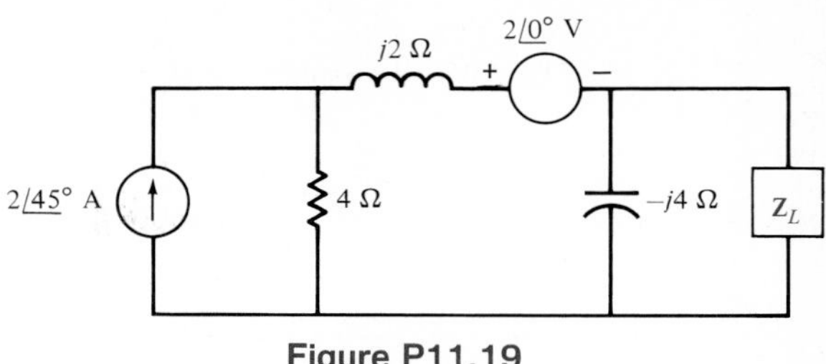

Figure P11.19

11.16. Determine the average power absorbed by a 2-Ω resistor connected at the output terminals of the network shown in Fig. P11.16.

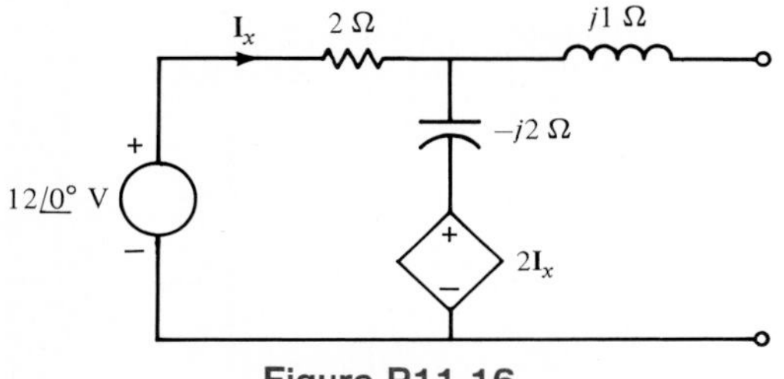

Figure P11.16

11.17. Show that the maximum average power transferred to a load is given by the following expression, where V_{oc} and R_{Th} are defined in the chapter.

$$P_L = \frac{V_{oc}^2}{8R_{Th}}$$

11.18. Determine the value of $\mathbf{Z}_L$ for maximum average power transfer in the circuit shown in Fig. P11.18.

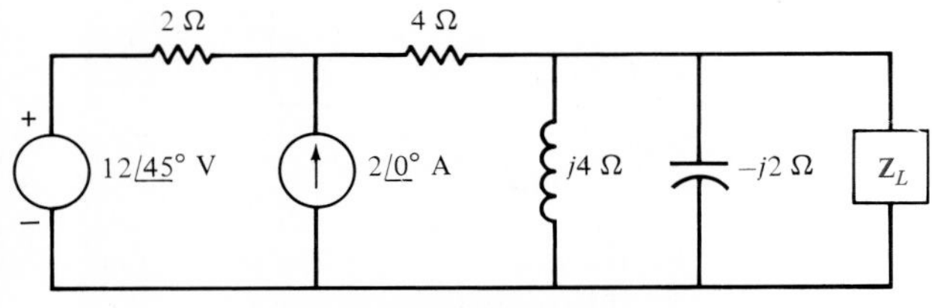

Figure P11.18

11.19. Determine the impedance $\mathbf{Z}_L$ for maximum average power transfer and the value of the maximum average power absorbed by the load in the network shown in Fig. P11.19.

11.20. Determine the impedance $\mathbf{Z}_L$ for maximum average power transfer and the value of the maximum average power transferred for the circuit shown in Fig. P11.20.

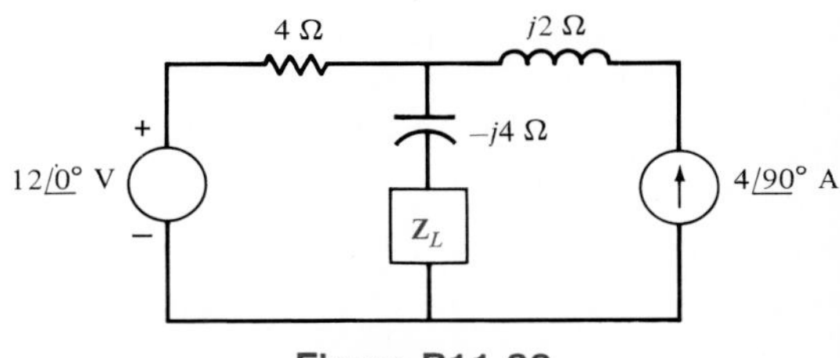

Figure P11.20

11.21. Find the impedance $\mathbf{Z}_L$ for maximum average power transfer and the value of the maximum average power transferred for the circuit shown in Fig. P11.21.

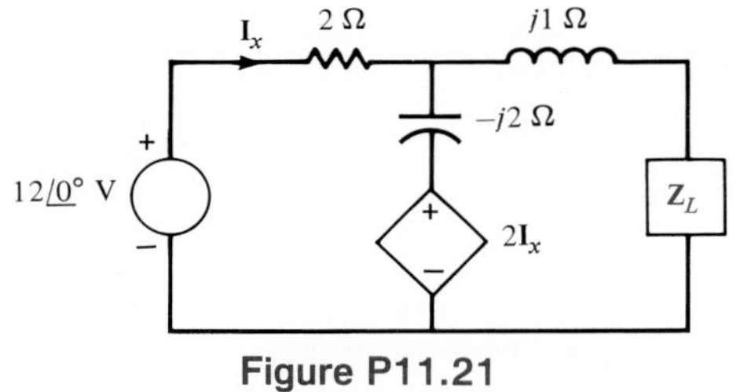

Figure P11.21

11.22. Find the value of $\mathbf{Z}_L$ for maximum average power transfer in the circuits shown in Fig. P11.22.

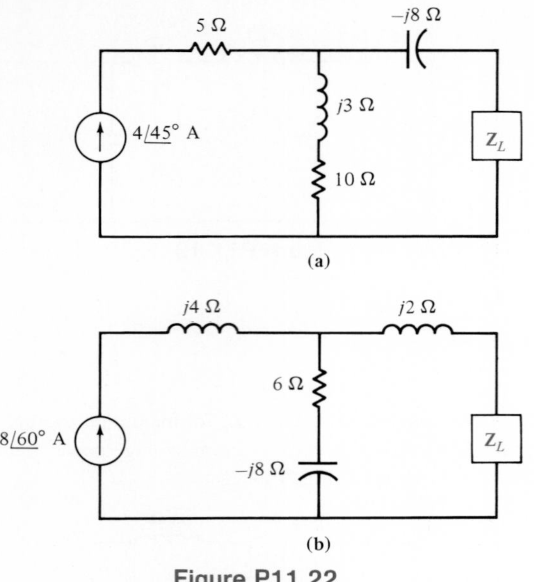

Figure P11.22

11.23. Determine the impedance $\mathbf{Z}_L$ for maximum average power transfer and the value of the maximum average power transferred for the circuit shown in Fig. P11.23.

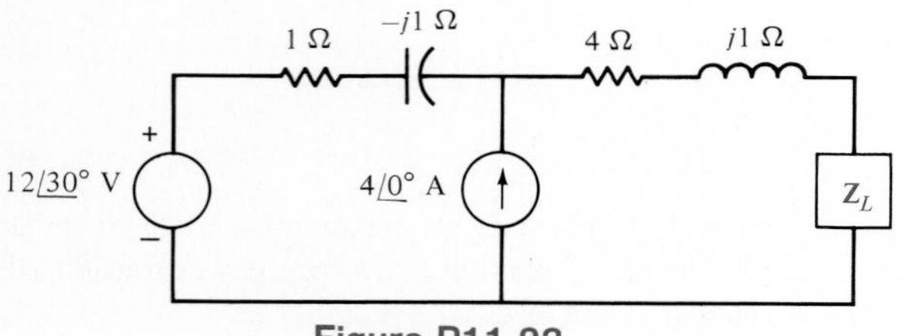

Figure P11.23

11.24. Determine the impedance $\mathbf{Z}_L$ for maximum average power transfer and the value of the maximum average power transferred for the network shown in Fig. P11.24.

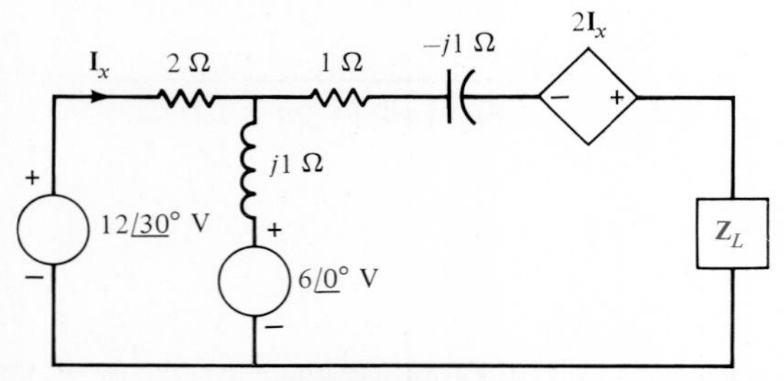

Figure P11.24

11.25. Determine $\mathbf{Z}_L$ for maximum average power transfer and the value of the maximum average power transferred for the network shown in Fig. P11.25.

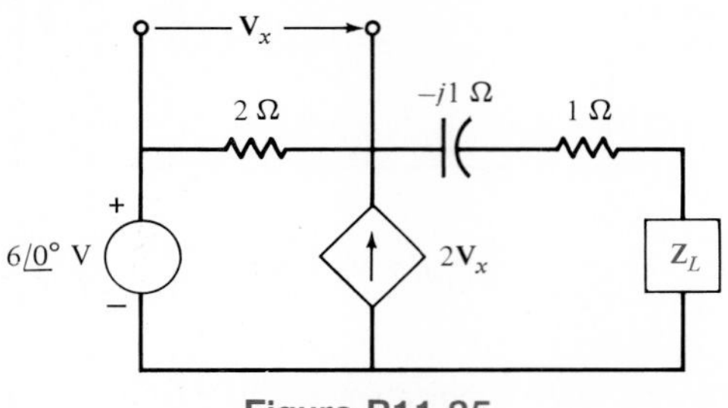

Figure P11.25

11.26. Determine the value of $\mathbf{Z}_L$ for maximum average power transfer in the circuit shown in Fig. P11.26.

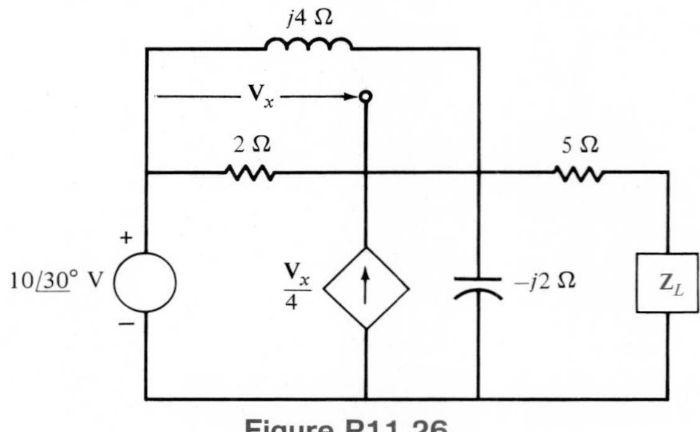

Figure P11.26

11.27. Calculate the rms value of the waveform shown in Fig. P11.27.

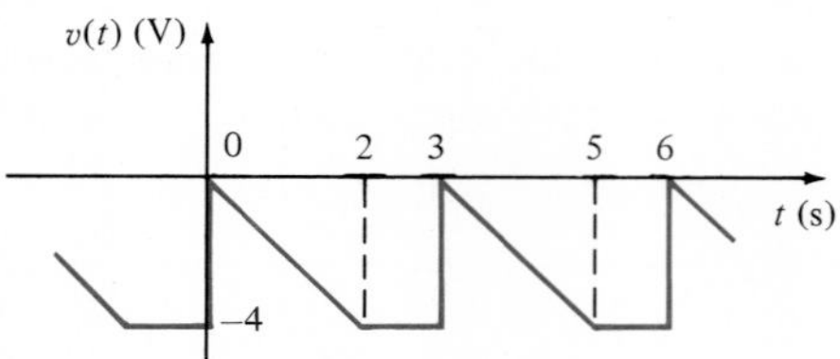

Figure P11.27

11.28. Compute the rms value of the voltage given by the waveform shown in Fig. P11.28.

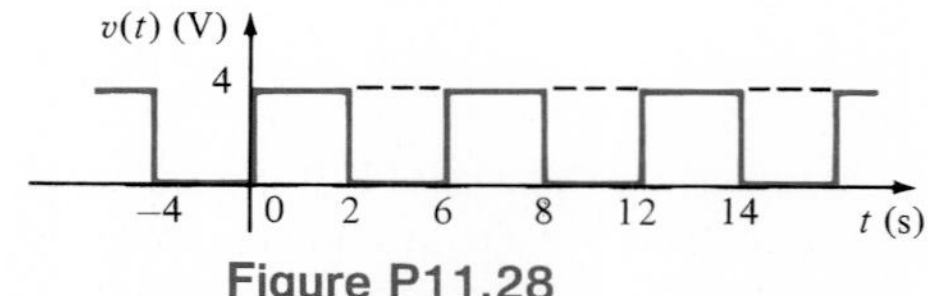

Figure P11.28

11.29. The current waveform in Fig. P11.29 is flowing through a 5-Ω resistor. Find the average power absorbed by the resistor.

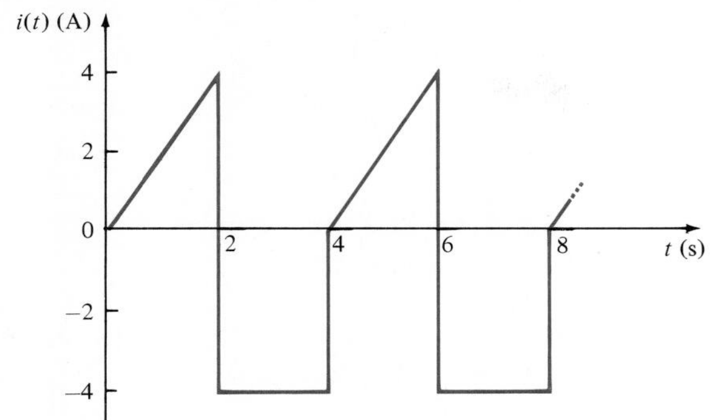

Figure P11.29

11.30. Calculate the rms value of the waveform shown in Fig. P11.30.

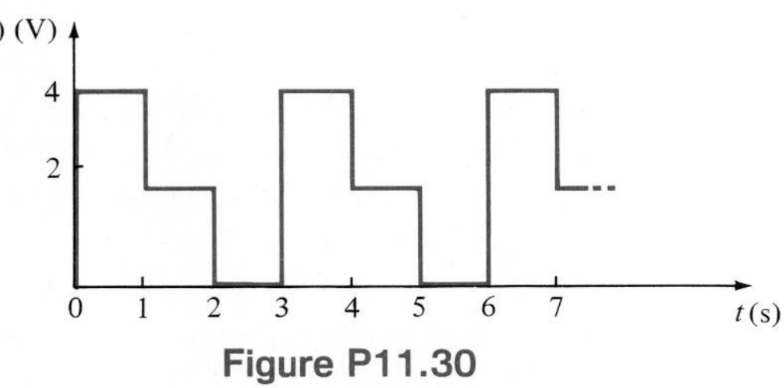

Figure P11.30

11.31. Calculate the rms value of the waveform shown in Fig. P11.31.

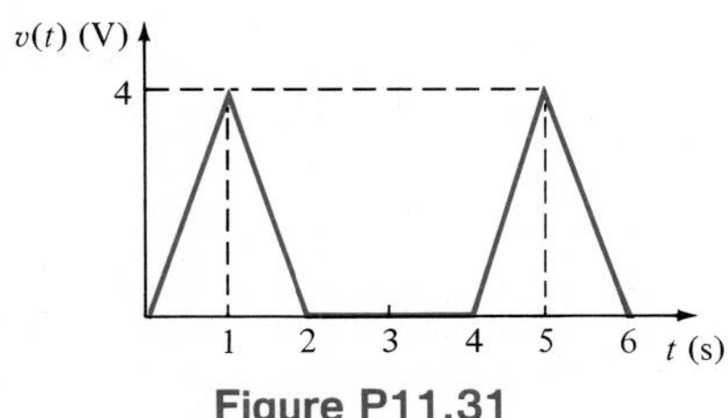

Figure P11.31

11.32. Find the rms value of the exponential waveform shown in Fig. P11.32.

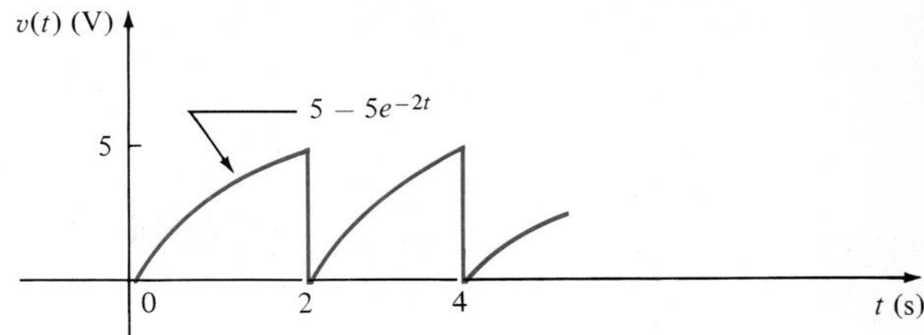

Figure P11.32

11.33. Calculate the rms value of the waveform shown in Fig. P11.33.

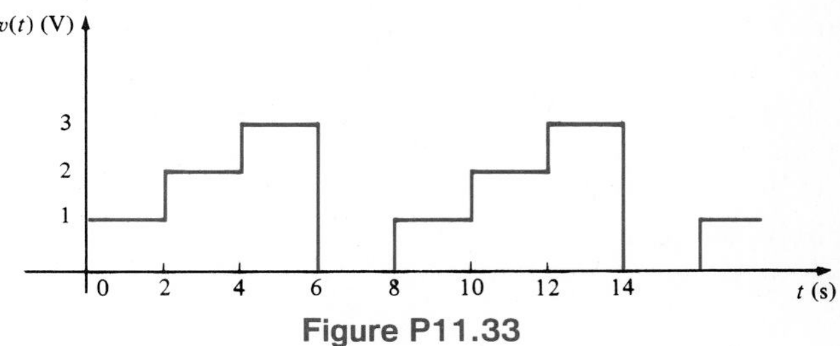

Figure P11.33

11.34. Determine the rms value of the waveform shown in Fig. 11.34.

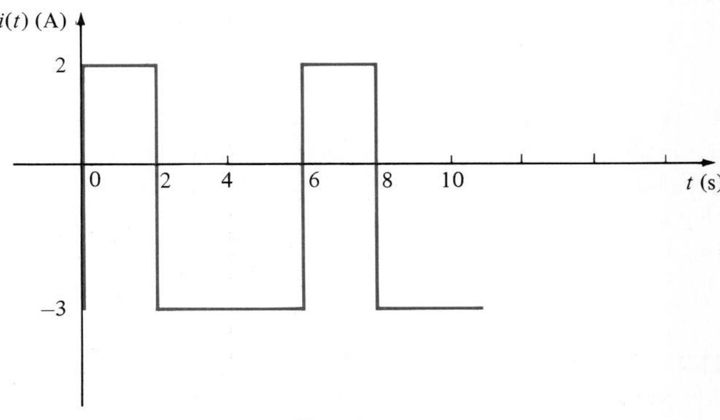

Figure P11.34

11.35. Determine the rms value of the rectified sine wave shown in Fig. P11.35. Use the fact that $\sin^2 \theta = \frac{1}{2} - \frac{1}{2} \cos 2\theta$.

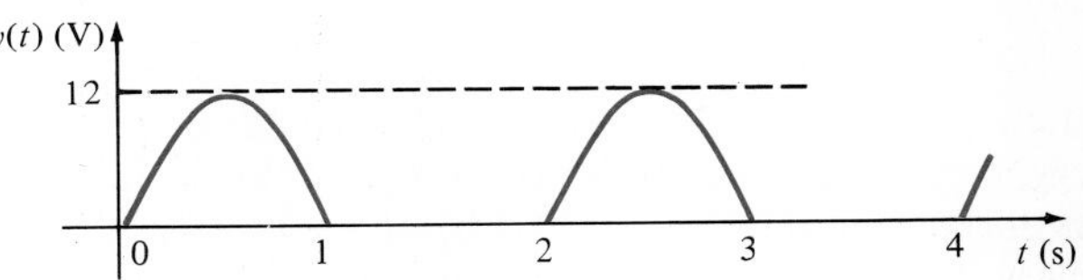

Figure P11.35

11.36. Compute the effective value of the voltage that is given by the expression

$$v(t) = 4 \sin(\omega t + 30°) + 6 \sin(2\omega t + 90°) \text{ V}$$

11.37. Find the rms value of the current defined by the expression

$$i(t) = 10 \cos(377t) + 12 \sin(754t + 80°) \text{ A}$$

11.38. For the following sinusoidal voltages, write the phasor voltage with rms magnitude in polar form.

(a) $v_1(t) = 50\sqrt{2} \cos(377t + 20°)$ V
(b) $v_2(t) = 32/\sqrt{2} \cos(377t - 30°)$ V
(c) $v_3(t) = 20 \sin(2\pi t + 10°)$ V

11.39. For the following rms voltages, write the voltage in the form of a sinusoid with $\omega = 377$ rad/s.

(a) $V_1 = 100 \angle 40°$ V rms
(b) $V_2 = 30\sqrt{2} \angle 30°$ V rms
(c) $V_3 = 16/\sqrt{2} \angle -100°$ V rms

11.40. Convert the following currents into their sinusoidal equivalents. Assume that $\omega = 377$ rad/s.

(a) $\mathbf{I}_a = 5 \angle 120°$ A rms
(b) $\mathbf{I}_b = 7.07 \angle 70°$ A rms
(c) $\mathbf{I}_c = 14\sqrt{2} \angle -30°$ A rms

11.41. Most voltmeters and ammeters display the rms values of the voltage or current being measured. If a sinusoidal waveform is being measured, determine its sinusoidal time-domain expression if the frequency is 60 Hz and the meter reads

(a) 120 V rms.
(b) 1.5 A rms.

11.42. A plant draws 250 A rms from a 440-V rms line to supply a load with 100 kW. What is the power factor of the load?

11.43. An industrial plant with an inductive load consumes 10 kW of power from a 220-V rms line. If the load power factor is 0.8, what is the angle by which the load voltage leads the load current?

11.44. As shown in Fig. P11.44, a plant consumes 50 kW at a lagging power factor of 0.85 from a 440 $\angle 0°$-V rms line. If the transmission line resistance is 0.09 Ω, what is the required supply voltage?

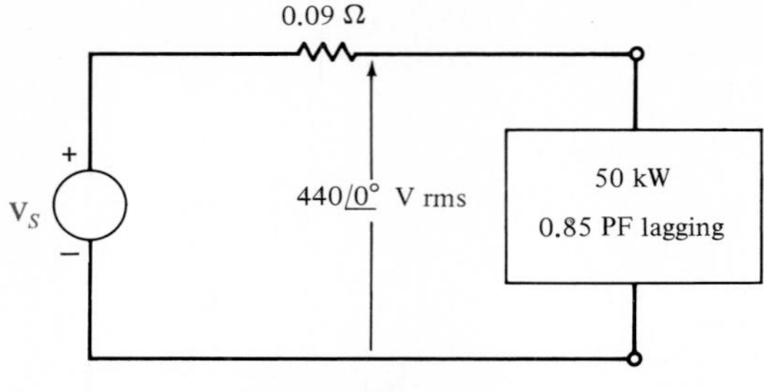

Figure P11.44

11.45. Determine the average power at the generator that is required to supply the load in Problem 11.44.

11.46. A plant consumes 60 kW at a power factor of 0.5 lagging from a 220-V rms 60-Hz line. If the transmission-line resistance from the power company generator to the load is 0.1 Ω, determine the average power that must be supplied by the generator.

11.47. Determine the value of the capacitor which when placed in parallel with the load of Problem 11.46 will change the load power factor to 0.9 lagging.

11.48. A particular load has a PF of 0.8 lagging. The power delivered to the load is 40 kW from a 220-V rms 60-Hz line. If the transmission-line resistance is 0.085 Ω, determine the real power that must be generated at the supply.

11.49. What value of capacitance, placed in parallel with the load in Problem 11.48 will raise the PF to 0.9 lagging?

11.50. A 15-kW load that operates at 0.75 PF lagging is fed from a transmission line with an impedance of $0.085 + j0.32$ Ω. If the load voltage is $200 \angle 0°$ V rms at 60 Hz, determine the real and quadrature power losses in the transmission line.

11.51. Determine the magnitude and phase angle of the input voltage to the transmission line in Problem 11.50.

11.52. An industrial load operates at 44 kW, 0.72 PF lagging. The load voltage is $220 \angle 0°$ V rms at 60 Hz. The line impedance is $0.1 + j0.3$ Ω. Determine the value of the capacitor required to change the load PF to 0.9 lagging. With the capacitor placed across the load, determine the complex power at the input to the line.

11.53. A large induction machine consumes 1 kW at a lagging power factor of 0.65 from a 220-V rms 60-Hz line.

(a) Find the value of the capacitance that must be placed across the machine's terminals to obtain a power factor of 0.90 lagging.
(b) Determine the difference in the power supplied to the terminals of the machine before and after the capacitor is added if the line resistance to the machine is 1 Ω.

11.54. Find the power factor of a load that has the following characteristics.
(a) $\mathbf{S} = 5000 + j4000$ VA
(b) $I = 5.2$ A rms, $V = 220$ V rms, $Q = 400$ var
(c) $|\mathbf{Z}| = 500$, $I = 4.8$ A rms, $P = +5000$ W, $Q_Z < 0$
(d) $I = 10\angle 40°$ A rms, $V = 400$ V rms, Re $\{Z\} = 25$ Ω, $\theta_Z > 0$

11.55. Calculate the voltage $\mathbf{V}_s$ that must be supplied to obtain 2 kW, 240 $\angle 0°$ V rms, and a power factor of 0.8 leading at the load $\mathbf{Z}_L$ in the network in Fig. P11.55.

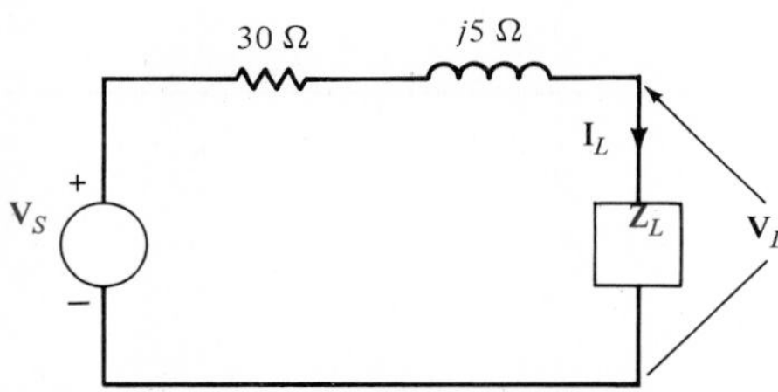

Figure P11.55

11.56. Determine the voltage $\mathbf{V}_s$ that must be supplied to obtain 1.5 kW, 0.75 $\angle 0°$ A rms, and a power factor of 0.85 leading at the load $\mathbf{Z}_L$ in the network in Fig. P11.56.

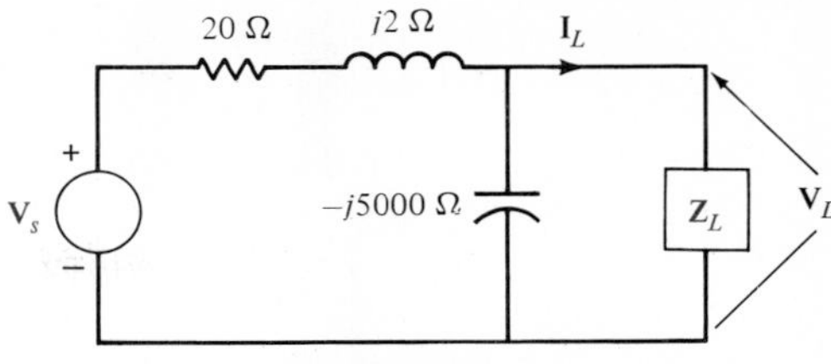

Figure P11.56

11.57. Determine the real power, reactive power, the complex power, and the power factor for a load having the following characteristics.
(a) $\mathbf{I} = 2\angle 40°$ A rms, $\mathbf{V} = 450\angle 70°$ V rms
(b) $\mathbf{I} = 1.5\angle -20°$ A rms, $\mathbf{Z} = 5000\angle 15°$ Ω
(c) $\mathbf{V} = 200\angle +35°$ V rms, $\mathbf{Z} = 1500\angle -15°$ Ω

11.58. A transmission line with impedance $0.08 + j0.25$ Ω is used to deliver power to a load. The load is inductive and the load voltage is 220 $\angle 0°$ V rms at 60 Hz. If the load requires 12 kW and the real power loss in the line is 560 W, determine the power factor angle of the load.

11.59. Determine the complex power input to the transmission line of Problem 11.58, and from this value compute the input voltage and power factor.

11.60. If the power factor of the load in Problem 11.58 is to be changed to 0.9 lagging, what value of capacitance placed in parallel with the load is required?

CHAPTER

12

Polyphase Circuits

12.1

Three-Phase Circuits

In this chapter we add a new dimension to our study of ac steady-state circuits. Up to this point we have dealt with what we refer to as single-phase circuits. Now we extend our analysis techniques to polyphase circuits, or more specifically, three-phase circuits, that is, circuits containing three voltage sources that are one-third of a cycle apart in time. There are a number of important reasons why we study three-phase circuits.

It is more advantageous and economical to generate and transmit electric power in the polyphase mode rather than with single-phase systems. As a result, essentially all electric power is transmitted in polyphase circuits. In the United States the power system frequency is 60 Hz, whereas in other parts of the world 50 Hz is common.

Power transmission is most effectively accomplished at very high voltage. Since this voltage can be extremely high in comparison to the level at which it is normally used (e.g., in the household), there is a need to raise and lower the voltage. This can be accomplished in ac systems using transformers, which we will study in Chapter 14.

As the name implies, three-phase circuits are those in which the forcing function is a three-phase system of voltages. If the three sinusoidal voltages have the same magnitude and frequency and each voltage is 120° out of phase with the other two, the voltages are said to be *balanced*. If the loads are such that the currents produced by the voltages are also balanced, the entire circuit is referred to as a *balanced three-phase circuit*.

A balanced set of three-phase voltages can be represented in the frequency domain as shown in Fig. 12.1a where we have assumed that their magnitudes are 120 V rms.

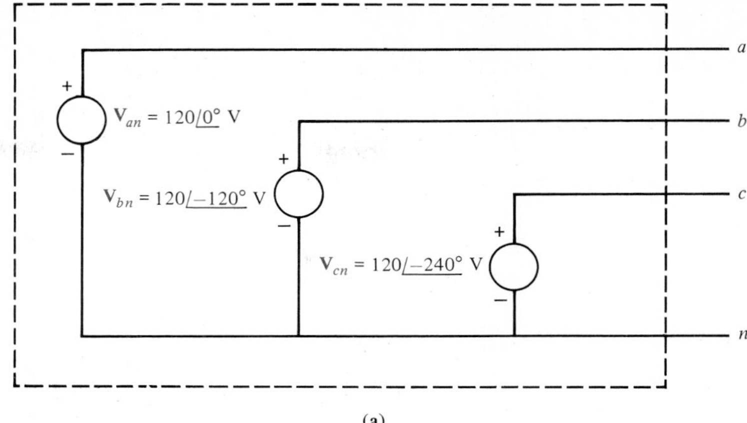

(a)

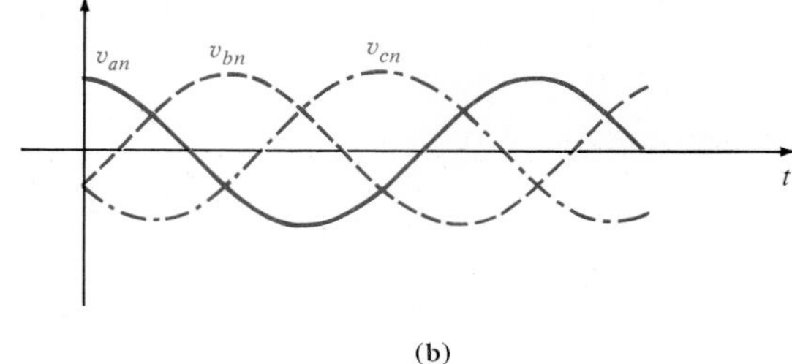

(b)

Figure 12.1 Balanced three-phase voltages.

From the figure we note that

$$\begin{aligned} \mathbf{V}_{an} &= 120 \underline{/0°} \text{ V} \\ \mathbf{V}_{bn} &= 120 \underline{/-120°} \text{ V} \\ \mathbf{V}_{cn} &= 120 \underline{/-240°} \text{ V} \\ &= 120 \underline{/120°} \text{ V} \end{aligned} \tag{12.1}$$

Note that our double-subscript notation is exactly the same as that employed in the earlier chapters; that is, $\mathbf{V}_{an}$ means the voltage at point a with respect to the point n. We will also employ the double-subscript notation for currents; that is, $\mathbf{I}_{an}$ is used to represent the current from a to n. However, we must be very careful in this case to describe the precise path, since in a circuit there will be more than one path between the two points. For example, in the case of a single loop the two currents in the two paths will be 180° out of phase with each other.

The phasor equations above can be expressed in the time domain as

$$\begin{aligned} v_{an}(t) &= 120\sqrt{2} \cos \omega t \text{ V} \\ v_{bn}(t) &= 120\sqrt{2} \cos (\omega t - 120°) \text{ V} \\ v_{cn}(t) &= 120\sqrt{2} \cos (\omega t - 240°) \text{ V} \end{aligned} \tag{12.2}$$

These time functions are shown in Fig. 12.1b.

Finally, let us examine the instantaneous power generated by a three-phase system. Assume that the voltages in Fig. 12.1 are

$$\begin{aligned} v_{an}(t) &= V_m \cos \omega t \text{ V} \\ v_{bn}(t) &= V_m \cos (\omega t - 120°) \text{ V} \\ v_{cn}(t) &= V_m \cos (\omega t - 240°) \text{ V} \end{aligned} \tag{12.3}$$

If the load is balanced, the currents produced by the sources are

$$\begin{aligned} i_a(t) &= I_m \cos (\omega t - \theta) \text{ A} \\ i_b(t) &= I_m \cos (\omega t - \theta - 120°) \text{ A} \\ i_c(t) &= I_m \cos (\omega t - \theta - 240°) \text{ A} \end{aligned} \tag{12.4}$$

The instantaneous power produced by the system is

$$\begin{aligned} p(t) &= p_a(t) + p_b(t) + p_c(t) \\ &= V_m I_m[\cos \omega t \cos (\omega t - \theta) + \cos (\omega t - 120°) \cos (\omega t - \theta - 120°) \\ &\quad + \cos (\omega t - 240°) \cos (\omega t - \theta - 240°)] \end{aligned} \tag{12.5}$$

Using the trigonometric identity

$$\cos \alpha \cos \beta = \tfrac{1}{2}[\cos (\alpha - \beta) + \cos (\alpha + \beta)] \tag{12.6}$$

Eq. (12.5) becomes

$$\begin{aligned} p(t) = \frac{V_m I_m}{2} &[\cos \theta + \cos (2\omega t - \theta) + \cos \theta \\ &+ \cos (2\omega t - \theta - 240°) + \cos \theta + \cos (2\omega t - \theta - 480°)] \end{aligned}$$

which can be written as

$$\begin{aligned} p(t) = \frac{V_m I_m}{2} &[3 \cos \theta + \cos (2\omega t - \theta) \\ &+ \cos (2\omega t - \theta - 120°) + \cos (2\omega t - \theta + 120°)] \end{aligned}$$

There exists a trigonometric identity which allows us to simplify the expression above. The identity, which we will prove later using phasors, is

$$\cos \phi + \cos (\phi - 120°) + \cos (\phi + 120°) = 0 \tag{12.7}$$

If we employ this identity, the expression for the power becomes

$$p(t) = 3 \frac{V_m I_m}{2} \cos \theta \text{ W} \tag{12.8}$$

Note that this equation indicates that the instantaneous power is always constant in time, rather than pulsating as in the single-phase case. Therefore, power delivery from a three-phase voltage source is very smooth, which is another important reason why power is generated in three-phase form.

In order to simplify the following analysis with no loss of generality, rms values will be used for all voltages and currents in the remainder of this chapter.

12.2 Single-Phase Three-Wire Systems

An important special case of polyphase circuits is the single-phase three-wire system shown in Fig. 12.2a. Its importance stems from the fact that it is the normal system found in households. Note that the voltage sources are equal (i.e., $\mathbf{V}_{an} = \mathbf{V}_{nb} = \mathbf{V}$), so that the magnitudes are equal and the phases are equal (single phase), and therefore the line-to-line voltage $\mathbf{V}_{ab} = 2\mathbf{V}_{an} = 2\mathbf{V}_{nb} = 2\,\mathbf{V}$. Typically, lights or small appliances are connected from one line to *neutral* n, and large appliances (e.g., hot water heaters) are connected line to line. Lights operate at about 120 V and the hot water heater operates at approximately 240 V.

Let us now attach two identical loads to the single-phase three-wire voltage system using perfect conductors as shown in Fig. 12.2b. From the figure we note that

$$\mathbf{I}_{aA} = \frac{\mathbf{V}}{\mathbf{Z}_L}$$

and

$$\mathbf{I}_{bB} = -\frac{\mathbf{V}}{\mathbf{Z}_L}$$

KCL at point N is

$$\begin{aligned}\mathbf{I}_{nN} &= -(\mathbf{I}_{aA} + \mathbf{I}_{bB})\\ &= -\left(\frac{\mathbf{V}}{\mathbf{Z}_L} - \frac{\mathbf{V}}{\mathbf{Z}_L}\right)\\ &= 0\end{aligned}$$

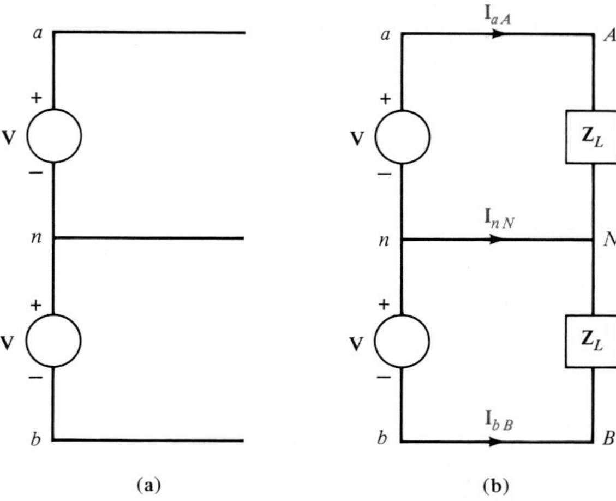

Figure 12.2 Single-phase three-wire system.

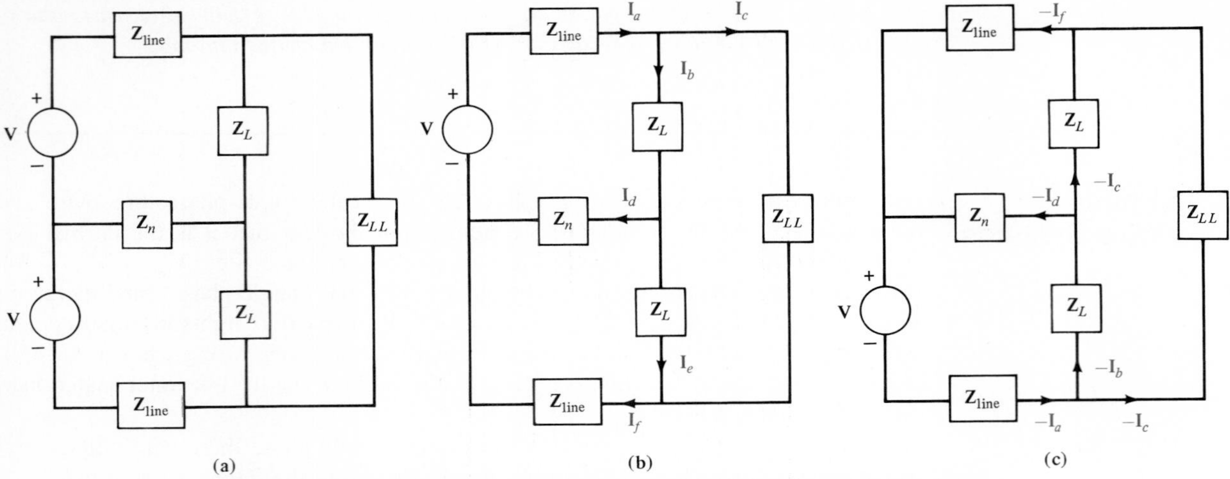

Figure 12.3 Circuits for analyzing a single-phase three-wire system.

Note that there is no current in the neutral wire, and therefore it could be removed without affecting the remainder of the system; that is, all the voltages and currents would be unchanged. One is naturally led to wonder just how far the simplicity exhibited by this system will extend. For example, what would happen if each line had a line impedance, if the neutral conductor had an impedance associated with it, and if there were a load tied from line to line? To explore these questions, consider the circuit in Fig. 12.3a. Although we could examine this circuit using many of the techniques we have employed in previous chapters, the symmetry of the network suggests that perhaps superposition may lead us to some conclusions without having to resort to a brute-force assault. Employing superposition, we consider the two circuits in Fig. 12.3b and c. The currents in Fig. 12.3b are labeled arbitrarily. Because of the symmetrical relationship between Fig. 12.3b and c, the currents in Fig. 12.3c correspond directly to those in Fig. 12.3b. If we add the two *phasor* currents in each branch, we find that the neutral current is again zero. A neutral current of zero is a direct result of the symmetrical nature of the network. If either the line impedances $\mathbf{Z}_{\text{line}}$ or the load impedances $\mathbf{Z}_L$ are unequal, the neutral current will be nonzero.

EXAMPLE 12.1

Consider the single-phase three-wire system shown in Fig. 12.4. For this network we wish to calculate all currents, the power loss in the various elements, and the power supplied by the two sources.

To begin, we compare this circuit with that in Fig. 12.3a. Since the loads are unequal, the system is unbalanced and therefore a neutral current should be present. The mesh equations for the network are

$$\begin{aligned} 11.5\mathbf{I}_1 - \mathbf{I}_2 - 10\mathbf{I}_3 &= 115\,\underline{/0^\circ} \\ -\mathbf{I}_1 + 21.5\mathbf{I}_2 - 20\mathbf{I}_3 &= 115\,\underline{/0^\circ} \\ -10\mathbf{I}_1 - 20\mathbf{I}_2 + (60 + j30)\mathbf{I}_3 &= 0 \end{aligned}$$

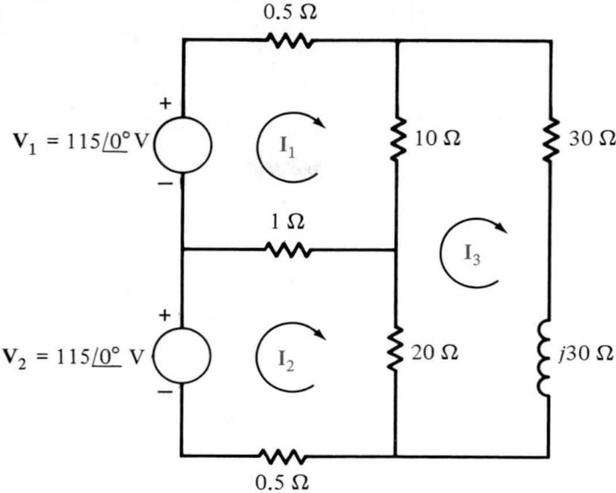

Figure 12.4 Unbalanced single-phase three-wire system.

Solving these equations via matrix analysis, we obtain

$$\mathbf{I}_1 = 14.44 \angle -13.68^\circ \text{ A}$$

$$\mathbf{I}_2 = 10.06 \angle -20.28^\circ \text{ A}$$

$$\mathbf{I}_3 = 5.15 \angle -44.09^\circ \text{ A}$$

These answers can be checked by substituting them into one of the mesh equations. In general, this is a wise move for complicated circuits, and this check assures the reader that it is safe to proceed.

The neutral current is

$$\mathbf{I}_1 - \mathbf{I}_2 = 14.44 \angle -13.68^\circ - 10.06 \angle -20.28^\circ$$

$$= 4.59 \angle 0.83^\circ \text{ A}$$

The other unknown branch currents for the loads are

$$\mathbf{I}_1 - \mathbf{I}_3 = 14.44 \angle -13.68^\circ - 5.15 \angle -44.09^\circ$$

$$= 10.33 \angle 0.89^\circ \text{ A}$$

and

$$\mathbf{I}_2 - \mathbf{I}_3 = 10.06 \angle -20.28^\circ - 5.15 \angle -44.09^\circ$$

$$= 5.74 \angle 0.90^\circ \text{ A}$$

The real power loss in the lines is

$$P_{\text{line}} = |\mathbf{I}_1|^2(0.5) + |\mathbf{I}_1 - \mathbf{I}_2|^2(1) + |\mathbf{I}_2|^2(0.5)$$

$$= 176 \text{ W}$$

The real power loss in the loads is

$$P_{\text{loads}} = |\mathbf{I}_1 - \mathbf{I}_3|^2(10) + |\mathbf{I}_2 - \mathbf{I}_3|^2(20) + |\mathbf{I}_3|^2(30)$$

$$= 2521 \text{ W}$$

The reactive power loss in the loads is

$$Q_{\text{loads}} = |\mathbf{I}_3|^2(30)$$
$$= 794 \text{ var}$$

The complex power supplied by the sources is

$$\mathbf{S}_S = \mathbf{V}_1\mathbf{I}_1^* + \mathbf{V}_2\mathbf{I}_2^*$$
$$= (115)(14.44)\,\underline{/13.68^\circ} + (115)(10.06)\,\underline{/20.28^\circ}$$
$$= 2697 + j794$$
$$= 2812\,\underline{/16.4^\circ} \text{ VA}$$

■

DRILL EXERCISE

D12.1. Given the network in Fig. D12.1, compute the line currents $\mathbf{I}_1$, $\mathbf{I}_2$, and $\mathbf{I}_3$.

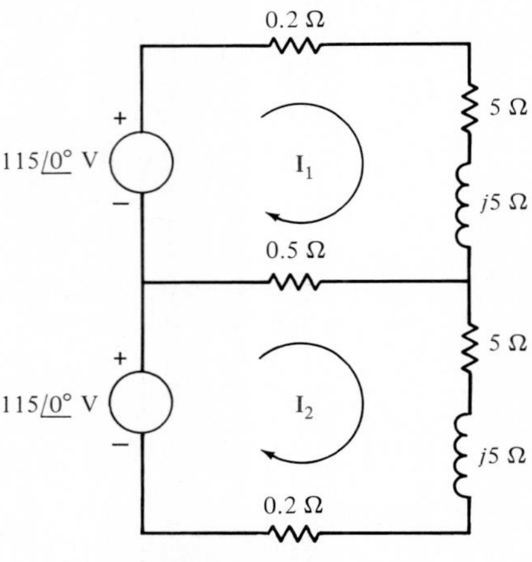

Figure D.12.1

12.3 Three-Phase Connections

By far the most important polyphase voltage source is the balanced three-phase source. This source, as illustrated by Fig. 12.5, has the following properties. The phase voltages, that is, the voltage from each line a, b, and c to the neutral n, are given by

$$\begin{aligned} \mathbf{V}_{an} &= V_p\,\underline{/0^\circ} \\ \mathbf{V}_{bn} &= V_p\,\underline{/-120^\circ} \\ \mathbf{V}_{cn} &= V_p\,\underline{/+120^\circ} \end{aligned} \tag{12.9}$$

The phasor diagram for these voltages is shown in Fig. 12.6a. The phase sequence of this set is said to be *abc*, meaning that $\mathbf{V}_{bn}$ *lags* $\mathbf{V}_{an}$ by 120°. The other possibility is for

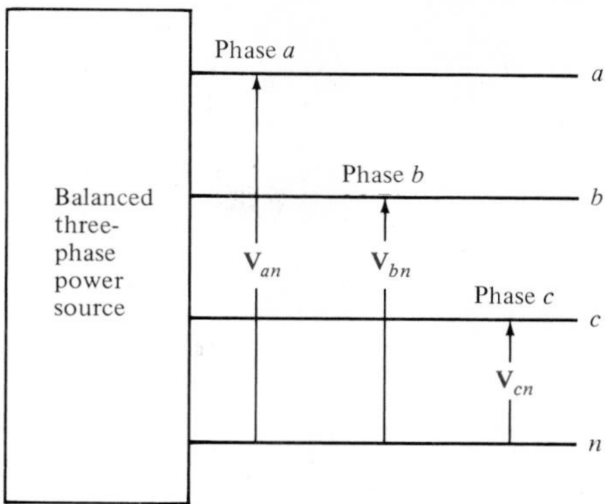

Figure 12.5 Balanced three-phase voltage source.

$\mathbf{V}_{bn}$ to *lead* $\mathbf{V}_{an}$ by 120°, described as a phase sequence of *cba*. This phase sequence is shown in Fig. 12.6b and is given by

$$\mathbf{V}_{an} = V_p \underline{/0^\circ}$$

$$\mathbf{V}_{bn} = V_p \underline{/+120^\circ}$$

$$\mathbf{V}_{cn} = V_p \underline{/-120^\circ}$$

We will standardize our notation so that we always label the voltages $\mathbf{V}_{an}$, $\mathbf{V}_{bn}$, and $\mathbf{V}_{cn}$ and observe them in the order *abc*. We will call this set of voltages a *positive phase sequence*.

An important property of the balanced voltage set is that

$$\mathbf{V}_{an} + \mathbf{V}_{bn} + \mathbf{V}_{cn} = 0 \tag{12.10}$$

This property can easily be seen by resolving the voltage phasors into components along the real and imaginary axes. It can also be demonstrated via Eq. (12.7).

From the standpoint of the user who connects a load to the balanced three-phase voltage source, it is not important how the voltages are generated. It is importnt to note, however, that if the load currents generated by connecting a load to the power source shown in Fig. 12.5 are also *balanced*, there are two possible equivalent configurations for the load. The equivalent load can be considered as being connected in either a *wye*

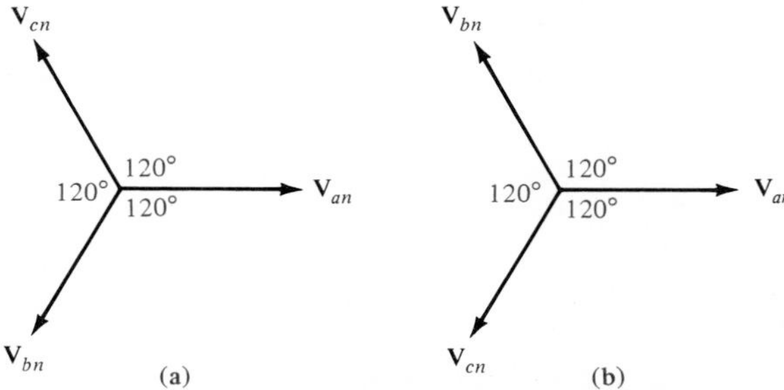

Figure 12.6 Phasor diagram for a balanced three-phase voltage source.

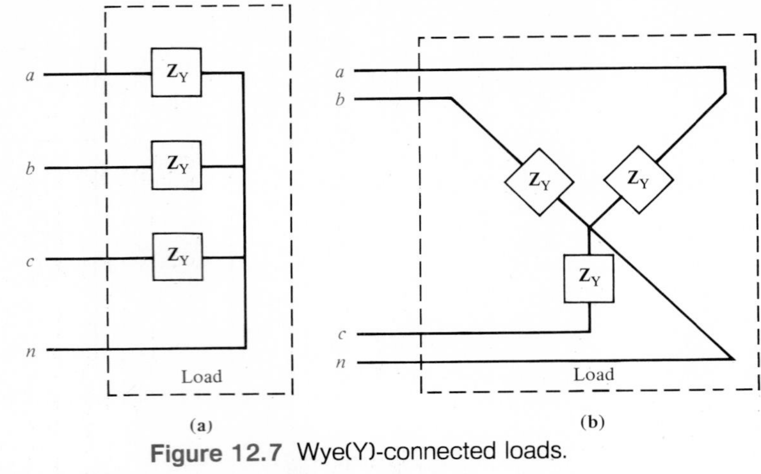

Figure 12.7 Wye(Y)-connected loads.

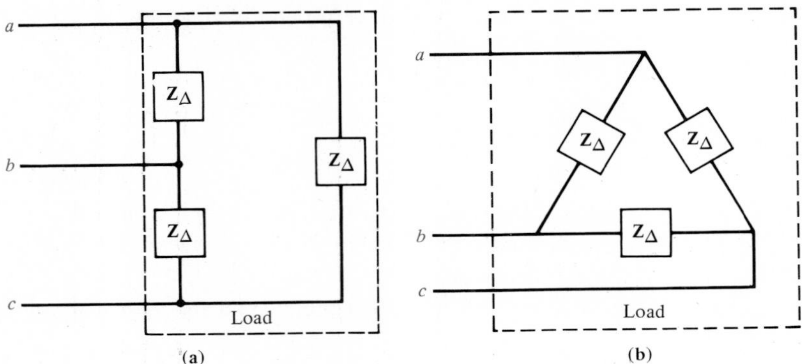

Figure 12.8 Delta (Δ)-connected loads.

(Y) or a *delta* (Δ) configuration. The balanced wye configuration is shown in Fig. 12.7a and equivalently in Fig. 12.7b. The delta configuration is shown in Fig. 12.8a and equivalently in Fig. 12.8b. Note that in the case of the delta connection, there is no neutral line. The actual function of the neutral line in the wye connection will be examined and it will be shown that in a balanced system the neutral line carries no current and therefore may be omitted.

Balanced Wye-Wye Connection

Suppose now that the source and load are both connected in a wye, as shown in Fig. 12.9. The phase voltages with positive phase sequence are

$$\begin{aligned} \mathbf{V}_{an} &= V_Y \angle 0^\circ \\ \mathbf{V}_{bn} &= V_Y \angle -120^\circ \\ \mathbf{V}_{cn} &= V_Y \angle -240^\circ \end{aligned} \tag{12.11}$$

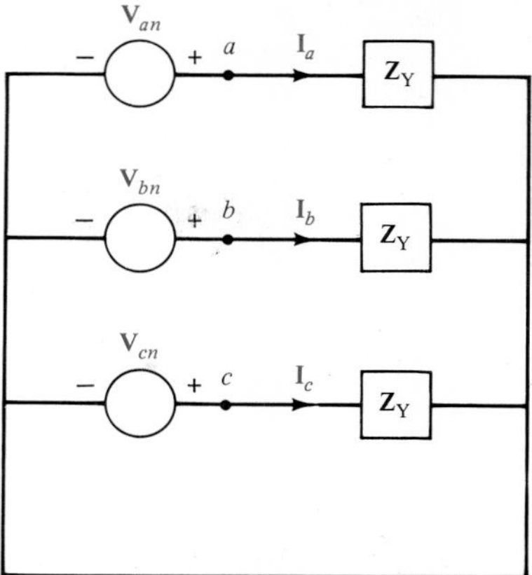

Figure 12.9 Balanced three-phase wye–wye connection.

where V_Y is the magnitude of the phase voltage in a wye-connected source. The *line-to-line* or, simply, *line voltages* can be calculated using KVL; for example,

$$\begin{aligned}
\mathbf{V}_{ab} &= \mathbf{V}_{an} - \mathbf{V}_{bn} \\
&= V_Y \angle 0^\circ - V_Y \angle -120^\circ \\
&= V_Y - V_Y\left[-\frac{1}{2} - j\frac{\sqrt{3}}{2}\right] \\
&= V_Y\left[\frac{3}{2} + j\frac{\sqrt{3}}{2}\right] \\
&= \sqrt{3}\, V_Y \angle 30^\circ
\end{aligned} \tag{12.12}$$

The phasor addition is shown in Fig. 12.10a. In a similar manner, we obtain the set of

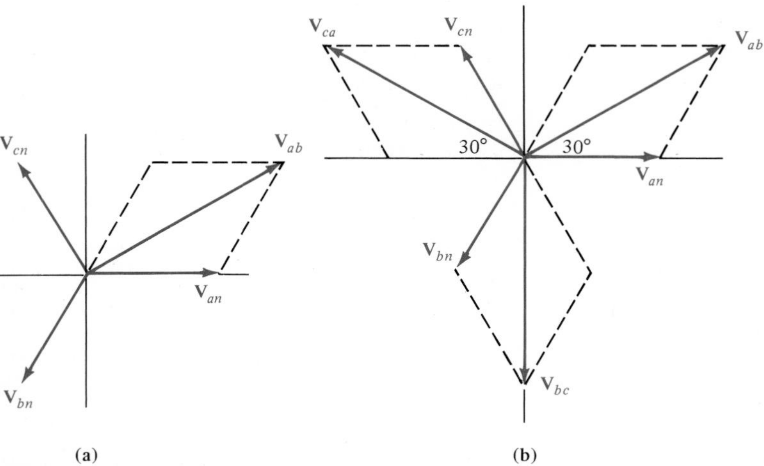

Figure 12.10 Phasor representation of phase and line voltages in a balanced wye–wye system.

line-to-line voltages as

$$\begin{aligned} \mathbf{V}_{ab} &= \sqrt{3}\, V_Y \underline{/30^\circ} \\ \mathbf{V}_{bc} &= \sqrt{3}\, V_Y \underline{/-90^\circ} \\ \mathbf{V}_{ca} &= \sqrt{3}\, V_Y \underline{/-210^\circ} \end{aligned} \tag{12.13}$$

All the line voltages together with the phase voltages are shown in Fig. 12.10b. We will denote the magnitude of the line voltages as V_L, and therefore for a balanced system,

$$V_L = \sqrt{3}\, V_Y \tag{12.14}$$

Hence in a wye-connected system the line voltage is equal to $\sqrt{3}$ times the phase voltage.

As shown in Fig. 12.9, the line currents are

$$\begin{aligned} \mathbf{I}_a &= \frac{\mathbf{V}_{an}}{\mathbf{Z}_Y} = \frac{V_Y \underline{/0^\circ}}{\mathbf{Z}_Y} \\ \mathbf{I}_b &= \frac{\mathbf{V}_{bn}}{\mathbf{Z}_Y} = \frac{V_Y \underline{/-120^\circ}}{\mathbf{Z}_Y} \\ \mathbf{I}_c &= \frac{\mathbf{V}_{cn}}{\mathbf{Z}_Y} = \frac{V_Y \underline{/-240^\circ}}{\mathbf{Z}_Y} \end{aligned} \tag{12.15}$$

The neutral current $\mathbf{I}_n$ is then

$$\mathbf{I}_n = (\mathbf{I}_a + \mathbf{I}_b + \mathbf{I}_c) = 0 \tag{12.16}$$

Since there is no current in the neutral line, this line could contain any impedance or it could be an open or a short circut.

As illustrated by the wye–wye connection in Fig. 12.9 the current in the line connecting the source to the load is the same as the phase current flowing through the impedance $\mathbf{Z}_Y$. Therefore, in a wye–wye connection,

$$I_L = I_Y \tag{12.17}$$

where I_L is the magnitude of the line current and I_Y is the magnitude of the phase current in a wye-connected load. Therefore, if $\mathbf{Z}_Y = Z_Y \underline{/\theta}$, the phasor line or phase currents can be written as

$$\begin{aligned} \mathbf{I}_a &= I_L \underline{/-\theta} = I_Y \underline{/-\theta} \\ \mathbf{I}_b &= I_L \underline{/-\theta - 120^\circ} = I_Y \underline{/-\theta - 120^\circ} \\ \mathbf{I}_c &= I_L \underline{/-\theta - 240^\circ} = I_Y \underline{/-\theta - 240^\circ} \end{aligned} \tag{12.18}$$

It is important to note that although we have a three-phase system composed of three sources and three loads, we can analyze this system on a "per phase" basis; that is, we treat each phase separately. This is, of course, a direct result of the balanced condition. We may even have impedances present in the lines; however, as long as the system remains balanced, we can analyze the circuit one phase at a time. If the line impedances in lines a, b, and c are equal, the system will be balanced. Recall that the balance of the system is unaffected by whatever appers in the netural line and since the netural line impedance is arbitrary, we assume that it is zero (i.e., a short circuit).

EXAMPLE 12.2

A posititive-sequence three-phase voltage source connected in a balanced wye has a line voltage of $\mathbf{V}_{ab} = 208 \angle -30^\circ$ V. We wish to determine the phase voltages.

The magnitude of the phase voltage is given by the expression

$$208 = \sqrt{3}\, V_Y$$

$$V_Y = 120 \text{ V}$$

The phase relationships between the line and phase voltages are shown in Fig. 12.10. From this figure we note that

$$\mathbf{V}_{an} = 120 \angle -60^\circ \text{ V}$$

$$\mathbf{V}_{bn} = 120 \angle -180^\circ \text{ V}$$

$$\mathbf{V}_{cn} = 120 \angle -300^\circ \text{ V}$$

The magnitudes of these voltages are quite common and one often hears that the power required in a building, for example, is three-phase 208/120 volts. ■

EXAMPLE 12.3

A three-phase wye-connected load is supplied by a positive-sequence balanced three-phase wye-connected source with a phase voltage of 120 V rms. If the line impedance and load impedance per phase are $1 + j1\ \Omega$ and $20 + j10\ \Omega$, respectively, we wish to determine the value of the line currents and the load voltages.

The phase voltages for the positive sequence are

$$\mathbf{V}_{an} = 120 \angle 0^\circ \text{ V}$$

$$\mathbf{V}_{bn} = 120 \angle -120^\circ \text{ V}$$

$$\mathbf{V}_{cn} = 120 \angle -240^\circ \text{ V}$$

The per phase circuit diagram is shown in Fig. 12.11. The line current for the *a* phase

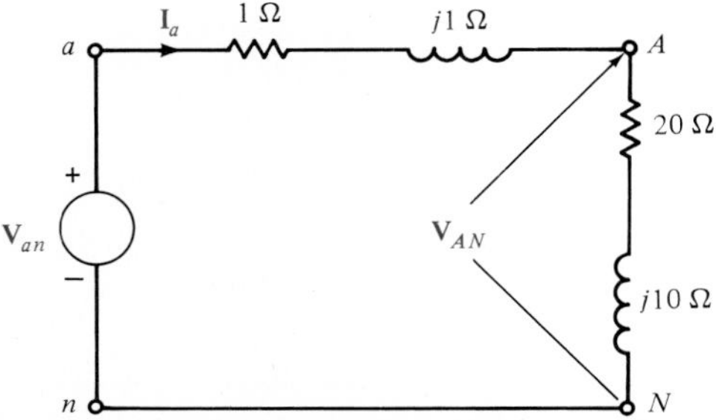

Figure 12.11 Per phase circuit diagram for the problem in Example 12.3.

is

$$\mathbf{I}_a = \frac{120 \underline{/0^\circ}}{21 + j11}$$

$$= 5.06 \underline{/-27.65^\circ} \text{ A}$$

The load voltage for the a phase, which we call $\mathbf{V}_{AN}$, is

$$\mathbf{V}_{AN} = (5.06 \underline{/-27.65^\circ})(20 + j10)$$

$$= 113.15 \underline{/-1.08^\circ} \text{ V}$$

The corresponding line currents and load voltages for the b and c phases are

$$\mathbf{I}_b = 5.06 \underline{/-147.65^\circ} \text{ A} \qquad \mathbf{V}_{BN} = 113.15 \underline{/-121.08^\circ} \text{ V}$$

$$\mathbf{I}_c = 5.06 \underline{/-267.65^\circ} \text{ A} \qquad \mathbf{V}_{CN} = 13.15 \underline{/-241.08^\circ} \text{ V}$$

■

DRILL EXERCISE

D12.2. The voltage for the a phase of a positive-phase-sequence balanced wye-connected source is $\mathbf{V}_{an} = 120 \underline{/90^\circ}$ V. Determine the line voltage for this source.

D12.3. A positive-phase-sequence three-phase voltage source connected in a balanced wye has a line voltage of $\mathbf{V}_{ab} = 208 \underline{/0^\circ}$ V. Determine the phase voltages of the source.

D12.4. A three-phase wye load is supplied by a positive-sequence-balanced three-phase wye-connected source through a transmission line with an impedance of $1 + j1$ ohms per phase. The load impedance is $8 + j3$ ohms per phase. If the load voltage for the a phase is $104.02 \underline{/26.6^\circ}$ V rms, determine the phase voltages of the source.

Balanced Wye-Delta Connection

Another very important connection for three-phase circuits is the balanced wye-delta system, that is, a wye-connected source and a delta-connected load, as shown in Fig. 12.12. From Fig. 12.12 we note that for this connection the line-to-line voltage is equal

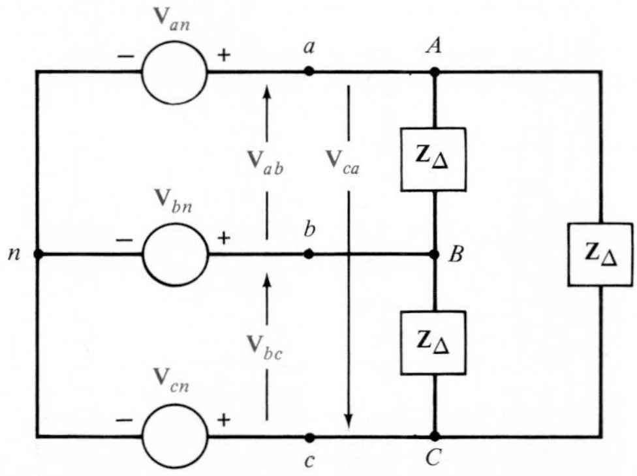

Figure 12.12 Balanced three-phase wye–delta system.

to the voltage across each load impedance in the delta-connected load. Therefore, if the phase voltages of the source are

$$\begin{aligned} \mathbf{V}_{an} &= V_Y \angle 0^\circ \\ \mathbf{V}_{bn} &= V_Y \angle -120^\circ \\ \mathbf{V}_{cn} &= V_Y \angle -240^\circ \end{aligned} \tag{12.19}$$

then the line voltages are

$$\begin{aligned} \mathbf{V}_{ab} &= \sqrt{3}\, V_Y \angle 30^\circ = V_\Delta \angle 30^\circ \\ \mathbf{V}_{bc} &= \sqrt{3}\, V_Y \angle -90^\circ = V_\Delta \angle -90^\circ \\ \mathbf{V}_{ca} &= \sqrt{3}\, V_Y \angle -210^\circ = V_\Delta \angle -210^\circ \end{aligned} \tag{12.20}$$

where V_Δ is the magnitude of the phase voltage at the delta-connected load. Hence the phase voltages at the load are

$$\begin{aligned} \mathbf{V}_{AB} &= V_\Delta \angle 30^\circ \\ \mathbf{V}_{BC} &= V_\Delta \angle -90^\circ \\ \mathbf{V}_{CA} &= V_\Delta \angle -210^\circ \end{aligned} \tag{12.21}$$

Note that the voltage across each of the phase impedances is the line voltage, and this voltage is related to the source voltages in the wye-connected source by the expression

$$V_\Delta = \sqrt{3}\, V_Y \tag{12.22}$$

From Fig. 12.12 we note that if $\mathbf{Z}_\Delta = Z_\Delta \angle \theta$, then the phase currents at the load are

$$\begin{aligned} \mathbf{I}_{AB} &= \frac{\mathbf{V}_{AB}}{\mathbf{Z}_\Delta} = I_\Delta \angle 30^\circ - \theta \\ \mathbf{I}_{BC} &= \frac{\mathbf{V}_{BC}}{\mathbf{Z}_\Delta} = I_\Delta \angle -90^\circ - \theta \\ \mathbf{I}_{CA} &= \frac{\mathbf{V}_{CA}}{\mathbf{Z}_\Delta} = I_\Delta \angle -210^\circ - \theta \end{aligned} \tag{12.23}$$

where I_Δ is the magnitude of these phase currents and is given by the expression

$$I_\Delta = \frac{V_\Delta}{Z_\Delta} \tag{12.24}$$

KCL can now be employed in conjunction with the phase currents to determine the line currents. For example,

$$\begin{aligned} \mathbf{I}_a &= \mathbf{I}_{AB} + \mathbf{I}_{AC} \\ &= \mathbf{I}_{AB} - \mathbf{I}_{CA} \end{aligned}$$

Using the same approach as that employed earlier to determine the relationship between the line voltages and the phase voltages in a wye-wye connection, we can show that

$$\mathbf{I}_a = \sqrt{3}\, I_\Delta \underline{/-\theta}$$
$$\mathbf{I}_b = \sqrt{3}\, I_\Delta \underline{/-120^\circ - \theta} \tag{12.25}$$
$$\mathbf{I}_c = \sqrt{3}\, I_\Delta \underline{/-240^\circ - \theta}$$

and therefore the relation between the magnitudes of the source currents and the load currents is

$$I_Y = \sqrt{3}\, I_\Delta \tag{12.26}$$

The phasor diagram in Fig. 12.13 illustrates all the important relationships between the currents and voltages for a delta-connected load.

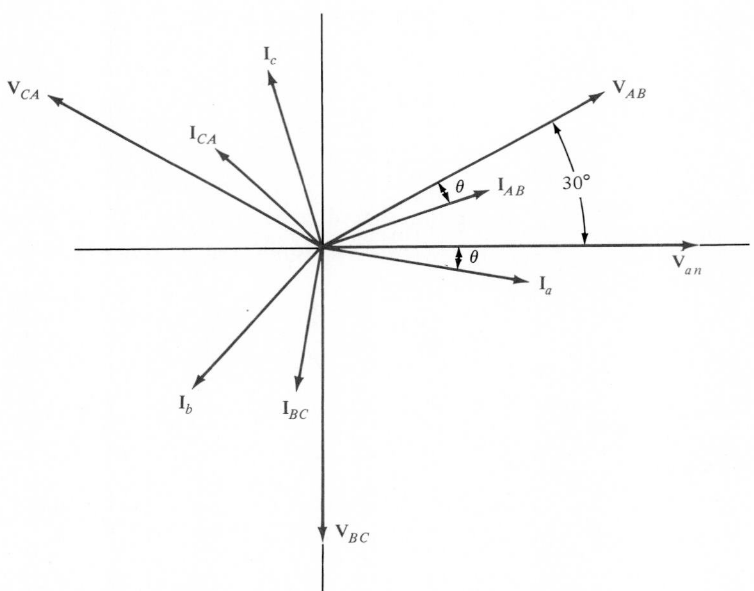

Figure 12.13 Current and voltage relationships for a delta-connected load.

EXAMPLE 12.4

A positive-sequence three-phase voltage source connected in a balanced wye supplies power to a balanced delta-connected load. The load current $\mathbf{I}_{AB} = 4 \underline{/20^\circ}$ A. We wish to determine the line currents.

From the relationships among the phase and line currents as specified in Eqs. (12.23) and (12.25) and illustrated in Fig. 12.13, we note that if

$$\mathbf{I}_{AB} = 4 \underline{/20^\circ} \text{ A}$$

then

$$\mathbf{I}_a = 4\sqrt{3}\,\underline{/-10°}\text{ A}$$

$$\mathbf{I}_b = 4\sqrt{3}\,\underline{/-130°}\text{ A}$$

$$\mathbf{I}_c = 4\sqrt{3}\,\underline{/-250°}\text{ A}$$

■

EXAMPLE 12.5

A balanced delta-connected load contains a 10-Ω resistor in series with a 20-mH inductor in each phase. The voltage source is a positive-sequence three-phase 60-Hz, balanced wye with a voltage $\mathbf{V}_{an} = 120\,\underline{/30°}$ V. We wish to determine all phase currents and line currents.

The impedance per phase in the delta load is $\mathbf{Z}_\Delta = 10 + j7.54\ \Omega$. The line voltage $\mathbf{V}_{ab} = 120\sqrt{3}\,\underline{/60°}$ V. Therefore, the phase voltage at the load is $\mathbf{V}_{AB} = 120\sqrt{3}\,\underline{/60°}$ V. Hence

$$\mathbf{I}_{AB} = \frac{120\sqrt{3}\,\underline{/60°}}{10 + j7.54}$$

$$= 16.60\,\underline{/+22.98°}\text{ A}$$

Then from the relationship specified in Eqs. (12.23) and (12.25), we have

$$\mathbf{I}_a = 16.60\sqrt{3}\,\underline{/-7.02°}$$

$$= 28.75\,\underline{/-7.02°}\text{ A}$$

Therefore, the remaining phase and line currents are

$$\mathbf{I}_{BC} = 16.60\,\underline{/-97.02°}\text{ A} \qquad \mathbf{I}_b = 28.75\,\underline{/-127.02°}\text{ A}$$

$$\mathbf{I}_{CA} = 16.60\,\underline{/-217.02°}\text{ A} \qquad \mathbf{I}_c = 28.75\,\underline{/-247.02°}\text{ A}$$

■

DRILL EXERCISE

D12.5. A positive sequence three-phase voltage source connected in a balanced wye supplies power to a balanced delta-connected load. The line current for the a phase is $\mathbf{I}_a = 12\,\underline{/40°}$ A. Find the phase currents in the delta-connected load.

D12.6. A positive-sequence balanced three-phase wye-connected source supplies power to a balanced delta-connected load. The load impedance per phase is $12 + j8\ \Omega$. If the current $\mathbf{I}_{AB}$ in one phase of the delta is $14.42\ \underline{/86.31°}$A, determine the line currents and phase voltages at the source.

Delta-Connected Source

Up to this point we have concentrated our discussion on circuits that have wye-connected sources. However, our analysis of the wye–wye and wye–delta connections provides us with the information necessary to handle a delta-connected source.

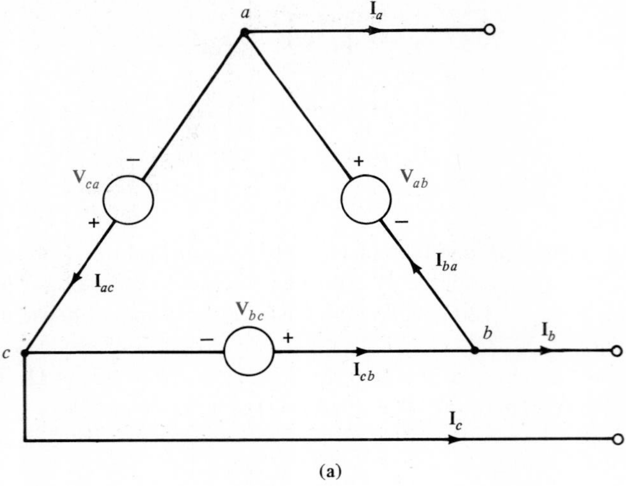

(a)

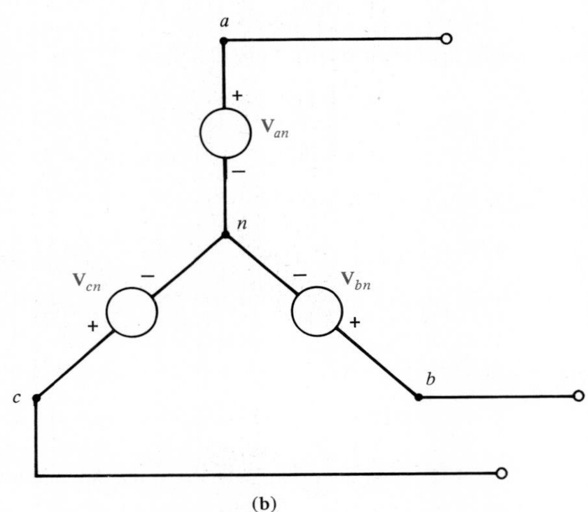

(b)

Figure 12.14 Sources connected in Δ and Y.

Consider the delta-connected source shown in Fig. 12.14a. Note that the sources are connected line to line. We found earlier that the relationship between line-to-line and line-to-neutral voltages was given by Eq. (12.13) and illustrated in Fig. 12.10 for a positive phase sequence of voltages. Therefore, if the delta sources are

$$\begin{aligned} \mathbf{V}_{ab} &= V_\Delta \underline{/0^\circ} \\ \mathbf{V}_{bc} &= V_\Delta \underline{/-120^\circ} \\ \mathbf{V}_{ca} &= V_\Delta \underline{/-240^\circ} \end{aligned} \tag{12.27}$$

where V_Δ is the magnitude of the phase voltage, then the equivalent wye sources shown in Fig. 12.14b are

$$\begin{aligned} \mathbf{V}_{an} &= \frac{V_\Delta}{\sqrt{3}} \underline{/-30^\circ} = V_Y \underline{/-30^\circ} \\ \mathbf{V}_{bn} &= \frac{V_\Delta}{\sqrt{3}} \underline{/-150^\circ} = V_Y \underline{/-150^\circ} \\ \mathbf{V}_{cn} &= \frac{V_\Delta}{\sqrt{3}} \underline{/-270^\circ} = V_Y \underline{/-270^\circ} \end{aligned} \tag{12.28}$$

where V_Y is the magnitude of the phase voltage of a wye-connected source. In addition, if the line currents are known to be

$$\begin{aligned} \mathbf{I}_a &= I_L \underline{/0^\circ} \\ \mathbf{I}_b &= I_L \underline{/-120^\circ} \\ \mathbf{I}_c &= I_L \underline{/-240^\circ} \end{aligned} \tag{12.29}$$

then the phase currents in the delta sources are

$$\begin{aligned} \mathbf{I}_{ba} &= \frac{I_L}{\sqrt{3}} \underline{/30^\circ} = I_\Delta \underline{/30^\circ} \\ \mathbf{I}_{cb} &= \frac{I_L}{\sqrt{3}} \underline{/-90^\circ} = I_\Delta \underline{/-90^\circ} \\ \mathbf{I}_{ac} &= \frac{I_L}{\sqrt{3}} \underline{/-210^\circ} = I_\Delta \underline{/-210^\circ} \end{aligned} \tag{12.30}$$

where I_Δ is the magnitude of the phase currents in a delta-connected source. Therefore, if we encounter a network containing a delta-connected source, we can easily convert the source from delta to wye so that all the techniques we have discussed previously can be applied in an analysis.

EXAMPLE 12.6

Consider the network shown in Fig. 12.15a. We wish to determine the line currents, the magnitude of the line voltage at the load, and the magnitude of the phase currents in the delta source.

The single-phase diagram for the network is shown in Fig. 12.15b. The line current $\mathbf{I}_{aA}$ is

$$\begin{aligned} \mathbf{I}_{aA} &= \frac{(208/\sqrt{3}) \underline{/-30^\circ}}{12.1 + j4.2} \\ &= 9.38 \underline{/-49.14^\circ} \text{ A} \end{aligned}$$

and thus $\mathbf{I}_{bB} = 9.38 \underline{/-169.14^\circ}$ A and $\mathbf{I}_{cC} = 9.38 \underline{/-289.14^\circ}$ A. The voltage $\mathbf{V}_{AN}$ is then

$$\begin{aligned} \mathbf{V}_{AN} &= (9.38 \underline{/-49.14^\circ})(12 + j4) \\ &= 118.66 \underline{/-30.71^\circ} \text{ V} \end{aligned}$$

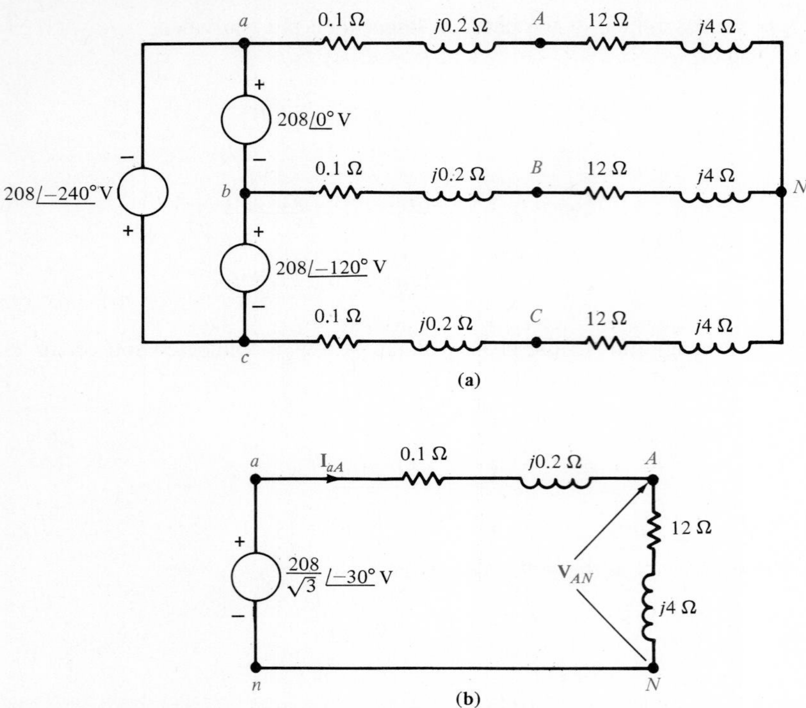

Figure 12.15 Delta–wye network and an equivalent single-phase diagram.

Therefore, the magnitude of the line voltage at the load is

$$|\mathbf{V}_L| = \sqrt{3}\,(118.66)$$
$$= 205.53 \text{ V}$$

The current $\mathbf{I}_{ba}$ is

$$\mathbf{I}_{ba} = \frac{9.38}{\sqrt{3}} \underline{/-49.14^\circ + 30^\circ}$$
$$= 5.42 \underline{/-19.14^\circ} \text{ A}$$

Therefore, the magnitude of the phase current in the Δ source is $|\mathbf{I}_{ba}| = 5.42$ A. ■

DRILL EXERCISE

D12.7. Consider the network shown in Fig. D12.7. Compute the magnitude of the line voltages at the load and the magnitude of the phase currents in the delta-connected source.

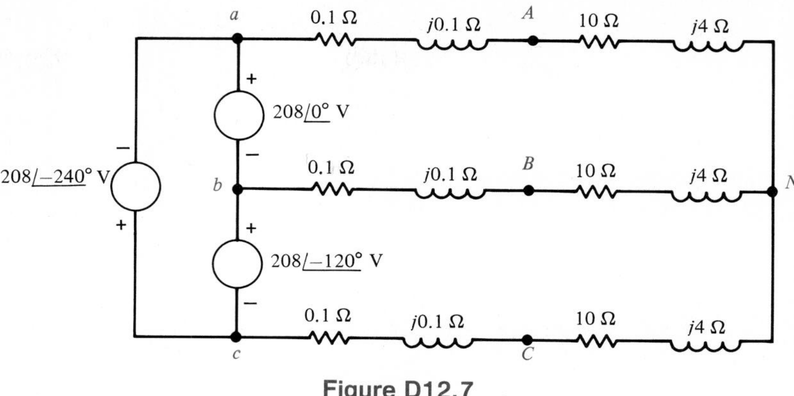

Figure D12.7

Finally, the magnitude relationships between the phase voltage and the line voltage and the phase current and the line current for the two geometrical configurations are presented in Table 12.1.

Wye⇄Delta Transformations

We have stated earlier that for a balanced system, the equivalent load configuration may be either wye or delta. If both of these configurations are connected at only three terminals, it would be very advantageous if an equivalence could be established between them. It is, in fact, possible to relate the impedances of one network to those of the other such that their terminal characteristics are the same. Consider, for example, the two networks shown in Fig. 12.16. For these two networks to be equivalent at each corresponding pair of terminals it is necessary that the input impedances at the corresponding terminals be equal, for example, if at terminals a and b, with c open-circuited, the impedance is the same for both configurations. Equating the impedances at each port yields

$$
\begin{aligned}
\mathbf{Z}_{ab} &= \mathbf{Z}_a + \mathbf{Z}_b = \frac{\mathbf{Z}_1(\mathbf{Z}_2 + \mathbf{Z}_3)}{\mathbf{Z}_1 + \mathbf{Z}_2 + \mathbf{Z}_3} \\
\mathbf{Z}_{bc} &= \mathbf{Z}_b + \mathbf{Z}_c = \frac{\mathbf{Z}_3(\mathbf{Z}_1 + \mathbf{Z}_2)}{\mathbf{Z}_1 + \mathbf{Z}_2 + \mathbf{Z}_3} \\
\mathbf{Z}_{ca} &= \mathbf{Z}_c + \mathbf{Z}_a = \frac{\mathbf{Z}_2(\mathbf{Z}_1 + \mathbf{Z}_3)}{\mathbf{Z}_1 + \mathbf{Z}_2 + \mathbf{Z}_3}
\end{aligned}
\tag{12.31}
$$

Table 12.1 Current-Voltage Relationships for the Wye and Delta Load Configurations

Parameter	Wye Configuration	Delta Configuration
Voltage	$\mathbf{V}_{\text{line to line}} = \sqrt{3}\, V_{Y}$	$\mathbf{V}_{\text{line to line}} = \mathbf{V}_{\Delta}$
Current	$\mathbf{I}_{\text{line}} = \mathbf{I}_{Y}$	$\mathbf{I}_{\text{line}} = \sqrt{3}\, \mathbf{I}_{\Delta}$

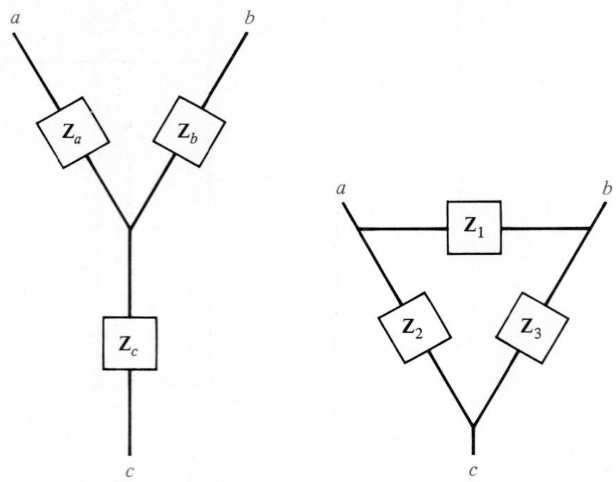

Figure 12.16 General wye- and delta-connected loads.

Solving this set of equations for $\mathbf{Z}_a$, $\mathbf{Z}_b$, and $\mathbf{Z}_c$ yields

$$\begin{aligned} \mathbf{Z}_a &= \frac{\mathbf{Z}_1\mathbf{Z}_2}{\mathbf{Z}_1 + \mathbf{Z}_2 + \mathbf{Z}_3} \\ \mathbf{Z}_b &= \frac{\mathbf{Z}_1\mathbf{Z}_3}{\mathbf{Z}_1 + \mathbf{Z}_2 + \mathbf{Z}_3} \\ \mathbf{Z}_c &= \frac{\mathbf{Z}_2\mathbf{Z}_3}{\mathbf{Z}_1 + \mathbf{Z}_2 + \mathbf{Z}_3} \end{aligned} \tag{12.32}$$

Similarly, if we solve Eq. (12.31) for $\mathbf{Z}_1$, $\mathbf{Z}_2$, and $\mathbf{Z}_3$, we obtain

$$\begin{aligned} \mathbf{Z}_1 &= \frac{\mathbf{Z}_a\mathbf{Z}_b + \mathbf{Z}_b\mathbf{Z}_c + \mathbf{Z}_c\mathbf{Z}_a}{\mathbf{Z}_c} \\ \mathbf{Z}_2 &= \frac{\mathbf{Z}_a\mathbf{Z}_b + \mathbf{Z}_b\mathbf{Z}_c + \mathbf{Z}_c\mathbf{Z}_a}{\mathbf{Z}_b} \\ \mathbf{Z}_3 &= \frac{\mathbf{Z}_a\mathbf{Z}_b + \mathbf{Z}_b\mathbf{Z}_c + \mathbf{Z}_c\mathbf{Z}_a}{\mathbf{Z}_a} \end{aligned} \tag{12.33}$$

Equations (12.32) and (12.33) are general relationships and apply to any set of impedances connected in a wye or delta configuration. For the balanced case where $\mathbf{Z}_a = \mathbf{Z}_b = \mathbf{Z}_c$ and $\mathbf{Z}_1 = \mathbf{Z}_2 = \mathbf{Z}_3$, the equations above reduce to

$$\mathbf{Z}_Y = \tfrac{1}{3}\mathbf{Z}_\Delta \tag{12.34}$$

and

$$\mathbf{Z}_\Delta = 3\mathbf{Z}_Y \tag{12.35}$$

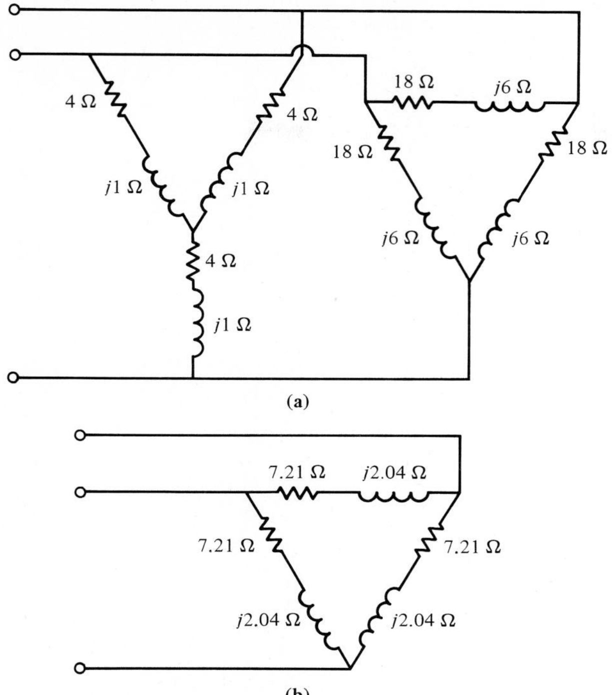

Figure 12.17 Three-phase load consisting of a balanced wye in parallel with a balanced delta converted to an equivalent delta load.

EXAMPLE 12.7

A three-phase load impedance consists of a balanced wye in parallel with a balanced delta, as shown in Fig. 12.17a. We wish to determine the equivalent delta load.

The equivalent delta for the given Y is

$$\begin{aligned}\mathbf{Z}_\Delta &= 3\mathbf{Z}_Y \\ &= 12 + j3\ \Omega\end{aligned}$$

This delta-connected impedance is in parallel with the original one and therefore the impedance per phase of the combined delta is

$$\begin{aligned}\mathbf{Z}'_\Delta &= \frac{(12 + j3)(18 + j6)}{12 + j3 + 18 + j6} \\ &= 7.21 + j2.04\ \Omega\end{aligned}$$

which is shown as the equivalent load in Fig. 12.17b. ■

EXAMPLE 12.8

A balanced three-phase system has a load consisting of a balanced wye in parallel with a balanced delta. The impedance per phase for the wye is $10 + j6\ \Omega$ and for the delta is $24 + j9\ \Omega$. The source is a balanced wye with a positive phase sequence and the line voltage $\mathbf{V}_{ab} = 208\,\underline{/30^\circ}$ V. If the line impedance per phase is $1 + j0.5\ \Omega$, we want to

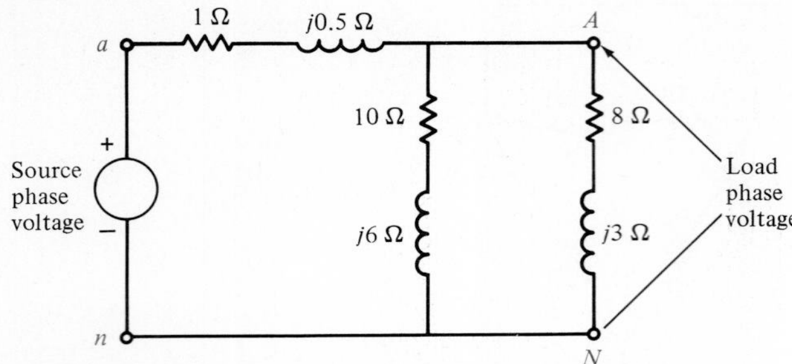

Figure 12.18 Per phase equivalent circuit for Example 12.8.

determine the line currents and the load phase voltages when the load is converted to an equivalent wye.

Converting the delta load to an equivalent wye load, we obtain

$$\mathbf{Z}_{Y1} = \tfrac{1}{3}\mathbf{Z}_{\Delta 1} = 8 + j3\ \Omega$$

The per phase equivalent circuit is shown in Fig. 12.18. Since the line voltage is $\mathbf{V}_{ab} = 208\,\underline{/30°}$ V, the phase voltage $\mathbf{V}_{an} = 120\,\underline{/0°}$ V. Recall that since the loads are balanced, the neutral connection can be a short circuit because no current is present in this line. The total load impedance is

$$\begin{aligned}\mathbf{Z}_Y &= \frac{\mathbf{Z}_{Y1}\mathbf{Z}_{Y2}}{\mathbf{Z}_{Y1} + \mathbf{Z}_{Y2}} \\ &= \frac{(10 + j6)(8 + j3)}{10 + j6 + 8 + j3} = 4.95\,\underline{/24.95°} \\ &= 4.49 + j2.09\ \Omega\end{aligned}$$

The line current $\mathbf{I}_a$ is then

$$\begin{aligned}\mathbf{I}_a &= \frac{\mathbf{V}_{an}}{\mathbf{Z}_{\text{Line}} + \mathbf{Z}_Y} \\ &= \frac{120\,\underline{/0°}}{1 + j0.5 + 4.49 + j2.09} \\ &= 19.77\,\underline{/-25.26°}\text{ A}\end{aligned}$$

The phase voltage at the load is then

$$\begin{aligned}\mathbf{V}_{AN} &= \mathbf{I}_a\mathbf{Z}_Y \\ &= (19.77\,\underline{/-25.26°})(4.95\,\underline{/24.95°}) \\ &= 97.86\,\underline{/-0.31°}\text{ V}\end{aligned}$$

Therefore, the line currents and load phase voltages are

$$\begin{aligned}\mathbf{I}_a &= 19.77\,\underline{/-25.26°}\text{ A} & \mathbf{V}_{AN} &= 97.86\,\underline{/-0.31°}\text{ V} \\ \mathbf{I}_b &= 19.77\,\underline{/-145.26°}\text{ A} & \mathbf{V}_{BN} &= 97.86\,\underline{/-120.31°}\text{ V} \\ \mathbf{I}_c &= 19.77\,\underline{/-265.26°}\text{ A} & \mathbf{V}_{CN} &= 97.86\,\underline{/-240.31°}\text{ V}\end{aligned}$$

■

DRILL EXERCISE

D12.8. In a balanced three-phase system the load consists of a balanced wye in parallel with a balanced delta. The impedance per phase for the wye is $8 + j4\ \Omega$ and for the delta is $18 + j6\ \Omega$. The source is a positive-phase-sequence balanced wye and $\mathbf{V}_{an} = 120\ \underline{/60^\circ}$ V. If the line impedance per phase is $1 + j1\ \Omega$, determine the magnitude of the phase currents in each load.

Power Relationships

Whether the load is connected in a wye or a delta, the real and reactive power per phase are

$$\begin{aligned} P_p &= V_p I_p \cos\theta \\ Q_p &= V_p I_p \sin\theta \end{aligned} \qquad (12.36)$$

where θ is the angle between the phase voltage and the phase current. From the information in Table 12.1 we note that Eq. (12.36) is equivalent to the following two expressions, which are valid for both the balanced wye and delta configurations:

$$\begin{aligned} P_p &= \frac{V_L I_L}{\sqrt{3}} \cos\theta \\ Q_p &= \frac{V_L I_L}{\sqrt{3}} \sin\theta \end{aligned} \qquad (12.37)$$

The total real and reactive power for all three phases is then

$$\begin{aligned} P_T &= \sqrt{3}\, V_L I_L \cos\theta \\ Q_T &= \sqrt{3}\, V_L I_L \sin\theta \end{aligned} \qquad (12.38)$$

and therefore the magnitude of the complex power is

$$\begin{aligned} |\mathbf{S}_T| &= \sqrt{P_T^2 + Q_T^2} \\ &= \sqrt{3}\, V_L I_L \end{aligned} \qquad (12.39)$$

EXAMPLE 12.9

A three-phase balanced wye–delta system has a line voltage of 208 V. The total real power absorbed by the load is 1200 W. If the power factor angle of the load is 20° lagging, we wish to determine the magnitude of the line current and the value of the load impedance per phase in the delta.

The line current can be obtained from Eq. (12.37). Since the real power per phase is 400 W,

$$400 = \frac{208 I_L}{\sqrt{3}} \cos 20^\circ$$

$$I_L = 3.5 \text{ A}$$

The phase current is

$$I_\Delta = \frac{I_L}{\sqrt{3}}$$

$$= 2.04 \text{ A}$$

Therefore, the magnitude of the impedance in each phase of the load is

$$|\mathbf{Z}| = \frac{V_L}{I_\Delta}$$

$$= \frac{208}{2.04}$$

$$= 101.77\ \Omega$$

Since the power factor angle is 20° lagging, the load impedance is

$$\mathbf{Z} = 101.77 \underline{/20^\circ}$$

$$= 95.63 + j34.81\ \Omega$$

■

EXAMPLE 12.10

For the circuit in Example 12.3 we wish to determine the real and reactive power per phase at the load and the total real power, reactive power, and the complex power at the source.

From the data in Example 12.3 the complex power per phase at the load is

$$\mathbf{S}_{\text{load}} = \mathbf{VI}^*$$

$$= (113.15 \underline{/-1.08^\circ})(5.06 \underline{/27.65^\circ})$$

$$= 572.54 \underline{/26.57^\circ}$$

$$= 512.07 + j256.1 \text{ VA}$$

Therefore, the real and reactive power per phase at the load are 512.07 W and 256.1 var, respectively.

The complex power per phase at the source is

$$\mathbf{S}_{\text{source}} = \mathbf{VI}^*$$

$$= (120 \underline{/0^\circ})(5.06 \underline{/27.65^\circ})$$

$$= 607.2 \underline{/27.65^\circ}$$

$$= 537.90 + j281.8 \text{ VA}$$

and therefore total real power, reactive power, and complex power at the source are 1613.7 W, 845.4 var, and 1821.6 VA, respectively. ■

EXAMPLE 12.11

A small shopping center contains three stores, which represent three balanced three-phase loads. The power lines into the shopping center represent a three-phase voltage source with a line voltage of 13.8 kV. The loads are as follows:

Load 1: 500 kVA, 0.8 PF lagging

Load 2: 400 kVA, 0.85 PF lagging

Load 3: 300 kVA, 0.90 PF lagging

We wish to determine the power line current.

$$\mathbf{S}_1 = 500 \underline{/\cos^{-1} 0.8°} = 400 + j300 \text{ kVA}$$

$$\mathbf{S}_2 = 400 \underline{/\cos^{-1} 0.85} = 340 + j210.71 \text{ kVA}$$

$$\mathbf{S}_2 = 300 \underline{/\cos^{-1} 0.9} = 270 + j130.77 \text{ kVA}$$

The total complex power is then

$$\mathbf{S}_{\text{total}} = 1010 + j641.48 = 1196.49 \underline{/32.42°} \text{ kVA}$$

Therefore,

$$I_L = \frac{|\mathbf{S}_{\text{total}}|}{V_L\sqrt{3}} = \frac{1196.49}{13.8\sqrt{3}} = 50.06 \text{ A}$$

■

DRILL EXERCISE

D12.9. A three-phase balanced wye–wye system has a line voltage of 208 V. The total real power absorbed by the load is 12 kW at 0.8 PF lagging. Determine the per phase impedance of the load.

D12.10. For the balanced wye–wye system desribed in Drill Exercise D12.4, determine the real and reactive power and the complex power at both the source and the load.

D12.11. A 480-V line feeds two balanced three-phase loads. If the two loads are rated as follows:

Load 1: 5 kVA at 0.8 PF lagging

Load 2: 10 kVA at 0.9 PF lagging

determine the magnitude of the line current from the 480-V source.

12.4 SPICE Analysis of Three-Phase Circuits

All the material required to apply SPICE to three-phase circuits has been covered in previous chapters. Therefore, our approach in this section will be to demonstrate the use of SPICE through several examples. In contrast to the earlier sections of this chapter, we will analyze the entire three-phase network and not rely on the balance condition to treat only a single phase. In addition, we will illustrate how to solve two problems with SPICE that naturally occur in a delta-connected system.

EXAMPLE 12.12

A balanced wye–wye three-phase system which is labeled for a SPICE analysis is shown in Fig. 12.19. The following SPICE program will calculate the line currents, the phase load voltages, and the line-to-line load voltages. $f = \frac{1}{2}\pi$ Hz. The program is

```
EXAMPLE 12.12
VAN 1 0 AC 120 0
VBN 5 0 AC 120 -120
VCN 6 0 AC 120 -240
RAN 1 2 10
RBN 5 4 10
RCN 6 7 10
LAN 2 3 2
LBN 4 3 2
LCN 7 3 2
.PRINT AC IM(VAN) IP(VAN) VM(1,3) VP(1,3)
.PRINT AC IM(VBN) IP(VBN) VM(5,3) VP(5,3)
.PRINT AC IM(VCN) IP(VCN) VM(6,3) VP(6,3)
.PRINT AC VM(1,5) VP(1,5) VM(5,6) VP(5,6)
+ VM(6,1) VP(6,1)
.AC LIN 1 .159155 .159155
.WIDTH OUT=80
.END
```

The computer output contains

```
I(VAN) = 11.77/168.7° A
V(1,3) = 120 /0° V
V(1,5) = 207.8 /30° V
```

■

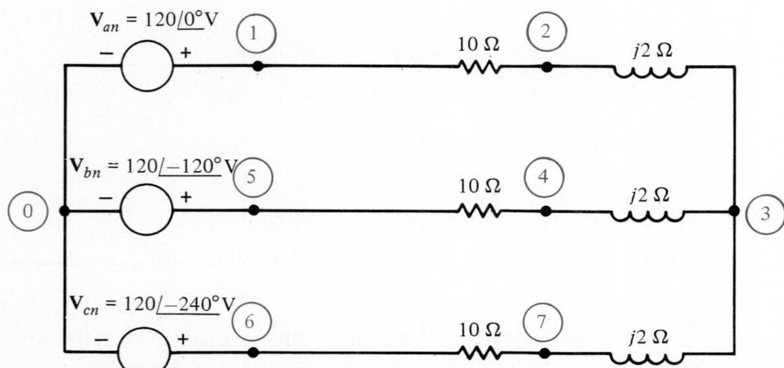

Figure 12.19 Balanced wye–wye three-phase system labelled for SPICE analysis.

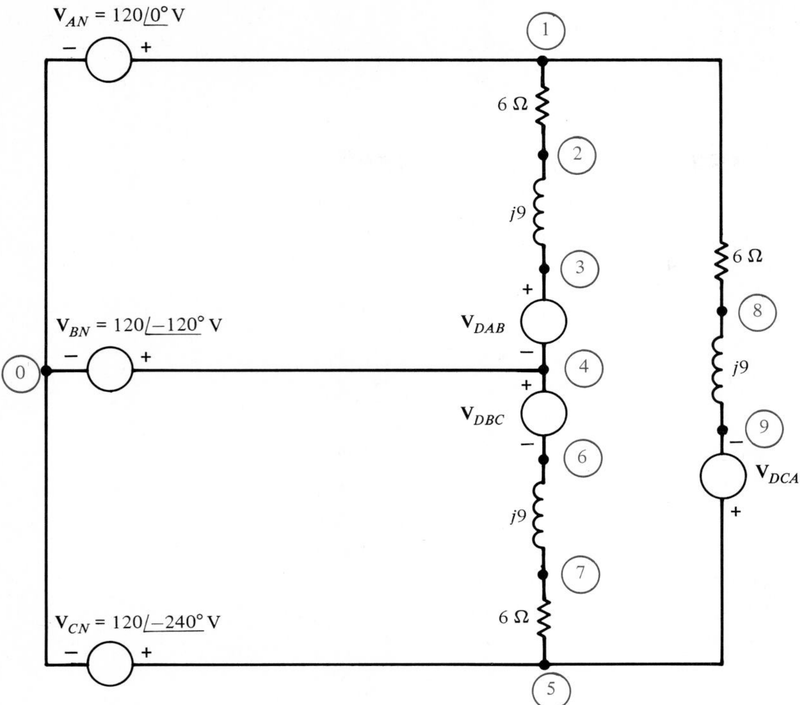

Figure 12.20 Balanced wye–delta three-phase system labelled for SPICE analysis.

EXAMPLE 12.13

A balanced wye–delta system labeled for a SPICE analysis is shown in Fig. 12.20. The following SPICE program will calculate the source currents, load currents, and load voltages. $f = \frac{1}{2}\pi$ Hz.

The computer program is

```
EXAMPLE 12.13
VAN 1 0 AC 120 0
VBN 4 0 AC 120 -120
VCN 5 0 AC 120 -240
RAB 1 2 6
LAB 2 3 9
VDAB 3 4
RBC 7 5 6
LBC 6 7 9
VDBC 4 6
RCA 1 8 6
LCA 8 9 9
VDCA 5 9
.PRINT AC IM(VAN) IP(VAN) IM(VBN) IP(VBN)
+ IM(VCN) IP(VCN)
.PRINT AC IM(VDAB) IP(VDAB) VM(1,4) VP(1,4)
```

```
.PRINT AC IM(VDBC) IP(VDBC) VM(4,5) VP(4,5)
.PRINT AC IM(VDCA) IP(VDCA) VM(5,1) VP(5,1)
.AC LIN 1 . 159155 . 159155
.WIDTH OUT=80
.END
```

The computer output contains

```
I(VAN) = 33.28 /123.7° A
I(VDAB) = 19.22 /-26.31° A
V(1,4) = 207.8 /30° V
```

■

EXAMPLE 12.14

A balanced three-phase delta-connected system is shown in Fig. 12.21a. We wish to use SPICE to calculate the currents and voltages in this network. From a SPICE standpoint this circuit has two problems: There is a loop of voltage sources, and there is no convenient ground node. Therefore, the circuit has been modified as shown in Fig. 12.21b. The 1-pΩ resistors placed in series around the voltage sources alleviates the problem of a voltage source loop. The I-TΩ resistors permit us to introduce the ground node required by SPICE. The position and value of these resistors are selected to maintain a balanced system and have a minimum impact on the operation of the original network.

The SPICE program, written for the network in Fig. 12.21b, which calculates the source currents, load currents, and load voltages, follows. $f = \frac{1}{2}\pi$ Hz.

```
EXAMPLE 12.14
VAB 2 3 AC 207.8 0
RSAB1 1 2 1P
RSAB2 3 4 1P
VBC 5 6 AC 207.8 -120
RSBC1 4 5 1P
RSBC2 6 7 1P
VCA 8 9 AC 207.8 -240
RSCA1 7 8 1P
RSCA2 9 1 1P
RSHZA 1 0 1T
RSHZB 4 0 1T
RSHZC 7 0 1T
RAB 10 11 6
CAB 1 10 .33333
VDAB 11 4
RBC 4 12 6
CCB 12 13 .33333
VDBC 13 7
RCA 1 14 6
CCA 14 15 .33333
VDCA 7 15
.PRINT AC IM(VAB) IP(VAB) VM(1,4) VP(1,4)
.PRINT AC IM(VBC) IP(VBC) VM(4,7) VP(4,7)
.PRINT AC IM(VCA) IP(VCA) VM(7,1) VP(7,1)
```

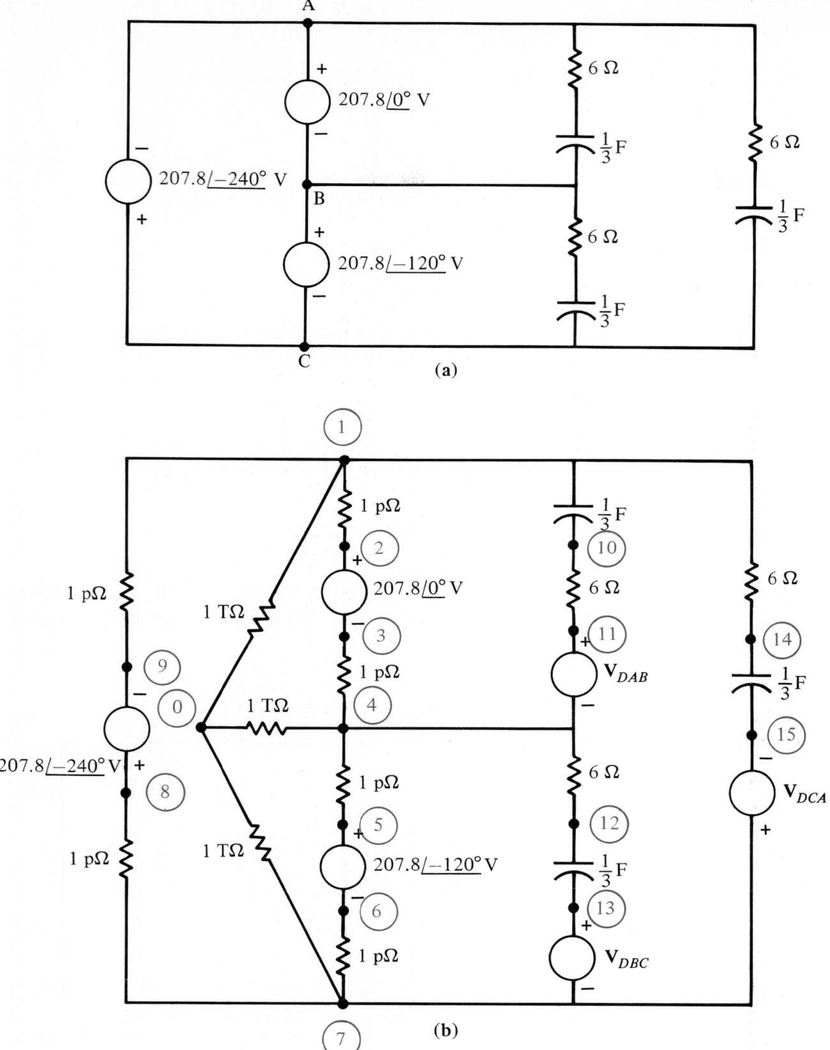

Figure 12.21 Three-phase delta-connected system analyzed using SPICE.

```
.PRINT AC IM(VDAB) IP(VDAB)
.PRINT AC IM(VDBC) IP(VDBC)
.PRINT AC IM(VDCA) IP(VDCA)
.AC LIN 1 .159155 .159155
.WIDTH OUT=80
.END
```

The computer output contains

```
I(VAB)  = 30.98 /-153.4° A
V(1,4)  = 207.8 /0° V
I(VDAB) = 30.98 /26.57° A
```

■

DRILL EXERCISE

D12.12. Given the network in Fig. D12.12, use SPICE to determine the line current, line voltage, and load voltage. $f = \frac{1}{2}\pi$ Hz.

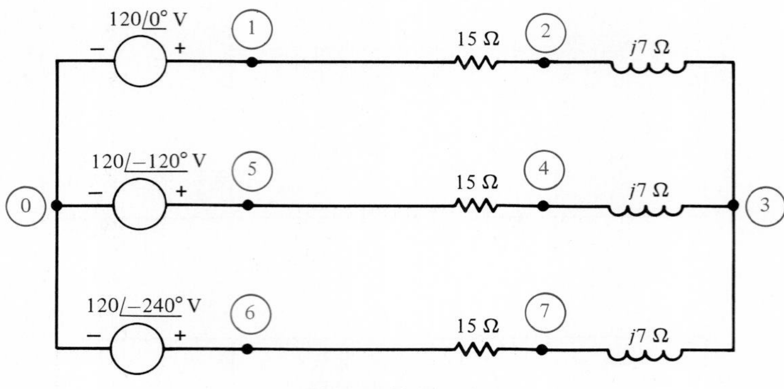

Figure D12.12

D12.13. The network in Fig. D12.13 is labeled for a SPICE analysis. Write a SPICE program for calculating the source currents, load currents, load voltage, and the voltage drop across the line impedance. $f = \frac{1}{2}\pi$ Hz.

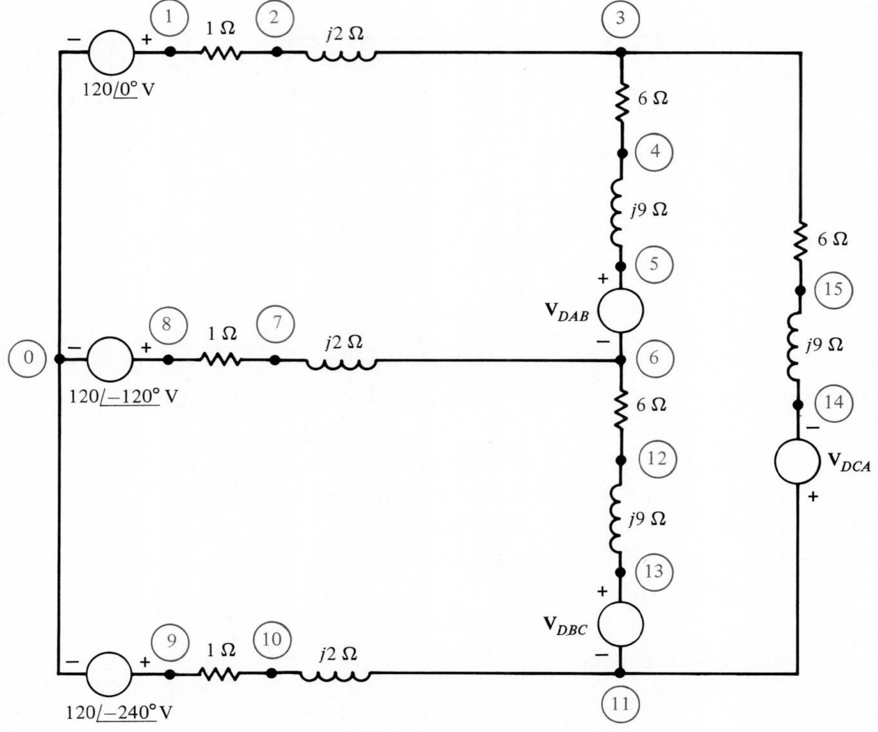

Figure D12.13

D12.14. The network in Fig. D12.14 is labeled for a SPICE analysis. Write a SPICE program for calculating the line currents, load currents, load voltages, and source currents. What is the purpose of the 1-TΩ resistors in the network? $f = \frac{1}{2}\pi$ Hz.

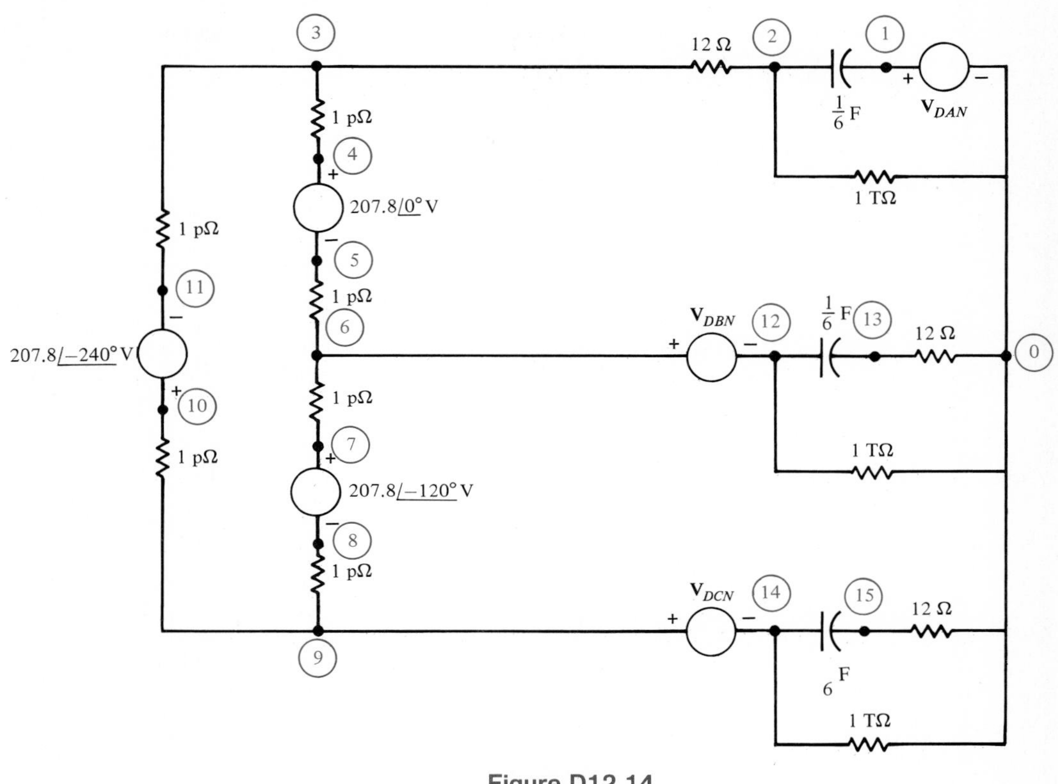

Figure D12.14

12.5 Power Measurements

In the preceding sections we have illustrated techniques for computing power. We will now show how power is actually measured in an electrical network. Our approach will be to treat the single-phase case first and then discuss the techniques used in three-phase systems.

Single-Phase Measurement

An instrument used to measure power is the wattmeter. This instrument contains a low-impedance current coil (which ideally has zero impedance) that is connected in series with the load, and a high-impedance voltage coil (which ideally has infinite impedance) that is connected across the load. If the voltage and current are periodic and the wattmeter

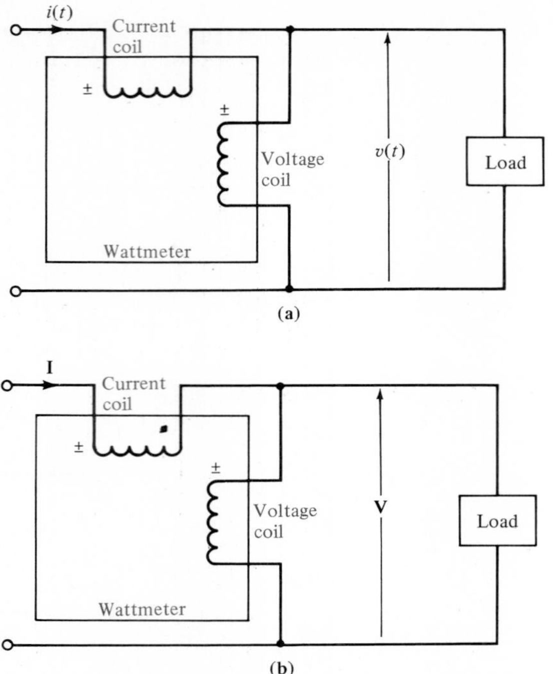

Figure 12.22 Wattmeter connections for power measurement.

is connected as shown in Fig. 12-22a, it will read

$$P = \frac{1}{T}\int_0^T v(t)i(t)\,dt$$

where $v(t)$ and $i(t)$ are defined on the figure. Note that $i(t)$ is referenced entering the $\pm$ terminal of the current coil and $v(t)$ is referenced positive at the $\pm$ terminal of the voltage coil. In the frequency domain the equivalent circuit is shown in Fig. 12.22b, where the current and voltage are referenced the same as in the time domain and the wattmeter reading will be

$$P = \text{Re}\,(\mathbf{VI}^*)$$

If $v(t)$ and $i(t)$ or $\mathbf{V}$ and $\mathbf{I}$ are correctly chosen, the reading will be average power. In Fig. 12-22 the connections will produce a reading of power delivered to the load. Since the two coils are completely isolated from one another, they could be connected anywhere in the circuit and the reading may or may not have meaning.

Note that if one of the coils on the wattmeter is reversed, the equations for the power are the negative of what they were before the coil reversal due to the change in the variable reference as related to the $\pm$ terminal. Due to the physical construction of wattmeters, the $\pm$ terminal of the potential coil should always be connected to the same line as the current coil, as shown in Fig. 12.23a. If it becomes necessary to reverse a winding to produce an upscale reading, the current coil should be reversed as shown in Fig.

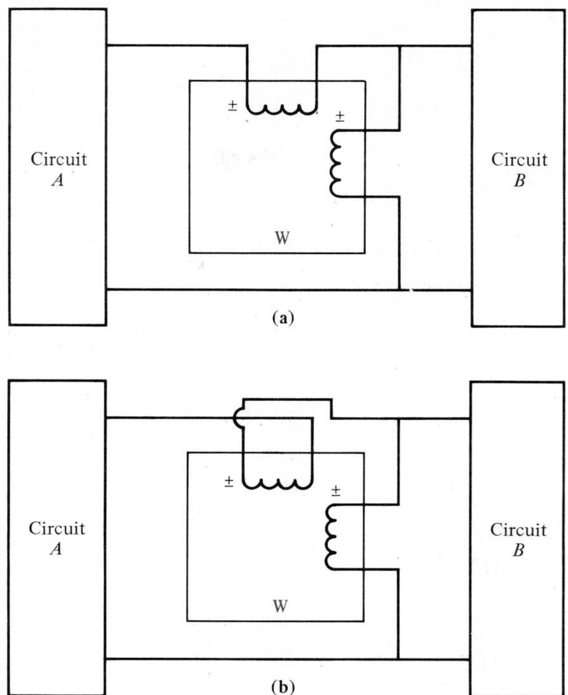

Figure 12.23 Wattmeter connections.

12-23b. Note that if the ± terminal of the potential coil is connected to the line containing the current coil and the meter is reading upscale, the power is flowing through the wattmeter from circuit A to circuit B.

EXAMPLE 12.15

Given the network shown in Fig. 12.24, we wish to determine the wattmeter reading.

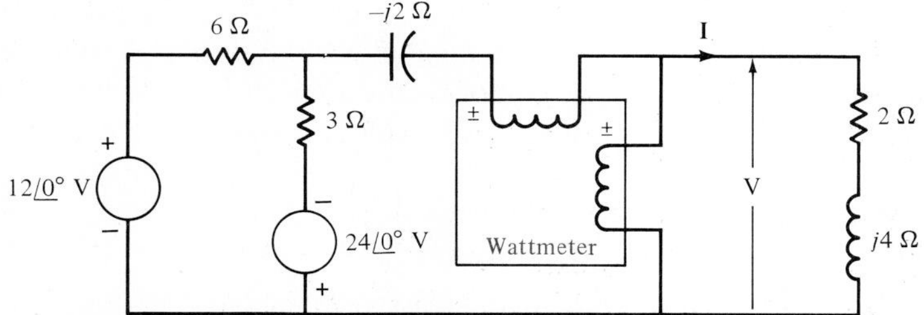

Figure 12.24 Use of the wattmeter in power measurement.

Using any one of a number of techniques (e.g., Thévenin's theorem), we can show that

$$\mathbf{I} = 2.68 \,\underline{/153.43^\circ}\text{ A}$$

$$\mathbf{V} = 11.99 \,\underline{/216.86^\circ}\text{ V}$$

and therefore the wattmeter reads

$$P = \text{Re}\,(\mathbf{VI}^*) = |\mathbf{V}||\mathbf{I}| \cos(\theta_v - \theta_i)$$

$$= (11.99)(2.68)\cos(216.86° - 153.43°)$$

$$= 14.37 \text{ W}$$

DRILL EXERCISE

D12.15. Determine the wattmeter reading in the network in Fig. D12.15.

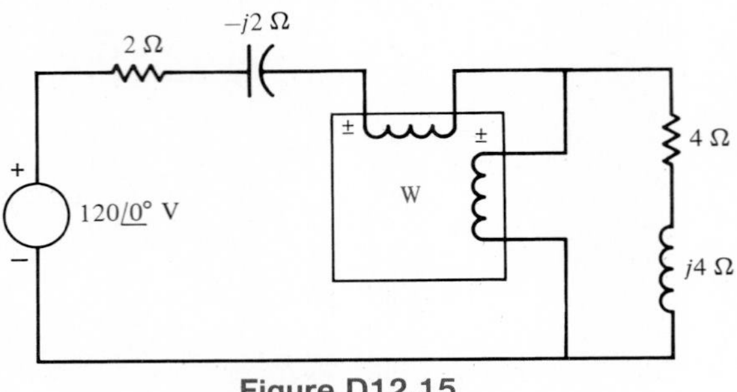

Figure D12.15

Three-Phase Measurement

The measurement of three-phase power would seem to be a simple task since it would appear that in order to accomplish the measurement we would simply replicate the single-phase measurement for each phase. For a balanced system this task would become even simpler since we need only measure the power in one phase and multiply the wattmeter reading by 3. However, as we pointed out earlier, the neutral terminal in the wye-connected load may be inaccessible, and of course, in a delta-connected load only the lines and not the phases are available in many cases. Our measurement technique must, therefore, deal only with the lines to the load.

The method we will describe is applicable for both wye and delta connections; however, we will present the measurement method using a wye-connected load. Consider the circuit shown in Fig. 12.25. The current coil of each wattmeter is connected in series with the line current, and the voltage coil of each wattmeter is connected between a line and what we will call the virtual neutral N^*, which is nothing more than an arbitrary point.

The average power measured by wattmeter A is

$$P_A = \frac{1}{T}\int_0^T v_{AN*} i_A \, dt \tag{12.40}$$

where T is the period of all voltages and currents in the system. In a similar manner,

$$P_B = \frac{1}{T}\int_0^T v_{BN*} i_B \, dt \tag{12.41}$$

$$P_C = \frac{1}{T}\int_0^T v_{CN*} i_C \, dt \tag{12.42}$$

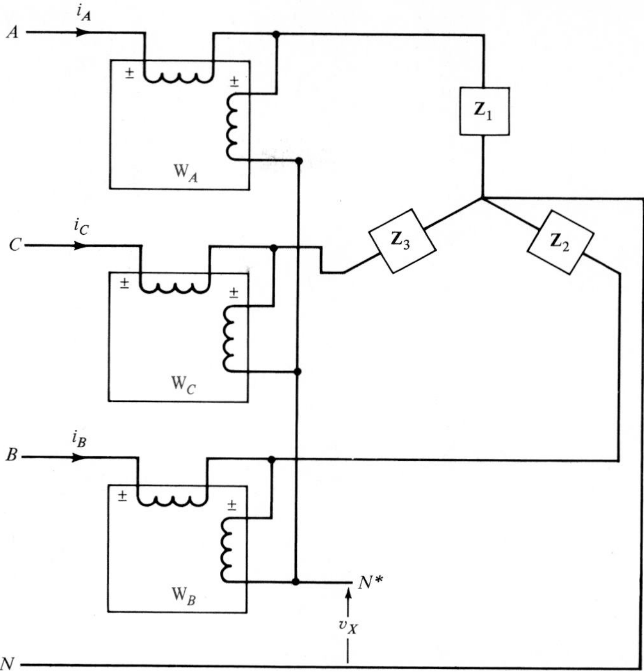

Figure 12.25 Wattmeter connections for power measurement.

The sum of all the wattmeter measurements is

$$P = \frac{1}{T}\int_0^T (v_{AN*}i_A + v_{BN*}i_B + v_{CN*}i_C)\,dt \tag{12.43}$$

As shown in Fig. 12.25, the voltages v_{AN*}, v_{BN*}, and v_{CN*} can be expressed as

$$\begin{aligned} v_{AN*} &= v_{AN} - v_X \\ v_{BN*} &= v_{BN} - v_X \\ v_{CN*} &= v_{CN} - v_X \end{aligned} \tag{12.44}$$

Substituting Eq. (12.44) into Eq. (12.43) yields

$$P = \frac{1}{T}\int_0^T (v_{AN}i_A + v_{BN}i_B + v_{CN}i_C)\,dt - \frac{1}{T}\int_0^T v_X(i_A + i_B + i_C)\,dt \tag{12.45}$$

However,

$$i_A + i_B + i_C = 0 \tag{12.46}$$

and hence the expression for the power reduces to

$$P = \frac{1}{T}\int_0^T v_{AN}i_A\,dt + \frac{1}{T}\int_0^T v_{BN}i_B\,dt + \frac{1}{T}\int_0^T v_{CN}i_C\,dt \tag{12.47}$$

which we recognize to be the total power absorbed by the three-phase Y load. Therefore,

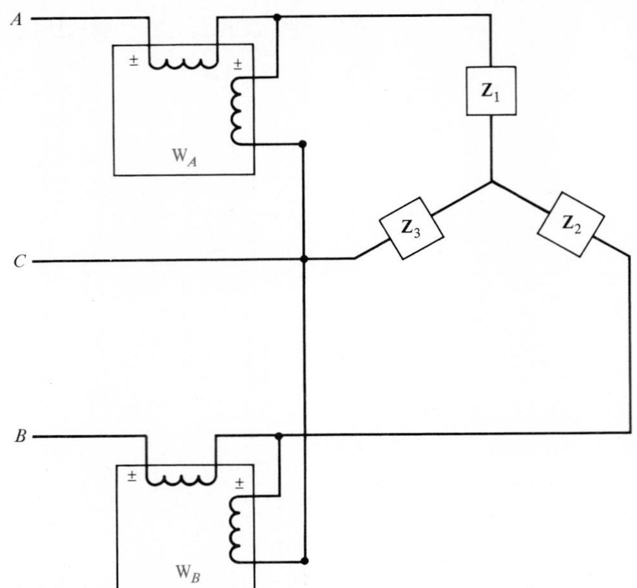

Figure 12.26 Two-wattmeter method for power measurement.

using the three wattmeters shown in Fig. 12.25, we can measure the power absorbed by a three-phase load, and in addition, this method applies whether the system is balanced or unbalanced, and whether the load is wye- or delta-connected.

If we connect the virtual neutral to line C, the voltage coil on wattmeter C will have zero volts across it; the wattmeter will read zero watts, and should be removed. Hence the two wattmeters A and B will then measure all the power in the three-phase load. This configuration is shown in Fig. 12.26 and is known as the *two-wattmeter method* for power measurement. The two-wattmeter method can always be used when the load is balanced, but in the unbalanced case, is valid only for a three-wire load. In general, if there are n wires from the source to the load, $n - 1$ wattmeters are required. As shown in Fig. 12.26, the total power measured by the two-wattmeter method is

$$P_T = P_A + P_B = |\mathbf{V}_{AC}||\mathbf{I}_A| \cos (\angle \mathbf{V}_{AC} - \angle \mathbf{I}_A) + |\mathbf{V}_{BC}||\mathbf{I}_B| \cos (\angle \mathbf{V}_{BC} - \angle \mathbf{I}_B) \quad (12.48)$$

where, for example, $\angle \mathbf{V}_{AC}$ represents the phase angle of the voltage $\mathbf{V}_{AC}$.

EXAMPLE 12.16

A balanced wye–delta system has a positive-sequence source with $\mathbf{V}_{an} = 120 \angle 0^\circ$ V. The balanced load has a phase impedance of $10 + j5\ \Omega$. We wish to find the power absorbed by the load using the two-wattmeter method.

If $\mathbf{V}_{an} = 120 \angle 0^\circ$ V, then

$$\mathbf{V}_{AB} = 208 \angle 30^\circ \text{ V}$$

$$\mathbf{V}_{BC} = 208 \angle -90^\circ \text{ V}$$

$$\mathbf{V}_{CA} = 208 \angle -210^\circ \text{ V}$$

and hence $\mathbf{V}_{AC} = 208 \underline{/-30°}$ V. Since the phase impedance is $10 + j5 = 11.18 \underline{/26.57°}\ \Omega$, the magnitude of the phase current is

$$I_\Delta = \frac{208}{11.18} = 18.60 \text{ A}$$

The power per phase is therefore

$$\begin{aligned} P_p &= (18.60)^2(10) \\ &= 3461 \text{ W} \end{aligned}$$

and hence the total power is

$$P_T = 10{,}383 \text{ W}$$

Using the two-wattmeter method, we note from Eq. (12.25) that

$$\mathbf{I}_a = 18.60\sqrt{3} \underline{/-26.57°} \text{ A}$$

$$\mathbf{I}_b = 18.60\sqrt{3} \underline{/-120 - 26.57°} \text{ A}$$

Therefore, using Eq. (12.48), we have

$$\begin{aligned} P_T &= (208)(32.22) \cos(-30 + 26.57°) \\ &\quad + (208)(32.22) \cos(-90 + 120 + 26.57°) \\ &= 10{,}383 \text{ W} \end{aligned}$$

■

DRILL EXERCISE

D12.16. Given the data in Drill Exercise D12.6, determine the power absorbed by the load using the two-wattmeter method.

Power Factor Measurement

The two-wattmeter method can also be used to compute the power factor angle if the load is balanced. To illustrate how this can be accomplished, consider the wye-connected load shown in Fig. 12.26, where $\mathbf{Z}_1 = \mathbf{Z}_2 = \mathbf{Z}_3 = Z_Y \underline{/\theta°}$. If the source is a positive-sequence balanced wye with $\underline{/v_{an}} = 0$, then as shown in Eq. (12.48),

$$P_A = |\mathbf{V}_{AC}||\mathbf{I}_A| \cos(\underline{/\mathbf{V}_{AC}} - \underline{/\mathbf{I}_A})$$

From Eq. (12.13), $|\mathbf{V}_{AC}| = V_L$ and $\underline{/\mathbf{V}_{AC}} = -30°$. Similarly, from Eq. (12.18), $|\mathbf{I}_A| = I_L$ and $\underline{/\mathbf{I}_A} = -\theta$. Therefore,

$$P_A = V_L I_L \cos(-30 + \theta)$$

In a similar manner we can show that

$$P_B = V_L I_L \cos(30 + \theta)$$

Then the ratio of the two wattmeter readings is

$$\frac{P_A}{P_B} = \frac{\cos(\theta - 30^\circ)}{\cos(\theta + 30^\circ)}$$

If we now employ the trigonometric identities in Eq. (9.11) and recall that $\cos 30^\circ = \sqrt{3}/2$ and $\sin 30^\circ = 1/2$, it is straightforward to show that the equation above can be reduced to

$$\tan\theta = \frac{(P_A - P_B)\sqrt{3}}{P_A + P_B}$$

and since $P_T = P_A + P_B$,

$$\theta = \tan^{-1}\frac{(P_A - P_B)\sqrt{3}}{P_T} \tag{12.49}$$

The equations above indicate that if $P_A = P_B$, the load is resistive; if $P_A > P_B$, the load is inductive; and if $P_A < P_B$, the load is capacitive. Finally, this technique is valid whether the load is wye- or delta-connected.

EXAMPLE 12.17

In a balanced wye-delta system two wattmeters are connected to measure the total power. We wish to determine the power factor of the load if the wattmeter readings are $P_A = 1200$ W and $P_B = 480$ W.

Using Eq. (12.49), we have

$$\begin{aligned}\theta &= \tan^{-1}\frac{(1200 - 480)\sqrt{3}}{1680}\\ &= 36.59^\circ\end{aligned}$$

Therefore,

$$\cos\theta = \text{PF} = 0.80 \text{ lagging}$$

■

EXAMPLE 12.18

If the source in Example 12.17 has a line voltage of 208 V, we wish to find the delta load impedance.

The total power $P_T = P_A + P_B = 1680$ W; therefore, the power per phase is $P_p = 1680/3 = 560$ W. Hence using the expression

$$V_\Delta I_\Delta \cos\theta = P_p$$

$$\begin{aligned}I_\Delta &= \frac{560}{(208)(0.8)}\\ &= 3.37 \text{ A}\end{aligned}$$

Therefore,

$$Z = \frac{208}{3.37} \underline{/36.59°} = 61.81 \underline{/36.59°}$$

$$= 49.63 + j36.84\ \Omega$$

■

EXAMPLE 12.19

In a balanced wye–delta system two wattmeters are used to measure total power. Wattmeter A reads 800 W and wattmeter B reads 400 W after the current coil terminals are reversed. If the line voltage is 208 V, we wish to determine the total power, the power factor, and the phase impedance of the load.

From the data given $P_A = 800$ and $P_B = -400$, therefore, the total power P_T is

$$P_T = P_A + P_B$$

$$= 400 \text{ W}$$

The power factor is computed from

$$\theta = \tan^{-1} \frac{[800 - (-400)]\sqrt{3}}{400}$$

$$\theta = 79.11°$$

and therefore

$$\cos \theta = \text{PF} = 0.19 \text{ lagging}$$

The phase impedance is determined from

$$V_\Delta I_\Delta \cos \theta = P_p$$

$$(208) I_\Delta (0.19) = \frac{400}{3}$$

$$I_\Delta = 3.37 \text{ A}$$

Therefore,

$$\mathbf{Z} = \frac{208}{3.37} \underline{/79.11°} = 61.65 \underline{/79.11°}\ \Omega$$

■

DRILL EXERCISE

D12.17. Two wattmeters are used to measure the total power in the load of a balanced wye–wye system. The line voltage is 208 V and the wattmeter readings are $P_A = 1600$ W and $P_B = 840$ W. Compute the impedance per phase of the load.

D12.18. Two wattmeters are used to measure the total power in the load of a balanced wye–wye system. The line voltage is 208 V. Wattmeter A reads 1280 W and wattmeter B reads 540 W when the current coil terminals are reversed. Determine the power factor of the load and the impedance per phase of the load.

Finally, the reader will recall that essentially our entire discussion in this chapter has focused on balanced systems. It is extremely important, however, to point out that an unbalanced three-phase system is nothing more than a circuit with three sources and, therefore, can be analyzed using the techniques we have presented earlier (e.g., loop equations, node equations, etc.).

12.6 Summary

As a prelude to our analysis of balanced three-phase circuits we have first introduced the single-phase three-wire system. We have shown that an important advantage of the balanced three-phase system is that it provides very smooth power delivery. Because of the balanced condition it is possible to analyze a circuit on a "per phase" basis, thereby providing a significant computational shortcut to a solution. The relationships between line voltages and phase voltages and line currents and phase currents for the balanced system in both a wye and a delta configuration were found to be very simple. The ability to perform wye-to-delta and delta-to-wye transformations was shown to be an important ingredient in our computational approach.

Techniques for power measurement in a three-phase environment were presented. The two-wattmeter method for the measurement of three-phase power was derived, and the use of this method for power factor measurement was presented.

KEY POINTS

- A balanced three-phase voltage source has three sinusoidal voltages of the same magnitude and frequency, and each voltage is 120° out of phase with the others.
- If the load currents generated by connecting a load to a balanced three-phase voltage source are also balanced, the load is connected in either a balanced wye or a balanced delta configuration.
- A positive-phase-sequence balanced voltage source is one in which $\mathbf{V}_{bn}$ lags $\mathbf{V}_{an}$ by 120° and $\mathbf{V}_{cn}$ lags $\mathbf{V}_{bn}$ by 120°.
- There is no current in the neutral line of a balanced wye–wye system.
- The phase voltage in a wye-connected system is the voltage from line to netural. The line voltage in a wye system is the voltage from line to line.
- The line current is the phase current in a wye-connected system.
- The phase current in a delta-connected system is the current in each leg of the delta. The line current in a delta-connected system is the current in the lines that connect the delta to the remainder of the network.
- The phase voltage is the line voltage in a delta-connected system.
- The total complex power in a balanced three-phase circuit is equal to three times the complex power per phase.
- The two-wattmeter method is a technique for measuring real power in a three-phase system using only two wattmeters.
- The power factor angle of a load in a balanced three-phase system can be computed using two wattmeters.

PROBLEMS

12.1. Determine the current in the neutral line of the single-phase three-wire system shown in Fig. P12.1.

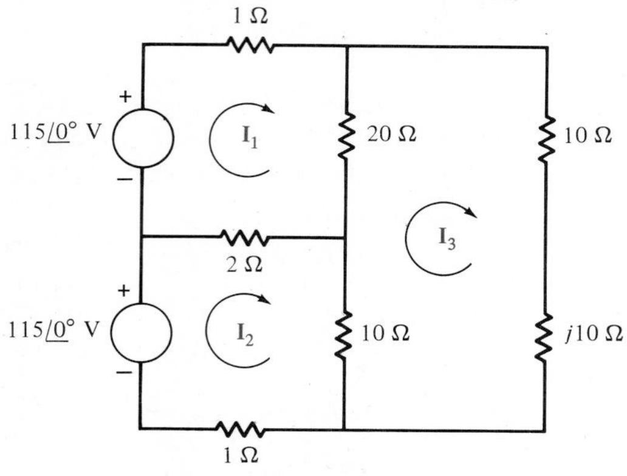

Figure P12.1

12.2. Find the real power losses in the line impedances in the circuit in Problem 12.1.

12.3. Find the real and reactive power losses in the load impedances in Problem 12.1.

12.4. Determine the magnitude of the complex power supplied by the sources in Problem 12.1.

12.5. For the single-phase three-wire system shown in Fig. P12.5, find the total power supplied by the voltage sources.

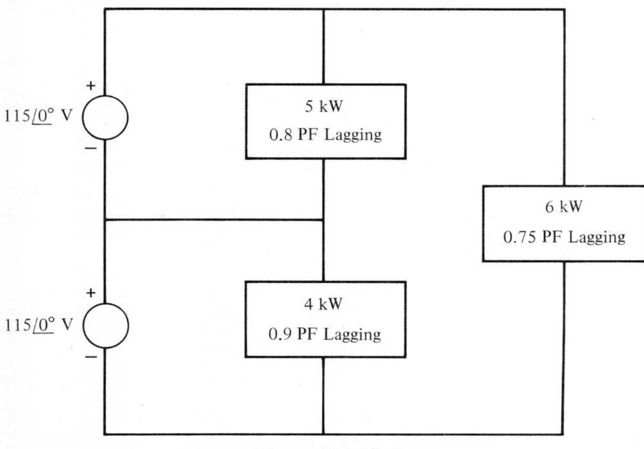

Figure P12.5

12.6. Determine the current in the neutral line of the single-phase, three-wire system shown in Fig. P12.5.

12.7. An air conditioner is connected to a single-phase, three-wire system. The compressor on the air conditioner draws 7.0 kW at a PF of 0.7 lagging and operates at 240 V. There are two fans inside the air conditioner which help circulate the air. Each operates at 120 V and consumes 150 W at a PF of 0.8 lagging. Determine the current flowing through each of these devices.

12.8. Sketch a phasor representation of a balanced three-phase system containing both phase voltages and line voltages if $\mathbf{V}_{an} = 100 \underline{/45^\circ}$ V. Label all magnitudes and assume a positive phase sequence.

12.9. For a balanced three-phase wye-wye connection with a load impedance of $20 + j40\ \Omega$ and a line impedance of $2 + j1\ \Omega$, calculate the phase voltages, the load voltages, the line currents, and the phase currents if $\mathbf{V}_{ab} = 220 \underline{/10^\circ}$ V. Assume a positive phase sequence.

12.10. A positive-sequence three-phase balanced wye voltage source has a phase voltage of $\mathbf{V}_{an} = 100 \underline{/20^\circ}$ V. Determine the line voltages of the source.

12.11. A positive-sequence balanced three-phase wye-connected source with a phase voltage of 100 V supplies power to a balanced wye-connected load. The per phase load impedance is $40 + j10\ \Omega$. Determine the line currents in the circuit if $\underline{/\mathbf{V}_{an}} = 0^\circ$.

12.12. A positive-sequence balanced three-phase wye-connected source supplies power to a balanced wye-connected load. The magnitude of the line voltages is 150 V. If the load impedance per phase is $36 + j12\ \Omega$, determine the line currents if $\underline{/\mathbf{V}_{an}} = 0^\circ$.

12.13. Calculate the phase voltages, line voltages, phase currents (at the load), and line currents for a balanced three-phase wye–delta connection with $\mathbf{Z}_\Delta = 60 \underline{/80^\circ}\ \Omega$ and $\mathbf{V}_{an} = 115 \underline{/-45^\circ}$ V.

12.14. A delta-connected load has an impedance of $54 + j27\ \Omega$ in each phase. If the wye-connected source is $150\sqrt{2}\cos(\omega t)$ V and the line impedance is $3 + j4\ \Omega$, determine the phase voltages (of the source), load voltages, line currents, and phase currents (at the load). (*Hint:* You may use delta-to-wye transformations.)

12.15. A positive-sequence balanced three-phase wye-connected source supplies power to a balanced wye-connected load. The line impedance per phase is 1 Ω, and the load impedance per phase is 20 + j20 Ω. If the source line voltage $\mathbf{V}_{ab}$ is 100 $\underline{/0°}$ V, find the line currents.

12.16. A positive-sequence balanced three-phase wye-connected source supplies power to a balanced wye-connected load of 50 + j20 Ω per phase. If the line impedance is 2 Ω per phase and the line voltage $\mathbf{V}_{ab}$ is 220 $\underline{/60°}$V, find the value of the load voltages.

12.17. A positive-sequence balanced three-phase wye-connected source supplies power to the load shown in Fig. P12.17. The phase voltage of the source is 120 V. If $\mathbf{Z}_1$ = 20 + j20 Ω and $\mathbf{Z}_2$ = 40 + j10 Ω, determine the line currents if $\underline{/\mathbf{V}_{an}}$ = 0°.

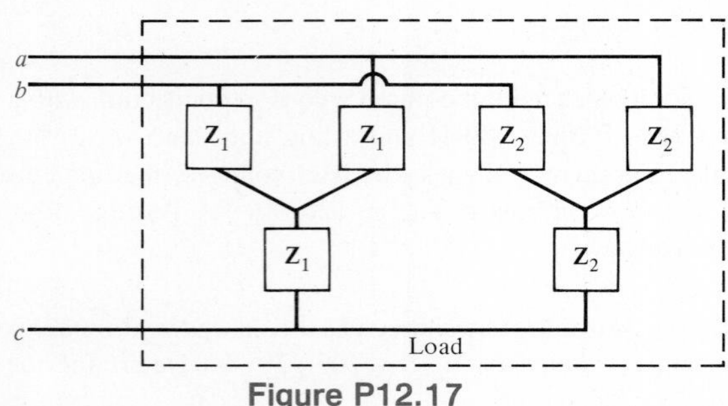

Figure P12.17

12.18. Compute the load voltages in Problem 12.17 if the line impedance per phase is 1 Ω.

12.19. Prove that the magnitude of the line currents feeding a balanced delta-connected load is equal to the phase currents multiplied by the square root of 3 (Fig. P12.19).

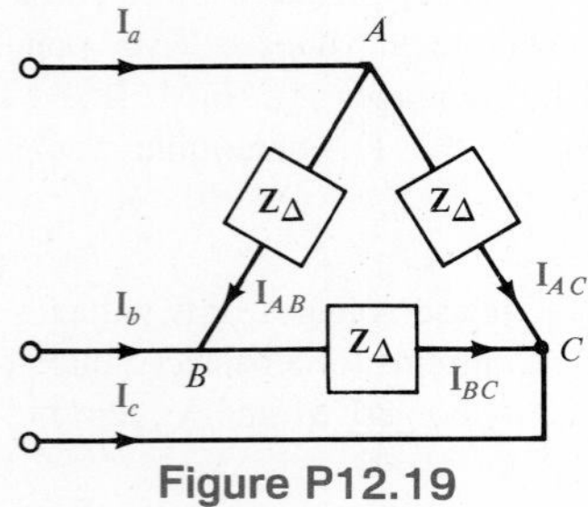

Figure P12.19

12.20. In a balanced wye–delta system the load per phase is 50 + j20 Ω. $\mathbf{V}_{an}$ for the positive-sequence balanced wye source is 120 $\underline{/-20°}$ V. Determine all the phase currents and line currents in the circuit.

12.21. A three-phase load consists of a balanced wye with an impedance of 9 + j6 Ω per phase in parallel with a balanced delta with an impedance of 24 + j9 Ω per phase. Calculate the equivalent balanced wye load.

12.22. Compute the equivalent balanced delta load for Problem 12.21.

12.23. In the preceding chapters we never encountered a circuit such as that shown in Fig. P12.23. Use the wye–delta transformation to determine the equivalent impedance at terminals *A-B*.

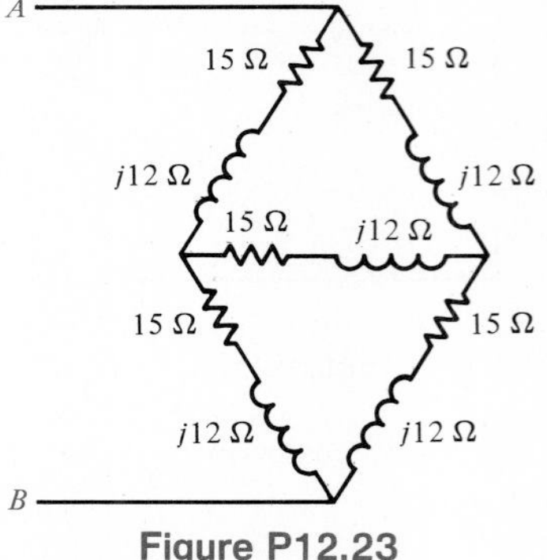

Figure P12.23

12.24. Use the wye–delta transformation to determine the equivalent impedance at terminals *A-B* in the network in Fig. P12.24.

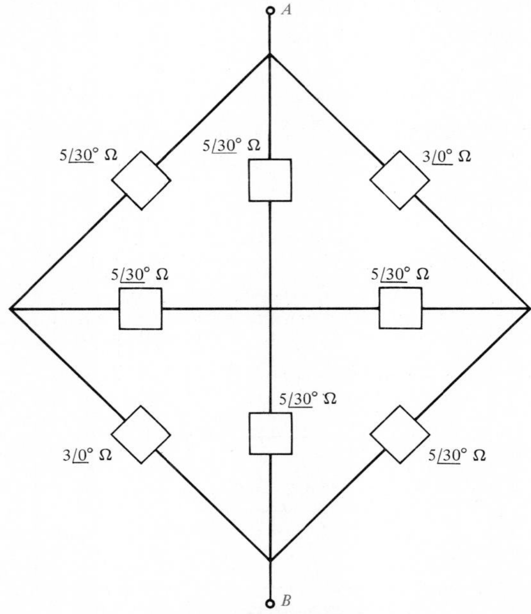

Figure P12.24

12.25. A balanced three-phase system has a load consisting of a delta in parallel with a wye. The impedance per phase for the delta is $15 + j6\ \Omega$ and for the wye is $9 + j3\ \Omega$. The source is a balanced wye with a positive phase sequence. If $\mathbf{V}_{an} = 120\ \underline{/-20^\circ}$ V, determine the line currents if the line impedance is $1 + j1\ \Omega$.

12.26. Determine the phase voltages of the equivalent wye-connected load in Problem 12.25.

12.27. Determine the line currents in Problem 12.25 if $\mathbf{V}_{ab} = 100\ \underline{/60^\circ}$ V.

12.28. Determine the phase voltages of the equivalent wye-connected load in Problem 12.27.

12.29. Determine the instantaneous power for a balanced three-phase load in which the load current is $1.5 \cos(377t + 20^\circ)$ A and the load voltage is $120 \cos(377t + 65^\circ)$ V.

12.30. A positive-sequence wye-connected source having a phase voltage of $120\ \underline{/0^\circ}$ V is attached to a wye-connected load having an impedance of $80\ \underline{/70^\circ}\ \Omega$. If the line impedance is $4\ \underline{/20^\circ}\ \Omega$, determine the total complex power produced by the voltage sources and the real and reactive power dissipated by the load.

12.31. A three-phase positive-sequence wye-connected source supplies 14 kVA with a power factor of 0.75 lagging to a parallel combination of a wye load and a delta load. If the wye load consumes 9 kVA at a power factor of 0.6 lagging and has a phase current of $10\ \underline{/-30^\circ}$ A, determine the phase impedance of the delta load.

12.32. A three-phase positive-sequence wye-connected source with $\mathbf{V}_{an} = 220\ \underline{/0^\circ}$ V supplies power to a wye-connected load that consumes 50 kW of power in each phase at a PF of 0.8 lagging. Three capacitors are found that have an impedance of $-j2.0\ \Omega$, and they are connected in parallel with the previous load in a wye configuration. Determine the power factor of the combined load as seen by the source.

12.33. If the three capacitors in the network in Problem 12.32 are connected in a delta configuration, determine the power factor of the combined load as seen by the source.

12.34. In a balanced three-phase system both the source and load are delta-connected. The transmission line has an impedance of $0.1 + j0.2\ \Omega$ per phase. The delta-connected source has a positive phase sequence and $\mathbf{V}_{ab} = 208\ \underline{/20^\circ}$ V. The impedance per phase at the load is $9 + j6\ \Omega$. Determine the line currents and the total power that must be supplied by the three-phase source.

12.35. A balanced three-phase delta-connected source supplies power to a load consisting of a balanced delta in parallel with a balanced wye. The phase impedance of the delta is $24 + j12\ \Omega$, and the phase impedance of the wye is $12 + j8\ \Omega$. The positive-phase-sequence source voltages are $\mathbf{V}_{ab} = 440\ \underline{/60^\circ}$ V, $\mathbf{V}_{bc} = 440\ \underline{/-60^\circ}$ V, and $\mathbf{V}_{ca} = 440\ \underline{/-180^\circ}$ V, and the line impedance per phase is $1 + j0.8\ \Omega$. Find the line currents and the power absorbed by the wye-connected load.

12.36. A balanced three-phase delta-connected source with a positive phase sequence is connected through a line impedance of $1 + j0.8\ \Omega$ to a balanced wye-connected load with a per phase impedance of $18 + j10\ \Omega$. If the source voltages are $\mathbf{V}_{ab} = 440\ \underline{/60^\circ}$ V, $\mathbf{V}_{bc} = 440\ \underline{/-60^\circ}$ V, and $\mathbf{V}_{ca} = 440\ \underline{/-180^\circ}$ V, determine the load voltages and the real power absorbed by the load.

12.37. A balanced three-phase wye-connected source supplies power to a load consisting of two delta-connected loads in parallel. The impedance of each load is $18 + j9\ \Omega$. The positive-phase-sequence source has a phase voltage of $\mathbf{V}_{an} = 120\ \underline{/30^\circ}$ V. Determine the line currents in the system.

12.38. A balanced three-phase wye-connected source supplies power to the load shown in Fig. P12.38. The positive-phase-sequence source has a line voltage $\mathbf{V}_{ab} = 440$ V $\underline{/30^\circ}$ V. In this network $\mathbf{Z}_{\text{line}} = 1 + j1\ \Omega$ and $\mathbf{Z}_{\text{load}} = 20 + j10\ \Omega$. Determine the current in each load and the total power losses in the lines.

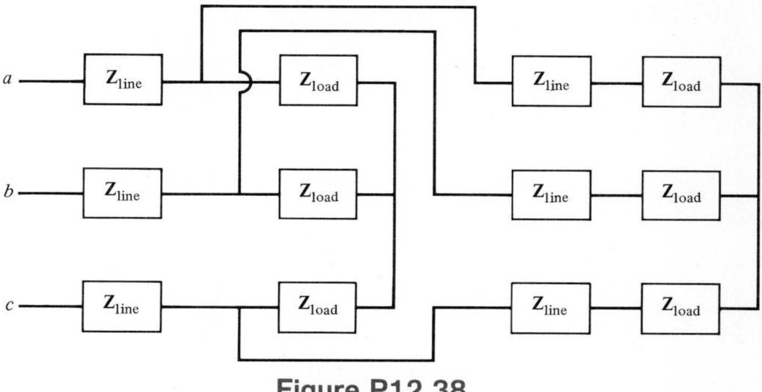

Figure P12.38

12.39. A balanced three-phase wye-connected source supplies power to a balanced delta load. The impedance per phase of the delta is $16 + j4\ \Omega$. The line impedance between the source and load is $\mathbf{Z}_{\text{line}} = 1 + j0.5\ \Omega$. Determine the phase voltage $\mathbf{V}_{an}$ of the positive-phase-sequence source if the phase current $\mathbf{I}_{ab}$ in the Δ load is $20\,\underline{/20°}$ A.

12.40. Two industrial plants represent balanced three-phase loads. The plants receive their power from a balanced three-phase source with a line voltage of 4.6 kV. Plant 1 is rated at 300 kVA, 0.8 PF lagging and plant 2 is rated at 350 kVA, 0.84 PF lagging. Determine the power-line current.

12.41. Two wattmeters are employed to measure the total power in a balanced three-phase wye-delta system. The phase voltage of the source is 120 V. Without reversing the meter terminals, the meter readings are $P_A = 1600$ W and $P_B = 800$ W. Determine the impedance per phase of the delta.

12.42. A three-phase balanced wye–delta system has a line voltage of 208 V and a line current of 9 A. If the reactive power per phase absorbed by the load is 300 var, find the total real power absorbed by the load.

12.43. A three-phase balanced wye–wye system has a line voltage of 208 V. The line current is 6 A and the total real power absorbed by the load is 1800 W. Determine the load impedance per phase.

12.44. A balanced three-phase wye–delta system has a line voltage of 208 V and a line current of 8.2 A. The total reactive power absorbed by the load is 600 var. Determine the resistive component of the load impedance.

12.45. A three-phase balanced positive-sequence wye–wye system has a load impedance per phase of $10 + j8\ \Omega$ and a line impedance per phase of $0.5 + j0.5\ \Omega$. If the source line voltage is 208 V, determine the total real power and the magnitude of the complex power supplied by the source.

12.46. The magnitude of the complex power supplied by a three-phase balanced wye–wye system is 3600 VA. The line voltage is 208 V. If the line impedance is negligible and the power factor angle of the load is 25°, determine the load impedance.

12.47. A balanced three-phase wye–wye system has two parallel loads. Load 1 is rated at 3000 VA, 0.7 PF lagging, and load 2 is rated at 2000 VA, 0.75 PF leading. If the line voltage is 208 V, find the magnitude of the line current.

12.48

(a) In the network in Fig. P12.48 find the line currents, phase currents, the power absorbed in each load, and the power supplied by the source.

(b) Repeat part (a) if the line has an impedance if $1 + j1\ \Omega$.

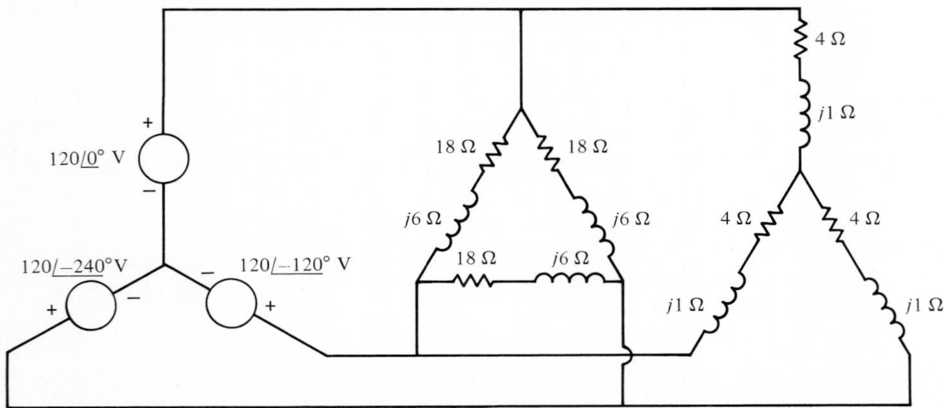

Figure P12.48

12.49. What is the wattmeter reading in the circuit shown in Fig. P.12.49?

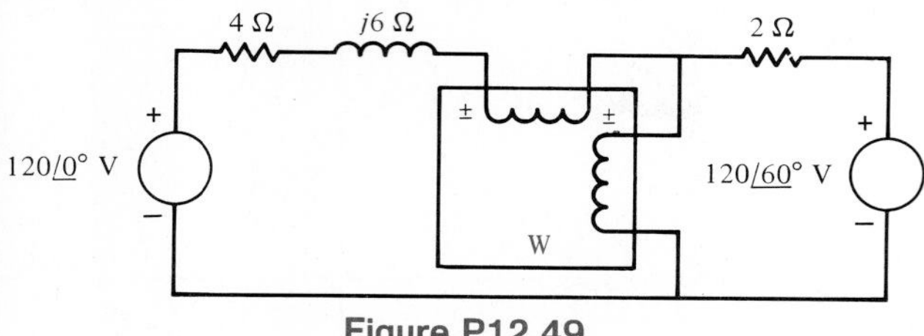

Figure P12.49

12.50 A balanced wye–wye three-phase system has a positive-sequence source with $\mathbf{V}_{an} = 120\,\underline{/0°}$ V. The balanced load has a phase impedance of $12 + j12\ \Omega$. Calculate the readings of the wattmeters if the two-wattmeter method is used to measure the total three-phase power.

12.51. A balanced wye–wye three-phase system has a positive-sequence source with $\mathbf{V}_{an} = 90\,\underline{/30°}$ V. The balanced load has a per phase impedance of $24 + j16\ \Omega$. Calculate the readings of the wattmeters if the two-wattmeter method is used to measure the total three-phase power.

12.52. A balanced wye–delta three-phase system employs two wattmeters to measure the total power. If the wattmeters read $P_A = 800$ W and $P_B = 400$ W without reversing the wattmeter terminals, and the line voltage is 208 V, find the value of the load impedance.

12.53. In a balanced wye–delta three-phase system two wattmeters are used to measure total power. The line voltage is 208 V. If wattmeter *A* reads 1200 W and wattmeter *B* reads 400 W when the current terminals are reversed, determine the power factor and the load impedance per phase.

12.54. If in Problem 12.53, wattmeter *A* reads 400 W when the current terminals are reversed, and wattmeter *B* reads 1200 W, determine the power factor and the load impedance per phase.

12.55. Calculate the line current, load current, and load voltages in the network in Fig. P12.55 using SPICE. $f = \frac{1}{2}\pi$ Hz.

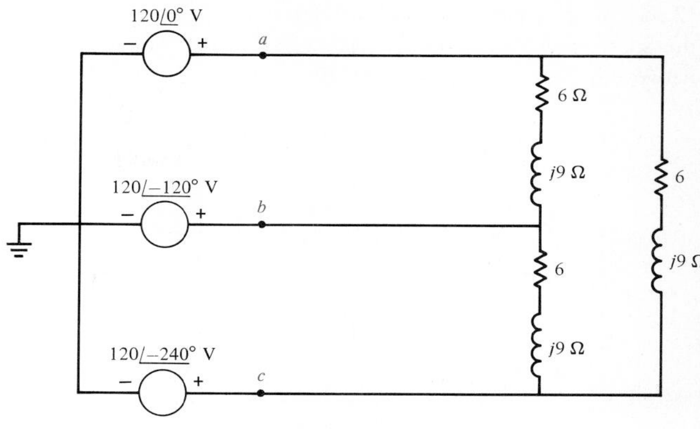

Figure P12.55

12.56. Find the source current, load current, and voltage across the line impedances in the network in Fig. P12.56 using SPICE. $f = \frac{1}{2}\pi$ Hz.

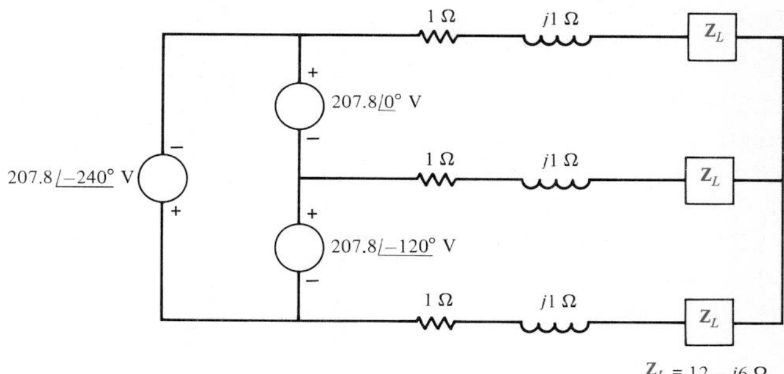

Figure P12.56

12.57. Use SPICE to calculate the source current, both load currents, and both load voltages in the circuit in Fig. P12.57. $f = \frac{1}{2}\pi$ Hz.

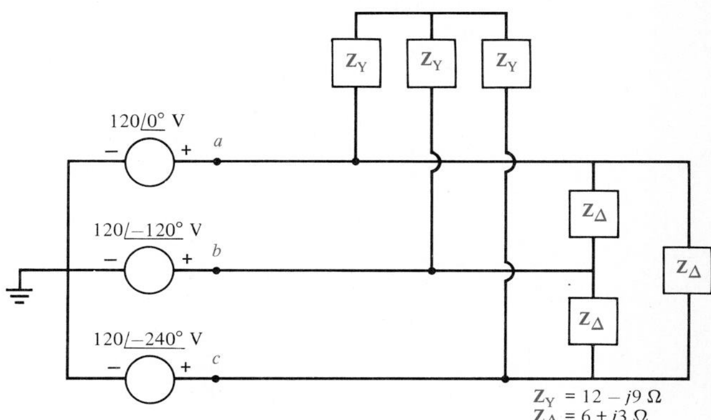

Figure P12.57

12.58. Write a SPICE program to determine the currents through the source, wye load, delta load, and the load voltages in the network in Fig. P12.58. $f = \frac{1}{2}\pi$ Hz.

12.60. Given the circuit in Fig. P12.60, use SPICE to calculate the source current, line-to-line voltage, power loss in the line impedance, and both load voltages. $f = \frac{1}{2}\pi$ Hz.

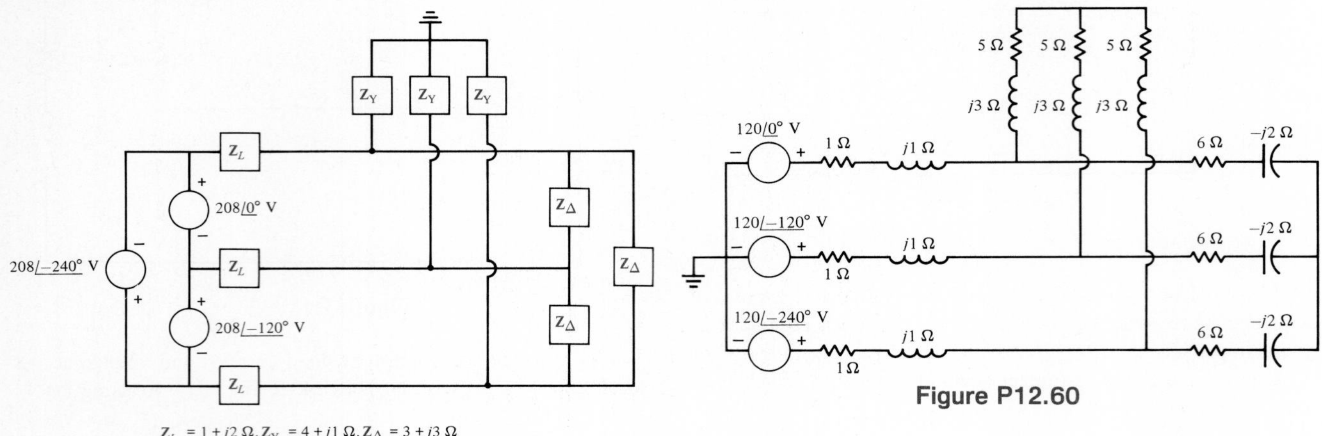

Figure P12.58

Figure P12.60

12.59. Write a SPICE program to determine the neutral wire current, load voltages, and load currents in the network in Fig. P12.59. $f = \frac{1}{2}\pi$ Hz.

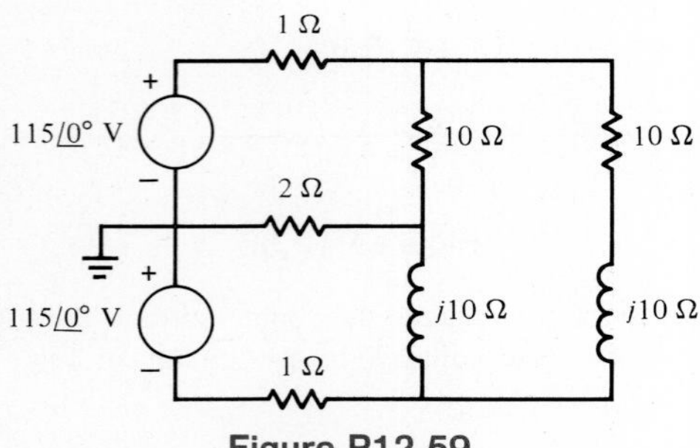

Figure P12.59

CHAPTER

13

Frequency Characteristics

In this chapter we examine the frequency characteristics of networks. Of particular interest to us will be the effect of the source frequency on a network. Techniques for determining frequency response will be presented, and networks with special filtering properties will be examined. In addition, SPICE will be employed to illustrate a network's frequency characteristics by means of plot routines. Networks that exhibit particular filtering characteristics have wide application in communications and control systems.

13.1 General Network Characteristics

In the earlier chapters we have investigated the response of networks to a number of different types of excitation. For example, we analyzed networks in which the forcing function was a constant (dc) or a sinusoid (ac). In the latter case the response of the network to a sinusoidal forcing function was obtained by determining the response due to the function $\mathbf{X}e^{j\omega t}$, where $\mathbf{X}$ represented either current or voltage, and then taking the real part of the resultant function. We demonstrated in Chapter 9 that the application of the forcing function $\mathbf{X}e^{j\omega t}$, which would appear to complicate the analysis, actually simplified the task, because it reduced the differential equation for the steady-state response to a set of algebraic equations with complex coefficients. At this point we change emphasis and examine the performance of a network as the frequency changes (i.e., frequency will become our variable). Networks in communication and control systems are designed to be frequency selective; in other words, they pass certain frequencies and reject others. For example, in a stereo system if the networks are designed so that each frequency in the audio range is amplified the same amount, perfect sound reproduction

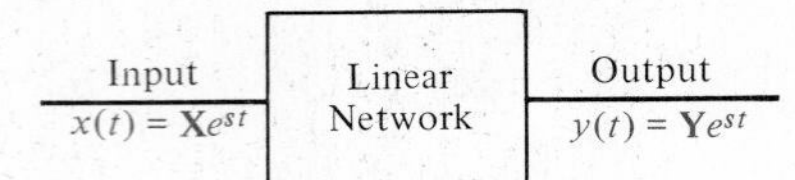

Figure 13.1 Representation of a linear network.

is achieved. If some frequencies are amplified more than others, distortion occurs and the output is not an exact duplicate of the input.

To simplify our notation, we can represent a linear network by a block diagram as shown in Fig. 13.1. The block represents the network, $x(t)$ represents the input or excitation, and $y(t)$ represents the output or response. Both $x(t)$ and $y(t)$ may be either a voltage or a current. As a further simplification we will substitute s for $j\omega$. At this point, we simply make the substitution for convenience in the algebraic manipulations. Later this variable will be called complex frequency and given more meaning. Since the network is linear, if $x(t) = \mathbf{X}e^{st}$, then $y(t) = \mathbf{Y}e^{st}$, and the ratio of these two functions is called the *network function*:

$$\mathbf{H}(s) = \frac{\mathbf{Y}e^{st}}{\mathbf{X}e^{st}} = \frac{\mathbf{Y}}{\mathbf{X}} \tag{13.1}$$

which is dependent on the complex frequency s and is independent of time.

Since both the input and output can generally be either voltage or current, there are four possible network functions. For example, if the input is current and the output is voltage, the network function is an impedance, more specifically a transfer impedance $\mathbf{Z}(s)$. Similarly, the other network functions are a transfer admittance $\mathbf{Y}(s)$, a voltage gain $\mathbf{G_v}(s)$, and a current gain $\mathbf{G_I}(s)$. These functions will be examined in more detail in Chapter 15 when we study two-port networks. We should also note that in addition to the *transfer functions* that we have just defined, there are *driving-point functions*. Driving-point functions are admittance or impedance, which are defined at a single pair of terminals. For example, the input impedance of a network is a driving-point impedance.

EXAMPLE 13.1

We wish to determine the transfer admittance and the voltage gain of the network shown in Fig. 13.2.

The mesh equations for the network are

$$(R_1 + sL)\mathbf{I}_1(s) - sL\mathbf{I}_2(s) = \mathbf{V}_1(s)$$

$$-sL\mathbf{I}_1(s) + \left(R_2 + sL + \frac{1}{sC}\right)\mathbf{I}_2(s) = 0$$

$$\mathbf{V}_o(s) = \mathbf{I}_2(s)R_2$$

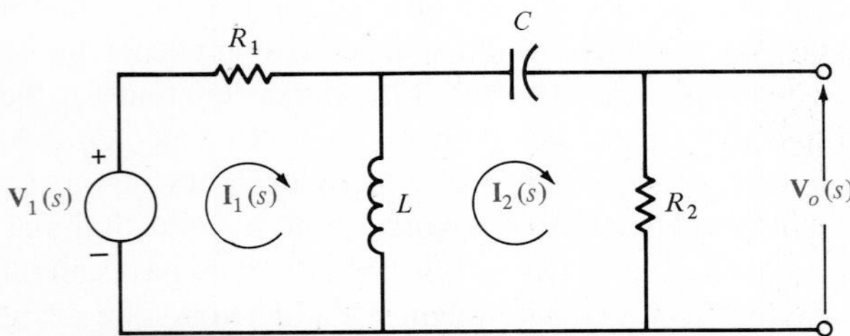

Figure 13.2 Circuit employed in Example 13.1.

Solving the equations for $\mathbf{I}_2(s)$ yields

$$\mathbf{I}_2(s) = \frac{sL\mathbf{V}_1(s)}{(R_1 + sL)(R_2 + sL + 1/sC) - s^2L^2}$$

Therefore, the transfer admittance is

$$\mathbf{Y}_T(s) = \frac{\mathbf{I}_2(s)}{\mathbf{V}_1(s)} = \frac{LCs^2}{(R_1 + R_2)LCs^2 + (L + R_1R_2C)s + R_1}$$

and the voltage gain is

$$\mathbf{G}_v(s) = \frac{\mathbf{V}_o(s)}{\mathbf{V}_1(s)} = \frac{LCR_2s^2}{(R_1 + R_2)LCs^2 + (L + R_1R_2C)s + R_1}$$ ■

DRILL EXERCISE

D13.1. Find the input impedance $\mathbf{Z}_i(s)$ for the network in Fig. D13.1.

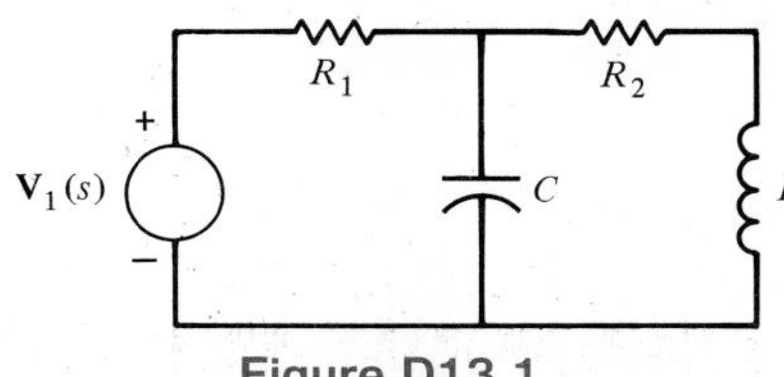

Figure D13.1

D13.2. Find the current gain $\mathbf{G}_I(s)$ for the network in Fig. D13.2.

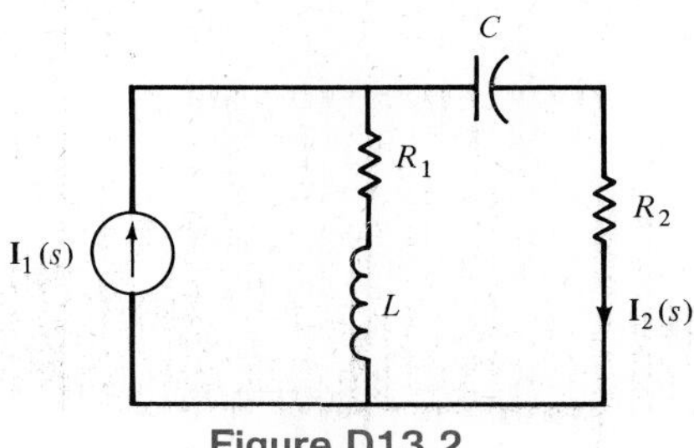

Figure D13.2

Poles and Zeros

In general the network functions can be expressed as the ratio of two polynomials in s. Example 13-1 is an illustration of this fact. In addition, we note that since the values of our circuit elements, or controlled sources, are real numbers, the coefficients of the two polynomials will be real. Therefore, we will express a network function in the form

$$\mathbf{H}(s) = \frac{N(s)}{D(s)} = \frac{a_ms^m + a_{m-1}s^{m-1} + \cdots + a_1s + a_0}{b_ns^n + b_{n-1}s^{n-1} + \cdots + b_1s + b_0} \tag{13.2}$$

where $N(s)$ is the numerator polynomial of degree m and $D(s)$ is the denominator polynomial of degree n. Equation (13.2) can also be written in the form

$$\mathbf{H}(s) = \frac{K_0(s - z_1)(s - z_2) \cdots (s - z_m)}{(s - p_1)(s - p_2) \cdots (s - p_n)} \tag{13.3}$$

where K_0 is a constant, $z_1, \ldots, z_m$ are the roots of $N(s)$, and $p_1, \ldots, p_n$ are the roots of $D(s)$. Note that if $s = z_1$ or $z_2, \ldots, z_m$, then $\mathbf{H}(s)$ becomes zero and hence $z_1, \ldots, z_m$ are called *zeros* of the function. Similarly, if $s = p_1$ or $p_2, \ldots, p_n$, then $\mathbf{H}(s)$ becomes infinite and therefore $p_1, \ldots, p_n$ are called *poles* of the function. The zeros or poles may actually be complex. However, if they are complex, they must occur in conjugate pairs since the coefficients of the polynomial are real. The representation of the network function specified in Eq. (13.3) is extremely important and is generally employed to represent any linear time-invariant dynamic system. The importance of this form stems from the fact that the dynamic properties of a system can be gleaned from an examination of the system poles.

13.2 Sinusoidal Frequency Analysis

Although there are specific cases in which a network operates at only one frequency (e.g., a power system network), in general we are interested in the behavior of a network as a function of frequency. In a sinusoidal steady-state analysis, the network function can be expressed as

$$\mathbf{H}(j\omega) = M(\omega)e^{j\phi(\omega)} \tag{13.4}$$

where $M(\omega) = |\mathbf{H}(j\omega)|$ and $\phi(\omega)$ is the phase. A plot of these two functions, which are commonly called the *magnitude* and *phase characteristics*, display the manner in which the response varies with the input frequency ω. We will now illustrate the manner in which to perform a frequency-domain analysis by simply evaluating the function at various frequencies within the range of interest or by using what is called a *Bode plot* (named after Hendrik W. Bode).

Basic Frequency Response Plots

If the network function is evaluated at some point $s = j\omega$, its value will be a complex quantity having a magnitude and phase. Therefore, in the case of a transfer function, two plots are used to describe the behavior of the network function: the amplitude plot and the phase plot. If the network function is a driving-point function, then separate graphs for the resistance and reactance or for the conductance and susceptance are normally used. In either case, however, we can evaluate the function $\mathbf{H}(j\omega)$ as a function of frequency by simply substituting values for ω into $\mathbf{H}(j\omega)$ which will yield both a magnitude and phase at each frequency.

Suppose that the network function is of the form

$$\mathbf{H}(s) = \frac{K_0(s - z_1)(s - z_2)}{(s - p_1)(s - p_2)} \tag{13.5}$$

If we evaluate this function at some point $s = j\omega_0$, then

$$\mathbf{H}(s = j\omega_0) = \frac{K_0(j\omega_0 - z_1)(j\omega_0 - z_2)}{(j\omega_0 - p_1)(j\omega_0 - p_2)} \tag{13.6}$$

and each term of the form $(j\omega_0 - z_i)$ and $(j\omega_0 - p_j)$ is a complex number having a magnitude and phase. Therefore, Eq. (13.6) can be written as

$$\mathbf{H}(j\omega_0) = \frac{K_0(Z_1 e^{j\theta_1})(Z_2 e^{j\theta_2})}{(P_1 e^{j\phi_1})(P_2 e^{j\phi_2})} \tag{13.7}$$

or

$$\mathbf{H}(j\omega_0) = \frac{K_0 Z_1 Z_2}{P_1 P_2} e^{j(\theta_1 + \theta_2 - \phi_1 - \phi_2)} \tag{13.8}$$

where, for example, $(j\omega_0 - z_1) = Z_1 e^{j\theta_1}$.

In general, the magnitude and phase of the network function can be expressed as

$$|\mathbf{H}(j\omega_0)| = \frac{K_0 \prod_{i=1}^{m} Z_i}{\prod_{j=1}^{n} P_j} \tag{13.9}$$

and

$$\underline{/\mathbf{H}(j\omega_0)} = \sum_{i=1}^{m} \theta_i - \sum_{j=1}^{n} \phi_j \tag{13.10}$$

where m is the number of finite zeros of $\mathbf{H}(j\omega)$ and n is the number of finite poles. If we evaluate these functions for all values of ω from 0 to ∞, we will obtain the magnitude and phase characteristics, or equivalently, the frequency response of $\mathbf{H}(j\omega)$.

EXAMPLE 13.2

We wish to determine the frequency response of a network with the transfer function

$$\mathbf{H}(s) = \frac{2s}{s^2 + 2s + 17}$$

$\mathbf{H}(s)$ can be written as

$$\mathbf{H}(s) = \frac{2s}{(s + 1 + j4)(s + 1 - j4)}$$

and hence

$$\mathbf{H}(j\omega) = \frac{2\omega \underline{/90^\circ}}{[1 + j(\omega + 4)][1 + j(\omega - 4)]}$$

The frequency response, which consists of the magnitude and phase of $\mathbf{H}(j\omega)$ as a function of frequency, can be obtained simply by evaluating $\mathbf{H}(j\omega)$ for various values

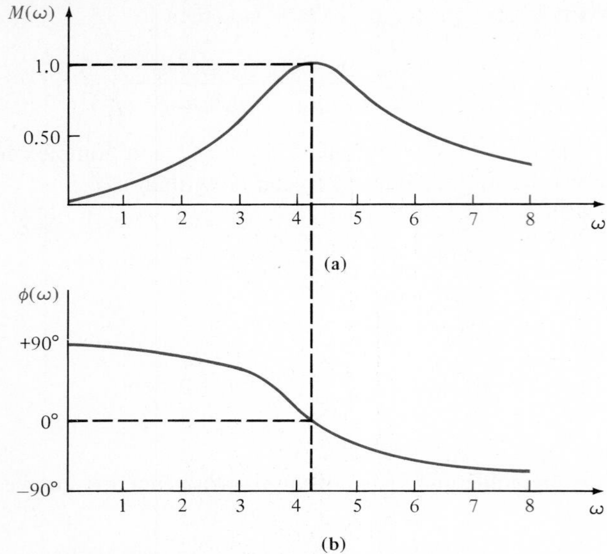

Figure 13.3 Graphs employed in Example 13.2.

of ω. The magnitude and phase of $\mathbf{H}(j\omega)$ for several values of ω are

ω	0	1	2	3	4	$\sqrt{17}$	5	6	7	8
$M(\omega)$	0	0.12	0.29	0.60	0.99	1.0	0.78	0.53	0.40	0.32
$\phi(\omega)$	90°	83°	73°	53°	7°	0°	−39°	−58°	−66°	−71°

Sketches of the amplitude and phase characteristics for the network function are shown in Fig. 13.3a and b. ■

Example 13.2 illustrates some rather subtle points which are true in general. Note that in the vicinity of a zero the network response approaches a minimum, and in the vicinity of a pole the network response approaches a maximum. The importance of this relationship stems from the fact that if we want to design a network filter to pass certain frequencies and reject others, we design the filter so that it has poles in the passband frequency range and zeros in the rejection frequency range.

DRILL EXERCISE

D13.3. Given the following network function $\mathbf{H}(j\omega)$, tabulate the values for $M(\omega)$ and $\phi(\omega)$ versus ω for $0 \leq \omega \leq 10$ rad/s in increments of 1 rad/s.

$$\mathbf{H}(j\omega) = \frac{j\omega + 2}{(j\omega + 6)(j\omega + 8)}$$

Frequency Response Using a Bode Plot

If the network characteristics are plotted on a log-log scale, they are known as *Bode plots*. In this case we plot $20 \log_{10} M(\omega)$ versus $\log_{10}(\omega)$ instead of $M(\omega)$ versus ω. The advantage of this technique is that rather than plotting the characteristic point by point, we can employ straight-line approximations to obtain the characteristic very efficiently. The ordinate for the magnitude plot is the decibel (dB). This unit was originally employed to measure the ratio of two powers; that is,

$$\text{number of dB} = 10 \log_{10} \frac{P_2}{P_1} \tag{13.11}$$

If the powers are absorbed by two equal resistors, then

$$\begin{aligned} \text{number of dB} &= 10 \log_{10} \frac{|\mathbf{V}_2|^2/R}{|\mathbf{V}_1|^2/R} = \frac{|\mathbf{I}_2|^2 R}{|\mathbf{I}_1|^2 R} \\ &= 20 \log_{10} \left|\frac{\mathbf{V}_2}{\mathbf{V}_1}\right| = 20 \log_{10} \left|\frac{\mathbf{I}_2}{\mathbf{I}_1}\right| \end{aligned} \tag{13.12}$$

The term "dB" has become so popular that it now is used for voltage and current ratios, as illustrated in Eq. (13.12), without regard to the impedance employed in each case.

In the sinusoidal steady-state case, $\mathbf{H}(j\omega)$ in Eq. (13.3) can be expressed in general as

$$\mathbf{H}(j\omega) = \frac{K_0(j\omega)^{\pm N}(1 + j\omega\tau_1)(1 + j\omega\tau_2)[1 + 2\zeta_3(j\omega\tau_3) + (j\omega\tau_3)^2] \cdots}{(1 + j\omega\tau_a)[1 + 2\zeta_b(j\omega\tau_b) + (j\omega\tau_b)^2] \cdots} \tag{13.13}$$

Note that this equation contains the following typical factors:

1. A frequency-independent factor, K_0.
2. Poles or zeros at the origin of the form $j\omega$, that is, $(j\omega)^{+N}$ for zeros and $(j\omega)^{-N}$ for poles.
3. Poles or zeros of the form $(1 + j\omega\tau)$.
4. Quadratic poles or zeros of the form $1 + 2\zeta(j\omega\tau) + (j\omega\tau)^2$.

Taking the logarithm of the magnitude of the function $\mathbf{H}(j\omega)$ in Eq. (13.13) yields

$$\begin{aligned} 20 \log_{10} |\mathbf{H}(j\omega)| &= 20 \log_{10} K_0 \pm 20N \log_{10} |j\omega| \\ &+ 20 \log_{10} |1 + j\omega\tau_1| \\ &+ 20 \log_{10} |1 + j\omega\tau_2| \\ &+ 20 \log_{10} |1 + 2\zeta_3(j\omega\tau_3) + (j\omega\tau_3)^2| \\ &+ \cdots - 20 \log_{10} |1 + j\omega\tau_a| - 20 \log_{10} |1 \\ &+ 2\zeta_b(j\omega\tau_b) + (j\omega\tau_b)^2| \cdots \end{aligned} \tag{13.14}$$

Note that we have used the fact that the log of the product of two or more terms is equal to the sum of the logs of the individual terms, the quotient of two terms is equal to the difference of the logs of the individual terms, and the fact that $\log_{10} A^n = n \log_{10} A$.

The phase angle for $\mathbf{H}(j\omega)$ is

$$\begin{aligned}\underline{/\mathbf{H}(j\omega)} &= 0 \pm N(90°) + \tan^{-1}\omega\tau_1 + \tan^{-1}\omega\tau_2 \\ &+ \tan^{-1}\left(\frac{2\zeta_3\omega\tau_3}{1-\omega^2\tau_3^2}\right) + \cdots \\ &- \tan^{-1}\omega\tau_a - \tan^{-1}\left(\frac{2\zeta_b\omega\tau_b}{1-\omega^2\tau_b^2}\right)\cdots\end{aligned} \tag{13.15}$$

As Eqs. (13.14) and (13.15) indicate, we will simply plot each factor individually on a common graph and then sum them algebraically to obtain the total characteristic.

The term $20\log_{10} K_0$ represents a constant magnitude with zero phase shift, as shown in Fig. 13.4a. Poles or zeros at the origin are of the form $(j\omega)^{\pm N}$, where + is used for a zero and − is used for a pole. The magnitude of this function is $\pm 20\ N\log_{10}\omega$, which is a straight line on log-log paper with a slope of $\pm 20\ N$ dB per decade; that is, the value will change by 20 each time the frequency is multiplied by 10, and the phase of this function is a constant $\pm N(90°)$. The magnitude and phase characteristics for poles and zeros at the origin are shown in Fig. 13.4b and c, respectively.

Linear approximations can be employed when a simple pole or zero of the form $(1 + j\omega\tau)$ is present in the network function. For $\omega\tau \ll 1$, $(1 + j\omega\tau) \approx 1$ and therefore $20\log_{10}|(1 + j\omega\tau)| = 20\log_{10} 1 = 0$ dB. Similarly, if $\omega\tau \gg 1$, then $(1 + j\omega\tau) \approx j\omega\tau$ and hence $20\log_{10}|(1 + j\omega\tau)| = 20\log_{10}\omega\tau$. Therefore, for $\omega\tau \ll 1$ the response is 0 dB and for $\omega\tau \gg 1$ the response has a slope that is the same as that of a simple pole or zero at the origin. The intersection of these two asymptotes, one for $\omega\tau \ll 1$ and one for $\omega\tau \gg 1$, is the point where $\omega\tau = 1$ or $\omega = 1/\tau$, which is called the *break frequency*. At this break frequency, where $\omega = 1/\tau$, $20\log_{10}|(1 + j1)| = 20\log_{10}(2)^{1/2} = 3$ dB. Therefore, the actual curve deviates from the asymptotes by 3 dB at the break frequency. It can be shown that at one-half and twice the break frequency, the deviations are 1 dB. The phase angle associated with a simple pole or zero is $\phi = \tan^{-1}\omega\tau$, which is a simple arc-tangent curve. Therefore, the phase shift is 45° at the break frequency and 26.5° and 63.5° at one-half and twice the break frequency, respectively. The actual magnitude curve for a pole of this form is shown in Fig. 13.5a. For a zero the magnitude curve and the asymptote for $\omega\tau \gg 1$ have a positive slope, and the phase curve extends from 0° to +90° as shown in Fig. 13.5b. If multiple poles or zeros of the form $(1 + j\omega\tau)^N$ are present, then the slope of the high frequency asymptote is multiplied by N, the deviation between the actual curve and the asymptote at the break frequency is $3N$ dB, and the phase curve extends from 0 to $N(90°)$ and is $N(45°)$ at the break frequency.

Quadratic poles or zeros are of the form $1 + 2\zeta(j\omega\tau) + (j\omega\tau)^2$. This term is a function not only of ω, but the dimensionless term ζ, which is called the *damping ratio*. If $\zeta > 1$ or $\zeta = 1$, the roots are real and unequal or real and equal, respectively, and these two cases have already been addressed. If $\zeta < 1$, the roots are complex conjugates, and it is this case which we will examine now. Following the argument above for a simple pole or zero, the log magnitude of the quadratic factor is 0 dB for $\omega\tau \ll 1$. For $\omega\tau \gg 1$,

$$20\log_{10}\left|1 - (\omega\tau)^2 + 2j\zeta(\omega\tau)\right| \approx 20\log_{10}\left|(\omega\tau)^2\right| = 40\log_{10}|\omega\tau|$$

and therefore, for $\omega\tau \gg 1$, the slope of the log magnitude curve is +40 dB/decade for

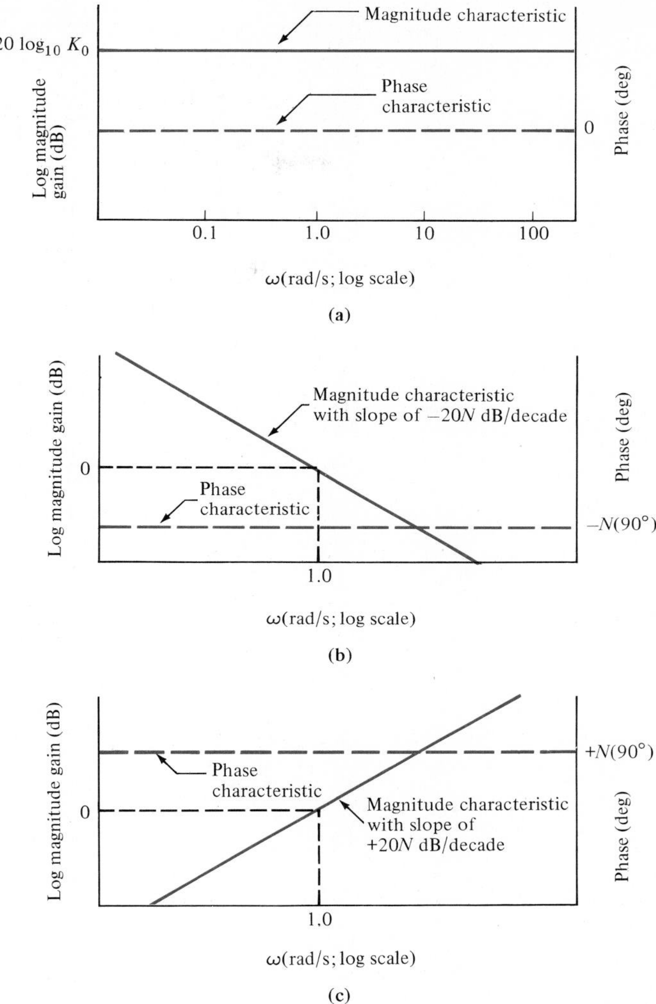

Figure 13.4 Magnitude and phase characteristics for a constant term and poles and zeros at the origin.

a quadratic zero and -40 dB/decade for a quadratic pole. Between the two extremes, $\omega\tau \ll 1$ and $\omega\tau \gg 1$, the behavior of the function is dependent on the damping ratio ζ. Figure 13.6a illustrates the manner in which the log magnitude curve for a quadratic *pole* changes as a function of the damping ratio. The phase shift for the quadratic factor is $\tan^{-1} 2\zeta\omega\tau/[1 - (\omega\tau)^2]$. The phase plot for quadratic *poles* is shown in Fig. 13.6b. Note that in this case the phase changes from 0° at frequencies for which $\omega\tau \ll 1$ to $-180°$ at frequencies for which $\omega\tau \gg 1$. For quadratic zeros the magnitude and phase curves are inverted; that is, the log magnitude curve has a slope of $+40$ dB/decade for $\omega\tau \gg 1$, and the phase curve is 0° for $\omega\tau \ll 1$ and $+180°$ for $\omega\tau \gg 1$.

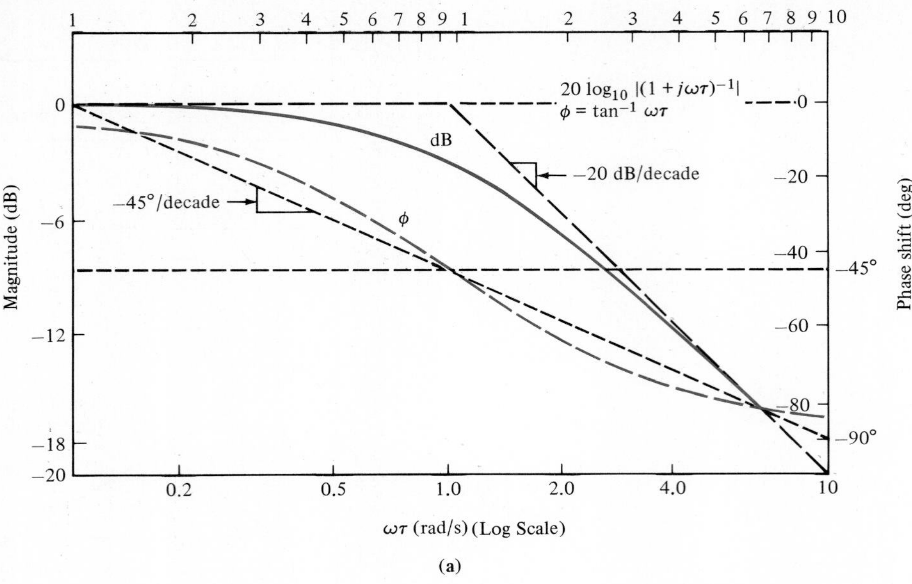

(a)

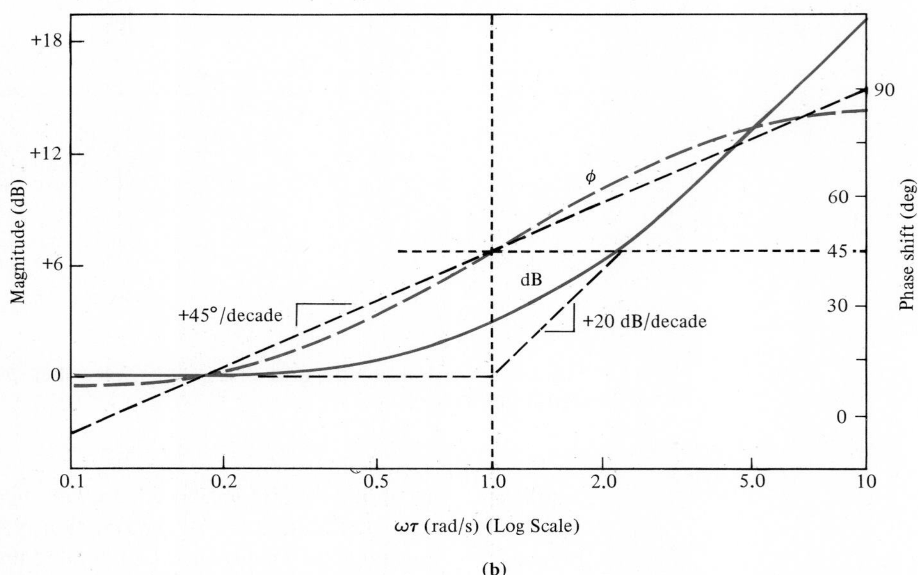

(b)

Figure 13.5 Magnitude and phase plot (a) for a simple pole, and (b) for a simple zero.

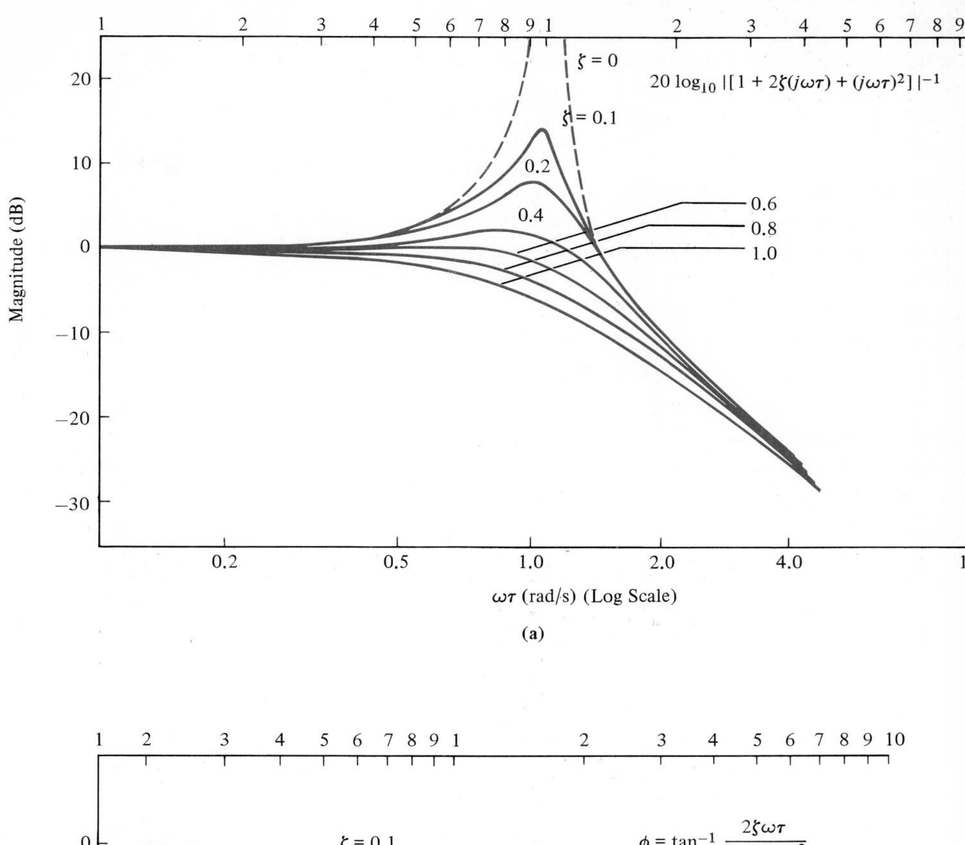

(a)

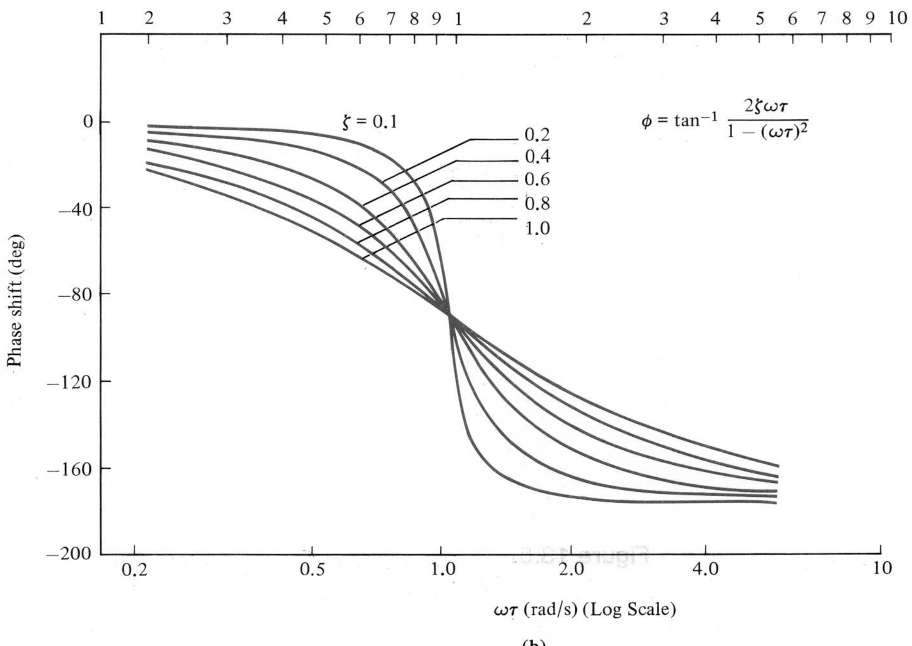

(b)

Figure 13.6 Magnitude and phase characteristics for quadratic poles.

EXAMPLE 13.3

We want to generate the magnitude and phase plots for the transfer function

$$\mathbf{G_V}(j\omega) = \frac{10(0.1j\omega + 1)}{(j\omega + 1)(0.02j\omega + 1)}$$

Note that this function is in standard form since every term is of the form $(j\omega\tau + 1)$. In order to determine the composite magnitude and phase characteristics, we will plot the individual terms and then add them as specified in Eqs. (13.14) and (13.15). Let us consider the magnitude plot first. Since $K_0 = 10$, $20 \log_{10} 10 = 20$ dB, which is a constant independent of frequency, as shown in Fig. 13.7a. The zero of the transfer function contributes a term of the form $+20 \log_{10} |1 + 0.1j\omega|$, which is 0 dB for $0.1\omega \ll 1$, has a slope of $+20$ dB/decade for $0.1\omega \gg 1$, and has a break frequency at $\omega = 10$ rad/s. The poles have break frequencies at $\omega = 1$ and $\omega = 50$ rad/s. The pole with break frequency at $\omega = 1$ rad/s contributes a term of the form $-20 \log_{10} |1 + j\omega|$, which is 0 dB for $\omega \ll 1$ and has a slope of -20 dB/decade for $\omega \gg 1$. A similar argument can be made for the pole that has a break frequency at $\omega = 50$ rad/s. These factors are all plotted individually in Fig. 13.7a.

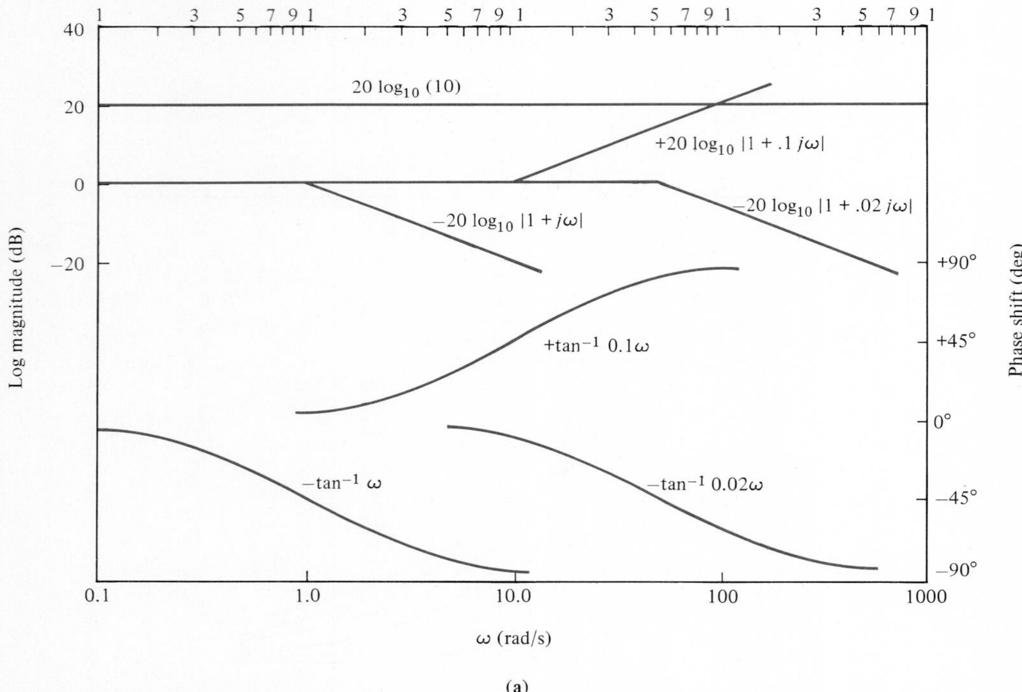

(a)

Figure 13.7a Magnitude and phase components for the poles and zeros of the transfer function in Example 13.3.

Consider now the individual phase curves. The term K_0 is not a function of ω and does not contribute to the phase of the transfer function. The phase curve for the zero is $+\tan^{-1} 0.1\,\omega$, which is an arctangent curve that extends from 0° for $0.1\,\omega \ll 1$ to $+90°$ for $0.1\omega \gg 1$ and has a phase of $+45°$ at the break frequency. The phase curve for the two poles are $-\tan^{-1} \omega$ and $-\tan^{-1} 0.02\omega$. The term $-\tan^{-1} \omega$ is 0° for $\omega \ll 1$, $-90°$ for $\omega \gg 1$ and $-45°$ at the break frequency $\omega = 1$. The phase curve for the remaining pole is plotted in a similar fashion. All the individual phase curves are shown in Fig. 13.7a.

As specified in Eqs. (13.14) and (13.15), the composite magnitude and phase of the transfer function are obtained simply by adding the individual terms. The composite curves are plotted in Fig. 13.7b. Note that the actual magnitude curve (solid line) differs from the straight-line approximation (dashed line) by 3 dB at the break frequencies and 1 dB at one-half and twice the break frequencies. ■

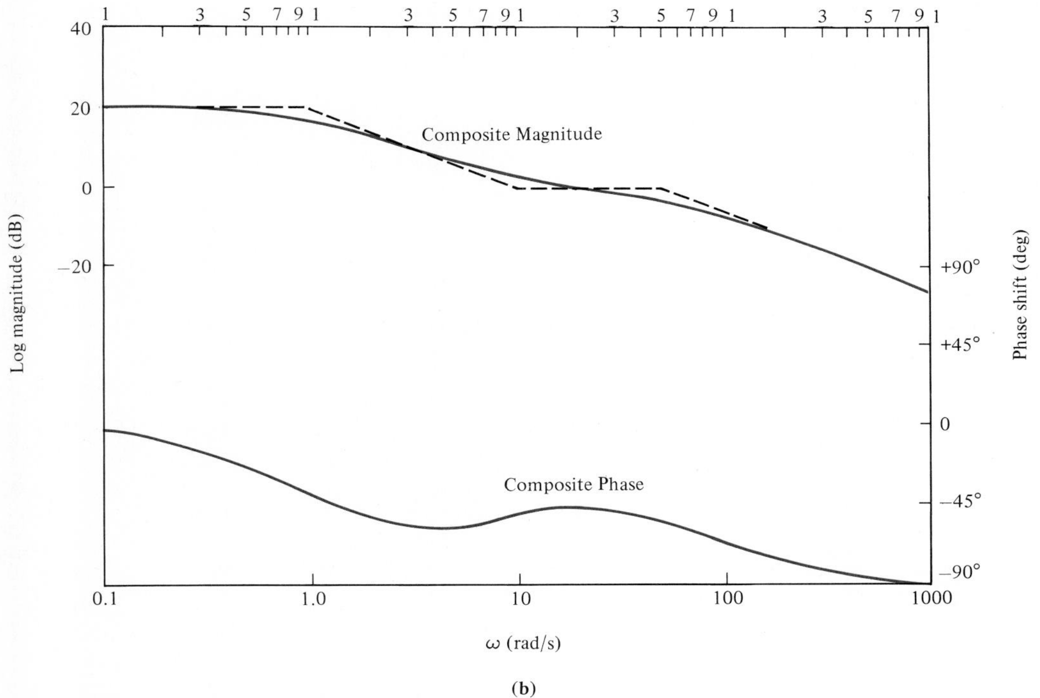

(b)

Figure 13.7b Bode plot for the transfer function in Example 13.3.

EXAMPLE 13.4

Let us draw the Bode plot for the following transfer function:

$$\mathbf{G_V}(j\omega) = \frac{25(j\omega + 1)}{(j\omega)^2(0.1j\omega + 1)}$$

Once again all the individual terms for both magnitude and phase are plotted in Fig. 13.8a. The straight line with a slope of -40 dB/decade is generated by the double pole at the origin. This line is a plot of $-40 \log_{10} \omega$ versus ω and therefore passes through 0 dB at $\omega = 1$ rad/s. The phase for the double pole is a constant $-180°$ for all frequencies. The remainder of the terms are plotted as illustrated in Example 13.3.

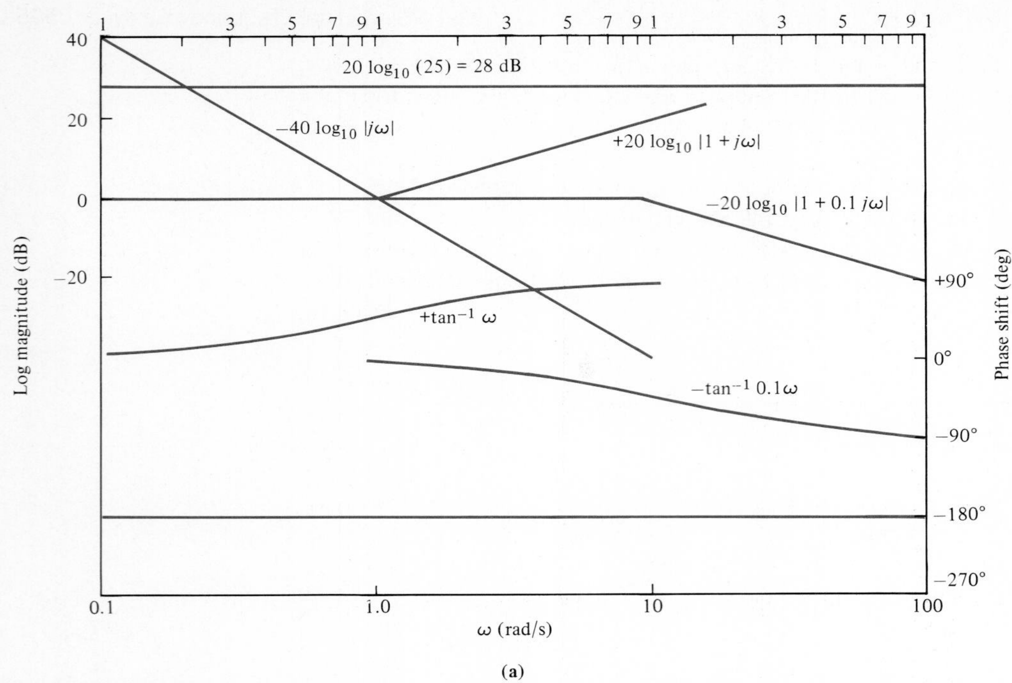

Figure 13.8a Magnitude and phase components for the poles and zeros of the transfer function in Example 13.4.

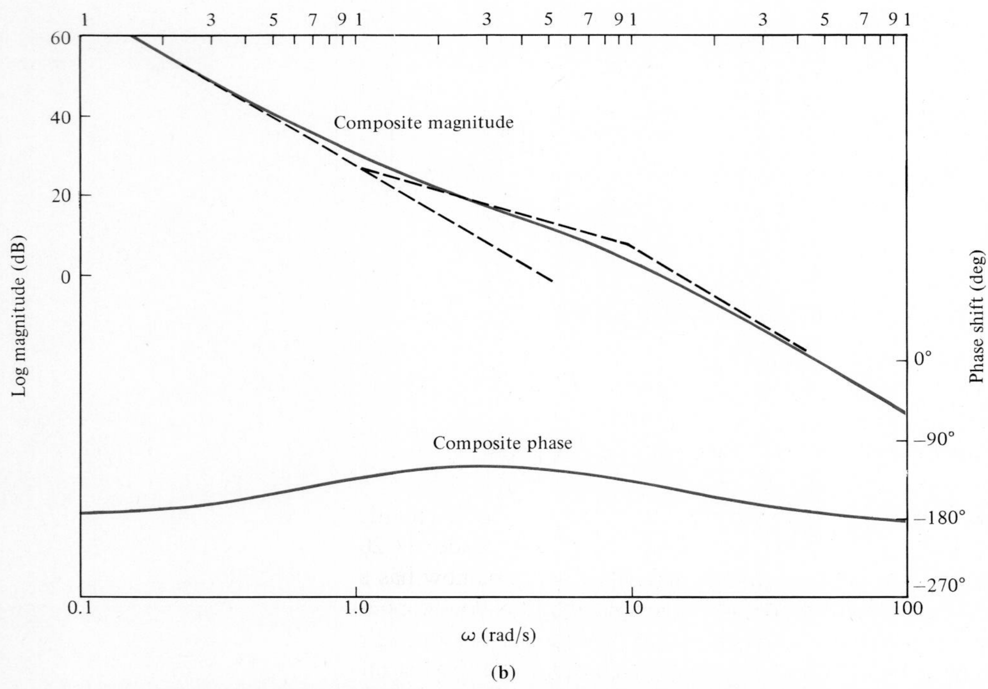

Figure 13.8b Bode plot for the transfer function in Example 13.4.

The composite plots are shown in Fig. 13.8b. Once again they are obtained simply by adding the individual terms in Fig. 13.8a. Note that for frequencies for which $\omega \ll 1$, the slope of the magnitude curve is -40 dB/decade. At $\omega = 1$ rad/s, which is the break frequency of the zero, the magnitude curve changes slope to -20 dB/decade. At $\omega = 10$ rad/s, which is the break frequency of the pole, the slope of the magnitude curve changes back to -40 dB/decade.

The composite phase curve starts at $-180°$ due to the double pole at the origin. Since the first break frequency encountered is a zero, the phase curve shifts toward $-90°$. However, before the composite phase reaches $-90°$, the pole with break frequency $\omega = 10$ rad/s begins to shift the composite curve back toward $-180°$. ■

Example 13.4 illustrates the manner in which to plot directly terms of the form $K_0/(j\omega)^N$. For terms of this form, the initial slope of $-20N$ dB/decade will intersect the 0-dB axis at a frequency of $(K_0)^{1/N}$ rad/s; that is, $-20 \log_{10} |K_0/(j\omega)^N| = 0$ dB implies that $K_0/(\omega)^N = 1$, and therefore $\omega = (K_0)^{1/N}$ rad/s. Note that the projected slope of the magnitude curve in Example 13.4 intersects the 0-dB axis at $\omega = (25)^{1/2} = 5$ rad/s.

Similarly, it can be shown that for terms of the form $K_0(j\omega)^N$, the initial slope of $+20N$ dB/decade will intersect the 0-dB axis at a frequency of $\omega = (1/K_0)^{1/N}$ rad/s; that is, $+20 \log_{10} |K_0 j\omega)^N| = 0$ dB implies that $K_0(\omega)^N = 1$, and therefore $\omega = (1/K_0)^{1/N}$ rad/s.

By applying the concepts we have just demonstrated, we can normally plot the log magnitude characteristic of a transfer function directly in one step.

EXAMPLE 13.5

We wish to generate the Bode plot for the following transfer function:

$$\mathbf{G}_\mathbf{V}(j\omega) = \frac{100(j\omega + 1)}{j\omega(j\omega + 10)}$$

This function can be written in standard form as

$$\mathbf{G}_\mathbf{V}(j\omega) = \frac{10(j\omega + 1)}{j\omega(0.1j\omega + 1)}$$

The Bode plot for $\mathbf{G}_\mathbf{V}(j\omega)$ is shown in Fig. 13.9. The magnitude curve can be plotted in one step as follows. Since the transfer function has a frequency-independent gain factor of 10 and a pole at the origin, the composite magnitude curve will have an initial slope of -20 dB/decade, which intersects the 0-dB axis at $\omega = 10$ rad/s. The initial slope of -20 dB/decade changes at the first break frequency, which, in this case, is a zero with break frequency $\omega = 1$ rad/s. Since the magnitude characteristic for the zero is 0 dB for $\omega \ll 1$ rad/s and $+20$ dB/decade for $\omega \gg 1$ rad/s, the composite magnitude curve changes slope from -20 dB/decade to 0 dB/decade (-20 dB/decade + 20 dB/decade) at $\omega = 1$ rad/s. The composite magnitude characteristic, which now has a slope of 0 dB/decade for $\omega > 1$ rad/s, continues until the next break frequency is encountered, which in this case is a pole with break frequency $\omega = 10$ rad/s. Since the magnitude characteristic for the pole is 0 dB/decade for $\omega < 10$ rad/s and -20 dB/decade for $\omega > 10$ rad/s, the composite magnitude characteristic changes slope from 0 dB/decade

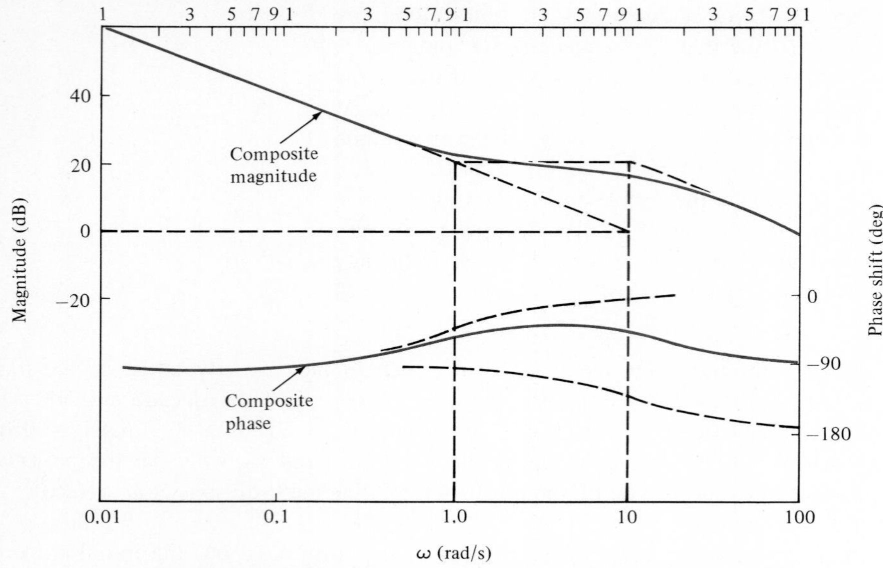

Figure 13.9 Bode plot for the function in Example 13.5.

to -20 dB/decade (0 dB/decade $-$ 20 dB/decade) at $\omega = 10$ rad/s. This final slope of -20 dB/decade continues for all values of $\omega > 10$ rad/s. Note that the composite magnitude curve in Fig. 13.9 has been adjusted for the 3-dB difference between the actual curve and the straight-line asymptotes at the break frequencies. At low frequencies the phase due to the pole at the origin is $-90°$. This curve, together with the two arctangent curves, one for the zero with break frequency $\omega = 1$ rad/s, and one for the pole with a break frequency of $\omega = 10$ rad/s, are summed to yield the commposite phase curve as shown in Fig. 13.9. ■

DRILL EXERCISE

D13.4. *Sketch* the magnitude characteristic of the Bode plot, labeling all critical slopes and points for the function

$$\mathbf{G}(j\omega) = \frac{10^4(j\omega + 2)}{(j\omega + 10)(j\omega + 100)}$$

D13.5. *Sketch* the magnitude characteristic of the Bode plot, labeling all critical slopes and points for the function

$$\mathbf{G}(j\omega) = \frac{10}{j\omega(0.05j\omega + 1)}$$

D13.6. *Sketch* the magnitude characteristic of the Bode plot, labeling all critical slopes and points for the function

$$\mathbf{G}(j\omega) = \frac{100(0.02j\omega + 1)}{(j\omega)^2}$$

D13.7. *Sketch* the magnitude characteristic of the Bode plot, labeling all critical slopes and points for the function

$$\mathbf{G}(j\omega) = \frac{10j\omega}{(j\omega + 1)(j\omega + 10)}$$

EXAMPLE 13.6

We wish to generate the Bode plot for the following transfer function:

$$\mathbf{G_V}(j\omega) = \frac{25j\omega}{(j\omega + 0.5)[(j\omega)^2 + 4j\omega + 100]}$$

Expressing this function in standard form, we obtain

$$\mathbf{G_V}(j\omega) = \frac{0.5j\omega}{(2j\omega + 1)[(j\omega/10)^2 + j\omega/25 + 1]}$$

The Bode plot is shown in Fig. 13.10. The initial low-frequency slope due to the zero at the origin is +20 dB/decade, and this slope intersects the 0-dB line at $\omega = \dfrac{1}{K_0} =$ 2 rad/s. At $\omega = 0.5$ rad/s the slope changes from +20 dB/decade to 0 dB/decade due to the presence of the pole with a break frequency at $\omega = 0.5$ rad/s. The quadratic term has a center frequency of $\omega = 10$ rad/s. (i.e., $\tau = 1/10$). Since

$$2\zeta\tau = \frac{1}{25}$$

and

$$\tau = 0.1$$

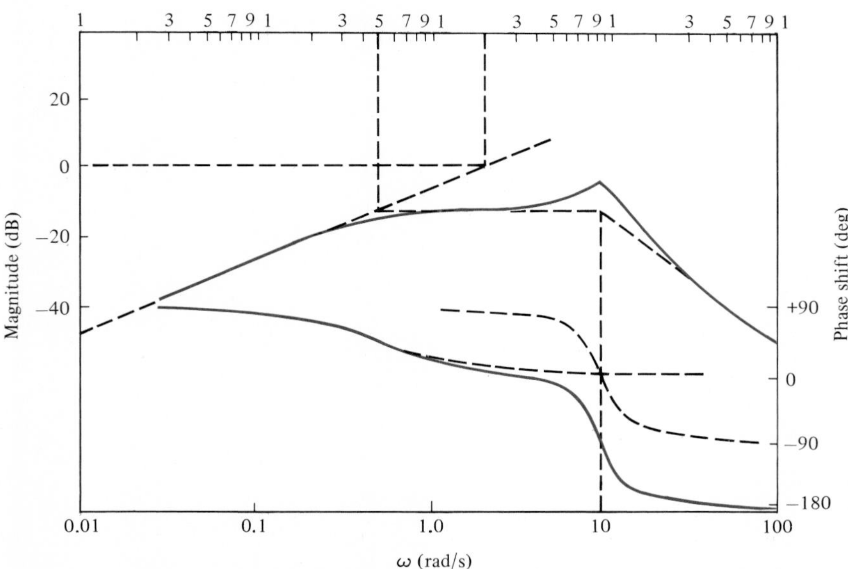

Figure 13.10 Bode plot for the transfer function in Example 13.6.

then

$$\zeta = 0.2$$

Plotting the curve in Fig. 13.6a with a damping ratio of $\zeta = 0.2$ at the center frequency $\omega = 10$ rad/s completes the composite magnitude curve for the transfer function.

The initial low-frequency phase curve is $+90°$, due to the zero at the origin. This curve, together with the phase curve for the simple pole and the phase curve for the quadratic term, as defined in Fig. 13.6b, are combined to yield the composite phase curve. ■

DRILL EXERCISE

D13.8. Given the following function $\mathbf{G}(j\omega)$, sketch the magnitude characteristic of the Bode plot, labeling all critical slopes and points.

$$\mathbf{G}(j\omega) = \frac{0.2(j\omega + 1)}{j\omega[(j\omega/12)^2 + j\omega/36 + 1]}$$

EXAMPLE 13.7

Given the straight-line magnitude characteristic shown in Fig. 13.11, we wish to determine the transfer function $\mathbf{G_V}(j\omega)$.

Since the initial slope is 0 dB/decade, and the level of the characteristic is 20 dB, the factor K_0 can be obtained from the expression

$$20 \text{ dB} = 20 \log_{10} K_0$$

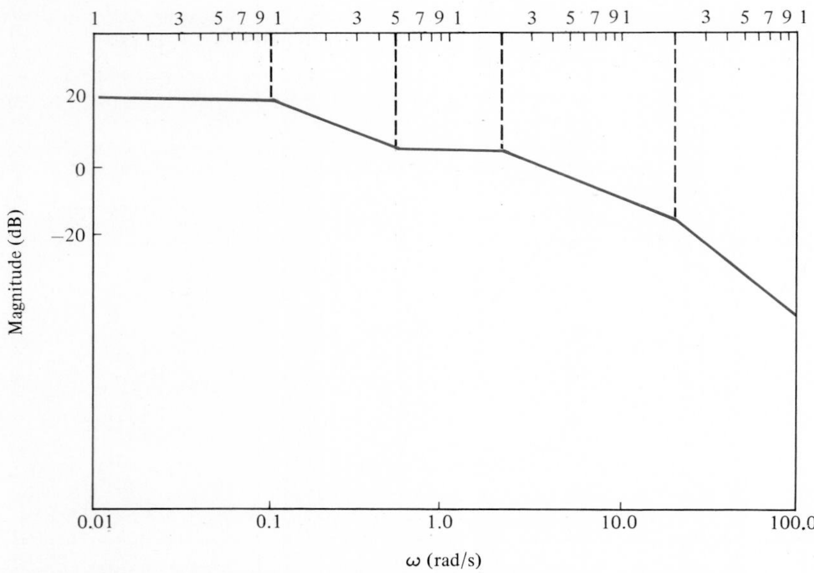

Figure 13.11 Straight-line magnitude plot employed in Example 13.7.

and hence

$$K_0 = 10$$

The -20-dB/decade slope starting at $\omega = 0.1$ rad/s indicates that the first pole has a break frequency at $\omega = 0.1$ rad/s, and therefore one of the factors in the denominator is $(10j\omega + 1)$. The slope changes by $+20$ dB/decade at $\omega = 0.5$ rad/s, indicating that there is a zero present with a break frequency at $\omega = 0.5$ rad/s, and therefore the numerator has a factor of $(2j\omega + 1)$. Two additional poles are present with break frequencies at $\omega = 2$ rad/s and $\omega = 20$ rad/s. Therefore, the composite transfer function is

$$\mathbf{G_V}(j\omega) = \frac{10(2j\omega + 1)}{(10j\omega + 1)(0.5j\omega + 1)(0.05j\omega + 1)}$$

The reader should note carefully the ramifications this example has with regard to network design. ■

DRILL EXERCISE

D13.9. Find the transfer function $\mathbf{G}(j\omega)$ if the straight-line magnitude characteristic approximation for this function is as shown in Fig. D13.9.

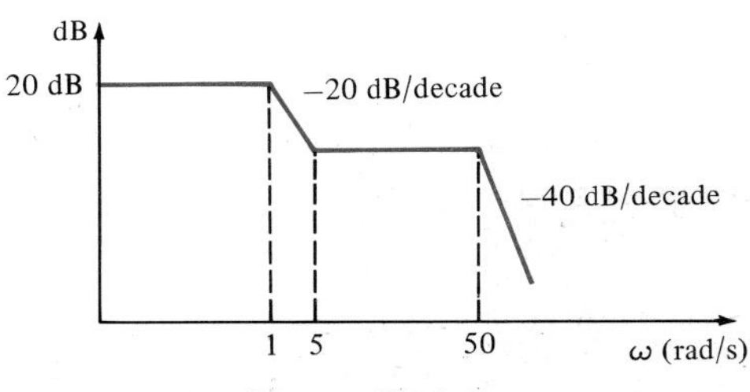

Figure D13.9

D13.10. Determine the transfer function $\mathbf{G}(j\omega)$ if the straight-line magnitude characteristic approximation for this function is as shown in Fig. D13.10.

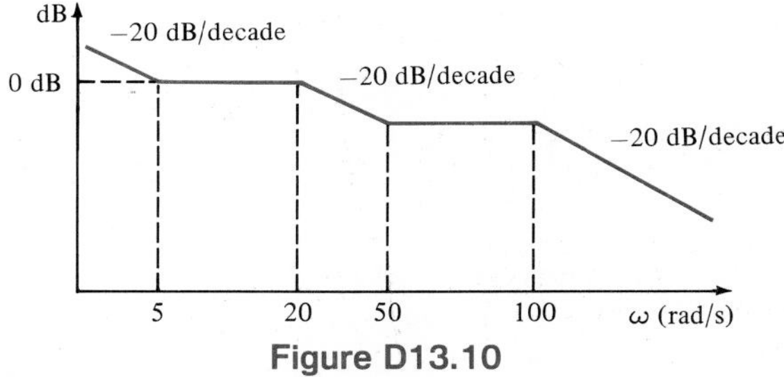

Figure D13.10

13.3 Filter Networks

Passive Filters

A filter network is generally designed to pass signals with a specific frequency range and reject or attenuate signals whose frequency spectrum is outside this passband. The most common filters are *low-pass* filters, which pass low frequencies and reject high frequencies; *high-pass* filters, which pass high frequencies and block low frequencies; *bandpass* filters, which pass some particular band of frequencies and reject all frequencies outside the range; and *band-rejection* filters, which are specifically designed to reject a particular band of frequencies and pass all other frequencies.

The ideal frequency characteristic for a low-pass filter is shown in Fig. 13.12a. Also shown is a typical or physically realizable characteristic. Ideally, we would like the low-pass filter to pass all frequencies to some frequency ω_0 and pass no frequency above that value; however, it is not possible to design such a filter with linear circuit elements. Hence we must be content to employ filters that we can actually build in the laboratory, and these filters have frequency characteristics that are simply not ideal.

A simple low-pass filter network is shown in Fig. 13.12b. The voltage gain for the network is

$$\mathbf{G}_\mathbf{V}(j\omega) = \frac{1}{1 + j\omega RC} \tag{13.16}$$

which can be written as

$$\mathbf{G}_\mathbf{V}(j\omega) = \frac{1}{1 + j\omega\tau} \tag{13.17}$$

where $\tau = RC$, the time constant. The amplitude characteristic is

$$M(\omega) = \frac{1}{[1 + (\omega\tau)^2]^{1/2}} \tag{13.18}$$

and the phase characteristic is

$$\phi(\omega) = -\tan^{-1}\omega\tau \tag{13.19}$$

Note that at the break frequency $\omega = 1/\tau$ and the amplitude is

$$M(\omega = 1/\tau) = \frac{1}{\sqrt{2}} \tag{13.20}$$

The break frequency is also commonly called the *half-power frequency*. This name is derived from the fact that if the voltage or current is $1/\sqrt{2}$ of its maximum value, then the power, which is proportional to the square of the voltage or current, is one-half its maximum value.

The magnitude and phase curves for this simple low-pass circuit are shown in Fig. 13.12c. Note that the magnitude curve is flat for low frequencies and rolls off at high frequencies. The phase shifts from 0° at low frequencies to −90° at high frequencies.

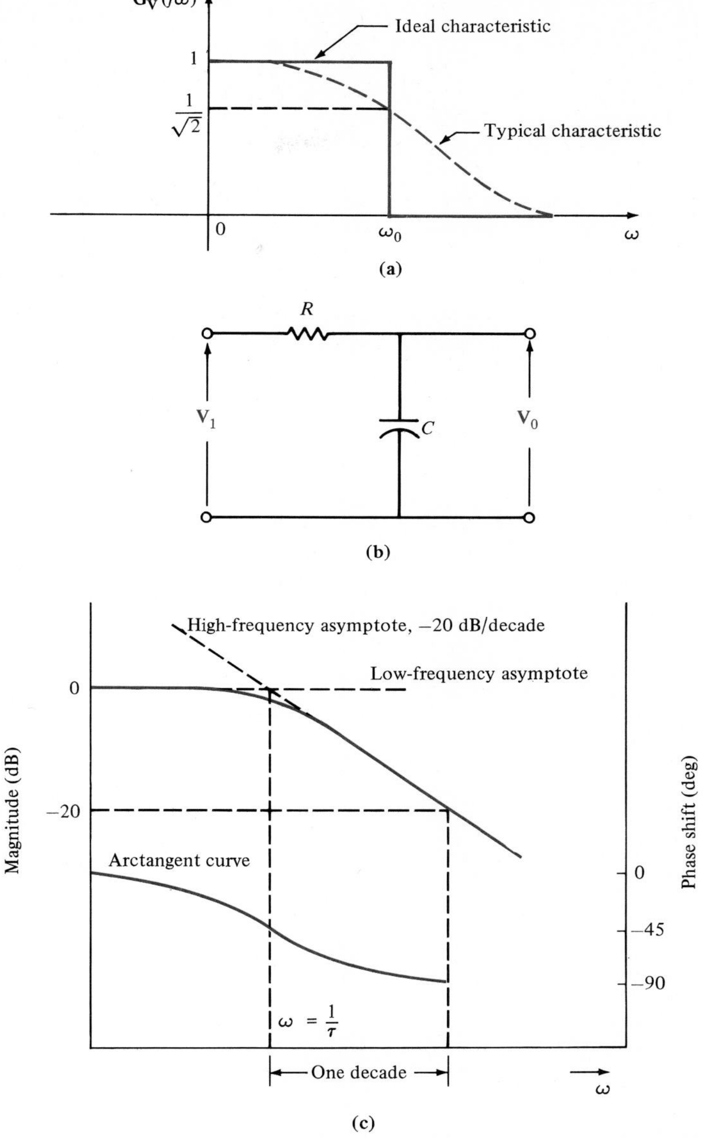

Figure 13.12 Low-pass filter circuit and its frequency characteristics.

The ideal frequency characteristic for a high-pass filter is shown in Fig. 13.13a, together with a typical characteristic that we could achieve with linear circuit components. Ideally, the high-pass filter passes all frequencies above some frequency ω_0 and no frequencies below that value.

A simple high-pass filter network is shown in Fig. 13.13b. This is the same network as shown in Fig. 13.12b except that the output voltage is taken across the resistor. The

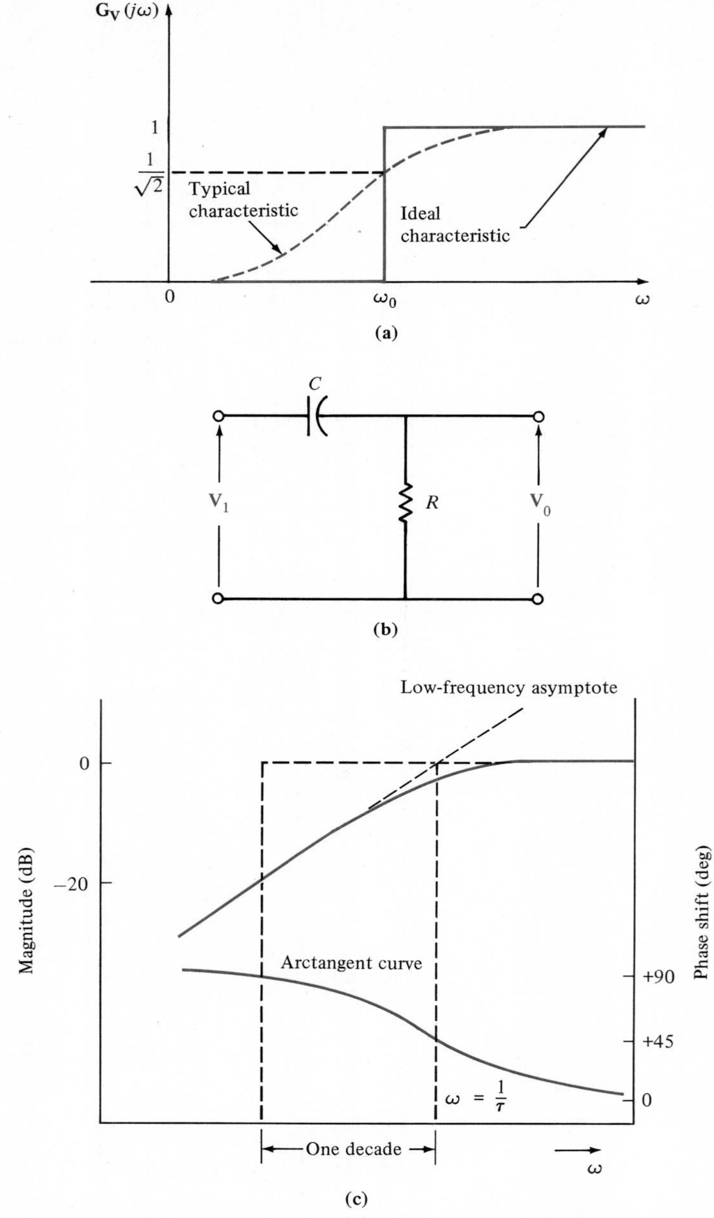

Figure 13.13 High-pass filter circuit and frequency characteristics.

voltage gain for this network is

$$\mathbf{G_V}(j\omega) = \frac{j\omega\tau}{1 + j\omega\tau} \tag{13.21}$$

where once again $\tau = RC$. The magnitude of this function is

$$M(\omega) = \frac{\omega\tau}{[1 + (\omega\tau)^2]^{1/2}} \tag{13.22}$$

and the phase is

$$\phi(\omega) = \frac{\pi}{2} - \tan^{-1} \omega\tau \tag{13.23}$$

The half-power frequency is $\omega = 1/\tau$, and the phase at this frequency is 45°.

The magnitude and phase curves for this high-pass filter are shown in Fig. 13.13c. At low frequencies the magnitude curve has a slope of +20 dB/decade due to the term $\omega\tau$ in the numerator of Eq. (13.22). Then at the break frequency the curve begins to flatten out. The phase curve is derived from Eq. (13.23).

Ideal and typical amplitude characteristics for simple bandpass and band rejection filters are shown in Fig. 13.14a and b, respectively. Simple networks that are capable of realizing the typical characteristics of each filter are shown below the characteristics in Fig. 13.14c and d. ω_0 is the center frequency of the pass or rejection band and the

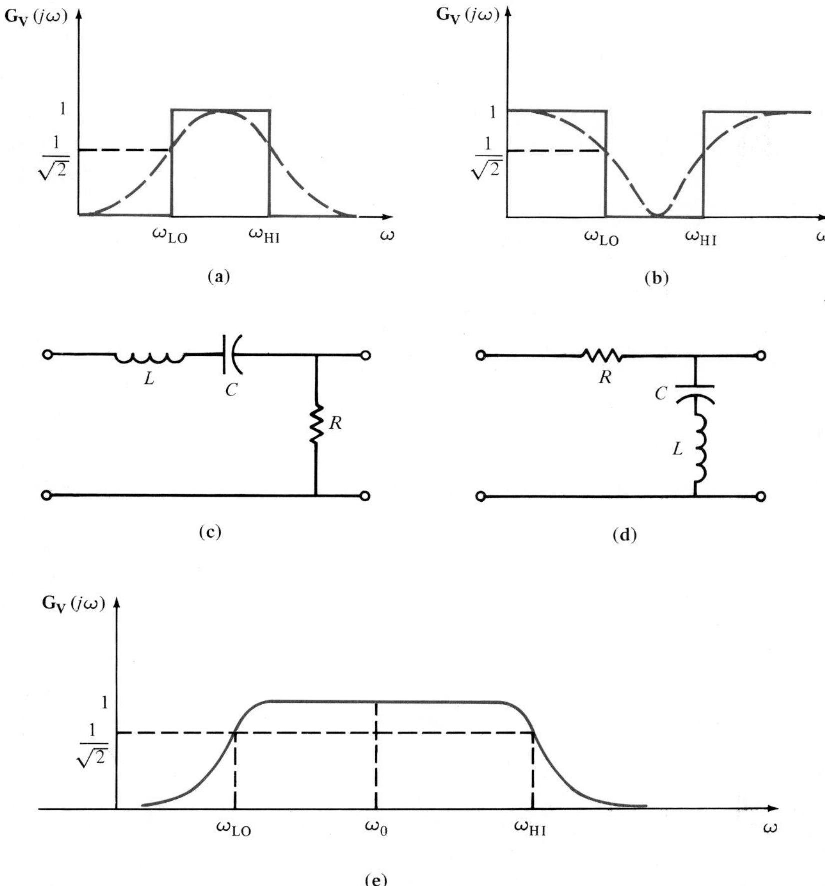

Figure 13.14 Bandpass and band-rejection filters and characteristics.

frequency at which the maximum or minimum amplitude occurs. ω_{LO} and ω_{HI} are the lower and upper break frequencies or *cutoff frequencies*, where the amplitude is $1/\sqrt{2}$ of the maximum value. The width of the pass or rejection band is called the *bandwidth*, and hence

$$BW = \omega_{HI} - \omega_{LO} \tag{13.24}$$

To illustrate these points, let us consider the bandpass filter. The voltage transfer function is

$$\mathbf{G_V}(j\omega) = \frac{R}{R + j(\omega L - 1/\omega C)}$$

and therefore the amplitude characteristic is

$$M(\omega) = \frac{RC\omega}{\sqrt{(RC\omega)^2 + (\omega^2 LC - 1)^2}}$$

At low frequencies

$$M(\omega) \approx \frac{RC\omega}{1} \approx 0$$

At high frequencies

$$M(\omega) \simeq \frac{RC\omega}{\omega^2 LC} \approx \frac{R}{\omega L} \approx 0$$

In the mid frequency range $(RC\omega)^2 \gg (\omega^2 LC - 1)^2$, and thus $M(\omega) \approx 1$. Therefore, the frequency characteristic for this filter is shown in Fig. 13.14e. The center frequency is $\omega_0 = 1/\sqrt{LC}$. At the lower cutoff frequency

$$\omega^2 LC - 1 = -RC\omega$$

or

$$\omega^2 + \frac{R\omega}{L} - \omega_0^2 = 0$$

Solving this expression for ω_{LO}, we obtain

$$\omega_{LO} = \frac{-(R/L) \pm \sqrt{(R/L)^2 + 4\omega_0^2}}{2}$$

At the upper cutoff frequency

$$\omega^2 LC - 1 = +RC\omega$$

or

$$\omega^2 - \frac{R}{L}\omega - \omega_0^2 = 0$$

Solving this expression for ω_{HI}, we obtain

$$\omega_{HI} = \frac{+(R/L) \pm \sqrt{(R/L)^2 + 4\omega_0^2}}{2}$$

Therefore, the bandwidth of the filter is

$$BW = \omega_{HI} - \omega_{LO} = \frac{R}{L}$$

DRILL EXERCISE

D13.11. Given the filter network shown in Fig. D13.11, sketch the magnitude characteristic of the Bode plot for $\mathbf{G_V}(j\omega)$.

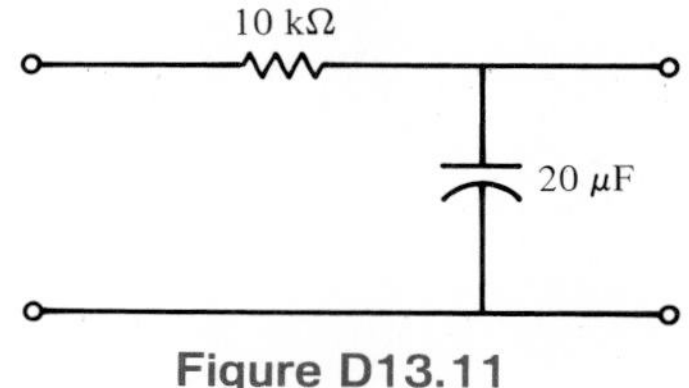

Figure D13.11

D13.12. Given the filter network in Fig. D13.12, sketch the magnitude characteristic of the Bode plot for $\mathbf{G_V}(j\omega)$.

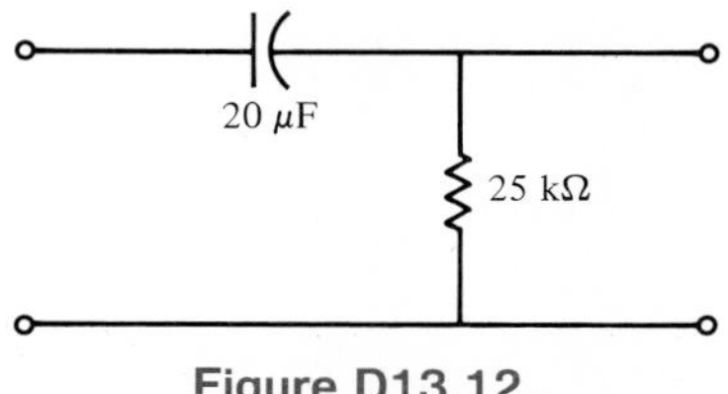

Figure D13.12

D13.13. A bandpass filter network is shown in Fig. D13.13. Sketch the magnitude characteristic of the Bode plot for $\mathbf{G_V}(j\omega)$.

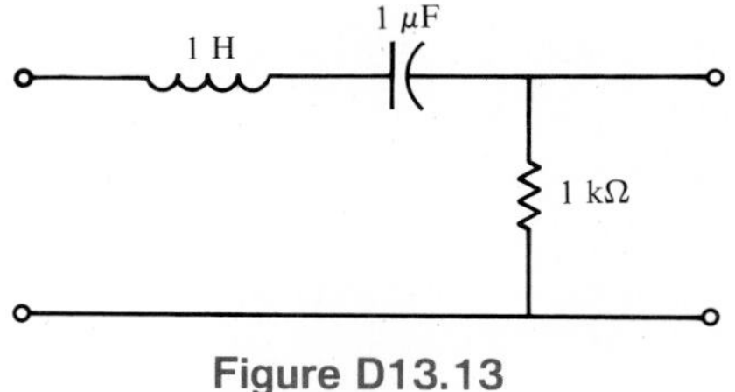

Figure D13.13

Active Filters

The equivalent circuits for the operational amplifiers derived in Chapter 2 are valid in the sinusoidal steady-state case also when we replace the attendant resistors with impedances. The equivalent circuits for the basic inverting and noninverting op amp circuits

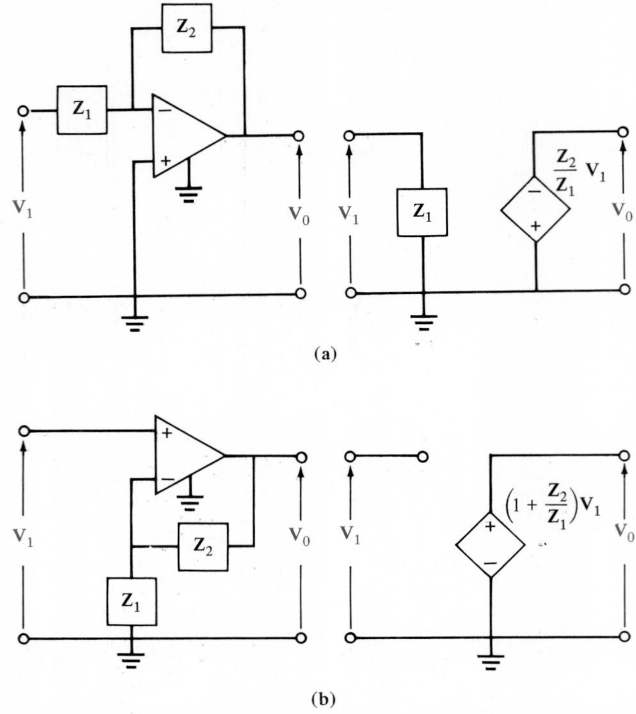

Figure 13.15 Equivalent circuits for the [a] inverting and [b] noninverting operational amplifier circuits.

are shown in Fig. 13.15a and b, respectively. Particular filter characteristics are obtained by judiciously selecting the impedances $\mathbf{Z}_1$ and $\mathbf{Z}_2$.

EXAMPLE 13.8

Let us determine the filter characteristics of the network shown in Fig. 13.16.

The impedances as illustrated in Fig. 13.15a are

$$\mathbf{Z}_1 = R_1$$

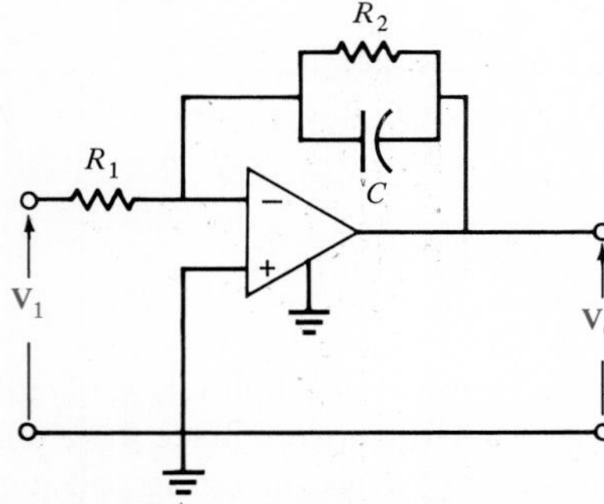

Figure 13.16 Operational amplifier filter circuit.

and

$$\mathbf{Z}_2 = \frac{R_2/j\omega C}{R_2 + 1/j\omega C} = \frac{R_2}{j\omega R_2 C + 1}$$

Therefore, the voltage gain of the network is

$$\mathbf{G_V}(j\omega) = \frac{-R_2/R_1}{j\omega R_2 C + 1}$$

Note that the transfer function is that of a low-pass filter. ■

EXAMPLE 13.9

We will show that the amplitude characteristic for the filter network in Fig. 13.17a is shown in Fig. 13.17b.

Comparing this network with that in Fig. 13.15b, we see that

$$\mathbf{Z}_1 = \frac{1}{j\omega C_1}$$

and

$$\mathbf{Z}_2 = \frac{R}{j\omega R C_2 + 1}$$

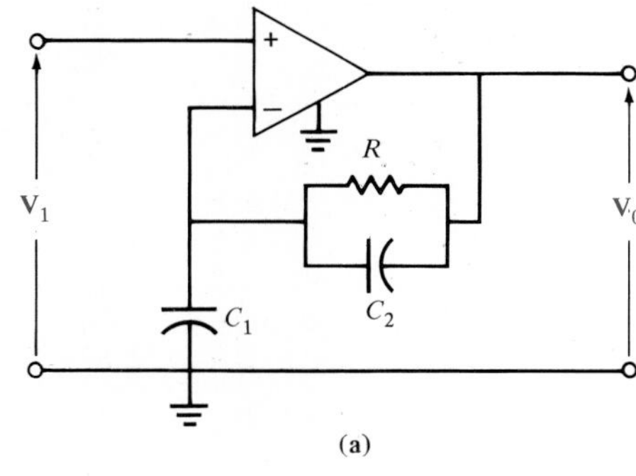

(a)

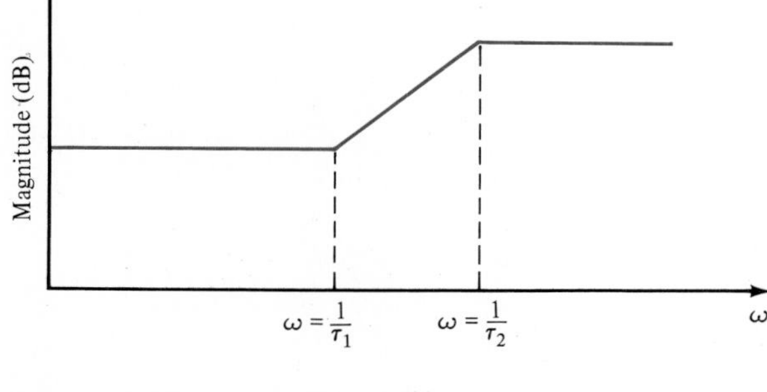

(b)

Figure 13.17 Operational amplifier circuit and its amplitude characteristic.

Therefore, the voltage gain for the network as a function of frequency is

$$\mathbf{G_V}(j\omega) = 1 + \frac{R/(j\omega RC_2 + 1)}{1/j\omega C_1}$$

$$= \frac{j\omega(RC_1 + RC_2) + 1}{j\omega RC_2 + 1}$$

$$= \frac{j\omega\tau_1 + 1}{j\omega\tau_2 + 1}$$

where $\tau_1 = R(C_1 + C_2)$ and $\tau_2 = RC_2$. Since $\tau_1 > \tau_2$, the amplitude characteristic is of the form shown in Fig. 13.17b. Note that the low frequencies have a gain of 1; however, the high frequencies are amplified. The exact amount of amplification is determined through selection of the circuit parameters. ■

DRILL EXERCISE

D13.14. Given the filter network shown in Fig. D13.14, determine the transfer function $\mathbf{G_V}(j\omega)$, sketch the magnitude characteristic of the Bode plot for $\mathbf{G_V}(j\omega)$, and identify the filter chracteristics of the network.

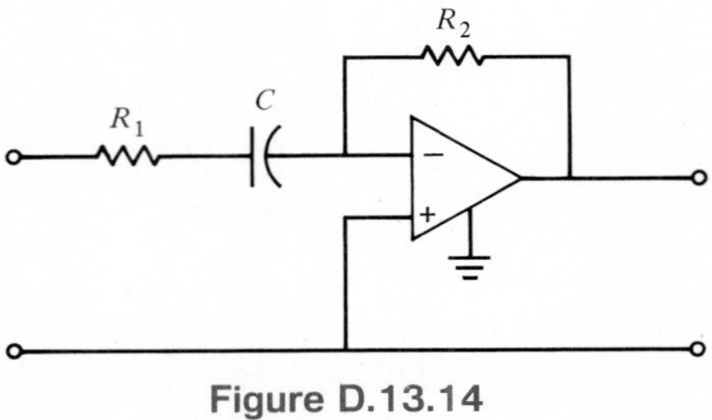

Figure D.13.14

The preceding examples indicate that if we wish to design a filter network, we simply select the resistor and capacitor values to provide the proper characteristics. It is then possible to adjust these values using frequency and magnitude scaling in order to obtain element values that are typical off-the-shelf items.

13.4 Resonant Circuits

Two circuits with extremely important frequency characteristics are shown in Fig. 13.18. The input impedance for the series *RLC* circuit is

$$Z(j\omega) = R + j\omega L + \frac{1}{j\omega C} \tag{13.25}$$

and the input admittance for the parallel *RLC* circuit is

$$\mathbf{Y}(j\omega) = G + j\omega C + \frac{1}{j\omega L} \tag{13.26}$$

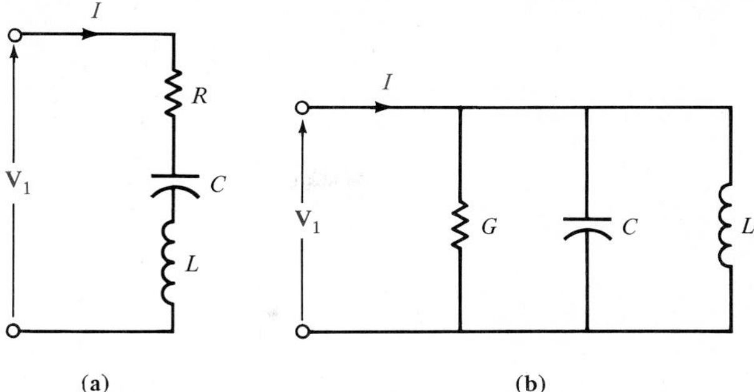

Figure 13.18 Series and parallel *RLC* circuits.

The fact that these two equations are similar is not surprising and, in fact, is expected as a direct consequence of duality. Note that the imaginary terms in both of the equations above will be zero if

$$\omega L = \frac{1}{\omega C}$$

The value of ω that satisfies this equation is

$$\omega_0 = \frac{1}{\sqrt{LC}} \tag{13.27}$$

and at this value of ω the impedance of the series circuit becomes

$$\mathbf{Z}(j\omega_0) = R \tag{13.28}$$

and the admittance of the parallel circuit is

$$\mathbf{Y}(j\omega_0) = G \tag{13.29}$$

This frequency ω_0, at which the impedance of the series circuit or the admittance of the parallel circuit is purely real, is called the *resonant frequency*, and the circuits themselves, at this frequency, are said to be *in resonance*. Resonance is a very important consideration in engineering design. For example, engineers designing the attitude control system for the Saturn vehicles had to ensure that the control system frequency did not excite the body bending (resonant) frequencies of the vehicle. Excitation of the bending frequencies would cause oscillation which, if continued unchecked, would result in a buildup of stress until the vehicle would finally break apart.

At resonance the voltage and current are in phase, and therefore the phase angle is zero. In the series case, at resonance the impedance is a minimum, and therefore the current is maximum for a given voltage. Figure 13.19 illustrates the frequency response of both the series and parallel *RLC* circuits. Note that at low frequencies the impedance of the series circuit is dominated by the capacitive term and the admittance of the parallel circuit is dominated by the inductive term. At high frequencies the impedance of the series circuit is dominated by the inductive term, and the admittance of the parallel circuit is dominated by the capacitive term.

Resonance can be viewed from another perspective—that of the phasor diagram. Once again we will consider the series and parallel cases together in order to illustrate

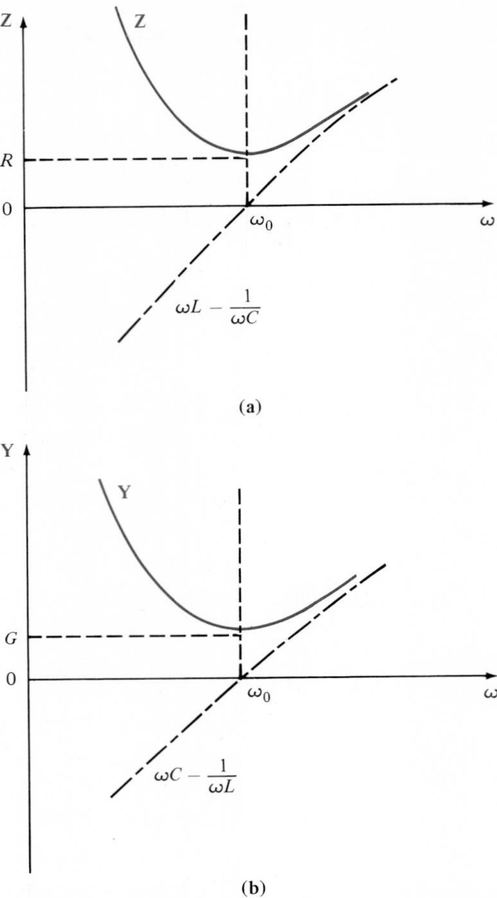

Figure 13.19 Frequency response of (a) a series and (b) a parallel *RLC* circuit.

the similarities between them. In the series case the current is common to every element, and in the parallel case the voltage is a common variable. Therefore, the current in the series circuit and the voltage in the parallel circuit are employed as references. Phasor diagrams for both circuits are shown in Fig. 13.20 for the three frequency values $\omega < \omega_0$, $\omega = \omega_0$, and $\omega > \omega_0$.

In the series case when $\omega < \omega_0$, $\mathbf{V}_C > \mathbf{V}_L$, θ_Z is negative and the voltage $\mathbf{V}_1$ lags the current. If $\omega = \omega_0$, $\mathbf{V}_L = \mathbf{V}_C$, θ_Z is zero, and the voltage $\mathbf{V}_1$ is in phase with the current. If $\omega > \omega_0$, $\mathbf{V}_L > \mathbf{V}_C$, θ_Z is positive, and the voltage $\mathbf{V}_1$ leads the current. Similar statements can be made for the parallel case in Fig. 13.20b. Because of the close relationship between series and parallel resonance, as illustrated by the preceding material, we will concentrate most of our discussion on the series case in the following developments.

For the series circuit we define what is commonly called the *quality factor* Q as

$$Q = \frac{\omega_0 L}{R} = \frac{1}{\omega_0 CR} = \frac{1}{R}\sqrt{\frac{L}{C}} \tag{13.30}$$

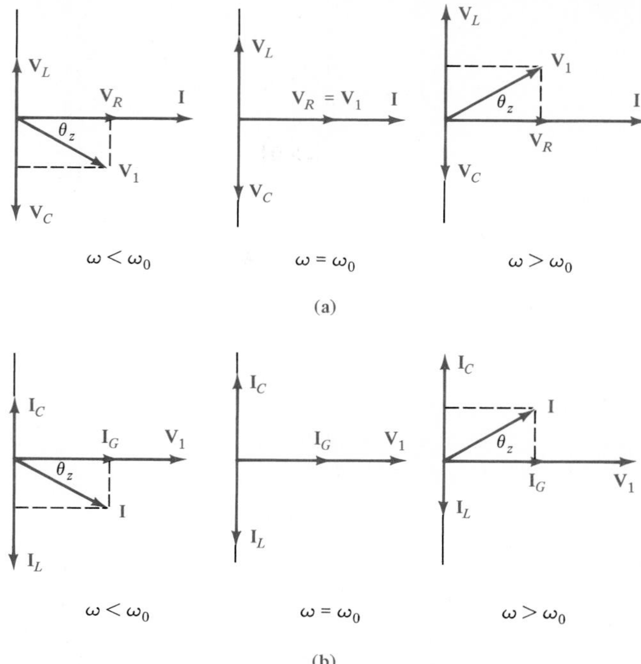

Figure 13.20 Phasor diagrams for (a) a series *RLC* circuit and (b) a parallel *GLC* circuit.

Q is a very important factor in resonant circuits, and its ramifications will be illustrated throughout the remainder of this section.

EXAMPLE 13.10

Consider the network shown in Fig. 13.21. Let us determine the resonant frequency, the voltages across each element at resonance, and the value of the quality factor.

The resonant frequency is obtained from the expression

$$\begin{aligned}\omega_0 &= \frac{1}{\sqrt{LC}}\\ &= \frac{1}{\sqrt{(25 \times 10^{-3})(10 \times 10^{-6})}}\\ &= 2000 \text{ rad/s}\end{aligned}$$

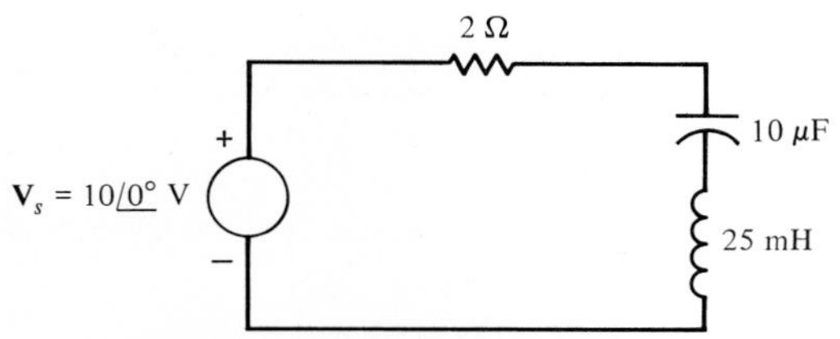

2 Ω

10 μF

$\mathbf{V}_s = 10\underline{/0°}$ V

25 mH

Figure 13.21 Series circuit.

At this resonant frequency

$$\mathbf{I} = \frac{\mathbf{V}}{\mathbf{Z}} = \frac{\mathbf{V}}{R} = 5 \angle 0^\circ \text{ A}$$

Therefore,

$$\mathbf{V}_R = (5 \angle 0^\circ)(2) = 10 \angle 0^\circ \text{ V}$$

$$\mathbf{V}_L = j\omega_0 L\mathbf{I} = 250 \angle 90^\circ \text{ V}$$

$$\mathbf{V}_C = \frac{\mathbf{I}}{j\omega_0 C} = 250 \angle -90^\circ \text{ V}$$

Note the magnitude of the voltages across the inductor and capacitor with respect to the input voltage. Note also that these voltages are equal and are 180° out of phase with one another. Therefore, the phasor diagram for this condition is shown in Fig. 13.20a for $\omega = \omega_0$. The quality factor Q derived from Eq. (13.30) is

$$Q = \frac{\omega_0 L}{R} = \frac{(2 \times 10^3)(25 \times 10^{-3})}{2} = 25$$

It is interesting to note that the voltages across the inductor and capacitor can be written in terms of Q as

$$|\mathbf{V}_L| = \omega_0 L|\mathbf{I}| = \frac{\omega_0 L}{R}\mathbf{V}_s = Q\mathbf{V}_s$$

and

$$|\mathbf{V}_C| = \frac{|\mathbf{I}|}{\omega_0 C} = \frac{1}{\omega_0 CR}\mathbf{V}_s = Q\mathbf{V}_s$$

This analysis indicates that for a given current there is a resonant voltage rise across the inductor and capacitor which is equal to the product of Q and the applied voltage. ■

DRILL EXERCISE

D13.15. Given the network in Fig. D13.15, find the value of C that will place the circuit in resonance at 1800 rad/s.

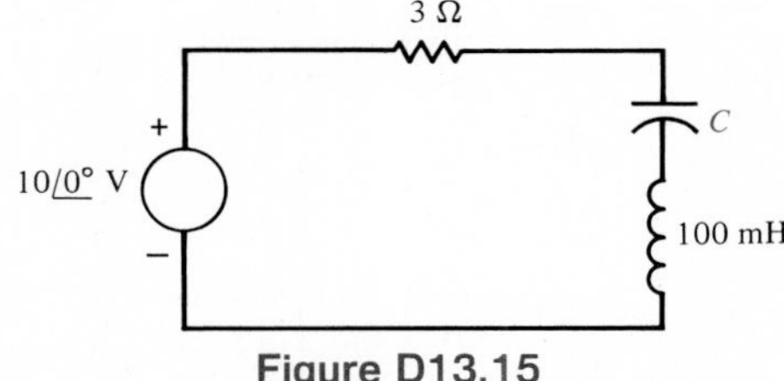

Figure D13.15

D13.16. Given the network in D13.15, determine the Q of the network and the magnitude of the voltage across the capacitor.

The impedance of the circuit in Fig. 13.18a is given by the Eq. (13.25), which can be expressed as an admittance,

$$\mathbf{Y}(j\omega) = \frac{1}{R[1 + j(1/R)(\omega L - 1/\omega C)]} \tag{13.31}$$

This equation can be expressed in a very convenient form in terms of the resonant frequency and the *quality factor*, Q, as

$$\mathbf{Y}(j\omega) = \frac{1}{R[1 + jQ(\omega/\omega_0 - \omega_0/\omega)]} \tag{13.32}$$

Since $\mathbf{I} = \mathbf{Y}\mathbf{V}_1$ and the voltage across the resistor is $\mathbf{V}_R = \mathbf{I}R$, then

$$\frac{\mathbf{V}_R}{\mathbf{V}_1} = \mathbf{G}_\mathbf{V}(j\omega) = \frac{1}{1 + jQ(\omega/\omega_0 - \omega_0/\omega)} \tag{13.33}$$

and the magnitude and phase are

$$M(\omega) = \frac{1}{[1 + Q^2(\omega/\omega_0 - \omega_0/\omega)^2]^{1/2}} \tag{13.34}$$

and

$$\phi(\omega) = -\tan^{-1} Q\left(\frac{\omega}{\omega_0} - \frac{\omega_0}{\omega}\right) \tag{13.35}$$

The sketches for these functions are shown in Fig. 13.22. Note that the circuit has the form of a bandpass filter. The bandwidth as shown is the difference between the two half-power frequencies. Since power is proportional to the square of the magnitude, these two frequencies may be derived by setting the magnitude $M(\omega) = 1/\sqrt{2}$, that is,

$$\left|\frac{1}{1 + jQ(\omega/\omega_0 - \omega_0/\omega)}\right| = \frac{1}{\sqrt{2}}$$

Therefore,

$$Q\left(\frac{\omega}{\omega_0} - \frac{\omega_0}{\omega}\right) = \pm 1 \tag{13.36}$$

Solving this equation, we obtain four frequencies,

$$\omega = \pm\frac{\omega_0}{2Q} \pm \omega_0\sqrt{\left(\frac{1}{2Q}\right)^2 + 1} \tag{13.37}$$

Taking only the positive values, we obtain

$$\omega_{LO} = \omega_0\left[-\frac{1}{2Q} + \sqrt{\left(\frac{1}{2Q}\right)^2 + 1}\right] \tag{13.38}$$

$$\omega_{HI} = \omega_0\left[\frac{1}{2Q} + \sqrt{\left(\frac{1}{2Q}\right)^2 + 1}\right] \tag{13.39}$$

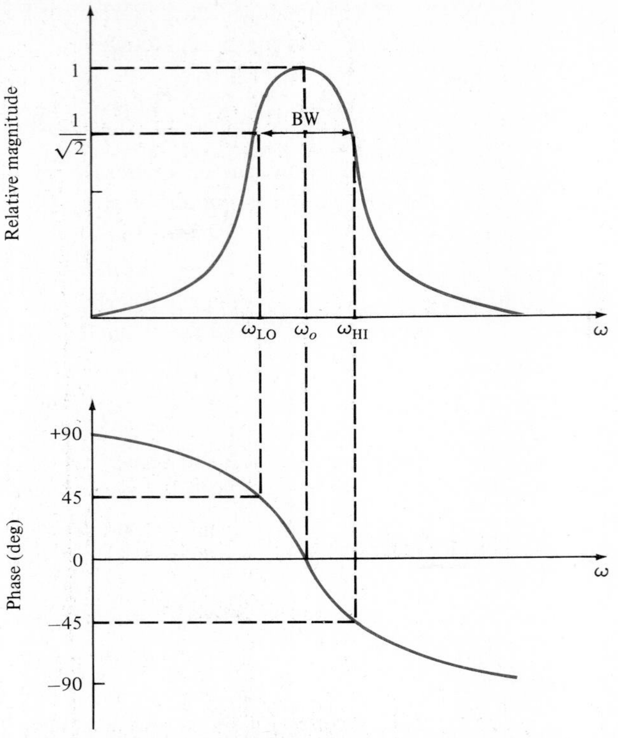

Figure 13.22 Magnitude and phase curves for Equation [13.33].

Subtracting these two equations yields the bandwidth as shown in Fig. 13.22:

$$\text{BW} = \omega_{\text{HI}} - \omega_{\text{LO}} = \frac{\omega_0}{Q} \tag{13.40}$$

and multiplying the two equations yields

$$\omega_0^2 = \omega_{\text{LO}}\omega_{\text{HI}} \tag{13.41}$$

which illustrates that the resonant frequency is the geometric mean of the two half-power frequencies.

DRILL EXERCISE

D13.17. For the network in Fig. D13.15, compute the two half-power frequencies and the bandwidth of the network.

Equation (13.30) indicates the dependence of Q on R. A high-Q series circuit has a small value of R, and as we will illustrate later, a high-Q parallel circuit has a large value of R.

Equation (13.40) illustrates that the bandwidth is inversely proportional to Q. Therefore, the frequency selectivity of the circuit is determined by the value of Q. A high-Q circuit has a small bandwith, and therefore the circuit is very selective. The manner in

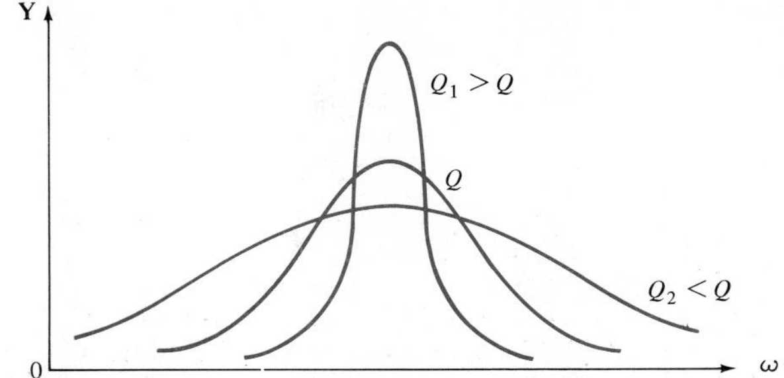

Figure 13.23 Network frequency response as a function of *Q*.

which Q affects the frequency selectivity of the network is graphically illustrated in Fig. 13-23. Hence, if we pass a signal with a wide frequency range through a high-Q circuit, only the frequency components within the bandwidth of the network will not be attenuated; that is, the network acts like a bandpass filter.

Q has a more general meaning which we can explore via an analysis of the series circuit. Suppose that the excitation is at the frequency ω_0 and that

$$i(t) = I_M \cos \omega_0 t$$

The energies stored in the inductor and capacitor are

$$w_L(t) = \tfrac{1}{2} L I_M^2 \cos^2 \omega_0 t$$

and

$$w_C(t) = \frac{1}{2} \frac{I_M^2}{\omega_0^2 C} \sin^2 \omega_0 t$$

The total energy stored is then

$$\begin{aligned} w_s(t) &= w_L(t) + w_C(t) \\ &= \tfrac{1}{2} I_M^2 \left(L \cos^2 \omega_0 t + \frac{1}{\omega_0^2 C} \sin^2 \omega_0 t \right) \end{aligned}$$

However, since $\omega_0 = 1/\sqrt{LC}$, the equation above reduces to

$$w_s(t) = \tfrac{1}{2} L I_M^2 \equiv W_s \tag{13.42}$$

which is a constant.

The energy dissipated per cycle can be derived by multiplying the average power absorbed by the resistor by the period of one cycle:

$$W_D = \left(\tfrac{1}{2} I_M^2 R\right) \frac{2\pi}{\omega_0} \tag{13.43}$$

Then

$$\begin{aligned} \frac{W_s}{W_D} &= \frac{\tfrac{1}{2} L I_M^2}{\tfrac{1}{2} I_M^2 R (2\pi/\omega_0)} \\ &= \frac{\omega_0 L}{R} \frac{1}{2\pi} \end{aligned} \tag{13.44}$$

Using Eq. (13.31), we find that this equation becomes

$$\frac{W_s}{W_D} = \frac{Q}{2\pi}$$

or

$$Q = 2\pi \frac{W_s}{W_D} \tag{13.45}$$

The importance of this definition of Q stems from the fact that this expression is applicable to acoustic, electrical, and mechanical systems, and therefore is generally considered to be the basic definition of Q.

EXAMPLE 13.11

Given a series circuit with $R = 2\ \Omega$, $L = 2$ mH, and $C = 5\ \mu F$, we wish to determine the resonant frequency, the quality factor, and the bandwidth for the circuit. Then we will determine the change in Q and the BW if R is changed from 2 to 0.2 Ω. Using Eq. (13.27), we have

$$\omega_0 = \frac{1}{\sqrt{LC}} = \frac{1}{[(2 \times 10^{-3})(5 \times 10^{-6})]^{1/2}}$$

$$= 10^4 \text{ rad/s}$$

and therefore the resonant frequency is $10^4/2\pi = 1592$ Hz. The quality factor is

$$Q = \frac{\omega_0 L}{R} = \frac{(10^4)(2 \times 10^{-3})}{2}$$

$$= 10$$

and the bandwidth is

$$BW = \frac{\omega_0}{Q} = \frac{10^4}{10}$$

$$= 10^3 \text{ rad/s}$$

If R is changed to $R = 0.2\ \Omega$, the new value of Q is 100, and therefore the new BW is 10^2 rad/s. ■

DRILL EXERCISE

D13.18. A series circuit is composed of $R = 2\ \Omega$, $L = 40$ mH, and $C = 100\ \mu$F. Determine the bandwidth of this circuit about its resonant frequency.

D13.19. A series RLC circuit has the following properties. $R = 4\ \Omega$, $\omega_0 = 4000$ rad/s, and the BW $= 100$ rad/s. Determine the values of L and C.

EXAMPLE 13.12

We wish to determine the parameters R, L, and C so that the circuit shown in Fig. 13.24 operates as a bandpass filter with an ω_0 of 1000 rad/s and a bandwidth of 100 rad/s.

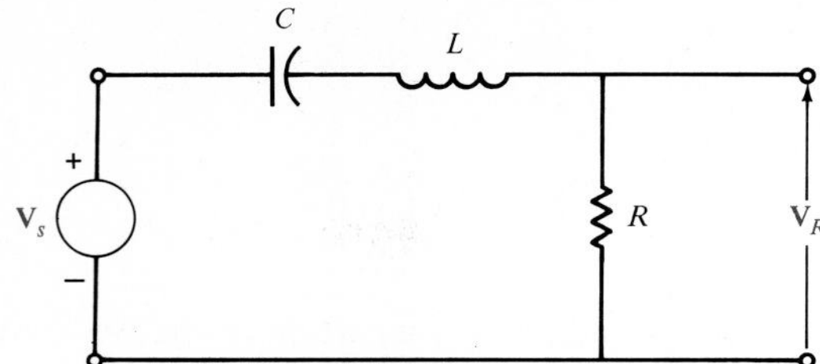

Figure 13.24 Series *RLC* circuit.

The voltage gain for the network is

$$\mathbf{G_V}(j\omega) = \frac{(R/L)j\omega}{(j\omega)^2 + (R/L)j\omega + 1/LC}$$

Hence

$$\omega_0 = \frac{1}{\sqrt{LC}}$$

and since $\omega_0 = 10^3$,

$$\frac{1}{LC} = 10^6$$

The bandwidth is

$$\text{BW} = \frac{\omega_0}{Q}$$

Then

$$Q = \frac{\omega_0}{\text{BW}} = \frac{1000}{100}$$
$$= 10$$

However,

$$Q = \frac{\omega_0 L}{R}$$

Therefore,

$$\frac{1000L}{R} = 10$$

Note that we have two equations in the three unknown circuit parameters R, L, and C. Hence if we select $C = 1\ \mu\text{F}$, then

$$L = \frac{1}{10^6 C} = 1\text{ H}$$

and

$$\frac{1000(1)}{R} = 10$$

yields

$$R = 100\ \Omega$$

Therefore, the parameters $R = 100\ \Omega$, $L = 1$ H, and $C = 1\ \mu$F will produce the proper filter characteristics. ■

In our presentation of resonance thus far, we have focused most of our discussion on the series resonant circuit. Although we should recall that the series and parallel resonant circuits are duals of one another, and therefore possess similar properties, it is nevertheless instructive to provide a couple of examples that deal directly with the parallel resonant case.

EXAMPLE 13.13

Given the parallel *RLC* circuit in Fig. 13.25,
(a) Derive the expression for the resonant frequency, the half-power frequencies, the bandwidth, and the quality factor for the transfer characteristic $\mathbf{V}_{out}/\mathbf{I}_{in}$ in terms of the circuit parameters R, L, and C.
(b) Compute the quantities in part (a) if $R = 1$ kΩ, $L = 10$ mH, and $C = 100\ \mu$F.

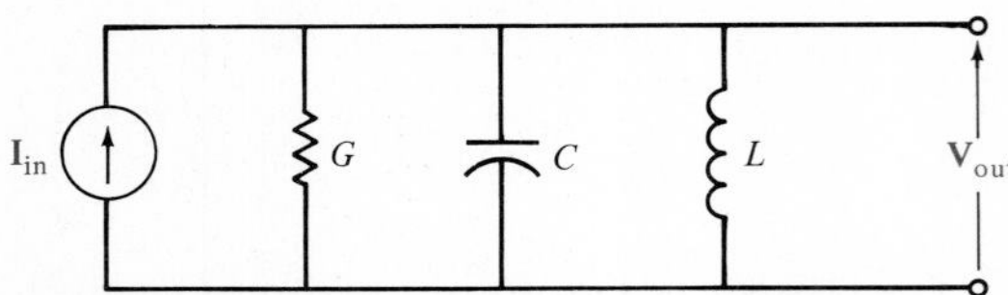

Figure 13.25 Circuit used in Example 13.13.

(a) The output voltage can be written as

$$\mathbf{V}_{out} = \frac{\mathbf{I}_{in}}{\mathbf{Y}_T}$$

and therefore the magnitude of the transfer characteristic can be expressed as

$$\left|\frac{\mathbf{V}_{out}}{\mathbf{I}_{in}}\right| = \frac{1}{\sqrt{(1/R^2) + (\omega C - 1/\omega L)^2}}$$

The transfer characteristic is a maximum at the resonant frequency

$$\omega_0 = \frac{1}{\sqrt{LC}} \tag{13.46}$$

and at this frequency

$$\left|\frac{\mathbf{V}_{out}}{\mathbf{I}_{in}}\right|_{max} = R \tag{13.47}$$

As demonstrated earlier, at the half-power frequencies the magnitude is equal to $1/\sqrt{2}$ of its maximum value, and hence the half-power frequencies can be obtained from the expression

$$\frac{1}{\sqrt{(1/R^2) + (\omega C - 1/\omega L)^2}} = \frac{R}{\sqrt{2}}$$

Solving this equation and taking only the positive values of ω yields

$$\omega_{\text{LO}} = -\frac{1}{2RC} + \sqrt{\frac{1}{(2RC)^2} + \frac{1}{LC}} \tag{13.48}$$

and

$$\omega_{\text{HI}} = \frac{1}{2RC} + \sqrt{\frac{1}{(2RC)^2} + \frac{1}{LC}} \tag{13.49}$$

Subtracting these two half-power frequencies yields the bandwidth,

$$\begin{aligned} \text{BW} &= \omega_{\text{HI}} - \omega_{\text{LO}} \\ &= \frac{1}{RC} \end{aligned} \tag{13.50}$$

Therefore, the quality factor is

$$\begin{aligned} Q &= \frac{\omega_0}{\text{BW}} \\ &= \frac{RC}{\sqrt{LC}} \\ &= R\sqrt{\frac{C}{L}} \end{aligned} \tag{13.51}$$

(b) Using the values given for the circuit components, we find that

$$\omega_0 = \frac{1}{\sqrt{(10^{-2})(10^{-4})}} = 10^3 \text{ rad/s}$$

The half-power frequencies are

$$\begin{aligned} \omega_{\text{LO}} &= \frac{-1}{(2)(10^3)(10^{-4})} + \sqrt{\frac{1}{(2)(10^{-1})} + 10^6} \\ &= 995 \text{ rad/s} \end{aligned}$$

and

$$\omega_{\text{HI}} = 1005 \text{ rad/s}$$

Therefore, the bandwidth is

$$\text{BW} = \omega_{\text{HI}} - \omega_{\text{LO}} = 10 \text{ rad/s}$$

and

$$\begin{aligned} Q &= 10^3\sqrt{\frac{10^{-4}}{10^{-2}}} \\ &= 100 \end{aligned}$$

■

DRILL EXERCISE

D13.20. A parallel *RLC* circuit has the following parameters: $R = 2\ \text{k}\Omega$, $L = 20$ mH, and $C = 150\ \mu$F. Determine the resonant frequency, the Q, and the bandwidth of the circuit.

D13.21. A parallel *RLC* circuit has the following parameters: $R = 6\ \text{k}\Omega$, BW = 1000 rad/s, and $Q = 120$. Determine the values of L, C, and ω_0.

EXAMPLE 13.14

A stereo receiver is tuned to 98 MHz on the FM band. The tuning knob controls a variable capacitor in a parallel resonant circuit. If the inductance of the circuit is 0.1 μH and the Q is 120, determine the values of C and G.

Using the expression for the resonant frequency, we obtain

$$C = \frac{1}{\omega_0^2 L}$$

$$= \frac{1}{(2\pi \times 98 \times 10^6)^2(0.1 \times 10^{-6})}$$

$$= 26.4\ \text{pF}$$

The conductance is

$$G = \frac{1}{\omega_0 LQ}$$

$$= \frac{1}{(2\pi \times 98 \times 10^6)(10^{-7})(120)}$$

$$= 135\ \mu\text{S}$$

■

EXAMPLE 13.15

Given the data in Example 13.14, suppose that another FM station in the vicinity is broadcasting at 98.1 MHz. Let us determine the relative value of the voltage across the resonant circuit at this frequency compared with that at 98 MHz, assuming that the current produced by both signals have the same amplitude.

At the resonant frequency of 98 MHz the voltage across the circuit is

$$\mathbf{V} = \frac{\mathbf{I}}{\mathbf{Y}} = \frac{\mathbf{I}}{135 \times 10^{-6}} = 7407\mathbf{I}$$

At 98.1 MHz the voltage is

$$\mathbf{V} = \frac{\mathbf{I}}{\sqrt{G^2 + (\omega C - 1/\omega L)^2}}$$

where

$$G = 135\ \mu\text{S}$$

$$\omega C = (2\pi \times 98.1 \times 10^6)(26.4 \times 10^{-12}) = 16{,}272 \times 10^{-6}\ \text{s}$$

$$\omega L = (2\pi \times 98.1 \times 10^6)(10^{-7}) = 61.6381\ \Omega$$

and therefore

$$\frac{1}{\omega L} = 1622 \times 10^{-6} \text{ s}$$

Hence

$$\mathbf{V} = \frac{\mathbf{I}}{14{,}650 \times 10^{-6}} = 68\mathbf{I}$$

This analysis indicates that the magnitude of the voltage across the resonant circuit produced by the 98-MHz signal is more than 100 times larger than that produced by the 98.1-MHz signal. Therefore, the network is very frequency selective. ■

In general, the resistance of the winding of an inductor cannot be neglected, and hence a more practical parallel resonant circuit is the one shown in Fig. 13.26. The input admittance of this circuit is

$$\begin{aligned}\mathbf{Y}(j\omega) &= j\omega C + \frac{1}{R + j\omega L} \\ &= j\omega C + \frac{R - j\omega L}{R^2 + \omega^2 L^2} \\ &= \frac{R}{R^2 + \omega^2 L^2} + j\left(\omega C - \frac{\omega L}{R^2 + \omega^2 L^2}\right)\end{aligned}$$

The resonant frequency at which the admittance is purely real is

$$\omega_r C - \frac{\omega_r L}{R^2 + \omega_r^2 L^2} = 0$$

$$\omega_r = \sqrt{\frac{1}{LC} - \frac{R^2}{L^2}} \tag{13.52}$$

Let us now try to relate some of the things we have learned about resonance to the Bode plots we presented earlier. The admittance for the series resonant circuit is

$$\begin{aligned}\mathbf{Y}(j\omega) &= \frac{1}{R + j\omega L + 1/j\omega C} \\ &= \frac{j\omega C}{(j\omega)^2 LC + j\omega CR + 1}\end{aligned} \tag{13.53}$$

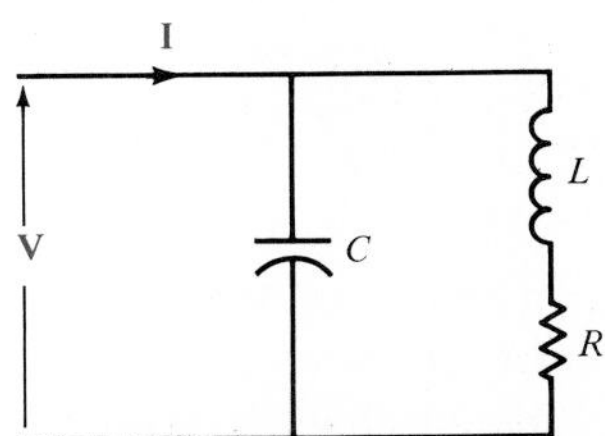

Figure 13.26 Practical parallel resonant circuit.

The standard form for the quadratic factor is

$$(j\omega\tau)^2 + 2\zeta\omega\tau j + 1$$

where $\tau = 1/\omega_0$, and hence in general the quadratic factor can be written as

$$\frac{(j\omega)^2}{\omega_0^2} + \frac{2\zeta\omega}{\omega_0} j + 1 \tag{13.54}$$

If we now compare this form of the quadratic factor with the denominator of $\mathbf{Y}(j\omega)$, we find that

$$\omega_0^2 = \frac{1}{LC}$$

$$\frac{2\zeta}{\omega_0} = CR$$

and therefore

$$\zeta = \frac{R}{2}\sqrt{\frac{C}{L}}$$

However, from Eq. (13.30),

$$Q = \frac{1}{R}\sqrt{\frac{L}{C}}$$

and hence

$$Q = \frac{1}{2\zeta} \tag{13.55}$$

To illustrate the significance of this equation, consider the Bode plot for the function $\mathbf{Y}(j\omega)$. The plot has an initial slope of $+20$ dB/decade due to the zero at the origin. If $\zeta > 1$ the poles, represented by the quadratic factor in the denominator, will simply roll off the frequency response as illustrated in Fig. 13.6a, and at high frequencies the slope of the composite characteristic will be -20 dB/dec. Note from Eq. (13.55) that if $\zeta > 1$, the Q of the circuit is very small. However, if $0 < \zeta < 1$, the frequency response will peak as shown in Fig. 13.6a, and the sharpness of the peak will be controlled by ζ. If ζ is very small, the peak of the frequency response is very narrow, the Q of the network is very large, and the circuit is very selective in filtering the input signal. Equation (13.55) together with Fig. 13.23 illustrate the connections among the frequency response, the Q, and the ζ of a network.

13.5 Scaling

Throughout this book we have employed a host of examples to illustrate the concepts being discussed. In many cases the actual values of the parameters were unrealistic in a practical sense, even though they may have simplified the presentation. In this section we illustrate how to *scale* the circuits to make them more realistic.

There are two ways to scale a circuit: *magnitude or impedance scaling* and *frequency scaling*. To magnitude-scale a circuit, we simply multiply the impedance of each element by a scale factor K_M. Therefore, a resistor R becomes $K_M R$. Multiplying the impedance of an inductor $j\omega L$ by K_M yields a new inductor $K_M L$, and multiplying the impedance of a capacitor $1/j\omega C$ by K_M yields a new capacitor C/K_M. Therefore, in magnitude scaling,

$$\begin{aligned} R' &\rightarrow K_M R \\ L' &\rightarrow K_M L \\ C' &\rightarrow \frac{C}{K_M} \end{aligned} \tag{13.56}$$

since

$$\omega_0' = \frac{1}{\sqrt{LC}} = \frac{1}{\sqrt{K_M L C/K_M}} = \omega_0$$

and Q' is

$$Q' = \frac{\omega_0 L}{R} = \frac{\omega_0 K_M L}{K_M R} = Q$$

The resonant frequency, the quality factor, and therefore the bandwidth are unaffected by magnitude scaling.

In frequency scaling the scale factor is denoted as K_F. The resistor is frequency independent and, therefore, unaffected by this scaling. The new inductor L', which has the same impedance at the scaled frequency ω_1', must satisfy the equation

$$j\omega_1 L = j\omega_1' L'$$

where $\omega_1' = K_F\omega_1$. Therefore,

$$j\omega_1 L = jK_F\omega_1 L'$$

Hence the new inductor value is

$$L' = \frac{L}{K_F}$$

Using a similar argument, we find that

$$C' = \frac{C}{K_F}$$

Therefore, to frequency-scale by a factor K_F,

$$\begin{aligned} R' &\rightarrow R \\ L' &\rightarrow \frac{L}{K_F} \\ C' &\rightarrow \frac{C}{K_F} \end{aligned} \tag{13.57}$$

Note that

$$\omega_0' = \frac{1}{\sqrt{(L/K_F)(C/K_F)}} = K_F\omega_0$$

and

$$Q' = \frac{\omega_0 L}{RK_F} = \frac{Q}{K_F}$$

and therefore

$$\text{BW}' = K_F(\text{BW})$$

Hence the resonant frequency, quality factor, and bandwidth of the circuit are affected by frequency scaling.

EXAMPLE 13.16

If the values of the circuit parameters in Fig. 13.26 are $R = 2\ \Omega$. $L = 1$ H, and $C = \frac{1}{2}$ F, let us determine the values of the elements if the circuit is magnitude-scaled by a factor $K_M = 10^2$ and frequency-scaled by a factor $K_F = 10^6$.

The magnitude scaling yields

$$R' = 2K_M = 200\ \Omega$$

$$L' = (1)K_M = 100\ \text{H}$$

$$C' = \frac{1}{2}\frac{1}{K_M} = \frac{1}{200}\ \text{F}$$

Applying frequency scaling to these values yields the final results:

$$R'' = 200\ \Omega$$

$$L'' = \frac{100}{K_F} = 100\ \mu\text{H}$$

$$C'' = \frac{1}{200}\frac{1}{K_F} = 0.005\ \mu\text{F}$$

■

DRILL EXERCISE

D13.22. An *RLC* network has the following parameter values: $R = 10\ \Omega$, $L = 1$ H, and $C = 2$ F. Determine the values of the circuit elements if the circuit is magnitude-scaled by a factor of 100 and frequency-scaled by a factor of 10,000.

13.6 SPICE Analysis Techniques

The SPICE circuit analysis program can be used to demonstrate network characteristics by means of the plotting routine. The following examples illustrate its use.

EXAMPLE 13.17

Consider the network in Fig. 13.27a, which is a low-pass filter prepared for SPICE analysis. Let us write a SPICE program to plot the output voltage using 10 points per decade over the frequency range from 1 Hz to 1 kHz. The program is listed below.

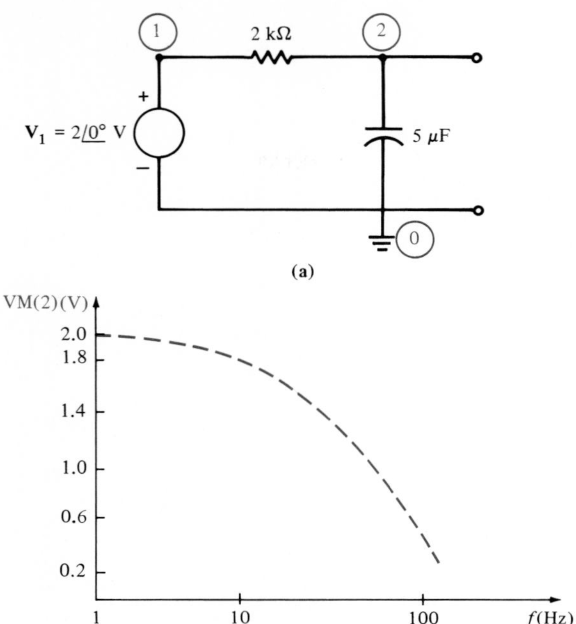

Figure 13.27 SPICE analysis of a low-pass filter.

```
EXAMPLE 13.14
V1    1    0    AC   2    0
R     1    2    2K
C     2    0    5UF
.AC   DEC  10   1      1000
.PLOT      AC   VM(2)
.END
```

The output plot is shown in Fig. 13.27b. ■

EXAMPLE 13.18

Let us examine the network in Fig. 13-28a, which has been labeled for SPICE analysis. We wish to write a SPICE program that will plot the output voltage using 20 points over the frequency range 100 to 200 Hz. The SPICE program is listed below.

```
EXAMPLE 13.15
V1    1    0    AC    100   0
C     1    2    1UF
R     3    0    100
L     2    3    1H
.AC   LIN  20   100    200
.PLOT      AC   VM(3)
.END
```

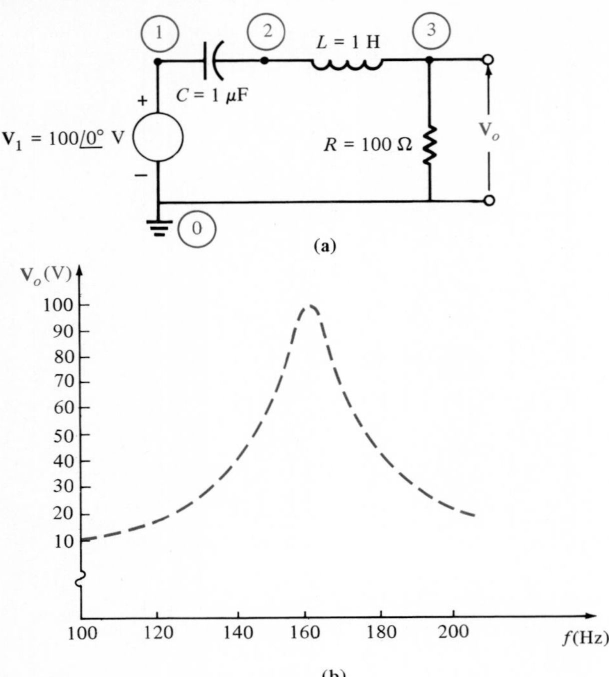

Figure 13.28 SPICE Analysis of a series *RLC* circuit.

The output plot is shown in Fig. 13.28b. Note that the resonant frequency of the circuit, calculated to be 160 Hz, is demonstrated on the plot. ■

DRILL EXERCISE

D13.23. Write a SPICE program to plot the frequency response of the voltage $\mathbf{V}_C$ in the network in Fig. D13.23 over the frequency range 100 to 1000 Hz using 50 points.

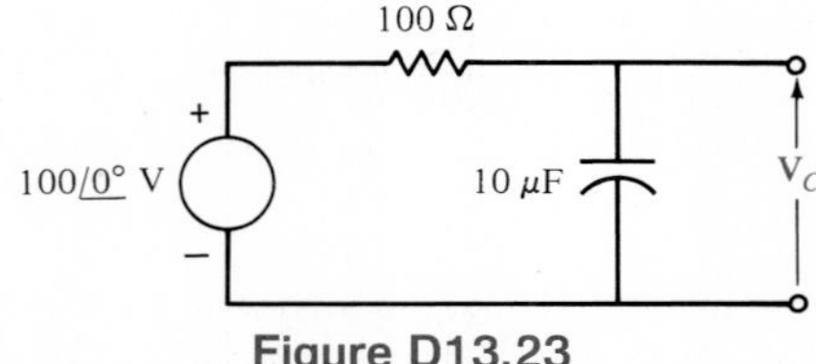

Figure D13.23

D13.24. Write a SPICE program to plot the frequency response of $\mathbf{V}_0$ in the network in Fig. D13.24 using 10 points per decade over the interval from 10 Hz to 10 kHz.

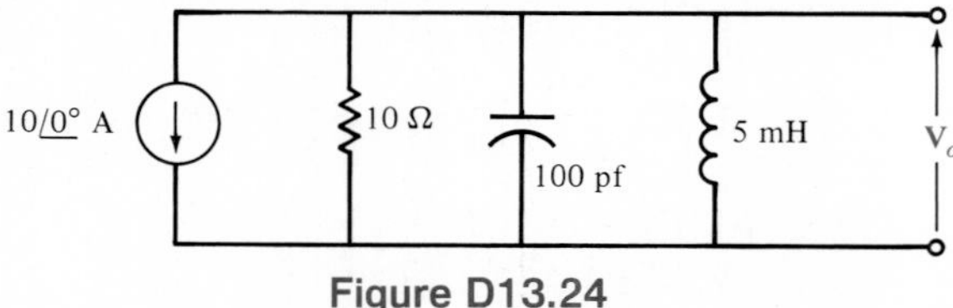

Figure D13.24

D13.25. Write a SPICE program to plot the voltage across the 1-Ω resistor in the network in Fig. D13.25 using two points per octave over the frequency range 1 Hz to 400 kHz.

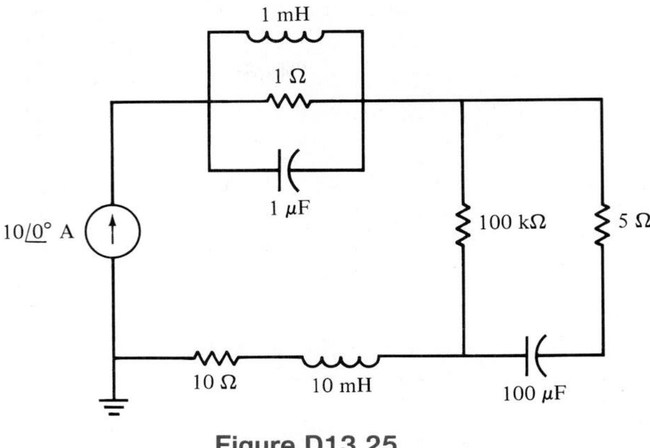

Figure D13.25

13.7 Summary

In this chapter we have examined the frequency characteristics of networks. The network function has been introduced and the frequency response of a network has been analyzed using both the pole-zero and Bode plots. These plots are very important because they indicate the behavior of the network as a function of frequency. Four types of passive network filters have been presented: low-pass, high-pass, bandpass, and band-rejection. Active network filters have been constructed using op amps.

Resonant circuits have been presented and discussed. At resonance the input voltage and current to a network are in phase. Bandwidth and half-power frequencies for resonant circuits have been introduced and network scaling has been presented as an aid to network design with practical circuit element values. Finally, the plot routines of SPICE are used to provide a vehicle for obtaining a frequency response of a network.

KEY POINTS

- There are four common network or transfer functions: $\mathbf{Z}(j\omega)$ is the ratio of the output voltage to the input current, $\mathbf{Y}(j\omega)$ is the ratio of the output current to the input voltage, $\mathbf{G}_V(j\omega)$ is the ratio of the output voltage to the input voltage, $\mathbf{G}_I(j\omega)$ is the ratio of the output current to the input current.
- The roots of the transfer function that cause it to become zero are called zeros of the function, and roots that cause it to become infinite are called poles.
- Bode plots are log-log plots of the magnitude and phase of transfer functions as a function of frequency.

- Straight-line approximations can be used to sketch quickly the magnitude characteristics of a Bode plot. The error between the actual curve and the straight-line approximation is well defined.
- The four most common types of filter networks are low pass, high pass, bandpass, and band rejection.
- The break, cutoff, or half-power frequencies are the frequencies at which the magnitude characteristic of a filter is $1/\sqrt{2}$ of its maximum value.
- The bandwidth of a bandpass or band-rejection filter is defined as the difference in frequency between the half-power points.
- Resonant frequency is defined as the frequency at which the impedance of a series *RLC* circuit or the admittance of a parallel *RLC* circuit is purely real.
- The quality factor, Q, is a measure of the sharpness of the resonance peak. The higher the Q, the sharper the peak.
- Unrealistic parameter values of passive circuit elements can be both magnitude and frequency scaled to produce realistic circuit element values.

PROBLEMS

13.1. Determine the driving-point impedance at the input terminals of the network shown in Fig. P13.1 as a function of s.

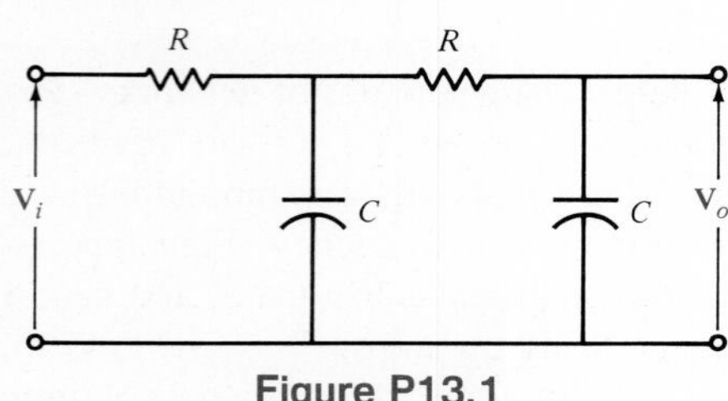

Figure P13.1

13.2. Determine the voltage transfer function as a function of s for the network in Problem 13.1.

13.3. Determine the voltage transfer function $\mathbf{V}_o/\mathbf{V}_i$ as a function of s for the network shown in Fig. P13.3.

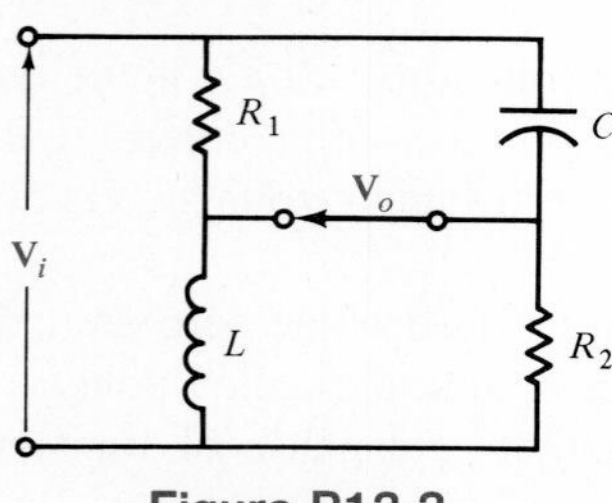

Figure P13.3

13.4. Determine the voltage transfer function as a function of frequency for the network in Fig. P13.4.

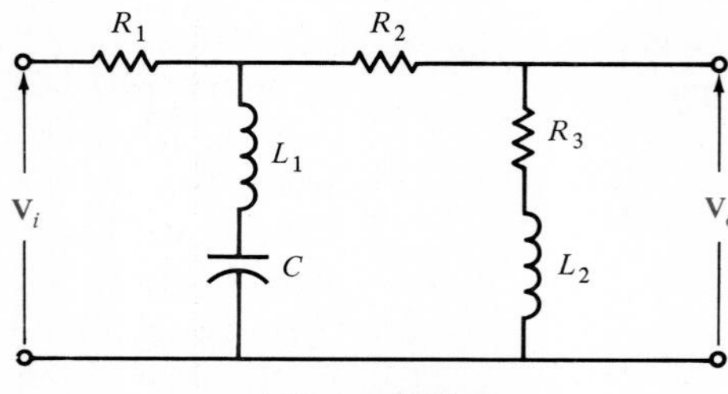

Figure P13.4

13.5. Draw the Bode plot for the network function

$$\mathbf{H}(j\omega) = \frac{j\omega 5 + 1}{j\omega 20 + 1}$$

13.6. Draw the Bode plot for the network function

$$\mathbf{H}(j\omega) = \frac{12}{j\omega(j\omega + 3)}$$

13.7. Draw the Bode plot for the network function

$$\mathbf{H}(j\omega) = \frac{10j\omega + 1}{j\omega(0.1j\omega + 1)}$$

13.8. Draw the Bode plot for the function

$$\mathbf{H}(j\omega) = \frac{16}{(j\omega)^2(j\omega 2 + 1)}$$

13.9. Draw the Bode plot for the network function

$$\mathbf{H}(j\omega) = \frac{-\omega^2}{(j\omega + 1)^3}$$

13.10. Sketch the magnitude characteristic of the Bode plot for the transfer function

$$\mathbf{G}(j\omega) = \frac{338(j\omega + 2)}{j\omega(j\omega + 5 - j12)(j\omega + 5 + j12)}$$

13.11. Sketch the magnitude characteristic of the Bode plot for the transfer function

$$\mathbf{G}(j\omega) = \frac{10j\omega}{(j\omega + 1)(j\omega + 10)^2}$$

13.12. Sketch the magnitude characteristic of the Bode plot for the transfer function

$$\mathbf{G}(j\omega) = \frac{5000(j\omega + 1)}{-\omega^2(j\omega + 10)(j\omega + 50)}$$

13.13. Sketch the magnitude characteristic of the Bode plot for the transfer function

$$\mathbf{G}(j\omega) = \frac{16(j\omega + 20)}{j\omega(-\omega^2 + 8j\omega + 32)}$$

13.14. Sketch the magnitude characteristic of the Bode plot for the transfer function

$$\mathbf{G}(j\omega) = \frac{800j\omega}{(j\omega + 2)(-\omega^2 + 4j\omega + 40)}$$

13.15. Sketch the magnitude characteristic of the Bode plot for the transfer function

$$\mathbf{G}(j\omega) = \frac{1000j\omega(-\omega^2 + 4j\omega + 16)}{(j\omega + 0.1)(j\omega + 40)^2(j\omega + 100)}$$

13.16. Draw the Bode plot for the network function

$$\mathbf{H}(j\omega) = \frac{72(j\omega + 2)}{j\omega[(j\omega)^2 + 2.4\,j\omega + 144]}$$

13.17. The magnitude characteristic for a network function is shown in Fig. P13.17. Determine the network function $\mathbf{H}(j\omega)$.

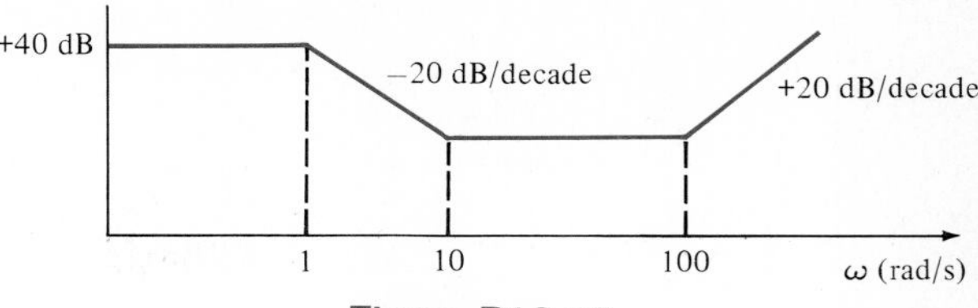

Figure P13.17

13.18. Determine $\mathbf{H}(j\omega)$ if the amplitude characteristic for $\mathbf{H}(j\omega)$ is shown in Fig. P13.18.

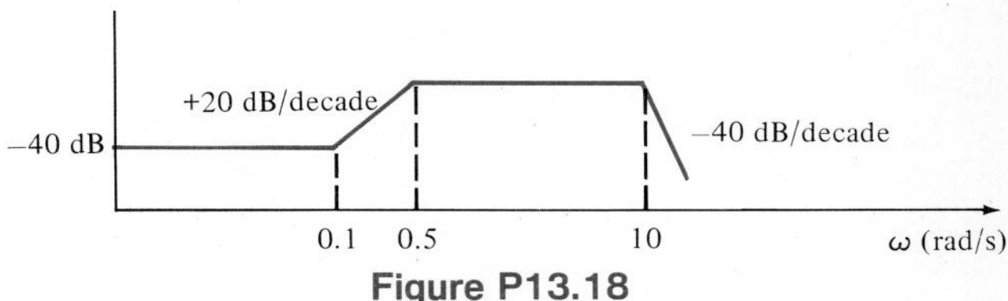

Figure P13.18

13.19. Find $\mathbf{H}(j\omega)$ if its magnitude characteristic is shown in Fig. P13.19.

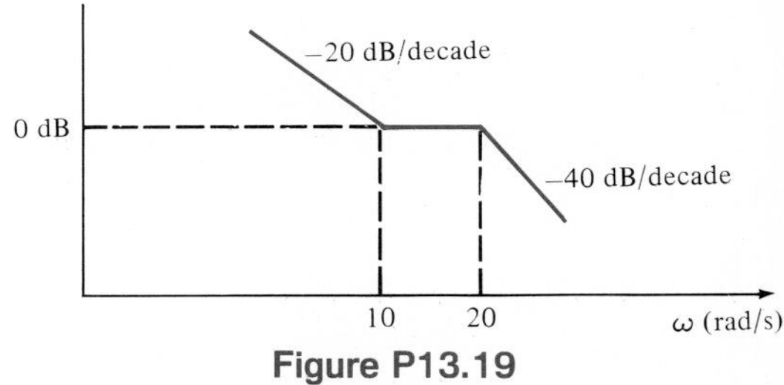

Figure P13.19

13.20. Determine $\mathbf{H}(j\omega)$ if its magnitude characteristic is shown in Fig. P13.20.

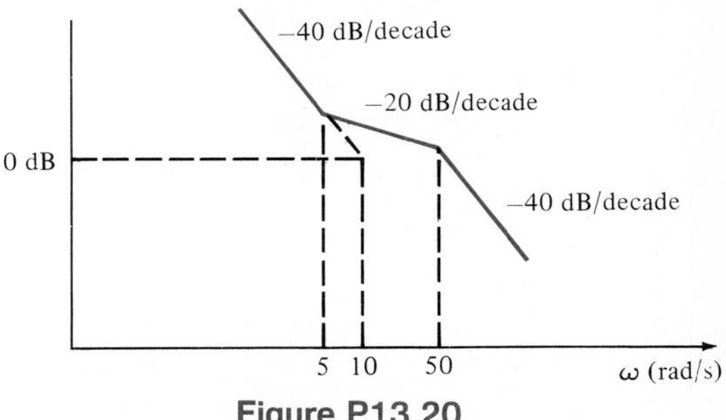

Figure P13.20

13.21. The magnitude characteristic of a band-elimination filter is shown in Fig. P13.21. Determine $\mathbf{H}(j\omega)$.

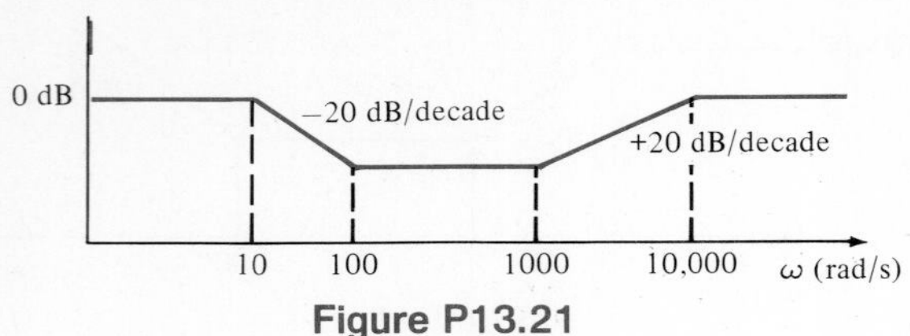

Figure P13.21

13.22. Find $\mathbf{H}(j\omega)$ for the magnitude characteristic shown in Fig. P13.22.

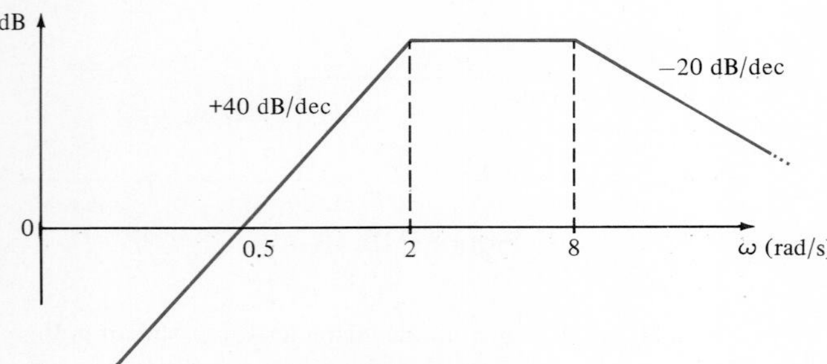

Figure P13.22

13.23. Given the magnitude characteristic for $\mathbf{G}(j\omega)$ in Fig. P13.23, determine the transfer function $\mathbf{G}(j\omega)$.

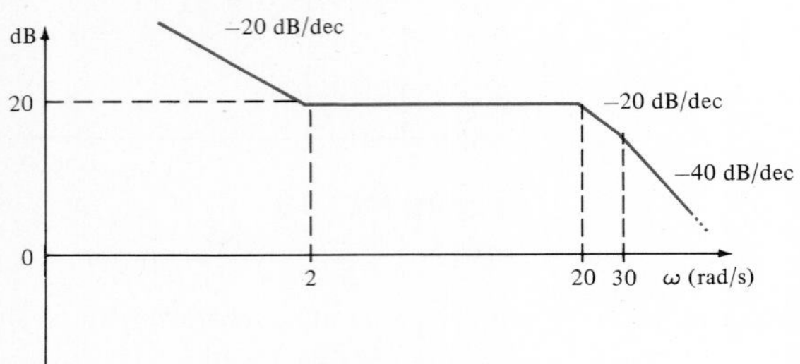

Figure P13.23

13.24. Given the magnitude characteristic for $\mathbf{G}(j\omega)$ in Fig. P13.24, determine the transfer function $\mathbf{G}(j\omega)$.

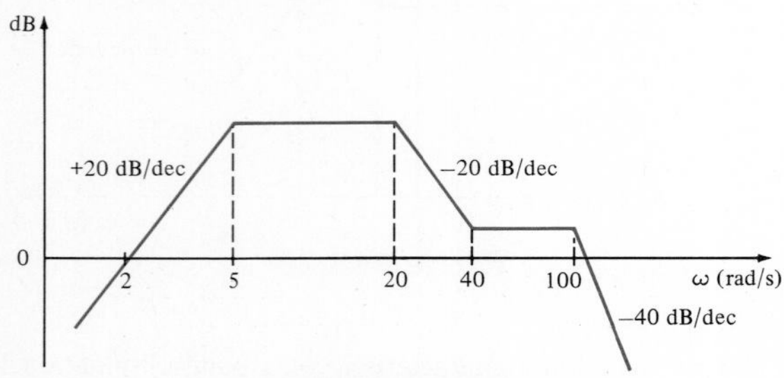

Figure P13.24

13.25. Given the magnitude characteristic for $\mathbf{G}(j\omega)$ in Fig. P13.25, determine the transfer function $\mathbf{G}(j\omega)$.

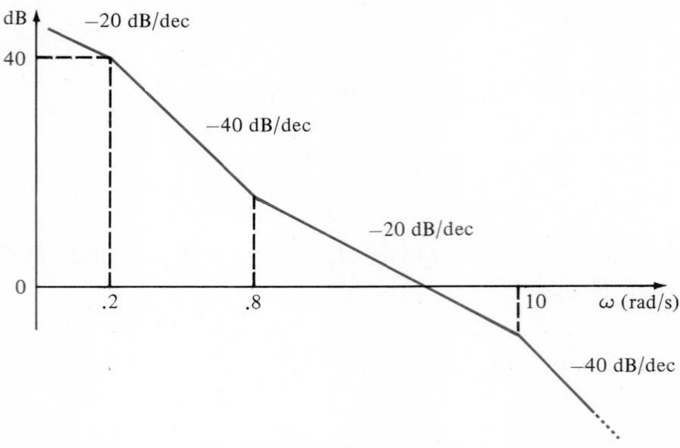

Figure P13.25

13.26. Determine what type of filter the network shown in Fig. P13.26 represents by determining the voltage transfer function.

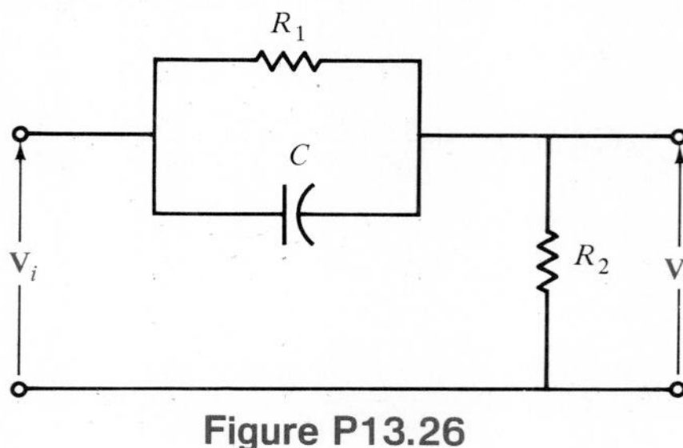

Figure P13.26

13.27. Given the lattice network shown in Fig. P13.27, determine what type of filter this network represents by determining the voltage transfer function.

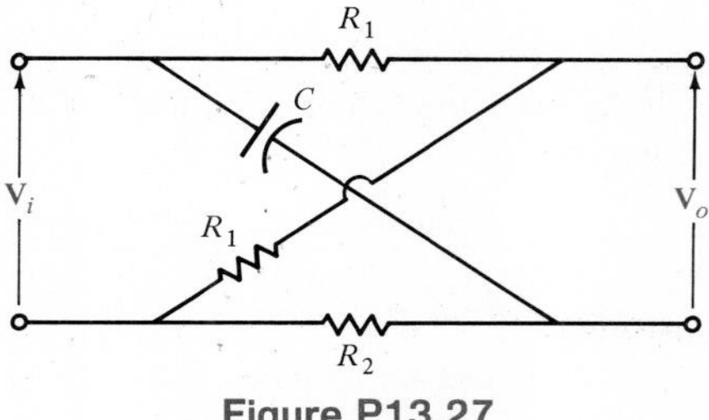

Figure P13.27

13.28. Given the network in Fig. 13.28, *sketch* the magnitude characteristic of the transfer function $\mathbf{G_V}(j\omega)$, labeling all critical values and identifying the type of filter.

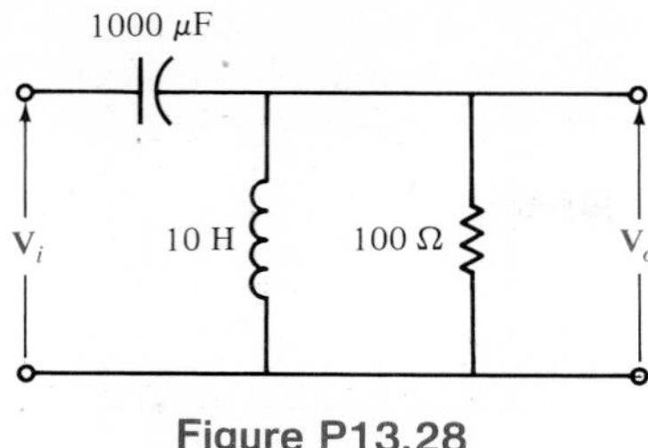

Figure P13.28

13.29. Draw the voltage transfer function magnitude and phase characteristics for the filter network shown in Fig. P13.29.

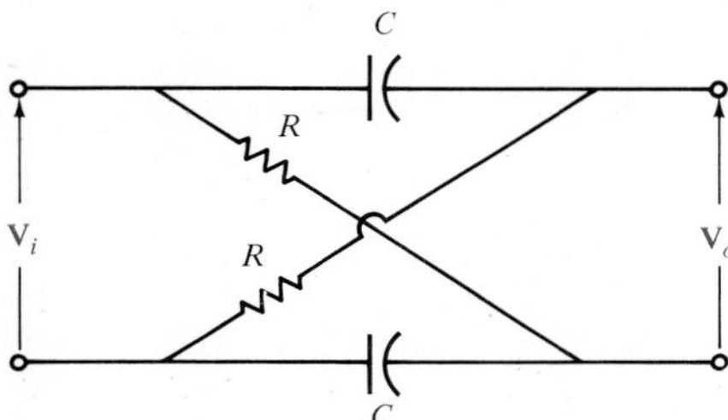

Figure P13.29

13.30. Determine the voltage transfer function and its magnitude characteristic for the network shown in Fig. P13.30 and identify the filter properties.

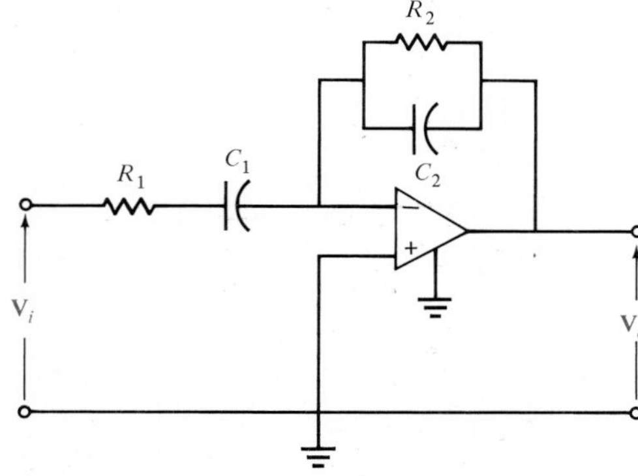

Figure P13.30

13.31. Given the network shown in Fig. P.13.31,
(a) Determine the voltage transfer function.
(b) Determine what type of filter the network represents.

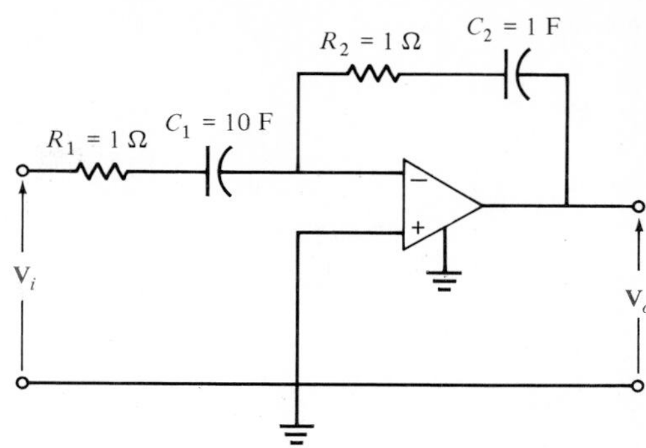

Figure P13.31

13.32. Repeat Problem 13.31 if $R_1 = R_2 = 1\ \Omega$, $C_1 = 1$ F, and $C_2 = 10$ F.

13.33. Given the network shown in Fig. P13.33.
(a) Determine the voltage transfer function.
(b) Determine what type of filter the network represents.

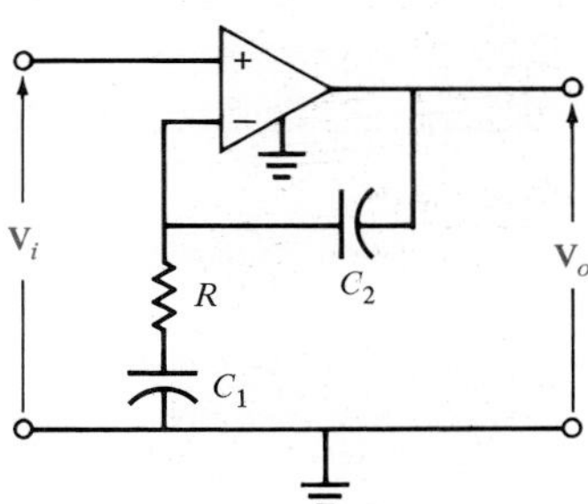

Figure P13.33

13.34. Repeat Problem 13.33 for the network shown in Fig. P13.34.

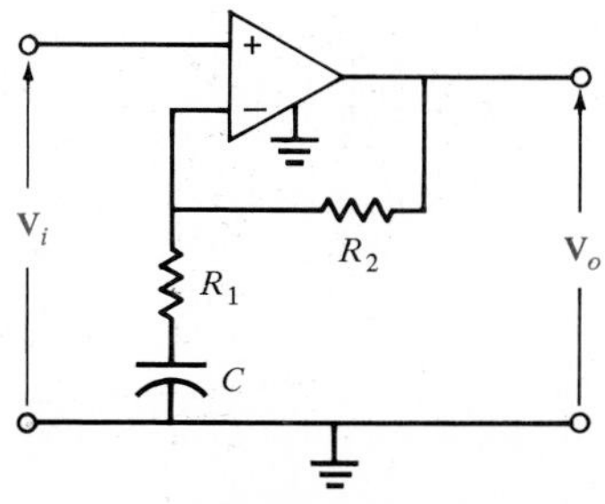

Figure P13.34

13.35. Repeat Problem 13.33 for the network shown in Fig. P13.35.

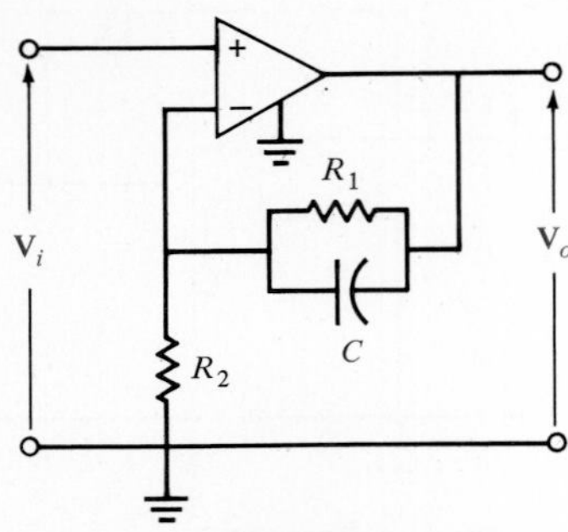

Figure P13.35

13.36. Given the network in Fig. P13.36, and employing the voltage follower analyzed in Chapter 2, determine the voltage transfer function, and its magnitude characteristic. What type of filter does the network represent?

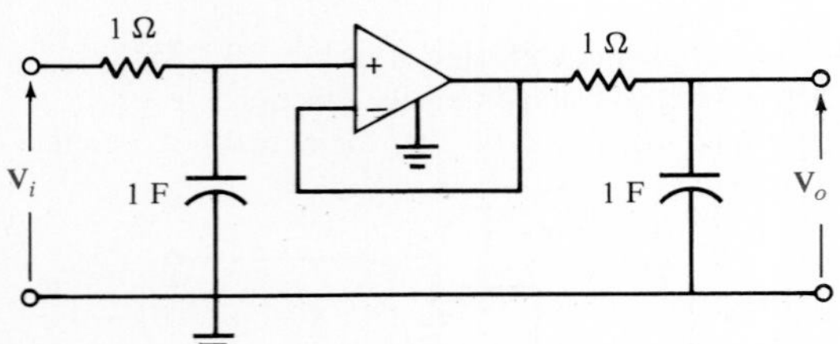

Figure P13.36

13.37. Given the series RLC circuit in Fig. P13.37, if $R = 10$ Ω, find the values of L and C such that the network will have a resonant frequency of 100 kHz and a bandwidth of 1 kHz.

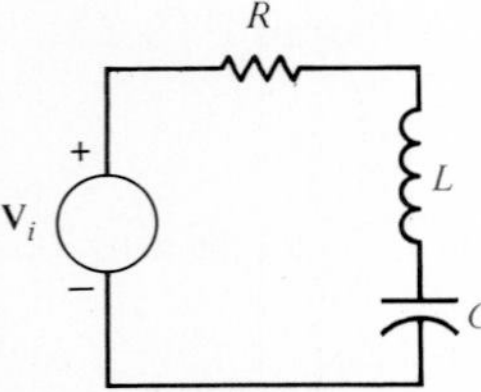

Figure P13.37

13.38. Show that for a parallel RLC resonant circuit, the total energy stored is $I_M^2 RC/2$ and that $Q = \omega_0 CR$.

13.39. Determine the parameters of a parallel resonant circuit which has the following properties: $\omega_0 = 2 \times 10^6$ rad/s, BW $= 20 \times 10^3$ rad/s, and an impedance at resonance of 2000 Ω.

13.40. Determine the value of C in the network shown in Fig. P13.40 in order for the circuit to be in resonance.

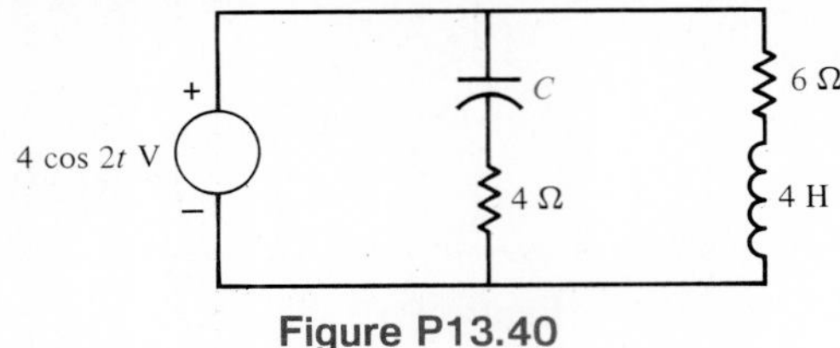

Figure P13.40

13.41. Determine the expression for the resonant frequency of the network shown in Fig. P13.41.

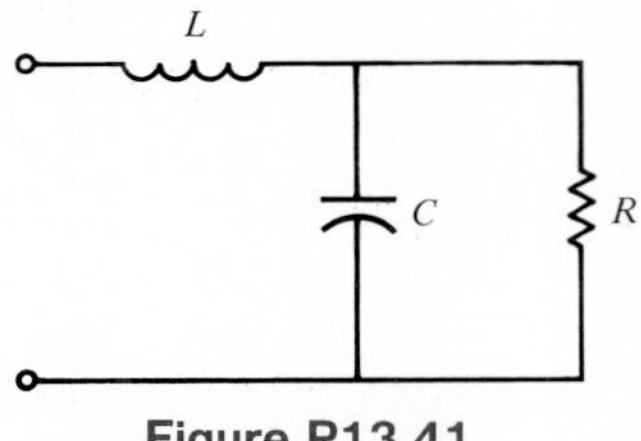

Figure P13.41

13.42. Given the series RLC circuit in Fig. P13.42,

(a) Derive the expression for the half-power frequencies, the resonant frequency, the bandwidth, and the quality factor for the transfer characteristic $\mathbf{I}/\mathbf{V}_{in}$ in terms of R, L, C.

(b) Compute the quantities in part (a) if $R = 10$ Ω, $L = 100$ mH, and $C = 10$ μF.

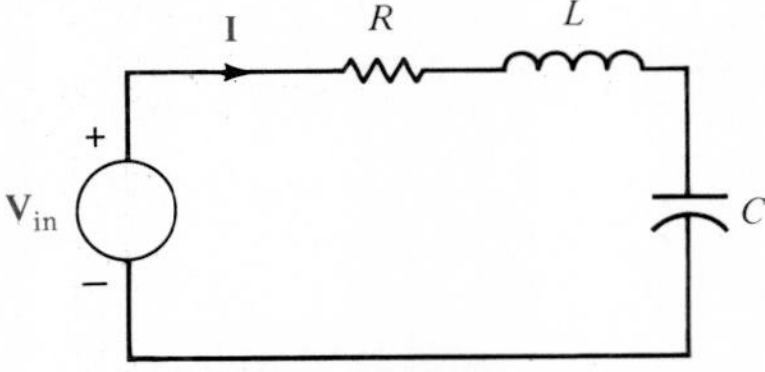

Figure P13.42

13.43. Consider the network in Fig. P13.43. If $R = 2$ kΩ, $L = 20$ mH, $C = 50$ μF, and $R_s = \infty$, determine the resonant frequency ω_0, the value of $\mathbf{V}_0$ at ω_0, the Q of the network, and the bandwidth of the network. What impact does an R_s of 10 k Ω have on the quantities determined?

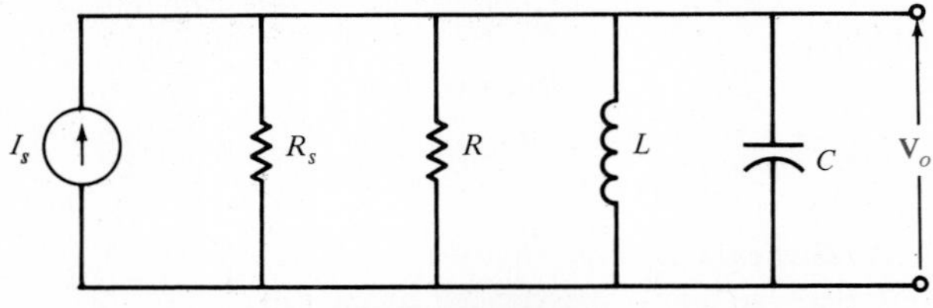

Figure P13.43

13.44. Given the network shown in Fig. P13.44, determine the minimum value of $\mathbf{Z}_1$ that will cause the circuit to be in resonance if $\mathbf{Z}_2 = 2 + j3\ \Omega$ and $\mathbf{Z}_3 = 3 - j1\ \Omega$.

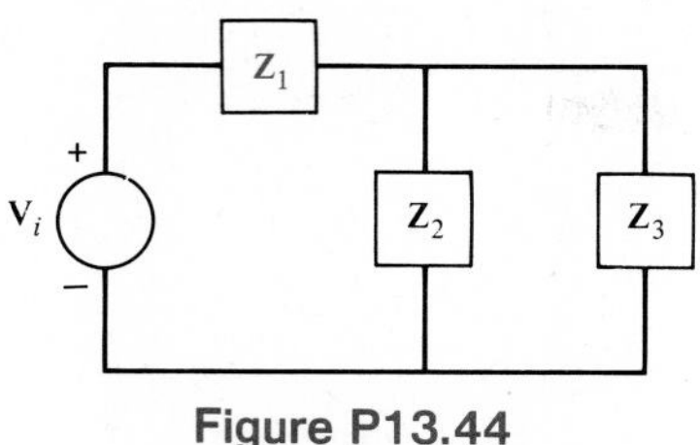

Figure P13.44

13.45. Determine the new parameters of the network shown in Fig. P13.45 if they are magnitude scaled by a factor of 1000.

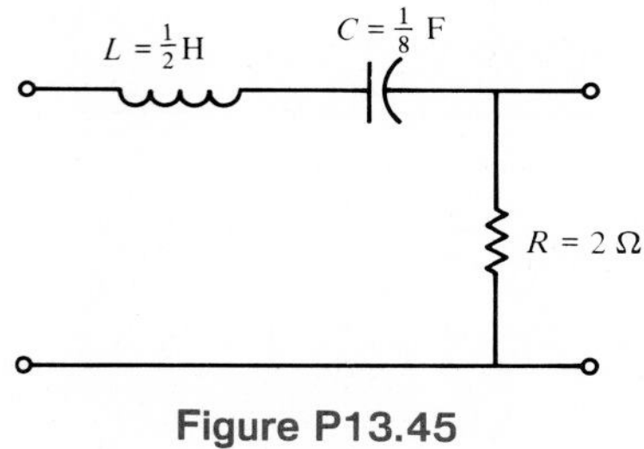

Figure P13.45

13.46. Determine the new parameters of the network in Problem 13.45 if they are frequency scaled by a factor of 10^5.

13.47. Determine the values of the circuit elements in Problem 13.31 if they are magnitude-scaled by a factor of 1000 and frequency-scaled by a factor of 10^8.

13.48. Write a SPICE program to plot the frequency response of $\mathbf{V}_0$ in Fig. P13.48. Plot over the range 1 to 50 Hz using 50 points.

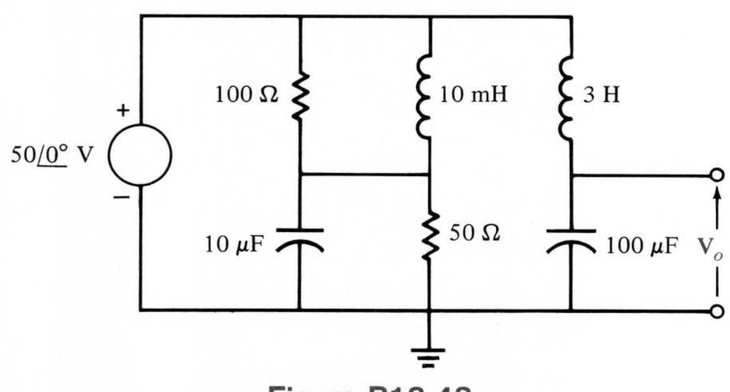

Figure P13.48

13.49. Given the network in Fig. P13.49, use 50 points to plot the frequency response of $\mathbf{V}$ over the range 1 to 100 Hz, using SPICE.

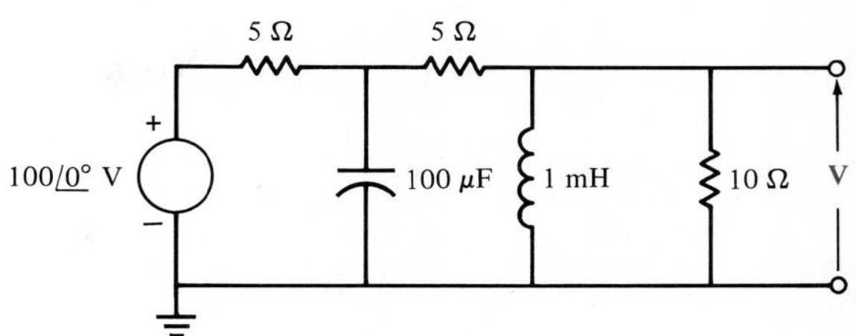

Figure P13.49

13.50. Write a SPICE program to plot $\mathbf{V}_x$ in the network in Fig. P13.50 using 10 points per decade over the range 10 kHz to 10 MHz.

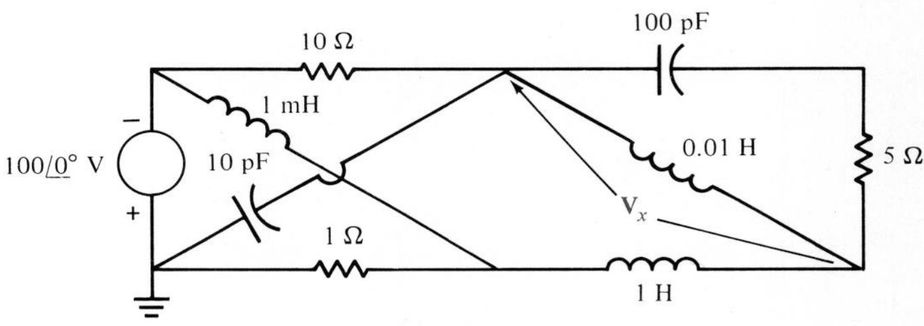

Figure P13.50

13.51. Given the network in Fig. P13.51, use SPICE to plot the frequency response of $\mathbf{V}_R$ using 50 points over the interval 0.01 to 5 Hz.

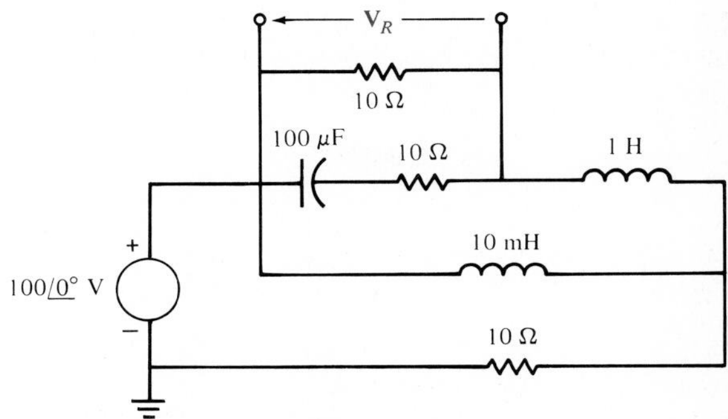

Figure P13.51

13.52. Write a SPICE program to plot $\mathbf{V}_0$, in the network shown in Fig. P13.52, over the interval 0.1 to 150 Hz using 50 points.

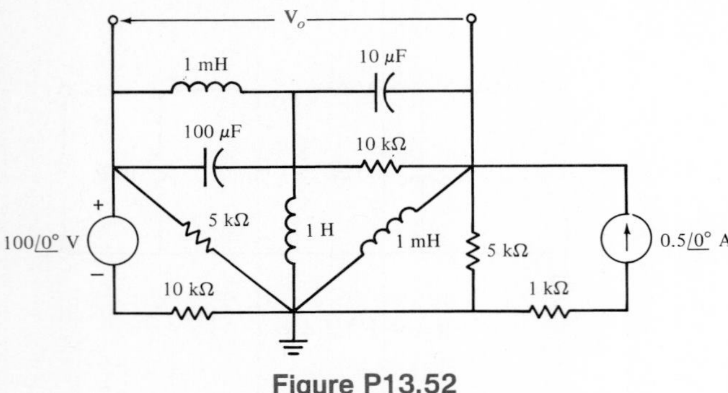

Figure P13.52

13.53. Given the network in Fig. P13.53, use SPICE to plot $\mathbf{V}_C$ and $\mathbf{V}_L$. Use 5 points per decade over the range 1 Hz to 1 MHz.

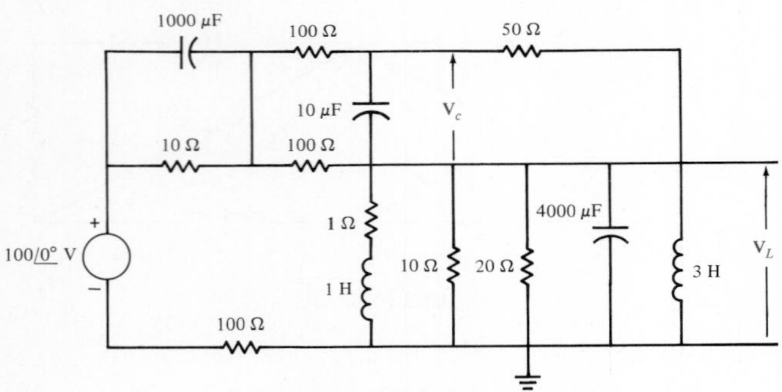

Figure P13.53

13.54. Write a SPICE program to plot $\mathbf{V}_C$ in the network in Fig. P13.54. Use 5 points per decade over the frequency range 0.1 to 1 MHz.

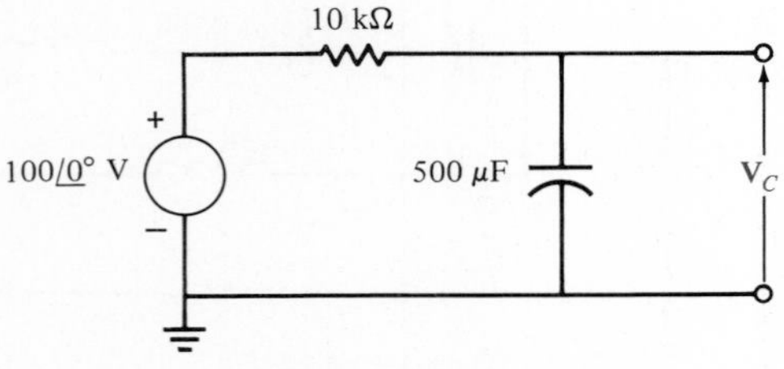

Figure P13.54

13.55. Plot $\mathbf{I}_x$ in the network in Fig. P13.55, using SPICE. Use 5 points per decade over the frequency range 0.1 to 10 MHz.

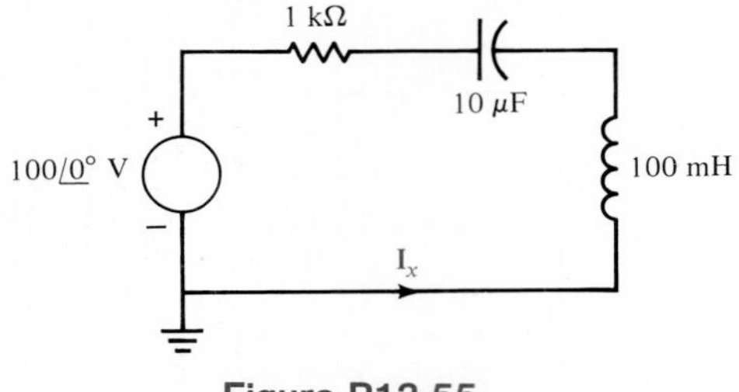

Figure P13.55

13.56. Repeat Problem 13.55 with $C = 1\ \mu\text{F}$ and $L = 10$ mH.

13.57. Given the network in Fig. P13.57, plot $\mathbf{V}_0$ using SPICE. Use 5 points per decade over the range 0.1 to 1 MHz.

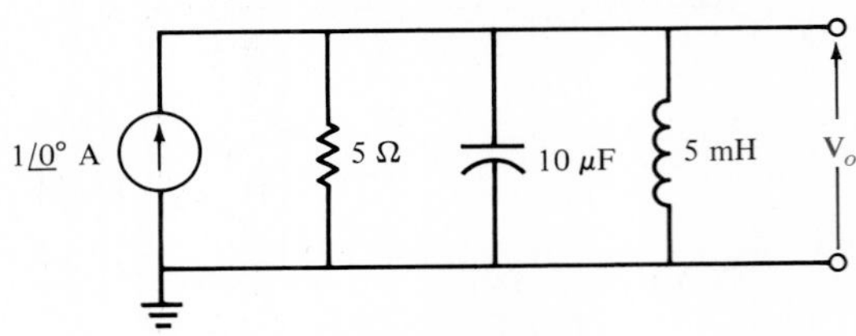

Figure P13.57

13.58. Repeat Problem 13.57 with $C = 100\ \mu\text{F}$.

13.59. Repeat Problem 13.57 with $L = 50$ mH.

13.60. Repeat Problem 13.57 with $C = 100\ \mu\text{F}$ and $L = 50$ mH.

CHAPTER

14

Magnetically Coupled Networks

We now introduce a new four-terminal element which is called a *transformer*. This circuit element consists of two inductors that are placed in close proximity to one another. Because of their close proximity, they share a common magnetic flux, and therefore the inductor coils are said to be mutually coupled.

We begin with a general description of two coupled coils and then show how circuit equations can be written for networks containing coupled inductors. We will then consider coils that are coupled with good magnetic material and then derive an approximation for ideal coupling, called the *ideal transformer*. Finally, we learn to solve magnetically coupled circuits using the SPICE circuit analysis program.

14.1 Mutual Inductance

To begin our description of two coupled coils, we employ *Faraday's law*, which can be stated as follows: The induced voltage in a coil is proportional to the rate of change of flux and the number of turns, N, in the coil. The two coupled coils are shown in Fig. 14.1 together with the following flux components.

ϕ_{l_1}	The flux in coil 1, which does not link coil 2, that is produced by the current in coil 1
ϕ_{l_2}	The flux in coil 2, which does not link coil 1, that is produced by the current in coil 2
ϕ_{12}	The flux in coil 1 produced by the current in coil 2
ϕ_{21}	The flux in coil 2 produced by the current in coil 1

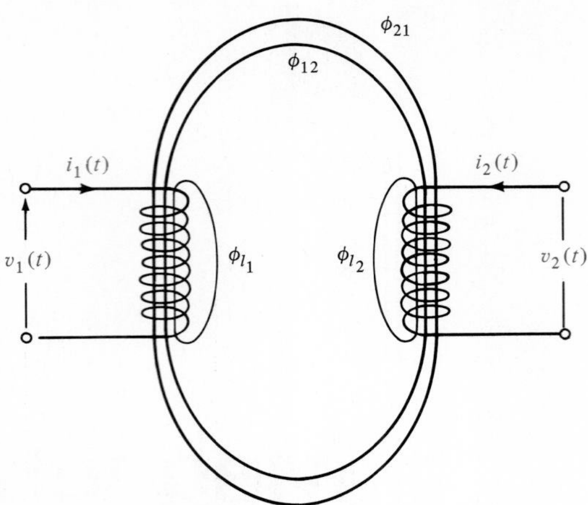

Figure 14.1 Flux relationships for mutually coupled coils.

$\phi_{11} = \phi_{l_1} + \phi_{21}$	The flux in coil 1 produced by the current in coil 1
$\phi_{22} = \phi_{l_2} + \phi_{12}$	The flux in coil 2 produced by the current in coil 2
ϕ_1	The total flux in coil 1
ϕ_2	The total flux in coil 2

In order to write the equations that describe the coupled coils, we define the voltages and currents, using the passive sign convention, at each pair of terminals as shown in Fig. 14.1.

Mathematically, Faraday's law can be written as

$$v_1(t) = N_1 \frac{d\phi_1}{dt} \tag{14.1}$$

The flux ϕ_1 will be equal to ϕ_{11}, the flux in coil 1 caused by current in coil 1, plus or minus the flux in coil 1 caused by current in coil 2; that is,

$$\phi_1 = \phi_{11} \pm \phi_{12} \tag{14.2}$$

If the current in coil 2 is such that the fluxes add, then the plus sign is used, and if the current in coil 2 is such that the fluxes oppose one another, the minus sign is used. The equation for the voltage can now be written

$$\begin{aligned} v_1(t) &= N_1 \frac{d\phi_1}{dt} \\ &= N_1 \frac{d\phi_{11}}{dt} \pm N_1 \frac{d\phi_{12}}{dt} \end{aligned} \tag{14.3}$$

From basic physics we know that

$$\begin{aligned} \phi_{11} &= N_1 i_1 \alpha_{11} \\ \phi_{12} &= N_2 i_2 \alpha_{12} \end{aligned} \tag{14.4}$$

where the α's are constants that depend on the magnetic paths taken by the flux components. The voltage equation can then be written

$$v_1(t) = N_1^2\alpha_{11}\frac{di_1}{dt} \pm N_1N_2\alpha_{12}\frac{di_2}{dt} \tag{14.5}$$

The constant $N_1^2\alpha_{11} = L_1$ (the same L that we used before) is now called the *self-inductance*, and the constant $N_1N_2\alpha_{12} = M_{12}$ is called the *mutual inductance*. Therefore,

$$v_1(t) = L_1\frac{di_1}{dt} \pm M_{12}\frac{di_2}{dt} \tag{14.6}$$

Using the same technique, we can write

$$v_2(t) = N_2^2\alpha_{22}\frac{di_2}{dt} \pm N_1N_2\alpha_{21}\frac{di_1}{dt} \tag{14.7}$$

which can be written

$$v_2(t) = L_2\frac{di_2}{dt} \pm M_{21}\frac{di_1}{dt} \tag{14.8}$$

We now need to examine the physical windings of the coupled coils to determine whether the "+" or "−" signs in Eq. (14.6) and (14.8) should be used. In basic physics we learned the *right-hand rule*, which states that if we curl the fingers of our right hand around the coil in the direction of the current, the flux produced by the current is in the direction of our thumb.

In order to indicate the physical relationship of the coils and, therefore, simplify the sign convention for the mutual terms, we employ what is commonly called the *dot convention*. Dots are placed beside each coil so that if currents are entering both dotted terminals or leaving both dotted terminals, the fluxes produced by these currents will add. In order to place the dots on a pair of coupled coils, we arbitrarily select one terminal of either coil and place a dot there. Using the right-hand rule, we determine the direction of the flux produced by this coil when current is entering the dotted terminal. We then examine the other coil to determine which terminal the current would have to enter to produce a flux that would add to the flux produced by the first coil. Place a dot on this terminal. The dots have been placed on the two coupled circuits in Fig. 14.2; verify that they are correct.

When the equations for the terminal voltages are written, the dots can be used to define the sign of the mutually induced voltages. If the currents $i_1(t)$ and $i_2(t)$ are both entering or leaving dots, the sign of the mutual voltage $M(di_2/dt)$ will be the same in an equation as the self-induced voltage $L_1(di_1/dt)$. If one current enters a dot and the other

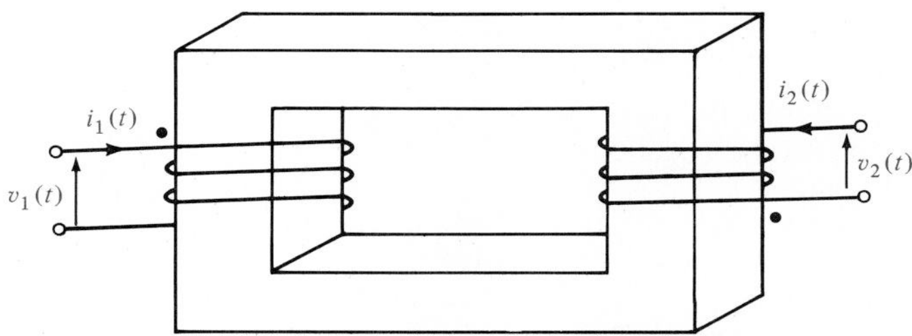

Figure 14.2 Illustration of the dot convention for coupled coils.

current leaves a dot, the mutual induced voltage and self-induced voltage terms will have opposite signs.

EXAMPLE 14.1

Determine the expressions for $v_1(t)$ and $v_2(t)$ in the circuits shown in Fig. 14.3.

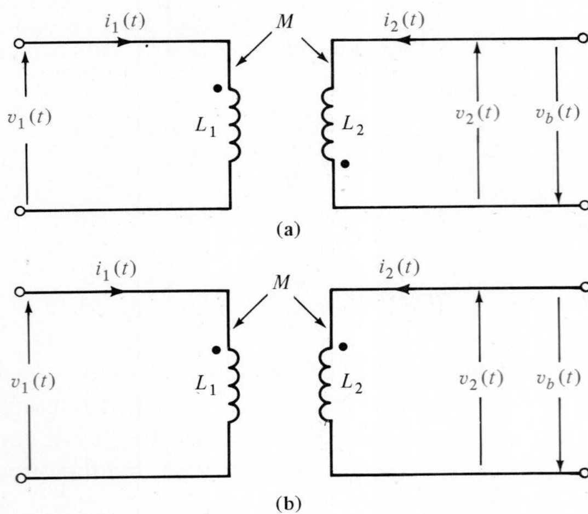

Figure 14.3 Two mutually coupled inductor circuits.

For the circuit in Fig. 14.3a the voltage equations for the variables as assigned on the figure are

$$v_1(t) = L_1 \frac{di_1}{dt} - M \frac{di_2}{dt}$$

$$v_2(t) = L_2 \frac{di_2}{dt} - M \frac{di_1}{dt}$$

In a similar manner the equations for the circuit in Fig. 14.3b are

$$v_1(t) = L_1 \frac{di_1}{dt} + M \frac{di_2}{dt}$$

$$v_2(t) = L_2 \frac{di_2}{dt} + M \frac{di_1}{dt}$$

■

EXAMPLE 14.2

Determine the equations for $v_1(t)$ and $v_b(t)$ in the circuits shown in Fig 14.3. For the network in Fig. 14.3a the equations are

$$v_1(t) = L_1 \frac{di_1}{dt} - M \frac{di_2}{dt}$$

$$v_b(t) = -L_2 \frac{di_2}{dt} + M \frac{di_1}{dt}$$

In a similar manner the equations for Fig. 14.3b are

$$v_1(t) = L_1 \frac{di_1}{dt} + M \frac{di_2}{dt}$$

$$v_b(t) = -L_2 \frac{di_2}{dt} - M \frac{di_1}{dt}$$

■

DRILL EXERCISE

D14.1. Write the equations for the $v_1(t)$ and $v_2(t)$ in the circuit in Fig. D14.1.

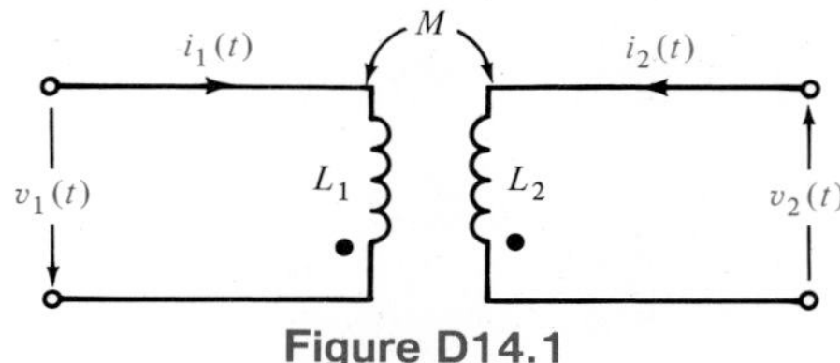

Figure D14.1

If the coupled circuit is a component in a network that is excited with a sinusoidal source, it can be drawn in the frequency domain. The voltages will be of the form $\mathbf{V}_1 e^{j\omega t}$ and $\mathbf{V}_2 e^{j\omega t}$ and the currents will be of the form $\mathbf{I}_1 e^{j\omega t}$ and $\mathbf{I}_2 e^{j\omega t}$, where $\mathbf{V}_1$, $\mathbf{V}_2$, $\mathbf{I}_1$, and $\mathbf{I}_2$ are phasors. Substituting these voltages and currents into Eq. (14.6) and (14.8) and assuming that $M_{12} = M_{21} = M$, we obtain

$$\begin{aligned} \mathbf{V}_1 &= j\omega L_1 \mathbf{I}_1 \pm j\omega M \mathbf{I}_2 \\ \mathbf{V}_2 &= j\omega L_2 \mathbf{I}_2 \pm j\omega M \mathbf{I}_1 \end{aligned} \tag{14.9}$$

The model of the coupled circuit in the frequency domain is identical to that in the time domain except for the way the elements are labeled. The sign on the mutual terms is handled in the same manner as is done in the time domain.

EXAMPLE 14.3

We wish to determine the output voltage $\mathbf{V}_o$ in the circuit in Fig. 14.4.

The two KVL equations for the network are

$$(2 + j4)\mathbf{I}_1 - j2\mathbf{I}_2 = 24 \angle 30^\circ$$

$$-j2\mathbf{I}_1 + (2 + j6 - j2)\mathbf{I}_2 = 0$$

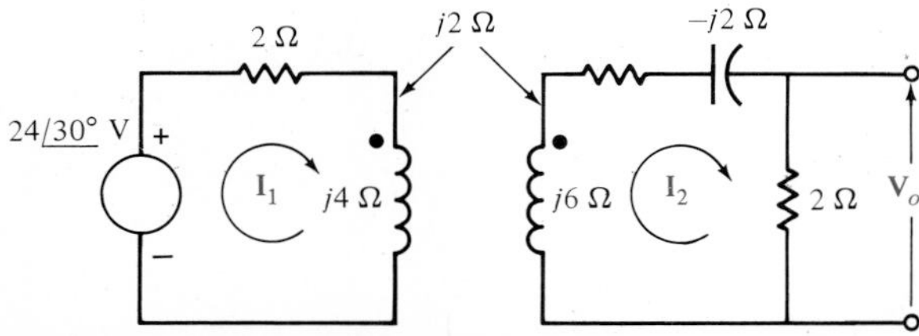

Figure 14.4 Example of a magnetically coupled circuit.

Solving the second equation for $\mathbf{I}_1$ and substituting it into the first equation yields

$$(2 - j1)(2 + j4)\mathbf{I}_2 - j2\mathbf{I}_2 = 24 \underline{/30^\circ}$$

$$\mathbf{I}_2 = \frac{24 \underline{/30^\circ}}{8 + j4}$$

$$= 2.68 \underline{/3.43^\circ} \text{ A}$$

Therefore,

$$\mathbf{V}_o = 2\mathbf{I}_2$$

$$= 5.37 \underline{/3.43^\circ} \text{ V}$$

■

Let us now consider a more complicated example involving mutual inductance.

EXAMPLE 14.4

Consider the circuit in Fig. 14.5. We wish to write the mesh equations for this network. Because of the multiple currents that are present in the coupled inductors, we must be very careful in writing the circuit equations.

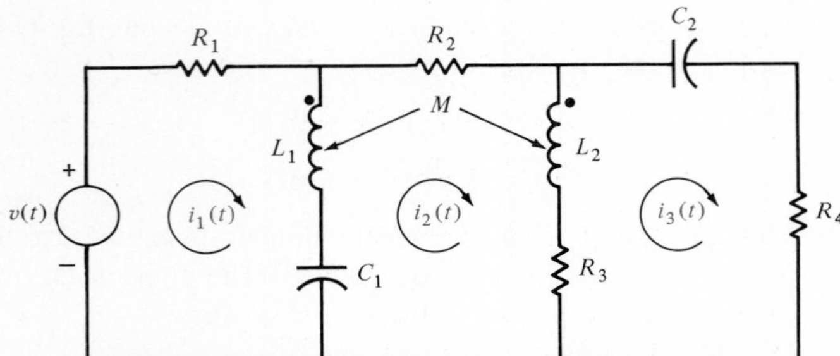

Figure 14.5 Example of a magnetically coupled circuit.

The mesh equations for the network are

$$\mathbf{I}_1 R_1 + j\omega L_1(\mathbf{I}_1 - \mathbf{I}_2) + j\omega M(\mathbf{I}_2 - \mathbf{I}_3) + \frac{1}{j\omega C_1}(\mathbf{I}_1 - \mathbf{I}_2) = \mathbf{V}$$

$$\frac{1}{j\omega C_1}(\mathbf{I}_2 - \mathbf{I}_1) + j\omega L_1(\mathbf{I}_2 - \mathbf{I}_1) + j\omega M(\mathbf{I}_3 - \mathbf{I}_2) + R_2\mathbf{I}_2 + j\omega L_2(\mathbf{I}_2 - \mathbf{I}_3)$$

$$- j\omega M(\mathbf{I}_2 - \mathbf{I}_1) + R_3(\mathbf{I}_2 - \mathbf{I}_3) = 0$$

$$R_3(\mathbf{I}_3 - \mathbf{I}_2) + j\omega L_2(\mathbf{I}_3 - \mathbf{I}_2) + j\omega M(\mathbf{I}_2 - \mathbf{I}_1) + \frac{1}{j\omega C_2}\mathbf{I}_3 + R_4\mathbf{I}_3 = 0$$

which can be rewritten in the form

$$\left(R_1 + j\omega L_1 + \frac{1}{j\omega C_1}\right)\mathbf{I}_1 - \left(j\omega L_1 + \frac{1}{j\omega C_1} - j\omega M\right)\mathbf{I}_2 - j\omega M\mathbf{I}_3 = \mathbf{V}$$

$$-\left(j\omega L_1 + \frac{1}{j\omega C_1} - j\omega M\right)\mathbf{I}_1 + \left(\frac{1}{j\omega C_1} + j\omega L_1 + R_2 + j\omega L_2 + R_3 - j2\omega M\right)\mathbf{I}_2 - (j\omega L_2 + R_3 - j\omega M)\mathbf{I}_3 = 0$$

$$-j\omega M\mathbf{I}_1 - (R_3 + j\omega L_2 - j\omega M)\mathbf{I}_2 + \left(R_3 + j\omega L_2 + \frac{1}{j\omega C_2} + R_4\right)\mathbf{I}_3 = 0$$

Note the symmetrical form of these equations. ■

DRILL EXERCISE

D14.2. Find the currents $\mathbf{I}_1$ and $\mathbf{I}_2$ and the output voltage $\mathbf{V}_0$ in the network in Fig. D14.2.

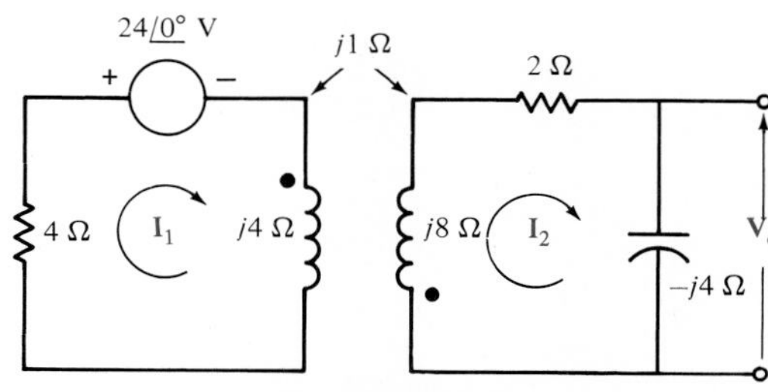

Figure D14.2

D14.3. Write the KVL equations in standard form for the network in Fig. D14.3.

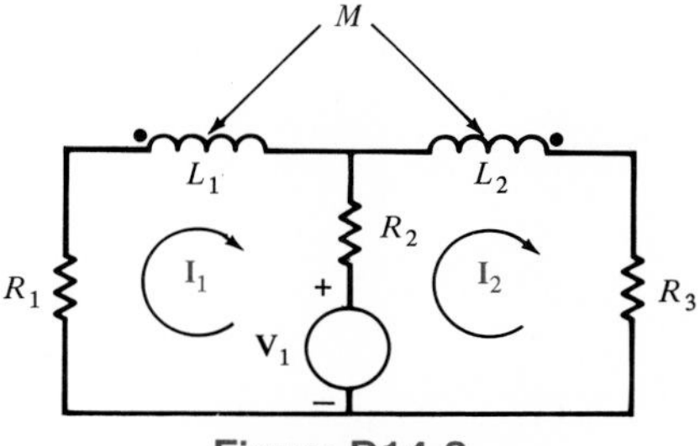

Figure D14.3

D14.4. Write the mesh equations for the network in Fig. D14.4.

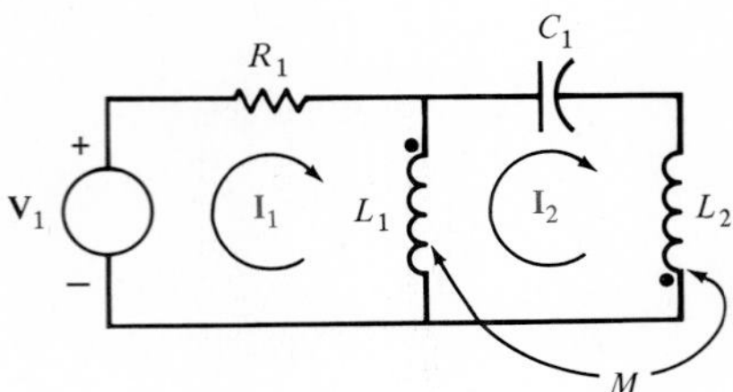

Figure D14.4

14.2 Energy Analysis

We now perform an energy analysis on a pair of mutually coupled inductors, which will yield some interesting relationships for the circuit elements. Our analysis will involve the performance of an experiment on the network shown in Fig. 14.6. Before beginning the experiment we set all voltages and currents in the circuit equal to zero. Once the circuit is quiescent, we begin by letting the current $i_1(t)$ increase from zero to some value I_1 with the right-side terminals open-circuited. Since the right-side terminals are open, $i_2(t) = 0$, and therefore the power entering these terminals is zero. The instantaneous power entering the left-side terminals is

$$p(t) = v_1(t)i_1(t) = \left[L_1 \frac{di_1(t)}{dt}\right] i_1(t)$$

The energy stored within the coupled circuit at t_1 when $i_1(t) = I_1$ is then

$$\int_0^{t_1} v_1(t)i_1(t)\,dt = \int_0^{I_1} L_1 i_1(t)\,di_1(t) = \tfrac{1}{2}L_1I_1^2$$

Continuing our experiment, starting at time t_1, we let the current $i_2(t)$ increase from zero to some value I_2 at time t_2 while holding $i_1(t)$ constant at I_1. The energy delivered through the right-side terminals is

$$\int_{t_1}^{t_2} v_2(t)i_2(t)\,dt = \int_0^{I_2} L_2 i_2(t)\,di_2(t) = \tfrac{1}{2}L_2I_2^2$$

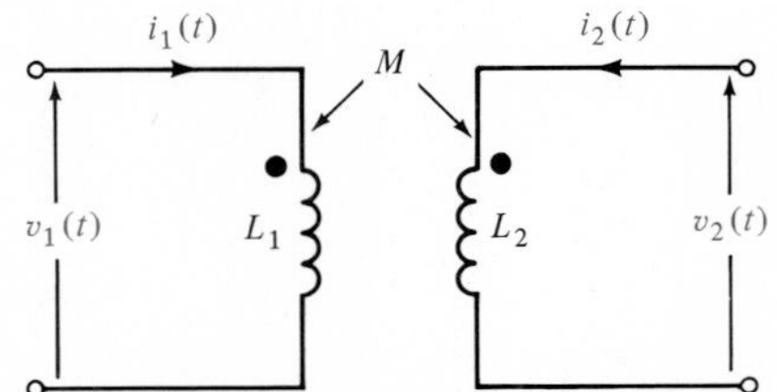

Figure 14.6 Magnetically coupled circuit.

However, during the interval t_1 to t_2 the voltage $v_1(t)$ is

$$v_1(t) = L_1 \frac{di_1(t)}{dt} + M_{12} \frac{di_2(t)}{dt}$$

where the subscript on M indicates a voltage induced in L_1 due to a current in L_2. Since $i_1(t)$ is a constant I_1, the energy delivered through the left-side terminals is

$$\int_{t_1}^{t_2} v_1(t)i_1(t)\, dt = \int_{t_1}^{t_2} M_{12} \frac{di_2(t)}{dt} I_1\, dt = M_{12}I_1 \int_0^{I_2} di_2(t)$$

$$= M_{12}I_1I_2$$

Therefore, the total energy stored in the network for $t > t_2$ is

$$w = \tfrac{1}{2}L_1I_1^2 + \tfrac{1}{2}L_2I_2^2 + M_{12}I_1I_2$$

Note that if in our experiment we had simply reversed the order in applying currents to the network [i.e., let $i_2(t)$ increase from zero to I_2 and then while holding $i_2(t)$ constant at I_2 let $i_1(t)$ increase from zero to I_1], the total energy delivered would be

$$w = \tfrac{1}{2}L_1I_1^2 + \tfrac{1}{2}L_2I_2^2 + M_{21}I_1I_2$$

Comparing the last two expressions, we find that the only difference between them is the mutual inductance terms. However, since the initial and final conditions in the network are identical, the energy stored in the network must be the same and hence

$$M_{12} = M_{21} = M \tag{14.10}$$

and therefore the expression for the total energy is

$$w = \tfrac{1}{2}L_1I_1^2 + \tfrac{1}{2}L_2I_2^2 + MI_1I_2$$

We could, of course, repeat our entire experiment with either the dot on L_1 or L_2, but not both reversed, and in this case the sign on the mutual inductance term would be negative. Therefore, in general, the expression for the total energy delivered or absorbed is

$$w = \tfrac{1}{2}L_1I_1^2 + \tfrac{1}{2}L_2I_2^2 \pm MI_1I_2$$

where the sign is chosen as + if the fluxes produced by I_1 and I_2 add, and − if the fluxes produced by I_1 and I_2 oppose one another (i.e., if both I_1 and I_2 enter or leave dotted terminals, the + sign is used).

It is very important for the reader to realize that in our derivation of the equation above, by means of the experiment, the values I_1 and I_2 could have been any values at any two *instants of time*; therefore, the energy stored in the magnetically coupled inductors at any instant of time is given by the expression

$$w(t) = \tfrac{1}{2}L_1[i_1(t)]^2 + \tfrac{1}{2}L_2[i_2(t)]^2 \pm Mi_1(t)i_2(t) \tag{14.11}$$

The two coupled inductors represent a passive network, and therefore the energy stored within this network must be nonnegative for any values of the inductances and currents. The equation for the instantaneous energy stored in the magnetic circuit can be written as

$$w(t) = \tfrac{1}{2}L_1i_1^2 + \tfrac{1}{2}L_2i_2^2 \pm Mi_1i_2$$

Adding and subtracting the term $\sqrt{L_1L_2}i_1i_2$, we obtain

$$w(t) = \tfrac{1}{2}L_1i_1^2 + \tfrac{1}{2}L_2i_2^2 \pm Mi_1i_2 + \sqrt{L_1L_2}i_1i_2 - \sqrt{L_1L_2}i_1i_2$$

Rearranging the order of the terms yields

$$w(t) = \tfrac{1}{2}L_1i_1^2 - \sqrt{L_1L_2}i_1i_2 + \tfrac{1}{2}L_2i_2^2 + \sqrt{L_1L_2}i_1i_2 \pm Mi_1i_2$$

which we recognize can be rewritten as

$$w(t) = \tfrac{1}{2}(\sqrt{L}i_1 - \sqrt{L_2}i_2)^2 + (\sqrt{L_1L_2} \pm M)i_1i_2$$

Since the instantaneous energy must be nonnegative for all values of i_1 and i_2,

$$M \leq \sqrt{L_1L_2} \tag{14.12}$$

Note that this equation specifies an upper limit on the value of the mutual inductance.

We define the coefficient of coupling between the two inductors L_1 and L_2 as

$$k = \frac{M}{\sqrt{L_1L_2}} \tag{14.13}$$

and we note from equation (14.12) that its range of values is

$$0 \leq k \leq 1 \tag{14.14}$$

This coefficient is an indication of how much flux in one coil is linked with the other coil; that is, if all the flux in one coil reaches the other coil, then we essentially have 100% coupling and $k = 1$. For large values of k (i.e., $k > 0.5$), the inductors are said to be tightly coupled, and for small values of k (i.e., $k < 0.5$), the coils are said to be loosely coupled. The previous equations indicate that the value for the mutual inductance is confined to the range

$$0 \leq M \leq \sqrt{L_1L_2} \tag{14.15}$$

and that the upper limit is the geometric mean of the inductances L_1 and L_2.

EXAMPLE 14.5

The coupled circuit in Fig. 14.7a has a coefficient of coupling of 1 (i.e., $k = 1$). We wish to determine the energy stored in the mutually coupled inductors at time $t = 5$ ms. $L_1 = 2.653$ mH and $L_2 = 10.61$ mH.

From the data the mutual inductance is

$$M = \sqrt{L_1L_2} = 5.31 \text{ mH}$$

The frequency domain equivalent circuit is shown in Fig. 14.7b, where the impedance values for X_{L_1}, X_{L_2}, and X_M are 1, 4, and 2, respectively. The loop equations for the network are then

$$(2 + j1)\mathbf{I}_1 - 2j\mathbf{I}_2 = 24 \underline{/0^\circ}$$

$$-j2\mathbf{I}_1 + (4 + 4j)\mathbf{I}_2 = 0$$

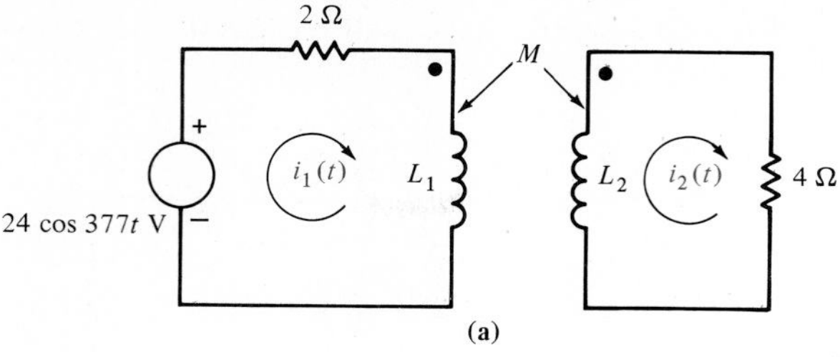

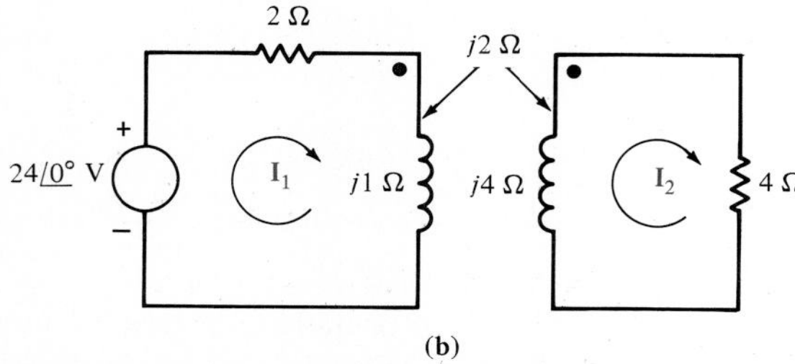

Figure 14.7 Example magnetically coupled circuit drawn in the time and frequency domains.

Solving these equations for the two loop currents yields

$$\mathbf{I}_1 = 9.41 \angle -11.31^\circ \text{ A} \quad \text{and} \quad \mathbf{I}_2 = 3.33 \angle +33.69^\circ \text{ A}$$

and therefore

$$i_1(t) = 9.41 \cos (377t - 11.31^\circ) \text{ A}$$

$$i_2(t) = 3.33 \cos (377t + 33.69^\circ) \text{ A}$$

At $t = 5$ ms. $377t = 1.885$ rad or 108°, and therefore

$$i_1(t = 5 \text{ ms}) = 9.41 \cos (108^\circ - 11.31^\circ) = -1.10 \text{ A}$$

$$i_2(t = 5 \text{ ms}) = 3.33 \cos (108^\circ + 33.69^\circ) = -2.61 \text{ A}$$

Therefore, the energy stored in the coupled inductors at $t = 5$ ms is

$$\begin{aligned} w(t)\big|_{t=0.005 \text{ s}} &= \tfrac{1}{2}(2.563 \times 10^{-3})(-1.10)^2 + \tfrac{1}{2}(10.61 \times 10^{-3})(-2.61)^2 \\ &\quad -(5.31 \times 10^{-3})(-1.10)(-2.61) \\ &= 1.55 \times 10^{-3} + 36.14 \times 10^{-3} - 15.25 \times 10^{-3} \\ &= 22.44 \text{ mJ} \end{aligned}$$

■

DRILL EXERCISE

D14.5. Two coils in a network are positioned such that there is 100% coupling between them. If the inductance of one coil is 10 mH and the mutual inductance is 6 mH, compute the inductance of the other coil.

D14.6. The network in Fig. D14.6 operates at 60 Hz. Compute the energy stored in the mutually coupled inductors at time $t = 10$ ms.

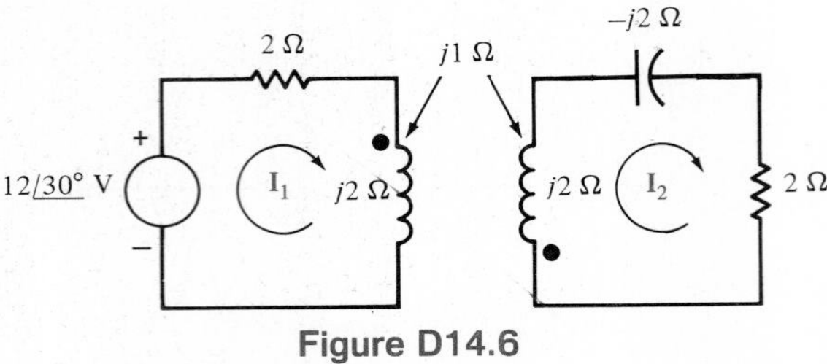

Figure D14.6

14.3 The Linear Transformer

A transformer is a device that contains two or more coils that are coupled magnetically. A typical transformer network is shown in Fig. 14.8. The source is connected to what is called the *primary* of the transformer, and the load is connected to the *secondary*. Thus R_1 and L_1 refer to the resistance and inductance of the primary, and R_2 and L_2 refer to the secondary's resistance and inductance. The transformer is said to be *linear* if no magnetic material is used to concentrate the flux that links the two inductors. Without the use of magnetic material, the coefficient of coupling, k, is typically very small. Transformers of this type find wide application in such products as radio and TV receivers.

There are a number of ways in which we can model the magnetically coupled inductor portion of the transformer. The model already developed is shown in Fig. 14.9a. All the models are developed so that the equations relating $\mathbf{V}_1$, $\mathbf{V}_2$, $\mathbf{I}_1$, and $\mathbf{I}_2$ in Fig. 14.9a are satisfied. These equations are

$$\begin{aligned} \mathbf{V}_1 &= j\omega L_1\mathbf{I}_1 + j\omega M\mathbf{I}_2 \\ \mathbf{V}_2 &= j\omega L_2\mathbf{I}_2 + j\omega M\mathbf{I}_1 \end{aligned} \tag{14.16}$$

One technique involves the use of dependent sources as shown in Fig. 14.9b. The

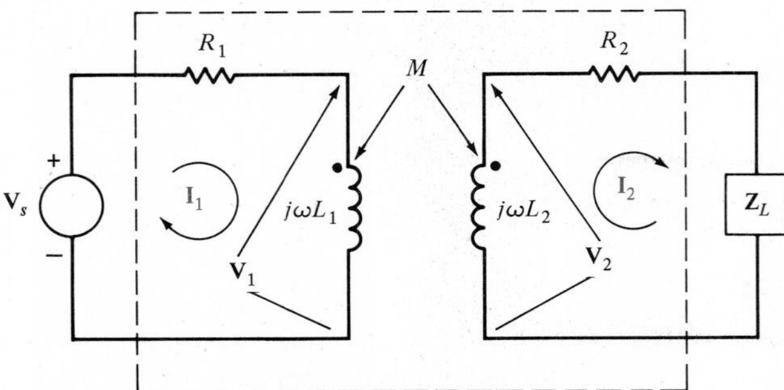

Figure 14.8 Transformer network.

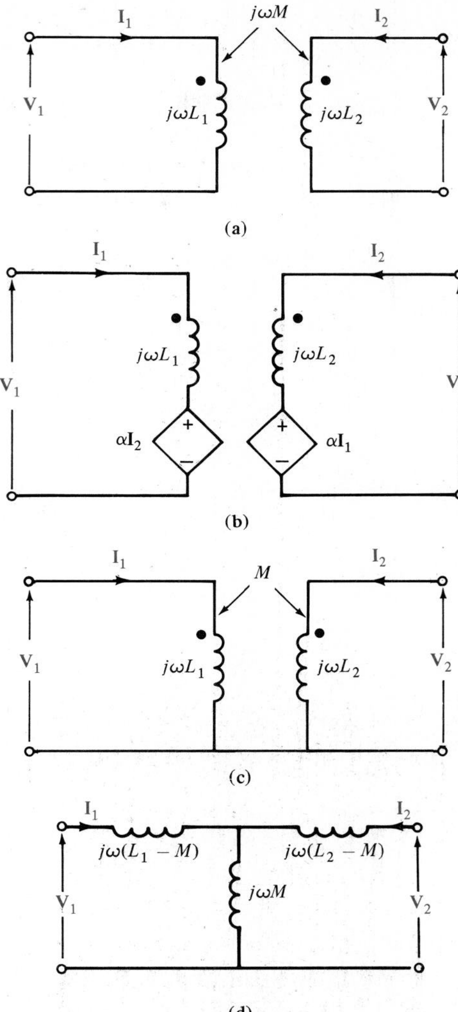

Figure 14.9 Transformer equivalent circuits.

circuit equations for this model are

$$\begin{aligned} \mathbf{V}_1 &= j\omega L_1\mathbf{I}_1 + \alpha\mathbf{I}_2 \\ \mathbf{V}_2 &= \alpha\mathbf{I}_1 + j\omega L_2\mathbf{I}_2 \end{aligned} \tag{14.17}$$

Comparing Eq. (14.16) and (14.17), we note that $\alpha = j\omega M$. Another network involves replacing the transformer with an equivalent T network. This equivalent circuit is valid only if the four-terminal network can be replaced by the three-terminal network shown in Fig. 14.9c. These circuit equations are

$$\begin{aligned} \mathbf{V}_1 &= j\omega L_1\mathbf{I}_1 + j\omega M\mathbf{I}_2 \\ \mathbf{V}_2 &= j\omega M\mathbf{I}_1 + j\omega L_2\mathbf{I}_2 \end{aligned} \tag{14.18}$$

If we simply examine these equations as they are given, they appear to represent two mesh equations in which the common element between the two loops is an inductor of value M. In order for the total inductance in the first mesh to be L_1 and the inductance of the common element to be M, as indicated in the first equation above, an inductor of value $L_1 - M$ must be placed in the first mesh of the equivalent circuit. A similar argument for the second mesh indicates that an inductor of value $L_2 - M$ must be placed in that mesh of the equivalent circuit. The equivalent circuit resulting from this analysis is shown in Fig. 14.9d. Note that this network satisfies the circuit Eqs. (14.18).

If the relationship between the currents and the dots are such that the mutual terms in Eq. (14.18) are negative, we may simply replace M by $-M$ in the equivalent network in Fig. 14.9d. From a mathematical modeling standpoint a negative inductance presents no problems; however, we must remember that such a physical element does not actually exist.

EXAMPLE 14.6

The coefficient of coupling for the linear transformer in the network in Fig. 14.10a is $k = 0.1$. We wish to determine the equivalent T network for the transformer and redraw the circuit using this equivalent model.

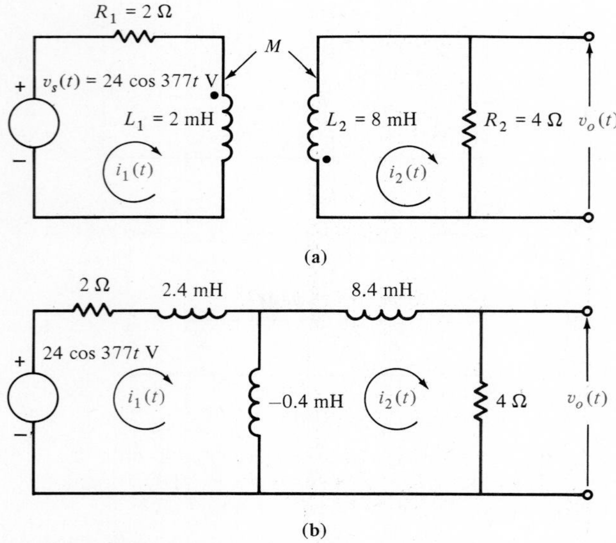

Figure 14.10 Example transformer circuit together with its T-equivalent circuit.

Using the relationship $k = M/\sqrt{L_1L_2}$, we find that

$$\begin{aligned} M &= k\sqrt{L_1L_2} \\ &= 0.1\sqrt{16 \times 10^{-6}} \\ &= 0.4 \text{ mH} \end{aligned}$$

Since both currents are into the dot, the equations for the original network are

$$\mathbf{V}_s = R_1\mathbf{I}_1 + j\omega L_1\mathbf{I}_1 + j\omega M\mathbf{I}_2$$

$$0 = j\omega M\mathbf{I}_1 + j\omega L_2\mathbf{I}_2 + R_2\mathbf{I}_2$$

Note that the mutual term is positive in these equations. If in the T network we use the same direction for the currents, the mutual term in the first equation will be positive only if the inductor in the center of the T network is negative. Therefore, the center inductor is chosen to have a value of $-M$, and the other two inductors have values $L_1 + M$ and $L_2 + M$, as shown in Fig. 14.10b. Write the loop equations for the network in Fig. 14.10b to verify that they are the same as the equations above. ■

With reference to Fig. 14.8, let us compute the input impedance to the transformer as seen by the source. The network equations are

$$\mathbf{V}_s = \mathbf{I}_1(R_1 + j\omega L_1) - j\omega M\mathbf{I}_2$$

$$0 = -j\omega M\mathbf{I}_1 + (R_2 + j\omega L_2 + \mathbf{Z}_L)\mathbf{I}_2$$

Solving the second equation for $\mathbf{I}_2$ and substituting it into the first equation yields

$$\mathbf{V}_s = \left(R_1 + j\omega L_1 + \frac{\omega^2 M^2}{R_2 + j\omega L_2 + \mathbf{Z}_L}\right)\mathbf{I}_1$$

Therefore, the input impedance is

$$\mathbf{Z}_i = \frac{\mathbf{V}_s}{\mathbf{I}_1} = R_1 + j\omega L_1 + \frac{\omega^2 M^2}{R_2 + j\omega L_2 + \mathbf{Z}_L} \tag{14.19}$$

As we look from the source into the network to determine $\mathbf{Z}_i$, we see the impedance of the primary (i.e., $R_1 + j\omega L_1$) plus an impedance that the secondary of the transformer reflects, due to mutual coupling, into the primary. This *reflected impedance* is

$$\mathbf{Z}_R = \frac{\omega^2 M^2}{R_2 + j\omega L_2 + \mathbf{Z}_L} \tag{14.20}$$

Note that this reflected impedance is independent of the dot locations, since M^2 will be positive regardless of the sign on M.

If $\mathbf{Z}_L$ in Eq. (14.20) is written as

$$\mathbf{Z}_L = R_L + jX_L \tag{14.21}$$

then

$$\mathbf{Z}_R = \frac{\omega^2 M^2}{R_2 + R_L + j(\omega L_2 + X_L)}$$

which can be written as

$$\mathbf{Z}_R = \frac{\omega^2 M^2[(R_2 + R_L) - j(\omega L_2 + X_L)]}{(R_2 + R_L)^2 + (\omega L_2 + X_L)^2} \tag{14.22}$$

This equation illustrates that if X_L is an inductive reactance, or if X_L is a capacitive reactance with $\omega L_2 > X_L$, then the reflected reactance is capacitive. In general, the reflected reactance is opposite in sign to that of the total reactance in the secondary. If $\omega L_2 + X_L = 0$ (i.e., the secondary is in resonance), $\mathbf{Z}_R$ is purely resistive and

$$\mathbf{Z}_R = \frac{\omega^2 M^2}{R_2 + R_L} \tag{14.23}$$

EXAMPLE 14.7

For the network shown in Fig. 14.11 we wish to determine the input impedance.

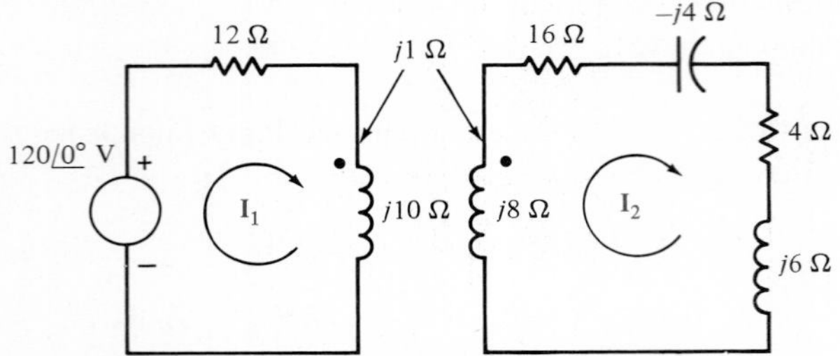

Figure 14.11 Example transformer circuit.

Following the development that led to Eq. (14.19), the input impedance is

$$\begin{aligned}\mathbf{Z}_i &= 12 + j10 + \frac{(1)^2}{16 + j8 - j4 + 4 + j6} \\ &= 12.04 + j9.98 \\ &= 15.64 \angle 39.65^\circ \ \Omega\end{aligned}$$

■

DRILL EXERCISE

D14.7. Given the network in Fig. D14.7, find the input impedance of the network and the current in the voltage source.

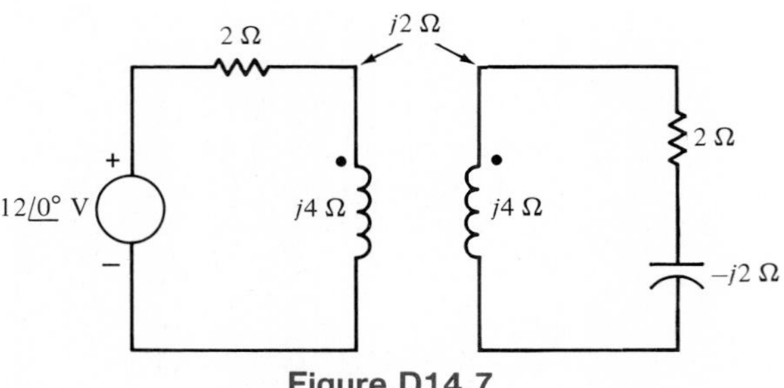

Figure D14.7

14.4 The Ideal Transformer

An ideal transformer is one in which the coupling coefficient is 1 and the inductive reactance of both the primary and secondary are extremely large in comparison with the load impedance. This model is a good approximation for a transformer that is constructed

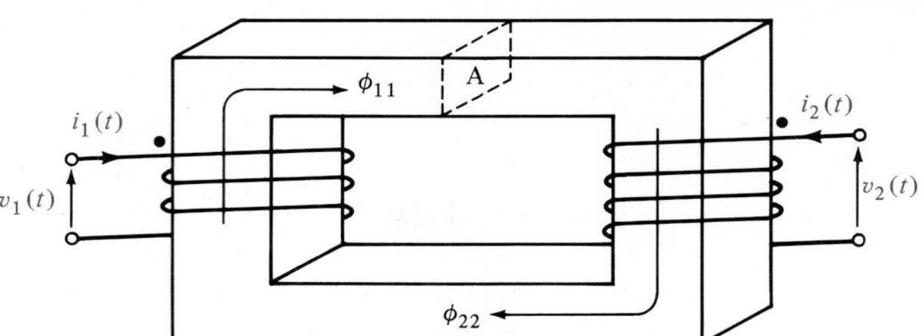

Figure 14.12 Transformer employing a magnetic core.

of two coils of wire wound on a single closed magnetic core as shown in Fig. 14.12. The magnetic core concentrates the flux so that all the flux links all the turns of both coils.

For the ideal case when good core material is used, let us examine the coupling equations under the condition that both coils are wrapped around the same core. Under this condition the same flux goes through each winding and therefore

$$v_1(t) = N_1 \frac{d\phi_1}{dt} = N_1 \frac{d\phi}{dt}$$

and

$$v_2(t) = N_2 \frac{d\phi_2}{dt} = N_2 \frac{d\phi}{dt}$$

and therefore

$$\frac{v_1}{v_2} = \frac{N_1}{N_2} \tag{14.24}$$

Another relationship can be developed between the currents $i_1(t)$ and $i_2(t)$ and the number of turns in each coil. In order to develop this relationship we employ, from electromagnetic field theory. Ampere's law, which is written in mathematical form as

$$\oint H \cdot dl = Hl = \frac{B}{\mu} l = \frac{\phi}{A\mu} l = N_1 i_1 \pm N_2 i_2 \tag{14.25}$$

Where H is the magnetic field intensity, the integral is over the closed path traveled by the flux around the transformer, B is the flux density, μ the permeability of the core material, l the mean length of the path traveled by the flux, ϕ the flux, and A the cross-sectional area of the core. The plus or minus sign is chosen depending on the relationship between the currents and the dots. For good transformer core material the μ is approximately 2000 times the μ of air, and in the ideal transformer we will assume that it is infinite (i.e., $\mu = \infty$). Therefore,

$$N_1 i_1 \pm N_2 I_2 = 0 \tag{14.26}$$

which for the case shown in Fig. 14.12 is (i.e., both currents entering the dots),

$$N_1 i_1 + N_2 i_2 = 0$$

or

$$\frac{i_1}{i_2} = -\frac{N_2}{N_1} \tag{14.27}$$

Therefore, in the case of the ideal transformer the positive sign in Eq. (14.26) is used if both currents enter or leave the dots, and the negative sign is used if one current enters the dot and the other current leaves the dot.

Therefore, to summarize the dot convention for an ideal transformer,

$$\frac{\mathbf{V}_1}{\mathbf{V}_2} = \frac{\pm N_1}{N_2} \tag{14.28}$$

where the + sign is used if both voltages are referenced to the dots and the − sign is used otherwise, and

$$\frac{\mathbf{I}_1}{\mathbf{I}_2} = \pm\frac{N_2}{N_1} \tag{14.29}$$

where the − sign is used if both currents are entering or leaving the dots, and the + sign is used otherwise.

Consider now the circuit shown in Fig. 14.13, where the symbol used for the transformer indicates that it is an iron-core transformer. Because of the relationship between the dots and both the assigned currents and voltages, the voltages $\mathbf{V}_1$ and $\mathbf{V}_2$ are related by the expression

$$\frac{\mathbf{V}_1}{\mathbf{V}_2} = \frac{N_1}{N_2}$$

and the currents, from Eq. (14.29), are related by

$$\frac{\mathbf{I}_1}{\mathbf{I}_2} = \frac{N_2}{N_1}$$

These two equations can be rewritten as

$$\mathbf{V}_1 = \frac{N_1}{N_2}\mathbf{V}_2$$

$$\mathbf{I}_1 = \frac{N_2}{N_1}\mathbf{I}_2$$

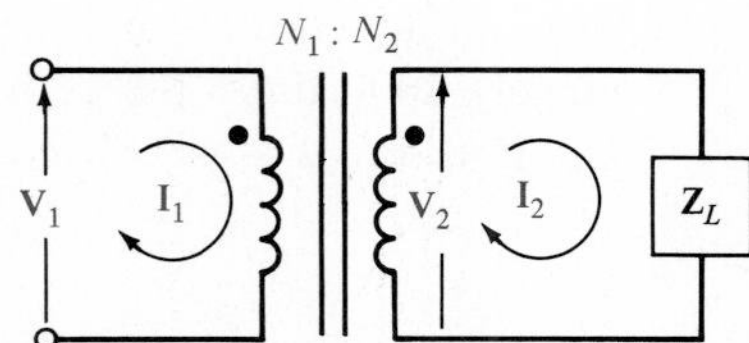

Figure 14.13 Ideal transformer circuit used to illustrate input impedance.

From the figure we note that $\mathbf{Z}_L = \mathbf{V}_2/\mathbf{I}_2$, and therefore

$$\mathbf{Z}_i = \frac{\mathbf{V}_1}{\mathbf{I}_1} = \left(\frac{N_1}{N_2}\right)^2 \mathbf{Z}_L \tag{14.30}$$

If we now define the turns ratio as

$$n = \frac{N_2}{N_1} \tag{14.31}$$

then the defining equations for the *ideal transformer* are

$$\begin{aligned} \mathbf{V}_1 &= \pm \frac{\mathbf{V}_2}{n} \\ \mathbf{I}_1 &= \pm n\mathbf{I}_2 \\ \mathbf{Z}_i &= \frac{\mathbf{Z}_L}{n^2} \end{aligned} \tag{14.32}$$

Equations (14.32) define the important relationships for an ideal transformer. Care must be exercised in using these relationships because the signs on the voltages and currents are dependent on the assigned references and how they are related to the dots as illustrated earlier.

EXAMPLE 14.8

We wish to determine the impedance seen by the source in the circuit in Fig. 14.14 if $\mathbf{Z}_L = 10 \angle 30^\circ \ \Omega$.

The reflected impedance is

$$\begin{aligned} \mathbf{Z}_F &= \frac{10 \angle 30^\circ}{(4)^2} \\ &= 0.54 + j0.313 \ \Omega \end{aligned}$$

Therefore, the input impedance is

$$\begin{aligned} \mathbf{Z}_i &= 4 - j2 + 0.54 + j0.313 \\ &= 4.54 - j1.69 \ \Omega \end{aligned}$$

■

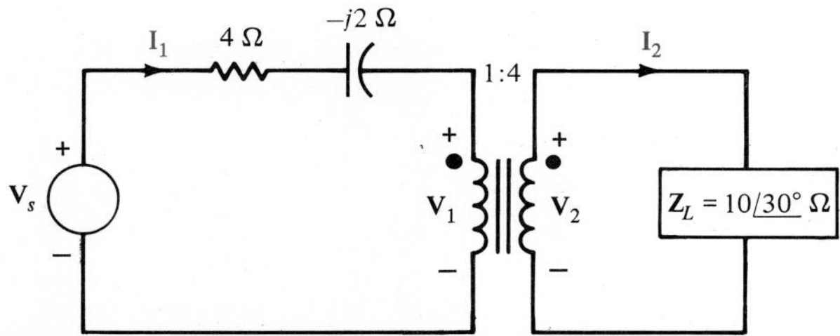

Figure 14.14 Example circuit containing an ideal transformer.

EXAMPLE 14.9

Given the circuit shown in Fig. 14.15, we wish to determine all voltages and currents.

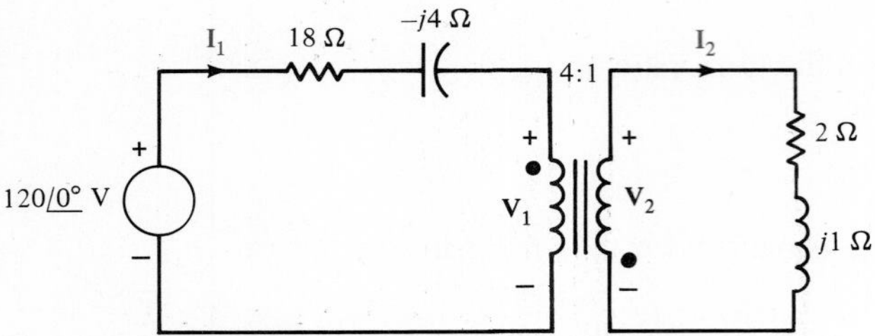

Figure 14.15 Ideal transformer circuit.

Because of the relationships between the dots and the currents and voltages, the defining equations for the transformer are

$$\mathbf{V}_1 = -\frac{\mathbf{V}_2}{n} \quad \text{and} \quad \mathbf{I}_1 = -n\mathbf{I}_2$$

where $n = \frac{1}{4}$. The reflected impedance at the input to the transformer is

$$\mathbf{Z}_F = 16(2 + j1) = 32 + j16\ \Omega$$

Therefore, the current in the source is

$$\mathbf{I}_1 = \frac{120\ \underline{/0^\circ}}{18 - j4 + 32 + j16} = 2.23\ \underline{/-13.5^\circ}\ \text{A}$$

The voltage across the input to the transformer is then

$$\begin{aligned}\mathbf{V}_1 &= \mathbf{I}_1\mathbf{Z}_F \\ &= (2.33\ \underline{/-13.5^\circ})(32 + j16) \\ &= 83.50\ \underline{/13.07^\circ}\ \text{V}\end{aligned}$$

Hence $\mathbf{V}_2$ is

$$\begin{aligned}\mathbf{V}_2 &= -n\mathbf{V}_1 \\ &= -\tfrac{1}{4}(83.50\ \underline{/13.07^\circ}) \\ &= 20.88\ \underline{/193.07^\circ}\ \text{V}\end{aligned}$$

The current $\mathbf{I}_2$ is

$$\begin{aligned}\mathbf{I}_2 &= -\frac{\mathbf{I}_1}{n} \\ &= -4(2.33\ \underline{/-13.5^\circ}) \\ &= 9.32\ \underline{/166.50^\circ}\ \text{A}\end{aligned}$$

■

DRILL EXERCISE

D14.8. Compute the current $\mathbf{I}_1$ in the network in Fig. D14.8.

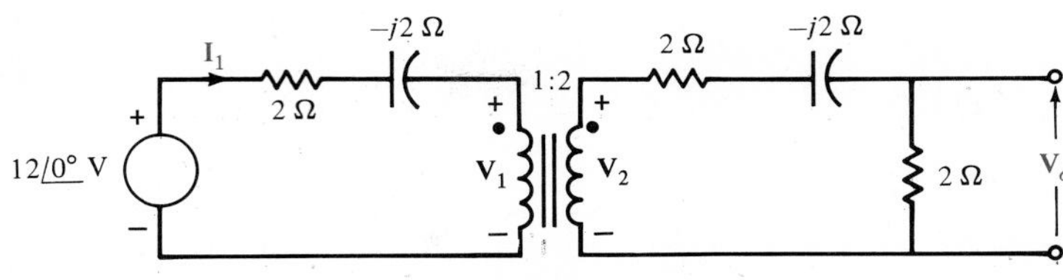

Figure D14.8

D14.9. Find $\mathbf{V}_o$ in the network in Fig. D14.8.

D14.10. In the network in Fig. D14.10 the voltage $\mathbf{V}_0 = 10\,\underline{/0^\circ}$ V. Find the input voltage $\mathbf{V}_s$.

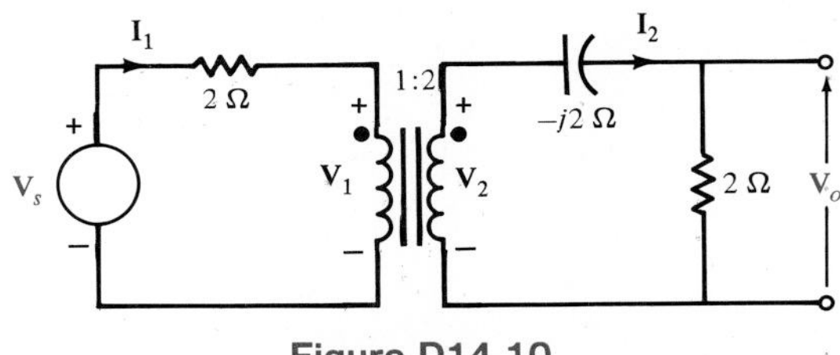

Figure D14.10

Another technique for simplifying the analysis of circuits containing an ideal transformer involves the use of either Thévenin's or Norton's theorems to obtain an equivalent circuit that replaces the transformer and either the primary or secondary circuit. Let us demonstrate this approach by employing Thévenin's theorem to derive an equivalent circuit for the transformer and primary circuit of the network shown in Fig. 14.16a. The equations for the transformer in view of the direction of the currents and voltages and the position of the dots are

$$\mathbf{I}_1 = n\mathbf{I}_2$$

$$\mathbf{V}_1 = \frac{\mathbf{V}_2}{n}$$

Forming a Thévenin equivalent at the secondary terminals as shown in Fig. 14.16b, we note that $\mathbf{I}_2 = 0$ and therefore $\mathbf{I}_1 = 0$. (Recall that the inductive reactance of both the primary and secondary is infinite.) Hence

$$\mathbf{V}_{oc} = \mathbf{V}_2 = n\mathbf{V}_1 = n\mathbf{V}_{s_1}$$

The Thévenin equivalent impedance obtained by the looking into the open-circuit terminals with $\mathbf{V}_{s_1}$ short-circuited is $\mathbf{Z}_1$, reflected into the secondary by the turns ratio, that is,

$$\mathbf{Z}_{Th} = n^2\mathbf{Z}_1$$

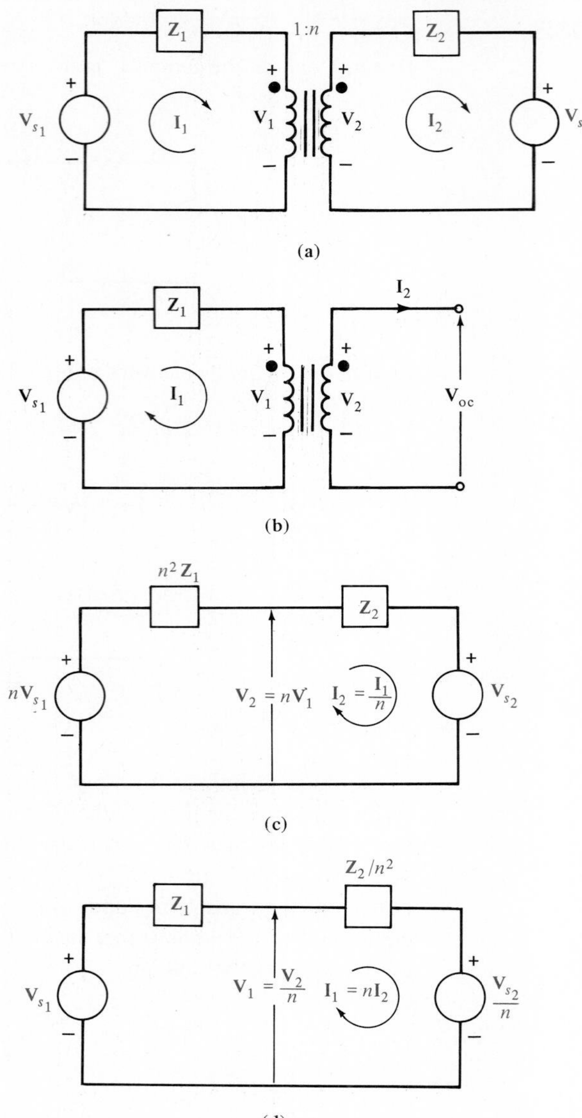

Figure 14.16 Circuit containing an ideal transformer and some of its equivalent networks.

Therefore, one of the resulting equivalent circuits for the network in Fig. 14.16a is as shown in Fig. 14.16c. In a similar manner we can show that replacing the transformer and its secondary circuit by an equivalent circuit results in the network shown in Fig. 14.16d.

It can be shown in general that when developing an equivalent circuit for the transformer and its primary circuit, each primary voltage is multiplied by n, each primary current is divided by n, and each primary impedance is multiplied by n^2. Similarly, when developing an equivalent circuit for the transformer and its secondary circuit, each secondary voltage is divided by n, each secondary current is multiplied by n, and each secondary impedance is divided by n^2.

The reader should recall from our previous analysis that if either dot on the transformer is reversed, then n is replaced by $-n$ in the equivalent circuits. In addition, it should be noted that the development of these equivalent circuits is predicated on the assumption that removing the transformer will divide the network into two parts; that is, there are no connections between the primary and secondary other than through the transformer. If any external connections exist, the equivalent circuit technique cannot in general be used. Finally, it should be pointed out that if the primary or secondary circuits are more complicated than those shown in Fig. 14.16a, Thévenin's theorem may be applied to reduce the network to that shown in Fig. 14.16a. The following examples should clarify these points.

EXAMPLE 14.10

Given the circuit in Fig. 14.17a, we wish to draw the two networks obtained by replacing the transformer and the primary, and the transformer and the secondary, with equivalent circuits.

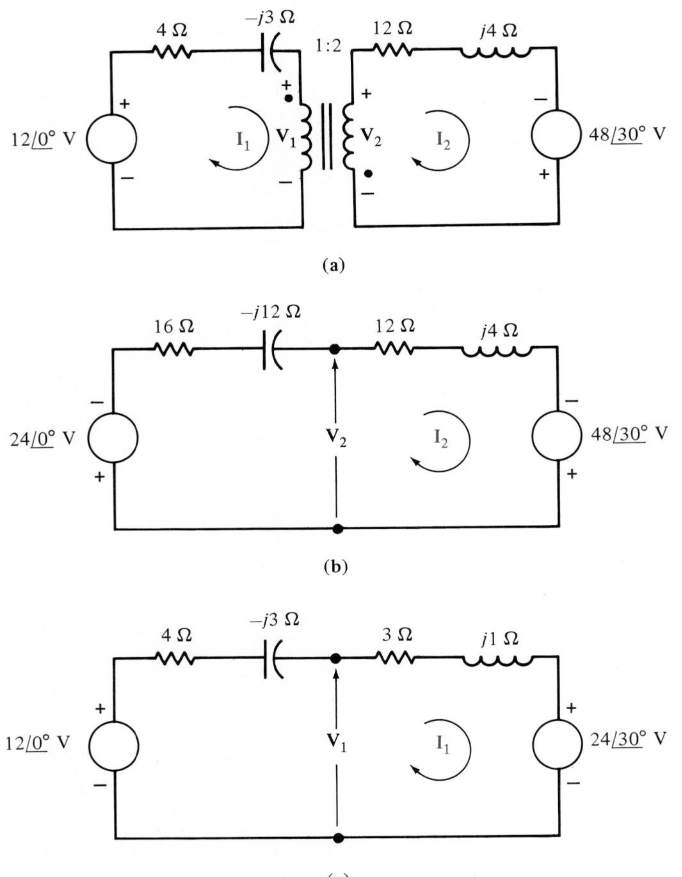

Figure 14.17 Example circuit and two equivalent circuits.

Due to the relationship between the assigned currents and voltages and the location of the dots, the network containing an equivalent circuit for the primary, and the network containing an equivalent circuit for the secondary, are shown in Fig. 14.17b and c, respectively. The reader should note carefully the polarity of the voltage sources in the equivalent networks. ■

EXAMPLE 14.11

We wish to derive an equivalent network for the circuit in Fig. 14.18a and use it to determine the current $\mathbf{I}_1$.

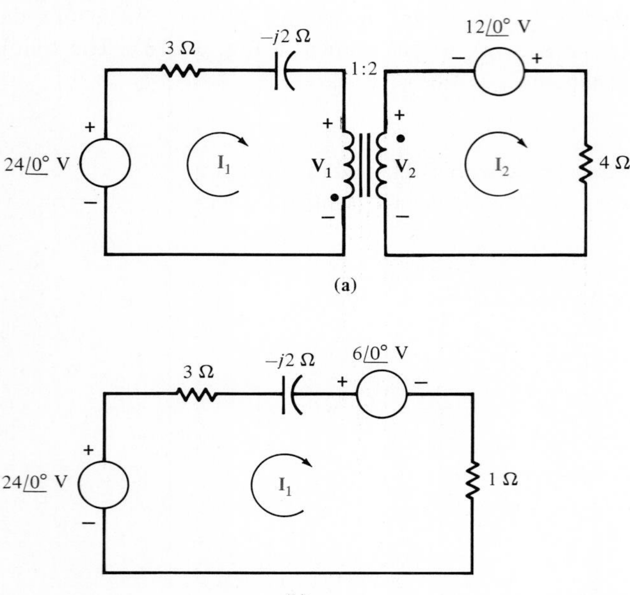

Figure 14.18 Example circuit containing an ideal transformer and an equivalent circuit.

For the given relationship between the assigned currents and voltages and the location of the dots, the network containing an equivalent circuit for the transformer and the secondary is shown in Fig. 14.18b. The current $\mathbf{I}_1$ is therefore

$$\begin{aligned}\mathbf{I}_1 &= \frac{24\,\underline{/0^\circ} - 6\,\underline{/0^\circ}}{3 - j2 + 1} \\ &= \frac{18\,\underline{/0^\circ}}{4 - j2} \\ &= 4.02\,\underline{/26.57^\circ}\text{ A}\end{aligned}$$

■

EXAMPLE 14.12

Let us determine the output voltage $\mathbf{V}_o$ in the circuit in Fig. 14.19a.

We begin our attack by forming a Thévenin equivalent for the primary circuit. From

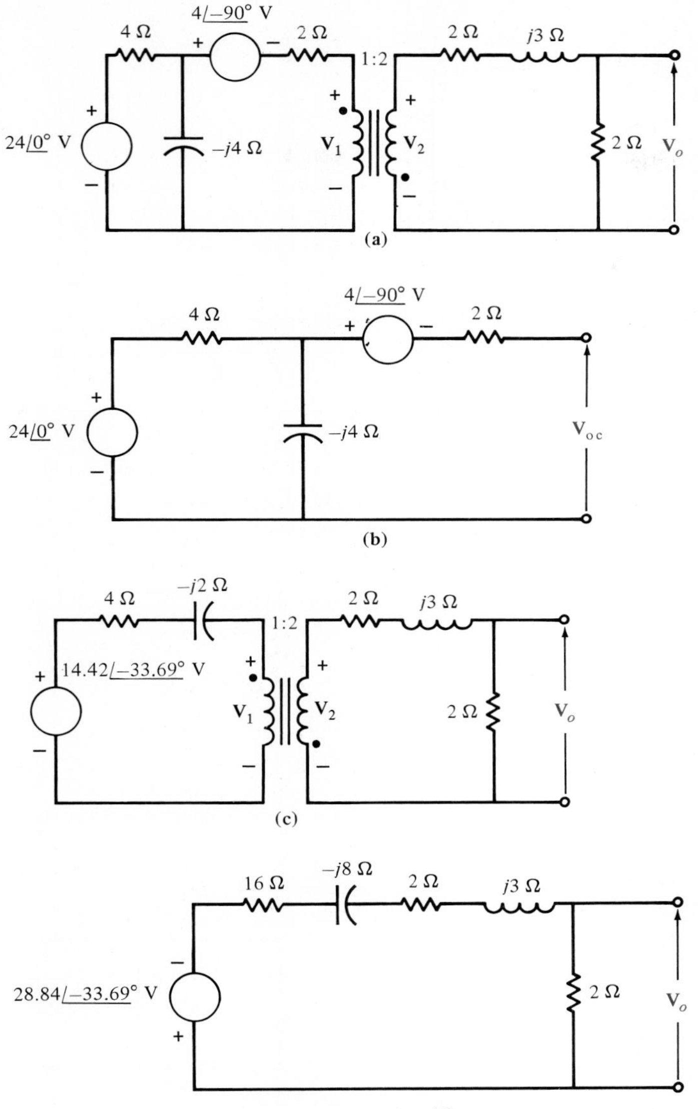

Figure 14.19 Example network and other circuits used to derive an equivalent network.

Fig. 14.19b we can show that the open-circuit voltage is

$$\mathbf{V}_{oc} = \frac{24 \underline{/0^\circ}}{4 - j4}(-j4) - 4 \underline{/-90^\circ}$$

$$= 12 - j8 = 14.42 \underline{/-33.69^\circ} \text{ V}$$

The Thévenin equivalent impedance looking into the open circuit terminals with the voltage sources short-circuited is

$$\mathbf{Z}_{\text{Th}} = \frac{(4)(-j4)}{4 - j4} + 2$$
$$= 4 - j2\ \Omega$$

The circuit in Fig. 14.19a thus reduces to that shown in Fig. 14.19c. Forming an equivalent circuit for the transformer and primary results in the network shown in Fig. 14.19d. Therefore, the voltage $\mathbf{V}_o$ is

$$\mathbf{V}_o = \frac{-28.84 \underline{/-33.69^\circ}}{20 - j5}(2)$$
$$= -2.80 \underline{/-19.65^\circ}\ \text{V}$$

■

DRILL EXERCISE

D14.11. Given the network in Fig. D14.11, form an equivalent circuit for the transformer and secondary, and use the resultant network to compute $\mathbf{I}_1$.

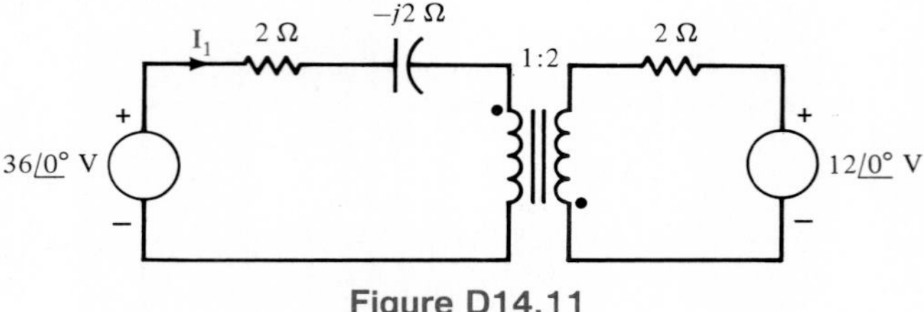

Figure D14.11

D14.12. Given the network in Fig. D14.12, form an equivalent circuit for the transformer and primary, and use the resultant network to find $\mathbf{V}_o$.

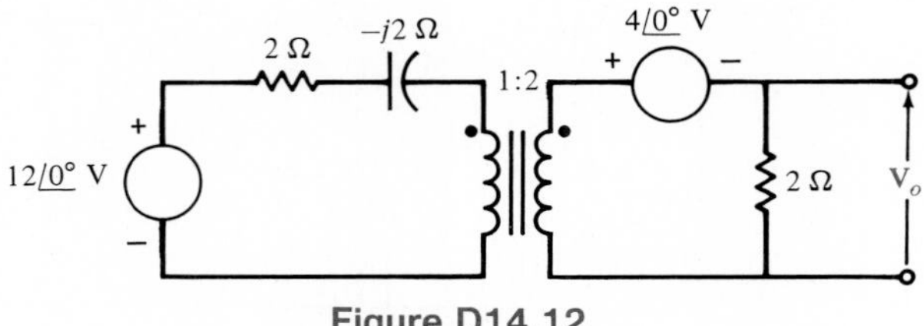

Figure D14.12

D14.13. Find $\mathbf{V}_o$ in the network in Fig. D14.13.

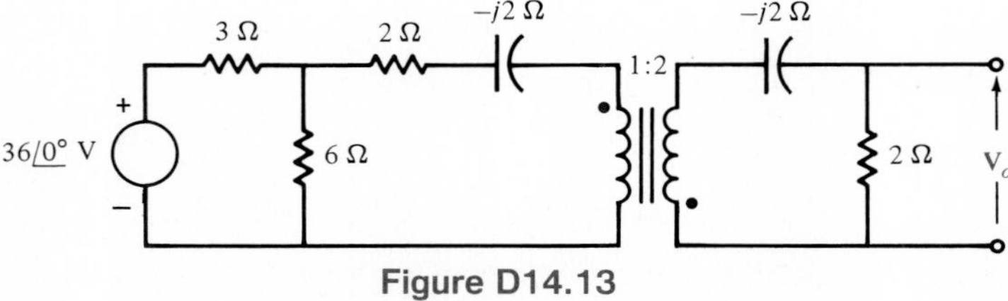

Figure D14.13

As a final point, consider the following example.

EXAMPLE 14.13

Given the circuit in Fig. 14.20a, we wish to find the value of the load R_o for maximum power transfer, and the value of the maximum power delivered to this load.

In order to determine the value of R_o for maximum power transfer, we will form a Thévenin equivalent at the terminals of R_o. The open-circuit voltage is derived from Fig. 14.20b as follows. The secondary resistance reflected into the primary is 16 Ω, and therefore

$$\mathbf{I}_1 = \frac{48 \underline{/0^\circ}}{6 + 2 + 16}$$
$$= 2 \underline{/0^\circ} \text{ A}$$

Therefore,

$$\mathbf{V}_1 = (2 \underline{/0^\circ})(16)$$
$$= 32 \underline{/0^\circ} \text{ V}$$

and by means of the turns ratio

$$\mathbf{I}_2 = 4 \underline{/0^\circ} \text{ A}$$
$$\mathbf{V}_2 = 16 \underline{/0^\circ} \text{ V}$$

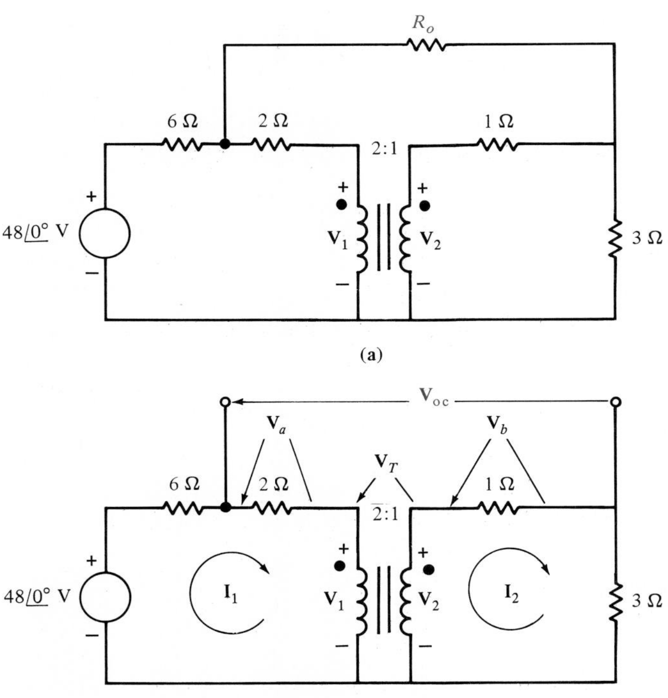

Figure 14.20 Maximum power transfer example involving an ideal transformer.

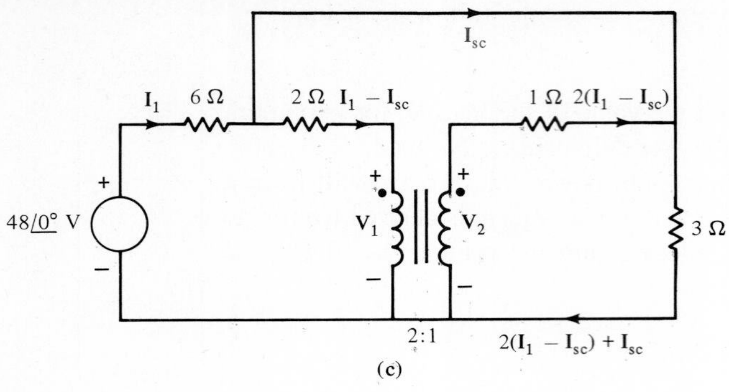

(c)

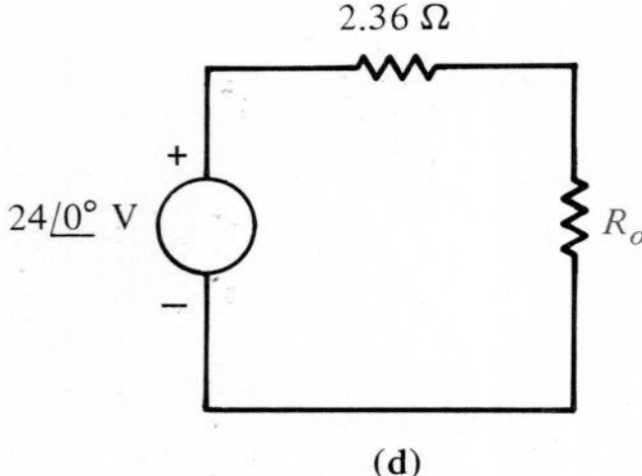

(d)

Figure 14.20 (Continued).

Now

$$\mathbf{V}_{oc} = \mathbf{V}_a + \mathbf{V}_T + \mathbf{V}_b = \mathbf{V}_a + \mathbf{V}_1 - \mathbf{V}_2 + \mathbf{V}_b = 24 \underline{/0^\circ} \text{ V}$$

The short-circuit current is obtained from Fig. 14.20c. The equations for this network are

$$48 \underline{/0^\circ} = 6\mathbf{I}_1 + 2(\mathbf{I}_1 - \mathbf{I}_{sc}) + \mathbf{V}_1$$

$$48 \underline{/0^\circ} = 6\mathbf{I}_1 + 3[2(\mathbf{I}_1 - \mathbf{I}_{sc}) + \mathbf{I}_{sc}]$$

$$\mathbf{V}_2 = (1)[2(\mathbf{I}_1 - \mathbf{I}_{sc})] + 3[2(\mathbf{I}_1 - \mathbf{I}_{sc}) + \mathbf{I}_{sc}]$$

$$\mathbf{V}_1 = 2\mathbf{V}_2$$

These equations reduce to

$$48 \underline{/0^\circ} = 12\mathbf{I}_1 - 3\mathbf{I}_{sc}$$

$$48 \underline{/0^\circ} = 26\mathbf{I}_1 - 12\mathbf{I}_{sc}$$

and therefore

$$\begin{bmatrix} \mathbf{I}_1 \\ \mathbf{I}_{sc} \end{bmatrix} = \frac{-1}{66} \begin{bmatrix} -12 & 3 \\ -26 & 12 \end{bmatrix} \begin{bmatrix} 48 \underline{/0^\circ} \\ 48 \underline{/0^\circ} \end{bmatrix}$$

$$= \frac{-1}{66} \begin{bmatrix} (-9)\ 48 \underline{/0^\circ} \\ (-14)\ 48 \underline{/0^\circ} \end{bmatrix} = \begin{bmatrix} 6.55 \underline{/0^\circ} \text{ A} \\ 10.18 \underline{/0^\circ} \text{ A} \end{bmatrix}$$

and

$$\mathbf{Z}_{Th} = \frac{24 \angle 0^\circ}{10.18 \angle 0^\circ} = 2.36 \ \Omega$$

Therefore, the equivalent circuit is shown in Fig. 14.20d, and $R_o = 2.36 \ \Omega$ for maximum power transfer. The maximum power delivered to R_o is

$$P_{max} = \left[\frac{24}{2(2.36)}\right]^2 (2.36) = 61.02 \text{ W}$$

■

14.5 SPICE Analysis Techniques

The SPICE circuit analysis program can be employed to solve magnetically coupled circuits. To use SPICE in this case we must present one additional statement that defines the coupling. The statement is

```
KXXXXXXX LYYYYYYY LZZZZZZZ value
```

LYYYYYYY and LZZZZZZZ are the names of the coupled inductors and value is the coefficient of coupling KXXXXXXX that must satisfy the relationship

$$0 < \text{KXXXXXXX} \leq 1$$

The "dot" convention applies in this analysis and the "dot" should be on the first node of each inductor.

The following examples illustrate the use of SPICE for coupled circuits.

EXAMPLE 14.14

Let us consider once again the network in Example 14.3, which is redrawn in Fig. 14.21 for SPICE analysis. Let us assume that the frequency is $f = 1/2\pi$ Hz. The coupling coefficient is then

$$k = \frac{2}{\sqrt{(4)(6)}} = 0.408248$$

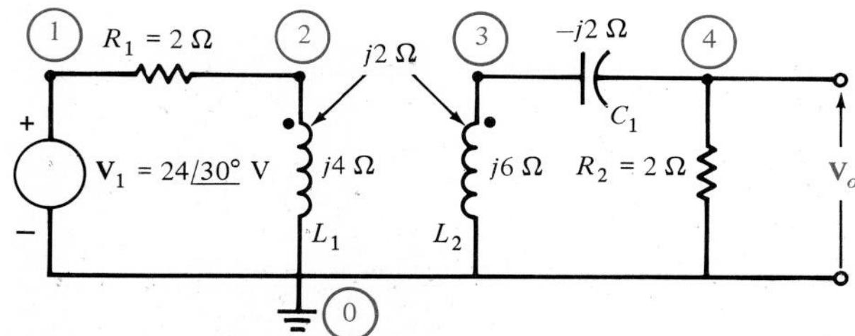

Figure 14.21 Figure 14.4 Prepared for SPICE analysis.

The SPICE program for determining the output voltage is

```
EXAMPLE 14.14
V1   1    0    AC       24     30
R1   1    2    2
L1   2    0    4
L2   3    0    6
K1   L1   L2   0.408248
C1   3    4    0.5
R2   4    0    2
.AC LIN   1    0.159155  0.159155
.PRINT    AC   VM(4)     VP(4)
.END
```

The computer output is

```
FREQ          VM(4)        VP(4)
1.592E-01   5.367E+00    3.435E+00
```

■

EXAMPLE 14.15

The network in Fig. 14.22a is redrawn in Fig. 14.22b for SPICE analysis. Let us write a program that will calculate the current in R_2 assuming that the frequency is $f = 1/2\pi$ Hz. The coefficient of coupling is

$$k = \frac{1}{\sqrt{(2)(3)}} = 0.408248$$

(a)

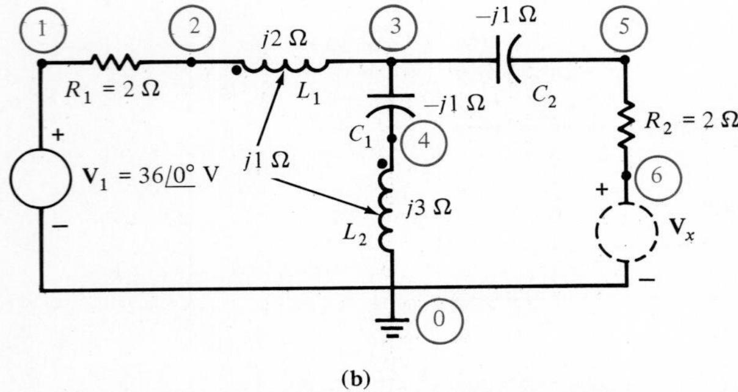

(b)

Figure 14.22 SPICE analysis example.

The program is listed below.

```
EXAMPLE 14.15
V1    1      0      AC     36     0
R1    1      2      2
L1    2      3      2
C1    3      4      1
L2    4      0      3
K1    L1     L2     0.408248
C2    3      5      1
R2    5      6      2
VX    6      0      AC     0      0
.AC LIN      1      0.159155      0.159155
.PRINT       AC     IM(VX)        IP(VX)
.END
```

The computer output is

```
FREQ          IM(VX)        IP(VX)
1.592E-01     6.900E+00     2.657E+01
```

DRILL EXERCISE

D14.14. Given the network in Fig. D14.14, with $f = 1/2\pi$ Hz and the coefficient of coupling $= 0.9$, write a SPICE program to compute $\mathbf{V}_o$.

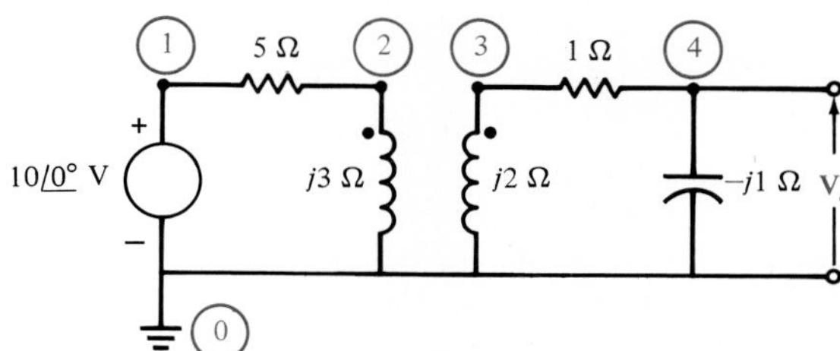

Figure D14.14

D14.15. Given the network in Fig. D14.15 with $f = 1/2\pi$ Hz and the coefficient of coupling $=$ 0.7071, write a SPICE program to compute $\mathbf{V}_o$.

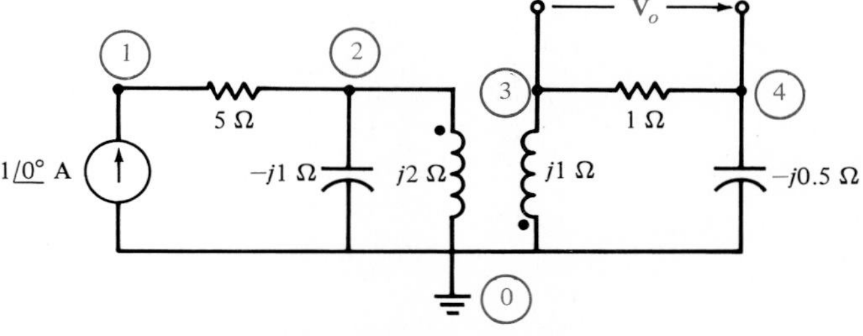

Figure D14.15

D14.16. Given the network in Fig. D14.16 with $f = 1/2\pi$ Hz and $k_{12} = 0.7071$, $k_{13} = 0.6325$, and $k_{23} = 0.4472$, write a SPICE program to compute the output voltage $\mathbf{V}_o$.

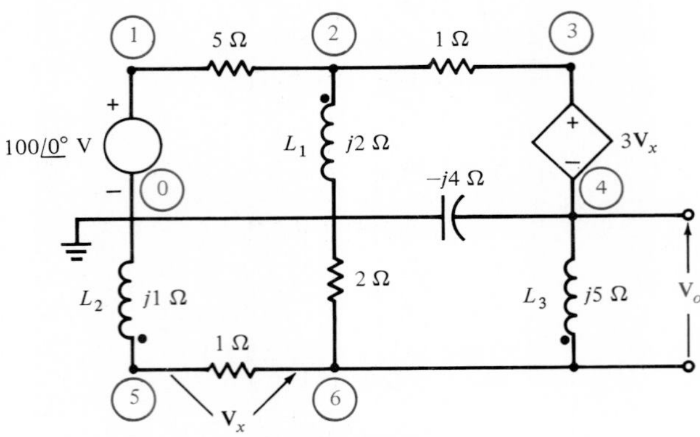

Figure D14.16

14.6 Summary

Magnetically coupled circuits have been presented and the circuit equations that describe these elements discussed. Mutual inductance has been defined and a dot convention adopted for indicating the physical relationship of the coils in order to simplify the sign convention for the mutual terms in the circuit equations. An energy analysis for mutually coupled coils has been performed which led to a coefficient of coupling between coils.

The linear transformer in which no magnetic material is used to couple the coils was discussed. Then coils coupled with good magnetic material were described and presented as an ideal transformer. Finally, Thévenin's theorem was employed to derive equivalent circuits for the transformer and either its primary or secondary circuit to simplify the analysis of circuits containing ideal transformers. Finally, the use of SPICE in analyzing circuits that contain mutual inductance was presented.

KEY POINTS

- A transformer is a device that contains two or more coils that are coupled magnetically.
- Inductor coils are said to be mutually coupled if they share a common magnetic flux.
- Faraday's law states that the induced voltage in a coil is proportional to the rate of change of flux and the number of turns in the coil.
- The dot convention is used to determine if the flux produced by current flowing through one inductor coil will add to or oppose the flux produced by another inductor coil.
- A transformer is said to be linear if no magnetic material is used to concentrate the flux that links the two inductors.
- The reflected impedance is the impedance on one side of the transformer that is reflected, due to mutual coupling, from the other side.

- An ideal transformer is one in which the coupling coefficient is 1 and the inductive reactance of both the primary and secondary are extremely large (assumed to be infinite) in comparison with the load impedance.
- The SPICE material presented throughout the text can be applied to mutually coupled circuits with the addition of one statement that defines the coupling.

PROBLEMS

14.1. Find the voltage $\mathbf{V}_o$ in the network shown in Fig. P14.1.

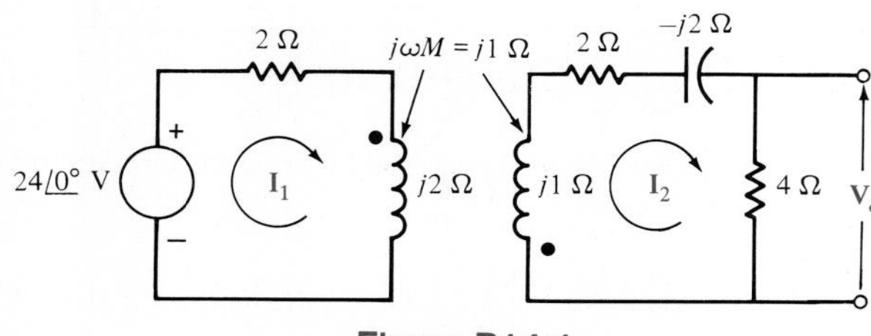

Figure P14.1

14.2. Find the voltage gain $\mathbf{V}_o/\mathbf{V}_s$ of the network shown in Fig. P14.2.

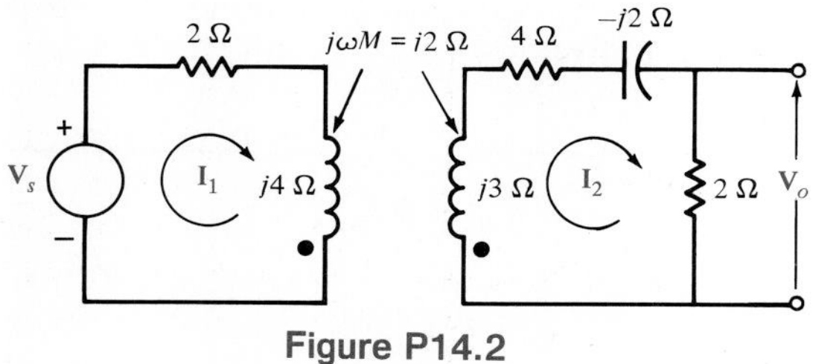

Figure P14.2

14.3. Write the mesh equations for the circuit shown in Fig. P14.3.

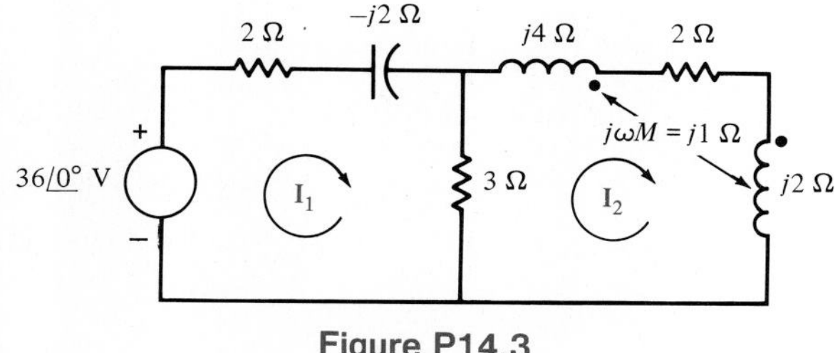

Figure P14.3

14.4. Write the mesh equations for the network shown in Fig. P14.4.

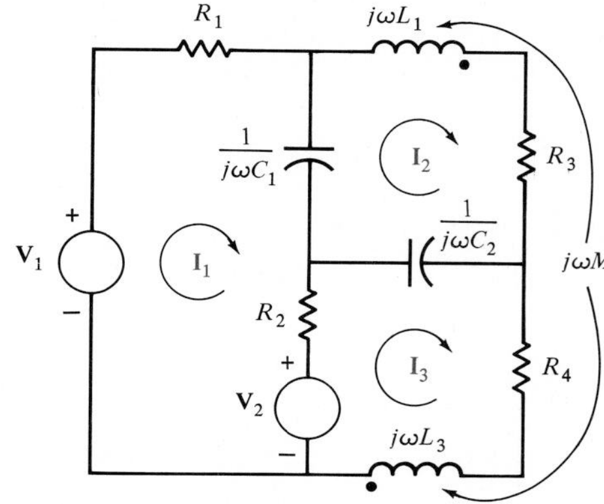

Figure P14.4

14.5. Write the set of linear independent phasor equations necessary to solve for the currents $i_1(t)$, $i_2(t)$, and $i_3(t)$ shown in Fig. P14.5.

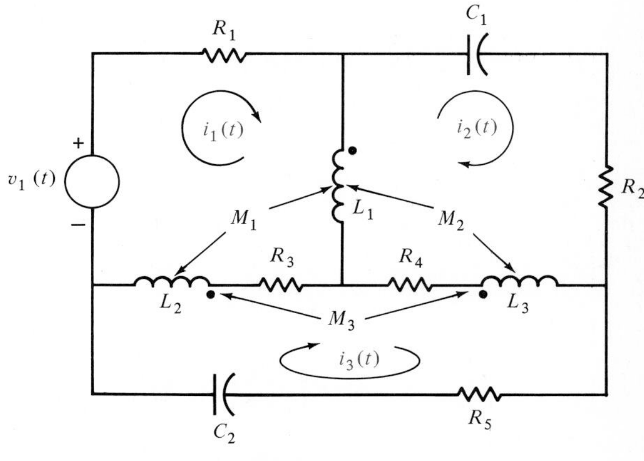

Figure P14.5

14.6. Write the mesh equations for the network shown in Fig. P14.6.

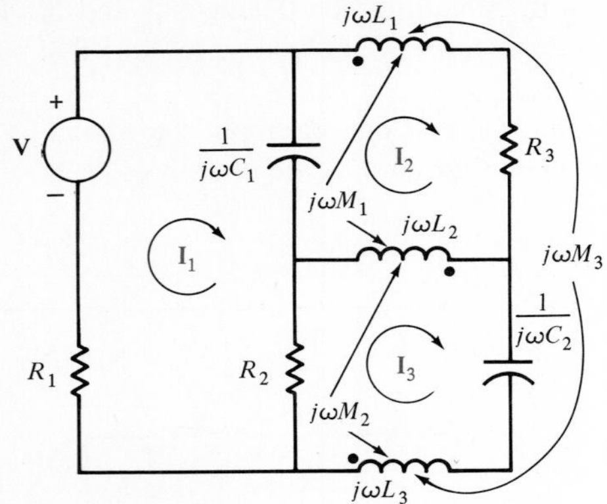

Figure P14.6

14.7. Write the mesh equations for the network shown in Fig. P14.7.

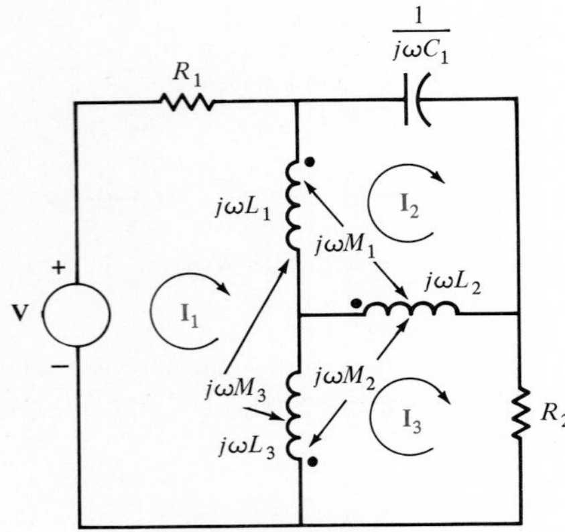

Figure P14.7

14.8. The currents in the network shown in Fig. P14.8 are

$$i_1(t) = 16 \cos (377t - 45°) \text{ mA}$$

$$i_2(t) = 3 \cos (377t - 45°) \text{ mA}$$

If the inductance values are $L_1 = 2$ H, $L_2 = 8$ H, and $M = 3$ H, determine the voltages $v_1(t)$ and $v_2(t)$.

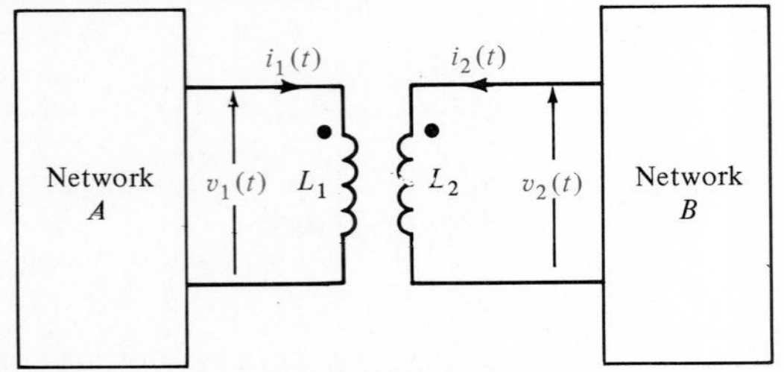

Figure P14.8

14.9. Determine the energy stored in the coupled inductors in Problem 14.8 at $t = 3$ ms.

14.10. Determine the expressions for $v_a(t)$ and $v_b(t)$ in the circuit shown in Fig. P14.10.

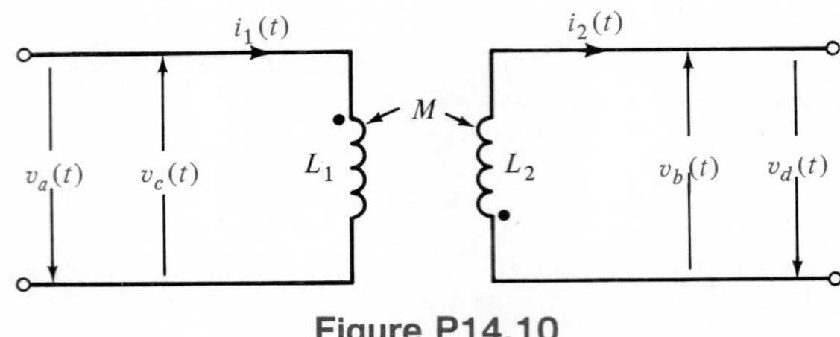

Figure P14.10

14.11. Find the expressions for $v_c(t)$ and $v_d(t)$ in the circuit shown in Fig. P14.10.

14.12. Find the expressions for $v_c(t)$ and $v_b(t)$ in the circuit shown in Fig. P14.10.

14.13. Find the expressions for $v_a(t)$ and $v_d(t)$ in the circuit shown in Fig. P14.10.

14.14. Determine the expressions for $v_a(t)$ and $v_b(t)$ in the circuit shown in Fig. P14.14.

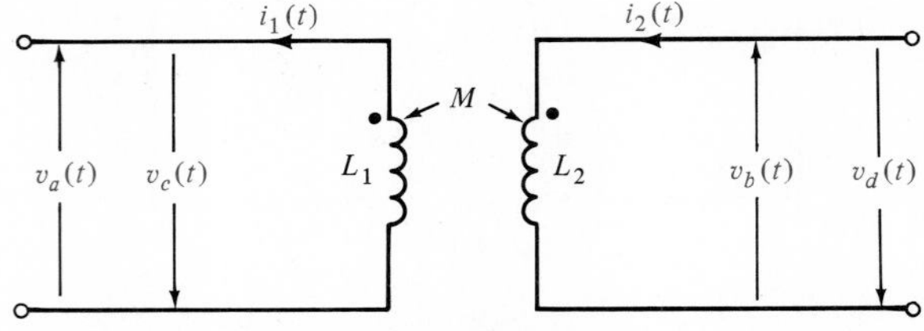

Figure P14.14

14.15. Find the expressions for $v_a(t)$ and $v_d(t)$ in the circuit shown in Fig. P14.14.

14.16. Find the expressions for $v_c(t)$ and $v_b(t)$ in the circuit shown in Fig. P14.14.

14.17. Find the expressions for $v_c(t)$ and $v_d(t)$ in the circuit shown in Fig. P14.14.

14.18. The currents in the magnetically coupled inductors shown in Fig. P14.18 are known to be $i_1(t) = 8 \cos (377t - 20°)$ mA and $i_2(t) = 4 \cos (377t - 50°)$ mA. The inductor values are $L_1 = 2$ H, $L_2 = 1$ H, and $k = 0.6$. Determine $v_1(t)$ and $v_2(t)$.

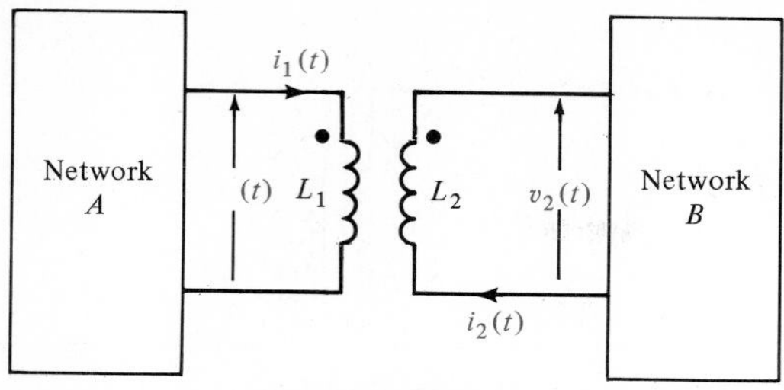

Figure P14.18

14.19. Determine the energy stored in the coupled inductors in Problem 14.18 at $t = 1$ ms.

14.20. In a coaxial cable there is always some amount of inductance in the wire, which usually allows for mutual coupling between the outer conductor and the center conductor. Shown in Fig. P14.20 is a representation of a coaxial cable which is used to transmit a signal to a load at 1 MHz. Determine the power delivered to the load if the coefficient of coupling is **(a)** 0.75 and **(b)** 1.0.

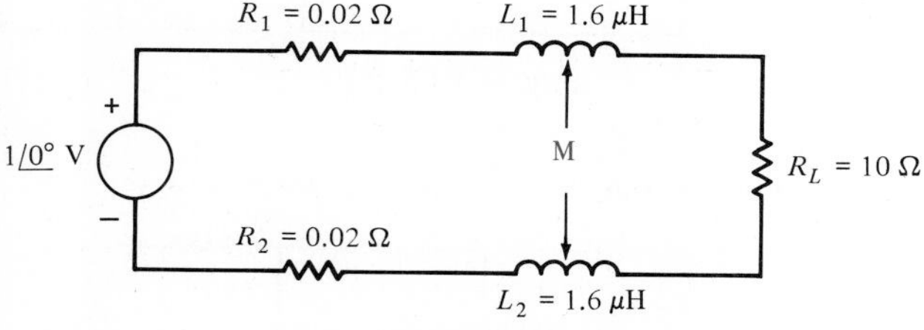

Figure P14.20

14.21. Repeat Problem 14.20 if the circuit operates at 10 kHz.

14.22. Determine the impedance seen by the source in the network shown in Fig. P14.22.

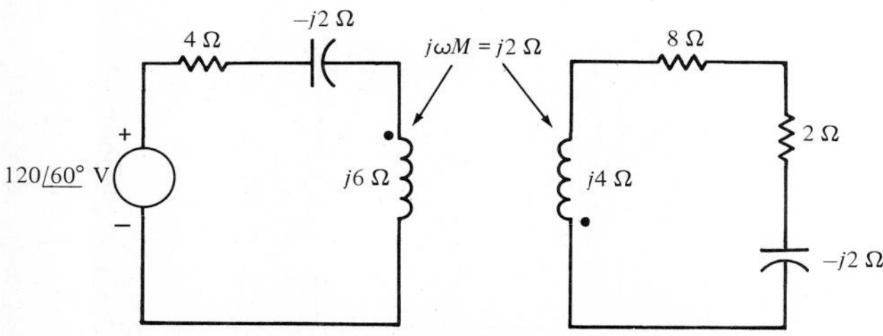

Figure P14.22

14.23. Compute the input impedance of the network in Fig. P14.23.

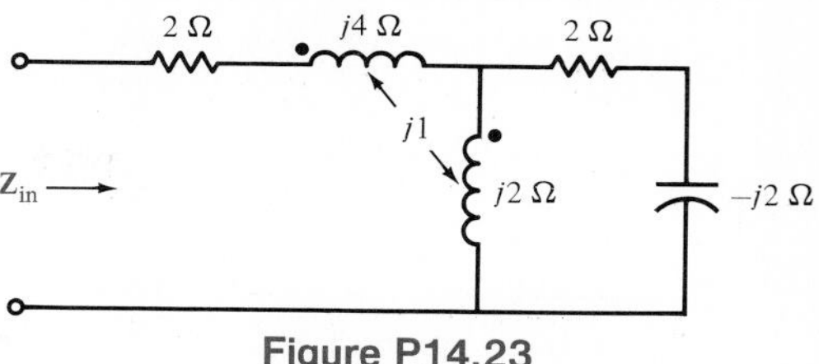

Figure P14.23

14.24. Determine the input impedance seen by the source in the network shown in Fig. P14.24.

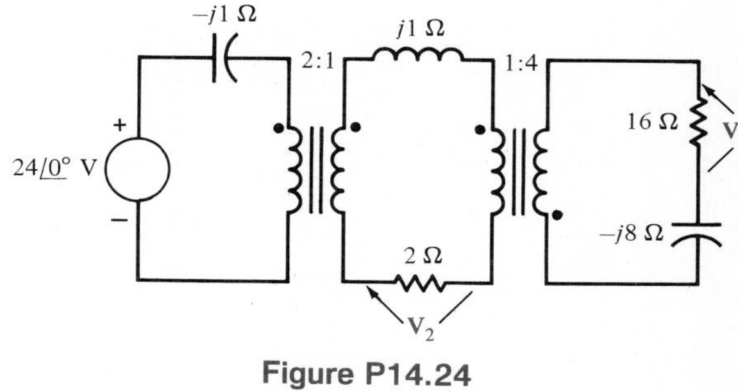

Figure P14.24

14.25. Determine the voltages $\mathbf{V}_1$ and $\mathbf{V}_2$ found in Figure P14.24.

14.26. Given the network shown in Fig. P14.26, determine the value of the capacitor C that will cause the reflected impedance to the primary to be purely resistive.

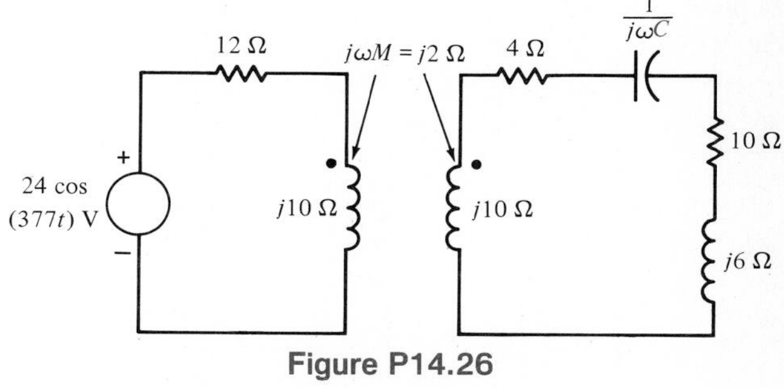

Figure P14.26

14.27. Analyze the network in Fig. P14.27 and determine if a value of $\mathbf{X}_c$ can be found such that the output voltage is equal to twice the input voltage.

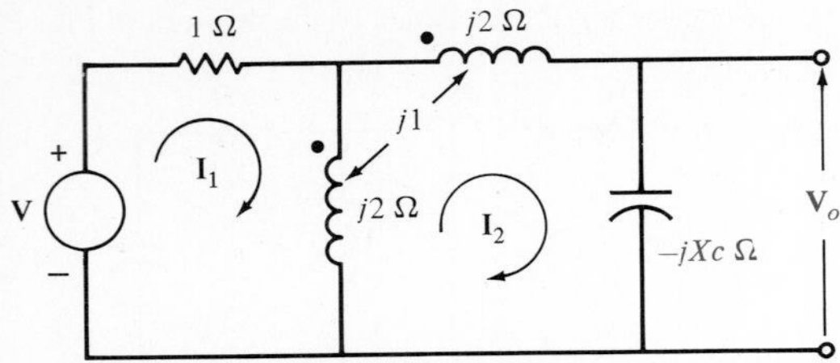

Figure P14.27

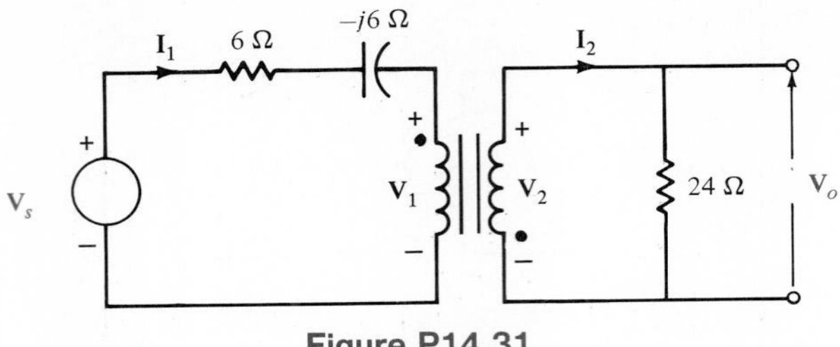

Figure P14.31

14.28. The equivalent circuit of a transformer is shown in Fig. P14.28. The resistance of the wire in the coils is represented by R_1 and R_2, while the flux which leaks from the coils is represented by L_1 and L_2. Also, the magnetization and core losses in the core of the transformer can be modeled by L_m and R_c. The load $\mathbf{Z}_L$ is a machine that consumes 1500 W of power at a power factor of 0.70 lagging. Determine the voltage $\mathbf{V}_s$ that must be supplied to provide $230 \underline{/0^\circ}$ V rms at the load $\mathbf{Z}_L$. Let $R_1 = 2\ \Omega$, $R_2 = 0.5\ \Omega$, $j\mathbf{X}_{L1} = j3\ \Omega$, $j\mathbf{X}_{L2} = j0.75\ \Omega$, $j\mathbf{X}_{Lm} = j500\ \Omega$, $R_c = 1000\ \Omega$, and $N_1{:}N_2 = 1{:}2$.

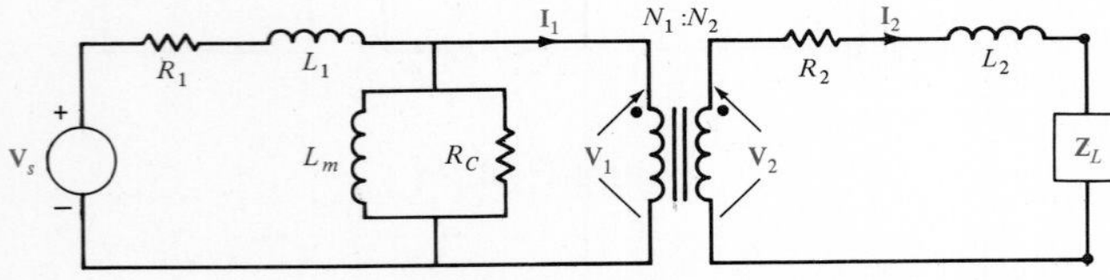

Figure P14.28

14.29. Using the values given in Problem 14.28, determine the power lost by the imperfections of the transformer shown in Fig. P14.28.

14.30. Calculate all currents and voltages in the circuit shown in Fig. P14.30.

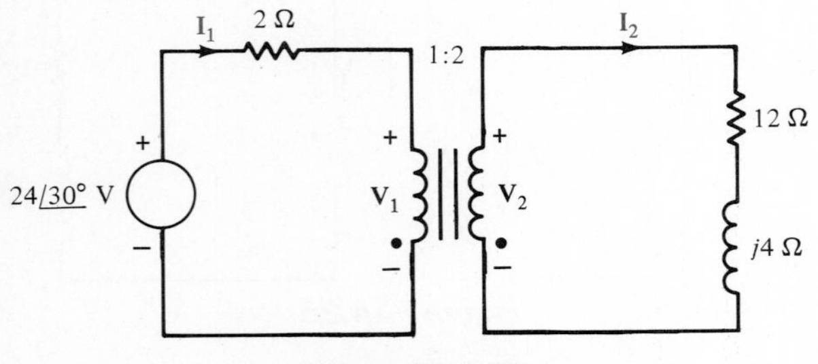

Figure P14.30

14.31. Given that $\mathbf{V}_o = 48 \underline{/30^\circ}$ V in the circuit shown in Fig. P14.31, determine $\mathbf{V}_s$.

14.32. If the voltage source $\mathbf{V}_s$ in the circuit of Problem 14.31 is $50 \underline{/0^\circ}$ V, determine $\mathbf{V}_o$.

14.33. In the network shown in Fig. P14.33 $\mathbf{I}_A$ is known to be $10 \underline{/30^\circ}$ A. Find the load voltage $\mathbf{V}_o$.

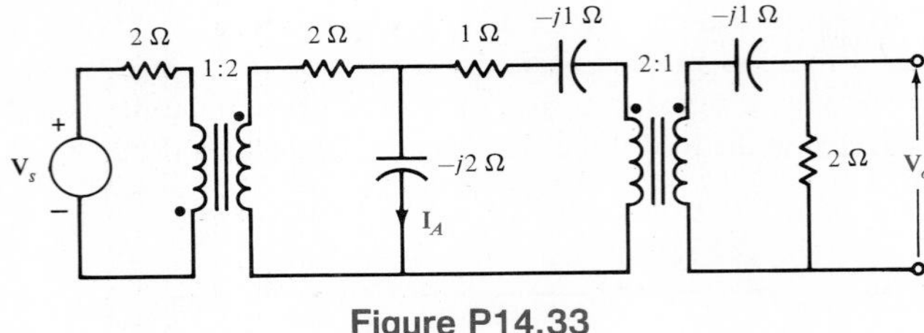

Figure P14.33

14.34. Find the current **I** in the network in Fig. P14.34.

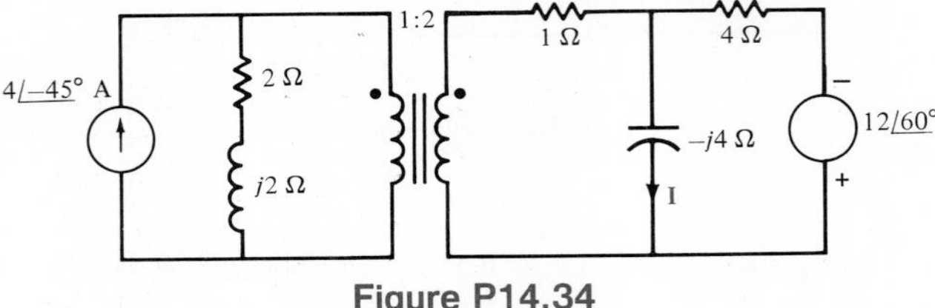

Figure P14.34

14.35. Determine the input impedance seen by the source in the network shown in Fig. P14.35.

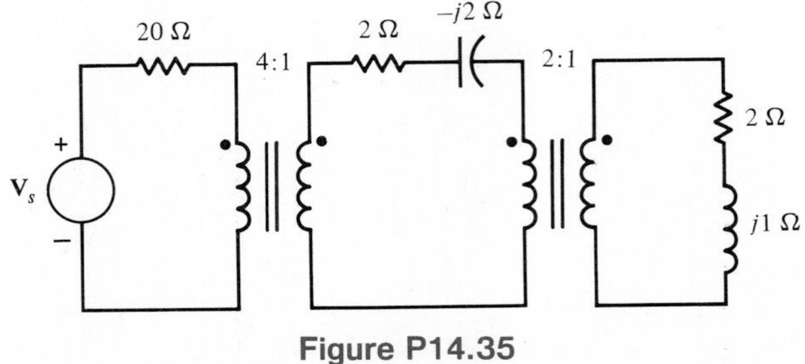

Figure P14.35

14.36. Determine the average power delivered to each resistor in the network shown in Fig. P14.36.

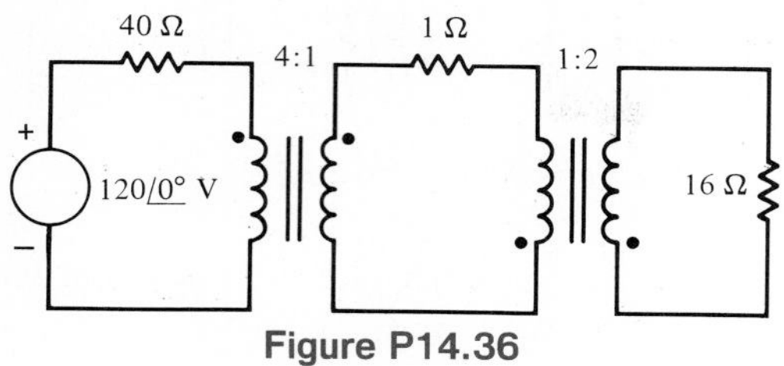

Figure P14.36

14.37. Find **I** in the network shown in Fig. P14.37.

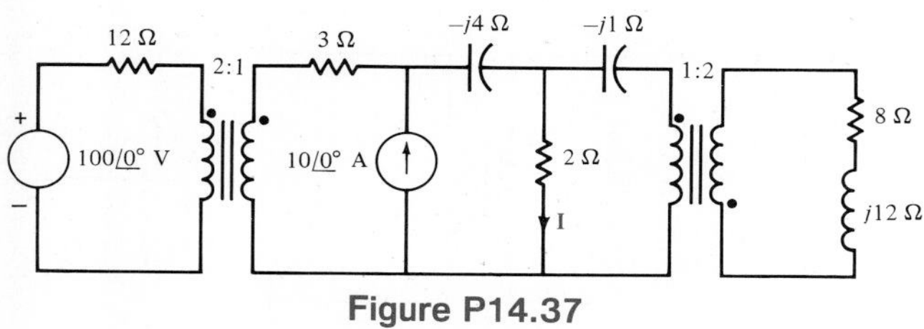

Figure P14.37

14.38. Determine the source current in the network shown in Fig. P14.38.

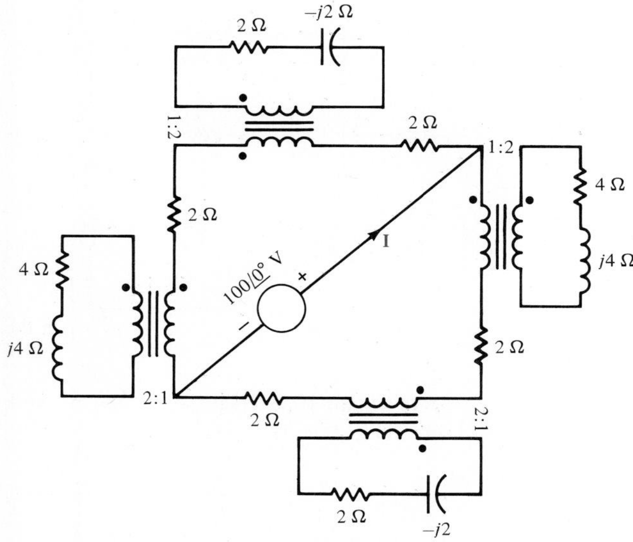

Figure P14.38

14.39. If the load voltage in the network shown in Fig. P14.39 is 100 /0° V rms, determine the input voltage $\mathbf{V}_s$.

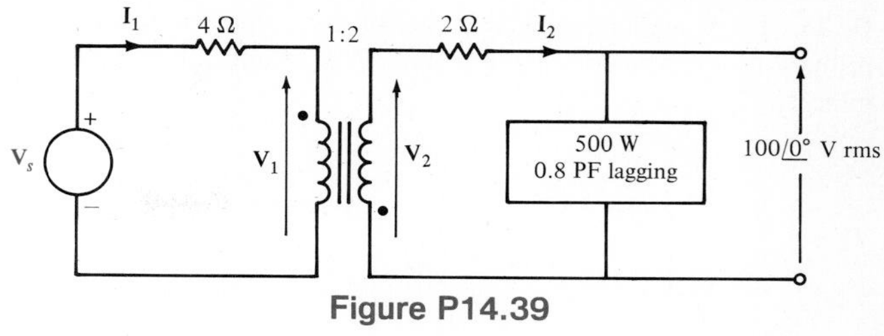

Figure P14.39

14.40. In the network shown in Fig. P14.40, if $\mathbf{V}_L = 120 \angle 0°$ V rms and the load $\mathbf{Z}_L$ absorbs 500 W at 0.85 PF lagging, compute the voltage $\mathbf{V}_s$.

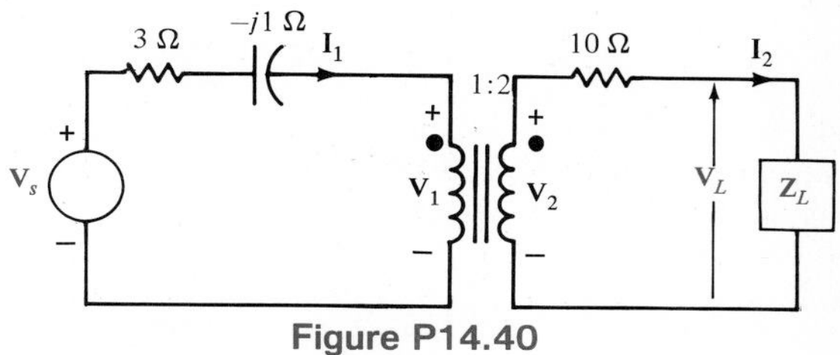

Figure P14.40

14.41. In the circuit shown in Fig. P14.41, if $\mathbf{V}_2 = 120 \angle 0°$ V rms and the load $\mathbf{Z}_L$ absorbs 400 W at 0.9 PF lagging, determine the wattmeter reading.

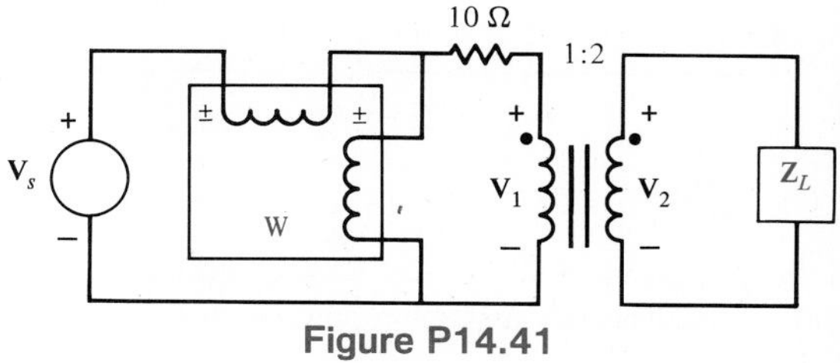

Figure P14.41

14.42. Given the network in Fig. P14.42, show that $\mathbf{V}_2$ can be written as

$$\mathbf{V}_2 = \frac{n\mathbf{Z}_L}{n^2\mathbf{Z}_s + \mathbf{Z}_L}\mathbf{V}_s$$

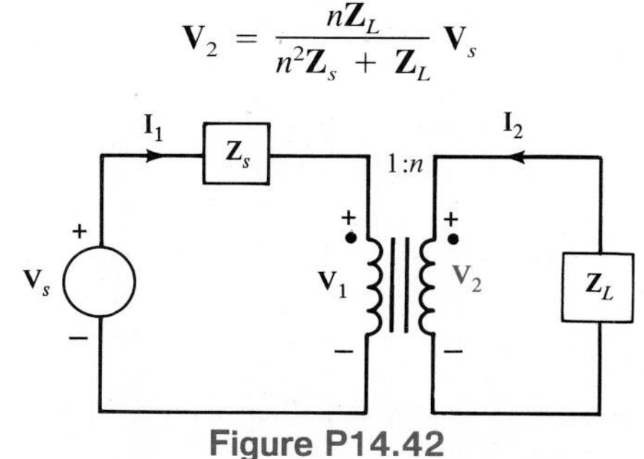

Figure P14.42

14.43. Form an equivalent circuit for the transformer and the primary in the network shown in Fig. P14.43 and use it to determine $\mathbf{I}_2$.

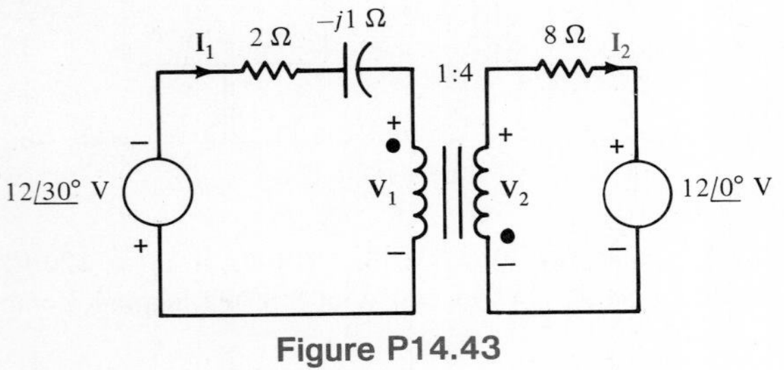

Figure P14.43

14.44. Form an equivalent circuit for the transformer and primary in the network shown in Fig. P14.44 and use this circuit to find the current $\mathbf{I}_2$.

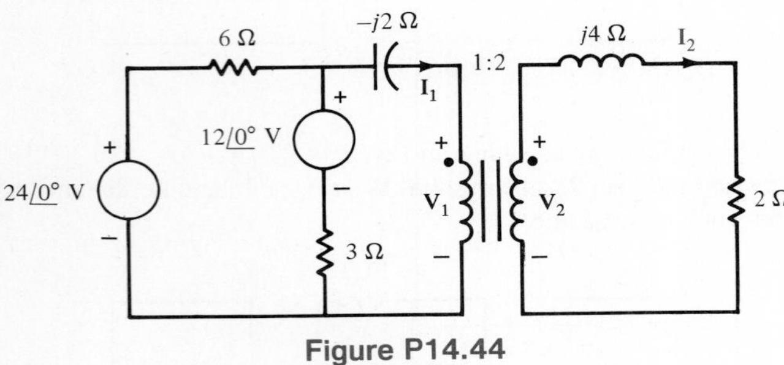

Figure P14.44

14.45. In the network shown in Fig. P14.45, determine the value of the load impedance for maximum power transfer.

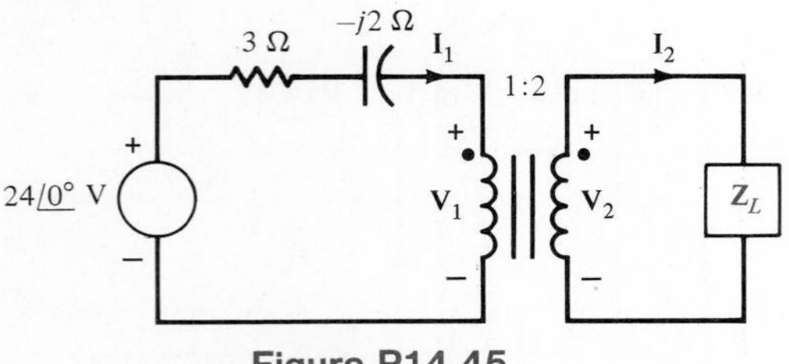

Figure P14.45

14.46. The output stage of an amplifier is to be matched to the impedance of a speaker as shown in Fig. P14.46. If the impedance of the speaker is 8 Ω and the amplifier requires a load impedance of 3.2 kΩ, determine the turns ratio of the ideal transformer.

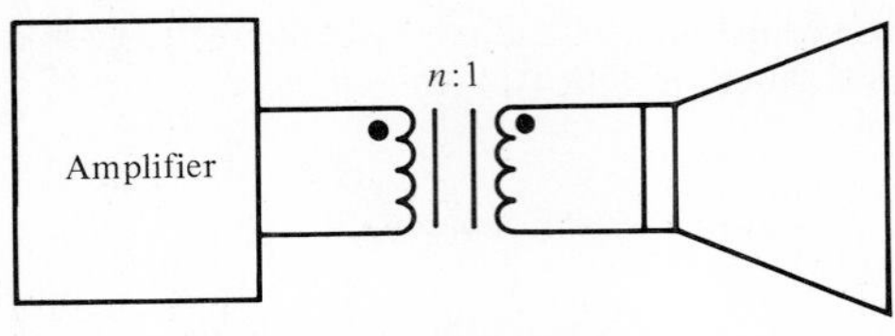

Figure P14.46

14.47. Determine the input impedance of the circuit shown in Fig. P14.47.

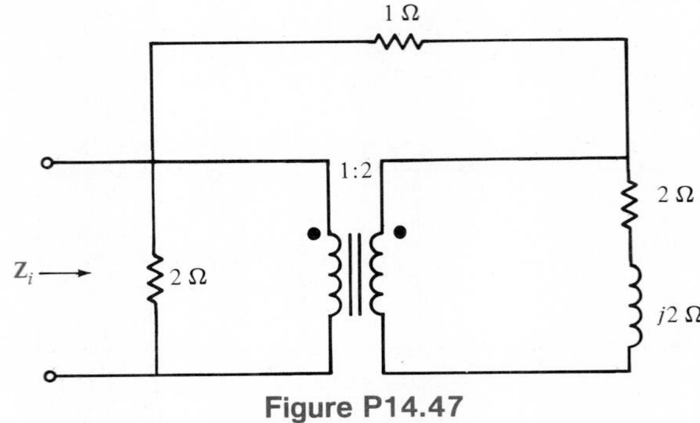

Figure P14.47

14.48. Find the value of the load resistance R_o in the network shown in Fig. P14.48 for maximum power transfer.

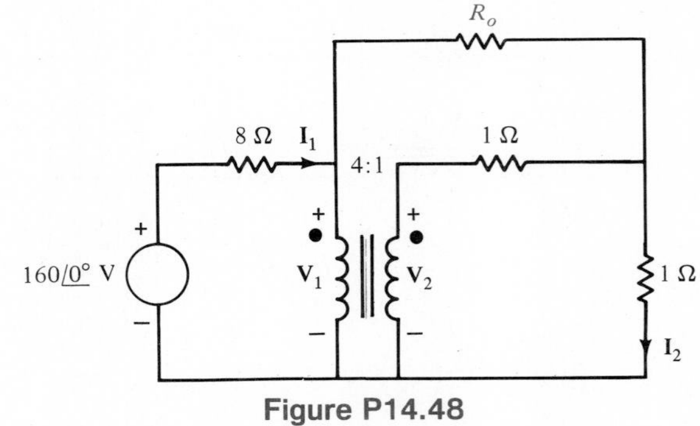

Figure P14.48

14.49. Write a SPICE program to calculate $\mathbf{V}_o$ in the network shown in Fig. P14.49, where $f = 1/2\pi$ Hz.

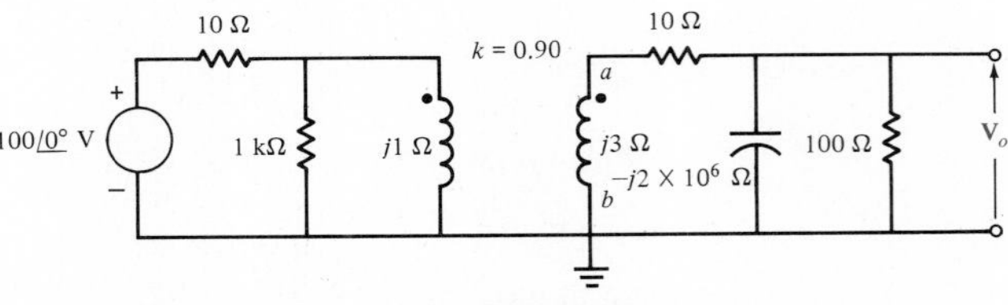

Figure P14.49

14.50. Repeat Problem 14.49 with the dot moved from position a to position b.

14.51. Given the network in Fig. P14.51, determine $\mathbf{V}_o$ using SPICE. $f = 1/2\pi$ Hz. Let

$$k_{12} = 0.995$$

$$k_{13} = .980$$

$$k_{23} = .950$$

Figure P14.51

14.52. Find $\mathbf{V}_o$ in the network in Fig. P14.52 using SPICE, where $f = 1/2\pi$ Hz.

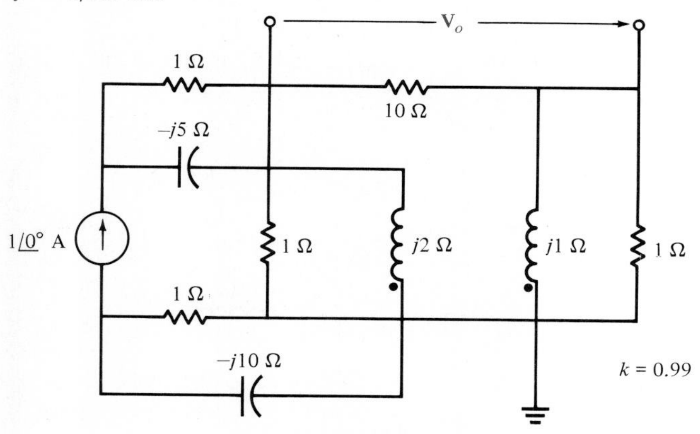

Figure P14.52

14.53. Write a SPICE program to calculate $\mathbf{V}_o$ in the network in Fig. P14.53, where $f = 1/2\pi$ Hz. Let

$k_{12} = 0.90$	$k_{23} = 0.99$
$k_{13} = 0.85$	$k_{24} = 0.95$
$k_{14} = 0.95$	$k_{34} = 0.50$

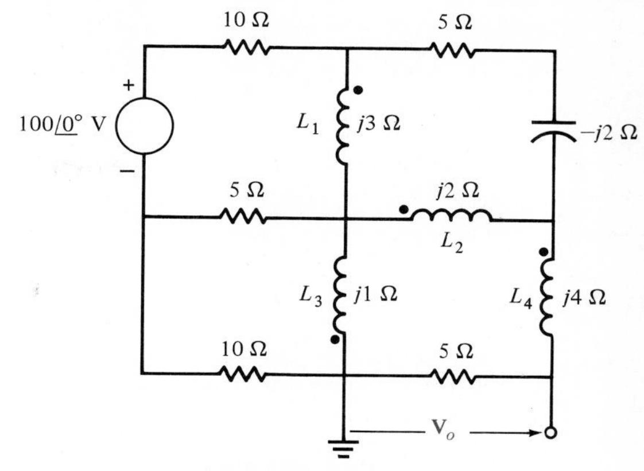

Figure P14.53

14.54. Use SPICE to find $\mathbf{V}_o$ in the network in Fig. P14.54, where $f = 1/2\pi$ Hz and

$$j\omega M_{12} = j2\ \Omega$$

$$j\omega M_{13} = j2\ \Omega$$

$$j\omega M_{23} = j1\ \Omega$$

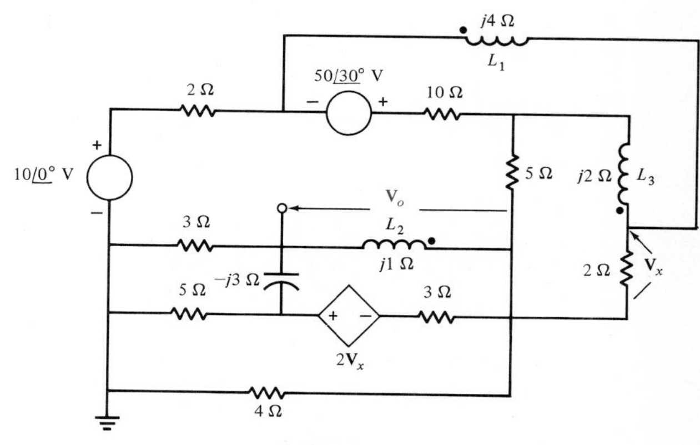

Figure P14.54

14.55. Write a SPICE program to find $\mathbf{V}_o$ in the circuit in Fig. P14.55, where $f = 1/2\pi$ Hz.

$j\omega M_{12} = j1\ \Omega$	$j\omega M_{23} = j1\ \Omega$
$j\omega M_{13} = j\sqrt{3}\ \Omega$	$j\omega M_{24} = j1\ \Omega$
$j\omega M_{14} = j1\ \Omega$	$j\omega M_{34} = j1\ \Omega$

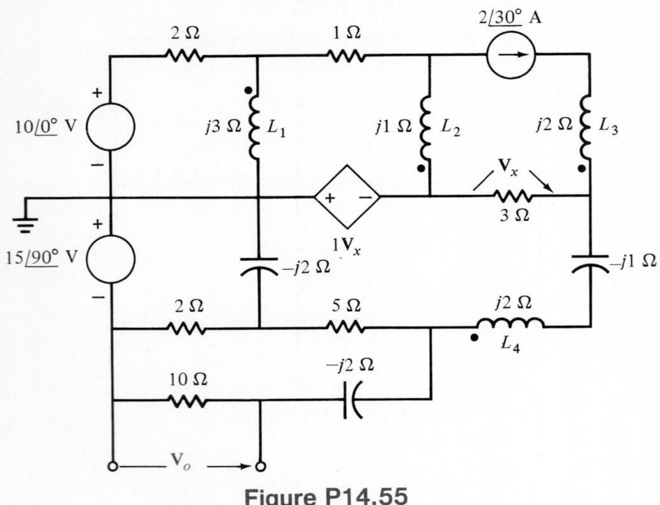
Figure P14.55

14.56. Given the circuit in Fig. P14.56, find $\mathbf{V}_o$ using SPICE. Let $f = 1/2\pi$ Hz and

$$j\omega M_{12} = j3\ \Omega$$

$$j\omega M_{13} = j4\ \Omega$$

$$j\omega M_{23} = j1\ \Omega$$

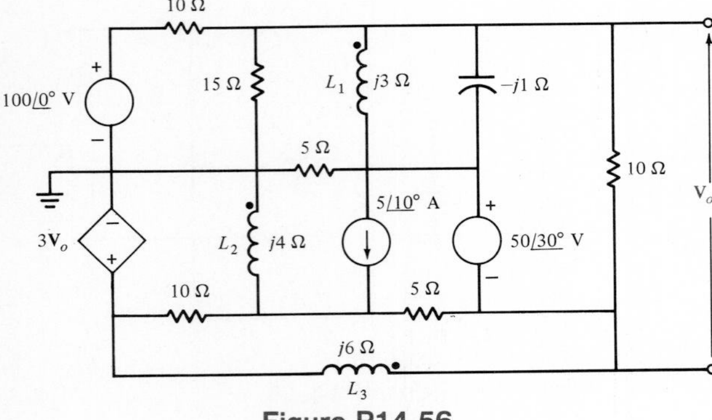
Figure P14.56

14.57. Use SPICE to find $\mathbf{V}_o$ in the network in Fig. P14.57, where $f = 1/2\pi$ Hz and

$$j\omega M_{12} = j2\ \Omega$$

$$j\omega M_{34} = j2\ \Omega$$

There is no other mutual inductance present in the network.

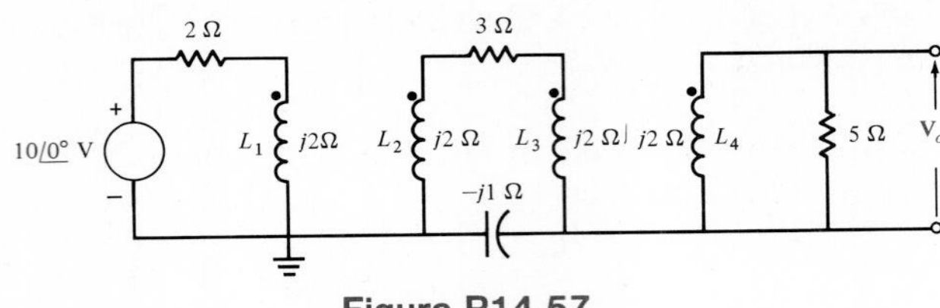
Figure P14.57

14.58. Write a SPICE program to find $\mathbf{V}_o$ in the network in Fig. P14.58, where all $j\omega M = j1\ \Omega$ and $f = 1/2\pi$ Hz.

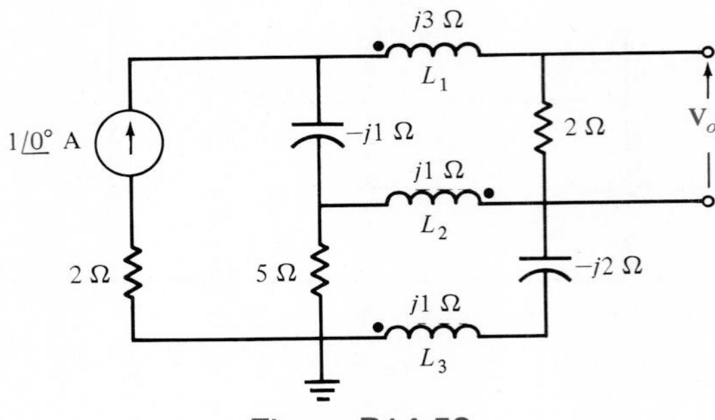
Figure P14.58

14.59. Write a SPICE program to find $\mathbf{V}_o$ in the circuit in Fig. P14.59, where $f = 1/2\pi$ Hz and

$$j\omega M_{12} = j1\ \Omega$$

$$j\omega M_{13} = j2\ \Omega$$

$$j\omega M_{23} = j1\ \Omega$$

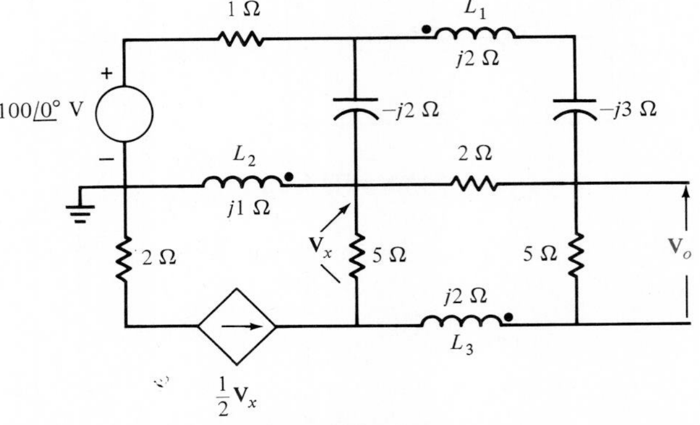
Figure P14.59

14.60. Find $\mathbf{V}_o$ in the circuit in Fig. P14.60 using SPICE. Let $f = 1/2\pi$ Hz and $j\omega M = j1\ \Omega$.

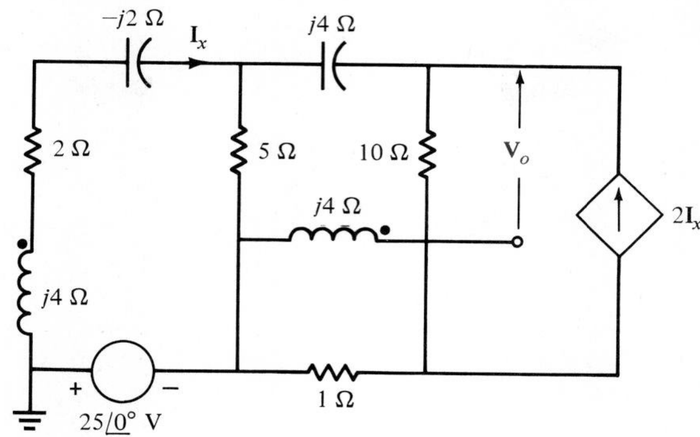

Figure P14.60

CHAPTER

15

Two-Port Networks

We say that the network shown in Fig. 15.1a has a single *port*. The pair of terminals that constitute this port could represent a single element (e.g., R, L, or C) or it could be some interconnection of these elements. The network in Fig. 15.1b is called a two-port, where once again each set of terminals represents a port. In our analyses thus far we have already examined some two-port networks (e.g., the operational amplifier and the transformer).

In general, we describe the two-port as a network consisting of R, L, and C elements, transformers, op amps, dependent sources, but no independent sources. The network has an input port and an output port, and as is the case with an op amp, one terminal may be common to both ports.

Although there are a number of ways to describe a two-port, we confine our discussion here to four types of parameters: admittance, impedance, hybrid, and transmission. We demonstrate the usefulness of each set of parameters, show how they are related to one another, and finally illustrate how two-ports can be interconnected in parallel, series, or cascade.

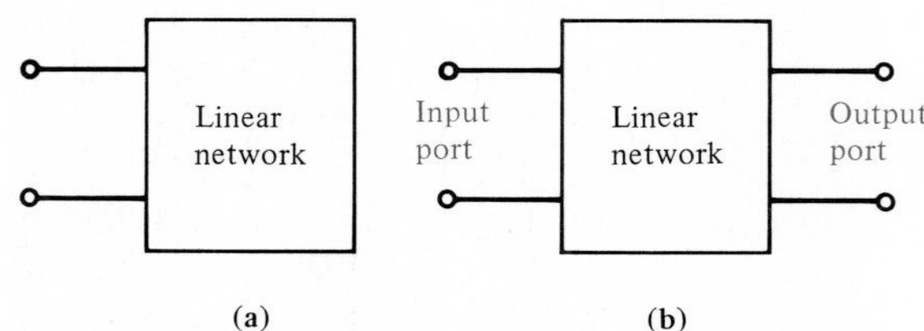

Figure 15.1 (a) Single-port network; (b) two-port network.

15.1 Admittance Parameters

In the two-port model shown in Fig. 15.2, it is customary to label the voltages and currents as shown; that is, the upper terminals are positive with respect to the lower terminals and the currents are into the two-port at the upper terminals and, because KCL must be satisfied at each port, the current is out of the two-port at the lower terminals. Since the network is linear and contains no independent sources, the principle of superposition can be applied to determine the current $\mathbf{I}_1$, which can be written as the sum of two components, one due to $\mathbf{V}_1$ and one due to $\mathbf{V}_2$. Using this principle we can write

$$\mathbf{I}_1 = \mathbf{y}_{11}\mathbf{V}_1 + \mathbf{y}_{12}\mathbf{V}_2$$

where $\mathbf{y}_{11}$ and $\mathbf{y}_{12}$ are essentially constants of proportionality and their units are siemens. In a similar manner $\mathbf{I}_2$ can be written as

$$\mathbf{I}_2 = \mathbf{y}_{21}\mathbf{V}_1 + \mathbf{y}_{22}\mathbf{V}_2$$

Therefore, the two equations that describe the two-port are

$$\begin{aligned} \mathbf{I}_1 &= \mathbf{y}_{11}\mathbf{V}_1 + \mathbf{y}_{12}\mathbf{V}_2 \\ \mathbf{I}_2 &= \mathbf{y}_{21}\mathbf{V}_1 + \mathbf{y}_{22}\mathbf{V}_2 \end{aligned} \tag{15.1}$$

or in matrix form,

$$\begin{bmatrix} \mathbf{I}_1 \\ \mathbf{I}_2 \end{bmatrix} = \begin{bmatrix} \mathbf{y}_{11} & \mathbf{y}_{12} \\ \mathbf{y}_{21} & \mathbf{y}_{22} \end{bmatrix} \begin{bmatrix} \mathbf{V}_1 \\ \mathbf{V}_2 \end{bmatrix}$$

Note that subscript 1 refers to the input port and subscript 2 refers to the output port, and the equations describe what we will call the *Y parameters* for a network. If these parameters $\mathbf{y}_{11}$, $\mathbf{y}_{12}$, $\mathbf{y}_{21}$, and $\mathbf{y}_{22}$ are known, the input/output operation of the two-port is completely defined.

From Eqs. (15.1) we can determine the Y parameters in the following manner. Note from the equations that $\mathbf{y}_{11}$ is equal to $\mathbf{I}_1$ divided by $\mathbf{V}_1$ with the output short-circuited (i.e., $\mathbf{V}_2 = 0$).

$$\mathbf{y}_{11} = \left.\frac{\mathbf{I}_1}{\mathbf{V}_1}\right|_{\mathbf{V}_2=0} \tag{15.2}$$

Since $\mathbf{y}_{11}$ is an admittance at the input measured in siemens with the output short-circuited, it is called the *short-circuit input admittance*. The equations indicate that the other Y parameters can be determined in a similar manner:

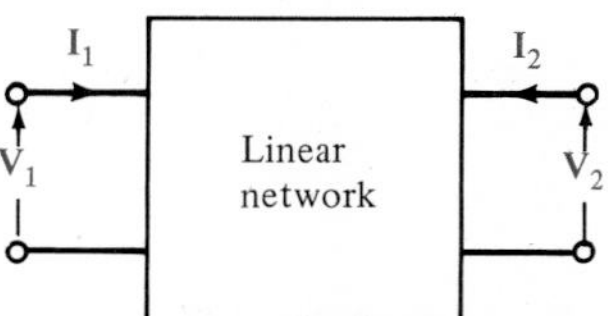

Figure 15.2 Two-port model.

$$\mathbf{y}_{12} = \left.\frac{\mathbf{I}_1}{\mathbf{V}_2}\right|_{\mathbf{V}_1=0}$$

$$\mathbf{y}_{21} = \left.\frac{\mathbf{I}_2}{\mathbf{V}_1}\right|_{\mathbf{V}_2=0} \qquad (15.3)$$

$$\mathbf{y}_{22} = \left.\frac{\mathbf{I}_2}{\mathbf{V}_2}\right|_{\mathbf{V}_1=0}$$

$\mathbf{y}_{12}$ and $\mathbf{y}_{21}$ are called the *short-circuit transfer admittances* and $\mathbf{y}_{22}$ is called the *short-circuit output admittance*. As a group the Y parameters are referred to as the *short-circuit admittance parameters*. Note that by applying the definitions above these parameters could be determined experimentally for a two-port whose actual configuration is unknown.

EXAMPLE 15.1

Let us determine the Y parameters for the network shown in Fig. 15.3a.

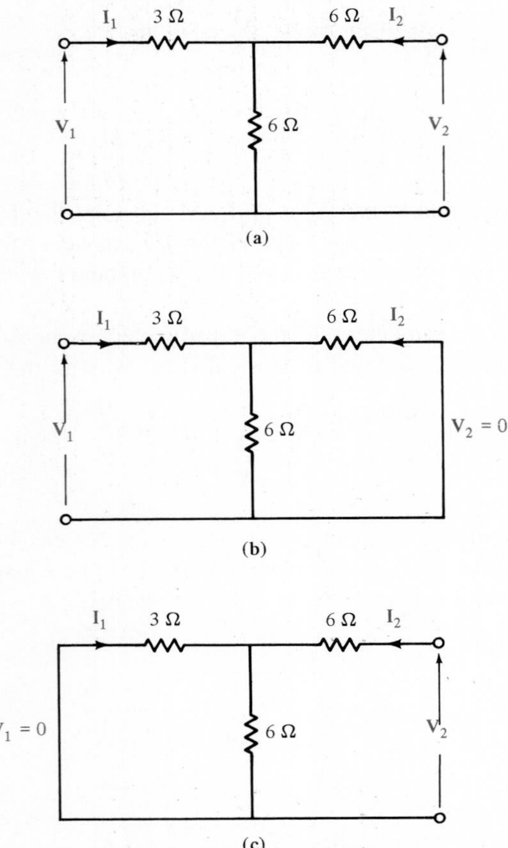

Figure 15.3 Two-port and the configurations used to compute the Y parameters.

From our definitions the admittance $\mathbf{y}_{11}$ is equal to $\mathbf{I}_1$ divided by $\mathbf{V}_1$, with $\mathbf{V}_2 = 0$ as shown in Fig. 15.3b. Therefore, with the output shorted, the two 6-Ω resistors are in parallel and $\mathbf{V}_1 = 6\mathbf{I}_1$, and hence

$$\mathbf{y}_{11} = \left.\frac{\mathbf{I}_1}{\mathbf{V}_1}\right|_{\mathbf{V}_2=0} = \frac{\mathbf{I}_1}{6\mathbf{I}_1} = \tfrac{1}{6}\text{ S}$$

The parameter $\mathbf{y}_{12}$ is evaluated using the circuit in Fig. 15.3c. Employing current division, we can write $\mathbf{I}_1$ for this circuit as

$$-\mathbf{I}_1 = \frac{\mathbf{V}_2}{6 + \dfrac{(3)(6)}{3 + 6}}\left(\frac{6}{3 + 6}\right)$$

Hence

$$\mathbf{y}_{12} = -\tfrac{1}{12}\text{ S}$$

In a similar manner, $\mathbf{y}_{21}$ can be determined from Fig. 15.3b.

$$-\mathbf{I}_2 = \frac{\mathbf{V}_1}{3 + \dfrac{(6)(6)}{6 + 6}}\left(\frac{6}{6 + 6}\right)$$

Therefore,

$$\mathbf{y}_{21} = -\tfrac{1}{12}\text{ S}$$

Finally, $\mathbf{y}_{22}$ is obtained from Fig. 15.3c as

$$\mathbf{I}_2 = \frac{\mathbf{V}_2}{6 + \dfrac{(6)(3)}{6 + 3}}$$

and hence

$$\mathbf{y}_{22} = \tfrac{1}{8}\text{ S}$$

From the values computed above we find that the equations that describe the two-port operation by means of admittance parameters are

$$\mathbf{I}_1 = \tfrac{1}{6}\mathbf{V}_1 - \tfrac{1}{12}\mathbf{V}_2$$

$$\mathbf{I}_2 = -\tfrac{1}{12}\mathbf{V}_1 + \tfrac{1}{8}\mathbf{V}_2$$

or in matrix form,

$$\begin{bmatrix}\mathbf{I}_1\\ \mathbf{I}_2\end{bmatrix} = \begin{bmatrix}\frac{1}{6} & -\frac{1}{12}\\ -\frac{1}{12} & \frac{1}{8}\end{bmatrix}\begin{bmatrix}\mathbf{V}_1\\ \mathbf{V}_2\end{bmatrix}$$ ■

EXAMPLE 15.2

We wish to determine the Y parameters for the two-port shown in Fig. 15.4a. Once these parameters are known, we will determine the current in a 4-Ω load which is connected to the two-port output when a 2-A current source is applied at the two-port input.

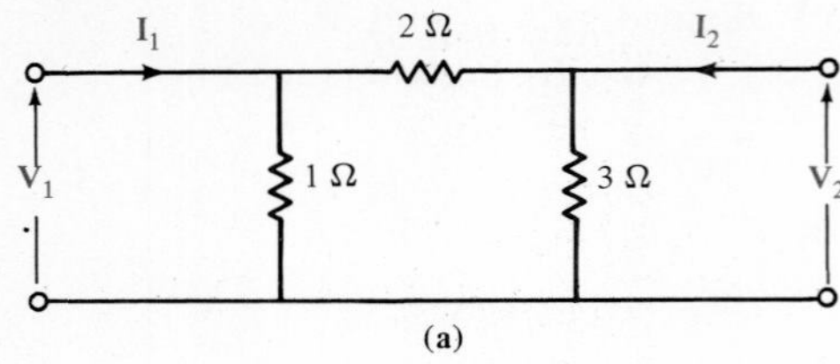

(a)

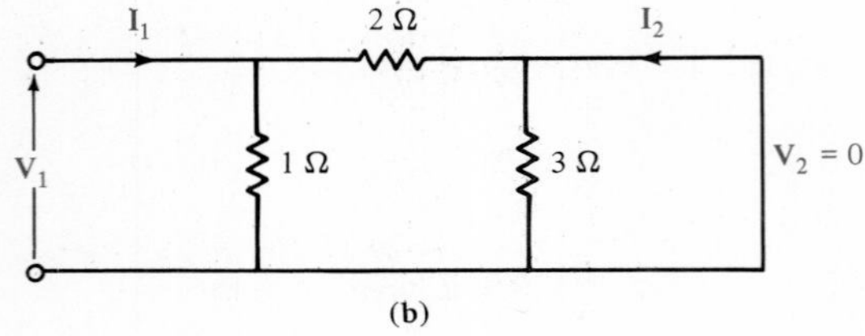

(b)

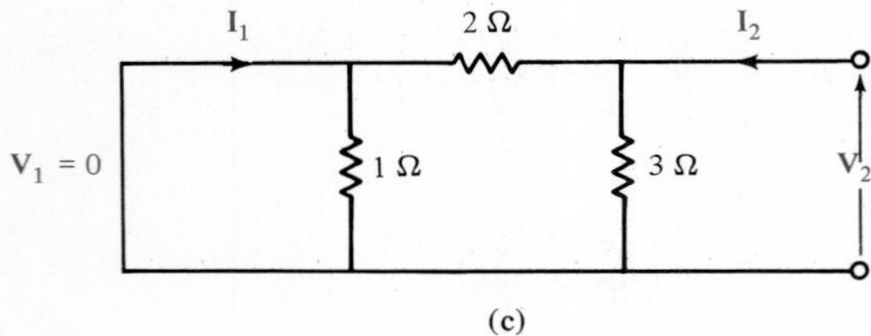

(c)

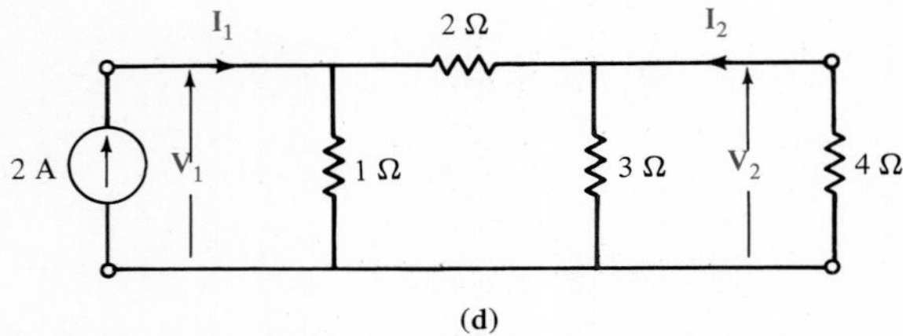

(d)

Figure 15.4 Networks employed in Example 15.2.

From Fig. 15.4b, we note that

$$\mathbf{I}_1 = \mathbf{V}_1(\tfrac{1}{1} + \tfrac{1}{2})$$

Therefore,

$$\mathbf{y}_{11} = \tfrac{3}{2}\text{ S}$$

As shown in Fig. 15.4c,

$$\mathbf{I}_1 = -\frac{\mathbf{V}_2}{2}$$

and hence

$$\mathbf{y}_{12} = -\tfrac{1}{2}\text{ S}$$

$\mathbf{y}_{21}$ is computed from Fig. 15.4b using the equation

$$\mathbf{I}_2 = -\frac{\mathbf{V}_1}{2}$$

and therefore

$$\mathbf{y}_{21} = -\tfrac{1}{2}\text{ S}$$

Finally, $\mathbf{y}_{22}$ can be derived from Fig. 15.4c using

$$\mathbf{I}_2 = \mathbf{V}_2(\tfrac{1}{3} + \tfrac{1}{2})$$

and

$$\mathbf{y}_{22} = \tfrac{5}{6}\text{ S}$$

Therefore, the equations that describe the two-port itself are

$$\mathbf{I}_1 = \tfrac{3}{2}\mathbf{V}_1 - \tfrac{1}{2}\mathbf{V}_2$$

$$\mathbf{I}_2 = -\tfrac{1}{2}\mathbf{V}_1 + \tfrac{5}{6}\mathbf{V}_2$$

These equations can now be employed to determine the operation of the two-port for some given set of terminal conditions. The terminal conditions we will examine are shown in Fig. 15.4d. From this figure we note that

$$\mathbf{I}_1 = 2\text{ A}$$

and

$$\mathbf{V}_2 = -4\mathbf{I}_2$$

Combining these with the two-port equations above yields

$$2 = \tfrac{3}{2}\mathbf{V}_1 - \tfrac{1}{2}\mathbf{V}_2$$

$$0 = -\tfrac{1}{2}\mathbf{V}_1 + \tfrac{13}{12}\mathbf{V}_2$$

or in matrix form

$$\begin{bmatrix} \frac{3}{2} & -\frac{1}{2} \\ -\frac{1}{2} & \frac{13}{12} \end{bmatrix} \begin{bmatrix} \mathbf{V}_1 \\ \mathbf{V}_2 \end{bmatrix} = \begin{bmatrix} 2 \\ 0 \end{bmatrix}$$

Note carefully that these equations are simply the nodal equations for the network in Fig. 15.4d. Solving the equations we obtain $\mathbf{V}_2 = \frac{8}{11}$ V and therefore $\mathbf{I}_2 = -\frac{2}{11}$ A. ■

DRILL EXERCISE

D15.1. Find the Y parameters for the two-port network shown in Fig. D15.1.

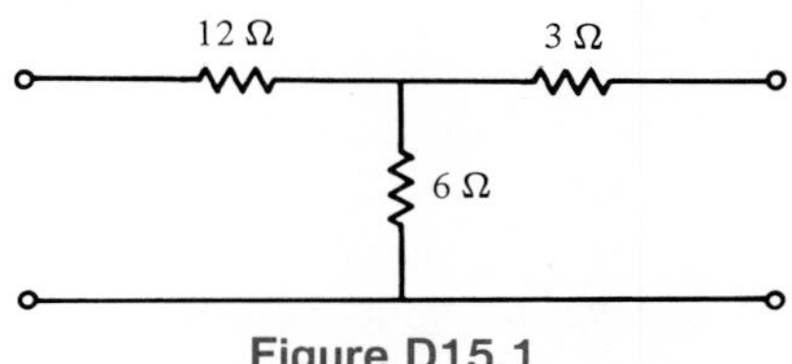

Figure D15.1

D15.2. Find the Y parameters for the two-port network shown in Fig. D15.2.

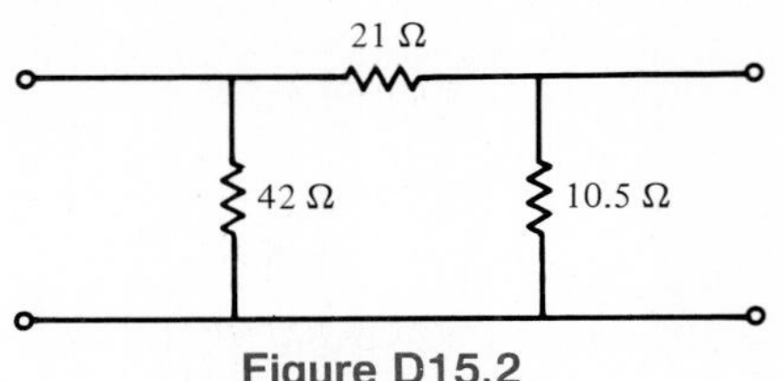

Figure D15.2

D15.3. If a 10-A source is connected to the input of the two-port in Fig. D15.2, find the current in a 5-Ω resistor connected to the output port.

15.2 Impedance Parameters

Once again if we assume that the two-port is a linear network that contains no independent sources, then by means of superposition we can write the input and output voltages as the sum of two components, one due to $\mathbf{I}_1$ and one due to $\mathbf{I}_2$:

$$\begin{aligned} \mathbf{V}_1 &= \mathbf{z}_{11}\mathbf{I}_1 + \mathbf{z}_{12}\mathbf{I}_2 \\ \mathbf{V}_2 &= \mathbf{z}_{21}\mathbf{I}_1 + \mathbf{z}_{22}\mathbf{I}_2 \end{aligned} \tag{15.4}$$

These equations which describe the two-port can also be written in matrix form as

$$\begin{bmatrix} \mathbf{V}_1 \\ \mathbf{V}_2 \end{bmatrix} = \begin{bmatrix} \mathbf{z}_{11} & \mathbf{z}_{12} \\ \mathbf{z}_{21} & \mathbf{z}_{22} \end{bmatrix} \begin{bmatrix} \mathbf{I}_1 \\ \mathbf{I}_2 \end{bmatrix} \tag{15.5}$$

Like the Y parameters, these *Z parameters* can be derived as follows:

$$\begin{aligned} \mathbf{z}_{11} &= \left.\frac{\mathbf{V}_1}{\mathbf{I}_1}\right|_{\mathbf{I}_2=0} \\ \mathbf{z}_{12} &= \left.\frac{\mathbf{V}_1}{\mathbf{I}_2}\right|_{\mathbf{I}_1=0} \\ \mathbf{z}_{21} &= \left.\frac{\mathbf{V}_2}{\mathbf{I}_1}\right|_{\mathbf{I}_2=0} \\ \mathbf{z}_{22} &= \left.\frac{\mathbf{V}_2}{\mathbf{I}_2}\right|_{\mathbf{I}_1=0} \end{aligned} \tag{15.6}$$

In the equations above, setting $\mathbf{I}_1$ or $\mathbf{I}_2 = 0$ is equivalent to open-circuiting the input or output port. Therefore, the Z parameters are called the *open-circuit impedance parameters*. $\mathbf{z}_{11}$ is called the *open-circuit input impedance*, $\mathbf{z}_{22}$ is called the *open-circuit output impedance*, and $\mathbf{z}_{12}$ and $\mathbf{z}_{21}$ are termed *open-circuit transfer impedances*.

EXAMPLE 15.3

We wish to find the Z parameters for the network in Fig. 15.5a. Once the parameters are known, we will use them to find the current in a 4-Ω resistor that is connected to the output terminals when a $12 \angle 0°$-V source with an internal impedance of $1 + j0\ \Omega$ is connected to the input.

From Fig. 15.5a we note that

$$\mathbf{z}_{11} = 2 - j4\ \Omega$$

$$\mathbf{z}_{12} = -j4\ \Omega$$

$$\mathbf{z}_{21} = -j4\ \Omega$$

$$\mathbf{z}_{22} = -j4 + j2 = -j2\ \Omega$$

The equations for the two-port are therefore

$$\mathbf{V}_1 = (2 - j4)\mathbf{I}_1 - j4\mathbf{I}_2$$

$$\mathbf{V}_2 = -j4\mathbf{I}_1 - j2\mathbf{I}_2$$

The terminal conditions for the network shown in Fig. 15.5b are

$$\mathbf{V}_1 = 12 \angle 0° - (1)\mathbf{I}_1$$

$$\mathbf{V}_2 = -4\mathbf{I}_2$$

Combining these with the two-port equations yields

$$12 \angle 0° = (3 - j4)\mathbf{I}_1 - j4\mathbf{I}_2$$

$$0 = -j4\mathbf{I}_1 + (4 - j2)\mathbf{I}_2$$

It is interesting to note that these equations are the mesh equations for the network. If we solve the equations for $\mathbf{I}_2$, we obtain $\mathbf{I}_2 = 1.61 \angle 137.73°$ A, which is the current in the 4-Ω load. ■

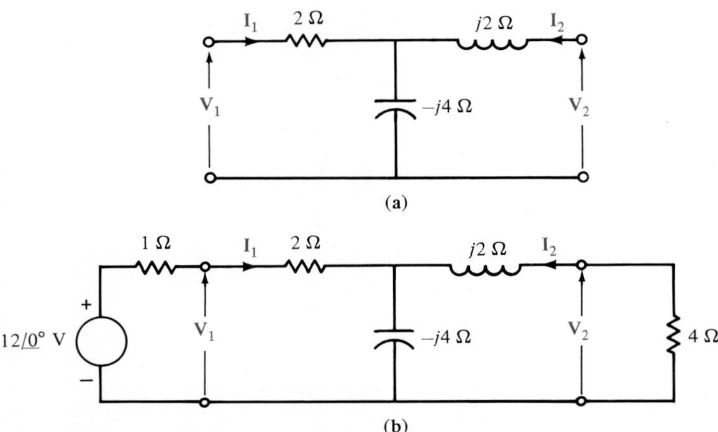

Figure 15.5 Circuits employed in Example 15.3.

EXAMPLE 15.4

Let us determine the Z parameters for the two-port network shown in Fig. 15.6a.

$\mathbf{z}_{11}$ is derived from Fig. 15.6b via the equation

$$\mathbf{I}_1 = \frac{\mathbf{V}_1}{R_1} + \beta\mathbf{V}_1$$

and hence

$$\mathbf{z}_{11} = \frac{R_1}{1 + \beta R_1}$$

$\mathbf{z}_{12}$ is obtained from Fig. 15.6c. Note that with $\mathbf{I}_1 = 0$, the current in resistors R_1 and R_2 is $\mathbf{I}_2 - \beta\mathbf{V}_1$. Therefore,

$$\mathbf{V}_1 = R_1(\mathbf{I}_2 - \beta\mathbf{V}_1)$$

and hence

$$\mathbf{z}_{12} = \frac{R_1}{1 + \beta R_1}$$

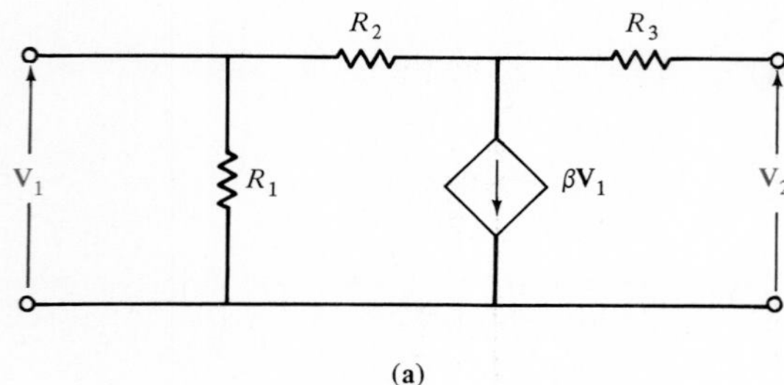

(a)

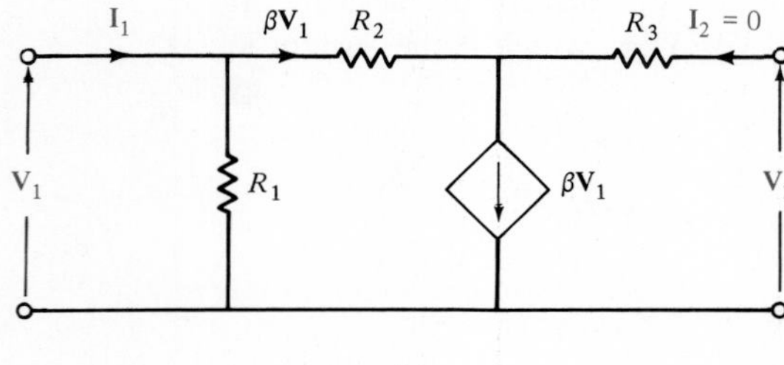

(b)

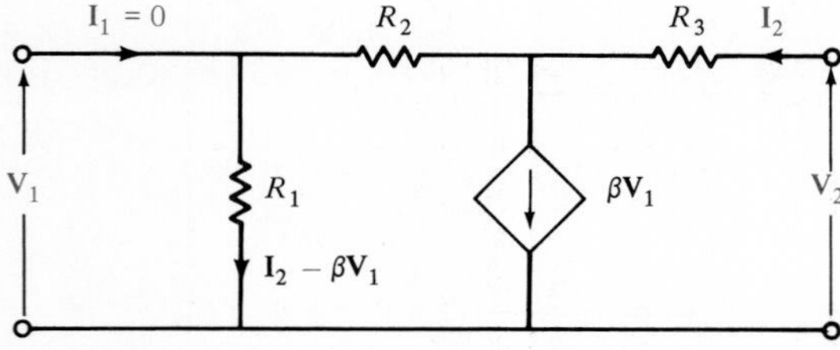

(c)

Figure 15.6 Circuits employed in Example 15.4.

The parameter $\mathbf{z}_{21}$ is derived from Fig. 15.6b. Since $\mathbf{I}_2 = 0$,

$$\mathbf{I}_1 = \frac{\mathbf{V}_1}{R_1} + \beta \mathbf{V}_1$$

In addition, KVL implies that

$$\mathbf{V}_1 - \beta \mathbf{V}_1 R_2 = \mathbf{V}_2$$

From these two equations we obtain

$$\mathbf{z}_{21} = \frac{R_1(1 - \beta R_2)}{1 + \beta R_1}$$

$\mathbf{z}_{22}$ can be obtained from Fig. 15.6c. Note that KVL around the outer loop is

$$\mathbf{V}_2 = \mathbf{I}_2 R_3 + (\mathbf{I}_2 - \beta \mathbf{V}_1) R_2 + \mathbf{V}_1$$

but

$$\mathbf{V}_1 = R_1(\mathbf{I}_2 - \beta \mathbf{V}_1)$$

Solving the two equations yields

$$\mathbf{z}_{22} = \frac{R_1 + R_2 + R_3 + \beta R_1 R_3}{1 + \beta R_1}$$

Therefore, the two-port equations for the network are

$$\mathbf{V}_1 = \frac{R_1}{1 + \beta R_1} \mathbf{I}_1 + \frac{R_1}{1 + \beta R_1} \mathbf{I}_2$$

$$\mathbf{V}_2 = \frac{R_1(1 - \beta R_2)}{1 + \beta R_1} \mathbf{I}_1 + \frac{R_1 + R_2 + R_3 + \beta R_1 R_3}{1 + \beta R_1} \mathbf{I}_2$$

■

At this point it is important for the reader to note that in all of our two-port analyses, the application of KVL and KCL will produce the equations needed to determine the parameters.

DRILL EXERCISE

D15.4. Find the Z parameters for the network in Fig. D15.1. Then compute the current in a 4-Ω load if a 24 $\underline{/0°}$-V source is connected at the input port.

D15.5. Find the Z parameters for the two-port in Fig. D15.5.

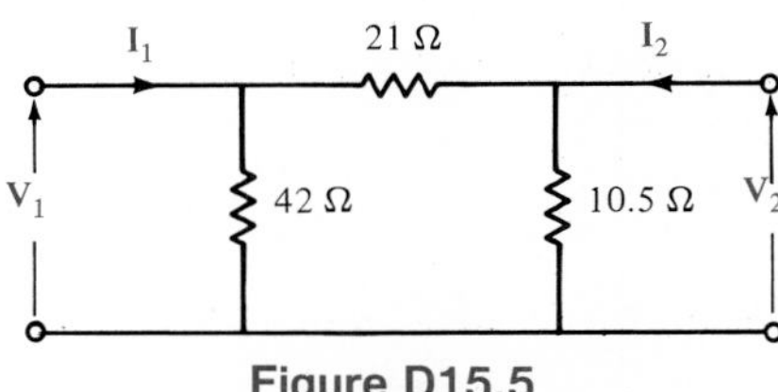

Figure D15.5

15.3

Hybrid Parameters

Under the assumptions used to develop the Y and Z parameters, we can obtain what are commonly called the *hybrid parameters*. In the pair of equations that define these parameters, $\mathbf{V}_1$ and $\mathbf{I}_2$ are the independent variables. Therefore, the two-port equations in terms of the hybrid parameters are

$$\begin{aligned}\mathbf{V}_1 &= \mathbf{h}_{11}\mathbf{I}_1 + \mathbf{h}_{12}\mathbf{V}_2 \\ \mathbf{I}_2 &= \mathbf{h}_{21}\mathbf{I}_1 + \mathbf{h}_{22}\mathbf{V}_2\end{aligned} \tag{15.7}$$

or in matrix form,

$$\begin{bmatrix}\mathbf{V}_1 \\ \mathbf{I}_2\end{bmatrix} = \begin{bmatrix}\mathbf{h}_{11} & \mathbf{h}_{12} \\ \mathbf{h}_{21} & \mathbf{h}_{22}\end{bmatrix}\begin{bmatrix}\mathbf{I}_1 \\ \mathbf{V}_2\end{bmatrix} \tag{15.8}$$

These parameters are especially important in transistor circuit analysis. The parameters are determined via the following equations:

$$\begin{aligned}\mathbf{h}_{11} &= \left.\frac{\mathbf{V}_1}{\mathbf{I}_1}\right|_{\mathbf{V}_2=0} \\ \mathbf{h}_{12} &= \left.\frac{\mathbf{V}_1}{\mathbf{V}_2}\right|_{\mathbf{I}_1=0} \\ \mathbf{h}_{21} &= \left.\frac{\mathbf{I}_2}{\mathbf{I}_1}\right|_{\mathbf{V}_2=0} \\ \mathbf{h}_{22} &= \left.\frac{\mathbf{I}_2}{\mathbf{V}_2}\right|_{\mathbf{I}_1=0}\end{aligned} \tag{15.9}$$

The parameters $\mathbf{h}_{11}$, $\mathbf{h}_{12}$, $\mathbf{h}_{21}$, and $\mathbf{h}_{22}$ represent the *short-circuit input impedance*, the *open-circuit reverse voltage gain*, the *short-circuit forward current gain*, and the *open-circuit output admittance*, respectively. Because of this mix of parameters, they are called *hybrid parameters*. In transistor circuit analysis the parameters $\mathbf{h}_{11}$, $\mathbf{h}_{12}$, $\mathbf{h}_{21}$, and $\mathbf{h}_{22}$ are normally labeled $\mathbf{h}_i$, $\mathbf{h}_r$, $\mathbf{h}_f$, and $\mathbf{h}_o$.

EXAMPLE 15.5

If the hybrid parameters for the two-port shown in Fig. 15.7 are known, we wish to find an expression for the input impedance of the network.

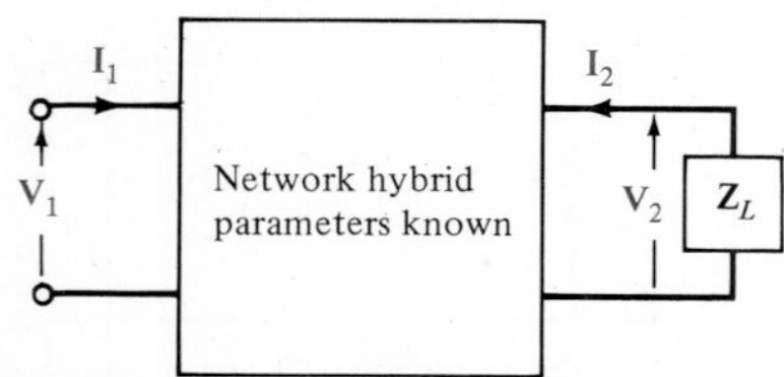

Figure 15.7 Network used in Example 15.5.

The equations for the two-port are

$$\mathbf{V}_1 = \mathbf{h}_{11}\mathbf{I}_1 + \mathbf{h}_{12}\mathbf{V}_2$$

$$\mathbf{I}_2 = \mathbf{h}_{21}\mathbf{I}_1 + \mathbf{h}_{22}\mathbf{V}_2$$

However,

$$\mathbf{V}_2 = -\mathbf{I}_2\mathbf{Z}_L$$

Therefore,

$$\mathbf{V}_1 = \mathbf{h}_{11}\mathbf{I}_1 - \mathbf{h}_{12}\mathbf{Z}_L\mathbf{I}_2$$

$$0 = \mathbf{h}_{21}\mathbf{I}_1 - (1 + \mathbf{h}_{22}\mathbf{Z}_L)\mathbf{I}_2$$

Solving the two equations for $\mathbf{I}_1$ and then forming the ratio $\mathbf{V}_1/\mathbf{I}_1$ yields

$$\mathbf{Z}_L = \frac{\mathbf{V}_1}{\mathbf{I}_1} = \mathbf{h}_{11} - \frac{\mathbf{h}_{12}\mathbf{h}_{21}\mathbf{Z}_L}{1 + \mathbf{h}_{22}\mathbf{Z}_L}$$ ■

EXAMPLE 15.6

Let us determine the hybrid parameters for the network shown in Fig. 15.8a.

$\mathbf{h}_{11}$ is determined using Fig. 15.8b. From KVL

$$\mathbf{V}_1 - \mathbf{I}_1 R_1 = \mathbf{V}$$

and using KCL we can write

$$\mathbf{I}_1 = \frac{\mathbf{V}}{R_2} + j\omega C\mathbf{V} + \alpha\mathbf{I}_1$$

Solving these two equations, we obtain

$$\mathbf{h}_{11} = R_1 + \frac{1 - \alpha}{1/R_2 + j\omega C}$$

The parameter $\mathbf{h}_{12}$ is derived from Fig. 15.8c. Since $\mathbf{I}_1 = 0$, the voltage across R_2 is $\mathbf{V}_1$ and therefore

$$\mathbf{V}_1 = \frac{\mathbf{V}_2}{R_2 + 1/j\omega C} R_2$$

from which we find that

$$\mathbf{h}_{12} = \frac{R_2}{R_2 + 1/j\omega C}$$

Figure 15.8b is used to compute $\mathbf{h}_{21}$. The two KCL equations for this network are

$$\mathbf{I}_1 = \frac{\mathbf{V}}{R_2} + j\omega C\mathbf{V} + \alpha\mathbf{I}_1$$

$$\mathbf{I}_2 = -(j\omega C\mathbf{V} + \alpha\mathbf{I}_1)$$

Solving these equations, we can show that

$$\mathbf{h}_{21} = -\frac{\alpha + j\omega CR_2}{1 + j\omega CR_2}$$

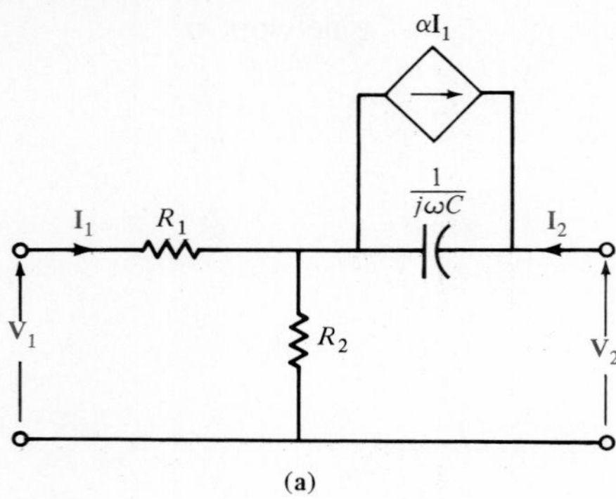

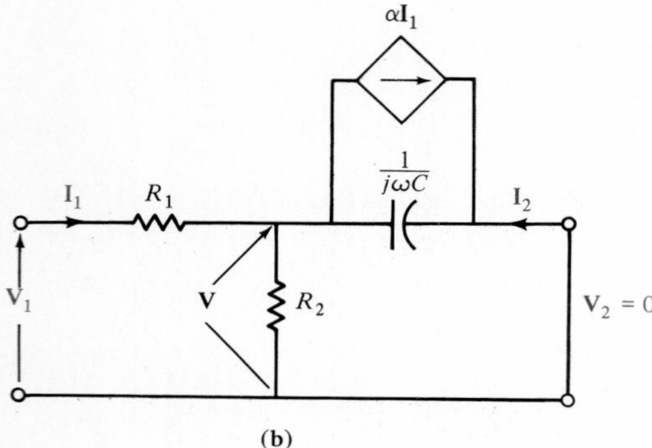

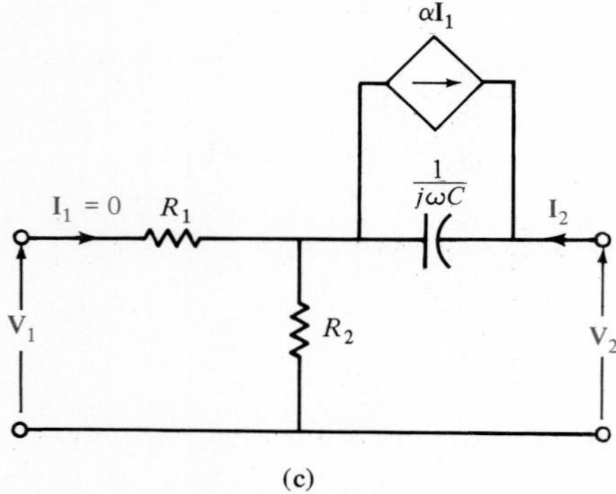

Figure 15.8 Circuits employed in Example 15.6.

Finally, $\mathbf{h}_{22}$ is obtained from Fig. 15.8c. The KVL equation for this network is

$$\mathbf{V}_2 = \mathbf{I}_2\left(R_2 + \frac{1}{j\omega C}\right)$$

and hence

$$\mathbf{h}_{22} = \frac{1}{R_2 + 1/j\omega C}$$

Therefore, the two-port equations in terms of the hybrid parameters for the network in Fig. 15.8a are

$$\mathbf{V}_1 = \left(R_1 + \frac{1 - \alpha}{1/R_2 + j\omega C}\right)\mathbf{I}_1 + \frac{R_2}{R_2 + 1/j\omega C}\mathbf{V}_2$$

$$\mathbf{I}_2 = -\frac{\alpha + j\omega CR_2}{1 + j\omega CR_2}\mathbf{I}_1 + \frac{1}{R_2 + 1/j\omega C}\mathbf{V}_2$$

■

EXAMPLE 15.7

An equivalent circuit for the op amp in Fig. 15.9a is shown in Fig. 15.9b. We will determine the hybrid parameters for this network.

Parameter $\mathbf{h}_{11}$ is derived from Fig. 15.9c. With the output shorted, $\mathbf{h}_{11}$ is a function of only R_i, R_1, and R_2 and

$$\mathbf{h}_{11} = R_i + \frac{R_1R_2}{R_1 + R_2}$$

Figure 15.9d is used to derive $\mathbf{h}_{12}$. Since $\mathbf{I}_1 = 0$, $\mathbf{V}_i = 0$ and the relationship between $\mathbf{V}_1$ and $\mathbf{V}_2$ is a simple voltage divider.

$$\mathbf{V}_1 = \frac{\mathbf{V}_2R_1}{R_1 + R_2}$$

Therefore,

$$\mathbf{h}_{12} = \frac{R_1}{R_1 + R_2}$$

KVL and KCL can be applied to Fig. 15.9c to determine $\mathbf{h}_{21}$. The two equations that relate $\mathbf{I}_2$ to $\mathbf{I}_1$ are

$$\mathbf{V}_i = \mathbf{I}_1R_i$$

$$\mathbf{I}_2 = \frac{-A\mathbf{V}_i}{R_o} - \frac{\mathbf{I}_1R_1}{R_1 + R_2}$$

Therefore,

$$\mathbf{h}_{21} = -\left(\frac{AR_i}{R_o} + \frac{R_1}{R_1 + R_2}\right)$$

Finally, the relationship between $\mathbf{I}_2$ and $\mathbf{V}_2$ in Fig. 15.9d is

$$\frac{\mathbf{V}_2}{\mathbf{I}_2} = \frac{\mathbf{R}_o(R_1 + R_2)}{R_o + R_1 + R_2}$$

and therefore

$$\mathbf{h}_{22} = \frac{R_o + R_1 + R_2}{R_o(R_1 + R_2)}$$

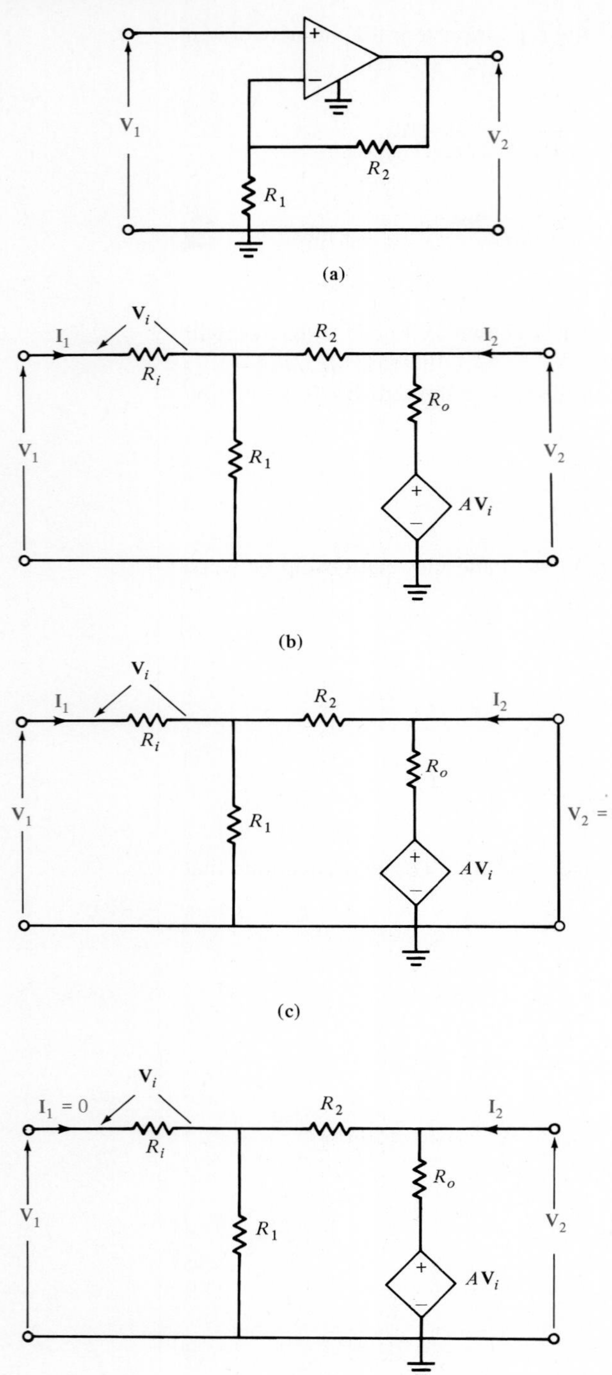

Figure 15.9 Circuits employed in Example 15.7.

The network equations are therefore

$$\mathbf{V}_1 = \left(R_i + \frac{R_1R_2}{R_1 + R_2}\right)\mathbf{I}_1 + \frac{R_1}{R_1 + R_2}\mathbf{V}_2$$

$$\mathbf{I}_2 = -\left(\frac{AR_i}{R_o} + \frac{R_1}{R_1 + R_2}\right)\mathbf{I}_1 + \frac{R_o + R_1 + R_2}{R_o(R_1 + R_2)}\mathbf{V}_2$$

■

DRILL EXERCISE

D15.6. Find the hybrid parameters for the network shown in Fig. D15.1.

D15.7. If a 4-Ω load is connected to the output port of the network examined in Drill Exercise D15.6, determine the input impedance of the two-port with the load connected.

D15.8. Compute the hybrid parameters for the network in Fig. D15.2.

D15.9. Consider the network in Fig. D15.9. The two-port is a hybrid model for a basic transistor. Determine the voltage gain of the entire network, $\mathbf{V}_2/\mathbf{V}_s$, if a source $\mathbf{V}_s$ with internal resistance R_1 is applied at the input to the two-port and a load R_L is connected at the output port.

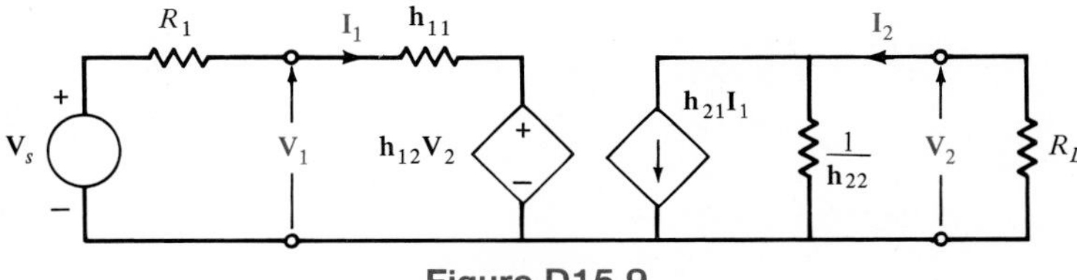

Figure D15.9

15.4 Transmission Parameters

The final parameters we will discuss are called the *transmission parameters*. They are defined by the equations

$$\begin{aligned}\mathbf{V}_1 &= \mathbf{A}\mathbf{V}_2 - \mathbf{B}\mathbf{I}_2\\ \mathbf{I}_1 &= \mathbf{C}\mathbf{V}_2 - \mathbf{D}\mathbf{I}_2\end{aligned} \tag{15.10}$$

or in matrix form,

$$\begin{bmatrix}\mathbf{V}_1\\ \mathbf{I}_1\end{bmatrix} = \begin{bmatrix}\mathbf{A} & \mathbf{B}\\ \mathbf{C} & \mathbf{D}\end{bmatrix}\begin{bmatrix}\mathbf{V}_2\\ -\mathbf{I}_2\end{bmatrix} \tag{15.11}$$

These parameters are very useful in the analysis of circuits connected in cascade as we will demonstrate later. The parameters are determined via the following equations:

$$\mathbf{A} = \left.\frac{\mathbf{V}_1}{\mathbf{V}_2}\right|_{\mathbf{I}_2=0}$$

$$\mathbf{B} = \left.\frac{\mathbf{V}_1}{-\mathbf{I}_2}\right|_{\mathbf{V}_2=0}$$

$$\mathbf{C} = \left.\frac{\mathbf{I}_1}{\mathbf{V}_2}\right|_{\mathbf{I}_2=0} \tag{15.12}$$

$$\mathbf{D} = \left.\frac{\mathbf{I}_1}{-\mathbf{I}_2}\right|_{\mathbf{V}_2=0}$$

A, B, C and **D** represent the *open-circuit voltage ratio*, the *negative short-circuit transfer impedance*, the *open-circuit transfer admittance*, and the *negative short-circuit current ratio*, respectively. For obvious reasons the transmission parameters are commonly referred to as the *ABCD parameters*.

EXAMPLE 15.8

Let us determine the transmission parameters for the network in Fig. 15.10a.

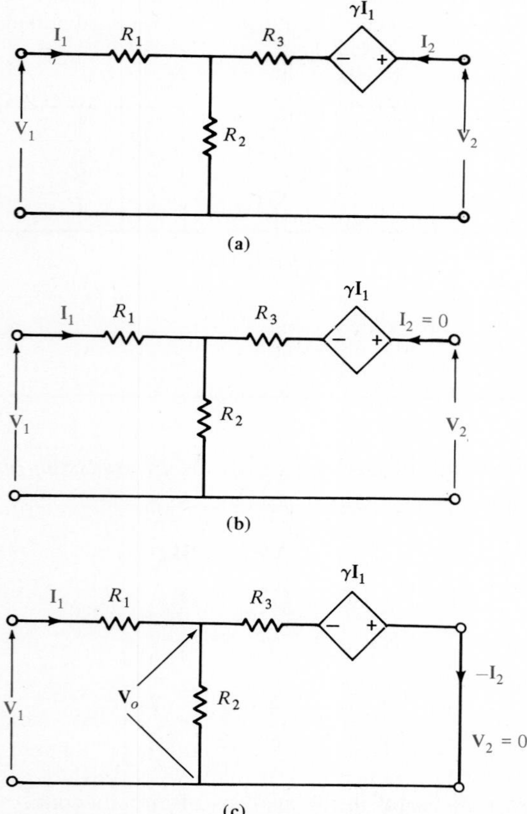

Figure 15.10 Circuits used in Example 15.8.

The parameter $\mathbf{A}$ is determined from Fig. 15.10b. Since $\mathbf{I}_2 = 0$, $\mathbf{V}_1$ and $\mathbf{V}_2$ are related via the two KVL equations

$$\mathbf{V}_2 = \gamma\mathbf{I}_1 + \mathbf{I}_1 R_2$$

$$\mathbf{V}_1 = \mathbf{I}_1(R_1 + R_2)$$

Solving these equations for $\mathbf{V}_1/\mathbf{V}_2$ yields

$$\mathbf{A} = \frac{R_1 + R_2}{\gamma + R_2}$$

Parameter $\mathbf{B}$ is obtained from Fig. 15.10c. In this case it is convenient to assume that $-\mathbf{I}_2 = 1$ A and solve for $\mathbf{V}_1$. Under this condition

$$\mathbf{V}_o = R_3 - \gamma\mathbf{I}_1$$

and

$$\mathbf{I}_1 = 1 + \frac{\mathbf{V}_o}{R_2}$$

Therefore,

$$\mathbf{I}_1 = \frac{R_2 + R_3}{R_2 + \gamma}$$

Then

$$\begin{aligned}\mathbf{V}_1 &= \mathbf{I}_1 R_1 + \mathbf{V}_o \\ &= \frac{R_2 + R_3}{R_2 + \gamma} R_1 + R_3 - \gamma \frac{R_2 + R_3}{R_2 + \gamma} \\ &= R_3 + \frac{(R_1 - \gamma)(R_2 + R_3)}{R_2 + \gamma}\end{aligned}$$

and hence

$$\mathbf{B} = R_3 + \frac{(R_1 - \gamma)(R_2 + R_3)}{R_2 + \gamma}$$

In Fig. 15.10b if we assume that $\mathbf{I}_1 = 1$ A, then $\mathbf{V}_2$ is

$$\mathbf{V}_2 = R_2 + \gamma$$

and therefore

$$\mathbf{C} = \frac{1}{R_2 + \gamma}$$

Figure 15.10c is employed to determine parameter $\mathbf{D}$. In our analysis above, in which we derived parameter $\mathbf{B}$, we found that if we assumed that $-\mathbf{I}_2 = 1$ A, then

$$\mathbf{I}_1 = \frac{R_2 + R_3}{R_2 + \gamma}$$

and therefore

$$\mathbf{D} = \frac{R_2 + R_3}{R_2 + \gamma}$$

The two-port equations for the network are therefore

$$\mathbf{V}_1 = \frac{R_1 + R_2}{R_2 + \gamma}\mathbf{V}_2 - \left[R_3 + \frac{(R_1 - \gamma)(R_2 + R_3)}{R_2 + \gamma}\right]\mathbf{I}_2$$

$$\mathbf{I}_1 = \frac{1}{R_2 + \gamma}\mathbf{V}_2 - \frac{R_2 + R_3}{R_2 + \gamma}\mathbf{I}_2$$

■

EXAMPLE 15.9

We will now determine the transmission parameters for the network in Fig. 15.11a. The most convenient way in which to attack this type of problem is to assume a value for one of the variables in a parameter and solve the network for the other variable using KVL and KCL. To simplify our analysis, we have also defined variables $\mathbf{V}_A$, $\mathbf{V}_B$, and $\mathbf{I}_o$ as shown in Fig. 15.11b and c.

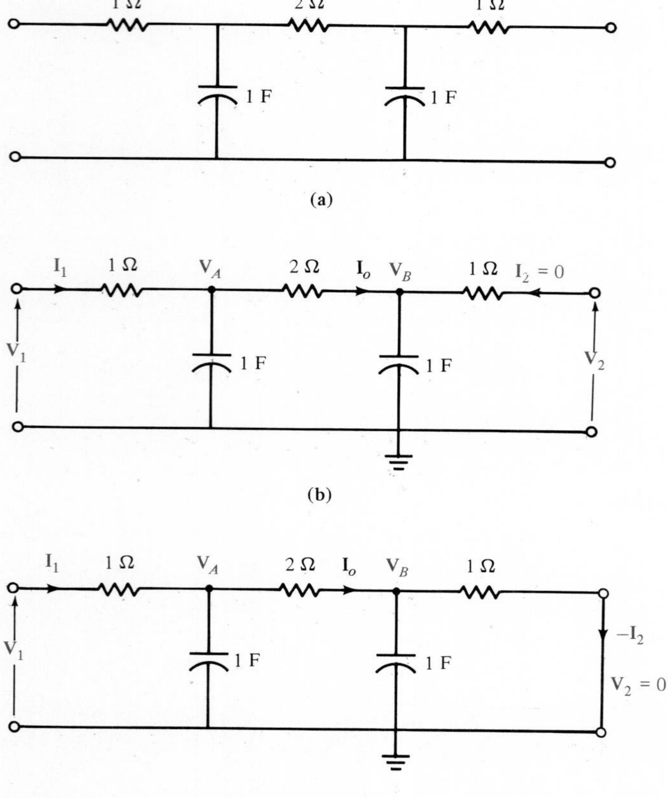

Figure 15.11 Circuits used in Example 15.9.

Parameter **A** is derived from Fig. 15.11b. If we assume that $\mathbf{V}_2 = 1$ V, then $\mathbf{V}_B = 1$ V and $\mathbf{I}_o = j\omega$ A, and hence

$$\mathbf{V}_A = 2\mathbf{I}_o + \mathbf{V}_B = 2j\omega + 1 \text{ V}$$

Then

$$\begin{aligned}\mathbf{I}_1 &= j\omega\mathbf{V}_A + \mathbf{I}_o \\ &= -2\omega^2 + 2j\omega \text{ A}\end{aligned}$$

Finally,

$$\begin{aligned}\mathbf{V}_1 &= (1)\mathbf{I}_1 + \mathbf{V}_A \\ &= -2\omega^2 + 4j\omega + 1 \text{ V}\end{aligned}$$

and therefore

$$\mathbf{A} = -2\omega^2 + 4j\omega + 1$$

Figure 15.11c is used to derive parameter **B**. Assuming that $-\mathbf{I}_2 = 1$ A, then $\mathbf{V}_B = 1$, $\mathbf{I}_o = 1 + j\omega$ A, and

$$\mathbf{V}_A = 2\mathbf{I}_o + \mathbf{V}_B = 3 + j2\omega \text{ V}$$

Then

$$\begin{aligned}\mathbf{I}_1 &= \mathbf{I}_o + j\omega\mathbf{V}_A \\ &= (1 + j\omega) + j\omega(3 + 2\omega) \\ &= 1 + 4j\omega - 2\omega^2 \text{ A}\end{aligned}$$

Finally,

$$\begin{aligned}\mathbf{V}_1 &= (1)\mathbf{I}_1 + \mathbf{V}_A \\ &= 4 + 6j\omega - 2\omega^2 \text{ V}\end{aligned}$$

Therefore,

$$\mathbf{B} = 4 + 6j\omega - 2\omega^2$$

From the analysis above used to derive parameter **A**, we find that

$$\mathbf{C} = -2\omega^2 + 2j\omega$$

In a similar manner it follows from the derivation of parameter **B** that

$$\mathbf{D} = -2\omega^2 + 4j\omega + 1$$

Therefore, the two-port equations are

$$\begin{aligned}\mathbf{V}_1 &= (-2\omega^2 + 4j\omega + 1)\mathbf{V}_2 - (-2\omega^2 + 6j\omega + 4)\mathbf{I}_2 \\ \mathbf{I}_1 &= (-2\omega^2 + 2j\omega)\mathbf{V}_2 - (-2\omega^2 + 4j\omega + 1)\mathbf{I}_2\end{aligned}$$

■

DRILL EXERCISE

D15.10. Find the transmission parameters for the network shown in Fig. D15.1.

D15.11. Compute the transmission parameters for the two-port in Fig. D15.2.

One final note concerning the derivation of two-port parameters is in order. It is entirely possible that some or all of the two-port parameters for a particular network may not exist. For example, consider the network in Fig. 15.12. The parameter $\mathbf{y}_{11}$ is determined by short-circuiting the output and computing the ratio $\mathbf{I}_1/\mathbf{V}_1$. However, shorting the output implies that $2\mathbf{I}_o = 0$, and therefore $\mathbf{I}_o = 0$. However, $\mathbf{I}_o$ cannot be zero if a nonzero voltage $\mathbf{V}_1$ is applied at the input. Therefore, this network does not have a set of Y parameters.

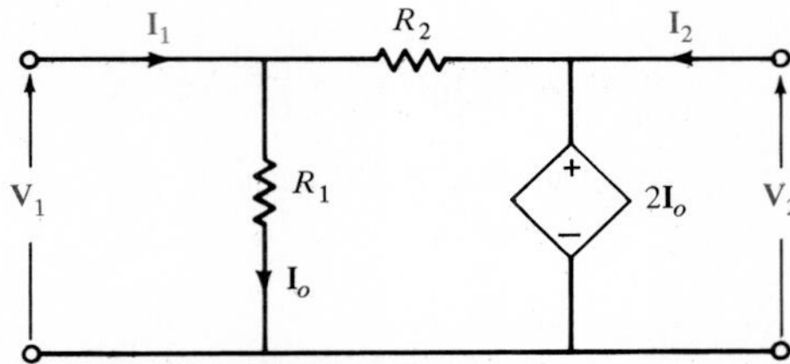

Figure 15.12 Example network that does not have a set of Y parameters unless the source has an internal impedance.

15.5 Equivalent Circuits

A number of standard equivalent circuits can be employed to represent a two-port network. These equivalent circuits have the same terminal characteristics as the original network, although they may not be physically realizable because they may possibly contain negative impedances. The equivalent circuits that are perhaps the most popular are shown in Fig. 15.13. Since there are four coefficients in the two-port equations, each of the circuits contains four elements. The reader can easily verify the validity of the circuits in Fig. 15.13 by writing the circuit equations that characterize each.

15.6 Parameter Conversions

If all the two-port parameters for a network exist, it is possible to relate one set of parameters to another since the parameters interrelate the variables $\mathbf{V}_1$, $\mathbf{I}_1$, $\mathbf{V}_2$, and $\mathbf{I}_2$. The following examples illustrate the manner in which to determine one set of parameters from another.

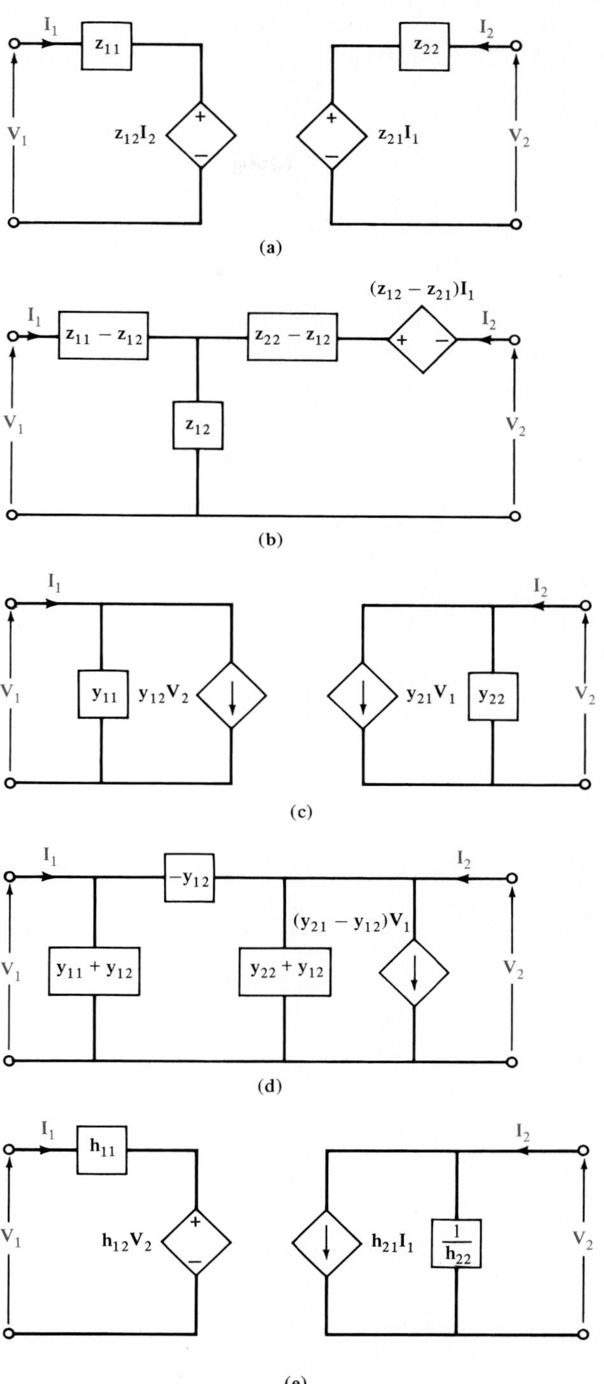

Figure 15.13 Some standard equivalent circuits for two-ports.

EXAMPLE 15.10

Let us show how to determine the hybrid parameters if the Z parameters are known. The two-port equations involving the Z parameters are

$$\mathbf{V}_1 = \mathbf{z}_{11}\mathbf{I}_1 + \mathbf{z}_{12}\mathbf{I}_2$$

$$\mathbf{V}_2 = \mathbf{z}_{21}\mathbf{I}_1 + \mathbf{z}_{22}\mathbf{I}_2$$

If we solve the second Z-parameter equation for $\mathbf{I}_2$, it will be directly in the form of the second hybrid-parameter equation. Therefore,

$$\mathbf{I}_2 = \frac{-\mathbf{z}_{21}}{\mathbf{z}_{22}}\mathbf{I}_1 + \frac{1}{\mathbf{z}_{22}}\mathbf{V}_2$$

Substituting this equation for $\mathbf{I}_2$ into the first Z-parameter equation, we obtain

$$\mathbf{V}_1 = \frac{\mathbf{z}_{11}\mathbf{z}_{22} - \mathbf{z}_{12}\mathbf{z}_{21}}{\mathbf{z}_{22}}\mathbf{I}_1 + \frac{\mathbf{z}_{12}}{\mathbf{z}_{22}}\mathbf{V}_2$$

Comparing the equations for $\mathbf{I}_2$ and $\mathbf{V}_1$ in terms of the Z parameters with the defining equations for the hybrid parameters shows that

$$\mathbf{h}_{11} = \frac{\Delta_z}{\mathbf{z}_{22}} \qquad \mathbf{h}_{12} = \frac{\mathbf{z}_{12}}{\mathbf{z}_{22}}$$

$$\mathbf{h}_{21} = \frac{-\mathbf{z}_{21}}{\mathbf{z}_{22}} \qquad \mathbf{h}_{22} = \frac{1}{\mathbf{z}_{22}}$$

where $\Delta_z = \mathbf{z}_{11}\mathbf{z}_{22} - \mathbf{z}_{12}\mathbf{z}_{21}$. ■

EXAMPLE 15.11

Let us determine the Z parameters from the Y parameters. Although this can be done very easily using determinants, we will illustrate the conversion using matrix analysis because of the nature of the two sets of equations. In matrix form we can write

$$\begin{bmatrix} \mathbf{z}_{11} & \mathbf{z}_{12} \\ \mathbf{z}_{21} & \mathbf{z}_{22} \end{bmatrix} \begin{bmatrix} \mathbf{I}_1 \\ \mathbf{I}_2 \end{bmatrix} = \begin{bmatrix} \mathbf{V}_1 \\ \mathbf{V}_2 \end{bmatrix}$$

and

$$\begin{bmatrix} \mathbf{y}_{11} & \mathbf{y}_{12} \\ \mathbf{y}_{21} & \mathbf{y}_{22} \end{bmatrix} \begin{bmatrix} \mathbf{V}_1 \\ \mathbf{V}_2 \end{bmatrix} = \begin{bmatrix} \mathbf{I}_1 \\ \mathbf{I}_2 \end{bmatrix}$$

From the second equation we note that

$$\begin{bmatrix} \mathbf{V}_1 \\ \mathbf{V}_2 \end{bmatrix} = \begin{bmatrix} \mathbf{y}_{11} & \mathbf{y}_{12} \\ \mathbf{y}_{21} & \mathbf{y}_{22} \end{bmatrix}^{-1} \begin{bmatrix} \mathbf{I}_1 \\ \mathbf{I}_2 \end{bmatrix}$$

and hence we can determine the Z parameters by inverting the matrix for the Y parameters (i.e., $[\mathbf{Z}] = [\mathbf{Y}]^{-1}$). The determininant for the matrix is

$$\Delta_Y = \mathbf{y}_{11}\mathbf{y}_{22} - \mathbf{y}_{12}\mathbf{y}_{21}$$

The adjoint of the matrix is

$$\begin{bmatrix} \mathbf{y}_{22} & -\mathbf{y}_{12} \\ -\mathbf{y}_{21} & \mathbf{y}_{11} \end{bmatrix}$$

Therefore, the inverse is

$$\frac{1}{\Delta_Y}\begin{bmatrix} \mathbf{y}_{22} & -\mathbf{y}_{12} \\ -\mathbf{y}_{21} & \mathbf{y}_{11} \end{bmatrix}$$

and hence the Z parameters in terms of the Y parameters are

$$\mathbf{z}_{11} = \frac{\mathbf{y}_{22}}{\Delta_Y} \qquad \mathbf{z}_{12} = \frac{-\mathbf{y}_{12}}{\Delta_Y}$$

$$\mathbf{z}_{21} = \frac{-\mathbf{y}_{21}}{\Delta_Y} \qquad \mathbf{z}_{22} = \frac{\mathbf{y}_{11}}{\Delta_Y}$$

■

Table 15.1 lists all the conversion formulas that relate one set of two-port parameters to another. Note that Δ_Z, Δ_Y, Δ_H, and Δ_T refer to the determinants of the matrices for the Z, Y, hybrid, and ABCD parameters, respectively. Therefore, given one set of parameters for a network, we can use Table 15.1 to find others.

Table 15.1 Two-Port Parameter Conversion Formulas

$\begin{bmatrix} \mathbf{z}_{11} & \mathbf{z}_{12} \\ \mathbf{z}_{21} & \mathbf{z}_{22} \end{bmatrix}$	$\begin{bmatrix} \frac{\mathbf{y}_{22}}{\Delta_Y} & \frac{-\mathbf{y}_{12}}{\Delta_Y} \\ \frac{-\mathbf{y}_{21}}{\Delta_Y} & \frac{\mathbf{y}_{11}}{\Delta_Y} \end{bmatrix}$	$\begin{bmatrix} \frac{\mathbf{A}}{\mathbf{C}} & \frac{\Delta_T}{\mathbf{C}} \\ \frac{1}{\mathbf{C}} & \frac{\mathbf{D}}{\mathbf{C}} \end{bmatrix}$	$\begin{bmatrix} \frac{\Delta_H}{\mathbf{h}_{22}} & \frac{\mathbf{h}_{12}}{\mathbf{h}_{22}} \\ \frac{-\mathbf{h}_{21}}{\mathbf{h}_{22}} & \frac{1}{\mathbf{h}_{22}} \end{bmatrix}$
$\begin{bmatrix} \frac{\mathbf{z}_{22}}{\Delta_Z} & \frac{-\mathbf{z}_{12}}{\Delta_Z} \\ \frac{-\mathbf{z}_{21}}{\Delta_Z} & \frac{\mathbf{z}_{11}}{\Delta_Z} \end{bmatrix}$	$\begin{bmatrix} \mathbf{y}_{11} & \mathbf{y}_{12} \\ \mathbf{y}_{21} & \mathbf{y}_{22} \end{bmatrix}$	$\begin{bmatrix} \frac{\mathbf{D}}{\mathbf{B}} & \frac{-\Delta_T}{\mathbf{B}} \\ -\frac{1}{\mathbf{B}} & \frac{\mathbf{A}}{\mathbf{B}} \end{bmatrix}$	$\begin{bmatrix} \frac{1}{\mathbf{h}_{11}} & \frac{-\mathbf{h}_{12}}{\mathbf{h}_{11}} \\ \frac{\mathbf{h}_{21}}{\mathbf{h}_{11}} & \frac{\Delta_H}{\mathbf{h}_{11}} \end{bmatrix}$
$\begin{bmatrix} \frac{\mathbf{z}_{11}}{\mathbf{z}_{21}} & \frac{\Delta_Z}{\mathbf{z}_{21}} \\ \frac{1}{\mathbf{z}_{21}} & \frac{\mathbf{z}_{22}}{\mathbf{z}_{21}} \end{bmatrix}$	$\begin{bmatrix} \frac{-\mathbf{y}_{22}}{\mathbf{y}_{21}} & \frac{-1}{\mathbf{y}_{21}} \\ \frac{-\Delta_Y}{\mathbf{y}_{21}} & \frac{-\mathbf{y}_{11}}{\mathbf{y}_{21}} \end{bmatrix}$	$\begin{bmatrix} \mathbf{A} & \mathbf{B} \\ \mathbf{C} & \mathbf{D} \end{bmatrix}$	$\begin{bmatrix} \frac{-\Delta_H}{\mathbf{h}_{21}} & \frac{-\mathbf{h}_{11}}{\mathbf{h}_{21}} \\ \frac{-\mathbf{h}_{22}}{\mathbf{h}_{21}} & \frac{-1}{\mathbf{h}_{21}} \end{bmatrix}$
$\begin{bmatrix} \frac{\Delta_Z}{\mathbf{z}_{22}} & \frac{\mathbf{z}_{12}}{\mathbf{z}_{22}} \\ \frac{-\mathbf{z}_{21}}{\mathbf{z}_{22}} & \frac{1}{\mathbf{z}_{22}} \end{bmatrix}$	$\begin{bmatrix} \frac{1}{\mathbf{y}_{11}} & \frac{-\mathbf{y}_{12}}{\mathbf{y}_{11}} \\ \frac{\mathbf{y}_{21}}{\mathbf{y}_{11}} & \frac{\Delta_Y}{\mathbf{y}_{11}} \end{bmatrix}$	$\begin{bmatrix} \frac{\mathbf{B}}{\mathbf{D}} & \frac{\Delta_T}{\mathbf{D}} \\ -\frac{1}{\mathbf{D}} & \frac{C}{D} \end{bmatrix}$	$\begin{bmatrix} \mathbf{h}_{11} & \mathbf{h}_{12} \\ \mathbf{h}_{21} & \mathbf{h}_{22} \end{bmatrix}$

DRILL EXERCISE

D15.12. Determine the Y parameters for a two-port if the Z parameters are

$$\mathbf{Z} = \begin{bmatrix} 18 & 6 \\ 6 & 9 \end{bmatrix}$$

D15.13. Given the Z parameters for a two-port in Drill Exercise D15.12, determine the hybrid parameters for this network.

15.7 Reciprocal Networks

Reciprocal networks are a special class of networks that possess a very interesting property. Let us examine this property by considering the network shown in Fig. 15.14a. This two-port network contains only R, L, and C elements. The total circuit, including the two-port, is assumed to contain n independent loops, and with no loss of generality the two loops shown can be labeled loop 1 and loop 2. Since $\mathbf{V}$ is the only source in the total network, the loop equations can be written as

$$\begin{aligned} \mathbf{z}_{11}\mathbf{I}_1 + \mathbf{z}_{12}\mathbf{I}_2 + \cdots + \mathbf{z}_{1n}\mathbf{I}_n &= \mathbf{V} \\ \mathbf{z}_{21}\mathbf{I}_1 + \mathbf{z}_{22}\mathbf{I}_2 + \cdots + \mathbf{z}_{2n}\mathbf{I}_n &= 0 \\ &\vdots \\ \mathbf{z}_{n1}\mathbf{I}_1 + \mathbf{z}_{n2}\mathbf{I}_2 + \cdots + \mathbf{z}_{nn}\mathbf{I}_n &= 0 \end{aligned} \tag{15.13}$$

The current $\mathbf{I}_2$ can then be written

$$\mathbf{I}_2 = \frac{\begin{vmatrix} \mathbf{z}_{11} & \mathbf{V} & \mathbf{z}_{13} & \cdots & \mathbf{z}_{1n} \\ \mathbf{z}_{21} & 0 & \mathbf{z}_{23} & \cdots & \mathbf{z}_{2n} \\ \vdots & \vdots & \vdots & & \vdots \\ \mathbf{z}_{n1} & 0 & \mathbf{z}_{n3} & \cdots & \mathbf{z}_{nn} \end{vmatrix}}{\Delta_Z}$$

or (15.14)

$$\mathbf{I}_2 = -\frac{\begin{vmatrix} \mathbf{z}_{21} & \mathbf{z}_{23} & \cdots & \mathbf{z}_{2n} \\ \mathbf{z}_{31} & \mathbf{z}_{33} & \cdots & \mathbf{z}_{3n} \\ \vdots & \vdots & & \vdots \\ \mathbf{z}_{n1} & \mathbf{z}_{n3} & \cdots & \mathbf{z}_{nn} \end{vmatrix}}{\Delta_Z}(\mathbf{V})$$

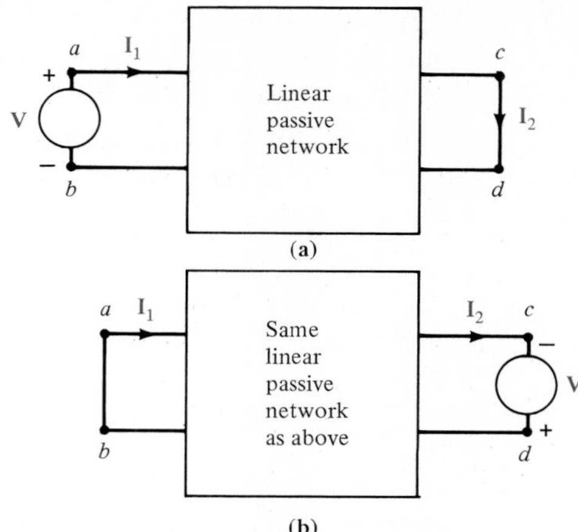

Figure 15.14 Networks used to illustrate reciprocity.

Now consider the circuit in Fig. 15.14b. The equations for this network are

$$\begin{aligned}
\mathbf{z}_{11}\mathbf{I}_1 + \mathbf{z}_{12}\mathbf{I}_2 + \cdots + \mathbf{z}_{1n}\mathbf{I}_n &= 0 \\
\mathbf{z}_{21}\mathbf{I}_1 + \mathbf{z}_{22}\mathbf{I}_2 + \cdots + \mathbf{z}_{2n}\mathbf{I}_n &= \mathbf{V} \\
\mathbf{z}_{31}\mathbf{I}_1 + \mathbf{z}_{32}\mathbf{I}_2 + \cdots + \mathbf{z}_{3n}\mathbf{I}_n &= 0 \\
&\vdots \\
\mathbf{z}_{n1}\mathbf{I}_1 + \mathbf{z}_{n2}\mathbf{I}_2 + \cdots + \mathbf{z}_{nn}\mathbf{I}_n &= 0
\end{aligned} \tag{15.15}$$

The current $\mathbf{I}_1$ is then

$$\mathbf{I}_1 = \frac{\begin{vmatrix} 0 & \mathbf{z}_{12} & \mathbf{z}_{13} & \cdots & \mathbf{z}_{1n} \\ \mathbf{V} & \mathbf{z}_{22} & \mathbf{z}_{23} & \cdots & \mathbf{z}_{2n} \\ 0 & \mathbf{z}_{32} & \mathbf{z}_{33} & \cdots & \mathbf{z}_{3n} \\ \vdots & \vdots & \vdots & & \vdots \\ 0 & \mathbf{z}_{n2} & \mathbf{z}_{n3} & \cdots & \mathbf{z}_{nn} \end{vmatrix}}{\Delta_Z}$$

or

$$\mathbf{I}_1 = -\frac{\begin{vmatrix} \mathbf{z}_{12} & \mathbf{z}_{13} & \cdots & \mathbf{z}_{1n} \\ \mathbf{z}_{32} & \mathbf{z}_{33} & \cdots & \mathbf{z}_{3n} \\ \vdots & \vdots & & \vdots \\ \mathbf{z}_{n2} & \mathbf{z}_{n3} & \cdots & \mathbf{z}_{nn} \end{vmatrix}}{\Delta_Z}(\mathbf{V}) \tag{15.16}$$

Let us now carefully compare Eqs. (15.14) and (15.16). Note that the difference in the two equations lies in the first row and the first column of the determinants in the numerators. Since we do not change the value of a determinant by interchanging all rows and columns, Eq. (15.16) can be written

$$\mathbf{I}_1 = -\frac{\begin{vmatrix} \mathbf{z}_{12} & \mathbf{z}_{32} & \cdots & \mathbf{z}_{n2} \\ \mathbf{z}_{13} & \mathbf{z}_{33} & \cdots & \mathbf{z}_{n3} \\ \vdots & \vdots & & \vdots \\ \mathbf{z}_{1n} & \mathbf{z}_{3n} & \cdots & \mathbf{z}_{nn} \end{vmatrix}}{\Delta_Z}(\mathbf{V}) \tag{5.17}$$

Note that Eqs. (15.14) and (15.17) would be identical if $\mathbf{z}_{ij} = \mathbf{z}_{ji}$ for all i and j. However, since $\mathbf{z}_{ij}$ is the impedance common to loops i and j and the network contains only R, L, and C elements, then $\mathbf{z}_{ij} = \mathbf{z}_{ji}$ for all i and j. From a two-port standpoint, Eqs. (15.14) and (15.17) can be written

$$\mathbf{I}_2 = \mathbf{y}_{21}\mathbf{V}$$

and

$$\mathbf{I}_1 = \mathbf{y}_{12}\mathbf{V}$$

and therefore

$$\mathbf{y}_{12} = \mathbf{y}_{21} \tag{15.18}$$

A network that possesses this property is called a *reciprocal network*. Using the conversion formulas for the two-port parameters, we can also show that for a reciprocal network.

$$\begin{aligned} \mathbf{z}_{12} &= \mathbf{z}_{21} \\ \mathbf{h}_{12} &= -\mathbf{h}_{21} \\ \mathbf{AD} - \mathbf{BC} &= 1 \end{aligned} \tag{15.19}$$

Equations (15.18) and (15.19) are valid not only for networks containing only R, L, and C elements but also for circuits containing mutual inductance and ideal transformers. It is more difficult to prove this more general case, however. The earlier examples have demonstrated that networks containing dependent sources are not reciprocal.

The results illustrated above are summarized by what is called the *reciprocity theorem*, which states that in a reciprocal network a single voltage source and a short circuit may be interchanged without affecting the value of the short-circuit current. The dual of this principle states that in a reciprocal network a single current source and an open circuit may be interchanged without affecting the value of the open-circuit voltage. In applying reciprocity we must ensure that interchanging the points of excitation and observation does not change the topology of the network.

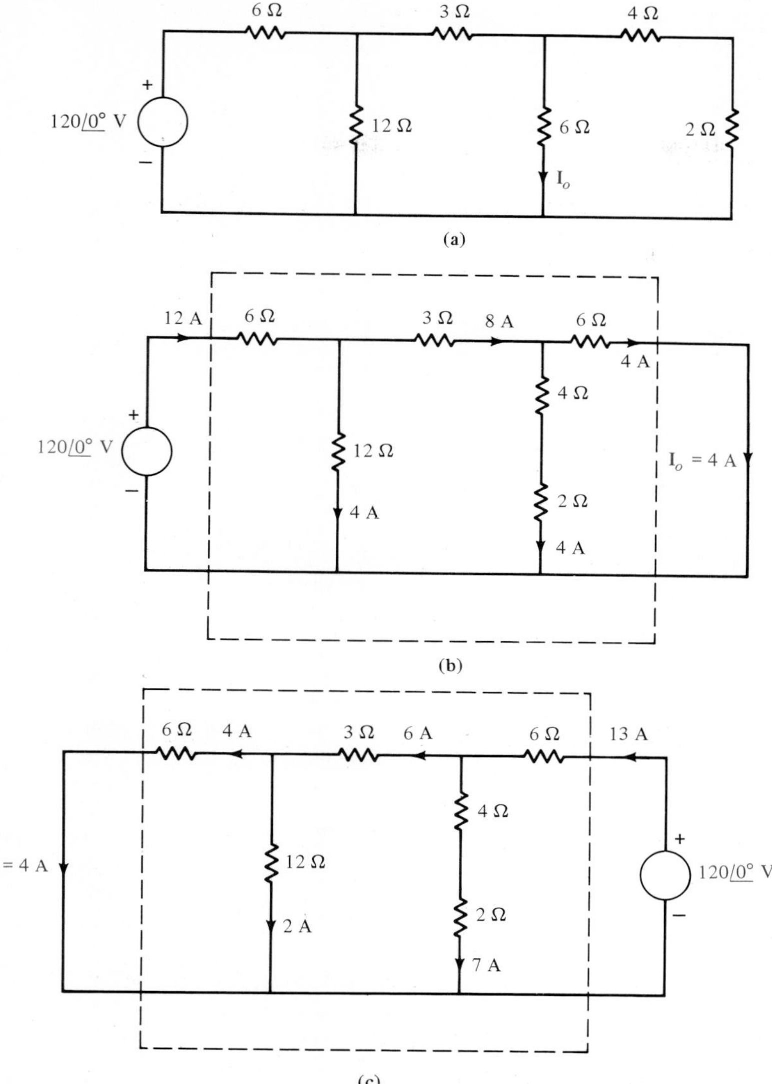

Figure 15.15 Networks used to illustrate reciprocity.

EXAMPLE 15.12

We will demonstrate reciprocity using the network in Fig. 15.15a. Redrawing the network as shown in Fig. 15.15b, we can easily compute $\mathbf{I}_o$ to be 4 A. Interchanging the voltage source and short circuit as shown in Fig. 15.15c, we also calculate $\mathbf{I}_o$ to be 4 A. ■

EXAMPLE 15.13

Let us demonstrate the reciprocity theorem by determining $\mathbf{V}_o$ in the network in Fig. 15.16a, and then interchanging the current source and open circuit to compute $\mathbf{V}_o$ in Fig. 15.16b.

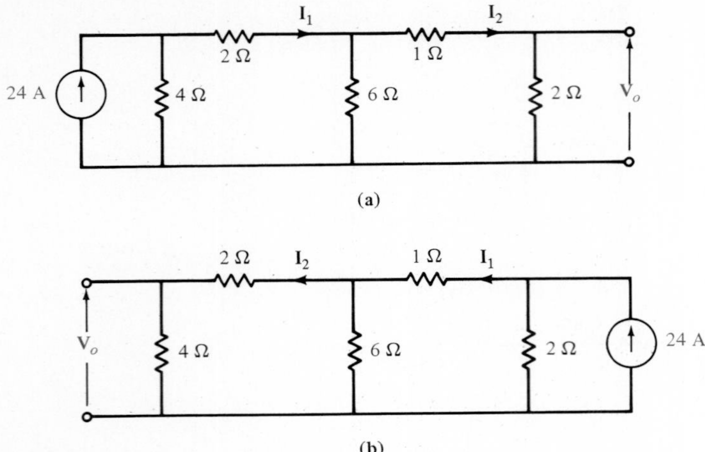

Figure 15.16 Networks used to illustrate the reciprocity theorem.

If we apply current division to the network in Fig. 15.16a, we find that $\mathbf{I}_1 = 12$ A and then $\mathbf{I}_2 = 8$ A, so that $\mathbf{V}_o = 16$ V. Using the same approach with the network in Fig. 15.16b, we compute $\mathbf{I}_1 = 8$ A, and then $\mathbf{I}_2 = 4$ A, so that $\mathbf{V}_o = 16$ V. ■

DRILL EXERCISE

D15.14. Using the network in Fig. D15.14, demonstrate reciprocity by computing $\mathbf{V}_o$ and then interchanging the current source and open circuit to compute $\mathbf{V}_o$ for this new circuit.

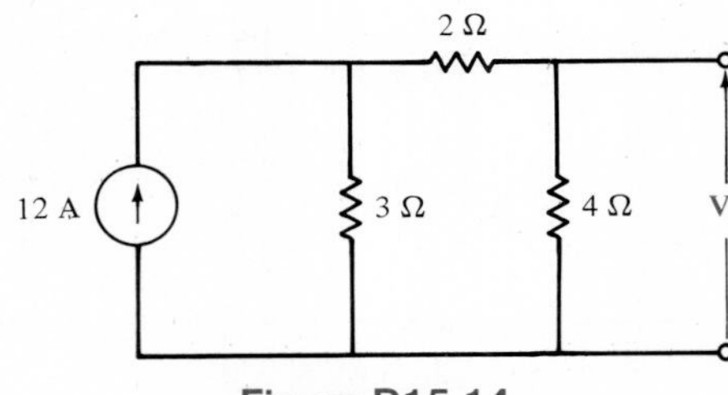
Figure D15.14

D15.15. Using the network in Fig. D15.15, illustrate the principle of reciprocity by determining $\mathbf{I}_o$, and then interchange the voltage source and short circuit to compute $\mathbf{I}_o$ for this new circuit.

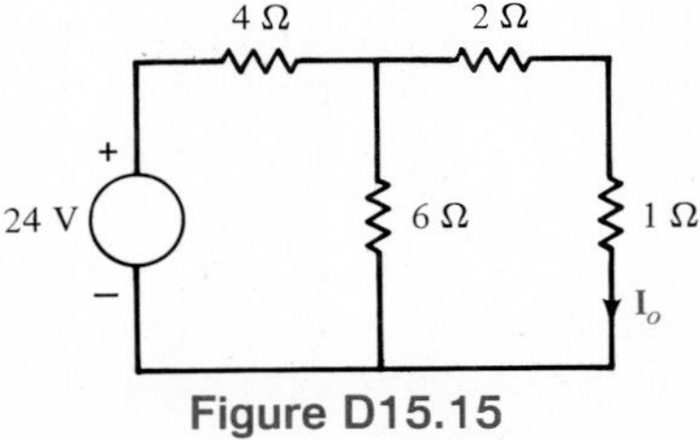
Figure D15.15

D15.16. Use the principle of reciprocity to find $\mathbf{I}_o$ in the network in Fig. D15.16.

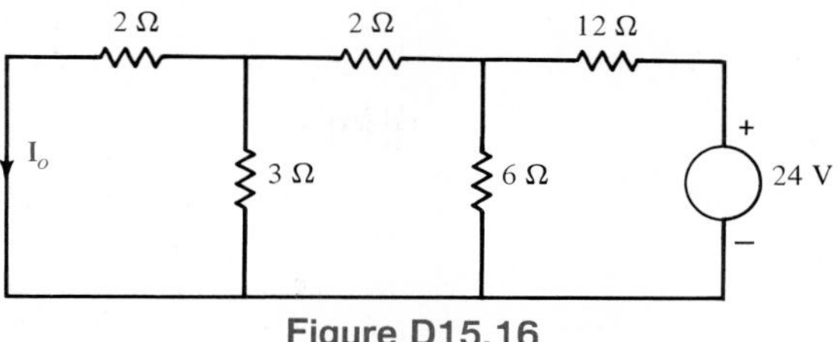

Figure D15.16

15.8 Interconnection of Two-Ports

In this section we will illustrate techniques for treating a network as a combination of subnetworks. We will, therefore, analyze a two-port network as an interconnection of simpler two-ports. Although two-ports can be interconnected in a variety of ways, we will treat only three types of connections: parallel, series, and cascade.

Parallel Interconnection

Suppose that a two-port N is composed of two two-ports N_a and N_b which are interconnected as shown in Fig. 15.17a. The defining equations for networks N_a and N_b are

$$\begin{aligned} \mathbf{I}_{1a} &= \mathbf{y}_{11a}\mathbf{V}_{1a} + \mathbf{y}_{12a}\mathbf{V}_{2a} \\ \mathbf{I}_{2a} &= \mathbf{y}_{21a}\mathbf{V}_{1a} + \mathbf{y}_{22a}\mathbf{V}_{2a} \end{aligned} \tag{15.20}$$

$$\begin{aligned} \mathbf{I}_{1b} &= \mathbf{y}_{11b}\mathbf{V}_{1b} + \mathbf{y}_{12b}\mathbf{V}_{2b} \\ \mathbf{I}_{2b} &= \mathbf{y}_{21b}\mathbf{V}_{1b} + \mathbf{y}_{22b}\mathbf{V}_{2b} \end{aligned} \tag{15.21}$$

Provided that the terminal characteristics of the two networks N_a and N_b are not altered by the interconnection illustrated in Fig. 15.17a, then

$$\begin{aligned} \mathbf{V}_1 &= \mathbf{V}_{1a} = \mathbf{V}_{1b} \\ \mathbf{V}_2 &= \mathbf{V}_{2a} = \mathbf{V}_{2b} \\ \mathbf{I}_1 &= \mathbf{I}_{1a} + \mathbf{I}_{1b} \\ \mathbf{I}_2 &= \mathbf{I}_{2a} + \mathbf{I}_{2b} \end{aligned} \tag{15.22}$$

$$\begin{aligned} \mathbf{I}_1 &= (\mathbf{y}_{11a} + \mathbf{y}_{11b})\mathbf{V}_1 + (\mathbf{y}_{12a} + \mathbf{y}_{12b})\mathbf{V}_2 \\ \mathbf{I}_2 &= (\mathbf{y}_{21a} + \mathbf{y}_{21b})\mathbf{V}_1 + (\mathbf{y}_{22a} + \mathbf{y}_{22b})\mathbf{V}_2 \end{aligned} \tag{15.23}$$

Therefore, the Y parameters for the total network are

$$\begin{bmatrix} \mathbf{y}_{11} & \mathbf{y}_{12} \\ \mathbf{y}_{21} & \mathbf{y}_{22} \end{bmatrix} = \begin{bmatrix} \mathbf{y}_{11a} + \mathbf{y}_{11b} & \mathbf{y}_{12a} + \mathbf{y}_{12b} \\ \mathbf{y}_{21a} + \mathbf{y}_{21b} & \mathbf{y}_{22a} + \mathbf{y}_{22b} \end{bmatrix} \tag{15.24}$$

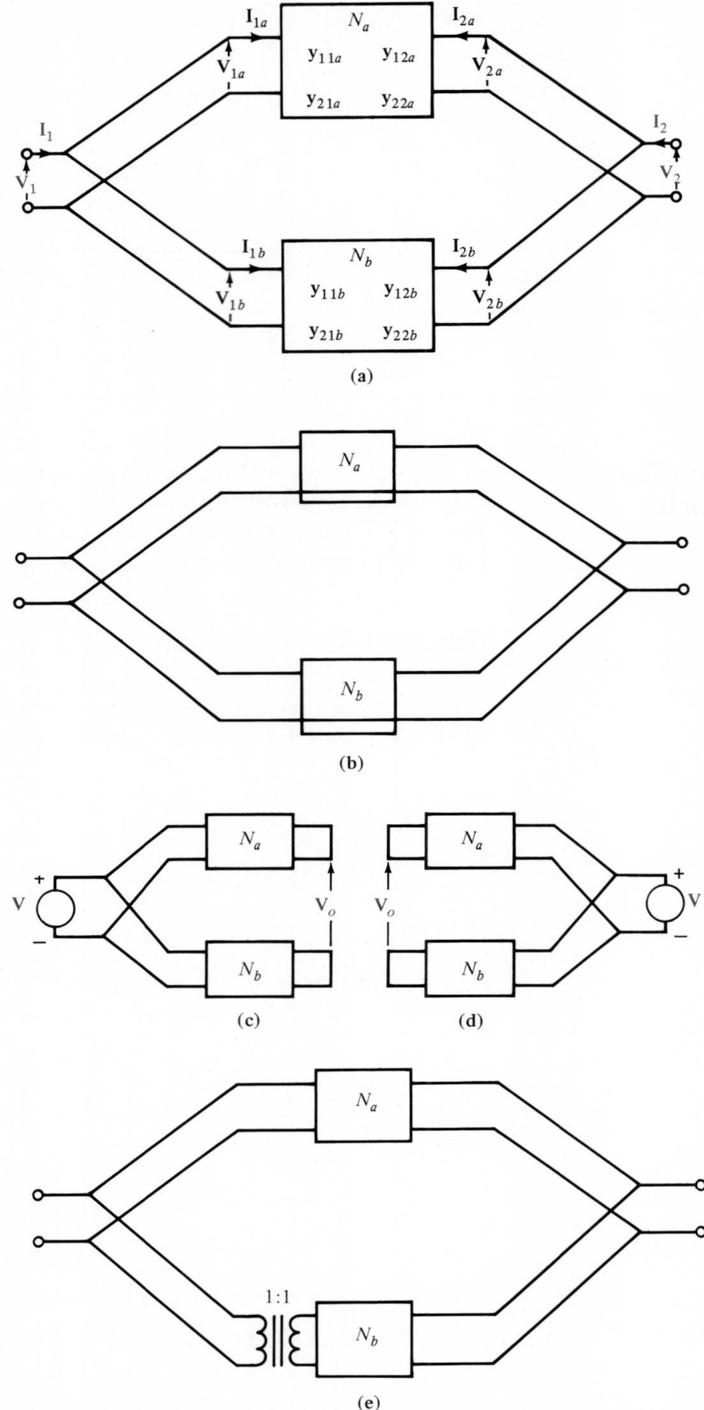

Figure 15.17 Some circuits describing the parallel interconnection of two-ports.

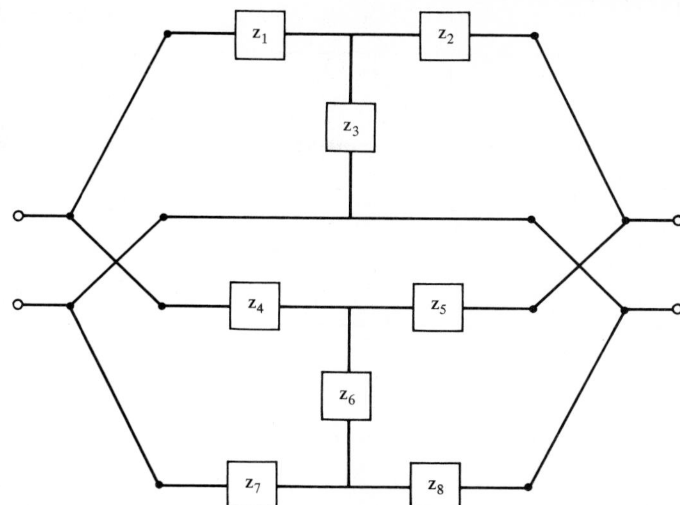

Figure 15.18 Example where parallel connection of two-ports cannot be performed without the use of an ideal transformer.

and hence to determine the Y parameters for the total network, we simply add the Y parameters of the two networks N_a and N_b. Equation (15.23) is always valid if the two networks are of the form shown in Fig. 15.17b. In the general case, Eq. (15.23) is valid if the networks satisfy the Brune tests (named after O. Brune, who developed these procedures). The Brune tests require that the voltage $\mathbf{V}_o$ be zero when the networks N_a and N_b are interconnected as shown in Fig. 15.17c and d. If networks N_a and N_b fail these tests, a one-to-one ideal transformer can be employed to ensure that Eqs. (15.23) are satisfied. One possible location of the transformer is shown in Fig. 15.17e; however, the transformer may actually be located at any one of the four ports shown in Fig. 15.17a. An example where an ideal transformer is required to provide the necessary isolation is shown in Fig. 15.18. Note that if the two two-ports are connected in parallel without the use of an ideal transformer, impedances $\mathbf{z}_7$ and $\mathbf{z}_8$ are short-circuited by the interconnection, and hence Eq. (15.23) does not apply.

EXAMPLE 15.14

We wish to determine the Y parameters for the network shown in Fig. 15.19a by considering it to be a parallel combination of two networks as shown in Figure 15.19b. The capacitive network will be referred to as N_a and the resistive network will be referred to as N_b.

The Y parameters for N_a are

$$\mathbf{y}_{11a} = j\tfrac{1}{2}\text{ S} \qquad \mathbf{y}_{12a} = -j\tfrac{1}{2}\text{ S}$$

$$\mathbf{y}_{21a} = -j\tfrac{1}{2}\text{ S} \qquad \mathbf{y}_{22a} = j\tfrac{1}{2}\text{ S}$$

and the Y parameters for N_b are

$$\mathbf{y}_{11b} = \tfrac{3}{5}\text{ S} \qquad \mathbf{y}_{12b} = -\tfrac{1}{5}\text{ S}$$

$$\mathbf{y}_{21b} = -\tfrac{1}{5}\text{ S} \qquad \mathbf{y}_{22b} = \tfrac{2}{5}\text{ S}$$

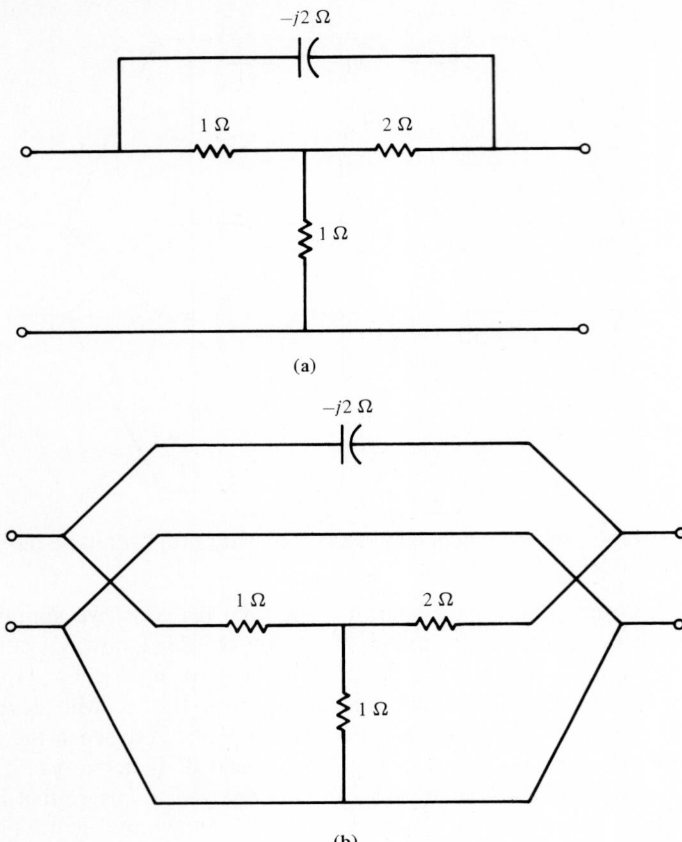

Figure 15.19 Network composed of the parallel combination of two subnetworks.

Hence the Y parameters for the network in Fig. 15.19 are

$$\mathbf{y}_{11} = \tfrac{3}{5} + j\tfrac{1}{2}\text{ S} \qquad \mathbf{y}_{12} = -(\tfrac{1}{5} + j\tfrac{1}{2})\text{ S}$$

$$\mathbf{y}_{21} = -(\tfrac{1}{5} + j\tfrac{1}{2})\text{ S} \qquad \mathbf{y}_{22} = \tfrac{2}{5} + j\tfrac{1}{2}\text{ S}$$

To gain an appreciation for the simplicity of this approach, the reader need only try to find the Y parameters for the network in Fig. 15.19a directly. ■

EXAMPLE 15.15

Consider the amplifier equivalent circuit shown in Fig. 15.20a. We will derive the Y parameters for this network via the parallel interconnection scheme shown in Fig. 15.20b.

We will refer to the upper network in Fig. 15.20b as N_a and the lower network as N_b. The Y parameters for N_a are

$$\mathbf{y}_{11a} = \frac{1}{R_F} \qquad \mathbf{y}_{12a} = -\frac{1}{R_F}$$

$$\mathbf{y}_{21a} = -\frac{1}{R_F} \qquad \mathbf{y}_{22a} = \frac{1}{R_F}$$

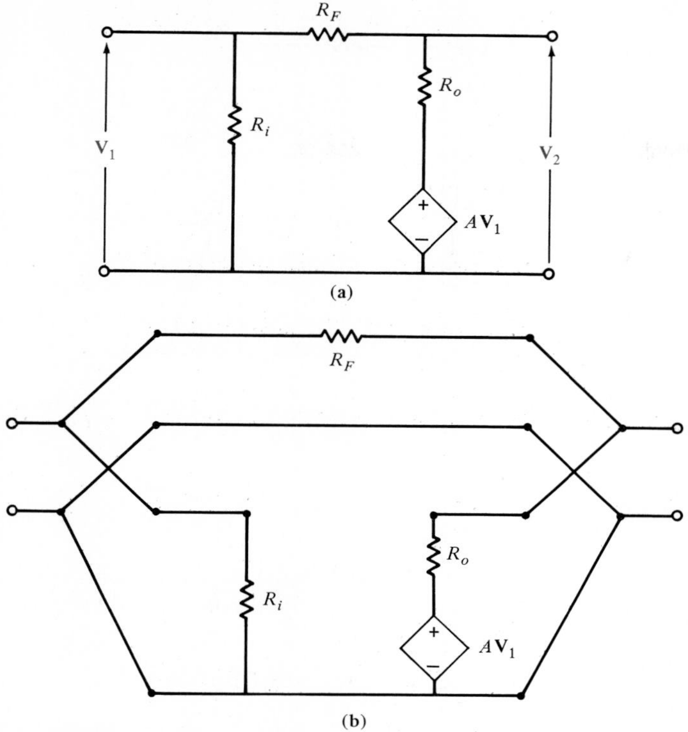

Figure 15.20 Amplifier circuit composed of two parallel networks.

and the Y parameters for N_b are

$$\mathbf{y}_{11b} = \frac{1}{R_i} \qquad \mathbf{y}_{12b} = 0$$

$$\mathbf{y}_{21b} = \frac{-A}{R_o} \qquad \mathbf{y}_{22b} = \frac{1}{R_o}$$

Hence the Y parameters for the network in Fig. 15.20a are

$$\mathbf{y}_{11} = \frac{1}{R_F} + \frac{1}{R_i} \qquad \mathbf{y}_{12} = -\frac{1}{R_F}$$

$$\mathbf{y}_{21} = -\left(\frac{1}{R_F} + \frac{A}{R_o}\right) \qquad \mathbf{y}_{22} = \frac{1}{R_F} + \frac{1}{R_o}$$

■

DRILL EXERCISE

D15.17. Find the Y parameters of the network in Fig. D15.2 by considering the network to be a parallel interconnection of two two-ports as shown in Fig. D15.17.

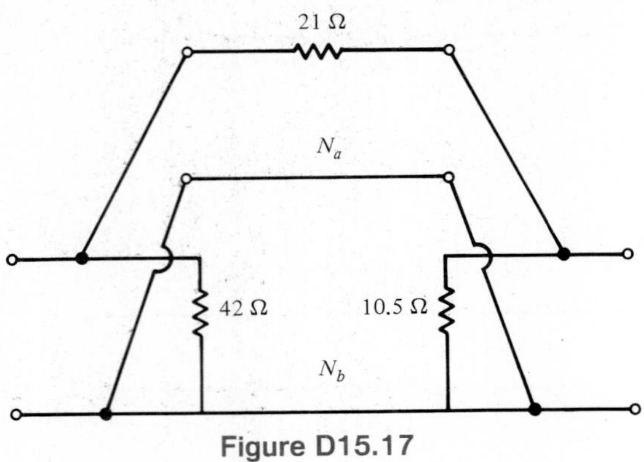

Figure D15.17

Series Interconnection

Consider the two-port N, which is composed of the series connection of N_a and N_b, as shown in Fig. 15.21a. The defining equations for the individual two-port networks N_a and N_b are

$$\begin{aligned}\mathbf{V}_{1a} &= \mathbf{z}_{11a}\mathbf{I}_{1a} + \mathbf{z}_{12a}\mathbf{I}_{2a}\\ \mathbf{V}_{2a} &= \mathbf{z}_{21a}\mathbf{I}_{1a} + \mathbf{z}_{22a}\mathbf{I}_{2a}\end{aligned} \tag{15.25}$$

$$\begin{aligned}\mathbf{V}_{1b} &= \mathbf{z}_{11b}\mathbf{I}_{1b} + \mathbf{z}_{12b}\mathbf{I}_{2b}\\ \mathbf{V}_{2b} &= \mathbf{z}_{21b}\mathbf{I}_{1b} + \mathbf{z}_{22b}\mathbf{I}_{2b}\end{aligned} \tag{15.26}$$

Once again, as long as the terminal characteristics of the two networks, N_a and N_b are not altered by the series interconnection, then

$$\begin{aligned}\mathbf{I}_1 &= \mathbf{I}_{1a} = \mathbf{I}_{1b}\\ \mathbf{I}_2 &= \mathbf{I}_{2a} = \mathbf{I}_{2b}\end{aligned} \tag{15.27}$$

and

$$\begin{aligned}\mathbf{V}_1 &= \mathbf{V}_{1a} + \mathbf{V}_{1b} = (\mathbf{z}_{11a} + \mathbf{z}_{11b})\mathbf{I}_1 + (\mathbf{z}_{12a} + \mathbf{z}_{12b})\mathbf{I}_2\\ \mathbf{V}_2 &= \mathbf{V}_{2a} + \mathbf{V}_{2b} = (\mathbf{z}_{21a} + \mathbf{z}_{21b})\mathbf{I}_1 + (\mathbf{z}_{22a} + \mathbf{z}_{22b})\mathbf{I}_2\end{aligned} \tag{15.28}$$

Therefore, the Z parameters for the total network are

$$\begin{bmatrix}\mathbf{z}_{11} & \mathbf{z}_{12}\\ \mathbf{z}_{21} & \mathbf{z}_{22}\end{bmatrix} = \begin{bmatrix}\mathbf{z}_{11a} + \mathbf{z}_{11b} & \mathbf{z}_{12a} + \mathbf{z}_{12b}\\ \mathbf{z}_{21a} + \mathbf{z}_{21b} & \mathbf{z}_{22a} + \mathbf{z}_{22b}\end{bmatrix} \tag{15.29}$$

Therefore, the Z parameters for the total network are equal to the sum of the Z parameters for the networks N_a and N_b. This procedure is always valid if the networks N_a and N_b are of the form shown in Fig. 15.21b. The Brune tests for the series interconnection of two-ports requires that $\mathbf{V}_o$ be zero when the networks are interconnected as shown in Fig. 15.21c and d. If these conditions are not satisfied, a one-to-one ideal transformer

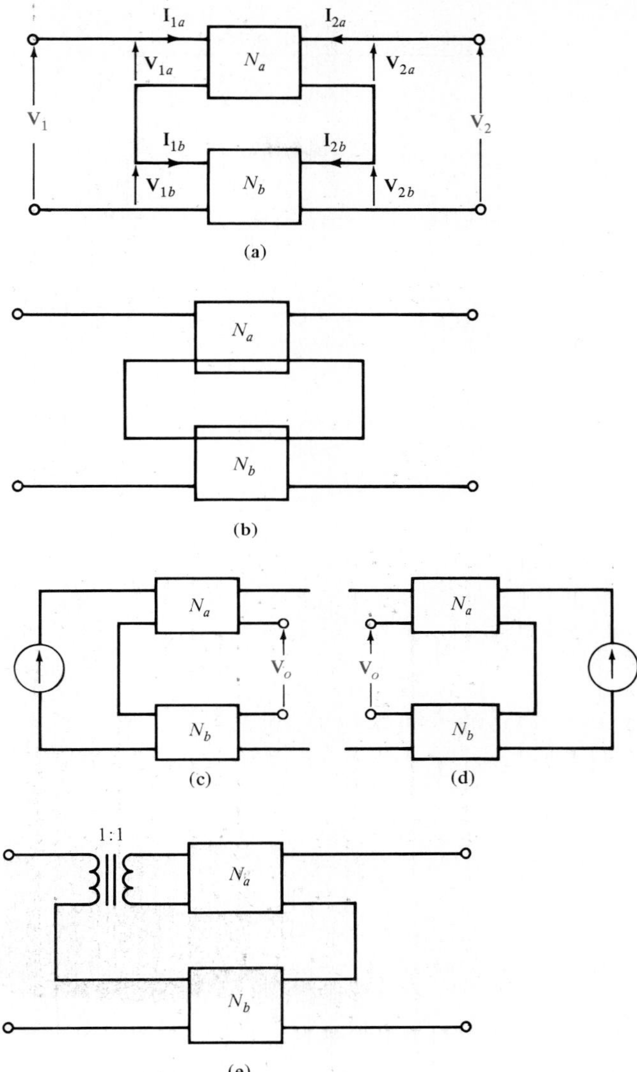

Figure 15.21 Some circuits describing the series interconnection of two-ports.

can be employed in the manner described for the parallel interconnection case and illustrated in Fig. 15.21e. For example, a transformer would be required if the two networks in Fig. 15.18 were interconnected in series, since $\mathbf{z}_4$ and $\mathbf{z}_5$ would be short-circuited.

EXAMPLE 15.16

Let us determine the Z parameters for the network shown in Fig. 15.19a. The circuit is redrawn in Fig. 15.22, illustrating a series interconnection. The upper network will be

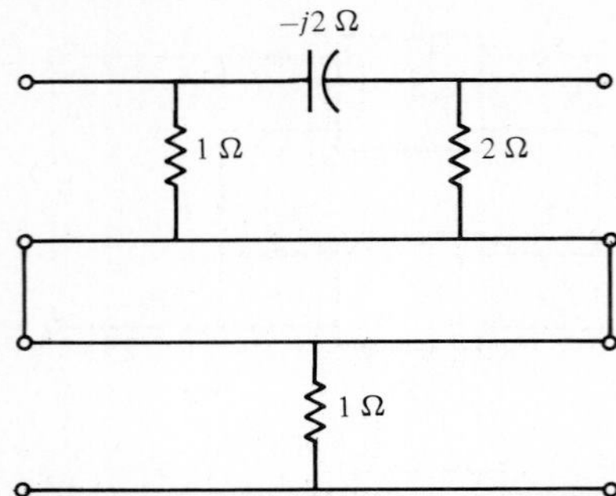

Figure 15.22 Network in Fig. 15.19a redrawn as a series interconnection of two networks.

referred to as N_a and the lower network as N_b. The Z parameters for N_a are

$$\mathbf{z}_{11a} = \frac{2 - 2j}{3 - 2j}\ \Omega \qquad \mathbf{z}_{12a} = \frac{2}{3 - 2j}\ \Omega$$

$$\mathbf{z}_{21a} = \frac{2}{3 - 2j}\ \Omega \qquad \mathbf{z}_{22a} = \frac{2 - 4j}{3 - 2j}\ \Omega$$

and the Z parameters for N_b are

$$\mathbf{z}_{11b} = \mathbf{z}_{12b} = \mathbf{z}_{21b} = \mathbf{z}_{22b} = 1\ \Omega$$

Hence the Z parameters for the total network are

$$\mathbf{z}_{11} = \frac{5 - 4j}{3 - 2j}\ \Omega \qquad \mathbf{z}_{12} = \frac{5 - 2j}{3 - 2j}\ \Omega$$

$$\mathbf{z}_{21} = \frac{5 - 2j}{3 - 2j}\ \Omega \qquad \mathbf{z}_{22} = \frac{5 - 6j}{3 - 2j}\ \Omega$$

We could easily check these results against those obtained in Example 15.14 by applying the conversion formulas in Table 15.1. ■

DRILL EXERCISE

D15.18. Find the Z parameters of the network in Fig. D15.1 by considering the circuit to be a series interconnection of two two-ports as shown in Fig. D15.18.

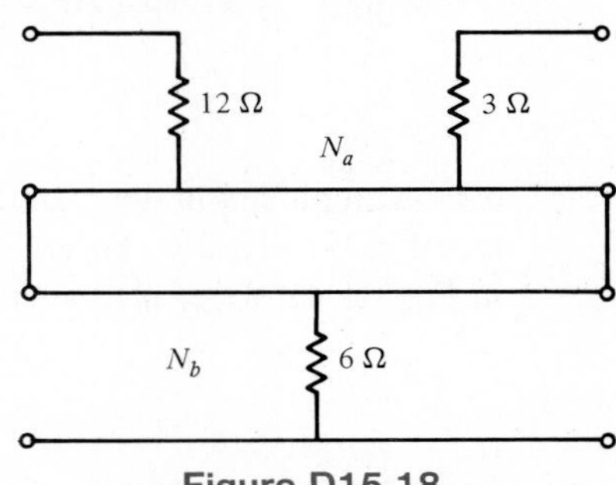

Figure D15.18

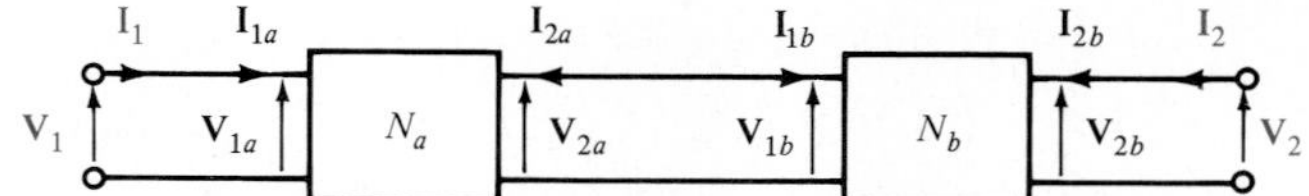

Figure 15.23 Cascade interconnection of networks.

Cascade Interconnection

A two-port N is composed of the cascade interconnection of N_a and N_b as shown in Fig. 15.23, where the parameter equations for N_a and N_b are

$$\begin{bmatrix} \mathbf{V}_{1a} \\ \mathbf{I}_{1a} \end{bmatrix} = \begin{bmatrix} \mathbf{A}_a & \mathbf{B}_a \\ \mathbf{C}_a & \mathbf{D}_a \end{bmatrix} \begin{bmatrix} \mathbf{V}_{2a} \\ -\mathbf{I}_{2a} \end{bmatrix} \tag{15.30}$$

$$\begin{bmatrix} \mathbf{V}_{1b} \\ \mathbf{I}_{1b} \end{bmatrix} = \begin{bmatrix} \mathbf{A}_b & \mathbf{B}_b \\ \mathbf{C}_b & \mathbf{D}_b \end{bmatrix} \begin{bmatrix} \mathbf{V}_{2b} \\ -\mathbf{I}_{2b} \end{bmatrix} \tag{15.31}$$

From Fig. 15.23, however, we note that

$$\begin{bmatrix} \mathbf{V}_1 \\ \mathbf{I}_1 \end{bmatrix} = \begin{bmatrix} \mathbf{V}_{1a} \\ \mathbf{I}_{1a} \end{bmatrix}, \qquad \begin{bmatrix} \mathbf{V}_{2a} \\ -\mathbf{I}_{2a} \end{bmatrix} = \begin{bmatrix} \mathbf{V}_{1b} \\ \mathbf{I}_{1b} \end{bmatrix}, \qquad \begin{bmatrix} \mathbf{V}_{2b} \\ -\mathbf{I}_{2b} \end{bmatrix} = \begin{bmatrix} \mathbf{V}_2 \\ -\mathbf{I}_2 \end{bmatrix}$$

and therefore the equations for the total network are

$$\begin{bmatrix} \mathbf{V}_1 \\ \mathbf{I}_1 \end{bmatrix} = \begin{bmatrix} \mathbf{A}_a & \mathbf{B}_a \\ \mathbf{C}_a & \mathbf{D}_a \end{bmatrix} \begin{bmatrix} \mathbf{A}_b & \mathbf{B}_b \\ \mathbf{C}_b & \mathbf{D}_b \end{bmatrix} \begin{bmatrix} \mathbf{V}_2 \\ -\mathbf{I}_2 \end{bmatrix} \tag{15.32}$$

Hence the transmission parameters for the total network are derived by matrix multiplication as indicated above. The order of the matrix multiplication is important and is performed in the order in which the networks are interconnected.

EXAMPLE 15.17

Let us reexamine the circuit that we considered in Example 15.9 and derive its two-port parameters by considering it to be a cascade connection of two networks as shown in Fig. 15.24. The ABCD parameters for the identical T networks can be easily calculated as

$$\mathbf{A} = 1 + j\omega \qquad \mathbf{B} = 2 + j\omega$$
$$\mathbf{C} = j\omega \qquad \mathbf{D} = 1 + j\omega$$

Figure 15.24 Two-port in Fig. 15.11 redrawn as a cascade connection of two networks.

Therefore, the transmission parameters for the total network are

$$\begin{bmatrix} \mathbf{A} & \mathbf{B} \\ \mathbf{C} & \mathbf{D} \end{bmatrix} = \begin{bmatrix} 1 + j\omega & 2 + j\omega \\ j\omega & 1 + j\omega \end{bmatrix} \begin{bmatrix} 1 + j\omega & 2 + j\omega \\ j\omega & 1 + j\omega \end{bmatrix}$$

Performing the matrix multiplication, we obtain

$$\begin{bmatrix} \mathbf{A} & \mathbf{B} \\ \mathbf{C} & \mathbf{D} \end{bmatrix} = \begin{bmatrix} 1 + 4j\omega - 2\omega^2 & 4 + 6j\omega - 2\omega^2 \\ 2j\omega - 2\omega^2 & 1 + 4j\omega - 2\omega^2 \end{bmatrix}$$ ■

DRILL EXERCISE

D15.19. Find the transmission parameters of the network in Fig. D15.1 by considering the circuit to be a cacade interconnection of three two-ports as shown in Fig. D15.19.

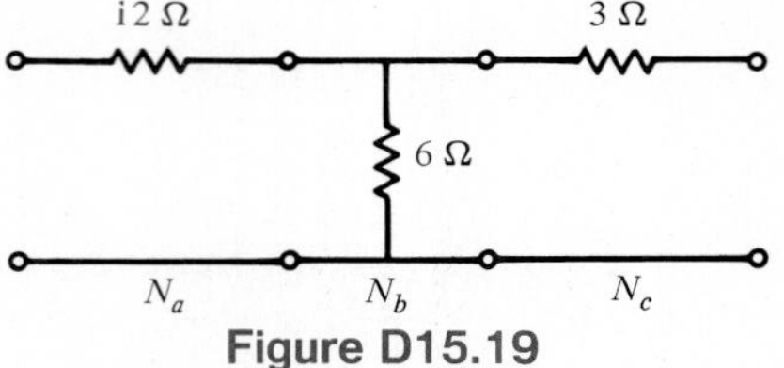

Figure D15.19

15.9 T–Π Equivalent Networks

Two very important two-ports are the T and Π networks shown in Fig. 15.25. Because we encounter these two geometrical forms often in two-port analyses, it is instructive to determine the conditions under which these two networks are equivalent. In order to determine the equivalence relationship, we will examine Z-parameter equations for the T network and the Y-parameter equations for the Π network.

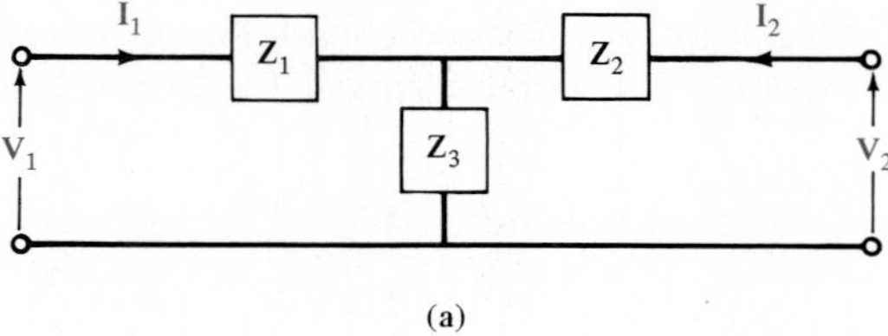

(a)

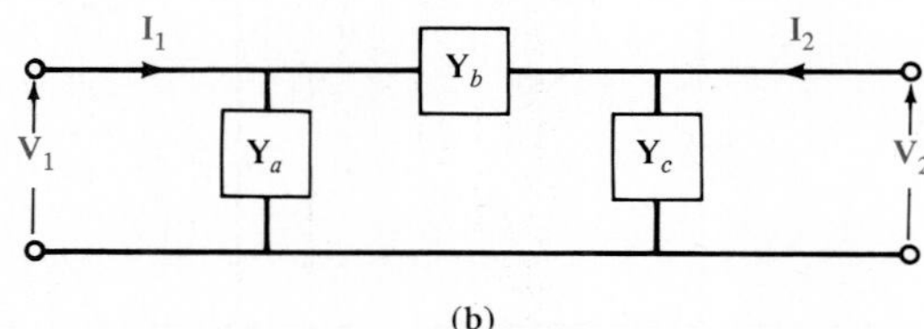

(b)

Figure 15.25 *T* and Π two-port networks.

For the T network the equations are

$$\mathbf{V}_1 = (\mathbf{Z}_1 + \mathbf{Z}_3)\mathbf{I}_1 + \mathbf{Z}_3\mathbf{I}_2$$

$$\mathbf{V}_2 = \mathbf{Z}_3\mathbf{I}_1 + (\mathbf{Z}_2 + \mathbf{Z}_3)\mathbf{I}_2$$

and for the Π network the equations are

$$\mathbf{I}_1 = (\mathbf{Y}_a + \mathbf{Y}_b)V_1 - \mathbf{Y}_bV_2$$

$$\mathbf{I}_2 = -\mathbf{Y}_bV_1 + (\mathbf{Y}_b + \mathbf{Y}_c)V_2$$

Solving the equations for the T network in terms of $\mathbf{I}_1$ and $\mathbf{I}_2$, we obtain

$$\mathbf{I}_1 = \left(\frac{\mathbf{Z}_2 + \mathbf{Z}_3}{\mathbf{D}_1}\right)V_1 - \frac{\mathbf{Z}_3V_2}{\mathbf{D}_1}$$

$$\mathbf{I}_2 = -\frac{\mathbf{Z}_3V_1}{\mathbf{D}_1} + \left(\frac{\mathbf{Z}_1 + \mathbf{Z}_3}{\mathbf{D}_1}\right)V_2$$

where $\mathbf{D}_1 = \mathbf{Z}_1\mathbf{Z}_2 + \mathbf{Z}_2\mathbf{Z}_3 + \mathbf{Z}_1\mathbf{Z}_3$. Comparing these equations with those for the Π network, we find that

$$\mathbf{Y}_a = \frac{\mathbf{Z}_2}{\mathbf{D}_1}$$

$$\mathbf{Y}_b = \frac{\mathbf{Z}_3}{\mathbf{D}_1} \tag{15.33}$$

$$\mathbf{Y}_c = \frac{\mathbf{Z}_1}{\mathbf{D}_1}$$

or in terms of the impedances of the Π network,

$$\mathbf{Z}_a = \frac{\mathbf{D}_1}{\mathbf{Z}_2}$$

$$\mathbf{Z}_b = \frac{\mathbf{D}_1}{\mathbf{Z}_3} \tag{15.34}$$

$$\mathbf{Z}_c = \frac{\mathbf{D}_1}{\mathbf{Z}_1}$$

If we reverse this procedure and solve the equations for the Π network in terms of $\mathbf{V}_1$ and $\mathbf{V}_2$ and then compare the resultant equations with those for the T network, we find that

$$\mathbf{Z}_1 = \frac{\mathbf{Y}_c}{\mathbf{D}_2}$$

$$\mathbf{Z}_2 = \frac{\mathbf{Y}_a}{\mathbf{D}_2} \tag{15.35}$$

$$\mathbf{Z}_3 = \frac{\mathbf{Y}_b}{\mathbf{D}_2}$$

where $\mathbf{D}_2 = \mathbf{Y}_a\mathbf{Y}_b + \mathbf{Y}_b\mathbf{Y}_c + \mathbf{Y}_a\mathbf{Y}_c$. Equation (15.35) can also be written in the form

$$\begin{aligned}\mathbf{Z}_1 &= \frac{\mathbf{Z}_a\mathbf{Z}_b}{\mathbf{Z}_a + \mathbf{Z}_b + \mathbf{Z}_c}\\ \mathbf{Z}_2 &= \frac{\mathbf{Z}_b\mathbf{Z}_c}{\mathbf{Z}_a + \mathbf{Z}_b + \mathbf{Z}_c}\\ \mathbf{Z}_3 &= \frac{\mathbf{Z}_a\mathbf{Z}_c}{\mathbf{Z}_a + \mathbf{Z}_b + \mathbf{Z}_c}\end{aligned} \tag{15.36}$$

Do Eqs. (15.34) and (15.36) look familiar? They should! They are simply the $\mathbf{Y} \leftrightarrows \Delta$ transformations employed in Chapter 12, since the T is a wye-connected network and the Π is a delta-connected network.

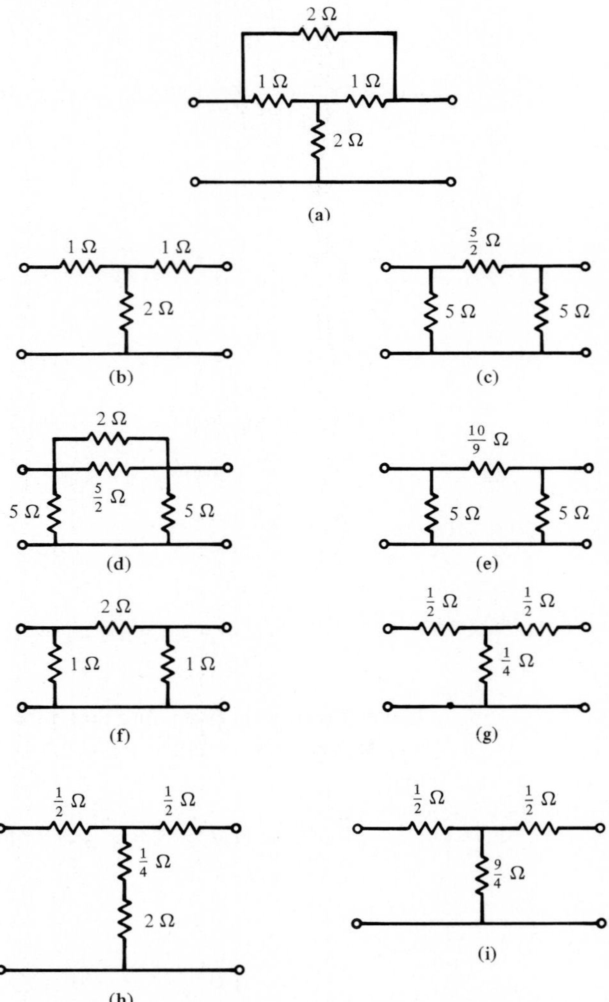

Figure 15.26 *T*–Π equivalent two-port analysis.

EXAMPLE 15.18

Let us determine both the equivalent-Π and equivalent-T networks for the two-port shown in Fig. 15.26a. Using the T- to-Π transformation equations, the T portion of the two-port shown in Fig. 15.26b is converted to a Π network in Fig. 15.26c. Combining the 2-Ω resistor in the original network with the Π network in Fig. 15.26c yields the network in Fig. 15.26d, which can immediately be converted to the final Π-equivalent network in Fig. 15.26e.

In a similar manner the Π portion of the original two-port shown in Fig. 15.26f is converted to the T network in Fig. 15.26g. Combining the remaining 2-Ω resistor in the original network with the T network in Fig. 15.26g yields the network in Fig. 15.26h, which can be converted immediately to the final T-equivalent network in Fig. 15.26i.■

DRILL EXERCISE

D15.20. Show that the two-ports in Figs. D15.1 and D15.2 are equivalent.

15.10

Two-Ports Embedded Within a Network

We may encounter situations in which a two-port is an integral part of a larger network. For example, the two-port may be some type of "black box" where all that is known concerning it is a set of two-port parameters. We will now try to illustrate that all the techniques we have described earlier can readily be applied in these situations also.

Consider the following example of a two-port embedded within a larger network.

EXAMPLE 15.19

Consider the network in Fig. 15.27a. The hybrid parameters of the two-port are known to be

$$\begin{bmatrix} \mathbf{h}_{11} & \mathbf{h}_{12} \\ \mathbf{h}_{21} & \mathbf{h}_{22} \end{bmatrix} = \begin{bmatrix} 100 & 0 \\ 1 & 10^{-3} \end{bmatrix}$$

We wish to determine the output voltage $\mathbf{V}_o$ of the entire network.

Our approach to the problem will be as follows. We will first find the input impedance of the two-port, add it to the 100-Ω resistor, and reflect this impedance to the left side of the transformer. We will then solve the two-loop circuit in which the transformer and everything to the right of it is represented by the reflected impedance. The voltage and current on the left side of the transformer will then be reflected to the right side. The latter voltage and current will be used in conjunction with the two-port parameters to determine the current in the load and therefore $\mathbf{V}_o$.

Using the results of Example 15.5, we find that the input impedance of the two-port is

$$\mathbf{Z}_{\text{in}} = \mathbf{h}_{11} - \frac{\mathbf{h}_{12}\mathbf{h}_{21}\mathbf{Z}_L}{1 + \mathbf{h}_{22}\mathbf{Z}_L}$$

where the hybrid parameters are given and $\mathbf{Z}_L = 100 + j100\ \Omega$. $\mathbf{Z}_{\text{in}}$ is therefore 100 Ω, and hence when this impedance is added to the 100 Ω in Fig. 15.27a and reflected to the left side of the transformer, we obtain the network in Fig. 15.27b. The mesh equations for this network are

$$\mathbf{I}_1 = \mathbf{I}_x$$

$$(2 - j2)\mathbf{I}_1 - (-j2)\mathbf{I}_2 = 12 \underline{/0^\circ} - 2\mathbf{I}_1$$

$$-(-j2)\mathbf{I}_1 + (2 - j1)\mathbf{I}_2 = 2\mathbf{I}_1$$

Solving these equations yields $\mathbf{I}_2 = 3.15 \underline{/-23.2^\circ}$ A, and therefore the voltage on the left side of the transformer is $6.30 \underline{/-23.2^\circ}$ V. Reflecting voltage and current to the right side of the transformer yields the results shown in Fig. 15.27c. Note that $\mathbf{I}_1 = 0.315 \underline{/-23.2^\circ}$ A and that $\mathbf{V}_2 = -\mathbf{I}_2(100 + j100)$. The equation for the two-port that relates the input and output currents is

$$\mathbf{I}_2 = \mathbf{h}_{21}\mathbf{I}_1 + \mathbf{h}_{22}\mathbf{V}_2$$

$$= \mathbf{h}_{21}\mathbf{I}_1 - \mathbf{I}_2(100 + j100)\,10^{-3}$$

$$= \frac{1(\mathbf{I}_1)}{1 + (100 + j100)}\,10^{-3}$$

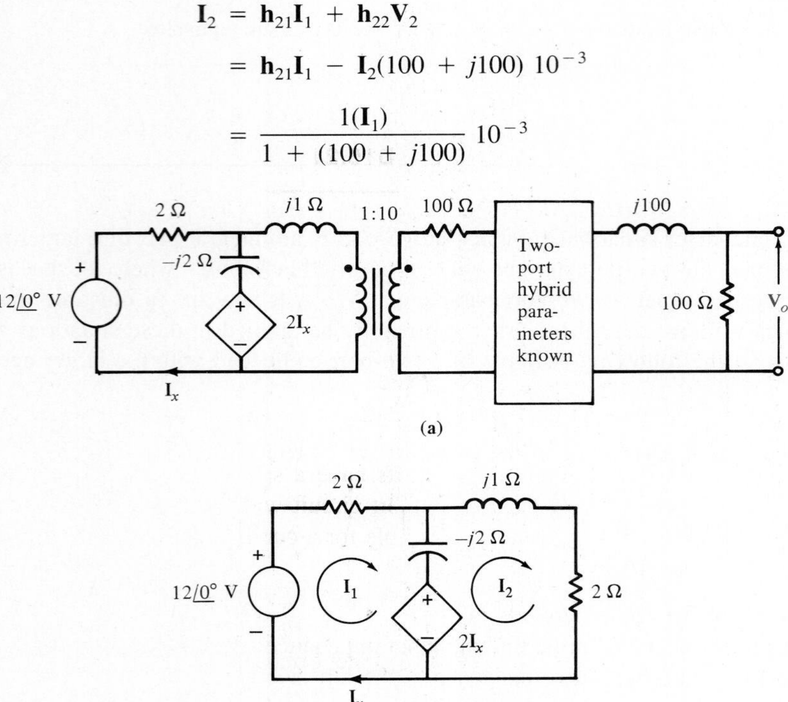

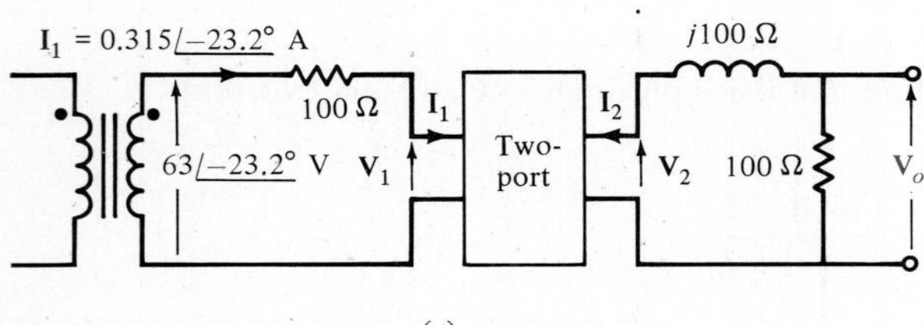

Figure 15.27 Circuits used in Example 15.19.

Therefore,

$$\mathbf{V}_o = -100\mathbf{I}_2$$
$$= -28.52 \underline{/-28.4°} \text{ V}$$

■

DRILL EXERCISE

D15.21. Given the network in Fig. D15.21, determine $\mathbf{V}_o$ if the hybrid parameters for the two-port are

$$\begin{bmatrix} \mathbf{h}_{11} & \mathbf{h}_{12} \\ \mathbf{h}_{21} & \mathbf{h}_{22} \end{bmatrix} = \begin{bmatrix} 14 & \frac{2}{3} \\ -\frac{2}{3} & \frac{1}{9} \end{bmatrix}$$

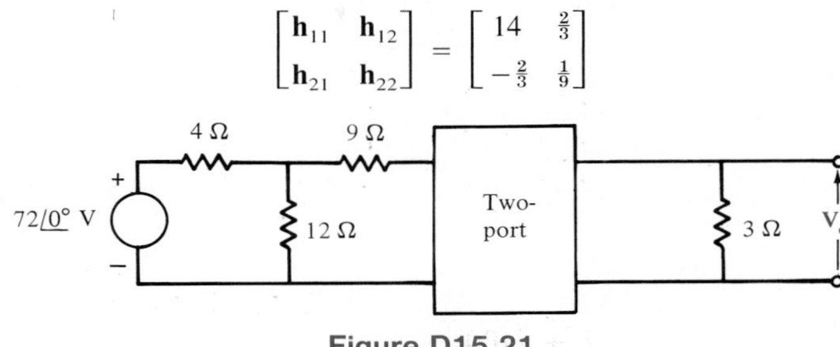

Figure D15.21

15.11 Summary

In this chapter two-port networks have been described using admittance, impedance, hybrid, and transmission parameters. Equivalent circuits for two-port networks have been described and conversion formulas for changing from one set of parameters to another have been presented.

We have shown that reciprocal networks have a special property that allows us to interchange a single voltage source and short circuit without affecting the value of the short-circuit current. The dual of this principle for a current source and open circuit also holds.

Finally, we have shown that a two-port may be treated as an interconnection of simpler networks. The interconnections described are parallel, series, and cascade. Finally, we have shown that the T and Π equivalent networks are the same as the Y and Δ networks described in Chapter 12.

KEY POINTS

- Four of the most common parameters used to describe a two-port network are the admittance, impedance, hybrid, and transmission parameters.
- Some or all of the two-port parameters for a network may not exist.
- If all the two-port parameters for a network exist, a set of conversion formulas can be used to relate one set of two-port parameters to another.
- A two-port network is said to be reciprocal if $\mathbf{y}_{12} = \mathbf{y}_{21}$, $\mathbf{z}_{12} = \mathbf{z}_{21}$, $\mathbf{h}_{12} = -H_{21}$, and $\mathbf{AB} - \mathbf{CD} = 1$.

- The reciprocity theorem states that in a reciprocal network a single voltage source and a short circuit may be interchanged without affecting the value of the short-circuit current, and similarly a single current source and an open circuit may be interchanged without affecting the value of the open-circuit voltage.
- The Brune tests must be satisfied in order to interconnect two-ports.
- When interconnecting two-ports, the Y parameters are added for a parallel connection, the Z parameters are added for a series connection, and the transmission parameters in matrix form are multiplied together for a cascade connection.
- T- to Π-transformations are shown to be the same as Y- to Δ-transformations.

PROBLEMS

15.1. Find the Y parameters for the two-port network shown in Fig. P15.1.

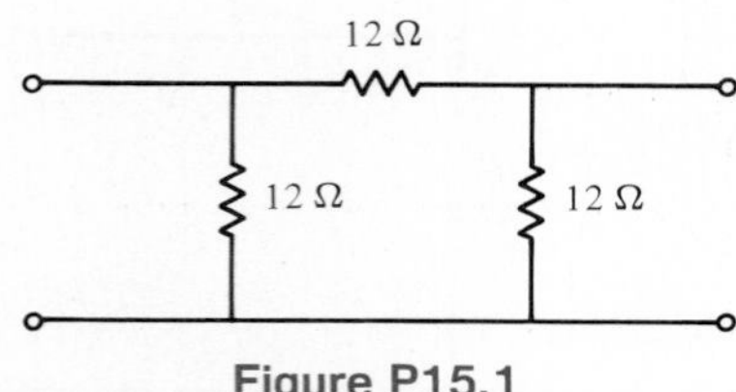

Figure P15.1

15.2. If a 12-A source is connected at the input port of the network shown in Fig. P15.1, find the current in a 4-Ω load resistor.

15.3. Find the Y parameters of the two-port network in Fig. P15.3.

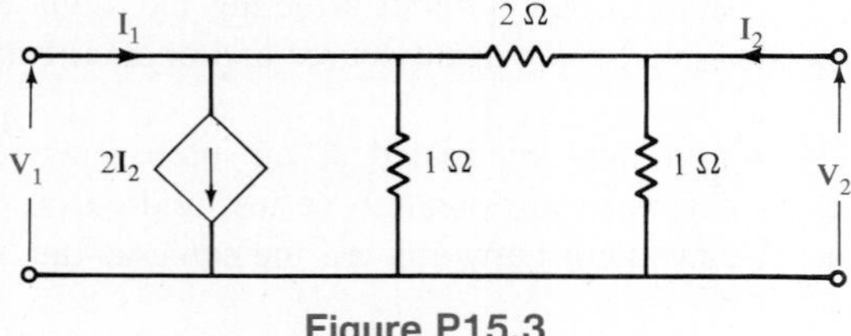

Figure P15.3

15.4. Find the Y parameters for the two-port network in Fig. P15.4.

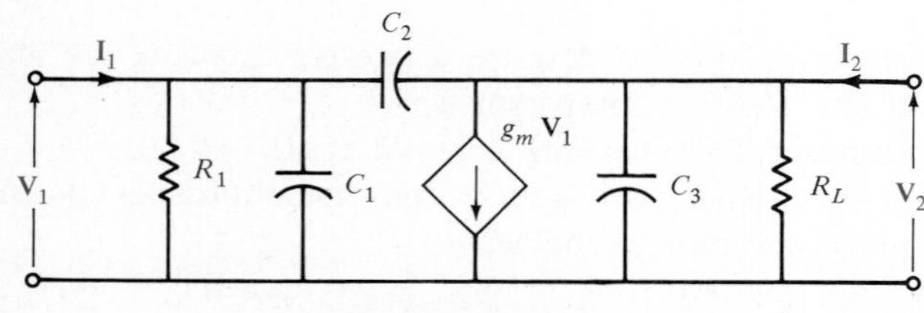

Figure P15.4

15.5. Find the Z parameters for the two-port in Problem 15.1.

15.6. Find the Z parameters for the two-port network in Fig. P15.6.

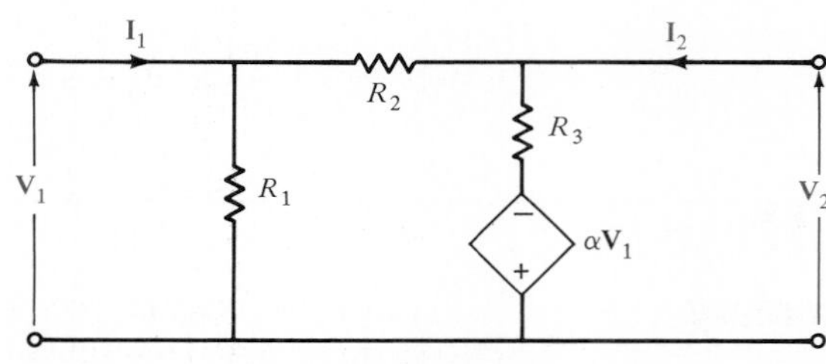

Figure P15.6

15.7. Find the Z parameters for the two-port shown in Fig. P15.7 and determine the voltage gain of the entire circuit with a 4-kΩ load attached to the output.

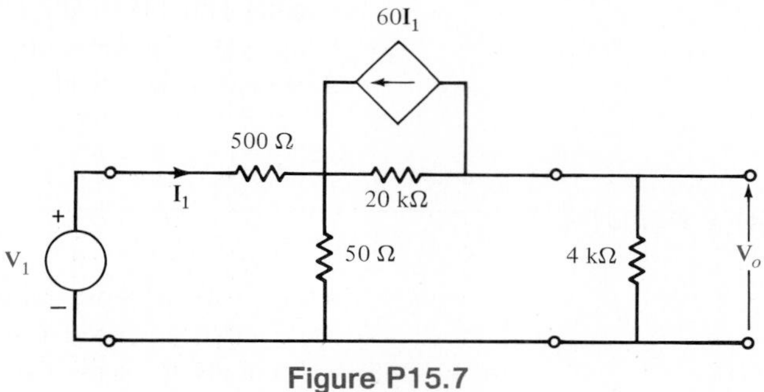

Figure P15.7

15.8. Calculate the input impedance of the network in Fig. P15.8.

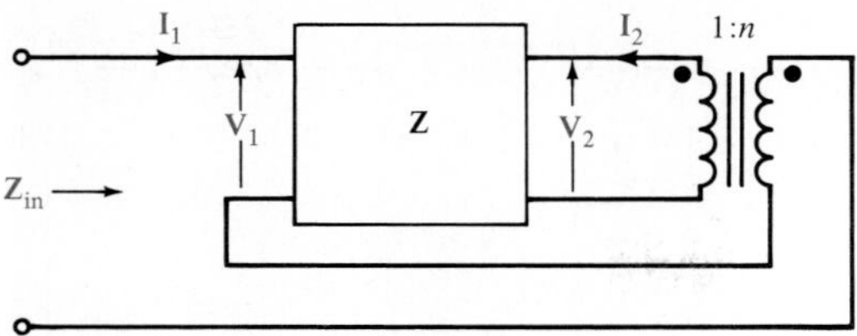

Figure P15.8

15.9. Compute the transmission parameters for the two-port in Problem 15.1.

15.10. Determine the admittance parameters for the network shown in Fig. P15.10.

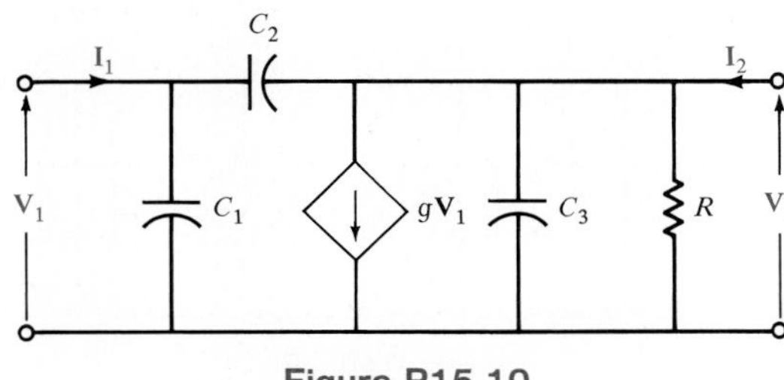

Figure P15.10

15.11. Derive the Z parameters for the network in Example 15.8.

15.12. Derive the hybrid parameters for the network in Fig. 15.10a.

15.13. Find the hybrid parameters for the network in Problem 15.1.

15.14. Derive the general conversion formulas for the Z parameters in terms of the hybrid parameters.

15.15. The G parameters are defined by the following equations:

$$\mathbf{I}_1 = \mathbf{g}_{11}\mathbf{V}_1 + \mathbf{g}_{12}\mathbf{I}_2$$

$$\mathbf{V}_2 = \mathbf{g}_{21}\mathbf{V}_1 + \mathbf{g}_{22}\mathbf{I}_2$$

Determine the Z parameters in terms of the G parameters.

15.16. Following are the hybrid parameters for a network.

$$\begin{bmatrix} \mathbf{h}_{11} & \mathbf{h}_{12} \\ \mathbf{h}_{21} & \mathbf{h}_{22} \end{bmatrix} = \begin{bmatrix} \frac{11}{5} & \frac{2}{5} \\ -\frac{2}{5} & \frac{1}{5} \end{bmatrix}$$

Determine the Y parameters for the network.

15.17. Derive the general conversion formulas for the admittance parameters in terms of the hybrid parameters.

15.18. Determine the input impedance of the network shown in Fig. P15.18, in terms of the hybrid parameters of the two-port and the two-port load $\mathbf{Z}_L$.

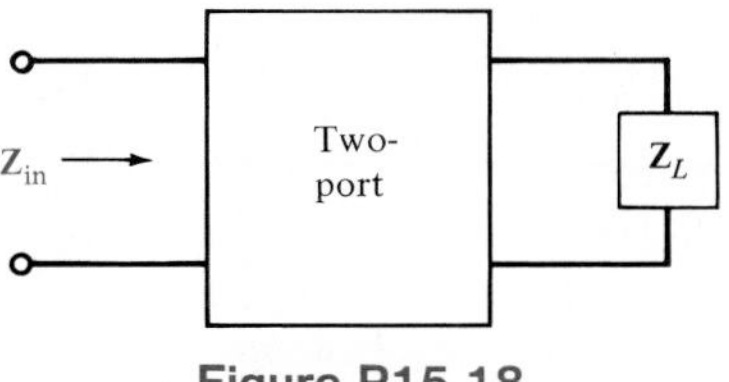

Figure P15.18

15.19. Find the Y parameters for the two-port network in Fig. P15.19.

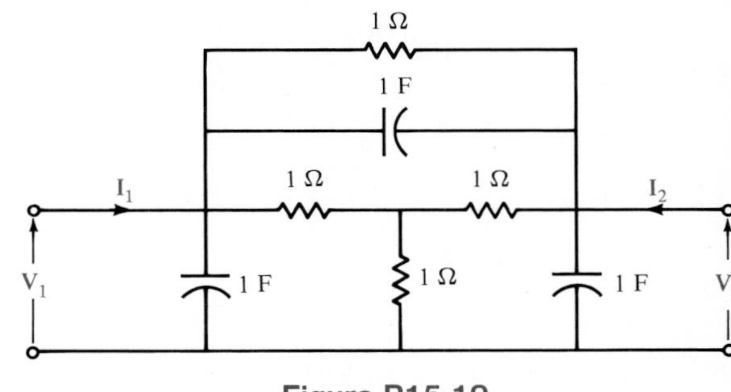

Figure P15.19

15.20. If the hybrid parameters for a network are as given in Problem 15.16, show that the parameters define a reciprocal network.

15.21. In the circuit shown in Fig. P15.21, illustrate the principle of reciprocity by interchanging the current source and the open circuit.

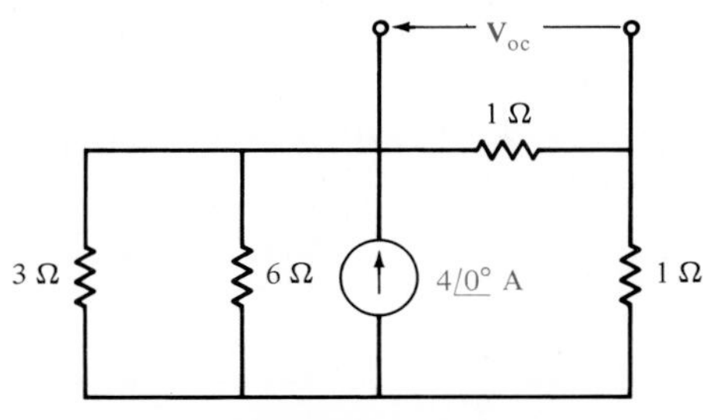

Figure P15.21

15.22. Use the principle of reciprocity to find $\mathbf{V}_o$ in the network in Fig. P15.22.

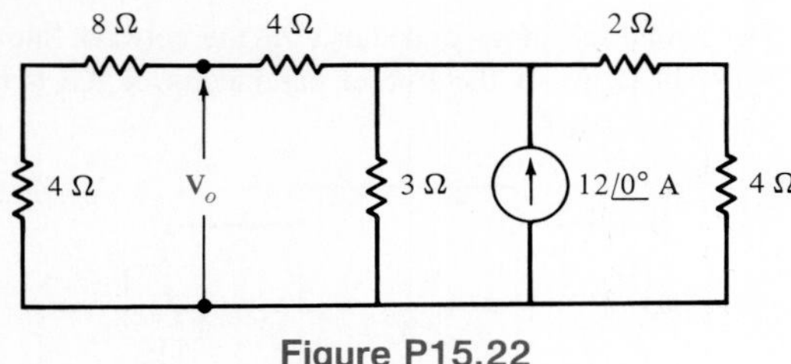

Figure P15.22

15.23. Use reciprocity to find $\mathbf{I}_o$ in the network in Fig. P15.23.

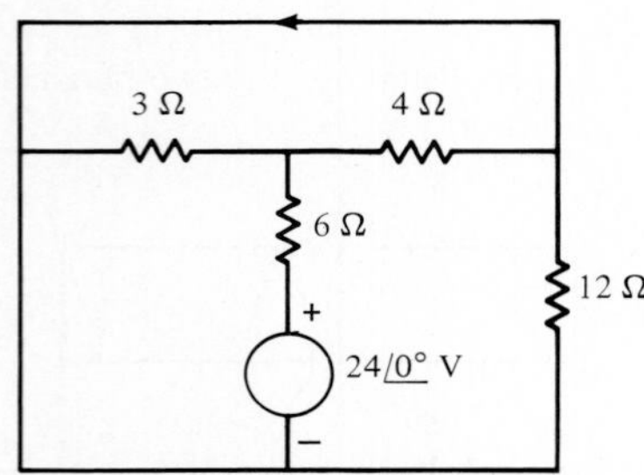

Figure P15.23

15.24. Use reciprocity to find $\mathbf{V}_o$ in the network in Fig. P15.24.

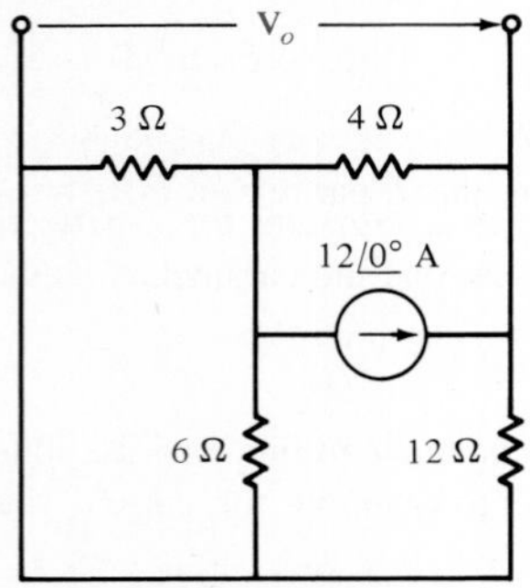

Figure P15.24

15.25. In Fig. P15.25, N is known to be a reciprocal network. If $\mathbf{V}_{oc} = 3\,\underline{/0°}$ V when $\mathbf{I}_o = 12\,\underline{/0°}$ A as shown in part (a), determine $\mathbf{V}_A$ when the network is connected as shown in part (b).

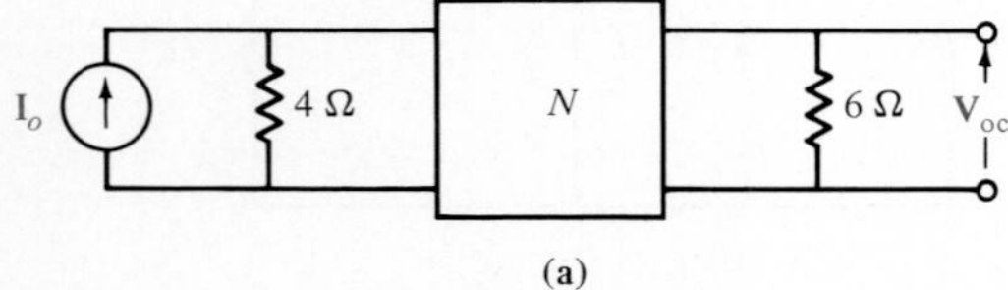

(a)

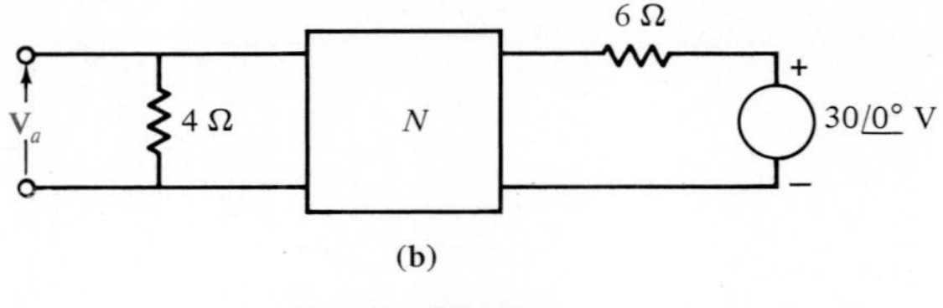

(b)

Figure P15.25

15.26. Compute the transmission parameters for the network in Fig. 15.3a.

15.27. Compute the transmission parameters for the network in Fig. 15.4a.

15.28. Determine the transmission parameters for the network shown in Fig. P15.28.

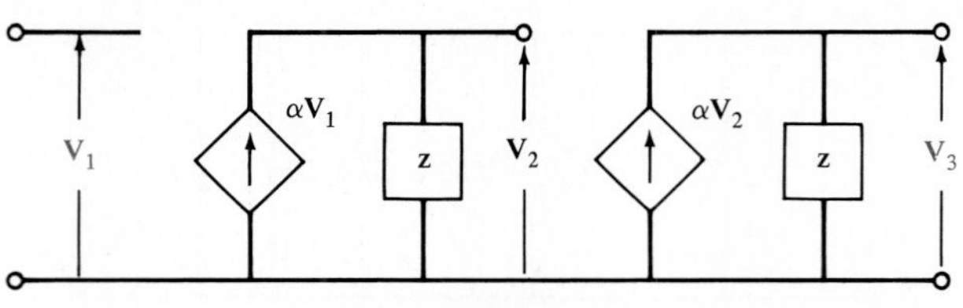

Figure P15.28

15.29. Determine the input impedance in the network shown in Fig. P15.29 assuming that the transmission parameters for the two identical two-ports are known.

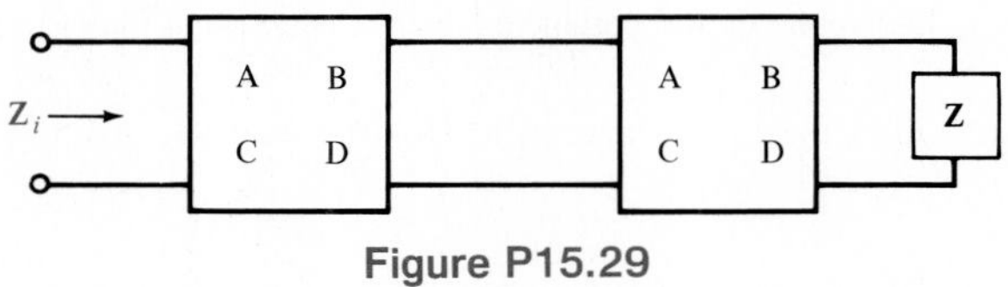

Figure P15.29

15.30. Determine the Y parameters for the network shown in Fig. P15.30.

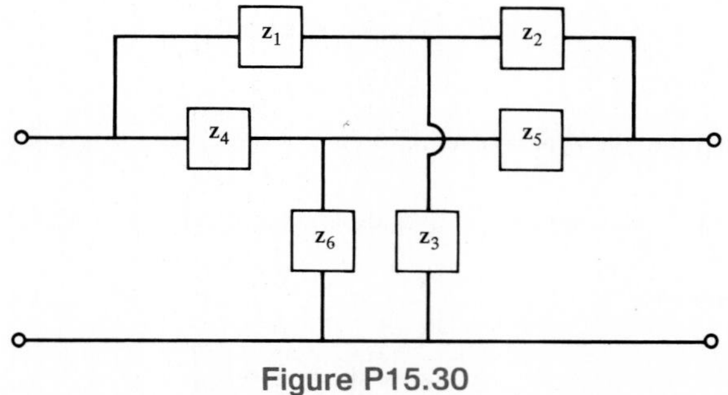

Figure P15.30

15.31. Determine the Y parameters for the network shown in Fig. P15.31.

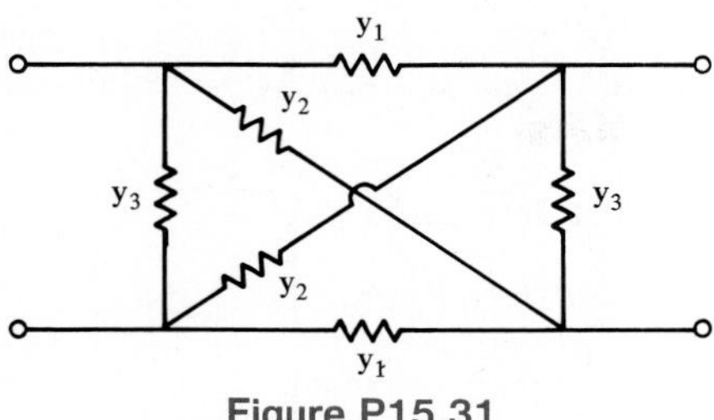

Figure P15.31

15.32. Find the Z parameters for the transistor model shown in Fig. P15.32.

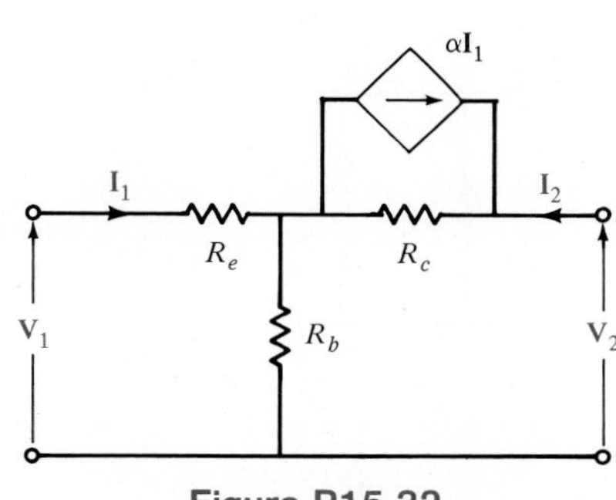

Figure P15.32

15.33. Find the Y parameters for the network in Fig. P15.33.

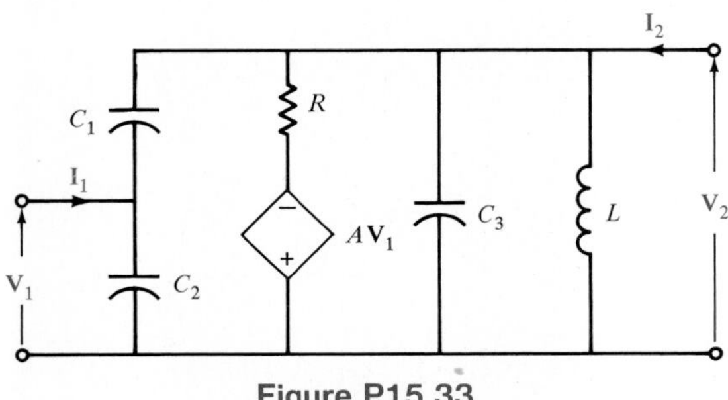

Figure P15.33

15.34. Find the Y parameters for the network in Fig. P15.34.

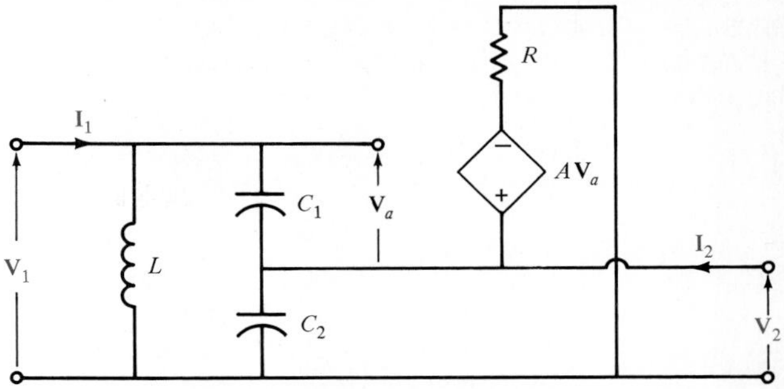

Figure P15.34

15.35. Determine the condition for reciprocity in the two-port network in Fig. P15.35.

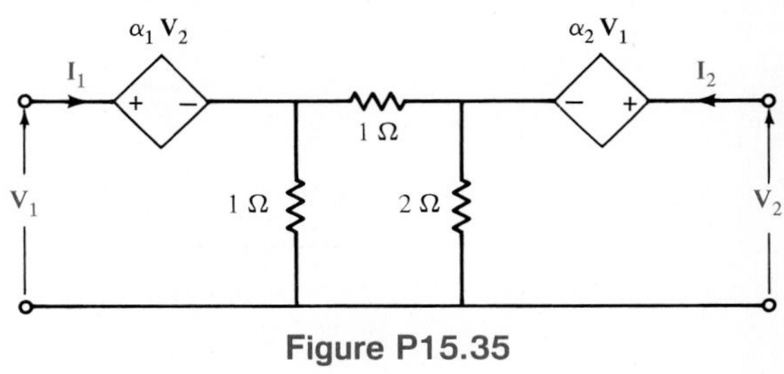

Figure P15.35

15.36. Determine the Z parameters for the two-port network in Fig. P15.36.

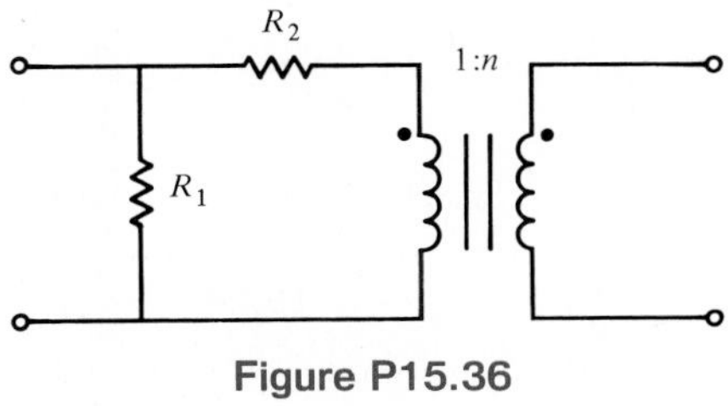

Figure P15.36

15.37. Given the network in Problem 15.36,

(a) Find the transmission parameters.

(b) Convert the transmission parameters to Z parameters and check them with the results of Problem 15.36.

15.38. The Z parameters for the two-port N are $\mathbf{z}_{11} = \mathbf{z}_{22} = \frac{5}{3}\ \Omega$ and $\mathbf{z}_{12} = \mathbf{z}_{21} = \frac{4}{3}\ \Omega$. Find the transmission parameters for the circuit in Fig. P15.38.

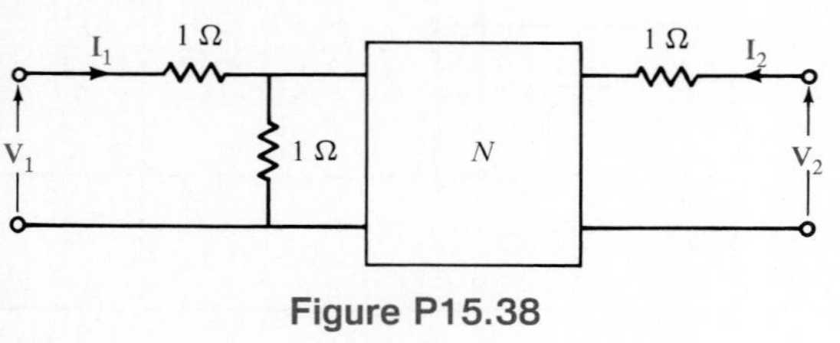

Figure P15.38

15.39. Determine the output voltage $\mathbf{V}_o$ in the network in Fig. P15.39 if the Z parameters for the two-port are

$$Z = \begin{bmatrix} 3 & 2 \\ 2 & 3 \end{bmatrix}$$

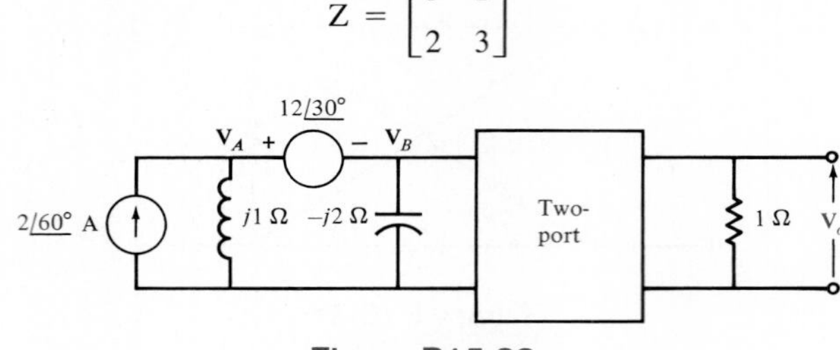

Figure P15.39

15.40. Determine the output voltage $\mathbf{V}_o$ in the network in Fig. P15.40 if the Z parameters of the two-port are

$$Z = \begin{bmatrix} 5 & 4 \\ 4 & 12 \end{bmatrix}$$

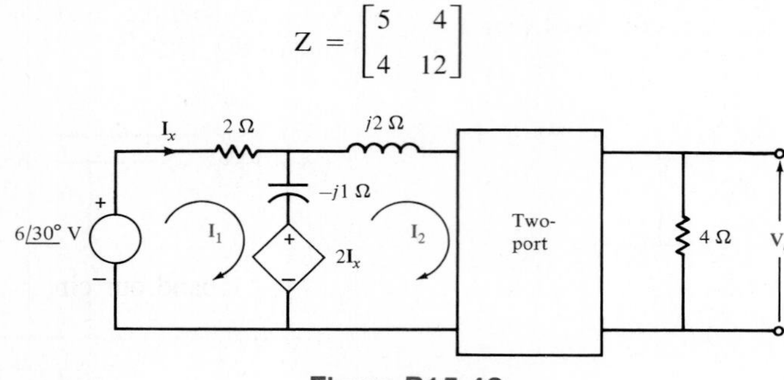

Figure P15.40

CHAPTER

16

Fourier Analysis Techniques

In this chapter we examine two very important topics: the Fourier series and the Fourier transform. These two techniques vastly expand our circuit analysis capabilities because they provide a means of effectively dealing with nonsinusoidal periodic signals and aperiodic signals. Using the Fourier series we show that we can determine the steady-state response of a network to a nonsinusoidal periodic input. The Fourier transform will allow us to analyze circuits with aperiodic inputs by transforming the problem to the frequency domain, solving it algebraically, and then transforming back to the time domain.

16.1 Fourier Series

A periodic function is one that satisfies the relationship

$$f(t) = f(t + nT_0), \qquad n = \pm 1, \pm 2, \pm 3, \ldots$$

for every value of t where T_0 is the period. As we have shown in previous chapters, the sinusoidal function is a very important periodic function. However, there are many other periodic functions that have wide applications. For example, laboratory signal generators produce the pulse-train and square-wave signals shown in Fig. 16.1a and b, respectively, which are used for testing circuits. The oscilloscope is another laboratory instrument and the sweep of its electron beam across the face of the cathode ray tube is controlled by a triangular signal of the form shown in Fig. 16.1c.

The techniques we will explore are based on the work of Jean Baptiste Joseph Fourier (1768–1830). Although our analyses will be confined to electric circuits, it is important

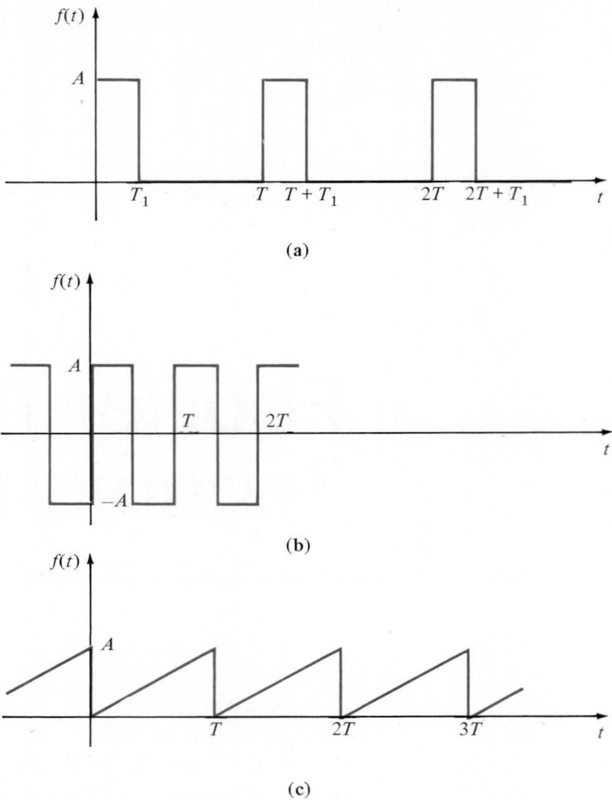

Figure 16.1 Some useful periodic signals.

to point out that the techniques are applicable to a wide range of engineering problems. In fact, it was Fourier's work in heat flow that led to the techniques which will be presented here.

In his work, Fourier demonstrated that a periodic function $f(t)$ could be expressed as a sum of sinusoidal functions. Therefore, given this fact and the fact that if a periodic function is expressed as a sum of linearly independent functions, each function in the sum must be periodic with the same period, and the function $f(t)$ can be expressed in the form

$$f(t) = a_0 + \sum_{n=1}^{\infty} D_n \cos(n\omega_0 t + \theta_n) \tag{16.1}$$

where $\omega_0 = 2\pi/T_0$ and a_0 is the average value of the waveform. An examination of this expression illustrates that all sinusoidal waveforms that are periodic with period T_0 have been included. For example, for $n = 1$, one cycle covers T_0 seconds and $D_1 \cos(\omega_0 t + \theta_1)$ is called the *fundamental*. For $n = 2$, two cycles fall within T_0 seconds and the term $D_2 \cos(2\omega_0 t + \theta_2)$ is called the *second harmonic*. In general, for $n = k$, k cycles fall within T_0 seconds and $D_k \cos(k\omega_0 t + \theta_k)$ is the *kth harmonic term*.

Since the function $\cos(n\omega_0 t + \theta_n)$ can be written in exponential form using Euler's identity or as a sum of cosine and sine terms of the form cos $n\omega_0 t$ and sin $n\omega_0 t$ as demonstrated in Chapter 9, the series in Eq. (16.1) can be written as

$$f(t) = a_0 + \sum_{\substack{n=-\infty \\ n \neq 0}}^{\infty} \mathbf{c}_n e^{jn\omega_0 t} = \sum_{n=-\infty}^{\infty} \mathbf{c}_n e^{jn\omega_0 t} \qquad (16.2)$$

$$= a_0 + \sum_{n=1}^{\infty} a_n \cos n\omega_0 t + b_n \sin n\omega_0 t \qquad (16.3)$$

Using the real-part relationship employed as a transformation between the time domain and the frequency domain, we can express $f(t)$ as

$$f(t) = a_0 + \sum_{n=1}^{\infty} \text{Re}\,[(D_n \angle \theta_n) e^{jn\omega_0 t}] \qquad (16.4)$$

$$= a_0 + \sum_{n=1}^{\infty} \text{Re}\,(2\mathbf{c}_n e^{jn\omega_0 t}) \qquad (16.5)$$

$$= a_0 + \sum_{n=1}^{\infty} \text{Re}\,[(a_n - jb_n) e^{jn\omega_0 t}] \qquad (16.6)$$

These equations allow us to write the Fourier series in a number of equivalent forms. Note that the *phasor* for the nth harmonic is

$$D_n \angle \theta_n = 2\mathbf{c}_n = a_n - jb_n \qquad (16.7)$$

The approach we will take will be to represent a nonsinusoidal periodic input by a sum of complex exponential functions, which because of Euler's identity is equivalent to a sum of sines and cosines. We will then use (1) the superposition property of linear systems and (2) our knowledge that the steady-state response of a time-invariant linear system to a sinusoidal input of frequency ω_0 is a sinusoidal function of the same frequency to determine the response of such a system.

In order to illustrate the manner in which a nonsinusoidal periodic signal can be represented by a Fourier series, consider the periodic function shown in Fig. 16.2a. In Fig. 16.2b–e we can see the impact of using a specific number of terms in the series to represent the original function. Note that the series more closely represents the original function as we employ more and more terms.

Exponential Fourier Series

Any physically realizable periodic signal may be represented over the interval $t_1 < t < t_1 + T_0$ by the *exponential Fourier series*

$$f(t) = \sum_{n=-\infty}^{\infty} \mathbf{c}_n e^{jn\omega_0 t} \qquad (16.8)$$

where the $\mathbf{c}_n$ are the complex (phasor) Fourier coefficients. These coefficients are derived as follows. Multiplying both sides of Eq. (16.8) by $e^{-jk\omega_0 t}$ and integrating over the interval t_1 to $t_1 + T_0$, we obtain

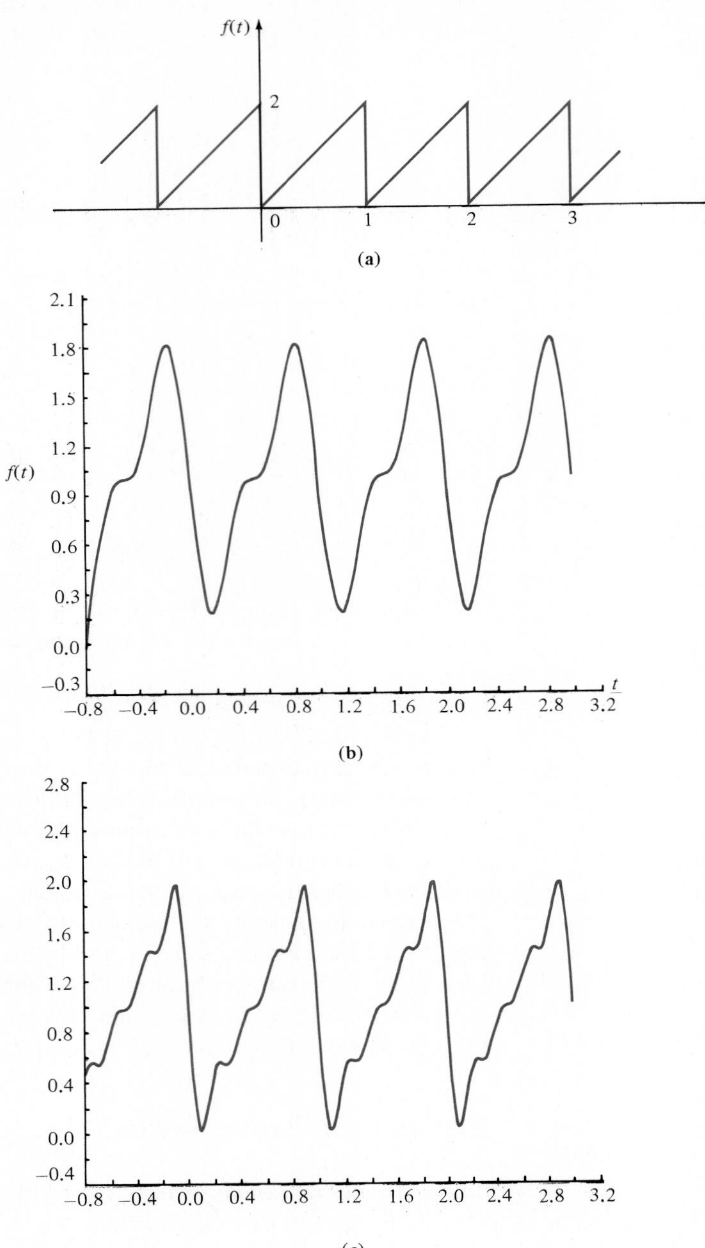

Figure 16.2 Periodic function (a) and its representation by a fixed number of Fourier series terms: (b) 2 terms; (c) 4 terms; (d) 100 terms.

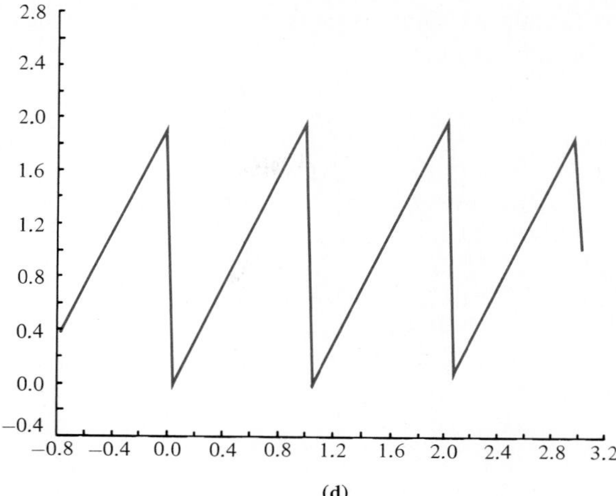

Figure 16.2 (Continued).

$$\int_{t_1}^{t_1+T_0} f(t)e^{-jk\omega_0 t}\,dt = \int_{t_1}^{t_1+T_0}\left(\sum_{n=-\infty}^{\infty} \mathbf{c}_n e^{jn\omega_0 t}\right)e^{-jk\omega_0 t}\,dt$$

$$= \mathbf{c}_k T_0$$

since

$$\int_{t_1}^{t_1+T_0} e^{j(n-k)\omega_0 t}\,dt = \begin{cases} 0 & \text{for } n \neq k \\ T_0 & \text{for } n = k \end{cases}$$

Therefore, the Fourier coefficients are defined by the equation

$$\mathbf{c}_n = \frac{1}{T_0}\int_{t_1}^{t_1+T_0} f(t)e^{-jn\omega_0 t}\,dt \tag{16.9}$$

The following example illustrates the manner in which we can represent a periodic signal by an exponential Fourier series.

Example 16.1

We wish to determine the exponential Fourier series for the periodic voltage waveform shown in Fig. 16.3.

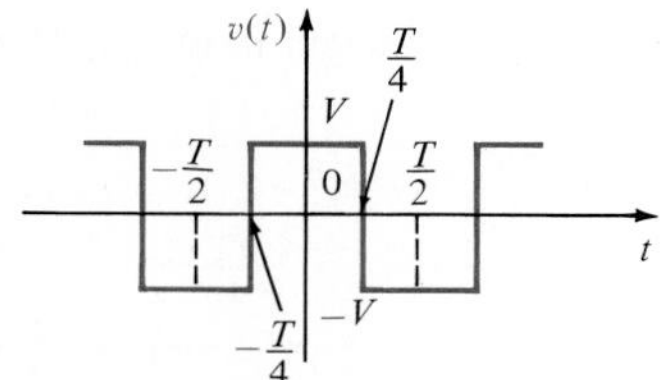

Figure 16.3 Periodic voltage waveform.

The Fourier coefficients are determined using Eq. (16.9) by integrating over one complete period of the waveform.

$$\begin{aligned}
\mathbf{c}_n &= \frac{1}{T}\int_{-T/2}^{T/2} f(t)e^{-jn\omega_0 t}\,dt \\
&= \frac{1}{T}\int_{-T/2}^{-T/4} -Ve^{-jn\omega_0 t}\,dt + \int_{-T/4}^{T/4} Ve^{-jn\omega_0 t}\,dt + \int_{T/4}^{T/2} -Ve^{-jn\omega_0 t}\,dt \\
&= \frac{V}{jn\omega_0 T}\left[+e^{-jn\omega_0 t}\Big|_{-T/2}^{-T/4} + e^{-jn\omega_0 t}\Big|_{-T/4}^{T/4} + e^{-jn\omega_0 t}\Big|_{T/4}^{T/2}\right] \\
&= \frac{V}{jn\omega_0 T}(2e^{jn\pi/2} - 2e^{-jn\pi/2} + e^{-jn\pi} - e^{+jn\pi}) \\
&= \frac{V}{n\omega_0 T}\left[4\sin\frac{n\pi}{2} - 2\sin(n\pi)\right] \\
&= 0 \qquad \text{for } n \text{ even} \\
&= \frac{2V}{n\pi}\sin\frac{n\pi}{2} \qquad \text{for } n \text{ odd}
\end{aligned}$$

Since $\mathbf{c}_0$ corresponds to the average value of the waveform $\mathbf{c}_0 = 0$. This term can also be evaluated using the original equation for $\mathbf{c}_n$. Therefore,

$$v(t) = \sum_{\substack{n=-\infty \\ n\neq 0 \\ n \text{ odd}}}^{\infty} \frac{2V}{n\pi}\sin\frac{n\pi}{2}\, e^{jn\omega_0 t}$$

Using Euler's identity, we can also write the equation as

$$v(t) = \sum_{\substack{n=1 \\ n \text{ odd}}}^{\infty} \frac{4V}{n\pi}\sin\frac{n\pi}{2}\cos n\omega_0 t$$

Note that this same result could have been obtained by integrating over the interval $-T/4$ to $3T/4$. ■

DRILL EXERCISE

D16.1. Find the Fourier coefficients for the waveform in Fig. D16.1.

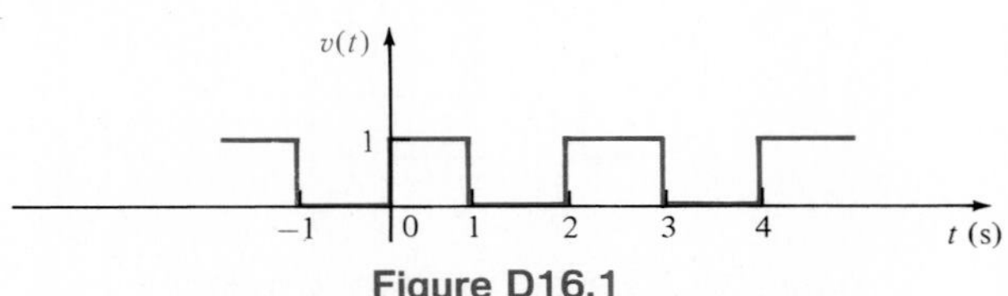

Figure D16.1

D16.2. Find the Fourier coefficients for the waveform in Fig. D16.2.

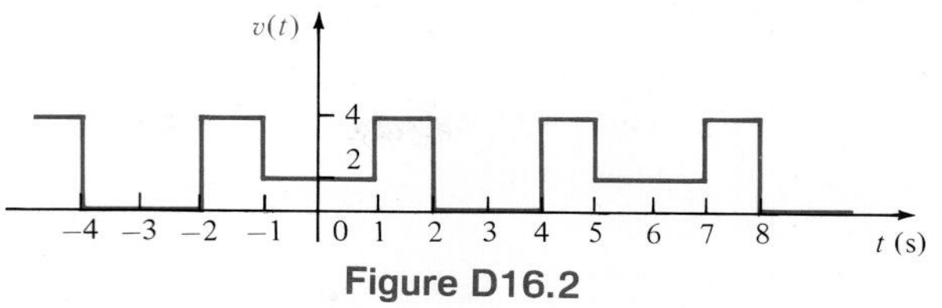

Figure D16.2

Trigonometric Fourier Series

Let us now examine another form of the Fourier series. Since

$$2\mathbf{c}_n = a_n - jb_n \tag{16.10}$$

We will examine this quantity $2\mathbf{c}_n$ and separate it into its real and imaginary parts. Using Eq. (16.9), we find that

$$2\mathbf{c}_n = \frac{2}{T_0}\int_{t_1}^{t_1+T_0} f(t)e^{-jn\omega_0 t}\,dt \tag{16.11}$$

Using Euler's identity, we can write this equation in the form

$$\begin{aligned}2\mathbf{c}_n &= \frac{2}{T_0}\int_{t_1}^{t_1+T_0} f(t)[\cos n\omega_0 t - j\sin n\omega_0 t]\,dt \\ &= \frac{2}{T_0}\int_{t_1}^{t_1+T_0} f(t)\cos n\omega_0 t\,dt - j\frac{2}{T_0}\int_{t_1}^{t_1+T_0} f(t)\sin n\omega_0 t\,dt\end{aligned}$$

From Eq. (16.10) we note then that

$$a_n = \frac{2}{T_0}\int_{t_1}^{t_1+T_0} f(t)\cos n\omega_0 t\,dt \tag{16.12}$$

$$b_n = \frac{2}{T_0}\int_{t_1}^{t_1+T_0} f(t)\sin n\omega_0 t\,dt \tag{16.13}$$

These are the coefficients of the Fourier series described by Eq. (16.3), which we call the *trigonometric Fourier series*. These equations are derived directly in most text books using the orthogonality properties of the cosine and sine functions. Note that we can now evaluate $\mathbf{c}_n$, a_n, b_n, and since

$$2\mathbf{c}_n = D_n \angle\theta_n \tag{16.14}$$

we can derive the coefficients for the *cosine Fourier series* described by Eq. (16.1). This form of the Fourier series is particularly useful because it allows us to represent each harmonic of the function as a phasor.

From Eq. (16.9) we note that $\mathbf{c}_0$, which is written as a_0 is

$$a_0 = \frac{1}{T_0}\int_{t_1}^{t_1+T_0} f(t)\,dt \tag{16.15}$$

This is the average value of the signal $f(t)$ and can often be evaluated directly from the waveform.

Symmetry and the Trigonometric Fourier Series

If a signal exhibits certain symmetrical properties, we can take advantage of these properties to simplify the calculations of the Fourier coefficients. There are three types of symmetry: (1) even-function symmetry, (2) odd-function symmetry, and (3) half-wave symmetry.

Even-Function Symmetry. A function is said to be even if

$$f(t) = f(-t) \tag{16.16}$$

An even function is symmetrical about the vertical axis and a notable example is the function cos $n\omega_0 t$. Note that the waveform in Fig. 16.3 also exhibits even-function symmetry. Let us now determine the expressions for the Fourier coefficients if the function satisfies Eq. (16.16).

If we let $t_1 = -T_0/2$ in Eq. (16.15), we obtain

$$a_0 = \frac{1}{T_0}\int_{-T_0/2}^{T_0/2} f(t)\,dt$$

which can be written as

$$a_0 = \frac{1}{T_0}\int_{-T_0/2}^{0} f(t)\,dt + \frac{1}{T_0}\int_{0}^{T_0/2} f(t)\,dt$$

If we now change the variable on the first integral (i.e., let $t = -x$), then $f(-x) = f(x)$, $dt = -dx$, and the range of integration is from $x = T_0/2$ to 0. Therefore, the equation above becomes

$$\begin{aligned} a_0 &= \frac{1}{T_0}\int_{T_0/2}^{0} f(x)(-dx) + \frac{1}{T_0}\int_{0}^{T_0/2} f(t)\,dt \\ &= \frac{1}{T_0}\int_{0}^{T_0/2} f(x)\,dx + \frac{1}{T_0}\int_{0}^{T_0/2} f(t)\,dt \\ &= \frac{2}{T_0}\int_{0}^{T_0/2} f(t)\,dt \end{aligned} \tag{16.17}$$

The other Fourier coefficients are derived in a similar manner. The a_n coefficient can be written

$$a_n = \frac{2}{T_0}\int_{-T_0/2}^{0} f(t)\cos n\omega_0 t\,dt + \frac{2}{T_0}\int_{0}^{T_0/2} f(t)\cos n\omega_0 t\,dt$$

Employing the change of variable that led to Eq. (16.17), we can express the equation above as

$$\begin{aligned} a_n &= \frac{2}{T_0}\int_{T_0/2}^{0} f(x)\cos(-n\omega_0 x)(-dx) + \frac{2}{T_0}\int_0^{T_0/2} f(t)\cos n\omega_0 t\,dt \\ &= \frac{2}{T_0}\int_0^{T_0/2} f(x)\cos n\omega_0 x\,dx + \frac{2}{T_0}\int_0^{T_0/2} f(t)\cos n\omega_0 t\,dt \\ &= \frac{4}{T_0}\int_0^{T_0/2} f(t)\cos n\omega_0 t\,dt \end{aligned} \tag{16.18}$$

Once again following the development above, we can write the equation for the b_n coefficient as

$$b_n = \frac{2}{T_0}\int_{-T_0/2}^{0} f(t)\sin n\omega_0 t\,dt + \int_0^{T_0/2} f(t)\sin n\omega_0 t\,dt$$

The variable change employed above yields

$$\begin{aligned} b_n &= \frac{2}{T_0}\int_{T_0/2}^{0} f(x)\sin(-n\omega_0 x)(-dx) + \frac{2}{T_0}\int_0^{T_0/2} f(t)\sin n\omega_0 t\,dt \\ &= \frac{-2}{T_0}\int_0^{T_0/2} f(x)\sin n\omega_0 x\,dx + \frac{2}{T_0}\int_0^{T_0/2} f(t)\sin n\omega_0\, t\,dt \\ &= 0 \end{aligned} \tag{16.19}$$

The preceding analysis indicates that the Fourier series for an even periodic function consists only of a constant term and cosine terms. Therefore, if $f(t)$ is even, $b_n = 0$ and from Eqs. (16.10) and (16.14), $\mathbf{c}_n$ are real and θ_n are multiples of 180°.

Odd-Function Symmetry. A function is said to be odd if

$$f(t) = -f(-t) \tag{16.20}$$

An example of an odd function is sin $n\omega_0 t$. Another example is the waveform in Fig. 16.4a. Following the mathematical development that led to Eqs. (16.17) to (16.19), we can show that for an odd function the Fourier coefficients are

$$a_0 = 0 \tag{16.21}$$

$$a_n = 0 \qquad \text{for all } n > 0 \tag{16.22}$$

$$b_n = \frac{4}{T_0}\int_0^{T_0/2} f(t)\sin n\omega_0 t\,dt \tag{16.23}$$

Therefore, if $f(t)$ is odd, $a_n = 0$ and, from Eqs. (16.10) and (16.14), $\mathbf{c}_n$ are pure imaginary and θ_n are odd multiples of 90°.

Half-Wave Symmetry. A function is said to possess *half-wave symmetry* if

$$f(t) = -f\left(t - \frac{T_0}{2}\right) \tag{16.24}$$

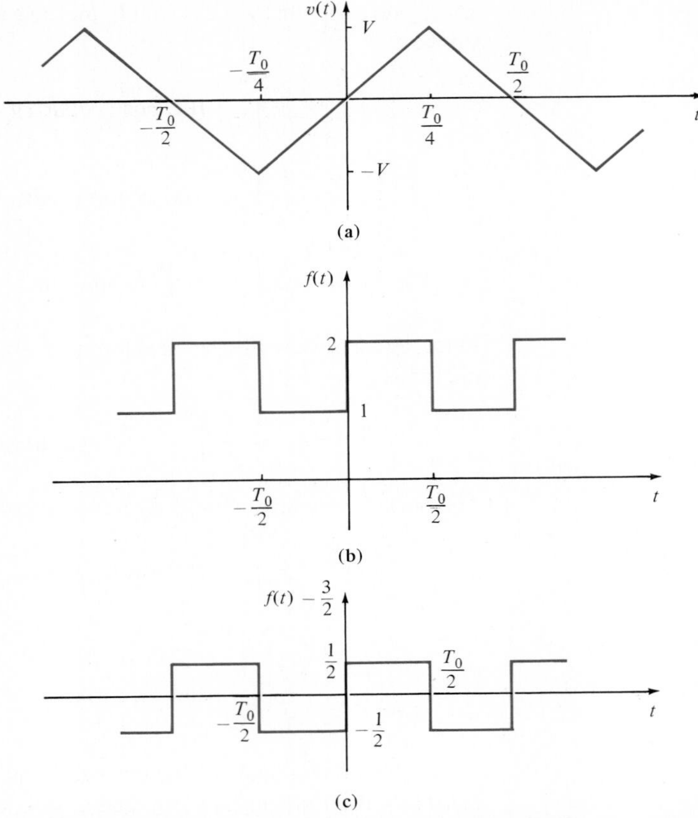

Figure 16.4 Three waveforms; (a) and (c) possess half-wave symmetry.

Basically, this equation states that each half-cycle is an inverted version of the adjacent half-cycle; that is, if the waveform from $-T_0/2$ to 0 is inverted, it is identical to the waveform from 0 to $T_0/2$. The waveform shown in Fig. 16.4a possesses half-wave symmetry.

Once again we can derive the expressions for the Fourier coefficients in this case by repeating the mathematical development that led to the equations for even-function symmetry using the change of variable $t = x + T_0/2$ and Eq. (16.24). The results of this development are the following equations:

$$a_0 = 0 \tag{16.25}$$

$$a_n = b_n = 0 \qquad \text{for } n \text{ even} \tag{16.26}$$

$$a_n = \frac{4}{T_0}\int_0^{T_0/2} f(t) \cos n\omega_0 t \, dt \qquad \text{for } n \text{ odd} \tag{16.27}$$

$$b_n = \frac{4}{T_0}\int_0^{T_0/2} f(t) \sin n\omega_0 t \, dt \qquad \text{for } n \text{ odd} \tag{16.28}$$

EXAMPLE 16.2

We wish to find the trigonometric Fourier series for the periodic signal in Fig. 16.3. The waveform exhibits even-function symmetry and therefore

$$a_0 = 0$$

$$b_n = 0 \qquad \text{for all } n$$

and

$$\begin{aligned}
a_n &= \frac{4}{T}\int_0^{T/2} f(t)\cos n\omega_0 t\,dt \qquad n \neq 0 \\
&= \frac{4}{T}\left[\int_0^{T/4} V\cos n\omega_0 t\,dt - \int_{T/4}^{T/2} V\cos n\omega_0 t\,dt\right] \\
&= \frac{4V}{n\omega_0 T}\left[\sin n\omega_0 t\Big|_0^{T/4} - \sin n\omega_0 t\Big|_{T/4}^{T/2}\right] \\
&= \frac{4V}{n\omega_0 T}\left(\sin\frac{n\pi}{2} - \sin n\pi + \sin\frac{n\pi}{2}\right) \\
&= \frac{8V}{n2\pi}\sin\frac{n\pi}{2} \qquad \text{for } n \text{ odd} \\
&= \frac{4V}{n\pi}\sin\frac{n\pi}{2} \qquad \text{for } n \text{ odd}
\end{aligned}$$

The reader should compare this result with that obtained in Example 16.1. Note that this function possesses half-wave symmetry. ■

EXAMPLE 16.3

Let us determine the trigonometric Fourier series expansion for the waveform shown in Fig. 16.4a. The function not only exhibits odd-function symmetry, but it possesses half-wave symmetry as well. Therefore, it is necessary to determine only the coefficients b_n for n odd. Note that

$$v(t) = \begin{cases} \dfrac{4Vt}{T_0} & 0 \leq t \leq T_0/4 \\[2ex] 2V - \dfrac{4Vt}{T_0} & T_0/4 < t \leq T_0/2 \end{cases}$$

The b_n coefficients are then

$$b_n = \frac{4}{T_0}\int_0^{T_0/4} \frac{4Vt}{T_0}\sin n\omega_0 t\,dt + \frac{4}{T_0}\int_{T_0/4}^{T_0/2}\left(2V - \frac{4Vt}{T_0}\right)\sin n\omega_0 t\,dt$$

The evaluation of these integrals is tedious but straightforward and yields

$$b_n = \frac{8V}{n^2\pi^2}\sin\frac{n\pi}{2} \qquad \text{for } n \text{ odd}$$

Hence the Fourier series expansion is

$$v(t) = \sum_{\substack{n=1 \\ n \text{ odd}}}^{\infty} \frac{8V}{n^2\pi^2} \sin\frac{n\pi}{2} \sin n\omega_0 t$$

EXAMPLE 16.4

We wish to find the trigonometric Fourier series expansion of the waveform in Fig. 16.4b. Note that this waveform has an average value of $\frac{3}{2}$. Therefore, instead of determining the Fourier series expansion of $f(t)$, we will determine the Fourier series for $f(t) - 3/2$, which is the waveform shown in Fig. 16.4c. The latter waveform possesses half-wave symmetry. The function is also odd and therefore

$$\begin{aligned} b_n &= \frac{4}{T_0}\int_0^{T_0/2} \tfrac{1}{2} \sin n\omega_0 t\, dt \\ &= \frac{2}{T_0}\left(\frac{-1}{n\omega_0} \cos n\omega_0 t \Big|_0^{T_0/2}\right) \\ &= \frac{-2}{n\omega_0 T_0}(\cos n\pi - 1) \\ &= \frac{2}{n\pi} \qquad n \text{ odd} \end{aligned}$$

Therefore, the Fourier series expansion for $f(t) - \frac{3}{2}$ is

$$f(t) - \tfrac{3}{2} = \sum_{\substack{n=1 \\ n \text{ odd}}}^{\infty} \frac{2}{n\pi} \sin n\omega_0 t$$

or

$$f(t) = \tfrac{3}{2} + \sum_{\substack{n=1 \\ n \text{ odd}}}^{\infty} \frac{2}{n\pi} \sin n\omega_0 t$$

DRILL EXERCISE

D16.3. Determine the type of symmetry exhibited by the waveforms in Figs. D16.2 and D16.3.

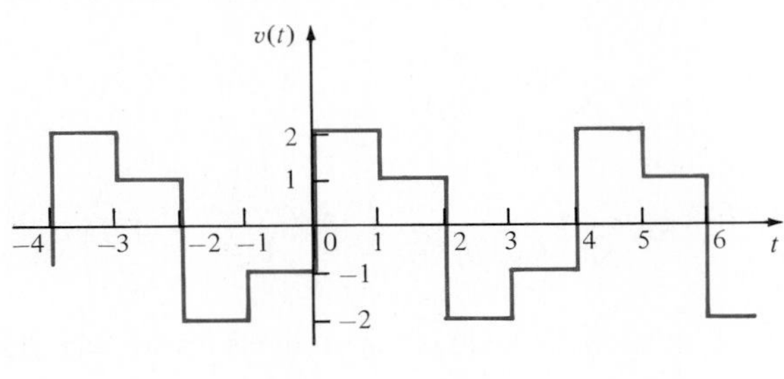

(a)

Figure D16.3a

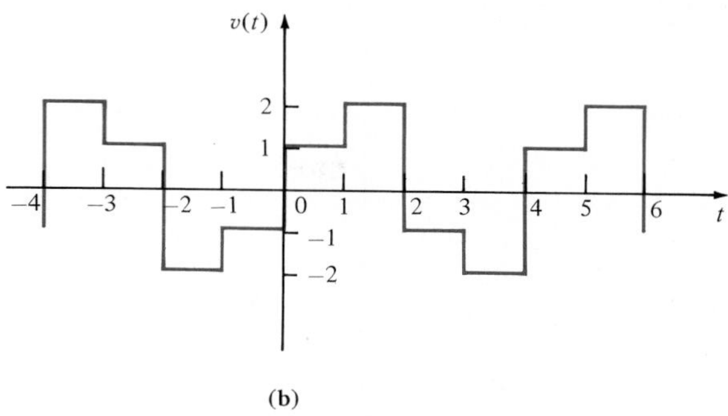

(b)

Figure D16.3b

D16.4. Find the trigonometric Fourier series for the voltage waveform in Fig. D16.2.

D16.5. Find the trigonometric Fourier series for the voltage waveform in Fig. D16.3a.

D16.6. Find the trigonometric Fourier series for the voltage waveform in Fig. D16.3b.

Time Shifting

Let us now examine the effect of time shifting a periodic waveform $f(t)$ defined by the equation

$$f(t) = \sum_{n=-\infty}^{\infty} \mathbf{c}_n e^{jn\omega_0 t}$$

Note that

$$f(t - t_0) = \sum_{n=-\infty}^{\infty} \mathbf{c}_n e^{jn\omega_0(t-t_0)}$$

$$= \sum_{n=-\infty}^{\infty} (\mathbf{c}_n e^{-jn\omega_0 t_0}) e^{jn\omega_0 t} \tag{16.29}$$

Since $e^{-jn\omega_0 t_0}$ corresponds to a phase shift, the Fourier coefficients of the time-shifted function are the Fourier coefficients of the original function with the angle shifted by an amount directly proportional to frequency. Therefore, time shift in the time domain corresponds to phase shift in the frequency domain.

EXAMPLE 16.5

Let us time delay the waveform in Fig. 16.3 by a quarter period and compute the Fourier series.

The waveform in Fig. 16.3 time delayed by $T_0/4$ is shown in Fig. 16.5. Since the time delay is $T_0/4$,

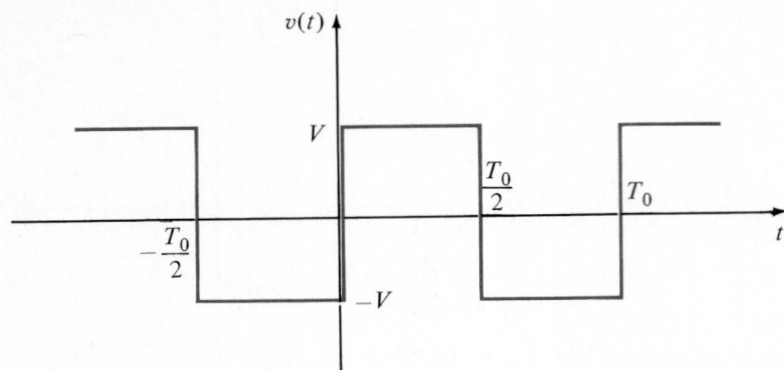

Figure 16.5 Waveform in Fig. 16.3 time shifted by $T_0/4$.

$$n\omega_0 t_d = n \frac{2\pi}{T_0}\frac{T_0}{4} = n\frac{\pi}{2} = n\ 90^\circ$$

Therefore, using Eq. (16.29) and the results of Example 16.1, the Fourier coefficients for the time-shifted waveform are

$$\mathbf{c}_n = \frac{2V}{n\pi}\sin\frac{n\pi}{2}\,\underline{/-n\ 90^\circ} \qquad n \text{ odd}$$

and therefore

$$v(t) = \sum_{\substack{n=1\\ n\text{ odd}}}^{\infty} \frac{4V}{n\pi}\sin\frac{n\pi}{2}\cos(n\omega_0 t - n\ 90^\circ)$$

However, this last equation can be expressed as

$$v(t) = \sum_{\substack{n=1\\ n\text{ odd}}}^{\infty} \frac{4V}{n\pi}\sin\frac{n\pi}{2}\sin n\omega_0 t$$

If we compute the Fourier coefficients for the time-shifted waveform in Fig. 16.5, we obtain

$$\begin{aligned}\mathbf{c}_n &= \frac{1}{T_0}\int_{-T_0/2}^{T_0/2} f(t)e^{-jn\omega_0 t}\,dt\\ &= \frac{1}{T_0}\int_{-T_0/2}^{0} -Ve^{-jn\omega_0 t}\,dt + \frac{1}{T_0}\int_{0}^{T_0/2} Ve^{-jn\omega_0 t}\,dt\\ &= \frac{2V}{jn\pi} \qquad \text{for } n \text{ odd}\end{aligned}$$

Therefore,

$$\mathbf{c}_n = \frac{2V}{n\pi}\,\underline{/-90^\circ} \qquad n \text{ odd}$$

Since n is odd, we can show that this expression is equivalent to the one obtained above. ■

In general, we can compute the phase shift in degrees using the expression

$$\text{phase shift (deg)} = \omega_0 t_d = (360°)\frac{t_d}{T_0} \tag{16.30}$$

so that a time shift of one-quarter period corresponds to a 90° phase shift.

As another interesting facet of the time shift, consider a function $f_1(t)$ which is nonzero in the interval $0 \le t \le T_0/2$ and is zero in the interval $T_0/2 < t \le T_0$. For purposes of illustration, let us assume that $f_1(t)$ is the triangular waveform shown in Fig. 16.6a. $f_1(t - T_0/2)$ is then shown in Fig. 16.6b. Then the function $f(t)$ defined as

$$f(t) = f_1(t) - f_1\left(t - \frac{T_0}{2}\right) \tag{16.31}$$

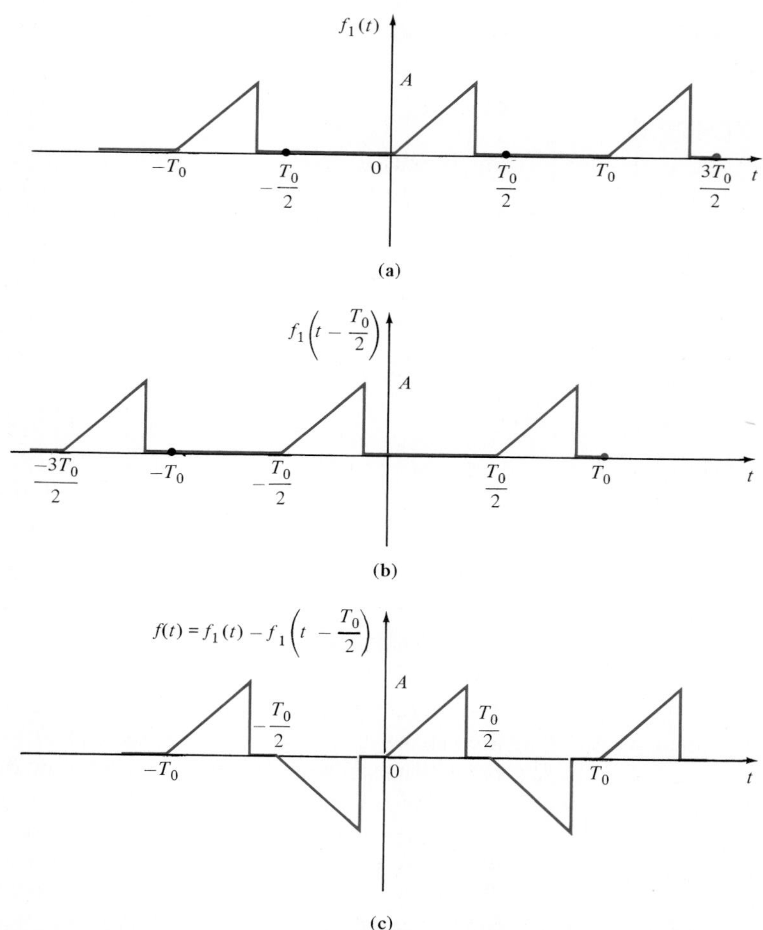

Figure 16.6 Waveforms that illustrate the generation of half-wave symmetry.

is shown in Fig. 16.6c. Note that $f(t)$ has half-wave symmetry. In addition, note that if

$$f_1(t) = \sum_{n=-\infty}^{\infty} \mathbf{c}_n e^{-jn\omega_0 t}$$

Then

$$f(t) = f_1(t) - f_1\left(t - \frac{T_0}{2}\right) = \sum_{n=-\infty}^{\infty} \mathbf{c}_n(1 - e^{-jn\pi})e^{jn\omega_0 t}$$

$$= \begin{cases} \sum_{n=-\infty}^{\infty} 2\mathbf{c}_n e^{jn\omega_0 t} & n \text{ odd} \\ 0 & n \text{ even} \end{cases} \tag{16.32}$$

Therefore, we see that any function with half-wave symmetry can be expressed in the form of Eq. (16.31), where the Fourier series is defined by Eq. (16.32), $\mathbf{c}_n$ is the Fourier coefficients for $f_1(t)$, and the integration to determine $\mathbf{c}_n$ is performed over half a period.

DRILL EXERCISE

D16.7. If the waveform in Fig. D16.1 is time-delayed 1 s, we obtain the waveform in Fig. D16.7. Compute the exponential Fourier coefficients for the waveform in Fig. D16.7 and show that they differ from the coefficients for the waveform in Fig. D16.1 by $n(180°)$.

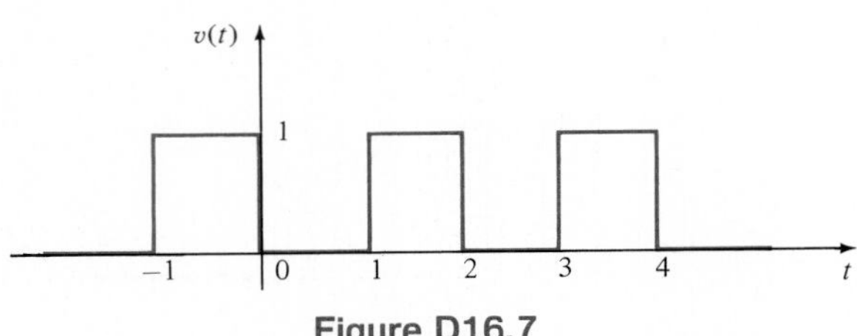

Figure D16.7

Waveform Generation

The magnitude of the harmonics in a Fourier series are independent of the time scale for a given wave shape. Therefore, the equations for a variety of waveforms can be given in tabular form without expressing a specific time scale. Table 16.1 is a set of commonly occurring periodic waves where the advantage of symmetry has been used to simplify the coefficients. These waveforms can be used to generate other waveforms. The level of a wave can be adjusted by changing the average value component; the time can be shifted by adjusting the angle of the harmonics; and two waveforms can be added to produce a third waveform; for example, the waveforms in Fig. 16.7a and b can be added to produce the waveform in Fig. 16.7c.

Table 16.1 Fourier Series for Some Common Waveforms

Waveform	Fourier series
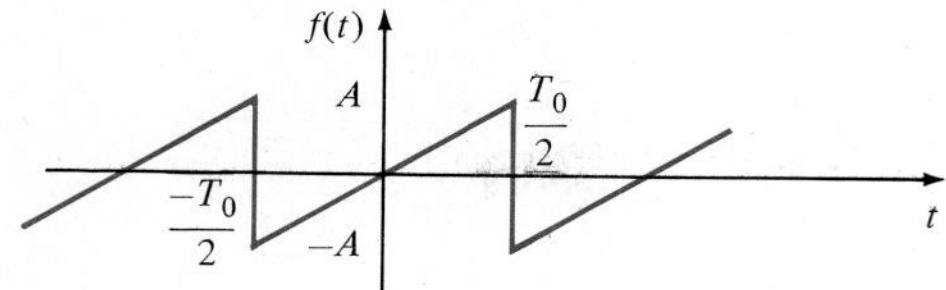	$f(t) = \sum_{n=1}^{\infty} (-1)^{n+1} \frac{2A}{n\pi} \sin n\omega_0 t$
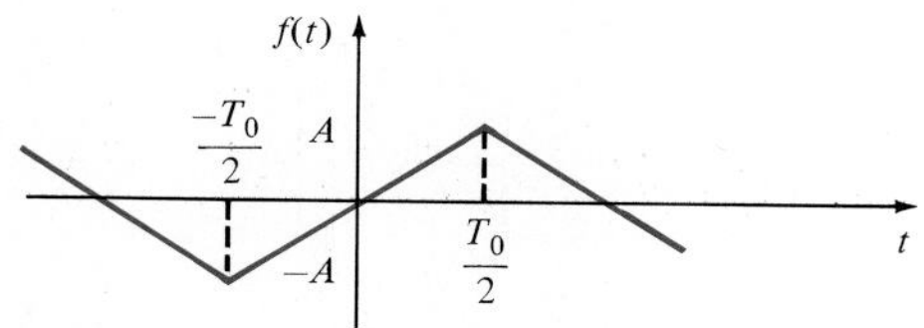	$f(t) = \sum_{\substack{n=1 \\ n \text{ odd}}}^{\infty} \frac{8A}{n^2\pi^2} \sin \frac{n\pi}{2} \sin n\omega_0 t$
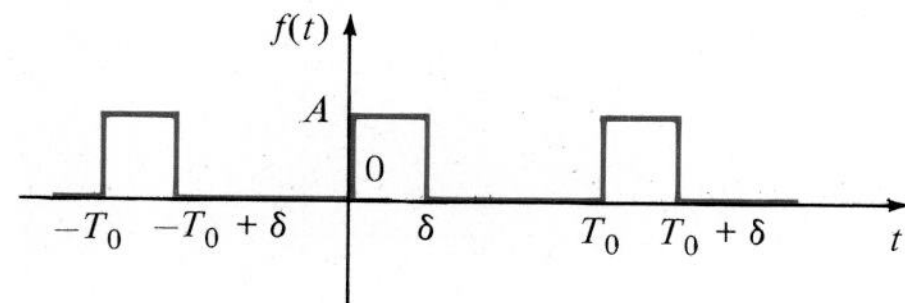	$f(t) = \sum_{n=-\infty}^{\infty} \frac{A}{n\pi} \sin \frac{n\pi\delta}{T_0} e^{jn\omega_0[t-(\delta/2)]}$
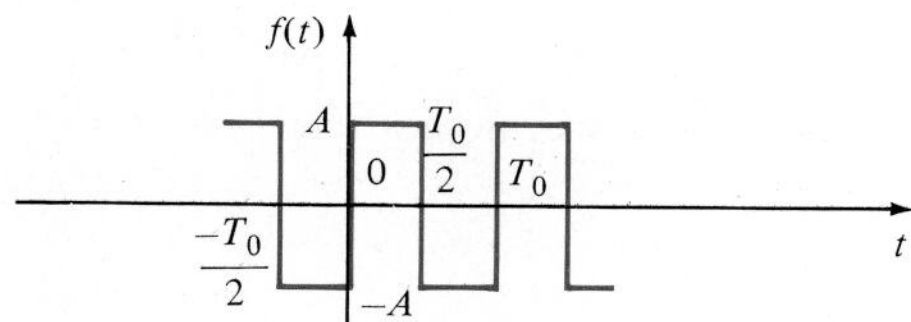	$f(t) = \sum_{\substack{n=1 \\ n \text{ odd}}}^{\infty} \frac{4A}{n\pi} \sin n\omega_0 t$
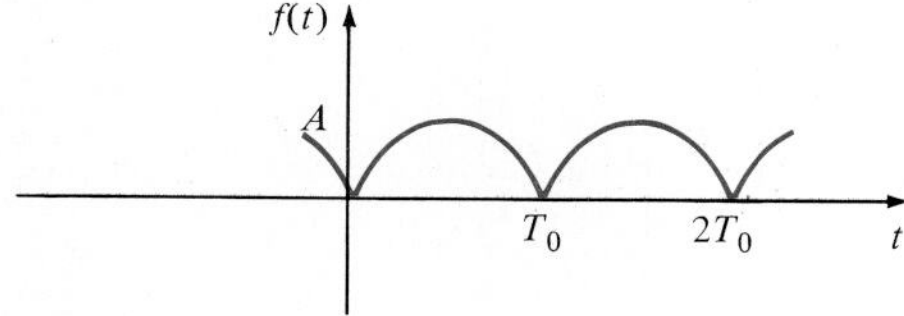	$f(t) = \frac{2A}{\pi} + \sum_{n=1}^{\infty} \frac{4A}{\pi(1 - 4n^2)} \cos n\omega_0 t$
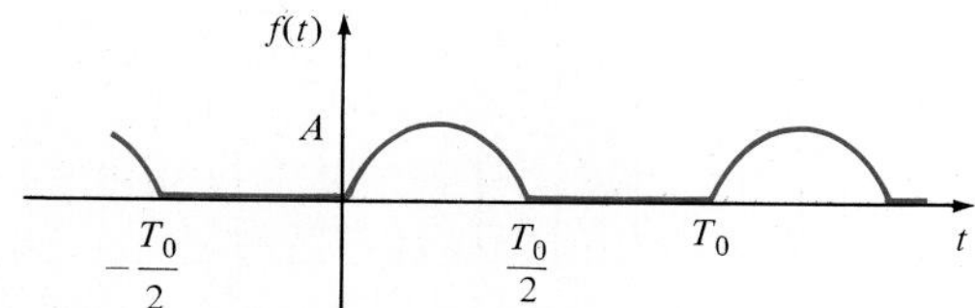	$f(t) = \frac{A}{\pi} + \frac{A}{2} \sin \omega_0 t + \sum_{\substack{n=2 \\ n \text{ even}}}^{\infty} \frac{2A}{\pi(1 - n^2)} \cos n\omega_0 t$

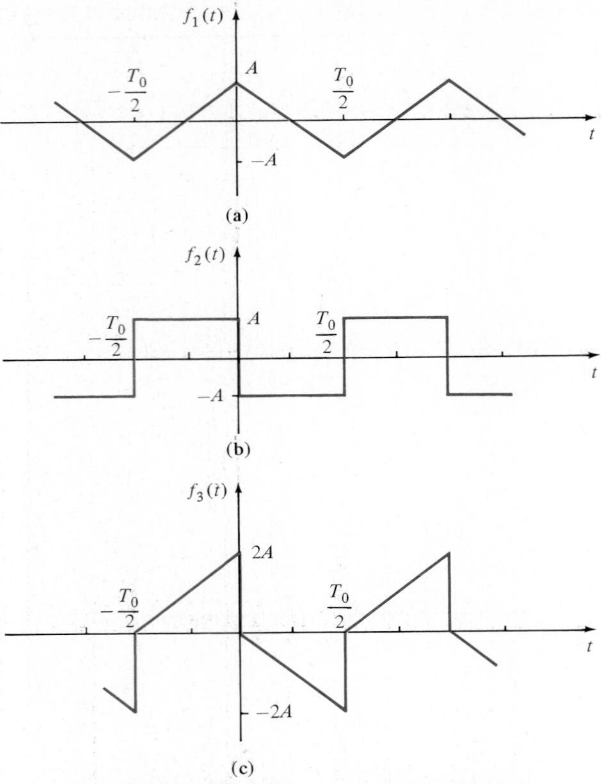

Figure 16.7 Example of waveform generation.

DRILL EXERCISE

D16.8. Two periodic waveforms are shown in Fig. D16.8. Compute the exponential Fourier series for each waveform, and then add the results to obtain the Fourier series for the waveform in Fig. D16.2.

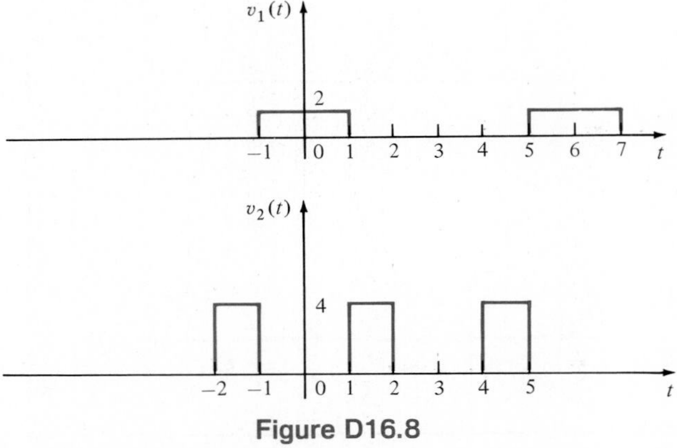

Figure D16.8

Frequency Spectrum

The *frequency spectrum* of the function $f(t)$ expressed as a Fourier series consists of a plot of the amplitude of the harmonics versus frequency, which we call the *amplitude spectrum,* and a plot of the phase of the harmonics versus frequency, which we call the *phase spectrum*. Since the frequency components are discrete, the spectra are called *line spectra*. Such spectra illustrate the frequency content of the signal. Plots of the amplitude and phase spectra are based on Eqs. (16.1), (16.3), and (16.7) and represent the amplitude and phase of the signal at specific frequencies.

EXAMPLE 16.6

The Fourier series for the triangular-type waveform shown in Fig. 16.7c with A = 5 is given by the equation

$$v(t) = \sum_{\substack{n=1 \\ n \text{ odd}}}^{\infty} \left[\frac{20}{n\pi} \sin n\omega_0 t - \frac{40}{n^2\pi^2} \cos n\omega_0 t \right]$$

We wish to plot the first four terms of the amplitude and phase spectra for this signal.

Since $D_n \underline{/\theta_n} = a_n - jb_n$, the first four terms for this signal are

$$D_1 \underline{/\theta_1} = -\frac{40}{\pi^2} - j\frac{20}{\pi} = 7.5 \underline{/-122^\circ}$$

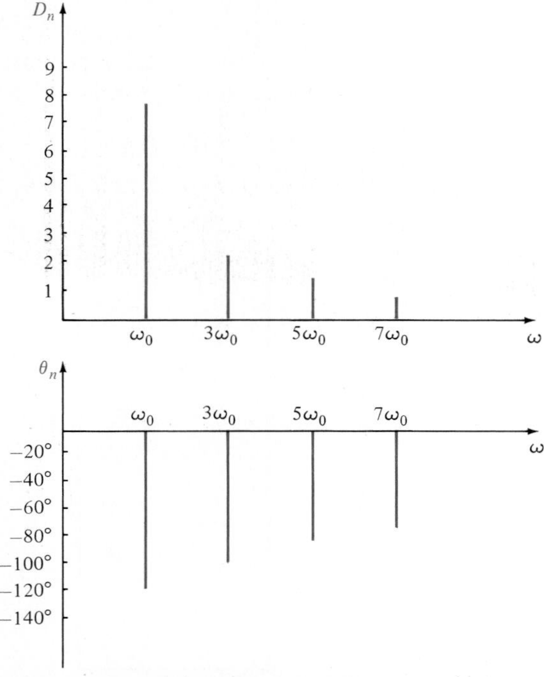

Figure 16.8 Amplitude and phase spectra.

$$D_3 \underline{/\theta_3} = -\frac{40}{9\pi^2} - j\frac{20}{3\pi} = 2.2 \underline{/-102^\circ}$$

$$D_5 \underline{/\theta_5} = -\frac{40}{25\pi^2} - j\frac{20}{5\pi} = 1.3 \underline{/-97^\circ}$$

$$D_7 \underline{/\theta_7} = -\frac{40}{49\pi^2} - j\frac{20}{7\pi} = 0.91 \underline{/-95^\circ}$$

Therefore, the plots of the amplitude and phase versus n are as shown in Fig. 16.8. ■

DRILL EXERCISE

D16.9. Determine the trigonometric Fourier series for the voltage waveform in Fig. D16.9 and plot the first four terms of the amplitude and phase spectra for this signal.

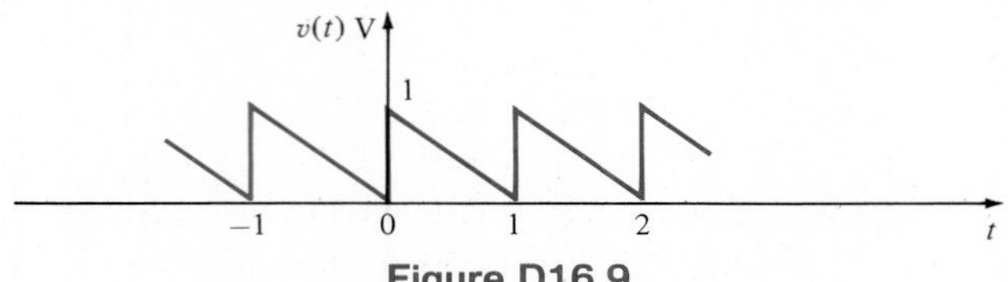

Figure D16.9

Steady-State Network Response

If a periodic signal is applied to a network, the steady-state voltage or current response at some point in the circuit can be found in the following manner. First, we represent the periodic forcing function by a Fourier series. If the input forcing function for a network is a voltage, the input can be expressed in the form

$$v(t) = v_0 + v_1(t) + v_2(t) + \cdots$$

and therefore represented in the time domain as shown in Fig. 16.9. Each source has its

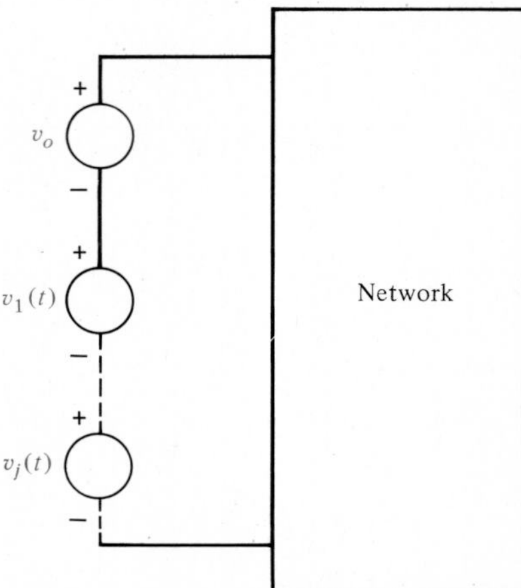

Figure 16.9 Network with a periodic voltage forcing function.

own amplitude and frequency. Next we determine the response due to each component of the input Fourier series; that is, we use phasor analysis in the frequency domain to determine the network response due to each source. The network response due to each source in the frequency domain is then transformed to the time domain. Finally, we add the time-domain solutions due to each source using the principle of superposition to obtain the Fourier series for the total *steady-state* network response.

EXAMPLE 16.7

We wish to determine the steady-state voltage $v_0(t)$ in Fig. 16.10 if the input voltage $v(t)$ is given by the expression

$$v(t) = \sum_{\substack{n=1 \\ n \text{ odd}}}^{\infty} \left[\frac{20}{n\pi} \sin 2nt - \frac{40}{n^2\pi^2} \cos 2nt \right]$$

Note that this source has no constant term, and therefore its dc value is zero. The amplitude and phase for the first four terms of this signal are given in Example 16.6, and therefore the signal $v(t)$ can be written as

$$\begin{aligned} v(t) &= 7.5 \cos (2t - 122°) + 2.2 \cos (6t - 102°) + 1.3 \cos (10t - 97°) \\ &\quad + 0.91 \cos (14t - 95°) + \cdots \end{aligned}$$

From the network we find that

$$I = \frac{\mathbf{V}}{2 + \dfrac{2/j\omega}{2 + 1/j\omega}} = \frac{\mathbf{V}(1 + 2j\omega)}{4 + 4j\omega}$$

$$\mathbf{I}_1 = \frac{\mathbf{I}(1/j\omega)}{2 + 1/j\omega} = \frac{\mathbf{I}}{1 + 2j\omega}$$

$$\mathbf{V}_0 = (1)\mathbf{I}_1 = 1 \cdot \frac{\mathbf{V}(1 + 2j\omega)}{4 + 4j\omega} \frac{1}{1 + 2j\omega} = \frac{\mathbf{V}}{4 + 4j\omega}$$

Therefore,

$$\mathbf{V}_o(\omega) = \frac{\mathbf{V}(\omega)}{4 + 4j\omega} \qquad \text{and hence} \qquad \mathbf{V}_o(n\omega_0) = \frac{\mathbf{V}(n\omega_0)}{4 + 4jn\omega_0}$$

or, since $\omega_0 = 2$,

$$\mathbf{V}_0(n\omega_0) = \frac{\mathbf{V}(n\omega_0)}{4 + j8n}$$

or

$$V_0(n) = \frac{V(n)}{4 + j8n}$$

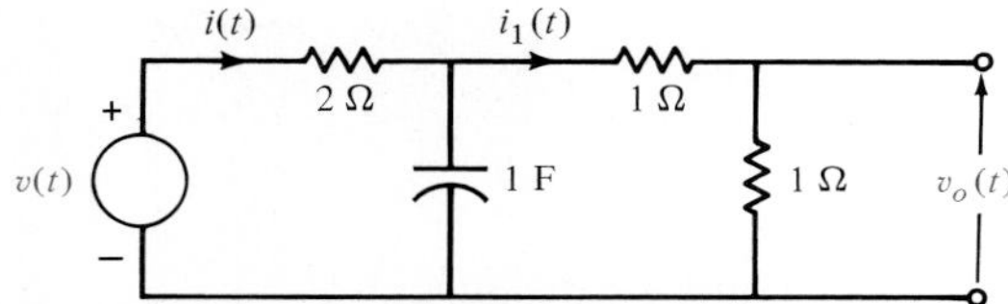

Figure 16.10 *RC* circuit employed in Example 16.7.

The individual components of the output due to the components of the input source are then

$$\mathbf{V}_0(\omega_0) = \frac{7.5\angle -122^\circ}{4 + j8} = 0.84\angle -185.4^\circ \text{ V}$$

$$\mathbf{V}_0(3\omega_0) = \frac{2.2\angle -102^\circ}{4 + j24} = 0.09\angle -182.5^\circ \text{ V}$$

$$\mathbf{V}_0(5\omega_0) = \frac{1.3\angle -97^\circ}{4 + j40} = 0.03\angle -181.3^\circ \text{ V}$$

$$\mathbf{V}_0(7\omega_0) = \frac{0.91\angle -95^\circ}{4 + j56} = 0.016\angle -181^\circ \text{ V}$$

Hence the steady-state output voltage $v_o(t)$ can be written as

$$v_o(t) = 0.84 \cos (2t - 185.4^\circ) + 0.09 \cos (6t - 182.5^\circ) + 0.03 \cos (10t - 181.3^\circ) + 0.016 \cos (14t - 181^\circ) + \cdots$$ ■

DRILL EXERCISE

D16.10 Determine the expression for the steady-state current $i(t)$ in Fig. D16.10 if the input voltge $v_s(t)$ is given by the expression

$$v_s(t) = \frac{20}{\pi} + \sum_{n=1}^{\infty} \frac{-40}{\pi(4n^2 - 1)} \cos 2nt$$

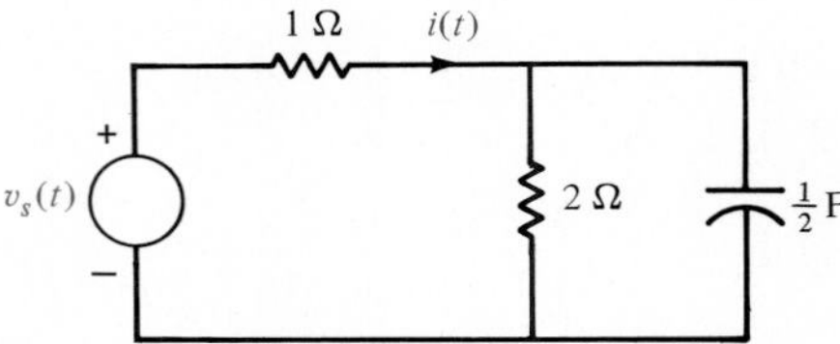

Figure D16.10

16.2 Fourier Transform

The preceding sections of this chapter have illustrated that the exponential Fourier series can be used to represent a periodic signal for all time. We will now consider a technique for representing an aperiodic signal for all values of time.

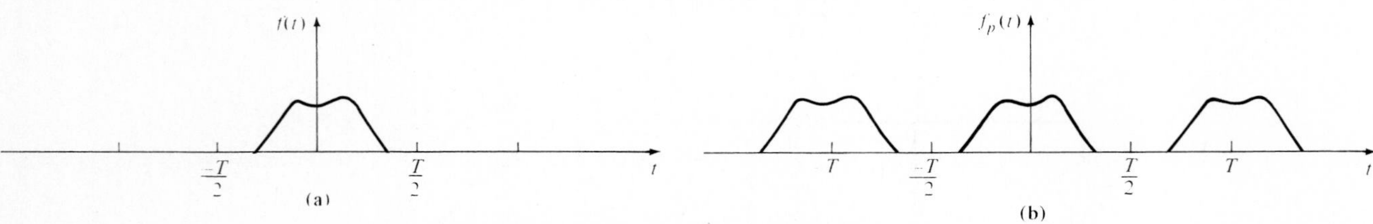

Figure 16.11 Aperiodic and periodic signals.

Suppose that an aperiodic signal $f(t)$ is as shown in Fig. 16.11a. We now construct a new signal $f_p(t)$ which is identical to $f(t)$ in the interval $-T/2$ to $T/2$ but is *periodic* with period T as shown in Fig. 16.11b. Since $f_p(t)$ is periodic, it can be represented in the interval $-\infty$ to ∞ by an exponential Fourier series.

$$f_p(t) = \sum_{n=-\infty}^{\infty} \mathbf{c}_n e^{jn\omega_0 t} \tag{16.33}$$

where

$$\mathbf{c}_n = \frac{1}{T}\int_{-T/2}^{T/2} f_p(t)e^{-jn\omega_0 t}\,dt \tag{16.34}$$

and

$$\omega_0 = \frac{2\pi}{T} \tag{16.35}$$

At this point we note that if we take the limit of the function $f_p(t)$ as $T \to \infty$, the periodic signal in Fig. 16.11b approaches the aperiodic signal in Fig. 16.11a; that is, the repetitious signals centered at $-T$ and $+T$ in Fig. 16.11b are moved to infinity.

The line spectrum for the periodic signal exists at harmonic frequencies $(n\omega_0)$ and the incremental spacing between the harmonics is

$$\Delta\omega = (n+1)\omega_0 - n\omega_0 = \omega_0 = \frac{2\pi}{T} \tag{16.36}$$

As $T \to \infty$ the lines in the frequency spectrum for $f_p(t)$ come closer and closer together, $\Delta\omega$ approaches the differential $d\omega$ and $n\omega_0$ can take on any value of ω. Under these conditions the line spectrum becomes a continuous spectrum. Since as $T \to \infty$, $\mathbf{c}_n \to 0$ in Eq. (16.34), we will examine the product $\mathbf{c}_n T$ where

$$\mathbf{c}_n T = \int_{-T/2}^{T/2} f_p(t)e^{-jn\omega_0 t}\,dt$$

In the limit as $T \to \infty$,

$$\lim_{T\to\infty} [\mathbf{c}_n T] = \lim_{T\to\infty} \int_{-T/2}^{T/2} f_p(t)e^{-jn\omega_0 t}\,dt$$

which in view of the previous discussion can be written as

$$\lim_{T\to\infty} [\mathbf{c}_n T] = \int_{-\infty}^{\infty} f(t)e^{-j\omega t}\,dt$$

This integral is the Fourier transform of $f(t)$, which we will denote as $\mathbf{F}(\omega)$, and hence

$$\mathbf{F}(\omega) = \int_{-\infty}^{\infty} f(t)e^{-j\omega t}\,dt \tag{16.37}$$

Similarly, $f_p(t)$ can be expressed as

$$f_p(t) = \sum_{n=-\infty}^{\infty} \mathbf{c}_n e^{jn\omega_0 t}$$

$$= \sum_{n=-\infty}^{\infty} (\mathbf{c}_n T) e^{jn\omega_0 t} \frac{1}{T}$$

$$= \sum_{n=-\infty}^{\infty} (\mathbf{c}_n T) e^{jn\omega_0 t} \frac{\Delta\omega}{2\pi}$$

which in the limit as $T \to \infty$ becomes

$$f(t) = \frac{1}{2\pi} \int_{-\infty}^{\infty} \mathbf{F}(\omega) e^{j\omega t}\, d\omega \tag{16.38}$$

Equations (16.37) and (16.38) constitute what is called the *Fourier transform pair*. Since $\mathbf{F}(\omega)$ is the Fourier transform of $f(t)$ and $f(t)$ is the inverse Fourier transform of $\mathbf{F}(\omega)$, they are normally expressed in the form

$$\mathbf{F}(\omega) = \mathscr{F}[f(t)] = \int_{-\infty}^{\infty} f(t) e^{-j\omega t}\, dt \tag{16.39}$$

$$f(t) = \mathscr{F}^{-1}[\mathbf{F}(\omega)] = \frac{1}{2\pi} \int_{-\infty}^{\infty} \mathbf{F}(\omega) e^{j\omega t}\, d\omega \tag{16.40}$$

Some Important Transform Pairs

There are a number of important Fourier transform pairs and in the following material we will derive a number of them and then list some of the more common ones in tabular form.

EXAMPLE 16.8

We wish to derive the Fourier transform for the voltage pulse shown in Fig. 16.12a.

Using Eq. (16.39), the Fourier transform is

$$\mathbf{F}(\omega) = \int_{-\delta/2}^{\delta/2} V e^{-j\omega t}\, dt$$

$$= \frac{V}{-j\omega} e^{-j\omega t} \Big|_{-\delta/2}^{\delta/2}$$

$$= V \frac{e^{-j\omega\delta/2} - e^{+j\omega\delta/2}}{-j\omega}$$

$$= V\delta \frac{\sin(\omega\delta/2)}{\omega\delta/2}$$

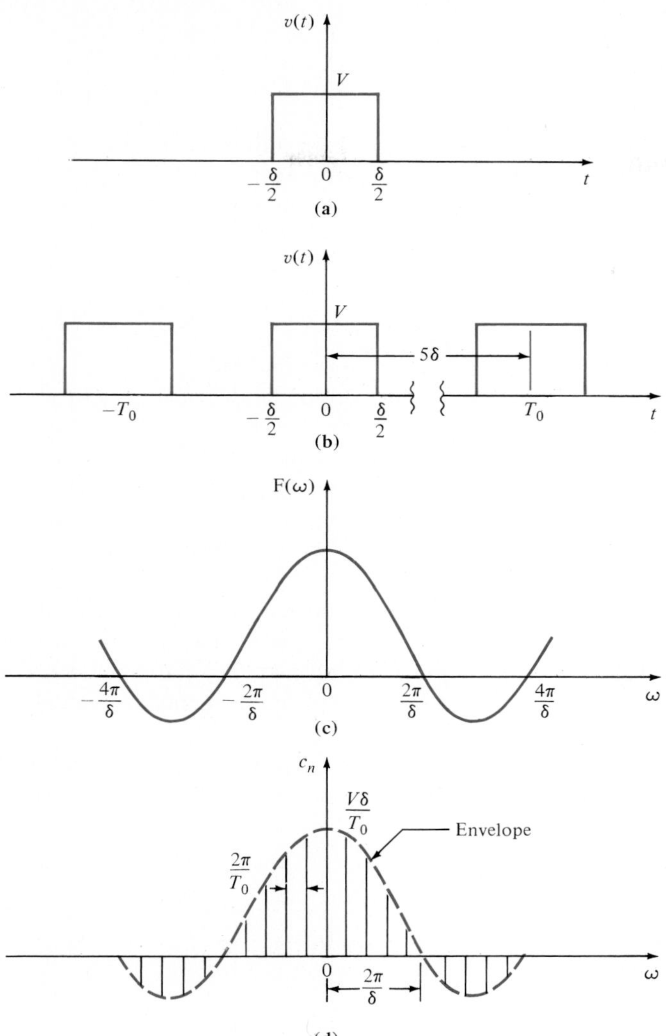

Figure 16.12 Pulses and their spectrums.

Therefore, the Fourier transform for the function

$$f(t) = \begin{cases} 0 & -\infty < t \leq -\dfrac{\delta}{2} \\ V & -\dfrac{\delta}{2} < t \leq \dfrac{\delta}{2} \\ 0 & \dfrac{\delta}{2} < t < \infty \end{cases}$$

is

$$\mathbf{F}(\omega) = V\delta \frac{\sin(\omega\delta/2)}{\omega\delta/2}$$

A plot of this function is shown in Fig. 16.12c. Let us explore this example even further. Consider now the pulse train shown in Fig. 16.12b. Using the techniques that have been demonstrated earlier, we can show that the Fourier coefficients for this waveform are

$$\mathbf{c}_n = \frac{V\delta}{T_0} \frac{\sin(n\omega_0\delta/2)}{n\omega_0\delta/2}$$

The line spectrum for $T_0 = 5\delta$ is shown in Fig. 16.12d.

What these equations and figures in this example indicate are: As $T_0 \to \infty$ and the periodic function becomes aperiodic, the lines in the discrete spectrum become denser and the amplitude gets smaller, and the amplitude spectrum changes from a line spectrum to a continuous spectrum. Note that the envelope for the discrete spectrum has the same shape as the continuous spectrum. Since the Fourier series represents the amplitude and phase of the signal at specific frequencies, the Fourier transform also specifies the frequency content of a signal. ■

There are two singularity functions that are very important in circuit analysis: (1) the unit step function, $u(t)$, discussed in Chapter 7, and (2) the unit impulse or delta function, $\delta(t)$. They are called *singularity functions* because they are either not finite or they do not possess finite derivatives everywhere. They are mathematical models for signals that we have and will employ in circuit analysis.

The unit impulse function can be represented in the limit by the rectangular pulse shown in Fig. 16.13a as $a \to 0$. The function is defined by the following:

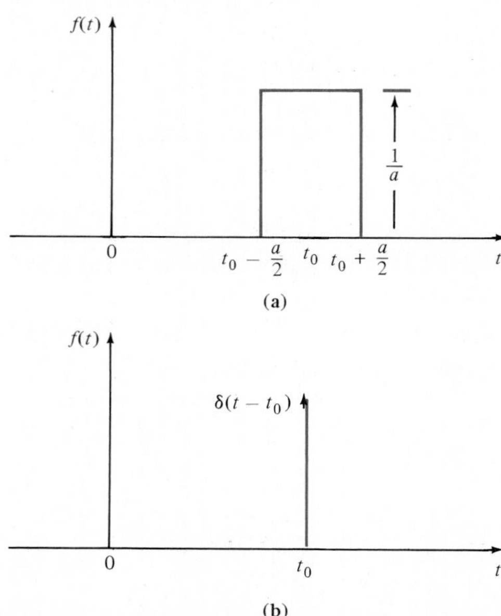

Figure 16.13 Representations of the unit impulse.

$$\delta(t - t_0) = 0 \qquad t \neq t_0 \tag{16.41}$$

$$\int_{t_0-\epsilon}^{t_0+\epsilon} \delta(t - t_0)\, dt = 1 \qquad \epsilon > 0 \tag{16.42}$$

The unit impulse is zero except at $t = t_0$, where it is undefined, but it has unit area (sometimes referred to as *strength*). We represent the unit impulse function on a plot as shown in Fig. 16.13b.

An important property of the unit impulse function is what is often called the *sampling property,* which is exhibited by the following integral:

$$\int_{t_1}^{t_2} f(t)\,\delta(t - t_0)\, dt = \begin{cases} f(t_0) & t_1 < t_0 < t_2 \\ 0 & t_0 < t_1, \quad t_0 > t_2 \end{cases} \tag{16.43}$$

for a finite t_0 and any $f(t)$ continuous at t_0. Note that the unit impulse function simply samples the value of $f(t)$ at $t = t_0$.

EXAMPLE 16.9

The Fourier transform of the unit impulse function $\delta(t - a)$ is

$$\mathbf{F}(\omega) = \int_{-\infty}^{\infty} \delta(t - a)e^{-j\omega t}\, dt$$

Using the sampling property of the unit impulse, we find that

$$\mathbf{F}(\omega) = e^{-j\omega a}$$

and if $a = 0$, then

$$\mathbf{F}(\omega) = 1$$

Note then that the $\mathbf{F}(\omega)$ for $f(t) = \delta(t)$ is *constant for all frequencies*. This is an important property, as we shall see later. ■

EXAMPLE 16.10

We wish to determine the Fourier transform of the function $f(t) = e^{j\omega_0 t}$. In this case we note that if $\mathbf{F}(\omega) = 2\pi\delta(\omega - \omega_0)$, then

$$\begin{aligned} f(t) &= \frac{1}{2\pi}\int_{-\infty}^{\infty} 2\pi\delta(\omega - \omega_0)e^{j\omega t}\, d\omega \\ &= e^{j\omega_0 t} \end{aligned}$$

Therefore, $f(t) = e^{j\omega_0 t}$ and $\mathbf{F}(\omega) = 2\pi\delta(\omega - \omega_0)$ represent a Fourier transform pair. ■

EXAMPLE 16.11

The Fourier transform of the function $f(t) = \cos \omega_0 t$ is

$$\begin{aligned} \mathbf{F}(\omega) &= \int_{-\infty}^{\infty} \cos \omega_0 t e^{-j\omega t}\, dt \\ &= \frac{1}{2}\int_{-\infty}^{\infty} e^{j\omega_0 t}e^{-j\omega t}\, dt + \frac{1}{2}\int_{-\infty}^{\infty} e^{-j\omega_0 t}e^{-j\omega t}\, dt \end{aligned}$$

Using the results of Example 16.10, we obtain

$$\mathbf{F}(\omega) = \pi\delta(\omega - \omega_0) + \pi\delta(\omega + \omega_0)$$

■

EXAMPLE 16.12

Let us determine the Fourier transform of the function $f(t) = e^{-at}u(t)$. The Fourier transform of this function is

$$\begin{aligned}\mathbf{F}(\omega) &= \int_{-\infty}^{\infty} e^{-at}u(t)e^{-j\omega t}\,dt \\ &= \int_{0}^{\infty} e^{-at}e^{-j\omega t}\,dt \\ &= \frac{-1}{a + j\omega}\,e^{-at}e^{-j\omega t}\Big|_0^{\infty} \\ &= \frac{1}{a + j\omega}\end{aligned}$$

for $a > 0$ since $e^{-at} \to 0$ as $t \to \infty$.

■

A number of useful Fourier transform pairs are shown in Table 16.2.

DRILL EXERCISE

D16.11. If $f(t) = \sin \omega_0 t$, find $\mathbf{F}(\omega)$.

Table 16.2 Fourier Transform Pairs

$f(t)$	$\mathbf{F}(\omega)$
$\delta(t)$	1
$\delta(t - a)$	$e^{-j\omega a}$
A	$2\pi A\,\delta(\omega)$
$e^{j\omega_0 t}$	$2\pi\delta(\omega - \omega_0)$
$\cos \omega_0 t$	$\pi\delta(\omega - \omega_0) + \pi\delta(\omega + \omega_0)$
$\sin \omega_0 t$	$j\pi\delta(\omega + \omega_0) - j\pi\delta(\omega - \omega_0)$
$e^{-at}u(t), \quad a > 0$	$\dfrac{1}{a + j\omega}$
$e^{-a\lvert t\rvert}, \quad a > 0$	$\dfrac{2a}{a^2 + \omega^2}$
$e^{-at} \cos \omega_0 t u(t), \quad a > 0$	$\dfrac{j\omega + a}{(j\omega + a)^2 + \omega_0^2}$
$e^{-at} \sin \omega_0 t u(t), \quad a > 0$	$\dfrac{\omega_0}{(j\omega + a)^2 + \omega_0^2}$

Some Properties of the Fourier Transform

Let us examine now some of the properties of the Fourier transform defined by the equation

$$\mathbf{F}(\omega) = \int_{-\infty}^{\infty} f(t)e^{-j\omega t}\, dt$$

Using Euler's identity, we can write this function as

$$\mathbf{F}(\omega) = A(\omega) + jB(\omega) \tag{16.44}$$

where

$$A(\omega) = \int_{-\infty}^{\infty} f(t) \cos \omega t\, dt \tag{16.45}$$

$$B(\omega) = -\int_{-\infty}^{\infty} f(t) \sin \omega t\, dt \tag{16.46}$$

From Eq. (16.44) we note that

$$\mathbf{F}(\omega) = |\mathbf{F}(\omega)|e^{j\theta(\omega)} \tag{16.47}$$

where

$$|\mathbf{F}(\omega)| = \sqrt{A^2(\omega) + B^2(\omega)} \tag{16.48}$$

$$\theta(\omega) = \tan^{-1} \frac{B(\omega)}{A(\omega)} \tag{16.49}$$

Given the definitions above, we can show that

$$\begin{aligned} A(\omega) &= A(-\omega) \\ B(\omega) &= -B(-\omega) \\ \mathbf{F}(-\omega) &= \mathbf{F}^*(\omega) \end{aligned} \tag{16.50}$$

Equations (16.50) simply state that $A(\omega)$ is an even function of ω, $B(\omega)$ is an odd function of ω, and $\mathbf{F}(\omega)$ evaluated at $-\omega$ is the complex conjugate of $\mathbf{F}(\omega)$. Therefore, $|\mathbf{F}(\omega)|$ is an even function of ω and $\theta(\omega)$ is an odd function of ω. Some additional properties of the Fourier transform are now presented in rapid succession.

Linearity. If $\mathcal{F}[f_1(t)] = \mathbf{F}_1(\omega)$ and $\mathcal{F}[f_2(t)] = \mathbf{F}_2(\omega)$, then

$$\begin{aligned} \mathcal{F}[af_1(t) + bf_2(t)] &= \int_{-\infty}^{\infty} [af_1(t) + bf_2(t)]e^{-j\omega t}\, dt \\ &= \int_{-\infty}^{\infty} af_1(t)e^{-j\omega t}\, dt + \int_{-\infty}^{\infty} bf_2(t)e^{-j\omega t}\, dt \\ &= a\mathbf{F}_1(\omega) + b\mathbf{F}_2(\omega) \end{aligned} \tag{16.51}$$

Time Scaling. If $\mathcal{F}[f(t)] = \mathbf{F}(\omega)$, then if $a > 0$,

$$\mathcal{F}[f(at)] = \int_{-\infty}^{\infty} f(at)e^{-j\omega t}\, dt$$

If we now let $x = at$, then

$$\begin{aligned}\mathcal{F}[f(at)] &= \int_{-\infty}^{\infty} f(x)e^{-j\omega x/a}\,\frac{dx}{a} \\ &= \frac{1}{a}\mathbf{F}\left(\frac{\omega}{a}\right)\end{aligned} \tag{16.52}$$

and if $a < 0$, then following the previous development,

$$\begin{aligned}\mathcal{F}[f(at)] &= \int_{+\infty}^{-\infty} f(x)e^{-j\omega x/a}\,\frac{dx}{a} \\ &= -\frac{1}{a}\mathbf{F}\left(\frac{\omega}{a}\right)\end{aligned} \tag{16.53}$$

Time Shifting. If $\mathcal{F}[f(t)] = \mathbf{F}(\omega)$, then

$$\mathcal{F}[f(t - t_0) = \int_{-\infty}^{\infty} f(t - t_0)e^{-j\omega t}\, dt$$

If we now let $x = t - t_0$, then

$$\begin{aligned}\mathcal{F}[f(t - t_0)] &= \int_{-\infty}^{\infty} f(x)e^{-j\omega(x + t_0)}\, dx \\ &= e^{-j\omega t_0}\,\mathbf{F}(\omega)\end{aligned} \tag{16.54}$$

Modulation. If $\mathcal{F}[f(t)] = \mathbf{F}(\omega)$, then

$$\begin{aligned}\mathcal{F}[e^{j\omega_0 t}f(t)] &= \int_{-\infty}^{\infty} f(t)e^{j\omega_0 t}e^{-j\omega t}\, dt \\ &= \int_{-\infty}^{\infty} f(t)e^{-j(\omega - \omega_0)t}\, dt \\ &= \mathbf{F}(\omega - \omega_0)\end{aligned} \tag{16.55}$$

Differentiation. If $f(t)$ is a function that is continuous in any finite interval, is absolutely integrable, and its first derivative is piecewise continuous and absolutely integrable, then

$$\mathcal{F}[f'(t)] = \int_{-\infty}^{\infty} f'(t)e^{-j\omega t}\, dt$$

Using integration by parts, we obtain

$$\mathscr{F}[f'(t)] = f(t)e^{-j\omega t}\Big|_{-\infty}^{\infty} + \int_{-\infty}^{\infty} j\omega f(t)e^{-j\omega t}\,dt$$

since $f(t)$ is absolutely integrable, $\lim_{t \to -\infty} |f(t)| = \lim_{t \to +\infty} |f(t)| = 0$, and therefore

$$\mathscr{F}[f'(t)] = j\omega \mathbf{F}(\omega) \tag{16.56}$$

Repeated application of this technique under the conditions listed above yields

$$\mathscr{F}[f^{(n)}(t)] = (j\omega)^n \mathbf{F}(\omega) \tag{16.57}$$

It is important to note that equation (16.56) does not guarantee the existence of $\mathbf{F}[f'(t)]$. It simply states that *if* the transform exists, it is given by Eq. (16.56).

Time Convolution. If $\mathscr{F}[f_1(t)] = \mathbf{F}_1(\omega)$ and $\mathscr{F}[f_2(t)] = \mathbf{F}_2(\omega)$, then

$$\begin{aligned}\mathscr{F}\left[\int_{-\infty}^{\infty} f_1(x)f_2(t-x)\,dx\right] &= \int_{t=-\infty}^{\infty}\int_{x=-\infty}^{\infty} f_1(x)f_2(t-x)\,dx\, e^{-j\omega t}\,dt \\ &= \int_{x=-\infty}^{\infty} f_1(x)\int_{t=-\infty}^{\infty} f_2(t-x)e^{-j\omega t}\,dt\,dx\end{aligned}$$

If we now let $u = t - x$, then

$$\begin{aligned}\mathscr{F}\left[\int_{-\infty}^{\infty} f_1(x)f_2(t-x)\,dx\right] &= \int_{x=-\infty}^{\infty} f_1(x)\int_{u=-\infty}^{\infty} f_2(u)e^{-j\omega(u+x)}\,du\,dx \\ &= \int_{x=-\infty}^{\infty} f_1(x)e^{-j\omega x}\int_{u=-\infty}^{\infty} f_2(u)e^{-j\omega u}\,du\,dx \\ &= \mathbf{F}_1(\omega)\mathbf{F}_2(\omega)\end{aligned} \tag{16.58}$$

We should note very carefully the time convolution property of the Fourier transform. With reference to Fig. 16.14, this property states that if $\mathbf{V}_i(j\omega) = \mathscr{F}[v_i(t)]$, $\mathbf{H}(j\omega) = \mathscr{F}[h(t)]$, and $\mathbf{V}_o(j\omega) = \mathscr{F}[v_o(t)]$, then

$$\mathbf{V}_o(j\omega) = \mathbf{H}(j\omega)\mathbf{V}_i(j\omega) \tag{16.59}$$

where $\mathbf{V}_i(j\omega)$ represents the input signal, $\mathbf{H}(j\omega)$ is the network transfer function, and $\mathbf{V}_o(j\omega)$ represents the output signal. Equation (16.59) tacitly assumes that the initial conditions of the network are zero.

Table 16.3 provides a short list of some of the Fourier transform properties.

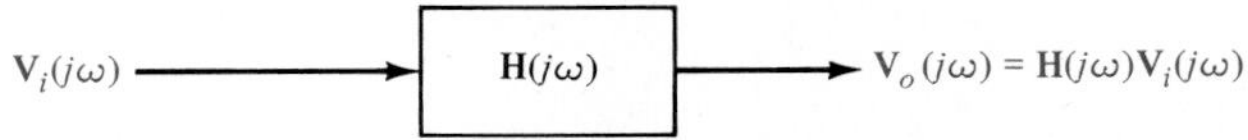

Figure 16.14 Representation of the time convolution property.

Table 16.3 Properties of the Fourier Transform

$f(t)$	$\mathbf{F}(\omega)$
$Af(t)$	$A\mathbf{F}(\omega)$
$f_1(t) \pm f_2(t)$	$\mathbf{F}_1(\omega) \pm \mathbf{F}_2(\omega)$
$f(at)$	$\frac{1}{a}\mathbf{F}\left(\frac{\omega}{a}\right),\ a > 0$
$f(t - t_0)$	$e^{-j\omega t_0}\mathbf{F}(\omega)$
$e^{j\omega t_0}f(t)$	$F(\omega - \omega_0)$
$\frac{d^n f(t)}{dt^n}$	$(j\omega)^n\mathbf{F}(\omega)$
$t^n f(t)$	$(j)^n \frac{d^n\mathbf{F}(\omega)}{d\omega^n}$
$f(t)\cos\omega_0 t$	$\frac{1}{2}[\mathbf{F}(\omega - \omega_0) + \mathbf{F}(\omega + \omega_0)]$
$\int_{-\infty}^{\infty} f_1(x)f_2(t - x)\,dx$	$\mathbf{F}_1(\omega)\mathbf{F}_2(\omega)$
$f_1(t)f_2(t)$	$\frac{1}{2\pi}\int_{-\infty}^{\infty} \mathbf{F}_1(x)\mathbf{F}_2(\omega - x)\,dx$

EXAMPLE 16.13

Let us determine the Fourier transform of the function $f(t - t_0) = e^{-(t - t_0)}u(t - t_0)$.

By definition,

$$\mathscr{F}[f(t - t_0)] = \int_{-\infty}^{\infty} f(t - t_0)e^{-j\omega t}\,dt$$

which can be written using the time-shifting property as

$$\begin{aligned}\mathscr{F}[f(t - t_0)] &= e^{-j\omega t_0}\int_{-\infty}^{\infty} f(t)e^{-j\omega t}\,dt \\ &= e^{-j\omega t_0}\int_{-\infty}^{\infty} e^{-t}u(t)e^{-j\omega t}\,dt\end{aligned}$$

which from the results of Example 16.12 is

$$\mathscr{F}[f(t - t_0)] = \frac{e^{-j\omega t_0}}{1 + j\omega}$$

■

EXAMPLE 16.14

Let us determine the Fourier transform of the function

$$f(t) = \frac{d}{dt}[e^{-at}u(t)] = -ae^{-at}u(t) + e^{-at}\,\delta(t)$$

then

$$\begin{aligned}\mathscr{F}[f(t)] &= \int_{-\infty}^{\infty} [-ae^{-at}u(t) + e^{-at}\,\delta(t)]e^{-j\omega t}\,dt \\ &= -\int_{-\infty}^{\infty} ae^{-at}u(t)e^{-j\omega t}\,dt + \int_{-\infty}^{\infty} e^{-at}\,\delta(t)e^{-j\omega t}\,dt \\ &= \frac{-a}{a + j\omega} + 1 \\ &= \frac{j\omega}{a + j\omega}\end{aligned}$$

The transform could also be evaluated using the differentiation property; that is, if

$$\mathscr{F}[e^{-at}u(t)] = \frac{1}{a + j\omega}$$

then

$$\mathscr{F}\left[\frac{d}{dt}\left(e^{-at}u(t)\right)\right] = j\omega\left(\frac{1}{a + j\omega}\right) = \frac{j\omega}{a + j\omega}$$ ■

DRILL EXERCISE

D16.12. If $\mathscr{F}[e^{-t}u(t)] = 1/(1 + j\omega)$, use the time-scaling property of the Fourier transform to find $\mathscr{F}[e^{-at}u(t)]$.

D16.13. Use the property $\mathscr{F}[t^n f(t)] = j^n[d^n\mathbf{F}(\omega)/d\omega^n]$ to determine the Fourier transform of $te^{-at}u(t)$.

D16.14. Determine the output $v_o(t)$ in Fig. D16.13 if the input signal $v_i(t) = e^{-t}u(t)$, the network transfer function $h(t) = e^{-2t}u(t)$, and all initial conditions are zero.

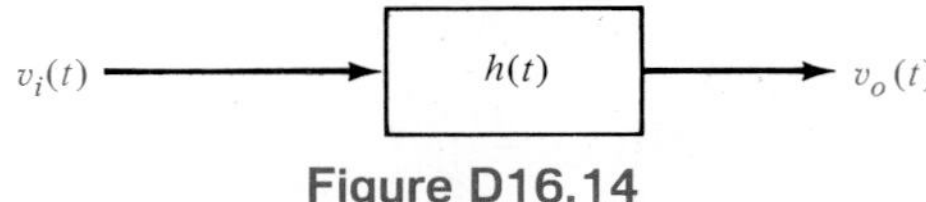

Figure D16.14

Parseval's Theorem

A mathematical statement of Parseval's theorem is

$$\int_{-\infty}^{\infty} f^2(t)\,dt = \frac{1}{2\pi}\int_{-\infty}^{\infty} |\mathbf{F}(\omega)|^2\,d\omega \qquad (16.60)$$

This relationship can be easily derived as follows:

$$\int_{-\infty}^{\infty} f^2(t)\,dt = \int_{-\infty}^{\infty} f(t)\frac{1}{2\pi}\int_{-\infty}^{\infty} \mathbf{F}(\omega)e^{j\omega t}\,d\omega\,dt$$

$$= \frac{1}{2\pi}\int_{-\infty}^{\infty} \mathbf{F}(\omega)\int_{-\infty}^{\infty} f(t)e^{-j(-\omega)t}\,dt\,d\omega$$

$$= \frac{1}{2\pi}\int_{-\infty}^{\infty} \mathbf{F}(\omega)\mathbf{F}(-\omega)\,d\omega$$

$$= \frac{1}{2\pi}\int_{-\infty}^{\infty} \mathbf{F}(\omega)\mathbf{F}^*(\omega)\,d\omega$$

$$= \frac{1}{2\pi}\int_{\infty}^{\infty} |\mathbf{F}(\omega)|^2\,d\omega$$

The importance of Parseval's theorem can be seen if we imagine that $f(t)$ represents the current in a 1-Ω resistor. Since $f^2(t)$ is power and the integral of power over time is energy, Eq. (16.60) shows that we can compute this 1-Ω energy or normalized energy in either the time domain or the frequency domain.

Applications

Let us now apply to circuit problems some of the things we have learned about the Fourier transform.

EXAMPLE 16.15

Using the transform technique, we wish to determine $v_0(t)$ in Fig. 16.15 if (a) $v_i(t) = 5e^{-2t}u(t)$ V, and (b) $v_i(t) = 5\cos 2t$ V.

(a) In this case since $v_i(t) = 5e^{-2t}u(t)$, V then

$$\mathbf{V}_i(\omega) = \frac{5}{2 + j\omega}\ \text{V}$$

$\mathbf{H}(\omega)$ for the network is

$$\mathbf{H}(\omega) = \frac{R}{R + j\omega L}$$

$$= \frac{10}{10 + j\omega}$$

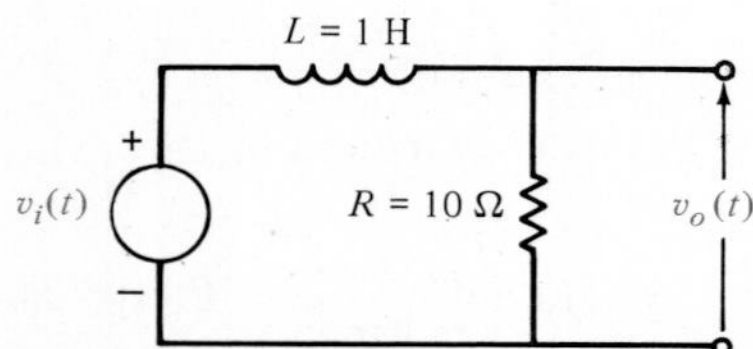

Figure 16.15 Simple *RL* circuit.

From Eq. (16.59).

$$\begin{aligned}\mathbf{V}_o(\omega) &= \mathbf{H}(\omega)\mathbf{V}_i(\omega)\\ &= \frac{50}{(2 + j\omega)(10 + j\omega)}\\ &= \frac{50}{8}\left(\frac{1}{2 + j\omega} - \frac{1}{10 + j\omega}\right) \text{ V}\end{aligned}$$

Hence, from Table 16.2, we see that

$$v_o(t) = 6.25[e^{-2t}u(t) - e^{-10t}u(t)] \text{ V}$$

(b) In this case, since $v_i(t) = 5 \cos 2t$

$$\mathbf{V}_i(\omega) = 5\pi\delta(\omega - 2) + 5\pi\delta(\omega + 2) \text{ V}$$

The output voltage in the frequency domain is then

$$\mathbf{V}_{a\to 0}(\omega) = \frac{50\pi[\delta(\omega - 2) + \delta(\omega + 2)]}{(10 + j\omega)}$$

Using the inverse Fourier transform gives us

$$v_o(t) = \mathcal{F}^{-1}[\mathbf{V}_o(\omega)] = \frac{1}{2\pi}\int_{-\infty}^{\infty} 50\pi \frac{\delta(\omega - 2) + \delta(\omega + 2)}{10 + j\omega} e^{j\omega t}\, d\omega$$

Employing the sampling property of the unit impulse function, we obtain

$$\begin{aligned}v_o(t) &= 25\left(\frac{e^{j2t}}{10 + j2} + \frac{e^{-j2t}}{10 - j2}\right)\\ &= 25\left(\frac{e^{j2t}}{10.2e^{j11.31^\circ}} + \frac{e^{-j2t}}{10.2e^{-j11.31^\circ}}\right)\\ &= 4.90 \cos (2t - 11.31^\circ) \text{ V}\end{aligned}$$

This result can be easily checked using phasor analysis. ■

Consider the network shown in Fig. 16.16a. This network represents a simple low-pass filter as shown in Chapter 13. We wish to illustrate the impact of this network on the input signal by examining the frequency characteristics of the output signal and the relationship between the 1-Ω or normalized energy at the input and output of the network.

The network transfer function is

$$\mathbf{H}(\omega) = \frac{1/RC}{1/RC + j\omega} = \frac{5}{5 + j\omega} = \frac{1}{1 + 0.2j\omega}$$

The Fourier transform of the input signal is

$$\mathbf{V}_i(\omega) = \frac{20}{20 + j\omega} = \frac{1}{1 + 0.05j\omega}$$

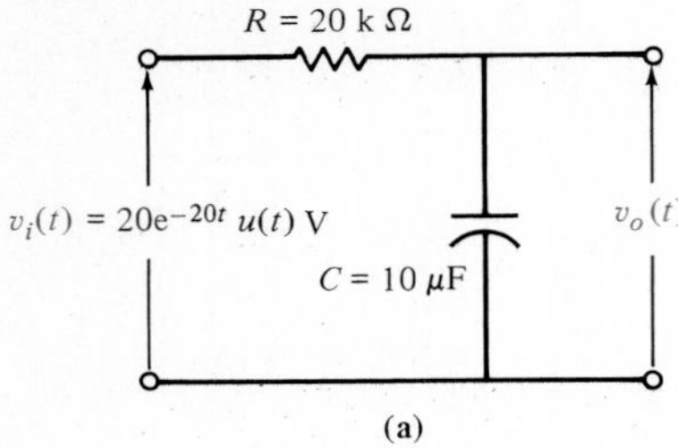

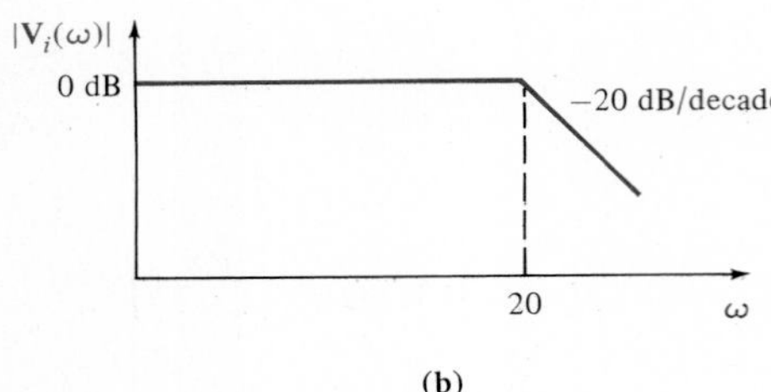

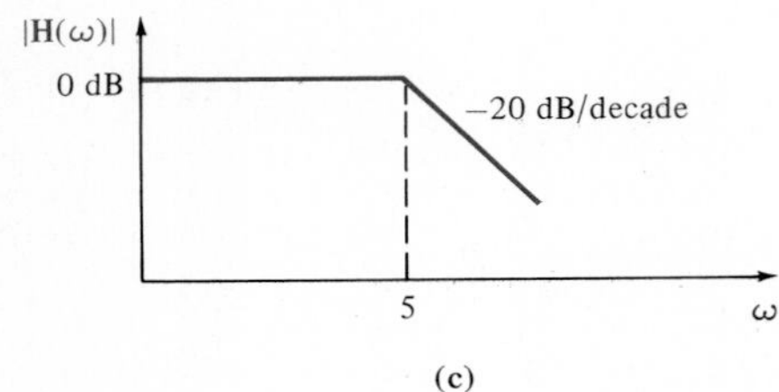

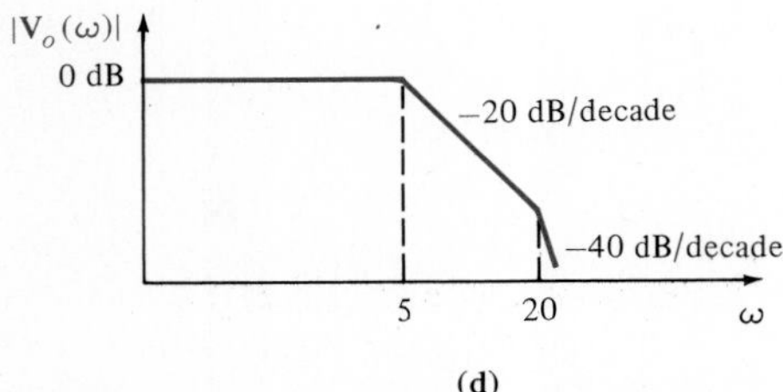

Figure 16.16 Low-pass filter, its frequency characteristic, and its input and output spectrums.

Then, using Eq. (16.59), the Fourier transform of the output is

$$\mathbf{V}_o(\omega) = \frac{1}{(1 + 0.2j\omega)(1 + 0.05j\omega)}$$

Using the techniques of Chapter 13, we note that the straight-line log-magnitude plot (frequency characteristic) for these functions is shown in Fig. 16.16b–d. Note that the low-pass filter passes the low frequencies of the input signal but attenuates the high frequencies.

The normalized energy at the filter input is

$$W_i = \int_0^\infty (20e^{-20t})^2\, dt$$
$$= \frac{400}{-40} e^{-40t}\Big|_0^\infty$$
$$= 10 \text{ J}$$

The normalized energy at the filter output can be computed using Parseval's theorem. Since

$$\mathbf{V}_o(\omega) = \frac{100}{(5 + j\omega)(20 + j\omega)}$$

$$|\mathbf{V}_o(\omega)|^2 = \frac{10^4}{(\omega^2 + 25)(\omega^2 + 400)}$$

Since $|\mathbf{V}_0(\omega)|^2$ is an even function,

$$W_o = 2\left(\frac{1}{2\pi}\right)\int_0^\infty \frac{10^4\, d\omega}{(\omega^2 + 25)(\omega^2 + 400)}$$

However, we can use the fact that

$$\frac{10^4}{(\omega^2 + 25)(\omega^2 + 400)} = \frac{10^4/375}{\omega^2 + 25} - \frac{10^4/375}{\omega^2 + 400}$$

Then

$$W_o = \frac{1}{\pi}\left(\int_0^\infty \frac{10^4/375}{\omega^2 + 25}\, d\omega - \int_0^\infty \frac{10^4/375}{\omega^2 + 400}\, d\omega\right)$$
$$= \frac{10^4}{375}\left(\frac{1}{\pi}\right)\left[\frac{1}{5}\left(\frac{\pi}{2}\right) - \frac{1}{20}\left(\frac{\pi}{2}\right)\right]$$
$$= 2.0 \text{ J}$$

■

Example 16.16 illustrates the effect that $\mathbf{H}(\omega)$ has on the frequency spectrum of the input signal. In general, $\mathbf{H}(\omega)$ can be selected to shape that spectrum in some prescribed manner. As an illustration of this effect, consider the *ideal* frequency spectrums shown in Fig. 16.17. In Fig. 16.17a is shown an ideal input magnitude spectrum $|\mathbf{V}_i(\omega)|$. $|\mathbf{H}(\omega)|$ and the output magnitude spectrum $|\mathbf{V}_o(\omega)|$, which are related by Eq. (16.59), are shown in Fig. 16.17b–e for an *ideal* low-pass, high-pass, bandpass, and band-elimination filter, respectively.

DRILL EXERCISE

D16.15. Compute the total 1-Ω energy content of the signal $v_i(t) = e^{-2t}u(t)$ using both the time-domain and frequency-domain approaches.

D16.16. Compute the 1-Ω energy content of the signal $v_1(t) = e^{-2t}u(t)$ in the frequency range from 0 to 1 rad/s.

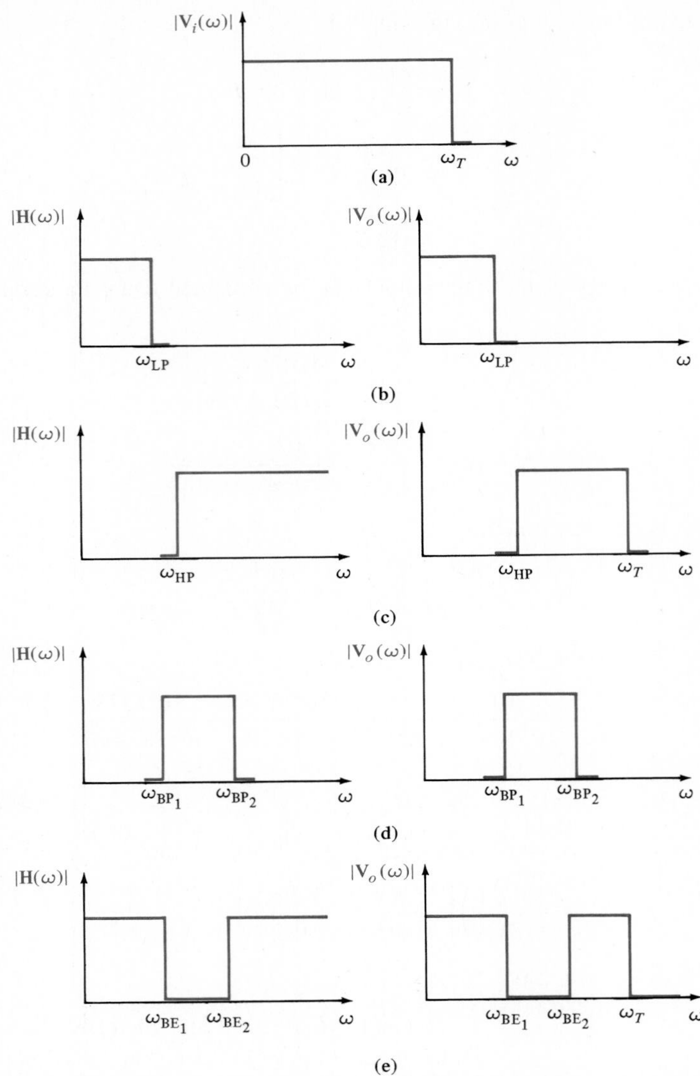

Figure 16.17 Frequency spectrums for the input and output of ideal low-pass, high-pass, bandpass, and band-elimination filters.

We note that by using Parseval's theorem we can compute the total energy content of a signal using either a time-domain or frequency-domain approach. However, the frequency domain is more flexible in that it permits us to determine the energy content of a signal within some specified frequency band.

16.3 Summary

We have shown that the Fourier series technique provides us with an effective means to deal with electric circuits which are forced with inputs that are periodic. The Fourier series for a particular signal can be expressed in several different equivalent forms. If the periodic signal displays some form of symmetry, this symmetry can be used to simplify the calculations involved in determining the Fourier coefficients employed to describe the waveform. We found that the steady-state response of a network to a periodic input involved expressing the input as a Fourier series, determining the circuit response to each component of the input, and summing the response due to each input component to determine the total network response.

The Fourier transform provides us with a technique for analyzing circuits with aperiodic forcing functions. A number of Fourier transform pairs have been presented together with some important properties of the Fourier transform. Two important relationships from a circuit analysis point of view are Parseval's theorem and the equation $\mathbf{Y}(\omega) = \mathbf{H}(\omega)\mathbf{X}(\omega)$, where $\mathbf{Y}(\omega)$ and $\mathbf{X}(\omega)$ represent the Fourier transform of the circuit output and input, respectively, and $\mathbf{H}(\omega)$ is the circuit transfer function. We have shown that Parseval's theorem allows us to compute the energy in a signal in the frequency domain and even determine the amount of signal energy that is contained within a certain range of frequencies. The input/output relationship in the frequency domain was shown to be an effective means of computing the circuit response to aperiodic inputs.

KEY POINTS

- A periodic function satisfies the relationship $f(t) = f(t + nT_0)$, $n = \pm 1, \pm 2, \pm 3, \ldots$
- A nonsinusoidal periodic function can be represented by a sum of sinusoidal functions.
- A Fourier series can be expressed in exponential or trigonometric form.
- A periodic function that exhibits even-function symmetry can be expressed by a Fourier series containing only a constant term and cosine terms.
- A periodic function that exhibits odd-function symmetry can be expressed by a Fourier series containing only sine terms.
- A time shift of a periodic signal in the time domain corresponds to a phase shift in the frequency domain.
- The frequency spectra of a function expressed as a Fourier series is called a line spectrum, since the frequency components are discrete.
- The network response of a nonsinusoidal periodic input can be obtained by expressing the input as a Fourier series, using phasor analysis to determine the response of each source in the series, then transforming the individual responses to the time domain and applying superposition.
- The Fourier transform pairs are expressed as

$$\mathbf{F}(\omega) = \mathcal{F}[f(t)] = \int_{-\infty}^{\infty} f(t)^{-j\omega t}\,dt$$

$$f(t) = \mathcal{F}^{-1}[\mathbf{F}(\omega)] = \frac{1}{2\pi}\int_{-\infty}^{\infty} \mathbf{F}(\omega)e^{j\omega t}\,d\omega$$

- $\mathbf{V}_0(j\omega) = \mathbf{H}(j\omega)\mathbf{V}_i(j\omega)$, where $\mathbf{V}_i(j\omega)$ represents the network input signal, $\mathbf{H}(j\omega)$ represents the network transfer function, and $\mathbf{V}_o(j\omega)$ represents the output signal of the network.

- Parseval's theorem can be used to compute the total energy content or the energy content over a specific frequency range of a signal using either a time-domain or frequency-domain approach.

PROBLEMS

16.1. Derive the exponential Fourier series for the periodic pulse train shown in Fig. P16.1.

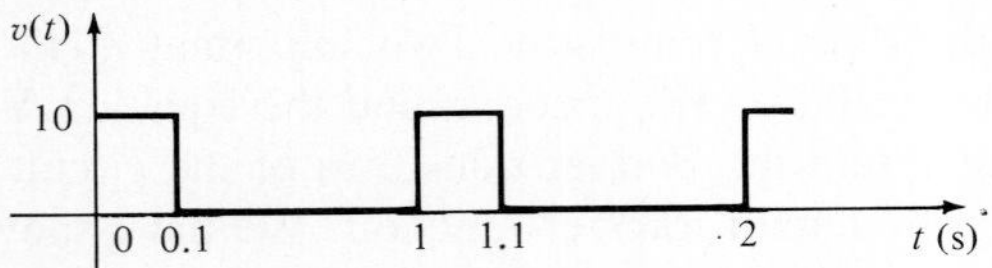

Figure P16.1

16.2. Derive the exponential Fourier series for the periodic signal shown in Fig. P16.2.

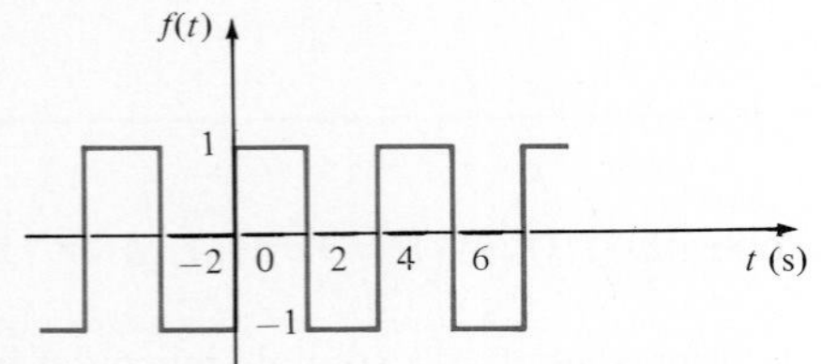

Figure P16.2

16.3. Find the exponential Fourier series for the periodic function shown in Fig. P16.3.

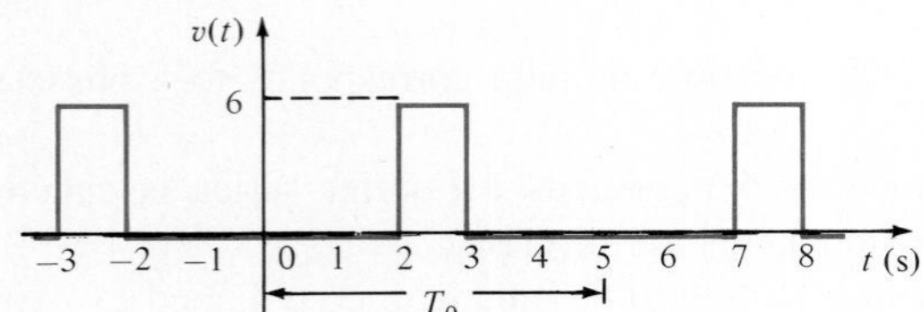

Figure P16.3

16.4. Find the exponential Fourier series for the waveform in Fig. P16.4.

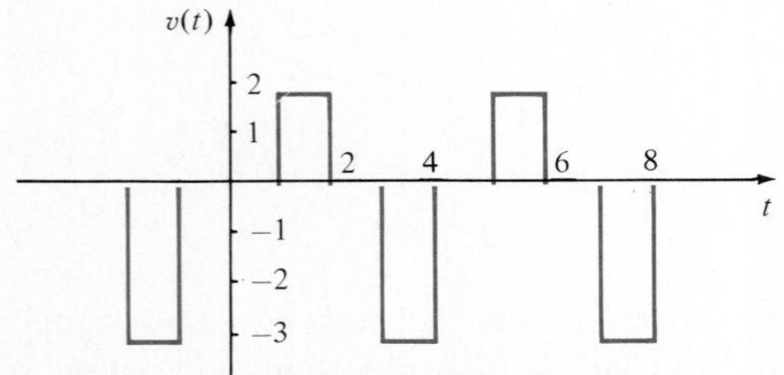

Figure P16.4

16.5. Compute the exponential Fourier series for the waveform that is the sum of the two waveforms in Fig. P16.5 by computing the exponential Fourier series of the two waveforms and adding them.

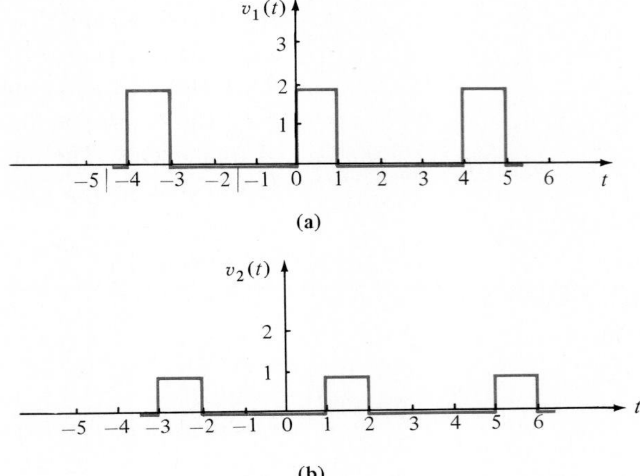

Figure P16.5

16.6. Derive the trigonometric Fourier series for the waveform shown in Fig. P16.6.

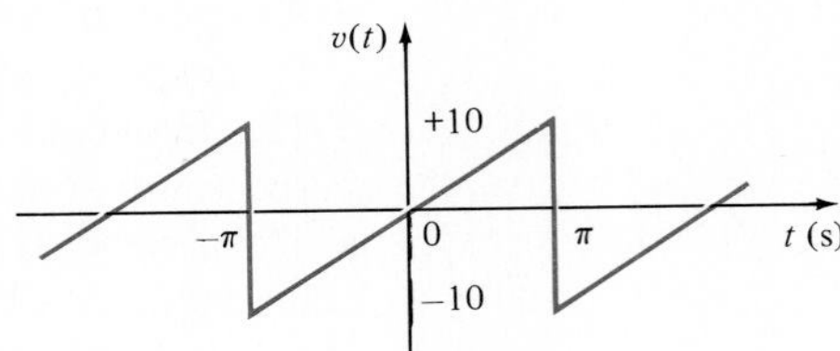

Figure P16.6

16.7. Derive the trigonometric Fourier series for the function shown in Fig. P16.7.

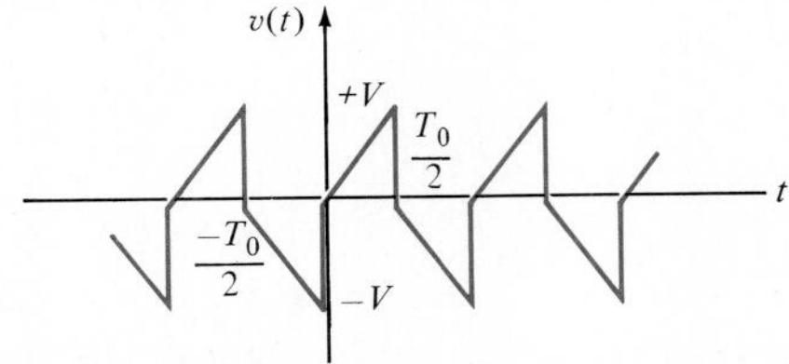

Figure P16.7

16.8. Derive the trigonometric Fourier series for the waveform shown in Fig. P16.8.

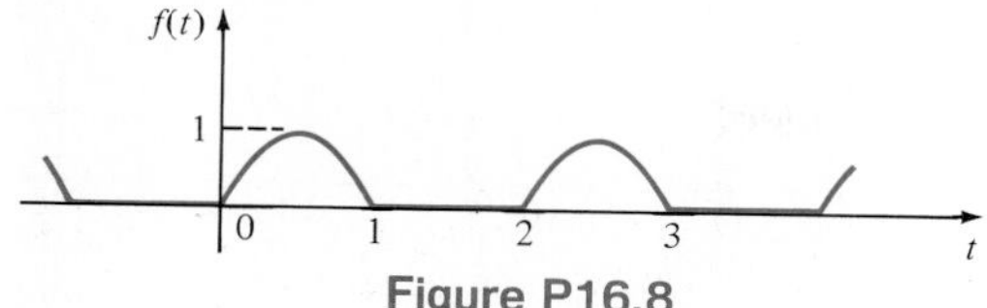

Figure P16.8

16.9. Find the trigonometric Fourier series for the waveform shown in Fig. P16.9.

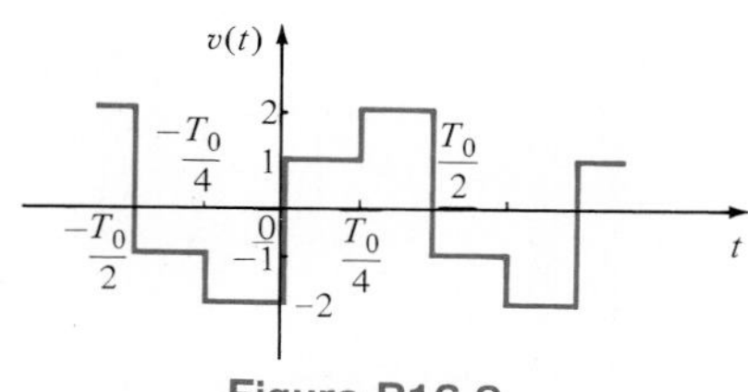

Figure P16.9

16.10. Find the Fourier series for the periodic waveform shown in Fig. P16.10.

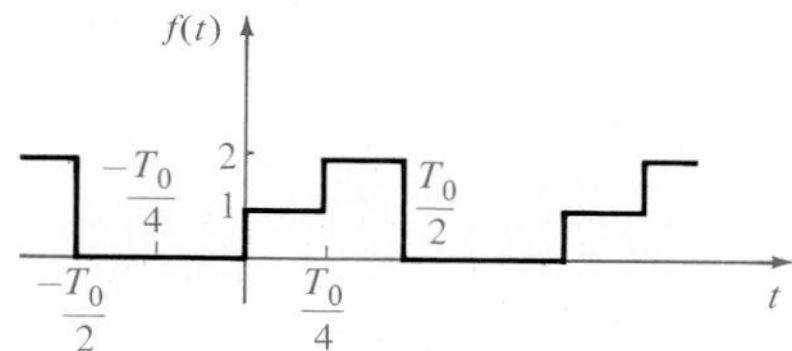

Figure P16.10

16.11. Sketch the missing section in the periodic function between $T/2$ and T in Fig. P16.11 that will make

(a) $a_0 = 0$ and $a_n = 0$

(b) $b_n = 0$

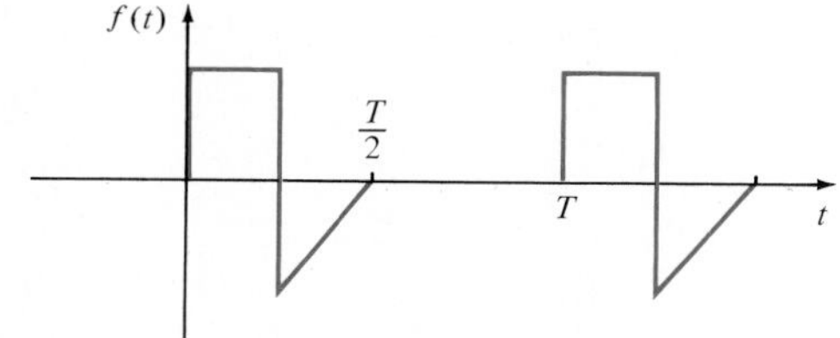

Figure P16.11

16.12. Determine which of the Fourier coefficients are zero for $f(t)$ shown in Fig. P16.12.

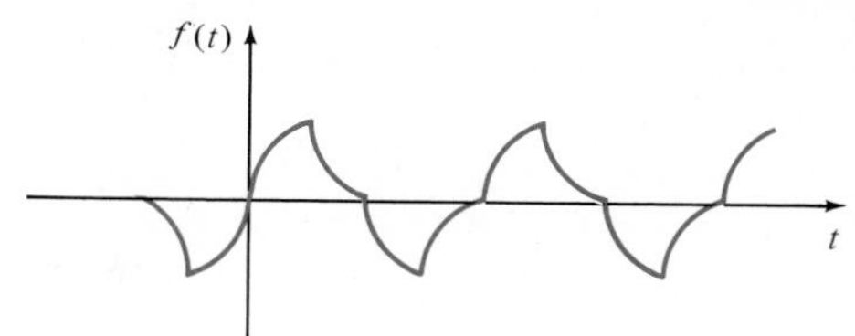

Figure P16.12

16.13. Find the trigonometric Fourier series for the waveform shown in Fig. P16.13.

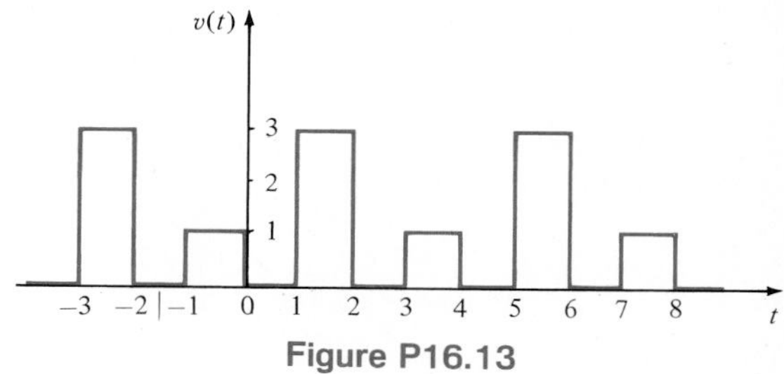

Figure P16.13

16.14. Find the trigonometric Fourier series for the waveform shown in Fig. P16.14.

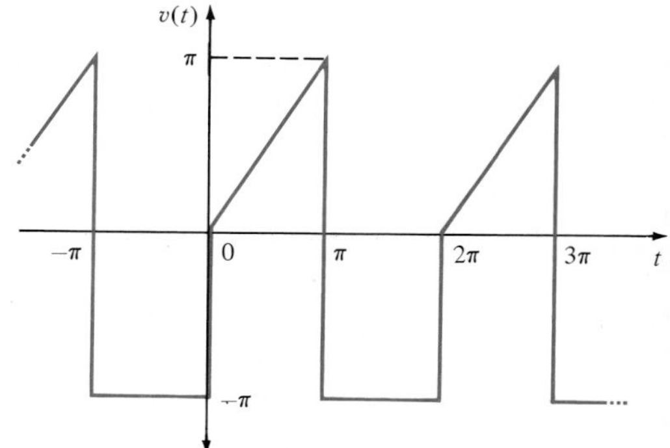

Figure P16.14

16.15. Find the trigonometric Fourier series for the voltage waveform that has a period of 2 and is defined as

$$v(t) = 1 - t^2 \qquad -1 \leq t \leq 1$$

16.16. Compute the trigonometric Fourier series for the waveform in Fig. P16.14 by determining the trigonometric Fourier series for the two waveforms in Fig. P16.16 of period 2π and adding them.

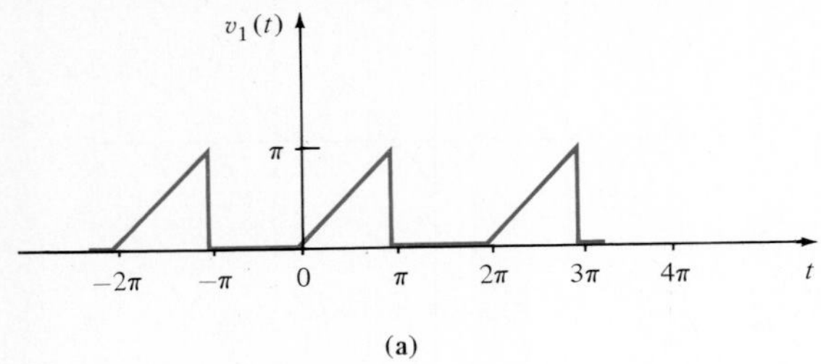

(a)

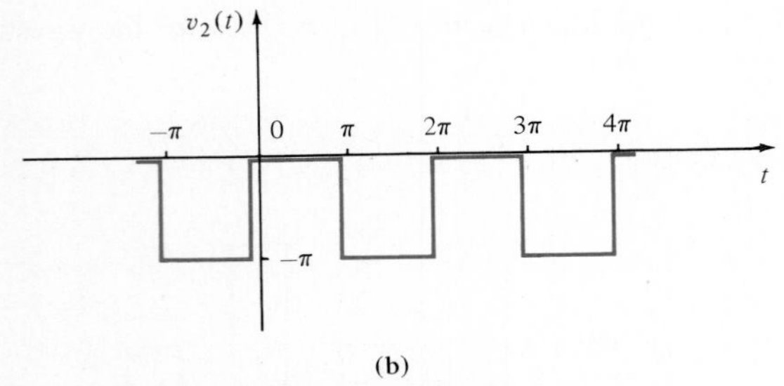

(b)

Figure P16.16

16.17. Plot the first four terms of the amplitude and phase spectra for the signal

$$v(t) = 1 - \frac{2}{\pi} \sum_{n=1}^{\infty} \frac{1}{n} \sin n\omega_0 t$$

16.18 The amplitude and phase spectra for a periodic function $v(t)$ which has only a small number of terms is shown in Fig. P16.18. Determine the expression for the $v(t)$ if $T_0 = 0.1$ s.

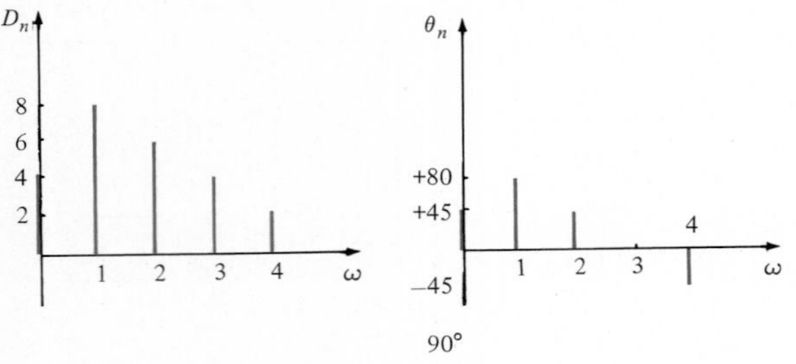

Figure P16.18

16.19. Determine the first four terms of the amplitude and phase spectra for the periodic signal defined in Problem 16.15.

16.20. The discrete line spectrum for a periodic function $f(t)$ is shown in Fig. P16.20. Determine the expression for $f(t)$.

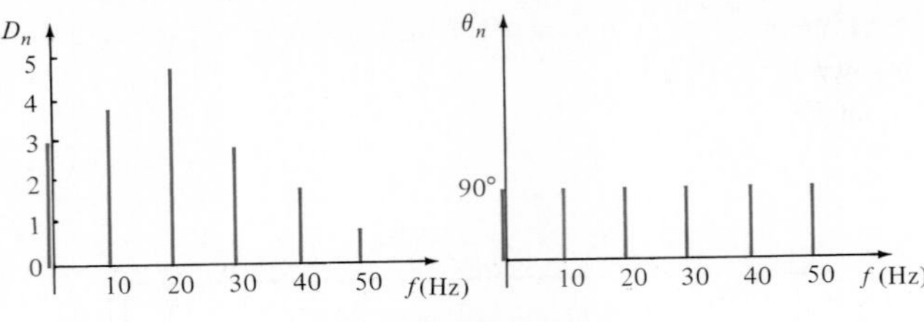

Figure P16.20

16.21. Shown in Fig. P16.21 is an amplitude and phase spectrum of a periodic function $f(t)$. Write $f(t)$ in the form of sine and cosine waveforms, making use of the Fourier series coefficients a_n and b_n.

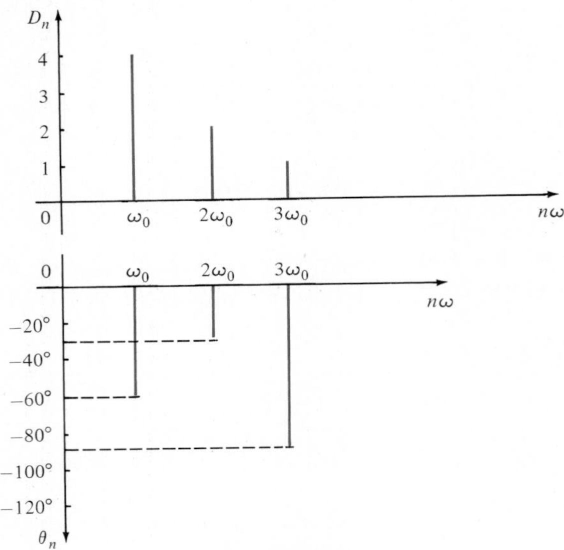

Figure P16.21

16.22. Write a function $f(t)$ in cosine form from the amplitude and phase spectra shown in Fig. P16.21.

16.23. Find the Fourier coefficient c_n for the three-phase full-wave rectified sinusoidal signal shown in Fig. P16.23. The maximum voltage is 100 V.

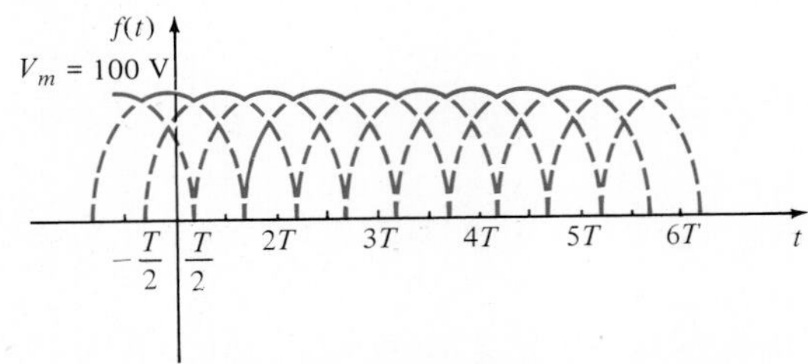

Figure P16.23

16.24. If the input voltage to the network shown in Fig. 16.24a is shown in Fig. P16.24b, find the expression for the steady-state current $i(t)$.

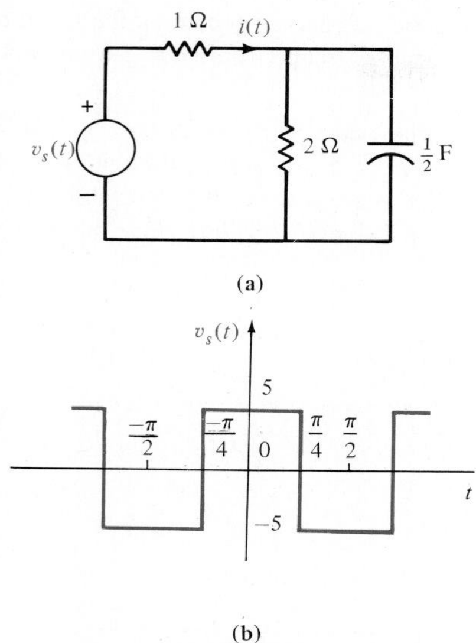

Figure P16.24

16.25. Determine the steady-state response of the current $i_0(t)$ in the circuit shown in Fig. P16.25 if the input voltage is described by the waveform shown in Problem 16.6.

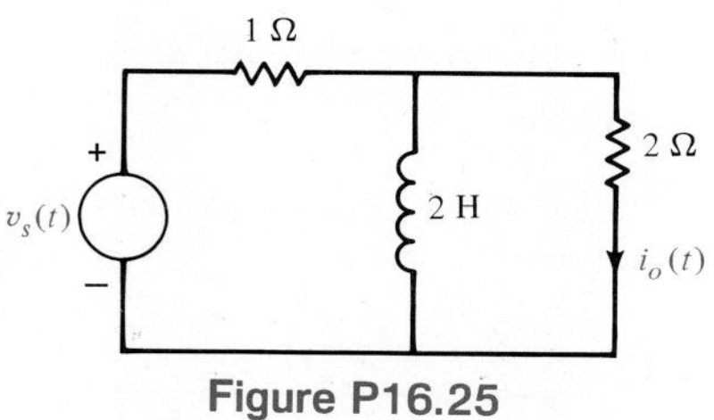

Figure P16.25

16.26. If the input voltage in Problem 16.25 is

$$v_s(t) = 1 - \frac{2}{\pi}\sum_{n=1}^{\infty}\frac{1}{n}\sin 0.2\pi nt \text{ V}$$

find the expression for the steady-state current $i_o(t)$.

16.27. Compute the first four terms of the steady-state voltage $v_o(t)$ in Fig. P16.27 if the input voltage is a periodic signal of the form

$$v(t) = \frac{2}{3} - \sum_{n=1}^{\infty}\frac{4}{\pi^2 n^2}(-1)^n \cos \pi nt \text{ V}$$

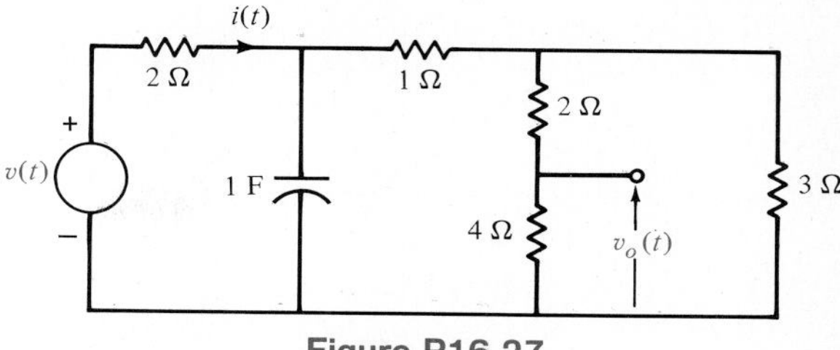

Figure P16.27

16.28. Determine the first three terms of the steady-state voltage $v_o(t)$ in Fig. P16.28 if the input voltage is a periodic signal of the form

$$v(t) = \frac{1}{2} + \sum_{n=1}^{\infty}\frac{1}{n\pi}(\cos n\pi - 1)\sin nt \text{ V}$$

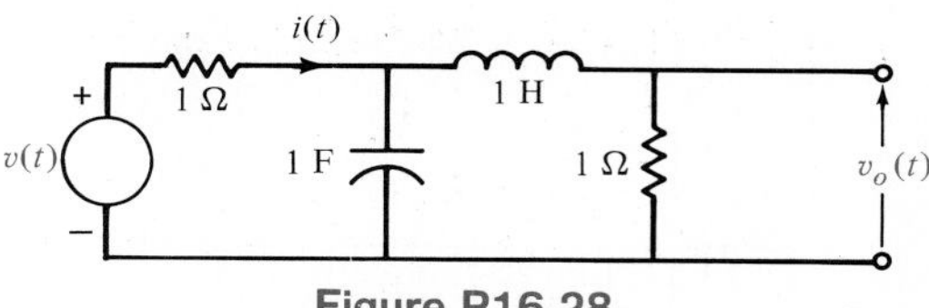

Figure P16.28

16.29. The current $i_s(t)$ shown in Fig. P16.29a is applied to the circuit shown in Fig. P16.29b. Determine the expression for the steady-state current $i_o(t)$ using the first four harmonics.

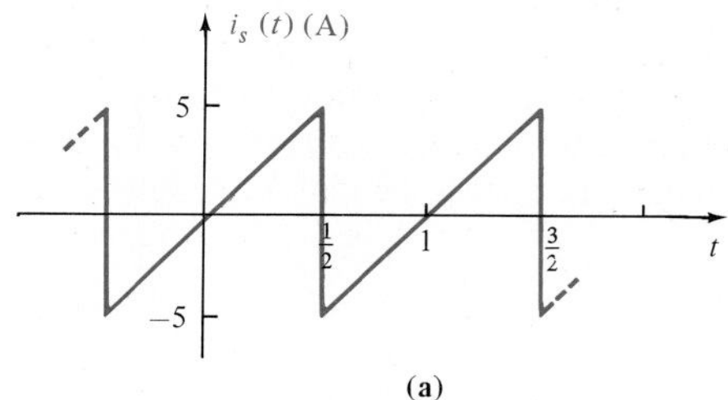

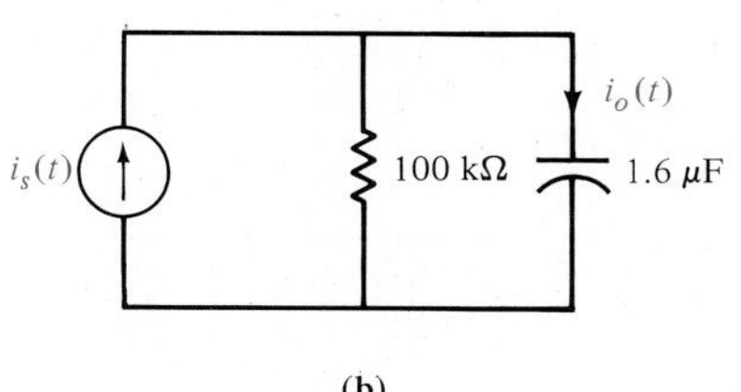

Figure P16.29

16.30. Find the Fourier transform of the function shown in Fig. P16.30.

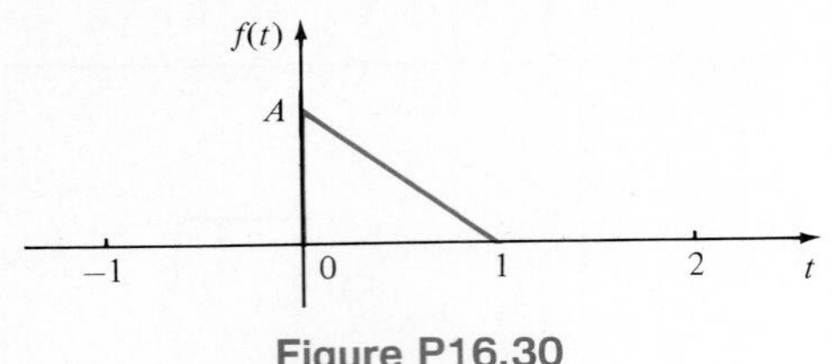

Figure P16.30

16.31. Find the Fourier transform of the function $f(t) = e^{-a|t|}$.

16.32. Find the Fourier transform of the function $f(t) = 12e^{-2|t|} \cos 4t$.

16.33. Find the Fourier transform of the waveform shown in Fig. P16.33.

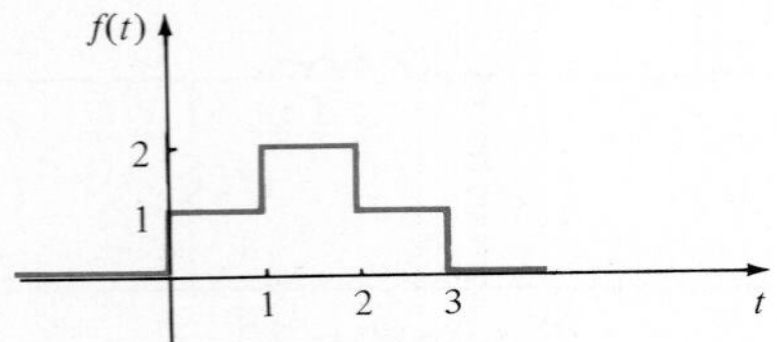

Figure P16.33

16.34. Determine the output signal $v_o(t)$ of a network with input signal $v_i(t) = 3e^{-t}u(t)$ and network transfer function $h(t) = e^{-2t}$. Assume that all initial conditions are zero.

16.35. The input signal to a network is $v_i(t) = e^{-3t}u(t)$. The transfer function of the network is $\mathbf{H}(j\omega) = 1/j\omega + 4$. Find the output of the network $v_o(t)$ if the initial conditions are zero.

16.36. Show that

$$\mathcal{F}[f_1(t)f_2(t)] = \frac{1}{2\pi}\int_{-\infty}^{\infty} \mathbf{F}_1(x)\mathbf{F}_2(\omega - x)\,dx$$

16.37. The input signal for the network in Fig. P16.37 is $v_i(t) = 10e^{-5t}u(t)$ V. Determine the total 1-Ω energy content of the output $v_o(t)$.

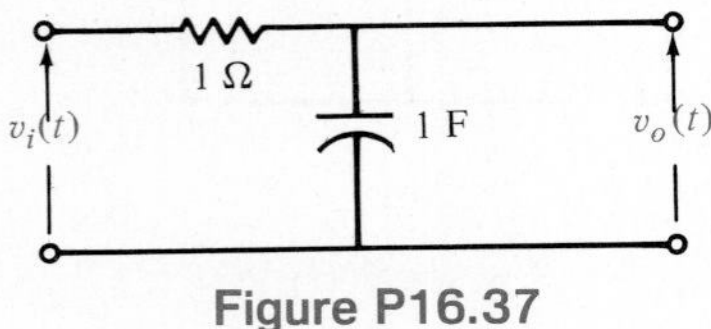

Figure P16.37

16.38. Determine the 1-Ω energy content of the signal $v_o(t)$ in Fig. P16.37 in the frequency range from 0 to 1 rad/s.

16.39. Compute the 1-Ω energy content of the signal $v_o(t)$ in Fig. P16.37 in the frequency range from $\omega = 2$ to $\omega = 4$ rad/s.

16.40. A triangular wave, $f(t)$ shown in Fig. P16.40(a), with $V_m = 10$ V and $T = 2$ s is passed through a bandpass filter, represented by $\mathbf{H}(\omega)$, as shown in Fig. P16.40(b). Use the information provided in Table 16.1 to determine the output $g(t)$.

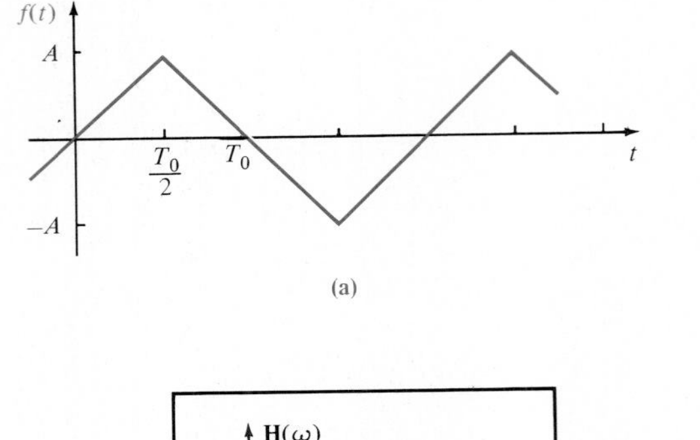

(a)

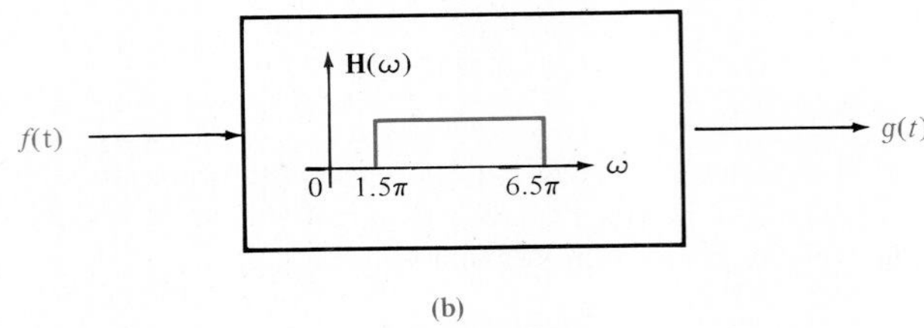

(b)

Figure P16.40

16.41. Determine the voltage $v_o(t)$ in the circuit shown in Fig. P16.41, using the Fourier transform if $i_i(t) = 2e^{-4t}u(t)$ A.

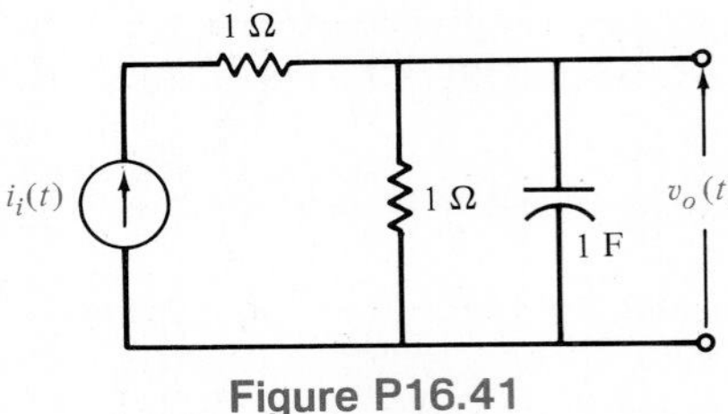

Figure P16.41

16.42. Compare the 1-Ω energy at both the input and output of the network in Problem 16.41 for the given input forcing function.

CHAPTER

17

The Laplace Transform

We will now introduce the Laplace transform. This is an extremely important technique in that for a given set of initial conditions, it will yield the total response of the circuit consisting of both the natural and forced responses in one operation.

Our use of the Laplace transform to solve circuit problems is analogous to that of using phasors in sinusoidal steady-state analysis. Using the Laplace transform we transform the circuit problem from the time domain to the frequency domain, solve the problem using algebra in the frequency domain, and then convert the solution in the complex frequency domain back to the time domain. Therefore, as we shall see, the Laplace transform is an integral transform that converts a set of linear simultaneous integrodifferential equations to a set of simultaneous algebraic equations.

Our approach is to define the Laplace transform, derive some of the transform pairs, consider some of the important properties of the transform, illustrate the inverse transform operation, introduce the convolution integral, and finally apply the transform to the solution of linear constant-coefficient integrodifferential equations.

17.1 Definition

The Laplace transform of a function $f(t)$ is defined by the equation

$$\mathscr{L}[f(t)] = \mathbf{F}(s) = \int_0^\infty f(t)e^{-st}\,dt \tag{17.1}$$

where s is the complex frequency

$$s = \sigma + j\omega \tag{17.2}$$

and the function $f(t)$ is assumed to possess the property that

$$f(t) = 0 \qquad \text{for } t < 0$$

Note that the Laplace transform is unilateral ($0 \leq t < \infty$), in contrast to the Fourier transform, which is bilateral ($-\infty < t < \infty$). In our analysis of circuits using the Laplace transform, we will focus our attention on the time interval $t \geq 0$. It is important to note that it is the initial conditions that account for the operation of the circuit prior to $t = 0$, and therefore our analyses will describe the circuit operation for $t \geq 0$.

In order for a function $f(t)$ to possess a Laplace transform, it must satisfy the condition

$$\int_0^\infty e^{-\sigma t}|f(t)|\,dt < \infty \tag{17.3}$$

for some real value of σ. Because of the convergence factor $e^{-\sigma t}$, there are a number of important functions that have Laplace transforms, even though Fourier transforms for these functions do not exist. All of the inputs we will apply to circuits possess Laplace transforms. Functions that do not have Laplace transforms (e.g., e^{t^2}) are of no interest to us in circuit analysis.

The inverse Laplace transform, which is analogous to the inverse Fourier transform, is defined by the relationship

$$\mathcal{L}^{-1}[\mathbf{F}(s)] = f(t) = \frac{1}{2\pi j}\int_{\sigma_1 - j\infty}^{\sigma_1 + j\infty} \mathbf{F}(s)e^{st}\,ds \tag{17.4}$$

where σ_1 is real and $\sigma_1 > \sigma$ in Eq. (17.3). The evaluation of this integral is based on complex variable theory, and therefore we will circumvent its use by developing and using a set of Laplace transform pairs.

17.2 Some Important Transform Pairs

We will now develop a number of basic transform pairs which are very useful in circuit analysis.

EXAMPLE 17.1

The Laplace transform of the impulse function defined in Chapter 16 is

$$\mathbf{F}(s) = \int_0^\infty \delta(t - t_0)e^{-st}\,dt$$

Using the sampling property of the delta function as outlined in Eq. (16.43), we obtain

$$\mathcal{L}[\delta(t - t_0)] = e^{-t_0 s}$$

In the limit as $t_0 \rightarrow 0$, $e^{-t_0 s} \rightarrow 1$, and therefore

$$\mathcal{L}[\delta(t)] = \mathbf{F}(s) = 1$$

■

EXAMPLE 17.2

The Laplace transform of the function

$$f(t) = te^{-at}\,\delta(t - 1)$$

is

$$\mathbf{F}(s) = \int_0^\infty te^{-at}\,\delta(t - 1)e^{-st}\,dt$$

$$= \int_0^\infty \delta(t - 1)te^{-(s+a)t}\,dt$$

Once again employing the sampling property of the impulse, we obtain

$$\mathbf{F}(s) = 1e^{-(s+a)} = \frac{e^{-s}}{e^a}$$

■

EXAMPLE 17.3

The Laplace transform of the unit step function defined in Chapter 7 is

$$\mathbf{F}(s) = \int_0^\infty u(t)e^{-st}\,dt$$

$$= \int_0^\infty 1e^{-st}\,dt$$

$$= -\frac{1}{s}e^{-st}\Big|_0^\infty$$

$$= \frac{1}{s} \qquad \sigma > 0$$

Therefore,

$$\mathcal{L}[u(t)] = \mathbf{F}(s) = \frac{1}{s}$$

■

EXAMPLE 17.4

The Laplace transform of the time-shifted unit step function shown in Fig. 17.1 is

$$\mathbf{F}(s) = \int_0^\infty u(t - a)e^{-st}\,dt$$

Note that

$$u(t - a) = \begin{cases} 1 & a < t < \infty \\ 0 & t < a \end{cases}$$

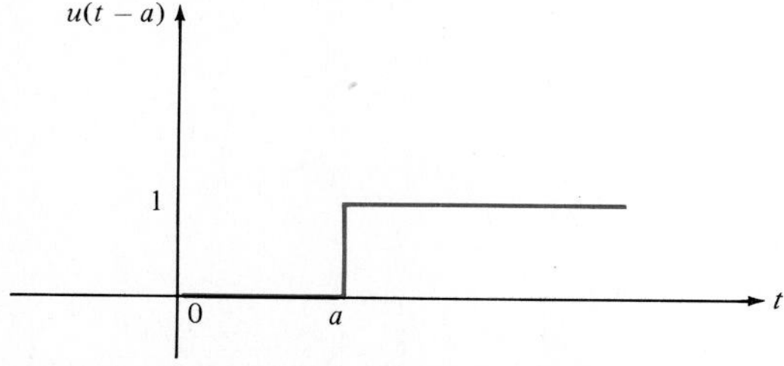

Figure 17.1 Time-shifted unit step function.

Therefore,

$$\mathbf{F}(s) = \int_a^{\infty} e^{-st}\, dt$$

$$= \frac{e^{-as}}{s}$$

■

EXAMPLE 17.5

The Laplace transform for the cosine function is

$$\mathbf{F}(s) = \int_0^{\infty} \cos \omega t \, e^{-st}\, dt$$

$$= \int_0^{\infty} \frac{e^{j\omega t} + e^{-j\omega t}}{2} e^{-st}\, dt$$

$$= \int_0^{\infty} \frac{e^{-(s-j\omega)t} + e^{-(s+j\omega)t}}{2}\, dt$$

$$= \frac{1}{2}\left(\frac{1}{s - j\omega} + \frac{1}{s + j\omega}\right) \qquad \sigma > 0$$

$$= \frac{s}{s^2 + \omega^2}$$

■

A short table of useful Laplace transform pairs is shown in Table 17.1.

DRILL EXERCISE

D17.1. If $f(t) = e^{-at} \cos \omega t \, \delta(t - 2)$, find $\mathbf{F}(s)$.

D17.2. Find $\mathbf{F}(s)$ if $f(t) = tu(t - 1)$.

D17.3. Show that if $f(t) = \sin \omega t$, then $\mathbf{F}(s) = \omega/(s^2 + \omega^2)$.

17.3 Some Useful Properties of the Transform

We now present a number of important properties of the Laplace transform and illustrate their usefulness by examples.

Theorem 1

$$\mathcal{L}[Af(t)] = A\mathbf{F}(s) \tag{17.5}$$

The transform is defined as

$$\mathcal{L}[Af(t)] = \int_0^{\infty} Af(t)e^{-st}\, dt$$

Table 17.1 Short Table of Laplace Transform Pairs

$f(t)$	$\mathbf{F}(s)$
$\delta(t)$	1
$u(t)$	$\frac{1}{s}$
e^{-at}	$\frac{1}{s+a}$
t	$\frac{1}{s^2}$
$\frac{t^n}{n!}$	$\frac{1}{S^{n+1}}$
te^{-at}	$\frac{1}{(s+a)^2}$
$\frac{t^n e^{-at}}{n!}$	$\frac{1}{(s+a)^{n+1}}$
$\sin \omega t$	$\frac{\omega}{s^2+\omega^2}$
$\cos \omega t$	$\frac{s}{s^2+\omega^2}$
$e^{-at} \sin \omega t$	$\frac{\omega}{(s+a)^2+\omega^2}$
$e^{-at} \cos \omega t$	$\frac{s+a}{(s+a)^2+\omega^2}$

Since A is not a function of time,

$$\mathscr{L}[Af(t)] = A \int_0^\infty f(t)e^{-st}\,dt$$

$$= A\mathbf{F}(s)$$

EXAMPLE 17.6

The Laplace transform of the function

$$f(t) = e^{-ax} \sin y \cos \omega t$$

can be found using the results of Example 17.5 as

$$\mathbf{F}(s) = e^{-ax} \sin y \frac{s}{s^2+\omega^2}$$

■

Theorem 2

$$\mathscr{L}[f_1(t) \pm f_2(t)] = \mathbf{F}_1(s) \pm \mathbf{F}_2(s) \tag{17.6}$$

by definition

$$\mathscr{L}[f_1(t) \pm f_2(t)] = \int_0^\infty [f_1(t) \pm f_2(t)]e^{-st}\,dt$$

$$= \int_0^\infty f_1(t)e^{-st}\,dt \pm \int_0^\infty f_2(t)e^{-st}\,dt$$

$$= \mathbf{F}_1(s) \pm \mathbf{F}_2(s)$$

EXAMPLE 17.7

The Laplace transform of the function

$$f(t) = e^{-t} + e^{-2t}$$

is

$$\mathbf{F}(s) = \int_0^\infty (e^{-t} + e^{-2t})e^{-st}\,dt$$

$$= \int_0^\infty e^{-(s+1)t}\,dt + \int_0^\infty e^{-(s+2)t}\,dt$$

$$= \frac{1}{s+1} + \frac{1}{s+2}$$

where $\mathscr{L}[e^{-t}] = 1/(s+1)$ and $\mathscr{L}[e^{-2t}] = 1/(s+2)$. ■

Theorem 3

$$\mathscr{L}[f(at)] = \frac{1}{a}\mathbf{F}\left(\frac{s}{a}\right) \qquad a > 0 \tag{17.7}$$

The Laplace transform of $f(at)$ is

$$\mathscr{L}[f(at)] = \int_0^\infty f(at)e^{-st}\,dt$$

Now let $\lambda = at$ and $d\lambda = a\,dt$. Then

$$\mathscr{L}[f(at)] = \int_0^\infty f(\lambda)e^{-(\lambda/a)s}\,\frac{d\lambda}{a}$$

$$= \frac{1}{a}\int_0^\infty f(\lambda)e^{-(s/a)\lambda}\,d\lambda$$

$$= \frac{1}{a}\mathbf{F}\left(\frac{s}{a}\right) \qquad a < 0$$

EXAMPLE 17.8

We wish to find the Laplace transform of $f(t) = \cos\omega(t/2)$. We have found that $\mathscr{L}[\cos\omega t] = s/(s^2 + \omega^2)$. Therefore, using Theorem 3 yields

$$\cos \omega\left(\frac{t}{2}\right) = \frac{4s}{(2s)^2 + \omega^2}$$

$$= \frac{s}{s^2 + \omega^2/4}$$

which should be fairly obvious, since $\mathcal{L}[\cos \alpha t] = s/(s^2 + \alpha^2)$, where in this case $\alpha = \omega/2$. ■

Theorem 4

$$\mathcal{L}[f(t - t_0)u(t - t_0)] = e^{-t_0 s}\mathbf{F}(s) \qquad t_0 \geq 0 \tag{17.8}$$

This theorem, commonly known as the shifting theorem, is illustrated as follows:

$$\mathcal{L}[f(t - t_0)u(t - t_0)] = \int_0^\infty f(t - t_0)u(t - t_0)e^{-st}\,dt$$

$$= \int_{t_0}^\infty f(t - t_0)e^{-st}\,dt$$

If we now let $\lambda = t - t_0$ *and* $d\lambda = dt$, *then*

$$\mathcal{L}[f(t - t_0)u(t - t_0)] = \int_0^\infty f(\lambda)e^{-s(\lambda + t_0)}\,d\lambda$$

$$= e^{-t_0 s}\int_0^\infty f(\lambda)e^{-s\lambda}\,d\lambda$$

$$= e^{-t_0 s}\mathbf{F}(s) \qquad t_0 \geq 0$$

EXAMPLE 17.9

If $f(t) = (t - 1)u(t - 1)$, then by employing Theorem 4, we find that

$$\mathcal{L}[(t - 1)u(t - 1)] = e^{-s}\mathcal{L}[t] = \frac{e^{-s}}{s^2}$$

■

Theorem 5

$$\mathcal{L}[f(t)u(t - t_0)] = e^{-t_0 s}\mathcal{L}[f(t + t_0)] \tag{17.9}$$

By definition,

$$\mathcal{L}[f(t)u(t - t_0)] = \int_0^\infty f(t)u(t - t_0)e^{-st}\,dt$$

Now let $\lambda = t - t_0$ *and* $d\lambda = dt$, *and therefore*

$$\int_0^\infty f(t)u(t - t_0)e^{-st}\,dt = \int_0^\infty f(\lambda + t_0)e^{-s(\lambda + t_0)}\,d\lambda$$

$$= e^{-t_0 s}\mathcal{L}[f(t + t_0)]$$

EXAMPLE 17.10

If $f(t) = tu(t-1)$, then $\mathcal{L}[f(t)]$ can be found from Theorem 5 as

$$\mathcal{L}[tu(t-1)] = e^{-s}\mathcal{L}[t+1]$$

$$= e^{-s}\left(\frac{1}{s^2} + \frac{1}{s}\right)$$

■

Theorem 6

$$\mathcal{L}[e^{-at}f(t)] = \mathbf{F}(s+a) \tag{17.10}$$

By definition,

$$\mathcal{L}[e^{-at}f(t)] = \int_0^\infty e^{-at}f(t)e^{-st}\,dt$$

$$= \int_0^\infty f(t)e^{-(s+a)}\,dt$$

$$= \mathbf{F}(s+a)$$

EXAMPLE 17.11

Since the Laplace transform of cos ωt is known to be

$$\mathcal{L}[\cos \omega t] = \frac{s}{s^2 + \omega^2}$$

then

$$\mathcal{L}[e^{-at}\cos \omega t] = \frac{s+a}{(s+a)^2 + \omega^2}$$

■

Theorem 7

$$\mathcal{L}\left[\frac{d^n f(t)}{dt^n}\right] = s^n\mathbf{F}(s) - s^{n-1}f(0) - s^{n-2}\frac{df(0)}{dt} \cdots s^0\frac{d^{n-1}f(0)}{dt^{n-1}} \tag{17.11}$$

Let us begin by examining $\mathcal{L}[df(t)/dt]$. By definition,

$$\mathcal{L}\left[\frac{df(t)}{dt}\right] = \int_0^\infty \frac{df(t)}{dt}e^{-st}\,dt$$

Using integration by parts gives us

$$u = e^{-st} \qquad dv = \frac{df(t)}{dt}\,dt = df(t)$$

$$du = -se^{-st} \qquad v = f(t)$$

Hence

$$\mathcal{L}\left[\frac{df(t)}{dt}\right] = f(t)e^{-st}\Big|_0^\infty + s\int_0^\infty f(t)e^{-st}\,dt$$

If we assume that the Laplace transform of the function $f(t)$ exists so that

$$\lim_{t\to\infty} e^{-st}f(t) = 0$$

then

$$\mathscr{L}\left[\frac{df(t)}{dt}\right] = -f(0) + s\mathbf{F}(s)$$

Using this result, we can write

$$\begin{aligned}\mathscr{L}\left[\frac{d^2f(t)}{dt^2}\right] &= s\mathscr{L}[f'(t)] - f'(0) \\ &= s[s\mathbf{F}(s) - f(0)] - f'(0) \\ &= s^2\mathbf{F}(s) - sf(0) - f'(0)\end{aligned}$$

By continuing in this manner we can demonstrate the original statement.

We will illustrate the usefulness of this theorem later in this chapter when we employ it to solve differential equations.

Theorem 8

$$\mathscr{L}[tf(t)] = \frac{-d\mathbf{F}(s)}{ds} \tag{17.12}$$

By definition,

$$\mathbf{F}(s) = \int_0^\infty f(t)e^{-st}\,dt$$

Then

$$\begin{aligned}\frac{d\mathbf{F}(s)}{ds} &= \int_0^\infty f(t)(-te^{-st})\,dt \\ &= -\int_0^\infty tf(t)e^{-st}\,dt \\ &= -\mathscr{L}[tf(t)]\end{aligned}$$

EXAMPLE 17.12

Let us demonstrate the use of Theorem 8. To begin, we know that

$$\mathscr{L}[u(t)] = \frac{1}{s}$$

Then

$$\mathscr{L}[tu(t)] = \frac{-d}{ds}\left(\frac{1}{s}\right) = \frac{1}{s^2}$$

Continuing, we note that

$$\mathscr{L}[t^2u(t)] = -\frac{d}{ds}\left(\frac{1}{s^2}\right) = \frac{2}{s^3}$$

and

$$\mathcal{L}[t^3u(t)] = \frac{-d}{ds}\left(\frac{2}{s^3}\right) = \frac{3!}{s^4}$$

and in general,

$$\mathcal{L}[t^nu(t)] = \frac{n!}{s^{n+1}}$$

■

Theorem 9

$$\mathcal{L}\left[\frac{f(t)}{t}\right] = \int_s^{\infty} \mathbf{F}(\lambda)\, d\lambda \tag{17.13}$$

By definition,

$$\int_0^{\infty} f(t)e^{-\lambda t}\, dt = \mathbf{F}(\lambda)$$

Therefore,

$$\int_s^{\infty}\int_0^{\infty} f(t)e^{-\lambda t}\, dt\, d\lambda = \int_s^{\infty} \mathbf{F}(\lambda)\, d\lambda$$

Since $f(t)$ is Laplace transformable, we can change the order of integration so that

$$\int_0^{\infty} f(t)\int_s^{\infty} e^{-\lambda t}\, d\lambda\, dt = \int_s^{\infty} \mathbf{F}(\lambda)\, d\lambda$$

and hence

$$\int_0^{\infty} \frac{f(t)}{t} e^{-st}\, dt = \int_s^{\infty} \mathbf{F}(\lambda)\, d\lambda$$

and therefore

$$\mathcal{L}\left[\frac{f(t)}{t}\right] = \int_s^{\infty} \mathbf{F}(\lambda)\, d\lambda$$

EXAMPLE 17.13

If $f(t) = te^{-at}$, then

$$\mathbf{F}(\lambda) = \frac{1}{(\lambda + a)^2}$$

Therefore,

$$\int_s^{\infty} \mathbf{F}(\lambda)\, d\lambda = \int_s^{\infty} \frac{1}{(\lambda + a)^2}\, d\lambda = \left.\frac{-1}{\lambda + a}\right|_s^{\infty} = \frac{1}{s + a}$$

Hence

$$f_1(t) = \frac{f(t)}{t} = \frac{te^{-at}}{t} = e^{-at} \quad \text{and} \quad \mathbf{F}_1(s) = \frac{1}{s + a}$$

■

Theorem 10

$$\mathcal{L}\left[\int_0^t f(\lambda)\, d\lambda\right] = \frac{1}{s}\mathbf{F}(s) \tag{17.14}$$

We begin with the expression

$$\mathcal{L}\left[\int_0^t f(\lambda)\, d\lambda\right] = \int_0^\infty \int_0^t f(\lambda)\, d\lambda e^{-st}\, dt$$

Integrating by parts yields

$$u = \int_0^t f(\lambda)\, d\lambda \qquad dv = e^{-st}\, dt$$

$$du = f(t)\, dt \qquad v = \frac{-1}{s} e^{-st}$$

Therefore,

$$\mathcal{L}\left[\int_0^t f(\lambda)\, d\lambda\right] = \frac{-e^{-st}}{s}\int_0^t f(\lambda)\, d\lambda\bigg|_0^\infty + \frac{1}{s}\int_0^\infty f(t)e^{-st}\, dt$$

$$= \frac{1}{s}\mathbf{F}(s)$$

We employ this theorem later in the chapter when we examine integrodifferential equations. The theorems we have presented are listed in Table 17.2 for quick reference.

DRILL EXERCISE

D17.4. Find $\mathbf{F}(s)$ if $f(t) = \frac{1}{2}(t - 4e^{-2t})$.

D17.5. If $f(t) = te^{-(t-1)}u(t - 1) - e^{-(t-1)}u(t - 1)$, determine $\mathbf{F}(s)$.

D17.6. Find $\mathbf{F}(s)$ if $f(t) = e^{-4t}(t - e^{-t})$.

D17.7. Find the Laplace transform of the function $te^{-4x}/(a^2 + 4)$.

D17.8. If $f(t) = \cos \omega t\, u(t - 1)$, find $\mathbf{F}(s)$.

D17.9. Use Theorem 8 to demonstrate that $\mathcal{L}[te^{-at}] = 1/(s + a)^2$.

17.4 The Gate Function

The *gate function* is defined as the difference of two unit step functions of the form

$$G_\tau(T_0) = u(t - \tau) - u(t - \tau - T_0) \tag{17.15}$$

The function is shown in Fig. 17.2. Note that the function is a rectangular pulse of unit height and width T_0 which starts at $t = \tau$. The importance of this gate is the fact that

Table 17.2 Some Useful Properties of the Laplace Transform

$f(t)$	$\mathbf{F}(s)$
$Af(t)$	$A\mathbf{F}(s)$
$f_1(t) \pm f_2(t)$	$\mathbf{F}_1(s) \pm \mathbf{F}_2(s)$
$f(at)$	$\frac{1}{a}\mathbf{F}\left(\frac{s}{a}\right),\ a > 0$
$f(t - t_0)u(t - t_0),\ t_0 \geq 0$	$e^{-t_0 s}\mathbf{F}(s)$
$f(t)u(t - t_0)$	$e^{-t_0 s}\mathcal{L}[f(t + t_0)]$
$e^{-at}f(t)$	$\mathbf{F}(s + a)$
$\frac{d^n f(t)}{dt^n}$	$s^n\mathbf{F}(s) - s^{n-1}f(0) - s^{n-2}f^1(0) \cdots s^0 f^{n-1}(0)$
$tf(t)$	$-\frac{d\mathbf{F}(s)}{ds}$
$\frac{f(t)}{t}$	$\int_s^\infty \mathbf{F}(\lambda)\, d\lambda$
$\int_0^t f(\lambda)\, d\lambda$	$\frac{1}{s}\mathbf{F}(s)$
$\int_0^t f_1(\lambda)f_2(t - \lambda)\, d\lambda$	$\mathbf{F}_1(s)\mathbf{F}_2(s)$

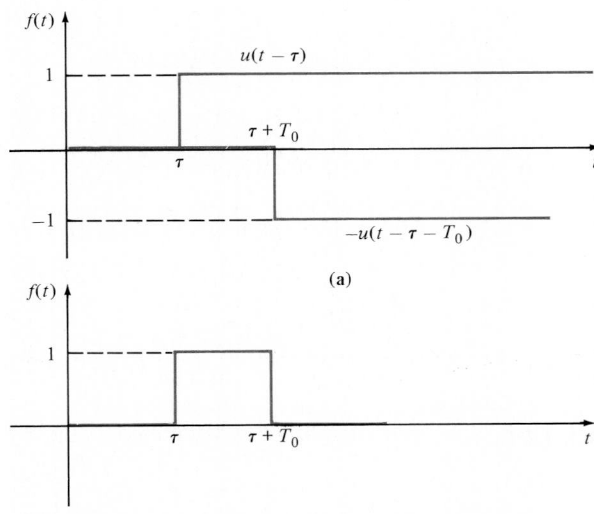

Figure 17.2 Gate function generated from two unit step functions.

any function $f(t)$ which is multiplied by the gate is equal to $f(t)$ within the gate interval $\tau < t < \tau + T_0$ and is zero everywhere else. Therefore, we can use the gate function to express a complex waveform by simply treating every different portion of the waveform separately using a gate. The following example illustrates the approach.

EXAMPLE 17.14

We wish to determine the Laplace transform of the sawtooth function shown in Fig. 17.3.

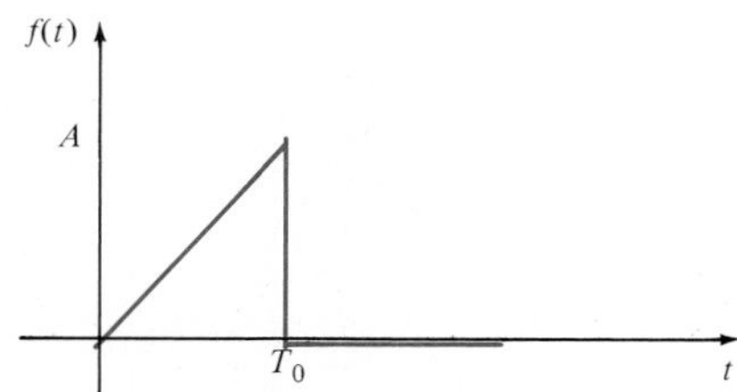

Figure 17.3 Sawtooth waveform.

Using the gate function, we can express the sawtooth as

$$\begin{aligned} f(t) &= \frac{A}{T_0} t[u(t) - u(t - T_0)] \\ &= \frac{A}{T_0} tu(t) - \frac{A}{T_0} tu(t - T_0) \\ &= \frac{A}{T_0} tu(t) - \frac{A}{T_0} (t - T_0 + T_0)u(t - T_0) \\ &= \frac{A}{T_0} tu(t) - \frac{A}{T_0} (t - T_0)u(t - T_0) - Au(t - T_0) \end{aligned}$$

Then $\mathbf{F}(s)$ is

$$\begin{aligned} \mathbf{F}(s) &= \frac{A}{T_0}\left(\frac{1}{s^2} - \frac{1}{s^2} e^{-T_0 s} - \frac{T_0}{s} e^{-T_0 s}\right) \\ &= \frac{A}{T_0 s^2} [1 - (1 + T_0 s)e^{-T_0 s}] \end{aligned}$$

■

Another technique for expressing a complex waveform involves the use of a combination of steps and ramps of the form $Au(t - t_0)$ and $B(t - t_0)u(t - t_0)$, respectively. The Laplace transforms of these functions are simply

$$\mathscr{L}[Au(t - t_0)] = \frac{A}{s} e^{-t_0 s}$$

and

$$\mathscr{L}[B(t - t_0)u(t - t_0)] = \frac{B}{s^2} e^{-t_0 s}$$

For each discontinuity in the waveform we will have a step in the expression and for each change in slope we will have a ramp in the expression. The following example illustrates the technique.

EXAMPLE 17.15

Let us determine the Laplace transform of the waveform shown in Fig. 17.4.

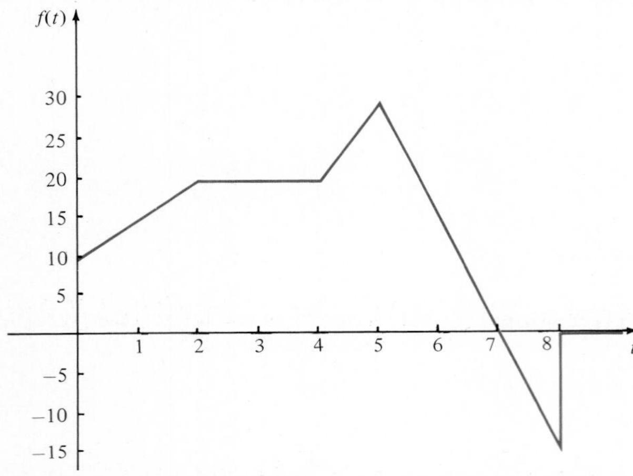

Figure 17.4 Piecewise linear function used in Example 17.15.

Starting at time $t = 0$, there is a discontinuity of 10; therefore, the function contains the term $10u(t)$. The slope changes from 0 to 5 at $t = 0$ and therefore the expression contains the term $5tu(t)$. At time $t = 2$ the slope changes from 5 to 0, and therefore the expression contains the term $-5(t - 2)u(t - 2)$. At time $t = 4$ the slope changes from 0 to 10, and hence the expression will contain the term $10(t - 4)u(t - 4)$. Continuing in this manner, we obtain

$$\begin{aligned} f(t) = {} & 10u(t) + 5tu(t) - 5(t - 2)u(t - 2) + 10(t - 4)u(t - 4) \\ & - 25(t - 5)u(t - 5) + 15(t - 8)u(t - 8) + 15u(t - 8) \end{aligned}$$

Therefore,

$$\mathbf{F}(s) = \frac{10}{s} + \frac{5}{s^2} - \frac{5}{s^2}e^{-2s} + \frac{10}{s^2}e^{-4s} - \frac{25}{s^2}e^{-5s} + \frac{15}{s^2}e^{-8s} + \frac{15}{s}e^{-8s}$$ ■

DRILL EXERCISE

D17.10. Find the Laplace transform of the waveform in Fig. D17.10.

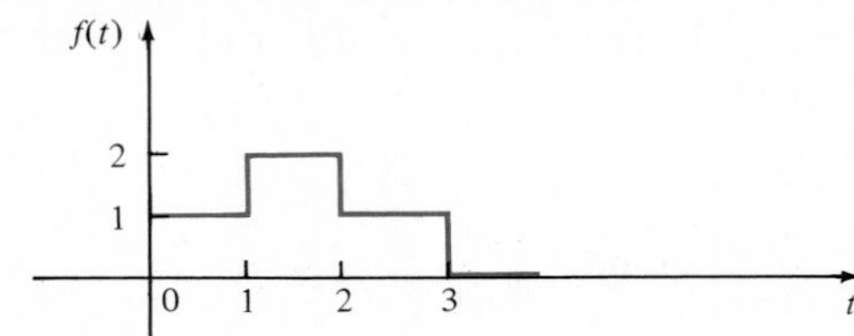

Figure D17.10

D17.11. Determine the Laplace transform of the waveform in Fig. D17.11.

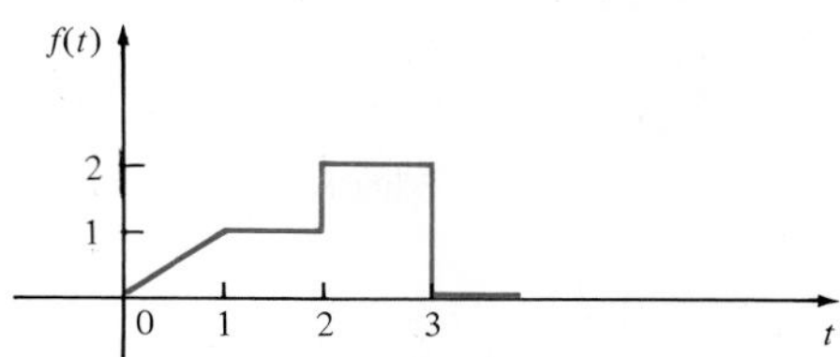

Figure D17.11

17.5 Periodic Functions

If $f(t)$ is a periodic function with period T for $t \geq 0$ and is zero for $t < 0$ then $f(t)$ can be expressed as

$$f(t) = f_1(t) + f_2(t) + f_3(t) + \cdots \tag{17.16}$$

where $f_1(t)$ is the function in the interval $0 \leq t < T$ and is zero elsewhere, $f_2(t)$ is the function in the interval $T \leq t < 2T$ and is zero elsewhere, and so on. However, since $f_2(t)$ in the interval $T \leq t < 2T$ is identical to $f_1(t)$ in the interval $0 \leq t < T$, and $f_3(t)$ in the interval $2T \leq t < 3T$ is identical to $f_1(t)$ in the interval $0 \leq t < T$, and so on, then $f(t)$ can be written as

$$f(t) = f_1(t) + f_1(t - T)u(t - T) + f_1(t - 2T)u(t - 2T) + \cdots \tag{17.17}$$

Hence, using Theorem 4, we can write

$$\begin{aligned} \mathcal{L}[f(t)] &= \mathbf{F}_1(s) + \mathbf{F}_1(s)e^{-Ts} + \mathbf{F}_1(s)e^{-2Ts} + \cdots \\ &= \mathbf{F}_1(s)(1 + e^{-Ts} + e^{-2Ts} + e^{-3Ts} + \cdots \end{aligned} \tag{17.18}$$

and since in general

$$\frac{1}{1 - x} = 1 + x + x^2 + x^3 + \cdots$$

then

$$\mathcal{L}[f(t)] = \frac{\mathbf{F}_1(s)}{1 - e^{-Ts}} \tag{17.19}$$

where $\mathbf{F}_1(s) = \mathcal{L}[f_1(t)]$, the Laplace transform of the first period of $f(t)$; and $\mathbf{F}_1(s) = \int_0^T f(t)e^{-st}\,dt$.

EXAMPLE 17.16

The Laplace transform of the periodic waveform shown in Fig. 17.5 is

$$\mathcal{L}[f(t)] = \frac{\mathbf{F}_1(s)}{1 - e^{-Ts}}$$

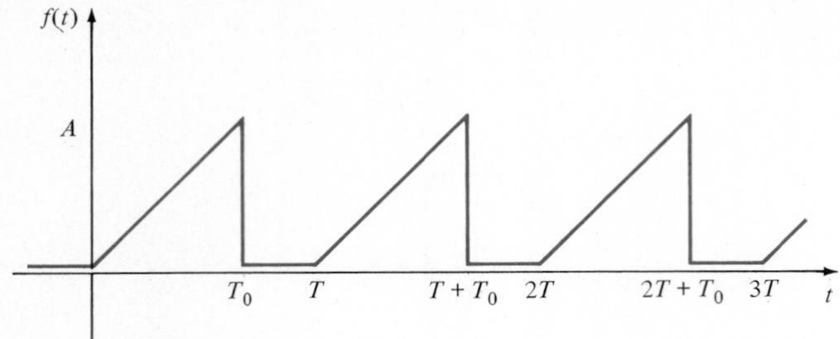

Figure 17.5 Periodic sawtooth waveform.

where from Example 17.14,

$$\mathbf{F}_1(s) = \frac{A}{T_0 s^2}[1 - (1 + T_0 s)e^{-T_0 s}]$$

■

DRILL EXERCISE

D17.12. Determine the Laplace transform of the periodic waveform in Fig. D17.12.

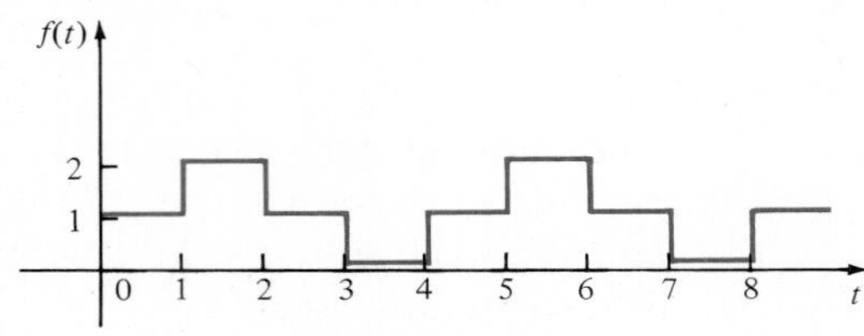

Figure D17.12

D17.13. Find the Laplace transform of the periodic waveform shown in Fig. D17.13.

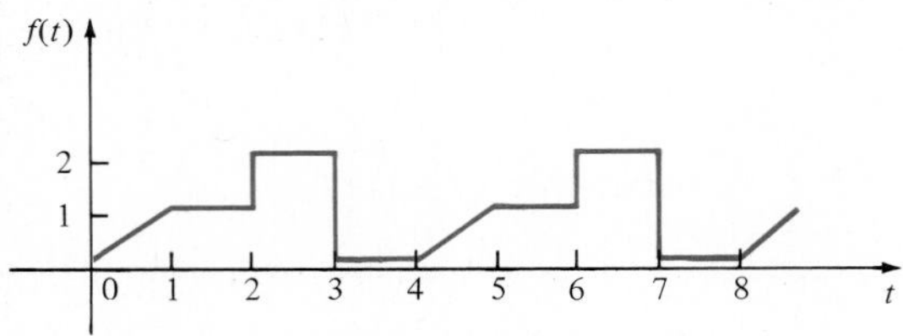

Figure D17.13

17.6

Performing the Inverse Transform

As we begin our discussion of this topic, let us review the procedure we will use in applying the Laplace transform to circuit analysis. First, we will transform the problem in the time domain to the complex frequency domain. Second, we will solve the circuit equations algebraically in the complex frequency domain. Finally, we will transform the solution in the frequency domain back to the time domain. It is this latter operation that we discuss now.

The algebraic solution of the circuit equations in the complex frequency domain results in a rational function of s of the form

$$\mathbf{F}(s) = \frac{\mathbf{P}(s)}{\mathbf{Q}(s)} = \frac{a_m s^m + a_{m-1} s^{m-1} + \cdots + a_1 s + a_0}{b_n s^n + b_{n-1} s^{n-1} + \cdots + b_1 s + b_0} \tag{17.20}$$

The roots of the polynomial $\mathbf{P}(s)$ are called the *zeros* of the function $\mathbf{F}(s)$ because at these values of s, $\mathbf{F}(s) = 0$. Similarly, the roots of the polynomial $\mathbf{Q}(s)$ are called *poles* of $\mathbf{F}(s)$, since at these values of s, $\mathbf{F}(s)$ becomes infinite.

If $\mathbf{F}(s)$ is a proper rational function of s, then $n > m$. However, if this is not the case, we simply divide $\mathbf{P}(s)$ by $\mathbf{Q}(s)$ to obtain a quotient and a remainder, that is,

$$\frac{\mathbf{P}(s)}{\mathbf{Q}(s)} = C_{m-n} s^{m-n} + \cdots + C_2 s^2 + C_1 s + C_0 + \frac{\mathbf{P}_1(s)}{\mathbf{Q}(s)} \tag{17.21}$$

Now $\mathbf{P}_1(s)/\mathbf{Q}(s)$ is a proper rational function of s. Let us examine the possible forms of the roots of $\mathbf{Q}(s)$.

1. If the roots are simple, $\mathbf{P}_1(s)/\mathbf{Q}(s)$ can be expressed in partial fraction form as

$$\frac{\mathbf{P}_1(s)}{\mathbf{Q}(s)} = \frac{K_1}{s + p_1} + \frac{K_2}{s + p_2} + \cdots + \frac{K_n}{s + p_n} \tag{17.22}$$

2. If $\mathbf{Q}(s)$ has simple complex roots, they will appear in complex-conjugate pairs and the partial fraction expansion of $\mathbf{P}_1(s)/\mathbf{Q}(s)$ for each pair of complex-conjugate roots will be of the form

$$\frac{\mathbf{P}_1(s)}{\mathbf{Q}_1(s)(s + \alpha - j\beta)(s + \alpha + j\beta)} = \frac{K_1}{s + \alpha - j\beta} + \frac{K_1^*}{s + \alpha + j\beta} + \cdots \tag{17.23}$$

where $\mathbf{Q}(s) = \mathbf{Q}_1(s)(s + \alpha - j\beta)(s + \alpha + j\beta)$ and K_1^* is the complex conjugate of K_1.

3. If $\mathbf{Q}(s)$ has a root of multiplicity r, the partial fraction expansion for each such root will be of the form

$$\frac{\mathbf{P}_1(s)}{\mathbf{Q}_1(s)(s + p_1)^r} = \frac{K_{11}}{(s + p_1)} + \frac{K_{12}}{(s + p_1)^2} + \cdots + \frac{K_{1r}}{(s + p_1)^r} + \cdots \tag{17.24}$$

The importance of these partial fraction expansions stems from the fact that once the function $\mathbf{F}(s)$ is expressed in this form, the individual inverse Laplace transforms can be obtained from known and tabulated transform pairs. The sum of these inverse Laplace transforms then yields the desired time function, $f(t) = \mathscr{L}^{-1}[\mathbf{F}(s)]$.

Simple Poles

Let us assume that all the poles of $\mathbf{F}(s)$ are simple, so that the partial fraction expansion of $\mathbf{F}(s)$ is of the form

$$\mathbf{F}(s) = \frac{\mathbf{P}(s)}{\mathbf{Q}(s)} = \frac{K_1}{s + p_1} + \frac{K_2}{s + p_2} + \cdots + \frac{K_n}{s + p_n} \tag{17.25}$$

Then the constant K_i can be computed by multiplying both sides of this equation by $(s + p_i)$ and evaluating the equation at $s = -p_i$, that is,

$$\left.\frac{(s + p_i)\mathbf{P}(s)}{\mathbf{Q}(s)}\right|_{s=-p_i} = 0 + \cdots + 0 + K_i + 0 + \cdots + 0 \qquad i = 1, 2, \ldots, n \tag{17.26}$$

Once all of the K_i terms are known, the time function $f(t) = \mathscr{L}^{-1}[\mathbf{F}(s)]$ can be obtained using the Laplace transform pair

$$\mathscr{L}^{-1}\left[\frac{1}{s + a}\right] = e^{-at} \tag{17.27}$$

EXAMPLE 17.17

Given that

$$\mathbf{F}(s) = \frac{12(s + 1)(s + 3)}{s(s + 2)(s + 4)(s + 5)}$$

let us find the function $f(t) = \mathscr{L}^{-1}[\mathbf{F}(s)]$.

Expressing $\mathbf{F}(s)$ in a partial fraction expansion, we obtain

$$\frac{12(s + 1)(s + 3)}{s(s + 2)(s + 4)(s + 5)} = \frac{K_0}{s} + \frac{K_1}{s + 2} + \frac{K_2}{s + 4} + \frac{K_3}{s + 5}$$

To determine K_0, we multiply both sides of the equation by s to obtain the equation

$$\frac{12(s + 1)(s + 3)}{(s + 2)(s + 4)(s + 5)} = K_0 + \frac{K_1 s}{s + 2} + \frac{K_2 s}{s + 4} + \frac{K_3 s}{s + 5}$$

Evaluating the equation at $s = 0$ yields

$$\frac{(12)(1)(3)}{(2)(4)(5)} = K_0 + 0 + 0 + 0$$

or

$$K_0 = \frac{36}{40}$$

Similarly,

$$\left.(s + 2)\mathbf{F}(s)\right|_{s=-2} = \left.\frac{12(s + 1)(s + 3)}{s(s + 4)(s + 5)}\right|_{s=-2} = K_1$$

or

$$K_1 = 1$$

Using the same approach, we find that $K_2 = \frac{36}{8}$ and $K_3 = -\frac{32}{5}$. Hence $\mathbf{F}(s)$ can be written as

$$\mathbf{F}(s) = \frac{36/40}{s} + \frac{1}{s + 2} + \frac{36/8}{s + 4} - \frac{32/5}{s + 5}$$

Then $f(t) = \mathcal{L}^{-1}[\mathbf{F}(s)]$ is

$$f(t) = \left(\frac{36}{40} + 1e^{-2t} + \frac{36}{8}e^{-4t} - \frac{32}{5}e^{-5t}\right)u(t)$$

■

DRILL EXERCISE

D17.14. Find $f(t)$ if $\mathbf{F}(s) = 10(s + 6)/(s + 1)(s + 3)$.

D17.15. If $\mathbf{F}(s) = 12(s + 2)/s(s + 1)$, find $f(t)$.

Complex-Conjugate Poles

Let us assume that $\mathbf{F}(s)$ has one pair of complex-conjugate poles. The partial fraction expansion of $\mathbf{F}(s)$ can then be written as

$$\mathbf{F}(s) = \frac{\mathbf{P}_1(s)}{\mathbf{Q}_1(s)(s + \alpha - j\beta)(s + \alpha + j\beta)} = \frac{K_1}{s + \alpha - j\beta} + \frac{K_1^*}{s + \alpha + j\beta} + \cdots \tag{17.28}$$

The constant K_1 can then be determined using the procedure employed for simple poles, that is,

$$(s + \alpha - j\beta)\mathbf{F}(s)\Big|_{s=-\alpha+j\beta} = K_1 \tag{17.29}$$

In this case K_1 is in general a complex number that can be expressed as $|K_1| \underline{/\theta}$. Then $K_1^* = |K_1| \underline{/-\theta}$. Hence, the partial fraction expansion can be expressed in the form

$$\begin{aligned}\mathbf{F}(s) &= \frac{|K_1| \underline{/\theta}}{s + \alpha - j\beta} + \frac{|K_1| \underline{/-\theta}}{s + \alpha + j\beta} + \cdots \\ &= \frac{|K_1|e^{j\theta}}{s + \alpha - j\beta} + \frac{|K_1|e^{-j\theta}}{s + \alpha + j\beta} + \cdots\end{aligned} \tag{17.30}$$

The corresponding time function is then of the form

$$\begin{aligned}f(t) = \mathcal{L}^{-1}[\mathbf{F}(s)] &= |K_1|e^{j\theta}e^{-(\alpha - j\beta)t} + |K_1|e^{-j\theta}e^{-(\alpha + j\beta)t} + \cdots \\ &= |K_1|e^{-\alpha t}[e^{j(\beta t + \theta)} + e^{-j(\beta t + \theta)}] + \cdots \\ &= 2|K_1|e^{-\alpha t}\cos(\beta t + \theta) + \cdots\end{aligned} \tag{17.31}$$

EXAMPLE 17.18

Let us determine the time function $f(t)$ for the function

$$\mathbf{F}(s) = \frac{10(s + 1)}{s(s^2 + 2s + 2)}$$

The function $\mathbf{F}(s)$ can be expressed as a partial fraction expansion of the form

$$\mathbf{F}(s) = \frac{10(s + 1)}{s(s + 1 - j1)(s + 1 + j1)} = \frac{K_0}{s} + \frac{K_1}{s + 1 - j1} + \frac{K_1^*}{s + 1 + j1}$$

Then

$$\mathbf{s F}(s)\Big|_{s=0} = K_0$$

$$\frac{(10)(1)}{(1 + j1)(1 - j1)} = 5 = K_0$$

In a similar manner,

$$(s + 1 - j1)\mathbf{F}(s)\Big|_{s=-1+j1} = K_1$$

$$\frac{(10)(-1 + j1 + 1)}{(-1 + j1)(-1 + j1 + 1 + j1)} = K_1$$

$$\frac{5}{\sqrt{2}}\underline{/-135^\circ} = K_1$$

Then, of course,

$$K_1^* = \frac{5}{\sqrt{2}}\underline{/135^\circ}$$

Hence

$$\mathbf{F}(s) = \frac{5}{s} + \frac{(5/\sqrt{2})\,\underline{/-135^\circ}}{s + 1 - j1} + \frac{(5/\sqrt{2})\,\underline{/135^\circ}}{s + 1 + j1}$$

and therefore

$$f(t) = [5 + 5\sqrt{2}e^{-t}\cos(t - 135^\circ)]u(t) \qquad ■$$

DRILL EXERCISE

D17.16. Determine $f(t)$ if $\mathbf{F}(s) = s/(s^2 + 4s + 8)$.

D17.17. If $\mathbf{F}(s) = (s + 1)/(s^2 + 6s + 13)$, find $f(t)$.

Multiple Poles

Let us suppose that $\mathbf{F}(s)$ has a pole of multiplicity r. Then $\mathbf{F}(s)$ can be written in a partial fraction expansion of the form

$$\mathbf{F}(s) = \frac{\mathbf{P}_1(s)}{\mathbf{Q}_1(s)(s + p_1)^r} = \frac{K_{11}}{s + p_1} + \frac{K_{12}}{(s + p_1)^2} + \cdots + \frac{K_{1r}}{(s + p_1)^r} + \cdots \tag{17.32}$$

Employing the approach for a simple pole, we can evaluate K_{1r} as

$$(s + p_1)^r\mathbf{F}(s)\Big|_{s=-p_1} = K_{1r} \tag{17.33}$$

In order to evaluate $K_{1_{r-1}}$ we multiply $\mathbf{F}(s)$ by $(s + p_1)^r$ as we did to determine K_{1_r}; however, prior to evaluating the equation at $s = -p_1$, we take the derivative with respect to s. The proof that this will yield $K_{1_{r-1}}$ can be obtained by multiplying both sides of equation (17.32) by $(s + p_1)^r$ and then taking the derivative with respect to s. Now when we evaluate the equation at $s = -p_1$, the only term remaining on the right side of the equation is $K_{1_{r-1}}$, and therefore

$$\frac{d}{ds}[(s + p_1)^r\mathbf{F}(s)]\Big|_{s=-p_1} = K_{1_{r-1}} \tag{17.34}$$

$K_{1_{r-2}}$ can be computed in a similar fashion and in that case the equation is

$$\frac{d^2}{ds^2}[(s + p_1)^r\mathbf{F}(s)\Big|_{s=-p_1} = (2!)K_{1_{r-2}} \tag{17.35}$$

The general expression for this case is

$$K_{1j} = \frac{1}{(r - j)!}\frac{d^{r-j}}{ds^{r-j}}[(s + p_1)^r\mathbf{F}(s)]\Big|_{s=-p_1} \tag{17.36}$$

Let us illustrate this procedure with an example.

EXAMPLE 17.19

Given the following function $\mathbf{F}(s)$, let us determine the corresponding time function $f(t) = \mathcal{L}^{-1}[\mathbf{F}(s)]$.

$$\mathbf{F}(s) = \frac{10(s + 3)}{(s + 1)^3(s + 2)}$$

Expressing $\mathbf{F}(s)$ as a partial fraction expansion,

$$\mathbf{F}(s) = \frac{10(s + 3)}{(s + 1)^3(s + 2)} = \frac{K_{11}}{s + 1} + \frac{K_{12}}{(s + 1)^2} + \frac{K_{13}}{(s + 1)^3} + \frac{K_2}{s + 2}$$

Then

$$(s + 1)^3\mathbf{F}(s)\Big|_{s=-1} = K_{13}$$

$$20 = K_{13}$$

K_{12} is now determined by the equation

$$\frac{d}{ds}[(s + 1)^3\mathbf{F}(s)]\Big|_{s=-1} = K_{12}$$

$$\frac{-10}{(s + 2)^2}\Big|_{s=-1} = -10 = K_{12}$$

In a similar fashion K_{11} is computed from the equation

$$\left.\frac{d^2}{ds^2}[(s+1)^3\mathbf{F}(s)]\right|_{s=-1} = 2K_{11}$$

$$\left.\frac{20}{(s+2)^3}\right|_{s=-1} = 20 = 2K_{11}$$

Therefore,

$$10 = K_{11}$$

In addition,

$$\left.(s+2)\mathbf{F}(s)\right|_{s=-2} = K_2$$

$$-10 = K_2$$

Hence $\mathbf{F}(s)$ can be expressed as

$$\mathbf{F}(s) = \frac{10}{s+1} - \frac{10}{(s+1)^2} + \frac{20}{(s+1)^3} - \frac{10}{s+2}$$

Now we employ the transform pair

$$\mathcal{L}^{-1}\left[\frac{1}{(s+a)^{n+1}}\right] = \frac{t^n}{n!}e^{-at}$$

and hence

$$f(t) = (10e^{-t} - 10te^{-t} + 10t^2e^{-t} - 10e^{-2t})u(t)$$ ■

DRILL EXERCISE

D17.18. Determine $f(t)$ if $\mathbf{F}(s) = s/(s+1)^2$.

D17.19. If $\mathbf{F}(s) = (s+2)/s^2(s+1)$, find $f(t)$.

17.7 Convolution Integral

Convolution is a very important concept and has wide application in circuit and systems analysis. We will use it in both this chapter and the next. In this chapter we employ it to determine the inverse transformation of the product of two functions $\mathbf{F}_1(s)$ and $\mathbf{F}_2(s)$ in terms of their inverse transforms $f_1(t)$ and $f_2(t)$. In Chapter 18 we use convolution to determine the response of a linear circuit to some arbitrary input in terms of its response to a unit impulse.

Theorem 11

If $f(t) = \int_0^t f_1(t - \lambda)f_2(\lambda)\,d\lambda = \int_0^t f_1(\lambda)f_2(t - \lambda)\,d\lambda$ (17.37)

and $\mathcal{L}[f(t)] = \mathbf{F}(s)$, $\mathcal{L}[f_1(t)] = \mathbf{F}_1(s)$, *and* $\mathcal{L}[f_2(t)] = \mathbf{F}_2(s)$, *then*

$$\mathbf{F}(s) = \mathbf{F}_1(s)\mathbf{F}_2(s) \tag{17.38}$$

Our demonstration begins with the definition

$$\mathcal{L}[f(t)] = \int_0^\infty \left[\int_0^t f_1(t - \lambda)f_2(\lambda)\,d\lambda\right] e^{-st}\,dt$$

We now force the function into the proper format by introducing into the integral within the brackets the unit step function $u(t - \lambda)$. We can do this because

$$u(t - \lambda) = \begin{cases} 1 & \text{for } \lambda < t \\ 0 & \text{for } \lambda > t \end{cases} \tag{17.39}$$

The first condition in Eq. (17.39) ensures that the insertion of the unit step function has no impact within the limits of integration. The second condition in Eq. (17.39) allows us to change the upper limit of integration from t to ∞. Therefore,

$$\mathcal{L}[f(t)] = \int_0^\infty \left[\int_0^\infty f_1(t - \lambda)u(t - \lambda)f_2(\lambda)\,d\lambda\right] e^{-st}\,dt$$

which can be written as

$$\mathcal{L}[f(t)] = \int_0^\infty f_2(\lambda)\left[\int_0^\infty f_1(t - \lambda)u(t - \lambda)e^{-st}\,dt\right] d\lambda$$

Note that the integral within the brackets is the shifting theorem illustrated in Example 17.9. Hence the equation can be written as

$$\begin{aligned} \mathcal{L}[f(t)] &= \int_0^\infty f_2(\lambda)\mathbf{F}_1(s)e^{-s\lambda}\,d\lambda \\ &= \mathbf{F}_1(s)\int_0^\infty f_2(\lambda)e^{-s\lambda}\,d\lambda \\ &= \mathbf{F}_1(s)\mathbf{F}_2(s) \end{aligned}$$

Convolution is often represented in shorthand notation as

$$f(t) = f_1(t) * f_2(t)$$

Therefore,

$$\begin{aligned} f(t) &= f_1(t) * f_2(t) \\ &= \int_0^t f_1(\lambda)f_2(t - \lambda)\,d\lambda \\ &= \int_0^t f_1(t - \lambda)f_2(\lambda)\,d\lambda \end{aligned} \tag{17.40}$$

and hence

$$\mathbf{F}(s) = \mathbf{F}_1(s)\mathbf{F}_2(s)$$

Note that the convolution in the time domain corresponds to multiplication in the frequency domain. Thus convolution affords us a method of obtaining the inverse Laplace transform by splitting $\mathbf{F}(s)$ up into the product of two factors $\mathbf{F}_1(s)$ and $\mathbf{F}_2(s)$ whose inverse transforms are known. Note that this technique is quite different from the partial-fraction expansion method, in which $\mathbf{F}(s)$ is split into a sum of factors whose inverse transforms are known.

Let us now illustrate the use of convolution in the evaluation of inverse Laplace transforms.

EXAMPLE 17.20

Let us determine the inverse Laplace transform of the function $\mathbf{F}(s)$ where

$$\mathbf{F}(s) = \left(\frac{s}{s+1}\right)^2$$

Using convolution, we let

$$\mathbf{F}_1(s) = \mathbf{F}_2(s) = \frac{s}{s+1} = 1 - \frac{1}{s+1}$$

Therefore,

$$f_1(t) = f_2(t) = \delta(t) - e^{-t}$$

Hence

$$\begin{aligned}
f(t) &= \int_0^t f_1(\lambda)f_2(t-\lambda)\,d\lambda \\
&= \int_0^t [\delta(\lambda) - e^{-\lambda}][\delta(t-\lambda) - e^{-(t-\lambda)}]\,d\lambda \\
&= \int_0^t \delta(\lambda)\delta(t-\lambda)\,d\lambda - \int_0^t \delta(\lambda)e^{-(t-\lambda)}\,d\lambda \\
&\quad - \int_0^t \delta(t-\lambda)e^{-\lambda}\,d\lambda + \int_0^t e^{-\lambda}e^{-(t-\lambda)}\,d\lambda
\end{aligned}$$

Employing the sampling property of the impulse function as illustrated in Eq. (16.43), we obtain

$$[f(t) = \delta(t) - 2e^{-t} + te^{-t}]\,u(t)$$

which is the inverse transform of $\mathbf{F}(s)$.

For comparison, let us determine $f(t)$ from $\mathbf{F}(s)$ using the partial-fraction expansion method described earlier. $\mathbf{F}(s)$ can be written as

$$\mathbf{F}(s) = \left(\frac{s}{s+1}\right)^2 = 1 - \frac{2s+1}{(s+1)^2}$$

The second term can then be expanded as

$$\frac{2s+1}{(s+1)^2} = \frac{K_{11}}{s+1} + \frac{K_{12}}{(s+1)^2}$$

Evaluating the constants, we obtain $K_{11} = 2$ and $K_{12} = -1$. Therefore,

$$\mathbf{F}(s) = 1 - \frac{2}{s+1} + \frac{1}{(s+1)^2}$$

and hence

$$[f(t) = \delta(t) - 2e^{-t} + te^{-t}]\, u(t)$$

■

EXAMPLE 17.21

Given the following $\mathbf{F}(s)$, let us determine $f(t)$.

$$\mathbf{F}(s) = \frac{s+1}{s(s^2+4)}$$

$\mathbf{F}(s)$ can be written as the following product:

$$\mathbf{F}(s) = \frac{1}{s}\left[\frac{s}{s^2+4} + \frac{2}{2(s^2+4)}\right]$$

If we let

$$\mathbf{F}_1(s) = \frac{1}{s}$$

$$\mathbf{F}_2(s) = \frac{s}{s^2+4} + \frac{1}{2}\left(\frac{2}{s^2+4}\right)$$

then

$$f_1(t) = u(t)$$

$$f_2(t) = \cos 2t + \tfrac{1}{2}\sin 2t$$

Using the convolution integral yields

$$f(t) = \int_0^t u(\lambda)[\cos 2(t-\lambda) + \tfrac{1}{2}\sin 2(t-\lambda)]\, d\lambda$$

$$= \int_0^t u(\lambda)\cos 2(t-\lambda)\, d\lambda + \tfrac{1}{2}\int_0^t u(\lambda)\sin 2(t-\lambda)\, d\lambda$$

Expanding the function using Eq. (9.11), we obtain

$$f(t) = \int_0^t (\cos 2t \cos 2\lambda + \sin 2t \sin 2\lambda)\, d\lambda$$

$$+ \tfrac{1}{2} \int_0^t (\sin 2t \cos 2\lambda - \cos 2t \sin 2\lambda) u(\lambda)\, d\lambda$$

$$= \cos 2t \int_0^t \cos 2\lambda u(\lambda)\, d\lambda + \sin 2t \int_0^t \sin 2\lambda u(\lambda)\, d\lambda$$

$$+ \tfrac{1}{2} \sin 2t \int_0^t \cos 2\lambda u(\lambda)\, d\lambda - \tfrac{1}{2} \cos 2t \int_0^t \sin 2\lambda u(\lambda)\, d\lambda$$

Performing the indicated integration and noting that $u(\lambda) = 1$ in the range of integration yields

$$f(t) = \left[\frac{1}{4} + \frac{\sin 2t}{2} - \frac{\cos 2t}{4}\right] u(t)$$

For comparison let us rework the problem and translate $f_1(t)$ rather than $f_2(t)$. In this case

$$f(t) = \int_0^t \cos 2t u(t - \lambda)\, d\lambda + \tfrac{1}{2} \int_0^t \sin 2\lambda u(t - \lambda)\, d\lambda$$

Once again $u(t - \lambda) = 1$ over the range of integration, and hence

$$f(t) = \tfrac{1}{2} \sin 2t \Big|_0^t - \tfrac{1}{4} \cos 2\lambda \Big|_0^t$$

$$= [\tfrac{1}{2} \sin 2t - \tfrac{1}{4} \cos 2t + \tfrac{1}{4}]\, u(t) \qquad \blacksquare$$

This example illustrates a very important point. When employing the convolution integral

$$f(t) = \int_0^t f_1(\lambda) f_2(t - \lambda)\, d\lambda$$

always translate the simpler function; that is, always select the simpler function as $f_2(t)$.

DRILL EXERCISE

D17.20. Given the function $\mathbf{F}(s) = 1/(s + 1)(s + 2)(s + 3)$, find $f(t)$ using the convolution integral. Let $\mathbf{F}_1(s) = 1/(s + 1)(s + 2)$ and $\mathbf{F}_2(s) = 1/(s + 3)$.

D17.21. Use the convolution integral to find $f(t)$ if $\mathbf{F}(s)$ is given by the expression $\mathbf{F}(s) = 1/s^2(s + 1)$.

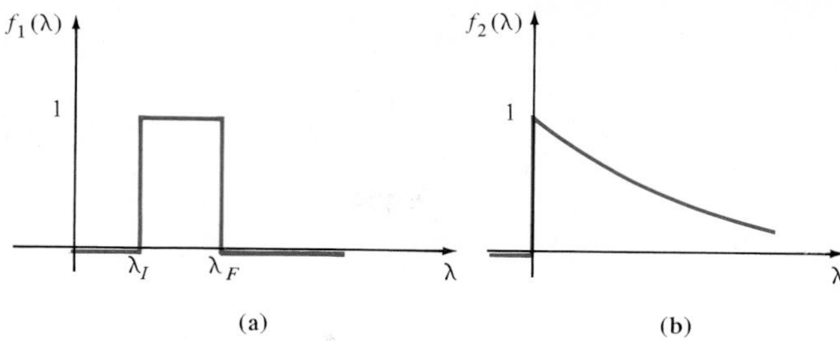

Figure 17.6 Two functions $f_1(\lambda)$ and $f_2(\lambda)$.

In an attempt to interpret the operation of the convolution integral, let us examine it from a graphical standpoint. Suppose that the functions $f_1(\lambda)$ and $f_2(\lambda)$ are as shown in Fig. 17.6. If $f_2(\lambda) = e^{-\alpha\lambda}u(\lambda)$, the time-shifted function

$$f_2(\lambda - t) = e^{-\alpha(\lambda - t)}u(\lambda - t)$$

is as shown in Fig. 17.7a. If we now change the sign of the argument so that the function is

$$f_2(t - \lambda) = e^{-\alpha(t-\lambda)}u(t - \lambda),$$

then the function is folded or reflected about the point $\lambda = t$ as shown in Fig. 17.7b. The integrand of the convolution integral, $f_1(\lambda)f_2(t - \lambda)$, will be nonzero only when the two functions overlap. The overlap, or product of the two functions, for several values of t is shown shaded in Fig. 17.8. The convolution integral for various values of t is equal to the shaded area shown in Fig. 17.8. A plot of these areas as t varies is shown in Fig. 17.9. If we had shifted and folded the function $f_1(\lambda)$, multiplied it by $f_2(\lambda)$ and integrated, we would again obtain the curve in Fig. 17.9.

This graphical explanation of convolution has hopefully provided some additional insight. It serves one other purpose, however. This graphical development indicates that convolution can be performed on two functions which may be available only as experimental curves and cannot be represented as analytical functions.

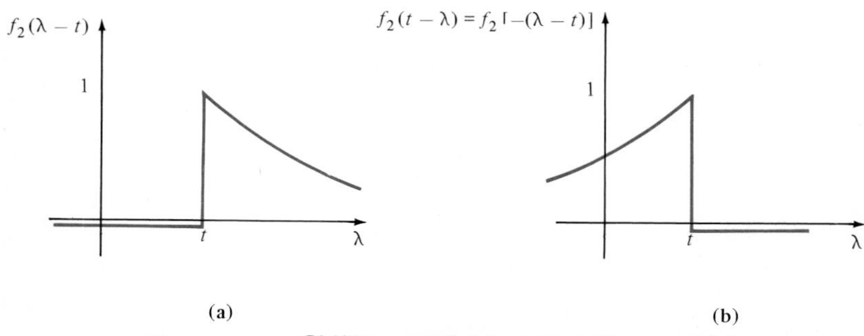

Figure 17.7 Shifting and folding operation on $f_2(\lambda)$.

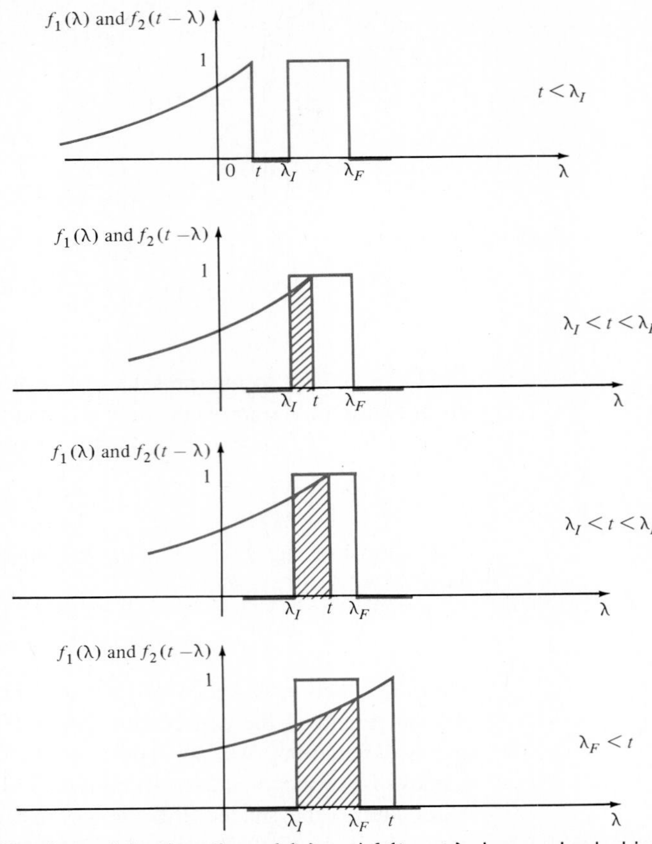

Figure 17.8 Product of the functions $f_1(\lambda)$ and $f_2(t - \lambda)$ shown shaded in the figures for different values of t.

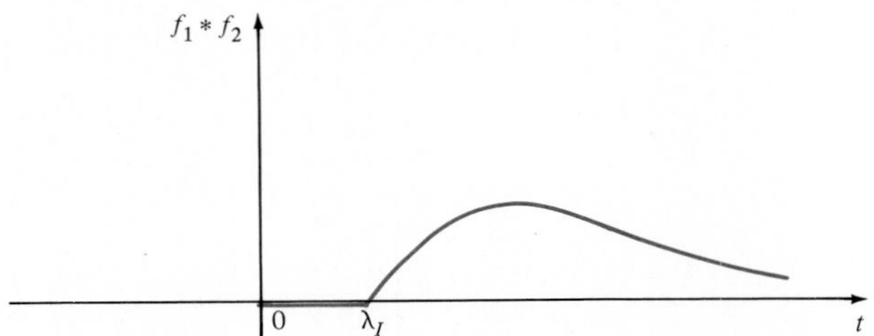

Figure 17.9 Plot of the convolution integral.

EXAMPLE 17.22

As a final example of the use of the convolution integral in the computation of an inverse Laplace transform, consider the following function:

$$\mathbf{F}(s) = \frac{10e^{-4s}}{s(s+2)}$$

We find it instructive to split this function into a product of the two following terms:

$$\mathbf{F}_1(s) = \frac{10e^{-s}}{s}$$

$$\mathbf{F}_2(s) = \frac{e^{-3s}}{s + 2}$$

The inverse transform of each term is

$$f_1(t) = 10u(t - 1)$$

$$f_2(t) = e^{-2(t-3)}u(t - 3)$$

If we translate the function $f_1(t)$, the convolution integral is

$$f(t) = 10 \int_0^t e^{-2(\lambda-3)}u(\lambda - 3)u(t - 1 - \lambda)\, d\lambda$$

Let us now examine the two unit step functions in the integrand. $u(\lambda - 3)$ is shown in Fig. 17.10a. $u(t - 1 - \lambda)$ is plotted in Fig. 17.10b. Note that the product $u(\lambda - 3)u(t - 1 - \lambda)$ will be nonzero, as shown in Fig. 17.10c, provided that $(t - 1) > 3$ or that $t > 4$. Hence the convolution integral is

$$f(t) = \begin{cases} 10e^6 \int_3^{t-1} e^{-2\lambda}\, d\lambda & \text{for } t > 4 \\ 0 & \text{for } t < 4 \end{cases}$$

Therefore,

$$f(t) = \frac{10e^6}{-2} e^{-2\lambda}\Big|_3^{t-1} = 5 - 5e^{-2(t-4)} \qquad \text{for } t > 4$$

or, in other words,

$$f(t) = [5 - 5e^{-2(t-4)}]u(t - 4)$$

Let us now translate the function $f_2(t)$ so that the convolution integral is

$$f(t) = \int_0^t 10u(\lambda - 1)e^{-2(t-3-\lambda)}u(t - 3 - \lambda)\, d\lambda$$

Once again the unit steps and their product are shown in Fig. 17.10d–f. The product term will be nonzero only if $(t - 3) > 1$ or $t > 4$. Hence the convolution integral becomes

$$f(t) = \begin{cases} 10e^{-2(t-3)} \int_1^{t-3} e^{2\lambda}\, d\lambda & \text{for } t > 4 \\ 0 & \text{for } t < 4 \end{cases}$$

Hence

$$f(t) = \frac{10e^{-2(t-3)}}{2} e^{2\lambda}\Big|_1^{t-3} \qquad \text{for } t > 4$$

$$5 - 5e^{-2(t-4)} \qquad \text{for } t > 4$$

or

$$f(t) = [5 - 5e^{-2(t-4)}]u(t - 4)$$

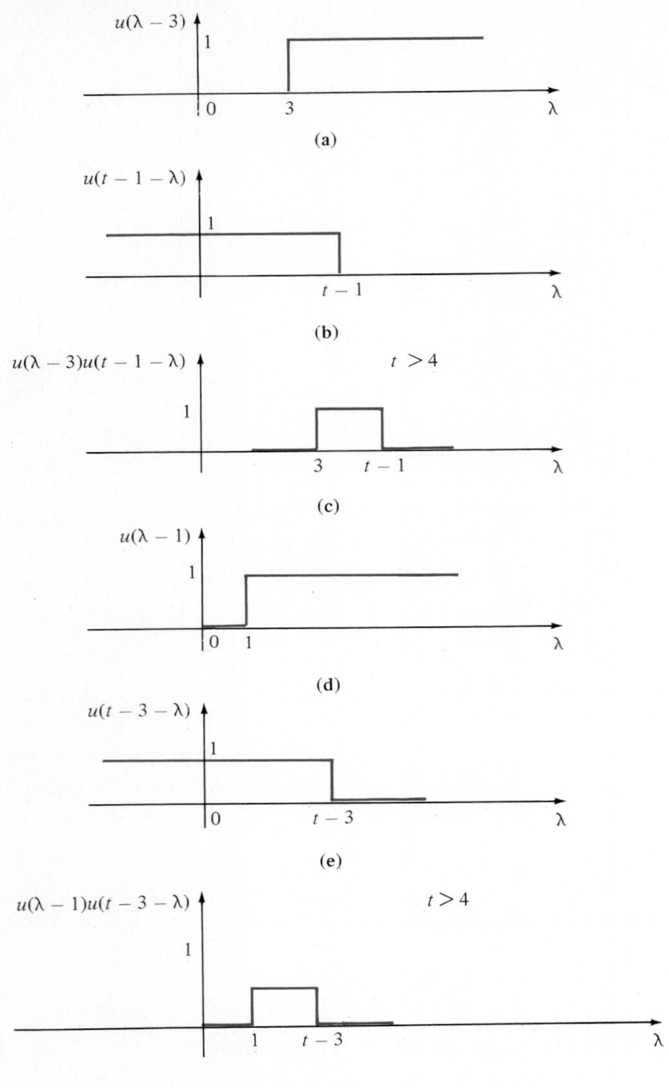

Figure 17.10 Plots employed in Example 17.22.

As a final check, we can simply take the inverse Laplace transform directly.

$$\mathbf{F}(s) = \frac{10e^{-4s}}{s(s+2)} = \left(\frac{5}{s} - \frac{5}{s+2}\right)e^{-4s}$$

Using the property

$$\mathcal{L}^{-1}[e^{-t_0 s}\mathbf{F}(s)] = f(t - t_0)u(t - t_0) \qquad t_0 \geq 0$$

$f(t)$ can be immediately written as

$$f(t) = (5 - 5e^{-2(t-4)})u(t - 4)$$

■

DRILL EXERCISE

D17.22. Use the convolutional integral to find $f(t)$ if

$$\mathbf{F}(s) = \frac{e^{-2s}}{(s+1)(s+3)}$$

17.8 Initial-Value and Final-Value Theorems

Suppose that we wish to determine the initial or final value of a circuit response in the time domain from the Laplace transform of the function in the s-domain without performing the inverse transform. If we determine the function $f(t) = \mathcal{L}^{-1}[\mathbf{F}(s)]$, we can find the initial value by evaluating $f(t)$ as $t \to 0$ and the final value by evaluating $f(t)$ as $t \to \infty$. It would be very convenient, however, if we could simply determine the initial and final values from $\mathbf{F}(s)$ without having to perform the inverse transform. The initial- and final-value theorems allow us to do just that.

The *initial-value theorem* states that

$$\lim_{t\to 0} f(t) = \lim_{s\to\infty} s\mathbf{F}(s) \tag{17.41}$$

provided that $f(t)$ and its first derivative are transformable.

The proof of this theorem employs the Laplace transform of the function $df(t)/dt$.

$$\int_0^\infty \frac{df(t)}{dt} e^{-st}\, dt = s\mathbf{F}(s) - f(0)$$

Taking the limit of both sides as $s \to \infty$,

$$\lim_{s\to\infty} \int_0^\infty \frac{df(t)}{dt} e^{-st}\, dt = \lim_{s\to\infty} [s\mathbf{F}(s) - f(0)]$$

and since

$$\int_0^\infty \frac{df(t)}{dt} \lim_{s\to\infty} e^{-st}\, dt = 0$$

then

$$f(0) = \lim_{s\to\infty} s\mathbf{F}(s)$$

which is, of course,

$$\lim_{t\to 0} f(t) = \lim_{s\to\infty} s\mathbf{F}(s)$$

The *final-value theorem* states that

$$\lim_{t\to\infty} f(t) = \lim_{s\to 0} s\mathbf{F}(s) \tag{17.42}$$

provided that $f(t)$ and its first derivative are transformable, and that $f(\infty)$ exists. This latter requirement means that all the poles of $\mathbf{F}(s)$ must be of the form $(s + \alpha)$ or

$(s + \alpha \pm j\beta)$ and cannot be of the form $(s - \alpha)$, since $\mathcal{L}^{-1}[1/(s - \alpha)] = e^{\alpha t} \rightarrow \infty$ as $t \rightarrow \infty$ for $\alpha > 0$.

The proof of this theorem also involves the Laplace transform of the function $df(t)/dt$.

$$\int_0^\infty \frac{df(t)}{dt} e^{-st}\, dt = s\mathbf{F}(s) - f(0)$$

Taking the limit of both sides as $s \rightarrow 0$ gives us

$$\lim_{s\rightarrow 0} \int_0^\infty \frac{df(t)}{dt} e^{-st}\, dt = \lim_{s\rightarrow 0} [s\mathbf{F}(s) - f(0)]$$

Therefore,

$$\int_0^\infty \frac{df(t)}{dt}\, dt = \lim_{s\rightarrow 0} [s\mathbf{F}(s) - f(0)]$$

and

$$f(\infty) - f(0) = \lim_{s\rightarrow 0} s\mathbf{F}(s) - f(0)$$

and hence

$$f(\infty) = \lim_{t\rightarrow\infty} f(t) = \lim_{s\rightarrow 0} s\mathbf{F}(s)$$

EXAMPLE 17.23

Let us determine the initial and final values for the function in Example 17.18.

The function is

$$\mathbf{F}(s) = \frac{10(s + 1)}{s(s^2 + 2s + 2)}$$

and the corresponding time function is

$$f(t) = 5 + 5\sqrt{2}\, e^{-t} \cos(t - 135°)$$

Applying the initial-value theorem, we have

$$\begin{aligned} f(0) &= \lim_{s\rightarrow\infty} s\mathbf{F}(s) \\ &= \lim_{s\rightarrow\infty} \frac{10(s + 1)}{s^2 + 2s + 2} \\ &= 0 \end{aligned}$$

The final value is obtained from the expression

$$\begin{aligned} f(\infty) &= \lim_{s\rightarrow 0} s\mathbf{F}(s) \\ &= \lim_{s\rightarrow 0} \frac{10(s + 1)}{s^2 + 2s + 2} \\ &= 5 \end{aligned}$$

Note that these values could be obtained directly from the time function $f(t)$. ■

DRILL EXERCISE

D17.23. Find the initial and final values of the function $f(t)$ if $\mathbf{F}(s) = \mathscr{L}[f(t)]$ is given by the expression

$$\mathbf{F}(s): = \frac{(s + 1)^2}{s(s + 2)(s^2 + 2s + 2)}$$

17.9 Applications to Differential Equations

Let us now begin to apply the power of the Laplace transform. In this section we demonstrate its use in the solution of linear, constant-coefficient integrodifferential equations. In Chapter 18 we illustrate the use of the Laplace transform to circuit analysis. In our approach we employ the transform pairs we have presented and discussed to convert the integrodifferential equation in the time domain to an algebraic equation in the s-domain. The solution in the s-domain will then be converted back to the time domain using the transform pairs. The solution we obtain will be the complete solution since the initial conditions are automatically included in the transform process.

EXAMPLE 17.24

Given the following differential equation, let us find $y(t)$.

$$\frac{d^2y(t)}{dt^2} + \frac{5dy(t)}{dt} + 6y(t) = e^{-t}$$

where $y(0) = 1$ and $y'(0) = 1$.

Transforming the equation, we obtain

$$s^2\mathbf{Y}(s) - sy(0) - y'(0) + 5[s\mathbf{Y}(s) - y(0)] + 6\mathbf{Y}(s) = \frac{1}{s + 1}$$

or

$$\mathbf{Y}(s)[s^2 + 5s + 6] = \frac{1}{s + 1} + s + 1 + 5$$

Therefore,

$$\mathbf{Y}(s) = \frac{1}{(s + 1)(s^2 + 5s + 6)} + \frac{s + 6}{s^2 + 5s + 6}$$

The first term on the right-hand side of the equation is the response for zero initial conditions. The second term on the right-hand side of the equation modifies the complementary term to give the required initial conditions. Note that the particular integral term is not affected by this second term. Using the techniques described earlier to determine the inverse transform, we obtain

$$y(t) = (\tfrac{1}{2}e^{-t} + 3e^{-2t} - \tfrac{5}{2}e^{-3t})u(t)$$ ■

Our approach can be applied equally well to a set of simultaneous linear constant-coefficient differential equations. The following example illustrates this use.

EXAMPLE 17.25

Let us use the Laplace transform technique to solve the equations

$$\frac{dx(t)}{dt} + \frac{dy(t)}{dt} + 2y(t) = 4u(t)$$

$$2\frac{dx(t)}{dt} + x(t) + \frac{dy(t)}{dt} = 2u(t)$$

where $y(0) = 1$ and $x(0) = -1$.

Transforming the equations, we obtain

$$s\mathbf{X}(s) - x(0) + s\mathbf{Y}(s) - y(0) + 2\mathbf{Y}(s) = \frac{4}{s}$$

$$2s\mathbf{X}(s) - 2x(0) + \mathbf{X}(s) + s\mathbf{Y}(s) - y(0) = \frac{2}{s}$$

These equations reduce to

$$s\mathbf{X}(s) + (s + 2)\mathbf{Y}(s) = \frac{4}{s}$$

$$(2s + 1)\mathbf{X}(s) + s\mathbf{Y}(s) = \frac{2}{s} - 1$$

Solving the algebraic equations, we obtain

$$\mathbf{X}(s) = \frac{-(s^2 + 4s - 4)}{s(s^2 + 5s + 2)}$$

$$\mathbf{Y}(s) = \frac{s^2 + 6s + 4}{s(s^2 + 5s + 2)}$$

from which we obtain

$$x(t) = (2 - 1.14e^{-0.44t} - 1.86e^{-4.56t})u(t)$$

$$y(t) = (2 - 0.86e^{-0.44t} - 0.14e^{-4.56t})u(t)$$

■

EXAMPLE 17.26

Finally, let us employ the Laplace transform to solve the equation

$$\frac{dy(t)}{dt} + 2y(t) + \int_0^t y(\lambda)e^{-2(t-\lambda)}\,d\lambda = 10u(t) \qquad y(0) = 0$$

Applying the transform, we obtain

$$s\mathbf{Y}(s) + 2\mathbf{Y}(s) + \frac{\mathbf{Y}(s)}{s + 2} = \frac{10}{s}$$

$$\mathbf{Y}(s)\left(s + 2 + \frac{1}{s + 2}\right) = \frac{10}{s}$$

$$\mathbf{Y}(s) = \frac{10(s + 2)}{s(s^2 + 4s + 5)}$$

Expressing the function in a partial-fraction expansion, we obtain

$$\frac{10(s+2)}{s(s+2-j1)(s+2+j1)} = \frac{K_0}{s} + \frac{K_1}{s+2-j1} + \frac{K_1^*}{s+2+j1}$$

$$\left.\frac{10(s+2)}{s^2+4s+5}\right|_{s=0} = K_0$$

$$4 = K_0$$

In a similar manner,

$$\left.\frac{10(s+2)}{s(s+2+j1)}\right|_{s=-2+j1} = K_1$$

$$2.236 \underline{/-153.43^\circ} = K_1$$

Therefore,

$$2.236 \underline{/153.43^\circ} = K_1^*$$

The partial-fraction expansion of $\mathbf{Y}(s)$ is then

$$\mathbf{Y}(s) = \frac{4}{s} + \frac{2.236 \underline{/-153.43^\circ}}{s+2-j1} + \frac{2.236 \underline{/153.43^\circ}}{s+2+j1}$$

and therefore,

$$y(t) = [4 + 4.472e^{-2t} \cos (t - 153.43^\circ)]u(t)$$

■

DRILL EXERCISE

D17.24. Solve the differential equation

$$\frac{d^2y(t)}{dt^2} + 11\frac{dy(t)}{dt} + 30y(t) = 4u(t) \qquad y(0) = y'(0) = 0$$

D17.25. Determine $y(t)$ if

$$\frac{dy(t)}{dt} + \int_0^t y(\lambda)e^{-2(t-\lambda)}\,d\lambda = u(t) \qquad y(0) = 0$$

17.10 Summary

In this chapter we have introduced a very powerful analysis technique called the Laplace transform. We have shown that by using this technique we can transform a set of linear, constant coefficient integrodifferential equations in the time domain into a set of algebraic equations in the s-domain. The equations in the s-domain can be solved algebraically and then the solution is transformed back to the time domain. The transformation from the s-domain to the time domain was accomplished by using a set of known transform pairs. The convolution integral was introduced and its use in the inversion of Laplace transforms was demonstrated. Finally, the transform technique was applied to the solution of integrodifferential equations.

KEY POINTS

- The Laplace transform allows us to convert a problem from the time domain to the frequency domain, solve the problem using algebra in the frequency domain, and then convert the solution in the frequency domain back to the time domain.
- The Laplace transform is defined as

$$\mathscr{L}[f(t)] = \mathbf{F}(s) = \int_0^\infty f(t)e^{-st}\,dt$$

- A set of transform pairs exist that permit us to transform a function from the time domain to the frequency domain, and vice versa.
- Ten important properties of the Laplace transform have been presented that facilitate its use.
- The gate function aids us in determining the Laplace transform of a waveform in the time domain.
- The partial-fraction expansion method can be applied to determine the inverse Laplace transform of functions containing simple, multiple, and complex conjugate poles.
- The convolution integral can be used to determine the inverse Laplace transform.
- The initial- and final-value theorems can be applied to determine initial and final values of time-domain functions when their Laplace transforms are known.
- The Laplace transform can be used to solve integrodifferential equations. The complete solution is obtained since the initial conditions are automatically included in the transform process.

PROBLEMS

17.1. Use the shifting theorem to determine $\mathscr{L}[f(t)]$ where

$$f(t) = [t - 1 + e^{-(t-1)}]u(t - 1)$$

17.2. Use the shifting theorem to determine $\mathscr{L}[f(t)]$ where

$$f(t) = [e^{-(t-2)} - e^{-2(t-2)}]u(t - 2)$$

17.3. Use Theorem 5 to find $\mathscr{L}[f(t)]$ if

$$f(t) = e^{-at}u(t - 1)$$

17.4. Use Eq. (17.1) to determine $\mathscr{L}[f(t)]$ where $f(t)$ is defined as in Problem 17.3.

17.5. Use Theorem 5 to find $\mathscr{L}[f(t)]$ if

$$f(t) = e^{-at} \sin \omega t \; u(t - 1)$$

17.6. Use Eq. (17.1) to determine $\mathscr{L}[f(t)]$ where $f(t)$ is as defined in Problem 17.5.

17.7. Use Theorem 6 to find $\mathscr{L}[f(t)]$ if

$$f(t) = te^{-at}u(t - 1)$$

17.8. Use Eq. (17.1) to determine $\mathscr{L}[f(t)]$ where $f(t)$ is as defined in Problem 17.7.

17.9. If $f(t) = t \cos \omega t \; u(t - 1)$, find $\mathbf{F}(s)$.

17.10. Find the Laplace transform of the waveform shown in Fig. P17.10.

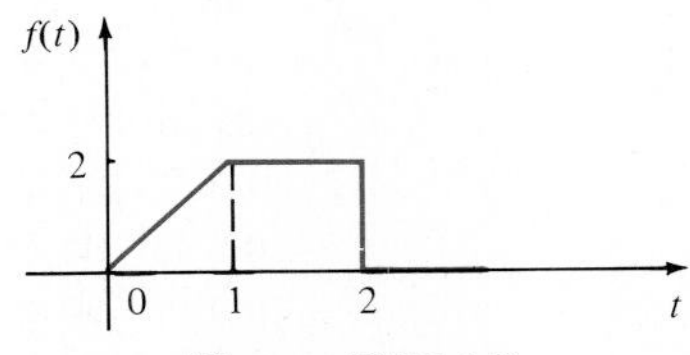

Figure P17.10

17.11. Determine the Laplace transform $\mathbf{F}(s)$ if $f(t)$ is as shown in Fig. P17.11.

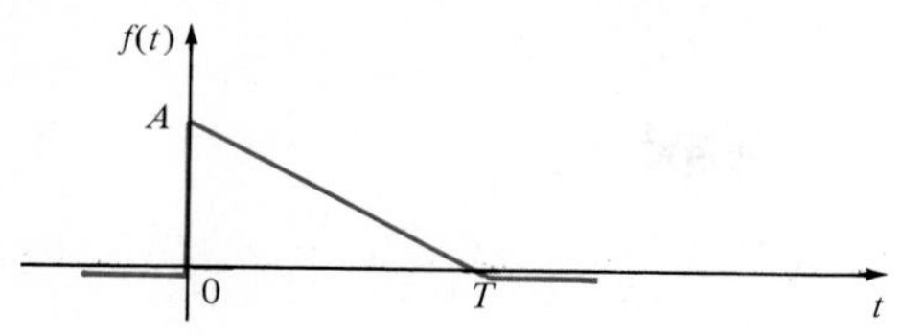

Figure P17.11

17.12. Determine the Laplace transform of the waveform shown in Fig. P17.12.

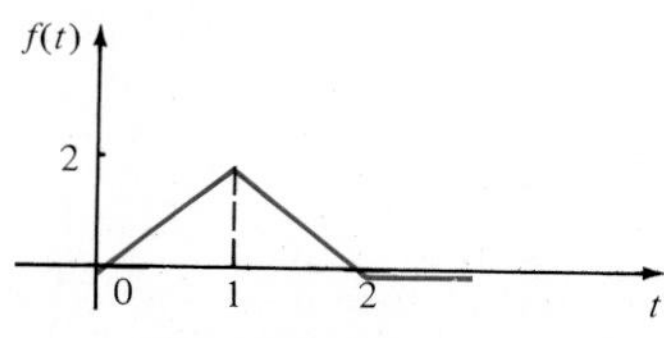

Figure P17.12

17.13. Determine the Laplace transform of the triangular waveform shown in Fig. P17.13.

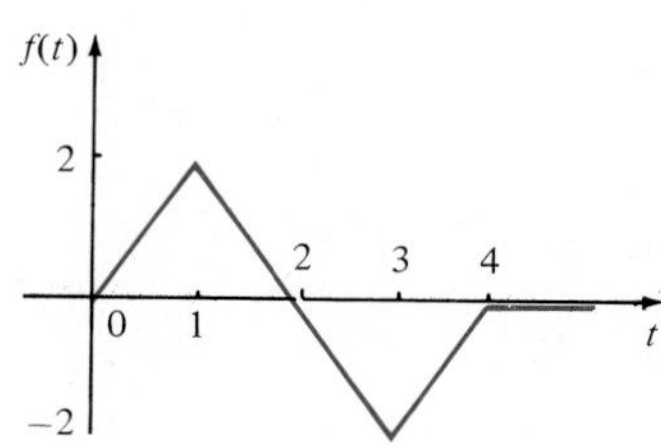

Figure P17.13

17.14. Find the Laplace transform of the periodic waveform shown in Fig. P17.14.

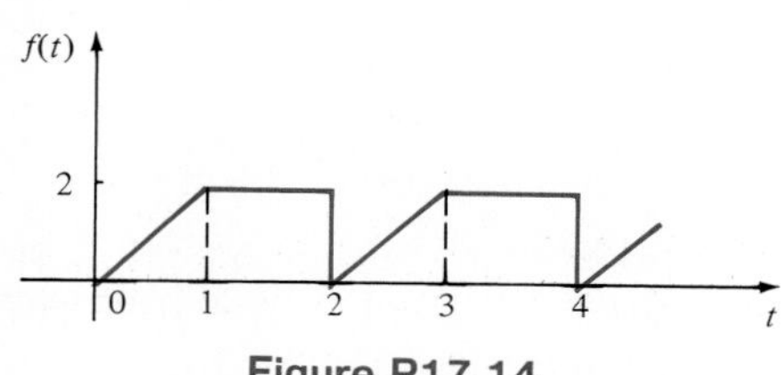

Figure P17.14

17.15. Determine the Laplace transform of the periodic waveform shown in Fig. P17.15.

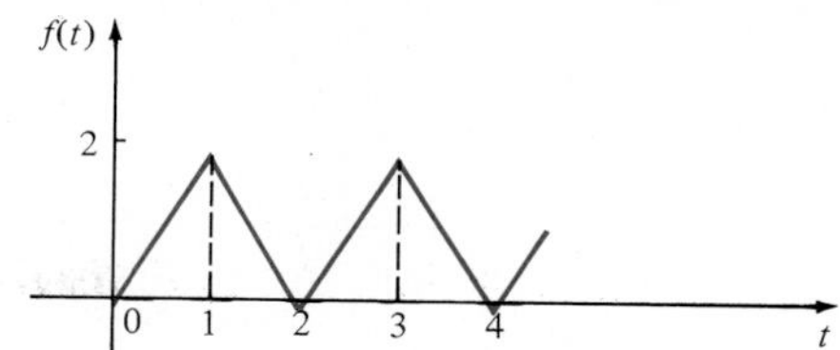

Figure P17.15

17.16. Find the Laplace transform of the waveform shown in Fig. P17.16.

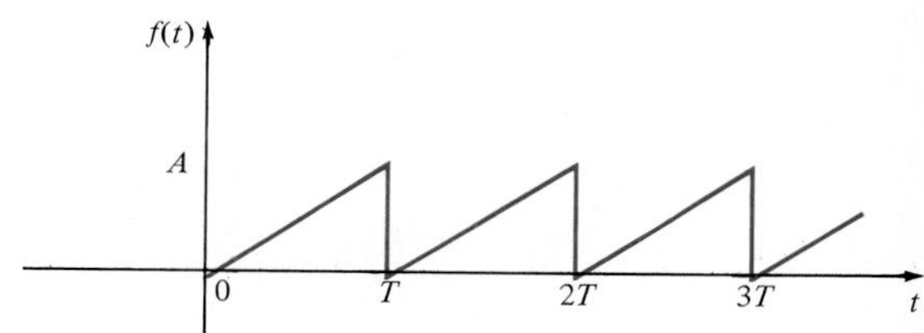

Figure P17.16

17.17. Find the Laplace transform of the waveform shown in Fig. 17.17.

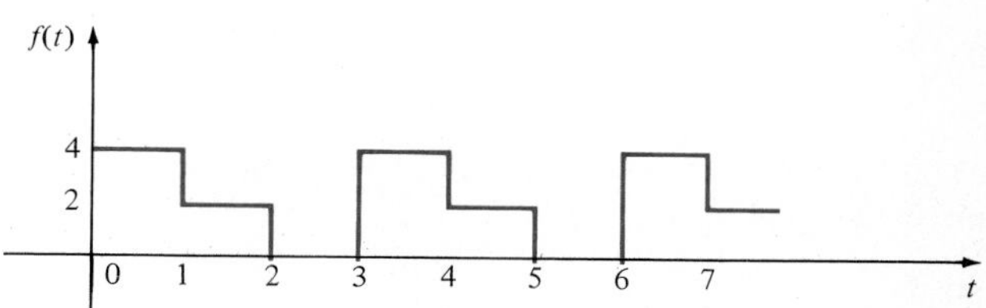

Figure P17.17

17.18. Determine the Laplace transform of the periodic waveform shown in Fig. P17.18.

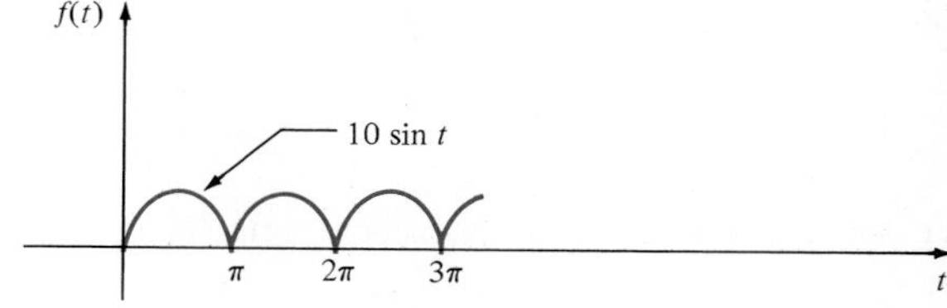

Figure P17.18

17.19. Determine the Laplace transform of the periodic waveform shown in Fig. P17.19.

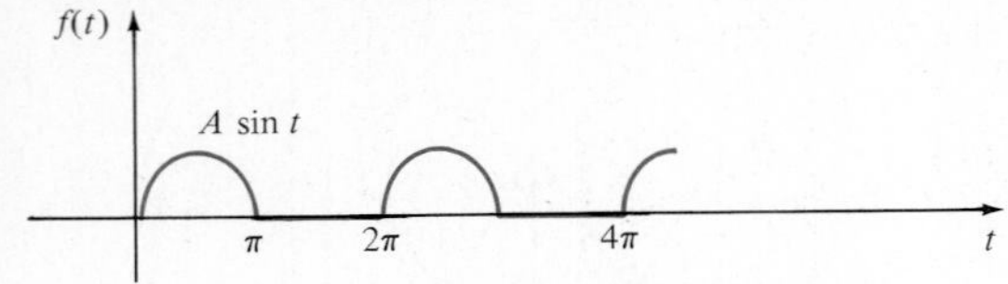

Figure P17.19

17.20. Find the Laplace transform of the periodic waveform shown in Fig. P17.20.

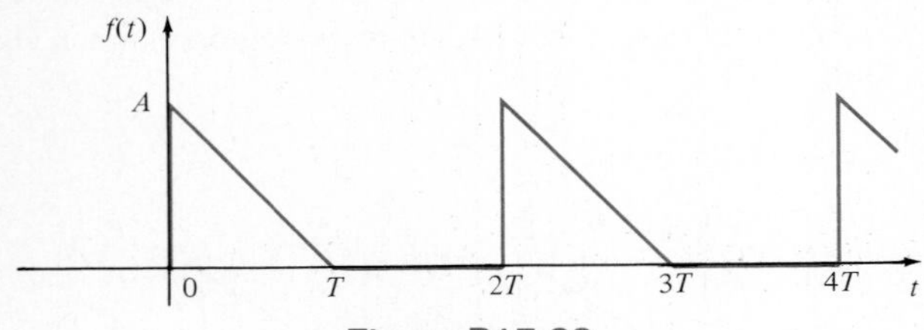

Figure P17.20

17.21. Given the following functions $\mathbf{F}(s)$, find the inverse Laplace transform of each function.

(a) $\mathbf{F}(s) = \dfrac{10(s+1)}{s^2+2s+2}$

(b) $\mathbf{F}(s) = \dfrac{s+1}{s(s^2+4s+5)}$

(c) $\mathbf{F}(s) = \dfrac{s^2+1}{s(s+2)(s^2+2s+2)}$

17.22. Given the following functions $\mathbf{F}(s)$, find the inverse Laplace transform of each function.

(a) $\mathbf{F}(s) = \dfrac{s+1}{s^2(s+2)}$

(b) $\mathbf{F}(s) = \dfrac{s+3}{(s+1)^2(s+4)}$

(c) $\mathbf{F}(s) = \dfrac{s(s+1)}{(s+2)^3(s+3)}$

17.23. Find $f(t)$ if $\mathbf{F}(s)$ is given by the expression

$$\mathbf{F}(s) = \frac{12(s+2)}{s^2(s+1)(s^2+4s+8)}$$

17.24. Find the inverse Laplace transform of the following functions.

(a) $\mathbf{F}(s) = \dfrac{e^{-s}}{s+1}$

(b) $\mathbf{F}(s) = \dfrac{1-e^{-2s}}{s}$

(c) $\mathbf{F}(s) = \dfrac{1-e^{-s}}{s+2}$

17.25. Find the inverse Laplace transform of the following functions.

(a) $\mathbf{F}(s) = \dfrac{(s+1)e^{-s}}{s(s+2)}$

(b) $\mathbf{F}(s) = \dfrac{10e^{-2s}}{(s+1)(s+3)}$

(c) $\mathbf{F}(s) = \dfrac{1-e^{-s}}{s^2(s+1)}$

17.26. Find the inverse Laplace transform $f(t)$ if $\mathbf{F}(s)$ is

$$\mathbf{F}(s) = \frac{se^{-s}}{(s+1)(s+2)}$$

17.27. Find the inverse Laplace transform of the function

$$\mathbf{F}(s) = \frac{(s+2)e^{-2s}}{(s^2+4s+8)}$$

17.28. Find the inverse Laplace transform of the function

$$\mathbf{F}(s) = \frac{s^2e^{-2s}}{(s^2+1)(s+1)(s^2+2s+2)}$$

17.29. Find the inverse Laplace transform of the function

$$\mathbf{F}(s) = \frac{10s(s+2)e^{-4s}}{(s+1)^2(s^2+2s+2)}$$

17.30. Find $f(t)$ using convolution if $\mathbf{F}(s)$ is

$$\mathbf{F}(s) = \frac{1}{(s+1)(s+2)}$$

17.31. Use convolution to find $f(t)$ if

$$\mathbf{F}(s) = \frac{1}{(s+1)(s+2)^2}$$

17.32. Given that

$$\mathbf{F}(s) = \frac{s}{(s+1)(s+2)(s+3)}$$

and that $\mathbf{F}_1(s) = 1/(s+1)$, $\mathbf{F}_2(s) = 1/(s+2)$, and $\mathbf{F}_3(s) = s/(s+3)$; find $f(t) = \mathcal{L}^{-1}[\mathbf{F}(s)]$ using the expression

$$f(t) = f_1(t) * f_2(t) * f_3(t)$$

17.33. Use convolution to determine $f(t)$ given that $\mathbf{F}(s) = 1/s^3$.

17.34. Find $f(t)$ using the convolution integral if $\mathbf{F}(s) = 1/s^4$.

17.35. Find $f(t)$ using the convolution integral if

$$\mathbf{F}(s) = \frac{1}{s(s^2 + 1)}$$

17.36. Use the convolution integral to find $f(t)$ if

$$\mathbf{F}(s) = \frac{1}{s^2(s^2 + 1)}$$

17.37. Find $f(t)$ using convolution, given that

$$\mathbf{F}(s) = \frac{1}{(s^2 + 4)^2}$$

17.38. Find $f(t)$ using the convolution integral if

$$\mathbf{F}(s) = \frac{e^{-2s}}{s(s^2 + 1)}$$

17.39. Find $f(t)$ using the convolution integral if

$$\mathbf{F}(s) = \frac{e^{-s}}{s^2(s + 1)}$$

17.40. Find $f(t)$ using the convolution integral if

$$\mathbf{F}(s) = \frac{e^{-2s}}{(s + 1)(s + 2)}$$

17.41. Find $f(t)$ using the convolution integral if

$$\mathbf{F}(s) = \frac{1 - e^{-2s}}{s(s + 2)}$$

17.42. Use convolution to find $f(t)$ given that $\mathbf{F}(s)$ is equal to

$$\mathbf{F}(s) = \left(\frac{1 - e^{-as}}{s}\right)^2$$

17.43. Find $f(t)$ using the convolution integral if

$$\mathbf{F}(s) = \frac{e^{-s}}{s^2(s^2 + 1)}$$

17.44. Find $f(t)$ using the convolution integral if

$$\mathbf{F}(s) = \frac{1 - e^{-4s}}{s^2(s + 1)}$$

17.45. Find $f(t)$ using the convolution integral if

$$\mathbf{F}(s) = \frac{e^{-s}}{(s + 2)(s^2 + 4)}$$

17.46. Find the initial and final values of the time function $f(t)$ if $\mathbf{F}(s)$ is given as

(a) $\mathbf{F}(s) = \dfrac{10(s + 2)}{(s + 1)(s + 3)}$

(b) $\mathbf{F}(s) = \dfrac{s^2 + 2s + 4}{(s + 6)(s^3 + 4s^2 + 8s + 10)}$

(c) $\mathbf{F}(s) = \dfrac{2s}{s^2 + 2s + 2}$

17.47. Given that $\mathbf{F}(s)$ is

$$\mathbf{F}(s) = \frac{10(s + 1)(s + 4)}{s(s + 2)(s + 3)}$$

find the initial and final values of $df(t)/dt$ if $f(t) = \mathscr{L}^{-1}[\mathbf{F}(s)]$.

17.48. Find the final values of the time function $f(t)$ given that

(a) $\mathbf{F}(s) = \dfrac{10(s + 1)}{(s + 2)(s - 3)}$

(b) $\mathbf{F}(s) = \dfrac{10}{s^2 + 4}$

17.49. Find the final value of the function $f(t)$ if $\mathbf{F}(s) = \mathscr{L}[f(t)]$ is given by the expression

$$\mathbf{F}(s) = \frac{s + 1}{s^3 + 3s^2 + 2s + 8}$$

17.50. Find the solution of the following differential equations using Laplace transforms.

(a) $\dfrac{d^2y(t)}{dt^2} + 3\dfrac{dy(t)}{dt} + 2y(t) = 2u(t),$
$y(0) = y'(0) = 0$

(b) $\dfrac{d^2y(t)}{dt^2} + 7\dfrac{dy(t)}{dt} + 12y(t) = e^{-t},$
$y(0) = 0,\ y'(0) = 1$

17.51. Solve the following differential equations using Laplace transforms.

(a) $\dfrac{d^2y(t)}{dt^2} + \dfrac{2dy(t)}{dt} + 2y(t) = e^{-t},$
$y(0) = y'(0) = 0$

(b) $\dfrac{d^2y(t)}{dt^2} + \dfrac{2dy(t)}{dt} + 5y(t) = u(t),$
$y(0) = 0,\ y'(0) = 1$

17.52. Solve the following differential equations using Laplace transforms.

(a) $\dfrac{d^2y(t)}{dt^2} + \dfrac{2dy(t)}{dt} + y(t) = e^{-2t},$
$y(0) = y'(0) = 0$

(b) $\frac{d^2y(t)}{dt^2} + \frac{4dy(t)}{dt} + 4y(t) = u(t),$
$y(0) = 0, y'(0) = 1$

17.53. Given the following differential equation, find $y(t)$ using Laplace transforms.

$$\frac{dy(t)}{dt} + 2y(t) = 2e^{-t}, \; y'(0) = -1$$

17.54. Solve the following integrodifferential equation using Laplace transforms.

$$\frac{dy(t)}{dt} + 5y(t) + 4\int_0^t y(x)\,dx = u(t) \qquad y(0) = 0$$

17.55. Determine $y(t)$ in the following equation if all initial conditions are zero.

$$\frac{d^3y(t)}{dt^3} + \frac{4d^2y(t)}{dt^2} + \frac{3dy(t)}{dt} = 10e^{-2t}$$

17.56. Solve the following integrodifferential equation using Laplace transforms.

$$\frac{dy(t)}{dt} + 2y(t) + \int_0^t y(\lambda)\,d\lambda = 1 - e^{-2t}\; y(0) = 0, \quad t > 0$$

17.57. Use Laplace transforms to solve the following integrodifferential equation.

$$\frac{dy(t)}{dt} + 2y(t) + \int_0^t y(\lambda)e^{-2(t-\lambda)}\,d\lambda = 4u(t),$$

$$y(0) = 1, \quad t > 0$$

17.58. Find $y(t)$ if

$$y(t) + \int_0^t y(\lambda)(t - \lambda)\,d\lambda = e^{-t}$$

17.59. Use Laplace transforms to solve the following equation.

$$y(t) + \int_0^t y(\lambda)e^{-(t-\lambda)}\,d\lambda = tu(t) \qquad t > 0$$

17.60. Solve the following equation using Laplace transforms.

$$y(t) + 2\int_0^t y(\lambda)e^{-(t-\lambda)}\sin(t - \lambda)\,d\lambda = u(t) \qquad t > 0$$

CHAPTER

18

Application of the Laplace Transform to Circuit Analysis

We have demonstrated in Chapter 17 how the Laplace transform can be used to solve linear constant-coefficient-differential equations. Since all of the linear networks with which we are dealing can be described by a set of linear constant-coefficient-differential equations, using the Laplace transform for circuit analysis would appear to be a viable scheme. The terminal characteristics for each circuit element can be described in the s-domain by transforming the appropriate time-domain equations. Kirchhoff's laws when applied to a circuit produce a set of integrodifferential equations in terms of the terminal characteristics of the network elements which when transformed yield a set of algebraic equations in the s-domain. Therefore, a complex frequency-domain analysis in which the passive network elements are represented by their transform impedance or admittance, and sources, whether independent or dependent, are represented in terms of their transform variables, can be performed. Such an analysis in the s-domain is algebraic and all of the techniques derived in dc analysis will apply. Therefore, the analysis is similar to the dc analysis of resistive networks and all of the network analysis techniques and network theorems that were applied in dc analysis are valid in the s-domain (e.g., node analysis, loop analysis, superposition, source transformation, Thévenin's theorem, Norton's theorem, and combinations of impedance or admittance). Therefore, our approach will be to transform each of the circuit elements, draw an s-domain equivalent circuit, and then using the transformed network, solve the circuit equations algebraically in the s-domain. Finally, a table of transform pairs can be employed to obtain a complete (transient plus steady-state) solution. Our approach in this chapter is to employ the circuit element models and demonstrate, using a variety of examples, many of the concepts and techniques that have been presented earlier.

18.1

Circuit Element Models

Let us examine the terminal characteristics of the network elements in the s-domain. The voltage-current relationship for a resistor in the time domain using the passive sign convention is

$$v(t) = Ri(t) \tag{18.1}$$

Using the Laplace transform, this relationship in the s-domain is

$$\mathbf{V}(s) = R\mathbf{I}(s) \tag{18.2}$$

Therefore, the time-domain and complex-frequency-domain representations of this element are as shown in Fig. 18.1a.

The time-domain relationships for a capacitor using the passive sign convention are

$$v(t) = \frac{1}{C}\int_0^t i(x)\, dx + v(0) \tag{18.3}$$

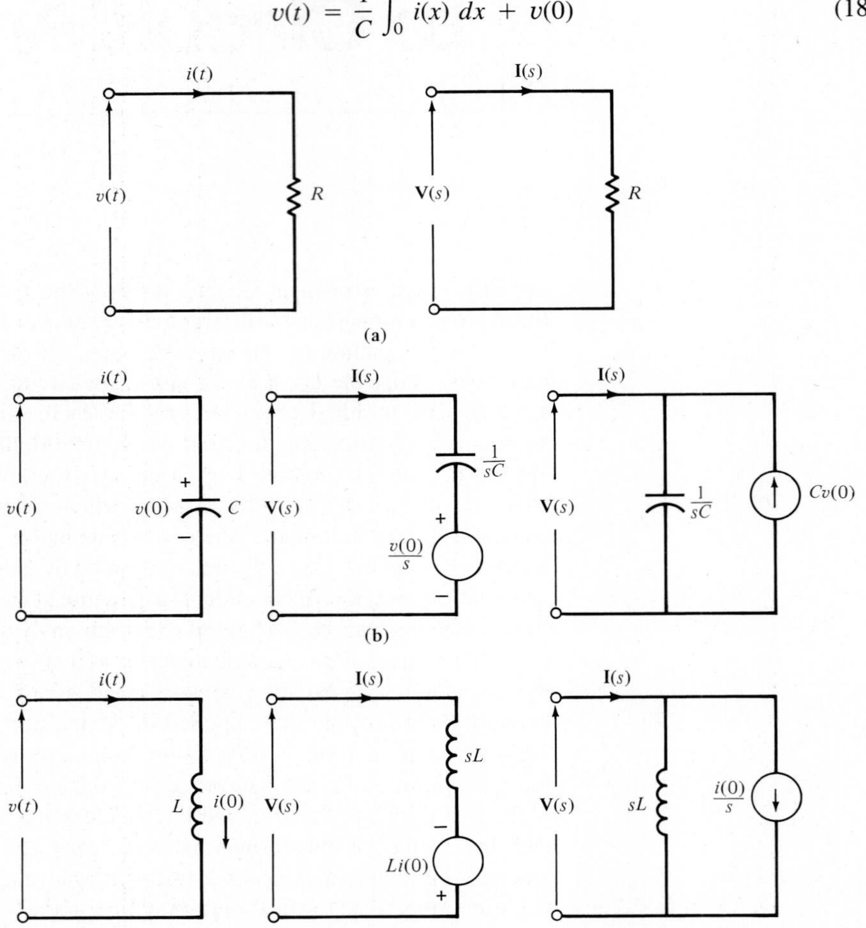

Figure 18.1 Time-domain and s-domain representations of circuit elements.

$$i(t) = C\frac{dv(t)}{dt} \tag{18.4}$$

The s-domain equations for the capacitor are then

$$\mathbf{V}(s) = \frac{\mathbf{I}(s)}{sC} + \frac{v(0)}{s} \tag{18.5}$$

$$\mathbf{I}(s) = sC\mathbf{V}(s) - Cv(0) \tag{18.6}$$

and hence the s-domain representation of this element is as shown in Fig. 18.1b.

For the inductor, the voltage-current relationships using the passive sign convention are

$$v(t) = L\frac{di(t)}{dt} \tag{18.7}$$

$$i(t) = \frac{1}{L}\int_0^t v(x)\,dx + i(0) \tag{18.8}$$

The relationships in the s-domain are then

$$\mathbf{V}(s) = sL\mathbf{I}(s) - Li(0) \tag{18.9}$$

$$\mathbf{I}(s) = \frac{\mathbf{V}(s)}{sL} + \frac{i(0)}{s} \tag{18.10}$$

The s-domain representation of this element is shown in Fig. 18.1c.

Independent and dependent voltage and current sources can also be represented by their transforms, that is,

$$\begin{aligned} \mathbf{V}_1(s) &= \mathcal{L}[v_1(t)] \\ \mathbf{I}_2(s) &= \mathcal{L}[i_2(t)] \end{aligned} \tag{18.11}$$

and if $v_1(t) = Ai_2(t)$, which represents a current-controlled voltage source, then

$$\mathbf{V}_1(s) = A\mathbf{I}_2(s) \tag{18.12}$$

The reader should note carefully the direction of the current sources and the polarity of the voltage sources in the transformed network which result from the initial conditions. If the polarity of the initial voltage or direction of the initial current is reversed, the sources in the transformed circuit that result from the initial condition are also reversed.

18.2 Analysis Techniques

Now that we have the s-domain representation for the circuit elements in terms of their complex frequency impedance or admittance, we are in a position to analyze networks using a transformed circuit.

EXAMPLE 18.1

Given the circuits in Fig. 18.2a and b, we wish to write the mesh equations in the s-domain for the network in Fig. 18.2a and the node equations in the s-domain for the network in Fig. 18.2b.

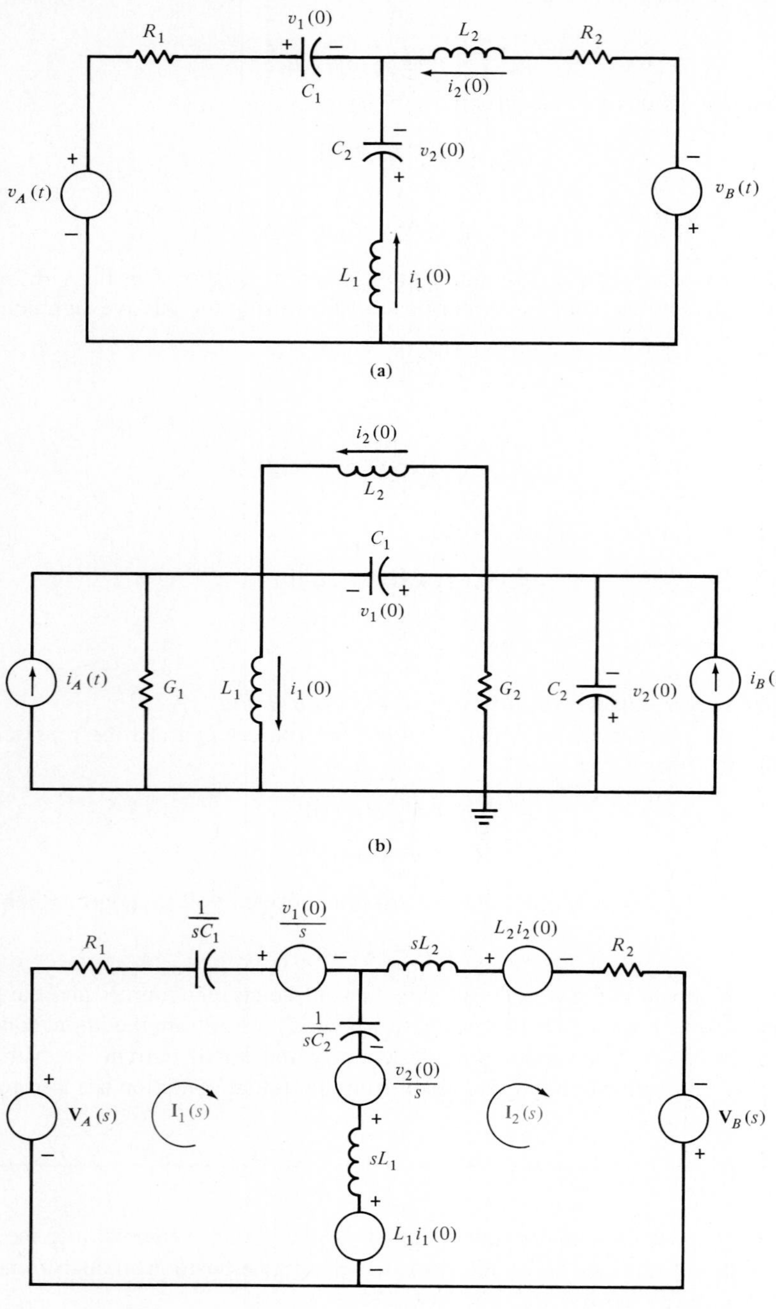

Figure 18.2 Circuits used in Example 18.1.

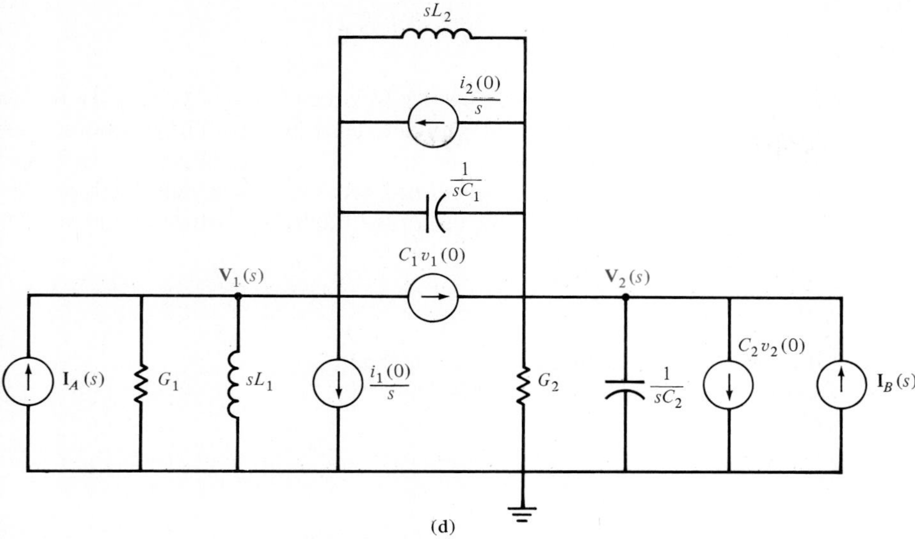

Figure 18.2 (Continued).

The transformed circuit using the impedance values for the network parameters in Fig. 18.2a is shown in Fig. 18.2c. The mesh equations for this network are

$$\left(R_1 + \frac{1}{sC_1} + \frac{1}{sC_2} + sL_1\right)\mathbf{I}_1(s) - \left(\frac{1}{sC_2} + sL_1\right)\mathbf{I}_2(s)$$

$$= \mathbf{V}_A(s) - \frac{v_1(0)}{s} + \frac{v_2(0)}{s} - L_1 i_1(0)$$

$$-\left(\frac{1}{sC_2} + sL_1\right)\mathbf{I}_1(s) + \left(\frac{1}{sC_2} + sL_1 + sL_2 + R_2\right)\mathbf{I}_2(s)$$

$$= L_1 i_1(0) - \frac{v_2(0)}{s} - L_2 i_2(0) + \mathbf{V}_B(s)$$

The transformed circuit using the impedance values for the network parameters in Fig. 18.2b is shown in Fig. 18.2d. The node equations for this network are

$$\left(G_1 + \frac{1}{sL_1} + sC_1 + \frac{1}{sL_2}\right)\mathbf{V}_1(s) - \left(\frac{1}{sL_2} + sC_1\right)\mathbf{V}_2(s)$$

$$= \mathbf{I}_A(s) - \frac{i_1(0)}{s} + \frac{i_2(0)}{s} - C_1 v_1(0)$$

$$-\left(\frac{1}{sL_2} + sC_1\right)\mathbf{V}_1(s) - \left(\frac{1}{sL_2} + sC_1 + G_2 + sC_2\right)\mathbf{V}_2(s)$$

$$= C_1 v_1(0) - \frac{i_2(0)}{s} - C_2 v_2(0) + \mathbf{I}_B(s)$$

Note that the two sets of equations above are of the same form. This phenomenon results from the fact that the networks in Fig. 18.2a and b are duals of one another. ■

Example 18.1 attempts to illustrate the manner in which to employ the two s-domain representations of the inductor and capacitor circuit elements when initial conditions are present. In the following examples we illustrate the use of a number of analysis techniques in obtaining the complete response of a transformed network. The circuits analyzed have been specifically chosen to demonstrate the use of the Laplace transform to circuits with a variety of passive and active elements.

EXAMPLE 18.2

Consider the network shown in Fig. 18.3a. Let us determine the node voltages $v_1(t)$ and $v_2(t)$ using both nodal analysis and superposition. The transformed network is shown in Fig. 18.3b. The nodal equations for the network are

$$\mathbf{V}_1(s)\left(\frac{1}{s} + \frac{1}{2} + \frac{s}{2}\right) - \mathbf{V}_2(s)\left(\frac{s}{2}\right) = \frac{12}{s^2}$$

$$-\mathbf{V}_1(s)\left(\frac{s}{2}\right) + \mathbf{V}_2(s)\left(\frac{s}{2} + \frac{1}{2}\right) = \frac{4}{s}$$

The equations in matrix form are

$$\begin{bmatrix} \dfrac{s^2 + s + 2}{2s} & \dfrac{-s}{2} \\[2ex] \dfrac{-s}{2} & \dfrac{s + 1}{2} \end{bmatrix} \begin{bmatrix} \mathbf{V}_1(s) \\ \mathbf{V}_2(s) \end{bmatrix} = \begin{bmatrix} \dfrac{12}{s^2} \\[2ex] \dfrac{4}{s} \end{bmatrix}$$

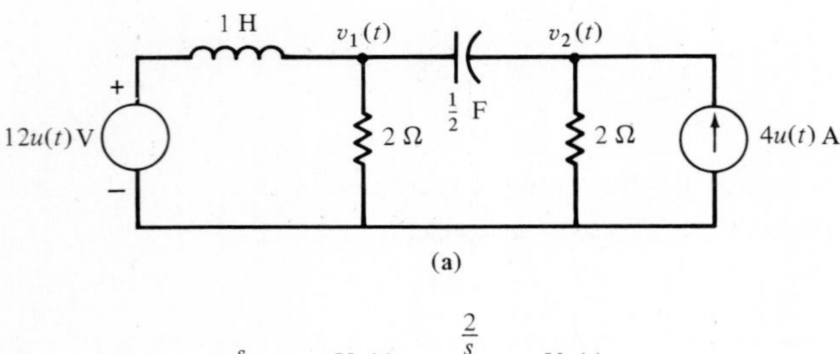

(a)

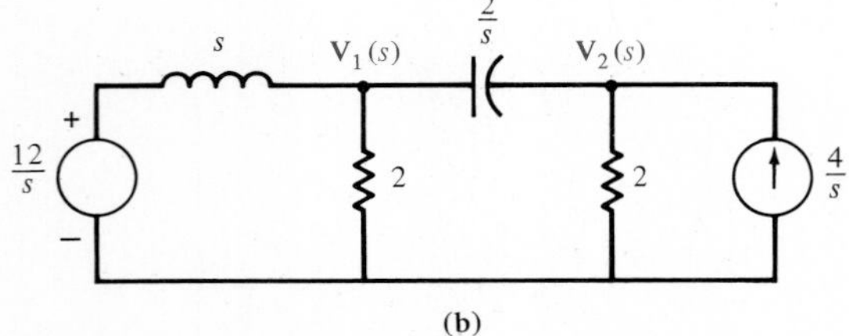

(b)

Figure 18.3 Circuits used in Example 18.2.

Solving for the node voltages, we obtain

$$\begin{bmatrix} \mathbf{V}_1(s) \\ \mathbf{V}_2(s) \end{bmatrix} = \begin{bmatrix} \dfrac{s^2 + s + 2}{2s} & \dfrac{-s}{2} \\ \dfrac{-s}{2} & \dfrac{s+1}{2} \end{bmatrix}^{-1} \begin{bmatrix} \dfrac{12}{s^2} \\ \dfrac{4}{s} \end{bmatrix}$$

$$= \frac{4s}{2s^2 + 3s + 2} \begin{bmatrix} \dfrac{s+1}{2} & \dfrac{s}{2} \\ \dfrac{s}{2} & \dfrac{s^2 + s + 2}{2s} \end{bmatrix} \begin{bmatrix} \dfrac{12}{s^2} \\ \dfrac{4}{s} \end{bmatrix}$$

$$= \begin{bmatrix} \dfrac{4(s^2 + 3s + 3)}{s(s^2 + 3s/2 + 1)} \\ \dfrac{4(s^2 + 4s + 2)}{s(s^2 + 3s/2 + 1)} \end{bmatrix}$$

Let us now solve for $\mathbf{V}_1(s)$ and $\mathbf{V}_2(s)$ using superposition. If we let $\mathbf{V}_1'(s)$ and $\mathbf{V}_2'(s)$ represent the node voltages caused by the 12-V source, then as shown in Fig. 18.4a, $\mathbf{V}_1'(s)$ can be calculated by multiplying the current in the 12-V source by the impedance, which consists of the parallel combination of the 2-Ω resistor and the series combination of the $\frac{1}{2}$-F capacitor and the other 2-Ω resistor. Therefore,

$$\mathbf{V}_1'(s) = \left[\frac{12/s}{s + \dfrac{2[(2/s) + 2]}{2 + 2/s + 2}} \right] \left[\frac{2[(2/s) + 2]}{2 + 2/s + 2} \right]$$

$$= \left[\frac{12/s}{s + (4s + 4)/(4s + 2)} \right] \left[\frac{4s + 4}{4s + 2} \right]$$

$$= \frac{12(s + 1)}{s(s^2 + \frac{3}{2}s + 1)}$$

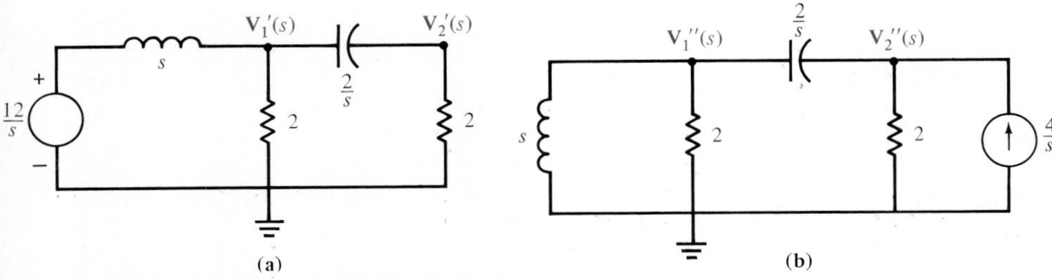

Figure 18.4 Circuits used in applying superposition in Example 18-2.

Once $\mathbf{V}_1'(s)$ is known, $\mathbf{V}_2'(s)$ can be calculated using voltage division as

$$\mathbf{V}_2'(s) = \mathbf{V}_1'(s)\left[\frac{2}{(2/s) + 2}\right] = \frac{12}{s^2 + \frac{3}{2}s + 1}$$

If we now let $\mathbf{V}_1''(s)$ and $\mathbf{V}_2''(s)$ represent the node voltages caused by the 4-A source, then as shown in Fig. 18.4b, $\mathbf{V}_2''(s)$ can be calculated by multiplying the current in the 4-A source by the impedance of the network as seen by the source. The impedance of the 1-H inductor in parallel with the 2-Ω resistor is $2s/(s + 2)$. The impedance of the $\frac{1}{2}$-F capacitor in series with this combination is $\dfrac{2}{s} + \dfrac{2s}{2 + s}$. Finally, the impedance seen by the source is the latter impedance in parallel with the 2-Ω resistor, and therefore

$$\mathbf{V}_2''(s) = \frac{4}{s}\left[\frac{\left(\dfrac{2s}{s + 2} + \dfrac{2}{s}\right)(2)}{\dfrac{2s}{s + 2} + \dfrac{2}{s} + 2}\right] = \frac{4}{s}\left[\frac{4s^2 + 4s + 8}{4s^2 + 6s + 4}\right] = \frac{4(s^2 + s + 2)}{s(s^2 + \frac{3}{2}s + 1)}$$

$\mathbf{V}_1''(s)$ can be computed from $\mathbf{V}_2''(s)$ using voltage division. Hence

$$\mathbf{V}_1''(s) = \mathbf{V}_2''(s)\left[\frac{\dfrac{2s}{s + 2}}{\dfrac{2s}{s + 2} + \dfrac{2}{s}}\right] = \mathbf{V}_2''(s)\left[\frac{s^2}{s^2 + s + 2}\right] = \frac{4s^2}{s(s^2 + \frac{3}{2}s + 1)}$$

Now using superposition, we have

$$\mathbf{V}_1(s) = \mathbf{V}_1'(s) + \mathbf{V}_1''(s) = \frac{12(s + 1)}{s(s^2 + \frac{3}{2}s + 1)} + \frac{4s^2}{s(s^2 + \frac{3}{2}s + 1)} = \frac{4(s^2 + 3s + 3)}{s(s^2 + \frac{3}{2}s + 1)}$$

and

$$\mathbf{V}_2(s) = \mathbf{V}_2'(s) + \mathbf{V}_2''(s)$$

$$= \frac{12}{s^2 + \frac{3}{2}s + 1} + \frac{4(s^2 + s + 2)}{s(s^2 + \frac{3}{2}s + 1)}$$

$$= \frac{4(s^2 + 4s + 2)}{s(s^2 + \frac{3}{2}s + 1)}$$

Note that the answers we have obtained via superposition are exactly the same as those obtained by nodal analysis. Let us now determine $v_1(t) = \mathcal{L}^{-1}[\mathbf{V}_1(s)]$ and $v_2(t) = \mathcal{L}^{-1}[\mathbf{V}_2(s)]$. $\mathbf{V}_1(s)$ can be written as

$$\mathbf{V}_1(s) = \frac{4(s^2 + 3s + 3)}{s(s^2 + \frac{3}{2}s + 1)} = \frac{4(s^2 + 3s + 3)}{s[s + \frac{3}{4} + j(\sqrt{7}/4)][s + \frac{3}{4} - j(\sqrt{7}/4)]}$$

Therefore,

$$\frac{4(s^2 + 3s + 3)}{s[s + \frac{3}{4} - j(\sqrt{7}/4)][s + \frac{3}{4} + j(\sqrt{7}/4)]}$$
$$= \frac{K_0}{s} + \frac{K_1}{s + \frac{3}{4} - j(\sqrt{7}/4)} + \frac{K_1^*}{s + \frac{3}{4} + j(\sqrt{7}/4)}$$

Evaluating the constants K_0 and K_1, we obtain

$$\left.\frac{4(s^2 + 3s + 3)}{s^2 + \frac{3}{2}s + 1}\right|_{s=0} = K_0$$

$$12 = K_0$$

and

$$\left.\frac{4(s^2 + 3s + 3)}{s[s + \frac{3}{4} + j(\sqrt{7}/4)]}\right|_{s=-3/4+j(\sqrt{7}/4)} = K_1$$

$$4\underline{/180^\circ} = K_1$$

Therefore,

$$v_1(t) = \left[12 + 8e^{-(3/4)t} \cos\left(\frac{\sqrt{7}}{4}t + 180^\circ\right)\right]u(t)\text{ V}$$

In a similar manner,

$$\mathbf{V}_2(s) = \frac{4(s^2 + 4s + 2)}{s(s^2 + \frac{3}{2}s + 1)} = \frac{K_0}{s} + \frac{K_1}{s + \frac{3}{4} - j(\sqrt{7}/4)} + \frac{K_1^*}{s + \frac{3}{4} + j(\sqrt{7}/4)}$$

Evaluating the constants K_0 and K_1, we obtain

$$\left.\frac{4(s^2 + 4s + 2)}{s^2 + \frac{3}{2}s + 1}\right|_{s=0} = K_0$$

$$8 = K_0$$

and

$$\left.\frac{4(s^2 + 4s + 2)}{s[s + \frac{3}{4} + j(\sqrt{7}/4)]}\right|_{s=-3/4+j(\sqrt{7}/4)} = K_1$$

$$5.66\underline{/-110.7^\circ} = K_1$$

Therefore,

$$v_2(t) = \left[8 + 11.32e^{-(3/4)t} \cos\left(\frac{\sqrt{7}}{4}t - 110.7^\circ\right)\right]u(t) \text{ V}$$

It is interesting to note that as $t \to \infty$, the sources in the circuit appear to be dc sources. In the dc case the inductor acts like a short circuit and the capacitor acts like an open circuit. Under these conditions the voltage $v_1(t) = 12$ V, and the voltage $v_2(t) = 8$ V. It is reassuring to realize that the equations for $v_1(t)$ and $v_2(t)$ yield these results also. ■

EXAMPLE 18.3

Consider the network shown in Fig. 18.5a. We wish to determine the output voltage $v_o(t)$. As we begin to attack the problem we note two things. First, because the source $12u(t)$ is connected between $v_1(t)$ and $v_2(t)$, we have a supernode. Second, if $v_2(t)$ is known, $v_0(t)$ can be easily obtained by voltage division. Hence we will use nodal analysis in conjunction with voltage division to obtain a solution.

The transformed network is shown in Fig. 18.5b. KCL for the supernode is

$$\frac{\mathbf{V}_1(s)}{2} + \mathbf{V}_1(s)\,\frac{s}{2} - 2\mathbf{I}(s) + \frac{\mathbf{V}_2(s)}{s + 1} = 0$$

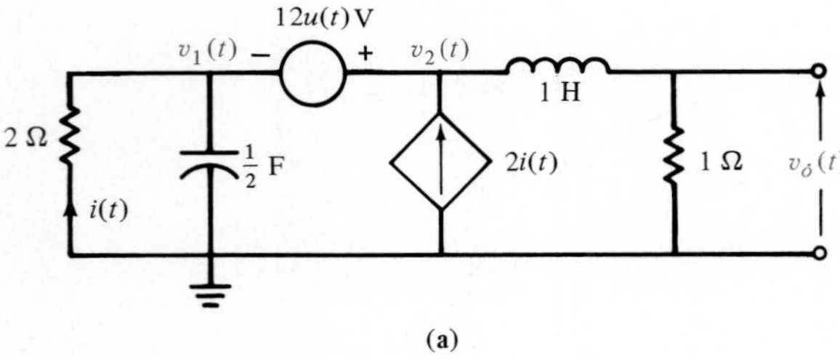

(a)

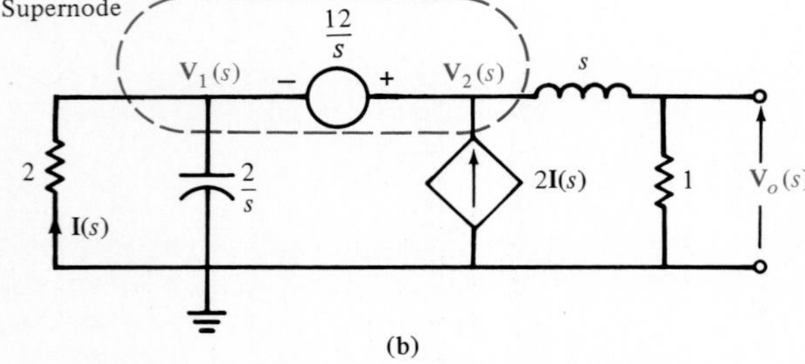

(b)

Figure 18.5 Circuits used in Example 18.3.

However,

$$\mathbf{I}(s) = -\frac{\mathbf{V}_1(s)}{2}$$

and

$$\mathbf{V}_1(s) = \mathbf{V}_2(s) - \frac{12}{s}$$

Substituting the last two equations into the first equation yields

$$\left[\mathbf{V}_2(s) - \frac{12}{s}\right]\frac{s + 3}{2} + \frac{\mathbf{V}_2(s)}{s + 1} = 0$$

or

$$\mathbf{V}_2(s) = \frac{12(s + 1)(s + 3)}{s(s^2 + 4s + 5)}$$

Employing a voltage divider, we obtain

$$\begin{aligned}\mathbf{V}_o(s) &= \mathbf{V}_2(s)\frac{1}{s + 1} \\ &= \frac{12(s + 3)}{s(s^2 + 4s + 5)} \\ &= \frac{12(s + 3)}{s(s + 2 - j1)(s + 2 + j1)}\end{aligned}$$

which can be written as

$$\frac{12(s + 3)}{s(s + 2 - j1)(s + 2 + j1)} = \frac{K_0}{s} + \frac{K_1}{s + 2 - j1} + \frac{K_1^*}{s + 2 + j1}$$

Evaluating the constants, we obtain

$$\left.\frac{12(s + 3)}{s^2 + 4s + 5}\right|_{s=0} = K_0$$

$$\frac{36}{5} = K_0$$

and

$$\left.\frac{12(s + 3)}{s(s + 2 + j1)}\right|_{s=-2+j1} = K_1$$

$$3.79\underline{/161.57^\circ} = K_1$$

Therefore,

$$v_o(t) = [7.2 + 7.58e^{-2t} \cos\ (t + 161.57^\circ)]u(t)\text{ V}$$ ■

EXAMPLE 18.4

Let us examine the network in Fig. 18.6a. We wish to determine the output voltage $v_o(t)$. We will solve this problem using both loop equations and Norton's theorem.

The transformed network is shown in Fig. 18.6b. KVL for the right-hand loop is

$$\frac{12}{s} - [\mathbf{I}_2(s) - \mathbf{I}_1(s)]s - \frac{\mathbf{I}_2(s)}{s} - 2\mathbf{I}_2(s) = 0$$

However, $\mathbf{I}_1(s) = 4/s$, and hence

$$\mathbf{I}_2(s) = \frac{4(s + 3)}{(s + 1)^2}$$

Therefore,

$$\mathbf{V}_o(s) = \frac{8(s + 3)}{(s + 1)^2}$$

Let us now apply Norton's theorem to determine $\mathbf{V}_o(s)$. The short-circuit current can be found from Fig. 18.7a. Although $\mathbf{I}_{sc}(s)$ can be determined in a number of ways, we will use superposition to find it. Let $\mathbf{I}'_{sc}(s)$ represent the value of the short-circuit current caused by the current source and let $\mathbf{I}''_{sc}(s)$ represent the value of $\mathbf{I}_{sc}(s)$ caused by the voltage source. These current values can be found from the circuits in Fig. 18.7b and c. Using current division

$$\mathbf{I}'_{sc}(s) = \frac{4}{s}\left(\frac{s}{s + 1/s}\right)$$
$$= \frac{4}{s}\left(\frac{s^2}{s^2 + 1}\right)$$

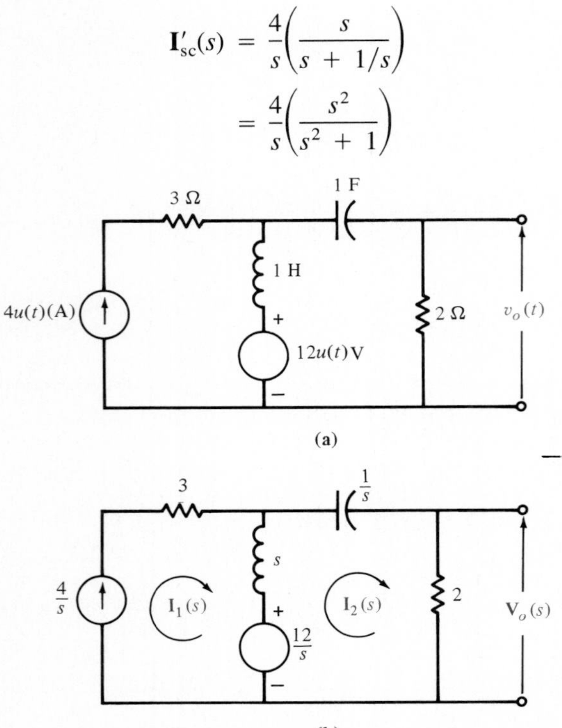

Figure 18.6 Circuits used in Example 18.4.

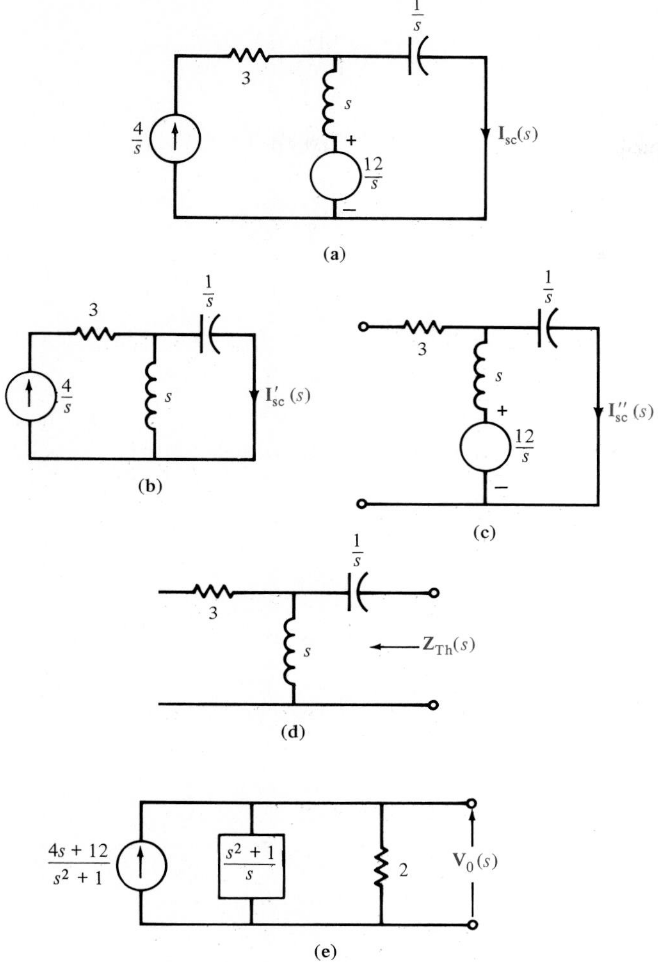

Figure 18.7 Circuits in Example 18.4.

From Fig. 18.7c we see that

$$\mathbf{I}''_{sc}(s) = \frac{12/s}{s + 1/s}$$

$$= \frac{12}{s^2 + 1}$$

Therefore,

$$\mathbf{I}_{sc}(s) = \mathbf{I}'_{sc}(s) + \mathbf{I}''_{sc}(s)$$

$$= \frac{4s + 12}{s^2 + 1}$$

The Thévenin equivalent impedance can be found from Fig. 18.7d as

$$\mathbf{Z}_{\text{Th}}(s) = s + \frac{1}{s} = \frac{s^2 + 1}{s}$$

Therefore, the Norton equivalent circuit together with the 2-Ω load is shown in Fig. 18.7e. Hence $\mathbf{V}_o(s)$ is

$$\mathbf{V}_o(s) = \frac{4s + 12}{s^2 + 1}\left[\frac{(2)\left(\dfrac{s^2 + 1}{s}\right)}{2 + \dfrac{s^2 + 1}{s}}\right]$$

$$= \frac{8(s + 3)}{(s + 1)^2}$$

which is identical to the value obtained earlier with loop equations. $\mathbf{V}_o(s)$ can be written as

$$\mathbf{V}_o(s) = \frac{8(s + 3)}{(s + 1)^2} = \frac{K_{11}}{(s + 1)^2} + \frac{K_{12}}{s + 1}$$

Evaluating the constants, we obtain

$$8(s + 3)\Big|_{s=-1} = K_{11}$$

$$16 = K_{11}$$

and

$$\frac{d}{ds}[8(s + 3)]\Big|_{s=-1} = K_{12}$$

$$8 = K_{12}$$

Therefore,

$$v_0(t) = (16te^{-t} + 8e^{-t})u(t) \text{ V}$$ ■

EXAMPLE 18.5

Let us examine the circuit shown in Fig. 18.8a. We will determine the output voltage $v_o(t)$ using both loop equations and Thévenin's theorem.

The transformed network is shown in Fig. 18.8b. The loop equations for this network are

$$2\mathbf{I}_1(s) + s[\mathbf{I}_1(s) - \mathbf{I}_2(s)] = \frac{12}{s} - 2\mathbf{V}_x(s)$$

$$-s\mathbf{I}_1(s) + \mathbf{I}_2(s)\left(s + \frac{1}{s} + 2\right) = 2\mathbf{V}_x(s)$$

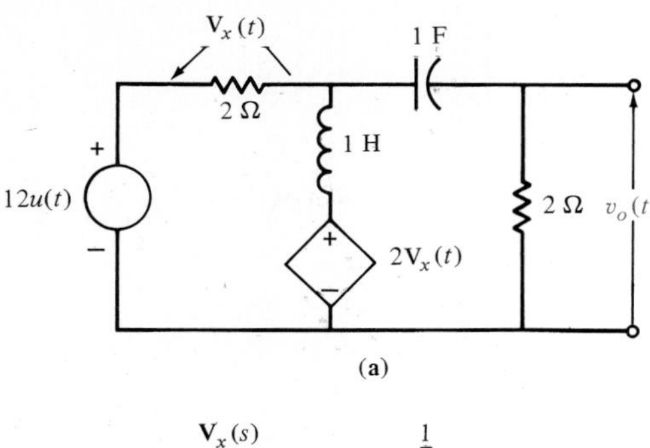

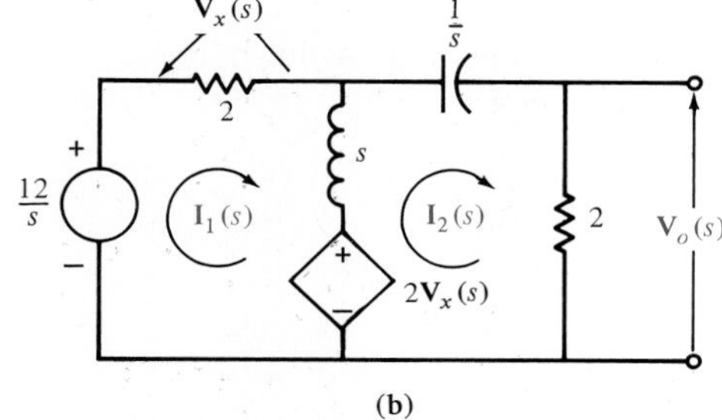

Figure 18.8 Circuits used in Example 18.5.

where

$$\mathbf{V}_x(s) = 2\mathbf{I}_1(s)$$

Substituting the last equation into the first two equations yields

$$(s + 6)\mathbf{I}_1(s) - s\mathbf{I}_2(s) = \frac{12}{s}$$

$$-(s + 4)\mathbf{I}_1(s) + \left(\frac{s^2 + 2s + 1}{s}\right)\mathbf{I}_2(s) = 0$$

which in matrix form is

$$\begin{bmatrix} s + 6 & -s \\ -(s + 4) & \dfrac{s^2 + 2s + 1}{s} \end{bmatrix} \begin{bmatrix} \mathbf{I}_1(s) \\ \mathbf{I}_2(s) \end{bmatrix} = \begin{bmatrix} \dfrac{12}{s} \\ 0 \end{bmatrix}$$

Solving this equation for the currents, we obtain

$$\begin{bmatrix} \mathbf{I}_1(s) \\ \mathbf{I}_2(s) \end{bmatrix} = \begin{bmatrix} s + 6 & -s \\ -(s + 4) & \dfrac{s^2 + 2s + 1}{s} \end{bmatrix}^{-1} \begin{bmatrix} \dfrac{12}{s} \\ 0 \end{bmatrix}$$

$$= \frac{s}{4s^2 + 13s + 6} \begin{bmatrix} \dfrac{s^2 + 2s + 1}{s} & s \\ s + 4 & s + 6 \end{bmatrix} \begin{bmatrix} \dfrac{12}{s} \\ 0 \end{bmatrix}$$

$$= \begin{bmatrix} \dfrac{12(s^2 + 2s + 1)}{s(4s^2 + 13s + 6)} \\ \dfrac{12(s + 4)}{4s^2 + 13s + 6} \end{bmatrix}$$

Therefore,

$$\mathbf{V}_0(s) = 2\mathbf{I}_2(s)$$

$$= \frac{24(s + 4)}{4s^2 + 13s + 6}$$

Let us now employ Thévenin's theorem to find $\mathbf{V}_o(s)$. The open-circuit voltage can be obtained using the circuit in Fig. 18.9a. KVL for the closed path is

$$(s + 2)\mathbf{I}_o(s) = \frac{12}{s} - 2\mathbf{V}_x(s)$$

where

$$\mathbf{V}_x(s) = 2\mathbf{I}_o(s)$$

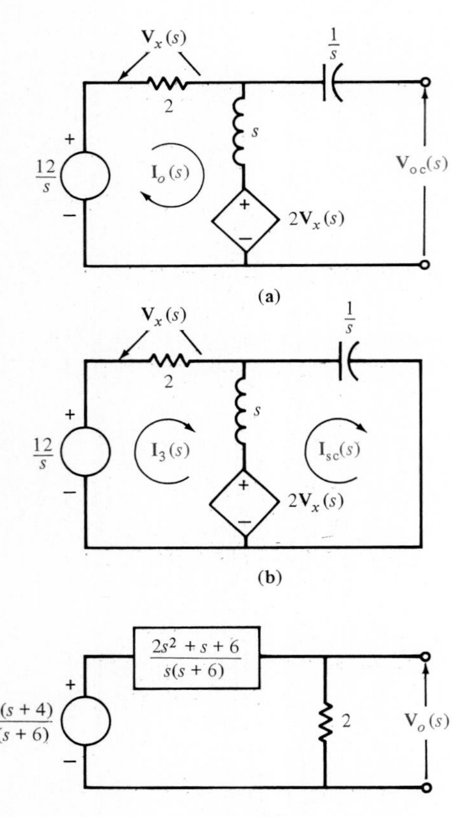

Figure 18.9 Circuits used to solve the circuit in Fig. 18.8, using Thévenin's theorem.

Solving these equations for $\mathbf{I}_o(s)$ yields

$$\mathbf{I}_o(s) = \frac{12}{s(s+6)}$$

Therefore,

$$\begin{aligned}\mathbf{V}_{oc}(s) &= s\mathbf{I}_o(s) + 2\mathbf{V}_x(s)\\ &= \frac{12(s+4)}{s(s+6)}\end{aligned}$$

The short-circuit current can be found from the circuit in Fig. 18.9b. The loop equations for this network are

$$(s+2)\mathbf{I}_3(s) - s\mathbf{I}_{sc}(s) = \frac{12}{s} - 2\mathbf{V}_x(s)$$

$$-s\mathbf{I}_3(s) + \left(s + \frac{1}{s}\right)\mathbf{I}_{sc}(s) = 2\mathbf{V}_x(s)$$

where

$$\mathbf{V}_x(s) = 2\mathbf{I}_3(s)$$

Substituting the expression for $\mathbf{V}_x(s)$ into the two loop equations yields

$$(s+6)\mathbf{I}_3(s) - s\mathbf{I}_{sc}(s) = \frac{12}{s}$$

$$-(s+4)\mathbf{I}_3(s) + \left(\frac{s^2+1}{s}\right)\mathbf{I}_{sc}(s) = 0$$

which in matrix form are

$$\begin{bmatrix} s+6 & -s \\ -(s+4) & \dfrac{s^2+1}{s}\end{bmatrix}\begin{bmatrix}\mathbf{I}_3(s)\\ \mathbf{I}_{sc}(s)\end{bmatrix} = \begin{bmatrix}\dfrac{12}{s}\\ 0\end{bmatrix}$$

The currents are

$$\begin{aligned}\begin{bmatrix}\mathbf{I}_3(s)\\ \mathbf{I}_{sc}(s)\end{bmatrix} &= \begin{bmatrix} s+6 & -s \\ -(s+4) & \dfrac{s^2+1}{s}\end{bmatrix}^{-1}\begin{bmatrix}\dfrac{12}{s}\\ 0\end{bmatrix}\\ &= \frac{s}{2s^2+s+6}\begin{bmatrix}\dfrac{s^2+1}{s} & s\\ s+4 & s+6\end{bmatrix}\begin{bmatrix}\dfrac{12}{s}\\ 0\end{bmatrix}\\ &= \begin{bmatrix}\dfrac{12(s^2+1)}{s(2s^2+s+6)}\\ \dfrac{12(s+4)}{2s^2+s+6}\end{bmatrix}\end{aligned}$$

The Thévenin equivalent impedance is then

$$\mathbf{Z}_{\mathrm{Th}}(s) = \frac{\mathbf{V}_{\mathrm{oc}}(s)}{\mathbf{I}_{\mathrm{sc}}(s)} = \frac{\dfrac{12(s+4)}{s(s+6)}}{\dfrac{12(s+4)}{2s^2+s+6}}$$

$$= \frac{2s^2+s+6}{s(s+6)}$$

The Thévenin equivalent circuit together with the 2-Ω load is shown in Fig. 18.9c. Using a simple voltage divider, we find that

$$\mathbf{V}_0(s) = \left[\frac{2}{\dfrac{2s^2+s+6}{s(s+6)} + 2}\right]\left[\frac{12(s+4)}{s(s+6)}\right]$$

$$= \frac{24(s+4)}{4s^2+13s+6}$$

which is, of course, the same answer as that obtained using loop equations.

Let us now find $v_o(t) = \mathcal{L}^{-1}[\mathbf{V}_o(s)]$. $\mathbf{V}_o(s)$ can be written as

$$\mathbf{V}_o(s) = \frac{6(s+4)}{s^2+\frac{13}{4}s+\frac{3}{2}} = \frac{6(s+4)}{(s+2.7)(s+0.56)}$$

Hence

$$\frac{6(s+4)}{(s+2.7)(s+0.56)} = \frac{K_1}{s+2.7} + \frac{K_2}{s+0.56}$$

Evaluating the constants, we obtain

$$\left.\frac{6(s+4)}{s+0.56}\right|_{s=-2.7} = K_1$$

$$-3.64 = K_1$$

and

$$\left.\frac{6(s+4)}{s+2.7}\right|_{s=-0.56} = K_2$$

$$9.64 = K_2$$

Therefore,

$$v_o(t) = (9.64e^{-0.56t} - 3.64e^{-2.7t})u(t) \text{ V}$$ ■

DRILL EXERCISE

D18.1. Find $i_o(t)$ in the network in Fig. D18.1 using node equations.

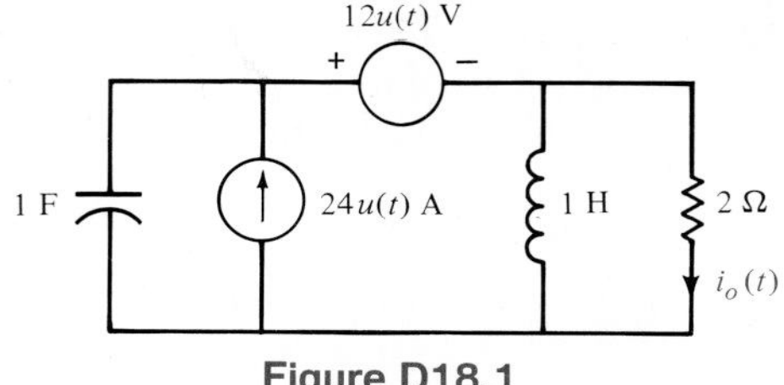

Figure D18.1

D18.2. Solve Drill Exercise D18.1 using superposition.

D18.3. Solve Drill Exercise D18.1 using source transformation.

D18.4. Solve Drill Exercise D18.1 using Norton's theorem.

D18.5. Find $v_o(t)$ in the network in Fig. D18.5 using loop equations.

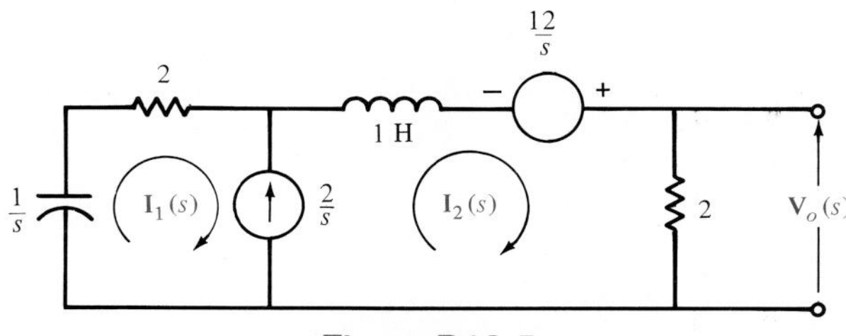

Figure D18.5

D18.6. Solve Drill Exercise D18.5 using superposition.

D18.7. Solve Drill Exercise D18.5 using source transformation.

D18.8. Solve Drill Exercise D18.5 using Thévenin's theorem.

18.3 Steady-State Response

In Section 18.2 we have demonstrated, using a variety of examples, the power of the Laplace transform technique in determining the complete response of a network. This complete response is composed of transient terms which disappear as $t \to \infty$ and steady-state terms which are present at all times. Let us now examine a method by which to determine the steady-state response of a network directly. Recall from previous examples that the network response can be written as

$$\mathbf{Y}(s) = \mathbf{H}(s)\mathbf{X}(s) \tag{18.13}$$

where $\mathbf{Y}(s)$ is the output or response, $\mathbf{X}(s)$ is the input or forcing function, and $\mathbf{H}(s)$ is the network function or transfer function defined in Section 13.1. The transient portion of the response $\mathbf{Y}(s)$ results from the poles of $\mathbf{H}(s)$, and the steady-state portion of the response results from the poles of the input or forcing function.

As a direct parallel to the sinusoidal response of a network as outlined in Section 9.2, we assume that the forcing function is of the form

$$x(t) = X_M e^{j\omega_0 t} \tag{18.14}$$

which by Euler's identity can be written as

$$x(t) = X_M \cos \omega_0 t + jX_M \sin \omega_0 t \tag{18.15}$$

The Laplace transform of Eq. (18.14) is

$$\mathbf{X}(s) = \frac{X_M}{s - j\omega_0} \tag{18.16}$$

and therefore

$$\mathbf{Y}(s) = \mathbf{H}(s)\left(\frac{X_M}{s - j\omega_0}\right) \tag{18.17}$$

At this point we tacitly assume that $\mathbf{H}(s)$ does not have any poles of the form $(s - j\omega_k)$. If, however, this is the case, we simply encounter difficulty in defining the steady-state response.

Performing a partial-fraction expansion of Eq. (18.17) yields

$$\mathbf{Y}(s) = \frac{X_M \mathbf{H}(j\omega_0)}{s - j\omega_0} + \text{terms that occur due to the poles of } \mathbf{H}(s) \tag{18.18}$$

The first term to the right of the equal sign can be expressed as

$$\mathbf{Y}(s) = \frac{X_M |\mathbf{H}(j\omega_0)| e^{j\phi(j\omega_0)}}{s - j\omega_0} + \cdots \tag{18.19}$$

since $\mathbf{H}(j\omega_0)$ is a complex quantity with a magnitude and phase that are a function of $j\omega_0$.

Performing the inverse transform of Eq. (18.19), we obtain

$$\begin{aligned} y(t) &= X_M |\mathbf{H}(j\omega_0)| e^{j\omega_0 t} e^{j\phi(j\omega_0)} + \cdots \\ &= X_M |\mathbf{H}(j\omega_0)| e^{j(\omega_0 t + \phi(j\omega_0))} + \cdots \end{aligned} \tag{18.20}$$

and hence the steady-state response is

$$y_{ss}(t) = X_M |\mathbf{H}(j\omega_0)| e^{j(\omega_0 t + \phi(j\omega_0))} \tag{18.21}$$

Since the actual forcing function is $X_M \cos \omega_0 t$, which is the real part of $X_M e^{j\omega_0 t}$, the steady-state response is the real part of Eq. (18.21).

$$y_{ss}(t) = X_M |\mathbf{H}(j\omega_0)| \cos [\omega_0 t + \phi(j\omega_0)] \tag{18.22}$$

In general, the forcing function may have a phase angle θ. In this case, θ is simply added to $\phi(j\omega_0)$ so that the resultant phase of the response is $\phi(j\omega_0) + \theta$.

EXAMPLE 18.6

For the circuit shown in Fig. 18.10a, we wish to determine the steady-state voltage $v_{oss}(t)$ for $t > 0$ if the initial conditions are zero.

As illustrated earlier, this problem could be solved using a variety of techniques, such as node equations, mesh equations, source transformation, and Thévenin's theorem. We will employ node equations to obtain the solution. The transformed network using the impedance values for the parameters is shown in Fig. 18.10b. The node equations for this network are

$$\left(\frac{1}{2} + \frac{1}{s} + \frac{s}{2}\right)\mathbf{V}_1(s) - \left(\frac{s}{2}\right)\mathbf{V}_o(s) = \frac{1}{2}\,\mathbf{V}_i(s)$$

$$-\left(\frac{s}{2}\right)\mathbf{V}_1(s) + \left(\frac{s}{2} + 1\right)\mathbf{V}_o(s) = 0$$

Solving these equations for $\mathbf{V}_o(s)$, we obtain

$$\mathbf{V}_o(s) = \frac{s^2}{3s^2 + 4s + 4}\,\mathbf{V}_i(s)$$

Note that this equation is in the form of Eq. (18.13), where $\mathbf{H}(s)$ is

$$\mathbf{H}(s) = \frac{s^2}{3s^2 + 4s + 4}$$

Since the forcing function is $10 \cos 2t\, u(t)$, then $V_M = 10$ and $\omega_0 = 2$. Hence

$$\mathbf{H}(j2) = \frac{(j2)^2}{3(j2)^2 + 4(j2) + 4}$$

$$= 0.354 \angle 45^\circ$$

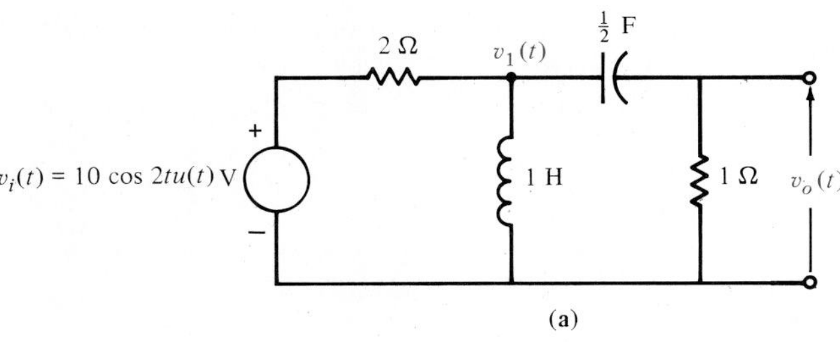

(a)

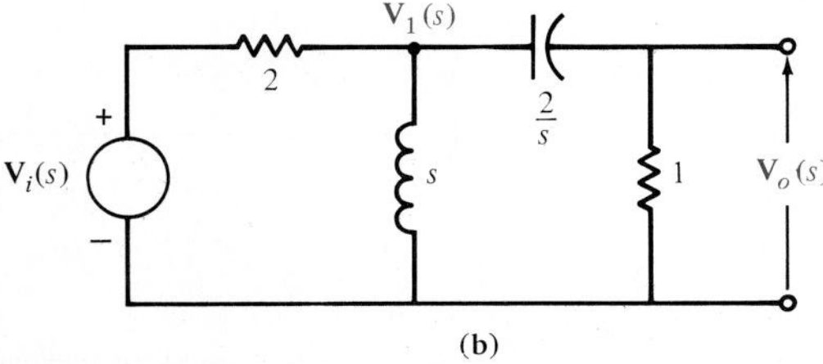

(b)

Figure 18.10 Circuit used in Example 18.6.

Therefore,

$$|\mathbf{H}(j2)| = 0.354$$

$$\phi(j2) = 45°$$

and hence the steady-state response is

$$v_{oss}(t) = V_M|\mathbf{H}(j2)| \cos [2t + \phi(j2)]$$

$$= 3.54 \cos (2t + 45°)$$

The complete (transient plus steady-state) response can be obtained from the expression

$$\mathbf{V}_o(s) = \frac{s^2}{3s^2 + 4s + 4} \mathbf{V}_i(s)$$

$$= \frac{s^2}{3s^2 + 4s + 4}\left(\frac{10s}{s^2 + 4}\right)$$

$$= \frac{10s^3}{(s^2 + 4)(3s^2 + 4s + 4)}$$

Determining the inverse Laplace transform of this function using the techniques of Chapter 17, we obtain

$$v_o(t) = 3.54 \cos (2t + 45°) + 1.44e^{-(2/3)t} \cos \left(\frac{2\sqrt{2}}{3} t - 55°\right) \text{ V}$$

Note that as $t \to \infty$ the second term approaches zero, and thus the steady-state response is

$$v_{oss}(t) = 3.54 \cos (2t + 45°) \text{ V}$$

which can easily be checked using a phasor analysis. ■

DRILL EXERCISE

D18.9. Determine the steady-state voltage $v_{oss}(t)$ in the network in Fig. D18.9 for $t > 0$ if the initial conditions in the network are zero.

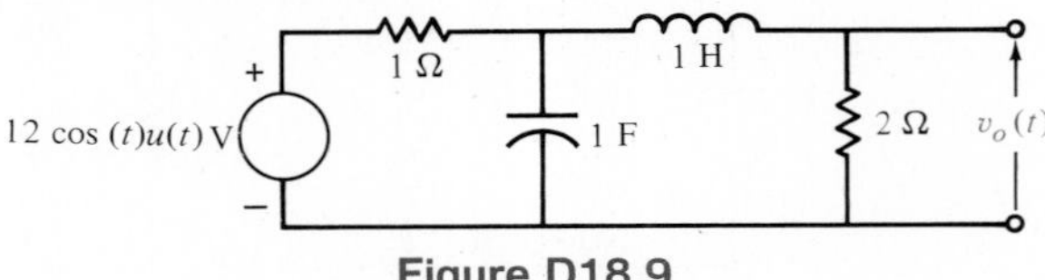

Figure D18.9

D18.10. Determine the steady-state voltage $v_{oss}(t)$ in the network in Fig. D18.10 for $t > 0$ if all initial conditions are zero.

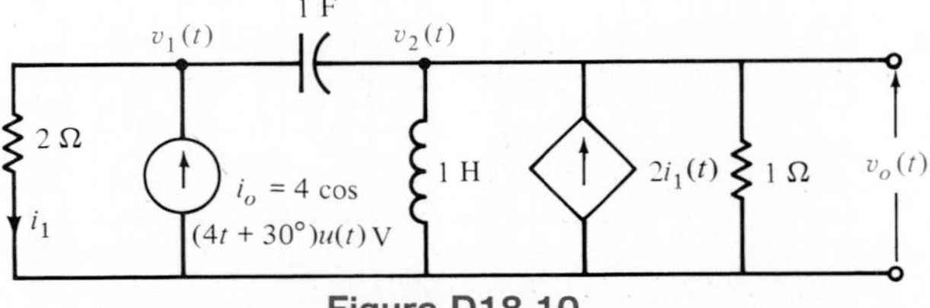

Figure D18.10.

18.4 Transient Response

We will now illustrate the use of the Laplace transform in the transient analysis of circuits. We will analyze networks such as those considered in Chapters 7 and 8. Our approach will first be to determine the initial conditions for the capacitors and inductors in the network, and then we will employ the element models that were specified at the beginning of this chapter together with the circuit analysis techniques to obtain a solution. The following examples will demonstrate the approach.

EXAMPLE 18.7

In the network shown in Fig. 18.11a, the switch moves from position 1 to 2 at $t = 0$. Let us determine the current $i(t)$ for $t > 0$.

In the steady-state condition prior to switch action, the capacitor appears as an open circuit and the voltage across it is the same as the voltage across the 3-kΩ resistor, which is

$$v_c(0-) = \frac{(12 \text{ V})(3 \text{ k}\Omega)}{6 \text{ k}\Omega + 3 \text{ k}\Omega} = 4 \text{ V}$$

The transformed circuit, valid for $t > 0$, is shown in Fig. 18.11b. The node equation for the voltage $\mathbf{V}(s)$ is

$$\mathbf{V}(s)\left(\frac{1}{6 \text{ k}} + 10^{-4}s + \frac{1}{3 \text{ k}}\right) = 4 \times 10^{-4}$$

or

$$\mathbf{V}(s) = \frac{4}{s + 5}$$

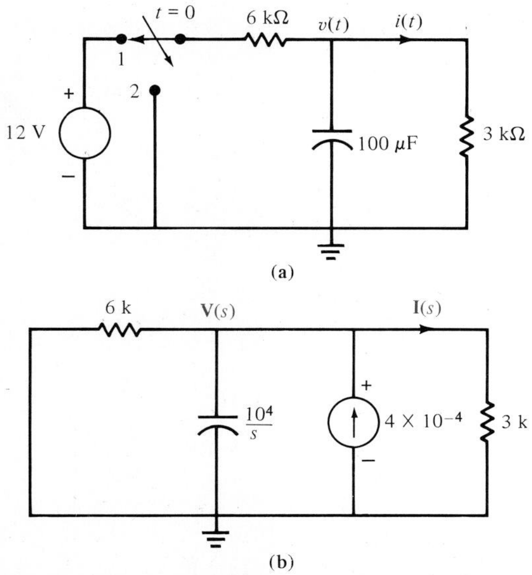

Figure 18.11 Example network in the time domain and s-domain.

Then

$$\mathbf{I}(s) = \frac{\frac{4}{3} \times 10^{-3}}{s + 5}$$

and hence

$$i(t) = \left(\frac{4}{3} e^{-5t}\right) u(t) \qquad \text{mA}$$

This is the same answer as that obtained in Example 7.1. ■

EXAMPLE 18.8

Given the network shown in Fig. 18.12a, let us find the current $i_2(t)$ for $t > 0$.

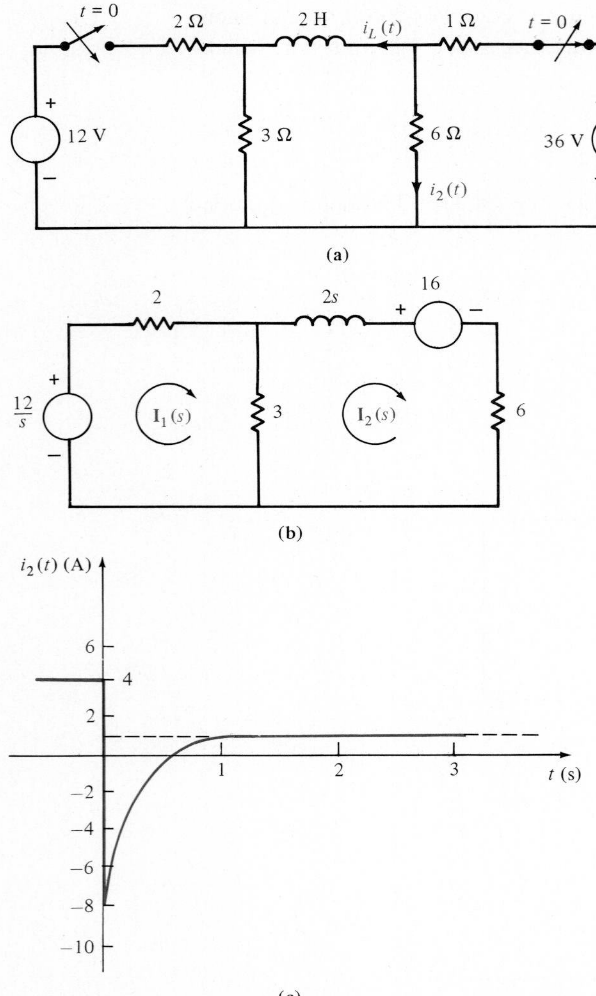

Figure 18.12 (a) Example circuit in the time domain; (b) the transformed circuit for $t > 0$; (c) the network response for $i_2(t)$.

In the steady state prior to switch action, the inductor appears as a short circuit and using dc analysis we find that

$$i_L(0-) = 8 \text{ A}$$

$$i_2(0-) = 4 \text{ A}$$

The transformed circuit using the impedance values of the parameters is shown in Fig. 18.12b. The mesh equations for this network are

$$5\mathbf{I}_1(s) - 3\mathbf{I}_2(s) = \frac{12}{s}$$

$$-3\mathbf{I}_1(s) + (2s + 9)\mathbf{I}_2(s) = -16$$

which can be written in matrix form as

$$\begin{bmatrix} 5 & -3 \\ -3 & 2s + 9 \end{bmatrix} \begin{bmatrix} \mathbf{I}_1(s) \\ \mathbf{I}_2(s) \end{bmatrix} = \begin{bmatrix} \dfrac{12}{s} \\ -16 \end{bmatrix}$$

Solving for the currents, we obtain

$$\begin{bmatrix} \mathbf{I}_1(s) \\ \mathbf{I}_2(s) \end{bmatrix} = \begin{bmatrix} 5 & -3 \\ -3 & 2s + 9 \end{bmatrix}^{-1} \begin{bmatrix} \dfrac{12}{s} \\ -16 \end{bmatrix}$$

$$= \frac{1}{10s + 36} \begin{bmatrix} 2s + 9 & 3 \\ 3 & 5 \end{bmatrix} \begin{bmatrix} \dfrac{12}{s} \\ -16 \end{bmatrix}$$

$$= \begin{bmatrix} \dfrac{-24s + 108}{s(10s + 36)} \\ \dfrac{-80s + 36}{s(10s + 36)} \end{bmatrix}$$

Therefore, $\mathbf{I}_2(s)$ is

$$\mathbf{I}_2(s) = \frac{-8s + 3.6}{s(s + 3.6)}$$

$$= \frac{1}{s} - \frac{9}{s + 3.6}$$

Therefore, $i_2(t)$ is

$$i_2(t) = (1 - 9e^{-3.6t})u(t) \text{ A}$$

A plot of this current is shown in Fig. 18.12c. ■

EXAMPLE 18.9

In this example we analyze the circuit in Example 8.7 using Laplace transforms. The circuit is redrawn in Fig. 18.13a. At $t = 0$ the initial voltage across the capacitor is

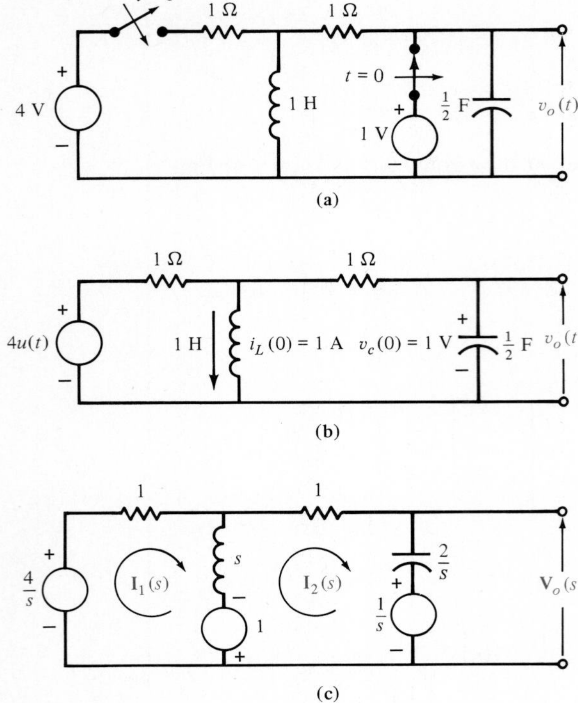

Figure 18.13 Circuits employed in Example 18.9.

1 V, and the initial current drawn through the inductor is 1 A. The circuit for $t > 0$ is shown in Fig. 18.13b with the initial conditions. The transformed network is shown in Fig. 18.13c.

The mesh equations for transformed network are

$$(s + 1)\mathbf{I}_1(s) - s\mathbf{I}_2(s) = \frac{4}{s} + 1$$

$$-s\mathbf{I}_1(s) + \left(s + \frac{2}{s} + 1\right)\mathbf{I}_2(s) = \frac{-1}{s} - 1$$

which can be written in matrix form as

$$\begin{bmatrix} s + 1 & -s \\ -s & \dfrac{s^2 + s + 2}{s} \end{bmatrix} \begin{bmatrix} \mathbf{I}_1(s) \\ \mathbf{I}_2(s) \end{bmatrix} = \begin{bmatrix} \dfrac{s + 4}{s} \\ \dfrac{-(s + 1)}{s} \end{bmatrix}$$

Solving for the currents, we obtain

$$\begin{bmatrix} \mathbf{I}_1(s) \\ \mathbf{I}_2(s) \end{bmatrix} = \begin{bmatrix} s + 1 & -s \\ -s & \dfrac{s^2 + s + 2}{s} \end{bmatrix}^{-1} \begin{bmatrix} \dfrac{s + 4}{s} \\ \dfrac{-(s + 1)}{s} \end{bmatrix}$$

$$= \frac{s}{2s^2 + 3s + 2}\begin{bmatrix} \frac{s^2+s+2}{s} & s \\ s & s+1 \end{bmatrix}\begin{bmatrix} \frac{s+4}{s} \\ \frac{-(s+1)}{s} \end{bmatrix}$$

$$= \begin{bmatrix} \frac{4s^2 + 6s + 8}{s(2s^2 + 3s + 2)} \\ \frac{2s - 1}{2s^2 + 3s + 2} \end{bmatrix}$$

The output voltage is then

$$\begin{aligned} \mathbf{V}_o(s) &= \frac{2}{s}\mathbf{I}_2(s) + \frac{1}{s} \\ &= \frac{2}{s}\left(\frac{2s - 1}{2s^2 + 3s + 2}\right) + \frac{1}{s} \\ &= \frac{s + \frac{7}{2}}{s^2 + \frac{3}{2}s + 1} \end{aligned}$$

This function can be written in a partial-fraction expansion as

$$\frac{s + \frac{7}{2}}{s^2 + \frac{3}{2}s + 1} = \frac{K_1}{s + \frac{3}{4} - j(\sqrt{7}/4)} + \frac{K_1^*}{s + \frac{3}{4} + j(\sqrt{7}/4)}$$

Evaluating the constants, we obtain

$$\left.\frac{s + \frac{7}{2}}{s + \frac{3}{4} + j(\sqrt{7}/4)}\right|_{s = -(3/4) + j(\sqrt{7}/4)} = K_1$$

$$2.14 \underline{/-76.5^\circ} = K_1$$

Therefore,

$$v_o(t) = \left[4.29e^{-(3/4)t} \cos\left(\frac{\sqrt{7}}{4}t - 76.5^\circ\right)\right]u(t) \text{ V}$$

This function can be rewritten using the trigonometric identity

$$\cos(A - B) = \cos A \cos B + \sin A \sin B$$

Then

$$\begin{aligned} v_o(t) &= 4.29e^{-(3/4)t}\left(\cos\frac{\sqrt{7}}{4}t \cos 76.5^\circ + \sin\frac{\sqrt{7}}{4}t \sin 76.5^\circ\right)\boldsymbol{u(t)} \\ &= \left(e^{-(3/4)t}\cos\frac{\sqrt{7}}{4}t + 4.16e^{-(3/4)t}\sin\frac{\sqrt{7}}{4}t\right)u(t) \text{ V} \end{aligned}$$

which is identical to Eq. (8.29). ■

EXAMPLE 18.10

In this example we analyze the circuit in Example 8.8 using Laplace transforms. The circuit is drawn again in Fig. 18.14a. The initial conditions can be found from the network in Fig. 18.14b. From this network we note that at $t = 0$ the capacitor voltage is 1 V and the inductor current is 0.5 A. The transformed network, valid for $t > 0$, is shown in Fig. 18.14c.

The node equations for the transformed network are

$$\frac{\mathbf{V}_1(s) - 4/s}{1} + \frac{\mathbf{V}_1(s)}{1/s} + \frac{\mathbf{V}_1(s) - \mathbf{V}_o(s)}{1} = 1$$

$$\frac{\mathbf{V}_o(s) - \mathbf{V}_1(s)}{1} + \frac{\mathbf{V}_o(s)}{2} + \frac{\mathbf{V}_o(s) - 4/s}{s/2} = \frac{1}{2s}$$

These equations can be written in the form

$$(s + 2)\mathbf{V}_1(s) - \mathbf{V}_o(s) = \frac{s + 4}{s}$$

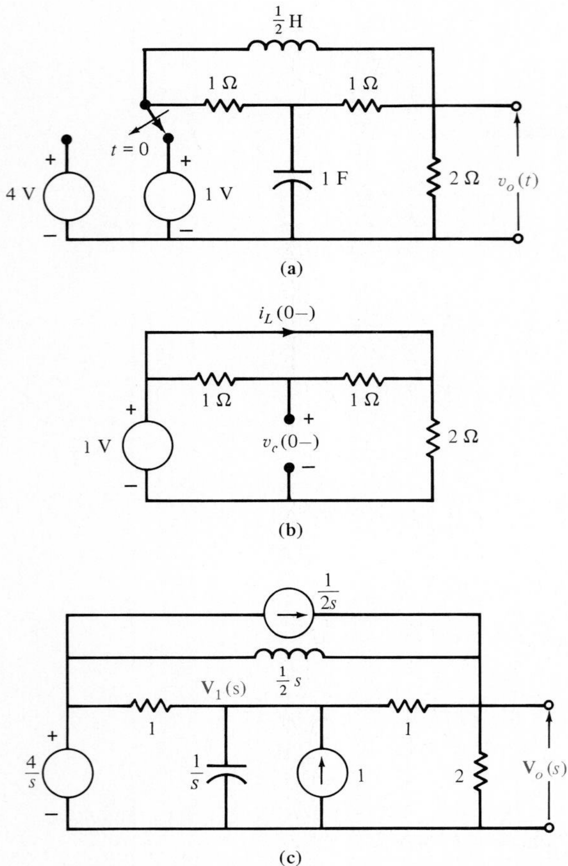

Figure 18.14 Circuits used in Example 18.10.

$$-\mathbf{V}_1(s) + \frac{3s + 4}{2s}\mathbf{V}_o(s) = \frac{s + 16}{2s^2}$$

which can be expressed in a matrix equation as

$$\begin{bmatrix} s + 2 & -1 \\ -1 & \dfrac{3s + 4}{2s} \end{bmatrix} \begin{bmatrix} \mathbf{V}_1(s) \\ \mathbf{V}_o(s) \end{bmatrix} = \begin{bmatrix} \dfrac{s + 4}{s} \\ \dfrac{s + 16}{2s^2} \end{bmatrix}$$

Solving for the voltages, we obtain

$$\begin{bmatrix} \mathbf{V}_1(s) \\ \mathbf{V}_o(t) \end{bmatrix} = \begin{bmatrix} s + 2 & -1 \\ -1 & \dfrac{3s + 4}{2s} \end{bmatrix}^{-1} \begin{bmatrix} \dfrac{s + 4}{s} \\ \dfrac{s + 16}{2s^2} \end{bmatrix}$$

$$= \frac{2s}{3s^2 + 8s + 8} \begin{bmatrix} \dfrac{3s + 4}{2s} & 1 \\ 1 & s + 2 \end{bmatrix} \begin{bmatrix} \dfrac{s + 4}{s} \\ \dfrac{s + 16}{2s^2} \end{bmatrix}$$

$$= \begin{bmatrix} \dfrac{3s^2 + 17s + 32}{s(3s^2 + 8s + 8)} \\ \dfrac{3s^2 + 26s + 32}{s(3s^2 + 8s + 8)} \end{bmatrix}$$

$\mathbf{V}_o(s)$ can be written as

$$\mathbf{V}_o(s) = \frac{s^2 + \frac{26}{3}s + \frac{32}{3}}{s(s^2 + \frac{8}{3}s + \frac{8}{3})}$$

The partial-fraction expansion of this function is

$$\frac{s^2 + \frac{26}{3}s + \frac{32}{3}}{s(s^2 + \frac{8}{3}s + \frac{8}{3})} = \frac{K_0}{s} + \frac{K_1}{s + \frac{4}{3} - j(2\sqrt{2}/3)} + \frac{K_1^*}{s + \frac{4}{3} + j(2\sqrt{2}/3)}$$

Evaluating the constants, we obtain

$$\left.\frac{s^2 + \frac{26}{3}s + \frac{32}{3}}{s(s^2 + \frac{8}{3}s + \frac{8}{3})}\right|_{s=0} = K_0$$

$$4 = K_0$$

and

$$\left.\frac{s^2 + \frac{26}{3}s + \frac{32}{3}}{s[s + \frac{4}{3} + j(2\sqrt{2}/3)]}\right|_{s=(-4/3)+j(2\sqrt{2}/3)} = K_1$$

$$1.84 \underline{/-144.74^\circ} = K_1$$

Therefore,

$$v_o(t) = \left[4 + 3.68e^{-(4/3)t} \cos\left(\frac{2\sqrt{2}}{3} t - 144.74°\right)\right] u(t) \text{ V}$$

Using the identity cos $(A - B) = \cos A \cos B + \sin A \sin B$, we obtain

$$v_o(t) = \left[4 + 3.68^{-(4/3)t}\left(\cos \frac{2\sqrt{2}}{3} t \cos 144.74° + \sin \frac{2\sqrt{2}}{3} t \sin 144.74°\right)\right] u(t)$$

$$= \left[4 - 3e^{-(4/3)t} \cos \frac{2\sqrt{2}}{3} t + 2.12e^{-(4/3)t} \sin \frac{2\sqrt{2}}{3} t\right] u(t) \text{ V}$$

which is identical to the answer obtained in Example 8.8. ■

DRILL EXERCISE

D18.11. Solve Drill Exercise D7.1 using Laplace transforms.

D18.12. Solve Drill Exercise D7.3 using Laplace transforms.

D18.13. Solve Drill Exercise D7.5 using Laplace transforms.

D18.14. Solve Drill Exercise D7.6 using Laplace transforms.

At this point it is informative to review briefly the natural response of both first-order and second-order networks. We have demonstrated in Chapter 7 that if only a single storage element is present, the natural response of a network to an initial condition is always of the form

$$x(t) = X_0 e^{-t/T_c}$$

where $X(t)$ can be either $v(t)$ or $i(t)$, X_0 is the initial value of $x(t)$, and T_c is the time constant of the network. Example 18.7 is an example of the use of the Laplace transform in determining the natural response of a first-order system.

As illustrated in Chapter 8, the natural response of a second-order network is controlled by the roots of the *characteristic equation,* which is of the form

$$s^2 + 2\xi\omega_0 s + \omega_0^2 = 0$$

where ξ is the *damping ratio* and ω_0 is the *undamped natural frequency*. These two key factors, ξ and ω_0, control the response, and there are basically three cases of interest.

Case 1, $\xi > 1$: Overdamped Network. The roots of the characteristic equation are $s_1, s_2 = -\xi\omega_0 \pm \omega_0\sqrt{\xi^2 - 1}$, and therefore the network response is of the form

$$x(t) = K_1 e^{-(\xi\omega_0 + \omega_0\sqrt{\xi^2-1})t} + K_2 e^{-(\xi\omega_0 - \omega_0\sqrt{\xi^2-1})t}$$

Case 2, $\xi < 1$: Underdamped Network. The roots of the characteristic equation are $s_1, s_2 = -\xi\omega_0 \pm j\omega_0\sqrt{1 - \xi^2}$, and therefore the network response is of the form

$$x(t) = Ke^{-\xi\omega_0 t} \cos(\omega_0\sqrt{1 - \xi^2}\, t + \Phi)$$

Case 3, $\xi = 1$: Critically Damped Network. The roots of the characteristic equation are $s_1, s_2 = -\omega_0$, and hence the response is of the form

$$x(t) = K_1 t e^{-\omega_0 t} + K_2 e^{-\omega_0 t}$$

It is very important that the reader note that the characteristic equation is the denominator of the response function $\mathbf{X}(s)$, and the roots of this equation, which are the poles of the network, determine the form of the network's natural response.

A convenient method for displaying the network's poles and zeros in graphical form is the use of a pole–zero plot. A pole–zero plot of a function can be accomplished using what is commonly called the *complex* or *s-plane*. In the complex plane the abscissa is σ and the ordinate is $j\omega$. Zeros are represented by 0's and poles are represented by $\times$'s. Although we are concerned only with the finite poles and zeros specified by the network or response function, we should point out that a rational function must have the same number of poles and zeros. Therefore, if $n > m$, there are $n - m$ zeros at the point at infinity; and if $n < m$, there are $m - n$ poles at the point at infinity. A systems engineer can tell a lot about the operation of a network or system by simply examining its pole–zero plot.

In order to correlate the natural response of a network to an initial condition with the network's pole locations, we have illustrated in Fig. 18.15 the correspondence for all

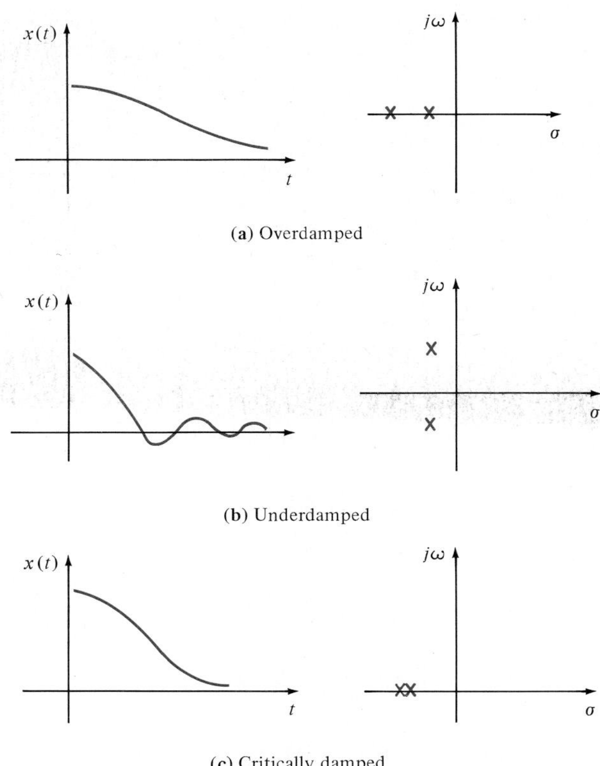

Figure 18.15 Natural response of a second-order network together with network pole locations for three cases.

three cases—overdamped, underdamped, and critically damped. Note that if the network poles are real and unequal, the response is slow, and therefore $x(t)$ takes a long time to reach zero. If the network poles are complex conjugates, the response is fast; however, it overshoots and is eventually damped out. The dividing line between the overdamped and underdamped cases is the critically damped case in which the roots are real and equal. In this case the transient response dies out as quickly as possible, with no overshoot. Example 18.10 illustrates an underdamped response with a steady-state nonzero component.

18.5 Initial and Final Values

The initial- and final-value theorems derived in Chapter 17 are, respectively,

$$\lim_{t\to 0} f(t) = \lim_{s\to\infty} s\mathbf{F}(s) \tag{18.23}$$

and

$$\lim_{t\to\infty} f(t) = \lim_{s\to 0} s\mathbf{F}(s) \tag{18.24}$$

Let us now demonstrate their use in circuit analysis by applying them to a couple of the examples presented in the preceding section.

EXAMPLE 18.11

The expression for $\mathbf{V}(s)$ in Example 18.7 is

$$\mathbf{V}(s) = \frac{4}{s+5}$$

Let us determine the initial and final values of this voltage using the theorems. The initial value is

$$\begin{aligned}\lim_{t\to 0} v(t) &= \lim_{s\to\infty} s\mathbf{V}(s)\\ &= \lim_{s\to\infty} \frac{4s}{s+5}\\ &= 4\text{ V}\end{aligned}$$

and the final value is

$$\begin{aligned}\lim_{t\to\infty} v(t) &= \lim_{s\to 0} s\mathbf{V}(s)\\ &= \lim_{s\to 0} \frac{4s}{s+5}\\ &= 0\text{ V}\end{aligned}$$

■

EXAMPLE 18.12

The expression for the voltage $\mathbf{V}_o(s)$ in Example 18.10 was found to be

$$\mathbf{V}_o(s) = \frac{3s^2 + 26s + 32}{s(3s^2 + 8s + 8)}$$

The initial and final values of this voltage are then

$$\begin{aligned}\lim_{t \to 0} v_o(t) &= \lim_{s \to \infty} s\mathbf{F}(s)\\ &= \lim_{s \to \infty} \frac{3s^2 + 26s + 32}{3s^2 + 8s + 8}\\ &= 1 \text{ V}\end{aligned}$$

and

$$\begin{aligned}\lim_{t \to \infty} v_o(t) &= \lim_{s \to 0} s\mathbf{F}(s)\\ &= \lim_{s \to 0} \frac{3s^2 + 26s + 32}{3s^2 + 8s + 8}\\ &= 4 \text{ V}\end{aligned}$$

These values can easily be checked using the equation for $v_o(t)$ in Example 18.10. ■

It is important to note that the final-value theorem applies only to cases in which the limit of $f(t)$ as $T \to \infty$ exists. Therefore, this theorem cannot be used if the forcing function is sinusoidal since sin ∞ does not have a precise value.

DRILL EXERCISE

D18.15. Find the initial and final values of the voltage

$$\mathbf{V}(s) = \frac{8(5s + 3)}{s(8s + 5)}$$

D18.16. Find the initial and final values of the voltage

$$\mathbf{V}(s) = \frac{8s + 36}{s(3s + 6)}$$

18.6 Response to Complex Waveform Inputs

In Chapter 17 we illustrated techniques for determining the Laplace transform of a complex waveform. By using these methods in conjunction with our analysis techniques, we can obtain the response of a network to any number of unusual input forcing functions. The following examples are provided to demonstrate the techniques.

EXAMPLE 18.13

In this example we analyze the circuit in Example 7.6 using Laplace transforms. The circuit and the input are redrawn in Fig. 18.16a. We wish to find the expression for the output voltage $v_o(t)$ for $t > 0$.

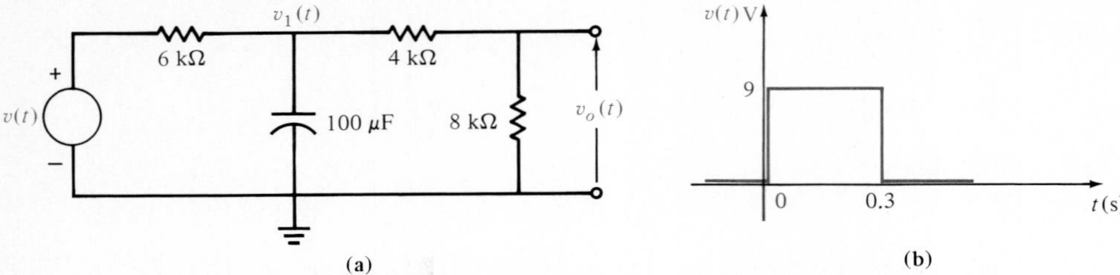

Figure 18.16 Network (a) and input (b) used in Example 18-13.

KCL can be used to compute the node voltage $\mathbf{V}_1(s)$ and then a simple voltage divider can be employed to find $\mathbf{V}_o(s)$. KCL yields

$$\frac{\mathbf{V}_1(s) - \mathbf{V}(s)}{6k} + \frac{\mathbf{V}_1(s)}{1/10^{-4}s} + \frac{\mathbf{V}_1(s)}{12k} = 0$$

Solving this equation for $\mathbf{V}_1(s)$, we obtain

$$\mathbf{V}_1(s) = \frac{\mathbf{V}(s)}{6k(10^{-4}s + 2.5 \times 10^{-4})}$$

Note that

$$\mathbf{V}_o(s) = \frac{8k}{4k + 8k}\mathbf{V}_1(s)$$

Therefore,

$$\mathbf{V}_o(s) = \frac{2\mathbf{V}(s)}{1.8s + 4.5}$$

$$= \frac{1}{0.9}\frac{\mathbf{V}(s)}{s + 2.5}$$

The expression for $\mathbf{V}(s) = \mathcal{L}[v(t)]$, where $v(t)$ is as shown in Fig. 18.16b, is

$$\mathbf{V}(s) = 9\left(\frac{1}{s} - \frac{e^{-0.3s}}{s}\right)$$

Substituting this equation into the expression for $\mathbf{V}_o(s)$ yields

$$\mathbf{V}_o(s) = \left(\frac{1}{0.9}\right)\frac{1}{s + 2.5}(9)\frac{1 - e^{-0.3s}}{s}$$

$$= \frac{10(1 - e^{-0.3s})}{s(s + 2.5)}$$

Since

$$\frac{10}{s(s + 2.5)} = \frac{4}{s} - \frac{4}{s + 2.5}$$

then

$$\mathbf{V}_o(s) = \left(\frac{4}{s} - \frac{4}{s + 2.5}\right)(1 - e^{-0.3s})$$

and therefore

$$v_o(t) = (4 - 4e^{-2.5t})u(t) - (4 - 4e^{-2.5(t-0.3)})u(t - 0.3) \text{ V}$$

This expression is of course identical to that obtained in Example 7.6. ■

EXAMPLE 18.14

Suppose that the input to the network of Example 18.13 is shown in Fig. 18.17. Let us find the network output for this new input. As shown in Example 17.14, the Laplace transform of this sawtooth waveform is

$$\mathbf{V}(s) = \frac{9}{s^2}[1 - (1 + s)e^{-s}]$$

and hence the output is

$$\begin{aligned}\mathbf{V}_0(s) &= \left(\frac{1}{0.9}\right)\frac{1}{s + 2.5}\left(\frac{9}{s^2}\right)[1 - (1 + s)e^{-s}] \\ &= \frac{10}{s^2(s + 2.5)}[1 - (1 + s)e^{-s}] \\ &= \frac{10}{s^2(s + 2.5)} - \frac{10(s + 1)}{s^2(s + 2.5)}e^{-s}\end{aligned}$$

Expanding each of the terms above in a partial-fraction expansion, we obtain

$$\frac{10}{s^2(s + 2.5)} = \frac{4}{s^2} - \frac{1.6}{s} + \frac{1.6}{s + 2.5}$$

and

$$\frac{10(s + 1)}{s^2(s + 2.5)} = \frac{4}{s^2} + \frac{2.4}{s} - \frac{2.4}{s + 2.5}$$

Therefore,

$$v_o(t) = (4t - 1.6 + 1.6e^{-2.5t})u(t) - [4(t - 1) + 2.4 - 2.4e^{-2.5(t-1)}]u(t - 1) \text{ V}$$

■

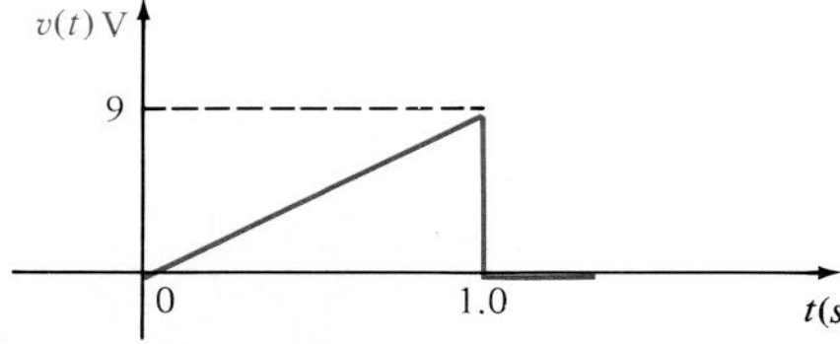

Figure 18.17 Input for the problem in Example 18.14.

DRILL EXERCISE

D18.17. Find the expression for the voltage $v_o(t)$ for $t > 0$ in the network in Fig. D18.17a if the input is shown in Fig. D18.17b.

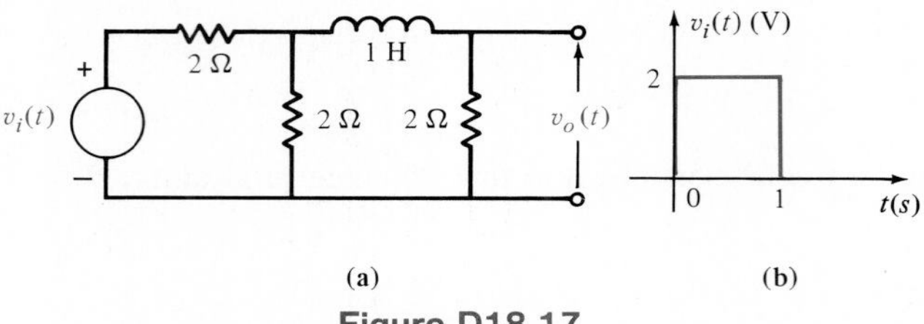

Figure D18.17

D18.18. Find the expression for the voltage $v_o(t)$ for $t > 0$ in the network in Fig. D18.18a if the input is shown in Fig. D18.18b.

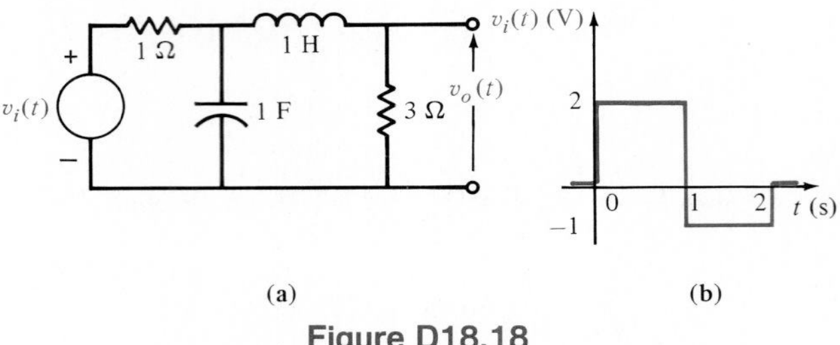

Figure D18.18

18.7 Transfer Function

In Chapter 13 we introduced the concept of network or transfer function. It is essentially nothing more than the ratio of some output variable to some input variable. If both variables are voltages, the transfer function is a voltage gain. If both variables are currents, the transfer function is a current gain. If one variable is a voltage and the other is a current, the transfer function becomes a transfer admittance or impedance.

In deriving a transfer function all initial conditions are set equal to zero. In addition, if the output is generated by more than one input source in a network, superposition can be employed in conjunction with the transfer function for each source.

To present this concept in a more formal manner, let us assume that the input/output relationship for a linear circuit is

$$b_n \frac{d^n y_o(t)}{dt^n} + b_{n-1} \frac{d^{n-1} y_o(t)}{dt^{n-1}} + \cdots + b_1 \frac{dy_o(t)}{dt} + b_0 y_o(t)$$
$$= a_m \frac{d^m x_i(t)}{dt^m} + a_{m-1} \frac{d^{m-1} x_i(t)}{dt^{m-1}} + \cdots + a_1 \frac{dx_i(t)}{dt} + a_0 x_i(t)$$

If all the initial conditions are zero, the transform of the equation is

$$(b_n s^n + b_{n-1} s^{n-1} + \cdots + b_1 s + b_0)\mathbf{Y}_o(s) = (a_m s^m + a_{m-1} s^{m-1} + \cdots + a_1 s + a_0)\mathbf{X}_i(s)$$

or

$$\frac{\mathbf{Y}_o(s)}{\mathbf{X}_i(s)} = \frac{a_m s^m + a_{m-1} s^{m-1} + \cdots + a_1 s + a_0}{b_n s^n + b_{n-1} s^{n-1} + \cdots + b_1 s + b_0}$$

This ratio of $\mathbf{Y}_o(s)$ to $\mathbf{X}_i(s)$ is called the *transfer* or *network function,* which we denote as $\mathbf{H}(s)$, that is,

$$\frac{\mathbf{Y}_o(s)}{\mathbf{X}_i(s)} = \mathbf{H}(s)$$

or

$$\mathbf{Y}_o(s) = \mathbf{H}(s)\mathbf{X}_i(s) \tag{18.25}$$

This equation states that the output response $\mathbf{Y}_o(s)$ is equal to the network function multiplied by the input $\mathbf{X}_i(s)$. Note that if $v_i(t) = \delta(t)$ and therefore $\mathbf{X}_i(s) = 1$, the impulse response is equal to the inverse Laplace transform of the network function. This is an extremely important concept because it illustrates that if we know the impulse response of a network, we can find the response due to some other forcing function using Eq. (18.25).

EXAMPLE 18.15

If the impulse response of a network is $h(t) = e^{-t}$, let us determine the response $v_o(t)$ to an input $v_i(t) = 10e^{-2t}u(t)$ V.

The transformed variables are

$$\mathbf{H}(s) = \frac{1}{s + 1}$$

$$\mathbf{V}_i(s) = \frac{10}{s + 2}$$

Therefore,

$$\mathbf{V}_o(s) = \mathbf{H}(s)\mathbf{V}_i(s) = \frac{10}{(s + 1)(s + 2)}$$

and hence

$$v_o(t) = 10(e^{-t} - e^{-2t})u(t) \text{ V}$$

■

The importance of the transfer function stems from the fact that it provides the systems engineer with a great deal of knowledge about the system's operation, since its dynamic properties are governed by the system poles.

EXAMPLE 18.16

Let us derive the transfer function $\mathbf{V}_o(s)/\mathbf{V}_i(s)$ for the network in Fig. 18.18a. Our output variable is the voltage across a variable capacitor and the input voltage is a unit step. The transformed network is shown in Fig. 18.18b. The mesh equations for the network are

$$2\mathbf{I}_1(s) - \mathbf{I}_2(s) = \mathbf{V}_i(s)$$

$$-\mathbf{I}_1(s) + \left(s + \frac{1}{sC} + 1\right)\mathbf{I}_2(s) = 0$$

and the output equation is

$$\mathbf{V}_o(s) = \frac{1}{sC}\mathbf{I}_2(s)$$

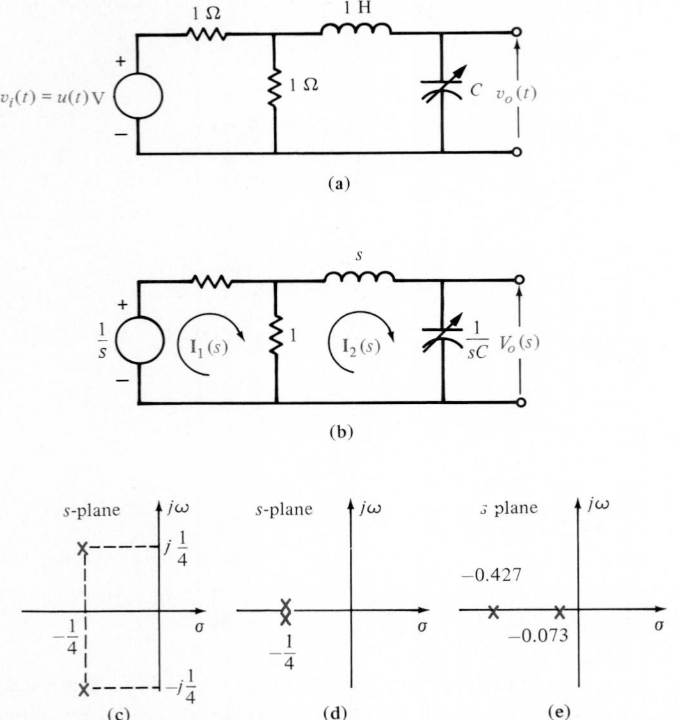

Figure 18.18 Networks and pole–zero plots used in Example 18.16.

From these equations we find that the transfer function is

$$\frac{\mathbf{V}_o(s)}{\mathbf{V}_i(s)} = \frac{1/2C}{s^2 + \frac{1}{2}s + 1/C}$$

Since the transfer function is dependent on the value of the capacitor, let us examine the tranfer function and the output response for three values of the capacitor.

(a) $C = 8$ F:

$$\frac{\mathbf{V}_o(s)}{\mathbf{V}_i(s)} = \frac{\frac{1}{16}}{(s^2 + \frac{1}{2}s + \frac{1}{8})}$$

$$= \frac{\frac{1}{16}}{(s + \frac{1}{4} - j\frac{1}{4})(s + \frac{1}{4} + j\frac{1}{4})}$$

The output response is

$$\mathbf{V}_o(s) = \frac{\frac{1}{16}}{s(s + \frac{1}{4} - j\frac{1}{4})(s + \frac{1}{4} + j\frac{1}{4})}$$

As illustrated in Chapter 8, the poles of the transfer function, which are the roots of the characteristic equation, are complex conjugates, as shown in Fig. 18.18c, and therefore the output response will be *underdamped*. The output response as a function of time is

$$v_o(t) = \left[\frac{1}{2} + \frac{1}{\sqrt{2}} e^{-t/4} \cos\left(\frac{t}{4} + 135°\right)\right] u(t) \text{ V}$$

Note that for large values of time the transient oscillations, represented by the second term in the response, become negligible and the output settles out to a value of $\frac{1}{2}$ V. This can also be seen directly from the circuit since for large values of time the input looks like a dc source, the inductor acts like a short circuit, the capacitor acts like an open circuit, and the resistors form a voltage divider.

(b) $C = 16$ F:

$$\frac{\mathbf{V}_o(s)}{\mathbf{V}_i(s)} = \frac{\frac{1}{32}}{s^2 + \frac{1}{2}s + \frac{1}{16}}$$

$$= \frac{\frac{1}{32}}{(s + \frac{1}{4})^2}$$

The output response is

$$\mathbf{V}_o(s) = \frac{\frac{1}{32}}{s(s + \frac{1}{4})^2}$$

Since the poles of the transfer function are real and equal as shown in Fig. 18.18d, the output response will be *critically damped*. $v_o(t) = \mathcal{L}^{-1}[\mathbf{V}_o(s)]$ is

$$v_o(t) = \left[\frac{1}{2} - \left(\frac{t}{8} + \frac{1}{2}\right) e^{-t/4}\right] u(t) \text{ V}$$

(c) $C = 32$ F:

$$\frac{\mathbf{V}_o(s)}{\mathbf{V}_i(s)} = \frac{\frac{1}{64}}{s^2 + \frac{1}{2}s + \frac{1}{32}}$$

$$= \frac{\frac{1}{64}}{(s + 0.427)(s + 0.073)}$$

The output response is

$$\mathbf{V}_o(s) = \frac{\frac{1}{64}}{s(s + 0.427)(s + 0.073)}$$

The poles of the transfer function are real and unequal as shown in Fig. 18.18e, and therefore the output response will be *overdamped*. The response as a function of time is

$$v_o(t) = (0.5 + 0.103e^{-0.427t} - 0.603e^{-0.073t})u(t) \text{ V}$$

Although the values selected for the network parameters are not very practical, the reader is reminded that both magnitude and frequency scaling, as outlined in Chapter 13, can be applied here also. ■

DRILL EXERCISE

D18.19. If the unit impulse response of a network is known to be $\frac{10}{9}(e^{-t} - e^{-10t})$, determine the unit step response.

D18.20. The transfer function for a network is

$$\mathbf{H}(s) = \frac{(s + 10)}{s^2 + 4s + 8}$$

Determine the pole–zero plot of $\mathbf{H}(s)$, the type of damping exhibited by the network, and the unit step response of the network.

Recall from our previous discussion that if a second-order network is underdamped, the characteristic equation of the network is of the form

$$s^2 + 2\xi\omega_0 s + \omega_0^2 = 0$$

and the roots of this equation, which are the network poles, are of the form

$$s_1, s_2 = -\xi\omega_0 \pm j\omega_0\sqrt{1 - \xi^2}$$

the roots s_1 and s_2, when plotted in the s-plane, generally appear as shown in Fig. 18.19, where

$$\xi = \text{damping ratio}$$
$$\omega_0 = \text{undamped natural frequency}$$

and as shown in Fig. 18.19,

$$\xi = \cos\theta$$

It is important for the reader to note that the damping ratio and the undamped natural frequency are exactly the same quantities as those employed in Chapter 13 when deter-

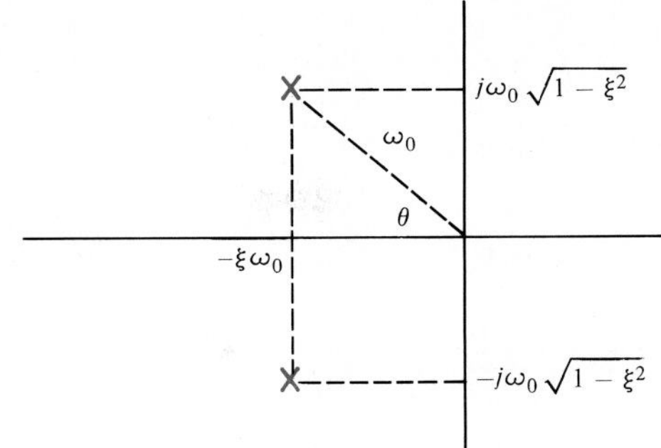

Figure 18.19 Pole locations for second-order underdamped network.

mining a network's frequency response. We find here that it is these same quantities that govern the network's transient response.

EXAMPLE 18.17

In the network in Fig. 18.20a we wish to select R to provide the desired step response of the network. Specifically, we wish to have a damping ratio of $\xi = 0.707$. We will find the value of R to provide this response, plot the pole–zero pattern for the network transfer function, and determine $v_o(t)$ for $t > 0$ if $v_i(t) = u(t)$ V.

The network transfer function is

$$\frac{\mathbf{V}_o(s)}{\mathbf{V}_i(s)} = \frac{Rs}{s^2 + Rs + 1}$$

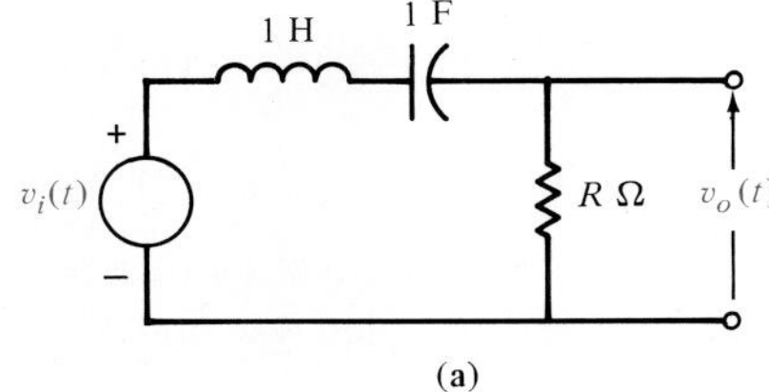

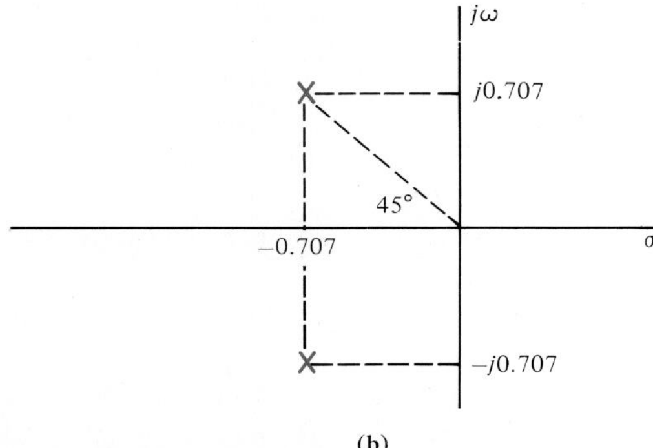

Figure 18.20 Network and pole locations for the problem in Example 18.17.

The characteristic equation is of the form $s^2 + 2\xi\omega_0 s + \omega_0^2$, where $\omega_0 = 1$ and $2\xi\omega_0 = R$. For $\xi = 0.707$, $R = 1.414$. The pole–zero plot for the transfer function is shown in Fig. 18.20b.

The output $\mathbf{V}_o(s)$ for a unit step input is

$$\mathbf{V}_o(s) = \frac{1.414s}{s(s + 0.707 - j0.707)(s + 0.707 + j0.707)}$$

$$= \frac{K_1}{s + 0.707 - j0.707} + \frac{K_1^*}{s + 0.707 + j0.707}$$

$$= \frac{1 \angle -90^\circ}{s + 0.707 - j0.707} + \frac{1 \angle +90^\circ}{s + 0.707 + j0.707}$$

Hence

$$v_o(t) = [2e^{-0.707t} \cos (0.707t - 90^\circ)]u(t) \text{ V}$$

■

18.8 Convolution Integral

In Chapter 17 we employed the convolution integral to determine the inverse Laplace transform of a product of two functions of s. At that time we split a function $\mathbf{F}(s)$ into a product of $\mathbf{F}_1(s)$ and $\mathbf{F}_2(s)$ such that

$$\mathbf{F}(s) = \mathbf{F}_1(s)\mathbf{F}_2(s) \tag{18.26}$$

Then we employed the convolution integral to find

$$f(t) = \int_0^t f_1(\lambda)f_2(t - \lambda)\, d\lambda \tag{18.27}$$

where $f(t) = \mathcal{L}^{-1}[\mathbf{F}(s)]$, $f_1(t) = \mathcal{L}^{-1}[\mathbf{F}_1(s)]$, and $f_2(t) = \mathcal{L}^{-1}[\mathbf{F}_2(s)]$.

At this point we note that if the transfer function of a network $\mathbf{H}(s)$ can be derived, and if the input function $\mathbf{X}(s)$ is known, then we find ourselves in exactly the same position; that is, the output response $\mathbf{Y}(s)$ is

$$\mathbf{Y}(s) = \mathbf{H}(s)\mathbf{X}(s) \tag{18.28}$$

and therefore

$$y(t) = \int_0^t h(\lambda)x(t - \lambda)\, d\lambda \tag{18.29}$$

Let us now employ the convolution integral to determine the output response of a network for various inputs.

EXAMPLE 18.18

Consider the network in Fig. 18.21a. We wish to determine the response $v_o(t)$ to the following set of inputs: (1) $v_i(t) = u(t)$ V, (2) $v_i(t) = e^{-t}u(t)$ V, and (3) $v_i(t) = 4[u(t) - u(t - 1)]$ V.

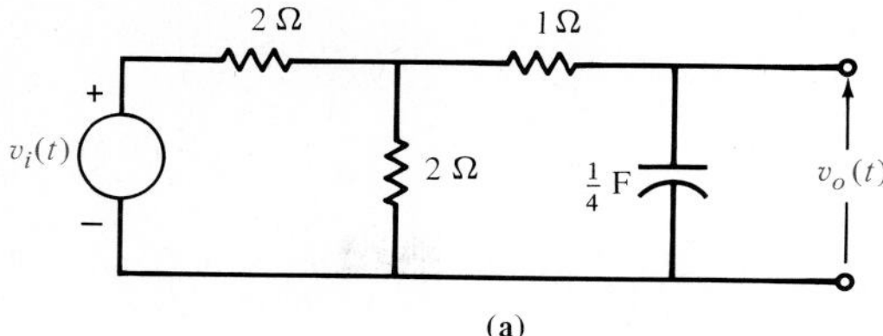

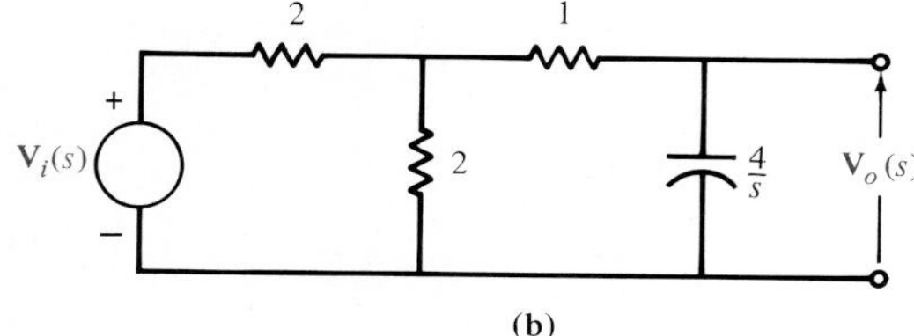

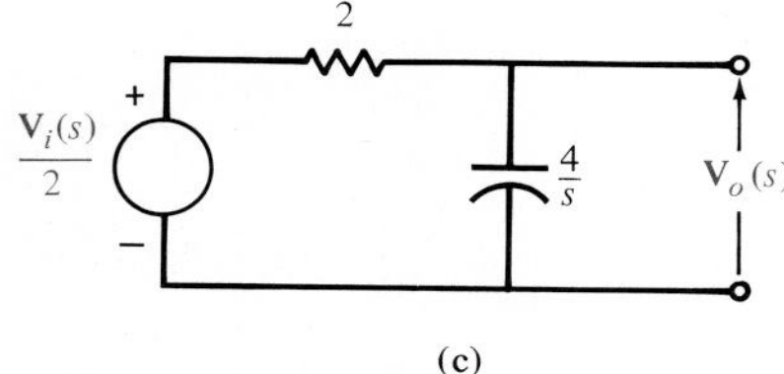

Figure 18.21 Circuits used in Example 18.18.

The transformed network is shown in Fig. 18.21b. The output voltage is

$$\mathbf{V}_o(s) = \frac{[\mathbf{V}_i(s)/2](4/s)}{2 + 4/s}$$

$$= \left(\frac{1}{s + 2}\right) \mathbf{V}_i(s)$$

and therefore the impulse response of the network is

$$h(t) = e^{-2t}$$

Now that the impulse response is known, convolution can be employed to determine the network's response to other inputs. If $v_i(t) = u(t)$, then

$$v_o(t) = \int_0^t u(\lambda)e^{-2(t-\lambda)}\, d\lambda$$

$$= e^{-2t} \int_0^t u(\lambda)e^{2\lambda}\, d\lambda$$

$$= e^{-2t} \int_0^t e^{2\lambda}\, d\lambda \qquad t > 0$$

$$= \tfrac{1}{2}(1 - e^{-2t})u(t) \text{ V}$$

If the input is $v_i(t) = e^{-t}u(t)$, then

$$v_o(t) = \int_0^t e^{-\lambda}u(\lambda)e^{-2(t-\lambda)}\, d\lambda$$

$$= e^{-2t}\int_0^t e^{-\lambda}e^{2\lambda}u(\lambda)\, d\lambda$$

$$= e^{-2t}\int_0^t e^{\lambda}\, d\lambda \qquad t > 0$$

$$= (e^{-t} - e^{-2t})u(t) \text{ V}$$

If the input is $v_i(t) = 4[u(t) - u(t - 1)]$, then

$$v_o(t) = \int_0^t 4[u(\lambda) - u(\lambda - 1)]e^{-2(t-\lambda)}\, d\lambda$$

$$= 4e^{-2t}\int_0^t [u(\lambda) - u(\lambda - 1)]e^{2\lambda}\, d\lambda$$

$$= 4e^{-2t}\int_0^t u(\lambda)e^{2\lambda}\, d\lambda - 4e^{-2t}\int_0^t u(\lambda - 1)e^{2\lambda}\, d\lambda$$

The first integral is

$$4e^{-2t}\int_0^t u(\lambda)e^{2\lambda}\, d\lambda = 4e^{-2t}\int_0^t e^{2\lambda}\, d\lambda \qquad t > 0$$

$$= 2(1 - e^{-2t})u(t)$$

The second integral above is

$$-4e^{-2t}\int_0^t u(\lambda - 1)e^{2\lambda}\, d\lambda = -4e^{-2t}\int_1^t e^{2\lambda}\, d\lambda \qquad t > 1$$

$$= -2(1 - e^{-2(t-1)})u(t - 1)$$

Therefore, the output response is

$$v_o(t) = 2(1 - e^{-2t})u(t) - 2(1 - e^{-2(t-1)})u(t - 1) \text{ V}$$

It is important to note that all of the outputs $v_o(t)$ for the different inputs $v_i(t)$ can be easily checked by computing

$$v_o(t) = \mathcal{L}^{-1}[\mathbf{V}_o(s)]$$

where

$$\mathbf{V}_o(s) = \mathbf{H}(s)\mathbf{V}_i(s)$$ ■

DRILL EXERCISE

D18.21. The output voltage of a network in response to an input of a unit impulse is $v_o(t) = \frac{10}{9}(e^{-t} - e^{10t})$ V. Using the convolution integral, find the output voltage if the input is $e^{-2t}u(t)$.

18.9

Summary

In this chapter we have applied the power of the Laplace transform to circuit analysis. We have demonstrated how to describe the circuit elements in the s-domain and we have used all the network analysis techniques and network theorems derived in dc analysis to obtain a solution. Both the steady-state and transient analyses have been performed. The ease with which one can determine the response of a network to complex inputs using the Laplace transform was shown. The transfer function was defined and its use with the convolution integral demonstrated.

KEY POINTS

- Circuit elements and their initial conditions can be represented in the s-domain by an equivalent circuit.
- All the dc network analysis techniques can be applied to networks represented in the s-domain.
- The response of a network obtained using the Laplace transform is the complete response (i.e., transient plus steady state).
- The steady-state response of a network can be obtained via an evaluation of the network function at the forcing function frequency.
- Once the initial conditions in a network are found, the Laplace transform yields the transient response of the network by solving a set of algebraic equations in the s-domain and then determining the inverse Laplace transform of the desired variable.
- In a network, the roots of the characteristic equation determine the network response.
- By representing complex waveforms using the Laplace transform, the response of a network to complex inputs can be obtained in a straightforward manner.
- The inverse Laplace transform of the network transfer function is the impulse response of the network.
- Convolution can be employed to obtain the response of a network to any input if the impulse response of the network is known.

PROBLEMS

18.1. Find the input impedance $\mathbf{Z}(s)$ of the network shown in Fig. P18.1.

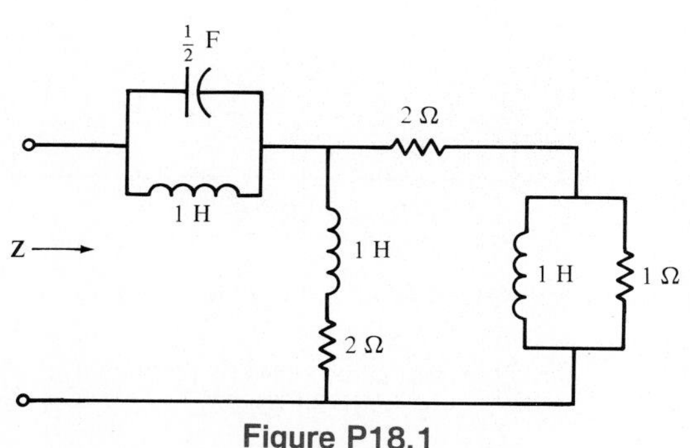

Figure P18.1

18.2. Find the input impedance $\mathbf{Z}(s)$ in the network shown in Fig. P18.2.

(a) When terminals B-B' are open-circuited.

(b) When terminals B-B' are short-circuited.

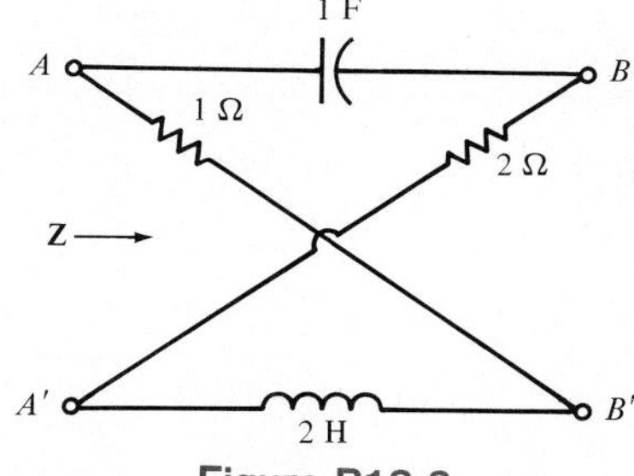

Figure P18.2

18.3. Find the input impedance $\mathbf{Z}(s)$ for the network shown in Fig. P18.3.

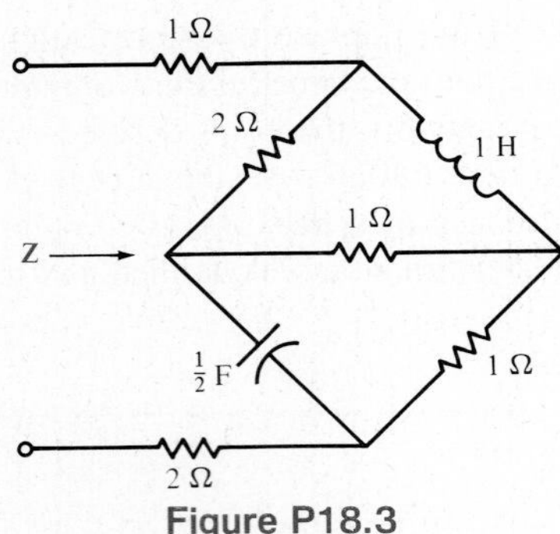

Figure P18.3

18.4. For the network shown in Fig. P18.4, determine the value of the output voltage as $t \to \infty$.

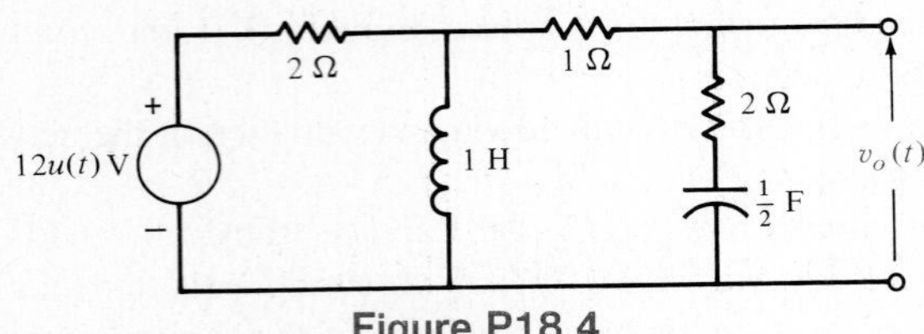

Figure P18.4

18.5. For the network shown in Fig. P18.5, determine the output voltage $v_o(t)$ as $t \to \infty$.

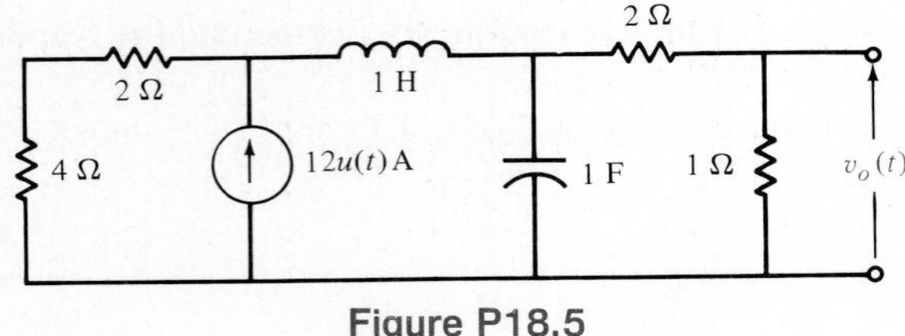

Figure P18.5

18.6. Use Laplace transforms and mesh analysis to find $i_1(t)$ for $t > 0$ in the network shown in Fig. P18.6. Assume zero initial conditions.

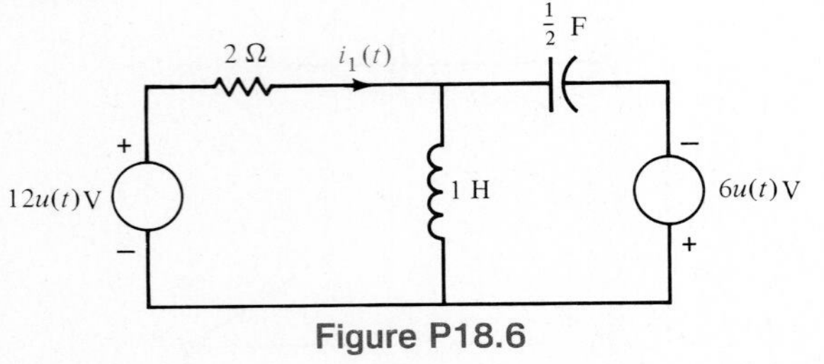

Figure P18.6

18.7. Solve Problem 18.6 using Laplace transforms and node analysis.

18.8. Solve Problem 18.6 using Laplace transforms and superposition.

18.9. Solve Problem 18.6 using Laplace transforms and source transformation.

18.10. Solve Problem 18.6 using Laplace transforms and Thévenin's theorem.

18.11. For the network shown in Fig. P18.11, find $v_o(t)$ using node equations.

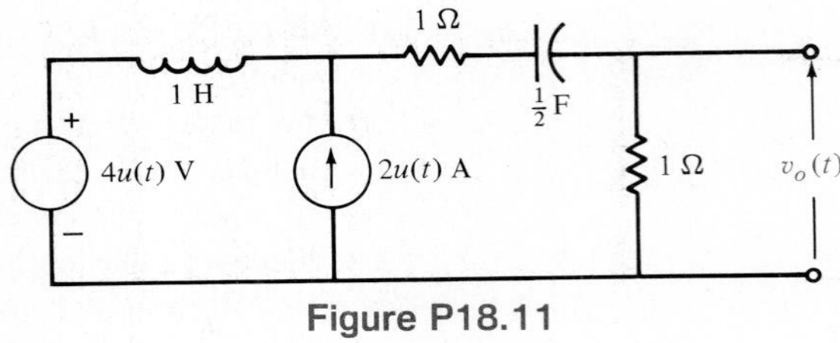

Figure P18.11

18.12. Use loop equations to solve Problem 18.11.

18.13. Use superposition to solve Problem 18.11.

18.14. Use source transformation to solve Problem 18.11.

18.15. Use Thévenin's theorem to solve Problem 18.11.

18.16. Use node equations to find $v_o(t)$ in the network shown in Fig. P18.16.

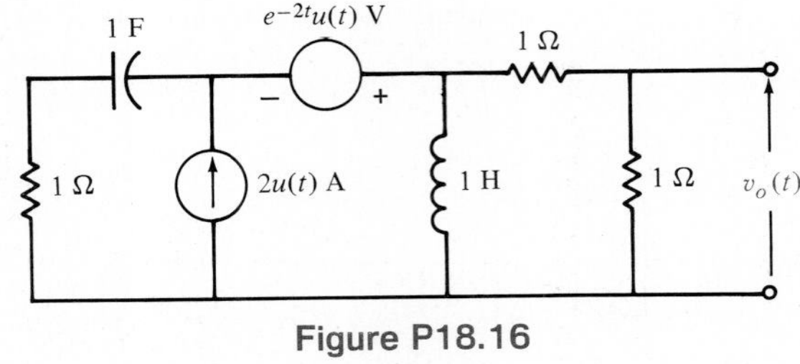

Figure P18.16

18.17. Use superposition to solve Problem 18.16.

18.18. Use source transformation to solve Problem 18.16.

18.19. Use Thévenin's theorem to solve Problem 18.16.

18.20. Find $v_o(t)$ in the network shown in Fig. P18.20 using node equations.

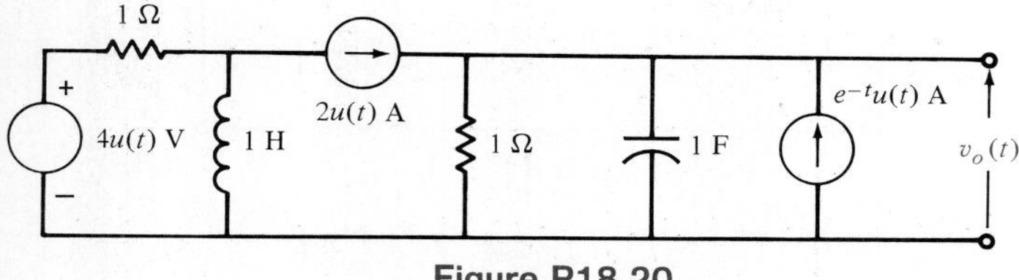

Figure P18.20

18.21. Use loop equations to find $v_o(t)$ in the network shown in Fig. P18.21.

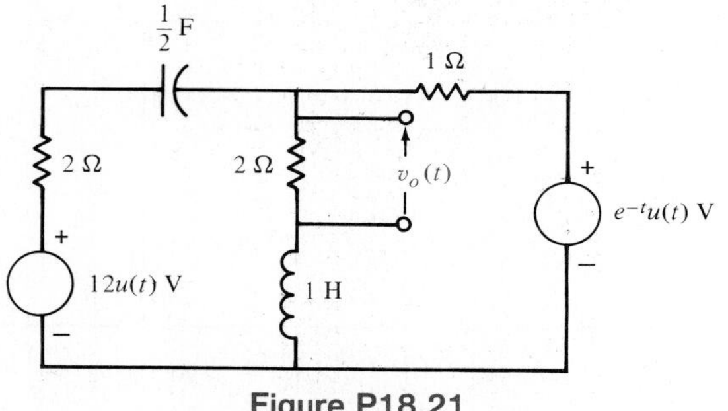

Figure P18.21

18.22. Use Thévenin's theorem to solve Problem 18.21.

18.23. Use loop equations to find $v_o(t)$ in the network shown in Fig. P18.23.

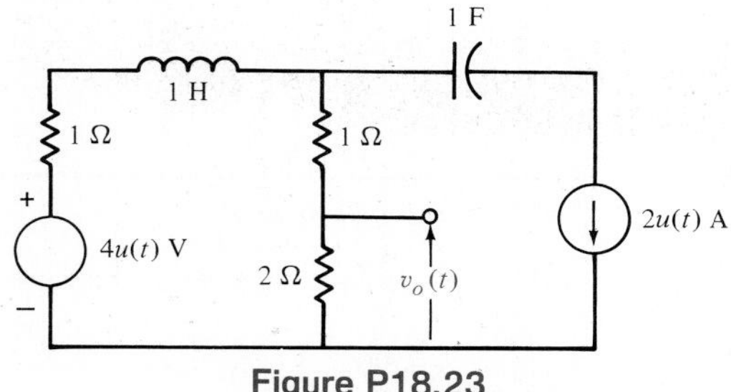

Figure P18.23

18.24. Use Thévenin's theorem to solve Problem 18.23.

18.25. Use nodal equations to find $i_o(t)$ in the network shown in Fig. P18.25.

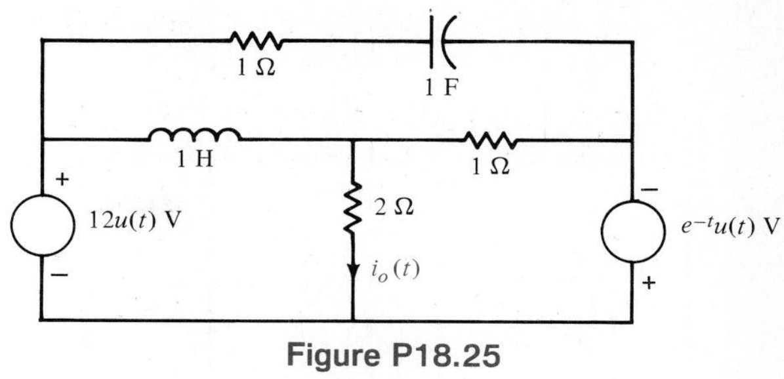

Figure P18.25

18.26. Use Norton's theorem to solve Problem 18.25.

18.27. Use loop equations to find $i_o(t)$ in the network shown in Fig. P18.27.

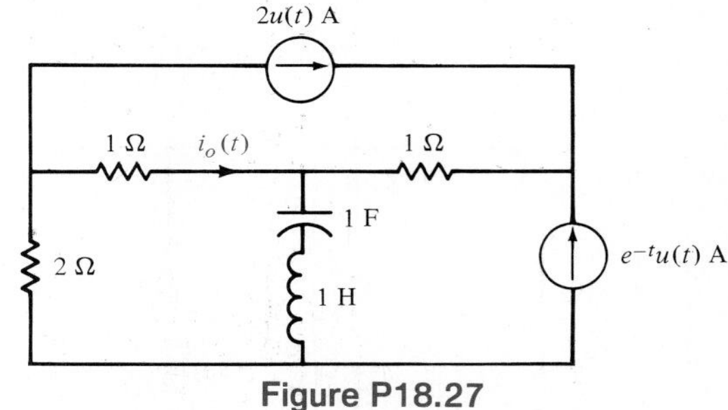

Figure P18.27

18.28. Use Thévenin's theorem to solve Problem 18.27.

18.29. Use loop equations to find $v_o(t)$ in the network shown in Fig. P18.29.

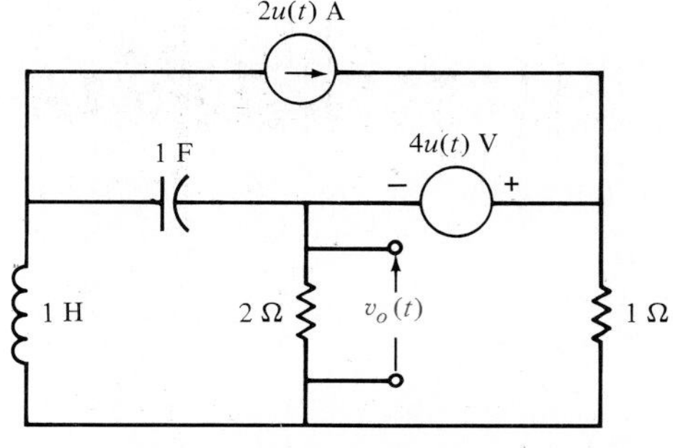

Figure P18.29

18.30. Use Thévenin's theorem to solve Problem 18.29.

18.31. Use loop equations to find $v_o(t)$ in the network shown in Fig. P18.31.

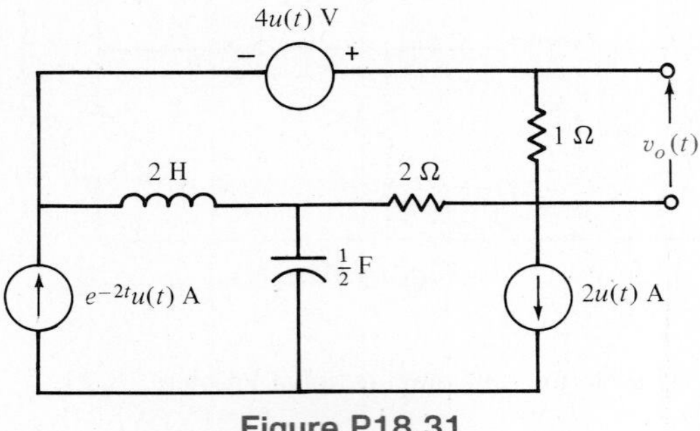

Figure P18.31

18.32. Use Thévenin's theorem to solve Problem 18.31.

18.33. Use nodal equations to find $i_o(t)$ in the network shown in Fig. P18.33.

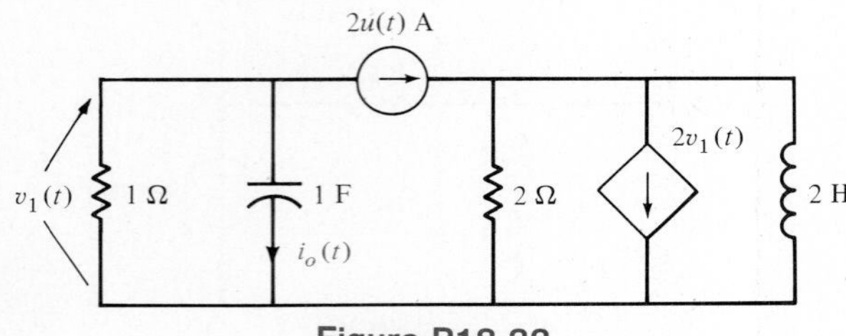

Figure P18.33

18.34. Use Thévenin's theorem to solve Problem 18.33.

18.35. Use nodal equations to find $v_o(t)$ in the network shown in Fig. P18.35.

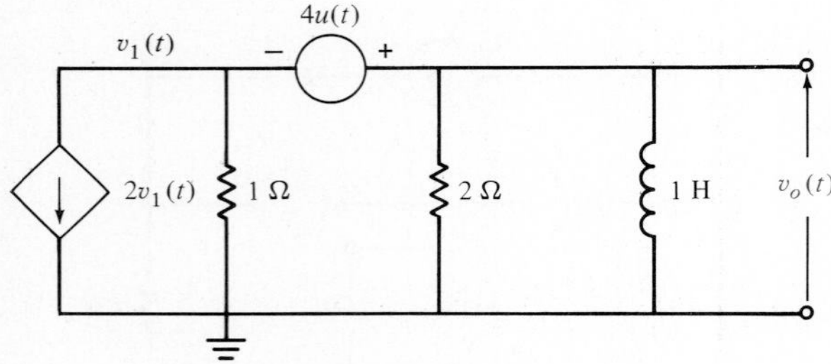

Figure P18.35

18.36. Use Thévenin's theorem to solve Problem 18.35.

18.37. Use loop equations to find $v_o(t)$ in the network shown in Fig. P18.37.

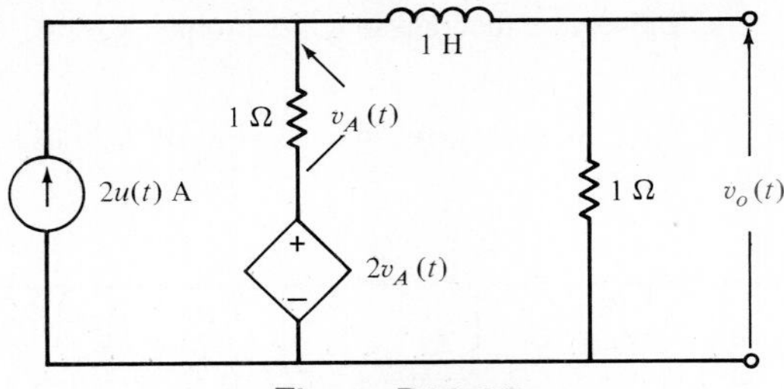

Figure P18.37

18.38. Use Thévenin's theorem to solve Problem 18.37.

18.39. Use loop equations to find $v_o(t)$ in the network shown in Fig. P18.39.

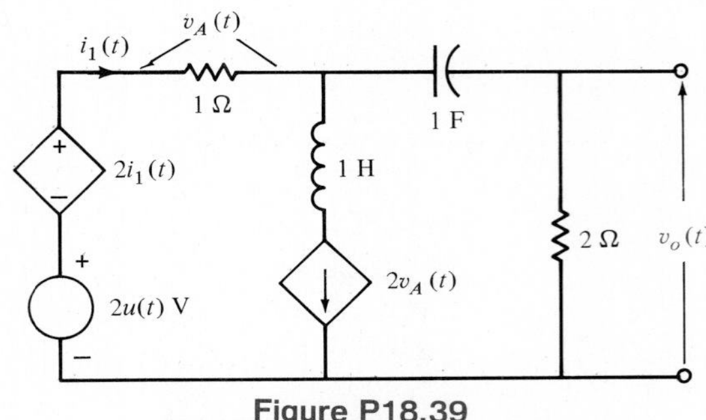

Figure P18.39

18.40. Use Thévenin's theorem to solve Problem 18.39.

18.41. Determine the steady-state response $i_o(t)$ in the network shown in Fig. P18.41.

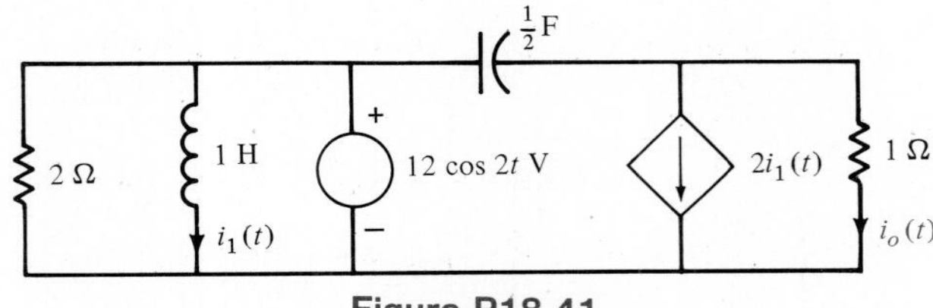

Figure P18.41

18.42. Find the steady-state response $v_o(t)$ in the network shown in Fig. P18.42.

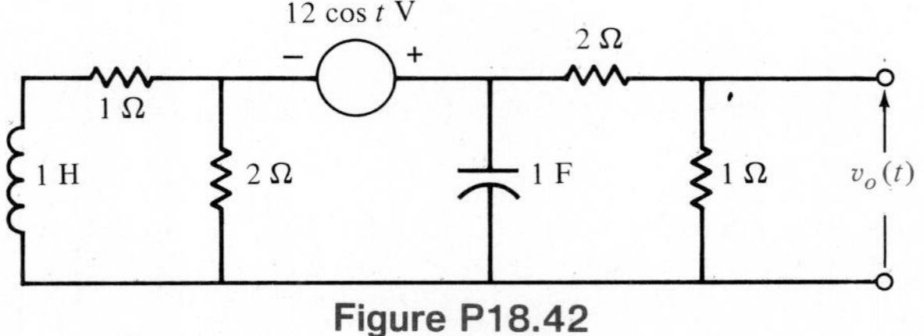

Figure P18.42

18.43. Find the steady-state response $v_o(t)$ in the network shown in Fig. P18.43.

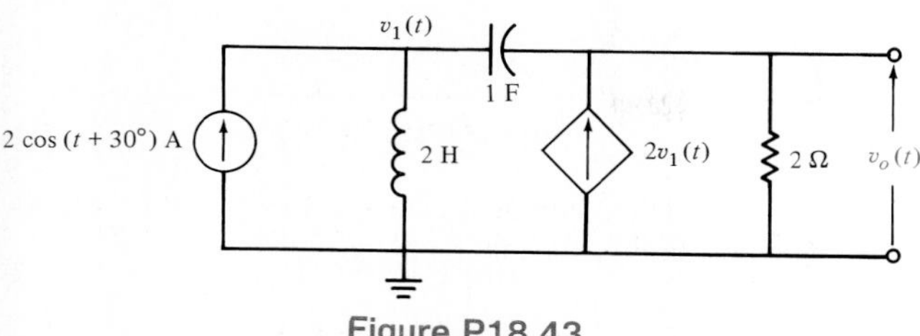

Figure P18.43

18.44. Find the steady-state response $v_o(t)$ in the network shown in Fig. P18.44.

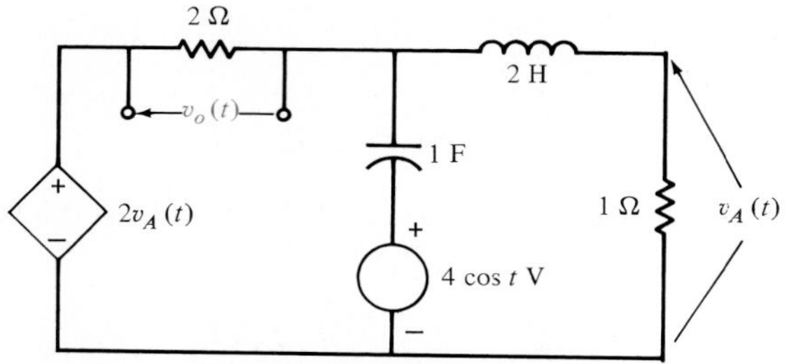

Figure P18.44

18.45. Find the steady-state response $i_o(t)$ in the network shown in Fig. P18.45.

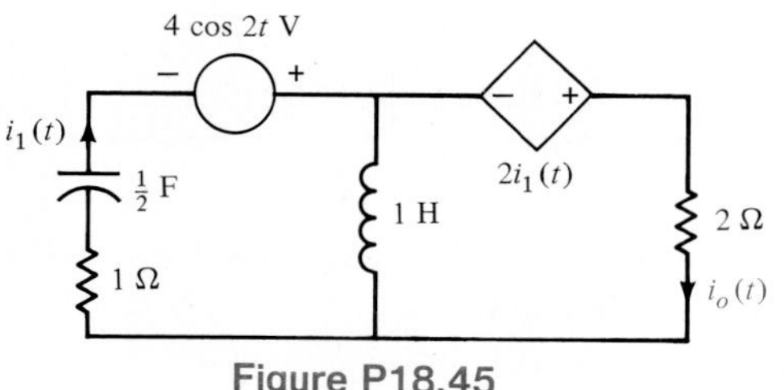

Figure P18.45

18.46. Find $i_o(t)$ for $t > 0$ in the network shown in Fig. P18.46.

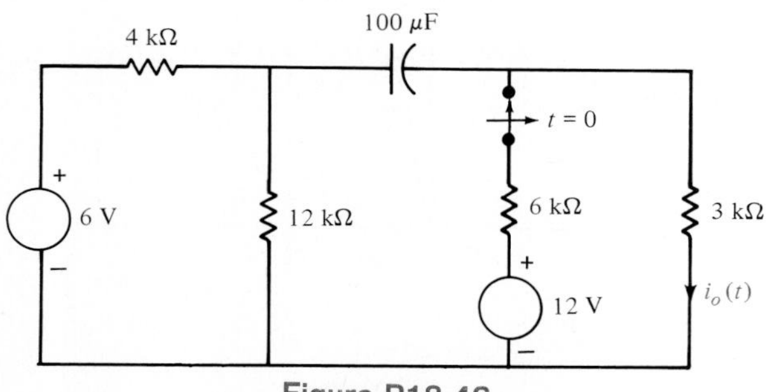

Figure P18.46

18.47. Find the current $i_o(t)$ for $t > 0$ in the network shown in Fig. P18.47.

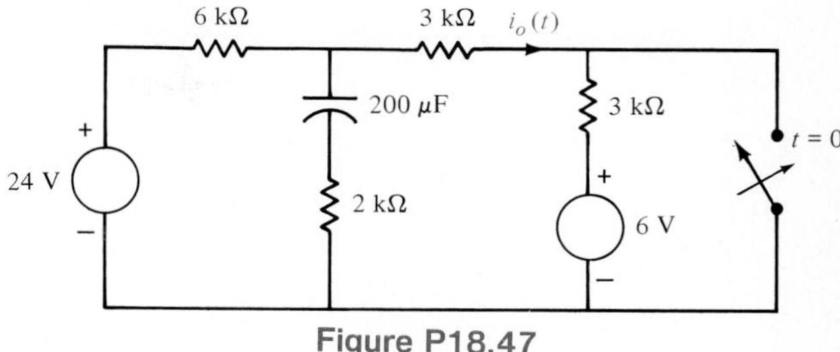

Figure P18.47

18.48. Find $i_o(t)$ for $t > 0$ in the network shown in Fig. P18.48.

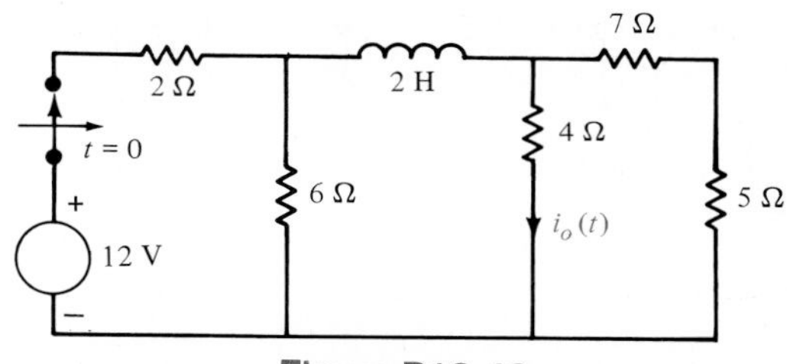

Figure P18.48

18.49. Find $v_o(t)$ for $t > 0$ in the network shown in Fig. P18.49 using Laplace transforms. Assume that the circuit has reached steady state at $t = 0-$.

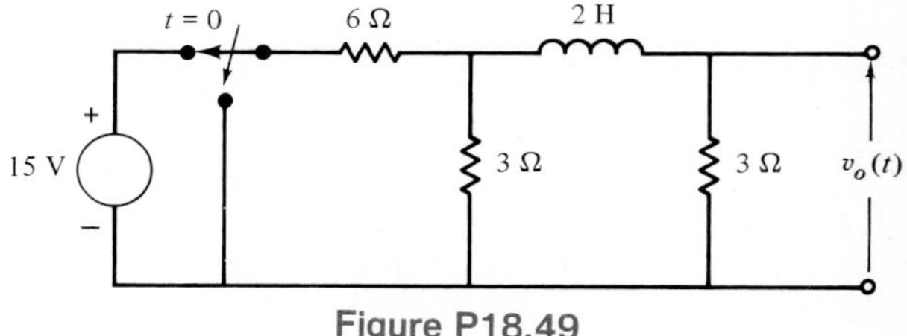

Figure P18.49

18.50. Use Laplace transforms and node analysis to find $v_o(t)$ for $t > 0$ in the network shown in Fig. P18.50. Assume that the circuit has reached steady state at $t = 0-$.

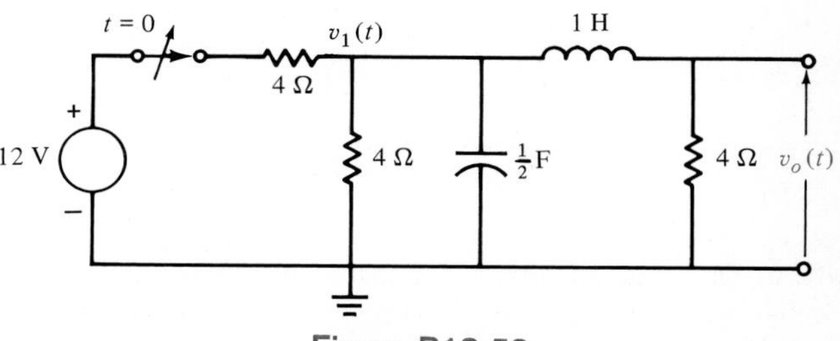

Figure P18.50

18.51. Find $i_o(t)$ for $t > 0$ in the network shown in Fig. P18.51.

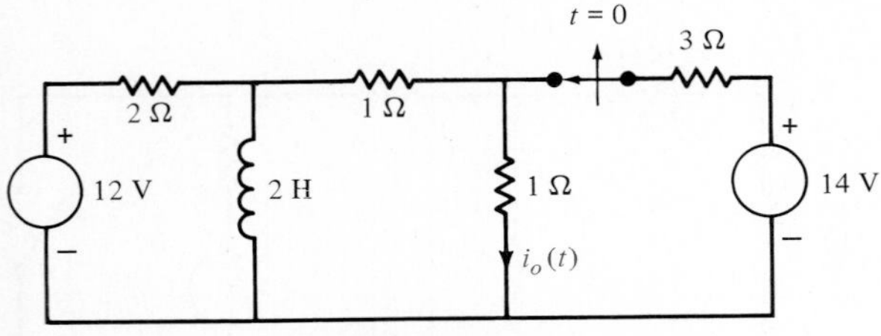

Figure P18.51

18.52. Find $v_o(t)$ for $t > 0$ in the network shown in Fig. P18.52.

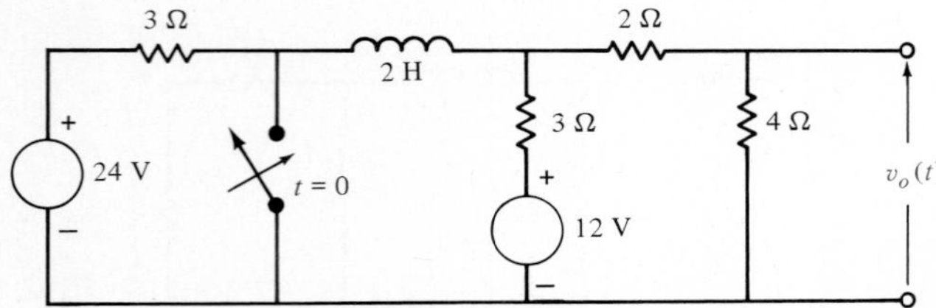

Figure P18.52

18.53. Find $i_o(t)$ for $t > 0$ in the network shown in Fig. P18.53.

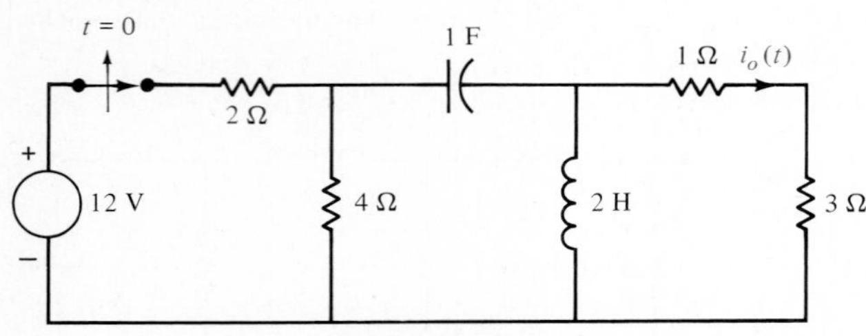

Figure P18.53

18.54. Find $i_o(t)$ for $t > 0$ in the network shown in Fig. P18.54.

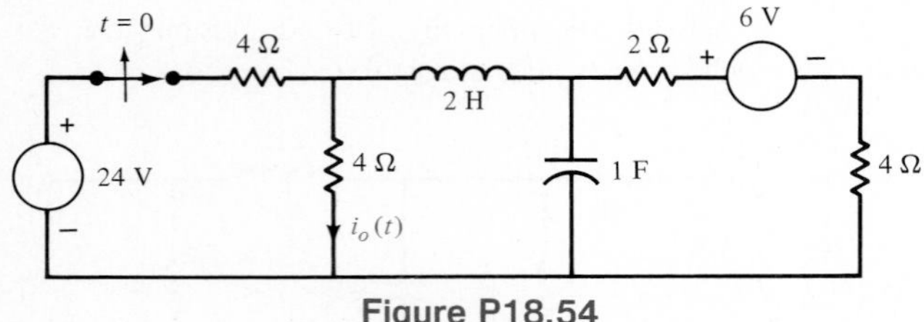

Figure P18.54

18.55. Determine the initial and final values of the voltage $v_o(t)$ in the network shown in Fig. P18.55.

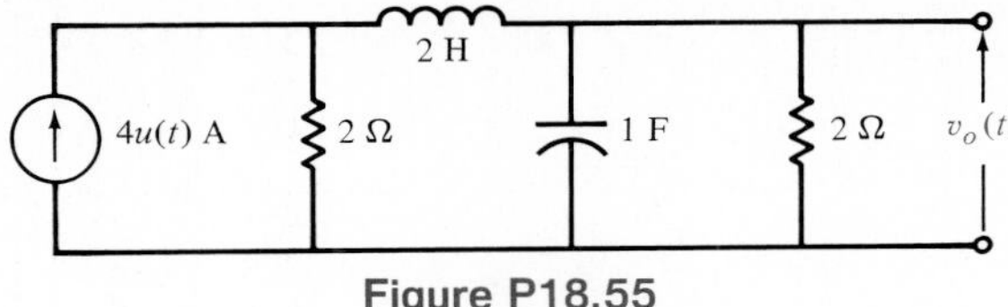

Figure P18.55

18.56. Find the initial and final values of the current $i_o(t)$ in the network shown in Fig. P18.56.

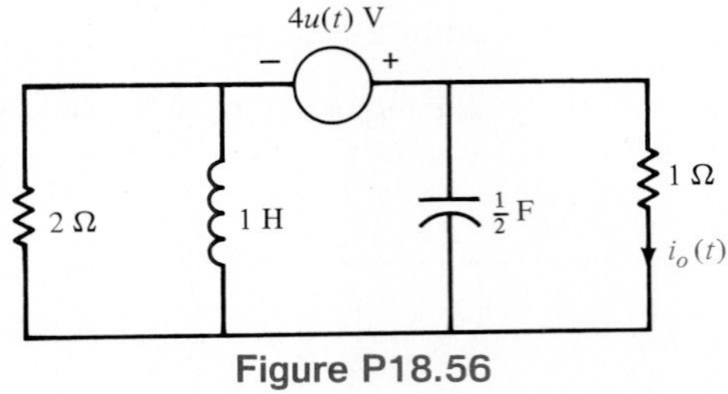

Figure P18.56

18.57. Determine the initial and final values of the voltage $v_o(t)$ in the network shown in Fig. P18.57.

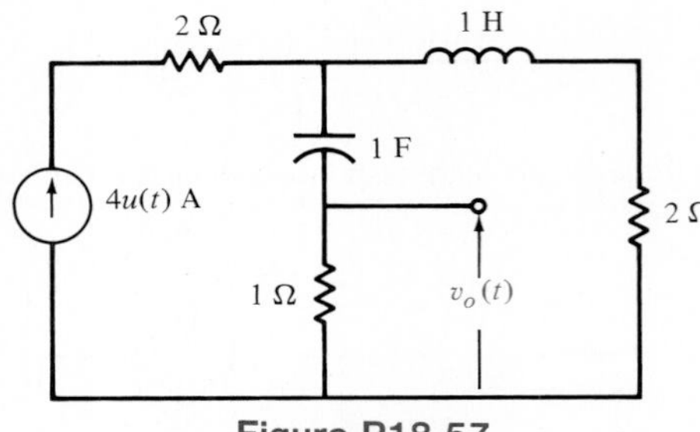

Figure P18.57

18.58. Find the initial and final values of the voltage $v_o(t)$ in the network shown in Fig. P18.58.

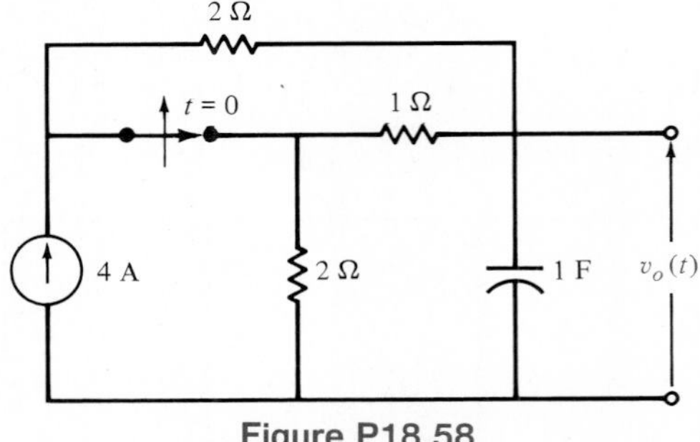

Figure P18.58

18.59. Find the initial and final values of the voltage $v_o(t)$ in the network shown in Fig. P18.59.

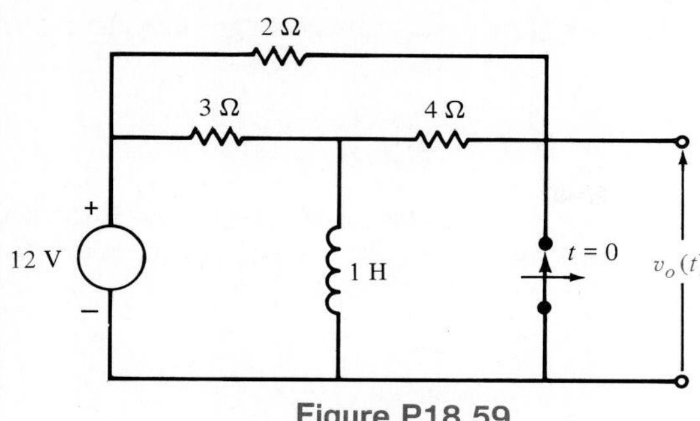

Figure P18.59

18.60. Determine the output voltage $v_o(t)$ in the network in Fig. P18.60a if the input is given by the source in Fig. P18.60b.

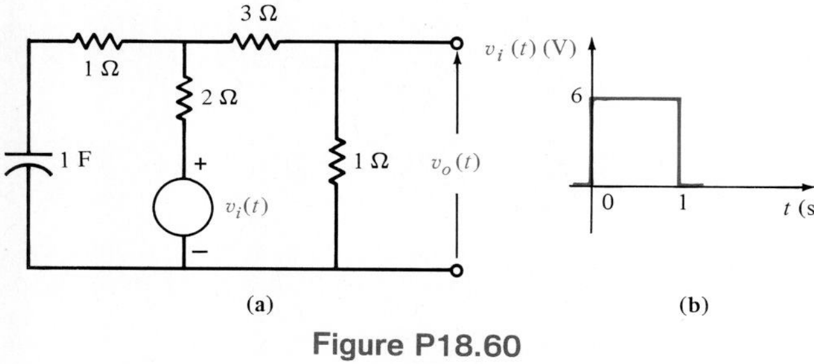

Figure P18.60

18.61. Determine the output voltage $v_o(t)$ in the network in Fig. P18.60a if the input is given by the source in Fig. P18.61.

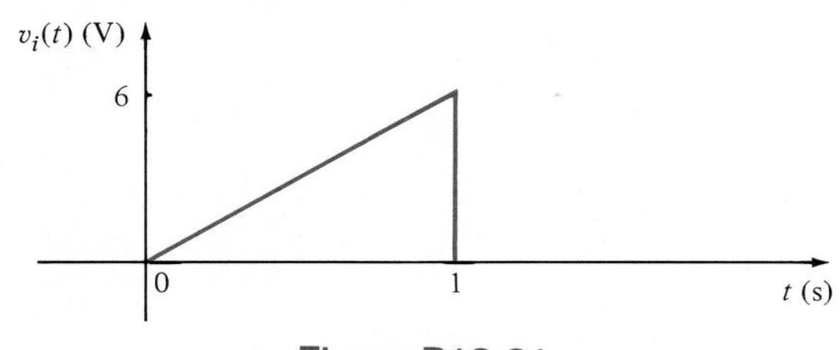

Figure P18.61

18.62. Find the output voltage $v_o(t)$ in the network shown in Fig. P18.60a if the input is given by the source in Fig. P18.62.

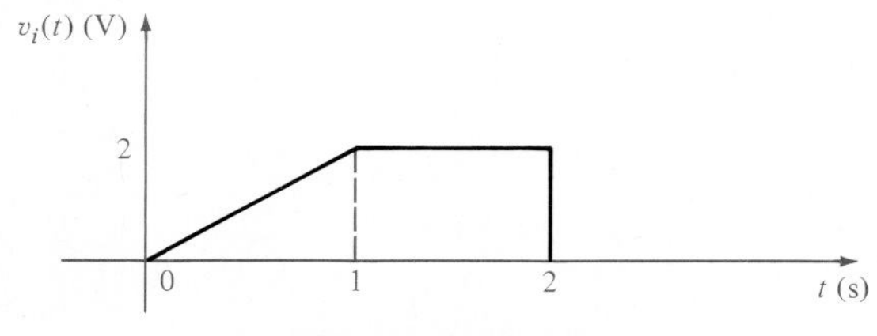

Figure P18.62

18.63. Determine the output voltage $v_o(t)$ in the network in Fig. P18.63 if the input is given by the source in Fig. P18.60b.

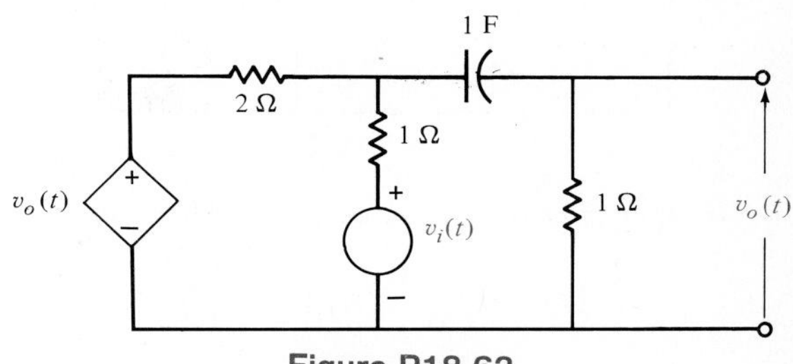

Figure P18.63

18.64. Find the output voltage $v_o(t)$ in the network in Fig. P18.63 if the input is given by the source in Fig. P18.61.

18.65. Find the output voltage $v_o(t)$ in the network in Fig. P18.63 if the input is given by the source in Fig. P18.62.

18.66. If the input to a network is $x(t) = (e^{-t} \cos 2t)u(t)$ and the network transfer function is $h(t) = 10e^{-4t}$, find the output $y(t)$.

18.67. If the input to a network is $x(t) = 10e^{-t}u(t)$ and the output is $y(t) = (5 - \frac{10}{3} e^{-t} - \frac{5}{3}t^{-4t})u(t)$, find the network transfer function $h(t)$.

18.68. The response of a network to the input $x(t) = (e^{-2t} - e^{-5t})u(t)$ is $y(t) = 4(1 - te^{-t} - e^{-t})$. Determine the impulse response of the network.

18.69. The input to a network is $e^{-t}u(t)$ and the network response is $(t - 1 + e^{-t})u(t)$. Find the network response to the input $(e^{-t} - e^{-2t})u(t)$.

18.70. Find the transfer function $\mathbf{V}_o(s)/\mathbf{V}_i(s)$ for the network shown in Fig. P18.70.

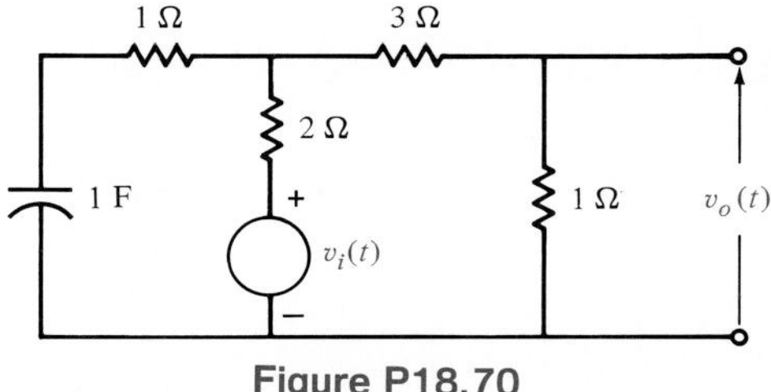

Figure P18.70

18.71. Determine the transfer function $\mathbf{I}_o(s)/\mathbf{I}_i(s)$ for the network shown in Fig. P18.71.

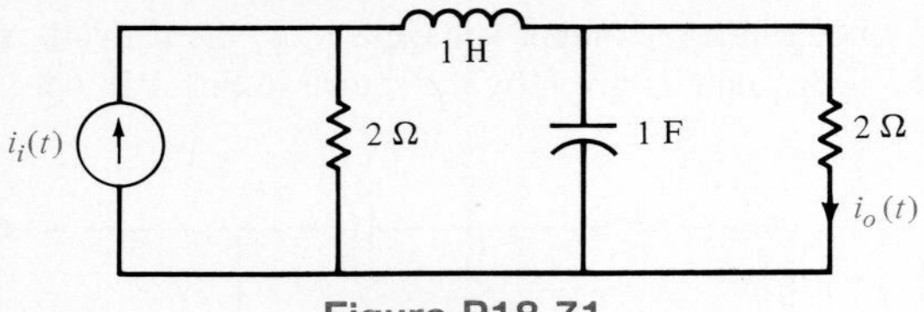

Figure P18.71

18.72. The current response of a network to a unit step input is

$$\mathbf{I}_o(s) = \frac{10(s + 2)}{s^2(s^2 + 11s + 30)}$$

Is the response underdamped?

18.73. The transfer function of a network is given by the expression

$$\mathbf{G}(s) = \frac{100s}{s^2 + 22s + 40}$$

Determine the damping ratio, the undamped natural frequency, and the type of response that will be exhibited by the network.

APPENDIX

Basic Elements of Determinants and Matrices

In the following material we present some of the basic concepts involved in the use of determinants and matrices in the solution of various circuit problems. The discussion will be very succinct, with examples to illustrate the points presented. Because of the concise nature of the presentation, the reader interested in a more general presentation of the subject is referred to the many mathematics texts on these subjects.

In the solution of various circuit problems we encounter a system of simultaneous equations of the form

$$\begin{aligned} a_{11}x_1 + a_{12}x_2 + \cdots + a_{1n}x_n &= b_1 \\ a_{21}x_1 + a_{22}x_2 + \cdots + a_{2n}x_n &= b_2 \\ \vdots \qquad \vdots \qquad\qquad \vdots \quad &\quad \vdots \\ a_{n1}x_1 + a_{n2}x_2 + \cdots + a_{nn}x_n &= b_n \end{aligned} \tag{A.1}$$

where the x's and b's are typically voltages and currents or currents and voltages, respectively. If the number of variables is small (e.g., two or three), we can solve the equations via a systematic elimination of variables. However, for a large number of variables, this technique can quickly become unwieldly.

A straightforward approach to the solution of a system of equations such as those shown above involves the use of *determinants* and *Cramer's rule*. This procedure also serves as a prelude to the more general methods involving *matrix analysis*.

A.1

Determinants

A *determinant* is a square array of numbers arranged in an equal number of rows and columns which are contained within vertical parallel lines. In its simplest form, the determinant is a 2 × 2 array:

$$\Delta = \begin{vmatrix} a_{11} & a_{12} \\ a_{21} & a_{22} \end{vmatrix}$$

where the symbol Δ is used to represent the value of the determinant. For the 2 × 2 array, Δ is defined to be

$$\Delta = a_{11}a_{22} - a_{21}a_{12}$$

that is, it is equal to the product $a_{11}a_{22}$ of elements *down* the diagonal from left to right minus the product $a_{21}a_{12}$ of elements *up* the diagonal from left to right.

EXAMPLE A.1

If

$$\Delta = \begin{vmatrix} a_{11} & a_{12} \\ a_{21} & a_{22} \end{vmatrix} = \begin{vmatrix} 4 & -1 \\ -1 & 3 \end{vmatrix}$$

$$= (4)(3) - (-1)(-1) = 11$$

The evaluation of larger determinants is accomplished by expanding the determinant in *cofactors*. The cofactor A_{ij} of the element a_{ij} of a determinant is equal to the product of $(-1)^{i+j}$ and the determinant remaining after row i and column j are deleted. ■

EXAMPLE A.2

Given the 3 × 3 determinant below, determine the cofactors A_{11}, A_{23}, and A_{31}.

$$\Delta = \begin{vmatrix} a_{11} & a_{12} & a_{13} \\ a_{21} & a_{22} & a_{23} \\ a_{31} & a_{32} & a_{33} \end{vmatrix}$$

$$A_{11} = (-1)^2 \begin{vmatrix} a_{22} & a_{23} \\ a_{32} & a_{33} \end{vmatrix} = a_{22}a_{33} - a_{32}a_{23}$$

$$A_{23} = (-1)^5 \begin{vmatrix} a_{11} & a_{12} \\ a_{31} & a_{32} \end{vmatrix} = -(a_{11}a_{32} - a_{31}a_{12})$$

and

$$A_{31} = (-1)^4 \begin{vmatrix} a_{12} & a_{13} \\ a_{22} & a_{23} \end{vmatrix} = a_{12}a_{23} - a_{22}a_{13}$$

■

Now in general the value of any determinant is equal to the sum of the products of the elements in any row or column and their corresponding cofactors. Consider, for example, the 3 × 3 determinant shown in Example A.2. Since we could use any row or column to evaluate the determinant, we can use the second column, so that

$$\Delta = a_{12}A_{12} + a_{22}A_{22} + a_{32}A_{32}$$

$$= a_{12}(-)^3\begin{vmatrix} a_{21} & a_{23} \\ a_{31} & a_{33} \end{vmatrix} + a_{22}(-1)^4\begin{vmatrix} a_{11} & a_{13} \\ a_{31} & a_{33} \end{vmatrix}$$

$$+ a_{32}(-1)^5\begin{vmatrix} a_{11} & a_{13} \\ a_{21} & a_{23} \end{vmatrix}$$

or the third row so that

$$\Delta = a_{31}A_{31} + a_{32}A_{32} + a_{33}A_{33}$$

$$= a_{31}(-1)^4\begin{vmatrix} a_{12} & a_{13} \\ a_{22} & a_{23} \end{vmatrix} + a_{32}(-1)^5\begin{vmatrix} a_{11} & a_{13} \\ a_{21} & a_{23} \end{vmatrix}$$

$$+ a_{33}(-1)^6\begin{vmatrix} a_{11} & a_{12} \\ a_{21} & a_{22} \end{vmatrix}$$

or some other row or column.

EXAMPLE A.3

Calculate the value of the following determinant.

$$\Delta = \begin{vmatrix} 7 & -4 & -1 \\ -4 & 7 & -2 \\ -1 & -2 & 3 \end{vmatrix}$$

Expanding the determinant using the first row, we obtain

$$\Delta = 7\begin{vmatrix} 7 & -2 \\ -2 & 3 \end{vmatrix} - (-4)\begin{vmatrix} -4 & -2 \\ -1 & 3 \end{vmatrix} - 1\begin{vmatrix} -4 & 7 \\ -1 & -2 \end{vmatrix}$$

$$= 7(17) + 4(-14) - (15)$$

$$= 48$$

■

With this background on determinants we are now in a position to illustrate the manner in which independent simultaneous equations are solved using *Cramer's rule*. For the system of equations given in Eq. (A.1), we define Δ, Δ_1, Δ_2, . . . , Δ_j as

$$\Delta = \begin{vmatrix} a_{11} & a_{12} & \cdots & a_{1n} \\ a_{21} & a_{22} & \cdots & a_{2n} \\ \vdots & \vdots & & \vdots \\ a_{n1} & a_{n2} & \cdots & a_{nn} \end{vmatrix}$$

$$\Delta_1 = \begin{vmatrix} b_1 & a_{12} & \cdots & a_{1n} \\ b_2 & a_{22} & \cdots & a_{2n} \\ \vdots & \vdots & & \vdots \\ b_n & a_{n2} & \cdots & a_{nn} \end{vmatrix}$$

$$\Delta_2 = \begin{vmatrix} a_{11} & b_1 & a_{13} & \cdots & a_{1n} \\ a_{21} & b_2 & a_{23} & \cdots & a_{2n} \\ \vdots & \vdots & \vdots & & \vdots \\ a_{n1} & b_n & a_{n3} & \cdots & a_{nn} \end{vmatrix}$$

$$\Delta_j = \begin{vmatrix} a_{11} & \cdots & a_{1j-1} & b_1 & a_{1j+1} & \cdots & a_{1n} \\ a_{21} & \cdots & a_{2j-1} & b_2 & a_{2j+1} & \cdots & a_{2n} \\ \vdots & & \vdots & \vdots & \vdots & & \vdots \\ a_{n1} & \cdots & a_{nj-1} & b_n & a_{nj+1} & \cdots & a_{nn} \end{vmatrix}$$

In other words, Δ_j is equal to the determinant Δ with the jth column replaced by the elements on the right side of the system of equations. Then Cramer's rule states that the variables $x_1, x_2, \ldots, x_n$ are determined by

$$\begin{aligned} x_1 &= \frac{\Delta_1}{\Delta} \\ x_2 &= \frac{\Delta_2}{\Delta} \\ &\vdots \\ x_n &= \frac{\Delta_n}{\Delta} \end{aligned} \tag{A.2}$$

EXAMPLE A.4

Given the system of equations below, determine the values of x_1, x_2, and x_3.

$$\begin{aligned} 7x_1 - 4x_2 - x_3 &= 4 \\ -4x_1 + 7x_2 - 2x_3 &= 0 \\ -x_1 - 2x_2 + 3x_3 &= -1 \end{aligned}$$

In Example A.3 we found that $\Delta = 48$. Δ_1 is

$$\Delta_1 = \begin{vmatrix} 4 & -4 & -1 \\ 0 & 7 & -2 \\ -1 & -2 & 3 \end{vmatrix}$$

Expanding using the first column, we obtain

$$\begin{aligned} \Delta_1 &= 4\begin{vmatrix} 7 & -2 \\ -2 & 3 \end{vmatrix} - 1\begin{vmatrix} -4 & -1 \\ 7 & -2 \end{vmatrix} \\ &= 4(17) - 1(15) = 53 \end{aligned}$$

$$\Delta_2 = \begin{vmatrix} 7 & 4 & -1 \\ -4 & 0 & -2 \\ -1 & -1 & 3 \end{vmatrix}$$

Expanding via the second column yields

$$\begin{aligned} \Delta_2 &= -4\begin{vmatrix} -4 & -2 \\ -1 & 3 \end{vmatrix} + 1\begin{vmatrix} 7 & -1 \\ -4 & -2 \end{vmatrix} \\ &= -4(-14)+1(-18) = 38 \end{aligned}$$

$$\Delta_3 = \begin{vmatrix} 7 & -4 & 4 \\ -4 & 7 & 0 \\ -1 & -2 & -1 \end{vmatrix}$$

An expansion via the third column yields

$$\begin{aligned} \Delta_3 &= 4\begin{vmatrix} -4 & 7 \\ -1 & -2 \end{vmatrix} - 1\begin{vmatrix} 7 & -4 \\ -4 & 7 \end{vmatrix} \\ &= 4(15) - 1(33) = 27 \end{aligned}$$

Then from Cramer's rule

$$x_1 = \frac{\Delta_1}{\Delta} = \frac{53}{48} = 1.10416$$

$$x_2 = \frac{\Delta_2}{\Delta} = \frac{38}{48} = 0.79166$$

$$x_3 = \frac{\Delta_3}{\Delta} = \frac{27}{48} = 0.5625$$

■

A.2 Matrices

A *matrix* is defined to be a rectangular array of numbers arranged in rows and columns and written in the form

$$\begin{bmatrix} a_{11} & a_{12} & \cdots & a_{1n} \\ a_{21} & a_{22} & \cdots & a_{2n} \\ \vdots & \vdots & & \vdots \\ a_{m1} & a_{m2} & \cdots & a_{mn} \end{bmatrix}$$

This array is called an m by n ($m \times n$) matrix because it has m rows and n columns. A matrix is a convenient way of representing arrays of numbers; however, one must remember that the matrix itself has no numerical value. In the array above the numbers or functions a_{ij} are called the *elements* of the matrix. Any matrix that has the same number of rows as columns is called a *square matrix*. The sum of the diagonal elements of a square matrix is called the *trace* of the matrix.

EXAMPLE A.5

The following are matrices:

$$\begin{bmatrix} a \\ b \\ c \\ d \end{bmatrix}, \quad \begin{bmatrix} 1 & 3 \\ 2 & 4 \end{bmatrix}, \quad \begin{bmatrix} 4 & 3 & 2 & 1 \\ 5 & 6 & 7 & 8 \end{bmatrix}, \quad [1 \quad 2 \quad 3]$$

However,

$$\begin{bmatrix} 1 & 3 \\ 2 & \end{bmatrix}, \quad \begin{bmatrix} & a & \\ b & & d \\ & c & \end{bmatrix}$$

are not matrices since they are not rectangular arrays arranged in rows and columns. ■

EXAMPLE A.6

The matrix $\boldsymbol{A}$ shown below is a square matrix:

$$\boldsymbol{A} = \begin{bmatrix} 1 & 2 & 3 \\ 4 & 5 & 6 \\ 7 & 8 & 9 \end{bmatrix}$$

and the trace (tr$\boldsymbol{A}$) of this square matrix is

$$\text{tr}\boldsymbol{A} = 1 + 5 + 9 = 15$$

■

A matrix whose elements are all zero is called a *null matrix* and is denoted by $\boldsymbol{0}$.

EXAMPLE A.7

The following are null matrices:

$$\boldsymbol{0} = \begin{bmatrix} 0 & 0 & 0 \\ 0 & 0 & 0 \end{bmatrix}, \quad \boldsymbol{0} = \begin{bmatrix} 0 \\ 0 \\ 0 \end{bmatrix}, \quad \boldsymbol{0} = \begin{bmatrix} 0 & 0 & 0 \\ 0 & 0 & 0 \\ 0 & 0 & 0 \end{bmatrix}$$

■

A square matrix that has nonzero elements along the main diagonal and zeros elsewhere is said to be a *diagonal matrix*.

EXAMPLE A.8

The following are diagonal matrices:

$$\begin{bmatrix} 1 & 0 \\ 0 & 2 \end{bmatrix}, \quad \begin{bmatrix} 1 & 0 & 0 \\ 0 & 2 & 0 \\ 0 & 0 & 3 \end{bmatrix}, \quad \begin{bmatrix} a & 0 & 0 & 0 \\ 0 & b & 0 & 0 \\ 0 & 0 & c & 0 \\ 0 & 0 & 0 & d \end{bmatrix}$$

■

The *identity matrix* is a diagonal matrix in which all diagonal elements are equal to one.

EXAMPLE A.9

The following are identity matrices:

$$\begin{bmatrix} 1 & 0 \\ 0 & 1 \end{bmatrix}, \quad \begin{bmatrix} 1 & 0 & 0 \\ 0 & 1 & 0 \\ 0 & 0 & 1 \end{bmatrix}, \quad \ldots, \quad \begin{bmatrix} 1 & 0 & \cdot & \cdot & \cdots & 0 \\ 0 & 1 & 0 & \cdot & \cdots & 0 \\ 0 & 0 & 1 & 0 & \cdots & 0 \\ \vdots & & & & & \\ 0 & 0 & \cdot & \cdot & \cdots & 1 \end{bmatrix}$$ ■

The matrices $\boldsymbol{A}$ and $\boldsymbol{B}$ are said to be *equal* if their corresponding elements are equal. In other words, $\boldsymbol{A} = \boldsymbol{B}$ if and only if $a_{ij} = b_{ij}$ for all i and j.

EXAMPLE A.10

$$\boldsymbol{A} = \begin{bmatrix} 3 & 2 & 1 \\ 4 & 5 & 6 \end{bmatrix} \quad \text{and} \quad \boldsymbol{B} = \begin{bmatrix} 3 & 2 & 1 \\ 4 & 5 & 6 \end{bmatrix}$$

then $\boldsymbol{A} = \boldsymbol{B}$ because $a_{ij} = b_{ij}$ for all i and j. However, if

$$\boldsymbol{A} = \begin{bmatrix} 2 & 4 \\ 6 & 8 \end{bmatrix}, \quad \boldsymbol{B} = \begin{bmatrix} 2 & 4 \\ 6 & 10 \end{bmatrix}, \quad \text{and} \quad \boldsymbol{C} = \begin{bmatrix} 2 & 4 & 0 \\ 6 & 8 & 0 \end{bmatrix}$$

then $\boldsymbol{A} \neq \boldsymbol{B}$ and $\boldsymbol{A} \neq \boldsymbol{C}$ and $\boldsymbol{B} \neq \boldsymbol{C}$. ■

The addition and subtraction of two matrices of the same order (i.e., $m \times n$) is accomplished as follows:

$$\boldsymbol{C} = \boldsymbol{A} \pm \boldsymbol{B} \tag{A.3}$$

or

$$c_{ij} = a_{ij} \pm b_{ij} \qquad \text{for all } i \text{ and } j$$

That is, the elements of $\boldsymbol{C}$ are the sum or difference of the corresponding elements of $\boldsymbol{A}$ and $\boldsymbol{B}$.

$$\boldsymbol{C} = \begin{bmatrix} c_{11} & \cdots & c_{1n} \\ \vdots & & \vdots \\ c_{m1} & \cdots & c_{mn} \end{bmatrix} = \begin{bmatrix} a_{11} & \cdots & a_{1n} \\ \vdots & & \vdots \\ a_{m1} & \cdots & a_{mn} \end{bmatrix} \pm \begin{bmatrix} b_{11} & \cdots & b_{1n} \\ \vdots & & \vdots \\ b_{m1} & \cdots & b_{mn} \end{bmatrix}$$

$$= \begin{bmatrix} a_{11} \pm b_{11} & \cdots & a_{1n} \pm b_{1n} \\ \vdots & & \vdots \\ a_{m1} \pm b_{m1} & \cdots & a_{mn} \pm b_{mn} \end{bmatrix}$$

EXAMPLE A.11

If

$$\boldsymbol{A} = \begin{bmatrix} 1 & 3 \\ 2 & 4 \end{bmatrix}, \quad \boldsymbol{B} = \begin{bmatrix} -1 & -2 \\ 3 & 4 \end{bmatrix}, \quad \text{and} \quad \boldsymbol{C} = \begin{bmatrix} 1 & 3 \\ 4 & 2 \end{bmatrix}$$

then

$$\boldsymbol{A} + \boldsymbol{B} = \begin{bmatrix} 0 & 1 \\ 5 & 8 \end{bmatrix}, \quad \boldsymbol{A} - \boldsymbol{B} = \begin{bmatrix} 2 & 5 \\ -1 & 0 \end{bmatrix}, \quad \boldsymbol{A} - \boldsymbol{B} + \boldsymbol{C} = \begin{bmatrix} 3 & 8 \\ 3 & 2 \end{bmatrix} \quad \blacksquare$$

Given a matrix $\boldsymbol{A}$ and a scalar λ, *the multiplication* of $\boldsymbol{A}$ by the scalar λ, written $\lambda\boldsymbol{A}$, is defined to be

$$\lambda\boldsymbol{A} = \begin{bmatrix} \lambda a_{11} & \cdots & \lambda a_{1n} \\ \vdots & & \\ \lambda a_{m1} & \cdots & \lambda a_{mn} \end{bmatrix} \tag{A.4}$$

EXAMPLE A.12

If $\lambda_1 = 2$, $\lambda_2 = -2$, $\lambda_3 = 3$, and

$$\boldsymbol{A} = \begin{bmatrix} 2 & 4 & 6 \\ 1 & 3 & 5 \end{bmatrix}$$

then

$$\lambda_1\boldsymbol{A} = \begin{bmatrix} 4 & 8 & 12 \\ 2 & 6 & 10 \end{bmatrix}$$

$$\lambda_2\boldsymbol{A} = \begin{bmatrix} -4 & -8 & -12 \\ -2 & -6 & -10 \end{bmatrix}$$

$$\lambda_3\boldsymbol{A} = \boldsymbol{A} + \boldsymbol{A} + \boldsymbol{A} = \begin{bmatrix} 6 & 12 & 18 \\ 3 & 9 & 15 \end{bmatrix} \quad \blacksquare$$

It is important to note that for matrices of the same order:

1. $\boldsymbol{A} + \boldsymbol{B} = \boldsymbol{B} + \boldsymbol{A}$
2. $\boldsymbol{A} + (\boldsymbol{B} + \boldsymbol{C}) = (\boldsymbol{A} + \boldsymbol{B}) + \boldsymbol{C}$
3. $\lambda(\boldsymbol{A} + \boldsymbol{B}) = \lambda\boldsymbol{A} + \lambda\boldsymbol{B} = (\boldsymbol{A} + \boldsymbol{B})\lambda$, for λ a scalar

Consider now the multiplication of two matrices. If we are given an $m \times n$ matrix $\boldsymbol{A}$ and an $n \times r$ matrix $\boldsymbol{B}$, the product $\boldsymbol{AB}$ is defined to be an $m \times r$ matrix $\boldsymbol{C}$ whose elements are given by the expression

$$c_{ij} = \sum_{k=1}^{n} a_{ik}b_{kj}, \qquad i = 1, \ldots, m, \quad j = 1, \ldots, r \tag{A.5}$$

Note that the product $\boldsymbol{AB}$ is defined only when the number of columns of $\boldsymbol{A}$ is equal to the number of rows of $\boldsymbol{B}$

EXAMPLE A.13

Suppose that the matrices $\boldsymbol{A}$ and $\boldsymbol{B}$ are defined as follows:

$$\boldsymbol{A} = \begin{bmatrix} a_{11} & a_{12} \\ a_{21} & a_{22} \\ a_{31} & a_{32} \end{bmatrix} \qquad \boldsymbol{B} = \begin{bmatrix} b_{11} & b_{12} \\ b_{21} & b_{22} \end{bmatrix}$$

Note that in the formula above, $m = 3$, $n = 2$, $r = 2$. Using this formula, we can calculate

$$c_{11} = \sum_{k=1}^{2} a_{1k}b_{k1} = a_{11}b_{11} + a_{12}b_{21}$$

$$c_{12} = \sum_{k=1}^{2} a_{1k}b_{k2} = a_{11}b_{12} + a_{12}b_{22}$$

$$\vdots$$

These elements form the array

$$\boldsymbol{C} = \boldsymbol{AB} = \begin{bmatrix} a_{11}b_{11} + a_{12}b_{21} & a_{11}b_{12} + a_{12}b_{22} \\ a_{21}b_{11} + a_{22}b_{21} & a_{21}b_{12} + a_{22}b_{22} \\ a_{31}b_{11} + a_{32}b_{21} & a_{31}b_{12} + a_{32}b_{22} \end{bmatrix}$$ ■

A close inspection of the product above illustrates that multiplication is a ''row-by-column'' operation. In other words, each element in a row of the first matrix is multiplied by the corresponding element in a column of the second matrix and then the products are summed. This operation is diagrammed as follows:

$$\begin{bmatrix} c_{11} & \cdots & c_{1p} \\ \vdots & \boxed{c_{ij}} & \vdots \\ c_{m1} & \cdots & c_{mp} \end{bmatrix} = \begin{bmatrix} a_{11} & \cdots & a_{1n} \\ \boxed{a_{i1} \quad \cdots \quad a_{in}} \\ a_{m1} & \cdots & a_{mn} \end{bmatrix} \begin{bmatrix} b_{11} & \cdots & \boxed{b_{1j}} & \cdots & b_{1p} \\ \vdots & & \vdots & & \vdots \\ b_{n1} & \cdots & \boxed{b_{nj}} & \cdots & b_{np} \end{bmatrix} \tag{A.6}$$

The following examples will illustrate the computational technique.

EXAMPLE A.14

If

$$\boldsymbol{A} = \begin{bmatrix} 1 & 3 \\ 2 & 4 \end{bmatrix} \quad \text{and} \quad \boldsymbol{B} = \begin{bmatrix} 2 & 1 \\ 3 & 5 \end{bmatrix}$$

then

$$\boldsymbol{AB} = \begin{bmatrix} (1)(2) + (3)(3) & (1)(1) + (3)(5) \\ (2)(2) + (4)(3) & (2)(1) + (4)(5) \end{bmatrix} = \begin{bmatrix} 11 & 16 \\ 16 & 22 \end{bmatrix}$$

If

$$\boldsymbol{C} = \begin{bmatrix} 1 & 2 \\ 3 & 4 \end{bmatrix} \quad \text{and} \quad \boldsymbol{D} = \begin{bmatrix} 1 \\ 2 \end{bmatrix}$$

then

$$\boldsymbol{CD} = \begin{bmatrix} (1)(1) + (2)(2) \\ (3)(1) + (4)(2) \end{bmatrix} = \begin{bmatrix} 5 \\ 11 \end{bmatrix}$$

and if

$$\boldsymbol{E} = \begin{bmatrix} 2 \\ 3 \\ -1 \end{bmatrix} \quad \text{and} \quad \boldsymbol{F} = [4 \quad 5 \quad 6]$$

then

$$\boldsymbol{EF} = \begin{bmatrix} 2 \\ 3 \\ -1 \end{bmatrix} [4 \quad 5 \quad 6] = \begin{bmatrix} (2)(4) & (2)(5) & (2)(6) \\ (3)(4) & (3)(5) & (3)(6) \\ (-1)(4) & (-1)(5) & (-1)(6) \end{bmatrix} = \begin{bmatrix} 8 & 10 & 12 \\ 12 & 15 & 18 \\ -4 & -5 & -6 \end{bmatrix}$$

Note that the product $\boldsymbol{DC}$ is not defined. ■

If the indicated sums and products are properly defined, then

1. $\boldsymbol{A}(\boldsymbol{B} + \boldsymbol{C}) = \boldsymbol{AB} + \boldsymbol{AC}$
2. $(\boldsymbol{A} + \boldsymbol{B})\boldsymbol{C} = \boldsymbol{AC} + \boldsymbol{BC}$
3. $\boldsymbol{A}(\boldsymbol{BC}) = (\boldsymbol{AB})\boldsymbol{C}$

However, it is important to note that in general:

4. $\boldsymbol{AB} \neq \boldsymbol{BA}$
5. $\boldsymbol{AB} = 0$ does not necessarily imply that $\boldsymbol{A} = 0$ or $\boldsymbol{B} = 0$
6. $\boldsymbol{AB} = \boldsymbol{AC}$ does not necessarily imply that $\boldsymbol{B} = \boldsymbol{C}$

The matrix of order $n \times m$ obtained by interchanging the rows and columns of an $m \times n$ matrix $\boldsymbol{A}$ is called the *transpose* of $\boldsymbol{A}$ and is denoted by $\boldsymbol{A}^T$.

EXAMPLE A.15

If

$$\boldsymbol{A} = \begin{bmatrix} 1 \\ 2 \\ 3 \end{bmatrix}$$

then

$$A^T = [1 \quad 2 \quad 3]$$

and if

$$A = \begin{bmatrix} 1 & 4 \\ 2 & 5 \\ 3 & 6 \end{bmatrix}$$

then

$$A^T = \begin{bmatrix} 1 & 2 & 3 \\ 4 & 5 & 6 \end{bmatrix}$$

■

The transpose of the product of two matrices is equal to the product of the transposes in reverse order. In other words,

$$(\boldsymbol{AB})^T = \boldsymbol{B}^T\boldsymbol{A}^T$$

and in general,

$$[\boldsymbol{A}_1\boldsymbol{A}_2 \cdots \boldsymbol{A}_n]^T = \boldsymbol{A}_n^T \cdots \boldsymbol{A}_2^T\boldsymbol{A}_1^T \tag{A.7}$$

A *symmetric matrix* is a square matrix $\boldsymbol{A}$ for which

$$\boldsymbol{A} = \boldsymbol{A}^T \tag{A.8}$$

EXAMPLE A.16

The following matrix is a symmetric matrix, since $\boldsymbol{A} = \boldsymbol{A}^T$:

$$\boldsymbol{A} = \begin{bmatrix} 1 & 0 & 3 \\ 0 & 6 & -4 \\ 3 & -4 & -8 \end{bmatrix}$$

■

As defined earlier for determinants, the *cofactor* A_{ij} of the element a_{ij} of any square matrix $\boldsymbol{A}$ is equal to the product $(-1)^{i+j}$ and the determinant of the submatrix obtained from $\boldsymbol{A}$ by deleting row i and column j.

EXAMPLE A.17

Given the matrix

$$\boldsymbol{A} = \begin{bmatrix} a_{11} & a_{12} & a_{13} \\ a_{21} & a_{22} & a_{23} \\ a_{31} & a_{32} & a_{33} \end{bmatrix}$$

the cofactors A_{11}, A_{12}, and A_{22} are

$$A_{11} = (-1)^2 \begin{vmatrix} a_{22} & a_{23} \\ a_{32} & a_{33} \end{vmatrix} = a_{22}a_{33} - a_{32}a_{23}$$

$$A_{12} = (-1)^3 \begin{vmatrix} a_{21} & a_{23} \\ a_{31} & a_{33} \end{vmatrix} = -(a_{21}a_{33} - a_{31}a_{23})$$

$$A_{22} = (-1)^2 \begin{vmatrix} a_{11} & a_{13} \\ a_{31} & a_{33} \end{vmatrix} = a_{11}a_{33} - a_{31}a_{13}$$

■

The *adjoint* of the matrix $\mathbf{A}$ (adj $\mathbf{A}$) is the transpose of the matrix obtained from $\mathbf{A}$ by replacing each element a_{ij} by its cofactor A_{ij}. In other words, if

$$\mathbf{A} = \begin{bmatrix} a_{11} & a_{12} & \cdots & a_{1n} \\ a_{21} & a_{22} & \cdots & \cdot \\ \vdots & \vdots & & \vdots \\ a_{n1} & \cdots & \cdots & a_{nn} \end{bmatrix}$$

then

$$\text{adj } \mathbf{A} = \begin{bmatrix} A_{11} & A_{21} & \cdots & A_{n1} \\ A_{12} & A_{22} & \cdots & \cdot \\ \vdots & & & \vdots \\ A_{1n} & \cdots & \cdots & A_{nn} \end{bmatrix}$$

EXAMPLE A.18

If

$$\mathbf{A} = \begin{bmatrix} 1 & 2 \\ 4 & 3 \end{bmatrix}$$

then

$$\text{adj } \mathbf{A} = \begin{bmatrix} 3 & -2 \\ -4 & 1 \end{bmatrix}$$

and if

$$\mathbf{B} = \begin{bmatrix} 1 & 2 & 3 \\ 2 & 3 & 2 \\ 3 & 3 & 4 \end{bmatrix}$$

then

$$\text{adj } \boldsymbol{B} = \begin{bmatrix} 6 & 1 & -5 \\ -2 & -5 & 4 \\ -3 & 3 & -1 \end{bmatrix}$$

■

If $\boldsymbol{A}$ is a square matrix and if there exists a square matrix $\boldsymbol{A}^{-1}$ such that

$$\boldsymbol{A}^{-1}\boldsymbol{A} = \boldsymbol{A}\boldsymbol{A}^{-1} = \boldsymbol{I} \tag{A.9}$$

then $\boldsymbol{A}^{-1}$ is called the *inverse* of $\boldsymbol{A}$. It can be shown that the inverse of the matrix $\boldsymbol{A}$ is equal to the adjoint divided by the determinant (written here as $|\boldsymbol{A}|$), that is,

$$\boldsymbol{A}^{-1} = \frac{\text{adj } \boldsymbol{A}}{|\boldsymbol{A}|} \tag{A.10}$$

EXAMPLE A.19
Given

$$\boldsymbol{A} = \begin{bmatrix} 2 & 3 \\ 1 & 4 \end{bmatrix}$$

then

$$|\boldsymbol{A}| = (2)(4) - (1)(3) = 5$$

and

$$\text{adj } \boldsymbol{A} = \begin{bmatrix} 4 & -3 \\ -1 & 2 \end{bmatrix}$$

Therefore,

$$\boldsymbol{A}^{-1} = \tfrac{1}{5}\begin{bmatrix} 4 & -3 \\ -1 & 2 \end{bmatrix}$$

Also, if

$$\boldsymbol{B} = \begin{bmatrix} 2 & 3 & 1 \\ 1 & 2 & 3 \\ 3 & 1 & 2 \end{bmatrix}$$

then

$$\begin{aligned} |\boldsymbol{B}| &= 2\begin{vmatrix} 2 & 3 \\ 1 & 2 \end{vmatrix} - 1\begin{vmatrix} 3 & 1 \\ 1 & 2 \end{vmatrix} + 3\begin{vmatrix} 3 & 1 \\ 2 & 3 \end{vmatrix} \\ &= 2 - 5 + 21 = 18 \end{aligned}$$

and

$$\text{adj } \boldsymbol{B} = \begin{bmatrix} 1 & -5 & 7 \\ 7 & 1 & -5 \\ -5 & 7 & 1 \end{bmatrix}$$

Therefore,

$$\boldsymbol{B}^{-1} = \frac{1}{18}\begin{bmatrix} 1 & -5 & 7 \\ 7 & 1 & -5 \\ -5 & 7 & 1 \end{bmatrix}$$ ■

It is important to note that the inverse of the product is equal to the product of the inverses in reverse order. In other words,

$$(\boldsymbol{AB})^{-1} = \boldsymbol{B}^{-1}\boldsymbol{A}^{-1}$$

and in general,

$$[\boldsymbol{A}_1\boldsymbol{A}_2 \cdots \boldsymbol{A}_n]^{-1} = \boldsymbol{A}_n^{-1} \cdots \boldsymbol{A}_2^{-1}\boldsymbol{A}_1^{-1} \tag{A.11}$$

EXAMPLE A.20

If $\boldsymbol{A}$ is a diagonal matrix of the form

$$\boldsymbol{A} = \begin{bmatrix} 1 & 0 & 0 \\ 0 & 3 & 0 \\ 0 & 0 & 4 \end{bmatrix}$$

then

$$\boldsymbol{A}^{-1} = \begin{bmatrix} 1 & 0 & 0 \\ 0 & \frac{1}{3} & 0 \\ 0 & 0 & \frac{1}{4} \end{bmatrix}$$ ■

An $m \times n$ matrix $\boldsymbol{A}$ is said to have *rank* r if at least one of its $r \times r$ determinants is nonzero, where every $(r + 1) \times (r + 1)$ determinant, if any, is zero.

EXAMPLE A.21

Given the matrix

$$\boldsymbol{B} = \begin{bmatrix} 1 & 2 & 3 \\ 2 & 3 & 4 \\ 3 & 5 & 7 \end{bmatrix}$$

the rank of $\boldsymbol{B}$ is $r = 2$, since

$$\begin{vmatrix} 1 & 2 \\ 2 & 3 \end{vmatrix} = -1 \neq 0, \text{ while } |\boldsymbol{B}| = 0.$$ ■

An $n \times n$ matrix $\boldsymbol{A}$ is called *nonsingular* if it has maximal rank (i.e., $r = n$), or in other words, if $|\boldsymbol{A}| \neq 0$. Otherwise, A is said to be a *singular* matrix. In Example A.21, $\boldsymbol{B}$ is singular. Note that we cannot take the inverse of a singular matrix.

Consider now a system of m *nonhomogeneous* linear equations in the n unknowns $x_1, \ldots, x_n$:

$$\begin{array}{ccccccccc} a_{11}x_1 & + & a_{12}x_2 & + & \cdots & + & a_{1n}x_n & = & b_1 \\ \vdots & & \vdots & & & & \vdots & & \vdots \\ a_{m1}x_1 & + & a_{m2}x_2 & + & \cdots & + & a_{mn}x_n & = & b_m \end{array}$$

When this system of equations has a solution, it is said to be *consistent;* otherwise, the equations are said to be *inconsistent*. If $m = n$, the linear equations above can be solved using matrix techniques, as illustrated in the following example.

EXAMPLE A.22

By employing matrix methods, we will solve the equations

$$\begin{aligned} 2x_1 + 3x_2 + x_3 &= 9 \\ x_1 + 2x_2 + 3x_3 &= 6 \\ 3x_1 + x_2 + 2x_3 &= 8 \end{aligned}$$

Note that this set of simultaneous equations can be written as a single matrix equation in the form

$$\begin{bmatrix} 2 & 3 & 1 \\ 1 & 2 & 3 \\ 3 & 1 & 2 \end{bmatrix} \begin{bmatrix} x_1 \\ x_2 \\ x_3 \end{bmatrix} = \begin{bmatrix} 9 \\ 6 \\ 8 \end{bmatrix}$$

or

$$\boldsymbol{AX} = \boldsymbol{B}$$

Multiplying both sides of the equation above through by $\boldsymbol{A}^{-1}$ yields

$$\boldsymbol{A}^{-1}\boldsymbol{AX} = \boldsymbol{A}^{-1}\boldsymbol{B}$$

or

$$\boldsymbol{X} = \boldsymbol{A}^{-1}\boldsymbol{B}$$

$\boldsymbol{A}^{-1}$ was calculated in Example A.19. Employing that inverse here, we obtain

$$\boldsymbol{X} = \tfrac{1}{18} \begin{bmatrix} 1 & -5 & 7 \\ 7 & 1 & -5 \\ -5 & 7 & 1 \end{bmatrix} \begin{bmatrix} 9 \\ 6 \\ 8 \end{bmatrix}$$

or

$$\begin{bmatrix} x_1 \\ x_2 \\ x_3 \end{bmatrix} = \tfrac{1}{18} \begin{bmatrix} 35 \\ 29 \\ 5 \end{bmatrix}$$

and hence

$$x_1 = \tfrac{36}{18},\ x_2 = \tfrac{29}{18},\ \text{and } x_3 = \tfrac{5}{18}$$

■

APPENDIX

Complex Numbers

The reader has normally already encountered complex numbers and their use in previous work, and therefore only a quick review of the elements employed in this text are presented here.

B.1 Complex Number Representation

Complex numbers are typically represented in three forms: *exponential, polar,* or *rectangular*. In the exponential form a complex number **A** is written as

$$\mathbf{A} = ze^{j\theta} \tag{B.1}$$

The real quantity z is known as the *amplitude* or *magnitude,* the real quantity θ is called the *angle,* and j is the *imaginary operator* $j = \sqrt{-1}$, where $j^2 = -1$, $j^3 = -\sqrt{-1} = -j$, and so on. As indicated in the main body of the text, θ is expressed in radians or degrees.

The *polar form* of a complex number **A**, which is symbolically equivalent to the exponential form, is written as

$$\mathbf{A} = z \underline{/\theta} \tag{B.2}$$

Note that in this case the expression $e^{j\theta}$ is replaced by the angle symbol $\underline{/\theta}$. The representation of a complex number **A** by a magnitude of z at a given angle θ suggests a representation using polar coordinates in a complex plane.

The rectangular representation of a complex number is written as

$$\mathbf{A} = x + jy \tag{B.3}$$

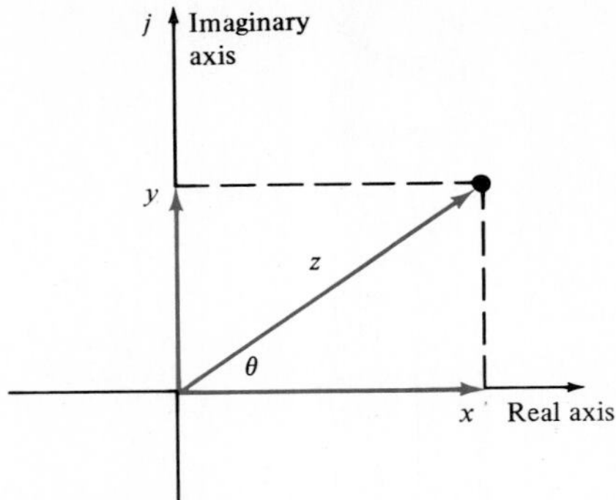

Figure B.1 Representation of a complex number in the complex plane.

where x is the real part of **A** and y is the imaginary part of **A**, which is usually expressed in the form

$$x = \text{Re}\,(\mathbf{A}) \qquad \text{(B.4)}$$
$$y = \text{Im}\,(\mathbf{A})$$

The complex number $\mathbf{A} = x + jy$ can be graphically represented in the complex plane as shown in Fig. B.1 where the abscissa is the real axis and the ordinate is the imaginary axis. Note that $x + jy$ uniquely locates a point in the complex plane which could also be specified by a magnitude z, representing the straight-line distance from the origin to the point, and an angle θ, which represents the angle between the positive real axis and the straight line connecting the point with the origin.

The connection between the various representations of **A** can be seen via Euler's identity, which is

$$e^{j\theta} = \cos\theta + j\sin\theta \qquad \text{(B.5)}$$

Using this identity, the complex number **A** can be written as

$$\mathbf{A} = ze^{j\theta} = z\cos\theta + jz\sin\theta \qquad \text{(B.6)}$$

which as shown in Fig. B.1 is equivalent to

$$\mathbf{A} = x + jy$$

Equating the real and imaginary parts of these two equations yields

$$x = z\cos\theta \qquad \text{(B.7)}$$
$$y = z\sin\theta$$

From these equations we obtain

$$x^2 + y^2 = z^2\cos^2\theta + z^2\sin^2\theta = z^2$$

Therefore,

$$z = \sqrt{x^2 + y^2} \qquad \text{(B.8)}$$

Additionally,

$$\frac{z \sin \theta}{z \cos \theta} = \tan \theta = \frac{y}{x}$$

and hence

$$\theta = \tan^{-1} \frac{y}{x} \tag{B.9}$$

The interrelationships among the three representations of a complex number are as follows.

Exponential	Polar	Rectangular
$ze^{j\theta}$	$z \underline{/\theta}$	$x + jy$
$\theta = \tan^{-1} \frac{y}{x}$	$\theta = \tan^{-1} \frac{y}{x}$	$x = z \cos \theta$
$z = \sqrt{x^2 + y^2}$	$z = \sqrt{x^2 + y^2}$	$y = z \sin \theta$

EXAMPLE B.1

If a complex number **A** in polar form is $\mathbf{A} = 10 \underline{/30^\circ}$, express **A** in both exponential and rectangular forms.

$$\mathbf{A} = 10 \underline{/30^\circ} = 10e^{j30^\circ} = 10 [\cos 30^\circ + j \sin 30^\circ] = 8.66 + j5.0 \qquad \blacksquare$$

EXAMPLE B.2

If $\mathbf{A} = 4 + j3$, express **A** in both exponential and polar forms. In addition, express $-\mathbf{A}$ in exponential and polar forms with a positive magnitude.

$$\mathbf{A} = 4 + j3 = \sqrt{4^2 + 3^2} \underline{/\tan^{-1} \frac{3}{4}}$$

$$= 5 \underline{/36.9^\circ}$$

Also,

$$-\mathbf{A} = -5 \underline{/36.9^\circ} = 5 \underline{/36.9^\circ + 180^\circ} = 5 \underline{/216.9^\circ} = 5e^{j216.9^\circ}$$

or

$$-\mathbf{A} = -5 \underline{/36.9^\circ} = 5 \underline{/36.9^\circ - 180^\circ} = 5 \underline{/-143.1^\circ} = 5e^{-j143.1^\circ} \qquad \blacksquare$$

B.2 Mathematical Operations

We will now show that the operations of addition, subtraction, multiplication, and division apply to complex numbers in the same manner that they apply to real numbers. Before proceeding with this illustration, however, let us examine two important definitions.

Two complex numbers **A** and **B** defined as

$$\mathbf{A} = z_1 e^{j\theta_1} = z_1 \underline{/\theta_1} = x_1 + jy_1$$

$$\mathbf{B} = z_2 e^{j\theta_2} = z_2 \underline{/\theta_2} = x_2 + jy_2$$

are *equal* if and only if $x_1 = x_2$ and $y_1 = y_2$ or $z_1 = z_2$ and $\theta_1 = \theta_2 \pm n360°$, where $n = 0, 1, 2, 3, \ldots$.

EXAMPLE B.3

If $\mathbf{A} = 2 + j3$, $\mathbf{B} = 2 - j3$, $\mathbf{C} = 4 \underline{/30°}$ and $\mathbf{D} = 4 \underline{/750°}$, then $\mathbf{A} \neq \mathbf{B}$, but $\mathbf{C} = \mathbf{D}$ since $30° = 30° + 2(360°)$. ■

The *conjugate,* $\mathbf{A}^*$, of a complex number $\mathbf{A} = x + jy$ is defined to be

$$\mathbf{A}^* = x - jy \tag{B.10}$$

that is, j is replaced by $-j$ to obtain the conjugate. Note that the magnitude of $\mathbf{A}^*$ is the same as that of **A**, since

$$z = \sqrt{x^2 + (-y)^2} = \sqrt{x^2 + y^2}$$

However, the angle is now

$$\tan^{-1} \frac{-y}{x} = -\theta$$

Therefore, the conjugate is written in exponential and polar form as

$$\mathbf{A}^* = ze^{-j\theta} = z \underline{/-\theta} \tag{B.11}$$

We also have the relationship

$$(\mathbf{A}^*)^* = \mathbf{A} \tag{B.12}$$

EXAMPLE B.4

If $\mathbf{A} = 10 \underline{/30°}$ and $\mathbf{B} = 4 + j3$, then $\mathbf{A}^* = 10 \underline{/-30°}$ and $\mathbf{B}^* = 4 - j3$. $(\mathbf{A}^*)^* = 10 \underline{/30°} = \mathbf{A}$ and $(\mathbf{B}^*)^* = 4 + j3 = \mathbf{B}$. ■

Addition

The sum of two complex numbers $\mathbf{A} = x_1 + jy_1$ and $\mathbf{B} = x_2 + jy_2$ is

$$\begin{aligned} \mathbf{A} + \mathbf{B} &= x_1 + jy_1 + x_2 + jy_2 \\ &= (x_1 + x_2) + j(y_1 + y_2) \end{aligned} \tag{B.13}$$

that is, we simply add the individual real parts, and we add the individual imaginary parts to obtain the components of the resultant complex number. This addition can be illustrated graphically by plotting each of the complex numbers as vectors and then performing the vector addition. This graphical approach is shown in Fig. B.2. Note that the vector addition is accomplished by plotting the vectors tail to head or simply completing the parallelogram.

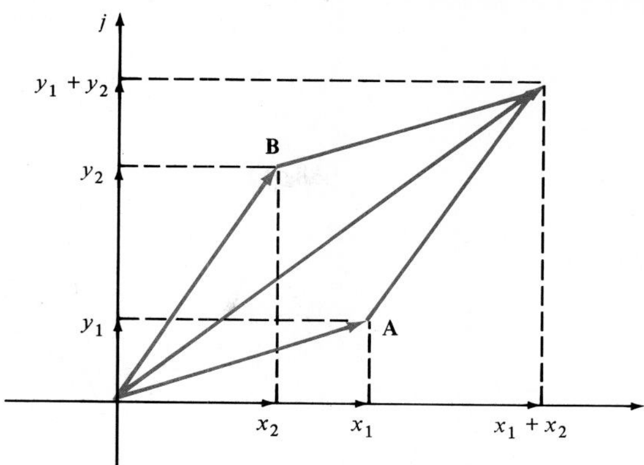

Figure B.2 Vector addition for complex numbers.

EXAMPLE B.5

Given the complex numbers $\mathbf{A} = 4 + j1$, $\mathbf{B} = 3 - j2$, and $\mathbf{C} = -2 - j4$, we wish to calculate $\mathbf{A} + \mathbf{B}$ and $\mathbf{A} + \mathbf{C}$ (Fig. B.3).

$$\begin{aligned}\mathbf{A} + \mathbf{B} &= (4 + j1) + (3 - j2) \\ &= 7 - j1 \\ \mathbf{A} + \mathbf{C} &= (4 + j1) + (-2 - j4) \\ &= 2 - j3\end{aligned}$$

■

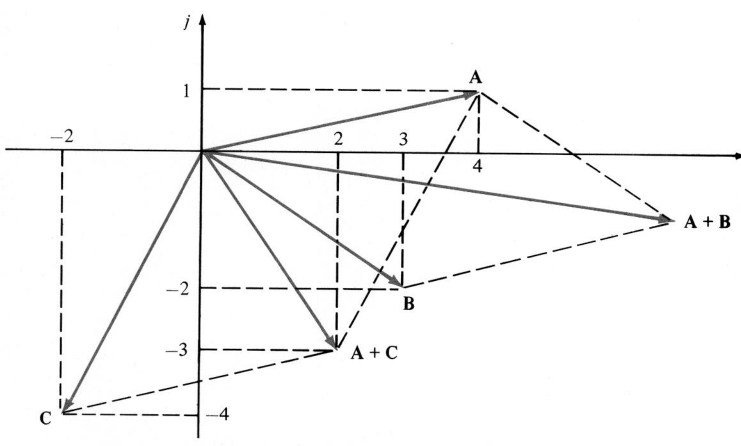

Figure B.3 Examples of complex number addition.

EXAMPLE B.6

We wish to calculate the sum $\mathbf{A} + \mathbf{B}$ if $\mathbf{A} = 5\,\underline{/36.9^\circ}$ and $\mathbf{B} = 5\,\underline{/53.1^\circ}$. We must first convert from polar to rectangular form.

$$\mathbf{A} = 5\,\underline{/36.9^\circ} = 4 + j3$$

$$\mathbf{B} = 5\,\underline{/53.1^\circ} = 3 + j4$$

Therefore,

$$\mathbf{A} + \mathbf{B} = 4 + j3 + 3 + j4 = 7 + j7$$
$$= 9.9 \angle 45°$$

■

Subtraction

The difference of two complex numbers $\mathbf{A} = x_1 + jy_1$ and $\mathbf{B} = x_2 + jy_2$ is

$$\begin{aligned}\mathbf{A} - \mathbf{B} &= (x_1 + jy_1) - (x_2 + jy_2) \\ &= (x_1 - x_2) + j(y_1 - y_2)\end{aligned} \tag{B.14}$$

that is, we simply subtract the individual real parts and we subtract the individual imaginary parts to obtain the components of the resultant complex number. Since a negative sign corresponds to a phase or angle change of 180°, the graphical technique for performing the subtraction ($\mathbf{A} - \mathbf{B}$) can be accomplished by drawing **A** and **B** as vectors, rotating the vector **B** 180° and then adding it to the vector **A**.

EXAMPLE B.7

Given $\mathbf{A} = 3 + j1$ and $\mathbf{B} = 2 - j2$, calculate the difference $\mathbf{A} - \mathbf{B}$.

$$\mathbf{A} - \mathbf{B} = (3 + j1) - (2 - j2)$$
$$= 1 + j3$$

■

The graphical solution is shown in Fig. B.4.

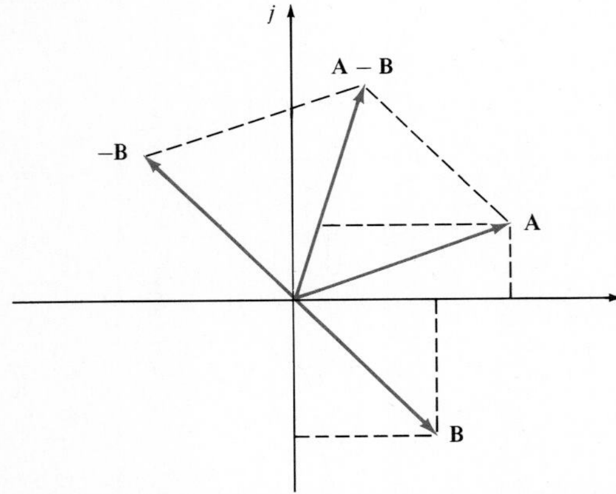

Figure B.4 Example of subtracting complex numbers.

EXAMPLE B.8

Let us calculate the difference $\mathbf{A} - \mathbf{B}$ if $\mathbf{A} = 5 \angle 36.9°$ and $\mathbf{B} = 5 \angle 53.1°$. Converting both numbers from polar to rectangular form,

$$\mathbf{A} = 5 \angle 36.9° = 4 + j3$$

$$\mathbf{B} = 5 \angle 53.1° = 3 + j4$$

then

$$\mathbf{A} - \mathbf{B} = (4 + j3) - (3 + j4) = 1 - j1 = \sqrt{2}\,\underline{/-45^\circ}$$ ■

EXAMPLE B.9

Given the complex number $\mathbf{A} = 5\,\underline{/36.9^\circ}$, calculate $\mathbf{A}^*$, $\mathbf{A} + \mathbf{A}^*$, and $\mathbf{A} - \mathbf{A}^*$. If $\mathbf{A} = 5\,\underline{/36.9^\circ} = 4 + j3$, then $\mathbf{A}^* = 5\,\underline{|-36.9^\circ} = 4 - j3$. Hence $\mathbf{A} + \mathbf{A}^* = 8$ and $\mathbf{A} - \mathbf{A}^* = j6$. ■

Note that addition and subtraction of complex numbers is a straightforward operation if the numbers are expressed in rectangular form. Note also that the sum of a complex number and its conjugate is a real number, and the difference of a complex number and its conjugate is an imaginary number.

Multiplication

The product of two complex numbers $\mathbf{A} = z_1 e^{j\theta_1} = z_1\,\underline{/\theta_1} = x_1 + jy_1$ and $\mathbf{B} = z_2 e^{j\theta_2} = z_2\,\underline{/\theta_2} = x_2 + jy_2$ is

$$\mathbf{AB} = (z_1 e^{j\theta_1})(z_2 e^{j\theta_2}) = z_1 z_2 e^{j(\theta_1 + \theta_2)} = z_1 z_2\,\underline{/\theta_1 + \theta_2} \tag{B.15}$$

or

$$\begin{aligned} \mathbf{AB} &= (x_1 + jy_1)(x_2 + jy_2) \\ &= x_1x_2 + jx_1y_2 + jx_2y_1 + j^2y_1y_2 \\ &= (x_1x_2 - y_1y_2) + j(x_1y_2 + x_2y_1) \end{aligned} \tag{B.16}$$

If the two complex numbers are in exponential or polar form, multiplication is readily accomplished by multiplying their magnitudes and adding their angles. Multiplication is straightforward, although slightly more complicated, if the numbers are expressed in rectangular form.

The product of a complex number and its conjugate is a real number, that is

$$\mathbf{AA}^* = (ze^{j\theta})(ze^{-j\theta}) = z^2e^{j0} = z^2\,\underline{/0^\circ} = z^2 \tag{B.17}$$

Note that this real number is the square of the magnitude of the complex number.

EXAMPLE B.10

If $\mathbf{A} = 10\,\underline{/30^\circ}$ and $\mathbf{B} = 5\,\underline{/15^\circ}$, the products $\mathbf{AB}$ and $\mathbf{AA}^*$ are

$$\mathbf{AB} = (10\,\underline{/30^\circ})(5\,\underline{/15^\circ}) = 50\,\underline{/45^\circ}$$

and

$$\mathbf{AA}^* = (10\,\underline{/30^\circ})(10\,\underline{/-30^\circ}) = 100\,\underline{/0^\circ} = 100$$ ■

EXAMPLE B.11

Given $\mathbf{A} = 5\,\underline{/36.9^\circ}$ and $\mathbf{B} = 5\,\underline{/53.1^\circ}$, we wish to calculate the product in both polar and rectangular forms.

$$\begin{aligned}\mathbf{AB} &= (5\,\underline{/36.9^\circ})(5\,\underline{/53.1^\circ}) = 25\,\underline{/90^\circ}\\ &= (4 + j3)(3 + j4)\\ &= 12 + j16 + j9 + j^2 12\\ &= 25j\\ &= 25\,\underline{/90^\circ}\end{aligned}$$

■

Division

The quotient of two complex numbers $\mathbf{A} = z_1 e^{j\theta_1} = z_1\,\underline{/\theta_1} = x_1 + jy_1$ and $\mathbf{B} = z_2 e^{j\theta_2} = z_2\,\underline{/\theta_2} = x_2 + jy_2$ is

$$\frac{\mathbf{A}}{\mathbf{B}} = \frac{z_1 e^{j\theta_1}}{z_2 e^{j\theta_2}} = \frac{z_1}{z_2} e^{j(\theta_1 - \theta_2)} = \frac{z_1}{z_2}\,\underline{/\theta_1 - \theta_2} \tag{B.18}$$

that is, if the numbers are in exponential or polar form, division is immediately accomplished by dividing their magnitudes and subtracting their angles as shown above. If the numbers are in rectangular form, or the answer is desired in rectangular form, then the following procedure can be used.

$$\frac{\mathbf{A}}{\mathbf{B}} = \frac{x_1 + jy_1}{x_2 + jy_2}$$

The denominator is rationalized by multiplying both numerator and denominator by $\mathbf{B}^*$:

$$\begin{aligned}\frac{\mathbf{AB}^*}{\mathbf{BB}^*} &= \frac{(x_1 + jy_1)(x_2 - jy_2)}{(x_2 + jy_2)(x_2 - jy_2)}\\ &= \frac{x_1x_2 + y_1y_2}{x_2^2 + y_2^2} + j\frac{x_2y_1 - x_1y_2}{x_2^2 + y_2^2}\end{aligned} \tag{B.19}$$

In this form the denominator is real and the quotient is given in rectangular form.

EXAMPLE B.12

If $\mathbf{A} = 10\,\underline{/30^\circ}$ and $\mathbf{B} = 5\,\underline{/53.1^\circ}$, determine the quotient $\mathbf{A}/\mathbf{B}$ in both polar and rectangular forms.

$$\begin{aligned}\frac{\mathbf{A}}{\mathbf{B}} = \frac{\mathbf{AB}^*}{\mathbf{BB}^*} &= \frac{8.66 + j5}{3 + j4}\,\frac{3 - j4}{3 - j4}\\ &= \frac{(8.66 + j5)(3 - j4)}{3^2 + 4^2}\\ &= \frac{45.98 - j19.64}{25}\\ &= 1.84 - j0.79\end{aligned}$$

or

$$\begin{aligned}\frac{\mathbf{A}}{\mathbf{B}} &= \frac{10\,\underline{/30^\circ}}{5\,\underline{/53.1^\circ}}\\ &= 2\,\underline{/-23.1^\circ}\\ &= 1.84 - j0.79\end{aligned}$$

■

As a final example, consider the following one, which requires the use of many of the techniques presented above.

EXAMPLE B.13

Given $\mathbf{A} = 10\,\underline{/30°}$, $\mathbf{B} = 2 + j2$, $\mathbf{C} = 4 + j3$, and $\mathbf{D} = 4\,\underline{/10°}$. Calculate the expression for $\mathbf{AB}/(\mathbf{C} + \mathbf{D})$ in rectangular form.

$$\frac{\mathbf{AB}}{\mathbf{C} + \mathbf{D}} = \frac{(10\,\underline{/30°})(2 + j2)}{(4 + j3) + (4\,\underline{/10°})}$$

$$= \frac{(10\,\underline{/30°})(2\sqrt{2}\,\underline{/45°})}{4 + j3 + 3.94 + j0.69}$$

$$= \frac{20\sqrt{2}\,\underline{/75°}}{7.94 + j3.69}$$

$$= \frac{20\sqrt{2}\,\underline{/75°}}{8.75\,\underline{/24.93°}}$$

$$= 3.23\,\underline{/50.07°}$$

$$= 2.07 + j2.48$$

■

References

ASELTINE, J. A., *Transform Methods in Linear System Analysis,* McGraw-Hill Book Company, New York, 1958.

BALABANIAN, N., *Fundamentals of Circuit Theory,* Allyn and Bacon, Inc., Boston, 1961.

BLACKWELL, W. A., and GRIGSBY, L. L., *Introductory Network Theory,* PWS Engineering, Boston, Mass., 1985.

BOBROW, L. S., *Elementary Linear Circuit Analysis,* Holt, Rinehart and Winston, New York, 1981.

CARLSON, A. B., and GISSER, D. G., *Electrical Engineering,* Addison-Wesley Publishing Co., Inc., Reading, Mass., 1981.

CHAN, SHU-PARK, CHAN, SHU-YAN, and CHAN, SHU-GAR, *Analysis of Linear Networks and Systems,* Addison-Wesley Publishing Co., Inc., Reading, Mass., 1972.

CHEN, WAI-KAI, *Linear Networks and Systems,* Brooks/Cole Engineering Division of Wadsworth, Inc., Monterey, Calif., 1983.

CHENG, D. K., *Analysis of Linear Systems,* Addison-Wesley Publishing Co., Inc., Reading, Mass., 1959.

FITZGERALD, A. E., HIGGINBOTHAM, E. E., and GRABEL, A., *Basic Electrical Engineering,* McGraw-Hill Book Company, New York, 1981.

FRIEDLAND, B., WING, O., and ASH, R. *Principles of Linear Networks,* McGraw-Hill Book Company, New York, 1961.

HAYT, W. H., JR., and KEMMERLY, JACK, E., *Engineering Circuit Analysis,* 4th ed., McGraw-Hill Book Company, New York, 1986.

JOHNSON, D. E., HILBURN, J. L., and JOHNSON, J. R., *Basic Electric Circuit Analysis,* Prentice-Hall, Inc., Englewood Cliffs, N.J., 1978.

KIRWIN, G. J., and GRODZINSKY, S. E., *Basic Circuit Analysis,* Houghton Mifflin Company, Boston, 1980.

KUO, F. F., *Network Analysis and Synthesis,* John Wiley & Sons, Inc., New York, 1962.

LEY, B. J., LUTZ, S. G., and REHBERG, C. F., *Linear Circuit Analysis,* McGraw-Hill Book Company, New York, 1959.

NAGLE, L. W., "SPICE 2: A Program to Simulate Semiconductor Circuits," Department of Electrical Engineering and Computer Science, University of California, Berkeley, Calif., ERL-M520, May 1975.

NILSSON, J. W., *Electric Circuits,* 2nd ed., Addison-Wesley Publishing Co., Inc., Reading, Mass., 1986.

STREMLER, F. G., *Introduction to Communication Systems,* Addison-Wesley Publishing Co., Inc., Reading, Mass., 1977.

SU, K. L., *Fundamentals of Circuit, Electronics and Signal Analysis,* Houghton Mifflin Company, Boston, 1978.

TRICK, T. N., *Introduction to Circuit Analysis,* John Wiley & Sons, Inc., New York, 1977.

VAN VALKENBURG, M. E., *Analog Filter Design,* Holt, Rinehart and Winston, New York, 1982.

VAN VALKENBURG, M. E., and KINARIWALA, B. K., *Linear Circuits,* Prentice-Hall, Inc., Englewood Cliffs, N.J., 1982.

VLADIMIRESCU, A., NEWTON, A. R., and PEDERSON, D. O., "SPICE Version 2F.1 User's Guide," Department of Electrical Engineering and Computer Science, University of California, Berkeley, Calif., February 1980.

VLADIMIRESCU, A., and LIU, SALLY, "The Simulation of MOS Integrated Circuits Using SPICE 2," Department of Electrical Engineering and Computer Science, University of California, Berkeley, Calif., UCB/ERL M80/7, February 1980.

WEINBERG, L., *Network Analysis and Synthesis,* McGraw-Hill Book Company, New York, 1962.

Answers to Drill Problems

D1.1. (a) $P = -48$ W; (b) $P = 8$ W

D1.2. (a) $V = -20$ V; (b) $I = -5$ A

D1.3. Current source supplies 36 W, element 1 absorbs 54 W and element 2 supplies 18 W

D1.4. (a) Power supplied = 80 W; (b) power supplied = 160 W

D2.1. $V = 8$ V, $P_{abs} = P_{sup} = 32$ W

D2.2. $R = 4\ \Omega$, $V = 32$ V, and $P_{sup} = P_{abs} = 256$ W

D2.3. $I_1 = -5$ A **D2.4.** $I_1 = 7$ A, $I_2 = 2$ A **D2.5.** $V_{R_2} = 14$ V, $V_{bd} = 8$ V

D2.6. $V_{ab} = 6$ V, $V_{cd} = -12$ V **D2.7.** $V_o = -9$ V

D2.8. $I = 2$ A, $P_{5\Omega} = 20$ W, $V_{bd} = 28$ V, and $V_{be} = 20$ V **D2.9.** $V_o = -4$ V

D2.10. $V_2 = 4$ V **D2.11.** $P_{6\Omega} = 24$ W **D2.12.** $I_o = -4$ A

D2.13. $I_o = 7.5$ A **D2.14.** $R_{eq} = 4.5\ \Omega$ **D2.15.** $R_T = 4\ \Omega$

D2.16. $I_1 = 3$ A, $I_2 = \frac{1}{2}$ A, $I_3 = \frac{3}{2}$ A, $I_4 = 2$ A, and $I_5 = 1$ A **D2.17.** $V_o = 92$ V

D2.18. $V_o = 4V$ **D2.19.** $V_o = 28.8$ V **D2.20.** Gain = 101, $V_o = 0.101$ V

D3.1.

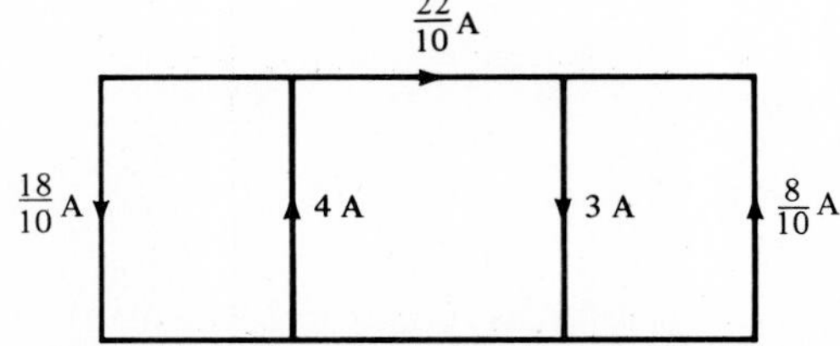

D3.2.

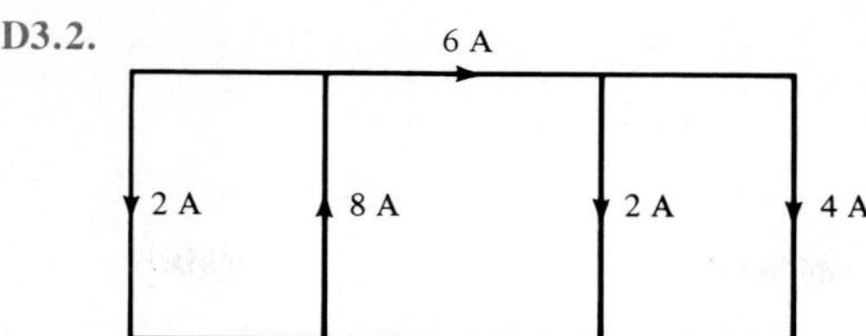

D3.3.

$$V_1\left(\frac{1}{R_1}+\frac{1}{R_2}\right) - V_2(0) - V_3\left(\frac{1}{R_1}\right) - V_4\left(\frac{1}{R_2}\right) = I_A$$

$$-V_1(0) + V_2\left(\frac{1}{R_3}+\frac{1}{R_5}\right) - V_3\left(\frac{1}{R_3}\right) - V_4(0) = -I_A$$

$$-V_1\left(\frac{1}{R_1}\right) - V_2\left(\frac{1}{R_3}\right) + V_3\left(\frac{1}{R_1}+\frac{1}{R_3}+\frac{1}{R_4}\right) - V_4\left(\frac{1}{R_4}\right) = -I_B$$

$$-V_1\left(\frac{1}{R_2}\right) - V_2(0) - V_3\left(\frac{1}{R_4}\right) + V_4\left(\frac{1}{R_2}+\frac{1}{R_4}+\frac{1}{R_6}\right) = 0$$

D3.4. $I_1 = \frac{20}{3}$ A, $I_2 = \frac{8}{6}$ A. Both currents flow top to bottom. **D3.5.** $I = 11$ A

D3.6. $V_1 = \frac{8}{5}$ V, $V_2 = \frac{-4}{5}$ V **D3.7.** $V_1 = 6$ V, $V_2 = 8$ V

D3.8. $I_0 = 8.4$ mA **D3.9.** $V_0 = -6.35$ V

D3.10.

$$\begin{bmatrix} R_1 + R_2 + R_4 & -R_2 & -R_4 \\ -R_2 & R_2 + R_3 & -R_3 \\ -R_4 & -R_3 & R_3 + R_4 + R_5 \end{bmatrix} \begin{bmatrix} I_1 \\ I_2 \\ I_3 \end{bmatrix} = \begin{bmatrix} 0 \\ V_1 + V_2 \\ -V_2 \end{bmatrix}$$

D3.11. $V_0 = 8$ V **D3.12.** $V_0 = 6$ V **D3.13.** $V_0 = 24$ V **D3.14.** $V_0 = -8$ V

D4.1. **(b)** and **(e)** No; **(c), (d),** and **(f)** Yes

D4.2. **(b)** and **(d)** Yes; **(c), (e),** and **(f)** No

D4.3. Fig. 4.1a, $L = 4$; Fig. 4.2a, $L = 5$

D4.4.

$$I_A = \frac{V_1 - V_2}{R_1} \qquad \frac{V_1 - V_2}{R_1} = \frac{V_2}{R_2} + \frac{V_2 - V_3}{R_3}$$

$$\frac{V_2 - V_3}{R_3} = \frac{V_3 - V_4}{R_4} + I_B \qquad \frac{V_3 - V_4}{R_4} = -I_C$$

D4.5.

$$\frac{V_1 - V_3}{R_1} + \frac{V_1 - V_2}{R_2} + \frac{V_1}{R_5} = 0 \qquad \frac{V_1 - V_3}{R_1} + \frac{V_2 - V_3}{R_3} - I_A + \frac{V_1}{R_5} = 0$$

$$\frac{V_1 - V_3}{R_1} + \frac{V_2 - V_3}{R_3} - \frac{V_3}{R_6} - \frac{V_4}{R_7} = 0 \qquad \frac{V_3 - V_4}{R_4} - \frac{V_4}{R_7} = 0$$

D4.6.

$$I_1(R_2 + R_3 + R_5) - I_2(R_3) - I_3(R_5) - I_4(0) = V_1$$

$$-I_1(R_3) + I_2(R_1 + R_3 + R_4) - I_3(0) - I_4(0) = -V_2$$

$$-I_1(R_5) - I_2(0) + I_3(R_5 + R_6 + R_8) - I_4(0) = -V_3$$

$$-I_1(0) - I_2(0) - I_3(0) + I_4(R_7 + R_9) = V_2 + V_3$$

D4.7.

$$V_1 - I_1R_1 - V_2 - (I_1 - I_3)R_4 - I_1R_6 = 0$$

$$V_2 - (I_2 + I_3)R_2 + V_3 - I_2R_3 = 0$$

$$-(I_3 - I_1)R_4 + I_2R_3 - I_3R_5 - I_3R_7 = 0$$

D4.8. $I_1 = \frac{20}{3}$ A **D4.9.** $I_0 = 2$ A **D4.10.** $I_0 = \frac{11}{4}$ A
D5.1. $V_0 = \frac{8}{7}$ V **D5.2.** $I_0 = 6$ A **D5.3.** $I_0 = 3$ A **D5.4.** $V_0 = -8$ V
D5.5. $V_0 = 4$ V **D5.6.** $V_0 = 15$ V **D5.7.** $V_0 = -8$ V **D5.8.** $V_0 = 4$ V
D5.9. $V_0 = 15$ V **D5.10.** $V_0 = -8$ V **D5.11.** $V_0 = 4$ V
D5.12. $V_0 = 15$ V **D5.13.** $V_0 = -8$ V **D5.14.** $V_0 = 4$ V
D5.15. $V_0 = \frac{24}{5}$ V **D5.16.** $V_0 = \frac{48}{5}$ V **D5.17.** $R_L = 6\ \Omega$, $P_{load} = \frac{2}{3}$ W
D5.18. $R_L = 3\ \Omega$, $P_{load} = \frac{4}{3}$ W

D5.19 $V_1 = 12.5$ V, $V_2 = 20$ V, $V_3 = 12.727$ V
$I_x = -4.318$ A, $I_0 = 1.818$ A

D5.20. $V_1 = -4.17$ V, $V_2 = 5.83$ V, $V_3 = 5$ V
$V_4 = 3.75$ V, $I_x = -0.83$ A
$I_Y = -1.46$ A, $I_0 = 0.625$ A

D5.21. $V_1 = 3.26$ V, $V_2 = -6.74$ V
$V_3 = 3.05$ V, $I_{10} = -0.54$ A
$I_x = 3.05$ A

D5.22. $V_1 = 10$ V, $V_2 = 5.27$ V, $V_3 = -0.182$ V
$V_4 = 3$ V, $I_{10} = -2.36$ A, $I_x = -2.73$ A
$I_a = -0.182$ A

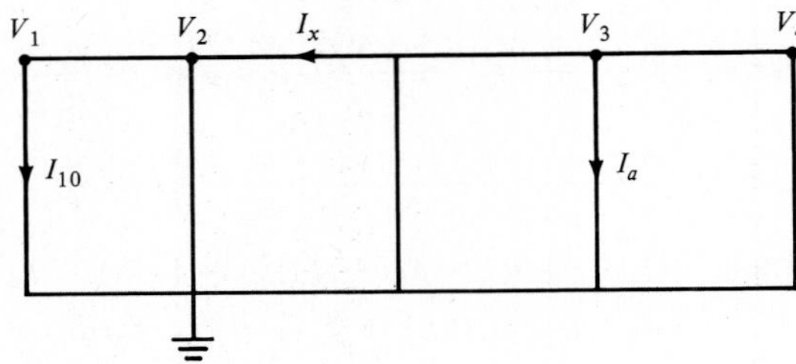

D6.1. $V = 0.05$ V
D6.2.

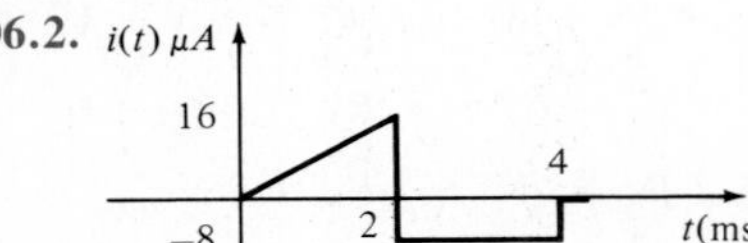

D6.3. $W(2\text{ ms}) = 32\text{ pJ}$

D6.4.

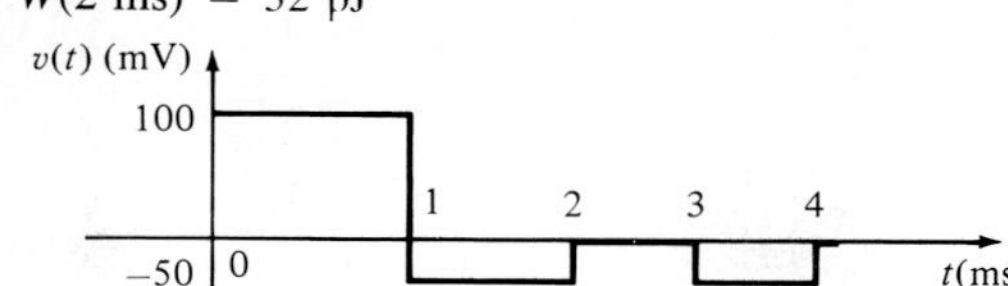

D6.5. $W(1.5\text{ ms}) = 562.5\text{ nJ}$ **D6.6.** $C_1 = 4\ \mu\text{F}$ **D6.7.** $C_{eq} = 1.5\ \mu\text{F}$

D6.8. $C_{eq} = \frac{32}{3}\ \mu\text{F}$ **D6.9.** $L_{eq} = \frac{44}{10}\text{ mH}$ **D6.10.** $L_{eq} = \frac{66}{7}\text{ mH}$

D6.11.

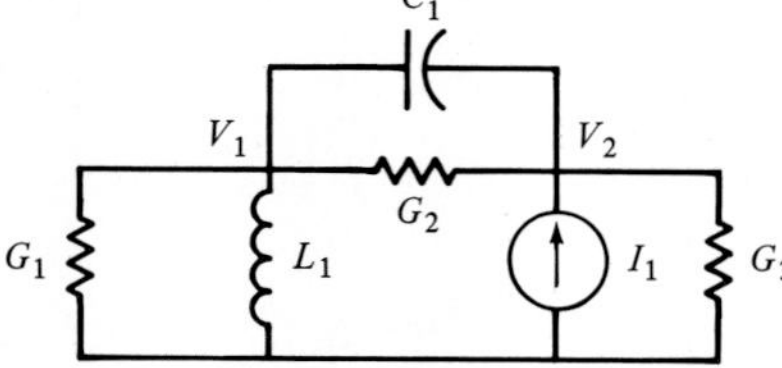

D6.12.

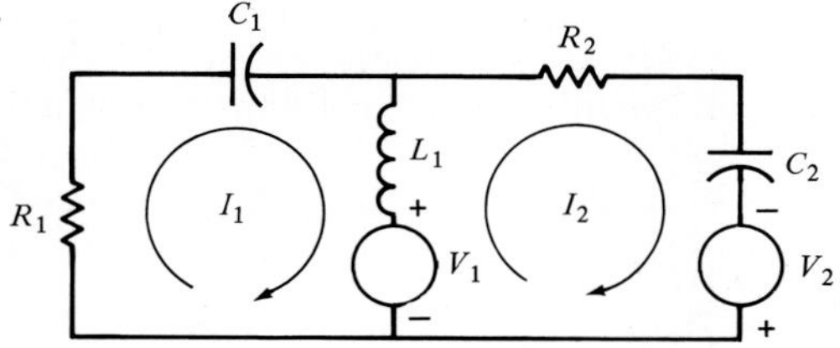

D7.1. $v_0(t) = \frac{8}{3}e^{-t/2}\text{ V}$ **D7.2.** $v_0(t) = -8e^{-t/8}\text{ V}$ **D7.3.** $v_0(t) = -6e^{-4t}\text{ V}$

D7.4. $i_0(t) = \frac{12}{5}e^{-18t/5}\text{ A}$ **D7.5.** $v_0(t) = \frac{24}{5} + \frac{1}{5}e^{-5t/8}\text{ V}$

D7.6. $v_0(t) = 6 - \frac{10}{3}e^{-2t}\text{ V}$ **D7.7.** $v_0(t) = 24 + 36e^{-t/12}\text{ V}$

D7.8.

$$v_0(t) = \begin{cases} 0 & t < 0 \\ 4(1 - e^{-3t/2})\text{ V} & 0 \le t \le 1 \\ 3.11e^{-3(t-1)/2}\text{ V} & 1 < t \end{cases}$$

D7.9.

$$i_0\,(t) = \begin{cases} 0 & t < 0 \\ 6 - 3e^{-t/8}\text{ A} & 0 \le t \le 5.6\text{ sec} \\ 5.29e^{-1(t-1)/8}\text{ A} & 4.5 < t \end{cases}$$

D7.10.

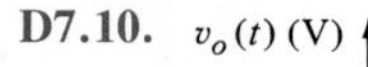

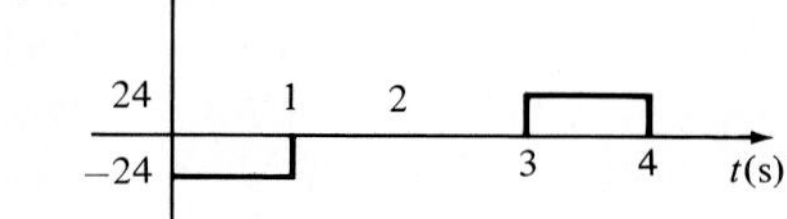

D7.11.

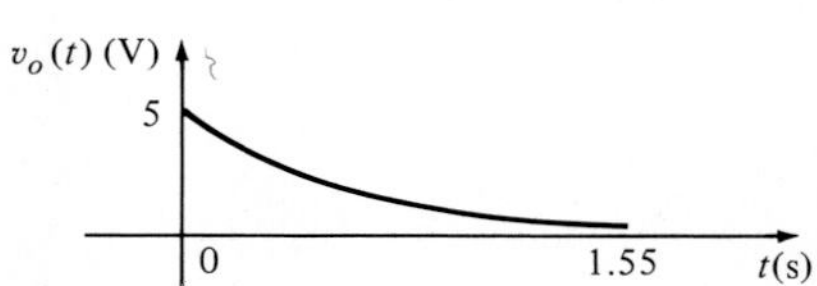

D7.12.

D7.13.

D7.14.

D8.1. $\alpha = 0.25$, $\omega_0 = 0.5$ rad/s
D8.2. **(a)** Underdamped; **(b)** critically damped; **(c)** overdamped
D8.3. $i(t) = -2e^{-t/2} + 4e^{-t}$ A **D8.4.** $v_0(t) = 2\,(e^{-t} - 3e^{-3t})$ V
D8.5. $t_s = 6.1$ s **D8.6.** $i_0(t) = -\frac{1}{2}e^{-3t} + e^{-6t}$ A **D8.7.** $v_c(t) = e^{-t} - te^{-t}$ V
D8.8.

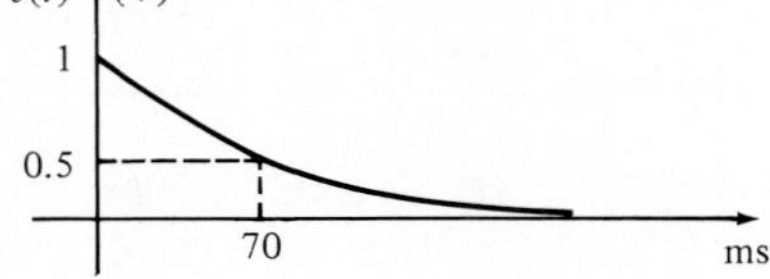

D8.9.

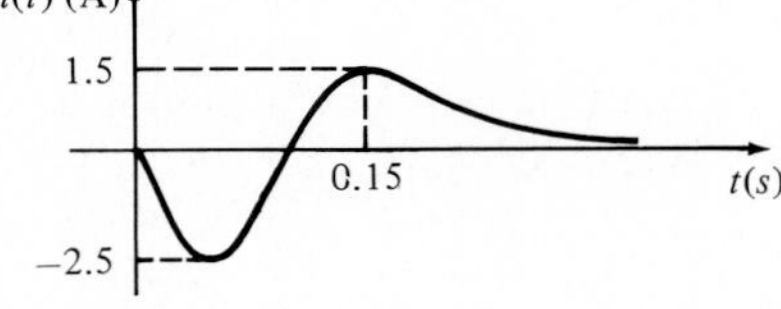

D9.1. $f = 400$ Hz, $v_1(t)$ leads $v_2(t)$ by $-40°$ **D9.2.** The two voltages are in phase.
D9.3. $i_1(t)$ leads $i_2(t)$ by $-15°$. **D9.4.** $\mathbf{V}_1 = 12\,\underline{/-425°}$ V, $\mathbf{V}_2 = 18\,\underline{/-85.8°}$ V
D9.5. $v_1(t) = 10\cos\,[2\pi(400)t + 20°]$ V, $v_2(t) = 12\cos\,[2\pi(400)t - 60°]$ V
D9.6. $v(t) = 48\cos\,(377t + 60°)$ V **D9.7.** $v_L(t) = 75.4\cos\,(377t + 60°)$ V
D9.8. $\mathbf{V}_c = 63.67\,\underline{/-235°}$ V **D9.9.** $v(t) = 50.1\cos\,(377t + 57°)$ V
D9.10. $i(t) = 3.88\cos\,(377t - 39.2°)$ A **D9.11.** $\mathbf{I} = 9\,\underline{/53.7°}$ A
D9.12. $\mathbf{V}_{cs} = 32.4\,\underline{/59.7°}$ V **D9.13.** $\mathbf{Z}_T = 3.38 + j1.08\ \Omega$
D9.14.

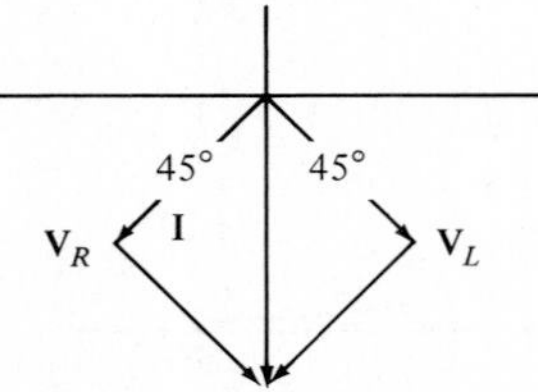

D9.15.

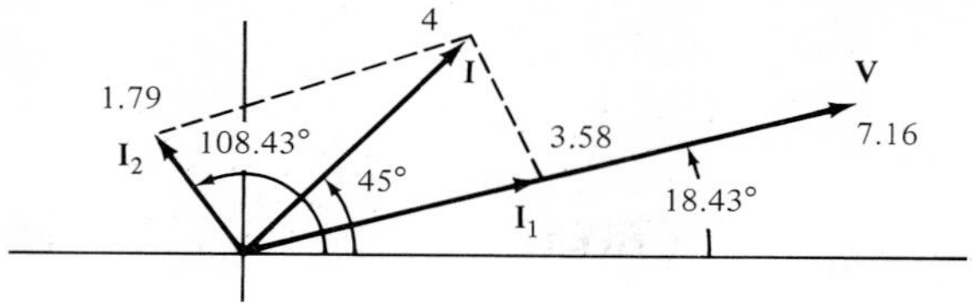

D9.16. $\mathbf{I}_1 = 3.38 \angle -17.7^\circ$ A, $\mathbf{I}_2 = 2.1 \angle -24.7^\circ$ A, $\mathbf{I}_3 = 1.33 \angle -6.3^\circ$ A
D9.17. $\mathbf{V}_s = 17.88 \angle -18.45^\circ$ V
D10.1. $\mathbf{V}_0 = 5.36 \angle -63.43^\circ$ V **D10.2.** $\mathbf{V}_0 = 2.12 \angle 75^\circ$ V
D10.3. $\mathbf{V}_1 = 2.77 \angle -58.4^\circ$ V, $\mathbf{V}_2 = -10.81 \angle 12.6^\circ$ V **D10.4.** $\mathbf{V}_0 = 10.8 \angle 36^\circ$ V
D10.5. $\mathbf{V}_0 = 24 \angle 53.1^\circ$ V **D10.6.** $\mathbf{V}_0 = 12 \angle 90^\circ$ V **D10.7.** $\mathbf{V}_0 = 5.66 \angle -45^\circ$ V
D10.8. $\mathbf{V}_0 = 12 \angle 90^\circ$ V **D10.9.** $\mathbf{V}_0 = 12 \angle 90^\circ$ V
D10.10. $\mathbf{V}_0 = 12 \angle 90^\circ$ V **D10.11.** $\mathbf{V}_0 = 5.66 \angle -45^\circ$ V
D10.12. $\mathbf{V}_0 = 24 \angle 53.1^\circ$ V **D10.13.** $\mathbf{V}_0 = 1.402 \angle -127.4^\circ$ V, $\mathbf{I}_0 = 1.402 \angle -37.4^\circ$ A
D10.14. $\mathbf{V}_0 = 0.6389 \angle -63.43^\circ$ V, $\mathbf{I}_0 = 0.3194 \angle -63.43^\circ$ A
D10.15. $\mathbf{V}_0 = 102.8 \angle 0.67^\circ$ V, $\mathbf{I}_0 = 6.856 \angle 0.67^\circ$ A
D11.1. $P_{2\Omega} = 7.18$ W, $P_{4\Omega} = 7.14$ W
D11.2. $P_{3\Omega} = 56.55$ W, $P_{4\Omega} = 33.95$ W, $P_{cs} = -90.50$ W
D11.3. $P_{cs} = -69.3$ W, $P_{vs} = 19.8$ W, $P_{4\Omega} = 49.5$ W
D11.4. $P_{24\angle 0^\circ} = -55.4$ W, $P_{12\angle 0^\circ} = 5.5$ W, $P_{2\Omega} = 22.1$ W, $P_{4\Omega} = 27.8$ W
D11.5. $\mathbf{Z}_L = 1 + j1\ \Omega$, $P_L = 44.94$ W **D11.6.** $\mathbf{Z}_L = 2 + j2\ \Omega$, $P_L = 180$ W
D11.7. $V_{rms} = 1.633$ V **D11.8.** $P = 32$ W **D11.9.** $P = 80$ W
D11.10. $P = 104$ W **D11.11.** Power saved = 3.77 kW **D11.12.** $C = 773\ \mu$F
D11.13. $P_{line} = 4.685$ kW, $Q_{line} = 11.713$ kVAR, $P_{source} = 44.685$ kW, $Q_{source} = 45.312$ kVAR
D11.14. $\mathbf{V}_{line} = 69.41 \angle 24.52^\circ$ V, $PF_{input} = 0.792$ lagging **D11.15.** $C = 773\ \mu$F
D12.1. $\mathbf{I}_1 = 34.5 - j34.5$ A, $\mathbf{I}_2 = 0$, $\mathbf{I}_3 = -(34.5 - j34.5)$ A
D12.2. $\mathbf{V}_{ab} = 208 \angle 120^\circ$ V, $\mathbf{V}_{bc} = 208 \angle 0^\circ$ V, $\mathbf{V}_{ca} = 208 \angle -120^\circ$ V
D12.3. $\mathbf{V}_{an} = 120 \angle -30^\circ$ V, $\mathbf{V}_{bn} = 120 \angle -150^\circ$ V, $\mathbf{V}_{cn} = 120 \angle -270^\circ$ V
D12.4. $\mathbf{V}_{an} = 120 \angle 30^\circ$ V, $\mathbf{V}_{bn} = 120 \angle -90^\circ$ V, $\mathbf{V}_{cn} = 120 \angle -210^\circ$ V
D12.5. $\mathbf{I}_{AB} = 6.93 \angle 70^\circ$ A, $\mathbf{I}_{BC} = 6.93 \angle -50^\circ$ A, $\mathbf{I}_{CA} = 6.93 \angle -170^\circ$ A
D12.6. $\mathbf{I}_a = 24.98 \angle 56.31^\circ$ A, $\mathbf{I}_b = 24.98 \angle -63.69^\circ$ A

$\mathbf{I}_c = 24.98 \angle -183.69^\circ$ A, $\mathbf{V}_{an} = 120 \angle 90^\circ$ V

$\mathbf{V}_{bn} = 120 \angle -30^\circ$ V, $\mathbf{V}_{cn} = 120 \angle -150^\circ$ V

D12.7. $I_\Delta = 6.36$ A, $V_L = 241$ V **D12.8.** $I_Y = 2.06$ A, $I_\Delta = 1.68$ A
D12.9. $\mathbf{Z} = 2.88 \angle 36.87^\circ\ \Omega$
D12.10. $\mathbf{S}_{source} = 1335.65 + j593.55$ VA, $\mathbf{S}_{load} = 1186.27 + j444.94$ VA
D12.11. $I_L = 17.97$ A
D12.12. For the a-phase $\mathbf{I}_{line} = 7.25 \angle 155^\circ$ A, $\mathbf{V}_{line} = 207.8 \angle 30^\circ$ V, $\mathbf{V}_{load} = 120 \angle 0^\circ$ V
D12.13. For the a-phase or AB phase, $I_{source} = 20.58 \angle 120^\circ$ A, $\mathbf{I}_{load} = 11.88 \angle -29.04^\circ$ A, $\mathbf{V}_{load} = 128.5 \angle 27.27^\circ$ V and the voltage across the line impedance is $46.02 \angle 4.4^\circ$ V.
D12.14. The output for the a-phase or AB phase is $\mathbf{I}_{source} = 5.162 \angle -153.4^\circ$ A, $\mathbf{I}_{line} = \mathbf{I}_{load} = 8.94 \angle -3.4^\circ$ A, $\mathbf{V}_{load} = 120 \angle -30^\circ$ V. The $1T\ \Omega$ resistors provide a dc path to ground.
D12.15. $P = 1440.24$ W **D12.16.** $P = 7488$ W **D12.17.** $\mathbf{Z} = 15.58 \angle 28.35^\circ\ \Omega$
D12.18. $PF = 0.23$ lagging, $\mathbf{Z} = 13.43 \angle 76.79^\circ\ \Omega$

D13.1. $$\mathbf{Z}_i(j\omega) = \frac{(R_1 + R_2) + j\omega(CR_1R_2 + L) + (j\omega)^2 LCR_1}{1 + j\omega CR_2 + (j\omega)^2 LC}$$

D13.2. $\mathbf{G}_I(j\omega) = \dfrac{j\omega C(R_1 + j\omega L)}{1 + j\omega C(R_1 + R_2) + (j\omega)^2 LC}$

D13.3.

ω	0	1	2	3	4	5	6	7	8	9	10
$M(\omega)$	0.042	0.046	0.054	0.063	0.069	0.073	0.074	0.074	0.073	0.071	0.68
$\Phi(\omega)$	0.0°	10°	12.6°	9.2°	3.2°	−3.6°	−10.3°	−16.5°	−22.2°	−27.2°	−31.7°

D13.4.

20 log₁₀ 20
+20 dB/dec
−20 dB/dec
2 10 100
ω (r/s)

D13.5.

0 dB
−20 dB/dec
−40 dB/dec
10 20
ω (r/s)

D13.6.

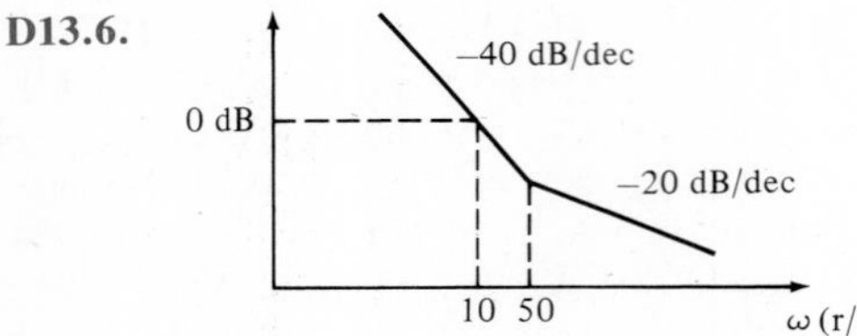

D13.7.

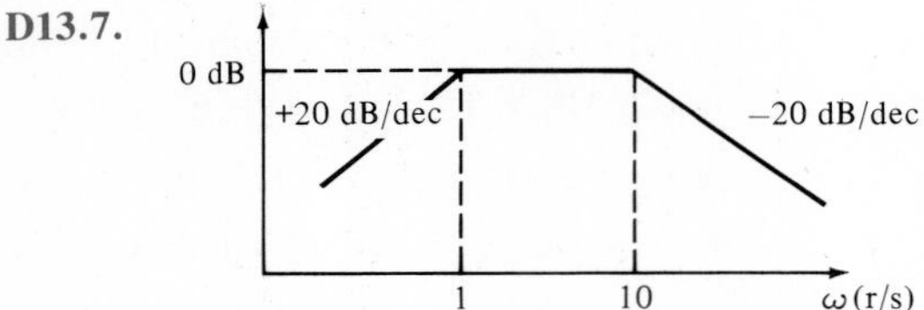

D13.8.

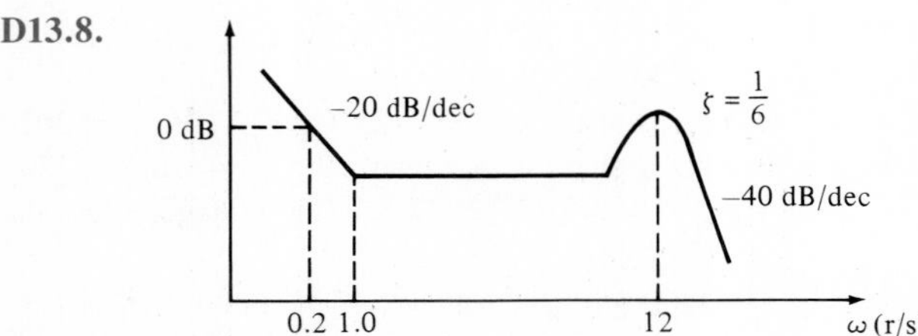

D13.9. $\mathbf{G}(j\omega) = \dfrac{10[(j\omega/5) + 1]}{(j\omega + 1)[(j\omega/50) + 1]^2}$

D13.10. $\mathbf{G}(j\omega) = \dfrac{5[(j\omega/5) + 1][(j\omega/50) + 1]}{j\omega[(j\omega/20) + 1][(j\omega/100) + 1]}$

D13.11.

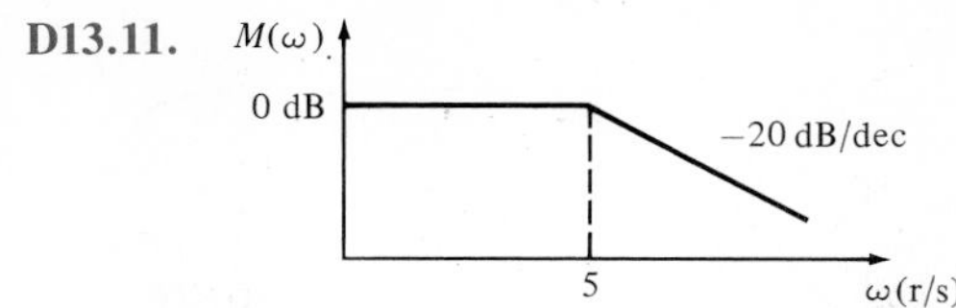

D13.12.

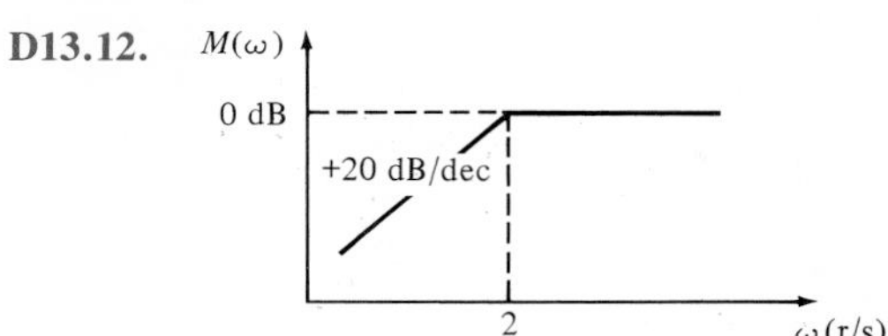

D13.13.

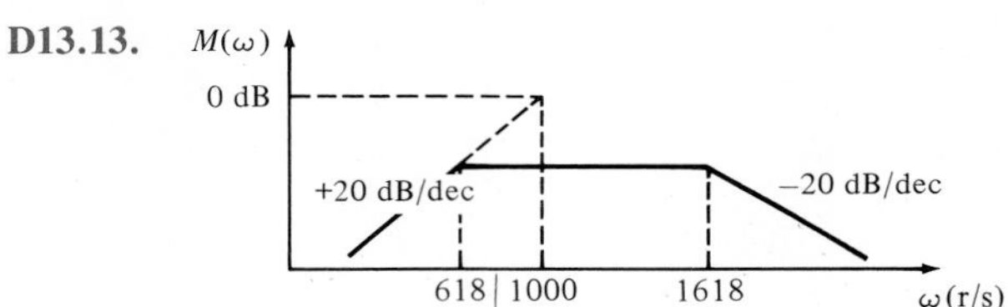

D13.14.

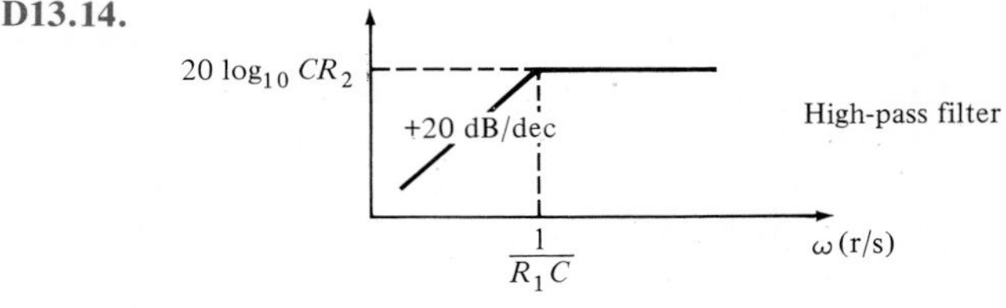

D13.15. $C = 3.09\ \mu\text{F}$ **D13.16.** $Q = 60$, $|\mathbf{V}_C| = 600$ V
D13.17. $\omega_{HI} = 1815$ rad/s, $\omega_{LO} = 1785$ rad/s, BW = 30 rad/s
D13.18. BW = 50 rad/s **D13.19.** $L = 40$ mH, $C = 1.56\ \mu\text{F}$
D13.20. $\omega_0 = 577$ rad/s, BW = 3.33 rad/s, $Q = 173$
D13.21. $C = 0.167\ \mu\text{F}$, $L = 417.5\ \mu\text{H}$, $\omega_0 = 119{,}760$ rad/s
D13.22. $R' = 1\ \text{k}\Omega$, $L' = 100$ H, $C' = \frac{1}{50}$ F, $R'' = 1\ \text{K}\ \Omega$, $L'' = 10$ mH, $C'' = 2\ \mu\text{F}$
D13.23.

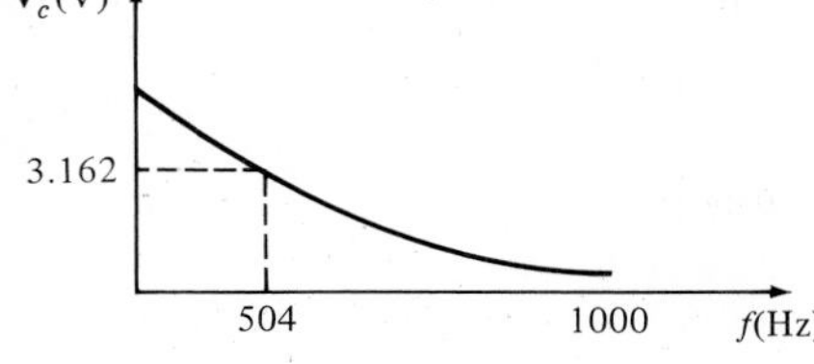

D13.24.

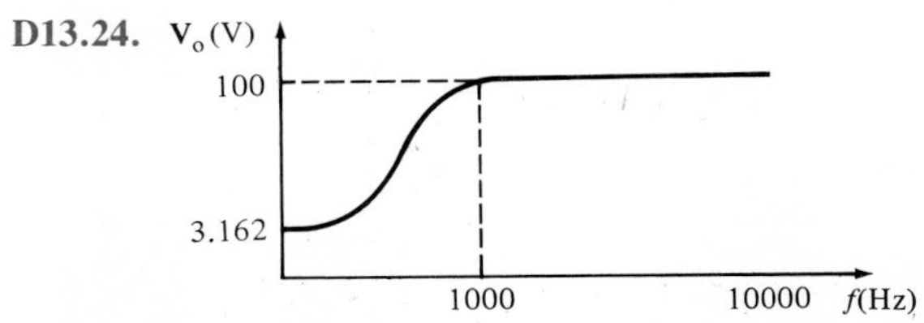

D13.25.

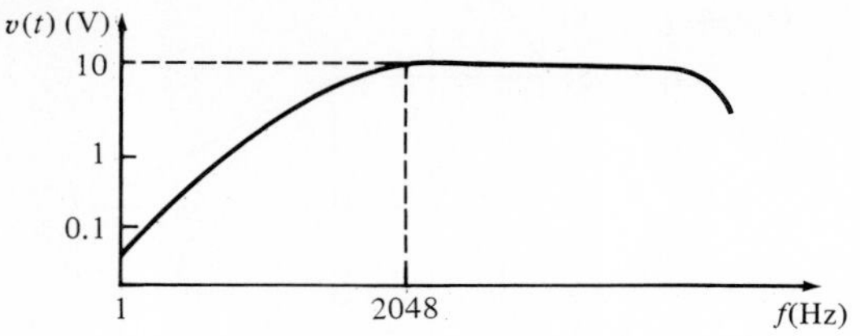

D14.1. $v_1(t) = -L_1 \dfrac{di_1}{dt} - M \dfrac{di_2}{dt}$

$v_2(t) = L_2 \dfrac{di_2}{dt} + M \dfrac{di_1}{dt}$

D14.2. $\mathbf{I}_1 = -4.29 \underline{/-42.8^\circ}$ A, $\mathbf{I}_2 = 0.96 \underline{/-16.26^\circ}$ A, $\mathbf{V}_0 = 3.84 \underline{/-106.26^\circ}$ V

D14.3. $(R_1 + j\omega L_1 + R_2)\mathbf{I}_1 - (R_2 + j\omega M)\mathbf{I}_2 = -\mathbf{V}_1$

$-(R_2 + j\omega M)\mathbf{I}_1 + (R_2 + j\omega L_2 + R_3)\mathbf{I}_2 = \mathbf{V}_1$

D14.4. $(R_1 + j\omega L_1)\mathbf{I}_1 - (j\omega L_1 + j\omega M)\mathbf{I}_2 = \mathbf{V}_1$

$-(j\omega L_1 + j\omega M)\mathbf{I}_1 + \left(j\omega L_1 + \dfrac{1}{j\omega C} + j\omega L_2 + 2j\omega M\right)\mathbf{I}_2 = 0$

D14.5. $L_2 = 3.6$ mH **D14.6.** $W(10 \text{ ms}) = 39$ mJ

D14.7. $\mathbf{Z}_i = 3 + j3\ \Omega$, $\mathbf{I}_{source} = 2 - j2$ A **D14.8.** $\mathbf{I}_1 = 3.07 \underline{/39.81^\circ}$ A

D14.9. $\mathbf{V}_0 = 3.07 \underline{/39.81^\circ}$ V **D14.10.** $\mathbf{V}_s = 25.5 \underline{/-11.31^\circ}$ V

D14.11. $\mathbf{I}_1 = 13.12 \underline{/38.66^\circ}$ A **D14.12.** $\mathbf{V}_0 = 3.12 \underline{/38.66^\circ}$ V

D14.13. $\mathbf{V}_0 = 4.66 \underline{/29.05^\circ}$ V **D14.14.** $\mathbf{V}_0 = 2.092 \underline{/-49.39^\circ}$ V

D14.15. $\mathbf{V}_0 = 0.8944 \underline{/-63.44^\circ}$ V **D14.16.** $\mathbf{V}_0 = 20.19 \underline{/103.40^\circ}$ V

D15.1. $\mathbf{y}_{11} = \frac{1}{14}$ S, $\mathbf{y}_{12} = \mathbf{y}_{21} = -\frac{1}{21}$ S, $\mathbf{y}_{22} = \frac{1}{7}$ S

D15.2. $\mathbf{y}_{11} = \frac{1}{14}$ S, $\mathbf{y}_{12} = \mathbf{y}_{21} = -\frac{1}{21}$ S, $\mathbf{y}_{22} = \frac{1}{7}$ S **D15.3.** $\mathbf{I}_2 = -4.29$ A

D15.4. $\mathbf{z}_{11} = 18\ \Omega$, $\mathbf{z}_{12} = \mathbf{z}_{21} = 6\ \Omega$, $\mathbf{z}_{22} = 9\ \Omega$ and $\mathbf{I}_2 = -0.73 \underline{/0^\circ}$ A

D15.5. $\mathbf{z}_{11} = 18\ \Omega$, $\mathbf{z}_{12} = \mathbf{z}_{21} = 6\ \Omega$, $\mathbf{z}_{22} = 9\ \Omega$

D15.6. $\mathbf{h}_{11} = 14\ \Omega$, $\mathbf{h}_{12} = \frac{2}{3}$, $\mathbf{h}_{21} = -\frac{2}{3}$, $\mathbf{h}_{22} = \frac{1}{9}$ S

D15.7. $\mathbf{Z}_i = 15.23\ \Omega$ **D15.8.** $\mathbf{h}_{11} = 14\ \Omega$, $\mathbf{h}_{12} = \frac{2}{3}$, $\mathbf{h}_{21} = -\frac{2}{3}$, $\mathbf{h}_{22} = \frac{1}{9}$ S

D15.9. $\dfrac{\mathbf{V}_2}{\mathbf{V}_s} = \dfrac{-\mathbf{h}_{21}}{(R_1 + \mathbf{h}_{11})[\mathbf{h}_{22} + (1/R_L)] - \mathbf{h}_{12}\mathbf{h}_{21}}$

D15.10. $\mathbf{A} = 3$, $\mathbf{B} = 21\ \Omega$, $\mathbf{C} = \frac{1}{6}$ S, $\mathbf{D} = \frac{3}{2}$

D15.11. $\mathbf{A} = 3$, $\mathbf{B} = 21\ \Omega$, $\mathbf{C} = \frac{1}{6}$ S, $\mathbf{D} = \frac{3}{2}$

D15.12. $\mathbf{y}_{11} = \frac{1}{14}$ S, $\mathbf{y}_{12} = \mathbf{y}_{21} = -\frac{1}{21}$ S, $\mathbf{y}_{22} = \frac{1}{7}$ S

D15.13. $\mathbf{h}_{11} = 14\ \Omega$, $\mathbf{h}_{12} = \frac{2}{3}$, $\mathbf{h}_{21} = -\frac{2}{3}$, $\mathbf{h}_{22} = \frac{1}{9}$ S

D15.14. $V_0 = 16$ V **D15.15.** $I_0 = \frac{8}{3}$ A **D15.16.** $I_0 = \frac{2}{3}$ A

D15.17. $\mathbf{Y}_T = \mathbf{Y}_a + \mathbf{Y}_b = \begin{bmatrix} \frac{1}{14} & -\frac{1}{21} \\ -\frac{1}{21} & \frac{1}{7} \end{bmatrix}$

D15.18. $\mathbf{Z}_T = \mathbf{Z}_a + \mathbf{Z}_b = \begin{bmatrix} 18 & 6 \\ 6 & 9 \end{bmatrix}$

D15.19. $\mathbf{ABCD}_T = \begin{bmatrix} 3 & 21 \\ \frac{1}{6} & \frac{3}{2} \end{bmatrix}$

D15.20. Their two port parameters are identical. **D15.21** $\mathbf{V}_o = 3 \underline{/0^\circ}$ V

D16.1. $\mathbf{c}_0 = \frac{1}{2},\ \mathbf{c}_n = \left[\dfrac{1 - e^{-jn\pi}}{j2\pi n}\right]$

D16.2. $\mathbf{c}_0 = 2,\ \mathbf{c}_n = \dfrac{2}{n\pi}\left[2 \sin \dfrac{2\pi n}{3} - \sin \dfrac{n\pi}{3}\right]$

D16.3. Figure 16.2 exhibits even symmetry; Fig. 16.3a and b exhibit half-wave and odd symmetry, respectively.

D16.4. $v(t) = 2 + \sum_{n=1}^{\infty} \dfrac{4}{n\pi}\left[2 \sin \dfrac{2n\pi}{3} - \sin \dfrac{n\pi}{3}\right] \cos \dfrac{n\pi}{3} t$ V

D16.5. $v(t) = \sum_{\substack{n=1 \\ n \text{ odd}}}^{\infty} \dfrac{2}{n\pi} \sin \dfrac{n\pi}{2} \cos \dfrac{n\pi}{2} t + \dfrac{2}{n\pi} [2 - \cos n\pi] \sin \dfrac{n\pi}{2} t$ V

D16.6. $v(t) = \sum_{n=1}^{\infty} \dfrac{2}{n\pi}\left[1 + \cos \dfrac{n\pi}{2} - 2 \cos n\pi\right] \sin \dfrac{n\pi}{2} t$ V

D16.7. For the waveform in Fig. D16.7, $\mathbf{c}_n = -\left(\dfrac{1 - e^{-j\pi n}}{j2\pi n}\right)$, which differs from that in Fig. D16.1 by $n(180^\circ)$

D16.8. $v_1(t) = \frac{2}{3} + \sum_{\substack{n=-\infty \\ n \neq 0}}^{\infty} \dfrac{2}{n\pi} \sin \dfrac{n\pi}{3} e^{jn\omega_0 t}$ V

$v_2(t) = \frac{4}{3} + \sum_{\substack{n=-\infty \\ n \neq 0}}^{\infty} \left(\dfrac{-4}{n\pi}\right)\left(\sin \dfrac{n\pi}{3} - \sin \dfrac{2n\pi}{3}\right) e^{jn\omega_0 t}$ V

Adding these two functions yields the one for D16.2.

D16.9. $v(t) = \dfrac{1}{2} + \sum_{n=1}^{\infty} \dfrac{1}{n\pi} \sin 2\pi nt$ V

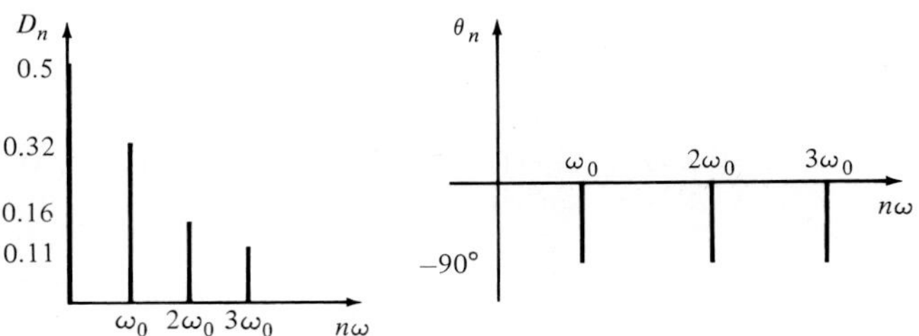

D16.10. $i(t) = 2.12 - 2.63 \cos(2t + 29.74^\circ) - 0.70 \cos(4t - 22.83^\circ)$ A

D16.11. $\mathbf{F}(\omega) = \pi j[\delta(\omega + \omega_0) - \delta(\omega - \omega_0)]$ **D16.12.** $\mathbf{F}(\omega) = \dfrac{1}{a + j\omega}$

D16.13. $\mathbf{F}(\omega) = \dfrac{1}{(a + j\omega)^2}$ **D16.14.** $v_0(t) = (e^{-t} - e^{-2t})\, u(t)$

D16.15. $W_T = \frac{1}{4}$ J **D16.16.** $W(0 \leftrightarrow 1) = 0.07$ J

D17.1. $\mathbf{F}(s) = e^{-2(s+a)} \cos 2\omega$ **D17.2.** $\mathbf{F}(s) = \dfrac{1}{s} e^{-s} + \dfrac{1}{s^2} e^{-s}$

D17.3. Use the definition of $\mathbf{F}(s)$ to show that $\mathbf{F}(s) = \omega/s^2 + \omega^2$.

D17.4. $\mathbf{F}(s) = \dfrac{1}{2s^2} - \dfrac{2}{s + 2}$ **D17.5.** $\mathbf{F}(s) = \dfrac{e^{-s}}{(s + a)^2}$

D17.6. $\mathbf{F}(s) = \dfrac{1}{(s+4)^2} - \dfrac{1}{s+5}$ **D17.7.** $\mathbf{F}(s) = \dfrac{e^{-4x}}{s^2(a^2+4)}$

D17.8. $\mathbf{F}(s) = e^{-s}\left(\dfrac{s\cos\omega}{s^2+\omega^2} - \dfrac{\omega\sin\omega}{s^2+\omega^2}\right)$ **D17.9.** $\mathbf{F}(s) = \dfrac{1}{(s+a)^2}$

D17.10. $\mathbf{F}(s) = \dfrac{1}{s} + \dfrac{e^{-s}}{s} - \dfrac{e^{-2s}}{s} - \dfrac{e^{-3s}}{s}$ **D17.11.** $\mathbf{F}(s) = \dfrac{1}{s^2} - \dfrac{e^{-s}}{s^2} + \dfrac{e^{-2s}}{s} - \dfrac{2e^{-3s}}{s}$

D17.12. $\mathbf{F}(s) = \dfrac{1}{1-e^{-4s}}\left(\dfrac{1}{s} + \dfrac{e^{-s}}{s} - \dfrac{e^{-2s}}{s} - \dfrac{e^{-3s}}{s}\right)$

D17.13. $\mathbf{F}(s) = \dfrac{1}{1-e^{-3s}}\left(\dfrac{1}{s^2} - \dfrac{e^{-s}}{s^2} + \dfrac{e^{-2s}}{s} - \dfrac{2e^{-3s}}{s}\right)$

D17.14. $f(t) = [25e^{-t} - 15e^{-3t}]u(t)$ **D17.15.** $f(t) = [24 - 12e^{-t}]u(t)$

D17.16. $f(t) = 1.41e^{-2t}\cos(2t + 45°)u(t)$ **D17.17.** $f(t) = 1.41e^{-3t}\cos(2t + 45°)u(t)$

D17.18. $f(t) = [e^{-t} - te^{-t}]u(t)$ **D17.19.** $f(t) = [-1 + 2t + e^{-t}]u(t)$

D17.20. $f(t) = (\frac{1}{2}e^{-t} + \frac{1}{2}e^{-3t} - e^{-2t})u(t)$ **D17.21.** $f(t) = [t - 1 + e^{-t}]u(t)$

D17.22. $f(t) = \frac{1}{2}(e^{-(t-2)} - e^{-3(t-2)})u(t-2)$ **D17.23.** $f(0) = 0,\ f(\infty) = \frac{1}{4}$

D17.24 $y(t) = (\frac{4}{30} - \frac{4}{5}e^{-5t} + \frac{2}{3}e^{-6t})u(t)$ **D17.25.** $y(t) = (2 - 2e^{-t} - te^{-t})u(t)$

D18.1. $i_0(t) = 6.53e^{-t/4}\cos\left(\dfrac{\sqrt{15}}{4}t - 156.72°\right)u(t)$ A

D18.2. Same as D18.1 **D18.3.** Same as D18.1 **D18.4.** Same as D18.1

D18.5. $v_0(t) = (4 - 8.93e^{-3.73t} + 4.93e^{-0.27t})u(t)$ V

D18.6. Same as D18.5 **D18.7.** Same as D18.5 **D18.8.** Same as D18.5

D18.9. $v_{0ss}(t) = 3.95\cos(2t - 99.46°)$ V **D18.10.** $v_{0ss}(t) = 6.86\cos(4t + 54.66°)$ V

D18.11. $v_0(t) = \frac{8}{3}e^{-t/2}u(t)$ V **D18.12.** $v_0(t) = -6e^{-4t}u(t)$ V

D18.13. $v_0(t) = (\frac{24}{5} + \frac{1}{5}e^{-5t/8})u(t)$ V **D18.14.** $v_0(t) = (6 - \frac{10}{3}e^{-2t})u(t)$ V

D18.15. $v(0) = 5$ V, $v(\infty) = \frac{24}{5}$ V **D18.16.** $v(0) = \frac{8}{3}$, $v(\infty) = 6$ V

D18.17. $v_0(t) = \frac{2}{3}[(1 - e^{-3t})u(t) - (1 - e^{-3(t-1)})u(t-1)]$ V

D18.18. $v_0(t) = 2(\frac{3}{4} - \frac{3}{4}e^{-2t} - \frac{3}{2}te^{-2t})u(t)$

$$-3(\tfrac{3}{4} - \tfrac{3}{4}e^{-2(t-1)} - \tfrac{3}{2}(t-1)e^{-2(t-1)})u(t-1)$$

$$+(\tfrac{3}{4} - \tfrac{3}{4}e^{-2(t-2)} - \tfrac{3}{2}(t-2)e^{-2(t-2)})u(t-2)\ \text{V}$$

D18.19. $x(t) = (1 - \frac{10}{9}e^{-t} + \frac{1}{9}e^{-10t})u(t)$

D18.20. $x(t) = [\frac{10}{8} + 1.46e^{-2t}\cos(2t - 210.96°)]u(t)$

D18.21. $v_0(t) = (\frac{10}{9}e^{-t} - \frac{90}{72}e^{-2t} + \frac{10}{72}e^{-10t})u(t)$ V

Answers to Selected Problems

Chapter 1

1.1. 0.801 pC **1.2.** 125 C **1.4.** **(a)** 30 W, **(b)** -16 W

1.8. **(a)** $I = 5$ A into bottom terminal; **(b)** $I = 5$ A into bottom terminal

1.12. $P_{10} = 60$ W supplied, $P_1 = 24$ W absorbed, $P_2 = 12$ W absorbed, $P_3 = 24$ W absorbed

1.14. $P_{24V} = 96$ W supplied, $P_1 = 32$ W absorbed, $P_{dep} = 64$ W absorbed

Chapter 2

2.1. $n = 5$, $B = 10$ **2.2.** The currents are $\overset{6A}{\rightarrow} \downarrow 3A \overset{3A}{\rightarrow}$, $P = 9$ W

2.6. $P_{4\Omega} = 16$ W **2.8.** $I_1 = -6$ A, $I_2 = -7$ A

2.11. $V_{R_1} = 3$ V, $V_{R_3} = 9$ V, $V_B = -12$ V

2.13.

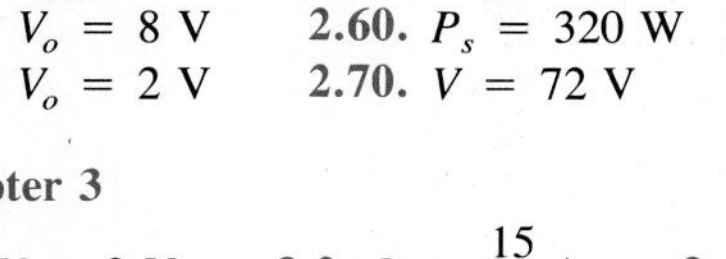

3 A

1.5 A

2.17. $V_1 = 5$ V, $V_2 = 6$ V, $V_3 = 6$ V **2.21.** $V = 4$ V, $I_o = \frac{4}{3}$ A

2.23. **(a)** 8 Ω, **(b)** 6 Ω

2.27. 12 Ω **2.34.** $I = 2$ A **2.37.** $V = 44$ V **2.41.** $R_o = 6$ Ω

2.44. $I_{total} = 4$ A **2.49.** $I_o = 2$ A **2.50.** $V = 64$ V **2.54.** $V_o = 16$ V

2.57. $V_o = 8$ V **2.60.** $P_s = 320$ W **2.63.** $V_A = -14$ V

2.66. $V_o = 2$ V **2.70.** $V = 72$ V **2.72.** $V_o = 200$ mV

Chapter 3

3.1. $V = 2$ V **3.2.** $I_o = \frac{15}{4}$ A **3.4.** $V_2 = \frac{208}{19}$ V **3.8.** $I_o = \frac{6}{7}$ A

3.11. $I_o = -\frac{13}{2}$ A **3.15.** $V_o = \frac{8}{5}$ V **3.19.** $V_1 = 12$ V, $V_2 = 8$ V

3.23. $V_1 = 73.1$ V, $V_2 = 16.2$ V, $I_o = 5.4$ A

3.26. $\frac{V_1 - 12}{1} + \frac{V_1}{2} + \frac{V_1}{2} - \frac{V_2}{2} = 0, \frac{-V_1}{2} + \frac{V_2}{2} + \frac{V_2}{1} = 2I_o, I_o = \frac{V_1}{2}$

$V_o = V_2 = 8$ V

3.29. $V_1\left(\frac{1}{R_1} + \frac{1}{R_2}\right) - V_2\left(\frac{1}{R_2}\right) = I_A + I_B, -V_1\left(\frac{1}{R_2}\right) + V_2\left(\frac{1}{R_2} + \frac{1}{R_3} + \frac{1}{R_4}\right) = -I_B$

3.31. $I_o = 0$ A **3.33.** $V_o = 5$ V **3.39.** $P_{1\Omega} = 1$ W

3.42 $I_1 = -I_A, -I_1(R_1) + I_2(R_1 + R_2 + R_3) - I_3(R_3) = V_1, -I_1(0) - I_2(R_3)$
$+ I_3(R_3 + R_4) = -V_2$

3.46. $I_o = 3$ A **3.50.** $V_{3\Omega} = 3$ V **3.53.** $V_o = 5$ V **3.58.** $V_o = \frac{24}{5}$ V

3.62. $v_o = \left(1 + \frac{R_4}{R_3}\right)v_2 - \frac{R_4}{R_3}\left(1 + \frac{R_2}{R_1}\right)v_1$

Chapter 4

4.1. **(a)** Planar; **(b)** Nonplanar

4.2. **(a)** 5 tree branches and 5 links; **(b)** 5 tree branches and 4 links

4.5. **(a)** 4 tree branches and 3 links; **(b)** 4 tree branches and 3 links

4.7. **(a)**
$$i_1 - i_2 = 0$$
$$i_4 + i_3 - i_2 = 0$$
$$i_4 + i_5 - i_7 = 0$$
$$i_6 - i_7 = 0$$
(b)
$$i_1 + i_2 = 0$$
$$i_2 - i_3 - i_7 = 0$$
$$i_2 + i_4 + i_6 = 0$$
$$i_7 - i_5 + i_6 = 0$$

4.12. **(a)**
$$-i_1 + i_2 + i_9 = 0$$
$$-i_9 - i_8 - i_7 = 0$$
$$-i_9 - i_8 - i_5 - i_6 = 0$$
$$-i_1 + i_4 + i_5 + i_8 + i_9 = 0$$
$$i_1 + i_3 - i_8 - i_9 = 0$$
(b)
$$i_3 + i_4 - i_7 = 0$$
$$-i_1 + i_2 - i_3 = 0$$
$$-i_1 + i_5 - i_7 = 0$$
$$i_7 - i_6 - i_8 = 0$$

4.13. **(a)**
$$V_2 = -V_1 - V_4 + V_7 - V_8$$
$$V_3 = -V_1 - V_4$$
$$V_5 = -V_4 + V_7$$
$$V_6 = V_7 - V_8$$
(b)
$$V_2 = V_1 + V_4$$
$$V_3 = V_1 + V_4 - V_7$$
$$V_5 = V_6 - V_4$$
$$V_8 = V_6 - V_7 + V_9$$

4.17. $I_1 = -6$ A, $I_2 = 4$ A, $I_3 = 2$ A, $I_4 = 2$ A, and $I_5 = -2$ A

4.18.
$$I_B + \frac{V_1}{R_1} + \frac{V_1 - V_2}{R_2} = 0$$
$$I_B + \frac{V_1}{R_1} + \frac{V_2}{R_3} + I_A = 0$$
$$-I_A + \frac{V_3}{R_4} + \frac{V_3}{R_5} = 0$$

4.20.
$$I_1R_2 + V_A + (I_1 - I_2 - I_3)R_3 = 0$$
$$(I_2 + I_3 - I_1)R_3 + I_2R_4 + V_B = 0$$
$$(I_2 + I_3 - I_1)R_3 - V_A + I_3R_1 + V_B = 0$$

4.25. $I_o = \frac{10}{3}$ A **4.30.** $I_o = 3$ A

Chapter 5

5.1. $I_o = 3$ A **5.3.** $V_o = 10$ V **5.7.** $V_o = \frac{80}{3}$ V **5.14.** $I_o = -1$ A
5.15. $I_o = -\frac{1}{2}$ A **5.17.** $I_o = 3$ A **5.21.** $I_o = 1.5$ A **5.24.** $V_o = 10$ V
5.25. $V_o = 10$ V **5.30.** $V_{RL} = 8$ V **5.31.** $I_o = 6$ A **5.35.** $V_o = 4$ V
5.38. $I_o = \frac{4}{3}$ A **5.41.** $I = \frac{4}{3}$ A **5.45.** $I = 2$ A **5.49.** $V_o = 3.43$ V
5.50. $R_L = 6\ \Omega$ **5.54.** $R_L = 4\ \Omega$

5.56.

Node	Voltage	Source	Current
1	-12V	V1	-2A
2	34.6V	V2	2A
3	29V	VD1	2A
4	39V		
5	35V		
6	10V		
7	27.4V		
8	35V		

5.58.

Node	Voltage	Source	Current
1	-5.5263V	V1	0.2368A
2	15.4737V	V2	2.434A
3	9.4737V	VD1	2.566A
4	1.1842V		
5	-24.4737V		
6	-14.7368V		
7	-9.7368V		
8	-24.4737V		

5.60.

Node	Voltage	Source	Current
1	25.5882V		
2	18.5294V	VD1	0.8824A
3	15V		
4	13.2353V		
5	4.4118V		
6	4.4118V		

Chapter 6

6.1. C parallel $= 9\ \mu$F, C series $= 2\ \mu$F **6.3.** $C_{eq} = 2\ \mu$F
6.7. $V_{12} = \frac{15}{2}$ V, $V_4 = \frac{45}{2}$ V **6.10.** $C_T = 0.8\ \mu$F **6.16.** $C_{total} = \frac{4}{5}\ \mu$F

6.21.

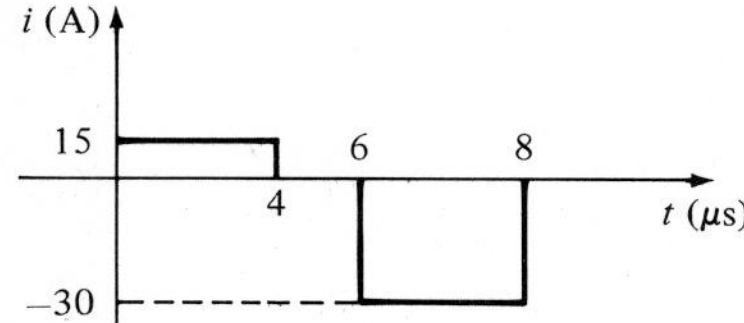

6.26.

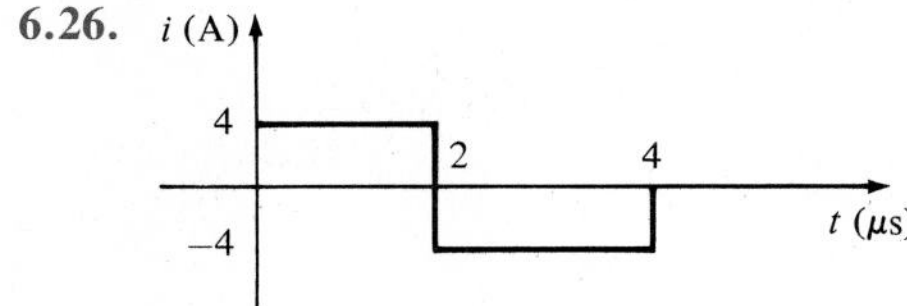

6.30. $W = 9$ J **6.31.** $L_{eq} = 9$ H **6.35.** $L_{eq} = 6$ mH **6.40.** $L_{eq} = 2$ H

6.43.

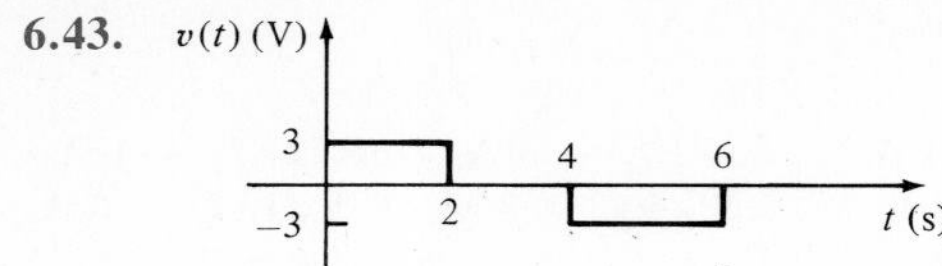

6.47.

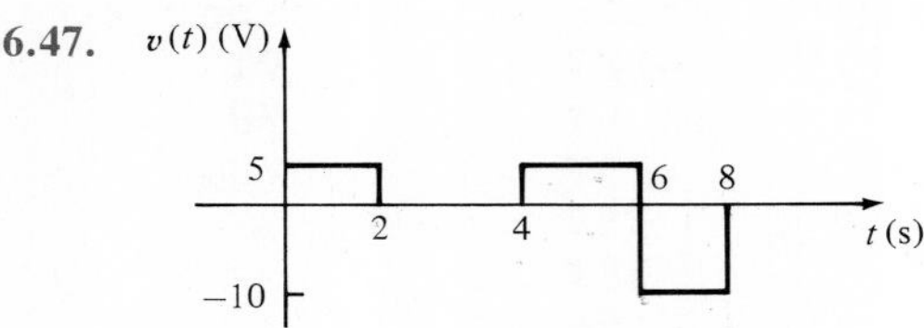

6.49.

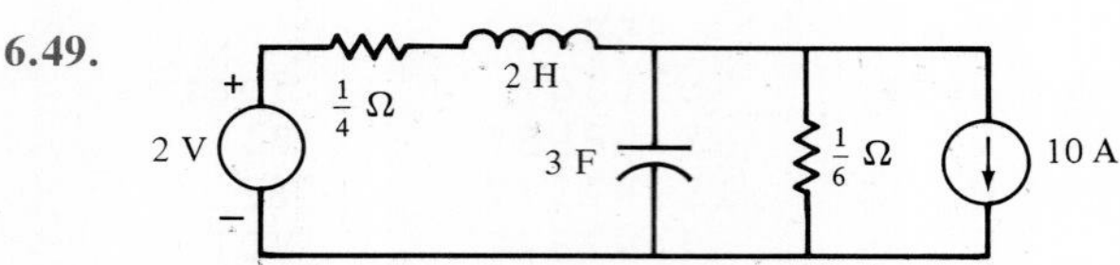

Chapter 7

7.1. $v_c\,(t) = 8e^{-t/0.6}$ V **7.3.** $i_o(t) = \frac{1}{2}\,e^{-t/0.8}$ mA
7.8. $i_o\,(t) = \frac{8}{3} + (\frac{23}{6} - \frac{8}{3})\,e^{-t/0.8}$ mA **7.11.** $v_o(t) = 2 + 12e^{-t/0.16}$ V
7.14. $v_o(t) = -\frac{48}{11}\,e^{-t/0.367}$ V **7.17.** $i_o(t) = 2 - e^{-t/12}$ A **7.20.** $i_1(t) = 1e^{-9t}$ A
7.23. $i_o(t) = -\frac{4}{11}\,e^{-12t/11}$ A

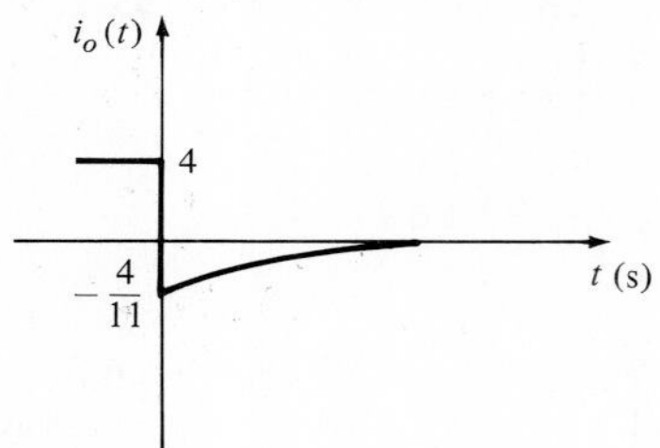

7.26. $i_o(t) = \frac{7}{6}e^{-t/3}$ A **7.32.** $i_o(t) = 2e^{-2t/3}$ A
7.35 **(a)** $i_1(t) = \frac{8}{3} + (3 - \frac{8}{3})e^{-t/0.2}$ mA; **(b)** $i_2(t) = 2e^{-t/\frac{2}{3}\times 10^{-3}}$ mA
7.38 **(a)** $2u(t-1) + u(t-3) - (t-4)u(t-4) + (t-6)u(t-6) - u(t-6)$;
(b) $2u(t) - tu(t) + 2u(t-3) - (t-4)u(t-4) + (t-6)u(t-6)$
7.40. $V_o(t) = \frac{48}{5}$ V **7.43.** $v_o(t) = \frac{96}{51} + \frac{40}{51}\,e^{-17t/6}$ V **7.48.** $i_o(t) = 2e^{-t/2}$ A
7.49. $v_o(t) = \frac{4}{5} - \frac{48}{35}\,e^{-5t/7}$ V **7.55.** $i(t) = \frac{4}{3} + (\frac{28}{9} - \frac{4}{3})e^{-2t}$ A
7.58.

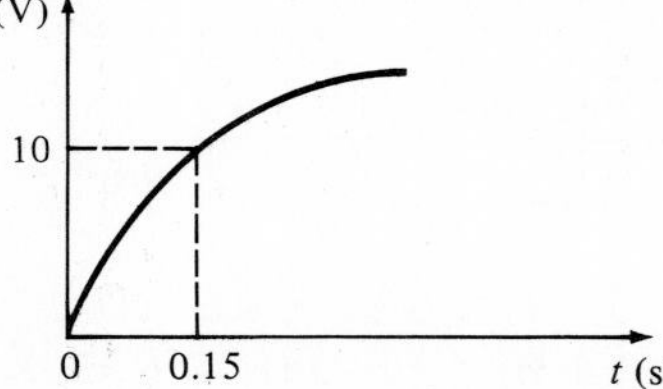

7.61.

7.64.

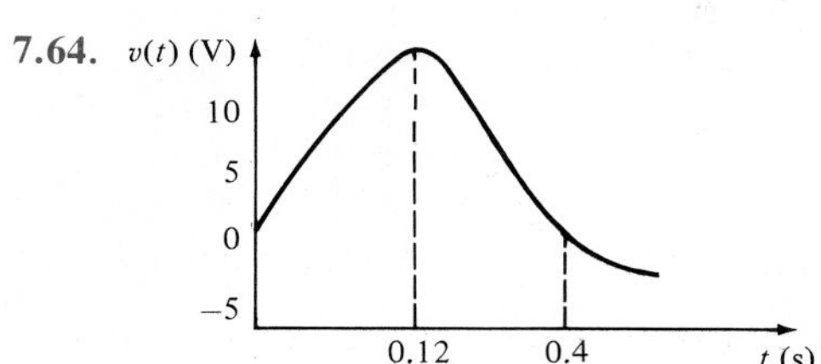

Chapter 8

8.1. Underdamped **8.2.** Overdamped **8.4.** $L = \frac{1}{8}$ H **8.6.** $t_s = 2.38$ s

8.10. $v(t) = 5e^{-5t} - 225te^{-5t}$ V **8.13.** $i(t) = -4e^{-t} \sin 2t$ A

8.16. $v(t) = 8 - 10e^{-t} + 2e^{-5t}$ V **8.21.** $v_o(t) = 8e^{-2t} - 4e^{-4t}$ V

8.24. **(a)** $s^2 + \dfrac{R_1R_2}{L(R_1 + R_2)} s + \dfrac{1}{LC} = 0$; **(b)** $v_o(t) = 8 - 28e^{-t} + 20e^{-3t}$ V

8.28.

8.32.

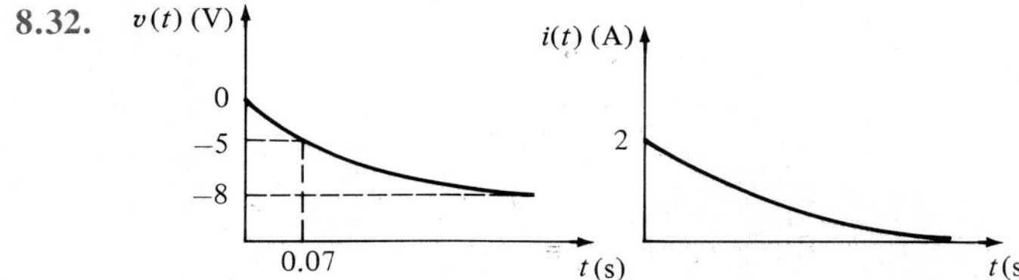

8.35.

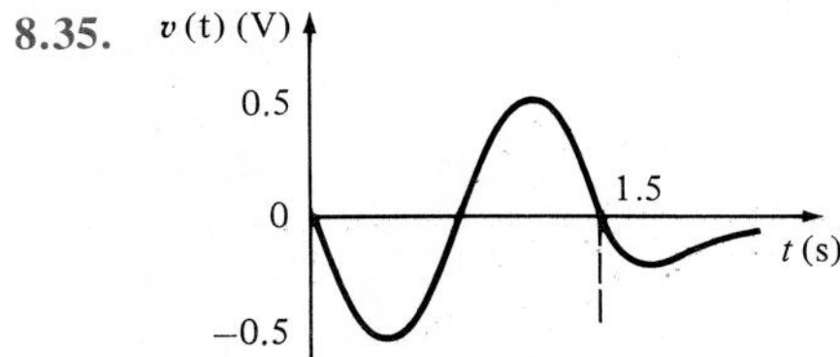

Chapter 9

9.1. 50 Hz, 45°

9.3.

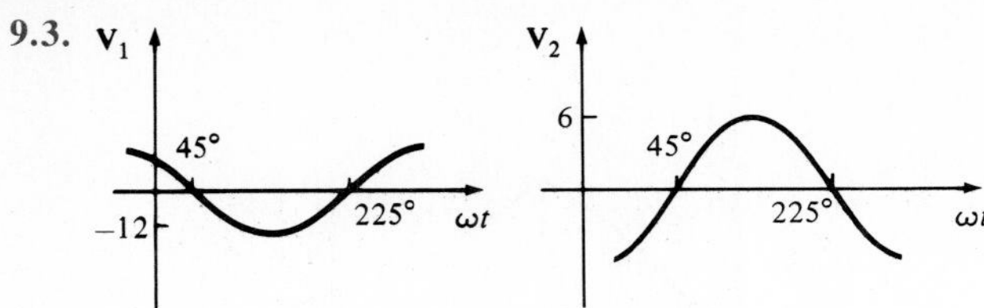

9.6. $14.42 \sin(\omega t - 33.69°)$
9.7. **(a)** $\mathbf{V}_1 = 12 \angle 60°$ V; **(b)** $\mathbf{V}_2 = 6 \angle -110°$ V
9.8. **(a)** $v_1(t) = 24 \cos(377t - 45°)$ V; **(b)** $v_2(t) = 10 \cos(377t + 120°)$ V
9.11. **(a)** $5 \cos 377t$; **(b)** $6 \cos(40t + 20°)$; **(c)** $2 \cos(377t - 180°)$; **(d)** $4 \cos(20t + 210°)$
9.14. **(a)** $\mathbf{I}_1 = 5 \angle 180°$ A; **(b)** $\mathbf{I}_2 = 6 \angle -45°$ A
9.21. **(a)** $\mathbf{Z} = 73.13 \angle 75.9°\ \Omega$; **(b)** $\mathbf{Z} = 18.75 \angle -62.1°\ \Omega$; **(c)** $\mathbf{Z} = 200 \angle 90°\ \Omega$
9.22. $\mathbf{Z}_{eq} = 4.75 - j0.61\ \Omega$ **9.26.** $\mathbf{Y}_{eq} = 0.18 \angle 17.01°$ S **9.29.** $\mathbf{V}_L = 78 \angle 119.9°$ V
9.34. $L = 0.276$H **9.38.** $\mathbf{I}_3 = 2.3 \angle 61.7°$ A **9.43.** $\mathbf{I}_s = 3 + j3$ A
9.47. $\mathbf{I}_0 = 6.51 \angle -49.4°$ A **9.52.** $\mathbf{V}_s = -8 + j3$ V **9.57.** $\mathbf{V}_s = 15.23 \angle 66.8°$ V
9.59. $\mathbf{V}_s = 12.39 \angle -2.75°$ V

Chapter 10

10.1. $I_0 = 4.72 \angle 18.31°$ A **10.3.** $\mathbf{V}_2 = 19.37 \angle -29.74°$ V **10.7.** $\mathbf{V}_x = 21.7 \angle 5.2°$ V
10.11. $\mathbf{I}_o = 3.93 \angle -30.47°$ A **10.14.** $\dfrac{\mathbf{V}_o}{\mathbf{V}_S} = \dfrac{-g_m R_1 R_2/(R_1 + R_s)}{R_2 + R_L + 1/j\omega C}$
10.17. $\mathbf{V}_x = 21.7 \angle 5.2°$ V **10.21.** $\mathbf{V}_o = 2.46 \angle -47.48°$ V
10.25. $\mathbf{I}_L = 2.95 \angle -38.03°$ A **10.29.** $\mathbf{I}_L = 2.95 \angle -38.03°$ A
10.33. $\mathbf{I}_o = 3.96 \angle -30.32°$ A **10.36.** $\mathbf{I}_L = 2.95 \angle -38.03°$ A
10.38. $\mathbf{V}_x = 21.7 \angle 5.2°$ V **10.41.** $\mathbf{V}_o = 6.30 \angle -23.2°$ V
10.47. Node voltage $= 41.49 \angle -42.52°$ V; current $= 13.83 \angle -42.52°$ A
10.49. Voltage $= 10 \angle 30°$ V; current $= 2.704 \angle 104.2°$ A
10.50. Voltage $= 3.199 \angle -46.70°$ V; current $= 1.599 \angle -136.7°$ A

Chapter 11

11.1. $P_{ave} = 0$ W **11.3.** $P = 1.58$ W **11.9.** $P_4 = 10.4$ W
11.12. $P_{supplied} = P_{absorbed} = 33.62$ W **11.16.** $P_{2\Omega} = 9.92$ W
11.18. $\mathbf{Z}_L = 1.85 + j2.77\ \Omega$ **11.20.** $P = 12.5$ W **11.23.** $P_L = 5.28$ W
11.27. $V_{rms} = \dfrac{4\sqrt{5}}{3}$ V **11.30.** $V_{rms} = \sqrt{\dfrac{20}{3}}$ V **11.36.** $V_{rms} = 5.10$ V
11.38. **(a)** $\mathbf{V}_1 = 50 \angle 20°$ V; **(b)** $\mathbf{V}_2 = 16 \angle -30°$ V; **(c)** $\mathbf{V}_3 = 14.14 \angle -80°$ V
11.42. PF $= 0.91$ **11.47.** $C = 4100\ \mu$F
11.50. $P_{line} = 702.48$ W, $Q_{line} = 2644.68$ VAR **11.55.** $\mathbf{V}_s = 512.8 \angle 26.55°$ V
11.58. $\theta = 49.31°$

Chapter 12

12.1. $\mathbf{I}_1 = 17.99 \angle -28.04°$ A, $\mathbf{I}_2 = 21.06 \angle -22.78°$ A, $\mathbf{I}_3 = 13.82 \angle -40.13°$ A and
$\mathbf{I}_n = 3.55 \angle -175.06°$ A
12.5. $\mathbf{S}_{total} = 18.58 \angle 36.20°$ kVA

12.8.

$|\mathbf{V}_{ab}| = 173\text{ V}$

$|\mathbf{V}_{an}| = 100\text{ V}$

$30°$

$45°$

$\mathbf{V}_{cn}$

$\mathbf{V}_{ca}$

$\mathbf{V}_{bn}$

$\mathbf{V}_{bc}$

12.11. $\mathbf{I}_{an} = 2.43\ \underline{/-14.04°}\text{ A}$, $\mathbf{I}_{bn} = 2.43\ \underline{/-134.04°}\text{ A}$, $\mathbf{I}_{cn} = 2.43\ \underline{/-254.04°}\text{ A}$
12.15. $\mathbf{I}_a = 1.99\ \underline{/-73.6°}\text{ A}$, $\mathbf{I}_b = 1.99\ \underline{/-193.6°}\text{ A}$, $\mathbf{I}_c = 1.99\ \underline{/-313.6°}\text{ A}$
12.20. $\mathbf{I}_a = 6.68\ \underline{/-41.8°}\text{ A}$ and $\mathbf{I}_{ab} = 3.86\ \underline{/-11.8°}\text{ A}$ **12.23.** $\mathbf{Z}_{eq} = 15 + j12\ \Omega$
12.25. $\mathbf{I}_a = 25.19\ \underline{/-47.64°}\text{ A}$ **12.30.** $P_{load} = 65.6\text{ W}$, $Q_{load} = 161.4\text{ VAR}$
12.34. $P_{total} = 9274\text{ W}$
12.37. $\mathbf{I}_a = 35.78\ \underline{/3.47°}\text{ A}$ **12.40.** $I_L = 244.6\text{ A}$ **12.43.** $\mathbf{Z} = 20\ \underline{/33.62°}\ \Omega$
12.47. $I_L = 10.25\text{ A}$ **12.49.** If one wattmeter coil is reversed, it will read 1239.06 W.
12.52. $\mathbf{Z}_\Delta = 94.1\ \underline{/30°}\ \Omega$
12.56. Source current $= 4.918\ \underline{/-159°}\text{ A}$, Load current $= 8.614\ \underline{/-8.963°}\text{ A}$, Voltage $= 12.81\ \underline{/36.04°}\text{ V}$

Chapter 13

13.1. $\mathbf{Z}_i(s) = \dfrac{s^2C^2R^2 + 3sCR + 1}{sC(sCR + 2)}$

13.4.
$$\frac{\mathbf{V}_o}{\mathbf{V}_i} = \frac{[(L_2/C) - \omega^2 L_1 L_2] + j(\omega L_1 R_3 - R_3/\omega C)}{[R_1(R_2 + R_3) - \omega^2 L_1 L_2 + L_2/C] + j[\omega L_1 R_1 + \omega L_1 R_2 + \omega L_1 R_3 + \omega L_2 R_1 - \dfrac{1}{\omega C}(R_1 + R_2 + R_3)]}$$

13.5.

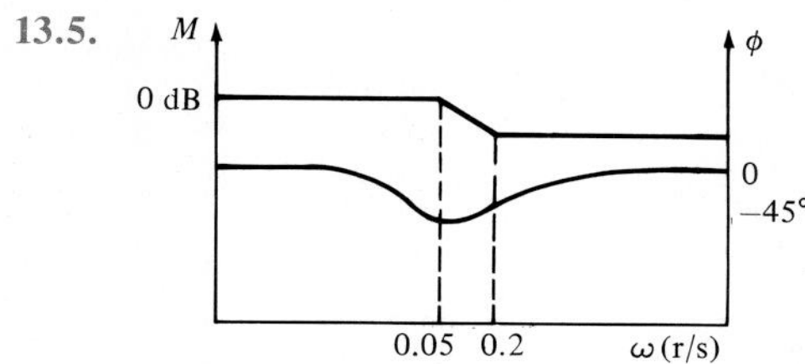

13.9.

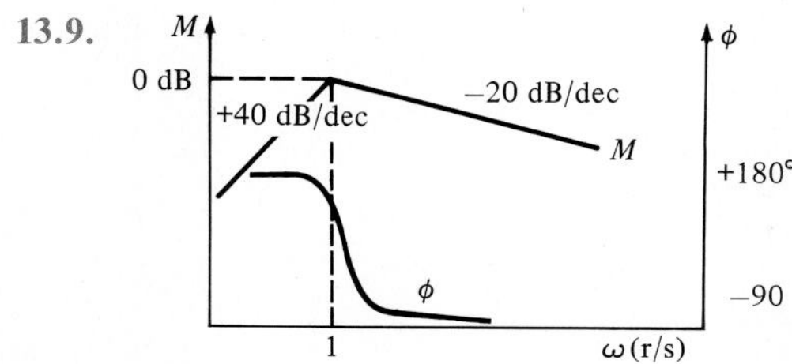

13.13.

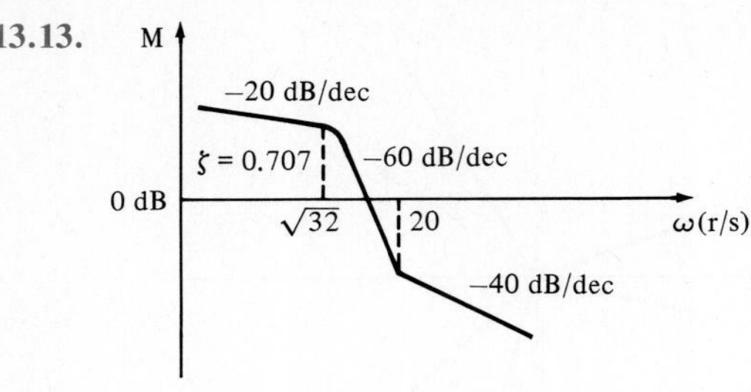

13.17. $$\mathbf{H}(j\omega) = \frac{100\left(\frac{j\omega}{10} + 1\right)\left(\frac{j\omega}{100} + 1\right)}{j\omega + 1}$$

13.19. $$\mathbf{H}(j\omega) = \frac{10\left(\frac{j\omega}{10} + 1\right)}{j\omega\left(\frac{j\omega}{20} + 1\right)^2}$$

13.23. $$\mathbf{G}(j\omega) = \frac{20\left(\frac{j\omega}{2} + 1\right)}{j\omega\left(\frac{j\omega}{20} + 1\right)\left(\frac{j\omega}{30} + 1\right)}$$

13.26. $\frac{\mathbf{V}_o}{\mathbf{V}_i} = \frac{j\omega + R_2/CR_1R_2}{j\omega + (R_1 + R_2)/CR_1R_2}$, high-pass filter

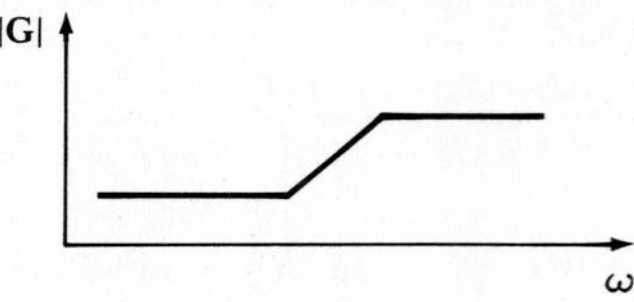

13.30. $\frac{\mathbf{V}_o}{\mathbf{V}_i} = \frac{-j\omega C_1R_2}{(j\omega C_1R_1 + 1)(j\omega C_2R_2 + 1)}$, bandpass filter

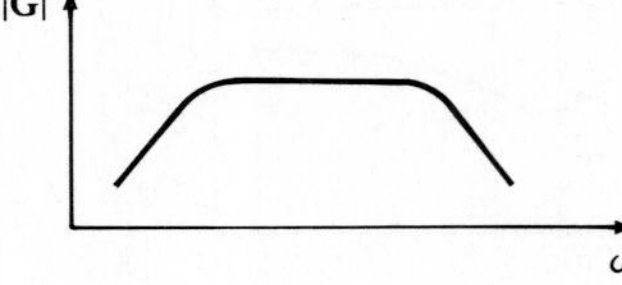

13.33. **(a)** $\frac{\mathbf{V}_o}{\mathbf{V}_i} = \frac{j\omega C_1R + 1 + C_1/C_2}{j\omega C_1R + 1}$; **(b)** low-pass filter

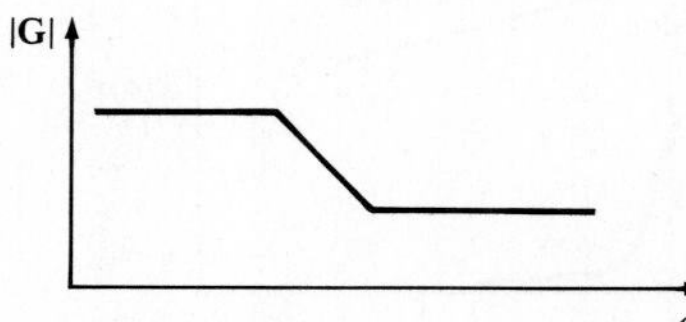

13.37. $Q = 100$, $L = 1.59$ mH, $C = 1.59$ nF **13.40.** $C = 0.345$ F or 0.045 F

13.44. $\mathbf{Z}_1 = -j\frac{17}{29}\ \Omega$ **13.45.** $R = 2000\ \Omega$, $L = 500$ H, $C = \frac{1}{8000}$ F

13.49.

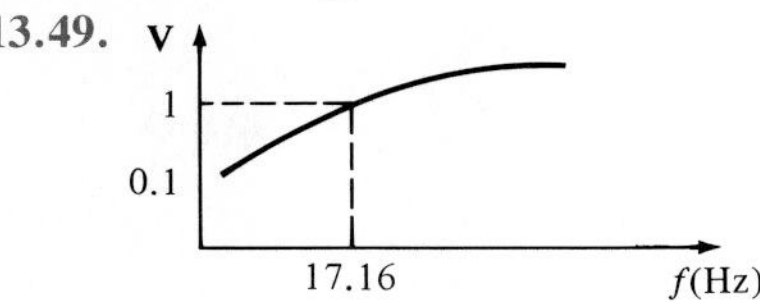

13.51.

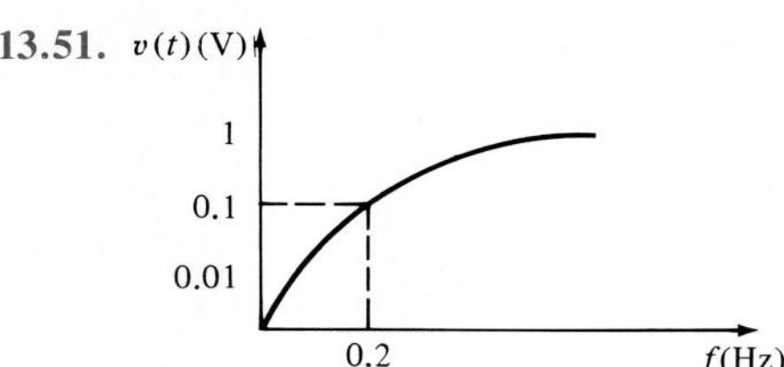

13.54. $v(t)$ (V)

100
30.33
1
10^{-2}
10^{-4}
4 kHz f

Chapter 14

14.1. $\mathbf{V}_o = 5.33\ \underline{/-123.69^\circ}$ V

14.3. $-36\ \underline{/0^\circ} + (2 - j2)\,\mathbf{I}_1 + 3(\mathbf{I}_1 - \mathbf{I}_2) = 0$, $3(\mathbf{I}_2 - \mathbf{I}_1) + j4(\mathbf{I}_2) - j1(\mathbf{I}_2) + 2(\mathbf{I}_2) + j2(\mathbf{I}_2) - j1(\mathbf{I}_2) = 0$

14.8. $v_1(t) = -15.45 \sin(377t - 45^\circ)$ V, $v_2(t) = -27.14 \sin(377t - 45^\circ)$ V

14.10.
$$v_a(t) = -L_1 \frac{di_1(t)}{dt} - M \frac{di_2(t)}{dt}$$
$$v_b(t) = -L_2 \frac{di_2(t)}{dt} - M \frac{di_1(t)}{dt}$$

14.14.
$$v_a(t) = -L_1 \frac{di_1(t)}{dt} + M \frac{di_2(t)}{dt}$$
$$v_b(t) = L_2 \frac{di_2(t)}{dt} - M \frac{di_1(t)}{dt}$$

14.20. **(a)** $P = 79.6$ mW; **(b)** $P = 99.6$ mW

14.22. $\mathbf{Z}_i = 4.38 + j3.92\ \Omega$ **14.26.** $C = 166\ \mu$F

14.30. $\mathbf{I}_1 = 4.71\ \underline{/18.69^\circ}$ A, $\mathbf{V}_1 = 14.89\ \underline{/37.12^\circ}$ V, $\mathbf{I}_2 = 2.36\ \underline{/18.69^\circ}$ A, $\mathbf{V}_2 = 29.79\ \underline{/37.12^\circ}$ V

14.32. $\mathbf{V}_o = -44.72\ \underline{/26.57^\circ}$ V **14.35.** $\mathbf{Z}_i = 180 + j32\ \Omega$

14.40. $\mathbf{V}_s = 107.15\ \underline{/-20.04^\circ}$ V **14.43.** $\mathbf{I}_2 = 0.88\ \underline{/60.86^\circ}$ A

14.45. $\mathbf{Z}_L = 12 + j8\ \Omega$ **14.49.** Voltage $= 14.06\ \underline{/82.69^\circ}$ V
14.52. Voltage $= 0.1899\ \underline{/-158.7^\circ}$ V **14.57.** Voltage $= 2.628\ \underline{/93.01^\circ}$ V

Chapter 15

15.1. $\mathbf{Y} = \begin{bmatrix} \frac{1}{6} & \frac{-1}{12} \\ \frac{-1}{12} & \frac{1}{6} \end{bmatrix}$

15.2. $\mathbf{V}_2 = 16$ V and $\mathbf{I}_2 = -4$ A **15.7.** $\dfrac{\mathbf{V}_o}{\mathbf{V}_1} = -65.6$

15.9. $\mathbf{A} = 2$, $\mathbf{B} = 12\ \Omega$, $\mathbf{C} = \frac{1}{4}$s, and $\mathbf{D} = 2$

15.14. $\mathbf{z}_{11} = \dfrac{\Delta H}{h_{22}}$, $\mathbf{z}_{12} = \dfrac{\mathbf{h}_{12}}{\mathbf{h}_{22}}$, $\mathbf{z}_{21} = \dfrac{-\mathbf{h}_{21}}{\mathbf{h}_{22}}$, $\mathbf{z}_{22} = \dfrac{1}{\mathbf{h}_{22}}$

15.18. $\mathbf{Z}_i = \dfrac{\mathbf{h}_{11} + \Delta H \mathbf{Z}_L}{1 + \mathbf{h}_{22}\mathbf{Z}_L}$ **15.22.** $\mathbf{V}_o = 16$ V

15.28.
$$\begin{bmatrix} \mathbf{A} & \mathbf{B} \\ \mathbf{C} & \mathbf{D} \end{bmatrix} = \begin{bmatrix} \frac{1}{\alpha^2 Z^2} & \frac{1}{\alpha^2 Z} \\ 0 & 0 \end{bmatrix}$$

15.30.
$$\begin{bmatrix} \mathbf{Y}_{11} & \mathbf{Y}_{12} \\ \mathbf{Y}_{21} & \mathbf{Y}_{22} \end{bmatrix} = \begin{bmatrix} \frac{Z_2 + Z_3}{\Delta Z_{123}} + \frac{Z_5 + Z_6}{\Delta Z_{456}} & \frac{-Z_3}{\Delta Z_{123}} - \frac{Z_6}{\Delta Z_{456}} \\ \frac{-Z_3}{\Delta Z_{123}} - \frac{Z_6}{\Delta Z_{456}} & \frac{Z_1 + Z_3}{\Delta Z_{123}} + \frac{Z_4 + Z_6}{\Delta Z_{456}} \end{bmatrix}$$

15.32. $\mathbf{z}_{11} = R_e + R_b$, $\mathbf{z}_{12} = \mathbf{R}_6$, $\mathbf{z}_{21} = R_c\alpha + R_b$, $\mathbf{z}_{22} = R_c + R_b$

15.36. $\mathbf{Z} = \begin{bmatrix} R_1 & nR_1 \\ nR_1 & n^2(R_1 + R_2) \end{bmatrix}$

15.39. $\mathbf{V}_o = 4.63\ \underline{/157.54^\circ}$ V

Chapter 16

16.1. $f(t) = \dfrac{10}{\pi} \sum_{n=-\infty}^{\infty} \dfrac{1}{n} \sin(0.1n\pi) e^{jn\pi(2t-0.1)}$

16.4. $v(t) = -\dfrac{1}{5} + \sum_{\substack{n=-\infty \\ n\neq 0}}^{\infty} \left[\dfrac{3e^{\frac{-j8\pi n}{5}} - 3e^{\frac{-j6\pi n}{5}} - 2e^{\frac{-j4\pi n}{5}} + 2e^{-j2\pi n/5}}{j2\pi n} \right] e^{\frac{j2\pi nt}{5}}$ V

16.6. $f(t) = \sum_{n=1}^{\infty} (-1)^{n+1} \dfrac{20}{n\pi} \sin nt$

16.12. $f(t)$ is odd and has half-wave symmetry, $a_0 = 0$, $a_n = 0$, $b_n = 0$ for n even

16.15. $v(t) = \dfrac{2}{3} - \sum_{n=1}^{\infty} \dfrac{4}{\pi^2 n^2} (-1)^n \cos \pi n t$ V

16.20. $f(t) = 3 - 4 \sin 20\pi t - 5 \sin 40\ \pi t - 3 \sin 60\pi t - 2 \sin 80\pi t - \sin 100\pi t$

16.24. $i_o(t) = \sum_{\substack{n=1 \\ n \text{ odd}}}^{\infty} \left(\dfrac{20}{n\pi} \sin \dfrac{n\pi}{2} \right) \dfrac{1}{A_n} \cos(2nt - \theta_n)$ A where $A_n \underline{/\theta_n} = \dfrac{3 + j2n}{1 + j2n}$

16.27. $v_o(t) = 0.18 + 0.028 \cos(\pi t - 75°) + 0.0036 \cos(2\pi t + 97.5°) + 0.0011 \cos(3\pi t - 85°) + \cdots$ V

16.30. $\mathbf{F}(\omega) = A\left[\frac{1}{j\omega} - \frac{1}{\omega^2}\left(e^{-j\omega} - 1\right)\right]$

16.35. $v_o(t) = (e^{-3t} - e^{-4t})\,u(t)$ V

16.37. $W_0 = \frac{5}{3}$ J

16.41. $v_o(t) = \frac{2}{3}(e^{-t} - e^{-4t})\,u(t)$ V

Chapter 17

17.1. $\mathbf{F}(s) = \frac{(s^2 + s + 1)e^{-s}}{s^2(s+1)}$ **17.5.** $\mathbf{F}(s) = e^{-(s+a)}\left[\frac{\omega \cos \omega + (s+a)\sin \omega}{(s+a)^2 + \omega^2}\right]$

17.9. $\mathbf{F}(s) = e^{-s}\left[\frac{2s^2 \cos \omega - \cos \omega (s^2 + \omega^2)}{(s^2 + \omega^2)^2} + \frac{2s\omega \sin \omega}{(s^2 + \omega^2)^2} + \frac{s \cos \omega}{s^2 + \omega^2} - \frac{\omega \sin \omega}{s^2 + \omega^2}\right]$

17.10. $\mathbf{F}(s) = \frac{2}{s^2} - \frac{2}{s^2}e^{-s} - \frac{2}{s}e^{-2s}$ **17.14.** $\mathbf{F}(s) = \frac{(2/s^2)(1 - e^{-s} - se^{-2s})}{1 - e^{-2s}}$

17.17. $\mathbf{F}(s) = \frac{4 - 2e^{-s} - 2e^{-2s}}{s(1 - e^{-3s})}$

17.21. **(a)** $f(t) = 10e^{-t}\cos t$; **(b)** $f(t) = \frac{1}{5} + 0.62e^{-2t}\cos(t - 108.43°)$; **(c)** $f(t) = \frac{1}{4} + \frac{1}{4}e^{-2t} + \frac{1}{2}e^{-t}\cos(t + 180°)$

17.24. **(a)** $f(t) = e^{-(t-1)}u(t-1)$; **(b)** $f(t) = u(t) - u(t-2)$; **(c)** $f(t) = e^{-2t}u(t) - e^{-2(t-1)}u(t-1)$

17.27. $f(t) = e^{-2(t-2)}\cos(t-2)u(t-2)$ **17.30.** $f(t) = e^{-t} - e^{-2t}$

17.33. $f(t) = \frac{t^2}{2}$ **17.37.** $f(t) = \frac{1}{16}(\sin 2t - 2t\cos 2t)$

17.41. $f(t) = \frac{1}{2}[(1 - e^{-2t})\,u(t) - (1 - e^{-2(t-2)})u(t-2)]$

17.46. **(a)** $f(0) = 10$, $f(\infty) = 0$; **(b)** $f(0) = 0$, $f(\infty) = 0$; **(c)** $f(0) = 2$, $f(\infty) = 0$

17.50. **(a)** $y(t) = (1 - 2e^{-t} + e^{-2t})u(t)$; **(b)** $y(t) = (\frac{1}{6}e^{-t} + \frac{1}{2}e^{-3t} - \frac{2}{3}e^{-4t})u(t)$

17.53. $y(t) = (2e^{-t} - \frac{1}{2}e^{-2t})u(t)$ **17.56.** $y(t) = (2te^{-t} - 2e^{-t} + 2e^{-2t})u(t)$

Chapter 18

18.1. $\mathbf{Z}(s) = \frac{3s^4 + 10s^3 + 22s^2 + 24s + 8}{(s^2+2)(s^2+6s+4)}$ **18.4.** $v_o(t \to \infty) = 0$ V

18.7. $i_1(t) = \left[6 + 6.42e^{-t/2}\cos\left(\frac{\sqrt{7}}{2}t + 62.16°\right)\right]u(t)$ A **18.8.** Same as Problem 18.7

18.12. $v_o(t) = [2\sqrt{2}e^{-t}\cos(t - 45°)]u(t)$ V **18.14.** Same as Problem 18.12

18.18. $v_o(t) = [\frac{1}{2}e^{-2t} + 1.264e^{-0.5t}\cos(0.646t - 142.2°)]u(t)$ V

18.19. Same as Problem 18.18 **18.23.** $v_o(t) = (1 - 5e^{-4t})u(t)$ V

18.27. $i_o(t) = (-2 + 2e^{-t} + 0.074e^{-0.382t} - 3.17e^{-2.62t})u(t)$ A

18.31. $v_o(t) = (\frac{8}{3} + 4e^{-2t} - \frac{17}{3}e^{-3t/2})u(t)$ V **18.33.** $i_o(t) = -2e^{-t}u(t)$ A

18.37. $v_o(t) = \frac{3}{2}(1 - e^{-4t})u(t)$ V **18.41.** $i_o(t) = 12\sqrt{2}\cos(2t + 45°)$ A

18.45. $i_o(t) = \frac{4}{\sqrt{2}}\cos(2t + 45°)$ A **18.46.** $i_o(t) = 0.66e^{-1.66t}u(t)$ mA

18.49. $v_o(t) = 3e^{-5t/2}u(t)$ V **18.53.** $i_o(t) = [1.61e^{-0.82t} - 0.61e^{-0.31t}]u(t)$ A

18.56. $i_o(0) = 0$ and $i_o(\infty) = 4$ A

18.60. $v_o(t) = (1 + \frac{4}{7}e^{-3t/7})u(t) - [1 + \frac{4}{7}e^{-3(t-1)/7}]u(t-1)$ V

18.62. $v_o(t) = (\frac{1}{3}t - \frac{4}{9} + \frac{4}{9}e^{-3t/7})u(t) - [\frac{1}{3}(t-1) - \frac{4}{9} + \frac{4}{9}e^{-3(t-1)/7}]u(t-1)$
$- (\frac{1}{3} - \frac{4}{21}e^{-3(t-2)/7})u(t-2)$ V

18.66. $y(t) = (\frac{30}{13}e^{-t}\cos 2t + \frac{20}{13}e^{-t}\sin 2t - \frac{30}{13}e^{-4t})u(t)$ **18.70.** $\dfrac{\mathbf{V}_o}{\mathbf{V}_i} = \dfrac{0.5(s+1)}{7s+3}$

Index

Commonly Used Prefixes

giga (G)	10^9
mega (M)	10^6
kilo (k)	10^3
milli (m)	10^{-3}
micro (μ)	10^{-6}
nano (n)	10^{-9}
pico (p)	10^{-12}